MW00845308

Risk Assessment in Geotechnical Engineering

Risk Assessment in Geotechnical Engineering

Gordon A. Fenton
Dalhousie University, Halifax, Nova Scotia

D. V. Griffiths
Colorado School of Mines, Golden, Colorado

John Wiley & Sons, Inc.

This book is printed on acid-free paper. ♾

Published by John Wiley & Sons, Inc., Hoboken, New Jersey.
Published simultaneously in Canada

Library of Congress Cataloging-in-Publication Data:

Fenton, Gordon A.
Risk assessment in geotechnical engineering / Gordon A. Fenton, D.V. Griffiths.
p. cm.
Includes index.
ISBN 978-0-470-17820-1 (cloth)
1. Soil mechanics. 2. Rock mechanics. 3. Risk assessment. I. Griffiths, D.V. II. Title.
TA710.F385 2008
624.1′51—dc22

2007044825

Printed in the United States of America

10 9 8 7 6 5 4 3 2 1

To Terry, Paul, Alex, Emily, Valerie, Will, and James

CONTENTS

PREFACE

Soils and rocks in their natural state are among the most variable of all engineering materials, and geotechnical engineers must often "make do" with materials that present themselves at a particular site. In a perfect world with no economic constraints, we would drill numerous boreholes and take multiple samples back to the laboratory for measurement of standard soil properties such as permeability, compressibility, and shear strength. Armed with all this information, we could then perform our design of a seepage problem, foundation, or slope and be very confident of our predictions. In reality we must usually deal with very limited site investigation data, and the traditional approach for dealing with this uncertainty in geotechnical design has been through the use of *characteristic values* of the soil properties coupled with a generous *factor of safety*.

If we were to plot the multitude of data from the hypothetical site investigation as a histogram for one of the properties, we would likely see a broad range of values in the form of a *bell-shaped curve*. The most likely values of the property would be somewhere in the middle, but a significant number of samples would display higher and lower values too. This variability inherent in soils and rocks suggests that geotechnical systems are highly amenable to a statistical interpretation. This is quite a different philosophy from the traditional approach mentioned above. In the probabilistic approach, we input soil properties characterized in terms of their means and variances (first and second moments) leading to estimates of the *probability of failure* or *reliability* of a design. Specific examples might involve estimation of the reliability of a slope design, the probability of excessive foundation settlement, or the probability of excessive leakage from a reservoir. When probabilities are coupled with consequences of design failure, we can then assess the *risk* associated with the design.

While the idea of using statistical concepts in geotechnical engineering is not new, the use of these methodologies has tended to be confined to *high-tech* projects, particularly relating to seismic design and offshore engineering. For example, the "hundred year" earthquake or wave is based on statistical analysis of historical records. In recent years, however, there has been a remarkable increase in activity and interest in the use of probabilistic methodologies applied to more traditional areas of geotechnical engineering. This growth has manifested itself in many forms and spans both academe and practice within the geotechnical engineering community, for example, more dedicated sessions at conferences, short courses for practitioners, and new journals and books.

The obvious question may then be, "why another book"? There is certainly no shortage of texts on structural reliability or general statistical methods for civil engineers, but there is only one other textbook to our knowledge, by Baecher and Christian (2003), specifically aimed at geotechnical engineers. In this rapidly evolving field, however, a number of important recent developments (in particular random-field simulation techniques) have reached a maturity and applicability that justify the current text. Our target audience therefore includes students and practitioners who wish to become acquainted with the theory and methodologies behind risk assessment in geotechnical engineering ranging from established *first-order* methods to the most recent numerical developments such as the random finite-element method (RFEM).

An additional unique feature of the current text is that the programs used in the geotechnical applications discussed in the second half of the book are made freely available for download from www.engmath.dal.ca/rfem.

The text is organized into two main parts with Part 1 devoted to theory and Part 2 to practice.

The first part of the book, (Chapters 1–7) describes the theory behind risk assessment techniques in geotechnical engineering. These chapters contain over 100 worked examples to help the reader gain a detailed understanding of the methods. Chapter 1 offers a review of probability theory intended as a gentle introduction to readers who may have forgotten most of their undergraduate "prob and stats." Chapters 2 and 3 offer a thorough description of both discrete and continuous random processes, leading into the theory of random fields used extensively in the practical applications described in Part 2. Chapter 4 describes how to make best estimates of uncertain parameters given observations (samples) at nearby locations along with some theory relating to how often we should expect to see exceptionally high (or low) soil properties. Chapter 5 describes the existing techniques available to statistically analyze spatially distributed soil data along with the shortcomings of each technique and to decide on a distribution to use in modeling soil variability. Chapter 6 discusses simulation and in particular lays out the underlying theory, associated algorithms, and accuracy of a variety of common methods of generating realizations of spatially variable random fields. Chapter 7 addresses reliability-based design in geotechnical engineering, which is currently an area of great activity both in North America and internationally. The chapter considers methods for choosing suitable load and resistance factors in the context of a target reliability in geotechnical design. The chapter also addresses some of the problems of implementing a reliability-based design, such as the fact that in frictional materials the load also contributes to the resistance, so that load and resistance are not independent as is commonly assumed in other reliability-based design codes.

The second part of the book, (Chapters 8–16) describes the use of advanced probabilistic tools to several classical geotechnical engineering applications. An emphasis in these chapters has been to study problems that will be familiar to all practicing geotechnical engineers. The examples use the RFEM as developed by the authors and made available through the website mentioned previously, in which random-field theory as described in Chapter 3 is combined with the finite-element method. Chapters 8 and 9 describe steady seepage with random permeability in both two and three dimensions. Both confined and unconfined flow examples are demonstrated. Chapter 10 considers settlements and differential settlements of strip and rectangular footings on soils with random compressibility. Chapters 11 (bearing capacity), 13 (slope stability), 14 (earth pressure), and 15 (mine pillar stability) describe limit analyses in geotechnical engineering in which the shear strength parameters are treated as being spatially variable and possibly cross-correlated. In all these cases, comparisons are made between the probability of failure and the traditional factor of safety that might be obtained from characteristic values of the shear strength parameters so that geotechnical engineers can get a sense for how traditional designs relate to failure probabilities. The limit analyses also highlight important deficiencies leading to unconservatism in some of the simpler probabilistic tools (e.g., first order) which are not able to properly account for spatial correlation structures. These chapters particularly draw attention to the important phenomenon of mechanisms of failure "seeking out" critical paths through the soil when weak spatially correlated zones dominate the solution. Chapter 12 considers probabilistic analysis of deep foundations such as piles in soils modeled with random t–z springs. Chapter 16 uses random-field models to quantify the probability of liquefaction and its extent at a particular site.

ACKNOWLEDGMENTS

Thanks are first of all due to Erik Vanmarcke, under whose guidance many of the thoughts presented in this book were born. We also wish to recognize Debbie Dupuis, who helped develop some of the introductory material presented in Chapter 1, and Ian Smith for his development of codes described in the text by Smith and Griffiths (2004) that underpin many of the random finite-element programs used in the second half of the book.

We are also aware of the contribution of many of our students and other colleagues over the years to the development and application of the random finite-element method. Notable mentions here should go to, in alphabetic order, Bill Cavers, Mark Denavit, Jason Goldsworthy, Jinsong Huang, Mark Jaksa, Carisa Lemons, Neil McCormick, Geoffrey Paice, Tom Szynakiewicz, Derena Tveten, Anthony Urquhart, Xianyue Zhang, Haiying Zhou, and Heidi Ziemann.

We greatly appreciate the efforts of those who reviewed the original proposals for this text and edited the book during the production stage. Thanks are also due to the Natural Sciences and Engineering Research Council of Canada for their financial support under grant OPG0105445; to the National Science Foundation under grants ECE-86-11521, CMS-0408150, and CMS-9877189; and to NCEER under grant 87-6003. The financial support of these agencies was an important component of the research and development that led to the publication of this text.

PART 1

Theory

CHAPTER 1

Review of Probability Theory

1.1 INTRODUCTION

Probability theory provides a rational and efficient means of characterizing the uncertainty which is prevalent in geotechnical engineering. This chapter summarizes the background, fundamental axioms, and main results constituting modern probability theory. Common discrete and continuous distributions are discussed in the last sections of the chapter.

1.2 BASIC SET THEORY

1.2.1 Sample Spaces and Events

When a system is random and is to be modeled as such, the first step in the model is to decide what all of the possible states (outcomes) of the system are. For example, if the load on a retaining wall is being modeled as being random, the possible load can range anywhere from zero to infinity, at least conceptually (while a zero load is entirely possible, albeit unlikely, an infinite load is unlikely—we shall see shortly that the likelihood of an infinite load can be set to be appropriately small). Once the complete set of possible states has been decided on, interest is generally focused on probabilities associated with certain portions of the possible states. For example, it may be of interest to determine the probability that the load on the wall exceeds the sliding resistance of the wall base, so that the wall slides outward. This translates into determining the probability associated with some portion, or *subset*, of the total range of possible wall loads (we are assuming, for the time being, that the base sliding resistance is *known*). These ideas motivate the following definitions:

Definitions

Experiment: Any process that generates a set of data. The experiment may be, for example, the monitoring of the volume of water passing through an earth dam in a unit time. The volume recorded becomes the data set.

Sample Space: The set of all possible outcomes of an experiment. The sample space is represented by the symbol S.

Sample Point: An outcome in the sample space. For example, if the experiment consists of monitoring the volume of water passing through an earth dam per hour, a sample point would be the observation 1.2 m^3/h. Another would be the observation 1.41 m^3/h.

Event: A subset of a sample space. Events will be denoted using uppercase letters, such as $A, B, \ldots$. For example, we might define A to be the event that the flow rate through an earth dam is greater than 0.01 m^3/h.

Null Set: The empty set, having no elements, is used to represent the impossible "event" and is denoted $\emptyset$. For example, the event that the flow rate through an earth dam is both less than 1 and greater than 5 m^3/h is impossible and so the event is the null set.

These ideas will be illustrated with some simple examples.

Example 1.1 Suppose an experiment consists of observing the results of two static pile capacity tests. Each test is considered to be a success (1) if the pile capacity exceeds a certain design criterion and a failure (0) if not. This is an *experiment* since a set of data is derived from it. The actual data derived depend on what is of interest. For example:

1. Suppose that only the number of successful pile tests is of interest. The *sample space* would then be $S = \{0, 1, 2\}$. The elements 0, 1, and 2 of the set S are *sample points*. From this sample space, the following events (which may be of interest) can be defined; $\emptyset$, $\{0\}$, $\{1\}$, $\{2\}$, $\{0, 1\}$, $\{0, 2\}$, $\{1, 2\}$, and $S = \{0, 1, 2\}$ are possible events. The *null set* is used to denote all impossible events (for example, the event that the number of successful tests, out of two tests, is greater than 2).
2. Suppose that the order of occurrence of the successes and failures is of interest. The sample space would then be $S = \{11, 10, 01, 00\}$. Each outcome is a doublet depicting the sequence. Thus, the elements 11, 10, 01, and 00 of S are sample points. The possible events are $\emptyset$, $\{11\}$, $\{10\}$, $\{01\}$, $\{00\}$, $\{11, 10\}$, $\{11, 01\}$, $\{11, 00\}$, $\{10, 01\}$, $\{10, 00\}$, $\{01, 00\}$, $\{11, 10, 01\}$, $\{11, 10, 00\}$, $\{11, 01, 00\}$, $\{10, 01, 00\}$, and $\{11, 10, 01, 00\}$.

Note that the information in 1 could be recovered from that in 2, but not vice versa, so it is often useful to define the experiment to be more general initially, when possible. Other types of events can then be derived after the experiment is completed.

Sample spaces may be either discrete or continuous:

Discrete Case: In this case, the sample space consists of a sequence of discrete values (e.g., 0, 1, . . .). For example, the number of blow counts in a standard penetration test (SPT). Conceptually, this could be any integer number from zero to infinity.

Continuous Case: In this case, the sample space is composed of a continuum of sample points and the number of sample points is effectively always infinite—for example, the elastic modulus of a soil sample. This could be any real number on the positive real line.

1.2.2 Basic Set Theory

The relationship between events and the corresponding sample space can often be illustrated graphically by means of a *Venn diagram*. In a Venn diagram the sample space is represented as a rectangle and events are (usually) drawn as circles inside the rectangle. For example, see Figure 1.1, where A_1, A_2, and A_3 are events in the sample space S.

We are often interested in probabilities associated with combinations of events; for example, the probability that a cone penetration test (CPT) sounding has tip resistance greater than x at the same time as the side friction is less that y. Such events will be formed as subsets of the sample space (and thus are sets themselves). We form these subsets using *set operators*. The union, intersection, and complement are set theory operators which are defined as follows:

The *union* of two events E and F is denoted $E \cup F$.

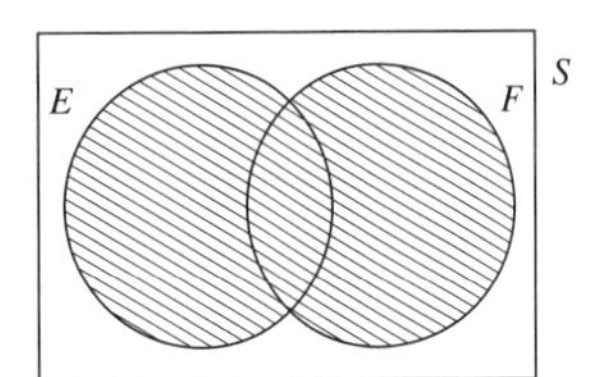

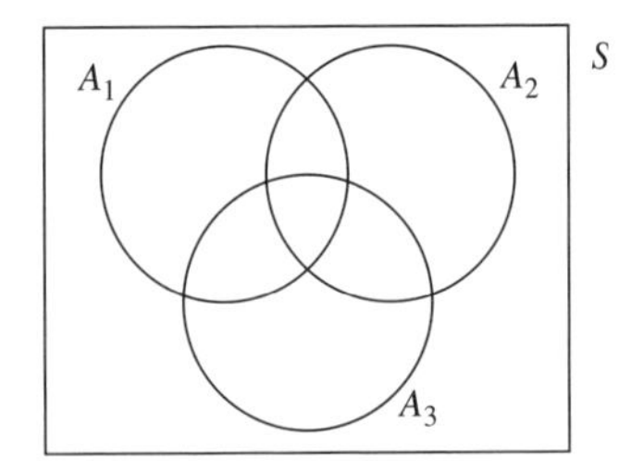

Figure 1.1 Simple Venn diagram.

The *intersection* of two events E and F is denoted $E \cap F$.

The *complement* of an event E is denoted E^c.

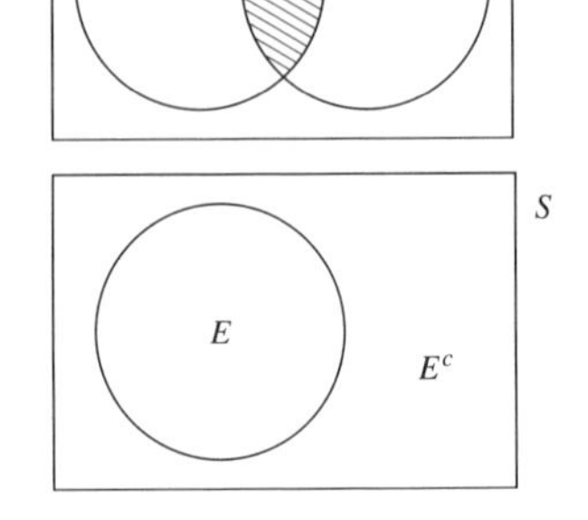

Two events E and F are said to be *mutually exclusive*, or *disjoint*, if $E \cap F = \emptyset$. For example, E and E^c are disjoint events.

Example 1.2 Three piles are being statically loaded to failure. Let A_i denote the event that the ith pile has a capacity exceeding specifications. Using only sets and set theory operators (i.e., using only $A_i, i = 1, 2, 3$, and $\cap$, $\cup$, and ^{c}), describe each of the following events. In each case, also draw a Venn diagram and shade the region corresponding to the event.

1. At least one pile has capacity exceeding the specification.
2. All three piles have capacities exceeding the specification.
3. Only the first pile has capacity exceeding the specification.
4. Exactly one pile has capacity exceeding the specification.
5. Either only the first pile or only both of the other piles have capacities exceeding the specification.

SOLUTION

1. $A_1 \cup A_2 \cup A_3$

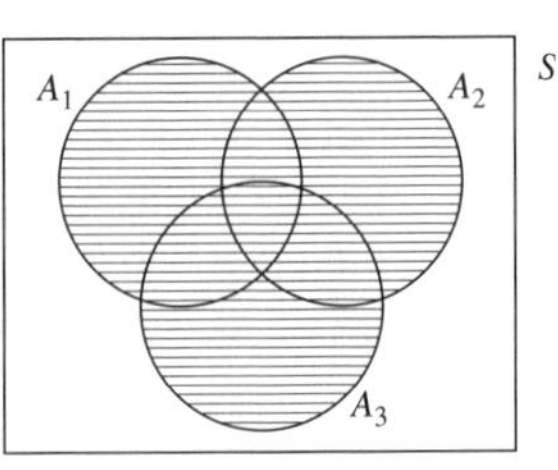

2. $A_1 \cap A_2 \cap A_3$

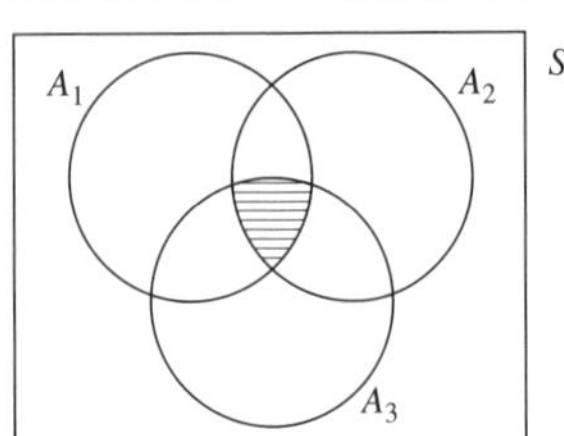

3. $A_1 \cap A_2^c \cap A_3^c$

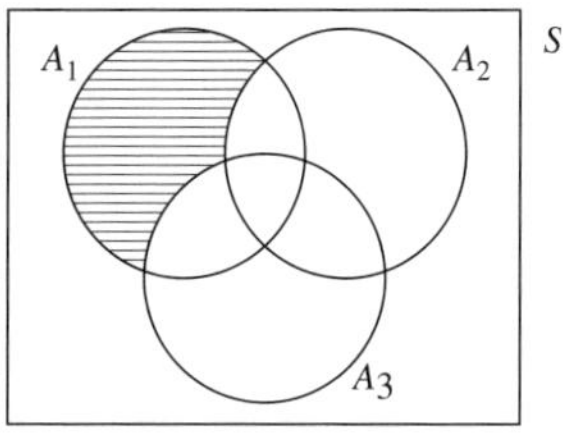

4. $(A_1 \cap A_2^c \cap A_3^c)$
$\cup (A_1^c \cap A_2 \cap A_3^c)$
$\cup (A_1^c \cap A_2^c \cap A_3)$

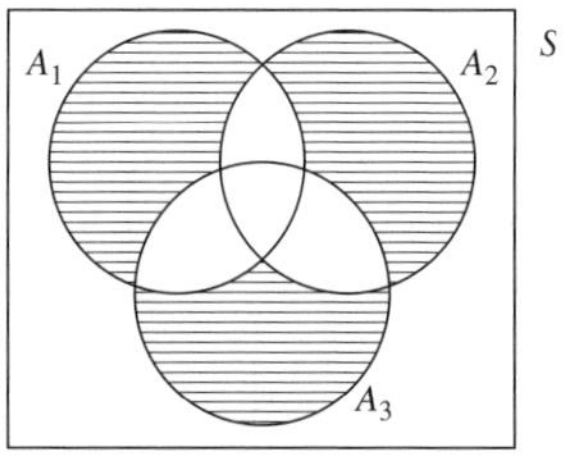

5. $(A_1 \cap A_2^c \cap A_3^c)$
$\cup (A_1^c \cap A_2 \cap A_3)$

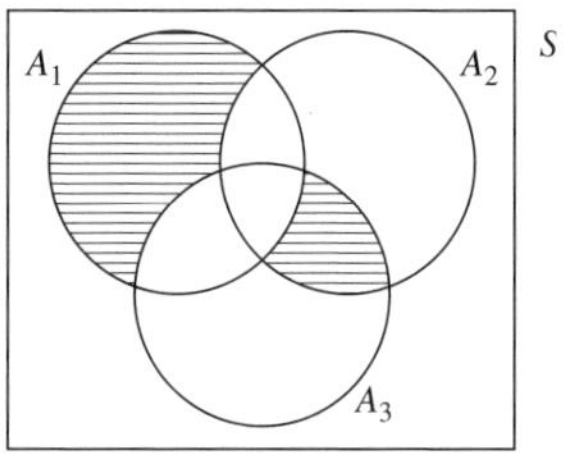

It is clear from the Venn diagram that, for example, $A_1 \cap A_2^c \cap A_3^c$ and $A_1^c \cap A_2 \cap A_3$ are disjoint events, that is, $(A_1 \cap A_2^c \cap A_3^c) \cap (A_1^c \cap A_2 \cap A_3) = \emptyset$.

1.2.3 Counting Sample Points

Consider experiments which have a finite number of possible outcomes. For example, out of a group of piles, we could have three failing to meet specifications but cannot have 3.24 piles failing to meet specifications. That is, the sample space, in this case, consists of only whole numbers. Such sample spaces are called *discrete* sample spaces. We are often interested in computing the probability associated with each possible value in the sample space. For example, we may want to be able to compute the probability that exactly three piles fail to meet specifications at a site. While it is not generally easy to assign probabilities to something like the number of soft soil lenses at a site, some discrete sample spaces consist of *equi-likely* outcomes, where all possible outcomes have the same probability of occurrence. In this case, we only need to know the total number of possible outcomes in order to assign probabilities to individual outcomes (i.e., the probability of each outcome is equal to 1 over the total number of possible outcomes). Knowing the total number of possible outcomes is often useful, so some basic *counting* rules will be considered here.

Multiplication Rule The fundamental principle of counting, often referred to as the multiplication rule, is:

If an operation can be performed in n_1 ways, and if for each of these, a second operation can be performed in n_2 ways, then the two operations can be performed together in $n_1 \times n_2$ different ways.

Example 1.3 How many possible outcomes are there when a soil's relative density is tested twice and the outcome of each test is either a pass or a fail? Assume that you are interested in the order in which the tests pass or fail.

SOLUTION On the first test, the test can proceed in any one of $n_1 = 2$ ways. For each of these, the second test can proceed in any one of $n_2 = 2$ ways. Therefore, by the multiplication rule, there are $n_1 \times n_2 = 2 \times 2 = 4$ possible test results. Consequently, there are four points in the sample space. These are (P,P), (P,F), (F,P), and (F,F) (see also Example 1.1).

The multiplication principle extends to k operations as follows:

If an operation can be performed in n_1 ways, and if for each of these a second operation can be performed in n_2 ways, and for each of the first two a third operation can be performed in n_3 ways, and so forth, then the sequence of k operations can be performed together in

$$n = n_1 \times n_2 \times \cdots \times n_k \tag{1.1}$$

different ways.

Example 1.4 Extending the previous example, suppose that a relative-density test classifies a soil into five possible states, ranging from "very loose" to "very dense." Then if four soil samples are tested, and the outcomes of the four tests are the ordered list of their states, how many possible ways can the tests proceed if the following conditions are assumed to hold?

1. The first sample is either very loose or loose, and all four tests are unique (i.e., all four tests result in different densities).
2. The first sample is either very loose or loose, and tests may yield the same results.
3. The first sample is anything but very loose, and tests may yield the same results.

SOLUTION

1. $2 \times 4 \times 3 \times 2 = 48$
2. $2 \times 5 \times 5 \times 5 = 250$
3. $4 \times 5 \times 5 \times 5 = 500$

Permutations Frequently, we are interested in sample spaces that contain, as elements, all possible orders or arrangements of a group of objects. For example, we may want to know the number of possible ways 6 CPT cones can be selected from a collection of 20 cones of various quality. Here are some examples demonstrating how this can be computed.

Example 1.5 Six piles are being driven to bedrock and the energy required to drive them will be recorded for each. That is, our experiment consists of recording the six measured energy levels. Suppose further that the pile results will be ranked from the one taking the highest energy to the one taking the lowest energy to drive. In how many different ways could this ranked list appear?

SOLUTION The counting process can be broken up into six simpler steps: (1) selecting the pile, out of the six, taking the highest energy to drive and placing it at the top of the list; (2) selecting the pile taking the next highest energy to drive from the remaining five piles and placing it next on the list, and so on for four more steps. Since we know in how many ways each of these operations can be done, we can apply the multiplication rule: $n = 6 \times 5 \times 4 \times 3 \times 2 \times 1 = 720$. Thus, there are 720 ways that the six piles could be ranked according to driving energy.

In the above example, the number of possible arrangements is 6!, where ! is the *factorial* operator. In general,

$$n! = n \times (n-1) \times \cdots \times 2 \times 1 \qquad (1.2)$$

if n is a nonzero integer. Also $0! = 1$ by definition. The reasoning of the above example will always prevail when counting the number of possible ways of arranging all objects in a sequence.

Definition A *permutation* is an arrangement, that is, an ordered sequence, of all or part of a set of objects. If we are looking for the number of possible ordered sequences of an entire set, then

The number of permutations of n distinct objects is n!.

If only part of the set of objects is to be ordered, the reasoning is similar to that proposed in Example 1.5, except that now the number of "operations" is reduced. Consider the following example.

Example 1.6 A company has six nuclear density meters, labeled A through F. Because the company wants to keep track of the hours of usage for each, they must each be signed out. A particular job requires three of the meters to be signed out for differing periods of time. In how many ways can three of the meters be selected from the six if the first is to be used the longest, the second for an intermediate amount of time, and the third for the shortest time?

SOLUTION We note that since the three meters to be signed out will be used for differing amounts of time, it will make a difference if A is selected first, rather than second, and so on. That is, the order in which the meters are selected is important. In this case, there are six possibilities for the first meter selected. Once this is selected, the second meter is select from the remaining five meters, and so on. So in total we have $6 \times 5 \times 4 = 120$ ways.

The product $6 \times 5 \times 4$ can be written as

$$\frac{6 \times 5 \times 4 \times 3 \times 2 \times 1}{3 \times 2 \times 1}$$

so that the solution to the above example can be written as

$$6 \times 5 \times 4 = \frac{6!}{(6-3)!}$$

In general, the number of permutations of r objects selected from n distinct objects, where order counts, is

$$P_r^n = \frac{n!}{(n-r)!} \qquad (1.3)$$

Combinations In other cases, interest is in the number of ways of selecting r objects from n distinct objects *without regard to order*.

Definition A *combination* is the number of ways that objects can be selected without regard to order.

Question: If there is no regard to order, are there going to be more or less ways of doing things?

Example 1.7 In how many ways can I select two letters from A, B, and C if I do it (a) with regard to order and (b) without regard to order?

SOLUTION

In Figure 1.2, we see that there are *fewer combinations than permutations*. The number of combinations is reduced

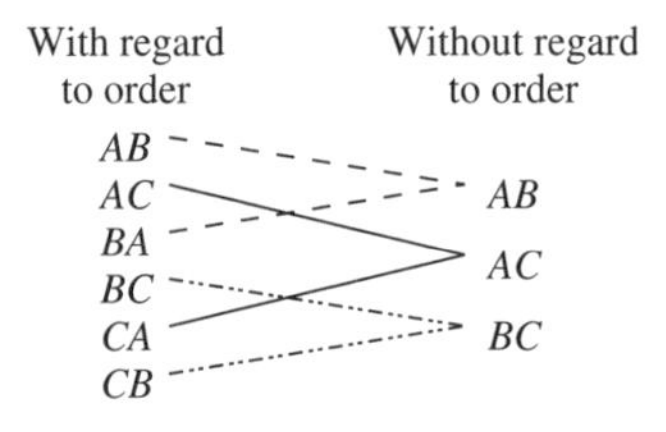

Figure 1.2 Selecting two letters from A, B, and C.

from the number of permutations by a factor of $2 \times 1 = 2$, which is the number of ways the two selected letters can be permuted among themselves.

In general we have:

The number of combinations of n distinct objects taken r at a time is written

$$\binom{n}{r} = \frac{n!}{r!(n-r)!} \tag{1.4}$$

Example 1.8 A geotechnical engineering firm keeps a list of eight consultants. Not all consultants are asked to provide a quote on a given request. Determine the number of ways three consultants can be chosen from the list.

SOLUTION

$$\binom{8}{3} = \frac{8!}{3!5!} = \frac{8 \times 7 \times 6}{3 \times 2 \times 1} = 56$$

Sometimes, the multiplication rule, permutations, and/or combinations must be used together to count the number of points in a sample space.

Example 1.9 A company has seven employees specializing in laboratory testing and five employees specializing in field testing. A job requires two employees from each area of specialization. In how many ways can the team of four be formed?

SOLUTION

$$\binom{7}{2} \times \binom{5}{2} = 210$$

1.3 PROBABILITY

1.3.1 Event Probabilities

The probability of an event A, denoted by $\mathrm{P}[A]$, is a number satisfying

$$0 \le \mathrm{P}[A] \le 1$$

Also, we assume that

$$\mathrm{P}[\emptyset] = 0, \qquad \mathrm{P}[S] = 1$$

Probabilities can sometimes be obtained using the counting rules discussed in the previous section. For example, if an experiment can result in any one of N different but equally likely outcomes, and if exactly m of these outcomes correspond to event A, then the probability of event A is $\mathrm{P}[A] = m/N$.

Example 1.10 Sixty soil samples have been taken at a site, 5 of which were taken of a liquefiable soil. If 2 of the samples are selected at random from the 60 samples, what is the probability that neither sample will be of the liquefiable soil?

SOLUTION We could solve this by looking at the number of ways of selecting the 2 samples from the 55 nonliquefiable soil and dividing by the total number of ways of selecting the 2 samples,

$$\mathrm{P}\left[0 \textit{ liquefiable}\right] = \frac{\binom{55}{2}}{\binom{60}{2}} = \frac{99}{118}$$

Alternatively, we could solve this by considering the probability of selecting the "first" sample from the 55 nonliquefiable samples and of selecting the second sample from the remaining 54 nonliquefiable samples,

$$\mathrm{P}\left[0 \textit{ liquefiable}\right] = \frac{55}{60} \times \frac{54}{59} = \frac{99}{118}$$

Note, however, that we have introduced an "ordering" in the second solution that was not asked for in the original question. This ordering needs to be carefully taken account of if we were to ask about the probability of having one of the samples being of a liquefiable soil. See the next example.

Example 1.11 Sixty soil samples have been taken at a site, 5 of which were taken of a liquefiable soil. If 2 of the samples are selected at random from the 60 samples, what is the probability that exactly 1 sample will be of the liquefiable soil?

SOLUTION We could solve this by looking at the number of ways of selecting one sample from the 5 liquefiable samples and 1 sample from the 55 nonliquefiable samples and dividing by the total number of ways of selecting the two samples:

$$\mathrm{P}\left[1 \textit{ liquefiable}\right] = \frac{\binom{5}{1}\binom{55}{1}}{\binom{60}{2}} = 2\left(\frac{5}{60}\right)\left(\frac{55}{59}\right) = \frac{55}{354}$$

We could also solve it by considering the probability of selecting the first sample from the 5 liquefiable samples and the second from the 55 nonliquefiable samples. However, since the question is only looking for the probability of one of the samples being liquefiable, we need to add in the probability that the first sample is nonliquefiable and the second is liquefiable:

$$\begin{aligned}\mathrm{P}\left[1 \textit{ liquefiable}\right] &= \frac{5}{60} \times \frac{55}{59} + \frac{55}{60} \times \frac{5}{59} \\ &= 2\left(\frac{5}{60}\right)\left(\frac{55}{59}\right) = \frac{55}{354}\end{aligned}$$

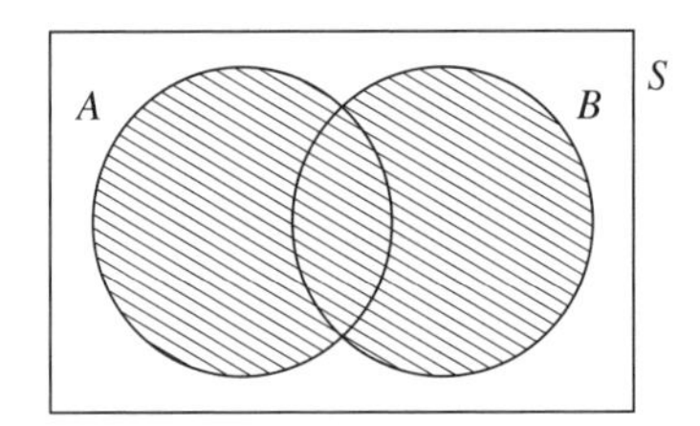

Figure 1.3 Venn diagram illustrating the union $A \cup B$.

1.3.2 Additive Rules

Often we must compute the probability of some event which is expressed in terms of other events. For example, if A is the event that company A requests your services and B is the event that company B requests your services, then the event that at least one of the two companies requests your services is $A \cup B$. The probability of this is given by the following relationship:

If A and B are any two events, then

$$\mathrm{P}[A \cup B] = \mathrm{P}[A] + \mathrm{P}[B] - \mathrm{P}[A \cap B] \tag{1.5}$$

This relationship can be illustrated by the Venn diagram in Figure 1.3. The desired quantity, $\mathrm{P}[A \cup B]$, is the area of $A \cup B$ which is shaded. If the shaded area is computed as the sum of the area of A, $\mathrm{P}[A]$, plus the area of B, $\mathrm{P}[B]$, then the intersection area, $\mathrm{P}[A \cap B]$, has been added twice. It must then be removed once to obtain the correct probability. Also,

If A and B are mutually exclusive, that is, are disjoint and so have no overlap, then

$$\mathrm{P}[A \cup B] = \mathrm{P}[A] + \mathrm{P}[B] \tag{1.6}$$

If $A_1, A_2, \ldots, A_n$ are mutually exclusive, then

$$\mathrm{P}[A_1 \cup \cdots \cup A_n] = \mathrm{P}[A_1] + \cdots + \mathrm{P}[A_n] \tag{1.7}$$

Definition We say that $A_1, A_2, \ldots, A_n$ is a *partition* of the sample space S if $A_1, A_2, \ldots, A_n$ are mutually exclusive and collectively exhaustive. Collectively exhaustive means that $A_1 \cup A_2 \cup \cdots\cdots \cup A_n = S$. If $A_1, A_2, \ldots, A_n$ is a partition of the sample space S, then

$$\mathrm{P}[A_1 \cup \cdots \cup A_n] = \mathrm{P}[A_1] + \cdots + \mathrm{P}[A_n] = \mathrm{P}[S] = 1 \tag{1.8}$$

The above ideas can be extended to the union of more than two events. For example:

For any three events A, B, and C, we have

$$\begin{aligned}\mathrm{P}[A \cup B \cup C] =& \mathrm{P}[A] + \mathrm{P}[B] + \mathrm{P}[C] - \mathrm{P}[A \cap B] \\ & - \mathrm{P}[A \cap C] - \mathrm{P}[B \cap C] \\ & + \mathrm{P}[A \cap B \cap C]\end{aligned} \tag{1.9}$$

This can be seen by drawing a Venn diagram and keeping track of the areas which must be added and removed in order to get $\mathrm{P}[A \cup B \cup C]$. Example 1.2 illustrates the union of three events.

For the complementary events A and A^c, $\mathrm{P}[A] + \mathrm{P}[A^c] = 1$. This is often used to compute $\mathrm{P}[A^c] = 1 - \mathrm{P}[A]$.

Example 1.12 A data-logging system contains two identical batteries, A and B. If one battery fails, the system will still operate. However, because of the added strain, the remaining battery is now more likely to fail than was originally the case. Suppose that the design life of a battery is three years. If at least one battery fails before the end of the battery design life in 7% of all systems and both batteries fail during that three-year period in only 1% of all systems, what is the probability that battery A will fail during the battery design life?

SOLUTION Let F_A be the event that battery A fails and F_B be the event that battery B fails. Then we are given that

$$\mathrm{P}[F_A \cup F_B] = 0.07, \qquad \mathrm{P}[F_A \cap F_B] = 0.01,$$
$$\mathrm{P}[F_A] = \mathrm{P}[F_B]$$

and we are looking for $\mathrm{P}[F_A]$. The Venn diagram in Figure 1.4 fills in the remaining probabilities. From this diagram, the following result is straightforward: $\mathrm{P}[F_A] = 0.03 + 0.01 = 0.04$.

Example 1.13 Based upon past evidence, it has been determined that in a particular region 15% of CPT soundings encounter soft clay layers, 12% encounter boulders, and 8% encounter both. If a sounding is selected at random:

1. What is the probability that it has encountered both a soft clay layer and a boulder?
2. What is the probability that it has encountered at least one of these two conditions?

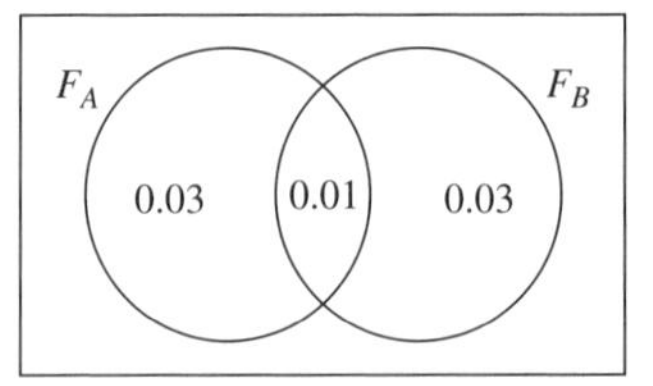

Figure 1.4 Venn diagram of battery failure events.

3. What is the probability that it has encountered neither of these two conditions?
4. What is the probability that it has *not* encountered a boulder?
5. What is the probability that it encounters a boulder but not a soft clay layer?

SOLUTION Let C be the event that the sounding encountered a soft clay layer. Let B be the event that the sounding encountered a boulder. We are given $\mathrm{P}[C] = 0.15$, $\mathrm{P}[B] = 0.12$, and $\mathrm{P}[C \cap B] = 0.08$, from which the Venn diagram in Figure 1.5 can be drawn:

1. $\mathrm{P}[C \cap B] = 0.08$
2. $\mathrm{P}[C \cup B] = \mathrm{P}[C] + \mathrm{P}[B] - \mathrm{P}[C \cap B] = 0.15 + 0.12 - 0.08 = 0.19$
3. $\mathrm{P}\left[C^c \cap B^c\right] = \mathrm{P}\left[(C \cup B)^c\right] = 1 - \mathrm{P}[C \cup B] = 1 - 0.19 = 0.81$
4. $\mathrm{P}[B^c] = 1 - \mathrm{P}[B] = 1 - 0.12 = 0.88$
5. $\mathrm{P}[B \cap C^c] = 0.04$ (see the Venn diagram)

1.4 CONDITIONAL PROBABILITY

The probability of an event is often affected by the occurrence of other events and/or the knowledge of information relevant to the event. Given two events, A and B, of an experiment, $\mathrm{P}[B \mid A]$ is called the conditional probability of B given that A has already occurred. It is defined by

$$\mathrm{P}[B \mid A] = \frac{\mathrm{P}[A \cap B]}{\mathrm{P}[A]} \tag{1.10}$$

That is, if we are given that event A has occurred, then A becomes our sample space. The probability that B has also occurred within this new sample space will be the ratio of the "area" of B within A to the "area" of A.

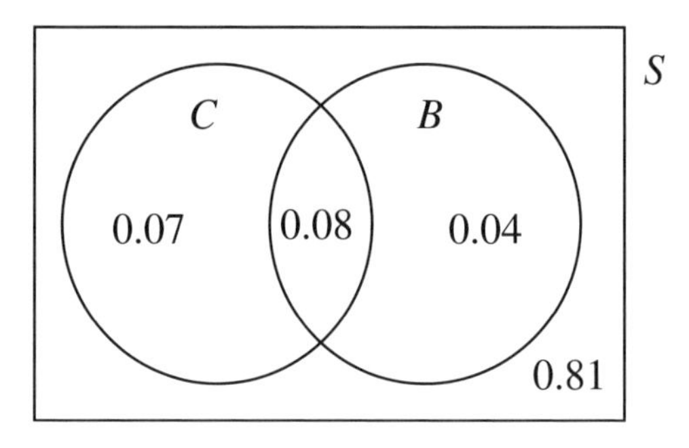

Figure 1.5 Venn diagram of CPT sounding events.

Example 1.14 Reconsidering Example 1.12, what is the probability that battery B will fail during the battery design life given that battery A has already failed?

SOLUTION We are told that F_A has occurred. This means that we are somewhere inside the F_A circle of Figure 1.4, which has "area" 0.04. We are asked to compute the conditional probability that F_B occurs given that F_A has occurred. This will be just the ratio of the area of F_B and F_A to the area of F_A,

$$\mathrm{P}[F_B|F_A] = \frac{\mathrm{P}[F_A \cap F_B]}{\mathrm{P}[F_A]} = \frac{0.01}{0.04} = 0.25$$

Example 1.15 A single soil sample is selected at random from a site. Three different toxic compounds, denoted A, B, and C, are known to occur in samples at this site with the following probabilities:

$$\mathrm{P}[A] = 0.01, \qquad \mathrm{P}[A \cap C] = 0.003,$$
$$\mathrm{P}[A \cap B] = 0.0025, \qquad \mathrm{P}[C] = 0.0075,$$
$$\mathrm{P}[A \cap B \cap C] = 0.001, \qquad \mathrm{P}[B \cap C] = 0.002,$$
$$\mathrm{P}[B] = 0.05$$

If both toxic compounds A and B occur in a soil sample, is the toxic compound C more likely to occur than if neither toxic compounds A nor B occur?

SOLUTION From the given information we can draw the Venn diagram in Figure 1.6.

We want to compare $\mathrm{P}[C|A \cap B]$ and $\mathrm{P}[C|A^c \cap B^c]$, where

$$\mathrm{P}[C \mid A \cap B] = \frac{\mathrm{P}[C \cap A \cap B]}{\mathrm{P}[A \cap B]} = \frac{0.001}{0.0025} = 0.4$$

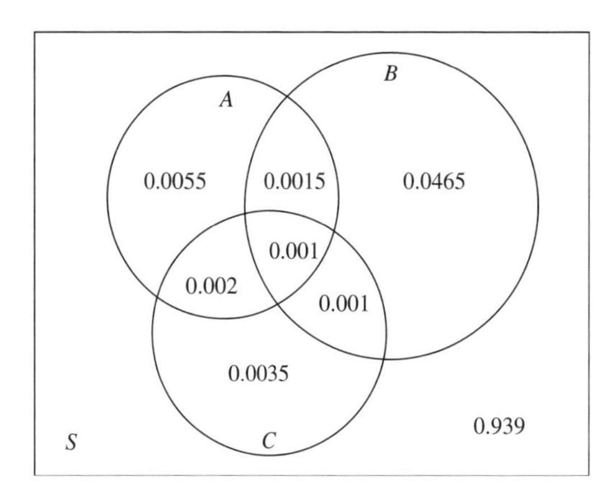

Figure 1.6 Venn diagram of toxic compound occurrence events.

$$P\left[C \mid A^c \cap B^c\right] = \frac{P[C \cap A^c \cap B^c]}{P[A^c \cap B^c]} = \frac{0.0035}{0.939 + 0.0035} = 0.0037$$

so the answer to the question is, yes, if both toxic compounds A and B occur in a soil sample, then toxic compound C is much more likely to also occur.

Sometimes we know $P[B \mid A]$ and wish to compute $P[A \cap B]$. If the events A and B can both occur, then

$$P[A \cap B] = P[B \mid A]\,P[A] \tag{1.11}$$

Example 1.16 A site is composed of 60% sand and 40% silt in separate layers and pockets. At this site, 10% of sand samples and 5% of silt samples are contaminated with trace amounts of arsenic. If a soil sample is selected at random, what is the probability that it is a sand sample and that it is contaminated with trace amounts of arsenic?

SOLUTION Let A be the event that the sample is sand. Let B be the event that the sample is silt. Let C be the event that the sample is contaminated with arsenic. Given $P[A] = 0.6$, $P[B] = 0.4$, $P[C \mid A] = 0.1$, and $P[C \mid B] = 0.05$. We want to find $P[A \cap C]$:

$$P[A \cap C] = P[A]\,P[C \mid A] = 0.6 \times 0.1 = 0.06$$

Two events A and B are independent *if and only if* $P[A \cap B] = P[A]\,P[B]$. This also implies that $P[A \mid B] = P[A]$, that is, if the two events are independent, then they do not affect the probability of the other occurring. Note that *independent events are not disjoint* and *disjoint events are not independent*! In fact, if two events are disjoint, then if one occurs, the other cannot have occurred. Thus, the occurrence of one of two disjoint events has a severe impact on the probability of occurrence of the other event (its probability of occurrence drops to zero).

If, in an experiment, the events $A_1, A_2, \ldots, A_k$ can all occur, then

$$\begin{aligned} P[A_1 \cap A_2 \cap \cdots \cap A_k] &= P[A_1]\,P[A_2 \mid A_1]\,P[A_3 \mid A_1 \cap A_2] \\ &\quad \cdots P\left[A_k \mid A_1 \cap \cdots \cap A_{k-1}\right] \\ &= P[A_k]\,P\left[A_{k-1} \mid A_k\right] \\ &\quad \cdots P[A_1 \mid A_k \cap \cdots \cap A_2] \end{aligned} \tag{1.12}$$

On the right-hand side, we could have any ordering of the A's. If the events $A_1, A_2, \ldots, A_k$ are independent, then this simplifies to

$$P[A_1 \cap A_2 \cap \cdots \cap A_k] = P[A_1]\,P[A_2] \cdots P[A_k] \tag{1.13}$$

Example 1.17 Four retaining walls, A, B, C, and D, are constructed independently. If their probabilities of sliding failure are estimated to be $P[A] = 0.01$, $P[B] = 0.008$, $P[C] = 0.005$, and $P[D] = 0.015$, what is the probability that none of them will fail by sliding?

SOLUTION Let A be the event that wall A will fail. Let B be the event that wall B will fail. Let C be the event that wall C will fail. Let D be the event that wall D will fail. Given $P[A] = 0.01$, $P[B] = 0.008$, $P[C] = 0.005$, $P[D] = 0.015$, and that the events A, B, C, and D are independent. We want to find $P[A^c \cap B^c \cap C^c \cap D^c]$:

$$\begin{aligned} P\left[A^c \cap B^c \cap C^c \cap D^c\right] &= P\left[A^c\right]P\left[B^c\right]P\left[C^c\right]P\left[D^c\right] \\ &\quad \text{(since } A, B, C, \text{ and } D \text{ are independent)} \\ &= (1 - P[A])(1 - P[B])(1 - P[C])(1 - P[D]) \\ &= (1 - 0.01)(1 - 0.008)(1 - 0.005)(1 - 0.015) \\ &= 0.9625 \end{aligned}$$

1.4.1 Total Probability

Sometimes we know the probability of an event in terms of the occurrence of other events and want to compute the *unconditional* probability of the event. For example, when we want to compute the *total* probability of failure of a bridge, we can start by computing a series of simpler problems such as:

1. Probability of bridge failure given a maximum static load
2. Probability of bridge failure given a maximum dynamic traffic load
3. Probability of bridge failure given an earthquake
4. Probability of bridge failure given a flood

The *total probability theorem* can be used to combine the above probabilities into the unconditional probability of bridge failure. We need to know the above conditional probabilities along with the probabilities that the "conditions" occur (e.g., the probability that the maximum static load will occur during the design life).

Example 1.18 A company manufactures cone penetration testing equipment. Of the piezocones they use, 50% are produced at plant A, 30% at plant B, and 20% at plant C. It is known that 1% of plant A's, 2% of plant B's, and 3% of plant C's output are defective. What is the probability that a piezocone chosen at random will be defective?

Setup

Let A be the event that the piezocone was produced at plant A. Let B be the event that the piezocone was produced at plant B. Let C be the event that the piezocone was produced at plant C. Let D be the event that the piezocone is defective. Given

$$\mathrm{P}[A] = 0.50, \qquad \mathrm{P}[D \mid A] = 0.01,$$
$$\mathrm{P}[B] = 0.30, \qquad \mathrm{P}[D \mid B] = 0.02,$$
$$\mathrm{P}[C] = 0.20, \qquad \mathrm{P}[D \mid C] = 0.03$$

We want to find $\mathrm{P}[D]$. There are at least two possible approaches.

Approach 1

A Venn diagram of the sample space is given in Figure 1.7. The information given in the problem does not allow the Venn diagram to be easily filled in. It is easy to see the event of interest, though, as it has been shaded in. Then

$$\begin{aligned}
\mathrm{P}[D] &= \mathrm{P}[(D \cap A) \cup (D \cap B) \cup (D \cap C)] \\
&= \mathrm{P}[D \cap A] + \mathrm{P}[D \cap B] + \mathrm{P}[D \cap C] \\
&\qquad \text{since } A \cap D,\ B \cap D, \text{ and } C \cap D \text{ are disjoint} \\
&= \mathrm{P}[D \mid A] \cdot \mathrm{P}[A] + \mathrm{P}[D \mid B] \cdot \mathrm{P}[B] \\
&\quad + \mathrm{P}[D \mid C] \cdot \mathrm{P}[C] \\
&= 0.01(0.5) + 0.02(0.3) + 0.03(0.2) \\
&= 0.017
\end{aligned}$$

Approach 2

Recall that when we only had probabilities like $\mathrm{P}[A]$, $\mathrm{P}[B]$, ..., that is, no conditional probabilities, we found it helpful to represent the probabilities in a Venn diagram. Unfortunately, there is no easy representation of the conditional probabilities in a Venn diagram: (In fact, conditional probabilities are ratios of probabilities that appear in the Venn diagram.) Conditional probabilities find a more natural home on *event trees*. Event trees must be constructed carefully and adhere to certain rules if they are going to be useful in calculations. Event trees consist of nodes and branches. There is a starting node from which two or more branches leave. At the end of each of these branches there is another node from which more branches may leave (and go to separate nodes). The idea is repeated from the newer nodes as often as required to completely depict all possibilities. A probability is associated with each branch and, for all branches except those leaving the starting node, the probabilities are conditional probabilities. Thus, the event tree is composed largely of conditional probabilities.

There is one other rule that event trees must obey: *Branches leaving any node must form a partition of the sample space.* That is, the events associated with each branch must be disjoint—you cannot be on more than one branch at a time—and must include all possibilities. The sum of probabilities of all branches leaving a node must be 1.0. Also keep in mind that an event tree will only be useful if all the branches can be filled with probabilities.

The event tree for this example is constructed as follows. The piezocone must *first be made at one of the three plants*, then *depending on where it was made*, it could be *defective* or not. The event tree for this problem is thus as given in Figure 1.8. Note that there are six "paths" on the tree. When a piezocone is selected at random, exactly one of these paths will have been followed—we will be on one of the branches. Recall that interest is in finding $\mathrm{P}[D]$. The event D will have occurred if either the first, third, or fifth path was followed. That is, the probability that the first, third, or fifth path was followed is sought. If the first path is followed, then the event $A \cap D$ has occurred. This has probability found by multiplying the probabilities along the path,

$$\mathrm{P}[A \cap D] = \mathrm{P}[D \mid A] \cdot \mathrm{P}[A] = 0.01(0.5) = 0.005$$

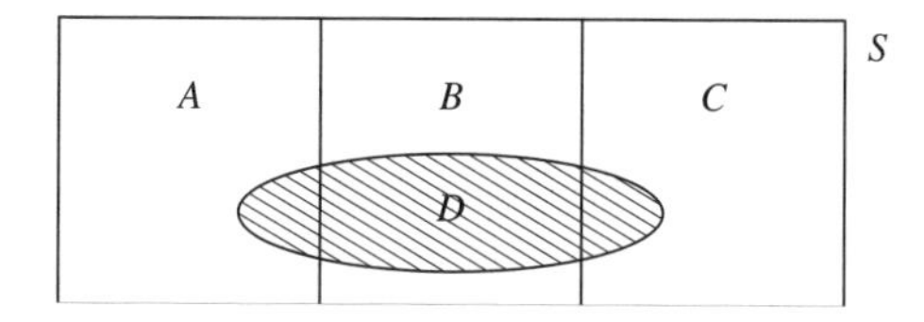

Figure 1.7 Venn diagram of piezocone events.

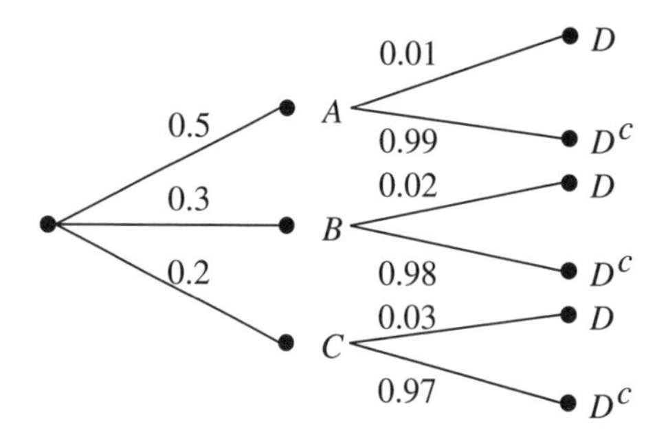

Figure 1.8 Event tree for piezocone events.

Looking back at the calculation performed in Approach 1, $P[D]$ was computed as

$$\begin{aligned} P[D] &= P[D \mid A] \cdot P[A] + P[D \mid B] \cdot P[B] \\ &\quad + P[D \mid C] \cdot P[C] \\ &= 0.01(0.5) + 0.02(0.3) + 0.03(0.2) \\ &= 0.017 \end{aligned}$$

which, in terms of the event tree, is just the sum of all the paths that lead to the outcome that you desire, D. Event trees make "total probability" problems much simpler. They give a "picture" of what is going on and allow the computation of some of the desired probabilities directly.

The above is an application of the total probability theorem, which is stated generally as follows:

Total Probability Theorem If the events $B_1, B_2, \ldots, B_k$ constitute a partition of the sample space S (i.e., are disjoint and collectively exhaustive), then for any event A in S

$$P[A] = \sum_{i=1}^{k} P[B_i \cap A] = \sum_{i=1}^{k} P[A \mid B_i] P[B_i] \tag{1.14}$$

1.4.2 Bayes' Theorem

Sometimes we want to improve an estimate of a probability in light of additional information. Bayes' theorem allows us to do this. It arises from the observation that $P[A \cap B]$ can be written in two ways:

$$\begin{aligned} P[A \cap B] &= P[A \mid B] \cdot P[B] \\ &= P[B \mid A] \cdot P[A] \end{aligned} \tag{1.15}$$

which implies that $P[B \mid A] \cdot P[A] = P[A \mid B] \cdot P[B]$, or

$$P[B \mid A] = \frac{P[A \mid B] \cdot P[B]}{P[A]} \tag{1.16}$$

Example 1.19 Return to the manufacturer of piezocones from above (Example 1.18). If a piezocone is selected at random and found to be defective, what is the probability that it came from plant A?

Setup
Same as before, except now the probability of interest is $P[A \mid D]$. Again, there are *two possible approaches.*

Approach 1
The relationship

$$P[A \mid D] = \frac{P[A \cap D]}{P[D]}$$

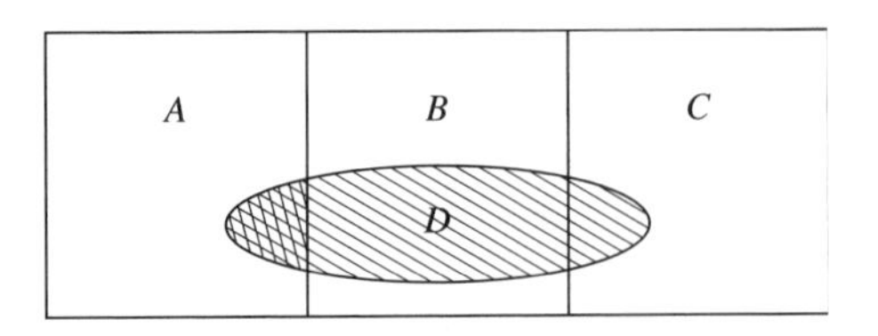

Figure 1.9 Venn diagram of conditional piezocone events.

can be seen as a ratio of areas in the Venn diagram in Figure 1.9, from which $P[A \mid D]$ can be computed as follows:

$$\begin{aligned} P[A \mid D] &= \frac{P[A \cap D]}{P[D]} \\ &= \frac{P[A \cap D]}{P[(A \cap D) \cup (B \cap D) \cup (C \cap D)]} \\ &= \frac{P[A \cap D]}{P[A \cap D] + P[B \cap D] + P[C \cap D]} \\ &\quad \text{since } A \cap D,\ B \cap D, \text{ and } C \cap D \text{ are disjoint} \\ &= \frac{P[D \mid A] P[A]}{P[D \mid A] P[A] + P[D \mid B] P[B] + P[D \mid C] P[C]} \\ &= \frac{0.01(0.5)}{(0.01)(0.5) + 0.02(0.3) + 0.03(0.2)} = \frac{0.005}{0.017} \\ &= 0.294 \end{aligned}$$

Note that the denominator had already been calculated in the previous question; however the computations have been reproduced here for illustrative purposes.

Approach 2
The probability $P[A \mid D]$ can be easily computed from the event tree. We are looking for the probability that A has occurred given that D has occurred. In terms of the paths on the tree, we know that (since D has occurred) one of the first, third, or fifth path has been taken. We want the probability that the first path was taken out of the three possible paths. Thus, we must compute the relative probability of taking path 1 out of the three paths:

$$\begin{aligned} P[A \mid D] &= \frac{P[D \mid A] P[A]}{P[D \mid A] P[A] + P[D \mid B] P[B] + P[D \mid C] P[C]} \\ &= \frac{0.01(0.5)}{(0.01)(0.5) + 0.02(0.3) + 0.03(0.2)} = \frac{0.005}{0.017} \\ &= 0.294 \end{aligned}$$

Event trees provide a simple graphical approach to solving problems involving conditional probabilities.

The above is an application of Bayes' Theorem, which is stated formally as follows.

Bayes' Theorem If the events $B_1, B_2, \ldots, B_k$ constitute a partition of the sample space S (i.e., are disjoint and collectively exhaustive), then for any event A of S such that $\mathrm{P}[A] \neq 0$

$$\mathrm{P}\left[B_j \mid A\right] = \frac{\mathrm{P}\left[B_j \cap A\right]}{\sum_{i=1}^{k} \mathrm{P}[B_i \cap A]} = \frac{\mathrm{P}\left[A \mid B_j\right]\mathrm{P}\left[B_j\right]}{\sum_{i=1}^{k} \mathrm{P}[A \mid B_i]\,\mathrm{P}[B_i]} = \frac{\mathrm{P}\left[A \mid B_j\right]\mathrm{P}\left[B_j\right]}{\mathrm{P}[A]} \tag{1.17}$$

for any $j = 1, 2, \ldots, k$.

Bayes' theorem is useful for revising or updating probabilities as more data and information become available. In the previous example on piezocones, there was an *initial* probability that a piezocone would have been manufactured at plant A: $\mathrm{P}[A] = 0.5$. This probability is referred to as the *prior* probability of A. That is, in the absence of any other information, a piezocone chosen at random has a probability of having been manufactured at plant A of 0.5. However, if a piezocone chosen at random is found to be defective (so that there is now more information on the piezocone), then the probability that it was manufactured at plant A reduces from 0.5 to 0.294. This latter probability is referred to as the *posterior* probability of A. *Bayesian* updating of probabilities is a very powerful tool in engineering reliability-based design.

For problems involving conditional probabilities, event trees are usually the easiest way to proceed. However, event trees are not always easy to draw, and the purely mathematical approach is sometimes necessary. As an example of a tree which is not quite straightforward, see if you can draw the event tree and answer the questions in the following exercise. Remember that you must set up the tree in such a way that you can fill in most of the probabilities on the branches. If you are left with too many empty branches and no other given information, you are likely to have confused the order of the events; try reorganizing your tree.

Exercise When contracting out a site investigation, an engineer will check companies A, B, and C in that sequence and will hire the first company which is available to do the work. From past experience, the engineer knows that the probability that company A will be available is 0.2. However, if company A is not available, then the probability that company B will be available is only 0.04. If neither company A nor B is available, then the probability that company C will be available is 0.4. If none of the companies are available, the engineer is forced to delay the investigation to a later time.

(a) What is the probability that one of the companies A or B will be available?
(b) What is the probability that the site investigation will take place on time?
(c) If the site investigation takes place on time, what is the probability that it was not investigated by company C?

Example 1.20 At a particular site, experience has shown that piles have a 20% probability of encountering a soft clay layer. Of those which encounter this clay layer, 60% fail a static load test. Of the piles which do not encounter the clay layer, only 10% fail a static load test.

1. What is the probability that a pile selected at random will fail a static load test?
2. Supposing that a pile has failed a static load test, what is the updated probability that it encountered the soft clay layer?

SOLUTION For a pile, let C be the event that a soft clay layer was encountered and let F be the event that the static load test was failed. We are given $\mathrm{P}[C] = 0.2$, $\mathrm{P}[F \mid C] = 0.6$, and $\mathrm{P}[F \mid C^c] = 0.1$.

1. We have the event tree in Figure 1.10 and thus $\mathrm{P}[F] = 0.2(0.6) + 0.8(0.1) = 0.2$.
2. From the above tree, we have

$$\mathrm{P}[C \mid F] = \frac{0.2 \times 0.6}{0.2} = 0.6$$

1.4.3 Problem-Solving Methodology

Solving real-life problems (i.e., "word problems") is not always easy. It is often not perfectly clear what is meant by a worded question. Two things improve one's chances of successfully solving problems which are expressed using words: (a) a systematic approach, and (b) *practice*. It is practice that allows you to identify those aspects of the question that need further clarification, if any. Below, a few basic recommendations are outlined.

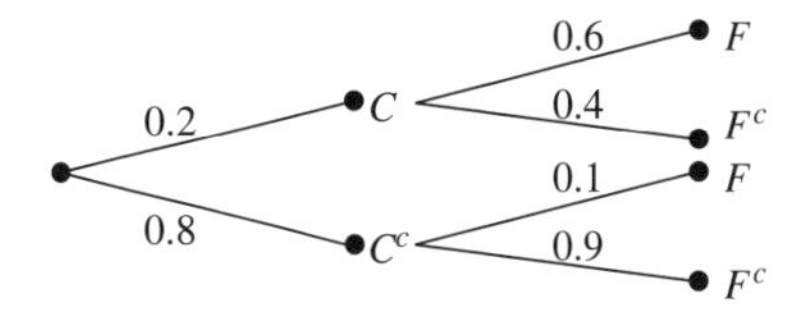

Figure 1.10 Event tree for pile encounter events.

1. Solving a word problem generally involves the computation of some quantity. Clearly identify this quantity at the beginning of the problem solution. Before starting any computations, it is good practice to write out your concluding sentence first. This forces you to concentrate on the essentials.
2. In any problem involving the probability of events, you should:

(a) *Clearly define your events.* Use the following guidelines:

   (i) Keep events as simple as possible.

   (ii) if your event definition includes the words *and, or, given, if, when,* and so on, then *it is NOT a good event definition.* Break your event into two (or more, if required) events and use the $\cap$, $\cup$, or $|$ operators to express what you had originally intended. The complement is also a helpful operator, see (iii).

   (iii) You do not need to define separate events for, for example, "an accident occurs" and "an accident does not occur". In fact, this will often lead to confusion. Simply define A to be one of the events and use A^c when you want to refer to the other. This may also give you some hints as to how to proceed since you know that $\mathrm{P}[A^c] = 1 - \mathrm{P}[A]$.

(b) Once your events are defined, you need to go through the worded problem to extract the given numerical information. Write this information down in the form of probabilities of the events that you defined above. For example, $\mathrm{P}[A] = 0.23$, $\mathrm{P}[B \mid A] = 0.6$, and so on. Note that the conditional probabilities, are often difficult to unravel. For example, the following phrases all translate into a probability statement of the form $\mathrm{P}[A \mid B]$:

If ... occurs, the probability of ... doubles....
In the event that ... occurs, the probability of ... becomes 0.6.
When ... occurs, the probability of ... becomes 0.43.
Given that ... occurs, the probability of ... is 0.3.

In this case, you will likely be using one of the conditional probability relationship ($\mathrm{P}[A \cap B] = \mathrm{P}[B \mid A]\,\mathrm{P}[A]$), the total probability theorem, or Bayes' Theorem.

(c) Now review the worded problem again and write down the probability that the question is asking for in terms of the events defined above. Although the question may be in worded form, you should be writing down something like $\mathrm{P}[A \cap B]$ or $\mathrm{P}[B \mid A]$. Make sure that you can express the desired probability in terms of the events you defined above. If you cannot, then you need to revise your original event definitions.

(d) Finally, use the rules of combining probabilities (e.g., probabilities of unions or intersections, Bayes' Theorem) to compute the desired probability.

1.5 RANDOM VARIABLES AND PROBABILITY DISTRIBUTIONS

Although probability theory is based on the idea of events and associated set theory, it becomes very unwieldy to treat random events like "time to failure" using explicit event definitions. One would conceivably have to define a separate event for each possible time of failure and so would soon run out of symbols for the various events. For this reason, and also because they allow the use of a wealth of mathematical tools, random variables are used to represent a suite of possible events. In addition, since most engineering problems are expressed in terms of numerical quantities, random variables are particularly appropriate.

Definition Consider a sample space S consisting of a set of outcomes $\{s_1, s_2, \ldots\}$. If X is a function that assigns a real number $X(s)$ to every outcome $s \in S$, then X is a *random variable*. Random variables will be denoted with uppercase letters.

Now what does this mean in plain English? Essentially a random variable is a means of identifying events in numerical terms. For example, if the outcome s_1 means that an apple was selected and s_2 means that an orange was selected, then $X(s_1)$ could be set equal to 1 and $X(s_2)$ could be set equal to 0. Then $X > 0$ means that an apple was selected. Now mathematics can be used on X, that is, if the fruit-picking experiment is repeated n times and $x_1 = X_1(s)$ is the outcome of the first experiment, $x_2 = X_2(s)$ the outcome of the second, and so on, then the total number of apples picked is $\sum_{i=1}^{n} x_i$. Note that mathematics could not be used on the actual outcomes themselves; for example, picking an apple is a real event which knows nothing about mathematics nor can it be used in a mathematical expression without first mapping the event to a number.

For each outcome s, there is exactly one value of $x = X(s)$, but different values of s may lead to the same x. We will see examples of this shortly.

The above discussion illustrates in a rather simple way one of the primary motivations for the use of random variables—simply so that mathematics can be used. One other thing might be noticed in the previous paragraph. After the "experiment" has taken place and the outcome is known, it is referred to using lowercase, x_i. That is x_i has a known fixed value while X is unknown. In other words

x is a realization of the random variable X. This is a rather subtle distinction, but it is important to remember that X is unknown. The most that we can say about X is to specify what its likelihoods of taking on certain values are—we cannot say exactly what the value of X is.

Example 1.21 Two piles are to be randomly selected for testing from a group of 60 piles. Five of the piles are 0.5 m in diameter, the rest are 0.3 m in diameter. If X is the number of 0.5-m-diameter piles selected for testing, then X is a random variable that assigns a number to each outcome in the sample space according to:

Sample Space	X
NN	0
NL	1
LN	1
LL	2

The sample space is made up of pairs of possible outcomes, where N represents a "normal" diameter pile (0.3 m) and L represents a large -diameter pile (0.5 m). For example, *LN* means that the first pile selected was large and the second pile selected was normal. Notice that the outcomes $\{NL\}$ and $\{LN\}$ both lead to $X = 1$.

Sample spaces corresponding to random variables may be discrete or continuous:

Discrete: A random variable is called a discrete random variable if its set of possible outcomes is countable. This usually occurs for any random variable which is a count of occurrences or of items, for example, the number of large-diameter piles selected in the previous example.

Continuous: A random variable is called a continuous random variable if it can take on values on a continuous scale. This is usually the case with measured data, such as cohesion.

Example 1.22 A few examples:

1. Let X be the number of blows in a standard penetration test—X is discrete.
2. Let Y be the number of piles driven in one day—Y is discrete.
3. Let Z be the time until consolidation settlement exceeds some threshold—Z is continuous.
4. Let W be the number of grains of sand involved in a sand cone test—W is discrete but is often approximated as continuous, particularly since W can be very large.

1.5.1 Discrete Random Variables

Discrete random variables are those that take on only discrete values $\{x_1, x_2, \ldots\}$, that is, have a countable number of outcomes. Note that countable just means that the outcomes can be numbered $1, 2, \ldots$, however there could still be an infinite number of them. For example, our experiment might be to count the number of soil tests performed before one yields a cohesion of 200 MPa. This is a discrete random variable since the outcome is one of $0, 1, \ldots$, but the number may be very large or even (in concept) infinite (implying that a soil sample with cohesion 200 MPa was never found).

Discrete Probability Distributions As mentioned previously, we can never know for certain what the value of a random variable is (if we do measure it, it becomes a *realization*—presumably the next measurement is again uncertain until it is measured, and so on). The most that we can say about a random variable is what its probability is of assuming each of its possible values. The set of probabilities assigned to each possible value of X is called a *probability distribution*. The sum of these probabilities over all possible values must be 1.0.

Definition The set of ordered pairs $(x, f_X(x))$ is the probability distribution of the discrete random variable X if, for each possible outcome x,

1. $0 \le f_X(x) \le 1$
2. $\sum_{\text{all } x} f_X(x) = 1$
3. $\mathrm{P}[X = x] = f_X(x)$

Here, $f_X(x)$ is called the *probability mass function* of X. The subscript is used to indicate what random variable is being governed by the distribution. We shall see when we consider continuous random variables why we call this a probability "mass" function.

Example 1.23 Recall Example 1.21. We can compute the probability mass function of the number of large piles selected by using the counting rules of Section 1.2.3. Specifically,

$$f_X(0) = \mathrm{P}[X = 0] = \frac{\binom{5}{0}\binom{55}{2}}{\binom{60}{2}} = 0.8390$$

$$f_X(1) = \mathrm{P}[X = 1] = \frac{\binom{5}{1}\binom{55}{1}}{\binom{60}{2}} = 0.1554$$

$$f_X(2) = \mathrm{P}[X = 2] = \frac{\binom{5}{2}\binom{55}{0}}{\binom{60}{2}} = 0.0056$$

and thus the probability mass function of the random variable X is

x	$f_X(x)$
0	0.8390
1	0.1554
2	0.0056

Discrete Cumulative Distributions An *equivalent* description of a random variable is the cumulative distribution function (cdf), which is defined as follows:

Definition The cumulative distribution function $F_X(x)$ of a discrete random variable X with probability mass function $f_X(x)$ is defined by

$$F_X(x) = \mathrm{P}[X \le x] = \sum_{t \le x} f_X(t) \tag{1.18}$$

We say that this is equivalent to the probability mass function because one can be obtained from the other,

$$f_X(x_i) = F_X(x_i) - F_X(x_{i-1}) \tag{1.19}$$

Example 1.24 In the case of an experiment involving tossing a fair coin three times we can count the number of heads which appear and assign that to the random variable X. The random variable X can assume four values 0, 1, 2, and 3 with probabilities $\frac{1}{8}$, $\frac{3}{8}$, $\frac{3}{8}$, and $\frac{1}{8}$ (do you know how these probabilities were computed?). Thus, $F_X(x)$ is defined as

$$F_X(x) = \begin{cases} 0 & \text{if } x < 0 \\ \frac{1}{8} & \text{if } 0 \le x < 1 \\ \frac{4}{8} & \text{if } 1 \le x < 2 \\ \frac{7}{8} & \text{if } 2 \le x < 3 \\ 1 & \text{if } 3 \le x \end{cases}$$

and a graph of $F_X(x)$ appears in Figure 1.11. The values of $F_X(x)$ at $x = 0, 1, \ldots$ are shown by the closed circles.

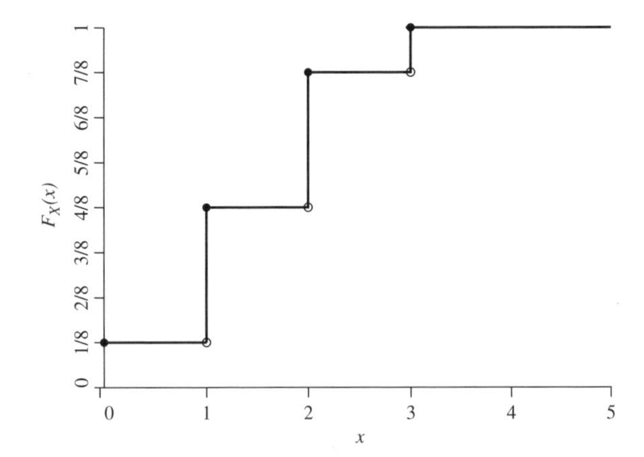

Figure 1.11 Cumulative distribution function for the three-coin toss.

Discrete probability mass functions are often represented using a bar plot, where the height of each bar is equal to the probability that the random variable takes that value. For example, the bar plot of the pile problem (Examples 1.21 and 1.23) would appear as in Figure 1.12.

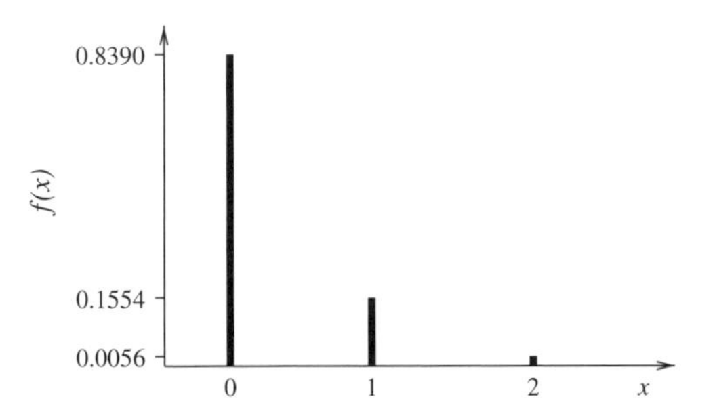

Figure 1.12 Bar plot of $f_X(x)$ for number of large piles selected, X.

1.5.2 Continuous Random Variables

Continuous random variables can take on an infinite number of possible outcomes—generally X takes values from the real line $\Re$. To illustrate the changes involved when we go from the discrete to the continuous case, consider the probability that a grain silo experiences a bearing capacity failure at *exactly* 4.3673458212... years from when it is installed. Clearly the probability that it fails at exactly that instant in time is essentially zero. In general the probability that it fails at any one instant in time is vanishingly small. In order to characterize probabilities for continuous random variables, we cannot use probabilities directly (since they are all essentially zero); we must use *relative likelihoods*. That is, we say that the probability that X lies in the small interval between x and $x + dx$ is $f_X(x)\,dx$, or

$$\mathrm{P}[x < X \le x + dx] = f_X(x)\,dx \tag{1.20}$$

where $f_X(x)$ is now called the *probability density function* (pdf) of the random variable X. The word *density* is used because "density" must be multiplied by a length measure in order to get a "mass." Note that the above probability is vanishingly small because dx is vanishingly small. The function $f_X(x)$ is now the relative likelihood that X lies in a very small interval near x. Roughly speaking, we can think of this as $\mathrm{P}[X = x] = f_X(x)\,dx$.

Continuous Probability Distributions

Definition The function $f_X(x)$ is a probability density function for the continuous random variable X defined over the set of real numbers if

1. $0 \le f_X(x) < \infty$ for all $-\infty < x < +\infty$,
2. $\int_{-\infty}^{\infty} f_X(x)\,dx = 1$ (i.e., the area under the pdf is 1.0), and
3. $P[a < X < b] = \int_a^b f_X(x)\,dx$ (i.e., the area under $f_X(x)$ between a and b).

Note: it is important to recognize that, in the continuous case, $f_X(x)$ is no longer a probability. It has units of probability per unit length. In order to get probabilities, we have to find *areas* under the pdf, that is, sum values of $f_X(x)\,dx$.

Example 1.25 Suppose that the time to failure, T in years, of a clay barrier has the probability density function

$$f_T(t) = \begin{cases} 0.02e^{-0.02t} & \text{if } t \ge 0 \\ 0 & \text{otherwise} \end{cases}$$

This is called an *exponential distribution* and distributions of this exponentially decaying form have been found to well represent many lifetime-type problems. What is the probability that T will exceed 100 years?

SOLUTION The distribution is shown in Figure 1.13. If we consider the more general case where

$$f_T(t) = \begin{cases} \lambda e^{-\lambda t} & \text{if } t \ge 0 \\ 0 & \text{otherwise} \end{cases}$$

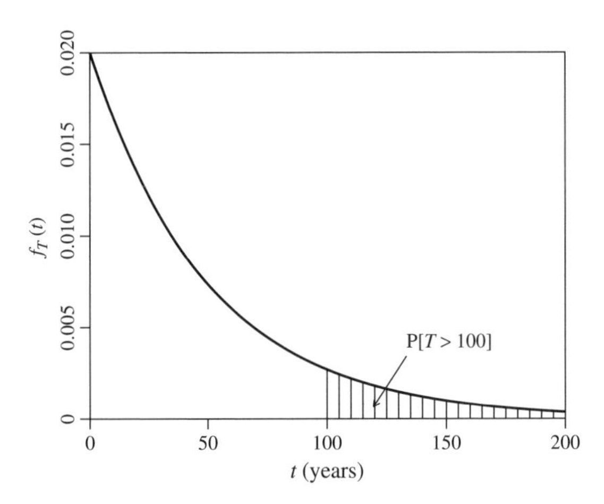

Figure 1.13 Exponential distribution illustrating $P[T > 100]$.

then we get

$$\begin{aligned} P[T > 100] = P[100 < T < \infty] &= \int_{100}^{\infty} \lambda e^{-\lambda t}\,dt \\ &= -e^{-\lambda t}\Big|_{100}^{\infty} = -e^{-\infty\lambda} + e^{-100\lambda} \\ &= e^{-100\lambda} \end{aligned}$$

For $\lambda = 0.02$, as is the case in this problem,

$$P[T > 100] = e^{-100\times 0.02} = e^{-2} = 0.1353$$

Continuous Cumulative Distribution The *cumulative distribution function* (cdf) for a continuous random variable is basically defined in the same way as it is for a discrete distribution (Figure 1.14).

Definition The cumulative distribution function $F_X(x)$ of a continuous random variable X having probability density function $f_X(x)$ is defined by the area under the density function to the left of x:

$$F_X(x) = P[X \le x] = \int_{-\infty}^{x} f_X(t)\,dt \tag{1.21}$$

As in the discrete case, the cdf is equivalent to the pdf in that one can be obtained from the other. It is simply another way of expressing the probabilities associated with a random variable. Since the cdf is an integral of the pdf, the pdf can be obtained from the cdf as a derivative:

$$f_X(x) = \frac{dF_X(x)}{dx} \tag{1.22}$$

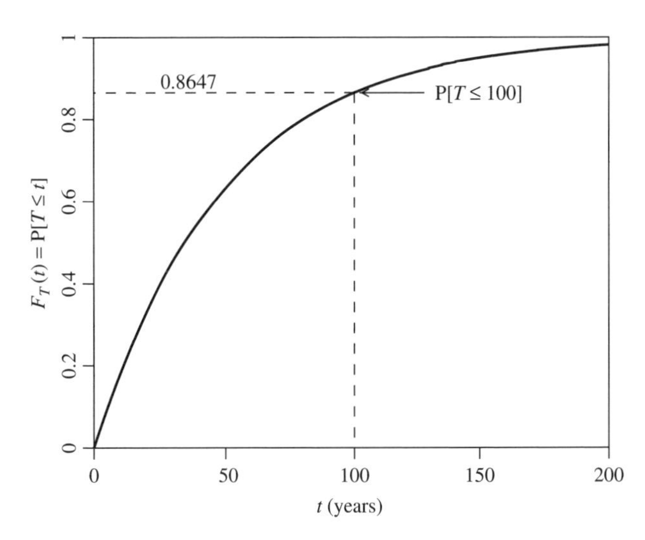

Figure 1.14 Cumulative distribution function for the exponential distribution.

Example 1.26 Note that we could also have used the cumulative distribution in Example 1.25. The cumulative distribution function of the exponential distribution is

$$F_T(t) = \mathrm{P}[T \le t] = \int_0^t \lambda e^{-\lambda t}\, dt = 1 - e^{-\lambda t}$$

and thus

$$\begin{aligned} \mathrm{P}[T > 100] &= 1 - \mathrm{P}[T \le 100] = 1 - F_T(100) \\ &= 1 - (1 - e^{-100\lambda}) = e^{-100\lambda} \end{aligned}$$

1.6 MEASURES OF CENTRAL TENDENCY, VARIABILITY, AND ASSOCIATION

A random variable is completely described, as well as can be, if its probability distribution is specified. However, we will never know the precise distribution of any natural phenomenon. Nature cares not at all about our mathematical models and the "truth" is usually far more complex than we are able to represent. So we very often have to describe a random variable using less complete but more easily estimated measures. The most important of these measures are *central tendency* and *variability*. Even if the complete probability distribution is known, these quantities remain useful because they convey information about the properties of the random variable that are of first importance in practical applications. Also, the parameters of the distribution are often derived as functions of these quantities or they may be the parameters themselves.

The most common measures of central tendency and variability are the *mean* and the *variance*, respectively. In engineering, the variability of a random quantity is often expressed using the dimensionless *coefficient of variation*, which is the ratio of the *standard deviation* over the mean. Also, when one has two random variables X and Y, it is frequently of interest to measure how strongly they are related (or associated) to one another. A typical measure of the strength of the relationship between two random variables is their *covariance*. As we shall see, covariance depends on the units of the random variables involved and their individual variabilities, and so a more intuitive measure of the strength of the relationship between two random variables is the *correlation coefficient*, which is both dimensionless and bounded. All of these characteristics will be covered in this section.

1.6.1 Mean

The *mean* is the most important characteristic of a random variable, in that it tells us about its central tendency. It is defined mathematically as follows:

Definition Let X be a random variable with probability density function $f(x)$. The mean, or *expected value*, of X, denoted μ_X, is defined by

$$\mu_X = \begin{cases} \mathrm{E}[X] = \sum_x x f(x) & \text{if } X \text{ is discrete} \quad (1.23\text{a}) \\ \mathrm{E}[X] = \int_{-\infty}^{\infty} x f(x)\, dx & \text{if } X \text{ is continuous} \quad (1.23\text{b}) \end{cases}$$

where the subscript on μ, when present, denotes what μ is the mean of.

Example 1.27 Let X be a discrete random variable which takes on the values listed in the table below with associated probabilities:

x	-2	-1	0	1	2
$f(x)$	$\frac{1}{12}$	$\frac{1}{6}$	k	$\frac{1}{3}$	$\frac{1}{4}$

1. Find the constant k such that $f_X(x)$ is a legitimate probability mass function for the random variable X.
2. Find the mean (expected value) of X.

SOLUTION

1. We know that the sum of all possible probabilities must be 1, so that $k = 1 - (\frac{1}{12} + \frac{1}{6} + \frac{1}{3} + \frac{1}{4}) = \frac{1}{6}$.
2. $\mathrm{E}[X] = (-2)(\frac{1}{12}) + (-1)(\frac{1}{6}) + 0(\frac{1}{6}) + 1(\frac{1}{3}) + 2(\frac{1}{4}) = \frac{1}{2}$.

Expectation The notation $\mathrm{E}[X]$ refers to a mathematical operation called *expectation*. The expectation of any random variable is a sum of all possible values of the random variable weighted by the probability of each value occurring. For example, if X is a random variable with probability (mass or density) function $f_X(x)$, then the expected value of the random variable $g(X)$, where g is any function of X, is

$$\mu_{g(X)} = \begin{cases} \mathrm{E}\left[g(X)\right] = \sum_x g(x) f_X(x) & \text{if } X \text{ is discrete} \\ \mathrm{E}\left[g(X)\right] = \int_{-\infty}^{\infty} g(x) f_X(x)\, dx & \text{if } X \text{ is continuous} \end{cases} \quad (1.24)$$

Example 1.28 A researcher is looking at fibers as a means of reinforcing soil. The fibers being investigated are nominally of radius 10 μm. However, they actually have random radius R with probability density function (in units of micrometers)

$$f_R(r) = \begin{cases} \frac{3}{4}\left[1 - (10-r)^2\right] & \text{if } 9 \le r \le 11 \\ 0 & \text{otherwise} \end{cases}$$

What is the expected area of a reinforcing fiber?

SOLUTION The area of a circle of radius R is πR^2. Thus,

$$\begin{aligned} \mathrm{E}\left[\pi R^2\right] &= \pi \mathrm{E}\left[R^2\right] = \pi \int_9^{11} r^2 \frac{3}{4}\left[1-(10-r)^2\right] dr \\ &= \frac{3}{4}\pi \int_9^{11} \left[-99r^2 + 20r^3 - r^4\right] dr \\ &= \frac{3}{4}\pi \left[-33r^3 + 5r^4 - \frac{r^5}{5}\right]_9^{11} \\ &= \frac{3}{4}\pi\left(\frac{668}{5}\right) = \frac{501}{5}\pi \\ &= 314.8 \quad \mu\text{m}^2 \end{aligned}$$

If we have a sample of observations $x_1, x_2, \ldots, x_n$ of some population X, then the population mean μ_X is estimated by the *sample mean* $\bar{x}$, defined as

$$\bar{x} = \frac{1}{n}\sum_{i=1}^{n} x_i \tag{1.25}$$

Example 1.29 Suppose $\mathbf{x} = \{x_1, x_2, \ldots, x_n\} = \{1, 3, 5, 7, 9\}$.

(a) What is $\bar{x}$?
(b) What happens to $\bar{x}$ if $\mathbf{x} = \{1, 3, 5, 7, 79\}$?

SOLUTION In both cases, the sample size is $n = 5$.

(a) $\bar{x} = \frac{1}{5}(1 + 3 + 5 + 7 + 9) = 5$

(b) $\bar{x} = \frac{1}{5}(1 + 3 + 5 + 7 + 79) = 19$

Notice that the one (possible erroneous) observation of 79 makes a big difference to the sample mean. An alternative measure of central tendency, which enthusiasts of robust statistics vastly prefer, is the median, discussed next.

1.6.2 Median

The *median* is another measure of central tendency. We shall denote the median as $\tilde{\mu}$. It is the point which divides the distribution into two equal halves. Most commonly, $\tilde{\mu}$ is found by solving

$$F_X(\tilde{\mu}) = \mathrm{P}\left[X \le \tilde{\mu}\right] = 0.5$$

for $\tilde{\mu}$. For example, if $f_X(x) = \lambda e^{-\lambda x}$, then $F_X(x) = 1 - e^{-\lambda x}$, and we get

$$1 - e^{-\lambda\tilde{\mu}} = 0.5 \qquad \Longrightarrow \qquad \tilde{\mu}_X = -\frac{\ln(0.5)}{\lambda} = \frac{0.693}{\lambda}$$

While the mean is strongly affected by extremes in the distribution, the median is largely unaffected.

In general, the mean and the median are not the same. If the distribution is positively skewed (or skewed right, which means a longer tail to the right than to the left), as are most soil properties, then the mean will be to the right of the median. Conversely, if the distribution is skewed left, then the mean will be to the left of the median. If the distribution is symmetric, then the mean and the median will coincide.

If we have a sample of observations $x_1, x_2, \ldots, x_n$ of some population X, then the population median $\tilde{\mu}_X$ is estimated by the *sample median* $\tilde{x}$. To define $\tilde{x}$, we must first order the observations from smallest to largest, $x_{(1)} \le x_{(2)} \le \cdots \le x_{(n)}$. When we have done so, the sample median is defined as

$$\tilde{x} = \begin{cases} x_{(n+1)/2} & \text{if } n \text{ is odd} \\ \frac{1}{2}\left(x_{(n/2)} + x_{(n+1)/2}\right) & \text{if } n \text{ is even} \end{cases}$$

Example 1.30 Suppose $\mathbf{x} = \{x_1, x_2, \ldots, x_n\} = \{1, 3, 5, 7, 9\}$.

(a) What is $\tilde{x}$?
(b) What happens to $\tilde{x}$ if $\mathbf{x} = \{1, 3, 5, 7, 79\}$?

SOLUTION In both cases, the sample size is odd with $n = 5$. The central value is that value having the same number of smaller values as larger values. In this case,

(a) $\tilde{x} = x_3 = 5$
(b) $\tilde{x} = x_3 = 5$

so that the (possibly erroneous) extreme value does not have any effect on this measure of the central tendency.

Example 1.31 Suppose that in 100 samples of a soil at a particular site, 99 have cohesion values of 1 kPa and 1 has a cohesion value of 3901 kPa (presumably this single sample was of a boulder or an error). What are the mean and median cohesion values at the site?

SOLUTION The mean cohesion is

$$\bar{x} = \tfrac{1}{100}(1 + 1 + \cdots + 1 + 3901) = 40 \quad \text{kPa}$$

The median cohesion is

$$\tilde{x} = 1 \quad \text{kPa}$$

Clearly, in this case, the median is a much better representation of the site. To design using the mean would almost certainly lead to failure.

1.6.3 Variance

The mean (expected value) or median of the random variable X tells where the probability distribution is "centered." The next most important characteristic of a random variable is whether the distribution is "wide," "narrow," or somewhere in between. This distribution "variability" is commonly measured by a quantity call the variance of X.

Definition Let X be a random variable with probability (mass or density) function $f_X(x)$ and mean μ_X. The *variance* σ_X^2 of X is defined by

$$\sigma_X^2 = \text{Var}[X] = \text{E}\left[(X - \mu_X)^2\right] = \begin{cases} \sum_x (x - \mu_X)^2 f_X(x) & \text{for discrete } X \\ \int_{-\infty}^{\infty} (x - \mu_X)^2 f_X(x)\, dx & \text{for continuous } X \end{cases} \tag{1.26}$$

The variance of the random variable X is sometimes more easily computed as

$$\sigma_X^2 = \text{E}\left[X^2\right] - \text{E}^2[X] = \text{E}\left[X^2\right] - \mu_X^2 \tag{1.27}$$

The variance σ_X^2 has units of X^2. The square root of the variance, σ_X, is called the *standard deviation* of X, which is illustrated in Figure 1.15. Since the standard deviation has the same units as X, it is often preferable to report the standard deviation as a measure of variability.

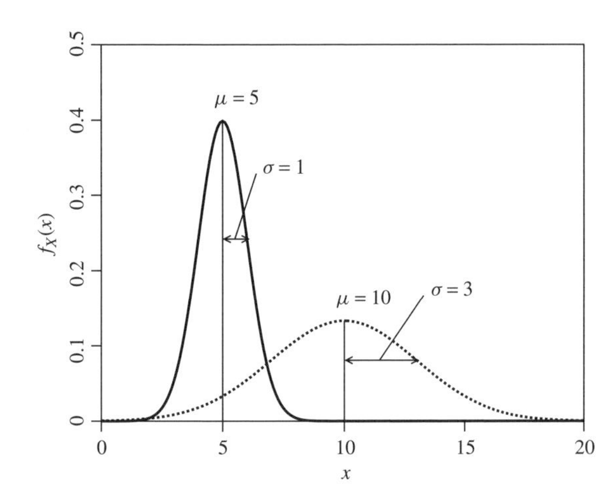

Figure 1.15 Two distributions illustrating how the position and shape change with changes in mean and variance.

Example 1.32 Recall Example 1.27. Find the variance and standard deviation of X.

SOLUTION $\text{Var}[X] = \text{E}\left[X^2\right] - \text{E}^2[X]$
where

$$\text{E}\left[X^2\right] = (-2)^2(\tfrac{1}{12}) + (-1)^2(\tfrac{1}{6}) + 0^2(\tfrac{1}{6}) + 1^2(\tfrac{1}{3}) + 2^2(\tfrac{1}{4}) = \tfrac{11}{6}$$

Thus, $\text{Var}[X] = \text{E}\left[X^2\right] - \text{E}^2[X] = \frac{11}{6} - (\frac{1}{2})^2 = \frac{19}{12}$ and

$$\sigma_X = \sqrt{\text{Var}[X]} = \sqrt{\tfrac{19}{12}} = 1.258$$

Even though the standard deviation has the same units as the mean, it is often still not particularly informative. For example, a standard deviation of 1.0 may indicate significant variability when the mean is 1.0 but indicates virtually deterministic behavior when the mean is one million. For example, an error of 1 m on a 1-m survey would be considered unacceptable, whereas an error of 1-m on a 1000-km survey might be considered quite accurate. A measure of variability which both is nondimensional and delivers a relative sense of the magnitude of variability is the *coefficient of variation*, defined as

$$v = \frac{\sigma}{\mu} \tag{1.28}$$

Example 1.33 Recall Examples 1.27 and 1.29. What is the coefficient of variation of X?

SOLUTION

$$v_X = \frac{\sqrt{19/12}}{1/2} = 2.52$$

or about 250%, which is a highly variable process.

Note that the coefficient of variation becomes undefined if the mean of X is zero. It is, however, quite popular as a way of expressing variability in engineering, particularly for material property and load variability, which generally have nonzero means.

1.6.4 Covariance

Often one must consider more than one random variable at a time. For example, the two components of a drained soil's shear strength, $\tan(\phi')$ and c', will vary randomly from location to location in a soil. These two quantities can be modeled by two random variables, and since they may influence one another (or they may be jointly influenced by some other factor), they are characterized by a *bivariate distribution*. See Figure 1.16.

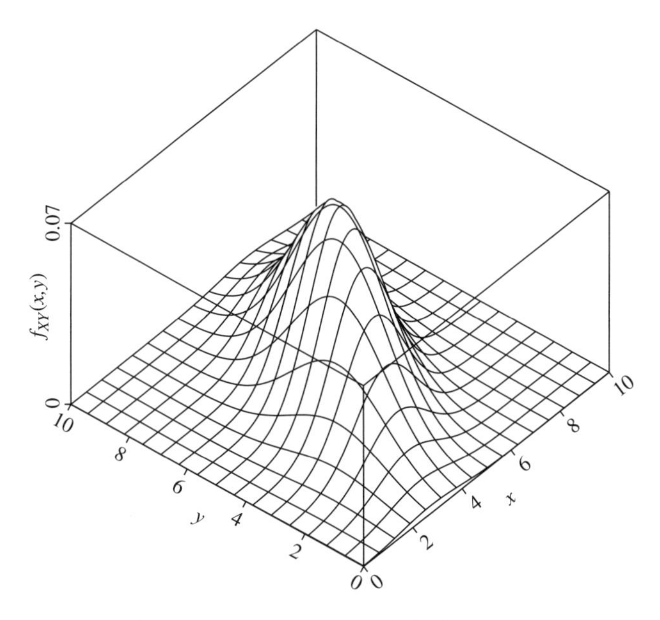

Figure 1.16 Example bivariate probability density function, $f_{XY}(x,y)$.

Properties of Bivariate Distribution

Discrete: $f_{XY}(x,y) = \mathrm{P}\left[X = x \cap Y = y\right]$

$$0 \le f_{XY}(x,y) \le 1$$

$$\sum_{\text{all } x} \sum_{\text{all } y} f_{XY}(x,y) = 1$$

Continuous: $f_{XY}(x,y)\;dx\;dy = \mathrm{P}\left[x < X \le x + dx \cap y < Y \le y + dy\right]$

$$f_{XY}(x,y) \ge 0 \text{ for all } (x,y) \in \Re^2$$

$$\int_{-\infty}^{\infty}\int_{-\infty}^{\infty} f_{XY}(x,y)\;dx\;dy = 1$$

$$\mathrm{P}\left[x_1 < X \le x_2 \cap y_1 < Y \le y_2\right] = \int_{y_1}^{y_2}\int_{x_1}^{x_2} f_{XY}(x,y)\;dx\;dy$$

Definition Let X and Y be random variables with joint probability distribution $f_{XY}(x,y)$. The *covariance* between X and Y is defined by

$$\begin{aligned}\mathrm{Cov}\,[X,Y] &= \mathrm{E}\,[(X-\mu_X)(Y-\mu_Y)] \quad (1.29\text{a})\\ &= \sum_x\sum_y (x-\mu_X)(y-\mu_Y)f_{XY}(x,y)\\ &\quad \text{(discrete case)}\\ &= \int_{-\infty}^{\infty}\int_{-\infty}^{\infty}(x-\mu_X)(y-\mu_Y)f_{XY}(x,y)\;dx\;dy\\ &\quad \text{(continuous case)} \quad (1.29\text{b})\end{aligned}$$

The covariance between two random variables X and Y, having means μ_X and μ_Y, respectively, may also be computed as

$$\mathrm{Cov}\,[X,Y] = \mathrm{E}\,[XY] - \mathrm{E}\,[X]\,\mathrm{E}\,[Y] = \mathrm{E}\,[XY] - \mu_X\mu_Y \quad (1.30)$$

Example 1.34 In order to determine the frequency of electrical signal transmission errors during a cone penetration test, a special cone penetrometer is constructed with redundant measuring and electrical systems. Using this penetrometer, the number of errors detected in the transmission of tip resistance during a typical cone penetration test can be measured and will be called X and the number of errors detected in the transmission of side friction will be called Y. Suppose that statistics are gathered using this penetrometer on a series of penetration tests and the following joint discrete probability mass function is estimated:

$f_{XY}(x,y)$		y (side)				
		0	1	2	3	4
	0	0.24	0.13	0.04	0.03	0.01
x	1	0.16	0.10	0.05	0.04	0.01
(tip)	2	0.08	0.05	0.01	0.00	0.00
	3	0.02	0.02	0.01	0.00	0.00

Assuming that these numbers are correct, compute

1. The expected number of errors in the transmission of the tip resistance
2. The expected number of errors in the transmission of the side friction
3. The variance of the number of errors in the transmission of the tip resistance
4. The variance of the number of errors in the transmission of the side friction
5. The covariance between the number of errors in the transmission of the tip resistance and the side friction

SOLUTION We expand the table by summing rows and columns to obtain the "marginal distributions" (i.e., unconditional distributions), $f_X(x)$ and $f_Y(y)$, of X and Y:

$f_{XY}(x,y)$		y (side)					$f_X(x)$
		0	1	2	3	4	
	0	0.24	0.13	0.04	0.03	0.01	0.45
x	1	0.16	0.10	0.05	0.04	0.01	0.36
(tip)	2	0.08	0.05	0.01	0.00	0.00	0.14
	3	0.02	0.02	0.01	0.00	0.00	0.05
$f_Y(y)$		0.50	0.30	0.11	0.07	0.02	1.00

so that

1. $\mathrm{E}[X] = \sum_x x f_X(x) = 0(0.45) + 1(0.36) + 2(0.14) + 3(0.05) = 0.79$
2. $\mathrm{E}[Y] = \sum_y y f_Y(y) = 0(0.50) + 1(0.30) + 2(0.11) + 3(0.07) + 4(0.02) = 0.81$
3. $\mathrm{E}\left[X^2\right] = \sum_x x^2 f_X(x) = 0^2(0.45) + 1^2(0.36) + 2^2(0.14) + 3^2(0.05) = 1.37$

   $\sigma_X^2 = \mathrm{E}\left[X^2\right] - \mathrm{E}^2[X] = 1.37 - 0.79^2 = 0.75$
4. $\mathrm{E}\left[Y^2\right] = \sum_y y^2 f_Y(y) = 0^2(0.50) + 1^2(0.30) + 2^2(0.11) + 3^2(0.07) + 4^2(0.02) = 1.69$

   $\sigma_Y^2 = \mathrm{E}\left[Y^2\right] - \mathrm{E}^2[Y] = 1.69 - 0.81^2 = 1.03$
5. $\mathrm{E}[XY] = \sum_x \sum_y xy f_{XY}(x, y) = (0)(0)(0.24) + (0)(1)(0.13) + \cdots + (3)(2)(0.01) = 0.62$

   $\mathrm{Cov}[X, Y] = \mathrm{E}[XY] - \mathrm{E}[X]\,\mathrm{E}[Y] = 0.62 - 0.79(0.81) = -0.02$

Although the covariance between two random variables does give information regarding the nature of the relationship, the magnitude of $\mathrm{Cov}[X, Y]$ does not indicate anything regarding the strength of the relationship. This is because $\mathrm{Cov}[X, Y]$ depends on the units and variability of X and Y. A quantity which is both normalized and nondimensional is the correlation coefficient, to be discussed next.

1.6.5 Correlation Coefficient

Definition Let X and Y be random variables with joint probability distribution $f_{XY}(x, y)$. The *correlation coefficient* between X and Y is defined to be

$$\rho_{XY} = \frac{\mathrm{Cov}[X, Y]}{\sigma_X \sigma_Y} \tag{1.31}$$

Figure 1.17 illustrates the effect that the correlation coefficient has on the shape of a bivariate probability density function, in this case for X and Y jointly normal. If $\rho_{XY} = 0$, then the contours form ovals with axes aligned with the cartesian axes (if the variances of X and Y are equal, then the ovals are circles). When $\rho_{XY} > 0$, the ovals become stretched and the major axis has a positive slope. What this means is that when Y is large X will also tend to be large. For example, when $\rho_{XY} = 0.6$, as shown on the right plot of Figure 1.17, then when $Y = 8$, the most likely value X will take is around 7, since this is the peak of the distribution along the line $Y = 8$. Similarly, if $\rho_{XY} < 0$, then the ovals will be oriented so that the major axis has a negative slope. In this case, large values of Y will tend to give small values of X.

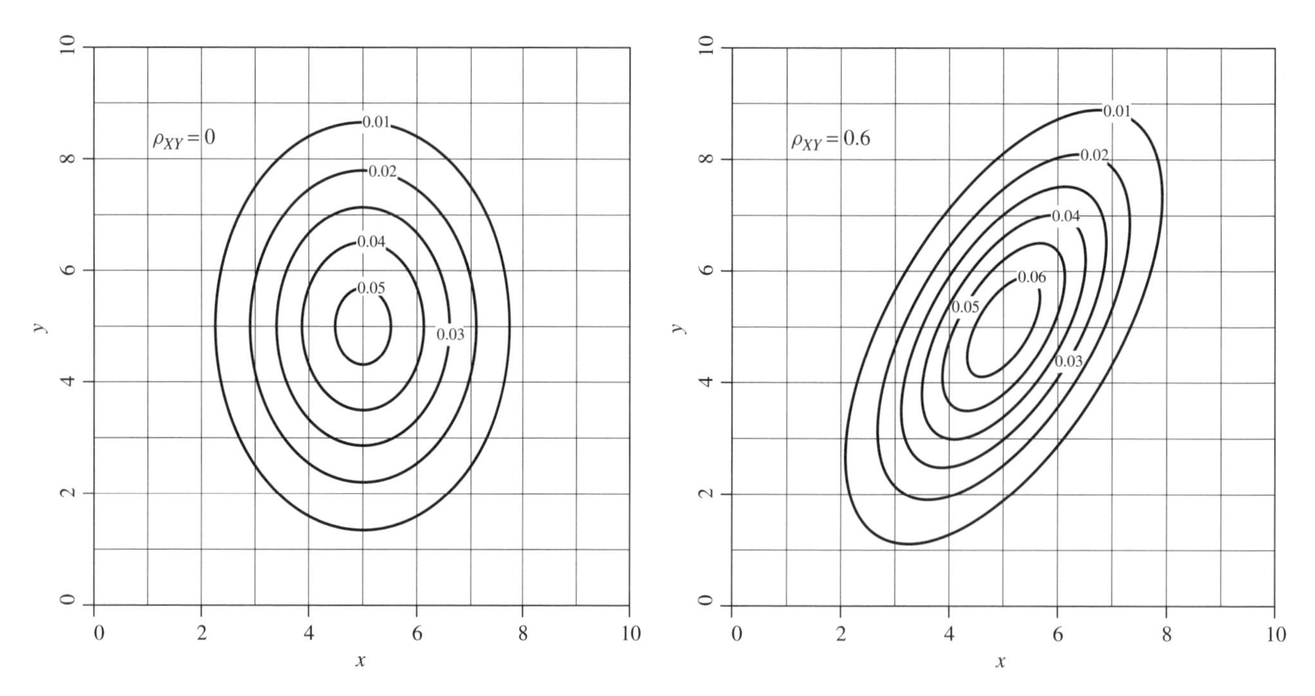

Figure 1.17 Effect of correlation coefficient ρ_{XY} on contours of a bivariate probability density function $f_{XY}(x, y)$ having $\mu_X = \mu_Y = 5$, $\sigma_X = 1.5$ and $\sigma_Y = 2.0$.

We can show that $-1 \le \rho_{XY} \le 1$ as follows: Consider two random variables X and Y having variances σ_X^2 and σ_Y^2, respectively, and correlation coefficient ρ_{XY}. Then

$$\begin{aligned}\text{Var}\left[\frac{X}{\sigma_X} + \frac{Y}{\sigma_Y}\right] &= \frac{\sigma_X^2}{\sigma_X^2} + \frac{\sigma_Y^2}{\sigma_Y^2} + 2\frac{\text{Cov}[X,Y]}{\sigma_X\sigma_Y} \\ &= 2\Big[1 + \rho_{XY}\Big] \\ &\ge 0\end{aligned}$$

which implies that $\rho_{XY} \ge -1$. Similarly,

$$\begin{aligned}\text{Var}\left[\frac{X}{\sigma_X} - \frac{Y}{\sigma_Y}\right] &= \frac{\sigma_X^2}{\sigma_X^2} + \frac{\sigma_Y^2}{\sigma_Y^2} - 2\frac{\text{Cov}[X,Y]}{\sigma_X\sigma_Y} \\ &= 2\Big[1 - \rho_{XY}\Big] \\ &\ge 0\end{aligned}$$

which implies that $\rho_{XY} \le 1$. Taken together, these imply that $-1 \le \rho_{XY} \le 1$.

The correlation coefficient is a direct measure of the degree of *linear* dependence between X and Y. When the two variables are perfectly linearly related, ρ_{XY} will be either $+1$ or -1 ($+1$ if Y increases with X and -1 if Y decreases when X increases). When $|\rho_{XY}| < 1$, the dependence between X and Y is not completely linear; however, there could still be a strong nonlinear dependence. If two random variables X and Y are independent, then their correlation coefficient will be zero. If the correlation coefficient between two random variables X and Y is 0, *it does not mean* that they are independent, only that they are uncorrelated. Independence is a much stronger statement than is $\rho_{XY} = 0$, since the latter only implies linear independence. For example, $Y = X^2$ may be linearly independent of X (this depends on the range of X), but clearly Y and X are completely (nonlinearly) dependent.

Example 1.35 Recall Example 1.30.

1. Compute the correlation coefficient between the number of errors in the transmission of tip resistance and the number of errors in the transmission of the side friction.
2. Interpret the value you found in 1.

SOLUTION

1. $\rho_{XY} = \dfrac{-0.02}{\sqrt{0.75}\sqrt{1.03}} = -0.023$
2. With ρ_{XY} as small as -0.023, there is essentially no linear dependence between the error counts.

1.7 LINEAR COMBINATIONS OF RANDOM VARIABLES

Consider the random variables $X_1, X_2, \ldots, X_n$ and the constants $a_1, a_2, \ldots, a_n$. If

$$Y = a_1X_1 + a_2X_2 + \cdots + a_nX_n = \sum_{i=1}^{n} a_iX_i \tag{1.32}$$

then Y is also a random variable, being a *linear combination* of the random variables $X_1, \ldots, X_n$. Linear combinations of random variables are common in engineering applications; any sum is a linear combination. For example, the weight of a soil mass is the sum of the weights of its constitutive particles. The bearing strength of a soil is due to the sum of the shear strengths along the potential failure surface. This section reviews the basic results associated with linear combinations.

1.7.1 Mean of Linear Combinations

The mean, or expectation, of a linear combination can be summarized by noting that *the expectation of a sum is the sum of the expectations*. Also, since constants can be brought out in front of an expectation, we have the following rules:

1. If a and b are constants, then
$$\text{E}[aX \pm b] = a\text{E}[X] \pm b \tag{1.33}$$
2. If g and h are functions of the random variable X, then
$$\text{E}\big[g(X) \pm h(X)\big] = \text{E}\big[g(X)\big] \pm \text{E}[h(X)] \tag{1.34}$$
3. Similarly, for any two random variables X and Y,
$$\text{E}\big[g(X) \pm h(Y)\big] = \text{E}\big[g(X)\big] \pm \text{E}[h(Y)] \tag{1.35}$$
Note that this means, for example, $\text{E}[X \pm Y] = \text{E}[X] \pm \text{E}[Y]$.
4. If X and Y are two *uncorrelated* random variables, then
$$\text{E}[XY] = \text{E}[X]\,\text{E}[Y] \tag{1.36}$$
by virtue of the fact that $\text{Cov}[X,Y] = \text{E}[XY] - \text{E}[X]\,\text{E}[Y] = 0$ when X and Y are uncorrelated. (This actually has nothing to do with linear combinations but often occurs in problems involving linear combinations.)

In general, if

$$Y = \sum_{i=1}^{n} a_iX_i \tag{1.37}$$

as in Eq. 1.32, then

$$\mathrm{E}[Y] = \sum_{i=1}^{n} a_i \mathrm{E}[X_i] \qquad (1.38)$$

1.7.2 Variance of Linear Combinations

The variance of a linear combination is complicated by the fact that the X_i's in the combination may or may not be correlated. If they are correlated, then the variance calculation will involve the covariances between the X_i's. In general, the following rules apply:

1. If a and b are constants, then

$$\begin{aligned}\mathrm{Var}[aX + b] &= \mathrm{Var}[aX] + \mathrm{Var}[b]\\ &= a^2\mathrm{Var}[X] = a^2\sigma_X^2 \qquad (1.39)\end{aligned}$$

   that is, the variance of a constant is zero, and since variance is defined in terms of *squared* deviations from the mean, all quantities, including constants, are squared. Variance has units of X^2 (which is why we often prefer the standard deviation in practice).

2. If X and Y are random variables with joint probability distribution $f_{XY}(x, y)$ and a and b are constants, then

$$\mathrm{Var}[aX + bY] = a^2\sigma_X^2 + b^2\sigma_Y^2 + 2ab\ \mathrm{Cov}[X, Y] \qquad (1.40)$$

   Note that the sign on the last term depends on the sign of a and b but that the variance terms are always positive. Note also that, if X and Y are uncorrelated, then $\mathrm{Cov}[X, Y] = 0$, so that, in this case, the above simplifies to

$$\mathrm{Var}[aX + bY] = a^2\sigma_X^2 + b^2\sigma_Y^2 \qquad (1.41)$$

If we consider the more general case where (as in Eq. 1.37)

$$Y = \sum_{i=1}^{n} a_i X_i$$

then we have the following results:

3. If $X_1, X_2, \ldots, X_n$ are correlated, then

$$\mathrm{Var}[Y] = \sum_{i=1}^{n}\sum_{j=1}^{n} a_i a_j\ \mathrm{Cov}\left[X_i, X_j\right] \qquad (1.42)$$

   where we note that $\mathrm{Cov}[X_i, X_i] = \mathrm{Var}[X_i]$. If $n = 2$, the equation given in item 2 is obtained by replacing X_1 with X and X_2 with Y.

4. If $X_1, X_2, \ldots, X_n$ are uncorrelated random variables, then

$$\begin{aligned}&\mathrm{Var}[a_1X_1 + \cdots + a_nX_n]\\ &= a_1^2\sigma_{X_1}^2 + \cdots + a_n^2\sigma_{X_n}^2 = \sum_{i=1}^{n} a_i^2\sigma_{X_i}^2 \qquad (1.43)\end{aligned}$$

   which follows from item 3 by noting that, if X_i and X_j are uncorrelated for all $i \neq j$, then $\mathrm{Cov}\left[X_i, X_j\right] = 0$ and we are left only with the $\mathrm{Cov}[X_i, X_i] = \sigma_{X_i}^2$ terms above. This means that, if the X's are uncorrelated, then *the variance of a sum is the sum of the variances*. (However, remember that this rule *only* applies if the X's are uncorrelated.)

Example 1.36 Let X and Y be independent random variables with $\mathrm{E}[X] = 2$, $\mathrm{E}\left[X^2\right] = 29$, $\mathrm{E}[Y] = 4$, and $\mathrm{E}\left[Y^2\right] = 52$. Consider the random variables $W = X + Y$ and $Z = 2X$. The random variables W and Z are clearly dependent since they both involve X. What is their covariance? What is their correlation coefficient?

SOLUTION Given $\mathrm{E}[X] = 2$, $\mathrm{E}\left[X^2\right] = 29$, $\mathrm{E}[Y] = 4$, and $\mathrm{E}\left[Y^2\right] = 52$; X and Y independent; and $W = X + Y$ and $Z = 2X$.

Thus,

$$\begin{aligned}
\mathrm{Var}[X] &= \mathrm{E}\left[X^2\right] - \mathrm{E}^2[X] = 29 - 2^2 = 25\\
\mathrm{Var}[Y] &= \mathrm{E}\left[Y^2\right] - \mathrm{E}^2[Y] = 52 - 4^2 = 36\\
\mathrm{E}[W] &= \mathrm{E}[X + Y] = 2 + 4 = 6\\
\mathrm{Var}[W] &= \mathrm{Var}[X + Y] = \mathrm{Var}[X] + \mathrm{Var}[Y]\\
&= 25 + 36 = 61\\
&\quad \text{(due to independence)}\\
\mathrm{E}[Z] &= \mathrm{E}[2X] = 2(2) = 4\\
\mathrm{Var}[Z] &= \mathrm{Var}[2X] = 4\mathrm{Var}[X] = 4(25) = 100\\
\mathrm{Cov}[W, Z] &= \mathrm{E}[WZ] - \mathrm{E}[W]\mathrm{E}[Z]\\
\mathrm{E}[WZ] &= \mathrm{E}[(X + Y)(2X)] = \mathrm{E}\left[2X^2 + 2XY\right]\\
&= 2\mathrm{E}\left[X^2\right] + 2\mathrm{E}[X]\mathrm{E}[Y]\\
&= 2(29) + 2(2)(4) = 74\\
\mathrm{Cov}[W, Z] &= 74 - 6(4) = 50\\
\rho_{WZ} &= \frac{50}{\sqrt{61}\sqrt{100}} = \frac{5}{\sqrt{61}} = 0.64
\end{aligned}$$

1.8 FUNCTIONS OF RANDOM VARIABLES

In general, deriving the distribution of a function of random variables [i.e., the distribution of Y where $Y = g(X_1, X_2, \ldots)$] can be quite a complex problem and exact solutions may be unknown or impractical to find.

In this section, we will cover only relatively simple cases (although even these can be difficult) and also look at some approximate approaches.

1.8.1 Functions of a Single Variable

Consider the function

$$Y = g(X) \tag{1.44}$$

and assume we know the distribution of X, that is, we know $f_X(x)$. When X takes on a specific value, that is, when $X = x$, we can compute $Y = y = g(x)$. If we assume, for now, that each value of x gives only one value of y and that each value of y arises from only one value of x (i.e., that $y = g(x)$ is a *one-to-one function*), then we must have the probability that $Y = y$ is just equal to the probability that $X = x$. That is, for discrete X,

$$\mathrm{P}\left[Y = y\right] = \mathrm{P}\left[X = x\right] = \mathrm{P}\left[X = g^{-1}(y)\right] \tag{1.45}$$

where $g^{-1}(y)$ is the inverse function, obtained by solving $y = g(x)$ for x, i.e. $x = g^{-1}(y)$. Eq. 1.45 implies that

$$f_Y(y) = f_X\left(g^{-1}(y)\right) \tag{1.46}$$

In terms of the discrete cumulative distribution function,

$$\begin{aligned} F_Y(y) = \mathrm{P}\left[Y \le y\right] &= F_X(g^{-1}(y)) = \mathrm{P}\left[X \le g^{-1}(y)\right] \\ &= \sum_{x_i \le g^{-1}(y)} f_X(x_i) \end{aligned} \tag{1.47}$$

In the continuous case, the distribution of Y is obtained in a similar fashion. Considering Figure 1.18, the probability that X lies in a neighborhood of x_1 is the area A_1. If X lies in the shown neighborhood of x_1, Y must lie in a corresponding neighborhood of y_1 and will do so with equal probability A_1. Since the two probabilities are equal, this defines the height of the distribution of Y in the neighborhood of y_1. Considering the situation in the neighborhood of x_2, we see that the height of the distribution of Y near y_2 depends not only on A_2, which is the probability that X is in the neighborhood of x_2, but also on the slope of $y = g(x)$ at the point x_2. As the slope flattens, the height of $f(y)$ increases; that is, $f(y)$ increases as the slope decreases.

We will develop the theory by first considering the continuous analog of the discrete cumulative distribution function developed above,

$$\begin{aligned} F_Y(y) &= \int_{-\infty}^{g^{-1}(y)} f_X(x)\, dx \\ &= \int_{-\infty}^{y} f_X(g^{-1}(y)) \left[\frac{d}{dy} g^{-1}(y)\right] dy \end{aligned} \tag{1.48}$$

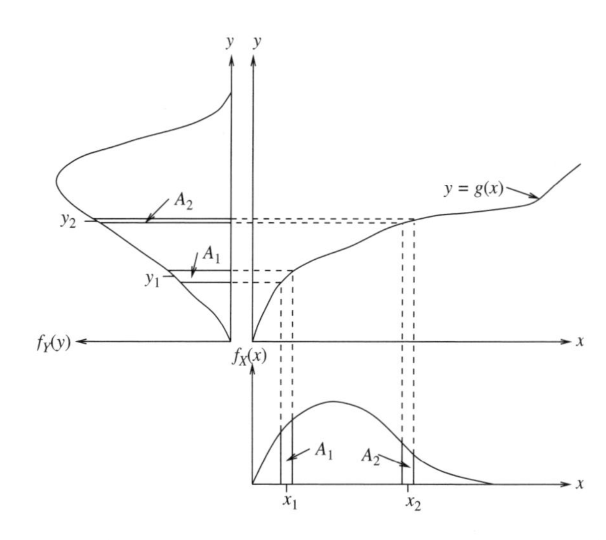

Figure 1.18 Deriving the distribution of $Y = g(X)$ from the distribution of X.

where we let $x = g^{-1}(y)$ to get the last result. To get the probability density function of Y, we can differentiate,

$$f_Y(y) = \frac{d}{dy} F_Y(y) = f_X(g^{-1}(y)) \left[\frac{d}{dy} g^{-1}(y)\right] \tag{1.49}$$

Note that the left-hand side here is found under the assumption that y always increases with increasing x. However, if y decreases with increasing x, then $\mathrm{P}\left[Y \le y\right]$ corresponds to $\mathrm{P}\left[X > x\right]$, leading to (see Eq. 1.47),

$$\begin{aligned} F_Y(y) &= 1 - F_X(g^{-1}(y)) \\ f_Y(y) &= f_X(g^{-1}(y)) \left[-\frac{d}{dy} g^{-1}(y)\right] \end{aligned}$$

To handle both possibilities (and since probabilities are always positive), we write

$$f_Y(y) = f_X\left(g^{-1}(y)\right) \left|\frac{d}{dy} g^{-1}(y)\right| \tag{1.50}$$

In terms of Figure 1.18 we can leave $x = g^{-1}(y)$ in the relationship and write our result as

$$f_Y(y) = f_X(x) \left|\frac{dx}{dy}\right| \tag{1.51}$$

which means that $f_Y(y)$ increases as the *inverse* of the slope, $|dx/dy|$, increases, which agrees with what is seen in Figure 1.18.

Example 1.37 Suppose that X has the following continuous distribution:

$$f_X(x) = \frac{1}{\sigma\sqrt{2\pi}} \exp\left\{-\frac{1}{2}\left(\frac{x-\mu}{\sigma}\right)^2\right\}$$

which is the normal distribution, which we will discuss further in Section 1.10.4. If $Z = (X - \mu)/\sigma$, then what is $f_Z(z)$? (Note, we use Z intentionally here, rather than Y, because as we shall see in Section 1.10.8, Z is the so-called standard normal.)

SOLUTION In order to determine $f_Z(z)$, we need to know both $f_X(x)$ and dx/dz. We know $f_X(x)$ is the normal distribution, as shown above. To compute dx/dz we need an expression for x, which we can get by inverting the given relationship for Z (note, for the computation of the slope, we assume that both X and Z are known, and are replaced by their lowercase equivalents):

$$x = g^{-1}(z) = \mu + \sigma z$$

which gives us

$$\left|\frac{dx}{dz}\right| = \left|\frac{dg^{-1}(z)}{dz}\right| = \sigma$$

Putting these results together gives us

$$f_Z(z) = f_X(x)\left|\frac{dx}{dz}\right| = f_X(\mu + \sigma z)\sigma$$
$$= \frac{1}{\sqrt{2\pi}} \exp\left\{-\frac{1}{2}z^2\right\}$$

Notice that the parameters μ and σ have now disappeared from the distribution of Z. As we shall see, Z is also normally distributed with $\mu_Z = 0$ and $\sigma_Z = 1$.

The question now arises as to what happens if the function $Y = g(X)$ is not one to one. The answer is that the probabilities of all the $X = x$ values which lead to each y are added into the probability that $Y = y$. That is, if $g(x_1), g(x_2), \ldots$ all lead to the same value of y, then

$$f_Y(y) = f_X(x_1)\left|\frac{dx_1}{dy}\right| + f_X(x_2)\left|\frac{dx_2}{dy}\right| + \cdots$$

The number of terms on the right-hand-side generally depends on y, so this computation over all y can be quite difficult. For example, the function $Y = a + bX + cX^2 + dX^3$ might have three values of x leading to the same value of y over some ranges in y but only one value of x leading to the same value of y on other ranges.

1.8.2 Functions of Two or More Random Variables

Here we consider functions of the form

$$\begin{aligned} Y_1 &= g_1(X_1, X_2, \ldots) \\ Y_2 &= g_2(X_1, X_2, \ldots) \\ &\vdots \end{aligned} \tag{1.52}$$

In the theory which follows, we require that the number of equations above equals the number of random variables $X_1, X_2, \ldots$ and that the equations be independent so that a unique inverse can be obtained. The theory will then give us the joint distribution of $Y_1, Y_2, \ldots$ in terms of the joint distribution of $X_1, X_2, \ldots$

More commonly, we only have a single function of the form

$$Y_1 = g_1(X_1, X_2, \ldots, X_n) \tag{1.53}$$

in which case an additional $n - 1$ *independent* equations, corresponding to $Y_2, \ldots, Y_n$, must be arbitrarily added to the problem in order to use the theory to follow. Once these equations have been added and the complete joint distribution has been found, the $n - 1$ arbitrarily added Y's can be integrated out to obtain the *marginal distribution* of Y_1. For example, if $Y_1 = X_1/X_2$ and we want the pdf of Y_1 given the joint pdf of (X_1, X_2), then we must

1. choose some function $Y_2 = g(X_1, X_2)$ which will allow us to find an inverse—for example, if we choose $Y_2 = X_2$, then we get $X_1 = Y_1Y_2$ and $X_2 = Y_2$ as our inverse;
2. obtain the joint pdf of (Y_1, Y_2) in terms of the joint pdf of (X_1, X_2); and
3. obtain the marginal pdf of Y_1 by integrating $f_{Y_1Y_2}$ over all possible values of Y_2.

In detail, suppose we start with the two-dimensional set of equations

$$\left.\begin{aligned} Y_1 &= g_1(X_1, X_2) \\ Y_2 &= g_2(X_1, X_2) \end{aligned}\right\} \iff \left\{\begin{aligned} X_1 &= h_1(Y_1, Y_2) \\ X_2 &= h_2(Y_1, Y_2) \end{aligned}\right. \tag{1.54}$$

where the right-hand equations are obtained by inverting the (given) left-hand equations. Recall that for one variable we had $f_Y(y) = f_X(x)\,|dx/dy|$. The generalization to multiple variables is

$$f_{Y_1Y_2}(y_1, y_2) = f_{X_1X_2}(h_1, h_2)\,|J| \tag{1.55}$$

where J is the *Jacobian* of the transformation,

$$J = \det\begin{bmatrix} \dfrac{\partial h_1}{\partial y_1} & \dfrac{\partial h_1}{\partial y_2} \\ \dfrac{\partial h_2}{\partial y_1} & \dfrac{\partial h_2}{\partial y_2} \end{bmatrix} \tag{1.56}$$

For more than two variables, the extension is

$$\left.\begin{aligned} Y_1 &= g_1(X_1, X_2, \ldots, X_n) \\ Y_2 &= g_2(X_1, X_2, \ldots, X_n) \\ &\vdots \\ Y_n &= g_n(X_1, X_2, \ldots, X_n) \end{aligned}\right\}$$

$$\Longleftrightarrow \begin{cases} X_1 = h_1(Y_1, Y_2, \ldots, Y_n) \\ X_2 = h_2(Y_1, Y_2, \ldots, Y_n) \\ \cdot \\ \cdot \\ \cdot \\ X_n = h_n(Y_1, Y_2, \ldots, Y_n) \end{cases} \tag{1.57}$$

$$J = \det \begin{bmatrix} \frac{\partial h_1}{\partial y_1} & \frac{\partial h_1}{\partial y_2} & \cdot \ \cdot \ \cdot & \frac{\partial h_1}{\partial y_n} \\ \frac{\partial h_2}{\partial y_1} & \frac{\partial h_2}{\partial y_2} & \cdot \ \cdot \ \cdot & \frac{\partial h_2}{\partial y_n} \\ \cdot & \cdot & \cdot & \cdot \\ \cdot & \cdot & \cdot & \cdot \\ \cdot & \cdot & \cdot & \cdot \\ \frac{\partial h_n}{\partial y_1} & \frac{\partial h_n}{\partial y_2} & \cdot \ \cdot \ \cdot & \frac{\partial h_n}{\partial y_n} \end{bmatrix} \tag{1.58}$$

and

$$f_{Y_1Y_2\cdots Y_n}(y_1, y_2, \ldots, y_n) = \begin{cases} f_{X_1X_2\cdots X_n}(h_1, h_2, \ldots, h_n)\,|J| \\ \qquad \text{for } (y_1, y_2, \ldots, y_n) \in T \\ 0 \qquad \text{otherwise} \end{cases} \tag{1.59}$$

where T is the region in $\mathbf{Y}$ space corresponding to possible values of $\mathbf{x}$, specifically

$$T = \{g_1, g_2, \ldots, g_n : (x_1, x_2, \ldots, x_n) \in S\} \tag{1.60}$$

and S is the region on which $f_{X_1X_2\cdots X_n}$ is nonzero.

Example 1.38 Assume X_1 and X_2 are jointly distributed according to

$$f_{X_1X_2}(x_1, x_2) = \begin{cases} 4x_1x_2 & \text{for } 0 < x_1 < 1 \text{ and } 0 < x_2 < 1 \\ 0 & \text{otherwise} \end{cases}$$

and that the following relationships exist between $\mathbf{Y}$ and $\mathbf{X}$:

$$\left.\begin{aligned} Y_1 &= \frac{X_1}{X_2} \\ Y_2 &= X_1X_2 \end{aligned}\right\} \Longleftrightarrow \begin{cases} X_1 = \sqrt{Y_1Y_2} \\ X_2 = \sqrt{\dfrac{Y_2}{Y_1}} \end{cases}$$

What is the joint pdf of (Y_1, Y_2)?

SOLUTION We first of all find the Jacobian,

$$\frac{\partial x_1}{\partial y_1} = \frac{1}{2}\sqrt{\frac{y_2}{y_1}}, \qquad \frac{\partial x_1}{\partial y_2} = \frac{1}{2}\sqrt{\frac{y_1}{y_2}}$$

$$\frac{\partial x_2}{\partial y_1} = -\frac{1}{2}\sqrt{\frac{y_2}{y_1^3}}, \qquad \frac{\partial x_2}{\partial y_2} = \frac{1}{2}\sqrt{\frac{1}{y_1y_2}}$$

so that

$$J = \det \begin{bmatrix} \frac{1}{2}\sqrt{\frac{y_2}{y_1}} & \frac{1}{2}\sqrt{\frac{y_1}{y_2}} \\ -\frac{1}{2}\sqrt{\frac{y_2}{y_1^3}} & \frac{1}{2}\sqrt{\frac{1}{y_1y_2}} \end{bmatrix} = \frac{1}{2y_1}$$

This gives us

$$\begin{aligned} f_{Y_1Y_2}(y_1, y_2) &= f_{X_1X_2}\left(\sqrt{y_1y_2}, \sqrt{\frac{y_2}{y_1}}\right)|J| \\ &= 4\sqrt{y_1y_2}\sqrt{\frac{y_2}{y_1}}\left(\frac{1}{2|y_1|}\right) \\ &= \frac{2y_2}{|y_1|} \end{aligned} \tag{1.61}$$

We must still determine the range of y_1 and y_2 over which this joint distribution is valid. We know that $0 < x_1 < 1$ and $0 < x_2 < 1$, so it must also be true that $0 < \sqrt{y_1y_2} < 1$ and $0 < \sqrt{y_2/y_1} < 1$. Now, if x_1 lies between 0 and 1, then x_1^2 must also lie between 0 and 1, so we can eliminate the square root signs and write our constraints on y_1 and y_2 as

$$0 < y_1y_2 < 1 \qquad \text{and} \qquad 0 < \frac{y_2}{y_1} < 1$$

If we consider the lines generated by replacing the inequalities above with equalities, we get the following bounding relationships:

$$y_1y_2 = 0, \qquad y_1y_2 = 1$$

$$\frac{y_2}{y_1} = 0, \qquad \frac{y_2}{y_1} = 1$$

If we plot these bounding relationships, the shape of the region, T, where $f_{Y_1Y_2}$ is defined by Eq. 1.61, becomes apparent. This is illustrated in Figure 1.19.

We see from Figure 1.19 that the range, T, is defined by

$$0 < y_2 < 1 \qquad \text{and} \qquad y_2 < y_1 < \frac{1}{y_2}$$

Our joint distribution can now be completely specified as

$$f_{Y_1Y_2}(y_1, y_2) = \begin{cases} \dfrac{2y_2}{y_1} & \text{for } 0 < y_2 < 1 \text{ and } y_2 < y_1 < \dfrac{1}{y_2} \\ 0 & \text{otherwise} \end{cases}$$

where we dropped the absolute value because y_1 is strictly positive.

Example 1.39 Consider the relationship

$$X = A \cos \Phi$$

where A and Φ are random variables with pdf $f_{A\Phi}(a, \phi)$. Assume that A and Φ are independent, that A follows a Rayleigh distribution with parameter s^2, and that Φ is uniformly distributed between 0 and 2π. What is the distribution of X?

SOLUTION First we must define a second function, Y, to give us a unique inverse relationship. Let us somewhat arbitrarily take

$$Y = A \sin \Phi$$

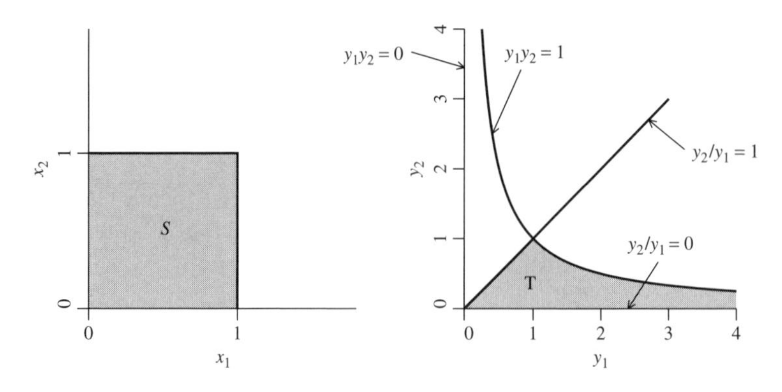

Figure 1.19 The ranges of (x_1, x_2) and (y_1, y_2) over which $f_{X_1X_2}$ and $f_{Y_1Y_2}$ are defined.

Note that there is no particular requirement for the choice of the second function so long as it leads to an inverse. This choice leads to the inverse relationships

$$\left.\begin{array}{l} X = A\cos\Phi \\ Y = A\sin\Phi \end{array}\right\} \iff \left\{\begin{array}{l} A = \sqrt{X^2+Y^2} \\ \Phi = \tan^{-1}\left(\dfrac{Y}{X}\right) \pm 2k\pi, \qquad k = 0, 1, \ldots \end{array}\right.$$

where we have assumed that $\tan^{-1}(Y/X)$ gives a unique value between 0 and 2π—for this, we must make use of the signs of Y and X in the determination of the angle. Notice that Φ is not single valued for each X and Y.

In determining the Jacobian, we will revert to lowercase letters to emphasize that the Jacobian is deterministic (despite the fact that J itself is uppercase),

$$\frac{\partial a}{\partial x} = \frac{x}{\sqrt{x^2+y^2}}, \qquad \frac{\partial a}{\partial y} = \frac{y}{\sqrt{x^2+y^2}}$$

$$\frac{\partial \phi}{\partial x} = -\frac{y}{x^2+y^2}, \qquad \frac{\partial \phi}{\partial y} = \frac{x}{x^2+y^2}$$

so that

$$J = \det\begin{bmatrix} \dfrac{\partial a}{\partial x} & \dfrac{\partial a}{\partial y} \\ \dfrac{\partial \phi}{\partial x} & \dfrac{\partial \phi}{\partial y} \end{bmatrix} = \frac{1}{\sqrt{x^2+y^2}}$$

Since A and Φ are independent, their joint distribution is just the product of their individual (marginal) distributions, namely $f_{A\Phi}(a,\phi) = f_A(a) f_\Phi(\phi)$. The joint distribution of X and Y is thus

$$f_{XY}(x,y) = \frac{f_A\left(\sqrt{x^2+y^2}\right)\sum_{k=-\infty}^{\infty} f_\Phi\left(\tan^{-1}(y/x) + 2k\pi\right)}{\sqrt{x^2+y^2}} \tag{1.62}$$

where the sum arises because Φ takes on an infinite number of possible values for each x and y—we must include the probability of each in the joint probability of X and Y.

The Rayleigh distribution, which is discussed further in Section 1.10.5, has probability density function

$$f_A(a) = \frac{a}{s^2}\exp\left\{-\frac{a^2}{2s^2}\right\}, \qquad a \ge 0$$

while the uniform distribution is

$$f_\Phi(\phi) = \frac{1}{2\pi}, \qquad 0 < \phi \le 2\pi$$

Since Φ has zero probability of being outside the interval $(0, 2\pi]$ and exactly one value of $[\tan^{-1}(y/x) + 2k\pi]$ will lie inside that interval, then only one term in the infinite sum is nonzero and the sum simplifies to

$$\sum_{k=-\infty}^{\infty} f_\Phi\left(\tan^{-1}\left(\frac{y}{x}\right) + 2k\pi\right) = \frac{1}{2\pi}$$

In this case, Eq. 1.62 becomes

$$\begin{aligned} f_{XY}(x,y) &= \frac{\sqrt{x^2+y^2}/s^2 \exp\{-(x^2+y^2)/2s^2\}}{2\pi\sqrt{x^2+y^2}} \\ &= \frac{1}{2\pi s^2}\exp\left\{-\frac{x^2+y^2}{2s^2}\right\} \end{aligned} \tag{1.63}$$

To find the marginal distribution of X (which was the original aim), we must integrate over all possible values of Y using the total probability theorem:

$$\begin{aligned} f_X(x) &= \int_{-\infty}^{\infty} \frac{1}{2\pi s^2}\exp\left\{-\frac{x^2+y^2}{2s^2}\right\} dy \\ &= \frac{e^{-x^2/(2s^2)}}{2\pi s^2}\int_{-\infty}^{\infty} e^{-y^2/(2s^2)}\, dy \\ &= \frac{1}{\sqrt{2\pi}\, s} e^{-x^2/(2s^2)} \end{aligned} \tag{1.64}$$

To get the final result, we needed to use the fact that

$$\int_{-\infty}^{\infty} e^{-y^2/(2s^2)}\, dy = s\sqrt{2\pi}$$

We note that Eq. 1.64 is just the normal distribution with mean zero and variance s^2. In addition, we see that Eq. 1.63 is separable and can be written as $f_{XY}(x,y) = f_X(x) \cdot f_Y(y)$, so that X and Y must be independent. A similar computation as was carried out above will show that $f_Y(y)$ is also a lognormal distribution with mean zero and variance s^2.

In summary, we see that if A is Rayleigh distributed with parameter s^2 and Φ is uniformly distributed between 0 and 2π, then

$$X = A \cos \Phi, \qquad Y = A \sin \Phi$$

will be a pair of identically normally distributed independent random variables, each with mean zero and variance s^2. As we shall see in Chapter 6, the above results suggest a very good approach to simulating normally distributed random variables.

1.8.2.1 Linear Transformations Say we have the simultaneous system of equations

$$\begin{aligned} Y_1 &= a_{11}X_1 + a_{12}X_2 + \cdots + a_{1n}X_n \\ Y_2 &= a_{21}X_1 + a_{22}X_2 + \cdots + a_{2n}X_n \\ &\;\;\vdots \\ Y_n &= a_{n1}X_1 + a_{n2}X_2 + \cdots + a_{nn}X_n \end{aligned}$$

which we can write using matrix notation as

$$\mathbf{Y} = A\mathbf{X} \tag{1.65}$$

If this relationship holds, then $\mathbf{X} = \mathbf{A}^{-1}\mathbf{Y}$ for nonsingular A (implies a one-to-one transformation). The joint distribution of $\mathbf{Y}$ is thus

$$f_Y(\mathbf{y}) = f_X(\mathbf{A}^{-1}\mathbf{y})\,|J| \tag{1.66}$$

where

$$J = \det\left[\mathbf{A}^{-1}\right] = \frac{1}{\det[\mathbf{A}]} \tag{1.67}$$

Example 1.40 Say that $Y_1 = X_1 + X_2$ and that the joint pdf of $\mathbf{X}$ is

$$f_{X_1X_2}(x_1,x_2) = \begin{cases} e^{-(x_1+x_2)} & \text{for } x_1,\ x_2 \ge 0 \\ 0 & \text{otherwise} \end{cases}$$

What is the distribution of Y_1?

SOLUTION Choose $Y_2 = X_2$ as our second equation. Then

$$\left.\begin{aligned} Y_1 &= X_1 + X_2 \\ Y_2 &= X_2 \end{aligned}\right\} \iff \left\{\begin{aligned} X_1 &= Y_1 - Y_2 \\ X_2 &= Y_2 \end{aligned}\right.$$

or

$$\begin{Bmatrix} X_1 \\ X_2 \end{Bmatrix} = \begin{bmatrix} 1 & -1 \\ 0 & 1 \end{bmatrix} \begin{Bmatrix} Y_1 \\ Y_2 \end{Bmatrix}$$

where we see from this that

$$\mathbf{A}^{-1} = \begin{bmatrix} 1 & -1 \\ 0 & 1 \end{bmatrix}$$

so that $J = \det A^{-1} = 1$. This gives us

$$\begin{aligned} f_{Y_1Y_2}(y_1,y_2) &= f_{X_1X_2}(y_1 - y_2, y_2)(1) \\ &= e^{-(y_1-y_2)-y_2}, \qquad y_1 - y_2 \ge 0 \text{ and } y_2 \ge 0 \\ &= e^{-y_1}, \qquad y_1 \ge 0 \text{ and } 0 \le y_2 \le y_1 \end{aligned}$$

To find the distribution of Y_2, we must integrate over all possible values of Y_1 using the total probability theorem,

$$\begin{aligned} f_{Y_1}(y_1) &= \int_{-\infty}^{\infty} f_{Y_1Y_2}(y_1,y_2)\, dy_2 = \int_0^{y_1} e^{-y_1}\, dy_2 \\ &= y_1 e^{-y_1}, \qquad y_1 \ge 0 \end{aligned}$$

In general, if $Y = X_1 + X_2$ and X_1 is independent of X_2 [so that their joint distribution can be written as the product $f_{X_1X_2}(x_1,x_2) = f_{X_1}(x_1)f_{X_2}(x_2)$], then the distribution of Y can be written as the convolution

$$\begin{aligned} f_Y(y) &= \int_{-\infty}^{\infty} f_{X_1}(y-x) f_{X_2}(x)\, dx \\ &= \int_{-\infty}^{\infty} f_{X_1}(x) f_{X_2}(y-x)\, dx \end{aligned} \tag{1.68}$$

1.8.3 Moments of Functions

In many cases the full distribution of a function of random variables is difficult to obtain. So we would like to be able to get at least the mean and variance (often the central limit theorem, discussed later, can be relied upon to suggest that the final distribution is either normal or lognormal). Obtaining just the mean and variance, at least approximately, is typically much easier than obtaining the complete distribution. In the following we will consider a function of the form $Y = g(X_1, X_2, \ldots, X_n)$ whose nth moment is defined by

$$\begin{aligned} \mathrm{E}\left[Y^n\right] = \int_{-\infty}^{\infty}\int_{-\infty}^{\infty}\cdots\int_{-\infty}^{\infty} & g^n(x_1,x_2,\ldots,x_n) \\ & \times f_{\mathbf{X}}(x_1,x_2,\ldots,x_n)\, dx_1\, dx_2 \cdots dx_n \end{aligned} \tag{1.69}$$

where $\mathbf{X}$ is the vector of $X's$; $\mathbf{X} = \{X_1, X_2, \ldots, X_n\}$.

1.8.3.1 Arbitrary Function of One Variable If g is an arbitrary function of one variable, $Y = g(X)$, then

$$\mathrm{E}\left[Y^n\right] = \int_{-\infty}^{\infty} g^n(x) f_X(x)\, dx \tag{1.70}$$

Various levels of approximations exist for this moment. Consider a Taylor's series expansion of $g(X)$ about μ_X,

$$Y = g(X) = g(\mu_X) + (X - \mu_X) \left.\frac{dg}{dx}\right|_{\mu_X} + \frac{1}{2}(X - \mu_X)^2 \left.\frac{d^2g}{dx^2}\right|_{\mu_X} + \cdots \tag{1.71}$$

A *first-order* approximation to the moments uses just the first two terms of the Taylor's series expansion:

$$\mathrm{E}[Y] \simeq \mathrm{E}\left[g(\mu_X) + (X - \mu_X) \left.\frac{dg}{dx}\right|_{\mu_X} \right] = g(\mu_X) \tag{1.72a}$$

$$\mathrm{Var}[Y] \simeq \mathrm{Var}\left[g(\mu_X) + (X - \mu_X) \left.\frac{dg}{dx}\right|_{\mu_X} \right] = \mathrm{Var}[X] \left(\left.\frac{dg}{dx}\right|_{\mu_X} \right)^2 \tag{1.72b}$$

This approximation is often referred to as the *first-order second-moment* (FOSM) method. Although it is generally only accurate for small variability and small nonlinearity, it is a widely used approximation because of its simplicity (see the next section).

The *second-order* approximation uses the first three terms of the Taylor's series expansion and so is potentially more accurate:

$$\mathrm{E}[Y] \simeq g(\mu_X) + \frac{1}{2}\,\mathrm{Var}[X] \left(\left.\frac{d^2g}{dx^2}\right|_{\mu_X} \right) \tag{1.73a}$$

$$\begin{aligned} \mathrm{Var}[Y] \simeq{}& \mathrm{Var}[X] \left(\left.\frac{dg}{dx}\right|_{\mu_X} \right)^2 - \left(\frac{1}{2}\,\mathrm{Var}[X] \left.\frac{d^2g}{dx^2}\right|_{\mu_X} \right)^2 \\ &+ \mathrm{E}\left[(X - \mu_X)^3\right] \left(\left.\frac{dg}{dx}\frac{d^2g}{dx^2}\right|_{\mu_X} \right) \\ &+ \frac{1}{4}\,\mathrm{E}\left[(X - \mu_X)^4\right] \left(\left.\frac{d^2g}{dx^2}\right|_{\mu_X} \right)^2 \end{aligned} \tag{1.73b}$$

Notice that the second-order approximation to the variance of Y involves knowledge of the third and fourth moments of X, which are generally difficult to estimate. Often, in practice, the second-order estimate of the mean is used along with the first-order estimate of the variance, since these both require no more than second-moment estimates of X.

1.8.3.2 Arbitrary Function of Several Variables If Y is an arbitrary function of several variables, $Y = g(X_1, X_2, \ldots, X_n)$, then the corresponding Taylor's series expansion is

$$\begin{aligned} Y ={}& g(\mu_{X_1}, \mu_{X_2}, \ldots, \mu_{X_n}) + \sum_{i=1}^{n} (X_i - \mu_{X_i}) \left.\frac{\partial g}{\partial x_i}\right|_{\boldsymbol{\mu}} \\ &+ \frac{1}{2} \sum_{i=1}^{n}\sum_{j=1}^{n} (X_i - \mu_{X_i})(X_j - \mu_{X_j}) \left.\frac{\partial^2 g}{\partial x_i\, \partial x_j}\right|_{\boldsymbol{\mu}} + \cdots \end{aligned} \tag{1.74}$$

where $\boldsymbol{\mu}$ is the vector of means, $\boldsymbol{\mu} = \{\mu_{X_1}, \mu_{X_2}, \ldots, \mu_{X_n}\}$. First-order approximations to the mean and variance of Y are then

$$\mathrm{E}[Y] \simeq g(\boldsymbol{\mu}) \tag{1.75a}$$

$$\mathrm{Var}[Y] \simeq \sum_{i=1}^{n}\sum_{j=1}^{n} \mathrm{Cov}\left[X_i, X_j\right] \left[\left.\frac{\partial g}{\partial x_i} \cdot \frac{\partial g}{\partial x_j}\right|_{\boldsymbol{\mu}} \right] \tag{1.75b}$$

Second-order approximations are

$$\mathrm{E}[Y] \simeq g(\boldsymbol{\mu}) + \frac{1}{2} \sum_{i=1}^{n}\sum_{j=1}^{n} \mathrm{Cov}\left[X_i, X_j\right] \left(\left.\frac{\partial^2 g}{\partial x_i\, \partial x_j}\right|_{\boldsymbol{\mu}} \right) \tag{1.76a}$$

$$\mathrm{Var}[Y] = (\text{involves quadruple sums and fourth-order moments}) \tag{1.76b}$$

Example 1.41 The average degree of consolidation, C, under combined vertical and radial drainage is given by the relationship (e.g., Craig, 2001)

$$C = R + V - RV \tag{1.77}$$

where R is the average degree of consolidation due to horizontal (radial) drainage only and V is the average degree of consolidation due to vertical drainage only. From observations of a particular experiment which was repeated many times, suppose that we have determined the following:

$$\begin{aligned} \mu_R &= \mathrm{E}[R] = 0.3, & \sigma_R^2 &= \mathrm{Var}[R] = 0.01 \\ \mu_V &= \mathrm{E}[V] = 0.5, & \sigma_V^2 &= \mathrm{Var}[V] = 0.04 \\ \mathrm{Cov}[R, V] &= 0.015, & (\rho_{RV} &= 0.75) \end{aligned}$$

Estimate the mean μ_C and variance σ_C^2 of the average degree of consolidation.

SOLUTION First, we will expand Eq. 1.77 in a Taylor's series about $\boldsymbol{\mu} = (\mu_R, \mu_V)$ as follows

$$C = (\mu_R + \mu_V - \mu_R\mu_V) + (R - \mu_R)\frac{\partial C}{\partial R}|_{\boldsymbol{\mu}} + \frac{1}{2}(R - \mu_R)^2\frac{\partial^2 C}{\partial R^2}|_{\boldsymbol{\mu}} + (V - \mu_V)\frac{\partial C}{\partial V}|_{\boldsymbol{\mu}} + \frac{1}{2}(V - \mu_V)^2\frac{\partial^2 C}{\partial V^2}|_{\boldsymbol{\mu}} + (R - \mu_R)(V - \mu_V)\frac{\partial^2 C}{\partial R \partial V}|_{\boldsymbol{\mu}} + \cdots$$

Truncating the approximation at second-order terms and taking the expectation result in a second-order approximation to the mean:

$$\mu_C \simeq (\mu_R + \mu_V - \mu_R\mu_V) + \mathrm{E}[R - \mu_R]\frac{\partial C}{\partial R}|_{\boldsymbol{\mu}} + \frac{1}{2}\mathrm{E}\left[(R - \mu_R)^2\right]\frac{\partial^2 C}{\partial R^2}|_{\boldsymbol{\mu}} + \mathrm{E}[V - \mu_V]\frac{\partial C}{\partial V}|_{\boldsymbol{\mu}} + \frac{1}{2}\mathrm{E}\left[(V - \mu_V)^2\right]\frac{\partial^2 C}{\partial V^2}|_{\boldsymbol{\mu}} + \mathrm{E}[(R - \mu_R)(V - \mu_V)]\frac{\partial^2 C}{\partial R \partial V}|_{\boldsymbol{\mu}}$$

$$= (\mu_R + \mu_V - \mu_R\mu_V) + \frac{1}{2}\sigma_R^2\frac{\partial^2 C}{\partial R^2}|_{\boldsymbol{\mu}} + \frac{1}{2}\sigma_V^2\frac{\partial^2 C}{\partial V^2}|_{\boldsymbol{\mu}} + \mathrm{Cov}[R, V]\frac{\partial^2 C}{\partial R \partial V}|_{\boldsymbol{\mu}}$$

The partial derivatives are

$$\frac{\partial^2 C}{\partial R^2} = 0, \qquad \frac{\partial^2 C}{\partial V^2} = 0, \qquad \frac{\partial^2 C}{\partial R \partial V} = -1$$

so that

$$\mu_C = (\mu_R + \mu_V - \mu_R\mu_V) - \mathrm{Cov}[R, V] = 0.3 + 0.5 - (0.3)(0.5) - 0.015 = 0.635$$

Note that since derivatives higher than second order disappear, this result is exact and could have been obtained directly:

$$\mathrm{E}[C] = \mathrm{E}[R + V - RV] = \mu_R + \mu_V - \mathrm{E}[RV] = \mu_R + \mu_V - \left(\mathrm{Cov}[R, V] + \mu_R\mu_V\right) = 0.3 + 0.5 - (0.015 + 0.3 \times 0.5) = 0.635$$

Can we also get an exact result for σ_C^2? If so, we would need to find

$$\mathrm{E}\left[C^2\right] = \mathrm{E}\left[(R + V - RV)^2\right] = \mathrm{E}\left[R^2 + V^2 + 2RV - 2R^2V - 2RV^2 + R^2V^2\right]$$

which involves third and fourth moments, which we do not know. We must therefore approximate σ_C^2. The first-order approximation involves just second-moment information, which we were given, and appears as follows:

$$\sigma_C^2 \simeq \mathrm{Cov}[R, R]\left(\frac{\partial C}{\partial R}\right)^2|_{\boldsymbol{\mu}} + 2\,\mathrm{Cov}[R, V]\left(\frac{\partial C}{\partial R}\right)\left(\frac{\partial C}{\partial V}\right)|_{\boldsymbol{\mu}} + \mathrm{Cov}[V, V]\left(\frac{\partial C}{\partial V}\right)^2|_{\boldsymbol{\mu}}$$

where

$$\frac{\partial C}{\partial R} = 1 - V|_{\boldsymbol{\mu}} = 1 - \mu_V = 1 - 0.5 = 0.5$$

$$\frac{\partial C}{\partial V} = 1 - R|_{\boldsymbol{\mu}} = 1 - \mu_R = 1 - 0.3 = 0.7$$

Recalling that $\mathrm{Cov}[R, R] = \sigma_R^2$ and $\mathrm{Cov}[V, V] = \sigma_V^2$, we get

$$\sigma_C^2 \simeq (0.01)(0.5)^2 + 2(0.015)(0.5)(0.7) + (0.04)(0.7)^2 = 0.0326$$

and $\sigma_C = 0.18$.

1.8.4 First-Order Second-Moment Method

The FOSM method is a relatively simple method of including the effects of variability of input variables on a resulting dependent variable. It is basically a formalized methodology based on a first-order Taylor series expansion, as discussed in the previous section. Since it is a commonly used method, it is worth describing it explicitly in this section.

The FOSM method uses a Taylor series expansion of the function to be evaluated. This expansion is truncated after the linear term (hence "first order"). The modified expansion is then used, along with the first two moments of the random variable(s), to determine the values of the first two moments of the dependent variable (hence "second moment").

Due to truncation of the Taylor series after first-order terms, the accuracy of the method deteriorates if second and higher derivatives of the function are significant. Furthermore, the method takes no account of the form of the probability density function describing the random variables, using only their mean and standard deviation. The skewness (third moment) and higher moments are ignored.

Another limitation of the traditional FOSM method is that explicit account of spatial correlation of the random variable is not typically done. For example, the soil properties at two geotechnical sites could have identical mean and standard deviations; however, at one site the properties could vary rapidly from point to point ("low" spatial correlation length) and at another they could vary gradually ("high spatial correlation length").

Consider a function $f(X, Y)$ of two random variables X and Y. The Taylor series expansion of the function about the mean values (μ_X, μ_Y), truncated after first-order terms from Eq. 1.74, gives

$$f(X,Y) \approx f(\mu_X, \mu_Y) + (X - \mu_X)\frac{\partial f}{\partial x} + (Y - \mu_Y)\frac{\partial f}{\partial y} \tag{1.78}$$

where derivatives are evaluated at (μ_X, μ_Y).

To a first order of accuracy, the expected value of the function is given by

$$\mathrm{E}\left[f(X,Y)\right] \approx f\left(\mathrm{E}[X], \mathrm{E}[Y]\right) \tag{1.79}$$

and the variance by

$$\mathrm{Var}\left[f(X,Y)\right] \approx \mathrm{Var}\left[(X - \mu_X)\frac{\partial f}{\partial x} + (Y - \mu_Y)\frac{\partial f}{\partial y}\right] \tag{1.80}$$

Hence,

$$\mathrm{Var}\left[f(X,Y)\right] \approx \left(\frac{\partial f}{\partial x}\right)^2 \mathrm{Var}[X] + \left(\frac{\partial f}{\partial y}\right)^2 \mathrm{Var}[Y] + 2\frac{\partial f}{\partial x}\frac{\partial f}{\partial y}\mathrm{Cov}[X,Y] \tag{1.81}$$

If X and Y are uncorrelated,

$$\mathrm{Var}\left[f(X,Y)\right] \approx \left(\frac{\partial f}{\partial x}\right)^2 \mathrm{Var}[X] + \left(\frac{\partial f}{\partial y}\right)^2 \mathrm{Var}[Y] \tag{1.82}$$

In general, for a function of n uncorrelated random variables, the FOSM method tells us that

$$\mathrm{Var}\left[f(X_1, X_2, \ldots, X_n)\right] \approx \sum_{i=1}^{n} \left(\frac{\partial f}{\partial x_i}\right)^2 \mathrm{Var}[X_i] \tag{1.83}$$

where the first derivatives are evaluated at the mean values $(\mu_{X_1}, \mu_{X_2}, \ldots, \mu_{X_n})$.

1.9 COMMON DISCRETE PROBABILITY DISTRIBUTIONS

Many engineered systems have the same statistical behavior: We generally only need a handful of probability distributions to characterize most naturally occurring phenomena. In this section, the most common discrete distribution will be reviewed (the next section looks at the most comment continuous distributions). These are the *Bernoulli family* of distributions, since they all derive from the first:

1. Bernoulli
2. Binomial
3. Geometric
4. Negative binomial
5. Poisson
6. Exponential
7. Gamma

The Poisson, exponential, and gamma are the continuous-time analogs of the binomial, geometric, and negative binomial, respectively, arising when each instant in time is viewed as an independent Bernoulli trial. In this section we consider the *discrete* members of the Bernoulli family, which are the first five members listed above, looking briefly at the main characteristics of each of these distributions and describing how they are most commonly used in practice. Included with the statistical properties of each distribution is the maximum-likelihood estimate (MLE) of their parameters. We do not formally cover the maximum-likelihood method until Section 5.2.1.2, but we present these results along with their distributions to keep everything together.

For a more complete description of these distributions, the interested reader should consult an introductory textbook on probability and statistics, such as Law and Kelton (1991) or Devore (2003).

1.9.1 Bernoulli Trials

All of the discrete distributions considered in this section (and the first two in the next section) are derived from the idea of *Bernoulli trials*. A Bernoulli trial is an experiment which has only two possible outcomes, *success* or *failure* (or $[1, 0]$, or [true, false], or $[< 5, \geq 5]$, etc). If a sequence of Bernoulli trials are mutually independent with constant (stationary) probability p of success, then the sequence is called a *Bernoulli process*. There are many examples of Bernoulli processes: One might model the failures of earth dams using a Bernoulli process. The success or failure of each of a sequence of bids made by a company might be a Bernoulli process. The failure of piles to support the load applied on them might be a Bernoulli process if it can be assumed that the piles fail (or survive) independently and with constant probability. However, if the failure of one pile is dependent on the failure of adjacent piles, as might be the case if the soil structures are similar and load transfer takes place, the Bernoulli model may not be appropriate and a more complex, "dependent," model may be required, for example, random field modeling of the soil and finite-element analysis of the structural response within a Monte Carlo simulation. Evidently, when we depart from satisfying the assumptions underlying the simple models, such as those required for the Bernoulli model, the required models rapidly become very much more complicated. In some cases, applying the simple model to the more complex problem will yield a ballpark estimate, or at least a bound on the probability, and so it may be appropriate to proceed with a Bernoulli model taking care to treat the results as approximate. The degree of approximation depends very much on the degree of dependence between "trials" and the "stationarity" of the probability of "success," p.

If we let

$$X_j = \begin{cases} 1 & \text{if the } j\text{th trial results in a success} \\ 0 & \text{if the } j\text{th trial results in a failure} \end{cases} \tag{1.84}$$

then the Bernoulli distribution, or probability mass function, is given by

$$\begin{aligned} \mathrm{P}\left[X_j = 1\right] &= p \\ \mathrm{P}\left[X_j = 0\right] &= 1 - p = q \end{aligned} \tag{1.85}$$

for all $j = 1, 2, \ldots$. Note that we commonly denote $1 - p$ as q for simplicity.

For a single Bernoulli trial the following results hold:

$$\begin{aligned} \mathrm{E}\left[X_j\right] &= \sum_{i=0}^{1} i \cdot \mathrm{P}\left[X_j = i\right] \\ &= 0(1-p) + 1(p) = p \end{aligned} \tag{1.86a}$$

$$\mathrm{E}\left[X_j^2\right] = \sum_{i=0}^{1} i^2 \cdot \mathrm{P}\left[X_j = i\right] = 0^2(1-p) + 1^2(p) = p$$

$$\mathrm{Var}\left[X_j\right] = \mathrm{E}\left[X_j^2\right] - \mathrm{E}^2\left[X_j\right] = p - p^2 = pq \tag{1.86b}$$

For a sequence of trials, the assumption of independence between the trials means that

$$\begin{aligned} &\mathrm{P}\left[X_1 = x_1 \cap X_2 = x_2 \cap \cdots \cap X_n = x_n\right] \\ &\quad = \mathrm{P}\left[X_1 = x_1\right] \mathrm{P}\left[X_2 = x_2\right] \cdots \mathrm{P}\left[X_n = x_n\right] \end{aligned} \tag{1.87}$$

The MLE of p is just the average of the set of observations, $x_1, x_2, \ldots, x_n$, of X,

$$\hat{p} = \frac{1}{n} \sum_{i=1}^{n} x_i \tag{1.88}$$

Notice that we use a hat to indicate that this is just an *estimate* of the true parameter p. Since the next set of observations will likely give a different value for $\hat{p}$, we see that $\hat{p}$ is actually a random variable itself, rather than the true population parameter, which is nonrandom. The mean and variance of the sequence of $\hat{p}$ can be found by considering the random $\hat{P}$,

$$\hat{P} = \frac{1}{n} \sum_{i=1}^{n} X_i \tag{1.89}$$

obtained *prior* to observing the results of our Bernoulli trials. We get

$$\begin{aligned} \mathrm{E}\left[\hat{P}\right] &= \mathrm{E}\left[\frac{1}{n} \sum_{i=1}^{n} X_i\right] \\ &= \frac{1}{n} \sum_{i=1}^{n} \mathrm{E}\left[X_i\right] = \frac{1}{n}(np) \\ &= p \end{aligned} \tag{1.90}$$

which means that the estimator given by Eq. 1.88 is *unbiased* (that is, the estimator is "aimed" at its desired target on average).

The estimator variance is

$$\begin{aligned} \mathrm{Var}\left[\hat{P}\right] &= \mathrm{Var}\left[\frac{1}{n} \sum_{i=1}^{n} X_i\right] \\ &= \frac{1}{n^2} \sum_{i=1}^{n} \mathrm{Var}\left[X_i\right] = \frac{1}{n^2}(npq) \\ &= \frac{pq}{n} \end{aligned} \tag{1.91}$$

where we made use of the fact that the variance of a sum is the sum of the variances *if the random variables are uncorrelated.* We are assuming that, since this is a Bernoulli process, not only are the random variables uncorrelated, but also they are completely independent (the probability of one occurring is not affected by the probability of other occurrences).

Note that the estimator variance depends on the true value of p on the right-hand-side of Eq. 1.91. Since we are estimating p, we obviously do not know the true value. The solution is to use our estimate of p to estimate its variance, so that

$$\sigma_{\hat{P}}^2 \simeq \frac{\hat{p}\hat{q}}{n} \tag{1.92}$$

Once we have determined the estimator variance, we can compute its *standard error*, which is commonly taken to be equal to the standard deviation and which gives an indication of how accurate our estimate is,

$$\sigma_{\hat{p}} \simeq \sqrt{\frac{\hat{p}\hat{q}}{n}} \tag{1.93}$$

For example, if $\hat{p} = 0.01$, then we would prefer $\sigma_{\hat{p}}$ to be quite a bit smaller than 0.01 and we can adjust the number of observations n to achieve this goal.

In Part 2 of this book, we will be estimating the probability of failure, p_f, of various classic geotechnical problems using a technique called *Monte Carlo simulation*. The standard error given by Eq. 1.93 will allow us to estimate the accuracy of our failure probability estimates, assuming that each "simulation" results in an independent failure/success trial.

Applications The classic Bernoulli trial is the toss of a coin, but many other experiments can lead to Bernoulli trials under the above conditions. Consider the following examples:

1. Soil anchors at a particular site have a 1% probability of pulling out. When an anchor is examined, it is classified as a success if it has not pulled out or a failure if it has. This is a Bernoulli trial with $p = 0.99$ if the anchors fail independently and if the probability of success remains constant from trial to trial.
2. Suppose that each sample of soil at a site has a 10% chance of containing significant amounts of chromium. A sample is analyzed and classified as a success if it does not contain significant amounts of chromium and a failure if it does. This is a Bernoulli trial with $p = 0.90$ if the samples are independent and if the probability of success remains constant from trial to trial.
3. A highway through a certain mountain range passes below a series of steep rock slopes. It is estimated that each rock slope has a 2% probability of failure (resulting in some amount of rock blocking the highway) over the next 10 years. If we define each rock slope as a trial which is a success if it does not fail in the next 10 years, then this can be modeled as a Bernoulli trial with $p = 0.98$ (assuming rock slopes fail independently, which might not be a good assumption if they generally fail due to, e.g., earthquakes).

1.9.2 Binomial Distribution

Let N_n be the number of successes in n Bernoulli trials, each with probability of success p. Then N_n follows a binomial distribution where

$$\mathrm{P}[N_n = k] = \binom{n}{k} p^k q^{n-k}, \qquad k = 0, 1, 2, \ldots, n \tag{1.94}$$

The quantity $p^k q^{n-k}$ is the probability of obtaining k successes and $n - k$ failures in n trials and $\binom{n}{k}$ is the number of possible ways of arranging the k successes over the n trials.

For example, consider eight trials which can be represented as a series of eight dashes:

___ ___ ___ ___ ___ ___ ___ ___

One possible realization of three successes in eight trials might be

F S F F S S F F

where successes are shown as S and failures as F. Another possible realization might be

S F F S F F F S

and so on. Clearly these involve three successes, which have probability p^3, and five failures, which have probability q^5. Combining these two probabilities with the fact that three successes in eight trials can be arranged in $\binom{8}{3}$ different ways leads to

$$\mathrm{P}[N_8 = 3] = \binom{8}{3} p^3 q^{8-3}$$

which generalizes to the binomial distribution for n trials and k successes given above. See Figure 1.20.

Properties In the following proofs, we make use of the binomial theorem, which states that

$$(\alpha + \beta)^n = \sum_{i=0}^{n} \binom{n}{i} \alpha^i \beta^{n-i} = \sum_{i=0}^{n} \frac{n!}{i!(n-i)!} \alpha^i \beta^{n-i} \tag{1.95}$$

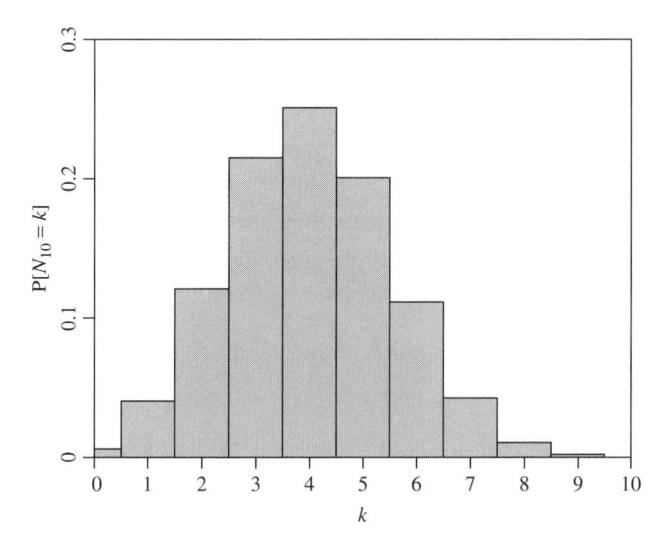

Figure 1.20 Binomial distribution for $n = 10$ and $p = 0.4$.

The expected number of successes in n trials can be found directly from the definition of the discrete-case expectation,

$$\begin{aligned}
\mathrm{E}[N_n] &= \sum_{i=0}^{n} i \binom{n}{i} p^i q^{n-i} \\
&= \sum_{i=0}^{n} i \left(\frac{n!}{i!(n-i)!} \right) p^i q^{n-i} \\
&= np \sum_{i=1}^{n} \frac{(n-1)!}{(i-1)!(n-i)!} p^{i-1} q^{n-i} \\
&= np \sum_{i=0}^{(n-1)} \frac{(n-1)!}{i!((n-1)-i)!} p^i q^{(n-1)-i} \\
&= np(p+q)^{n-1} \\
&= np
\end{aligned} \tag{1.96}$$

since $p + q = 1$.

Alternatively, we could write

$$\begin{aligned}
\mathrm{E}[N_n] &= \mathrm{E}[X_1 + X_2 + \cdots + X_n] \\
&= \mathrm{E}[X_1] + \mathrm{E}[X_2] + \cdots + \mathrm{E}[X_n] \\
&= np
\end{aligned}$$

where X_i is a Bernoulli random variable having expectation p.

To find the variance of N_n, we first need to find

$$\begin{aligned}
\mathrm{E}\left[N_n^2\right] &= \sum_{i=0}^{n} i^2 \binom{n}{i} p^i q^{n-i} = \sum_{i=1}^{n} i^2 \left(\frac{n!}{i!(n-i)!} \right) p^i q^{n-i} \\
&= np \sum_{i=1}^{n} i \left(\frac{(n-1)!}{(i-1)!(n-i)!} \right) p^{i-1} q^{n-i} \\
&= np \sum_{i=0}^{n-1} (i+1) \left(\frac{(n-1)!}{i!(n-1-i)!} \right) p^i q^{n-1-i} \\
&= np \left\{ \sum_{i=0}^{n-1} i \left(\frac{(n-1)!}{i!(n-1-i)!} \right) p^i q^{n-1-i} \right. \\
&\quad \left. + \sum_{i=0}^{n-1} \left(\frac{(n-1)!}{i!(n-1-i)!} \right) p^i q^{n-1-i} \right\} \\
&= np \{(n-1)p + 1\} \\
&= npq + n^2 p^2
\end{aligned}$$

where for the first sum we made use of the result given by Eq. 1.96. The variance is thus

$$\mathrm{Var}[N_n] = \mathrm{E}\left[N_n^2\right] - \mathrm{E}^2[N_n] = npq + n^2p^2 - n^2p^2 = npq \tag{1.97}$$

The same result could have been obtained much more easily by considering the variance of a sum of independent random variables, since in this case the variance of a sum is the sum of the variances:

$$\mathrm{Var}[N_n] = \mathrm{Var}\left[\sum_{i=1}^{n} X_i \right] = \sum_{i=1}^{n} \mathrm{Var}\, X_i = npq$$

The MLE of p is

$$\hat{p} = \frac{\bar{N}_n}{n} \tag{1.98}$$

if n is known, where $\bar{N}_n$ is the average of the observed values of N_n. If both n and p are unknown, see Law and Kelton (2000) for the MLE. This estimator is precisely the same as that given by Eq. 1.89 since $N_n = \sum_{i=1}^{n} X_i$, and so its mean and standard error are discussed in the previous section (with n replaced by the total number of trials making up $\bar{N}_n$).

Example 1.42 A manufacturer of geotextile sheets wishes to control the quality of its product by rejecting any lot in which the proportion of textile sheets having unacceptably low tensile strength appears to be too high. To this end, out of each large lot (1000 sheets), 25 will be selected and tested. If 5 or more of these sheets have an unacceptably low tensile strength, the entire lot will be rejected. What is the probability that a lot will be rejected if

1. 5% of the sheets in the lot have unacceptably low tensile strength?
2. 10% of the sheets in the lot have unacceptably low tensile strength?

SOLUTION

1. Let N_{25} be the number of sheets that have unacceptably low tensile strengths out of the 25 sampled.
   If the sheets fail the tension test independently with constant probability of failure, then N_{25} follows a binomial distribution with $p = 0.05$. We note that since the number of low-strength sheets in a lot is fixed, the probability of failure will change as sheets are tested. For example, if 50 out of 1000 sheets are low strength, then the probability of failure of the first sheet tested is 0.05. The probability of failure of the second sheet tested is either 49/999 or 50/999, depending on whether the first sheet tested was low strength or not. However, if the lot size (1000 in this case) is large relative to the number selected for testing (25 in this case), then the approximation that p is constant is reasonable and will lead to fairly accurate results. We will make this assumption here, so

that

$$\begin{aligned}
\mathrm{P}[N_{25} \geq 5] &= 1 - \mathrm{P}[N_{25} \leq 4] \\
&= 1 - \mathrm{P}[N_{25} = 0] - \mathrm{P}[N_{25} = 1] \\
&\quad - \mathrm{P}[N_{25} = 2] - \mathrm{P}[N_{25} = 3] \\
&\quad - \mathrm{P}[N_{25} = 4] \\
&= 1 - \binom{25}{0}(0.05)^0(0.95)^{25} \\
&\quad - \binom{25}{1}(0.05)^1(0.95)^{24} \\
&\quad - \binom{25}{2}(0.05)^2(0.95)^{23} \\
&\quad - \binom{25}{3}(0.05)^3(0.95)^{22} \\
&\quad - \binom{25}{4}(0.05)^4(0.95)^{21} \\
&= 0.00716
\end{aligned}$$

Thus, there is a very small probability of rejecting a lot where 5% of the sheets have an unacceptably low tensile strength.

2. Let N_{25} be the number of sheets that have unacceptably low tensile strengths out of the 25 sampled. Then N_{25} follows a binomial distribution with $p = 0.10$ (we will again assume sheets fail the test independently and that the probability of this happening remains constant from sheet to sheet):

$$\begin{aligned}
\mathrm{P}[N_{25} \geq 5] &= 1 - \mathrm{P}[N_{25} \leq 4] \\
&= 1 - \mathrm{P}[N_{25} = 0] - \mathrm{P}[N_{25} = 1] \\
&\quad - \mathrm{P}[N_{25} = 2] - \mathrm{P}[N_{25} = 3] \\
&\quad - \mathrm{P}[N_{25} = 4] \\
&= 1 - \binom{25}{0}(0.10)^0(0.90)^{25} \\
&\quad - \binom{25}{1}(0.10)^1(0.90)^{24} \\
&\quad - \binom{25}{2}(0.10)^2(0.90)^{23} \\
&\quad - \binom{25}{3}(0.10)^3(0.90)^{22} \\
&\quad - \binom{25}{4}(0.10)^4(0.90)^{21} \\
&= 0.098
\end{aligned}$$

There is now a reasonably high probability (about 10%) that a lot will be rejected if 10% of the sheets have an unacceptably low tensile strength.

1.9.3 Geometric Distribution

Consider a Bernoulli process in which T_1 is the number of trials required to achieve the first success. Thus, if $T_1 = 3$, then we must have had two failures followed by a success (the value of T_1 fully prescribes the sequence of trials). This has probability

$$\mathrm{P}[T_1 = 3] = \mathrm{P}[\{\text{failure, failure, success}\}] = q^2 p$$

In general

$$\mathrm{P}[T_1 = k] = q^{k-1}p, \qquad k = 1, 2, \ldots \tag{1.99}$$

Note that this is a valid probability mass function since

$$\sum_{k=1}^{\infty} q^{k-1}p = p\sum_{k=0}^{\infty} q^k = \frac{p}{1-q} = 1$$

where we used the fact that for any $\alpha < 1$ (see, e.g., Gradshteyn and Ryzhik, 1980)

$$\sum_{k=0}^{\infty} \alpha^k = \frac{1}{1-\alpha} \tag{1.100}$$

As an example, in terms of the actual sequence of trials, the event that the first success occurs on the eighth trial appears as

F F F F F F F S

That is, the single success always occurs on the last trial. If $T_1 = 8$, then we have had seven failures, having probability q^7, and one success, having probability p. Thus

$$\mathrm{P}[T_1 = 8] = q^7 p$$

Generalizing this for $T_1 = k$ leads to the geometric distribution shown in Figure 1.21.

Because trials are assumed independent, the geometric distribution also models the number of trials *between* successes in a Bernoulli process. That is, suppose we observe the result of the Bernoulli process at trial number 1032. We will observe either a success or failure, but whichever is observed, it is now *known*. We can then ask a question such as: What is the probability that the next success occurs on trial 1040? To determine this, we start with trial 1032. Because we have observed that there is no uncertainty associated with trial 1032, it does not enter into the probability problem. However, trials 1033, 1034, ..., 1040 are unknown. We are asking for the probability that trial 1040 is the first success after 1032. In order for this *event* to occur, trials 1033–1039 must be failures. Thus, the eight trials,

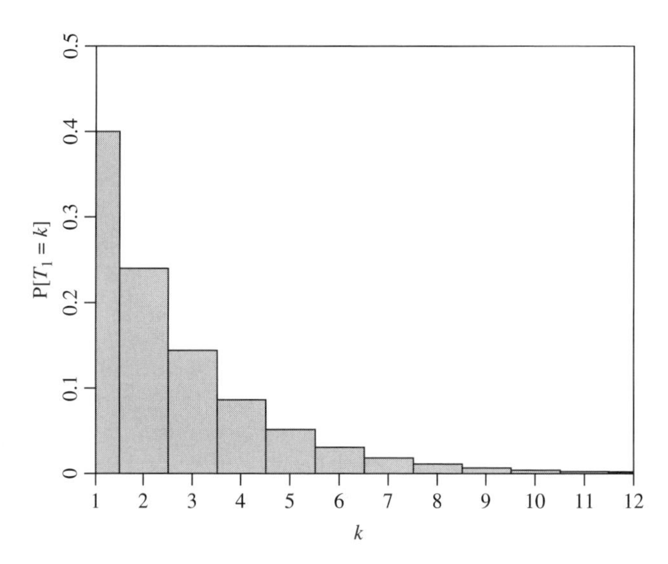

Figure 1.21 Geometric distribution for $p = 0.4$.

1033–1040, must involve seven failures (q^7) followed by one success (p). The required probability is just the product

$$\mathrm{P}\left[T_1 = 8\right] = q^7 p$$

What this means is that the geometric distribution, by virtue of the independence between trials, is *memoryless*. It does not matter when you start looking at a Bernoulli process, the number of trials to the next success is given by the geometric distribution (and is independent of the trial number).

Properties The mean of T_1, which is also sometimes referred to as the *return period* or the *mean recurrence time*, is determined as

$$\begin{aligned}\mathrm{E}\left[T_1\right] &= \sum_{k=1}^{\infty} kpq^{k-1} = p\sum_{k=1}^{\infty} kq^{k-1} \\ &= p\frac{d}{dq}\sum_{k=1}^{\infty} q^k = p\frac{d}{dq}\left(\frac{q}{1-q}\right) \\ &= p\left(\frac{1}{(1-q)^2}\right) = \frac{1}{p} \qquad (1.101)\end{aligned}$$

where we used Eq. 1.100 to evaluate the final sum above. We will use the second to last sum in the following proof.

The variance of T_1 is obtained from $\mathrm{Var}\left[T_1\right] = \mathrm{E}\left[T_1^2\right] - \mathrm{E}^2[T_1]$ as

$$\begin{aligned}\mathrm{E}\left[T_1^2\right] &= \sum_{k=1}^{\infty} k^2 pq^{k-1} = p\sum_{k=1}^{\infty} k^2 q^{k-1} = p\frac{d}{dq}\sum_{k=1}^{\infty} kq^k \\ &= p\frac{d}{dq}\left(\frac{q}{(1-q)^2}\right) \\ &= \frac{1}{p} + \frac{2q}{p^2}\end{aligned}$$

Thus

$$\begin{aligned}\mathrm{Var}\left[T_1\right] &= \mathrm{E}\left[T_1^2\right] - \mathrm{E}^2[T_1] \\ &= \frac{1}{p} + \frac{2q}{p^2} - \frac{1}{p^2} \\ &= \frac{q}{p^2} \qquad (1.102)\end{aligned}$$

As an aside, in engineering problems, we often reverse the meaning of success and failure and use the geometric distribution to model time to failure, where time is measured in discrete steps (trials).

The MLE of p is

$$\hat{p} = \frac{n}{\sum_{i=1}^{n} t_i} = \frac{1}{\bar{t}} \qquad (1.103)$$

where $t_1, t_2, \ldots, t_n$ are n independent observations of T_1.

Example 1.43 Recall the previous example where a manufacturer of geotextile sheets wishes to control the quality of its product by rejecting any lot in which the proportion of textile sheets having unacceptably low tensile strength appears to be too high. Suppose now that the sampling scheme is changed and the manufacturer decides to only sample geotextile sheets until one is encountered having an unacceptably low tensile strength. If this occurs on or before the eighth sheet tested, the entire lot will be rejected. What is the probability that a lot will be rejected if

1. 5% of the sheets in the lot have unacceptably low tensile strengths?
2. 10% of the sheets in the lot have unacceptably low tensile strengths?

If having 5% of the sheets in a lot with unacceptably low tensile strength is detrimental to the manufacturer's image and such a lot should not be sent to market, it appears that this control approach would work better than that of Example 1.39. However, if the manufacturer is more concerned with profit, this control approach is definitely not to their advantage. What might be the disadvantage of this approach from the point of view of the manufacturer? Explain with the help of a numerical example.

SOLUTION

1. Let T_1 be the trial number of the first sheet to have an unacceptably low tensile strength. Then, assuming independence between sheets and constant probability of success, T_1 follows a geometric distribution with

$p = 0.05$ and

$$\begin{aligned} P[T_1 \le 8] &= P[T_1 = 1] + P[T_1 = 2] \\ &\quad + \cdots + P[T_1 = 7] + P[T_1 = 8] \\ &= 0.05 + 0.95(0.05) \\ &\quad + \cdots + 0.95^6(0.05) + 0.95^7(0.05) \\ &= 0.337 \end{aligned}$$

2. Let T_1 be the trial number of the first sheet to have an unacceptably low tensile strength. Then, under the same assumptions as in item 1, T_1 follows a geometric distribution with $p = 0.10$ and

$$\begin{aligned} P[T_1 \le 8] &= P[T_1 = 1] + P[T_1 = 2] \\ &\quad + \cdots + P[T_1 = 7] + P[T_1 = 8] \\ &= 0.10 + 0.90(0.10) \\ &\quad + \cdots + 0.90^6(0.10) + 0.90^7(0.10) \\ &= 0.570 \end{aligned}$$

3. The problem with this approach, from the point of view of the manufacturer, is that a significant proportion of lots with less than 5% unacceptably low-strength sheets would be rejected (e.g., about a third). In addition, consider what happens under this quality control approach when only 2% of the sheets in the lot have unacceptably low tensile strength. (We will assume here that this is actually fairly good quality control, although, in practice, the acceptable risks can certainly vary.)
Let T_1 be the trial number of the first sheet to have an unacceptably low tensile strength. Then T_1 follows a geometric distribution, under the above assumptions, with $p = 0.02$ and

$$\begin{aligned} P[T_1 \le 8] &= P[T_1 = 1] + P[T_1 = 2] \\ &\quad + \cdots + P[T_1 = 7] + P[T_1 = 8] \\ &= 0.02 + 0.98(0.02) + \cdots + 0.98^6(0.02) \\ &\quad + 0.98^7(0.02) \\ &= 0.149 \end{aligned}$$

so that there is still approximately a 15% chance that such a "good" lot would be rejected. This test does not sufficiently "resolve" the critical fraction of defectives.

1.9.4 Negative Binomial Distribution

Suppose we wish to know the number of trials (time) in a Bernoulli process until the mth success. Letting T_m be the number of trials until the mth success,

$$P[T_m = k] = \binom{k-1}{m-1} p^m q^{k-m} \quad \text{for } k = m, m+1, \ldots \tag{1.104}$$

which is the *negative binomial distribution*. Whereas a binomial distributed random variable is the number of successes in a fixed number of trials, a negative binomial distributed random variable is the number of trials for a fixed number of successes. We note that the negative binomial is also often used to model the number of failures before the mth success, which results in a somewhat different distribution. We prefer the interpretation that the negative binomial distribution governs the number of trials until the mth success because it is a natural generalization of the geometric distribution and because it is then a discrete analog of the gamma distribution considered in Section 1.10.2.

The name of the negative binomial distribution arises from the *negative binomial series*

$$(1-q)^{-m} = \sum_{k=m}^{\infty} \binom{k-1}{m-1} q^{k-m} \tag{1.105}$$

which converges for $|q| < 1$. This series can be used to show that the negative binomial distribution is a valid distribution, since

$$\begin{aligned} \sum_{k=m}^{\infty} P[T_m = k] &= \sum_{k=m}^{\infty} \binom{k-1}{m-1} p^m q^{k-m} \\ &= p^m \sum_{k=m}^{\infty} \binom{k-1}{m-1} q^{k-m} \\ &= p^m (1-q)^{-m} \\ &= 1 \end{aligned} \tag{1.106}$$

as expected.

We see that the geometric distribution is a special case of the negative binomial distribution with $m = 1$. The negative binomial distribution is often used to model 'time to the mth failure, where time is measured in discrete steps, or trials. Consider one possible realization which has the third success on the eighth trial:

F S S F F F F S

Another possible realization might be

F F F S F S F S

In both cases, the number of successes is 3, having probability p^3, and the number of failures is 5, having probability q^5. In terms of ordering, if $T_3 = 8$, then the third success must occur on the eighth trial (as shown above). Thus, the only other uncertainty is the ordering of the other two successes. This can occur in $\binom{7}{2}$ ways. The probability that the third success occurs on the eighth trial is therefore given by

$$P[T_3 = 8] = \binom{7}{2} p^3 q^5$$

Generalizing this for m successes and k trials leads to the negative binomial distribution shown in Eq. 1.104.

Properties The mean is determined as

$$\begin{aligned}
E[T_m] &= \sum_{j=m}^{\infty} j P[T_m = j] = \sum_{j=m}^{\infty} j \binom{j-1}{m-1} p^m q^{j-m} \\
&= \sum_{j=m}^{\infty} j \left(\frac{(j-1)!}{(m-1)!(j-m)!} \right) p^m q^{j-m} \\
&= m p^m \sum_{j=m}^{\infty} \left(\frac{j!}{m!(j-m)!} \right) q^{j-m} \\
&= m p^m \left[1 + (m+1)q + \frac{(m+2)(m+1)}{2!} q^2 \right. \\
&\quad \left. + \frac{(m+3)(m+2)(m+1)}{3!} q^3 + \cdots \right] \\
&= \frac{m p^m}{(1-q)^{m+1}} \\
&= \frac{m}{p}
\end{aligned} \tag{1.107}$$

which is just m times the mean of a single geometrically distributed random variable T_1, as expected, since the number of trials between successes follows a geometric distribution. In fact, this observation leads to the following alternative representation of T_m,

$$T_m = T_{1,1} + T_{1,2} + \cdots + T_{1,m} \tag{1.108}$$

where $T_{1,1}$ is the number of trials until the first success, $T_{1,2}$ is the number of trials after the first success until the second success, and so on. That is, the $T_{1,i}$ terms are just the times between successes. Since all trials are independent, each of the $T_{1,i}$ terms will be independent geometrically distributed random variables, all having common probability of success, p. This leads to the following much simpler computation:

$$E[T_m] = E[T_{1,1}] + E[T_{1,2}] + \cdots + E[T_{1,m}] = \frac{m}{p} \tag{1.109}$$

since $E[T_{1,i}] = 1/p$ for all $i = 1, 2, \ldots, m$. The mean in Figure 1.22 is $3/0.4 = 7.5$.

To get the variance, $\text{Var}[T_m]$, we again use Eq. 1.108. Due to independence of the $T_{1,i}$ terms, the variance of the sum is the sum of the variances,

$$\begin{aligned}
\text{Var}[T_m] &= \text{Var}[T_{1,1}] + \text{Var}[T_{1,2}] + \cdots + \text{Var}[T_{1,m}] \\
&= m \,\text{Var}[T_1] \\
&= \frac{mq}{p^2}
\end{aligned} \tag{1.110}$$

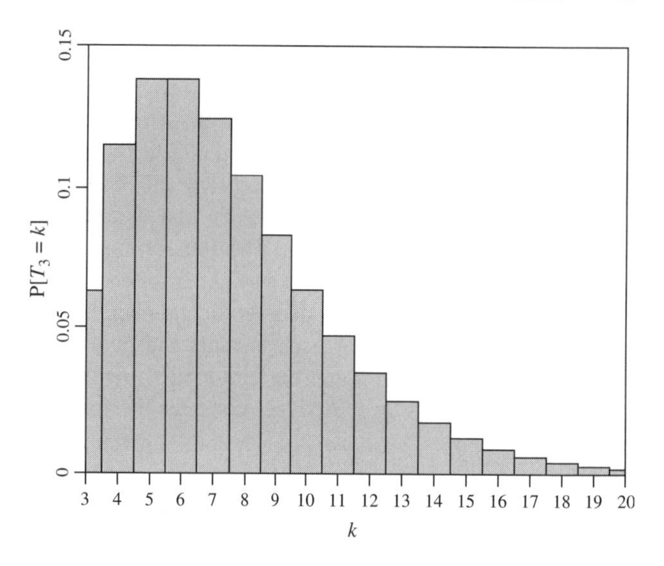

Figure 1.22 Negative binomial distribution for T_3 (i.e., $m = 3$) and $p = 0.4$.

which is just m times the variance of a single geometrically distributed random variable T_1, as expected.

If m is known, then the MLE of p is

$$\hat{p} = \frac{mn}{\sum_{i=1}^{n} x_i} = \frac{m}{\bar{x}} \tag{1.111}$$

where $x_1, x_2, \ldots, x_n$ are n independent observations of T_m. If m is unknown, see Law and Kelton (2000), although beware of the fact that Law and Kelton define their negative binomial as governing the number of failures prior to the mth success, not as the number of trials until the mth success, as is done here.

Example 1.44 Consider again the problem of the tensile strength of geotextile sheets of the previous two examples. If 10% of the sheets have unacceptably low tensile strengths, what is the probability that on the next series of tests the third sheet to fail the tensile test is the eighth sheet tested?

SOLUTION Let T_3 be the number of sheets tested when the third sheet to fail the tensile test is encountered (note, this *includes* the sheet being tested). Then we are looking for

$$P[T_3 = 8] = \binom{7}{2} (0.10)^3 (0.9)^{8-3} = 0.0124$$

1.9.5 Poisson Distribution

If we now allow every instant in time (or space) to be a Bernoulli trial, we get a family of three distributions: the Poisson distribution, the exponential distribution, and the gamma distribution. The latter two are continuous distributions governing the time between trial successes and are discussed in the next section. The Poisson distribution is analogous to the binomial distribution: It is derived from the binomial distribution by letting the number of trials go to infinity (one trial for each instant) and governs the number of successes in some time interval t. To see how the Poisson distribution is derived, consider the following example.

Example 1.45 Derivation from Binomial Distribution Suppose that it is known that along a certain long highway stretch an average of 1 slope subsidence occurs per year. What is the probability that exactly 10 slope subsidences will occur in the next 10-year interval?

SOLUTION If we attempt to model this using the binomial distribution, we must first divide time up into a series of intervals within each of which a slope can either subside (success) or not (failure). As a starting point, let us assume that at most one slope can subside in any half-year interval. We make this assumption because a Bernoulli trial can only have two outcomes, and if we wish to be able to count the number of subsidences, we must make these two possible outcomes either 1 (a single slope subsides) or 0 (no slopes subside). If our trials are a half-year in duration, then we have 20 trials in 10 years and the probability of a success (a slope subsides) in each trial is the rate per year divided by the number of trials per year: $p = \frac{1}{2}$. In our 10-year interval the probability we are looking for is

$$\begin{aligned} \mathrm{P}\left[10 \text{ subsidences in 10 years}\right] \\ \simeq \binom{20}{10}(0.5)^{10}(0.5)^{20-10} = 0.176 \end{aligned}$$

Of course, we know that two or more slope subsidences could easily occur within any half-year interval. An improved solution is obtained by using a shorter trial interval. If 2-month intervals were to be used then we now have six trials per year and the probability of a slope subsidence in any interval becomes $p = \frac{1}{6}$. The number of trials in 10 years (120 months) becomes $n = \frac{120}{2} = 60$

$$\begin{aligned} \mathrm{P}\left[10 \text{ subsidences in 10 years}\right] \\ \simeq \binom{60}{10}\left(\frac{1}{6}\right)^{10}\left(\frac{5}{6}\right)^{50} = 0.137 \end{aligned}$$

which is quite a bit more accurate.

In general, if time interval t is divided into n intervals and the mean arrival rate is λ, then

$$p = \frac{\lambda t}{n} \tag{1.112}$$

and if N_t is the number of subsidences in t years,

$$\mathrm{P}\left[N_t = k\right] = \binom{n}{k}\left(\frac{\lambda t}{n}\right)^k\left(1 - \frac{\lambda t}{n}\right)^{n-k}$$

where λt is the mean number of subsidences ("arrivals") occurring in time interval t. If arrivals are instantaneous (so that no more than one can occur in any instant with probability 1) and can occur at any instant in time, so that each instant in time becomes a Bernoulli trial, then

$$\begin{aligned} \mathrm{P}\left[N_t = k\right] &= \lim_{n\to\infty}\binom{n}{k}\left(\frac{\lambda t}{n}\right)^k\left(1 - \frac{\lambda t}{n}\right)^{n-k} \\ &= \lim_{n\to\infty}\left[\left\{\frac{n}{n}\cdot\frac{n-1}{n}\cdots\frac{n-k+1}{n}\right\}\right. \\ &\quad \left.\times\frac{(\lambda t)^k}{k!}\left(1 - \frac{\lambda t}{n}\right)^n\left(1 - \frac{\lambda t}{n}\right)^{-k}\right] \end{aligned}$$

but since

$$\lim_{n\to\infty}\left\{\frac{n}{n}\cdot\frac{n-1}{n}\cdots\frac{n-k+1}{n}\right\} = 1$$

$$\lim_{n\to\infty}\left(1 - \frac{\lambda t}{n}\right)^{-k} = 1$$

$$\lim_{n\to\infty}\left(1 - \frac{\lambda t}{n}\right)^{n} = e^{-\lambda t}$$

then our distribution simplifies to

$$\mathrm{P}\left[N_t = k\right] = \frac{(\lambda t)^k}{k!}e^{-\lambda t}$$

which is the Poisson distribution. In other words, the Poisson distribution is a limiting case of the binomial distribution, obtained when the number of trials goes to infinity, one for each instant in time, and p is replaced by the mean rate λ.

For our problem $\lambda = 1$ subsidence per year and $t = 10$ years. The probability of exactly 10 subsidences in 10 years using the Poisson distribution is

$$\mathrm{P}\left[N_{10} = 10\right] = \frac{(10)^{10}}{10!}e^{-10} = 0.125$$

and we see that the binomial model using 2-month trial intervals gives a reasonably close result (with a relative error of less than 10%).

We note that the Poisson model assumes independence between arrivals. In the subsidence problem mentioned above, there may be significant dependence between occurrences, if, for example, they are initiated by spatially extended rainfall or freeze/thaw action. When dependence exists between trials and some common outside influence (e.g., weather), the model is complicated by the fact that the rate of occurrence becomes dependent on time. One possible solution is to apply different Poisson models for different time periods (e.g., wet season vs. dry season) or to investigate nonstationary Poisson models.

The Poisson distribution is often used to model arrival processes. We shall see in Chapter 4 that it is also useful to model "excursion" processes, for example, the number of weak pockets in a soil mass. For simplicity, we will talk about Poisson processes in time, but recognize that they can be equivalently applied over space simply by replacing t with a distance (or area, volume, etc.) measure.

For any nonzero time interval we have an infinite number of Bernoulli trials, since any time interval is made up of an infinite number of instants. Thus, the probability of success, p, in any one instant must go to zero (see Eq. 1.112); otherwise we would have an infinite number of successes in each time interval ($np \to \infty$ as $n \to \infty$). This means that we must abandon the probability of success, p, in favor of a *mean rate of success*, λ, which quantifies the mean number of successes per unit time.

The basic assumption on which the Poisson distribution rests is that each instant in time is a Bernoulli trial. Since Bernoulli trials are independent and have constant probability of success and only two possible outcomes, the Poisson process enjoys the following properties:

1. Successes (arrivals) are independently and can occur at any instant in time.
2. The mean arrival rate is constant.
3. Waiting times between arrivals are independent and exponentially distributed.
4. The time to the kth arrival is gamma distributed.

In fact, if the first two or either of the last two properties are known to hold for a sequence of arrivals, then the arrival process belongs to the Poisson family.

As in the previous example, we will define N_t to be the number of successes (arrivals or "occurrences") occurring in time t. If the above assumptions hold, then N_t is governed by the following distribution:

$$\mathrm{P}[N_t = k] = \frac{(\lambda t)^k}{k!} e^{-\lambda t}, \qquad k = 0, 1, 2, \ldots \tag{1.113}$$

where λ is the mean rate of occurrence (λ has units of reciprocal time). This distribution is illustrated in Figure 1.23

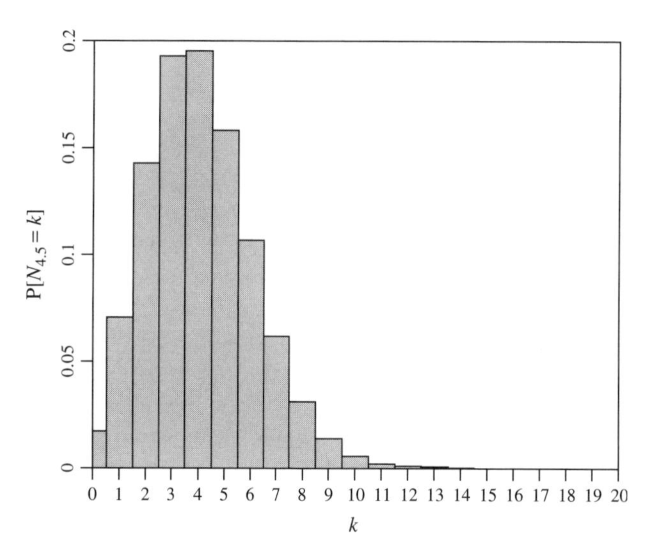

Figure 1.23 Poisson distribution for $t = 4.5$ and $\lambda = 0.9$.

Properties The mean is determined as

$$\begin{aligned} \mathrm{E}[N_t] &= \sum_{j=0}^{\infty} j \frac{(\lambda t)^j}{j!} e^{-\lambda t} = \lambda t e^{-\lambda t} \sum_{j=1}^{\infty} \frac{(\lambda t)^{j-1}}{(j-1)!} \\ &= \lambda t e^{-\lambda t} \sum_{j=0}^{\infty} \frac{(\lambda t)^j}{j!} \\ &= \lambda t \end{aligned} \tag{1.114}$$

The mean of the distribution shown in Figure 1.23 is $\mathrm{E}[N_{4.5}] = 0.9(4.5) = 4.05$. To determine the variance, we first need to find

$$\begin{aligned} \mathrm{E}\left[N_t^2\right] &= \sum_{j=0}^{\infty} j^2 \frac{(\lambda t)^j}{j!} e^{-\lambda t} = \lambda t e^{-\lambda t} \sum_{j=0}^{\infty} (j+1) \frac{(\lambda t)^j}{j!} \\ &= \lambda t e^{-\lambda t} \left[\sum_{j=0}^{\infty} j \frac{(\lambda t)^j}{j!} + \sum_{j=0}^{\infty} \frac{(\lambda t)^j}{j!} \right] \\ &= (\lambda t)^2 + (\lambda t) \end{aligned}$$

Thus

$$\mathrm{Var}[N_t] = \mathrm{E}\left[N_t^2\right] - \mathrm{E}^2[N_t] = \lambda t \tag{1.115}$$

That is, the mean and variance of a Poisson process are the same.

The Poisson distribution is also often written in terms of the single parameter $\nu = \lambda t$,

$$\mathrm{P}[N_t = k] = \frac{\nu^k}{k!} e^{-\nu}, \qquad k = 0, 1, 2, \ldots \tag{1.116}$$

If $x_1, x_2, \ldots, x_n$ are n independent observations of N_t, then the MLE of ν is

$$\hat{\nu} = \frac{1}{n}\sum_{i=1}^{n} x_i = \bar{x} \tag{1.117}$$

If t is known, then $\hat{\lambda} = \hat{\nu}/t$.

Example 1.46 Many research papers suggest that the arrivals of earthquakes follow a Poisson process over time. Suppose that the mean time between earthquakes is 50 years at a particular location.

1. How many earthquakes can be expected to occur during a 100-year period?
2. What is the probability that more than three earthquakes occur in a 100-year period?
3. How long must the time period be so that the probability that no earthquakes occur during that period is at most 0.1?
4. Suppose that 50 years pass without any earthquakes occurring. What is the probability that another 50 years will pass without any earthquakes occurring?

SOLUTION

1. Let N_t be the number of earthquakes occurring over t years. Then

$$\mathrm{P}[N_t = k] = \frac{(\lambda t)^k}{k!} e^{-\lambda t}$$

where $\lambda = \frac{1}{50} = 0.02$ per year is the mean rate of occurrence of earthquakes and $t = 100$ years. Using this, we have $\mathrm{E}[N_{100}] = 100\lambda = 100(0.02) = 2$. Thus, we can expect two earthquakes to occur during a 100-year period, which makes sense since the mean time between earthquakes is 50 years.

2. Since $\lambda t = 0.02 \times 100 = 2$, we have

$$\begin{aligned}\mathrm{P}[N_{100} > 3] &= 1 - \mathrm{P}[N_{100} \le 3] \\ &= 1 - \Big(\mathrm{P}[N_{100} = 0] + \mathrm{P}[N_{100} = 1] \\ &\quad + \mathrm{P}[N_{100} = 2] + \mathrm{P}[N_{100} = 3]\Big) \\ &= 1 - e^{-2}\left[1 + 2 + \frac{2^2}{2} + \frac{2^3}{3!}\right] \\ &= 0.143\end{aligned}$$

3. Let N_t be the number of occurrences over the time interval t. We want to find t such that $\mathrm{P}[N_t = 0] = e^{-\lambda t} \le 0.1$. This gives us $t \ge -\ln(0.1)/\lambda = -\ln(0.1)/0.02 = 115$ years.

4. Let N_{50} be the number of occurrences over the first 50 years and N_{100} be the number of occurrences over the first 100 years. Then, we have

$$\begin{aligned}\mathrm{P}[N_{100} = 0 \mid N_{50} = 0] &= \frac{\mathrm{P}[N_{100} = 0 \cap N_{50} = 0]}{\mathrm{P}[N_{50} = 0]} \\ &= \frac{\mathrm{P}[N_{100} = 0]}{\mathrm{P}[N_{50} = 0]} = \frac{e^{-100\lambda}}{e^{-50\lambda}} \\ &= e^{-50\lambda} = e^{-1} \\ &= 0.368\end{aligned}$$

We note that due to the memorylessness of the Poisson process (which is in turn due to the independence between trials) this result is identical to the probability of having no earthquakes in any 50-year period,

$$\mathrm{P}[N_{50} = 0] = e^{-50\lambda} = e^{-1} = 0.368$$

Now consider a Poisson process with arrival rate λ. If arrivals are retained randomly from this process with probability p and rejected with probability $q = 1 - p$, then the resulting process of retained arrivals is also Poisson with arrival rate $p\lambda$ [see Cinlar (1975) for a proof]. This is illustrated by the following example.

Example 1.47 Earthquakes in a particular region occur as a Poisson process with mean rate $\lambda = 3$ per year. In addition, it has been observed that every third earthquake, on average, has magnitude exceeding 5.

(a) What is the probability of having two or more earthquakes of magnitude in excess of 5 in the next one year?
(b) What is the probability that the next earthquake of magnitude in excess of 5 will occur within the next 2 months?

SOLUTION We are told that earthquakes occur as a Poisson process with $\lambda = 3$ per year. This means that an earthquake can occur at any instant in time but that on average there are three "successes" each year. We are also told that on average one in three of these earthquakes has a higher magnitude (i.e., exceeding 5). The "on average" part of this statement implies that each earthquake that does occur has a $\frac{1}{3}$ chance of having a higher magnitude. The mean rate of occurrence of higher magnitude earthquakes is thus $\lambda' = 1$ per year.

(a) Let N_t be the number of higher magnitude earthquakes which occur in t years. Under the above conditions,

N_t follows a Poisson distribution and the desired probability is

$$\begin{aligned} \mathrm{P}[N_1 \geq 2] &= 1 - \mathrm{P}[N_1 = 0] - \mathrm{P}[N_1 = 1] \\ &= 1 - e^{-\lambda' t}[1 + \lambda' t] \\ &= 1 - e^{-1(1)}[1 + 1(1)] \\ &= 0.2643 \end{aligned}$$

(b) The number of higher magnitude earthquakes which might occur in the next two months is $N_{1/6}$. The question is "What is the probability that one or more higher magnitude earthquakes will occur in the next two months?" which can be solved as follows:

$$\begin{aligned} \mathrm{P}\left[N_{1/6} \geq 1\right] &= 1 - \mathrm{P}\left[N_{1/6} = 0\right] = 1 - e^{-\lambda' t} \\ &= 1 - e^{-1/6} = 0.1535 \end{aligned}$$

As mentioned above, and as we will see more of shortly, the time to the next occurrence of a Poisson process is exponentially distributed (compare the above result to the exponential distribution presented in Section 1.10.1).

The previous example seems to suggest that the distribution of every third occurrence is also Poisson, which is *not* correct. This raises a rather subtle issue, but the distinction lies between whether we are selecting every third occurrence or whether we are selecting occurrences randomly with probability $\frac{1}{3}$ of success. Here are the rules and the reasoning for a process in which we are selecting every kth occurrence on average or deterministically:

1. If we are selecting every kth occurrence on average, and so randomly (i.e., the probability of selecting an occurrence is $1/k$), then the *time* until the next selection follows an exponential distribution (see Section 1.10.1) with mean rate $\lambda' = \lambda/k$, where λ is the mean occurrence rate of the original process. In this case, the likelihood of having success in the next instant is $1/k$, and the likelihood decreases exponentially thereafter. The resulting process is a Poisson process.
2. If we are selecting every kth occurrence nonrandomly (e.g., every kth customer arriving at a website is asked to fill out a survey), then the time between selections follows a gamma distribution (see Section 1.10.2). The main implication of having to have exactly $k - 1$ occurrences of the original process before a selection is that the likelihood of a selection in the next $k - 1$ instants is zero. In other words, we expect the gamma distribution to start at zero when $t = 0$. The resulting process is not Poisson.

In the above the word "likelihood" is used loosely to denote the relative probability of an occurrence in a vanishingly small time interval (i.e., an instant), dp/dt.

1.10 COMMON CONTINUOUS PROBABILITY DISTRIBUTIONS

Many naturally occurring and continuous random phenomena can be well modeled by a relatively small number of distributions. The following six continuous distributions are particularly common in engineering applications:

1. Exponential
2. Gamma
3. Uniform
4. Weibull
5. Rayleigh
6. Normal
7. Lognormal

As mentioned in the previous section, the exponential and gamma distributions are members of the *Bernoulli family*, deriving from the idea that each instant in time constitutes an independent Bernoulli trial. These are the continuous-time analogs of the geometric and negative binomial distributions.

Aside from the above, there are certainly other continuous distributions which may be considered. Distributions which involve more than two parameters are generally difficult to justify because we rarely have enough data to estimate even two parameters with much accuracy. From a practical point of view what this means is that even if a geotechnical researcher has large volumes of data at a particular site and can accurately estimate, for example, a modified six-parameter beta distribution, it is unlikely that anyone else will be able to do so at other sites. Thus, complex distributions, such as a six-parameter beta distribution, are of questionable value at any site other than the site at which it was estimated (see Chapter 4 for further discussion of this issue).

As with the common discrete distributions, this section looks briefly at the main characteristics of each of these continuous distributions and describes how they are most commonly used in practice. For a more complete description of these distributions, the interested reader should consult an introductory textbook on probability and statistics, such as Law and Kelton (1991) or Devore (2003).

1.10.1 Exponential Distribution

The exponential distribution is yet another distribution derived from the Bernoulli family: It is the continuous analog of the geometric distribution. Recall that the geometric distribution governs the number of trials until the first success (or to the next success). If we imagine that each instant in time is now an independent trial, then the time until the first (or next) success is given by the exponential distribution (the mathematics associated with this transition from the geometric distribution involving "discrete" trials to a "continuous" sequence of trials is similar to that shown previously for the transition from the binomial to the Poisson distribution and will not be repeated here).

As with the geometric distribution, the exponential distribution is often used to describe "time-to-failure" problems. It also governs the time between arrivals of a Poisson process. If T_1 is the time to the occurrence (or failure) in question and T_1 is exponentially distributed, then its probability density function is (see Figure 1.24)

$$f_{T_1}(t) = \lambda e^{-\lambda t}, \qquad t \geq 0 \tag{1.118}$$

where λ is the *mean rate* of occurrence (or failure). Its cumulative distribution function is

$$F_{T_1}(t) = \mathrm{P}[T_1 \leq t] = 1 - e^{-\lambda t}, \qquad t \geq 0 \tag{1.119}$$

Properties

$$\mathrm{E}[T_1] = \frac{1}{\lambda} \tag{1.120a}$$

$$\mathrm{Var}[T_1] = \frac{1}{\lambda^2} \tag{1.120b}$$

That is, the mean and standard deviation of an exponentially distributed random variable are equal.

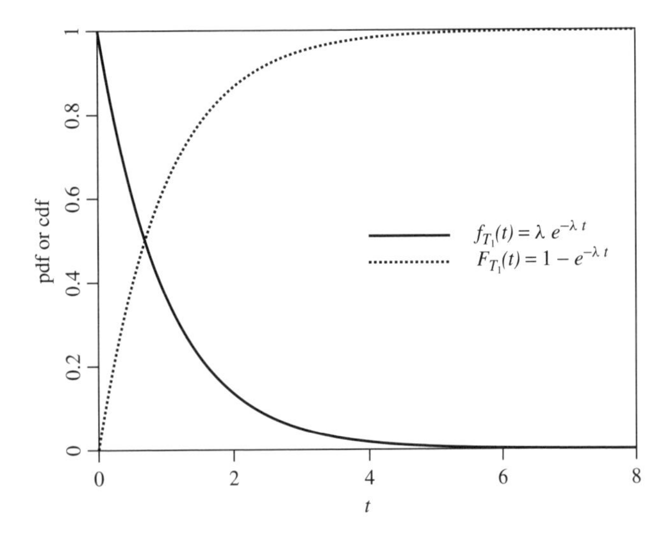

Figure 1.24 Exponential distribution for $\lambda = 1$.

Memoryless Property We will illustrate this property with an example: Let T_1 denote the time between occurrences of earthquakes in a particular region. Assume that T_1 has an exponential distribution with a mean of 4 months (i.e., on average, earthquakes in this region occur once every 4 months). Thus, T_1 has mean arrival rate of $\lambda = \frac{1}{4} = 0.25$ earthquakes per month. The probability that an earthquake occurs within the next 2 weeks (half-month) is thus

$$\begin{aligned}\mathrm{P}[T_1 < 2 \text{ weeks}] &= \mathrm{P}[T_1 < 0.5 \text{ months}]\\ &= 1 - e^{-0.5\times 0.25} = 0.1175\end{aligned}$$

Now, suppose that we set up a ground motion accelerometer in this region and 8 months pass without an earthquake occurring. What is the probability that an earthquake will occur in the next half-month (i.e., between 8 and 8.5 months from our setup time)? Because 8 months have gone by without an earthquake occurring, you might feel that an occurrence is overdue and therefore more likely. That is, that the probability of an occurrence in the next half-month should be greater than 0.1175. However, for the exponential distribution, this is not the case, which is one of the features of the exponential distribution—the past is ignored. Each instant in time constitutes a trial which is independent of all other trials. In fact,

$$\begin{aligned}\mathrm{P}[T_1 < 8.5 \mid T_1 > 8] &= \frac{\mathrm{P}[8 < T_1 < 8.5]}{\mathrm{P}[T_1 > 8]}\\ &= \frac{(1 - e^{-8.5\times 0.25}) - (1 - e^{-8\times 0.25})}{e^{-8\times 0.25}}\\ &= 0.1175\end{aligned}$$

Thus, after 8 months without an occurrence, the probability of an occurrence in the next half-month is the same as the probability of an occurrence in any half-month interval. We found this same property existed in the Poisson process; indeed, the times between arrivals in the Poisson process are exponentially distributed.

More generally, if T_1 is exponentially distributed with mean rate λ, then the *memoryless* property means that the probability that T_1 is greater than $t + s$, given that $T_1 > t$, is the same as the probability that T_1 is greater than s with no past history knowledge. In other words,

$$\begin{aligned}\mathrm{P}[T_1 > t + s \mid T_1 > t] &= \frac{\mathrm{P}[T_1 > t + s \,\cap\, T_1 > t]}{\mathrm{P}[T_1 > t]}\\ &= \frac{\mathrm{P}[T_1 > t + s]}{\mathrm{P}[T_1 > t]} = \frac{e^{-\lambda(t+s)}}{e^{-\lambda t}}\\ &= e^{-\lambda s}\\ &= \mathrm{P}[T_1 > s]\end{aligned} \tag{1.121}$$

Link to Poisson It was mentioned above that the exponential distribution governs the time between the occurrences of a Poisson process. This can be clearly seen through the following argument: Let N_t be a Poisson distributed random variable with mean arrival rate λ. We wish to know the distribution of the time until the first arrival. Let T_1 be the time to the first arrival. Then,

$$\mathrm{P}[T_1 > t] = \mathrm{P}[N_t = 0] = \frac{(\lambda t)^0}{0!} e^{\lambda t} = e^{-\lambda t}$$

and so

$$\mathrm{P}[T_1 \leq t] = F_{T_1}(t) = 1 - e^{-\lambda t}$$

But $1 - e^{-\lambda t}$ is the cumulative distribution for the exponential probability density function $\lambda e^{-\lambda t}$. Consequently, T_1 must follow an exponential distribution with mean rate λ; that is, the time to the first occurrence in a Poisson process follows an exponential distribution with parameter λ which is equal to the Poisson rate λ. The same holds for the time *between* any occurrences of a Poisson process.

In many cases, the assumption of "independence" between trials at every instant in time makes sense (e.g., arrivals of customers at a bank, cars traveling along a highway). However, earthquakes tend to occur only once sufficient strain levels have developed between adjacent tectonic plates, and that generally takes some time. Thus, the times between measurable earthquake occurrences depend on tectonic movement rates and interplate friction, which will not generally lead to a constant probability of occurrence at each instant in time. The Poisson model is usually more reasonable for moderate to high earthquake magnitudes (in Chapter 4 we discuss the fact that higher level excursions tend to a Poisson process).

If $x_1, x_2, \ldots x_n$ are n independent observations of T_1, then the MLE of λ is

$$\hat{\lambda} = \frac{1}{n} \sum_{i=1}^{n} x_i = \bar{x} \tag{1.122}$$

Example 1.48 Suppose the lifetime of a particular type of nuclear density meter has an exponential distribution with a mean of 28,700 h. Compute the probability of a density meter of this type failing during its 8000-h warranty?

SOLUTION Let T_1 be the lifetime of this type of density meter. Then T_1 is exponentially distributed with $\lambda = 1/28{,}700$ per hour, and

$$\begin{aligned} \mathrm{P}[T_1 < 8000] &= F_{T_1}(8000) \\ &= 1 - \exp\left\{-\frac{8000}{28{,}700}\right\} = 0.243 \end{aligned}$$

Example 1.49 Let us assume that earthquakes in a certain region occur on average once every 50 years and that the number of earthquakes in any time interval follows a Poisson distribution. Under these conditions, what is the probability that less than 30 years will pass before the next earthquake occurs?

SOLUTION Let T_1 be the time to the next earthquake. Then, since the number of earthquakes follow a Poisson distribution, the time between earthquakes follows an exponential distribution. Thus, T_1 follows an exponential distribution with $\lambda = 1/50 = 0.02$ earthquakes per year (on average), and

$$\mathrm{P}\left[T_1 < 30 \text{ years}\right] = 1 - e^{-0.02\times 30} = 0.549$$

We could also solve this using the Poisson distribution. Let N_{30} be the number of earthquakes to occur in the next 30 years. Then the event that less than 30 years will pass before the next earthquake is equivalent to the event that one or more earthquakes will occur in the next 30 years. That is,

$$\begin{aligned} \mathrm{P}\left[T_1 < 30 \text{ years}\right] &= \mathrm{P}[N_{30} \geq 1] = 1 - \mathrm{P}[N_{30} < 1] \\ &= 1 - \mathrm{P}[N_{30} = 0] = 1 - e^{-0.02\times 30} \\ &= 0.549 \end{aligned}$$

1.10.2 Gamma Distribution

We consider here a particular form of the gamma distribution which is a member of the Bernoulli family and is the continuous-time analog of the negative binomial distribution. It derives from an infinite sequence of Bernoulli trials, one at each instant in time, with mean rate of success λ, and governs the time between every kth occurrence of successes in a Poisson process. Specifically, if T_k is defined as the time to the kth success in a Poisson process, then T_k is the sum of k independent exponentially distributed random variables E_i each with parameter λ. That is, $T_k = E_1 + E_2 + \cdots + E_k$ and T_k has the probability density function

$$f_{T_k}(t) = \frac{\lambda\,(\lambda t)^{k-1}}{(k-1)!} e^{-\lambda t}, \qquad t \geq 0 \tag{1.123}$$

which is called the gamma distribution (Figure 1.25). This form of the gamma distribution (having integer k) is also referred to as the *k-Erlang distribution*. Note that $k = 1$ gives the exponential distribution, as expected. The above distribution can be generalized to noninteger k if $(k-1)!$ is replaced by $\Gamma(k)$, which is the gamma function; see Law and Kelton (2000) for more information on the general gamma distribution. We also give a brief discussion of noninteger k at the end of this section.

To derive the cumulative distribution function, we integrate the above probability density function (by parts) to

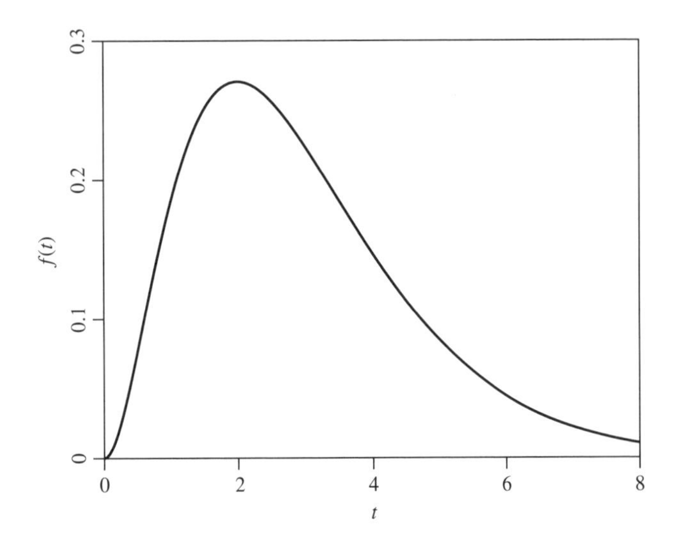

Figure 1.25 Gamma probability density function for $\lambda = 1$ and $k = 3$.

obtain, for integer k,

$$F_{T_k}(t) = \mathrm{P}[T_k \le t] = 1 - e^{-\lambda t} \sum_{j=0}^{k-1} \frac{(\lambda t)^j}{j!} \qquad (1.124)$$

The cumulative distribution function can also be found by recognizing that the event that the kth arrival occurs within time t (i.e., $T_k < t$) is equivalent to the event that there are k or more arrivals within time t (i.e., $N_t \ge k$). In other words,

$$\begin{aligned} F_{T_k}(t) &= \mathrm{P}[T_k \le t] = \mathrm{P}[N_t \ge k] = 1 - \mathrm{P}[N_t < k] \\ &= 1 - e^{-\lambda t} \sum_{j=0}^{k-1} \frac{(\lambda t)^j}{j!} \end{aligned}$$

Properties

$$\mathrm{E}[T_k] = \frac{k}{\lambda} \qquad \left(= k\ \mathrm{E}[E_i]\right) \qquad (1.125a)$$

$$\mathrm{Var}[Y] = \frac{k}{\lambda^2} \qquad \left(= k\ \mathrm{Var}[E_i]\right) \qquad (1.125b)$$

If k is known and $x_1, x_2, \ldots, x_n$ are n independent observations of T_k, then the MLE of λ is

$$\hat{\lambda} = \frac{nk}{\sum_{i=1}^{n} x_i} = \frac{k}{\bar{x}} \qquad (1.126)$$

Example 1.50 As in the previous example, let us assume that earthquakes in a certain region occur on average once every 50 years and that the number of earthquakes in any time interval follows a Poisson distribution. Under these conditions, what is the probability that less than 150 years will pass before two or more earthquakes occur?

SOLUTION Let T_2 be the time to the occurrence of the second earthquake. Then, since earthquakes occur according to a Poisson process, T_2 must follow a gamma distribution with $k = 2$ and $\lambda = \frac{1}{50}$ and

$$\begin{aligned} \mathrm{P}[T_2 < 150] &= F_{T_2}(150) \\ &= 1 - e^{-150/50}\left(1 + \frac{150/50}{1!}\right) = 0.801 \end{aligned}$$

Note that the same result is obtained by computing

$$\begin{aligned} \mathrm{P}[N_{150} \ge 2] &= 1 - \mathrm{P}[N_{150} < 2] \\ &= 1 - \mathrm{P}[N_{150} = 0] - \mathrm{P}[N_{150} = 1] \\ &= 1 - e^{-150/50} - \frac{150/50}{1!} e^{-150/50} \\ &= 0.801 \end{aligned}$$

The gamma distribution presented above is specialized to the sum of k independent and identically exponentially distributed random variables. It can be extended to other types of problems, so long as k is (at least approximately) a positive integer.

Example 1.51 Suppose that for clay type A the length of time in years until achieving 80% of consolidation settlement follows a gamma distribution with a mean of 4 and a variance of 8. Suppose also that for clay type B the time required to achieve the same fraction of consolidation settlement also follows a gamma distribution but with mean 4 and variance 16. Which clay type has a higher probability of reaching 80% consolidation in less than one year?

SOLUTION Let X be the time required to achieve 80% consolidation settlement for clay type A. Then X follows a gamma distribution with $\mu = k/\lambda = 4$ and $\sigma^2 = k/\lambda^2 = 8$. Solving these two equations for k and λ gives us $k = 2$ and $\lambda = \frac{1}{2}$.
Now let Y be the time required to achieve 80% consolidation settlement for clay type B. Then Y follows a gamma distribution with $\mu = k/\lambda = 4$ and $\sigma^2 = k/\lambda^2 = 16$. Solving these two equations for k and λ gives us $k = 1$ and $\lambda = \frac{1}{4}$. For clay type A we then have

$$\begin{aligned} \mathrm{P}[X < 1] &= F_{T_2}(1) \\ &= 1 - e^{-\lambda}(1 + \lambda) \\ &= 1 - e^{-1/2}(1 + \tfrac{1}{2}) \\ &= 0.0902 \end{aligned}$$

while for clay type B we have

$$\begin{aligned} P[Y < 1] &= F_{T_1}(1) \\ &= 1 - e^{-\lambda} \\ &= 1 - e^{-1/4} \\ &= 0.2212 \end{aligned}$$

Thus, we are more likely to achieve 80% consolidation in under one year with clay type B.

Although the gamma distribution is not limited to integer values of k, the interpretation of the gamma PDF as the distribution of a sum of independent and identically exponentially distributed random variables is lost if k is not an integer. The more general gamma distribution has the form

$$f_X(x) = \frac{\lambda\,(\lambda x)^{k-1}}{\Gamma(k)} e^{-\lambda x}, \qquad x \geq 0 \tag{1.127}$$

which is valid for any $k > 0$ and $\lambda > 0$. The *gamma function* $\Gamma(k)$ for $k > 0$ is defined by the integral

$$\Gamma(k) = \int_0^\infty x^{k-1} e^{-x}\, dx \tag{1.128}$$

Tabulations of the gamma function can be found in Abramowitz and Stegun (1970), for example. When k is an integer, $\Gamma(k) = (k-1)!$.

1.10.3 Uniform Distribution

The continuous uniform distribution is the simplest of all continuous distributions since its density function is constant (over a range) (Figure 1.26). Its general definition is

$$f(x) = \frac{1}{\beta - \alpha}, \qquad \alpha \leq x \leq \beta$$

and its cumulative distribution is

$$F(x) = P[X \leq x] = \frac{x - \alpha}{\beta - \alpha}, \qquad \alpha \leq x \leq \beta \tag{1.129}$$

The uniform distribution is useful in representing random variables which have known *upper* and *lower* bounds and which have equal likelihood of occurring anywhere between these bounds. Another way of looking at the uniform distribution is that it is *noninformative* or *nonpresumptive*. That is, if you know nothing else about the relative likelihood of a random variable, aside from its upper and lower bounds, then the uniform distribution is appropriate—it makes no assumptions regarding preferential likelihood of the random variable since all possible values are equally likely.

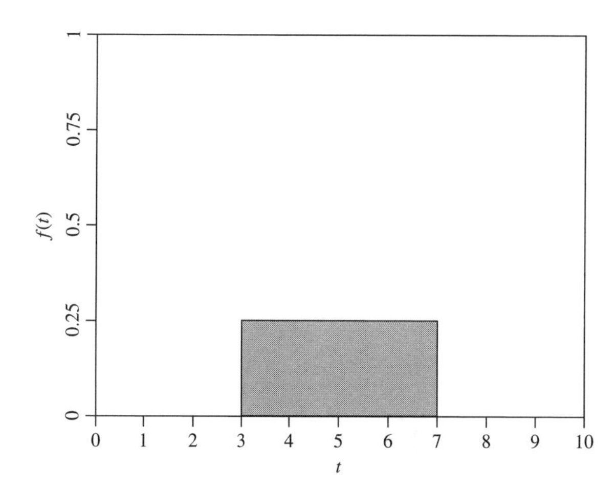

Figure 1.26 Uniform distribution for $\alpha = 3$ and $\beta = 7$.

Properties

$$E[X] = \int_\alpha^\beta \frac{x\,dx}{\beta - \alpha} = \frac{\alpha + \beta}{2}$$

(this is the midpoint) (1.130a)

$$\text{Var}[X] = \int_\alpha^\beta \frac{x^2}{\beta - \alpha} dx - E^2[X] = \frac{(\beta - \alpha)^2}{12} \tag{1.130b}$$

If $x_1, x_2, \ldots, x_n$ are n independent observations of uniformly distributed X with minimum value $x_{\min}$ and maximum value $x_{\max}$, then the MLEs of α and β are

$$\hat{\alpha} = x_{\min}, \qquad \hat{\beta} = x_{\max}$$

That is, the MLEs of the lower and upper bounds of the uniform distribution are just equal to the observed minimum and maximum values.

Example 1.52 The C function `rand()` returns numbers uniformly distributed on the interval [0, `RAND_MAX`), which includes zero but excludes `RAND_MAX`. If X_i is assigned subsequent values returned by `rand()/RAND_MAX`, then each X_i is uniformly distributed on the interval [0, 1). If we further define

$$Y = \alpha \left[\sum_{i=1}^{12} X_i - 6 \right]$$

then what is the mean and variance of Y?

SOLUTION

$$\begin{aligned} E[Y] &= \alpha \left[\sum_{i=1}^{12} E[X_i] - 6 \right] = \alpha\,[12\,E[X_i] - 6] \\ &= \alpha\left[12(\tfrac{1}{2}) - 6\right] \\ &= 0 \end{aligned}$$

$$\begin{aligned}\mathrm{Var}[Y] &= \mathrm{Var}\left[\alpha\left(\sum_{i=1}^{12} X_i - 6\right)\right] = \mathrm{Var}\left[\alpha\sum_{i=1}^{12} X_i\right] \\ &= \alpha^2\,\mathrm{Var}\left[\sum_{i=1}^{12} X_i\right] \\ &= \alpha^2 \sum_{i=1}^{12}\mathrm{Var}[X_i] = \alpha^2(12)(\tfrac{1}{12}) \\ &= \alpha^2\end{aligned}$$

1.10.4 Weibull Distribution

Often, engineers are concerned with the strength properties of materials and the lifetimes of manufactured devices. The Weibull distribution has become very popular in describing these types of problems (Figure 1.27). One of the attractive features of the Weibull distribution is that its cumulative distribution function is quite simple.

If a continuous random variable X has a Weibull distribution, then it has probability density function

$$f(x) = \frac{\beta}{x}(\lambda x)^\beta e^{-(\lambda x)^\beta} \quad \text{for } x > 0 \tag{1.131}$$

having parameters $\lambda > 0$ and $\beta > 0$. The Weibull has a particularly simple cumulative distribution function

$$F(x) = 1 - e^{-(\lambda x)^\beta} \quad \text{if } x \geq 0 \tag{1.132}$$

Note that the exponential distribution is a special case of the Weibull distribution (simply set $\beta = 1$). While the exponential distribution has constant, memoryless failure rate, the Weibull allows for a failure rate that decreases with time ($\beta < 1$) or a failure rate that increases with time ($\beta > 1$). This gives increased flexibility for modeling lifetimes of systems that improve with time (e.g., a good red wine might have $\beta < 1$) or degrade with time (e.g., reinforced concrete bridge decks subjected to salt might have $\beta > 1$).

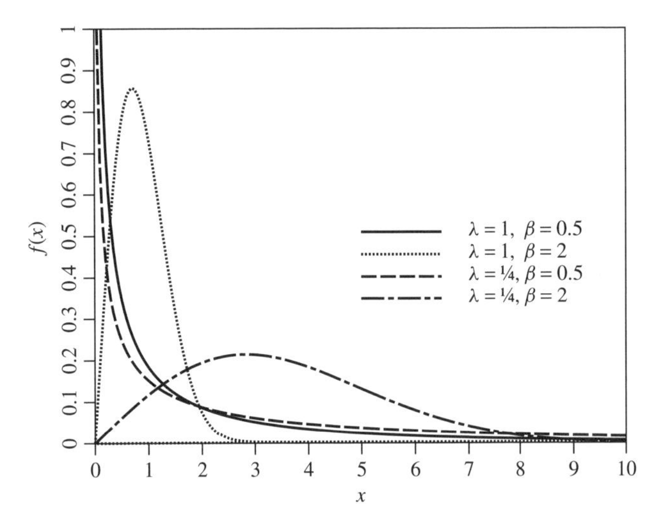

Figure 1.27 Weibull distribution.

The mean and variance of a Weibull distributed random variable are

$$\mu = \frac{1}{\lambda\beta}\Gamma\left(\frac{1}{\beta}\right) \tag{1.133a}$$

$$\sigma^2 = \frac{1}{\lambda^2\beta}\left\{2\Gamma\left(\frac{2}{\beta}\right) - \frac{1}{\beta}\left[\Gamma\left(\frac{1}{\beta}\right)\right]^2\right\} \tag{1.133b}$$

where Γ is the gamma function, which is commonly tabulated in math tables.

To find MLEs of λ and β, we must solve the following two equations for the estimators $\hat{\lambda}$ and $\hat{\beta}$ (Law and Kelton, 2000):

$$\frac{\sum_{i=1}^n x_i^{\hat{\beta}} \ln x_i}{\sum_{i=1}^n x_i^{\hat{\beta}}} - \frac{1}{\hat{\beta}} = \frac{1}{n}\sum_{i=1}^n \ln x_i \qquad \hat{\lambda} = \left(\frac{1}{n}\sum_{i=1}^n x_i^{\hat{\beta}}\right)^{-1/\hat{\beta}} \tag{1.134}$$

The first equation involves only $\hat{\beta}$, which can be solved for numerically. Once $\hat{\beta}$ has been obtained, the second equation can be solved directly for $\hat{\lambda}$. Thomas et al. (1969) provide an efficient general recursive formula using Newton's root-finding method,

$$\hat{\beta}_{k+1} = \hat{\beta}_k + \frac{A + (1/\hat{\beta}_k) - C_k/B_k}{(1/\hat{\beta}_k^2) + (B_k H_k - C_k^2)/B_k^2} \tag{1.135}$$

where

$$A = \frac{1}{n}\sum_{i=1}^n \ln x_i$$

$$B_k = \sum_{i=1}^n x_i^{\hat{\beta}_k}$$

$$C_k = \sum_{i=1}^n x_i^{\hat{\beta}_k}\ln x_i$$

$$H_k = \sum_{i=1}^n x_i^{\hat{\beta}_k}(\ln x_i)^2$$

An appropriate initial starting point is given by Menon (1963) and Thoman et al. (1969) to be

$$\hat{\beta}_0 = \left\{\frac{6}{(n-1)\pi^2}\left[\sum_{i=1}^n(\ln x_i)^2 - \frac{1}{n}\left(\sum_{i=1}^n \ln x_i\right)^2\right]\right\}^{-1/2} \tag{1.136}$$

See also Thoman et al. (1969) for confidence intervals on the true λ and β.

Example 1.53 The time to 90% consolidation of a sample of a certain clay has a Weibull distribution with $\beta = \frac{1}{2}$. A significant number of tests have shown that 81% of clay samples reach 90% consolidation in under 5516 h. What is the median time to attain 90% consolidation?

SOLUTION Let X be the time until a clay sample reaches 90% consolidation. Then we are told that X follows a Weibull distribution with $\beta = 0.5$. We first need to compute the other Weibull parameter, λ. To do this we make use of the fact that we know $\text{P}[X < 5516] = 0.81$, and since $\text{P}[X < 5516] = F(5516)$, we have

$$F(5516) = 1 - \exp\left\{-(5516\lambda)^{0.5}\right\} = 0.81$$

$$\exp\left\{-(5516\lambda)^{0.5}\right\} = 0.19$$

$$\lambda = \frac{1}{2000}$$

We are now looking for the median, $\tilde{x}$, which is the point which divides the distribution into half. That is, we want to find $\tilde{x}$ such that $F(\tilde{x}) = 0.5$,

$$1 - \exp\left\{-\left(\frac{\tilde{x}}{2000}\right)^{0.5}\right\} = 0.5$$

$$\exp\left\{-\left(\frac{\tilde{x}}{2000}\right)^{0.5}\right\} = 0.5$$

$$\tilde{x} = 960.9 \text{ h}$$

1.10.5 Rayleigh Distribution

The Rayleigh distribution (Figure 1.28) is a nonnegative distribution which finds application in the simulation of normally distributed random processes (see Section 3.3 and Chapter 6). In particular, consider the two orthogonal components τ_1 and τ_2 of the vector $\boldsymbol{\tau}$ in two-dimensional space. If the two components are independent and identically normally distributed random variables with zero means and common variance s^2, then the vector length $|\tau| = \sqrt{\tau_1^2 + \tau_2^2}$ will be Rayleigh distributed with probability density function

$$f(x) = \frac{x}{s^2} \exp\left\{-\frac{x^2}{2s^2}\right\}, \qquad x \geq 0 \tag{1.137}$$

and cumulative distribution function

$$F(x) = 1 - e^{-\frac{1}{2}(x/s)^2} \quad \text{if } x \geq 0 \tag{1.138}$$

which is actually a special case of the Weibull distribution ($\beta = 2$ and $\lambda = 1/(s\sqrt{2})$).

The mean and variance of a Rayleigh distributed random variable are

$$\mu = s\sqrt{\tfrac{1}{2}\pi} \qquad \sigma^2 = (2 - \tfrac{1}{2}\pi)s^2$$

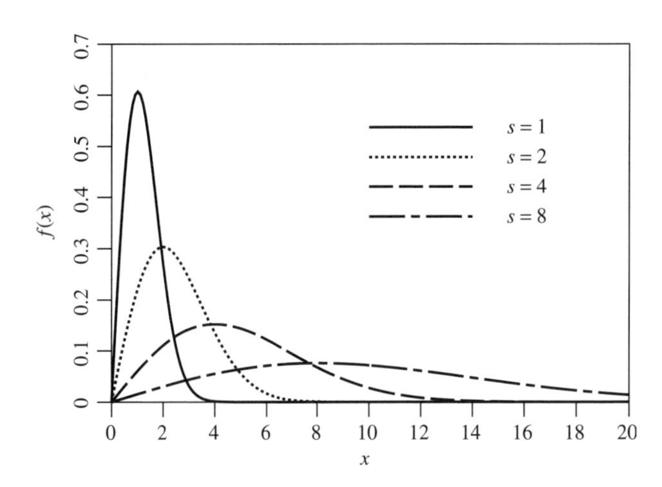

Figure 1.28 Rayleigh distribution.

1.10.6 Student *t*-Distribution

If Z is a standard normal random variable, V is a chi-square random variable with ν degrees of freedom, and Z and V are independent, then the random variable T defined by

$$T = \frac{Z}{\sqrt{V/\nu}} \tag{1.139}$$

follows the Student *t*-distribution with probability function

$$f(t) = \frac{\Gamma[(\nu+1)/2]}{\sqrt{\pi\nu}\,\Gamma(\nu/2)} \left(1 + \frac{t^2}{\nu}\right)^{-(\nu+1)/2}, \qquad -\infty < t < \infty \tag{1.140}$$

This distribution was discovered in 1908 by William Gossett, who was working for the Guinness Brewing Company in Dublin, Ireland. The company considered the discovery to be proprietary information and told Gossett he could not publish it. Gossett published it anyway under the pseudonym "Student."

Table A.2 shows values of $t_{\alpha,\nu}$ such that $\text{P}\left[T > t_{\alpha,\nu}\right] = \alpha$ for commonly used values of α. We shall see more of this distribution in Chapters 2 and 3. Figure 1.29 shows some of the family of *t*-distributions. Notice that the *t*-distribution becomes wider in the tails as the number of degrees of freedom ν decreases. Conversely, as ν increases, the distribution narrows, becoming the standard normal distribution as $\nu \to \infty$. Thus, the last line of Table A.2 corresponds to the standard normal distribution, which is useful when finding z for given cumulative probability. (Note that Table A.2 is in terms of *areas to the right*.)

The mean and variance of a Student *t*-distributed random variable are

$$\mu = 0, \qquad \sigma^2 = \frac{\nu}{\nu - 2} \quad \text{for } \nu > 2$$

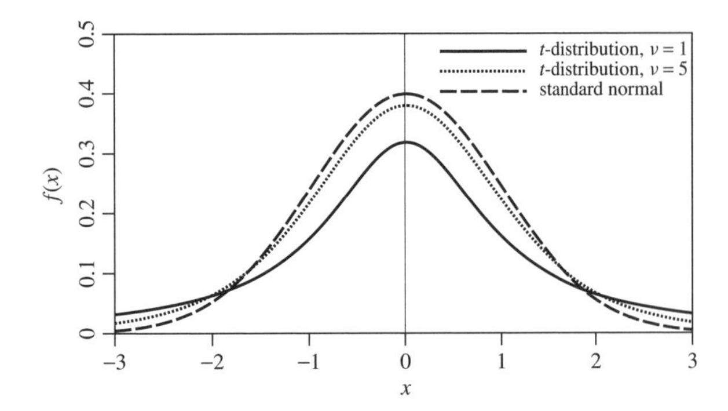

Figure 1.29 Student t-distribution.

1.10.7 Chi-Square Distribution

If $Z_1, Z_2, \ldots, Z_\nu$ are independent *standard* normal random variables [i.e., each $N(0,1)$], then the sum

$$\chi_k^2 = Z_1^2 + Z_2^2 + \cdots + Z_\nu^2 \tag{1.141}$$

has the probability density function

$$f(x) = \frac{1}{2^{\nu/2}\Gamma(\frac{\nu}{2})} x^{\nu/2-1} e^{-x/2} \quad \text{for } x > 0 \tag{1.142}$$

which is called a chi-square distribution with ν degrees of freedom (Figure 1.30). This is actually a special case of the gamma distribution with $k = \nu/2$ and $\lambda = \frac{1}{2}$. To get probabilities, we write

$$\mathrm{P}\left[\chi_k^2 \geq \chi_{\alpha,k}^2\right] = \alpha = \int_{\chi_{\alpha,k}^2}^{\infty} f(u)\, du \tag{1.143}$$

and use standard chi-square tables. See Table A.3. For example, $\mathrm{P}\left[\chi_{10}^2 \geq 15.99\right] = 0.10$, which is found by entering the table with $\nu = 10$ degrees of freedom, looking across for 15.99, and then reading up at the top of the table for the associated probability. Note that both Tables A.2 and A.3 are in terms of area to the right and are used with *inverse* problems where we want values on the horizontal axis having area to the right specified by a given α.

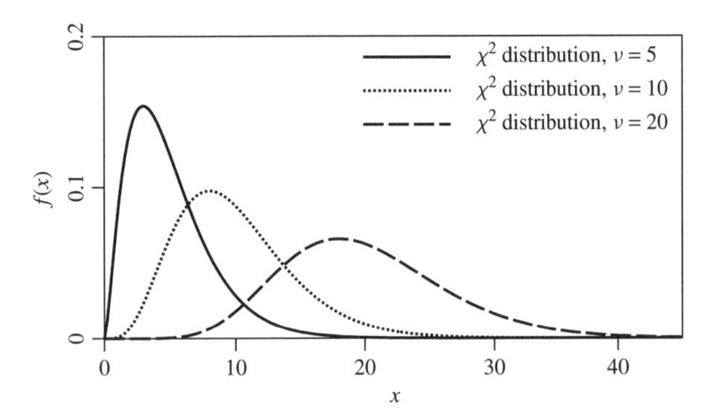

Figure 1.30 Chi-square distribution.

The mean and variance of a chi-square distributed random variable are

$$\mu = \nu, \qquad \sigma^2 = 2\nu$$

1.10.8 Normal Distribution

The *normal distribution* is probably the single most important distribution in use today (Figure 1.31). This is largely because sums of random variables tend to a normal distribution, as was proven by the *central limit theorem*—a theorem to be discussed shortly. Many natural "additive" type phenomena, or phenomena involving many accumulating factors, therefore tend to have a normal distribution. For example, the cohesive strength of a soil is due to the sum of a very large number of electrochemical interactions taking place at the molecular level; thus, the normal distribution has been widely used to represent the distribution of cohesion (its main competitor as a representative distribution is the lognormal distribution, discussed next).

A random variable X follows a normal (or *Gaussian*) distribution if its probability density function has the form

$$f(x) = \frac{1}{\sigma\sqrt{2\pi}} \exp\left[-\frac{1}{2}\left(\frac{x-\mu}{\sigma}\right)^2\right] \quad \text{for } -\infty < x < \infty \tag{1.144}$$

The notation $X \sim N(\mu, \sigma^2)$ will be used to mean that X follows a normal distribution with mean μ and variance σ^2.

Properties

1. The distribution is symmetric about the mean μ (which means that μ is also equal to the median).
2. The maximum point, or *mode*, of the distribution occurs at μ.
3. The inflection points of $f(x)$ occur at $x = \mu \pm \sigma$.

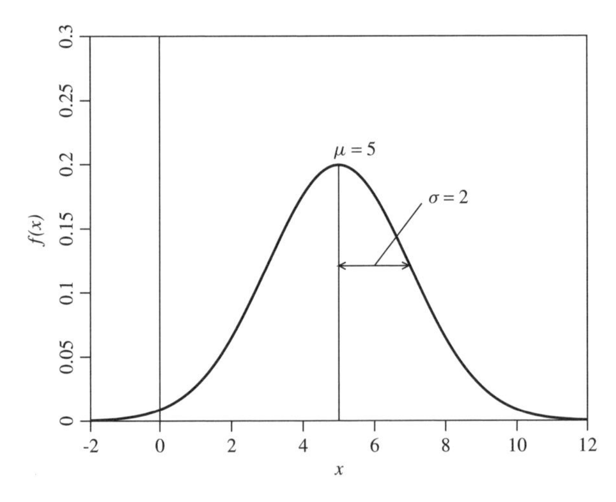

Figure 1.31 Normal distribution with $\mu = 5$ and $\sigma = 2$.

The mean and variance are given as

$$\mathrm{E}[X] = \mu, \qquad \mathrm{Var}[X] = \sigma^2$$

If $x_1, x_2, \ldots, x_n$ are n independent observations of normally distributed X, then the MLEs of μ and σ^2 are

$$\hat{\mu} = \frac{1}{n}\sum_{i=1}^{n} x_i = \bar{x} \tag{1.145a}$$

$$\hat{\sigma}^2 = \frac{1}{n}\sum_{i=1}^{n} (x_i - \hat{\mu})^2 \tag{1.145b}$$

The more common estimator for σ^2 is slightly different, having the form

$$s^2 = \frac{1}{n-1}\sum_{i=1}^{n} (x_i - \hat{\mu})^2 \tag{1.146}$$

The latter is an *unbiased* estimator (see Section 5.2.1), which is generally more popular, especially for smaller n.

Standard Normal Unfortunately, no closed-form solution exists for the integral of the normal probability density function. Probabilities associated with the normal distribution must be obtained by numerical integration. Traditionally, this has meant that normal probabilities have had to be obtained by consulting tables presented in manuals and textbooks. Of course, no book is big enough to contain the complete set of tables necessary for all possible values of μ and σ, so some way of encapsulating the tables is necessary. As it turns out, if the random variable X is transformed by subtracting its mean and dividing by its standard deviation,

$$Z = \frac{X - \mu}{\sigma} \tag{1.147}$$

then the resulting random variable Z has mean zero and unit variance (Figure 1.32). If a probability table is developed for Z, which is called the *standard normal* variate, then probabilities for all other normally distributed random variables can be obtained by performing the above normalizing transformation. That is, probabilities for any normally distributed random variable can be obtained by performing the above transformation and then consulting the single standard normal probability table.

The distribution of the standard normal Z is given the special symbol $\phi(z)$, rather than $f(z)$, because of its importance in probability modeling and is defined by

$$\phi(z) = \frac{1}{\sqrt{2\pi}} e^{-\frac{1}{2}z^2} \quad \text{for } -\infty < z < \infty \tag{1.148}$$

The cumulative distribution function of the standard normal also has a special symbol, $\Phi(z)$, rather than $F(z)$, again because of its importance. Tables of $\Phi(z)$ are commonly included in textbooks, and one appears in Appendix A. Computing probabilities for any normally distributed random variables proceeds by *standardization*, that is, by subtracting the mean and dividing by the standard deviation on both sides of the inequality in the following:

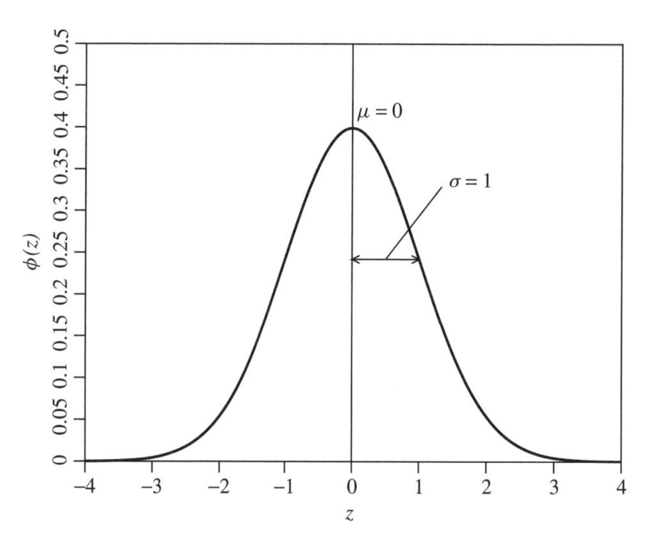

Figure 1.32 Standard normal distribution.

$$\begin{aligned}\mathrm{P}[X < x] &= \mathrm{P}\left[\frac{X-\mu}{\sigma} < \frac{x-\mu}{\sigma}\right] \\ &= \mathrm{P}\left[Z < \frac{x-\mu}{\sigma}\right] \\ &= \Phi\left(\frac{x-\mu}{\sigma}\right) \\ &= \Phi(z)\end{aligned} \tag{1.149}$$

at which point, Table A.1 can be consulted, with $z = (x-\mu)/\sigma$, to obtain the desired probability.

Example 1.54 Suppose X is normally distributed with mean 5 and standard deviation 2. Then, what is $\mathrm{P}[X < 2.0]$?

SOLUTION In order to use Table A.1, we *standardize* on *both* sides of the inequality by subtracting the mean and dividing by the standard deviation:

$$\begin{aligned}\mathrm{P}[X < 2.0] &= \mathrm{P}\left[\frac{X-\mu}{\sigma} < \frac{2-\mu}{\sigma}\right] \\ &= \mathrm{P}\left[Z < \frac{2-5}{2}\right] = \mathrm{P}[Z < -1.5] \\ &= \Phi(-1.5)\end{aligned}$$

Table A.1 does not include negative values, so we make use of the symmetry of the standard normal. That is, the area under the distribution to the left of $z = -1.5$ (see the figures below) is equal to the area under the distribution to the right of $z = 1.5$. The table only gives areas to the left; it is a cumulative distribution. This means that the area to the right of a point must be obtained by subtracting the area to the left from 1.0. This leaves us with

$$P[X < 2.0] = 1 - \Phi(1.5) = 1 - 0.93319 = 0.06681$$

Note, for increased precision, interpolation can be used between table values, for example, if you are trying to determine $\Phi(\frac{2}{3})$. However, given the typical uncertainty in the estimates of the mean and standard deviation, there is probably little point in trying to obtain the final probability too precisely.

The probability areas involved in this question are shown below. The plot on the left illustrates the original $P[X < 2]$ while the plot on the right illustrates the transformed standardized problem, $P[Z < -1.5]$. The shaded areas are of equal size.

Example 1.55 The reliability of soil anchor cables against tensile failure is to be assessed. Suppose that a particular brand of cable has normally distributed tensile strength with mean 35 kN and a standard deviation of 2 kN.

1. What is the probability that the tensile strength of a randomly selected cable is less than 40 kN?
2. Approximately 10% of all sampled cables will have a tensile strength stronger than which value?
3. Can you see any problems with modeling tensile strength using a normal distribution?

SOLUTION Let X be the tensile strength of the cable. Then X is normally distributed with mean $\mu = 35$ kN and standard deviation $\sigma = 2$ kN.

1. $P[X < 40] = P\left[\frac{X-\mu}{\sigma} < \frac{40-35}{2}\right] = P[Z < 2.5]$ $= 0.9938.$
2. $P[X > x] = 0.10 \quad \rightarrow \quad P\left[\frac{X-\mu}{\sigma} > \frac{x-35}{2}\right]$ $= 0.10.$
   Since $P[Z > 1.28] = 0.10$, we have
   $$\tfrac{1}{2}(x - 35) = 1.28 \qquad \Longrightarrow \qquad x = 37.56$$
   so that 10% of all samples are stronger than 37.56 kN. Note that in this solution we had to search through Table A.1 for the probability as close as possible to $1 - 0.10 = 0.9$ and then read "outwards" to see what value of z it corresponded to. A much simpler solution is to look at the last line of Table A.2 under the heading $\alpha = 0.10$. As we saw previously, Table A.2 is the inverse t-distribution, and the t-distribution collapsed to the standard normal when $\nu \rightarrow \infty$.
3. The normal distribution allows negative tensile strengths, which are not physically meaningful. This is a strong motivation for the *lognormal* distribution covered in Section 1.10.9.

1.10.8.1 Central Limit Theorem If $X_1, X_2, \ldots, X_n$ are independent random variables having arbitrary distributions, then the random variable

$$Y = X_1 + X_2 + \cdots + X_n \tag{1.150}$$

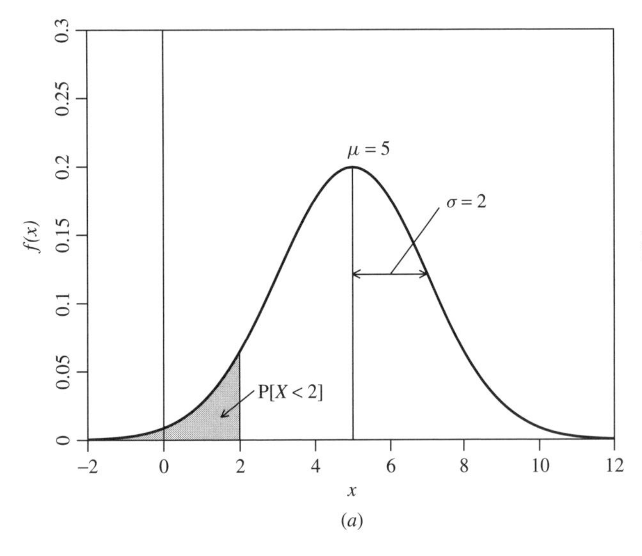

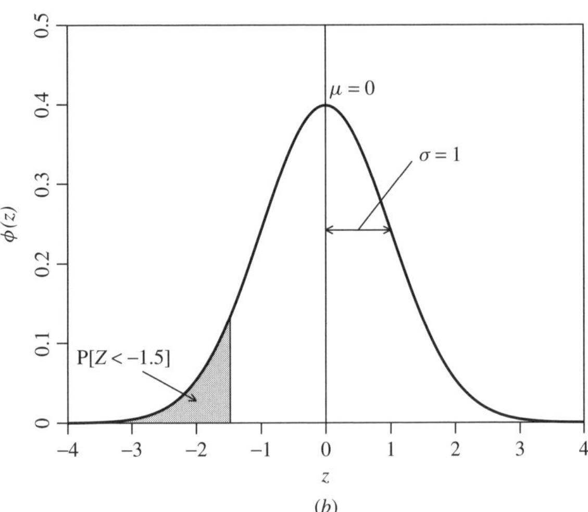

Probability plots for Example 1.54

has a normal distribution as $n \to \infty$ if all the X's have about the same mean and variance (i.e., none is dominant). See Papoulis (1991) for a proof of this theorem. In addition, if the X's are all normally distributed then Y is normally distributed for any n.

Specifically we will find the following result useful. If

$$\bar{X}_n = \frac{1}{n}\sum_{i=1}^{n} X_i$$

where $X_1, X_2, \ldots, X_n$ are independent samples taken from population X having mean μ and variance σ^2 (any distribution), then

$$\lim_{n\to\infty} \mathrm{P}\left[\frac{(\bar{X}_n - \mu)}{\sigma/\sqrt{n}} \leq x\right] = \Phi(x) \tag{1.151}$$

Implications

1. The sum of normal variates is normal (for any n) as mentioned above.
2. If the distributions of the X's are well behaved (almost normal), Then $n \geq 4$ gives a good approximation to the normal distribution.
3. If the distributions of the X's are uniform (or almost so), then $n \geq 6$ yields a reasonably good approximation to the normal distribution (out to at least about three standard deviations from the mean).
4. For poorly behaved distributions, you may need $n > 100$ before the distribution begins to look reasonably normal. This happens, for example, with distributions whose tails fall off very slowly.

Thus for n sufficiently large and $X_1, X_2, \ldots, X_n$ independent and identically distributed (iid)

$$Y = X_1 + X_2 + \cdots + X_n$$

is approximately normally distributed with

$$\mu_Y = \mathrm{E}[Y] = n\ \mathrm{E}[X_i] \tag{1.152a}$$

$$\sigma_Y^2 = \mathrm{Var}[Y] = n\ \mathrm{Var}[X_i] \tag{1.152b}$$

If the X's are *not* identically distributed but are still independent, then

$$\mu_Y = \sum_{i=1}^{n} \mathrm{E}[X_i] \tag{1.153a}$$

$$\sigma_Y^2 = \sum_{i=1}^{n} \mathrm{Var}[X_i] \tag{1.153b}$$

1.10.8.2 Normal Approximation to Binomial By virtue of the central limit theorem, the binomial distribution, which as you will recall arises from the sum of a sequence of Bernoulli random variables, can be approximated by the normal distribution (Figure 1.33). Specifically, if N_n is the number of successes in n trials, then

$$N_n = \sum_{i=1}^{n} X_i \tag{1.154}$$

where X_i is the outcome of a Bernoulli trial ($X_i = 1$ with probability p, $X_i = 0$ with probability $q = 1 - p$). Since N_n is the sum of identically distributed random variables, which are assumed independent, if n is large enough, the central limit theorem says that N_n can be approximated by a normal distribution. We generally consider this approximation to be reasonably accurate when both $np \geq 5$ and $nq \geq 5$. In this case, the normal distribution approximation has mean and standard deviation

$$\mu = np \tag{1.155a}$$

$$\sigma = \sqrt{npq} \tag{1.155b}$$

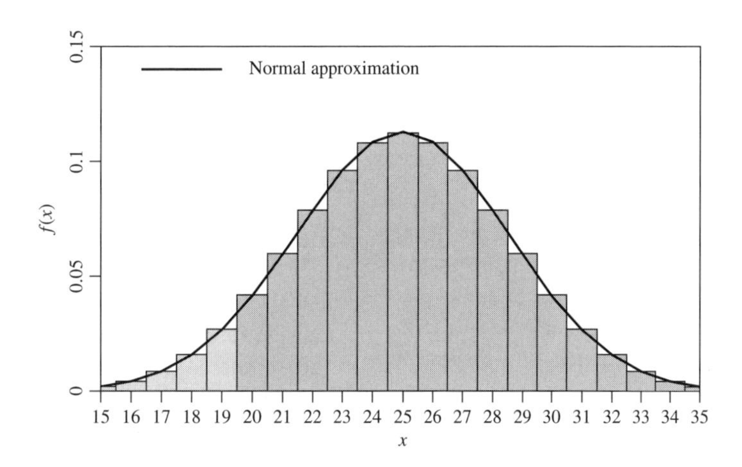

Figure 1.33 Normal approximation to binomial distribution.

Of course, we know that N_n is discrete while the normal distribution governs a continuous random variable. When we want to find the approximate probability that N_n is greater than or equal to, say, k, using the normal distribution, we should include all of the binomial *mass* at k. This means that we should look at the normal probability that $(N_n > k - \frac{1}{2})$. For example, in Figure 1.33, the probability that $N_n \geq 20$ is better captured by the area under the normal distribution above 19.5.

In general, the following corrections apply. Similar corrections apply for two-sided probability calculations.

$$\mathrm{P}[N_n \geq k] \simeq 1 - \Phi\left(\frac{k - 0.5 - \mu}{\sigma}\right) \tag{1.156a}$$

$$\mathrm{P}[N_n > k] \simeq 1 - \Phi\left(\frac{k + 0.5 - \mu}{\sigma}\right) \tag{1.156b}$$

$$\mathrm{P}[N_n \leq k] \simeq \Phi\left(\frac{k + 0.5 - \mu}{\sigma}\right) \tag{1.156c}$$

$$\mathrm{P}[N_n < k] \simeq \Phi\left(\frac{k - 0.5 - \mu}{\sigma}\right) \tag{1.156d}$$

Example 1.56 Suppose that in a certain region it is equally likely for a soil sample to pass a particular soil test as it is to fail it. If this is true, what is the probability that more than 495 samples pass the test over the next 900 tests?

SOLUTION If we assume that soil tests pass or fail independently with constant probability of passing the test, then the number of tests passing, out of n tests, is N_n, which follows a binomial distribution. The exact probability is then given by

$$\begin{aligned}\mathrm{P}[N_{900} > 495] &= \mathrm{P}[N_{900} = 496] + \mathrm{P}[N_{900} = 497] \\ &\quad + \cdots + \mathrm{P}[N_{900} = 900] \\ &= \binom{900}{496} p^{496} q^{404} + \binom{900}{497} p^{497} q^{403} \\ &\quad + \cdots + \binom{900}{900} p^{900} q^{0}\end{aligned}$$

It is not practical to compute this with a simple hand calculator, and even with a computer the calculations are prone to numerical roundoff and overflow errors. The normal approximation will give a very accurate result with a fraction of the effort. We start by computing the mean and variance of N_{900}:

$$\mathrm{E}[N_{900}] = \mu = np = (900)(0.5) = 450$$

$$\mathrm{Var}[N_{900}] = \sigma^2 = npq = (900)(0.5)(0.5) = 225$$

so that $\sigma = \sqrt{225} = 15$. We now make the following approximation:

$$\begin{aligned}\mathrm{P}[N_{900} > 495] &= \mathrm{P}\left[\frac{N_{900} - \mu}{\sigma} > \frac{495 - \mu}{\sigma}\right] \\ &\simeq \mathrm{P}\left[Z > \frac{495 + 0.5 - 450}{15}\right] \\ &= 1 - \Phi(3.03) \\ &= 0.00122\end{aligned}$$

where, in the second line of the equation, we say that $(N_{900} - \mu)/\sigma$ is approximately a standard normal, and, at the same time, apply the half-interval correction for increased accuracy. (Note that without the half-interval correction we would get $\mathrm{P}[N_{900} > 495] \simeq 0.00135$, a small absolute difference but a 10% relative difference.)

1.10.8.3 Multivariate Normal Distribution The normal distribution is also popular as a distribution governing multiple random variables because it is simply defined knowing only the mean and variance of each random variable and the covariances acting between them. Consider two random variables, X and Y; these follow a bivariate normal distribution if their joint distribution has the form

$$\begin{aligned}f_{XY}(x,y) &= \frac{1}{2\pi\sigma_X\sigma_Y\sqrt{1-\rho^2}} \exp\left\{\frac{-1}{2(1-\rho^2)}\left[\left(\frac{x-\mu_X}{\sigma_X}\right)^2\right.\right. \\ &\quad \left.\left. -2\rho\left(\frac{x-\mu_X}{\sigma_X}\right)\left(\frac{y-\mu_Y}{\sigma_Y}\right) + \left(\frac{y-\mu_Y}{\sigma_Y}\right)^2\right]\right\}\end{aligned} \tag{1.157}$$

for $-\infty < x, y < \infty$, where ρ is the correlation coefficient between X and Y and μ_X, μ_Y and σ_X, σ_Y are the means and standard deviations of X and Y, respectively. Figures 1.16 and 1.17 illustrate the bivariate normal distribution.

If X and Y follow a bivariate normal distribution, then their marginal probability density functions, defined as

$$f_X(x) = \int_{-\infty}^{\infty} f_{XY}(x,y)\,dy \tag{1.158a}$$

$$f_Y(y) = \int_{-\infty}^{\infty} f_{XY}(x,y)\,dx \tag{1.158b}$$

are also normal distributions. For example, the marginal distribution of X is a normal distribution with mean μ_X and standard deviation σ_X, and similarly for the marginal distribution of Y. That is,

$$f_X(x) = \frac{1}{\sigma_X\sqrt{2\pi}} \exp\left\{-\frac{1}{2}\left(\frac{x-\mu_X}{\sigma_X}\right)^2\right\} \tag{1.159a}$$

$$f_Y(y) = \frac{1}{\sigma_Y\sqrt{2\pi}} \exp\left\{-\frac{1}{2}\left(\frac{y-\mu_Y}{\sigma_Y}\right)^2\right\} \tag{1.159b}$$

Recall that the conditional probability of A given B is

$$\text{P}[A \mid B] = \frac{\text{P}[A \cap B]}{\text{P}[B]}$$

From this, we get the following result for *conditional distributions*:

$$f_{X \mid Y}(x \mid y) = \frac{f_{XY}(x, y)}{f_Y(y)} \tag{1.160}$$

In particular, if X and Y follow a bivariate normal distribution, then it can be shown that

$$f_{X \mid Y}(x \mid y) = \frac{1}{\sigma_X\sqrt{1-\rho^2}\sqrt{2\pi}} \times \exp\left\{-\frac{1}{2}\left[\frac{x-\mu_X-\rho(y-\mu_Y)\sigma_X/\sigma_Y}{\sigma_X\sqrt{1-\rho^2}}\right]^2\right\} \tag{1.161}$$

It can be seen from this that the conditional distribution of X for a given $Y = y$ also follows a normal distribution with mean and standard deviation

$$\mu_{X \mid Y} = \mu_X + \frac{\rho(y-\mu_Y)\sigma_X}{\sigma_Y} \tag{1.162a}$$

$$\sigma_{X \mid Y} = \sigma_X\sqrt{(1-\rho^2)} \tag{1.162b}$$

Example 1.57 Suppose that the load capacities of two neighboring piles, X and Y, are jointly normally distributed with correlation coefficient $\rho = 0.7$. Based on similar pile capacities in the area, the following statistics have been determined:

$$\mu_X = \mu_Y = 2000, \quad \sigma_X = \sigma_Y = 500$$

What is the probability that the load capacity of pile X is less than 1700 if nothing is known about the load capacity of pile Y? Alternatively, if the load capacity of pile Y has been measured to be 1800, what is the probability that X is less than 1700 in light of this information?

SOLUTION If nothing is known about the load capacity of Y, then the probability that X is less than 1700 depends only on the marginal distribution of X. That is,

$$\begin{aligned}
\text{P}[X < 1700] &= \text{P}\left[Z < \frac{1700-\mu_X}{\sigma_X}\right] \\
&= \text{P}\left[Z < \frac{1700-2000}{500}\right] \\
&= \Phi(-0.6) \\
&= 0.274
\end{aligned}$$

If, however, we know that $Y = 1800$, then we are looking for the probability that pile $X < 1700$ conditioned on the fact that pile $Y = 1800$. The conditional mean of X given $Y = 1800$ is

$$\begin{aligned}
\mu_{X \mid Y} &= \mu_X + \frac{\rho(y-\mu_Y)\sigma_X}{\sigma_Y} \\
&= 2000 + \frac{0.7(1800-2000)(500)}{500} \\
&= 1860
\end{aligned}$$

This is saying, as expected, that the conditional mean of pile X is substantially reduced as a result of the fact that the neighboring pile had a relatively low load capacity. The conditional standard deviation of X given $Y = 1800$ is

$$\begin{aligned}
\sigma_{X \mid Y} &= \sigma_X\sqrt{1-\rho^2} \\
&= 500\sqrt{1-0.7^2} \\
&= 357.07
\end{aligned}$$

This is reduced from the unconditional standard deviation of 500 because the relatively high correlation with the neighboring pile constrains the possible values of pile X. For example, if the correlation between pile capacities were 1.0, then we would know that $X = Y$. In this case, once we know Y, we would know X with certainty. That is, when $\rho = 1$, the variance of $X \mid Y$ falls to zero. When $\rho = 0$, X and Y will be uncorrelated, and thus independent, since they are normally distributed, and the observation of Y will then make no difference to the variability (and distribution) of X.

For our question, the desired conditional probability is now

$$\begin{aligned}
\text{P}[X < 1700 \mid Y = 1800] &= \Phi\left(\frac{1700-\mu_{X \mid Y}}{\sigma_{X \mid Y}}\right) \\
&= \Phi\left(\frac{1700-1860}{357.07}\right) \\
&= \Phi(-0.45) \\
&= 0.326
\end{aligned}$$

As expected, the observation of a low load capacity at a neighboring pile has increased the probability of a low load capacity at the pile of interest.

To extend the multivariate normal distribution to more than two random variables, it is useful to use vector–matrix notation. Define

$$\boldsymbol{\mu} = \begin{Bmatrix} \mu_1 \\ \mu_2 \\ \cdot \\ \cdot \\ \cdot \\ \mu_n \end{Bmatrix} \tag{1.163}$$

to be the vector of means of the sequence of n random variables $\boldsymbol{X} = \{X_1, X_2, \ldots, X_n\}$ and

$$\boldsymbol{C} = \begin{bmatrix} C_{11} & C_{12} & \cdot & \cdot & \cdot & C_{1n} \\ C_{21} & C_{22} & \cdot & \cdot & \cdot & C_{2n} \\ \cdot & \cdot & \cdot & & & \cdot \\ \cdot & \cdot & & \cdot & & \cdot \\ \cdot & \cdot & & & \cdot & \cdot \\ C_{n1} & C_{n2} & \cdot & \cdot & \cdot & C_{nn} \end{bmatrix} \tag{1.164}$$

to be the matrix of covariances between X_i and X_j, $i = 1, 2, \ldots, n$ and $j = 1, 2, \ldots, n$. Each element of the covariance matrix is defined as

$$\begin{aligned} C_{ij} &= \text{Cov}\left[X_i, X_j\right] = \rho_{ij}\sigma_i\sigma_j && \text{if } i \neq j \\ &= \text{Var}\left[X_i\right] = \sigma_i^2 && \text{if } i = j \end{aligned}$$

Note that if the X_i's are uncorrelated, then the covariance matrix is diagonal:

$$\boldsymbol{C} = \begin{bmatrix} \sigma_1^2 & 0 & \cdot & \cdot & \cdot & 0 \\ 0 & \sigma_2^2 & \cdot & \cdot & \cdot & 0 \\ \cdot & \cdot & \cdot & & & \cdot \\ \cdot & \cdot & & \cdot & & \cdot \\ \cdot & \cdot & & & \cdot & \cdot \\ 0 & 0 & \cdot & \cdot & \cdot & \sigma_n^2 \end{bmatrix}$$

Using these definitions, the joint normal distribution of $\mathbf{X} = \{X_1, X_2, \ldots, X_n\}$ is

$$f_{\mathbf{X}}(\mathbf{x}) = \frac{1}{(2\pi)^{n/2}\sqrt{|\boldsymbol{C}|}} \times \exp\left\{-\tfrac{1}{2}(\mathbf{x} - \boldsymbol{\mu})^{\mathrm{T}}\boldsymbol{C}^{-1}(\mathbf{x} - \boldsymbol{\mu})\right\} \tag{1.165}$$

where $|\boldsymbol{C}|$ is the determinant of $\boldsymbol{C}$ and superscript T means the transpose.

As in the bivariate case, all marginal distributions are also normally distributed:

$$f_{X_i}(x_i) = \frac{1}{\sigma_i\sqrt{2\pi}} \exp\left\{-\frac{1}{2}\left(\frac{x_i - \mu_i}{\sigma_i}\right)^2\right\} \tag{1.166}$$

The conditional distributions may be obtained by partitioning the vector $\mathbf{X}$ into two parts (Vanmarcke, 1984): $\mathbf{X}_a$ and $\mathbf{X}_b$ of size n_a and n_b, where $n_a + n_b = n$, that is,

$$\mathbf{X} = \left\{\begin{matrix} X_1 \\ \cdot \\ \cdot \\ \cdot \\ X_{n_a} \\ X_{n_a+1} \\ \cdot \\ \cdot \\ \cdot \\ X_n \end{matrix}\right\} = \left\{\begin{matrix} \mathbf{X}_a \\ \mathbf{X}_b \end{matrix}\right\} \tag{1.167}$$

having mean vectors

$$\boldsymbol{\mu}_a = \left\{\begin{matrix} \mu_1 \\ \cdot \\ \cdot \\ \cdot \\ \mu_{n_a} \end{matrix}\right\}, \qquad \boldsymbol{\mu}_b = \left\{\begin{matrix} \mu_{n_a+1} \\ \cdot \\ \cdot \\ \cdot \\ \mu_n \end{matrix}\right\} \tag{1.168}$$

Using this partition, the covariance matrix can be split into four submatrices:

$$\boldsymbol{C} = \begin{pmatrix} \boldsymbol{C}_{aa} & \boldsymbol{C}_{ab} \\ \boldsymbol{C}_{ba} & \boldsymbol{C}_{bb} \end{pmatrix} \tag{1.169}$$

where $\boldsymbol{C}_{ba} = \boldsymbol{C}_{ab}^{\mathrm{T}}$. Using these partitions, the conditional mean of the vector $\mathbf{X}_a$ given the vector $\mathbf{X}_b$ can be obtained from

$$\boldsymbol{\mu}_{a\,|\,b} = \boldsymbol{\mu}_a + \boldsymbol{C}_{ab}\boldsymbol{C}_{bb}^{-1}(\mathbf{X}_b - \boldsymbol{\mu}_b) \tag{1.170}$$

Similarly, the conditional covariance matrix is

$$\boldsymbol{C}_{a\,|\,b} = \boldsymbol{C}_{aa} - \boldsymbol{C}_{ab}\boldsymbol{C}_{bb}^{-1}\boldsymbol{C}_{ab}^{\mathrm{T}} \tag{1.171}$$

With these results, the conditional distribution of $\mathbf{X}_a$ given $\mathbf{X}_b$ is

$$f_{\mathbf{X}_a\,|\,\mathbf{X}_b}(\mathbf{x}_a \,|\, \mathbf{x}_b) = \frac{1}{(2\pi)^{n_a/2}\sqrt{|\boldsymbol{C}_{a\,|\,b}|}} \times \exp\left\{-\tfrac{1}{2}(\mathbf{x}_a - \boldsymbol{\mu}_{a\,|\,b})^T\boldsymbol{C}_{a\,|\,b}^{-1}(\mathbf{x}_a - \boldsymbol{\mu}_{a\,|\,b})\right\} \tag{1.172}$$

1.10.9 Lognormal Distribution

From the point of view of modeling material properties and loads in engineering, which are generally nonnegative, the normal distribution suffers from the disadvantage of allowing negative values. For example, if a soil's elastic modulus were to be modeled using a normal distribution, then there would be a nonzero probability of obtaining a negative elastic modulus. Since a negative elastic modulus does not occur in practice, the normal cannot be its true distribution.

As an approximation, the normal is nevertheless often used to represent material properties. The error incurred may be slight when the coefficient of variation v is small. For example, if $v \leq 0.3$, then $\mathrm{P}[X < 0] \leq 0.0004$, which may be fine unless it is at these extremes that failure is initiated. A simple way to avoid such problems is to fit a nonnegative distribution to the population in question, and one such candidate is the *lognormal distribution* (Figure 1.34). The lognormal distribution arises from the normal distribution through a simple, albeit nonlinear, transformation. In particular, if G is a normally distributed random variable, having range $-\infty < g < +\infty$, then $X = \exp\{G\}$ will have range $0 \leq x < \infty$. We say that the resulting random variable X is *lognormally* distributed—note that its natural logarithm is normally distributed.

The random variable X is lognormally distributed if $\ln(X)$ is normally distributed. If this is true, then X has probability density function

$$f(x) = \frac{1}{x\sigma_{\ln X}\sqrt{2\pi}} \exp\left\{-\frac{1}{2}\left(\frac{\ln x - \mu_{\ln X}}{\sigma_{\ln X}}\right)^2\right\}, \quad 0 \leq x < \infty \tag{1.173}$$

Note that this distribution is strictly nonnegative and so is popular as a distribution of nonnegative engineering properties, such as cohesion, elastic modulus, the tangent of the friction angle, and so on. The two parameters of the distribution,

$$\mu_{\ln X} = \mathrm{E}[\ln X], \qquad \sigma^2_{\ln X} = \mathrm{Var}[\ln X]$$

are the mean and variance of the underlying normally distributed random variable, $\ln X$.

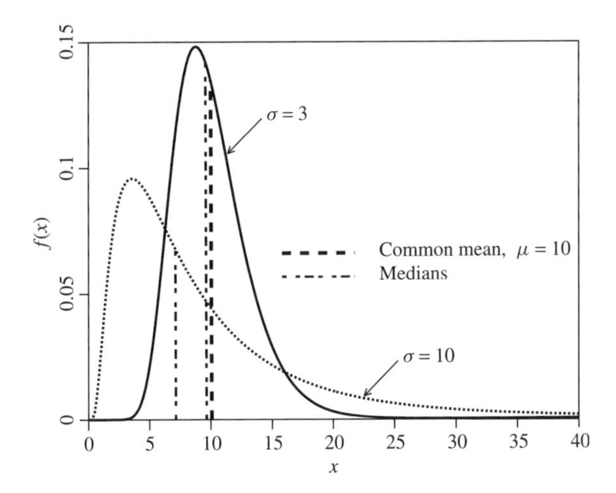

Figure 1.34 Two lognormal distributions illustrating effect of changing variance.

Computing Probabilities In order to compute probabilities from the lognormal distribution, we must make use of the fact that $\ln(X)$ is normally distributed so that we can use the standard normal table. That is, in a probability expression, we take logarithms on both sides of the inequality, then standardize by subtracting the mean and dividing by the standard deviation of $\ln X$,

$$\begin{aligned}
\mathrm{P}[X \leq a] &= \mathrm{P}[\ln(X) < \ln(a)] \\
&= \mathrm{P}\left[\frac{\ln(X) - \mu_{\ln X}}{\sigma_{\ln X}} < \frac{\ln(a) - \mu_{\ln X}}{\sigma_{\ln X}}\right] \\
&= \mathrm{P}\left[Z < \frac{\ln(a) - \mu_{\ln X}}{\sigma_{\ln X}}\right] \\
&= \Phi\left(\frac{\ln(a) - \mu_{\ln X}}{\sigma_{\ln X}}\right)
\end{aligned} \tag{1.174}$$

where, as before, Z is the standard normal random variate.

Mean and Variance The mean and variance of X are obtained by transforming the two parameters of the lognormal distribution,

$$\mu_X = \mathrm{E}[X] = e^{\mu_{\ln X} + \frac{1}{2}\sigma^2_{\ln X}} \tag{1.175a}$$

$$\sigma^2_X = \mathrm{Var}[X] = \mu^2_X\left(e^{\sigma^2_{\ln X}} - 1\right) \tag{1.175b}$$

Alternatively, if you are given μ_X and σ^2_X, you can obtain the parameters $\mu_{\ln X}$ and $\sigma^2_{\ln X}$ as follows:

$$\sigma^2_{\ln X} = \ln\left(1 + \frac{\sigma^2_X}{\mu^2_X}\right) \tag{1.176a}$$

$$\mu_{\ln X} = \ln(\mu_X) - \tfrac{1}{2}\sigma^2_{\ln X} \tag{1.176b}$$

Characteristics and Moments

$$\text{Mode} = e^{\mu_{\ln X} - \sigma^2_{\ln X}} \tag{1.177a}$$

$$\text{Median} = e^{\mu_{\ln X}} \tag{1.177b}$$

$$\text{Mean} = e^{\mu_{\ln X} + \frac{1}{2}\sigma^2_{\ln X}} \tag{1.177c}$$

$$\mathrm{E}\left[X^k\right] = e^{k\mu_{\ln X} + \frac{1}{2}k^2\sigma^2_{\ln X}} \tag{1.177d}$$

Note that the mode < median < mean, and thus the lognormal distribution has *positive skew*. A distribution is skewed if one of its tails is longer than the other, and, by tradition, the sign of the skew indicates the direction of the longer tail.

Figure 1.35 illustrates the relative locations of the mode, median, and mean for the nonsymmetric lognormal distribution. Because of the positive-skewed, or "skewed-right," shape of the distribution, with the long distribution tail to the right, realizations from the lognormal distribution will

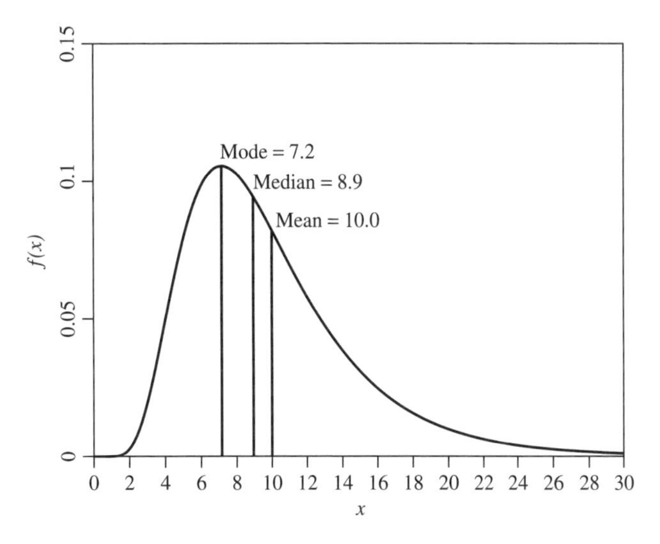

Figure 1.35 Location of mode, median, and mean in lognormal distribution for $\mu_X = 10$ and $\sigma_X = 5$.

have very large values every now and then. This results in the mean being drawn to the right (e.g., the arithmetic average is affected by very large values in the sum). Often, for the lognormal distribution, the median is actually viewed as the primary characteristic of the distribution, since it divides the distribution into equal halves.

It is worth digressing slightly at this point and consider the median of a lognormal distribution in a bit more detail, especially with respect to its estimation. Suppose that we have taken several observations $x_1, x_2, \ldots, x_n$ of a lognormally distributed random variable X. An estimate of the mean of $\ln(X)$ is just the average of $\ln(x_1), \ln(x_2), \ldots, \ln(x_n)$,

$$\hat{\mu}_{\ln X} = \frac{1}{n}\sum_{i=1}^{n} \ln(x_i) \tag{1.178}$$

where the hat denotes that this is an estimate of $\mu_{\ln X}$. From this, an estimate of the median, $\tilde{x}$, is

$$\tilde{x} = \exp\{\hat{\mu}_{\ln X}\} = \exp\left\{\frac{1}{n}\sum_{i=1}^{n} \ln(x_i)\right\} \tag{1.179}$$

Alternatively, the geometric average x_g of a sequence of nonnegative numbers is defined as the nth root of the product of the n observations,

$$\begin{aligned} x_g &= (x_1 x_2 \cdots x_n)^{1/n} \\ &= \exp\left\{\ln\left((x_1 x_2 \cdots x_n)^{1/n}\right)\right\} \\ &= \exp\left\{\frac{1}{n}\sum_{i=1}^{n} \ln(x_i)\right\} \end{aligned} \tag{1.180}$$

which is identical to the equation for $\tilde{x}$, so we see that the geometric average is an estimate of the median of a lognormally distributed random variable. As we shall see in Section 4.4, this also means that the median of a lognormal distribution is preserved under geometric averaging.

Multiplicative Property If $X = Y_1 Y_2 \cdots Y_n$ and each Y_i are (positive) independent random variables of any distribution having about the same "weight," then

$$\ln X = \ln Y_1 + \ln Y_2 + \cdots + \ln Y_n \tag{1.181}$$

and by the central limit theorem $\ln X$ tends to a normal distribution with

$$\mu_{\ln X} = \mu_{\ln Y_1} + \mu_{\ln Y_2} + \cdots + \mu_{\ln Y_n} \tag{1.182a}$$

$$\sigma^2_{\ln X} = \sigma^2_{\ln Y_1} + \sigma^2_{\ln Y_2} + \cdots + \sigma^2_{\ln Y_n} \tag{1.182b}$$

Thus X tends to a lognormal distribution with parameters $\mu_{\ln X}$ and $\sigma^2_{\ln X}$. This is a useful property since it can be used to approximate the distribution of many multiplicative functions.

In particular, if X is any multiplicative function, say

$$X = \frac{AB}{C} \quad \Longrightarrow \quad \ln X = \ln A + \ln B - \ln C \tag{1.183}$$

and A, B, and C are independent and lognormally distributed, then X is also lognormally distributed with

$$\mu_{\ln X} = \mu_{\ln A} + \mu_{\ln B} - \mu_{\ln C}$$

$$\sigma^2_{\ln X} = \sigma^2_{\ln A} + \sigma^2_{\ln B} + \sigma^2_{\ln C}$$

Recall that for variances the coefficient of -1 appearing before the last term in Eq. 1.183 is squared, so that, in the case of independence, the variance of a sum is literally the sum of the variances. (If A, B, and C were correlated, then the covariance terms which would have to be added in to find $\sigma^2_{\ln X}$ would have sign dependent on the signs appearing in the original sum.)

Consider again the geometric average, this time for random observations (i.e., before we have observed them),

$$X_g = (X_1 X_2 \cdots X_n)^{1/n} = X_1^{1/n} \times X_2^{1/n} \times \cdots \times X_n^{1/n}$$

which is a product of n random variables. By the central limit theorem, X_g will tend to a lognormal distribution so that

$$\ln(X_g) = \ln\left((X_1 X_2 \cdots X_n)^{1/n}\right) = \frac{1}{n}\sum_{i=1}^{n} \ln(X_i)$$

is normally distributed. As mentioned above, X_g is an estimate of the median of X if X is lognormally distributed. However, even if X is not lognormally distributed, X_g will tend to have a lognormal distribution, by the central limit theorem, if the X_i's are nonnegative. We shall see more of this in Chapter 4 where we suggest that in a variety of cases the lognormal distribution is a natural distribution for soil properties according to the central limit theorem.

The MLEs for $\mu_{\ln X}$ and $\sigma^2_{\ln X}$ are the same as for the normal distribution except that $\ln(X)$ is used in the estimate. If $x_1, x_2, \ldots, x_n$ are n independent observations of a lognormally distributed random variable, then the MLEs are

$$\hat{\mu}_{\ln X} = \frac{1}{n}\sum_{i=1}^{n} \ln\, x_i \tag{1.184a}$$

$$\hat{\sigma}^2_{\ln X} = \frac{1}{n}\sum_{i=1}^{n} (\ln\, x_i - \hat{\mu}_{\ln X})^2 \tag{1.184b}$$

The more common estimator for $\sigma^2_{\ln X}$ is slightly different, having the form

$$\hat{\sigma}^2_{\ln X} = \frac{1}{n-1}\sum_{i=1}^{n} (\ln\, x_i - \hat{\mu}_{\ln X})^2 \tag{1.185}$$

which is an *unbiased* estimator (see Section 5.2.1).

Example 1.58 The settlement δ of a shallow foundation, in meters, can be computed as

$$\delta = c\,\frac{L}{E}$$

where L is the footing load, E is the soil's effective elastic modulus, and c is a constant which accounts for geometry (footing area and aspect ratio, depth to bedrock, etc.) and Poisson's ratio. Assume that c is nonrandom and equal to $0.15\ \text{m}^{-1}$ and that the load and elastic modulus are both lognormally distributed with

$$\mu_E = 20,000.0\ \text{kN/m}^2, \qquad \sigma_E = 4000.0\ \text{kN/m}^2$$

$$\mu_L = 1200.0\ \text{kN}, \qquad \sigma_L = 300.0\ \text{kN}$$

What is the probability that the footing settlement exceeds 0.025 m?

SOLUTION First write $\ln(\delta) = \ln(c) + \ln(L) - \ln(E)$, so that

$$\mu_{\ln\delta} = \ln(c) + \mu_{\ln L} - \mu_{\ln E}, \qquad \sigma^2_{\ln\delta} = \sigma^2_{\ln L} + \sigma^2_{\ln E}$$

where we assumed independence between $\ln(L)$ and $\ln(E)$ when computing the variance of $\ln(\delta)$ (so that the covariance terms can be dropped). To compute the above, we must first find the means and variances of $\ln(L)$ and $\ln(E)$:

$$\begin{aligned}\sigma^2_{\ln L} &= \ln\left(1 + \frac{\sigma_L^2}{\mu_L^2}\right) = \ln\left(1 + \frac{300^2}{1200^2}\right)\\ &= 0.060625\\ \mu_{\ln L} &= \ln(\mu_L) - \tfrac{1}{2}\sigma^2_{\ln L} = \ln(1200) - \tfrac{1}{2}(0.060625)\\ &= 7.059765\end{aligned}$$

$$\begin{aligned}\sigma^2_{\ln E} &= \ln\left(1 + \frac{\sigma_E^2}{\mu_E^2}\right) = \ln\left(1 + \frac{4000^2}{20{,}000^2}\right) = 0.039221\\ \mu_{\ln E} &= \ln(\mu_E) - \tfrac{1}{2}\sigma^2_{\ln E} = \ln(20{,}000) - \tfrac{1}{2}(0.039221)\\ &= 9.883877\end{aligned}$$

Thus,

$$\begin{aligned}\mu_{\ln\delta} &= \ln(0.15) + 7.059765 - 9.883877 = -4.721232\\ \sigma^2_{\ln\delta} &= 0.060625 + 0.039221 = 0.099846\\ \sigma_{\ln\delta} &= \sqrt{0.099846} = 0.315984\end{aligned}$$

and

$$\begin{aligned}\text{P}[\delta > 0.025] &= 1 - \text{P}[\delta \le 0.025]\\ &= 1 - \text{P}\left[Z \le \frac{\ln(0.025) - \mu_{\ln\delta}}{\sigma_{\ln\delta}}\right]\\ &= 1 - \text{P}[Z \le 3.27]\\ &= 1 - \Phi(3.27) = 1 - 0.9994622\\ &= 0.00054\end{aligned}$$

Most foundations are designed to have probability of failure ranging from 0.001 to 0.0001 against ultimate limit states (e.g., bearing capacity failure). This foundation would be considered very safe with respect to settlement failure, especially since excessive settlement is generally considered to be only a serviceability limit state issue.

1.10.9.1 Bivariate Lognormal Distribution Generally, the multivariate lognormal distribution is handled by directly considering the underlying multivariate normal distribution. That is, rather than considering the joint distribution between the lognormally distributed variates $X_1, X_2, \ldots$, we consider the joint distribution between $\ln X_1, \ln X_2, \ldots$ since these are all normally distributed and the results presented in the previous section can be used. However, we sometimes need to consider the lognormally distributed variates directly. Here we will present some results for two lognormally distributed random variables X_1 and X_2.

If X_1 and X_2 are jointly lognormally distributed, then their bivariate distribution is

$$f_{X_1X_2}(x,y) = \frac{1}{2\pi\sigma_{\ln X_1}\sigma_{\ln X_2} rxy} \times \exp\left\{-\frac{1}{2r^2}\left[\Psi_1^2 - 2\rho_{\ln 12}\Psi_1\Psi_2 + \Psi_2^2\right]\right\}, \quad x \ge 0, y \ge 0 \tag{1.186}$$

where $\Psi_1 = (\ln x - \mu_{\ln X_1})/\sigma_{\ln X_1}$, $\Psi_2 = (\ln y - \mu_{\ln X_2})/\sigma_{\ln X_2}$, $r^2 = 1 - \rho^2_{\ln 12}$, and $\rho_{\ln 12}$ is the correlation coefficient between $\ln X_1$ and $\ln X_2$.

In general, the parameters $\mu_{\ln X_1}$, $\sigma_{\ln X_1}$ can be obtained using the transformation equations given in the previous section from the parameters μ_{X_1}, σ_{X_1}, and so on. If we happen to have an estimate for the correlation coefficient ρ_{12} acting between X_1 and X_2, we can get $\rho_{\ln 12}$ from

$$\rho_{\ln 12} = \frac{\ln(1 + \rho_{12} v_{X_1} v_{X_2})}{\sqrt{\ln(1 + v_{X_1}^2)\ln(1 + v_{X_2}^2)}} \tag{1.187}$$

where $v_{X_i} = \sigma_{X_i}/\mu_{X_i}$ is the coefficient of variation of X_i. We can also invert this relationship to obtain an expression for ρ_{12},

$$\rho_{12} = \frac{\exp\{\rho_{\ln 12}\sigma_{\ln X_1}\sigma_{\ln X_2}\} - 1}{\sqrt{\left(\exp\{\sigma_{\ln X_1}^2\} - 1\right)\left(\exp\{\sigma_{\ln X_2}^2\} - 1\right)}} \tag{1.188}$$

1.10.10 Bounded tanh Distribution

The second half of this book is devoted to a variety of traditional geotechnical problems which are approached in a nontraditional way. In particular, the soil or rock is treated as a spatially variable random field. We shall see in Chapter 3 that a random field with a multivariate normal distribution has the advantage of being fully specified by only its mean and covariance structure. In addition, the simulation of random fields is relatively straightforward when the random field is normally distributed and more complicated when it is not.

Unfortunately, the normal distribution is not appropriate for many soil and rock properties. In particular, most material properties are strictly nonnegative (e.g., elastic modulus). Since the normal distribution has range $(-\infty, +\infty)$, it will always admit some possibility of negative values. When one is simulating possibly millions of realizations of a soil or rock property using the normal distribution, some realizations will inevitably involve negative soil/rock properties, unless the coefficient of variation is quite small and chance is on your side. The occurrence of negative properties often leads to fundamental modeling difficulties (e.g., what happens when Poisson's ratio or the elastic modulus becomes negative?).

In cases where the normal distribution is not appropriate, there are usually two options: (1) choose a distribution on the interval $(0, +\infty)$ (e.g., the lognormal distribution) or (2) choose a distribution which is bounded both above and below on some interval (a, b). The latter would be appropriate for properties such as friction angle, Poisson's ratio, and void ratio.

As we saw above, the lognormal transformation $X = e^G$, where G is normally distributed, leads to a random variable X which takes values on the interval $(0, +\infty)$. Thus, the lognormal distribution derives from a simple transformation of a normally distributed random variable or field. In the case of a bounded distribution, using the transformation

$$X = a + \tfrac{1}{2}(b - a)\left[1 + \tanh\left(\frac{m + sG}{2\pi}\right)\right] \tag{1.189}$$

leads to the random variable X being bounded on the interval (a, b) if G is a standard normally distributed random variable (or at least bounded distribution—we shall assume that G is a standard normal here). The parameter m is a location parameter. If $m = 0$, then the distribution of X is symmetric about the midpoint of the interval, $\frac{1}{2}(a + b)$. The parameter s is a scale parameter—the larger s is, the more variable X is. The function tanh is defined as

$$\tanh(z) = \frac{e^z - e^{-z}}{e^z + e^{-z}} \tag{1.190}$$

In essence, Eq. 1.189 can be used to produce a random variable with a distribution bounded on the interval (a, b), which is a simple transformation of a normally distributed random variable. Thus, a bounded property is easily simulated by first simulating the normally distributed random variable G and then applying Eq. 1.189. Such a simulation would require that the mean and covariance structure of the simulated normally distributed random process be known. To this end, Eq. 1.189 can be inverted to yield

$$m + sG = \pi \ln\left(\frac{X - a}{b - X}\right) \tag{1.191}$$

Since G is a standard normal (having mean zero and unit variance), the parameters m and s are now seen as the mean and standard deviation of the normally distributed random process $(m + sG)$. These two parameters can be estimated by observing a sequence of realizations of X, that is, $x_1, x_2, \ldots, x_n$, transforming each according to

$$y_i = \pi \ln\left(\frac{x_i - a}{b - x_i}\right) \tag{1.192}$$

and then estimating the mean m and standard deviation s using the traditional estimators,

$$m = \frac{1}{n}\sum y_i \tag{1.193a}$$

$$s = \sqrt{\frac{1}{n-1}\sum_{i=1}^{n}(y_i - m)^2} \tag{1.193b}$$

In order to estimate the correlation structure, the spatial location, $\mathbf{x}$, of each observation must also be known, so that our observations become $x(\mathbf{x}_i)$, $i = 1, 2, \ldots, n$, and y_i also becomes a function of $\mathbf{x}_i$. The methods of estimating the correlation function discussed in Sections 5.3.6 and 5.4.1.1 can then be applied to the transformed observations, $y(\mathbf{x}_i)$.

The probability density function of X is

$$f_X(x) = \frac{\sqrt{\pi}(b-a)}{\sqrt{2}s(x-a)(b-x)} \times \exp\left\{-\frac{1}{2s^2}\left[\pi \ln\left(\frac{x-a}{b-x}\right) - m\right]^2\right\} \quad (1.194)$$

If $m = 0$, then the mean of X is at the midpoint, $\mu_X = \frac{1}{2}(a+b)$. Since most bounded distributions are symmetric about their midpoints, the remainder of this discussion will be for $m = 0$.

Figure 1.36 illustrates how the distribution of X changes as s changes for $m = 0$, $a = 0$, and $b = 1$. The distribution shapes are identical for different choices in a and b, the only change being that the horizontal axis scales with $b - a$ and the vertical axis scales with $1/(b-a)$. For example, if $a = 10$ and $b = 30$, the $s = 2$ curve looks identical to that shown in Figure 1.36 except that the horizontal axis runs from 10 to 30 while the vertical axis runs from 0 to 0.3. When $s > 5$, the distribution becomes U shaped, which is not a realistic material property shape. Practically speaking, values ranging from $s = 0$, which is nonrandom and equal to the mean, to $s = 5$, which is almost uniformly distributed between a and b, are reasonable.

The relationship between the parameter s and the standard deviation σ_X of X is also of interest. In the limit as $s \to \infty$, the transformation given by Eq. 1.189 becomes a Bernoulli distribution with $p = 0.5$ and X taking possible values a or b. The standard deviation of X for $s \to \infty$ must therefore be $0.5(b-a)$. At the other extreme, as $s \to 0$, we end up with $X = \frac{1}{2}(a+b)$, which is nonrandom. Thus, when $s \to 0$ the standard deviation of X is zero and when $s \to \infty$ the standard deviation of X is $0.5(b-a)$. We suggest, therefore, that σ_X increases from zero when $s = 0$ to $0.5(b-a)$ when $s \to \infty$.

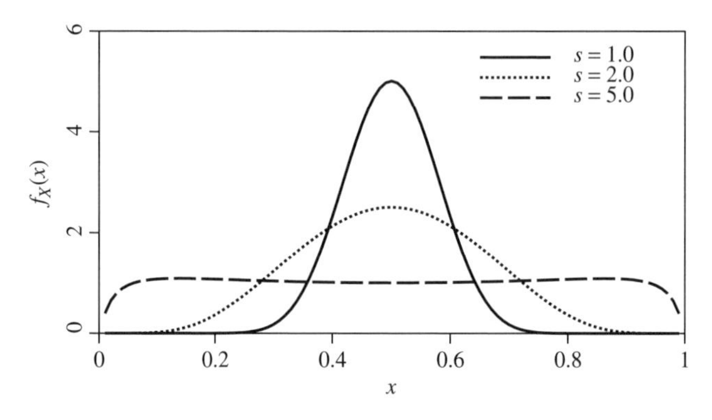

Figure 1.36 Probability density function of X determined as bounded transformation (Eq. 1.189) of normally distributed random variable $(m + sG)$ for $m = 0$ and various values of s.

The following relationship between s and the variance of X derives from a third-order Taylor series approximation to tanh and a first-order approximation to the expectation:

$$\begin{aligned} \sigma_X^2 &= (0.5)^2(b-a)^2 \, \mathrm{E}\left[\tanh^2\left(\frac{sG}{2\pi}\right)\right] \\ &\simeq (0.5)^2(b-a)^2 \, \mathrm{E}\left[\frac{[sG/(2\pi)]^2}{1+[sG/(2\pi)]^2}\right] \\ &\simeq (0.5)^2(b-a)^2 \frac{s^2}{4\pi^2+s^2} \end{aligned} \quad (1.195)$$

where $\mathrm{E}[G^2] = 1$ since G is a standard normal random variable. Equation 1.195 slightly overestimates the true standard deviation of X by 0% when $s = 0$ to 11% when $s = 5$. A much closer approximation over the entire range $0 \le s \le 5$ is obtained by slightly decreasing the 0.5 factor to 0.46 (this is an empirical adjustment),

$$\sigma_X \simeq \frac{0.46(b-a)s}{\sqrt{4\pi^2+s^2}} \quad (1.196)$$

The close agreement between Eq. 1.196 and a simulation-based estimate is illustrated in Figure 1.37.

Equation 1.195 can be generalized to yield an approximation to the covariance between two random variables X_i and X_j, each derived as tanh transformations of two standard normal variables G_i and G_j according to Eq. 1.189. If G_i and G_j are correlated, with correlation coefficient ρ_{ij}, then

$$\mathrm{Cov}[X_i, X_j] = (0.5)^2(b-a)^2 \times \mathrm{E}\left[\tanh\left(\frac{sG_i}{2\pi}\right)\tanh\left(\frac{sG_j}{2\pi}\right)\right]$$

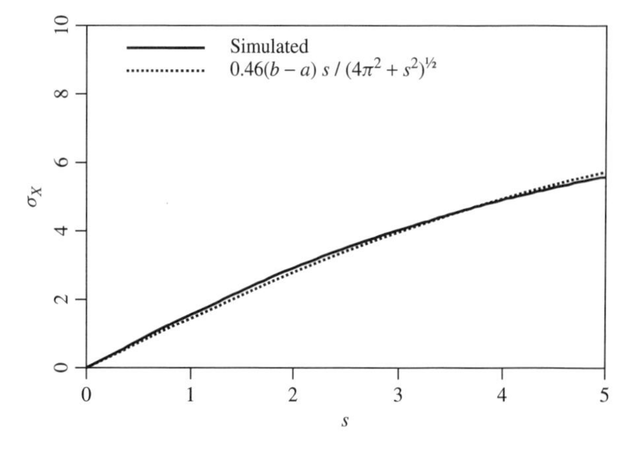

Figure 1.37 Relationship between σ_X and s derived from simulation (100,000 realizations for each s) and Taylor's series derived approximation given by Eq. 1.196. The vertical scale corresponds to $b - a = 20°$.

$$\simeq (0.5)^2(b-a)^2$$

$$\times \mathrm{E}\left[\frac{[sG_i/(2\pi)][sG_j/(2\pi)]}{1+(1/2)\left\{[sG_i/(2\pi)]^2+[sG_j/(2\pi)]^2\right\}}\right]$$

$$\simeq (0.46)^2(b-a)^2\frac{s^2\rho_{ij}}{4\pi^2+s^2}$$

where the empirical correction given in Eq. 1.196 was introduced in the last step.

1.11 EXTREME-VALUE DISTRIBUTIONS

Most engineering systems fail only when extreme loads occur and failure tends to initiate at the weakest point. Thus, it is of considerable interest to investigate the distribution of extreme values. Consider a sequence of n random variables $X_1, X_2, \ldots, X_n$. This could, for example, be the sequence of tensile strengths of individual links in a chain, or the sequence of daily average soil moisture levels, or earthquake intensities, and so on. Now define the extremes of this set of random variables as

$$Y_n = \max(X_1, X_2, \ldots, X_n) \tag{1.197a}$$

$$Y_1 = \min(X_1, X_2, \ldots, X_n) \tag{1.197b}$$

so that if X_i is the daily average soil moisture level, then Y_n is the maximum daily average soil moisture level over n days. Similarly, if X_i is the tensile strength of the ith link in a chain, then Y_1 is the tensile strength of a chain composed of n links.

1.11.1 Exact Extreme-Value Distributions

Let us first examine the behavior of the maximum, Y_n. We know that if the maximum is less than some number y, then each X_i must also be less than y. That is, the event $(Y_n \le y)$ must be equivalent to the event $(X_1 \le y \cap X_2 \le y \cap \cdots \cap X_n \le y)$. In other words the *exact* distribution of Y_n is

$$\mathrm{P}[Y_n \le y] = \mathrm{P}[X_1 \le y \cap X_2 \le y \cap \cdots \cap X_n \le y] \tag{1.198}$$

If it can be further assumed that the X's are *independent and identically distributed* (iid) (if this is not the case, the problem becomes very complex and usually only solved via simulation), then

$$\begin{aligned} F_{Y_n}(y) &= \mathrm{P}[Y_n \le y] \\ &= \mathrm{P}[X_1 \le y]\,\mathrm{P}[X_2 \le y]\cdots\mathrm{P}[X_n \le y] \\ &= [F_X(y)]^n \end{aligned} \tag{1.199}$$

where F_X is the cumulative distribution function of X. Taking the derivative gives us the probability density function

$$\begin{aligned} f_{Y_n}(y) &= \frac{dF_{Y_n}(y)}{dy} = n[F_X(y)]^{n-1}\frac{dF_X(y)}{dy} \\ &= n[F_X(y)]^{n-1}f_X(y) \end{aligned} \tag{1.200}$$

Example 1.59 Suppose that fissure lengths X in a rock mass have an exponential distribution with $f_X(x) = e^{-x}$. What, then, does the distribution of the maximum fissure length Y_n look like for $n = 1, 5, 50$ fissures?

SOLUTION If $n = 1$, then Y_n is the maximum of one observed fissure, which of course is just the distribution of the single fissure length. Thus, when $n = 1$, the distribution of Y_n is just the exponential distribution

$$f_{Y_1}(y) = f_X(y) = e^{-y}$$

When $n = 5$, we have

$$\begin{aligned} F_{Y_5}(y) &= \mathrm{P}[Y_5 \le y] \\ &= \mathrm{P}[X_1 \le y]\,\mathrm{P}[X_2 \le y]\cdots\mathrm{P}[X_5 \le y] \\ &= [F_X(y)]^5 \\ &= [1-e^{-y}]^5 \end{aligned}$$

where we used the fact that $F_X(x) = 1 - e^{-x}$. To find the probability density function (which is usually more informative graphically), we must differentiate:

$$f_{Y_5}(y) = \frac{dF_{Y_5}(y)}{dy} = 5e^{-y}[1-e^{-y}]^4$$

Similarly, when $n = 50$, we have

$$\begin{aligned} F_{Y_{50}}(y) &= \mathrm{P}[Y_{50} \le y] \\ &= \mathrm{P}[X_1 \le y]\,\mathrm{P}[X_2 \le y]\cdots\mathrm{P}[X_{50} \le y] \\ &= [F_X(y)]^{50} \\ &= [1-e^{-y}]^{50} \end{aligned}$$

and

$$f_{Y_{50}}(y) = \frac{dF_{Y_{50}}(y)}{dy} = 50e^{-y}[1-e^{-y}]^{49}$$

Plots of these three distributions appear as in Figure 1.38.

Example 1.60 Suppose that X follows an exponential distribution with

$$f_X(x) = \lambda e^{-\lambda x}, \qquad x \ge 0$$

Then what is the probability that the largest from a sample of five observations of X will exceed 3 times the mean?

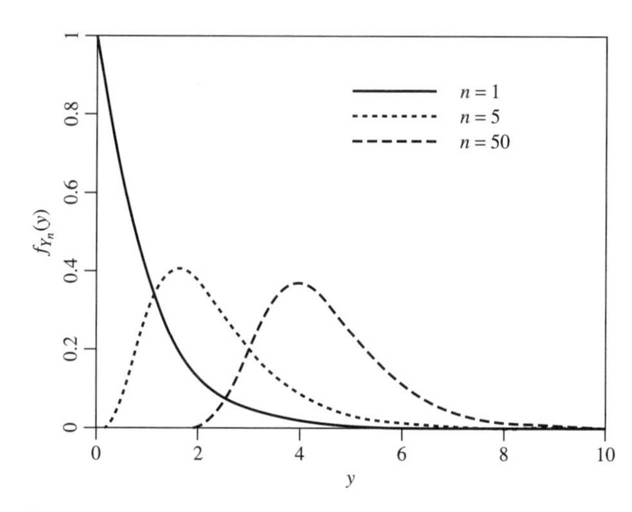

Figure 1.38 Distributions of maximum value of n observations of exponentially distributed random variable.

SOLUTION For $n = 5$, we have

$$\begin{aligned} F_{Y_5}(y) &= \mathrm{P}\left[Y_5 \le y\right] \\ &= \mathrm{P}\left[X_1 \le y\right]\mathrm{P}\left[X_2 \le y\right]\cdots\mathrm{P}\left[X_5 \le y\right] \\ &= \left[F_X(y)\right]^5 \\ &= \left[1 - e^{-\lambda y}\right]^5 \end{aligned}$$

so that

$$\mathrm{P}\left[Y_5 > y\right] = 1 - \left[1 - e^{-\lambda y}\right]^5$$

The mean of X is $1/\lambda$ (see Eq. 1.120), so the probability that Y_5 exceeds 3 times the mean is

$$\begin{aligned} \mathrm{P}\left[Y_5 > \frac{3}{\lambda}\right] &= 1 - \left[1 - e^{-\lambda(3/\lambda)}\right]^5 \\ &= 1 - \left[1 - e^{-3}\right]^5 \\ &= 0.14205 \end{aligned}$$

Now consider the distribution of the minimum out of n samples, Y_1. If we proceed as we did for Y_n, then we would look at the event $Y_1 \le y$. This event just means that $X_1 \le y$ or $X_2 \le y$ or ..., that is,

$$\mathrm{P}\left[Y_1 \le y\right] = \mathrm{P}\left[X_1 \le y \,\cup\, X_2 \le y \,\cup\, \cdots \,\cup\, X_n \le y\right] \tag{1.201}$$

The union on the right expands into $\binom{n}{1} + \binom{n}{2} + \binom{n}{3} + \cdots + \binom{n}{n}$ terms—in other words potentially a *lot* of terms. A better way to work out this distribution is to look at the complement:

$$\begin{aligned} \mathrm{P}\left[Y_1 > y\right] &= \mathrm{P}\left[X_1 > y \,\cap\, X_2 > y \,\cap\, \cdots \,\cap\, X_n > y\right] \\ &= \mathrm{P}\left[X_1 > y\right]\mathrm{P}\left[X_2 > y\right]\cdots\mathrm{P}\left[X_n > y\right] \\ &= \left[1 - F_X(y)\right]^n \end{aligned} \tag{1.202}$$

and since $\mathrm{P}\left[Y_1 > y\right] = 1 - F_{Y_1}(y)$ we get

$$F_{Y_1}(y) = 1 - \left[1 - F_X(y)\right]^n \tag{1.203}$$

and, taking the derivative,

$$f_{Y_1}(y) = n\left[1 - F_X(y)\right]^{n-1} f_X(y) \tag{1.204}$$

Example 1.61 A series of five soil samples are taken at a site and their shear strengths determined. Suppose that a subsequent design is going to be based on the minimum shear strength observed out of the five samples. If the shear strengths of the individual samples are exponentially distributed with parameter $\lambda = 0.025$ m^2/kN, then what is the distribution of the design shear strength?

SOLUTION If we let Y_1 be the design shear strength, where Y_1 is the minimum shear strength observed from the $n = 5$ samples, then

$$F_{Y_1}(y) = 1 - \left[1 - F_X(y)\right]^5$$

where, for the exponential distribution, $F_X(x) = 1 - e^{-\lambda x}$. Thus,

$$\begin{aligned} F_{Y_1}(y) &= 1 - \left[1 - \left(1 - e^{-\lambda y}\right)\right]^5 \\ &= 1 - e^{-5\lambda y} \end{aligned}$$

From this we see that the extreme-value distribution of the minimum of samples from an exponential distribution is also exponentially distributed with new parameter $\lambda' = n\lambda = 5(0.025) = 0.125$. Notice that while the individual samples have mean shear strength equal to $1/\lambda = 1/0.025 = 40$ kN/m^2, the mean design shear strength is one-fifth this value, $1/\lambda' = 1/0.125 = 8$ kN/m^2.

1.11.2 Asymptotic Extreme-Value Distributions

In cases where the cumulative distribution function $F_X(x)$ is not known explicitly (e.g., the normal or lognormal), the exact distributions given above are of questionable value. It turns out that if n is large enough and the sample is random (i.e., composed of independent observations), then the distribution of an extreme value tends toward one of three "asymptotic" forms, which are explained as follows. Thus, even if you do not know the precise form of the distribution of X, the distribution of the extreme value of $X_1, X_2, \ldots, X_n$ can often be deduced, since there are only three possibilities. The results presented below were developed by Gumbel (1958).

1.11.2.1 Type I Asymptotic Form If X has a distribution with an *unlimited exponentially* decaying tail in the direction of the extreme under consideration, then the distribution of the extreme will tend to the *type I asymptotic form.* Examples of such distributions are the normal (in either direction) and the exponential (in the positive direction).

In the case of the maximum, the type I extreme-value distribution has the form

$$F_{Y_n}(y) = \exp\left\{-e^{-\alpha_n(y-u_n)}\right\} \tag{1.205a}$$

$$f_{Y_n}(y) = \alpha_n\, e^{-\alpha_n(y-u_n)} \exp\left\{-e^{-\alpha_n(y-u_n)}\right\} \tag{1.205b}$$

where

$$\begin{aligned} u_n &= \text{characteristic largest value of } X \\ &= F_X^{-1}\left(1 - \frac{1}{n}\right) \\ &= \text{mode of } Y_n \end{aligned} \tag{1.206a}$$

$$\begin{aligned} \alpha_n &= \text{inverse measure of variance of } Y_n \\ &= nf_X(u_n) \end{aligned} \tag{1.206b}$$

In particular, u_n is defined as the value that X exceeds with probability $1/n$. It is found by solving $\mathrm{P}[X > u_n] = 1/n$ for u_n, giving the result shown above. If $F_X^{-1}(p)$ is not readily available, you will either have to consult the literature or determine this extreme-value distribution via simulation.

The mean and variance of the type I maximum asymptotic distribution are as follows:

$$\mathrm{E}[Y_n] = u_n + \frac{\gamma}{\alpha_n} \tag{1.207a}$$

$$\mathrm{Var}[Y_n] = \frac{\pi^2}{6\alpha_n^2} \tag{1.207b}$$

where $\gamma = 0.577216\ldots$ is Euler's number.

Example 1.62 Suppose that a structure is supported by $n = 20$ piles and that long-term pile settlements are distributed according to $f_X(x) = \lambda e^{-\lambda x}$ for $x \geq 0$ being the settlement, where $\lambda = 0.2$ mm^{-1}. If we make the assumption that the piles settle independently (probably a questionable assumption, so that the following results should only be considered approximate), then find the asymptotic parameters of the largest pile settlement, Y_n, out of the n piles, assuming that n is large enough that the asymptotic extreme-value distribution holds.

SOLUTION To find u_n, we solve $\mathrm{P}[X > u_n] = 1/n$ for u_n. For the exponential distribution,

$$\mathrm{P}[X > u_n] = e^{-\lambda u_n} = \frac{1}{n}$$

$$\begin{aligned} -\lambda u_n &= -\ln(n) \\ u_n &= \frac{\ln(n)}{\lambda} = \frac{\ln(20)}{0.2} \\ &= 14.98 \text{ mm} \end{aligned}$$

and

$$\alpha_n = nf_X(u_n) = n\lambda e^{-\lambda \ln(n)/\lambda} = \lambda$$

The parameter $u_n = 14.98$ is the most probable largest settlement out of the 20 piles (e.g., the mode of the distribution).

The asymptotic extreme-value distribution is then

$$F_{Y_n}(y) = \exp\left\{-e^{-\lambda y - \ln(n)}\right\} = \exp\left\{\frac{-e^{-\lambda y}}{n}\right\}$$

The distribution of the minimum value, where the distribution of X is exponentially decaying and unlimited in the direction of the minimum, has the form

$$F_{Y_1}(y) = 1 - \exp\left\{-e^{-\alpha_1(y-u_1)}\right\} \tag{1.208a}$$

$$f_{Y_1}(y) = \alpha_1\, e^{-\alpha_1(y-u_1)} \exp\left\{-e^{-\alpha_1(y-u_1)}\right\} \tag{1.208b}$$

where

$$\begin{aligned} u_1 &= \text{characteristic smallest value of } X \\ &= F_X^{-1}\left(\frac{1}{n}\right) \\ &= \text{mode of } Y_1 \end{aligned} \tag{1.209a}$$

$$\begin{aligned} \alpha_1 &= \text{inverse measure of variance of } Y_1 \\ &= nf_X(u_1) \end{aligned} \tag{1.209b}$$

In particular, u_1 is defined as the value that X has probability $1/n$ of being below. It is found by solving $\mathrm{P}[X \leq u_1] = 1/n$ for u_1. The mean and variance of Y_1 are as follows:

$$\mathrm{E}[Y_1] = u_1 - \frac{\gamma}{\alpha_1} \tag{1.210a}$$

$$\mathrm{Var}[Y_1] = \frac{\pi^2}{6\alpha_1^2} \tag{1.210b}$$

Because of the mirror symmetry of the minimum and maximum type I extreme-value distributions, the skewness coefficient of Y_n is 1.1414 whereas the skewness coefficient of Y_1 is -1.1414. That is, the two distributions are mirror images of one another.

1.11.2.2 Type II Asymptotic Form If X has a distribution with an unlimited polynomial tail, in the direction of the extreme, then its extreme value will have a type II distribution. Examples of distributions with polynomial tails are the lognormal (in the positive direction) and the Pareto (in the positive direction) distributions, the latter of which has the form

$$F_X(x) = 1 - \left(\frac{b}{x}\right)^\alpha \qquad \text{for } x \ge b$$

If the coefficient b is replaced by $u_n/n^{1/\alpha}$, then we get

$$F_X(x) = 1 - \frac{1}{n}\left(\frac{u_n}{x}\right)^\alpha \qquad \text{for } x \ge u_n/n^{1/\alpha}$$

The corresponding extreme-value distribution for the maximum, in the limit as $n \to \infty$, is

$$F_{Y_n}(y) = \exp\left\{-\left(\frac{u_n}{y}\right)^\alpha\right\} \qquad \text{for } y \ge 0 \qquad (1.211a)$$

$$f_{Y_n}(y) = \left(\frac{\alpha}{u_n}\right)\left(\frac{u_n}{y}\right)^{\alpha+1} \exp\left\{-\left(\frac{u_n}{y}\right)^\alpha\right\} \qquad (1.211b)$$

$$\begin{aligned} u_n &= \text{characteristic largest value of } X \\ &= F_X^{-1}\left(1 - \frac{1}{n}\right) \\ &= \text{mode of } Y_n && (1.212a) \\ \alpha &= \text{shape parameter} \\ &= \text{order of polynomial decay of } F_X(x) \\ &\quad\ \text{in direction of extreme} && (1.212b) \end{aligned}$$

Note that although the lognormal distribution seems to have an exponentially decaying tail in the direction of the maximum, the distribution is actually a function of the form $a\exp\{-b(\ln x)^2\}$, which has a polynomial decay. Thus, the extreme-value distribution of n lognormally distributed random variables follows a type II distribution with

$$\alpha = \frac{\sqrt{2\ \ln n}}{\sigma_{\ln X}}$$

$$u_n = \exp\{u_n'\}$$

$$u_n' = \sigma_{\ln X}\sqrt{2\ln n} - \frac{\sigma_{\ln X}\left[\ln(\ln n) + \ln(4\pi)\right]}{2\sqrt{2\ln n}} + \mu_{\ln X}$$

The mean and variance of the type II maximum asymptotic distribution are as follows:

$$\mathrm{E}[Y_n] = u_n\Gamma\left(1 - \frac{1}{\alpha}\right) \qquad \text{if } \alpha > 1 \qquad (1.213a)$$

$$\mathrm{Var}[Y_n] = u_n^2\Gamma\left(1 - \frac{2}{\alpha}\right) - \mathrm{E}^2[Y_n] \qquad \text{if } \alpha > 2 \qquad (1.213b)$$

where Γ is the gamma function (see Eq. 1.128).

The distribution of the minimum for an unbounded polynomial decaying tail can be found as the negative "reflection" of the maximum, namely as

$$F_{Y_1}(y) = 1 - \exp\left\{-\left(\frac{u_1}{y}\right)^\alpha\right\}, \quad y \le 0, \quad u_1 < 0 \qquad (1.214a)$$

$$f_{Y_1}(y) = -\left(\frac{\alpha}{u_1}\right)\left(\frac{u_1}{y}\right)^{\alpha+1} \exp\left\{-\left(\frac{u_1}{y}\right)^\alpha\right\} \qquad (1.214b)$$

where

$$\begin{aligned} u_1 &= \text{characteristic smallest value of } X \\ &= F_X^{-1}\left(\frac{1}{n}\right) \\ &= \text{mode of } Y_1 && (1.215a) \\ \alpha &= \text{shape parameter} \\ &= \text{order of polynomial decay of } F_X(x) \\ &\quad\ \text{in direction of extreme} && (1.215b) \end{aligned}$$

The mean and variance of the type II minimum asymptotic distribution are as follows:

$$\mathrm{E}[Y_1] = u_1\Gamma\left(1 - \frac{1}{\alpha}\right) \qquad \text{if } \alpha > 1 \qquad (1.216a)$$

$$\mathrm{Var}[Y_1] = u_1^2\Gamma\left(1 - \frac{2}{\alpha}\right) - \mathrm{E}^2[Y_1] \qquad \text{if } \alpha > 2 \qquad (1.216b)$$

Example 1.63 Suppose that the pile settlements, X, discussed in the last example actually have the distribution

$$f_X(x) = \frac{1}{x^2} \qquad \text{for } x \ge 1 \text{ mm}$$

Determine the exact distribution of the maximum of a random sample of size n and the asymptotic distribution of the maximum.

SOLUTION We first need to find the cumulative distribution function of X,

$$F_X(x) = \int_1^x \frac{1}{t^2}\,dt = 1 - \frac{1}{x}, \qquad x \ge 1$$

The exact cumulative distribution function of the maximum pile settlement, Y_n, is thus

$$F_{Y_n}(y) = \left[F_X(y)\right]^n = \left[1 - \frac{1}{y}\right]^n \qquad \text{for } y \ge 1$$

and the exact probability density function of Y_n is the derivative of $F_{Y_n}(y)$,

$$f_{Y_n}(y) = \frac{n}{y^2}\left[1 - \frac{1}{y}\right]^{n-1} \qquad \text{for } y \ge 1$$

For the asymptotic distribution, we need to find u_n such that $F_X(u_n) = 1 - 1/n$,

$$F_X(u_n) = 1 - \frac{1}{u_n} = 1 - \frac{1}{n}$$

so that $u_n = n$. The order of polynomial decay of $F_X(x)$ in the direction of the extreme (positive direction) is $\alpha = 1$, so that the asymptotic extreme-value distribution of the maximum, Y_n, is

$$F_{Y_n}(y) = \exp\left\{-\frac{n}{y}\right\} \qquad \text{for } y \geq 0$$

$$f_{Y_n}(y) = \frac{n}{y^2}\exp\left\{-\frac{n}{y}\right\} \qquad \text{for } y \geq 0$$

We see immediately that one result of the approximation is that the lower bound of the asymptotic approximations is $y \geq 0$, rather than $y \geq 1$ found in the exact distributions. However, for $n = 10$, Figure 1.39 compares the exact and asymptotic distributions, and they are seen to be very similar.

1.11.2.3 Type III Asymptotic Form If the distribution of X is bounded by a value, u, in the direction of the extreme, then the asymptotic extreme-value distribution (as $n \to \infty$) is the type III form. Examples are the lognormal and exponential distributions toward the left and the beta and uniform distributions in either direction. For the maximum, the type III asymptotic form is

$$F_{Y_n}(y) = \exp\left\{-\left(\frac{u-y}{u-u_n}\right)^\alpha\right\} \qquad \text{for } y \leq u \qquad (1.217a)$$

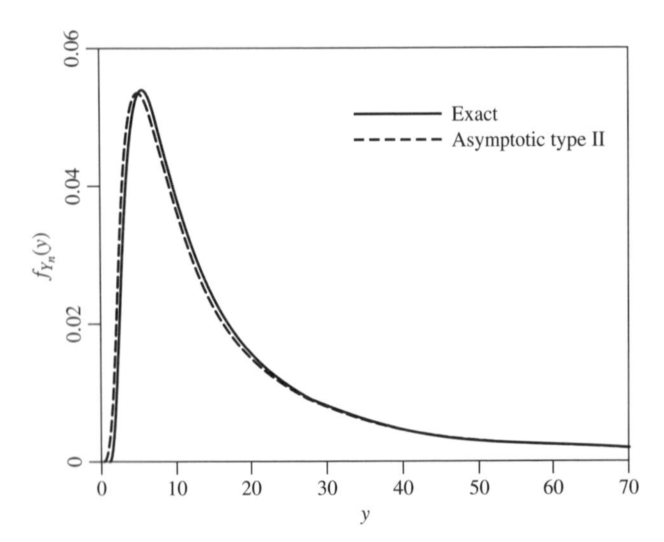

Figure 1.39 Comparison of exact and asymptotic (type II) extreme-value distributions for $n = 10$.

$$f_{Y_n}(y) = \frac{\alpha(u-y)^{\alpha-1}}{(u-u_n)^\alpha}\exp\left\{-\left(\frac{u-y}{u-u_n}\right)^\alpha\right\} \qquad \text{for } y \leq u \qquad (1.217b)$$

where

$$\begin{aligned} u_n &= \text{characteristic largest value of } X \\ &= F_X^{-1}\left(1 - \frac{1}{n}\right) \qquad (1.218a) \\ &= \text{mode of } Y_n \\ \alpha &= \text{shape parameter} \\ &= \text{order of polynomial decay of } F_X(x) \\ &\qquad \text{in direction of extreme} \qquad (1.218b) \end{aligned}$$

The mean and variance of the type III maximum asymptotic distribution are as follows:

$$\mathrm{E}[Y_n] = u - (u - u_n)\Gamma\left(1 + \frac{1}{\alpha}\right) \qquad (1.219a)$$

$$\mathrm{Var}[Y_n] = (u - u_n)^2 \times \left[\Gamma\left(1 + \frac{2}{\alpha}\right) - \Gamma^2\left(1 + \frac{1}{\alpha}\right)\right] \qquad (1.219b)$$

In the case of the minimum, the asymptotic extreme-value distribution is

$$F_{Y_1}(y) = 1 - \exp\left\{-\left(\frac{y-u}{u_1-u}\right)^\alpha\right\} \qquad \text{for } y \geq u \qquad (1.220a)$$

$$f_{Y_1}(y) = \frac{\alpha(y-u)^{\alpha-1}}{(u_1-u)^\alpha}\exp\left\{-\left(\frac{y-u}{u_1-u}\right)^\alpha\right\} \qquad (1.220b)$$

where

$$\begin{aligned} u_1 &= \text{characteristic smallest value of } X \\ &= F_X^{-1}\left(\frac{1}{n}\right) \\ &= \text{mode of } Y_1 \qquad (1.221a) \\ \alpha &= \text{shape parameter} \\ &= \text{order of polynomial decay of } F_X(x) \\ &\qquad \text{in direction of extreme} \qquad (1.221b) \end{aligned}$$

and u is the minimum bound on X. This distribution is also a form of the Weibull distribution. The shape parameter α is, as mentioned, the order of the polynomial $F_X(x)$ in the direction of the extreme. For example, if X is exponentially distributed and we are looking at the distribution of the minimum, then $F_X(x)$ has Taylor's series expansion for small x of

$$F_X(x) = 1 - e^{-\lambda x} \simeq 1 - (1 - \lambda x) = \lambda x \qquad (1.222)$$

which has order 1 as $x \to 0$. Thus, for the minimum of an exponential distribution, $\alpha = 1$.

The mean and variance of the type III maximum asymptotic distribution are as follows:

$$E[Y_n] = u + (u_1 - u)\Gamma\left(1 + \frac{1}{\alpha}\right) \tag{1.223a}$$

$$\text{Var}[Y_n] = (u_1 - u)^2 \times \left[\Gamma\left(1 + \frac{2}{\alpha}\right) - \Gamma^2\left(1 + \frac{1}{\alpha}\right)\right] \tag{1.223b}$$

Example 1.64 A series of 50 soil samples are taken at a site and their shear strengths determined. Suppose that a subsequent design is going to be based on the minimum shear strength observed out of the 50 samples. If the shear strengths of the individual samples are exponentially distributed with parameter $\lambda = 0.025$ m^2/kN, then what is the asymptotic distribution of the design shear strength (i.e., their minimum)? Assume that n is large enough that the asymptotic extreme-value distributions hold.

SOLUTION If we let Y_1 be the design shear strength, then Y_1 is the minimum shear strength observed among the $n = 50$ samples. Since the shear strengths are exponentially distributed, they are bounded by $u = 0$ in the direction of the minimum (to the left). This means that the asymptotic extreme-value distribution of Y_1 is type III. For this distribution, we first need to find u_1 such that $F_X(u_1) = 1/n$,

$$F_X(u_1) = 1 - e^{-\lambda u_1} = 1/n$$

$$\Longrightarrow \qquad u_1 = -(1/\lambda)\ \ln(1 - 1/n)$$

so that $u_1 = -\ln(0.98)/0.025 = 0.8081$.

The order of polynomial decay of $F_X(x)$ in the direction of the extreme (toward $X = 0$) is $\alpha = 1$, as determined by Eq. 1.222, so that the asymptotic extreme-value distribution of the minimum, Y_1, is

$$F_{Y_1}(y) = 1 - \exp\left\{-\left(\frac{y}{0.8081}\right)\right\}, \quad \text{for } y \ge 0$$

$$f_{Y_1}(y) = \frac{1}{0.8081}\exp\left\{-\left(\frac{y}{0.8081}\right)\right\} \quad \text{for } y \ge 0$$

which is just an exponential distribution with parameter $\lambda' = 1/0.8081 = 1.237$. Note that the exact distribution of the minimum is exponential with parameter $\lambda' = n\lambda = 50(0.025) = 1.25$, so the asymptotic approximation is reasonably close to the exact. Figure 1.40 illustrates the close agreement between the two distributions.

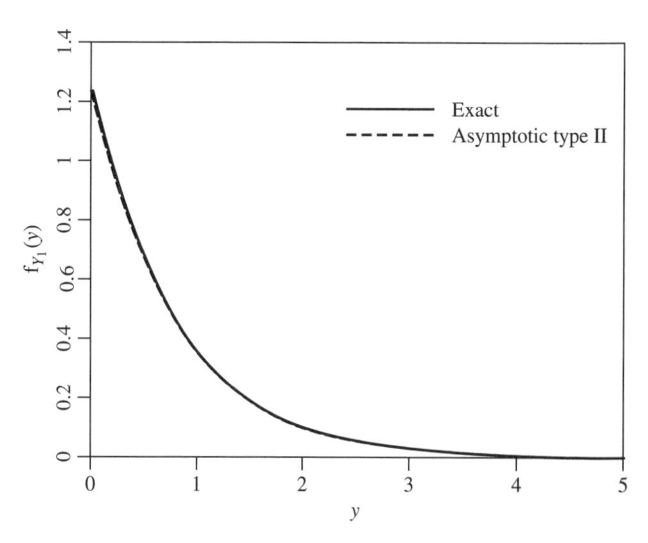

Figure 1.40 Comparison of exact and asymptotic (type III) extreme-value distributions for $n = 50$.

1.12 SUMMARY

De Morgan	$(A \cup B)^c = A^c \cap B^c, \quad (A \cap B)^c = A^c \cup B^c$
Probability	$P[A \cup B] = P[A] + P[B] - P[A \cap B]$
	$P[A \cap B] = P[A\|B] \cdot P[B] = P[B\|A] \cdot P[A]$
Bayes' theorem	$P[A_j \mid E] = \frac{P[E\|A_j] \cdot P[A_j]}{P[E]} = \frac{P[E\|A_j] \cdot P[A_j]}{\sum_{i=1}^{n} P[E\|A_i] \cdot P[A_i]}$
PDFs and CDFs	$F(x) = \int_{-\infty}^{x} f(\xi)\, d\xi \Longleftrightarrow f(x) = \frac{d}{dx} F(x)$

Expectations	$\mathrm{E}[X] = \int_{-\infty}^{\infty} x f_X \; dx$, $\quad \mathrm{E}\left[X^2\right] = \int_{-\infty}^{\infty} x^2 f_X \; dx$
	$\mathrm{E}\left[g(X)\right] = \int_{-\infty}^{\infty} g(x) f_X \; dx$, $\quad \mathrm{E}[a + bX] = a + b\mathrm{E}[X]$
	$\mathrm{E}[XY] = \int_{-\infty}^{\infty}\int_{-\infty}^{\infty} xy\, f_{XY}(x,y) \; dx \; dy$
Variance	$\mathrm{Var}[X] = \mathrm{E}\left[(X-\mu)^2\right] = \mathrm{E}\left[X^2\right] - \mathrm{E}^2[X] = \sigma^2$
	$\mathrm{Var}[a + bX] = b^2 \; \mathrm{Var}[X]$
Covariance	$\mathrm{Cov}[X,Y] = \mathrm{E}[(X-\mu_X)(Y-\mu_Y)] = \mathrm{E}[XY] - \mathrm{E}[X]\mathrm{E}[Y]$, $\quad \rho_{XY} = \dfrac{\mathrm{Cov}[X,Y]}{\sigma_X \, \sigma_Y}$
Taylor's series	$Y = g(X) = g(\mu_X) + (X-\mu_X) \left.\dfrac{dg}{dx}\right\rvert_{\mu_X} + \dfrac{1}{2!}(X-\mu_X)^2 \left.\dfrac{d^2 g}{dx^2}\right\rvert_{\mu_X} + \cdots$
Linear functions	If $Y = \sum_{i=1}^{n} a_i X_i$ and $Z = \sum_{i=1}^{n} b_i X_i$, then $\mathrm{E}[Y] = \sum_{i=1}^{n} a_i \mathrm{E}[X_i]$
	$\mathrm{Var}[Y] = \sum_{i=1}^{n}\sum_{j=1}^{n} a_i a_j \mathrm{Cov}\left[X_i, X_j\right]$, $\quad \mathrm{Cov}[Y,Z] = \sum_{i=1}^{n}\sum_{j=1}^{n} a_i b_j \; \mathrm{Cov}\left[X_i, X_j\right]$
Functions	If $Y = g(X)$ is one to one, then $f_Y(y) = f_X(x)\left\lvert \dfrac{dx}{dy} \right\rvert$
Miscellaneous	$\bar{X} = \dfrac{1}{n}\sum_{i=1}^{n} X_i$, $\quad S^2 = \dfrac{1}{n-1}\sum_{i=1}^{n}(X_i - \bar{X})^2 = \dfrac{1}{n-1}\left\{\sum_{i=1}^{n} X_i^2 - n\bar{X}^2\right\}$
	$\dbinom{n}{k} = \dfrac{n!}{k!(n-k)!}$, $\quad \Gamma(r) = (r-1)!$ (r integer)
Binomial	$\mathrm{P}[N_n = k] = \dbinom{n}{k} p^k q^{n-k}$ $\quad$ for $0 \le k \le n$
	$\mathrm{E}[N_n] = np$ $\quad \mathrm{Var}[N_n] = npq$
Geometric	$\mathrm{P}[T_1 = k] = pq^{k-1}$ $\quad$ for $k \ge 1$
	$\mathrm{E}[T_1] = \dfrac{1}{p}$ $\quad \mathrm{Var}[T_1] = \dfrac{q}{p^2}$
Negative binomial	$\mathrm{P}[T_k = m] = \dbinom{m-1}{k-1} p^k \, q^{m-k}$ $\quad$ for $m \ge k$
	$\mathrm{E}[T_k] = \dfrac{k}{p}$ $\quad \mathrm{Var}[T_k] = \dfrac{kq}{p^2}$
Poisson	$\mathrm{P}[N_t = k] = \dfrac{(\lambda t)^k}{k!} e^{-\lambda t}$ $\quad$ for $k \ge 0$
	$\mathrm{E}[N_t] = \lambda t$ $\quad \mathrm{Var}[N_t] = \lambda t$
Uniform	$f(x) = \dfrac{1}{\beta - \alpha}$ $\quad F(x) = \dfrac{x-\alpha}{\beta-\alpha}$ $\quad$ for $\alpha \le x \le \beta$
	$\mathrm{E}[X] = \frac{1}{2}(\alpha + \beta)$ $\quad \mathrm{Var}[X] = \frac{1}{12}(\beta - \alpha)^2$
Exponential	$f(t) = \lambda e^{-\lambda t}$ $\quad F(t) = 1 - e^{-\lambda t}$ $\quad$ for $t \ge 0$
	$\mathrm{E}[T] = \dfrac{1}{\lambda}$ $\quad \mathrm{Var}[T] = \dfrac{1}{\lambda^2}$

Gamma	$f(x) = \frac{\lambda}{(k-1)!}(\lambda x)^{k-1} e^{-\lambda x}$	$F(x) = 1 - e^{-\lambda x} \sum_{j=0}^{k-1} \frac{(\lambda x)^j}{j!}$	k integer
	$\mathrm{E}[X] = \frac{k}{\lambda}$	$\mathrm{Var}[X] = \frac{k}{\lambda^2}$	
Normal	$f(x) = \frac{1}{\sigma\sqrt{2\pi}} \exp\left\{-\frac{1}{2}\left(\frac{x-\mu}{\sigma}\right)^2\right\}$	$F(x) = \Phi\left(\frac{x-\mu}{\sigma}\right)$	for $-\infty < x < \infty$
	$\mathrm{E}[X] = \mu$	$\mathrm{Var}[X] = \sigma^2$	
	$\mathrm{P}[X \le x] = \mathrm{P}\left[Z \le \frac{x-\mu}{\sigma}\right] = \Phi\left(\frac{x-\mu}{\sigma}\right)$		
Lognormal	$f(x) = \frac{1}{x\sigma_{\ln X}\sqrt{2\pi}} \exp\left[-\frac{1}{2}\left(\frac{\ln x - \mu_{\ln X}}{\sigma_{\ln X}}\right)^2\right]$	$F(x) = \Phi\left(\frac{\ln x - \mu_{\ln X}}{\sigma_{\ln X}}\right)$	for $0 \le x < \infty$
	$\mathrm{E}[X] = \mu_X = e^{\mu_{\ln X} + \frac{1}{2}\sigma_{\ln X}^2}$	$\mathrm{Var}[X] = \sigma_X^2 = \mu_X^2\left(e^{\sigma_{\ln X}^2} - 1\right)$	
	$\sigma_{\ln X}^2 = \ln\left(1 + \frac{\sigma_X^2}{\mu_X^2}\right)$	$\mu_{\ln X} = \ln(\mu_X) - \frac{1}{2}\sigma_{\ln X}^2$	
Weibull	$f(x) = \frac{\beta}{x}(\lambda x)^\beta e^{-(\lambda x)^\beta}$	$F(x) = 1 - e^{-(\lambda x)^\beta}$	for $x \ge 0$
	$\mathrm{E}[X] = \frac{1}{\lambda\beta}\Gamma\left(\frac{1}{\beta}\right)$	$\mathrm{Var}[X] = \frac{1}{\lambda^2\beta}\left\{2\Gamma\left(\frac{2}{\beta}\right) - \frac{1}{\beta}\left[\Gamma\left(\frac{1}{\beta}\right)\right]^2\right\}$	
Extreme Value Distributions:			
Type I	$F_{Y_n}(y) = \exp\{-e^{-\alpha_n(y-u_n)}\}$	$u_n = F_X^{-1}\left(1 - \frac{1}{n}\right)$	$\alpha_n = nf_X(u_n)$
	$F_{Y_1}(y) = 1 - \exp\{-e^{-\alpha_1(y-u_1)}\}$	$u_1 = F_X^{-1}\left(\frac{1}{n}\right)$	$\alpha_1 = nf_X(u_1)$
Type II	$F_{Y_n}(y) = \exp\left\{-\left(\frac{u_n}{y}\right)^\alpha\right\}$	$u_n = F_X^{-1}\left(1 - \frac{1}{n}\right)$	α = polynomial order
	$F_{Y_1}(y) = 1 - \exp\left\{-\left(\frac{u_1}{y}\right)^\alpha\right\}$	$u_1 = F_X^{-1}\left(\frac{1}{n}\right)$	
Type III	$F_{Y_n}(y) = \exp\left\{-\left(\frac{u-y}{u-u_n}\right)^\alpha\right\}$	$u_n = F_X^{-1}\left(1 - \frac{1}{n}\right)$	α = polynomial order
	$F_{Y_1}(y) = 1 - \exp\left\{-\left(\frac{y-u}{u_1-u}\right)^\alpha\right\}$	$u_1 = F_X^{-1}\left(\frac{1}{n}\right)$	u = bound value

CHAPTER 2

Discrete Random Processes

2.1 INTRODUCTION

We are surrounded in nature by spatially and temporally varying phenomena, be it the height of the ocean's surface, the temperature of the air, the number of computational cycles demanded of a CPU per second, or the cohesion of a soil or rock. In this and subsequent chapters models which allow the quantification of natural variability along with our uncertainty about spatially varying processes will be investigated. The models considered are called *random processes*, or, more generally, *random fields*. To illustrate the basic theory, processes which vary in discrete steps (in either time or space) will be presented in this chapter. For example, Figure 2.1 illustrates a SPT where the number of blowcounts at each depth is a discrete number (i.e., $0, 1, 2, \ldots$). This particular soil test can be modeled using a discrete random process.

In theory a *random process* $X(t)$, for all t on the real line, is a collection of random variables whose randomness reflects our uncertainty. Once we have taken a sample of $X(t)$, such as we have done in Figure 2.1, there is no longer any uncertainty in the observation, and our sample is denoted $x(t)$. However, the blowcounts encountered by a SPT at an adjacent location will not be the same as seen in Figure 2.1, although it may be similar if the adjacent test location is nearby. Before performing the test, the test results will be uncertain: $X(1)$ will be a discrete random variable, as will $X(2)$, $X(3)$, and so on. The index t refers to a spatial position or time and we will often refer to $X(t)$ as the *state* of the process at position or time t. For example, $X(t)$ might equal the number of piles which have failed a static load test by time t during the course of substructure construction, where t is measured in time. Alternatively, $X(t)$ might be the depth to the water table, rounded to the nearest meter, at the tth boring, where t is now an index (might also be measured in distance if borings are arranged along a line) giving the boring number. For ease of understanding, the theory developed in this chapter will largely interpret t as time—the bulk of the theory presented in this chapter has been developed for time-varying random processes—but it is emphasized that t can be measured along any one-dimensional line. For geotechnical engineering, taking t along a line in space (e.g., depth) would probably be the most common application.

When the index t takes only a finite (or a *countable*) set of possible values, for example $t = 0, 1, 2, \ldots$, the process $X(t)$ is a *discrete-time* random process. In such cases, the notation X_k, $k = 0, 1, \ldots$, will be used to denote the random process at each discrete time. Alternatively, if the index t varies continuously along the real line, then the random process is said to be a *continuous-time* process. In this case, each instant in time can lead to a new random variable.

The *state space* of a random process is defined as the set of all possible values that the random variable $X(t)$ can assume. For example, we could have $X(1) = 3$, in which case 3 is an element of the *state space*. In general, the state space can be discrete or continuous. For example, if $X(t)$ is the number of SPT blows at depth t, then $X(t)$ has state space $1, 2, \ldots$. Alternatively, if $X(t)$ is the soil's cohesion at depth t, then $X(t)$ can take any nonnegative real value; in this case the state space is continuous. Continuous state spaces are somewhat more complicated to deal with mathematically, so we will start by considering discrete state spaces and save the continuous state spaces until the next chapter.

Thus, a random process is a sequence of random variables that describe the evolution through time (or space) of some (physical) process which, for the observer at least, is uncertain or inherently random.

2.2 DISCRETE-TIME, DISCRETE-STATE MARKOV CHAINS

2.2.1 Transition Probabilities

We will first consider a random process $X_n = X(t_n)$ which steps through time discretely. For example, a CPT sounding will take readings at discrete depth intervals $X_0 = X(0.000), X_1 = X(0.005), \ldots, X_n = X(n\Delta z)$, and so on. In addition, we will assume in this section that X_n can only assume a finite number of possible states (e.g., $X_n = 100, 200, \ldots$ kPa). Unless otherwise noted, the set of possible states (i.e., $a, b, \ldots$) of the random process X_n will be denoted by the positive integers $i_n = \{1, \ldots, m\}$. If $X_n = i_n$, then the process is said to be in state i_n at time n.

Furthermore, suppose now that whenever the process X_n is in state i_n, there is a *fixed* probability that it will go to

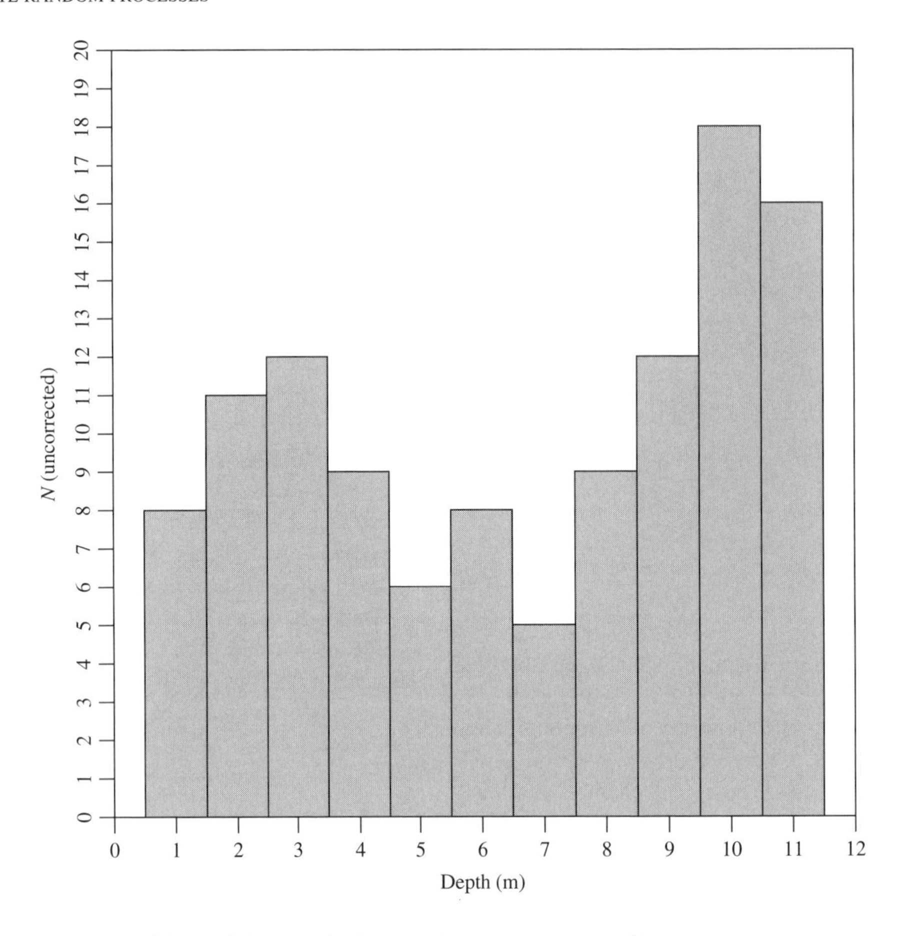

Figure 2.1 Standard penetration test: uncorrected blowcounts.

state j when we move to the next time step $(n+1)$. This probability is denoted as p_{ij}. Specifically, it is supposed that

$$p_{ij} = \mathbf{P}\big[X_{n+1} = j \mid X_n = i_n,\ X_{n-1} = i_{n-1}, \ldots, X_1 = i_1,\ X_0 = i_0\big] \tag{2.1}$$

where the symbol | means "given that." This equation says the following: Given that the random process starts in state i_0 and then progresses through states $i_1, i_2, \ldots$ and is currently in state i_n, the probability that it will be in state j in the next time step is given by p_{ij}.

A closer look at the right-hand-side of Eq. 2.1 indicates that there is a dependence not only on the current state $X_n = i_n$ but also on past states $X_{n-1}, \ldots$. Models of the future which depend on not only the present but also the past history are not uncommon. An example is creep strain in concrete. However, such models are typically rather complex and difficult to deal with, particularly mathematically. As a result, almost all random process theories make use of a simplifying assumption, namely that the future is dependent *only* on the present and is not dependent on the past. This is called the *Markovian* assumption or the *Markov property* and this property allows Eq. 2.1 to be written much more simply as

$$p_{ij} = \mathrm{P}\big[X_{n+1} = j | X_n = i\big] \tag{2.2}$$

The probability p_{ij} is called the *one-step transition probability*.

The Markov property results in simple and thus popular models. There are a great number of physical models in which the future is predicted using only the current state; for example, the future spatial position of a baseball can be accurately predicted given its current position, velocity, wind velocity, drag coefficient, mass, center of gravity, spin, rotational inertia, local gravity, relative location of the sun and moon, and so on. In fact, it can be argued that all mathematical models of the physical universe can be represented as Markov models, dependent only on knowledge of the complete current state to predict the future. Of course, sometimes the level of detail required about the current state, in order to accurately predict the future, is impractical (weather prediction being a classic example). This lack of complete knowledge about the current state leads to uncertainties in predictions, so that future states are most naturally characterized using probabilities.

In addition to the assumption that the future depends only on the present, a further simplification is often introduced, namely that the one-step transition probabilities are *stationary*. This means that probabilities remain constant from step to step. For example, the probability of going from state i in step 3 to state j in step 4 is the same as the probability of going from state i in step 276 to state j in step 277, and so on. Mathematically, this can be realized by stating that Eq. 2.2 remains true for any $n = 0, 1, \ldots$, that is, p_{ij} is independent of the step n under consideration.

Furthermore, since probabilities are nonnegative and since the process must make a transition into some state, the following must also be true;

$$0 \leq p_{ij} \leq 1 \quad \text{for all } 1 \leq i,\ j \leq m$$

$$\sum_{j=1}^{m} p_{ij} = 1 \quad \text{for all } i = 1, 2, \ldots, m$$

which is to say that the sum of probabilities of going from state i to any other state (including i) must be 1.0.

The probability p_{ij} is really just an element of a one-step transition probability *matrix*, which will be denoted as $\boldsymbol{P}$. Specifically, $\boldsymbol{P}$ is a nonsymmetric matrix whose rows sum to 1.0. The one-step transition matrix for a random process with m possible states appears as follows:

$$\boldsymbol{P} = \begin{bmatrix} p_{11} & p_{12} & \cdot & \cdot & \cdot & p_{1m} \\ p_{21} & p_{22} & \cdot & \cdot & \cdot & p_{2m} \\ \cdot & \cdot & \cdot & & & \cdot \\ \cdot & \cdot & & \cdot & & \cdot \\ \cdot & \cdot & & & \cdot & \cdot \\ p_{m1} & p_{m2} & \cdot & \cdot & \cdot & p_{mm} \end{bmatrix}$$

Example 2.1 Consider a sequence of piles arranged along a line. The piles are load tested sequentially. Because of the proximity of the piles, if one fails the load test there is a 40% probability that the next one in the sequence will also fail the test. Conversely, if a pile passes the load test, the probability that the next will also pass the load test is 70%. What is the one-step transition probability matrix for this problem?

SOLUTION Let state 1 be that the pile passes the load test and state 2 be that the pile fails the load test. Then p_{11} is the probability of going from state 1 to state 1, which is to say the probability that the next pile will pass the load test given that the current pile has passed the load test. We are told that $p_{11} = 0.7$. Similarly $p_{22} = 0.4$. Because rows sum to 1.0, we must have $p_{12} = 0.3$ and $p_{21} = 0.6$, which gives us

$$\boldsymbol{P} = \begin{bmatrix} p & 1-p \\ 1-p & p \end{bmatrix} = \begin{bmatrix} 0.7 & 0.3 \\ 0.6 & 0.4 \end{bmatrix}$$

Having established the probabilities associated with the state in the next time step, it is natural to ask what the probability of going from state i to state j in two time steps will be? What about three time steps? In general, the k-step transition probabilities $p_{ij}^{(k)}$ can be defined to be the probability that a process which is currently in state i will be in state j in exactly k time steps. In this definition, the intervening states assumed by the random process are of no interest, so long as it arrives in state j after k time steps. Mathematically, this k-step transition probability is defined as

$$p_{ij}^{(k)} = \mathrm{P}\left[X_{n+k} = j \,|\, X_n = i\right], \quad n, k \geq 0, \quad 0 \leq i, j \leq m \tag{2.3}$$

Again, only stationary k-step transition probabilities are considered, in which $p_{ij}^{(k)}$ is independent of the starting step number, n. As with the one-step transition probabilities, the k-step transition probabilities can be assembled into an $m \times m$ matrix

$$\boldsymbol{P}^{(k)} = \left[p_{ij}^{(k)}\right] \tag{2.4}$$

where

$$0 \leq p_{ij}^{(k)} \leq 1, \qquad k = 0, 1, \ldots, \qquad i = 1, 2, \ldots, m, \quad j = 1, 2, \ldots, m$$

and

$$\sum_{j=1}^{m} p_{ij}^{(k)} = 1 \quad k = 0, 1, \ldots, \qquad i = 1, 2, \ldots, m \tag{2.5}$$

Note that the zero-step transition matrix $\boldsymbol{P}^{(0)}$ is just the identity matrix while the one-step transition matrix $\boldsymbol{P}^{(1)} = \boldsymbol{P}$ so that $p_{ij}^{(1)} = p_{ij}$.

Example 2.2 In any given day, a company undertakes either zero, one, or two site investigations. The next day the number of sites investigated can be either zero, one, or two again, but there is some dependence from day to day. This is a simple three-state, discrete-time Markov chain. Suppose that the one-step transition matrix for this problem appears as follows;

$$\boldsymbol{P} = \begin{bmatrix} 0.7 & 0.3 & 0.0 \\ 0.2 & 0.3 & 0.5 \\ 0.0 & 0.4 & 0.6 \end{bmatrix}$$

(Notice that rows must sum to 1.0 but columns need not.) Figure 2.2 is called a transition diagram, which is a useful graphical depiction of a Markov Chain. What is the two-step transition matrix for this problem?

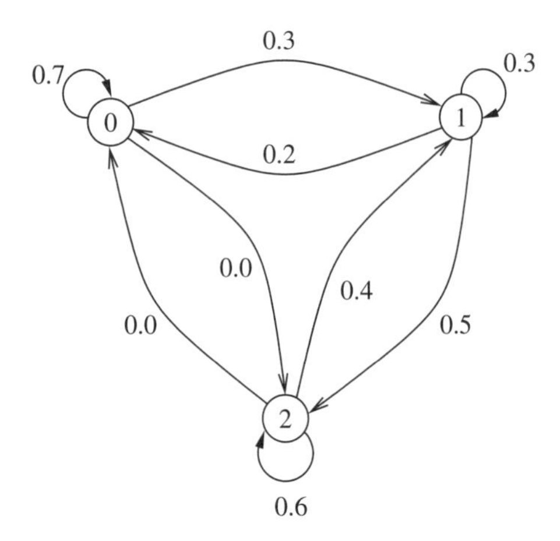

Figure 2.2 Three-state transition diagram.

SOLUTION We will number our possible states 0, 1, and 2 to correspond with the number of sites investigated. Thus, p_{01} is the probability of going from no sites investigated on day 1 to one site investigated on day 2 ($p_{01} = 0.3$ from the above matrix).

Note that the numbering of the states is arbitrary. We normally index the first state with a 1, the second with a 2, and so on, up to m states (which is the usual matrix/vector convention). However, when the first state is 0, the second is 1, and so on, up to $m-1$, it makes more sense to index the states starting at 0 rather than at 1.

Let us start by computing the probability of going from no sites investigated on day 1 to two sites investigated on day 3. Clearly, since $p_{02} = 0$, the company cannot go from zero to two sites in a single day (presumably this never happens for the company, unfortunately). Thus, the probability that the company goes from zero to two sites in two days is just the probability that the company goes from zero to one site in the next day times the probability that the company goes from one to two sites in the second day:

$$p_{02}^{(2)} = p_{01} \cdot p_{12} = (0.3)(0.5) = 0.15$$

The probability that the company goes from zero to one site in the next two days is a bit more complicated. In this case, two paths can be followed: (a) the company starts with zero sites, remains at zero sites in the next day, then investigates one site in the second day or (b) the company starts with zero sites, moves to one site in the next day, then remains at one site in the second day. The desired probability now involves a sum:

$$p_{01}^{(2)} = p_{00}p_{01} + p_{01}p_{11} = (0.7)(0.3) + (0.3)(0.3) = 0.3$$

Similarly, going from one site investigated to one site investigated in two steps now involves three paths: (a) one to zero and back to one, (b) one to one to one, or (c) one to two and back to one. The probability of this is

$$p_{11}^{(2)} = p_{10}p_{01} + p_{11}p_{11} + p_{12}p_{21} = (0.2)(0.3) + (0.3)(0.3) + (0.5)(0.4) = 0.35$$

A closer look at the above equations reveals that in general we can compute

$$p_{ij}^{(2)} = \sum_{k=0}^{2} p_{ik}p_{kj}$$

Using matrix notation, this can be expressed as

$$\boldsymbol{P}^{(2)} = \boldsymbol{P} \cdot \boldsymbol{P}$$

so that

$$\boldsymbol{P}^{(2)} = \begin{bmatrix} 0.7 & 0.3 & 0.0 \\ 0.2 & 0.3 & 0.5 \\ 0.0 & 0.4 & 0.6 \end{bmatrix} \begin{bmatrix} 0.7 & 0.3 & 0.0 \\ 0.2 & 0.3 & 0.5 \\ 0.0 & 0.4 & 0.6 \end{bmatrix} = \begin{bmatrix} 0.55 & 0.30 & 0.15 \\ 0.20 & 0.35 & 0.45 \\ 0.08 & 0.36 & 0.56 \end{bmatrix}$$

More generally, the *Chapman–Kolmogorov equations* provide a method for computing the k-step transition probabilities from the intermediate-step probabilities. These equations are (reverting to the usual matrix indexing starting from 1)

$$p_{ij}^{(k)} = \sum_{\ell=1}^{m} p_{i\ell}^{(\nu)} p_{\ell j}^{(k-\nu)}, \qquad \begin{array}{l} i = 1, 2, \ldots, m, \\ j = 1, 2, \ldots, m \end{array} \tag{2.6}$$

for any $\nu = 0, \ldots, k$. These equations are most easily understood by noting that $p_{i\ell}^{(\nu)} p_{\ell j}^{(k-\nu)}$ represents the probability that starting in state i the process will go to state j in k transitions through a path that takes it into state ℓ at the νth transition. Hence, summing over all possible intermediate states ℓ yields the probability that the process will be in state j after k transitions. In terms of the transition matrices, Eq. 2.6 is equivalent to

$$\boldsymbol{P}^{(k)} = \boldsymbol{P}^{(\nu)} \cdot \boldsymbol{P}^{(k-\nu)} \tag{2.7}$$

where $\cdot$ represents matrix multiplication (see Eq. 2.6). Hence, in particular,

$$\boldsymbol{P}^{(2)} = \boldsymbol{P} \cdot \boldsymbol{P} = \boldsymbol{P}^2$$

and by induction

$$\boldsymbol{P}^{(k)} = \boldsymbol{P}^{k-1} \cdot \boldsymbol{P} = \boldsymbol{P}^k \tag{2.8}$$

That is, *the k-step transition matrix may be obtained by multiplying the matrix* $\boldsymbol{P}$ *by itself k times*. Note that although

$\boldsymbol{P}^{(k)}$ is the same as $\boldsymbol{P}^k$, the superscript (k) is retained to ensure that the matrix components $p_{ij}^{(k)}$ are not interpreted as the component raised to the power k.

2.2.2 Unconditional Probabilities

So far, all of the probabilities we have considered are conditional probabilities. For instance, $p_{ij}^{(k)}$ is the probability that the state at time step k is j *given* that the initial state at time step 0 is i. If the *unconditional* distribution of the state at time step k is desired, we will first need to know the probability distribution of the initial state. Let us denote the initial probabilities by the row vector

$$\boldsymbol{\pi}(0) = \left\{\pi_1(0) \quad \pi_2(0) \quad \cdots \quad \pi_m(0)\right\} \tag{2.9}$$

where the ith element in this vector, $\pi_i(0)$, is the probability that the initial state is i, namely

$$\pi_i(0) = \mathrm{P}\left[X_0 = i\right] \tag{2.10}$$

for all $1 \leq i \leq m$. Also, since the initial state must be one of the possible states, the following must also be true:

$$\sum_{i=1}^{m} \pi_i(0) = 1 \tag{2.11}$$

The desired unconditional probabilities at time step n may be computed by using the total probability theorem (which combines all possible ways of getting to a certain state), that is,

$$\begin{aligned} \mathrm{P}\left[X_n = j\right] &= \sum_{i=1}^{m} \mathrm{P}\left[X_n = j \mid X_0 = i\right] \mathrm{P}\left[X_0 = i\right] \\ &= \sum_{i=1}^{m} p_{ij}^{(n)} \pi_i(0) \end{aligned} \tag{2.12}$$

If we define the n-step unconditional probabilities

$$\boldsymbol{\pi}(n) = \{\pi_1(n), \ldots, \pi_m(n)\} \tag{2.13}$$

with $\pi_i(n) = \mathrm{P}\left[X_n = i\right]$ being the probability of being in state i at time step n, then $\boldsymbol{\pi}(n)$ can be found from

$$\boldsymbol{\pi}(n) = \boldsymbol{\pi}(0) \cdot \boldsymbol{P}^n \tag{2.14}$$

Example 2.3 In an electronic load-measuring system, under certain adverse conditions, the probability of an error on each sampling cycle depends on whether or not it was preceded by an error. We will define 1 as the error state and 2 as the nonerror state. Suppose the probability of an error if preceded by an error is 0.75, the probability of an error if preceded by a nonerror is 0.50, and thus the probability of a nonerror if preceded by an error is 0.25, and the probability of a nonerror if preceded by a nonerror is 0.50. This gives the one-step transition matrix

$$\boldsymbol{P} = \begin{bmatrix} 0.75 & 0.25 \\ 0.50 & 0.50 \end{bmatrix}$$

The two-step, three-step, ..., seven-step transition matrices are shown below:

$$\boldsymbol{P}^2 = \begin{bmatrix} 0.688 & 0.312 \\ 0.625 & 0.375 \end{bmatrix}, \qquad \boldsymbol{P}^3 = \begin{bmatrix} 0.672 & 0.328 \\ 0.656 & 0.344 \end{bmatrix},$$

$$\boldsymbol{P}^4 = \begin{bmatrix} 0.668 & 0.332 \\ 0.664 & 0.336 \end{bmatrix}, \qquad \boldsymbol{P}^5 = \begin{bmatrix} 0.667 & 0.333 \\ 0.666 & 0.334 \end{bmatrix},$$

$$\boldsymbol{P}^6 = \begin{bmatrix} 0.667 & 0.333 \\ 0.667 & 0.333 \end{bmatrix}, \qquad \boldsymbol{P}^7 = \begin{bmatrix} 0.667 & 0.333 \\ 0.667 & 0.333 \end{bmatrix}$$

If we know that initially the system is in the nonerror state, then $\pi_1(0) = 0$, $\pi_2(0) = 1$, and $\boldsymbol{\pi}(n) = \boldsymbol{\pi}(0) \cdot \boldsymbol{P}^{(n)}$. Thus, for example, $\boldsymbol{\pi}(7) = \{0.667, 0.333\}$. Clearly either the load-measuring system and/or the adverse conditions should be avoided since the system is spending two-thirds of its time in the error state.

Notice also that the above powers of $\boldsymbol{P}$ are tending towards a "steady state." These are called the steady-state probabilities, which we will see more of later.

2.2.3 First Passage Times

The length of time (i.e., in this discrete case, the *number of steps*) for the process to go from state i to state j for the first time is called the *first passage time* N_{ij}. This is important in engineering problems as it can represent the recurrence time for a loading event, the time to (first) failure of a system, and so on. If $i = j$, then this is the number of steps needed for the process to *return* to state i for the first time, and this is called the *first return time* or the *recurrence time* for state i.

First passage times are random variables and thus have an associated probability distribution function. The probability that n steps will be needed to go from state i to j will be denoted by $f_{ij}^{(n)}$. It can be shown (using simple results on the union of two or more events) that

$$\begin{aligned} f_{ij}^{(1)} &= p_{ij}^{(1)} = p_{ij} \\ f_{ij}^{(2)} &= p_{ij}^{(2)} - f_{ij}^{(1)} \cdot p_{jj} \\ &\;\;\vdots \\ f_{ij}^{(n)} &= p_{ij}^{(n)} - f_{ij}^{(1)} \cdot p_{jj}^{(n-1)} - f_{ij}^{(2)} \cdot p_{jj}^{(n-2)} - \cdots - f_{ij}^{(n-1)} p_{jj} \end{aligned} \tag{2.15}$$

The first equation, $f_{ij}^{(1)}$, is just the one-step transition probability. The second equation, $f_{ij}^{(2)}$, is the probability of going from state i to state j in two steps minus the probability of going to state j in one time step—$f_{ij}^{(2)}$ is the probability of going from state i to j in two time steps *for the first time*, so we must remove the probabilities associated with going to state j prior to time step 2. Similarly, $f_{ij}^{(3)}$ is the probability of going from state i to state j in three time steps minus all probabilities which involve entering state j prior to the third time step.

Equations 2.15 are solved recursively starting from the one-step probabilities to finally obtain the probability of taking n steps to go from state i to state j. The computations are quite laborious; they are best solved using a computer program.

Example 2.4 Using the one-step transition probabilities presented in Example 2.3, the probability distribution governing the passage time n to go from state $i = 1$ to state $j = 2$ is

$$f_{12}^{(1)} = p_{12} = 0.25$$
$$f_{12}^{(2)} = 0.312 - (0.25)(0.5) = 0.187$$
$$f_{12}^{(3)} = 0.328 - (0.25)(0.375) - (0.187)(0.5) = 0.141$$
$$f_{12}^{(4)} = 0.332 - (0.25)(0.344) - (0.187)(0.375) - (0.141)(0.5) = 0.105$$

.
.
.

There are four such distributions, one for each (i, j) pair: $(1, 1)$, $(1, 2)$ (as above), $(2, 1)$, and $(2, 2)$.

Starting out in state i, it is not always guaranteed that state j will be reached at some time in the future. If it is guaranteed, then the following must be true:

$$\sum_{n=1}^{\infty} f_{ij}^{(n)} = 1$$

Alternatively, if there exists a possibility that state j will never be reached when starting from state i, then

$$\sum_{n=1}^{\infty} f_{ij}^{(n)} < 1$$

This observation leads to two possible cases:

1. If the sum (above) is equal to 1, then the values $f_{ij}^{(n)}$ for $n = 1, 2, \ldots$ represent the probability distribution of the first passage time for specific states i and j, and this passage will occur sooner or later. If $i = j$, then the state i is called a *recurrent state* since, starting in state i, the process will always return to i sooner or later.
2. If the sum (above) is less than 1, a process in state i may never reach state j. If $i = j$, then the state i is called a *transient state* since there is a chance that the process will never return to its starting state. (This means that, sooner or later, the process will leave state i forever.)

If $p_{ii} = 1$ for some state i, then the state i is called an *absorbing state*. Once this state is entered, it is never left.

Example 2.5 Is state 0 of the three-state Example 2.2 transient or recurrent?

SOLUTION Since all states in Example 2.2 "communicate," that is, state 0 can get to state 1, state 1 can get to state 2, and vice versa, all of the states in Example 2.2 are recurrent—they will all recur over and over with time.

Example 2.6 Considering the transition diagram in Figure 2.3 for a three-state discrete-time Markov chain answer the following questions:

(a) Is state 2 transient or recurrent?
(b) Compute the probabilities that, starting in state 0, state 2 is reached for the first time in one, two, or three time steps.
(c) Estimate (or make a reasonable guess at) the probability that state 2 is reached from state 0.

SOLUTION

(a) Since states 0 or 2 will eventually transit to state 1 (both have nonzero probabilities of going to state 1) and since

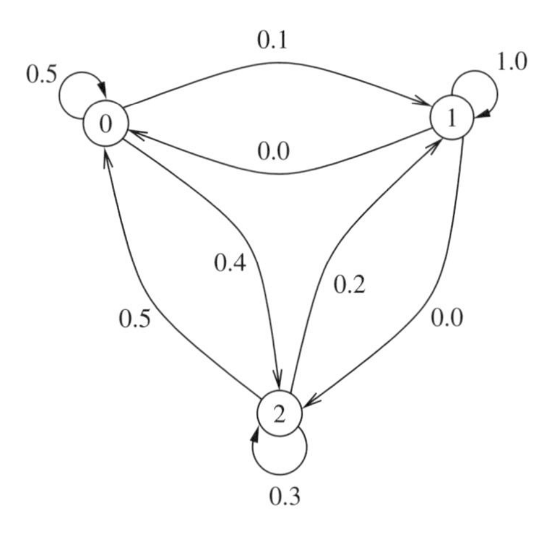

Figure 2.3 Three-state discrete-time Markov chain.

state 1 is absorbing (stays there forever), both states 0 and 2 are transient. In other words, no matter where this Markov chain starts, it will eventually end up in state 1 forever.

(b) If we start in state 0, the probability of going to state 2 in the next time step is 0.4. The probability of going to state 2 in two time steps is equal to the probability of staying in state 0 for the first time step (if we ever go to state 1, we will never get to state 2) times the probability of going to state 2 in the second time step,

$$f_{02}^{(2)} = p_{00}\,p_{02} = (0.5)(0.4) = 0.2$$

The probability of going from state 0 to state 2 in three time steps is equal to the probability of remaining in state 0 for two time steps times the probability of going to state 2,

$$f_{02}^{(3)} = p_{00}^2\,p_{02} = (0.5)^2(0.4) = 0.1$$

(c) The probability that state 2 is reached from state 0 is equal to the probability that starting in state 0, state 2 is reached in any of time steps $1, 2, \ldots$. This is a union of the events that we reach state 2 in any time step. Unfortunately, these events are not disjoint (i.e., we could reach state 2 in both steps 2 and 4). It is easier to compute this probability as 1 minus the probability that we reach state 1 prior to reaching state 2,

$$\begin{aligned} &\mathrm{P}\,[\text{state 2 is reached}] \\ &= 1 - \left[p_{01} + p_{00}p_{01} + p_{00}^2 p_{01} + \cdots\right] \\ &= 1 - p_{01}\sum_{k=0}^{\infty} p_{00}^k = 1 - \frac{p_{01}}{1 - p_{00}} \\ &= 1 - \frac{0.1}{1 - 0.5} = 0.8 \end{aligned}$$

where each term represents the probability of remaining in state 0 for k time steps and then going to state 1 in the $(k+1)$th time step.

2.2.4 Expected First Passage Time

It is usually very difficult to calculate the first passage time probabilities $f_{ij}^{(n)}$ for all n, especially considering the fact that n goes to infinity. If one succeeds in calculating them in some sort of functional form, then one could speak of the *expected first passage time* of the process from state i to state j, which is denoted μ_{ij}. In terms of the probabilities $f_{ij}^{(n)}$, the expected first passage time is given by

$$\mathrm{E}\left[N_{ij}\right] = \mu_{ij} = \begin{cases} \infty, & \sum_{n=1}^{\infty} f_{ij}^{(n)} < 1 \\ \sum_{n=1}^{\infty} n f_{ij}^{(n)}, & \sum_{n=1}^{\infty} f_{ij}^{(n)} = 1 \end{cases} \tag{2.16}$$

and if $\sum_{n=1}^{\infty} f_{ij}^{(n)} = 1$, which is to say state j will eventually be reached from state i, it can be shown that

$$\mu_{ij} = 1 + \sum_{k \neq j} p_{ik}\,\mu_{kj} \tag{2.17}$$

If $i = j$, then the expected first passage time is called the *expected recurrence time* (see Example 2.6). If $\mu_{ii} = \infty$ for a recurrent state, it is called *null*; however, this can only occur if there are an infinite number of possible states. If $\mu_{ii} < \infty$, then the state i is called *nonnull* or *positive recurrent*. There are no null recurrent states in a finite-state Markov chain. All of the states in such chains are either positive recurrent or transient. Note that expected recurrence times, μ_{ii}, are easily computed from the steady-state probabilities, as discussed next.

2.2.5 Steady-State Probabilities

As seen in Example 2.3, some Markov chains settle down quite quickly into a steady state, where the unconditional probability of being in a state becomes a constant. Only certain types of Markov chains have this property. Fortunately, they are the most commonly occurring types of Markov chains. To investigate the properties of such Markov chains, a few more definitions are required, as follows:

1. A state is called *periodic* with period $\tau > 1$ if a return is possible only in $\tau, 2\tau, 3\tau, \ldots$ steps; this means that $p_{ii}^{(n)} = 0$ for all values of n that are not divisible by $\tau > 1$, and τ is the smallest integer having this property. Clearly the Markov chain in Figure 2.4 is periodic (and not very interesting!).
2. State j is said to be *accessible* from state i if $p_{ij}^{(n)} > 0$ for some $n \geq 0$. What this means is that the process can get to state j from state i sooner or later.
3. Two states i and j that are *accessible to each other* are said to *communicate*, and this is denoted $i \leftrightarrow j$. Note that, by definition, any state communicates with itself since $p_{ii}^{(0)} = 1$. Also if state i communicates with state j and state j communicates with state k, then state i communicates with state k.
4. Two states that communicate are said to be in the same *class*. Note that as a consequence of 1–3 above *any two classes of states are either identical or disjoint.*

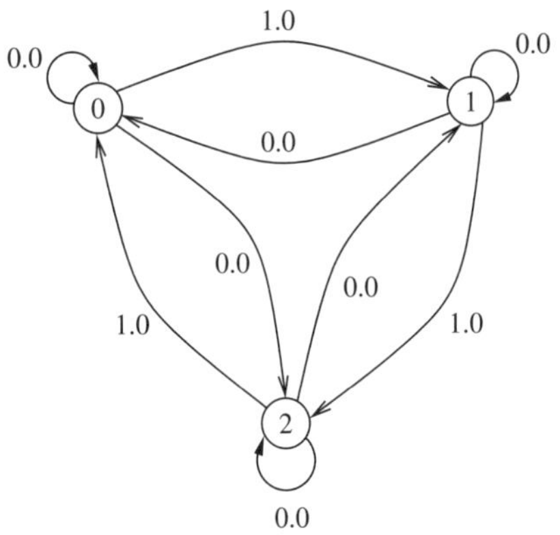

0.0 1.0 0.0 0 1 0.0 0.0 0.0 1.0 1.0 2 0.0

Figure 2.4 Example of a periodic Markov chain.

5. A Markov chain is said to be *irreducible* if it contains only one class, that is, if all states communicate with each other.
6. If the state i in a class is *aperiodic* (i.e., not periodic) and if the state is also *positive recurrent*, then the state is said to be *ergodic*.
7. An irreducible Markov chain is ergodic if all of its states are ergodic.

It is the *irreducible ergodic Markov chain* which settles down to a steady state. For such Markov chains, the unconditional state distribution

$$\boldsymbol{\pi}(n) = \boldsymbol{\pi}(0) \cdot \boldsymbol{P}^n \tag{2.18}$$

converges as $n \to \infty$ to a constant vector, and the resulting *limiting distribution* is independent of the initial probabilities $\boldsymbol{\pi}(0)$. In general, for irreducible ergodic Markov chains,

$$\lim_{n\to\infty} p_{ij}^{(n)} = \lim_{n\to\infty} \pi_j(n) = \pi_j$$

and the π_j's are independent of i. The π_j's are called the *steady-state probabilities* and they satisfy the following *state equations*:

1. $0 < \pi_j < 1$ for all $j = 1, 2, \ldots, m$
2. $\sum_{j=1}^{m} \pi_j = 1$
3. $\pi_j = \sum_{i=1}^{m} \pi_i \cdot p_{ij}$, $j = 1, 2, \ldots, m$

Using $m = 3$ as an example, item 3 can be reexpressed using vector–matrix notation as

$$\{\pi_1 \ \pi_2 \ \pi_3\} = \{\pi_1 \ \pi_2 \ \pi_3\} \begin{bmatrix} p_{11} & p_{12} & p_{13} \\ p_{21} & p_{22} & p_{23} \\ p_{31} & p_{32} & p_{33} \end{bmatrix}$$

Since there are $m + 1$ equations in items 2 and 3 above and there are m unknowns, one of the equations is redundant. The redundancy arises because the rows of $\boldsymbol{P}$ sum to 1 and are thus not independent. Choose $m - 1$ of the m equations in 3 along with the equation in 2 to solve for the steady-state probabilities.

Example 2.7 In the case of the electronic load-measuring system presented in Example 2.3, the state equations above become

$$\pi_1 = 0.75\pi_1 + 0.50\pi_2, \qquad 1 = \pi_1 + \pi_2$$

Solving these for the steady-state probabilities yields

$$\pi_1 = \tfrac{2}{3}, \qquad \pi_2 = \tfrac{1}{3}$$

which agrees with the emerging results of Example 2.3 as n increases above about 5.

Note that steady-state probabilities and the mean recurrence times for irreducible ergodic Markov chains have a reciprocal relationship:

$$\mu_{jj} = \frac{1}{\pi_j}, \qquad j = 1, 2, \ldots, m \tag{2.19}$$

Thus, the mean recurrence time can be computed without knowing the probability distribution of the first passage time.

Example 2.8 A sequence of soil samples are taken along the line of a railway. The samples are tested and classified into three states:

1. Good
2. Fair (needs some remediation)
3. Poor (needs to be replaced)

After taking samples over a considerable distance, the geotechnical engineer in charge notices that the soil classifications are well modeled by a three-state stationary Markov chain with the transition probabilities

$$\boldsymbol{P} = \begin{bmatrix} 0.6 & 0.2 & 0.2 \\ 0.3 & 0.4 & 0.3 \\ 0.0 & 0.3 & 0.7 \end{bmatrix}$$

and the transition diagram in Figure 2.5.

(a) What are the steady-state probabilities?
(b) On average, how many samples must be taken until the next sample to be classified as poor is encountered?

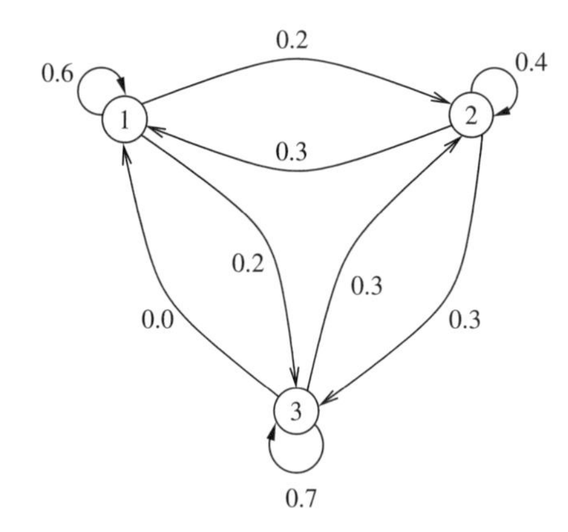

Figure 2.5 Transition diagram for railway example.

SOLUTION

(a) The following equations can be solved simultaneously:

$$\pi_1 = 0.6\pi_1 + 0.3\pi_2 + 0.0\pi_3$$
$$\pi_2 = 0.2\pi_1 + 0.4\pi_2 + 0.3\pi_3$$
$$1 = \pi_1 + \pi_2 + \pi_3$$

to yield the steady-state probabilities

$$\pi_1 = \frac{3}{13}, \qquad \pi_2 = \frac{4}{13}, \qquad \pi_3 = \frac{6}{13}$$

It appears that soil samples are most likely to be classified as poor (twice as likely as being classified as good).

(b) The mean number of samples required to return to state 3 (poor) is μ_{33} (see Eq. 2.19) where

$$\mu_{33} = \frac{1}{\pi_3} = 2.17 \ \textit{samples}$$

Example 2.9 The water table at a particular site may be idealized into three states: low, moderate, and high. Because of the probabilistic nature of rainfall patterns, irrigation pumping, and evaporation, the water table level may shift from one state to another between seasons as a Markov chain. Suppose that the transition probabilities from one state to another are as indicated in Figure 2.6, where low, moderate, and high water table levels are denoted by states 1, 2, and 3, respectively.

(a) Derive the one-step transition matrix for this problem.

(b) Suppose that for season 1 you predict that there is an 80% probability the water table will be high at the beginning of season 1 on the basis of extended weather reports. Also if it is not high, the water table will be three times as likely to be moderate as it is to be low. On the basis of this prediction, what is the probability that the water table will be high at the beginning of season 2?

(c) What is the steady-state probability that the water table will be high in any one season?

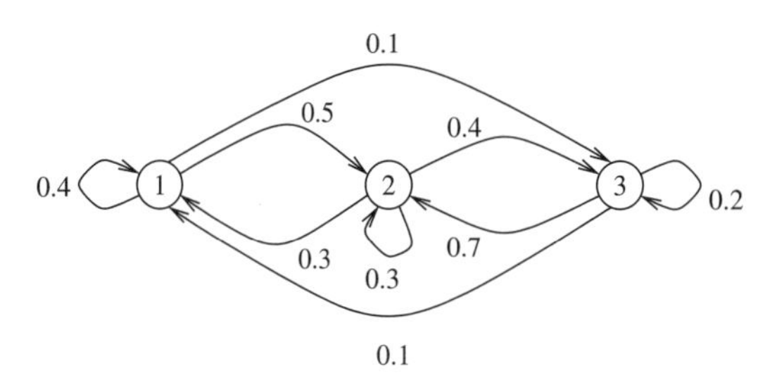

Figure 2.6 Transition diagram for water table problem.

SOLUTION

(a) From Figure 2.6, the probability of going from low to low (state 1 to 1) is 0.4, going from low to moderate (1 to 2) is 0.5, and so on, leading to the following one-step transition matrix:

$$\boldsymbol{P} = \begin{bmatrix} 0.4 & 0.5 & 0.1 \\ 0.3 & 0.3 & 0.4 \\ 0.1 & 0.7 & 0.2 \end{bmatrix}$$

(b) In this case, the initial state probabilities are $\boldsymbol{\pi}(0) = \{0.05 \ 0.15 \ 0.8\}$. Thus, at the beginning of season 2, the unconditional state probabilities become

$$\boldsymbol{\pi}(1) = \{0.05 \quad 0.15 \quad 0.8\} \begin{bmatrix} 0.4 & 0.5 & 0.1 \\ 0.3 & 0.3 & 0.4 \\ 0.1 & 0.7 & 0.2 \end{bmatrix}$$
$$= \{0.145 \quad 0.63 \quad 0.225\}$$

Thus, the probability that the water table will be high at the beginning of season 2 is 22.5%.

(c) To find the steady-state probabilities, we need to find $\{\pi_1 \ \pi_2 \ \pi_3\}$ such that

$$\{\pi_1 \quad \pi_2 \quad \pi_3\} = \{\pi_1 \quad \pi_2 \quad \pi_3\} \begin{bmatrix} 0.4 & 0.5 & 0.1 \\ 0.3 & 0.3 & 0.4 \\ 0.1 & 0.7 & 0.2 \end{bmatrix}$$

and $\pi_1 + \pi_2 + \pi_3 = 1.0$. Using this and the first two equations from above (since the third equation is linearly dependent on the other two), we have

$$\pi_1 = 0.4\pi_1 + 0.3\pi_2 + 0.1\pi_3$$
$$\pi_2 = 0.5\pi_1 + 0.3\pi_2 + 0.7\pi_3$$
$$1.0 = 1.0\pi_1 + 1.0\pi_2 + 1.0\pi_3$$

or

$$\begin{bmatrix} -0.6 & 0.3 & 0.1 \\ 0.5 & -0.7 & 0.7 \\ 1.0 & 1.0 & 1.0 \end{bmatrix} \begin{Bmatrix} \pi_1 \\ \pi_2 \\ \pi_3 \end{Bmatrix} = \begin{Bmatrix} 0.0 \\ 0.0 \\ 1.0 \end{Bmatrix}$$

which has solution

$$\begin{Bmatrix} \pi_1 \\ \pi_2 \\ \pi_3 \end{Bmatrix} = \begin{Bmatrix} 0.275 \\ 0.461 \\ 0.265 \end{Bmatrix}$$

so that the steady-state probability that the water table will be high at the start of any one season is 26.5%.

Example 2.10 A bus arrives at its stops either early, on time, or late. If the bus is late at a stop, its probabilities of being early, on time, and late at the next stop are $\frac{1}{6}$, $\frac{2}{6}$, and $\frac{3}{6}$, respectively. If the bus is on time at a stop, it is equi-likely to be early, on time, or late at the next stop. If it is early at a stop, it is twice as likely to be on time at the next stop as either early or late, which are equi-likely.

(a) Why can this sequence of bus stops be modeled using a Markov chain?
(b) Find the one-step transition matrix $\boldsymbol{P}$.
(c) If the bus is early at the first stop, what is the probability that it is still early at the third stop? What is this probability at the fourth stop?
(d) If the controller estimates the bus to have probabilities of 0.1, 0.7, and 0.2 of being early, on time, or late at the first stop, what now is the probability that the bus is early at the third stop?
(e) After many stops, at what fraction of stops is the bus early on average?

SOLUTION

(a) Since the probabilities of being early, on time, or late at any stop depend only on the state at the previous stop, the sequence of stops can be modeled as a three-state Markov chain.
(b) Define the states as 1 for early, 2 for on time, and 3 for late. Then from the given information and making use of the fact that each row of the transition matrix must sum to 1.0, we get

$$\boldsymbol{P} = \begin{bmatrix} \frac{1}{4} & \frac{2}{4} & \frac{1}{4} \\ \frac{1}{3} & \frac{1}{3} & \frac{1}{3} \\ \frac{1}{6} & \frac{2}{6} & \frac{3}{6} \end{bmatrix}$$

(c) For this question we need the two-step transition matrix:

$$\boldsymbol{P}^2 = \boldsymbol{P} \cdot \boldsymbol{P} = \begin{bmatrix} \frac{1}{4} & \frac{2}{4} & \frac{1}{4} \\ \frac{1}{3} & \frac{1}{3} & \frac{1}{3} \\ \frac{1}{6} & \frac{2}{6} & \frac{3}{6} \end{bmatrix} \begin{bmatrix} \frac{1}{4} & \frac{2}{4} & \frac{1}{4} \\ \frac{1}{3} & \frac{1}{3} & \frac{1}{3} \\ \frac{1}{6} & \frac{2}{6} & \frac{3}{6} \end{bmatrix} = \begin{bmatrix} \boxed{0.271} & 0.375 & 0.354 \\ 0.250 & 0.389 & 0.361 \\ 0.236 & 0.361 & 0.403 \end{bmatrix}$$

(note that all rows sum to 1.0, OK). This gives the probability that it is still early at the third stop to be 0.271. For the next stop, we need to compute the three-step transition matrix:

$$\boldsymbol{P}^3 = \boldsymbol{P}^2 \cdot \boldsymbol{P} = \begin{bmatrix} 0.271 & 0.375 & 0.354 \\ 0.250 & 0.389 & 0.361 \\ 0.236 & 0.361 & 0.403 \end{bmatrix} \begin{bmatrix} \frac{1}{4} & \frac{2}{4} & \frac{1}{4} \\ \frac{1}{3} & \frac{1}{3} & \frac{1}{3} \\ \frac{1}{6} & \frac{2}{6} & \frac{3}{6} \end{bmatrix} = \begin{bmatrix} \boxed{0.252} & \cdot & \cdot \\ \cdot & \cdot & \cdot \\ \cdot & \cdot & \cdot \end{bmatrix}$$

so that the probability that the bus is early at the fourth stop is 0.252.

(d) Now we have uncertainty about the initial state, and we must multiply

$$\{0.1 \quad 0.7 \quad 0.2\} \begin{bmatrix} 0.271 & 0.375 & 0.354 \\ 0.250 & 0.389 & 0.361 \\ 0.236 & 0.361 & 0.403 \end{bmatrix} = \{0.249 \quad 0.382 \quad 0.369\}$$

so that the probability that the bus is early at the third stop is now 0.249.

(e) For the steady-state probabilities, we solve

$$\pi_1 = \tfrac{1}{4}\pi_1 + \tfrac{1}{3}\pi_2 + \tfrac{1}{6}\pi_3$$
$$\pi_2 = \tfrac{1}{2}\pi_1 + \tfrac{1}{3}\pi_2 + \tfrac{1}{3}\pi_3$$
$$1.0 = \pi_1 + \pi_2 + \pi_3$$

which gives us

$$\begin{Bmatrix} \pi_1 \\ \pi_2 \\ \pi_3 \end{Bmatrix} = \begin{Bmatrix} 0.250 \\ 0.375 \\ 0.375 \end{Bmatrix}$$

so that the steady-state probability of being early at a stop is 0.25 (as suggested by the first column of the matrices in part (c), which appear to be tending toward 0.25).

2.3 CONTINUOUS-TIME MARKOV CHAINS

The transition from the discrete-time Markov chain to the continuous-time Markov chain is entirely analogous to the transition from the binomial (number of "successes" in n discrete trials) to the Poisson (number of successes in time interval t). In fact, the Markov chain is really simply a generalization of the binomial and Poisson random variables—rather than just success and "failure" as possible outcomes, the Markov chain allows any number of possible "states" (m states have been considered in the examples so far, where m can be any integer). In addition, the Markov chain allows for statistical dependence between states from step to step. Nevertheless, a deeper understanding of both discrete and continuous-time Markov chains is possible through a more careful study of the binomial and Poisson random variables. Recall that the binomial random variable is characterized by p, the probability of success. The Markov analogs are the set of state probabilities $\boldsymbol{\pi}(0)$ and transition probabilities p_{ij}. However, when time becomes continuous, the number of "trials" becomes infinite (one at each instant in time), and it is no longer meaningful to talk about the probability of success on an individual trial. Rather, the Poisson distribution becomes characterized by a *mean rate of occurrence* λ. The mean rate of occurrence can also be described as a mean *intensity* which encourages "occurrences" over time. Higher intensities result in a larger number of occurrences over any time interval.

In the continuous-time Markov chain, occurrences translate into "state changes," and each state has associated with it an intensity which expresses the rate at which changes into the state are likely to occur. State changes can be characterized either by transition probabilities, which vary with elapsed time and are difficult to compute, or by constant intensities. The transition probability approach will be discussed first.

Continuous-time Markov chains are denoted $X(t)$, $t \geq 0$, and the transition probability is now a function of elapsed time (t) since time zero. The continuous-time analog to the Chapman–Kolmogorov equations are

$$p_{ij}(t) = \sum_{k=1}^{m} p_{ik}(\nu) p_{kj}(t - \nu), \qquad t \geq 0 \tag{2.20}$$

for any $0 \leq \nu \leq t$, where

$$p_{ij}(t) = \mathrm{P}\left[X(t) = j \mid X(0) = i\right], \qquad i = 1, 2, \ldots, m, \quad j = 1, 2, \ldots, m$$

Only stationary Markov chains are considered here, which means that $p_{ij}(t)$ depends only on the elapsed time, not on the starting time, which was assumed to be zero above.

The property of stationarity has some implications that are worth investigating further. Suppose that a continuous-time Markov chain enters state i at some time, say time $t = 0$, and suppose that the process does not leave state i (that is, a transition does not occur) during the next 10 min. What, then, is the probability that the process will not leave state i during the following 5 min? Well, since the process is in state i at time $t = 10$, it follows, by the Markov and stationarity properties, that the probability that it remains in that state during the interval [10, 15] is just the same as the probability that it stays in state i for at least 5 min to start with. This is because the probabilities relating to states in the future from $t = 10$ are identical to those from $t = 0$, given that the current state is known at time t. That is, if T_i denotes the amount of time that the process stays in state i before making the transition into a different state, then

$$\mathrm{P}\left[T_i > 15 | T_i > 10\right] = \mathrm{P}\left[T_i > 5\right]$$

or, in general, and by the same reasoning,

$$\mathrm{P}\left[T_i > t + s | T_i > t\right] = \mathrm{P}\left[T_i > s\right]$$

for all $s \geq 0$, $t \geq 0$. Hence, the random variable T_i is *memoryless* and must thus (by results seen in Section 1.10.1) be *exponentially distributed.* This is entirely analogous to the Poisson process, as stated earlier in this section.

In other words, a continuous-time Markov chain is a random process that moves from state to state in accordance with a (discrete-time) Markov chain but is such that the amount of *time spent in each state*, before proceeding to the next state, is exponentially distributed. In addition, the times the process spends in state i and in the next state visited must be independent random variables.

In analogy with discrete-time Markov chains, the probability that a continuous-time Markov chain will be in state j at time t sometimes converges to a limiting value which is independent of the initial state (see the discrete-time Markov chain discussion for conditions under which this occurs). The resulting π_j's are once again called the steady-state probabilities and are defined by

$$\lim_{t \to \infty} p_{ij}(t) = \pi_j$$

where π_j exists. Each π_j is independent of the initial state probability vector $\boldsymbol{\pi}(0)$ and the steady-state probabilities satisfy

1. $0 < \pi_j < 1$ for all $j = 1, 2, \ldots, m$
2. $\sum_{j=1}^{m} \pi_j = 1$
3. $\pi_j = \sum_{i=1}^{m} \pi_i \cdot p_{ij}(t)$ for $j = 1, 2, \ldots, m$ and $t \geq 0$

As an alternative to transition probabilities, the transition intensities may be used. Intensities may be interpreted as the mean rate of transition from one state to another. In this sense, the intensity u_{ij} may be defined as the mean rate of transition from state i into state j for any i not equal to j. This has the following formal definition in terms of the transition probability:

$$u_{ij} = \frac{d}{dt} p_{ij}(t)|_{t=0} \tag{2.21}$$

where the derivative exists.

The definition for u_{jj} is special—it is the intensity of transition *out* of state j, with formal definition

$$u_{jj} = -\frac{d}{dt} p_{jj}(t)|_{t=0} \tag{2.22}$$

Armed with these two definitions, the steady-state equations can be rewritten as

$$\pi_j \cdot u_{jj} = \sum_{i \neq j} \pi_i \cdot u_{ij}, \qquad j = 1, 2, \ldots, m \tag{2.23}$$

The above is a *balance* equation, that is, the "tendency" to enter state j is equal to the tendency to exit state j, where tendency is probability times mean transition rate.

Commonly the intensities u_{ij} are easier to find than are the corresponding transition probabilities. An example where this is the case follows.

Example 2.11 A university has two triaxial test machines. Only one technician is available, so the test facility only ever conducts one test at a time. If at least one test machine is in proper repair, the test can proceed. If one of the machines is out of order, it is sent to the university machine shop for repair. The university machine shop has the capacity to repair both machines simultaneously, if necessary, although the actual repair time depends on the problem. If both triaxial test machines are out of order, the test facility becomes unavailable until one or the other test machines have been repaired. The "system" here consists of the two triaxial test machines and the repair shop. The system states $X(t)$ are defined as:

1. Both test machines operating
2. One test machine operating and one test machine in repair
3. Two test machines in repair (testing unavailable)

The time to failure of a triaxial test machine has been found to follow an exponential distribution,

$$f_T(t) = \begin{cases} \lambda e^{-\lambda t}, & t \geq 0 \\ 0, & t < 0 \end{cases}$$

as does the repair time at the machine shop,

$$r_T(t) = \begin{cases} \mu e^{-\mu t}, & t \geq 0 \\ 0, & t < 0 \end{cases}$$

Assuming that interfailure and interrepair times are independent, then $X(t)$ is a continuous-time, irreducible (i.e., no-absorbing-state) discrete-state Markov chain with transitions only from a state to its neighbor states: $1 \to 2$, $2 \to 1$, $2 \to 3$, and $3 \to 2$. Of course, there may be no state change as well. (This chain is similar to that in Example 2.2 in that state changes from first $\to$ last and from last $\to$ first are not possible.)

In this problem, transition intensities can be obtained directly from the mean rates of the exponential distributions. For example, the intensity u_{11} is just the transition rate out of state 1. This is 2λ since there are two machines "waiting" to fail. The intensity u_{22}, the transition rate out of state 2, is $\lambda + \mu$, since either an additional failure or a repair results in a move from state 2 (to either state 3 or state 1, respectively). Altogether, the transition intensities are

$$\begin{aligned} u_{11} &= 2\lambda, & u_{22} &= (\lambda + \mu) \\ u_{12} &= 2\lambda, & u_{23} &= \lambda \\ u_{13} &= 0, & u_{31} &= 0 \\ u_{21} &= \mu, & u_{32} &= 2\mu \\ & & u_{33} &= 2\mu \end{aligned}$$

The simplest way to view this system is to draw a transition diagram (Figure 2.7), where, for the continuous-time problem, the arrows are labeled with the mean transition rates rather than probabilities. Using this diagram, the balance equations can be derived as follows:

1. For state 1 (both test machines operating), the tendency to leave state 1 is $2\lambda \times \pi_1$ and the tendency to

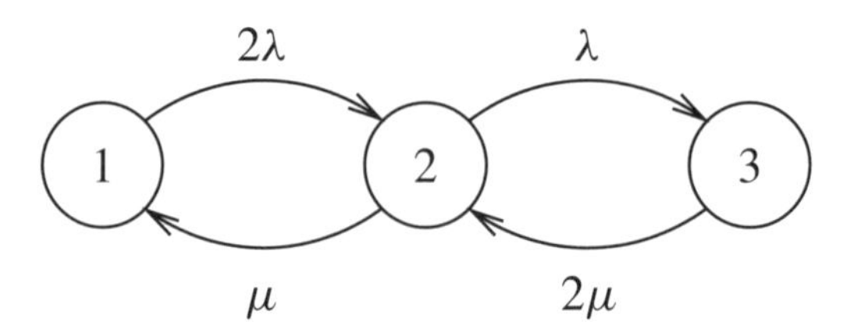

Figure 2.7 Continuous-time transition diagram for triaxial test example.

enter state 1 is $\mu \times \pi_2$, so the corresponding balance equation is

$$2\lambda\pi_1 = \mu\pi_2$$

2. For state 2 (one test machine operating and one in the repair shop), the tendency to leave state 2 is $(\lambda + \mu) \times \pi_2$ and the tendency to enter state 2 is $2\lambda \times \pi_1 + 2\mu \times \pi_3$, so the corresponding balance equation is

$$(\lambda + \mu)\pi_2 = 2\lambda\pi_1 + 2\mu\pi_3$$

3. For state 3 (both test machines in the repair shop), the tendency to leave state 3 is $2\mu \times \pi_3$ and the tendency to enter state 3 is $\lambda \times \pi_2$, which leads to the balance equation

$$2\mu\pi_3 = \lambda\pi_2$$

4. Finally, we know that the sum of the steady-state probabilities must be equal to 1, so we have an additional equation,

$$\pi_1 + \pi_2 + \pi_3 = 1$$

We cannot use all four of the above equations, since we have only three unknown steady-state probabilities. In fact, the fourth equation tells us that the first three equations are not linearly independent. Thus, using two of the first, second, or third equations along with the fourth equation leads to the solution

$$\pi_1 = \frac{\mu^2}{(\lambda + \mu)^2}, \quad \pi_2 = \frac{2\lambda\mu}{(\lambda + \mu)^2}, \quad \pi_3 = \frac{\lambda^2}{(\lambda + \mu)^2}$$

The probability of test availability (that triaxial tests can be performed) under steady-state conditions is thus 1 minus the probability that the two machines are both in the repair shop, namely,

$$Availability = 1 - \frac{\lambda^2}{(\lambda + \mu)^2}$$

2.3.1 Birth-and-Death Processes

As suggested by the name, birth-and-death processes have traditionally been used to model population dynamics. The basic idea is that the number of births and deaths in a population depends on the current population. Small populations encourage high birth rates, while large populations tend to have high death rates due to depletion of available resources, competition, and so on. Birth-and-death processes have found application in other areas. For example, economists have traditionally used birth-and-death processes to model the number of customers in a bank or a store at any instant in time. The process can also be successfully used to model the number of paying clients that any business has at any point in time.

Consider a system whose state at any time is the number of customers in a store or bank. Suppose that if there are j customers in the system, then:

1. New customer arrivals enter the system at a mean rate λ_j.
2. Customers depart the system at a mean rate μ_j.

That is, if there are j customers in the system, then:

1. The time until the next arrival is exponentially distributed with mean $1/\lambda_j$.
2. The time until the next arrival is independent of the time until the next departure.
3. The time until the next departure is exponentially distributed with mean $1/\mu_j$.

Such a system is called a *birth-and-death process*. The parameters λ_j for $j = 0, 1, \ldots$ and μ_j for $j = 1, 2, \ldots$ are called, respectively, the *arrival* (or *birth*) and *departure* (or *death*) rates. Of course, $\mu_0 = 0$ since the departure rate when the population is zero must also be zero. In this model, the arrival and departure rates are allowed to depend on j, the number of people currently in the system. As mentioned above, the dependence of birth and death rates on the population size is quite realistic (e.g., limited food resources mean higher death rates when the population becomes too large).

In essence, a birth-and-death process is a *continuous-time Markov chain* with states $\{0, 1, \ldots\}$ for which transitions from state j may go only to either state $j - 1$ or state $j + 1$. Thus, $p_{ij}(\Delta t) = 0$ for $j < i - 1$ or $j > i + 1$, where Δt is the (infinitesimally small) time step increment. The matrix of transition probabilities for time increment Δt may be expressed as a tridiagonal matrix with the following form:

$$P = \begin{bmatrix} 1-\lambda_0\Delta t & \lambda_0\Delta t & \cdots & 0 & 0 \\ \mu_1\Delta t & 1-(\lambda_1+\mu_1)\Delta t & \cdots & 0 & 0 \\ \cdot & \cdot & \cdot & \cdot & \cdot \\ \cdot & \cdot & \cdot & \cdot & \cdot \\ \cdot & \cdot & \cdot & \cdot & \cdot \\ 0 & 0 & \cdots & 1-(\lambda_{j-1}+\mu_{j-1})\Delta t & \lambda_{j-1}\Delta t \\ 0 & 0 & \cdots & \mu_j\Delta t & 1-(\lambda_j+\mu_j)\Delta t \end{bmatrix}$$

The transition diagram for the birth-and-death process appears as in Figure 2.8. The steady-state equations are obtained by applying Eq. 2.23. These give

$$\mu_1\pi_1 = \lambda_0\pi_0$$

$$\lambda_0\pi_0 + \mu_2\pi_2 = (\lambda_1 + \mu_1) \cdot \pi_1$$

$$\lambda_1\pi_1 + \mu_3\pi_3 = (\lambda_2 + \mu_2) \cdot \pi_2$$

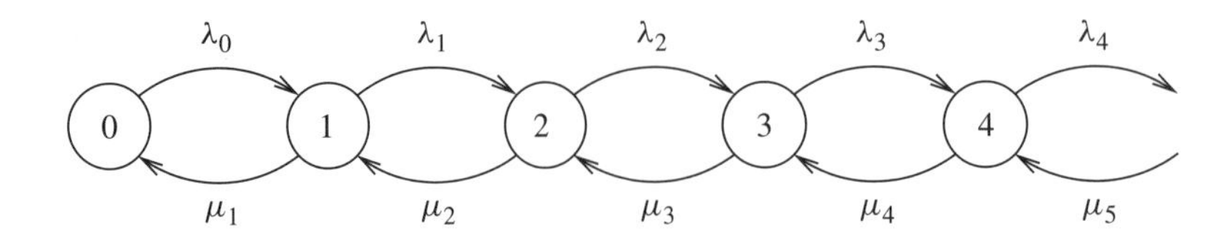

Figure 2.8 Transition diagram for birth-and-death processes.

$$\vdots$$

$$\lambda_{j-2} \cdot \pi_{j-2} + \mu_j \cdot \pi_j = (\lambda_{j-1} + \mu_{j-1}) \cdot \pi_{j-1}$$

$$\lambda_{j-1}\pi_{j-1} + \mu_{j+1}\pi_{j+1} = (\lambda_j + \mu_j) \cdot \pi_j$$

$$\vdots$$

and $\sum_{j=0}^{\infty} \pi_j = 1$. Solving these equations yield

$$\pi_1 = \frac{\lambda_0}{\mu_1} \cdot \pi_0$$

$$\pi_2 = \frac{\lambda_1}{\mu_2} \cdot \pi_1 = \frac{\lambda_1 \lambda_0}{\mu_2 \mu_1} \cdot \pi_0$$

$$\pi_3 = \frac{\lambda_2}{\mu_3} \cdot \pi_2 = \frac{\lambda_2 \lambda_1 \lambda_0}{\mu_3 \mu_2 \mu_1} \cdot \pi_0$$

and so on. In general,

$$\pi_{j+1} = \frac{\lambda_j}{\mu_{j+1}} \cdot \pi_j = \frac{\lambda_j \lambda_{j-1} \cdots \lambda_0}{\mu_{j+1} \mu_j \cdots \mu_1} \cdot \pi_0 \tag{2.24}$$

If the following is defined,

$$C_j = \frac{\lambda_{j-1} \lambda_{j-2} \cdots \lambda_0}{\mu_j \mu_{j-1} \cdots \mu_1} \tag{2.25}$$

then $\pi_j = C_j \cdot \pi_0$, $j = 1, 2, \ldots$, and since

$$\sum_{j=0}^{\infty} \pi_j = 1 \quad \text{or} \quad \pi_0 + \sum_{j=1}^{\infty} \pi_j = 1$$

$$\text{or} \quad \pi_0 + \pi_0 \sum_{j=1}^{\infty} C_j = 1$$

the final result becomes

$$\pi_0 = \frac{1}{1 + \sum_{j=1}^{\infty} C_j} \tag{2.26}$$

from which all the other steady-state probabilities can be obtained using the equations shown above.

Note that the steady-state equations (and the solutions that are derived from them) assume that the λ_j and μ_j values are such that a steady state can be reached. This will be true if

1. $\lambda_j = 0$ for $j > k$, so that there is a finite number of states, or
2. the mean arrival rate λ_j is less than the mean service rate μ_j for all j.

Example 2.12 A very simple example of a birth-and-death process (without a steady state) is the Poisson process. The Poisson process has the following parameters:

$$\mu_n = 0 \quad \text{for all } n \geq 0$$

$$\lambda_n = \lambda \quad \text{for all } n \geq 0$$

This is a process in which departures never occur, and the time between successive arrivals is exponential with mean $1/\lambda$. Hence, this is just a Poisson process which counts the total number of arrivals.

Example 2.13 Suppose that a geotechnical engineer receives jobs at a mean rate of one every three days and takes two days to complete each job, on average. What fraction of the time does the engineer have two jobs waiting (i.e., three jobs "in the system")?

SOLUTION This is a special kind of birth-and-death process, where the birth (job arrival) rates and the death (job completion) rates are constant. Specifically

$$\lambda_0 = \lambda_1 = \cdots = \lambda, \qquad \mu_1 = \mu_2 = \cdots = \mu$$

where $\lambda = \frac{1}{3}$ arrival per day and $\mu = \frac{1}{2}$ job completed per day. In this case, because all birth and death rates are constant, we have

$$C_j = \left(\frac{\lambda}{\mu}\right)^j$$

and

$$1 + \sum_{j=1}^{\infty} C_j = \sum_{j=0}^{\infty} \left(\frac{\lambda}{\mu}\right)^j = \begin{cases} \infty & \text{if } \lambda \geq \mu \\ \dfrac{1}{1 - \frac{\lambda}{\mu}} & \text{if } \lambda < \mu \end{cases}$$

since $(\lambda/\mu)^0 = 1$. Clearly, if the mean arrival rate of jobs exceeds the mean rate at which jobs can be completed, then the number of jobs waiting in the system will almost certainly (i.e., sooner or later) grow to infinity. This result gives, when $\lambda < \mu$,

$$\pi_0 = \frac{1}{1 + \sum_{j=1}^{\infty} c_j} = 1 - \frac{\lambda}{\mu}$$

$$\pi_1 = \left(\frac{\lambda}{\mu}\right)\left(1 - \frac{\lambda}{\mu}\right)$$

$$\pi_2 = \left(\frac{\lambda}{\mu}\right)^2\left(1 - \frac{\lambda}{\mu}\right)$$

$$\vdots$$

$$\pi_j = \left(\frac{\lambda}{\mu}\right)^j\left(1 - \frac{\lambda}{\mu}\right)$$

$$\vdots$$

From this, we see that the probability that three jobs are in the system (two waiting to be started) at any one time is just

$$\pi_3 = \left(\frac{\lambda}{\mu}\right)^3\left(1 - \frac{\lambda}{\mu}\right) = \left(\frac{1/3}{1/2}\right)^3\left(1 - \frac{1/3}{1/2}\right) = \left(\frac{2}{3}\right)^3\left(\frac{1}{3}\right) = 0.0988$$

so that the engineer spends just under 10% of the time with two more jobs waiting.

Example 2.14 Now suppose that the geotechnical engineer of the last example has developed a policy of refusing jobs once she has three jobs waiting (i.e., once she has four jobs in the system—three waiting plus the one she is working on). Again the job arrival rate is one every three days and jobs are completed in two days on average. What now is the fraction of time that the engineer has two jobs waiting? Also, what fraction of incoming jobs is the engineer having to refuse (this is a measure of lost economic potential)?

SOLUTION In this case, the population (number of jobs) is limited in size to 4. The states, 0–4, denote the number of jobs she needs to accomplish. The transition diagram for this problem appears as in Figure 2.9. For a limited population size, the solution is only slightly more complicated. Our arrival and departure rates are now

$$\lambda_0 = \lambda_1 = \cdots = \lambda_{m-1} = \lambda, \qquad \lambda_m = \lambda_{m+1} = \cdots = 0$$

$$\mu_1 = \mu_2 = \cdots = \mu$$

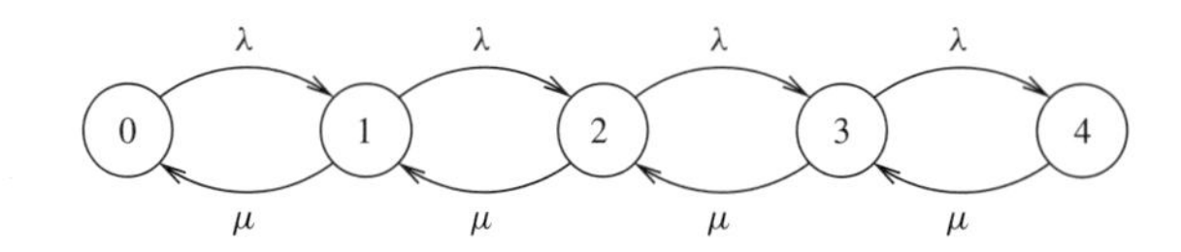

Figure 2.9 Transition diagram for limited-queue problem.

where $m = 4$. This gives us

$$C_1 = \left(\frac{\lambda}{\mu}\right), \quad C_2 = \left(\frac{\lambda}{\mu}\right)^2, \quad \ldots, \quad C_m = \left(\frac{\lambda}{\mu}\right)^m$$

$$C_{m+1} = C_{m+2} = \cdots = 0$$

so that

$$1 + \sum_{j=1}^{\infty} C_j = \sum_{j=0}^{m} \left(\frac{\lambda}{\mu}\right)^j$$

and

$$\pi_0 = \frac{1}{\sum_{j=0}^{m} \left(\frac{\lambda}{\mu}\right)^j}$$

$$\pi_1 = \left(\frac{\lambda}{\mu}\right)\pi_0$$

$$\pi_2 = \left(\frac{\lambda}{\mu}\right)^2\pi_0$$

$$\vdots$$

Using these results with $m = 4$, $\lambda = \frac{1}{3}$, $\mu = \frac{1}{2}$, and $\lambda/\mu = \frac{2}{3}$ gives

$$\pi_0 = \frac{1}{1 + (2/3) + (2/3)^2 + (2/3)^3 + (2/3)^4} = 0.3839$$

$$\pi_1 = \left(\tfrac{2}{3}\right)(0.3839) = 0.2559$$

$$\pi_2 = \left(\tfrac{2}{3}\right)^2(0.3839) = 0.1706$$

$$\pi_3 = \left(\tfrac{2}{3}\right)^3(0.3839) = 0.1137$$

$$\pi_4 = \left(\tfrac{2}{3}\right)^4(0.3839) = 0.0758$$

So the probability of having three jobs in the system increases when the engineer has a limit to the number of jobs waiting. This is perhaps as expected since the engineer no longer spends any of her time in states $5, 6, \ldots$, those times are now divided among the states 0 to 4. We also see that 7.58% of her time is spent in state 4. During this fraction of time, incoming jobs are rejected. Thus, over the course of, say, a year, the engineer loses $0.0758 \times (\frac{1}{3}) \times 365 = 9.2$ jobs on average. It does not appear that it would be worthwhile hiring another engineer to handle the lost jobs in this case (although this does depend on the value of lost jobs).

2.4 QUEUEING MODELS

In this section, Markov chains are extended to include models in which customers arrive in some random manner at a service facility. In these models, arriving customers are made to wait in a queue until it is their turn to be served by one of s servers. When a server becomes free, the next customer in the queue moves out of the queue to the server and the server then takes a random amount of time (often exponentially distributed) to serve the customer. Once served the customer is generally assumed to leave the system.

For queueing problems such as these, interest is usually focused on one or more of the following quantities:

1. $L = \sum_{j=0}^{\infty} j \cdot \pi_j$ = expected number of customers in system (including both queue and those being served)
2. $L_q = \sum_{j=s}^{\infty} (j - s) \cdot \pi_j$ = expected queue length (not including those customers currently being served)
3. W = expected waiting time in system
4. W_q = expected waiting time in queue (excluding service time)

Several relationships between the above quantities exist. For instance, if λ is the mean arrival rate of customers and λ is constant (independent of the number of customers in the system), then the expected number of customers in the system is just the mean arrival rate times the expected waiting time in the system:

$$L = \lambda_a W \tag{2.27}$$

that is, by the time the first customer in the system is leaving the system, at time W on average, the number of customers in the system has grown to $\lambda_a W$, *on average*. Note that when there is a limit N on the number of customers in the system, the arrival rate is the effective arrival rate $\lambda_a = \lambda(1 - \pi_N)$, otherwise $\lambda_a = \lambda$.

Similarly, the expected number of customers in the queue itself is just the mean arrival rate times the expected waiting time in the queue (again, using the effective arrival rate if the queue size is limited):

$$L_q = \lambda_a W_q \tag{2.28}$$

As with the birth-and-death model, queueing models may be characterized by arrival rates λ_j and departure rates μ_j, which are dependent on how many customers there are in a queue (e.g., customers entering a bank with long queues often decide to do their banking later). The major difference from the birth-and-death model is that queueing models allow for more than one server.

Queueing models differ from one another by the number of servers and by the manner in which λ_j and μ_j vary as a function j. Here are two different common queueing models:

[M/M/1] Suppose that customers arrive at a single-server service station according to a Poisson process with mean arrival rate λ. That is, the times between successive arrivals are independent exponentially distributed random variables having mean $1/\lambda$. Upon arrival, each customer goes directly into service if the server is free, and if not, then the customer joins the queue (i.e., waits in line) and there is no limit to the size of the queue. When the server finishes serving a customer, the customer leaves the system and the next customer in line, if any are waiting, enters the service. The successive service times are assumed to be independent exponentially distributed random variables having mean $1/\mu$.

This is called a M/M/1 queueing system because:

(a) The first M refers to the fact that the interarrival process is Markovian (and thus times between successive arrivals are independent and exponentially distributed).

(b) The second M refers to the fact that the service process is Markovian (and thus service times are independent and exponentially distributed).

(c) The 1 refers to the fact that there is a single server.

[M/M/s] Suppose that customers arrive at a multiple-server service station, having s servers, according to a Poisson process with mean arrival rate λ. That is, the times between successive arrivals are independent exponentially distributed random variables having mean $1/\lambda$. Upon arrival, each customer goes directly into service if one or more of the s servers is free, and if not, then the customer joins the single queue (i.e., waits in a single line with everybody else not being served). When one of the servers finishes serving a customer, the customer leaves the system and the next customer in line, if any are waiting, enters the service of the free server. For each server, the successive service times are assumed to be independent exponentially distributed random variables having mean $1/\mu$. Also servers operate independently.

Table 2.1 presents mathematical results for four different queueing models. Of note is that a closed-form

Table 2.1 Quantities of Interest for Four Queueing Models

	Model 1	Model 2	Model 3	Model 4
Birth rates	$\lambda_0 = \lambda_1 = \cdots = \lambda$	$\lambda_0 = \cdots = \lambda_{N-1} = \lambda$ $\lambda_N = \lambda_{N+1} = \cdots = 0$	$\lambda_0 = \lambda_1 = \cdots = \lambda$	$\lambda_0 = \lambda_1 = \cdots = \lambda$
Death rates	$\mu_1 = \mu_2 = \cdots = \mu$	$\mu_0 = \cdots = \mu_N = \mu$	$\mu_j = \begin{cases} j \cdot \mu & \text{for } j \le s \\ s \cdot \mu & \text{for } j > s \end{cases}$	Arbitrary with mean $1/\mu$ and variance σ^2
Steady-state probabilities	$\pi_j = (1-\rho)\rho^j$ $\rho = \lambda/\mu,\ j = 0, 1, \ldots$	$\pi_j = \rho^j \left[\frac{1-\rho}{1-\rho^{N+1}}\right]$ $j = 0, 1, \ldots, N$ $\rho = \frac{\lambda}{\mu}$	$\pi_j = \begin{cases} \frac{\pi_0 \rho^j}{j!} & \text{for } j \le s \\ \frac{\pi_0 \rho^j}{s! s^{j-s}} & \text{for } j > s \end{cases}$ $\pi_0 = \left[\frac{\rho^s}{s!}\left(\frac{1}{1-\phi}\right) + \sum_{j=0}^{s-1} \frac{\rho^j}{j!}\right]^{-1}$ $\rho = \frac{\lambda}{\mu} \quad \phi = \frac{\rho}{s}$	$\pi_0 = 1 - \rho$ $\rho = \frac{\lambda}{\mu}$
L	$\frac{\rho}{1-\rho}$	$\rho\left[\frac{1+\rho^N(N\rho-N-1)}{(1-\rho)(1-\rho^{N+1})}\right]$	$L_q + \rho$	$\rho + L_q$
L_q	$\frac{\lambda^2}{\mu(\mu-\lambda)}$	$L - (1 - \pi_0)$	$\frac{\phi \pi_0 \rho^s}{s!(1-\phi)^2}$	$\frac{\lambda^2\sigma^2+\rho^2}{2(1-\rho)}$
W	$\frac{1}{\mu-\lambda}$	$\frac{L}{\lambda(1-\pi_N)}$	$W_q + \frac{1}{\mu}$	$W_q + \frac{1}{\mu}$
W_q	$\frac{\lambda}{\mu(\mu-\lambda)}$	$\frac{L_q}{\lambda(1-\pi_N)}$	$\frac{L_q}{\lambda}$	$\frac{L_q}{\lambda}$

Note:
Model 1 has a single server with constant birth-and-death rates and unlimited queue size (this is an M/M/1 model). If $\lambda > \mu$, then the queue grows to infinite size on average.
Model 2 has a single server with no more than N in the system.
Model 3 has s servers with unlimited queue size (this is an M/M/s model).
Model 4 has a single server, but service time has an arbitrary distribution with mean $1/\mu$ and variance σ^2 (arrival times still exponentially distributed with mean λ)

expression for the quantities of interest (e.g., steady-state probabilities and L, L_q, W, and W_q) could be obtained because these are really quite simple models. If one deviates from these (and this is often necessary in practice), closed-form solutions may be very difficult to find. So how does one get a solution in these cases? One must *simulate* the queueing process. Thus, simulation methods are essential for a practical treatment of queueing models. They are studied in the next chapter.

Example 2.15 Two laboratory technicians independently process incoming soil samples. The samples arrive at a mean rate of 40 per hour, during working hours, and each technician takes approximately 2 min, on average, to perform the soil test for which they are responsible. Assume that both the arrival and testing sequences are Poisson in nature.

(a) For a soil sample arriving during working hours, what is the chance that it will be immediately tested?
(b) What is the expected number of soil samples not yet completed testing ahead of an arriving soil sample?
(c) Suppose that one technician is off sick and the other is consequently having to work harder, processing arriving soil samples at a rate of 50 per hour. In this case, what is the expected time that an arriving sample will take from the time of its arrival until it has been processed?

SOLUTION Assume that the mean arrival rate is not affected by the number of soil samples in the system. Assume also that the interarrival and interservice times are independent and exponentially distributed so that this is a birth-and-death queueing process. If there are two or less soil samples in the queueing system, the mean testing rate will be proportional to the number of soil samples in the system, whereas if there are more than two soil samples in the system, both of the technicians will be busy and the processing rate is limited to 60 per hour. Thus,

$$\lambda_0 = \lambda_1 = \cdots = \lambda_\infty = \lambda = 40$$

$$\mu_0 = 0, \quad \mu_1 = \mu = 30,$$

$$\mu_2 = \mu_3 = \cdots = \mu_\infty = 2\mu = 60$$

The transition diagram for this problem appears as in Figure 2.10.

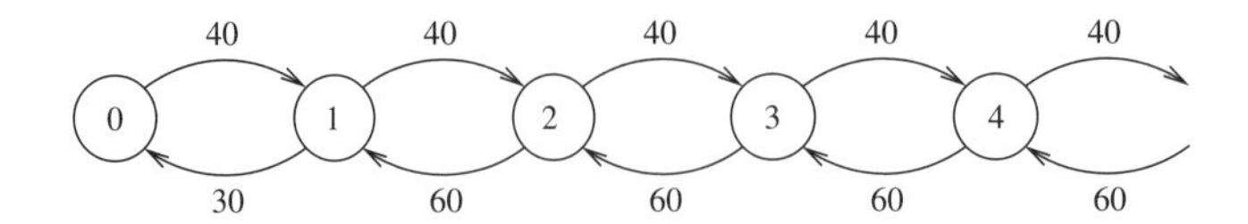

Figure 2.10 Transition diagram for laboratory technician problem.

Using these parameters we get

$$\begin{aligned}
\pi_0 &= \frac{1}{1+\sum_{j=1}^{\infty} C_j} \\
&= \left[1+\frac{\lambda}{\mu}+\frac{\lambda}{\mu}\left(\frac{\lambda}{2\mu}\right)+\frac{\lambda}{\mu}\left(\frac{\lambda}{2\mu}\right)^2+\cdots\right]^{-1} \\
&= \left[1+\left(\frac{\lambda}{\mu}\right)\left\{1+\left(\frac{\lambda}{2\mu}\right)+\left(\frac{\lambda}{2\mu}\right)^2+\cdots\right\}\right]^{-1} \\
&= \left[1+\left(\frac{\lambda}{\mu}\right)\frac{1}{1-(\lambda/2\mu)}\right]^{-1} \\
&= \frac{2\mu-\lambda}{2\mu+\lambda} \\
&= 0.2
\end{aligned}$$

and

$$\pi_1 = \frac{\lambda_0}{\mu_1}\pi_0 = \frac{40}{30}0.2 = 0.267$$

(a) The event that an arriving soil sample is immediately tested is equivalent to the event that there is zero or one soil sample in the system (since if there are already two soil samples in the system, the arriving sample will have to wait). Hence, the pertinent probability is

$$\pi_0 + \pi_1 = 0.2 + 0.267 = 0.467$$

(b) The expected number of unfinished soil samples ahead of an arriving soil sample is given by

$$\begin{aligned}
L &= 1\cdot\frac{\lambda}{\mu}\pi_0 + 2\cdot\frac{\lambda}{\mu}\left(\frac{\lambda}{2\mu}\right)\pi_0 \\
&\quad + 3\cdot\frac{\lambda}{\mu}\left(\frac{\lambda}{2\mu}\right)^2\pi_0+\cdots \\
&= \left(\frac{\lambda}{\mu}\right)\frac{\pi_0}{\left(1-\dfrac{\lambda}{(2\mu)}\right)^2} \\
&= \left(\frac{40}{30}\right)\frac{0.2}{\left(1-\dfrac{40}{60}\right)^2} \\
&= 2.403
\end{aligned}$$

That is, on average, an arriving soil sample can expect two or three soil samples ahead of it in the system. Note that the wording here is somewhat delicate. It is assumed here that the arriving soil sample is not yet in the system, so that the expected number of samples in the system are what the arriving sample can expect to "see."

(c) This problem corresponds to model 1 of Table 2.1, so that the expected waiting time W, including time in the queue, is given by

$$W = \frac{1}{\mu-\lambda} = \frac{1}{50-40} = 0.1 \quad \text{h (i.e., 6 min)}$$

Example 2.16 Consider a geotechnical firm which employs four engineers. Jobs arrive once per day, on average. Suppose that each engineer takes an average of two days to complete a job. If all four engineers are busy, newly arriving jobs are turned down.

(a) What fraction of time are all four engineers busy?
(b) What is the expected number of jobs being worked on on any given day?
(c) By how much does the result of part (a) change if arriving jobs are allowed/willing to wait in a queue (i.e., are not turned down if all four engineers are busy)?

SOLUTION This is a four-server model with limited queue size (more specifically, the queue size cannot be greater than zero), so it does not correspond to any of our simplified models shown in Table 2.1. We must use the basic equations with rates

$$\lambda_0=\lambda_1=\lambda_2=\lambda_3=1, \qquad \lambda_4=\lambda_5=\cdots=0$$
$$\mu_0=0,\quad \mu_1=\tfrac{1}{2},\quad \mu_2=\tfrac{2}{2},\quad \mu_3=\tfrac{3}{2},\quad \mu_4=\tfrac{4}{2}$$

which has the transition diagram shown in Figure 2.11:

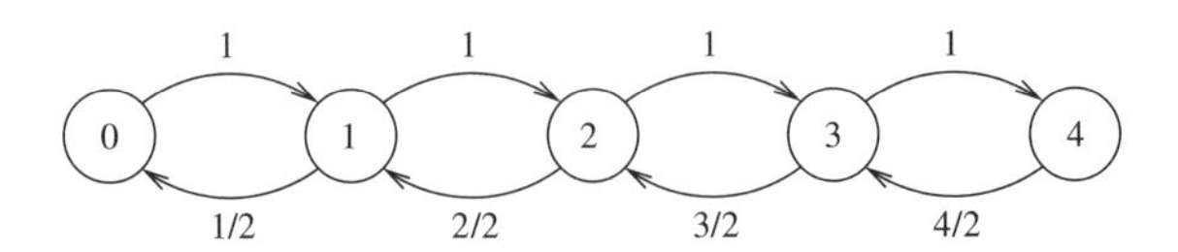

Figure 2.11 Transition diagram for four-engineer problem.

(a) This gives us

$$C_1 = \frac{\lambda_0}{\mu_1} = 2$$
$$C_2 = \frac{\lambda_1}{\mu_2} C_1 = 2$$
$$C_3 = \frac{\lambda_2}{\mu_3} C_2 = \frac{4}{3}$$
$$C_4 = \frac{\lambda_3}{\mu_4} C_3 = \frac{2}{3}$$
$$C_5 = C_6 = \cdots = 0$$

which yields probabilities

$$\pi_0 = \frac{1}{1+2+2+4/3+2/3} = 0.1428$$
$$\pi_1 = C_1\pi_0 = 0.2857$$
$$\pi_2 = C_2\pi_0 = 0.2857$$
$$\pi_3 = C_3\pi_0 = 0.1905$$
$$\pi_4 = C_4\pi_0 = 0.0952$$

so that the four engineers are fully occupied $\pi_4 = 0.095$, or 9.5%, of the time.

(b) If N is the number of jobs being worked on on any day, then the expected number of jobs on any one day is

$$\mathrm{E}[N] = 0\pi_0 + 1\pi_1 + 2\pi_2 + 3\pi_3 + 4\pi_4 = 1.81$$

(c) Now we have queueing model 3, since the queue size is unlimited, with $\rho = 2$, $s = 4$, and $\phi = \frac{1}{2}$. Then

$$\pi_0 = \frac{1}{1+2+2+4/3+(2^4/4!)[1/(1-1/2)]}$$
$$= 0.1304$$

The probability that the firm is fully occupied corresponds to the probability that the number of jobs in the system is $4, 5, \ldots$. The probability of this is $1 - (\pi_0 + \pi_1 + \pi_2 + \pi_3)$, where

$$\pi_1 = \frac{2^1}{1!}\pi_0 = 2\pi_0, \qquad \pi_2 = \frac{2^2}{2!}\pi_0 = 2\pi_0,$$
$$\pi_3 = \frac{2^3}{3!}\pi_0 = \frac{4}{3}\pi_0$$

so that the desired probability is

$$1 - \left(1+2+2+\tfrac{4}{3}\right)(0.1304) = 0.174$$

which is greater than the limited queue result of part (a).

CHAPTER 3

Random Fields

3.1 INTRODUCTION

In the previous chapter, we considered only discrete-state Markov chains (with both discrete and continuous time). We turn our attention in this chapter to continuous-state processes where the random process $X(t)$ can now take on an infinite number of possible values at each point t. As an example of a continuous-state random process, Figure 3.1 illustrates the tip resistance measured during a CPT. Aside from soil disturbance, measurement errors, and problems with extracting engineering properties from CPT data, Figure 3.1 presumably gives a reasonably good idea about the soil properties at the location at which the CPT was taken. However, what can be said about the soil properties 10 (or 50) m away from the CPT sounding? The data presented in Figure 3.1 could be used to characterize the randomness (uncertainty) at locations which have not been sampled. But how can the variability at one location be used to represent the variability at other locations? Some considerations involved in characterizing spatial variability are as follows:

1. *Variability at a Point:* Pick a specific position t^*. At this point the process has a random value $X(t^*) = X^*$ which is governed by a probability density function $f_{X^*}(x)$. If we picked another position, say t', then $X(t') = X'$ would have another, possibly different pdf, $f_{X'}(x)$. That is, the pdf's could evolve with position. In practice, evolving pdf's become quite difficult to estimate for anything beyond a simple trend in the mean or variance. An example where the point, or marginal, distribution evolves with time is earthquake ground motion where the motion variance increases drastically during the strong motion portion of the record.
2. *Spatial Dependence:* Consider again two positions t^* and t' separated by distance $\tau = t' - t^*$. Presumably, the two random variables $X(t')$ and $X(t^*)$ will exhibit some dependence on each other. For example, if X is cohesion, then we would expect $X(t')$ and $X(t^*)$ to be quite similar (i.e., highly dependent) when τ is small (e.g., a few centimeters) and possibly quite dissimilar (i.e., largely independent) when τ is large (e.g., tens, hundreds, or thousands of meters). If $X(t^*)$ and $X(t')$ are independent for any two positions with separation $\tau = t' - t^* \neq 0$, then the process would be infinitely rough—points separated by vanishingly small lags could have quite different values. This is not physically realistic for most natural phenomena. Thus, $X(t^*)$ and $X(t')$ generally have some sort of dependence that often decreases with separation distance. This interdependence results in a smoothing of the random process. That is, for small τ, nearby states of X are preferential—the random field is constrained by its neighbors to be similar. We characterize this interdependence using the joint bivariate distribution $f_{X^*X'}(x^*, x')$ which specifies the probability that $X^* = x^*$ and $X' = x'$ at the same time. If we extend this idea to the consideration of any three, or four, or five, ..., points, then the complete probabilistic description of a random process is the infinite-dimensional probability density function

$$f_{X_1 X_2 \dots}(x_1, x_2, \dots)$$

Such an infinite-dimensional pdf is difficult to use in practice, not only mathematically, but also because its parameters are difficult to estimate from real data.

To simplify the characterization problem, we introduce a number of assumptions which are commonly made:

1. *Gaussian Process:* The joint pdf is a multivariate *normally distributed* random process. Such a process is also commonly referred to as a *Gaussian process*. The great advantage to the multivariate normal distribution is that the complete distribution can be specified by just the mean vector and the covariance matrix. As we saw in Section 1.10.8, the multivariate normal pdf has the form

$$f_{X_1 X_2 \cdots X_k}(x_1, x_2, \dots, x_k) = \frac{1}{(2\pi)^{k/2}} \frac{1}{|\boldsymbol{C}|^{1/2}} \times \exp\left\{-\tfrac{1}{2}(\mathbf{x} - \boldsymbol{\mu})^{\mathrm{T}} \boldsymbol{C}^{-1} (\mathbf{x} - \boldsymbol{\mu})\right\}$$

where $\boldsymbol{\mu}$ is the vector of mean values, one for each X_i, $\boldsymbol{C}$ is the covariance matrix between the X's, and

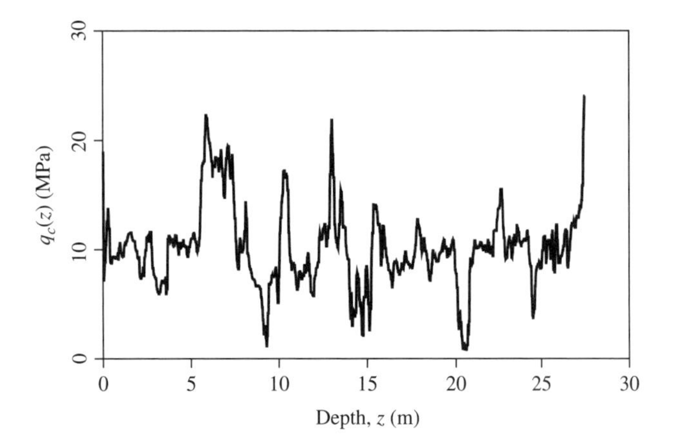

Figure 3.1 Tip resistance $q_c(z)$ measured over depth z by a cone penetrometer.

$|\boldsymbol{C}|$ is its determinant. Specifically,

$$\boldsymbol{\mu} = \mathrm{E}[\mathbf{X}]$$

$$\boldsymbol{C} = \mathrm{E}\left[(\mathbf{X} - \boldsymbol{\mu})(\mathbf{X} - \boldsymbol{\mu})^{\mathrm{T}}\right]$$

where the superscript T means the transpose. The covariance matrix $\boldsymbol{C}$ is a $k \times k$ symmetric, positive-definite matrix. For a continuous random field, the dimensions of $\boldsymbol{\mu}$ and $\boldsymbol{C}$ are still infinite, since the random field is composed of an infinite number of X's, one for each point. To simplify things, we often quantify $\boldsymbol{\mu}$ and $\boldsymbol{C}$ using continuous functions of space based on just a few parameters. For example, in a one-dimensional random field (or random process), the mean may vary linearly:

$$\mu(t) = a + bt$$

and the covariance matrix can be expressed in terms of the standard deviations, which may vary with t, and the correlation function ρ as in

$$C(t_1, t_2) = \sigma(t_1)\sigma(t_2)\rho(t_1, t_2)$$

which specifies the covariance between $X(t_1)$ and $X(t_2)$. Because the mean and covariance can vary with position, the resulting joint pdf is still difficult to use in practice, both mathematically and to estimate from real data, which motivates the following further simplifications.

2. *Stationarity or Statistical Homogeneity:* The joint pdf is, independent of spatial position, that is, it depends just on *relative* positions of the points. This assumption implies that the mean, covariance, and higher order moments are constant in time (or space) and thus that the marginal, or point, pdf is also constant in time (or space). So-called *weak stationarity* or *second-order stationarity* just implies that the mean and variance are constant in space.
3. *Isotropy:* In two- and higher dimensional random fields, isotropy implies that the joint pdf is invariant under rotation. This condition implies stationarity (although stationarity does not imply isotropy). Isotropy means that the correlation between two points only depends on the distance between the two points, not on their orientation relative to one another.

A random field $X(t)$ having nonstationary mean and variance can be converted to a random field which is stationary in its mean and variance by the following transformation:

$$X'(t) = \frac{X(t) - \mu(t)}{\sigma(t)} \tag{3.1}$$

The random field $X'(t)$ will now have zero mean and unit variance everywhere. Also a nonstationary random field can be produced from a stationary random field. For example, if $X(t)$ is a standard Gaussian random field (having zero mean and unit variance) and

$$Y(t) = 2 + \tfrac{1}{2}t + \tfrac{1}{4}\sqrt{t}X(t)$$

then $Y(t)$ is a nonstationary Gaussian random field with

$$\mathrm{E}[Y(t)] = \mu_Y(t) = 2 + \tfrac{1}{2}t$$

$$\mathrm{Var}[Y(t)] = \sigma_Y^2(t) = \tfrac{1}{2}t$$

in which both the mean and variance increase with t.

Note that a nonstationary *correlation* structure, where the correlation coefficient between $X(t)$ and $X(t+\tau)$ depends on t, is *not* rendered stationary by Eq. 3.1. Equation 3.1 only renders the mean and variance stationary, not correlation. At the moment, nonstationary correlation structures are uncommon in geotechnical engineering because of the prohibitive volumes of data required to estimate their parameters. Random-field models in geotechnical engineering are generally at most nonstationary in the mean. The variance and covariance structure will almost always be assumed to be stationary. We shall see more about why this is so in Chapter 5 when we talk about *ergodicity*. The practical implications are that Eq. 3.1 can almost always be used to transform a geotechnical random-field model into one which is stationary.

Quite often soil properties are not well modeled by the Gaussian (normal) distribution. For example, a normally distributed elastic modulus is admitting that some fraction of the soil has a negative elastic modulus, which is not physically meaningful. For such nonnegative soil properties the normal distribution is not appropriate and a non-Gaussian random field would be desired, such as the lognormal distribution. Nevertheless, Gaussian random fields are desirable because of their simple characterization and simple probabilistic nature. Fortunately, we can retain a lot of these desirable features, at least at some level, by using non-Gaussian random fields which are derived as simple transformations of a Gaussian random field. For example, the random field $Y(t)$ defined by the transformation

$$Y(t) = e^{X(t)} \tag{3.2}$$

will have a lognormal distribution if $X(t)$ is normally distributed. A note of caution here, however, is that the covariance structure of the resulting field is also nonlinearly transformed. For example, if $X(1)$ has correlation coefficient 0.2 with $X(2)$, the same is no longer true of $Y(1)$ and $Y(2)$. In fact, the correlation function of Y is now given by (Vanmarcke, 1984)

$$\rho_Y(\tau) = \frac{\exp\{\sigma_X^2 \rho_X(\tau)\} - 1}{\exp\{\sigma_X^2\} - 1} \tag{3.3}$$

for stationary processes, where $\rho_X(\tau)$ is the correlation coefficient between $X(t)$ and $X(t+\tau)$.

In this book, we will largely restrict ourselves to stationary Gaussian random fields and to fields derived through simple transformations from Gaussian random fields (e.g., lognormally distributed random fields). Gaussian random fields are completely specified by their mean and covariance structure, that is, their first two moments. In practice, we are sometimes able to reasonably accurately estimate the mean, and sometimes a mean trend, of a soil property at a site. Estimating the variance and covariance requires considerably more data—we often need to resort to information provided by the literature in order to specify the variance and covariance structure. Because of this uncertainty in the basic parameters of even the covariance, there is often little point in adopting other joint distributions, which are more complicated and depend on higher moments, to govern the random fields representing soil properties, unless these distributions are suggested by mechanical or physical theory.

Under the simplifying assumptions that the random field is Gaussian and stationary, we need to know three things in order to characterize the field:

1. The field mean μ_X
2. The field variance σ_X^2
3. How rapidly the field varies in space

The last is characterized by the second moment of the field's joint distribution, which is captured equivalently by the covariance function, the spectral density function, or the variance function. These functions are discussed in the next few sections.

3.2 COVARIANCE FUNCTION

The second-moment nature of a Gaussian random field can be expressed by the *covariance function*,

$$\begin{aligned} C(t', t^*) &= \text{Cov}\left[X(t'), X(t^*)\right] \\ &= \text{E}\left[\left(X(t') - \mu_X(t')\right)\left(X(t^*) - \mu_X(t^*)\right)\right] \\ &= \text{E}\left[X(t')X(t^*)\right] - \mu_X(t')\mu_X(t^*) \end{aligned} \tag{3.4}$$

where $\mu_X(t)$ is the mean of X at the position t. Since the magnitude of the covariance depends on the size of the variance of $X(t')$ and $X(t^*)$, it tells us little about the degree of linear dependence between $X(t')$ and $X(t^*)$. A more meaningful measure, in this sense, is the *correlation function*,

$$\rho(t', t^*) = \frac{C(t', t^*)}{\sigma_X(t')\sigma_X(t^*)} \tag{3.5}$$

where $\sigma_X(t)$ is the standard deviation of X at the position t. As seen in Chapter 1, $-1 \le \rho(t', t^*) \le 1$, and when $\rho(t', t^*) = 0$, we say that $X(t')$ and $X(t^*)$ are uncorrelated. When X is Gaussian, being uncorrelated also implies independence. If $\rho(t', t^*) = \pm 1$, then $X(t')$ and $X(t^*)$ are perfectly linearly correlated, that is, $X(t')$ can be expressed in terms of $X(t^*)$ as

$$X(t') = a \pm bX(t^*)$$

Furthermore, if $X(t')$ and $X(t^*)$ are perfectly correlated and the random field is stationary, then $X(t') = \pm X(t^*)$. The sign to use is the same as the sign of $\rho(t', t^*)$.

For stationary random fields, the mean and covariance are independent of position, so that

$$\begin{aligned} C(t', t^*) &= C(t' - t^*) = C(\tau) = \text{Cov}\left[X(t), X(t+\tau)\right] \\ &= \text{Cov}\left[X(0), X(\tau)\right] = \text{E}\left[X(0)X(\tau)\right] - \mu_X^2 \end{aligned} \tag{3.6}$$

and the correlation function becomes

$$\rho(\tau) = \frac{C(\tau)}{C(0)} = \frac{C(\tau)}{\sigma_X^2}$$

Because $C(t', t^*) = C(t^*, t')$, we must have $C(\tau) = C(-\tau)$ when the field is stationary, and similarly $\rho(\tau) = \rho(-\tau)$.

At this point, we can, in principle, describe a Gaussian random field and ask probabilistic questions of it.

Example 3.1 Suppose that the total amount Q of toxic waste which flows through a clay barrier of thickness D in an interval of time is proportional to the average hydraulic conductivity K_{ave} through the barrier (note that the harmonic average is probably a better model for this problem, but the arithmetic average is much easier to deal with and so will be used here in this simple illustration).

That is,

$$Q = cK_{ave}$$

where c is a constant. A one-dimensional idealization for K_{ave} is

$$K_{ave} = \frac{1}{D}\int_0^D K(x)\,dx$$

where $K(x)$ is the *point* hydraulic conductivity, meaning it expresses the hydraulic conductivity of the clay at the point x. Assume that $K(x)$ is a continuous-state stationary Gaussian random process with mean 1, coefficient of variation 0.20, and correlation function $\rho(\tau) = \exp\{-|\tau|/4\}$. One possible realization of $K(x)$ appears in Figure 3.2.

(a) Give expressions for the mean and variance of Q in terms of the mean and variance of $K(x)$.
(b) If the correlation function is actually $\rho(\tau) = \exp\{-|\tau|\}$, will the variance of Q increase or decrease? Explain your reasoning.

SOLUTION

(a) Since $Q = cK_{ave}$, we must have

$$Q = \frac{c}{D}\int_0^D K(x)\,dx$$

Taking expectations of both sides gives the mean of Q,

$$\mathrm{E}[Q] = \mathrm{E}\left[\frac{c}{D}\int_0^D K(x)\,dx\right] = \frac{c}{D}\int_0^D \mathrm{E}[K(x)]\,dx$$
$$= \frac{c}{D}\int_0^D (1)\,dx = c$$

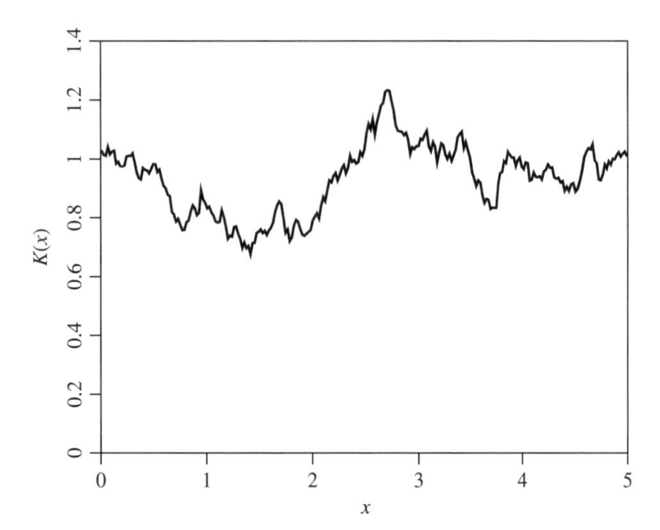

Figure 3.2 One possible realization of $K(x)$.

while the variance is obtained as (recalling that the square of a sum becomes a double sum)

$$\begin{aligned}\mathrm{Var}[Q] &= \mathrm{E}\left[(Q-\mu_Q)^2\right] \\ &= \mathrm{E}\left[(Q-c)^2\right] = \mathrm{E}\left[c^2(K_{ave}-1)^2\right] \\ &= c^2\,\mathrm{E}\left[\left(\frac{1}{D}\int_0^D (K(x)-1)\,dx\right)^2\right] \\ &= \frac{c^2}{D^2}\mathrm{E}\left[\int_0^D\int_0^D (K(\xi)-1)(K(\eta)-1)\,d\xi\,d\eta\right] \\ &= \frac{c^2}{D^2}\int_0^D\int_0^D \mathrm{E}\left[(K(\xi)-1)(K(\eta)-1)\right]\,d\xi d\eta\end{aligned}$$

Recognizing that $\mathrm{E}[(K(\xi)-1)(K(\eta)-1)] = \sigma_K^2\,\rho(\xi-\eta)$ is just the covariance between the hydraulic conductivities at the two points ξ and η, we get

$$\begin{aligned}\mathrm{Var}[Q] &= \frac{c^2\sigma_K^2}{D^2}\int_0^D\int_0^D \rho(\xi-\eta)\,d\xi\,d\eta \\ &= \frac{c^2\sigma_K^2}{D^2}\int_0^D\int_0^D \exp\left\{-\frac{|\xi-\eta|}{4}\right\}\,d\xi\,d\eta \\ &= \frac{2c^2\sigma_K^2}{D^2}\int_0^D (D-\tau)\,\exp\left\{-\frac{\tau}{4}\right\}\,d\tau\end{aligned}$$

which can be solved with the aid of a good integral table. Note that the collapse of a two-dimensional integral to a one-dimensional integral in the last step was accomplished by taking advantage of the fact that $\rho(\xi-\eta)$ is constant along diagonal lines through the integration space. That is, we need only integrate along a line perpendicular to these diagonally constant values and multiply by the length of the diagonals (there are some $\sqrt{2}$ factors that cancel out). We shall illustrate the details of this integration reduction in Section 3.4 (see, e.g., Figure 3.7). The end result is

$$\mathrm{Var}[Q] = c^2\sigma_K^2\left(\frac{32}{D^2}\right)\left[\frac{D}{4} + \exp\left\{-\frac{D}{4}\right\} - 1\right]$$

where $\sigma_K = 0.2\mu_K = 0.2$.

(b) Since the correlation function $\rho(\tau) = \exp\{-|\tau|\}$ falls more rapidly with τ than does the correlation function used in part (a), the conductivity values become more independent of one another through the clay barrier. Since Q is an average of the conductivity values, increasing independence between values serves to decrease the variance of Q. That is, the variability in $K(x)$ now tends to cancel out to a greater degree. We note that in order to understand this somewhat counterintuitive result that the variance of Q decreases as $K(x)$ becomes more independent (and thus more random), we need to remember that we are talking about variability over the ensemble of possible realizations. For strongly correlated random fields, there is less variability within each realization but more variability from realization to realization. Conversely, for weakly correlated random fields, there is more variability within each realization but less variability between realizations (e.g., all realizations look similar). In the latter case, averages of each realization are very similar (small variability).

This discussion illustrates the contrast between characterizing an entire population (which is what we are doing in this example) and characterizing a particular realization (if we happen to know things about the particular realization). We shall discuss this issue at greater length in Chapter 5.

Another property of the covariance function is that it is *positive definite*. To illustrate this property, consider a linear combination of n of the random variables in the process $X(t)$, say $X_i = X(t_i)$ for any sequence of times $t_1, t_2, \dots, t_n$,

$$Y = a_1 X_1 + a_2 X_2 + \cdots + a_n X_n = \sum_{i=1}^{n} a_i X_i$$

where $a_1, a_2, \dots, a_n$ are any set of coefficients. We saw in Chapter 1 that the variance of a linear combination is

$$\text{Var}[Y] = \sum_{i=1}^{n} \sum_{j=1}^{n} a_i a_j \, \text{Cov}\left[X_i, X_j\right]$$

Since $\text{Var}[Y]$ is also defined as $\text{E}\left[(Y - \mu_Y)^2\right]$, it cannot be negative. This means that the covariances between the X's must satisfy the following inequality for any a_i:

$$\sum_{i=1}^{n} \sum_{j=1}^{n} a_i a_j \, \text{Cov}\left[X_i, X_j\right] \geq 0 \tag{3.7}$$

which is the statement of positive definiteness. In the case of a stationary process where $\text{Cov}\left[X_i, X_j\right] = \sigma_X^2 \rho(t_i - t_j) = \sigma_X^2 \rho_{ij}$, we see that the correlation function is also positive definite,

$$\sum_{i=1}^{n} \sum_{j=1}^{n} a_i a_j \rho_{ij} \geq 0 \tag{3.8}$$

since $\sigma_X^2 \geq 0$.

One of the points of Eqs. 3.7 and 3.8 is that not just any covariance and correlation function can be used to characterize the second moment of a random field. In particular, the following properties of the covariance function must be satisfied:

1. $|\text{Cov}\left[X_i, X_j\right]| \leq \sigma_{X_i} \sigma_{X_j}$, which ensures that $-1 \leq \rho_{ij} \leq 1$
2. $\text{Cov}\left[X_i, X_j\right] = \text{Cov}\left[X_j, X_i\right]$
3. $\sum_{i=1}^{n} \sum_{j=1}^{n} a_i a_j \, \text{Cov}\left[X_i, X_j\right] \geq 0$

For isotropic covariance functions in two and higher dimensions, see also Section 3.7.6. If two covariance functions $C_1(X_i, X_j)$ and $C_2(X_i, X_j)$ each satisfy the above conditions, then their sum $C(X_i, X_j) = C_1(X_i, X_j) + C_2(X_i, X_j)$ will also satisfy the above conditions and be a valid covariance function.

If the set of covariances $\text{Cov}\left[X_i, X_j\right]$ is viewed as a matrix $\boldsymbol{C} = [C_{ij}]$, with elements $C_{ij} = \text{Cov}\left[X_i, X_j\right]$, then one of the results of positive definiteness is that the *square root* of $\boldsymbol{C}$ will be real. The square root will be defined here as the lower triangular matrix $\boldsymbol{L}$ such that $\boldsymbol{L}\boldsymbol{L}^{\mathrm{T}} = \boldsymbol{C}$, where the superscript T denotes the matrix transpose. The lower triangular matrix $\boldsymbol{L}$ has the form

$$\boldsymbol{L} = \begin{bmatrix} \ell_{11} & 0 & 0 & \cdot & \cdot & \cdot & 0 \\ \ell_{21} & \ell_{22} & 0 & \cdot & \cdot & \cdot & 0 \\ \ell_{31} & \ell_{32} & \ell_{33} & \cdot & \cdot & \cdot & 0 \\ \cdot & \cdot & \cdot & \cdot & & & \cdot \\ \cdot & \cdot & \cdot & & \cdot & & \cdot \\ \cdot & \cdot & \cdot & & & \cdot & \cdot \\ \ell_{n1} & \ell_{n2} & \ell_{n3} & \cdot & \cdot & \cdot & \ell_{nn} \end{bmatrix} \tag{3.9}$$

which is generally obtained by Cholesky decomposition. We shall see how this matrix can be used to simulate a random field in Section 6.4.2.

A positive-definite covariance matrix can also be decomposed into a matrix of eigenvectors $\boldsymbol{Q}$ and positive eigenvalues $\boldsymbol{\Psi}$ such that

$$\boldsymbol{C} = \boldsymbol{Q}^{\mathrm{T}} \boldsymbol{\Psi} \boldsymbol{Q} \tag{3.10}$$

where $\boldsymbol{\Psi}$ is a diagonal matrix whose elements are the eigenvalues $\psi_1, \psi_2, \dots, \psi_n$ of the covariance matrix $\boldsymbol{C}$.

The eigenvectors composing each column of the matrix $\boldsymbol{Q}$ make up an *orthonormal basis*, which is a set of unit vectors which are mutually perpendicular. A property of orthonormal vectors is that $\boldsymbol{Q}^{\mathrm{T}} = \boldsymbol{Q}^{-1}$. If we premultiply and postmultiply Eq. 3.10 by $\boldsymbol{Q}$ and $\boldsymbol{Q}^{\mathrm{T}}$, respectively, we get

$$\boldsymbol{Q}\boldsymbol{C}\boldsymbol{Q}^{\mathrm{T}} = \boldsymbol{\Psi} = \begin{bmatrix} \psi_1 & 0 & 0 & \cdot & \cdot & \cdot & 0 \\ 0 & \psi_2 & 0 & \cdot & \cdot & \cdot & 0 \\ 0 & 0 & \psi_3 & \cdot & \cdot & \cdot & 0 \\ \cdot & \cdot & \cdot & \cdot & & & \cdot \\ \cdot & \cdot & \cdot & & \cdot & & \cdot \\ \cdot & \cdot & \cdot & & & \cdot & \cdot \\ 0 & 0 & 0 & \cdot & \cdot & \cdot & \psi_n \end{bmatrix} \tag{3.11}$$

Now let us define the vector $\mathbf{X} = \{X_1, X_2, \dots, X_n\}^{\mathrm{T}}$ which contains the sequence of $X(t)$ values discussed above, having covariance matrix $\boldsymbol{C} = \text{E}\left[(\mathbf{X} - \mu_X)(\mathbf{X} - \mu_X)^{\mathrm{T}}\right]$. If we let

$$\mathbf{Z} = \boldsymbol{Q}\mathbf{X} \tag{3.12}$$

be a sequence of random variables obtained by rotating the vector $\mathbf{X}$ by the orthonormal basis $\boldsymbol{Q}$, then $\mathbf{Z}$ is composed of uncorrelated random variables having variances $\psi_1, \psi_2, \ldots, \psi_n$. We can show this by computing the covariance matrix of $\mathbf{Z}$. For this we will assume, without loss of generality and merely for simplicity, that $\mathrm{E}[X(t)] = 0$ so that $\mathrm{E}[\mathbf{Z}] = \mathbf{0}$. (The end result for a nonzero mean is exactly the same—it is just more complicated getting there.) The covariance matrix of $\mathbf{Z}$, in this case, is given by

$$\begin{aligned} \boldsymbol{C}_Z &= \mathrm{E}\left[\mathbf{Z}\mathbf{Z}^{\mathrm{T}}\right] = \mathrm{E}\left[(\boldsymbol{Q}\mathbf{X})(\boldsymbol{Q}\mathbf{X})^{\mathrm{T}}\right] = \mathrm{E}\left[\boldsymbol{Q}\mathbf{X}\mathbf{X}^{\mathrm{T}}\boldsymbol{Q}^{\mathrm{T}}\right] \\ &= \boldsymbol{Q}\ \mathrm{E}\left[\mathbf{X}\mathbf{X}^{\mathrm{T}}\right]\boldsymbol{Q}^{\mathrm{T}} \\ &= \boldsymbol{Q}\boldsymbol{C}\boldsymbol{Q}^{\mathrm{T}} \\ &= \boldsymbol{\Psi} \end{aligned}$$

so that the matrix of eigenvectors $\boldsymbol{Q}$ can be viewed as a rotation matrix which transforms the set of correlated random variables $X_1, X_2, \ldots, X_n$ into a set of uncorrelated random variables $\mathbf{Z} = \{Z_1, Z_2, \ldots, Z_n\}^{\mathrm{T}}$ having variances $\psi_1, \psi_2, \ldots, \psi_n$, respectively.

3.2.1 Conditional Probabilities

We are often interested in conditional probabilities of the form: Given that $X(t)$ has been observed to have some value x at position t, what is the probability distribution of $X(t+s)$? For example, if the cohesion at $t = 4$ m is known, what is the conditional distribution of the cohesion at $t = 6$ m (assuming that the cohesion field is stationary and that we know the correlation coefficient between the cohesion at $t = 4$ and the cohesion at $t = 6$ m)? If $X(t)$ is a stationary Gaussian process, then the conditional distribution of $X(t+s)$ given $X(t) = x$ is also normally distributed with mean and variance

$$\mathrm{E}[X(t+s) \,|\, X(t) = x] = \mu_X + (x - \mu_X)\rho(s) \quad (3.13a)$$

$$\mathrm{Var}[X(t+s) \,|\, X(t) = x] = \sigma_X^2(1 - \rho^2(s)) \quad (3.13b)$$

where $\rho(s)$ is the correlation coefficient between $X(t+s)$ and $X(t)$.

3.3 SPECTRAL DENSITY FUNCTION

We now turn our attention to an equivalent second-moment description of a stationary random process, namely its *spectral representation*. We say "equivalent" because the spectral representation, in the form of a *spectral density function*, contains the same information as the covariance function, just expressed in a different way. As we shall see, the spectral density function can be obtained from the covariance function and vice versa. The two forms are merely transforms of one another.

Priestley (1981) shows that if $X(t)$ is a stationary random process, with $\rho(\tau)$ continuous at $\tau = 0$, then it can be expressed as a sum of sinusoids with mutually independent random amplitudes and phase angles,

$$\begin{aligned} X(t) &= \mu_X + \sum_{k=-N}^{N} C_k \cos(\omega_k t + \Phi_k) \\ &= \mu_X + \sum_{k=-N}^{N} \left[A_k \cos(\omega_k t) + B_k \sin(\omega_k t)\right] \end{aligned} \quad (3.14)$$

where μ_X is the process mean, C_k is a random amplitude, and Φ_k is a random phase angle. The equivalent form involving A_k and B_k is obtained by setting $A_k = C_k \cos(\Phi_k)$ and $B_k = -C_k \sin(\Phi_k)$. If the random amplitudes A_k and B_k are normally distributed with zero means, then $X(t)$ will also be normally distributed with mean μ_X. For this to be true, C_k must be Raleigh distributed and Φ_k must be uniformly distributed on the interval $[0, 2\pi]$. Note that $X(t)$ will tend to a normal distribution anyhow, by virtue of the central limit theorem, for wide-band processes, so we will assume that $X(t)$ is normally distributed.

Consider the kth component of $X(t)$ and ignore μ_X for the time being,

$$X_k(t) = C_k \cos(\omega_k t + \Phi_k) \quad (3.15)$$

If C_k is independent of Φ_k, then $X_k(t)$ has mean

$$\begin{aligned} \mathrm{E}[X_k(t)] &= \mathrm{E}[C_k \cos(\omega_k t + \Phi_k)] \\ &= \mathrm{E}[C_k]\,\mathrm{E}[\cos(\omega_k t + \Phi_k)] = 0 \end{aligned}$$

due to independence and the fact that, for any t, $\mathrm{E}[\cos(\omega_k t + \Phi_k)] = 0$ since Φ_k is uniformly distributed on $[0, 2\pi]$. The variance of $X_k(t)$ is thus

$$\begin{aligned} \mathrm{Var}[X_k(t)] &= \mathrm{E}\left[X_k^2(t)\right] = \mathrm{E}\left[C_k^2\right]\mathrm{E}\left[\cos^2(\omega_k t + \Phi_k)\right] \\ &= \tfrac{1}{2}\ \mathrm{E}\left[C_k^2\right] \end{aligned} \quad (3.16)$$

Note that $\mathrm{E}\left[\cos^2(\omega_k t + \Phi_k)\right] = \frac{1}{2}$, which again uses the fact that Φ_k is uniformly distributed between 0 and 2π.

Priestley also shows that the component sinusoids are independent of one another, that is, that $X_k(t)$ is independent of $X_j(t)$ for all $k \neq j$. Using this property, we can put the components back together to find the mean and variance of $X(t)$,

$$\mathrm{E}[X(t)] = \mu_X + \sum_{k=-N}^{N} \mathrm{E}[X_k(t)] = \mu_X \quad (3.17a)$$

$$\mathrm{Var}[X(t)] = \sum_{k=-N}^{N} \mathrm{Var}[X_k(t)] = \sum_{k=-N}^{N} \tfrac{1}{2}\ \mathrm{E}\left[C_k^2\right] \quad (3.17b)$$

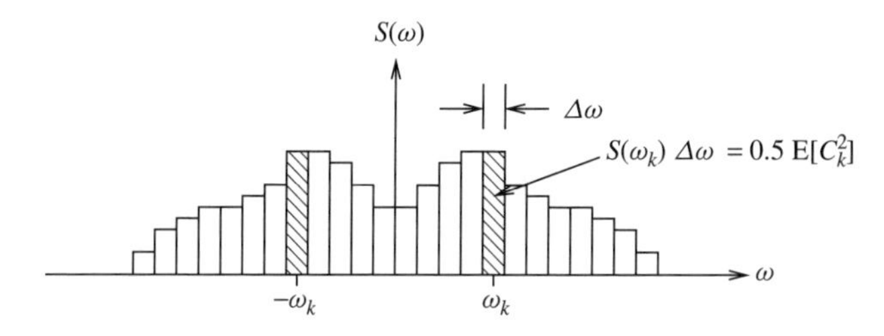

Figure 3.3 Two-sided spectral density function $S(\omega)$.

In other words, the prescribed mean of $X(t)$ is preserved by the spectral representation and the variance of the sum is the sum of the variances of each component frequency, since the component sinusoids are independent. The amount that each component frequency contributes to the overall variance of $X(t)$ depends on the "power" in the sinusoid amplitude, $\frac{1}{2}\,\mathrm{E}\left[C_k^2\right]$.

Now define the *two-sided spectral density function* $S(\omega)$ such that

$$S(\omega_k)\ \Delta\omega = \mathrm{Var}\left[X_k(t)\right] = \mathrm{E}\left[X_k^2(t)\right] = \tfrac{1}{2}\ \mathrm{E}\left[C_k^2\right] \quad (3.18)$$

Then the variance of $X(t)$ can be written as

$$\mathrm{Var}\left[X(t)\right] = \sum_{k=-N}^{N} S(\omega_k)\ \Delta\omega \quad (3.19)$$

In the limit as $\Delta\omega \to 0$ and $N \to \infty$, we get

$$\mathrm{Var}\left[X(t)\right] = \sigma_X^2 = \int_{-\infty}^{\infty} S(\omega)\ d\omega \quad (3.20)$$

which is to say the variance of $X(t)$ is just the area under the two-sided spectral density function (Figure 3.3).

3.3.1 Wiener–Khinchine Relations

We can use the spectral representation to express the covariance function $C(\tau)$. Assuming that $\mu_X = 0$ for the time being to simplify the algebra (this is not a restriction, the end results are the same even if $\mu_X \neq 0$), we have

$$\begin{aligned}
C(\tau) &= \mathrm{Cov}\left[X(0), X(\tau)\right] \quad \text{(due to stationarity)} \\
&= \mathrm{E}\left[\sum_k X_k(0) \sum_j X_j(\tau)\right] \\
&= \sum_k \sum_j \mathrm{E}\left[X_k(0) X_j(\tau)\right] \\
&= \sum_k \mathrm{E}\left[X_k(0) X_k(\tau)\right] \quad \text{(due to independence)}
\end{aligned}$$

Now, since $X_k(0) = C_k \cos(\Phi_k)$ and $X_k(\tau) = C_k \cos(\omega_k \tau + \Phi_k)$, we get

$$\begin{aligned}
C(\tau) &= \sum_k \mathrm{E}\left[C_k^2\right] \mathrm{E}\left[\cos(\Phi_k)\cos(\omega_k\tau + \Phi_k)\right] \\
&= \sum_k \mathrm{E}\left[C_k^2\right] \mathrm{E}\left[\tfrac{1}{2}\{\cos(\omega_k\tau + 2\Phi_k) + \cos(\omega_k\tau)\}\right] \\
&= \sum_k \tfrac{1}{2}\ \mathrm{E}\left[C_k^2\right]\cos(\omega_k\tau) \\
&= \sum_k S(\omega_k)\cos(\omega_k\tau)\ \Delta\omega
\end{aligned}$$

which in the limit as $\Delta\omega \to 0$ gives

$$C(\tau) = \int_{-\infty}^{\infty} S(\omega)\cos(\omega\tau)\ d\omega \quad (3.21)$$

Thus, the covariance function $C(\tau)$ is the Fourier transform of the spectral density function $S(\omega)$. The inverse transform can be applied to find $S(\omega)$ in terms of $C(\tau)$,

$$S(\omega) = \frac{1}{2\pi}\int_{-\infty}^{\infty} C(\tau)\cos(\omega\tau)\ d\tau \quad (3.22)$$

so that knowing either $C(\tau)$ or $S(\omega)$ allows the other to be found (and hence these are equivalent in terms of information). Also, since $C(\tau) = C(-\tau)$, that is, the covariance between one point and another is the same regardless of which point you consider first, and since $\cos(x) = \cos(-x)$, we see that

$$S(\omega) = S(-\omega) \quad (3.23)$$

In other words, the two-sided spectral density function is an even function (see Figure 3.3). The fact that $S(\omega)$ is symmetric about $\omega = 0$ means that we need only know the positive half in order to know the entire function. This motivates the introduction of the *one-sided spectral density function* $G(\omega)$ defined as

$$G(\omega) = 2S(\omega), \qquad \omega \geq 0 \quad (3.24)$$

(See Figure 3.4). The factor of 2 is included to preserve the total variance when only positive frequencies are considered. Now the Wiener–Khinchine relations become

$$C(\tau) = \int_0^{\infty} G(\omega)\cos(\omega\tau)\ d\omega \quad (3.25a)$$

$$G(\omega) = \frac{1}{\pi}\int_{-\infty}^{\infty} C(\tau)\cos(\omega\tau)\ d\tau \quad (3.25b)$$

$$= \frac{2}{\pi}\int_0^{\infty} C(\tau)\cos(\omega\tau)\ d\tau \quad (3.25c)$$

and the variance of $X(t)$ is the area under $G(\omega)$ (set $\tau = 0$ in Eq. 3.25a to see this),

$$\sigma_X^2 = C(0) = \int_0^{\infty} G(\omega)\ d\omega \quad (3.26)$$

The spectral representation of a stationary Gaussian process is primarily used in situations where the frequency

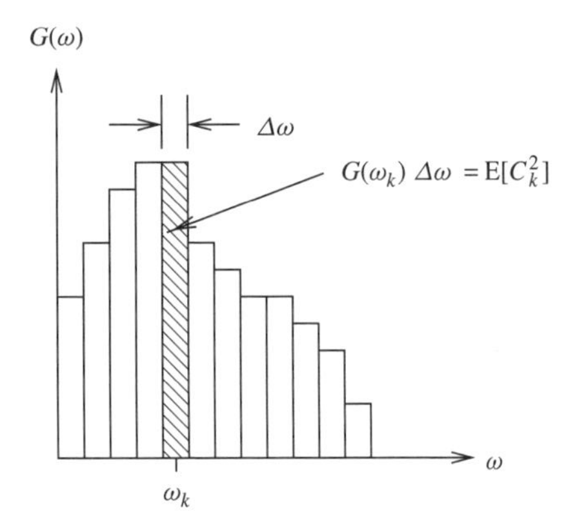

Figure 3.4 One-sided spectral density function $G(\omega) = 2S(\omega)$ corresponding to Figure 3.3.

domain is an integral part of the problem being considered. For example, earthquake ground motions are often represented using the spectral density function because the motions are largely sinusoidal with frequency content dictated by resonance in the soil or rock through which the earthquake waves are traveling. In addition, the response of structures to earthquake motion is often performed using Fourier response "modes," each having its own resonance frequency. Thus, if a structure has a 1-Hz primary response mode (single mass-and-spring oscillation), then it is of interest to see what power the input ground motion has at 1 Hz. This is given by $G(\omega_k)\,\Delta\omega$ at $\omega_k = 1$ Hz.

In addition, the spectral representation provides a means to simulate a stationary Gaussian process, namely to simulate independent realizations of C_k and Φ_k for $k = 0, 1, \ldots, N$ and then recombine using the spectral representation. We shall see more of this in Chapter 6.

3.3.2 Spectral Density Function of Linear Systems

Let us consider a system which is excited by an *input* $X(t)$ and which has a *response* $Y(t)$. If the system is *linear*, then doubling the input $X(t)$ will double the response $Y(t)$. More generally, when the input is a sum, $X(t) = X_1(t) + X_2(t) + \cdots$, and $Y_i(t)$ is the response of the system to each individual $X_i(t)$, the total response of a linear system will be the sum $Y(t) = Y_1(t) + Y_2(t) + \cdots$. This is often referred to as the principle of *superposition*, which is one of the main features of a linear system.

Although there are many different types of linear systems, those described by linear differential equations are most easily represented using the spectral density function, as we shall see. A linear differential equation is one in which a linear combination of derivatives of $Y(t)$ is set equal to a linear combination of derivatives of $X(t)$,

$$c_n \frac{dy^n}{dt^n} + c_{n-1}\frac{dy^{n-1}}{dt^{n-1}} + \cdots + c_1\frac{dy}{dt} + c_0 y = d_m\frac{dx^m}{dt^m} + d_{m-1}\frac{dx^{m-1}}{dt^{m-1}} + \cdots + d_1\frac{dx}{dt} + d_0 x \qquad (3.27)$$

In particular, the coefficients c_i and d_j are independent of x, y, and t in a linear differential equation.

One of the features of a linear system is that when excited by a sinusoidal input at a specific frequency ω the response will also be at the frequency ω, possibly phase shifted and amplified. That is, if the input is $X(t) = \cos(\omega t)$, then the response will have the form $Y(t) = a_\omega \cos(\omega t + \phi_\omega)$, where a_ω is the output amplitude and ϕ_ω is a phase shift between input and response, both at frequency ω. We can also write the response as

$$\begin{aligned} Y(t) &= a_\omega \cos(\omega t + \phi_\omega) \\ &= a_\omega(\cos\omega t \cos\phi_\omega - \sin\omega t \sin\phi_\omega) \\ &= A_\omega \cos\omega t - B_\omega \sin\omega t \end{aligned} \qquad (3.28)$$

where $A_\omega = a_\omega \cos\phi_\omega$ and $B_\omega = a_\omega \sin\phi_\omega$.

It is convenient to solve linear differential equations in the complex domain. To this end, we define the complex input

$$X_c(t) = e^{i\omega t} = \cos\omega t + i\ \sin\omega t \qquad (3.29)$$

where $i = \sqrt{-1}$. Our actual input is $X(t) = \text{Re}\big(X_c(t)\big)$, where Re(·) means "real part of." Also, let us define the *transfer function*

$$H(\omega) = A_\omega + iB_\omega \qquad (3.30)$$

The complex response $Y_c(t)$ to the complex input $X_c(t)$ can now be written as

$$\begin{aligned} Y_c(t) &= H(\omega)X_c(t) = [A_\omega + iB_\omega][\cos\omega t + i\sin\omega t] \\ &= A_\omega \cos\omega t - B_\omega \sin\omega t + i[A_\omega \sin\omega t \\ &\quad + B_\omega \cos\omega t] \end{aligned} \qquad (3.31)$$

from which we can see that $Y(t) = \text{Re}\big(Y_c(t)\big)$. To see how these results are used to solve a linear differential equation, consider the following example.

Example 3.2 Suppose a system obeys the linear differential equation

$$c\dot{y} + \alpha y = x$$

where the overdot implies differentiation with respect to t. If $x(t) = \cos\omega_k t$, what is the response $y(t)$?

SOLUTION We will first derive the complex response of the system to complex input, then take the real part for the solution. The complex response of the system to the frequency ω_k is obtained by setting the input $x_c(t)$ and output $y_c(t)$ as follows:

$$x_c(t) = e^{i\omega_k t}$$

$$y_c(t) = H(\omega_k)x_c(t) = H(\omega_k)e^{i\omega_k t}$$

Substitution of these into the system differential equation gives

$$c\frac{d}{dt}H(\omega_k)e^{i\omega_k t} + \alpha H(\omega_k)e^{i\omega_k t} = e^{i\omega_k t}$$

or

$$(ic\omega_k + \alpha)H(\omega_k)e^{i\omega_k t} = e^{i\omega_k t}$$

which can be solved for the transfer function to give

$$\begin{aligned} H(\omega_k) &= \frac{1}{ic\omega_k + \alpha} \\ &= \frac{\alpha - ic\omega_k}{\alpha^2 + c^2\omega_k^2} \\ &= \left(\frac{\alpha}{\alpha^2 + c^2\omega_k^2}\right) - i\left(\frac{c\omega_k}{\alpha^2 + c^2\omega_k^2}\right) \end{aligned}$$

The magnitude of the transfer function tells us how much the input signal is amplified,

$$|H(\omega_k)| = \frac{\sqrt{\alpha^2 + c^2\omega_k^2}}{\alpha^2 + c^2\omega_k^2} = \frac{1}{\sqrt{\alpha^2 + c^2\omega_k^2}}$$

Recalling that $H(\omega) = A_\omega + iB_\omega$, we must have

$$A_{\omega_k} = \frac{\alpha}{\alpha^2 + c^2\omega_k^2}, \qquad B_{\omega_k} = -\frac{c\omega_k}{\alpha^2 + c^2\omega_k^2}$$

The complex response $y_c(t)$ to the complex input $x_c(t) = e^{i\omega_k t}$ is thus $y_c(t) = H(\omega_k)e^{i\omega_k t}$, which expands into

$$\begin{aligned} y_c(t) = \frac{1}{\alpha^2 + c^2\omega_k^2}\Big[&\alpha \cos\omega_k t + c\omega_k \sin\omega_k t \\ &+ i(\alpha \sin\omega_k t - c\omega_k \cos\omega_k t)\Big] \end{aligned}$$

The real response to the real input $x(t) = \cos\omega_k t$ is therefore

$$\begin{aligned} y(t) &= \mathrm{Re}\Big(y_c(t)\Big) \\ &= \frac{1}{\alpha^2 + c^2\omega_k^2}\Big[\alpha \cos\omega_k t + c\omega_k \sin\omega_k t\Big] \\ &= \left(\frac{\sqrt{\alpha^2 + c^2\omega_k^2}}{\alpha^2 + c^2\omega_k^2}\right)\cos\left(\omega_k t + \tan^{-1}\left(\frac{c\omega_k}{\alpha}\right)\right) \\ &= \frac{1}{\sqrt{\alpha^2 + c^2\omega_k^2}}\cos(\omega_k t + \phi_k) \\ &= |H(\omega_k)| \cos(\omega_k t + \phi_k) \end{aligned} \tag{3.32}$$

where $\phi_k = \tan^{-1}(c\omega_k/\alpha)$ is the phase shift.

The transfer function $H(\omega)$ gives the steady-state response of a linear system to a sinusoidal input at frequency ω. If we make use of the superposition principle of linear systems, then we could compute a series of transfer functions $H(\omega_1)$, $H(\omega_2)$, ... corresponding to sinusoidal excitations at frequencies ω_1, ω_2, ... The overall system response would be the sum of all the individual responses.

To determine the spectral density function $S_Y(\omega)$ of the system response $Y(t)$, we start by assuming that the input $X(t)$ is equal to the sinusoidal component given by Eq. 3.15,

$$X_k(t) = C_k \cos(\omega_k t + \Phi_k) \tag{3.33}$$

where Φ_k is uniformly distributed between 0 and 2π and independent of C_k. Assuming that the spectral density function of $X(t)$, $S_X(\omega)$, is known, we select C_k to be random with

$$\mathrm{E}\left[C_k^2\right] = 2S_X(\omega_k)\Delta\omega$$

so that Eq. 3.18 holds. Equation 3.32 tells us that the random response $Y_k(t)$ will be amplified by $|H(\omega_k)|$ and phase shifted by ϕ_k from the random input $X_k(t)$,

$$Y_k(t) = |H(\omega_k)|C_k \cos(\omega_k t + \Phi_k + \phi_k)$$

The spectral density of $Y_k(t)$ is obtained in exactly the same way as the spectral density of $X_k(t)$ was found in Eq. 3.18,

$$\begin{aligned} S_Y(\omega_k)\ \Delta\omega &= \mathrm{Var}\left[Y_k(t)\right] = \mathrm{E}\left[Y_k^2(t)\right] \\ &= \mathrm{E}\left[|H(\omega_k)|^2 C_k^2 \cos^2(\omega_k t + \Phi_k + \phi_k)\right] \\ &= |H(\omega_k)|^2\ \mathrm{E}\left[C_k^2\right]\mathrm{E}\left[\cos^2(\omega_k t + \Phi_k + \phi_k)\right] \\ &= |H(\omega_k)|^2(2S_X(\omega_k)\Delta\omega)\left(\tfrac{1}{2}\right) \\ &= |H(\omega_k)|^2 S_X(\omega_k)\Delta\omega \end{aligned}$$

Generalizing this to any input frequency leads to one of the most important results in random vibration theory, namely that the response spectrum is a simple function of the input spectrum,

$$S_Y(\omega) = |H(\omega)|^2 S_X(\omega)$$

3.3.3 Discrete Random Processes

So far in the discussion of spectral representation we have been considering only processes that vary continuously in time. Consider now a process which varies continuously but which we have only sampled at discrete points in time. The upper plot of Figure 3.5 illustrates what we might observe if we sample $X(t)$ at a series of points separated by Δt. When we go to represent $X(t)$ as a sum of sinusoids, we need to know which component sinusoids to use and what

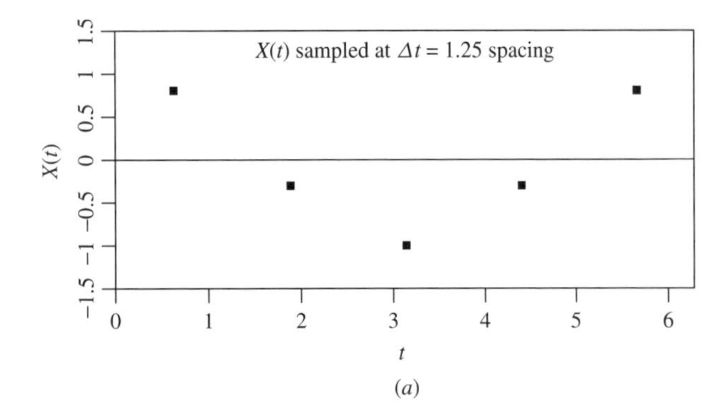

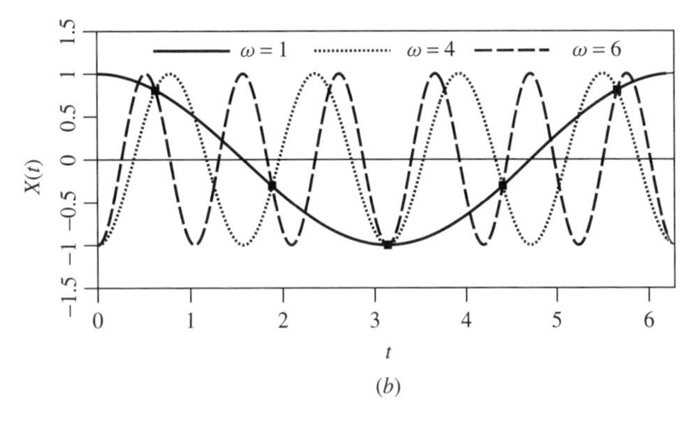

Figure 3.5 (*a*) Observations of $X(t)$ at spacing Δt. (*b*) Several frequencies each of which could result in the same sequence of observations.

their amplitudes are. When $X(t)$ varies continuously and is known for all time, there is a unique set of sinusoids composing $X(t)$. However, as seen in the lower plot of Figure 3.5, there exist many sinusoidal waves each of which could have produced the sampled values. Thus, when $X(t)$ is only known discretely, we can no longer uniquely determine its frequency components.

Frequencies which are indistinguishable from each other when sampled discretely are called *aliases* of one another. In fact, all frequencies having wavelength shorter than $2\Delta t$ will have an alias with a frequency which is longer than $2\Delta t$. We call the frequency corresponding to this critical wavelength the *Nyquist frequency* ω_N, where

$$\omega_N = \frac{\pi}{\Delta t} \tag{3.34}$$

Each frequency in the range $0 \le \omega \le \omega_N$ has aliases at $2\omega_N - \omega$, $2\omega_N + \omega$, $4\omega_N - \omega$, $4\omega_N + \omega$, and so on. We call the low-frequency (long-wavelength) components, where $0 \le \omega \le \omega_N$, the *principal aliases*. In Figure 3.5, $\omega_N = \pi/\Delta t = \pi/1.25 = 2.5$, and two aliases of the principal alias $\omega = 1$ are $2\omega_N - \omega = 2(2.5) - 1 = 4$ and $2\omega_N + \omega = 2(2.5) + 1 = 6$.

Just as a bicycle wheel appears to be turning more slowly when "sampled" by a stroboscope, the high-frequency aliases appear to the viewer to be the low-frequency principal alias. For example, if $X(t)$ consists of just a single sinusoidal component having frequency $2.5\omega_N$, it will appear after sampling to be a sinusoid having frequency $0.5\omega_N$. That is, the power of the frequencies above ω_N are folded into the power of the frequencies below ω_N. This complicates the estimation of $G(\omega)$ whenever $X(t)$ has significant power above ω_N. We shall see more of this in Chapter 5.

The discrete observations, $X_i = X(t_i) = X(i\,\Delta t)$ for $i = 0, 1, \ldots, n$ can be fully represented by sinusoids having frequencies between zero and the Nyquist frequency ω_N. That is, frequencies above ω_N are not needed to reproduce X_i. In fact, only the frequencies below ω_N are uniquely defined by X_i. This means that the spectral density function of X_i should be taken as zero beyond $\omega_N = \pi/\Delta t$. For such discrete processes, the covariance function can be obtained from the spectral density function through a slight modification of the Wiener–Khinchine relationship as follows:

$$C(\tau) = \int_0^{\pi/\Delta t} G(\omega)\cos(\omega\tau)\, d\omega \tag{3.35}$$

for $|\tau| = k\,\Delta t$, $k = 0, 1, \ldots, n$.

3.4 VARIANCE FUNCTION

Virtually all engineering properties are actually properties of a local average of some sort. For example, the hydraulic conductivity of a soil is rarely measured at a point since, at the point level, we are either in a void having infinite conductivity or in a solid having negligible conductivity. Just as we rarely model soils at the microscopic, or particle, level for use in designs at the macroscopic level, the hydraulic conductivity is generally estimated using a laboratory sample of some volume, supplying a differential total head, and measuring the quantity of water which passes through the sample in some time interval. The paths that the water takes to migrate through the sample are not considered individually; rather it is the sum of these paths that are measured. This is a "local average" over the laboratory sample. (As we shall see later there is more than one possible type of average to take, but for now we shall concentrate on the more common arithmetic average.)

Similarly, when the compressive strength of a material is determined, a load is applied to a finite-sized sample until failure occurs. Failure takes place when the shear/tensile resistances of a large number of bonds are broken—the failure load is then a function of the average bond strength throughout the failure region.

Thus, it is of considerable engineering interest to investigate how averages of random fields behave. Consider the local average defined as

$$X_T(t) = \frac{1}{T}\int_{t-T/2}^{t+T/2} X(\xi)\, d\xi \tag{3.36}$$

which is a "moving" local average. That is, $X_T(t)$ is the local average of $X(t)$ over a window of width T centered at t. As this window is moved along in time, the local average $X_T(t)$ changes more slowly (see Figure 3.6).

For example, consider the boat-in-the-water example: If the motion of a piece of sawdust on the surface of the ocean is tracked, it is seen to have considerable variability in its elevation. In fact, it will have as much variability as the waves themselves. Now, replace the sawdust with an ocean liner. The liner does not bounce around with every wave, but rather it "averages" out the wave motion over the area of the liner. Its vertical variability is drastically reduced.

In this example, it is also worth thinking about the spectral representation of the ocean waves. The piece of sawdust sees all of the waves, big and small, whereas the local averaging taking place over the ocean liner damps out the high-frequency components leaving just the long-wavelength components (wavelengths of the order of the size of the ship and longer). Thus, local averaging is a low-pass filter. If the ocean waves on the day that the sawdust and ocean liner are being observed are composed of just long-wavelength swells, then the variability of the sawdust and liner will be the same. Conversely, if the ocean surface is just choppy without any swells, then the ocean liner may hardly move up and down at all. Both the sawdust and the ocean liner will have the same mean elevation in all cases.

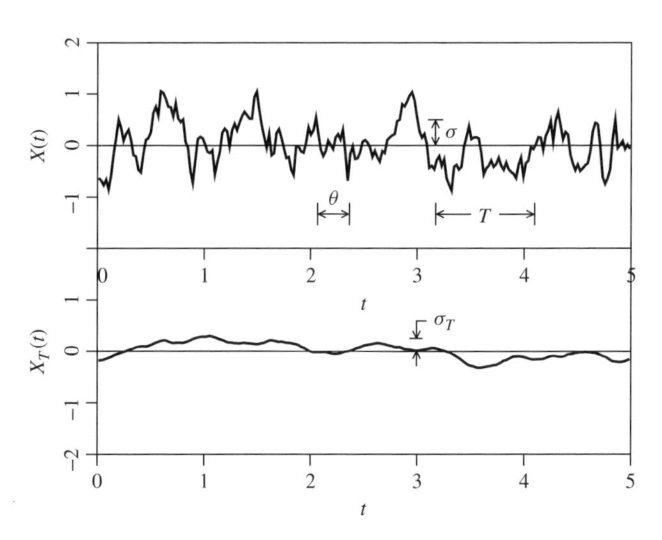

Figure 3.6 Effect of local averaging on variance; T is the moving window length over which the top plot is averaged to get the lower plot.

The two main effects of local averaging are to reduce the variance and to damp the contribution from the high-frequency components. The amount of variance reduction increases with increasing high-frequency content in the random field. An increased high-frequency content corresponds to increasing independence in the random field, so that another way of putting this is that variance reduction increases when the random field consists of more "independence." This is illustrated in Figure 3.6. A random process is shown in the upper plot, which is then averaged within a moving window of width T to obtain the lower plot. Notice that averaging both smooths the process and reduces its variance.

Let us look in more detail at the moments of $X_T(t)$. Its mean is

$$\begin{aligned} \mathrm{E}[X_T(t)] &= \mathrm{E}\left[\frac{1}{T}\int_{t-T/2}^{t+T/2} X(\xi)\, d\xi\right] \\ &= \frac{1}{T}\int_{t-T/2}^{t+T/2} \mathrm{E}[X(\xi)]\, d\xi \\ &= \mathrm{E}[X] \end{aligned} \tag{3.37}$$

for stationary $X(t)$. That is, local arithmetic averaging preserves the mean of the random field (the mean of an arithmetic average is just the mean of the process). Now consider the variance,

$$\mathrm{Var}[X_T(t)] = \mathrm{E}\left[(X_T(t) - \mu_{X_T})^2\right] \tag{3.38}$$

where, since $\mu_{X_T} = \mu_X$,

$$\begin{aligned} X_T - \mu_{X_T} &= \frac{1}{T}\int_{t-T/2}^{t+T/2} X(\xi)\, d\xi - \mu_X \\ &= \frac{1}{T}\int_{t-T/2}^{t+T/2} [X(\xi) - \mu_X]\, d\xi \end{aligned}$$

so that (due to stationarity, the bounds of the integral can be changed to any domain of length T without changing the expectation; we will use the domain $[0, T]$ for simplicity)

$$\begin{aligned} &\mathrm{Var}[X_T(t)] \\ &= \mathrm{E}\left[\frac{1}{T}\int_0^T [X(\xi) - \mu_X]\, d\xi\; \frac{1}{T}\int_0^T [X(\eta) - \mu_X]\, d\eta\right] \\ &= \frac{1}{T^2}\int_0^T\int_0^T \mathrm{E}[(X(\xi) - \mu_X)(X(\eta) - \mu_X)]\, d\xi\, d\eta \\ &= \frac{1}{T^2}\int_0^T\int_0^T C_X(\xi - \eta)\, d\xi\, d\eta \\ &= \frac{\sigma_X^2}{T^2}\int_0^T\int_0^T \rho_X(\xi - \eta)\, d\xi\, d\eta \\ &= \sigma_X^2\, \gamma(T) \end{aligned} \tag{3.39}$$

where $C_X(\tau)$ is the covariance function of $X(t)$ and $\rho_X(\tau)$ is the correlation function of $X(t)$ such that $C_X(\tau) = \sigma_X^2 \rho_X(\tau)$. In the final expression, $\gamma(T)$ is the so-called *variance function*, which gives the amount that the variance is reduced when $X(t)$ is averaged over the length T. The variance function has value 1.0 when $T = 0$, which is to say that $X_T(t) = X(t)$ when $T = 0$, and so the variance is not at all reduced. As T increases, the variance function decreases toward zero. It has the mathematical definition

$$\gamma(T) = \frac{1}{T^2} \int_0^T \int_0^T \rho_X(\xi - \eta)\, d\xi\, d\eta \tag{3.40}$$

The variance function can be seen, in Eq. 3.40, to be an average of the correlation coefficient between every pair of points on the interval $[0, T]$. If the correlation function falls off rapidly, so that the correlation between pairs of points becomes rapidly smaller with separation distance, then $\gamma(T)$ will be small. On the other hand, if all points on the interval $[0, T]$ are perfectly correlated, having $\rho(\tau) = 1$ for all τ, then $\gamma(T)$ will be 1.0. Such a field displays no variance reduction under local averaging. [In fact, if the field is stationary, all points will have the same random value, $X(t) = X$.]

The integral in Eq. 3.40 is over the square region $[0, T] \times [0, T]$ in (ξ, η) space. Considering Figure 3.7, one sees that $\rho_X(\xi - \eta)$ is constant along diagonal lines where $\xi - \eta =$ const. The length of the main diagonal, where $\xi = \eta$, is $\sqrt{2}T$, and the other diagonal lines decrease linearly in length to zero in the corners. The double integral can be collapsed to a single integral by integrating in a direction perpendicular to the diagonals; each diagonal differential area has length $\sqrt{2}(T - |\tau|)$, width $d\tau/\sqrt{2}$, and height equal to $\rho_X(\xi - \eta) = \rho_X(\tau)$. The integral can therefore be written as

$$\begin{aligned}\gamma(T) &= \frac{1}{T^2} \int_0^T \int_0^T \rho_X(\xi - \eta)\, d\xi\, d\eta \\ &= \frac{1}{T^2} \left[\int_{-T}^0 \sqrt{2}(T - |\tau_1|)\rho_X(\tau_1) \frac{d\tau_1}{\sqrt{2}} \right. \\ &\quad \left. + \int_0^T \sqrt{2}(T - |\tau_2|)\rho_X(\tau_2) \frac{d\tau_2}{\sqrt{2}} \right] \\ &= \frac{1}{T^2} \int_{-T}^T (T - |\tau|)\rho_X(\tau)\, d\tau \end{aligned} \tag{3.41}$$

Furthermore, since $\rho_X(\tau) = \rho_X(-\tau)$, the integrand is even, which results in the additional simplification

$$\gamma(T) = \frac{2}{T^2} \int_0^T (T - \tau)\rho_X(\tau)\, d\tau \tag{3.42}$$

Figure 3.8 shows two typical variance functions, the solid line corresponding to an exponentially decaying correlation function (the Markov model, see Section 3.6.5) and the dashed line corresponding to the Gaussian correlation function (Section 3.6.6). The variance function is another equivalent second-moment description of a random field, since it can be obtained through knowledge of the correlation function, which in turn can be obtained from the spectral density function. The inverse relationship between $\gamma(T)$ and $\rho(\tau)$ is obtained by differentiation:

$$\rho(\tau) = \frac{1}{2} \frac{d^2}{d\tau^2} [\tau^2 \gamma(\tau)] \tag{3.43}$$

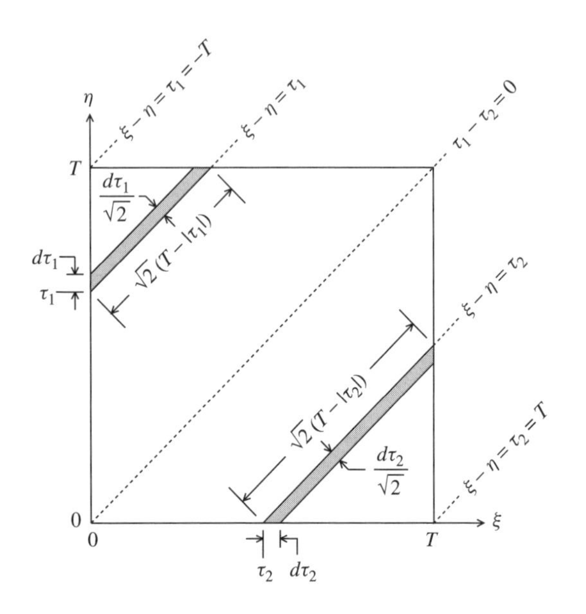

Figure 3.7 Reduction of two-dimensional integral of $\rho(\xi - \eta)$ to a one-dimensional integral.

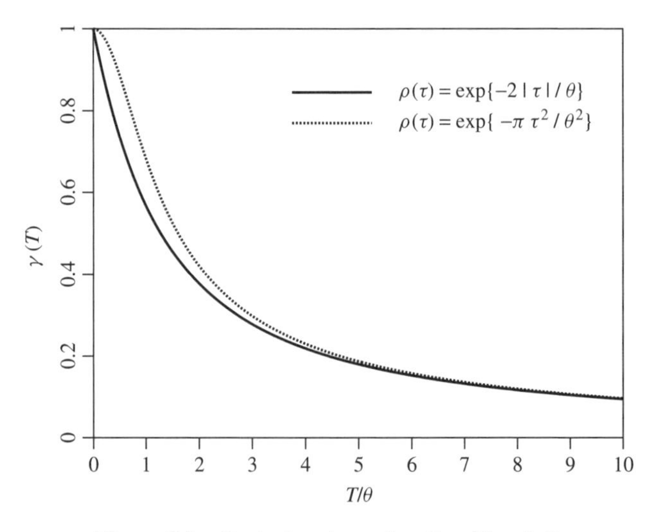

Figure 3.8 Typical variance function ($\theta = 0.4$).

The variance function can also be obtained from the spectral density function (Vanmarcke, 1984):

$$\gamma(T) = \int_0^\infty \frac{G(\omega)}{\sigma_X^2} \left[\frac{\sin(\omega T/2)}{\omega T/2} \right]^2 d\omega \tag{3.44}$$

Example 3.3 In Figure 3.6, a process having the Markov covariance function

$$C(\tau) = \sigma^2 \exp\left\{ -\frac{2|\tau|}{\theta} \right\}$$

has been observed (upper plot). For this process, $\sigma = 0.5$ and the correlation length (to be discussed in the next section) is $\theta = 0.3$. The process $X(t)$ is averaged over the length $T = 0.93$ at each t, that is,

$$X_T(t) = \frac{1}{T} \int_{t-T/2}^{t+T/2} X(\xi)\, d\xi$$

and this is shown in the lower plot of Figure 3.6. What is the standard deviation of $X_T(t)$?

SOLUTION Let σ_T be the standard deviation of $X_T(t)$. We know that

$$\sigma_T^2 = \sigma^2 \gamma(T) \quad \Longrightarrow \quad \sigma_T = \sigma \sqrt{\gamma(T)}$$

where

$$\begin{aligned}
\gamma(T) &= \frac{2}{T^2} \int_0^T (T - \tau) \rho_X(\tau)\, d\tau \\
&= \frac{2}{T^2} \int_0^T (T - \tau) \exp\left\{ -\frac{2|\tau|}{\theta} \right\} d\tau \\
&= \frac{\theta^2}{2T^2} \left[\frac{2|T|}{\theta} + \exp\left\{ -\frac{2|T|}{\theta} \right\} - 1 \right]
\end{aligned}$$

So, for $T = 0.93$ and $\theta = 0.3$, we get

$$\gamma(0.93) = 0.2707$$

The standard deviation of $X_T(t)$ is therefore

$$\sigma_T = 0.5\sqrt{0.2707} = 0.26$$

The averaging in this case approximately halves the standard deviation of the original field.

3.5 CORRELATION LENGTH

A convenient measure of the variability of a random field is the *correlation length* θ, also sometimes referred to as the *scale of fluctuation*. Loosely speaking θ is the distance within which points are significantly correlated (i.e., by more than about 10%). Conversely, two points separated by a distance more than θ will be largely uncorrelated. Mathematically, θ is defined here as the area under the correlation function (Vanmarcke, 1984),

$$\theta = \int_{-\infty}^{\infty} \rho(\tau)\, d\tau = 2 \int_0^\infty \rho(\tau)\, d\tau \tag{3.45}$$

The correlation length is sometimes defined without the factor of 2 shown on the right-hand side of Eq. 3.45 (see, e.g., Journel and Huijbregts, 1978)

Equation 3.45 implies that if θ is to be finite then $\rho(\tau)$ must decrease sufficiently quickly to zero as τ increases. Not all correlation functions will satisfy this criterion, and for such random processes, $\theta = \infty$. An example of a process with infinite correlation length is a *fractal* process (see Section 3.6.7).

In addition, the correlation length is really only meaningful for strictly nonnegative correlation functions. Since $-1 \le \rho \le 1$, one could conceivably have an oscillatory correlation function whose integrated area is zero but which has significant correlations (positive or negative) over significant distances. An example of such a correlation function might be that governing wave heights in a body of water.

The correlation length can also be defined in terms of the spectral density function,

$$G(\omega) = \frac{2\sigma^2}{\pi} \int_0^\infty \rho(\tau) \cos(\omega\tau)\, d\tau \tag{3.46}$$

since, when $\omega = 0$,

$$G(0) = \frac{2\sigma^2}{\pi} \int_0^\infty \rho(\tau)\, d\tau = \frac{\sigma^2}{\pi}\theta \tag{3.47}$$

which means that

$$\theta = \frac{\pi G(0)}{\sigma^2} \tag{3.48}$$

What this means is that if the spectral density function is finite at the origin, then θ will also be finite. In practice $G(0)$ is quite difficult to estimate, since it requires data over an infinite distance ($\omega = 0$ corresponds to an infinite wavelength). Thus, Eq. 3.48 is of limited value in estimating the correlation length from real data. This is our first hint that θ is fundamentally difficult to estimate and we will explore this further in Chapter 5.

The correlation length can also be defined in terms of the variance function as a limit (Vanmarcke, 1984):

$$\theta = \lim_{T\to\infty} T\gamma(T) \tag{3.49}$$

This implies that if the correlation length is finite, then the variance function has the following limiting form as the averaging region grows very large:

$$\lim_{T\to\infty} \gamma(T) = \frac{\theta}{T} \tag{3.50}$$

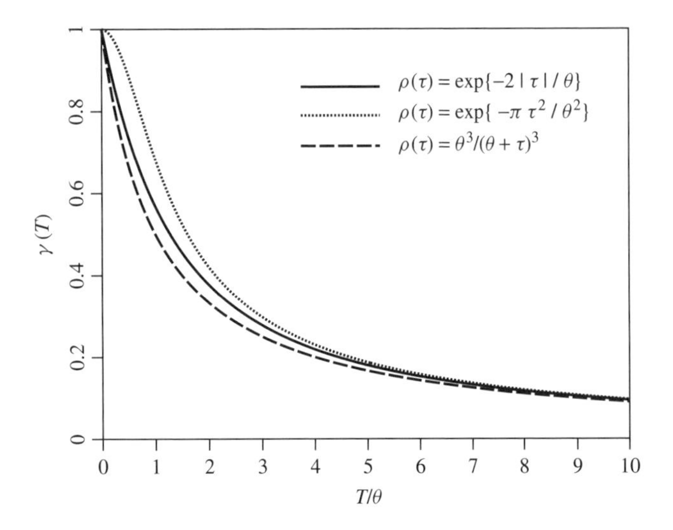

Figure 3.9 Variance function corresponding to three different correlation models.

which in turn means that θ/T can be used as an approximation for $\gamma(T)$ when $T >> \theta$. A more extensive approximation for $\gamma(T)$, useful when the precise correlation structure of a random field is unknown but for which θ is known (or estimated), is

$$\gamma(T) \simeq \frac{\theta}{\theta + |T|} \tag{3.51}$$

which has the correct limiting form for $T >> \theta$ and which has value 1.0 when $T = 0$, as expected. The correlation function corresponding to Eq. 3.51 is

$$\rho(\tau) = \frac{\theta^3}{(\theta + \tau)^3} \tag{3.52}$$

which is illustrated in Figure 3.9.

Some comments about what effect the correlation length has on a random field are in order. When the correlation length is small, the field tends to be somewhat "rough." In the limit, when $\theta \to 0$, all points in the field become uncorrelated and the field becomes infinitely rough, which is physically unrealizable. Such a field is called *white noise* (see Section 3.6.1). Conversely, when the correlation length becomes large, the field becomes smoother. In certain cases, such as under the Markov correlation function (see Section 3.6.5), the random field becomes completely uniform when $\theta \to \infty$—different from realization to realization but each realization is composed of a single random value. Traditional soil variability models, where the entire soil mass is represented by a single random variable, are essentially assuming $\theta = \infty$.

Figure 3.10 shows two random-field realizations. The field on the left has a small correlation length ($\theta = 0.04$) and can be seen to be quite rough. The field on the right has a large correlation length ($\theta = 2$) and can be seen to be more slowly varying.

3.6 SOME COMMON MODELS

3.6.1 Ideal White Noise

The simplest type of random field is one in which $X(t)$ is composed of an infinite sequence of iid random variables, one for each t. That is, $X_1 = X(t_1)$, $X_2 = X(t_2)$, ..., each have marginal distribution $f_X(x)$, and, since they are independent, their joint distribution is just the product of their marginal distributions,

$$f_{X_1X_2\ldots}(x_1, x_2, \ldots) = f_X(x_1)f_X(x_2)\cdots$$

The covariance between any two points, $X(t_1)$ and $X(t_2)$, is

$$C(t_1, t_2) = C(|t_1 - t_2|) = C(\tau) = \begin{cases} \sigma^2 & \text{if } \tau = 0 \\ 0 & \text{if } \tau \neq 0 \end{cases}$$

In practice, the simulation of white noise processes proceeds using the above results; that is, simply simulate a sequence of iid random variables. However, the above also implies that two points arbitrarily close to one another will have independent values, which is not very realistic—the field would be infinitely rough at the microscale.

The nature of *ideal white noise* for continuous t can be illustrated by considering two equispaced sequences of

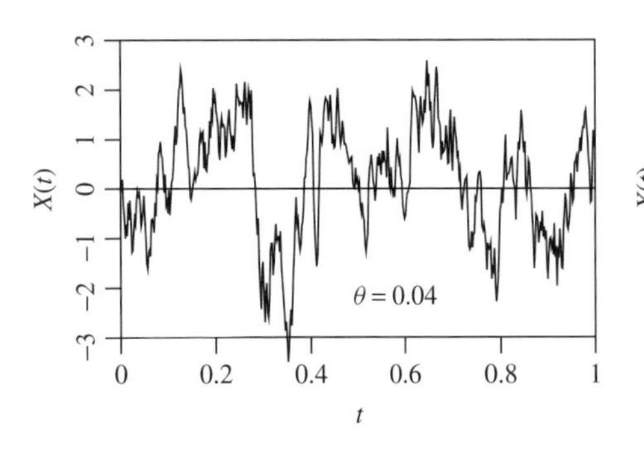

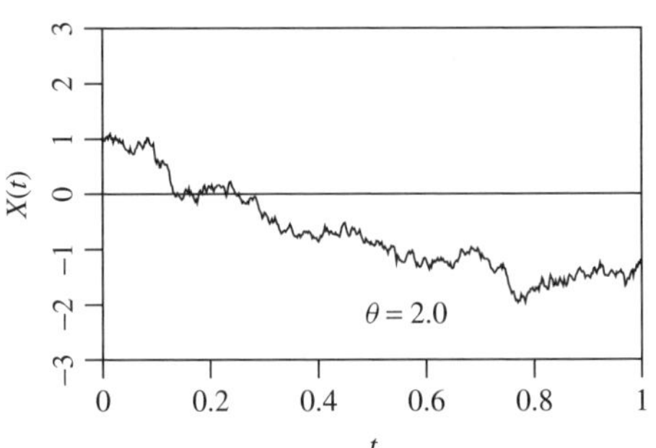

Figure 3.10 Sample realizations of $X(t)$ for two different correlation lengths.

observations of averages of an ideal white noise process. The first sequence, $X(0)$, $X(\Delta t)$, $X(2\Delta t)$, ..., is taken by averaging the white noise process over adjacent intervals of width Δt. Now, suppose that n successive values of the series $X(t)$ are averaged to produce another sequence $X_a(t)$. That is $X_a(0)$ is an average of $X(0)$, $X(\Delta t)$, ..., $X((n-1)\Delta t)$, and $X_a(\Delta t_a)$ is an average of $X(n\Delta t)$, $X((n+1)\Delta t)$, ..., $X((2n-1)\Delta t)$, and so on,

$$X_a(0) = \frac{1}{n}\sum_{i=0}^{n-1} X(i\Delta t)$$

$$X_a(\Delta t_a) = \frac{1}{n}\sum_{i=n}^{2n-1} X(i\Delta t)$$

$$\vdots$$

where $\Delta t_a = n\Delta t$. Because averaging preserves the mean, the mean of both sequences is identical. However, if σ^2 is the variance of the sequence $X(t)$ and σ_a^2 is the variance of the sequence $X_a(t)$, then classical statistics tells us that the average of n independent observations will have variance

$$\sigma_a^2 = \frac{\sigma^2}{n} \tag{3.53}$$

Noting that $n = \Delta t_a/\Delta t$, Eq. 3.53 can be reexpressed as

$$\sigma_a^2 \Delta t_a = \sigma^2 \Delta t = \pi G_o \tag{3.54}$$

That is, the product $\sigma^2\Delta t$ is a constant which we will set equal to πG_o, where G_o is the *white noise intensity*. The factor of π arises here so that we can let the white noise spectral density function $G(\omega)$ equal G_o, as we shall see shortly. Equation 3.54 can also be rearranged to give the variance of local averages of white noise in terms of the white noise intensity,

$$\sigma^2 = \frac{\pi G_o}{\Delta t} \tag{3.55}$$

For ideal white noise, Δt goes to zero so that σ^2 goes to infinity. Another way of understanding why the variance of white noise must be infinite is to reconsider Eq. 3.53. For the continuous white noise case, any interval Δt will consist of an infinite number of independent random variables ($n = \infty$). Thus, if the white noise variance σ^2 were finite, then $\sigma_a^2 = \sigma^2/n$ would be zero for any nonzero averaging region. That is, a white noise having finite variance would appear, at all practical averaging resolutions, to be a deterministic constant equal to the mean.

As the name suggests, white noise has spectral density function which is constant, implying equal power in all frequencies (and hence the analogy with "white" light), as shown in Figure 3.11,

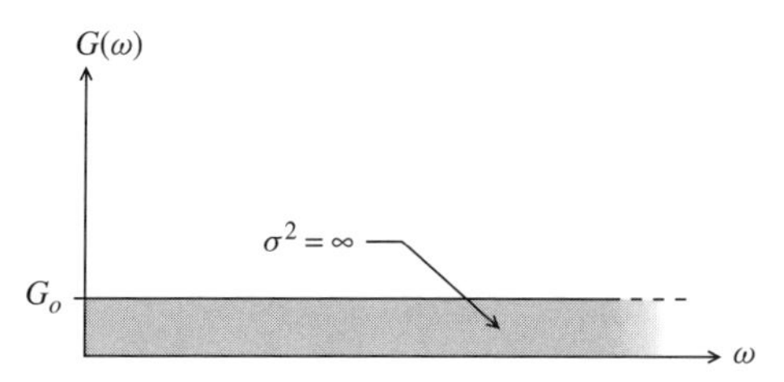

Figure 3.11 One-sided spectral density function for white noise.

$$G(\omega) = G_o \tag{3.56}$$

The primary, and attractive, feature of a white noise random process is that all points in the field are uncorrelated,

$$\rho(\tau) = \begin{cases} 1 & \text{if } \tau = 0 \\ 0 & \text{otherwise} \end{cases} \tag{3.57}$$

If the random field is also Gaussian, then all points are also independent, which makes probability calculations easier. White noise is often used as input to systems to simplify the computation of probabilities relating to the system response.

The covariance function corresponding to white noise is

$$C(\tau) = \pi G_o \delta(\tau) \tag{3.58}$$

where $\delta(\tau)$ is *Dirac delta function*, which is zero everywhere except at $\tau = 0$, where it assumes *infinite height*, *zero width*, but *unit area*. The Dirac delta function has the following useful property in integrals:

$$\int_{-\infty}^{\infty} f(x)\delta(x-a)\,dx = f(a)$$

That is, the delta function acts to extract a single value of the integrand at the point where the delta function argument becomes zero. We can use this property to test if Eq. 3.58 is in fact the covariance function corresponding to white noise, since we know that white noise should have constant spectrum, Eq. 3.56. Considering Eq. 3.25b,

$$\begin{aligned} G(\omega) &= \frac{1}{\pi}\int_{-\infty}^{\infty} C(\tau)\cos(\omega\tau)\,d\tau \\ &= \frac{\pi G_o}{\pi}\int_{-\infty}^{\infty} \delta(\tau)\cos(\omega\tau)\,d\tau \\ &= G_o\cos(0) \\ &= G_o \end{aligned}$$

as expected. This test also illustrates why the constant π appears in Eq. 3.58. We could not directly use the one-sided Eq. 3.25c in the above test, since the doubling of the area from Eq. 3.25b assumes only a vanishingly small contribution from $C(\tau)$ at $\tau = 0$, which is not the case for white noise. To double the contribution of $C(\tau)$ at $\tau = 0$

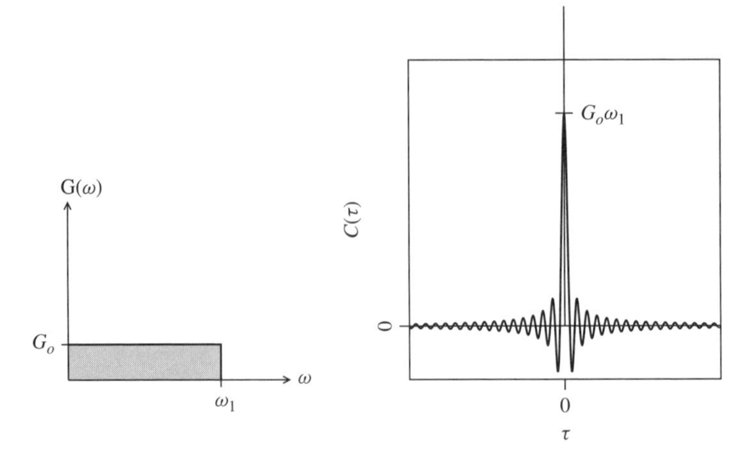

Figure 3.12 One-sided spectral density function and corresponding covariance function of band-limited white noise.

would be an error (which is one example of why white noise can be mathematically difficult).

The troublesome thing about white noise is that the area under the spectral density function is infinite,

$$\sigma^2 = \int_0^\infty G(\omega)\, d\omega = \int_0^\infty G_o\, d\omega = \infty$$

so that the process has infinite variance. Ideal white noise is "infinitely rough," which is *physically unrealizable*. For problems where a continuous white noise process must actually be simulated, it is usually a band-limited form of the white noise that is actually employed. The *band-limited white noise* has a flat spectral density function which is truncated at some upper frequency, ω_1,

$$G(\omega) = \begin{cases} G_o & \text{for } 0 \le \omega \le \omega_1 \\ 0 & \text{otherwise} \end{cases} \tag{3.59}$$

where G_o is some intensity constant. In this case, the variance of the process is finite and equal to $G_o\omega_1$. The covariance and correlation functions corresponding to band-limited white noise are

$$C(\tau) = G_o \frac{\sin \omega_1 \tau}{\tau} \tag{3.60a}$$

$$\rho(\tau) = \frac{\sin \omega_1 \tau}{\omega_1 \tau} \tag{3.60b}$$

Figure 3.12 illustrates the fact that, as $\omega_1 \to \infty$, $C(\tau)$ approaches the infinite-height Dirac delta function of Eq. 3.58.

The variance function can be obtained by integrating the correlation function, Eq. 3.60b (see also Eq. 3.42),

$$\gamma(T) = \frac{2}{\omega_1^2 T^2}\left[\omega_1 T \operatorname{Si}(\omega_1 T) + \cos(\omega_1 T) - 1\right] \tag{3.61}$$

where Si is the *sine integral*, defined by

$$\operatorname{Si}(\omega_1 T) = \int_0^{\omega_1 T} \frac{\sin t}{t}\, dt$$

See Abramowitz and Stegun (1970) for more details. For large $\omega_1 T$,

$$\gamma(T) \to \frac{\pi}{\omega_1 T} + \frac{2\cos(\omega_1 T)}{\omega_1^2 T^2}$$

since $\lim_{\omega_1 T \to \infty} \operatorname{Si}(\omega_1 T) \to \pi/2$.

The correlation length of band-limited white noise may be obtained by using Eq. 3.48. Since $G(0) = G_o$ and $\sigma^2 = G_o\omega_1$, we get

$$\theta = \frac{\pi G(0)}{\sigma^2} = \frac{\pi G_o}{G_o \omega_1} = \frac{\pi}{\omega_1}$$

3.6.2 Triangular Correlation Function

One of the simplest correlation functions is triangular, as illustrated in Figure 3.13,

$$\rho(\tau) = \begin{cases} 1 - |\tau|/\theta & \text{if } |\tau| \le \theta \\ 0 & \text{if } |\tau| > \theta \end{cases} \tag{3.62}$$

where θ is the correlation length.

One common process having a triangular correlation function is the moving average of white noise. Suppose that $W(t)$ is an ideal white noise process with intensity G_o (see the previous section) and we define

$$X(t) = \frac{1}{\theta}\int_{t-\theta/2}^{t+\theta/2} W(\xi)\, d\xi \tag{3.63}$$

to be a moving average of the white noise. Then $X(t)$ will be stationary with variance (we can take $\mu_X = \mu_W = 0$ and $t = \theta/2$ in the following for simplicity)

$$\sigma_X^2 = \mathrm{E}\left[X^2\right] = \mathrm{E}\left[\frac{1}{\theta^2}\int_0^\theta\int_0^\theta W(s)W(t)\,ds\,dt\right]$$
$$= \frac{1}{\theta^2}\int_0^\theta\int_0^\theta \mathrm{E}\left[W(s)W(t)\right]\,ds\,dt$$
$$= \frac{1}{\theta^2}\int_0^\theta\int_0^\theta \pi G_o\delta(t-s)\,ds\,dt$$
$$= \frac{\pi G_o}{\theta^2}\int_0^\theta 1\,dt$$
$$= \frac{\pi G_o}{\theta} \tag{3.64}$$

Alternatively, if σ_X^2 is known, we can use this to compute the required white noise intensity, $G_o = \theta\sigma_X^2/\pi$.

The covariance function of $X(t)$ is

$$C_X(\tau) = \begin{cases} \sigma_X^2(1 - |\tau|/\theta) & \text{if } |\tau| \le \theta \\ 0 & \text{if } |\tau| > \theta \end{cases} \tag{3.65}$$

The spectral density function of $X(t)$ is the spectral density of an average of white noise and so reflects the transfer function of a low-pass filter,

$$G_X(\omega) = G_o\left[\frac{\sin(\omega\theta/2)}{\omega\theta/2}\right]^2, \qquad \omega \ge 0 \tag{3.66}$$

where the filter transfer function amplitude is

$$|H(\omega)| = \frac{\sin(\omega\theta/2)}{\omega\theta/2}$$

Finally, the variance function of $X(t)$ is

$$\gamma(T) = \begin{cases} 1 - \dfrac{T}{3\theta} & \text{if } T \le \theta \\ \dfrac{\theta}{T}\left[1 - \dfrac{\theta}{3T}\right] & \text{if } T > \theta \end{cases} \tag{3.67}$$

3.6.3 Polynomial Decaying Correlation Function

A simple correlation function which may be useful if little is known about the characteristics of a random field's spatial variability is

$$\rho(\tau) = \frac{\theta^3}{(\theta+\tau)^3} \tag{3.68}$$

which has the variance function

$$\gamma(T) = \frac{\theta}{\theta + T} \tag{3.69}$$

This variance function has the correct theoretical limiting values, namely $\gamma(0) = 1$ and $\lim_{T\to\infty} \gamma(T) = \theta/T$.

The correlation function of Eq. 3.68 is compared to two other correlation functions in Figure 3.9.

3.6.4 Autoregressive Processes

A class of popular one-dimensional random fields are the *autoregressive* processes. These are simple to simulate and, because they derive from linear differential equations excited by white noise, represent a wide variety of engineering problems. Consider a first-order linear differential equation of the form discussed in Example 3.2,

$$c\frac{dX(t)}{dt} + \alpha X(t) = W(t) \tag{3.70}$$

where c and α are constants and $W(t)$ is an ideal white noise input with mean zero and intensity G_o. In physics, the steady-state solution, $X(t)$, to this equation is called the *Ornstein–Uhlenbeck* process, which is a classical Brownian motion problem.

The numerical *finite-difference* approximation to the derivative in Eq. 3.70 is

$$\frac{dX(t)}{dt} \simeq \frac{X(t+\Delta t) - X(t)}{\Delta t} \tag{3.71}$$

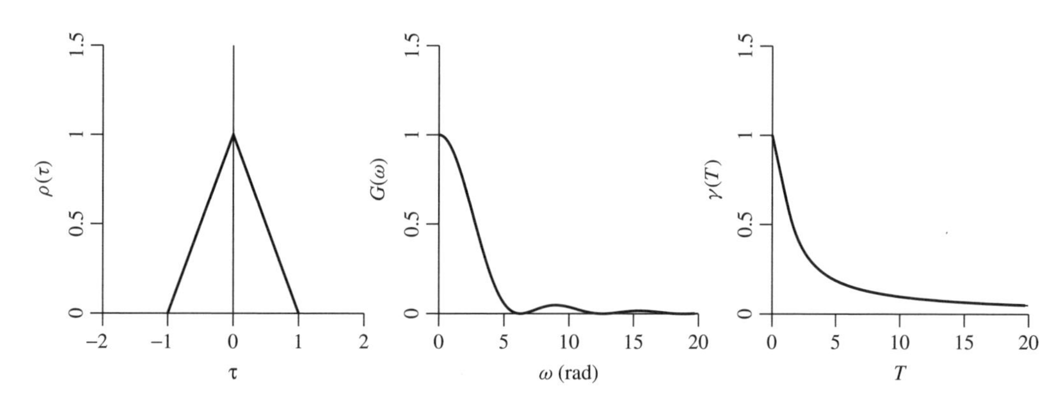

Figure 3.13 Triangular correlation function for $\theta = 1.0$, corresponding spectral density function $G(\omega)$ for $G_o = 1$, and variance function $\gamma(T)$.

If we let $\Delta t = 1$, then $dX(t)/dt \simeq X(t+1) - X(t)$ and Eq. 3.70 can be approximated by the finite-difference equation

$$c[X(t+1) - X(t)] + \alpha X(t) = W_b(t) \qquad (3.72)$$

where, since $X(t)$ is now a discrete process, $W_b(t)$ is a band-limited white noise process having constant intensity G_o up to the Nyquist frequency, $\pi/\Delta t$,

$$G_{W_b}(\omega) = \begin{cases} G_o & \text{if } 0 \le \omega \le \pi/\Delta t \\ 0 & \text{otherwise} \end{cases} \qquad (3.73)$$

Equation 3.72 can now be rearranged to allow the computation of the *future*, $X(t+1)$, given the present, $X(t)$, and the band-limited white noise input, $W_b(t)$,

$$X(t+1) = \frac{(c-\alpha)X(t) + W_b(t)}{c} = \left(\frac{c-\alpha}{c}\right)X(t) + \left(\frac{1}{c}\right)W_b(t) \qquad (3.74)$$

This is a *first-order autoregressive process* in which the future, $X(t+1)$, is expressed as a linear regression on the present, $X(t)$, with $W_b(t)$ playing the role of the regression error. We can simulate a first-order autoregressive process in one dimension using Eq. 3.74. We need only assume an initial value, $X(0)$, which can be taken to be the process mean. Subsequent values of X are obtained by generating a series of realizations of the random white noise, $W_b(0)$, $W_b(1), \ldots$, and then repeatedly applying Eq. 3.74,

$$X(1) = \left(1 - \frac{\alpha}{c}\right)X(0) + \left(\frac{1}{c}\right)W_b(0)$$

$$X(2) = \left(1 - \frac{\alpha}{c}\right)X(1) + \left(\frac{1}{c}\right)W_b(1)$$

$$\vdots$$

As indicated in Example 3.2, the transfer function corresponding to the continuous $X(t)$, Eq. 3.70, is

$$H(\omega) = \frac{1}{ic\omega + \alpha} \qquad (3.75)$$

so that the spectral density function corresponding to the solution of Eq. 3.70 is

$$G_X(\omega) = |H(\omega)|^2 G_W(\omega) = \frac{G_o}{c^2\omega^2 + \alpha^2} \qquad (3.76)$$

The covariance function of the continuous $X(t)$ can be obtained by using Eq. 3.25a, giving

$$C_X(\tau) = \sigma_X^2 e^{-\alpha|\tau|/c} \qquad (3.77)$$

where the variance σ_X^2 is the area under $G_X(w)$,

$$\sigma_X^2 = \int_0^\infty \frac{G_o}{c^2\omega^2 + \alpha^2}\, d\omega = \frac{\pi G_o}{2\alpha c} \qquad (3.78)$$

Note that Eq. 3.77 is a *Markov* correlation function, which will be covered in more detail in the next section.

Although Eq. 3.72, via Eq. 3.74, is popular as a means of simulating the response of a linear differential equation to white noise input, it is nevertheless only an approximation to its defining differential equation, Eq. 3.70. The approximation can be improved by taking Δt to be smaller; however, $\Delta t = 1$ is commonly used and so will be used here. Figures 3.14 and 3.15 compare the spectral density functions and covariance functions of the exact differential equation (Eq. 3.70) and its finite difference approximation

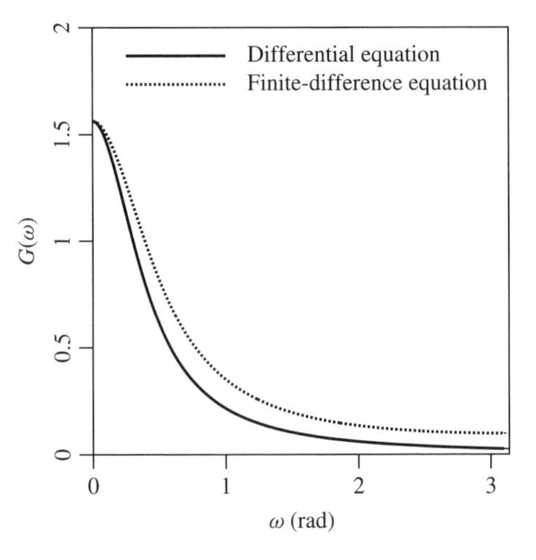

Figure 3.14 Comparison of spectral density functions of exact differential equation, Eq. 3.70, and its finite-difference approximation, Eq. 3.72, for $c = 2$, $\alpha = 0.8$, $G_o = 1$, and $\Delta t = 1$.

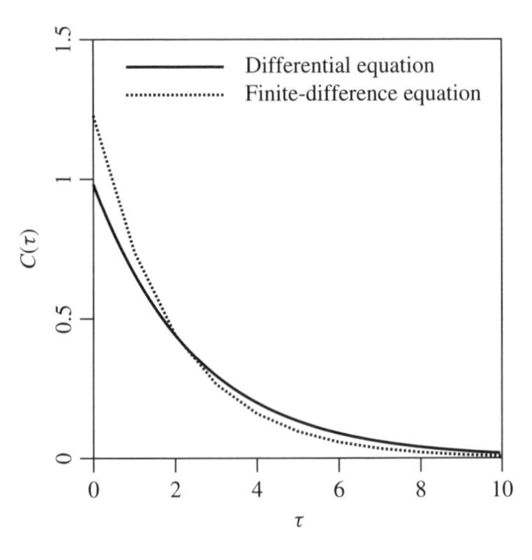

Figure 3.15 Comparison of covariance functions of exact differential equation, Eq. 3.70, and its finite-difference approximation, Eq. 3.72, for $c = 2$, $\alpha = 0.8$, $G_o = 1$, and $\Delta t = 1$.

(Eq. 3.72). As we shall see next, the mean and covariance structure of the discrete process (Eq. 3.74) can be found, so that the coefficients c and α can always be adjusted to get the desired discrete behavior.

It is informative to compare the second-moment characteristics of the differential equation and its finite-difference approximation. So long as $\mathrm{E}[W(t)] = \mathrm{E}[W_b(t)] = 0$, the mean (first moment) of both the differential equation response and the finite difference response is zero.

The actual spectral density function of Eq. 3.72 can be obtained in a number of ways, but one approach is to first obtain its transfer function. Letting $W_b(t) = e^{i\omega t}$ and the steady-state response $X(t) = H(\omega)W_b(t) = H(\omega)e^{i\omega t}$, Eq. 3.72 becomes

$$c\left[H(\omega)e^{i\omega(t+1)} - H(\omega)e^{i\omega t}\right] + \alpha H(\omega)e^{i\omega t} = e^{i\omega t}$$

which we can solve for $H(\omega)$,

$$H(\omega) = \frac{1}{\alpha + c(e^{i\omega} - 1)} = \frac{1}{-(c-\alpha) + ce^{i\omega}} \tag{3.79}$$

The squared magnitude of $H(\omega)$ is

$$|H(\omega)|^2 = \frac{1}{(c-\alpha)^2 - 2c(c-\alpha)\cos\omega + c^2} \tag{3.80}$$

The spectral density function of Eq. 3.72 is therefore (for $\Delta t = 1$)

$$\begin{aligned} G_X(\omega) &= |H(\omega)|^2 G_{W_b}(\omega) \\ &= \frac{G_o}{(\alpha - c)^2 + 2c(\alpha - c)\cos\omega + c^2}, \\ &\qquad 0 \le \omega \le \pi \end{aligned} \tag{3.81}$$

Note that these results assume that a steady state exists for the response, $X(t)$. The system will reach a steady state if $\alpha < c$ and we will assume this to be the case.

The variance of the approximate discrete process, Eq. 3.72, is

$$\begin{aligned} \sigma_X^2 &= \int_0^\pi \frac{G_o}{(\alpha - c)^2 + 2c(\alpha - c)\cos\omega + c^2}\, d\omega \\ &= \frac{\pi G_o}{2\alpha c - \alpha^2} \end{aligned} \tag{3.82}$$

Note that the integral has been truncated at $\omega_N = \pi/\Delta t = \pi$ because Eq. 3.72 is a discrete process with $\Delta t = 1$. The covariance function and correlation functions are

$$C(\tau) = \frac{\pi G_o}{2\alpha c - \alpha^2}\left(\frac{c-\alpha}{c}\right)^{|\tau|} \tag{3.83}$$

$$\rho(\tau) = \left(\frac{c-\alpha}{c}\right)^{|\tau|} \tag{3.84}$$

for $|\tau| = 0, 1, \ldots$ and $\alpha < c$.

Autoregressive models can be extended to higher order processes. Consider, for example, the second-order differential equation

$$\frac{d^2X(t)}{dt^2} + \alpha\frac{dX(t)}{dt} + \beta X(t) = W(t) \tag{3.85}$$

The spectral density function of $X(t)$ can be found by setting

$$\begin{aligned} W(t) &= e^{i\omega t} \\ X(t) &= H(\omega)e^{i\omega t} \\ \dot{X}(t) &= H(\omega)i\omega e^{i\omega t} \\ \ddot{X}(t) &= -H(\omega)\omega^2 e^{i\omega t} \end{aligned}$$

where the overdots indicate differentiation with respect to time. Substituting these into Eq. 3.85 gives

$$H(\omega)e^{i\omega t}[-\omega^2 + i\alpha\omega + \beta] = e^{i\omega t}$$

which yields

$$H(\omega) = \frac{1}{(\beta - \omega^2) + i\alpha\omega} \tag{3.86}$$

The spectral density function corresponding to Eq. 3.85 is thus

$$G_X(\omega) = |H(\omega)|^2 G_W(\omega) = \frac{G_o}{(\beta - \omega^2)^2 + \alpha^2\omega^2} \tag{3.87}$$

Making use of the following numerical approximations to the derivatives,

$$\begin{aligned} \frac{d^2X(t)}{dt^2} &\simeq \frac{X(t+\Delta t) - 2X(t) + X(t-\Delta t)}{\Delta t^2} \\ \frac{dX(t)}{dt} &\simeq \frac{X(t+\Delta t) - X(t-\Delta t)}{2\Delta t} \end{aligned}$$

where we used the more accurate central difference approximation for the first derivative, allows Eq. 3.85 to be approximated (and simulated) as the regression

$$X(t+1) = a_1 X(t) + a_2 X(t-1) + \epsilon(t)$$

where

$$\begin{aligned} a_1 &= \frac{2-\beta}{1+\alpha/2} \\ a_2 &= -\frac{1-\alpha/2}{1+\alpha/2} \\ \epsilon(t) &= \frac{W_b(t)}{1+\alpha/2} \end{aligned}$$

The latter means that $\epsilon(t)$ is a band-limited white noise process, from $\omega = 0$ to $\omega = \pi$, having intensity $G_o/(1 + \alpha/2)^2$.

Because higher dimensions do not have a well-defined "direction" (e.g., future), the autoregressive processes are not commonly used in two and higher dimensions.

3.6.5 Markov Correlation Function

The Markov correlation function is very commonly used because of its simplicity. Part of its simplicity is due to the fact that it renders a process where the "future" is dependent only on the "present" and not on the past. Engineering models which depend on the entire past history are relatively rare, but creep strains in concrete and masonry are one example. Most engineering models, however, allow the future to be predicted given only knowledge of the present state, and so the Markov property is quite applicable to such models. In terms of probabilities, the Markov property states that the conditional probability of the future state depends only on the current state (see Chapter 2), that is,

$$\begin{aligned} &\mathrm{P}\left[X(t_{n+1}) \le x \mid X(t_n), X(t_{n-1}), X(t_{n-2}, \ldots\right] \\ &\quad = \mathrm{P}\left[X(t_{n+1}) \le x \mid X(t_n)\right] \end{aligned}$$

which generally leads to simplified probabilistic models. More generally, the Markov property states that the future depends only on the most recently known state. So, for example, if we want to know a conditional probability relating to $X(t_{n+1})$ and we only know $X(t_{n-3}), X(t_{n-4}), \ldots$, then

$$\begin{aligned} &\mathrm{P}\left[X(t_{n+1}) \le x \mid X(t_{n-3}), X(t_{n-4}), \ldots\right] \\ &\quad = \mathrm{P}\left[X(t_{n+1}) \le x \mid X(t_{n-3})\right] \end{aligned}$$

The Markov correlation function has the form

$$\rho(\tau) = \exp\left\{-\frac{2|\tau|}{\theta}\right\} \tag{3.88}$$

where θ is the correlation length. This correlation function governs the solution to the first-order differential equation 3.70, the Ornstein–Uhlenbeck process. The parameter θ can be interpreted as the separation distance beyond which the random field is largely uncorrelated. For example, Eq. 3.88 says that when two points in the field are separated by $\tau = \theta$, their correlation has dropped to $e^{-2} = 0.13$.

The Markov process has variance function

$$\gamma(T) = \frac{\theta^2}{2T^2}\left[\frac{2|T|}{\theta} + \exp\left\{-\frac{2|T|}{\theta}\right\} - 1\right] \tag{3.89}$$

and "one-sided" spectral density function

$$G(\omega) = \frac{\sigma^2\theta}{\pi\left[1 + (\theta\omega/2)^2\right]} \tag{3.90}$$

which are illustrated in Figure 3.16. Although simple, the Markov correlation function is not *mean square differentiable*, which means that its derivative is discontinuous and infinitely variable, a matter which is discussed in more detail in Chapter 4. The lack of a finite variance derivative tends to complicate some things, such as the computation of level excursion statistics.

3.6.6 Gaussian Correlation Function

If a random process $X(t)$ has a *Gaussian* correlation function, then its correlation function has the form

$$\rho(\tau) = \exp\left\{-\pi\left(\frac{\tau}{\theta}\right)^2\right\} \tag{3.91}$$

where θ is the correlation length. The corresponding variance function is

$$\gamma(T) = \frac{\theta^2}{\pi T^2}\left[\frac{\pi|T|}{\theta}\operatorname{erf}\left\{\frac{\sqrt{\pi}|T|}{\theta}\right\} + \exp\left\{-\frac{\pi T^2}{\theta^2}\right\} - 1\right] \tag{3.92}$$

where $\operatorname{erf}(x) = 2\Phi(\sqrt{2}x) - 1$ is the error function and $\Phi(z)$ is the standard normal cumulative distribution function. The

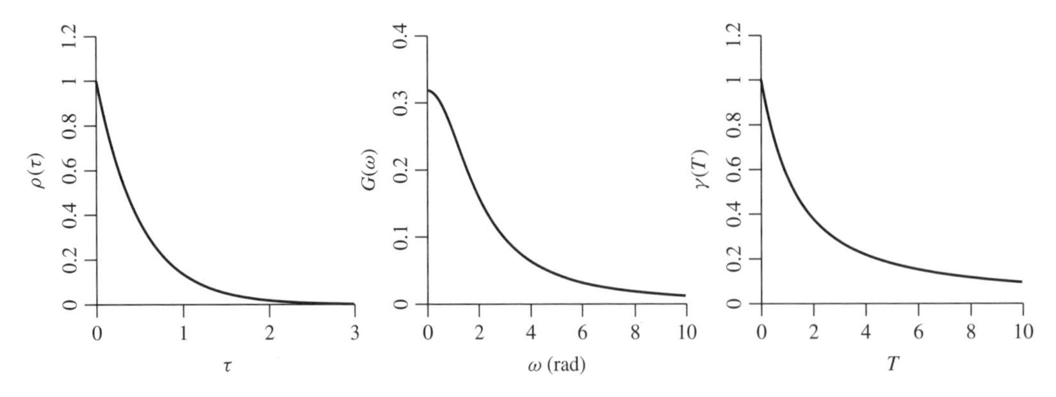

Figure 3.16 Markov correlation function for $\theta = 1.0$, corresponding spectral density function $G(\omega)$ for $\sigma_X = 1$, and variance function $\gamma(T)$.

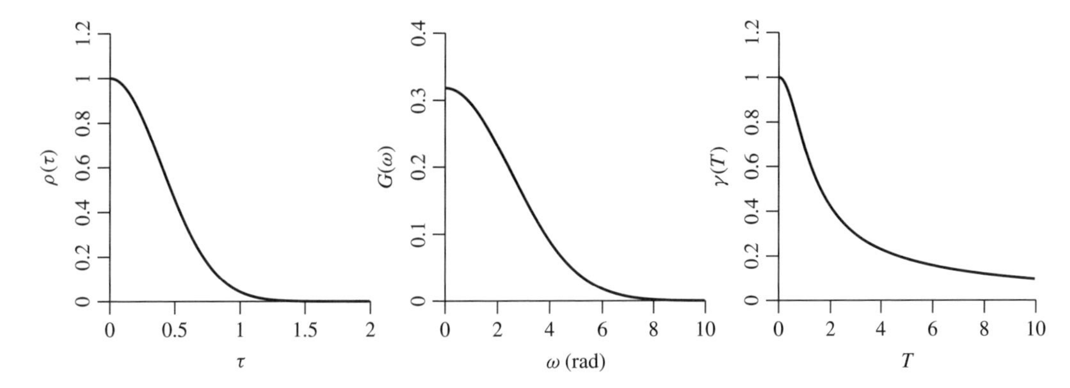

Figure 3.17 Gaussian correlation function for $\theta = 1.0$, corresponding spectral density function $G(\omega)$ for $\sigma_X = 1$, and variance function $\gamma(T)$.

spectral density function is exponentially decaying,

$$G(\omega) = \sigma_X^2 \left(\frac{\theta}{\pi}\right) \exp\left\{-\frac{\theta^2\omega^2}{4\pi}\right\} \tag{3.93}$$

as illustrated in Figure 3.17.

One advantage, at least mathematically, to the Gaussian correlation function is that it is mean square differentiable. That is, its derivative has finite variance and so level excursion statistics are more easily computed, as will be seen in Chapter 4. Mean square differentiable processes have correlation function with slope zero at the origin, and we can see that for this process $\rho(\tau)$ flattens out at the origin. From the point of view of simulation, one potential disadvantage to the Gaussian correlation function is that at larger correlation lengths the correlation between nearby points can become very close to 1 and so difficult to deal with numerically. If any off-diagonal value becomes 1.0, the correlation matrix loses its positive definiteness. A correlation matrix with all 1's off diagonal becomes singular. So, although the zero slope at $\tau = 0$ leads to mean square differentiable processes, it can also lead to numerical difficulties in simulation for large correlation lengths.

3.6.7 Fractal Processes

A random-field model which has gained some acceptance in a wide variety of applications is the *fractal model*, also known as statistically *self-similar*, *long memory*, or *1/f noise*. This model has an infinite correlation length and correlations remain high over very large distances. An example of such a process is shown in Figure 3.18.

Notice, in Figure 3.18, that the samples remain statistically similar, regardless of viewing resolution, under suitable scaling of the vertical axis. Such processes are often described by the (one-sided) spectral density function

$$G(\omega) = \frac{G_o}{\omega^\gamma} \tag{3.94}$$

in which the parameter γ controls how the spectral power is partitioned from the low to the high frequencies and G_o can be viewed as a spectral intensity (white noise intensity when $\gamma = 0$). In particular, the case where $0 \le \gamma < 1$ corresponds to infinite high-frequency power and results in a stationary random process called fractional Gaussian noise (Mandelbrot and van Ness, 1968), assuming a normal marginal distribution. When $\gamma > 1$, the spectral density falls off more rapidly at high frequencies, but grows more rapidly at low frequencies so that the infinite power is now in the low frequencies. This then corresponds to a nonstationary random process called fractional Brownian motion. Both cases are infinite-variance processes which are physically unrealizable. Their spectral densities must be truncated in some fashion to render them stationary with finite variance.

Self-similarity for fractional Gaussian noise is expressed by saying that the process $X(z)$ has the same distribution as the scaled process $a^{1-H}X(az)$ for some $a > 0$ and some H lying between 0.5 and 1. Alternatively, self-similarity for fractional Brownian motion means that $X(z)$ has the same distribution as $a^{-H}X(az)$, where the different exponent on a is due to the fact that fractional Gaussian noise is the derivative of fractional Brownian motion. Figure 3.18 shows a realization of fractional Gaussian noise with $H = 0.95$ produced using the local average subdivision method (Fenton, 1990). The uppermost plot is of length $n = 65{,}536$. Each plot in Figure 3.18 zooms in by a factor of $a = 8$, so that each lower plot has its vertical axis stretched by a factor of $8^{0.05} = 1.11$ to appear statistically similar to the next higher plot. The reason the scale expands as we zoom in is because less averaging is being performed. The variance is increasing without bound.

Probably the best way to envisage the spectral density interpretation of a random process is to think of the random process as being composed of a number of sinusoids each with random amplitude (power). The fractal model is saying

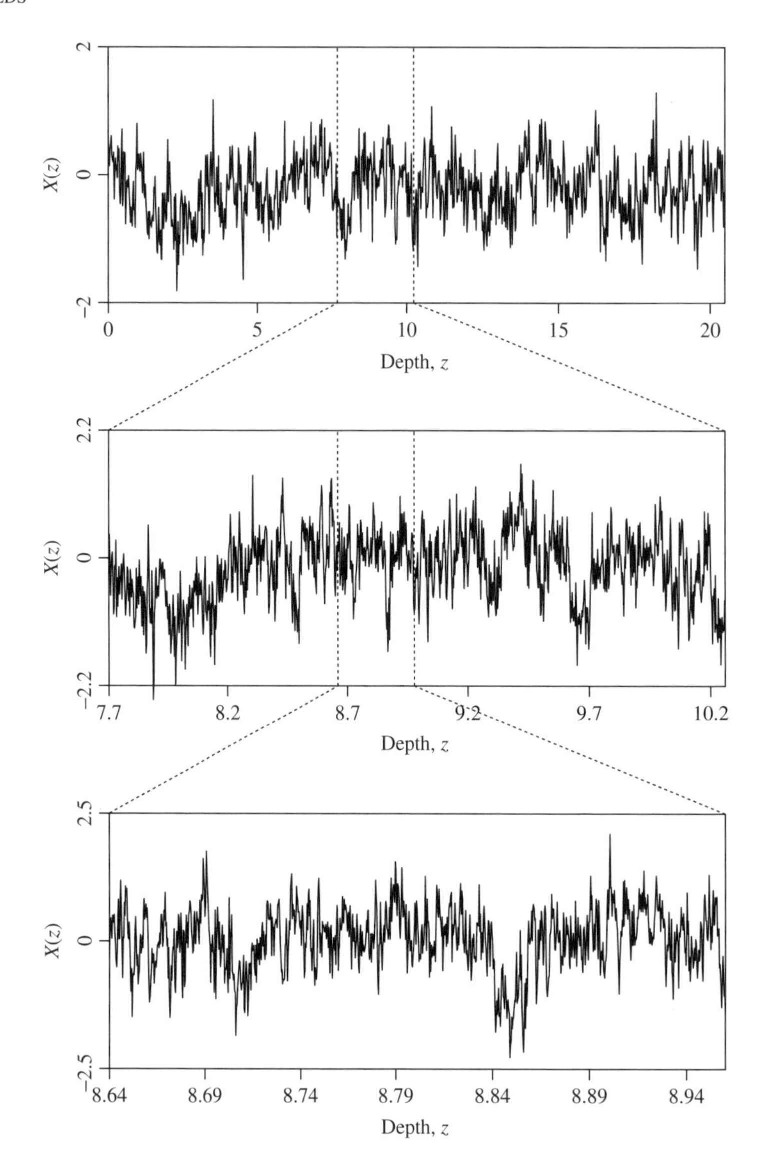

Figure 3.18 Example of a fractal process (fractional Gaussian noise with $H = 0.95$) at three resolutions.

that these random processes are made up of high-amplitude long-wavelength (low-frequency) sinusoids added to successively less powerful short-wavelength sinusoids. The long-wavelength components provide for what are seen as *trends* when viewed over a finite interval. As one "zooms" out and views progressively more of the random process, even longer wavelength (scale) sinusoids become apparent. Conversely, as one zooms in, the short-wavelength components dominate the (local) picture. This is the nature of self-similarity attributed to fractal processes—realizations of the process look the same (statistically) at any viewing scale.

By locally averaging the fractional Gaussian noise ($0 < \gamma < 1$) process over some distance δ, Mandelbrot and van Ness (1968) render fractional Gaussian noise (fGn) physically realizable (i.e., having finite variance). The resulting correlation function is

$$\rho(\tau) = \frac{1}{2\delta^{2H}}\left[|\tau + \delta|^{2H} - 2|\tau|^{2H} + |\tau - \delta|^{2H}\right] \quad (3.95)$$

where $H = \frac{1}{2}(\gamma + 1)$ is called the Hurst or self-similarity coefficient with $\frac{1}{2} \leq H < 1$. The case $H = \frac{1}{2}$ gives white noise, while $H = 1$ corresponds to perfect correlation [all $X(z) = X$ in the stationary case]. The spectral density function corresponding to fractional Gaussian noise is approximately (Mandelbrot and van Ness, 1968)

$$G(\omega) = \frac{G_o}{\omega^{2H-1}} \quad (3.96)$$

where

$$G_o = \frac{\sigma_X^2 H(2H-1)(2\pi\delta)^{2-2H}}{\Gamma(2-2H)\cos[\pi(1-H)]} \tag{3.97}$$

which is valid for small $\delta\omega$ and where $\Gamma(x)$ is the gamma function tabulated in, for example, Abramowitz and Stegun (1970). If we know the spectral density function, Eq. 3.97 can be inverted to determine the process variance

$$\sigma_X^2 = \frac{G_o\Gamma(2-2H)\cos[\pi(1-H)]}{H(2H-1)(2\pi\delta)^{2-2H}}$$

which goes to infinity as the local averaging distance δ goes to zero, as expected for a fractal process. Local averaging is effectively a low-pass filter, damping out high-frequency contributions, so that Mandelbrot's approach essentially truncates the spectral density function at the high end. Both the tail behavior of the spectral density function and the variance of the process thus depends on the choice of δ, which makes it a quite important parameter even though it is largely ignored in the literature (it is generally taken to equal 1 arbitrarily). Because of the local averaging, Eq. 3.94 can only be considered approximate for fractional Gaussian noise, the accuracy improving as $\delta \to 0$.

The variance function corresponding to fractional Gaussian noise is given by

$$\gamma(T) = \frac{|T+\delta|^{2H+2} - 2|T|^{2H+2} + |T-\delta|^{2H+2} - 2\delta^{2H+2}}{T^2(2H+1)(2H+2)\delta^{2H}} \tag{3.98}$$

Because the fractional Gaussian noise has, for $\delta \to 0$, an infinite variance, its use in practice is limited (any desired variance can be obtained simply by modifying δ). The nature of the process is critically dependent on H and δ, and these parameters are quite difficult to estimate from real data (for δ we need to know the behavior at the microscale while for H we need to know the behavior at the macroscale).

Notice in Figure 3.19 that the correlation function remains very high (and, hence, so does the variance function since highly correlated random variables do not provide much variance reduction when averaged). This is one of the main features of fractal processes and one of the reasons they are also called long-memory processes.

3.7 RANDOM FIELDS IN HIGHER DIMENSIONS

Figure 3.20 illustrates a two-dimensional random field $X(t_1, t_2)$ where X varies randomly in two directions, rather than just along a line. The elevation of a soil's surface and the thickness of a soil layer at any point on the plan area of a site are examples of two-dimensional random fields. The cohesion of the soil at plan location (t_1, t_2) and depth t_3 is an example of a three-dimensional random field $X(t_1, t_2, t_3)$. The coordinate labels t_1, t_2, and t_3 are often replaced by the more common Cartesian coordinates x, y, and z. We shall keep the current notation to remain consistent with that developed in the one-dimensional case.

In this section, we will concentrate predominately on two-dimensional random fields, the three-dimensional case generally just involving adding another coordinate. As in the one-dimensional case, a random field is characterized by the following:

1. Its first moment, or mean, $\mu(t_1, t_2)$, which may vary in space. If the random field is *stationary*, then the mean does not change with position; $\mu(t_1, t_2) = \mu$.
2. Its second moment, or covariance structure, $C(t_1', t_1^*, t_2', t_2^*)$, which gives the covariance between two points in the field, $X(t_1', t_2')$ and $X(t_1^*, t_2^*)$. If the field is stationary, then the covariance structure remains the same regardless of where the axis origin is located, that is, the covariance function becomes a function of just the difference, $(\mathbf{t}' - \mathbf{t}^*)$, that is, $C(t_1' - t_1^*, t_2' - t_2^*)$.
3. Its higher order moments. If the field is Gaussian, it is completely characterized by its first two moments.

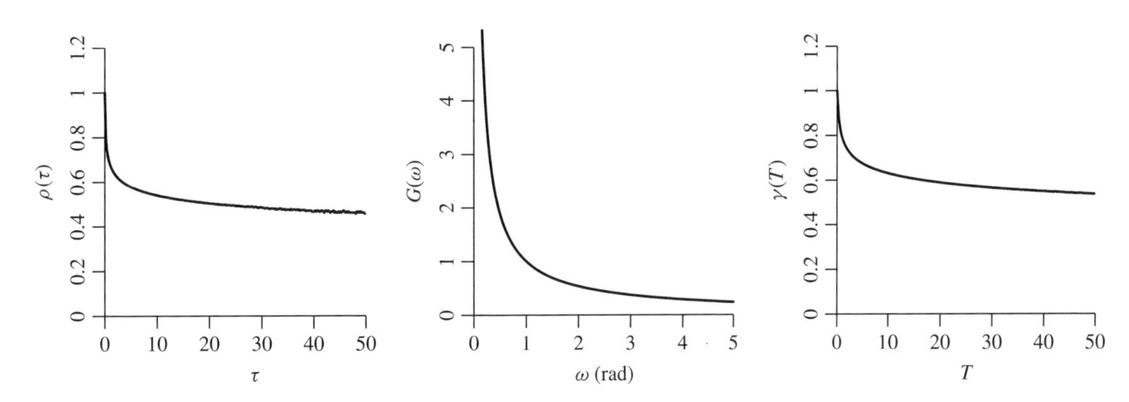

Figure 3.19 Correlation function, approximate spectral density function, and variance function for fractional Gaussian noise (with $H = 0.95$, $\delta = 0.1$).

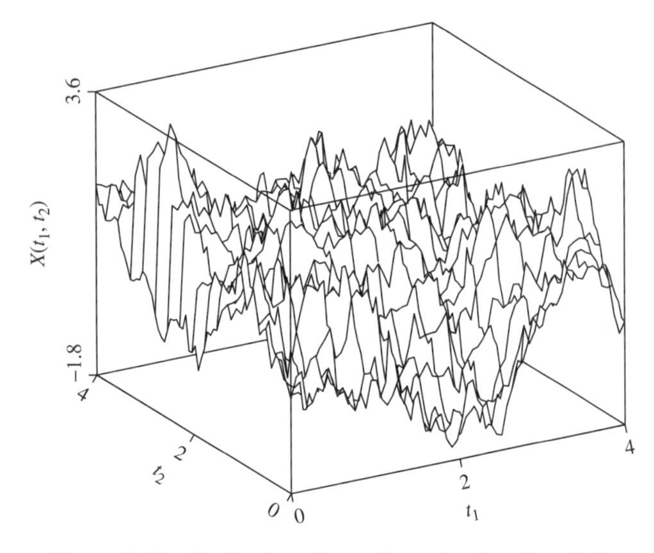

Figure 3.20 Realization of two-dimensional random field.

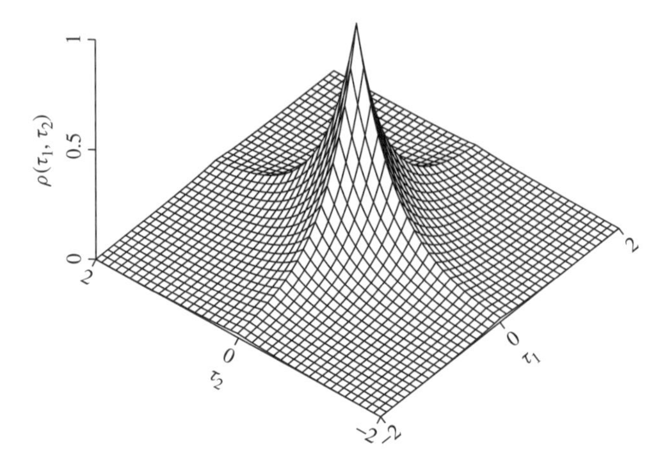

Figure 3.21 Two-dimensional correlation function $\rho(\tau_1, \tau_2)$ given by Eq. 3.102 for $\theta_1 = \theta_2 = 1$.

We will restrict our attention to just the first two moments of a random field. For simplicity, we will mostly concentrate on stationary random fields since any random field X' can be converted to a random field which is stationary in its mean and variance, X (with zero mean and unit variance), through the transformation

$$X(\mathbf{t}) = \frac{X'(\mathbf{t}) - \mu'(\mathbf{t})}{\sigma'(\mathbf{t})} \tag{3.99}$$

where $\mathbf{t}$ is a vector denoting spatial position (in two dimensions, $\mathbf{t}$ has components t_1 and t_2) and $\mu'(\mathbf{t})$ and $\sigma'(\mathbf{t})$ are the mean and standard deviation of X' at the spatial location $\mathbf{t}$.

In the following sections we investigate various ways that the second-moment characteristics of a random field can be expressed.

3.7.1 Covariance Function in Higher Dimensions

The covariance function gives the covariance between two points in the field, $X' = X(\mathbf{t}')$ and $X^* = X(\mathbf{t}^*)$. Since the covariance between X' and X^* is the same as the covariance between X^* and X' (i.e., it does not matter which way you look at the pair), then $C(t_1', t_1^*, t_2', t_2^*) = C(t_2', t_2^*, t_1', t_1^*)$. If the random field is stationary, this translates into the requirement that $C(\boldsymbol{\tau}) = C(-\boldsymbol{\tau})$, where $\boldsymbol{\tau} = \mathbf{t}' - \mathbf{t}^*$ is the spatial *lag* vector having components $\tau_1 = t_1' - t_1^*$, $\tau_2 = t_2' - t_2^*$. For example, for a two-dimensional stationary random field $C(t_1' - t_1^*, t_2' - t_2^*) = C(t_1^* - t_1', t_2^* - t_2')$, or $C(\tau_1, \tau_2) = C(-\tau_1, -\tau_2)$.

In two dimensions, the correlation function is defined as

$$\rho(\tau_1, \tau_2) = \frac{\text{Cov}\left[X', X^*\right]}{\sigma'\sigma^*} = \frac{C(\tau_1, \tau_2)}{\sigma'\sigma^*} \tag{3.100}$$

where σ' and σ^* are the standard deviations of $X' = X(\mathbf{t}')$ and $X^* = X(\mathbf{t}^*)$, respectively. Since we are assuming the random field is stationary, then $\sigma' = \sigma^* = \sigma$, and the correlation function becomes

$$\rho(\tau_1, \tau_2) = \frac{C(\tau_1, \tau_2)}{\sigma^2} \tag{3.101}$$

Figure 3.21 illustrates the two-dimensional correlation function

$$\begin{aligned}\rho(\tau_1, \tau_2) &= \exp\left\{-\frac{2}{\theta}(|\tau_1| + |\tau_2|)\right\} \\ &= \exp\left\{-\frac{2|\tau_1|}{\theta_1}\right\} \exp\left\{-\frac{2|\tau_2|}{\theta_2}\right\}\end{aligned} \tag{3.102}$$

which is Markovian in each coordinate direction. Note that even if the directional correlation lengths θ_1 and θ_2 are equal, this function is not isotropic, as seen Figure 3.21.

3.7.2 Spectral Density Function in Higher Dimensions

In two dimensions, the spectral representation of a stationary random field, $X(t_1, t_2)$, is the double sum

$$X(t_1, t_2) = \mu_X + \sum_{i=-N_1}^{N_1} \sum_{j=-N_2}^{N_2} C_{ij} \cos(\omega_{1_i} t_1 + \omega_{2_j} t_2 + \Phi_{ij}) \tag{3.103}$$

where, as in the one-dimensional case, C_{ij} is a random amplitude and Φ_{ij} a random phase angle. The variance of $X(t_1, t_2)$ is obtained by assuming the random variables C_{ij} and Φ_{ij} are all mutually independent,

$$\sigma_X^2 = \text{E}\left[(X(t_1, t_2) - \mu_X)^2\right] = \sum_{i=-N_1}^{N_1} \sum_{j=-N_2}^{N_2} \frac{1}{2}\,\text{E}\left[C_{ij}^2\right] \tag{3.104}$$

We define the two-dimensional spectral density function $S(\omega_1, \omega_2)$ such that

$$S(\omega_{1_i}, \omega_{2_j})\ \Delta\omega_1\ \Delta\omega_2 = \tfrac{1}{2}\ \mathrm{E}\left[C_{ij}^2\right] \tag{3.105}$$

Figure 3.22 illustrates a two-dimensional spectral density function. Note that if the correlation function is separable, as is Eq. 3.102, then both the spectral density and the variance functions will also be of separable form (although in the case of the spectral density function the variance does not appear more than once in the product). In the case of Figure 3.22 the spectral density function is obtained directly from Eq. 3.90 as

$$\begin{aligned} G(\omega_1, \omega_2) &= \frac{\sigma^2\theta_1\theta_2}{\pi^2\left[1 + (\theta_1\omega_1/2)^2\right]\left[1 + (\theta_2\omega_2/2)^2\right]} \\ &= 4S(\omega_1, \omega_2) \end{aligned} \tag{3.106}$$

In the limit as both $\Delta\omega_1$ and $\Delta\omega_2$ go to zero, we can express the variance of X as the volume under the spectral density function,

$$\sigma_X^2 = \int_{-\infty}^{\infty}\int_{-\infty}^{\infty} S(\omega_1, \omega_2)\ d\omega_1\ d\omega_2 \tag{3.107}$$

In the two-dimensional case, the Wiener–Khinchine relationships become

$$\begin{aligned} C(\tau_1, \tau_2) = \int_{-\infty}^{\infty}\int_{-\infty}^{\infty} & S(\omega_1, \omega_2) \\ & \times \cos(\omega_1\tau_1 + \omega_2\tau_2)\ d\omega_1\ d\omega_2 \end{aligned} \tag{3.108a}$$

$$\begin{aligned} S(\omega_1, \omega_2) = \frac{1}{(2\pi)^2}\int_{-\infty}^{\infty}\int_{-\infty}^{\infty} & C(\tau_1, \tau_2) \\ & \times \cos(\omega_1\tau_1 + \omega_2\tau_2)\ d\tau_1\ d\tau_2 \end{aligned} \tag{3.108b}$$

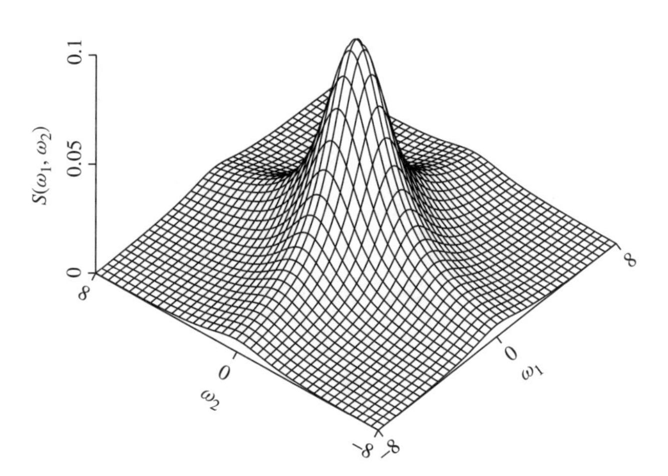

Figure 3.22 Two-dimensional spectral density function $S(\omega_1, \omega_2)$ corresponding to Eq. 3.102 for $\theta = 1$ and $\sigma = 1$.

If we express the components of spatial lag and frequency using vectors, $\boldsymbol{\tau} = \{\tau_1, \tau_2\}^{\mathrm{T}}$ and $\boldsymbol{\omega} = \{\omega_1, \omega_2\}^{\mathrm{T}}$, where superscript T denotes the transpose, then the Wiener–Khinchine relationships can be written for n dimensions succinctly as

$$C(\boldsymbol{\tau}) = \int_{-\infty}^{\infty} S(\boldsymbol{\omega})\cos(\boldsymbol{\omega}\cdot\boldsymbol{\tau})\ d\boldsymbol{\omega} \tag{3.109a}$$

$$S(\boldsymbol{\omega}) = \frac{1}{(2\pi)^n}\int_{-\infty}^{\infty} C(\boldsymbol{\tau})\cos(\boldsymbol{\omega}\cdot\boldsymbol{\tau})\ d\boldsymbol{\tau} \tag{3.109b}$$

where it is understood that we have a double integral for two dimensions, a triple integral for three dimensions, and so forth. The centered dot denotes the *vector dot product*, for example, $\boldsymbol{\omega}\cdot\boldsymbol{\tau} = \omega_1\tau_1 + \omega_2\tau_2$.

3.7.3 Variance Function in Higher Dimensions

In two dimensions, we can define the moving local average of a random field, $X(t_1, t_2)$, over an area of dimension $A = T_1 \times T_2$ to be

$$X_A(t_1, t_2) = \frac{1}{A}\int_{t_1 - T_1/2}^{t_1 + T_1/2}\int_{t_2 - T_2/2}^{t_2 + T_2/2} X(\xi_1, \xi_2)\ d\xi_2\ d\xi_1 \tag{3.110}$$

Figure 3.23 illustrates a moving local average field for $T_1 \times T_2 = 2 \times 2$. To determine the statistics of X_A, we will first assume that the random field $X(\mathbf{t})$ is stationary, so that we can choose to find the mean and variance of $X_A = X_A(T_1/2, T_2/2)$ as representative,

$$X_A = \frac{1}{A}\int_0^{T_1}\int_0^{T_2} X(t_1, t_2)\ dt_2\ dt_1 \tag{3.111}$$

The mean of X_A is

$$\mu_{X_A} = \frac{1}{A}\int_0^{T_1}\int_0^{T_2} \mathrm{E}\left[X(t_1, t_2)\right]\ dt_2\ dt_1 = \mu_X$$

Assuming that the random field $X(t_1, t_2)$ has "point" mean $\mu_X = 0$ and variance σ_X, then the variance of X_A is

$$\begin{aligned} \mathrm{Var}\left[X_A\right] &= \sigma_A^2 = \mathrm{E}\left[X_A^2\right] \\ &=^2 \frac{1}{A^2}\int_0^{T_1}\int_0^{T_1}\int_0^{T_2}\int_0^{T_2} \mathrm{E}\left[X(t_1, t_2)X(\xi_1, \xi_2)\right] \\ &\quad \times d\xi_2\ dt_2\ d\xi_1\ dt_1 \\ &= \frac{1}{A^2}\int_0^{T_1}\int_0^{T_1}\int_0^{T_2}\int_0^{T_2} \mathrm{Cov}\left[X(t_1, t_2), X(\xi_1, \xi_2)\right] \\ &\quad \times d\xi_2\ dt_2\ d\xi_1\ dt_1 \\ &= \frac{\sigma_X^2}{A^2}\int_0^{T_1}\int_0^{T_1}\int_0^{T_2}\int_0^{T_2} \rho(t_1 - \xi_1, t_2 - \xi_2) \\ &\quad \times d\xi_2\ dt_2\ d\xi_1\ dt_1 \end{aligned}$$

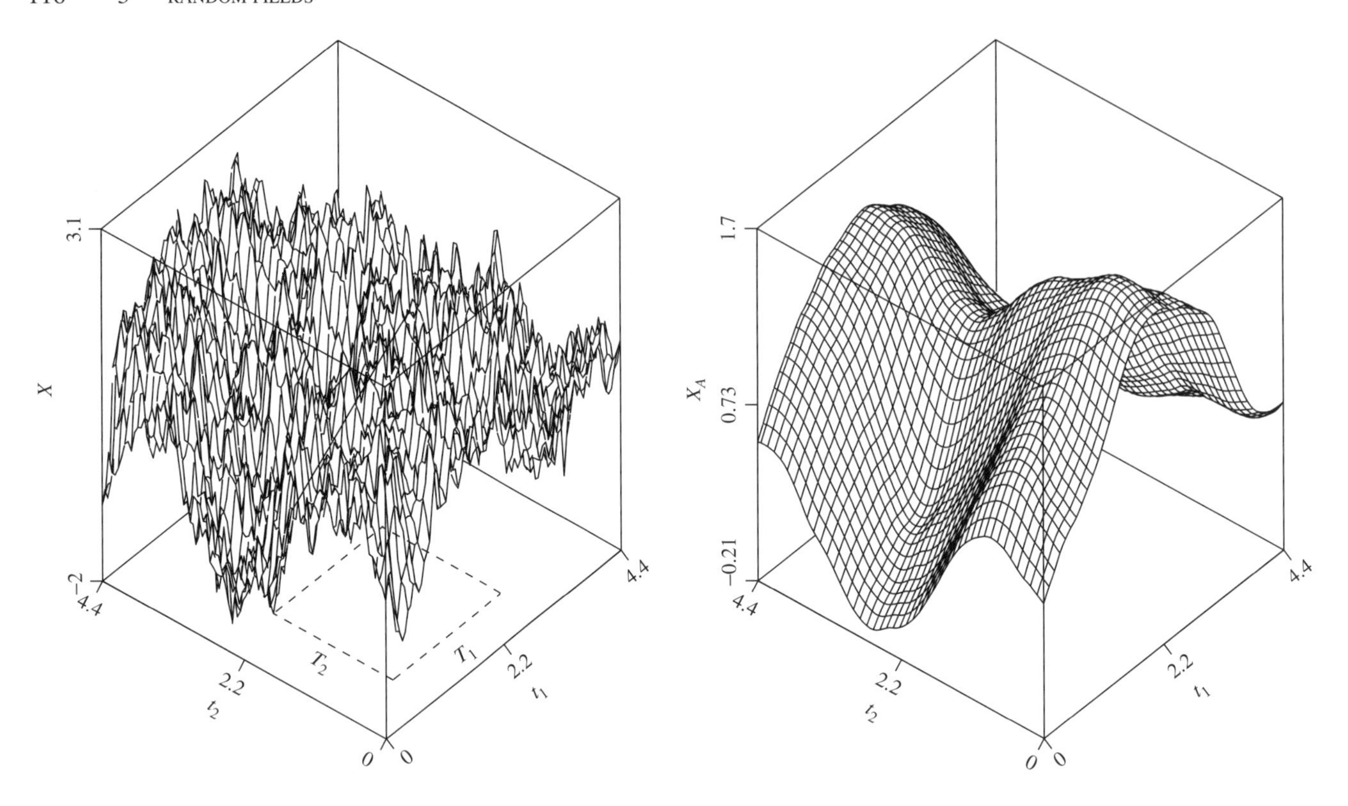

Figure 3.23 The field X_A on the right is a moving local average over a window of size $T_1 \times T_2$ of the field X on the left.

The same result would have been obtained even if $\mu_X \neq 0$ (at the expense of somewhat more complicated algebra).

Making use of the fact that, for stationary random fields, ρ is constant along diagonal lines where $t_1 - \xi_1$ and $t_2 - \xi_2$ are constant, we can reduce the fourfold integral to a double integral (see Eq. 3.41 and Figure 3.7), so that

$$\begin{aligned}\text{Var}\,[X_A] &= \frac{\sigma_X^2}{A^2}\int_{-T_1}^{T_1}\int_{-T_2}^{T_2}(|T_1| - |\tau_1|)(|T_2| - |\tau_2|) \\ &\quad\times \rho(\tau_1, \tau_2)\, d\tau_2\, d\tau_1 \\ &= \sigma_X^2 \gamma(T_1, T_2)\end{aligned}$$

where, since $A = T_1 T_2$, the variance function is defined by

$$\begin{aligned}\gamma(T_1, T_2) = \frac{1}{T_1^2 T_2^2}\int_{-T_1}^{T_1}\int_{-T_2}^{T_2}(|T_1| - |\tau_1|)(|T_2| - |\tau_2|) \\ \times \rho(\tau_1, \tau_2)\, d\tau_2\, d\tau_1 \qquad (3.112)\end{aligned}$$

Some additional simplification is possible if $\rho(\tau_1, \tau_2) = \rho(-\tau_1, \tau_2) = \rho(\tau_1, -\tau_2) = \rho(-\tau_1, -\tau_2)$ (this is called quadrant symmetry, which will be discussed shortly), in which case

$$\begin{aligned}\gamma(T_1, T_2) = \frac{4}{T_1^2 T_2^2}\int_0^{T_1}\int_0^{T_2}(|T_1| - \tau_1)(|T_2| - \tau_2) \\ \times \rho(\tau_1, \tau_2)\, d\tau_2\, d\tau_1 \qquad (3.113)\end{aligned}$$

The variance function corresponding to the separable Markov correlation function of Eq. 3.102 is shown in Figure 3.24. Although $\gamma(T_1, T_2)$ is perhaps questionably defined when T_1 or T_2 is negative, we shall assume that an averaging area of size -2×3 is the same as an averaging

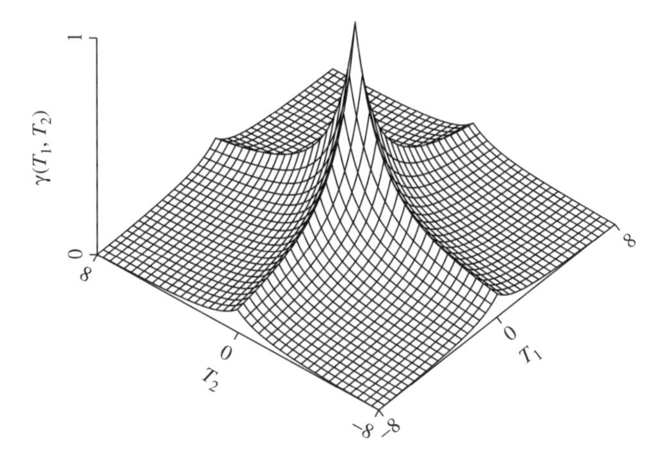

Figure 3.24 Two-dimensional variance function $\gamma(T_1, T_2)$ corresponding to Eq. 3.102 for $\theta = 1$.

area of size 2 × 3, the sign only arising because T_1 is measured in opposite directions. By this assumption, $\gamma(T_1, T_2)$ is automatically quadrant symmetric, as will be discussed next.

Figure 3.24 illustrates the separable two-dimensional variance function corresponding to Eq. 3.102, which is

$$\gamma(T_1, T_2) = \frac{\theta_1^2\theta_2^2}{4T_1^2T_2^2}\left[\frac{2|T_1|}{\theta_1} + \exp\left\{-\frac{2|T_1|}{\theta_1}\right\} - 1\right] \times \left[\frac{2|T_2|}{\theta_2} + \exp\left\{-\frac{2|T_2|}{\theta_2}\right\} - 1\right] \tag{3.114}$$

3.7.4 Quadrant Symmetric Correlation Structure

Figure 3.25 shows three points in a two-dimensional plane. If we say that $X^* = X(0, 0)$, then, when $X' = X(2, 4)$, the covariance between X^* and X' is $C(t_1' - t_1^*, t_2' - t_2^*) = C(2, 4)$. Alternatively, when $X' = X(-2, 4)$, the covariance between X^* and X' is $C(-2, 4)$. If these two covariances are equal, then we say that the random field is *quadrant symmetric* (Vanmarcke, 1984). Since also $C(\boldsymbol{\tau}) = C(-\boldsymbol{\tau})$, quadrant symmetry implies that $C(2, 4) = C(-2, 4) = C(-2, -4) = C(2, -4)$. One of the simplifications that arises from this condition is that we only need to know the covariances in the first quadrant ($t_1 \geq 0$ and $t_2 \geq 0$) in order to know the entire covariance structure. A quadrant-symmetric random process is also stationary, at least up to the second moment.

If the covariance function $C(\boldsymbol{\tau})$ is quadrant symmetric, then its spectral density function $S(\boldsymbol{\omega})$ will also be quadrant symmetric. In this case, we need only know the spectral power over the first quadrant and can define

$$G(\boldsymbol{\omega}) = 2^n S(\boldsymbol{\omega}), \qquad \boldsymbol{\omega} > 0 \tag{3.115}$$

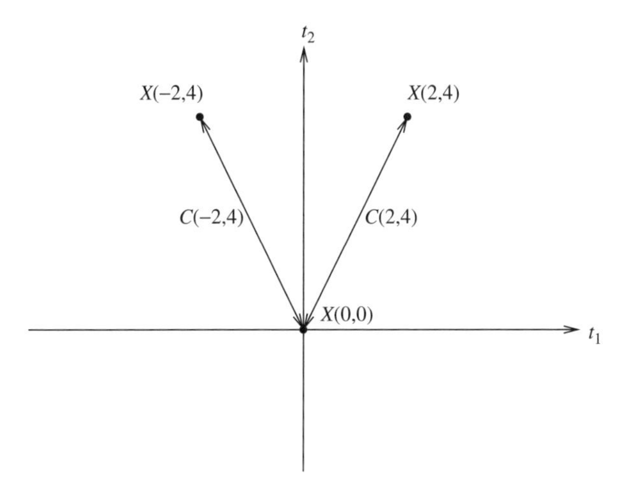

Figure 3.25 Three points on a plane and their covariances.

where n is the number of dimensions. For example, if $n = 2$, then $G(\boldsymbol{\omega})$ is defined as

$$G(\omega_1, \omega_2) = 4S(\omega_1, \omega_2) \tag{3.116}$$

and the two-dimensional Wiener–Khinchine relationships, defined in terms of the first quadrant only, become

$$C(\tau_1, \tau_2) = \int_0^\infty \int_0^\infty G(\omega_1, \omega_2)\cos\omega_1\tau_1 \times \cos\omega_2\tau_2 \, d\omega_1 \, d\omega_2 \tag{3.117a}$$

$$G(\omega_1, \omega_2) = \left(\frac{2}{\pi}\right)^2 \int_0^\infty \int_0^\infty C(\tau_1, \tau_2)\cos\omega_1\tau_1 \times \cos\,\omega_2\tau_2 \, d\tau_1 \, d\tau_2 \tag{3.117b}$$

and similarly in higher dimensions. We get Eq. 3.117b by starting with Eq. 3.108b,

$$\begin{aligned} S(\omega_1, \omega_2) &= \frac{1}{(2\pi)^2}\int_{-\infty}^\infty \int_{-\infty}^\infty C(\tau_1, \tau_2)\cos(\omega_1\tau_1 + \omega_2\tau_2)\, d\tau_1 \, d\tau_2 \\ &= \frac{1}{(2\pi)^2}\int_0^\infty \int_0^\infty \Big\{C(\tau_1, \tau_2)\Psi(\tau_1, \tau_2) \\ &\quad + C(\tau_1, -\tau_2)\Psi(\tau_1, -\tau_2) + C(-\tau_1, \tau_2)\Psi(-\tau_1, -\tau_2) \\ &\quad + C(-\tau_1, -\tau_2)\Psi(-\tau_1, -\tau_2)\Big\} d\tau_1 \, d\tau_2 \end{aligned}$$

where we introduced and used the short form $\Psi(\tau_1, \tau_2) = \cos(\omega_1\tau_1 + \omega_2\tau_2)$. Since C is quadrant symmetric, so that $C(\tau_1, \tau_2) = C(\tau_1, -\tau_2) = C(-\tau_1, \tau_2) = C(-\tau_1, -\tau_2)$, then the above simplifies to

$$\begin{aligned} S(\omega_1, \omega_2) &= \frac{1}{(2\pi)^2}\int_0^\infty \int_0^\infty C(\tau_1, \tau_2)\Big\{\Psi(\tau_1, \tau_2) + \Psi(\tau_1, -\tau_2) \\ &\quad + \Psi(-\tau_1, -\tau_2) + \Psi(-\tau_1, -\tau_2)\Big\}\, d\tau_1 \, d\tau_2 \\ &= \frac{4}{(2\pi)^2}\int_0^\infty \int_0^\infty C(\tau_1, \tau_2)\cos\omega_1\tau_1 \cos\omega_2\tau_2 d\tau_1 d\tau_2 \end{aligned}$$

In the last step we used the trigonometric identities relating to cosines of sums of angles to simplify the expression. Writing $G(\omega_1, \omega_2) = 2^2 S(\omega_1, \omega_2)$ gives us Eq. 3.117b.

The n-dimensional quadrant-symmetric Wiener–Khinchine relationships are

$$C(\boldsymbol{\tau}) = \int_0^\infty G(\boldsymbol{\omega})\cos\omega_1\tau_1 \cdots \cos\omega_n\tau_n \, d\boldsymbol{\tau}$$

$$G(\boldsymbol{\omega}) = \left(\frac{2}{\pi}\right)^n \int_0^\infty C(\boldsymbol{\tau})\cos\omega_1\tau_1 \cdots \cos\omega_n\tau_n \, d\boldsymbol{\omega}$$

Since the variance function $\gamma(T_1, T_2)$ is a function of $|T_1|$ and $|T_2|$, it is automatically quadrant symmetric.

3.7.5 Separable Correlation Structure

One of the simplest forms that the multidimensional correlation function can take is as the product of the directional one-dimensional correlation functions, that is,

$$\rho(t_1', t_1^*, t_2', t_2^*) = \rho_1(t_1', t_1^*)\rho_2(t_2', t_2^*) \tag{3.118}$$

If the random field is also stationary, then only the differences in position are important, so that the separable correlation function becomes

$$\rho(t_1', t_1^*, t_2', t_2^*) = \rho_1(t_1' - t_1^*)\rho_2(t_2' - t_2^*) = \rho_1(\tau_1)\rho_2(\tau_2) \tag{3.119}$$

Because $\rho(\tau) = \rho(-\tau)$, a separable correlation function is also quadrant symmetric and thus also at least second-moment stationary. That is, $\rho(\tau_1, \tau_2) = \rho(-\tau_1, \tau_2) = \rho(\tau_1, -\tau_2) = \rho(-\tau_1, -\tau_2)$. Figures 3.21, 3.22, and 3.24 are illustrations of a separable Markov process having $\theta_1 = \theta_2 = 1$ and $\sigma_X^2 = 1$. Clearly, the processes shown in Figures 3.21, 3.22, and 3.24 are not isotropic, even though their directional correlation lengths are equal. As we shall see in the next section, it is only when $\rho(\tau_1, \tau_2)$ can be written as a function of $\sqrt{\tau_1^2 + \tau_2^2}$ that we can have an isotropic correlation structure.

The covariance function corresponding to a separable process is

$$C(\tau_1, \tau_2) = \sigma^2\rho_1(\tau_1)\rho_2(\tau_2) \tag{3.120}$$

If the correlation structure is separable, then the spectral density and variance functions will also be separable. The variance function can be written as

$$\gamma(T_1, T_2) = \gamma_1(T_1)\gamma_2(T_2) \tag{3.121}$$

The separable spectral density must be written in terms of the product of the variance and unit-area (i.e., unit-variance) density functions,

$$G(\omega_1, \omega_2) = \sigma^2 g_1(\omega_1) g_2(\omega_2)$$

The unit-area spectral density functions $g_1(\omega_1)$ and $g_2(\omega_2)$ are analogous to the normalized correlation functions $\rho_1(\tau)$ and $\rho_2(\tau_2)$. That is, $g_1(\omega_1) = G_1(\omega_1)/\sigma^2$ and $g_2(\omega_2) = G_2(\omega_2)/\sigma^2$. They can also be defined by replacing $C(\tau)$ with $\rho(\tau)$ in the Wiener–Khinchine relationship,

$$g_1(\omega_1) = \frac{2}{\pi}\int_0^\infty \rho_1(\tau_1)\, \cos\omega_1\tau_1 \, d\tau_1 \tag{3.122}$$

Example 3.4 If the covariance function of a two-dimensional random field $X(t_1, t_2)$ is given by

$$C(\tau_1, \tau_2) = \sigma_X^2 \, \exp\left\{-2\left(\frac{|\tau_1|}{\theta_1} + \frac{|\tau_2|}{\theta_2}\right)\right\}$$

then what are the corresponding spectral density and variance functions?

SOLUTION We note that $C(\tau_1, \tau_2)$ can be written as

$$\begin{aligned} C(\tau_1, \tau_2) &= \sigma_X^2 \, \exp\left\{-2\frac{|\tau_1|}{\theta_1}\right\} \, \exp\left\{-2\frac{|\tau_2|}{\theta_2}\right\} \\ &= \sigma_X^2\rho_1(\tau_1)\rho_2(\tau_2) \end{aligned}$$

where

$$\rho_i(\tau_i) = \exp\left\{-\frac{2\tau_i}{\theta_i}\right\} \tag{3.123}$$

Evidently, the correlation structure is separable and each directional correlation function is Markovian (see Section 3.6.5). The spectral density function corresponding to a (directional) Markov process is given by Eq. 3.90,

$$G_i(\omega_i) = \frac{\sigma_i^2\theta_i}{\pi\left[1 + (\theta_i\omega_i/2)^2\right]}$$

so that the directional unit-area spectral density functions are obtained from $G_i(\omega_i)/\sigma_i^2$ as

$$g_1(\omega_1) = \frac{\theta_1}{\pi\left[1 + (\theta_1\omega_1/2)^2\right]}$$

$$g_2(\omega_2) = \frac{\theta_2}{\pi\left[1 + (\theta_2\omega_2/2)^2\right]}$$

The desired spectral density function is thus

$$\begin{aligned} G(\omega_1, \omega_2) &= \sigma_X^2 g_1(\omega_1) g_2(\omega_2) \\ &= \frac{\sigma_X^2\theta_1\theta_2}{\pi^2\left[1 + (\theta_1\omega_1/2)^2\right]\left[1 + (\theta_2\omega_2/2)^2\right]} \end{aligned}$$

The variance function corresponding to a (directional) Markov process is given by Eq. 3.89 as

$$\gamma_i(T_i) = \frac{\theta_i^2}{2T_i^2}\left[\frac{2|T_i|}{\theta_i} + \exp\left\{-\frac{2|T_i|}{\theta_i}\right\} - 1\right]$$

so that $\gamma(T_1, T_2) = \gamma_1(T_1)\gamma(T_2)$ is

$$\begin{aligned} \gamma(T_1, T_2) = \frac{\theta_1^2\theta_2^2}{4T_1^2T_2^2}&\left[\frac{2|T_1|}{\theta_1} + \exp\left\{-\frac{2|T_1|}{\theta_1}\right\} - 1\right] \\ \times &\left[\frac{2|T_2|}{\theta_2} + \exp\left\{-\frac{2|T_2|}{\theta_2}\right\} - 1\right] \end{aligned}$$

3.7.6 Isotropic Correlation Structure

If the correlation between two points depends only on the absolute distance between the points, and not on their orientation, then we say that the correlation structure is *isotropic*. In this case, the correlation between $X(1, 1)$ and $X(2, 1)$ is the same as the correlation coefficient between $X(1, 1)$ and any of $X(1, 2)$, $X(0, 1)$, and $X(1, 0)$ or, for that matter, any of the other points on the circle shown

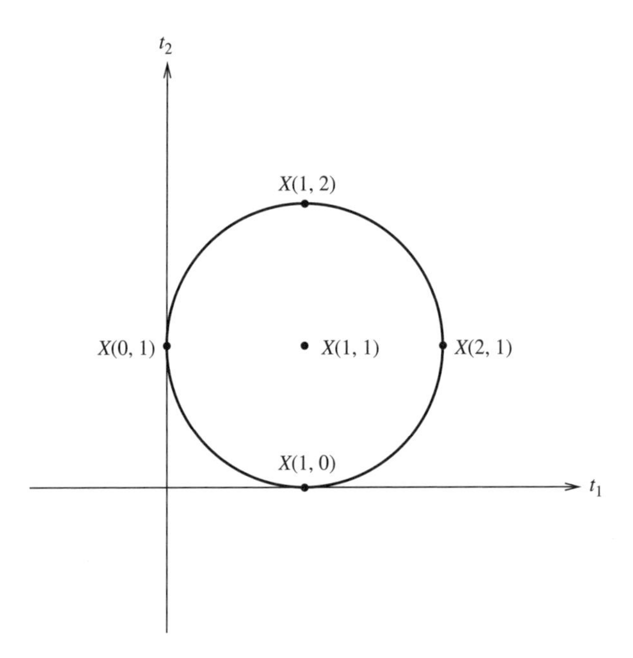

Figure 3.26 Isotropy implies that the correlation coefficient between $X(1,1)$ and any point on the circle are all the same.

in Figure 3.26. If a process is isotropic, it must also be quadrant symmetric and thus also at least second-moment stationary. The dependence only on distance implies that

$$\rho(\tau_1, \tau_2) = \rho\left(\sqrt{\tau_1^2 + \tau_2^2}\right) \tag{3.124}$$

For example, the isotropic two-dimensional Markov correlation function is given by

$$\rho(\tau) = \exp\left\{-\frac{2}{\theta}\sqrt{\tau_1^2 + \tau_2^2}\right\} = \exp\left\{-\frac{2|\tau|}{\theta}\right\} \tag{3.125}$$

which is illustrated in Figure 3.27, where $|\tau| = \sqrt{\tau_1^2 + \tau_2^2}$. The Gaussian correlation function can be both isotropic, if $\theta_1 = \theta_2 = \theta$, and separable, for example,

$$\begin{aligned}
\rho(\tau_1, \tau_2) &= \exp\left\{-\pi\left(\frac{\tau_1}{\theta_1}\right)^2\right\} \exp\left\{-\pi\left(\frac{\tau_2}{\theta_2}\right)^2\right\} \\
&= \exp\left\{-\frac{\pi}{\theta^2}\left(\tau_1^2 + \tau_2^2\right)\right\} \\
&= \exp\left\{-\frac{\pi}{\theta^2}\left(\sqrt{\tau_1^2 + \tau_2^2}\right)^2\right\} \\
&= \exp\left\{-\pi\left(\frac{|\tau|}{\theta}\right)^2\right\}
\end{aligned}$$

which is isotropic since it is a function of $|\tau| = \sqrt{\tau_1^2 + \tau_2^2}$.

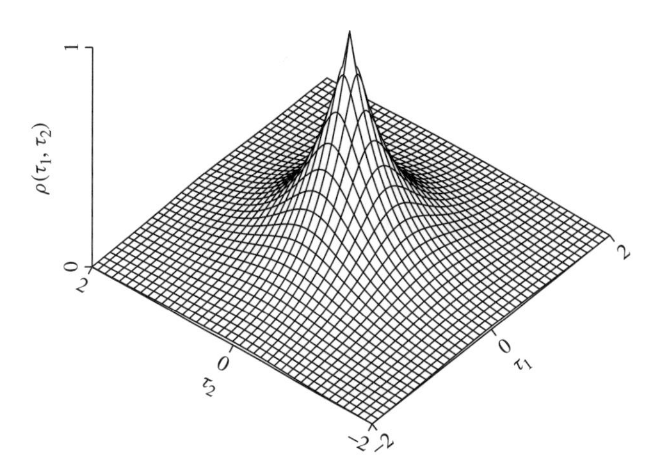

Figure 3.27 Isotropic Markov process in two dimensions.

Not all functions of $|\tau| = \sqrt{\tau_1^2 + \tau_2^2}$ are acceptable as isotropic correlation functions. Matern (1960) showed that for an n-dimensional isotropic field the correlation function must satisfy

$$\rho(\tau) \geq -\frac{1}{n} \tag{3.126}$$

which can be shown by considering $n + 1$ *equidistant* points, for example, an equilateral triangle in $n = 2$ dimensions or a tetrahedron in $n = 3$ dimensions, combined with the requirement that the correlation function be *positive definite* (see Eq. 3.8),

$$\sum_{i=1}^{n+1}\sum_{j=1}^{n+1} a_i a_j \rho_{ij} \geq 0 \tag{3.127}$$

where ρ_{ij} is the correlation coefficient between the ith and jth points. Since the points are equidistant and the field is isotropic, we must have

$$\rho_{ij} = \begin{cases} \rho(\tau) & \text{if } i \neq j \\ 1.0 & \text{if } i = j \end{cases}$$

where τ is the distance between points. If we also set the coefficients a_i to 1.0, that is, $a_1 = a_2 = \cdots = a_{n+1} = 1$, then Eq. 3.127 becomes

$$(n + 1) + [(n + 1)^2 - (n + 1)]\rho(\tau) \geq 0$$

which leads to Eq. 3.126.

Example 3.5 Suppose

$$\rho(\tau) = \exp\{-\tfrac{1}{2}\tau\}\cos(2\tau)$$

for $\tau \geq 0$. Can this function be used as an isotropic correlation function in $n = 3$ dimensions?

SOLUTION For $n = 3$ dimensions we require that $\rho(\tau) \geq -\frac{1}{3}$. The minimum value of ρ occurs when τ reaches the first root of $d\rho/d\tau = 0$ in the positive direction. For generality, we write

$$\rho(\tau) = \exp\{-a\tau\} \, \cos(\omega\tau)$$

where, in our problem, $a = \frac{1}{2}$ and $\omega = 2$. The derivative is

$$\frac{d\rho}{d\tau} = \exp\{-a\tau\}\Big[-a \,\cos\, \omega\tau - \omega \,\sin\, \omega\tau\Big]$$

so that setting $d\rho/d\tau = 0$ leads to the root

$$\tau_{min} = \frac{1}{\omega}\tan^{-1}\left(-\frac{a}{\omega}\right) = -\frac{1}{\omega}\tan^{-1}\left(\frac{a}{\omega}\right)$$

But we want the first positive root, so we shift to the right by π, that is,

$$\tau_{min} = \frac{\pi - \tan^{-1}(a/\omega)}{\omega}$$

Substituting this into our correlation function gives us the minimum value the correlation function will take on

$$\rho(\tau_{min}) = \exp\left\{-\frac{a}{\omega}\left(\pi - \tan^{-1}\frac{a}{\omega}\right)\right\} \, \cos\left(\pi - \tan^{-1}\frac{a}{\omega}\right)$$

For $a/\omega = 0.5/2 = 0.25$ we get

$$\rho(\tau_{min}) = \exp\left\{-0.25(\pi - \tan^{-1} 0.25)\right\} \times \cos\left(\pi - \tan^{-1} 0.25\right) = -0.47$$

But $-0.47 < -\frac{1}{3}$, so that this is not an acceptable isotropic correlation function in three dimensions. It would lead to a covariance structure which is not positive definite. We require the ratio $a/\omega \geq 0.37114$ in order for this function to be used as an isotropic correlation function in three dimensions.

If the random field is isotropic, then its spectral density function can be specified by a radial function (Vanmarcke, 1984). In two dimensions, the isotropic radial spectral density function has the form

$$G(\omega_1, \omega_2) = G^r\left(\sqrt{\omega_1^2 + \omega_2^2}\right) = G^r(\omega) \tag{3.128}$$

where $\omega = \sqrt{\omega_1^2 + \omega_2^2}$ is the absolute distance between the origin and any point in the frequency domain. A complication with the radial spectral density function is that the area beneath it is no longer equal to the variance of the random field, σ^2. To obtain the variance from the radial spectral density function, we must integrate over the original (ω_1, ω_2) space both radially and circumferentially. For $n = 2$ dimensions, the end result is

$$\sigma^2 = \frac{\pi}{2}\int_0^\infty \omega G^r(\omega)\, d\omega \tag{3.129}$$

while for $n = 3$ dimensions

$$\sigma^2 = \frac{\pi}{2}\int_0^\infty \omega^2 G^r(\omega)\, d\omega \tag{3.130}$$

The variance function is defined as the variance reduction factor after averaging the field over a *rectangle* of size $T_1 \times T_2$ (or $T_1 \times T_2 \times T_3$ in three dimensions). Since the rectangle is not isotropic, even if the random field being averaged is isotropic, the variance function does not have an isotropic form. An isotropic form would be possible if the variance function was defined using a circular averaging window, but this option will not be pursued further here.

3.7.7 Ellipsoidal Correlation Structure

If an isotropic random field is stretched in either or both coordinate directions, then the resulting field will have an *ellipsoidal* correlation structure. Stretching the axes results in a scaling of the distances τ_1 and τ_2 to, for example, τ_1/a_1 and τ_2/a_2 so that the correlation becomes a function of the effective distance τ,

$$\tau = \sqrt{\left(\frac{\tau_1}{a_1}\right)^2 + \cdots + \left(\frac{\tau_n}{a_n}\right)^2} \tag{3.131}$$

in n dimensions.

Example 3.6 Suppose we have simulated a random field X' in two dimensions which has isotropic correlation function

$$\rho'(\tau') = \exp\left\{-\frac{2|\tau'|}{4}\right\}$$

where $|\tau'| = \sqrt{(\tau_1')^2 + (\tau_2')^2}$. We wish to transform the simulated field into one which has correlation lengths in the t_1 (horizontal) and t_2 (vertical) directions of $\theta_1 = 8$ and $\theta_2 = 2$, respectively. How can we transform our simulation to achieve the desired directional correlation lengths?

SOLUTION The simulated random field is isotropic with $\theta_1' = \theta_2' = 4$ (see, e.g., Eq. 3.88). What this means is that the X' random field is such that when the distance between points in the t_1 or t_2 direction exceeds 4 the correlation between the two points becomes negligible. If we first consider the t_1 direction, we desire a correlation length of 8 in the t_1 (horizontal) direction. If we stretch the distance in the horizontal direction by a factor of 2, then, in the stretched field, it is only when points are separated by more than 8 that their correlation becomes negligible. Similarly, if we "stretch" the field in the vertical direction by a factor of $\frac{1}{2}$, then, in the stretched field it is only when points are separated by more than $\frac{1}{2}(4) = 2$ that their correlation becomes negligible. In other words, an isotropic field with correlation length $\theta = 4$ can be converted into an ellipsoidally correlated field with scales $\theta_1 = 8$ and $\theta_2 = 2$ by stretching the field in the t_1 by a factor of 2 and shrinking the field in the t_2 direction by a factor of $\frac{1}{2}$. This is illustrated in Figure 3.28.

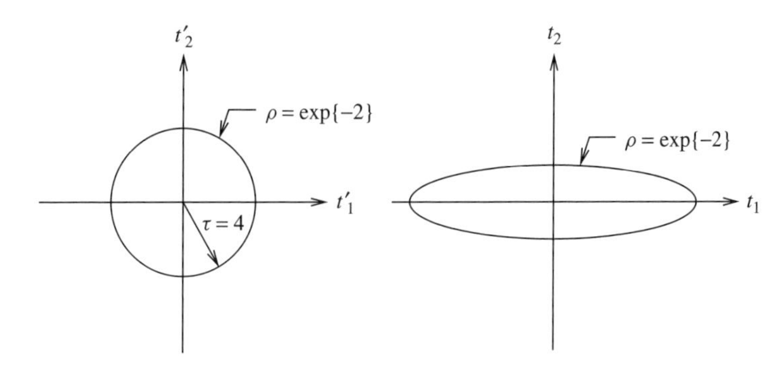

Figure 3.28 Ellipsoidal correlation function: after stretching the t' axes to the t axes, all points on the ellipse have equal correlation with the origin.

The resulting correlation function of the stretched field is

$$\rho(\tau) = \exp\left\{-2\sqrt{\left(\tfrac{1}{8}\tau_1\right)^2 + \left(\tfrac{1}{2}\tau_2\right)^2}\right\}$$

When the effective distance between points is ellipsoidal, that is,

$$\tau = \sqrt{\left(\frac{\tau_1}{a_1}\right)^2 + \cdots + \left(\frac{\tau_n}{a_n}\right)^2}$$

then the spectral density function will be a function of the effective frequency

$$\omega = \sqrt{(a_1\omega_1)^2 + \cdots + (a_n\omega_n)^2} \tag{3.132}$$

We shall give specific examples of the ellipsoidal spectral density function in Section 3.7.10.

3.7.8 Anisotropic Correlation Structure

If the correlation function depends on direction, we say that the field is anisotropic. The ellipsoidal correlation function just discussed is a special case of anisotropy, as were most of the separable correlation functions considered in Section 3.7.5. Figures 3.21, 3.22, and 3.24 are illustrations of separable Markov processes having $\theta_1 = \theta_2 = 1$ and $\sigma_X^2 = 1$. Despite the equivalence in the directional correlation lengths, the anisotropy arising from the separability is clearly evident.

Another possible form for an anisotropic correlation function is to express ρ as a function of the interpoint distance, $|\tau| = \sqrt{\tau_1^2 + \tau_2^2}$, and the angle of the vector from point 1 to point 2, ϕ. (In three and higher dimensions, additional angles are needed.) For example, Ewing (1969) suggests an anisotropic correlation function of the form

$$\rho(\tau, \phi) = \rho(\tau)\, \cos^2(\phi - \phi_o) \tag{3.133}$$

to model the spatial dependence of ocean wave heights, where ϕ_o gives the orientation of the waves. Notice that this model assumes correlation along an individual wave crest to decay with $\rho(\tau)$ but gives zero correlation in a direction perpendicular to the waves, that is, from crest to crest.

Most commonly, anisotropic random fields are of either separable or ellipsoidal form. This may be due largely to simplicity, since the separable and ellipsoidal forms are generally parameterized by directional correlation lengths.

3.7.9 Cross-Correlated Random Fields

Often different soil properties will be correlated with one another. For example, Holtz and Krizek (1971) suggest that liquid limit and water content have a cross-correlation coefficient of 0.67 [they present a much more extensive list of soil property cross-correlations; see also Baecher and Christian (2003) for a summary]. As another example, both Cherubini (2000) and Wolff (1985) suggest that cohesion and friction angle are reasonably strongly negatively correlated, with cross-correlation coefficients as large as -0.7.

Consider two soil properties, $X(\mathbf{t})$ and $Y(\mathbf{t})$, both of which are spatially random fields, where $\mathbf{t} = \{t_1, t_2, \ldots, t_n\}$ is the spatial position in n dimensions. If X and Y are cross-correlated, then the complete specification of the correlation structure involves three correlation functions:

$\rho_X(\mathbf{t}, \mathbf{t}') =$ correlations between $X(\mathbf{t})$ and $X(\mathbf{t}')$ for all $\mathbf{t}$ and $\mathbf{t}'$

$\rho_Y(\mathbf{t}, \mathbf{t}') =$ correlations between $Y(\mathbf{t})$ and $Y(\mathbf{t}')$ for all $\mathbf{t}$ and $\mathbf{t}'$

$\rho_{XY}(\mathbf{t}, \mathbf{t}') =$ cross-correlations between $X(\mathbf{t})$ and $Y(\mathbf{t}')$ for all $\mathbf{t}$ and $\mathbf{t}'$

The corresponding covariance structures are

$$C_X(\mathbf{t}, \mathbf{t}') = \sigma_X \sigma_{X'} \rho_X(\mathbf{t}, \mathbf{t}')$$

$$C_Y(\mathbf{t},\mathbf{t}') = \sigma_Y \sigma_{Y'} \rho_Y(\mathbf{t},\mathbf{t}')$$

$$C_{XY}(\mathbf{t},\mathbf{t}') = \sigma_X \sigma_{Y'} \rho_{XY}(\mathbf{t},\mathbf{t}')$$

where $\sigma_X^2 = \text{Var}[X(\mathbf{t})]$, $\sigma_{X'}^2 = \text{Var}\left[X(\mathbf{t}')\right]$, and similarly for Y. If the fields are both stationary, the correlation and covariance structures simplify to

$$C_X(\boldsymbol{\tau}) = \sigma_X^2 \rho_X(\boldsymbol{\tau})$$

$$C_Y(\boldsymbol{\tau}) = \sigma_Y^2 \rho_Y(\boldsymbol{\tau})$$

$$C_{XY}(\boldsymbol{\tau}) = \sigma_X \sigma_Y \rho_{XY}(\boldsymbol{\tau})$$

where $\boldsymbol{\tau} = \mathbf{t} - \mathbf{t}'$. When $\boldsymbol{\tau} = 0$, $C_{XY}(\mathbf{0})$ gives the covariance between X and Y at a point.

The covariance structure can be expressed in matrix form as

$$\boldsymbol{C} = \begin{bmatrix} \boldsymbol{C}_X & \boldsymbol{C}_{XY} \\ \boldsymbol{C}_{YX} & \boldsymbol{C}_Y \end{bmatrix}$$

where $\boldsymbol{C}_{YX}$ is the transpose of $\boldsymbol{C}_{XY}$ (equal, except that X and Y are interchanged).

The cross-spectral density function can be derived from the following transform pair (for stationary fields):

$$C_{XY}(\boldsymbol{\tau}) = \int_{-\infty}^{\infty} S_{XY}(\boldsymbol{\omega}) \cos(\boldsymbol{\omega} \cdot \boldsymbol{\tau}) \, d\boldsymbol{\omega} \quad (3.134a)$$

$$S_{XY}(\boldsymbol{\omega}) = \frac{1}{(2\pi)^n} \int_{-\infty}^{\infty} C_{XY}(\boldsymbol{\tau}) \cos(\boldsymbol{\omega} \cdot \boldsymbol{\tau}) \, d\boldsymbol{\tau} \quad (3.134b)$$

The estimation of the complete cross-correlation structure between soil properties requires a large amount of spatially distributed data and, preferably, multiple statistically identical realizations of a soil site. It is unlikely for the latter to happen, since each soil site is unique, and the former can be quite expensive. In practice, the complete cross-correlation structure will rarely be known, although it may be assumed. For example, the cross-correlation between cohesion and friction angle given by Cherubini (2000) ranges from −0.24 to −0.7, indicating a high degree of uncertainty in this particular cross-correlation. (As an aside, this uncertainty may, in large part, be due to difficulty in discerning between the cohesion and friction angle contributions to the measured shear strength.)

In general, the cross-correlation between soil properties is estimated by taking a number of samples assumed to be from the same population and statistically comparing their properties by pairs (the formula used to estimate the cross-correlation is given in Chapter 5). Any resulting estimate is then assumed to be the correlation between a pair of properties in any one sample or between a pair of properties at any point in the soil site. We will refer to this as a *pointwise* cross-correlation between pairs of properties, ignoring for the time being the distinction between a "point" and a finite-volume lab sample (we will investigate this distinction a little more closely in Chapter 5).

If we consider the available published information on soil property cross-correlations appearing in geotechnical engineering journals, we find that only pointwise cross-correlations are reported, that is, the correlation between $X(\mathbf{t})$ and $Y(\mathbf{t})$. See, for example, Holtz and Krizek (1971), Cherubini (2000), and Wolff (1985). In the event that only pointwise cross-correlations are known, the cross-correlation function $\rho_{XY}(\mathbf{t},\mathbf{t}')$ becomes a function only of $\mathbf{t}$, $\rho_{XY}(\mathbf{t})$. If the field is stationary, the cross-correlation function simplifies further to just the constant ρ_{XY}. Stationary pointwise correlated fields are thus specified by the correlation functions $\rho_X(\boldsymbol{\tau})$ and $\rho_Y(\boldsymbol{\tau})$ and by the cross-correlation ρ_{XY}. We shall see how this information can be used to simulate pointwise cross-correlated random fields in Chapter 6.

3.7.10 Common Higher Dimensional Models

3.7.10.1 White Noise and Triangular Correlation Function Consider an n-dimensional stationary white noise process $W(\mathbf{t})$ having spectral density function

$$C_W(\boldsymbol{\tau}) = \pi^n G_o \, \delta(\boldsymbol{\tau}) \quad (3.135)$$

where $\delta(\boldsymbol{\tau})$ is the n-dimensional *Dirac delta function* defined by

$$\delta(\boldsymbol{\tau}) = \delta(\tau_1)\delta(\tau_2)\cdots\delta(\tau_n) \quad (3.136)$$

The Dirac delta function $\delta(\boldsymbol{\tau})$ has value zero everywhere except at $\boldsymbol{\tau} = \mathbf{0}$ ($\tau_1 = \tau_2 = \cdots = 0$), where it assumes infinite height but unit (n-dimensional) volume.

Because $\delta(\tau_i) = \delta(-\tau_i)$ for all $i = 1, 2, \ldots, n$, white noise is quadrant symmetric and its spectral density function can be expressed in terms of the "one-sided" spectral density function

$$\begin{aligned} G_W(\boldsymbol{\omega}) = 2^n S(\boldsymbol{\omega}) &= \frac{2^n}{(2\pi)^n} \int_{-\infty}^{\infty} C_W(\boldsymbol{\tau}) \cos(\boldsymbol{\omega} \cdot \boldsymbol{\tau}) \, d\boldsymbol{\tau} \\ &= \frac{1}{\pi^n} \int_{-\infty}^{\infty} \pi^n G_o \delta(\boldsymbol{\tau}) \cos(\boldsymbol{\omega} \cdot \boldsymbol{\tau}) \, d\boldsymbol{\tau} \\ &= G_o \int_{\infty}^{\infty} \delta(\tau_n) \cdots \int_{\infty}^{\infty} \delta(\tau_1) \cos(\omega_1 \tau_1 + \cdots \\ &\quad + \omega_n \tau_n) \, d\tau_1 \cdots d\tau_n \\ &= G_o \end{aligned}$$

as expected.

If we average $W(\mathbf{t})$ over an n-dimensional "rectangular" region of size $\theta_1 \times \theta_2 \times \cdots \times \theta_n$,

$$X(\mathbf{t}) = \frac{1}{\theta_1 \theta_2 \cdots \theta_n} \int_{\mathbf{t}-\boldsymbol{\theta}/2}^{\mathbf{t}+\boldsymbol{\theta}/2} W(\boldsymbol{\xi}) \, d\boldsymbol{\xi} \quad (3.137)$$

and assume, for simplicity, that the white noise has mean zero, then $X(\mathbf{t})$ will also have mean zero. The variance of $X(\mathbf{t})$ is then computed as

$$\begin{aligned}
\sigma_X^2 &= \mathrm{E}\left[X^2\right] \\
&= \frac{1}{(\theta_1\theta_2\cdots\theta_n)^2}\int_{\mathbf{t}-\boldsymbol{\theta}/2}^{\mathbf{t}+\boldsymbol{\theta}/2}\int_{\mathbf{t}-\boldsymbol{\theta}/2}^{\mathbf{t}+\boldsymbol{\theta}/2} \mathrm{E}\left[W(\boldsymbol{\xi})W(\boldsymbol{\eta})\right]\, d\boldsymbol{\xi}\, d\boldsymbol{\eta} \\
&= \frac{\pi^n G_o}{(\theta_1\theta_2\cdots\theta_n)^2}\int_{\mathbf{t}-\boldsymbol{\theta}/2}^{\mathbf{t}+\boldsymbol{\theta}/2}\int_{\mathbf{t}-\boldsymbol{\theta}/2}^{\mathbf{t}+\boldsymbol{\theta}/2} \delta(\boldsymbol{\xi}-\boldsymbol{\eta})\, d\boldsymbol{\xi}\, d\boldsymbol{\eta} \\
&= \frac{\pi^n G_o}{(\theta_1\theta_2\cdots\theta_n)^2}\int_{\mathbf{t}-\boldsymbol{\theta}/2}^{\mathbf{t}+\boldsymbol{\theta}/2} 1\, d\boldsymbol{\eta} \\
&= \frac{\pi^n G_o}{\theta_1\theta_2\cdots\theta_n}
\end{aligned}$$

So far as the variance is concerned, the end result is identical even if the mean of W is not zero. Assuming the mean is zero just simplifies the algebra.

The covariance function of $X(\mathbf{t})$ is triangular in shape, but in multiple dimensions

$$C_X(\boldsymbol{\tau}) = \begin{cases} \sigma_X^2\prod_{i=1}^{n}\left(1-\dfrac{|\tau_i|}{\theta_i}\right), & |\tau_i|\le\theta_i, \quad i=1,\ldots n \\ 0, & \text{otherwise.} \end{cases} \tag{3.138}$$

where $\prod$ means product of (analogous to $\sum$ meaning sum of). If we write

$$\rho_i(\tau_i) = \begin{cases} \left(1-\dfrac{|\tau_i|}{\theta_i}\right) & \text{for } |\tau_i|\le\theta_i \\ 0 & \text{otherwise} \end{cases} \tag{3.139}$$

then Eq. 3.138 can be expressed as

$$C_X(\boldsymbol{\tau}) = \sigma_X^2\prod_{i=1}^{n}\rho_i(\tau_i) = \sigma_X^2\left[\rho_1(\tau_1)\rho_2(\tau_2)\cdots\rho_n(\tau_n)\right] \tag{3.140}$$

which demonstrates that $C_X(\boldsymbol{\tau})$ is separable and thus also quadrant symmetric. The latter allows the associated spectral density function to be expressed using the one-sided $G_X(\boldsymbol{\omega})$, which is also separable,

$$G_X(\boldsymbol{\omega}) = G_o\prod_{i=1}^{n}\left[\frac{\sin(\omega_i\theta_i/2)}{\omega_i\theta_i/2}\right]^2, \qquad \boldsymbol{\omega}\ge 0 \tag{3.141}$$

The relationship $G_X(\boldsymbol{\omega}) = 2^n S_X(\boldsymbol{\omega})$ can be used if the two-sided spectral density function $S_X(\boldsymbol{\omega})$ is desired.

The variance function, which gives the variance reduction factor when $X(\mathbf{t})$ is itself averaged over an "area" of size $T_1\times T_2\times\cdots\times T_n$, is also separable,

$$\gamma(T_1,T_2,\ldots,T_n) = \prod_{i=1}^{n}\gamma_i(T_i) \tag{3.142}$$

where the individual "directional" variance functions come from Eq. 3.67:

$$\gamma_i(T_i) = \begin{cases} 1-\dfrac{T_i}{3\theta_i} & \text{if } T_i\le\theta_i \\ \dfrac{\theta_i}{T_i}\left[1-\dfrac{\theta_i}{3T_i}\right] & \text{if } T_i>\theta_i \end{cases}$$

Note that $X(\mathbf{t})$ does not have an isotropic correlation structure even if $\theta_1=\theta_2=\cdots=\theta_n$. This is because the averaging region is an n-dimensional rectangle, which is an anisotropic shape. If an n-dimensional *spherical* averaging region were used, then the resulting correlation function would be isotropic.

3.7.10.2 Markov Correlation Function In higher dimensions, the Markovian property—where the future is dependent only on the most recently known past—is lost because higher dimensions do not have a clear definition of "past." As a result, the two models presented here are not strictly Markovian, but we shall refer to them as such since they derive from one-dimensional Markov models.

Separable Markov Model If the correlation function is separable and equal to the product of directional Markovian correlation functions, for example,

$$\rho(\boldsymbol{\tau}) = \rho_1(\tau_1)\rho_2(\tau_2)\cdots\rho_n(\tau_n) \tag{3.143}$$

where, according to Eq. 3.88,

$$\rho_i(\tau_i) = \exp\left\{-\frac{2|\tau_i|}{\theta_i}\right\} \tag{3.144}$$

then the spectral density function is also separable and thus quadrant symmetric,

$$G(\boldsymbol{\omega}) = \sigma^2 g_1(\omega_1)g_2(\omega_2)\cdots g_n(\omega_n), \quad \boldsymbol{\omega}\ge 0 \tag{3.145}$$

The individual unit-variance spectral density functions are obtained by dividing Eq. 3.90 by σ^2,

$$g_i(\omega_i) = \frac{\theta_i}{\pi\left[1+(\theta_i\omega_i/2)^2\right]} \tag{3.146}$$

The variance function associated with the separable Markov model of Eq. 3.143 is also separable and is given by

$$\gamma(\mathbf{T}) = \gamma_1(T_1)\gamma_2(T_2)\cdots\gamma_n(T_n) \tag{3.147}$$

where the one-dimensional variance functions are given by Eq. 3.89,

$$\gamma_i(T_i) = \frac{\theta_i^2}{2T_i^2}\left[\frac{2|T_i|}{\theta_i}+\exp\left\{-\frac{2|T_i|}{\theta_i}\right\}-1\right]$$

Ellipsoidal Markov Model The Markov correlation function, Eq. 3.88, can be extended to multiple dimensions by replacing τ by the lag $|\boldsymbol{\tau}|$,

$$\rho(\boldsymbol{\tau}) = \exp\left\{-\frac{2|\boldsymbol{\tau}|}{\theta}\right\} \tag{3.148}$$

where $|\boldsymbol{\tau}| = \sqrt{\tau_1^2 + \cdots + \tau_n^2}$. In this case, the field is isotropic with correlation length equal to θ in any direction. Equation 3.148 can be further generalized to an ellipsoidal correlation structure by expressing $|\tau|$ as the scaled lag

$$|\tau| = \sqrt{\left(\frac{2\tau_1}{\theta_1}\right)^2 + \cdots + \left(\frac{2\tau_n}{\theta_n}\right)^2} \tag{3.149}$$

so that the correlation function becomes ellipsoidal,

$$\rho(\boldsymbol{\tau}) = \exp\left\{-\sqrt{\left(\frac{2\tau_1}{\theta_1}\right)^2 + \cdots + \left(\frac{2\tau_n}{\theta_n}\right)^2}\right\} \tag{3.150}$$

If $\theta_1 = \theta_2 = \cdots = \theta_n$, we regain the isotropic model of Eq. 3.148.

According to Eq. 3.132, the ellipsoidal Markov spectral density function is a function of $\sqrt{(\theta_1\omega_1/2)^2 + \cdots + (\theta_n\omega_n/2)^2}$. In particular, for $n = 2$ dimensions

$$G(\boldsymbol{\omega}) = \frac{\sigma^2\theta_1\theta_2}{2\pi\left[1 + (\theta_1\omega_1/2)^2 + (\theta_2\omega_2/2)^2\right]^{3/2}} \tag{3.151}$$

while for $n = 3$ dimensions

$$G(\boldsymbol{\omega}) = \frac{\sigma^2\theta_1\theta_2\theta_3}{\pi^2\left[1 + (\theta_1\omega_1/2)^2 + (\theta_2\omega_2/2)^2 + (\theta_3\omega_3/2)^2\right]^2} \tag{3.152}$$

A closed-form expression for the variance function does not exist for the higher dimensional ellipsoidal Markov model. If needed, it can be obtained by numerically integrating Eq. 3.112,

$$\gamma(T_1, T_2) = \frac{1}{T_1^2 T_2^2}\int_{-T_1}^{T_1}\int_{-T_2}^{T_2}(|T_1| - |\tau_1|)(|T_2| - |\tau_2|) \times \rho(\tau_1, \tau_2)\, d\tau_2\, d\tau_1$$

in the two-dimensional case or

$$\gamma(T_1, T_2, T_3) = \frac{1}{T_1^2 T_2^2 T_3^2}\int_{-T_1}^{T_1}\int_{-T_2}^{T_2}\int_{-T_3}^{T_3}(|T_1| - |\tau_1|)(|T_2| - |\tau_2|) \times (|T_3| - |\tau_3|)\rho(\tau_1, \tau_2, \tau_3)\, d\tau_3\, d\tau_2\, d\tau_1 \tag{3.153}$$

in three dimensions.

Example 3.7 Suppose a three-dimensional soil mass has a random elastic modulus field with mean 30 kPa, standard deviation 6 kPa, and correlation function

$$\rho(\boldsymbol{\tau}) = \exp\left\{-\sqrt{\left(\frac{2\tau_1}{\theta_1}\right)^2 + \left(\frac{2\tau_2}{\theta_2}\right)^2 + \left(\frac{2\tau_3}{\theta_3}\right)^2}\right\} \tag{3.154}$$

where $\theta_1 = \theta_2 = 4$ and $\theta_3 = 1$ (assume that θ_3 is the correlation length in the vertical direction). Suppose further that settlement of a foundation on this soil has been found to depend on the average elastic modulus over a volume of size $V = T_1 \times T_2 \times T_3 = 2 \times 3 \times 8$. What is the mean and standard deviation of this average?

SOLUTION Suppose that the elastic modulus field is denoted by $X(t_1, t_2, t_3)$, where (t_1, t_2, t_3) is the spatial position with t_3 measured vertically. Let our average elastic modulus be X_V, defined by

$$X_V = \frac{1}{2(3)(8)}\int_0^8\int_0^3\int_0^2 X(t_1, t_2, t_3)\, dt_1\, dt_2\, dt_3 \tag{3.155}$$

where we have placed our origin at one corner of the averaging domain, with t_3 positive downward. (Since the field is stationary, we can place the origin wherever we want—stationarity is suggested by the fact that the mean and variance are not dependent on position, and Eq. 3.154 is a function of τ rather than position.)

The mean of X_V is obtained by taking expectations:

$$\begin{aligned}
\mathrm{E}[X_V] &= \mathrm{E}\left[\frac{1}{2(3)(8)}\int_0^8\int_0^3\int_0^2 X(t_1, t_2, t_3)\, dt_1\, dt_2\, dt_3\right] \\
&= \frac{1}{2(3)(8)}\int_0^8\int_0^3\int_0^2 \mathrm{E}[X(t_1, t_2, t_3)]\, dt_1\, dt_2\, dt_3 \\
&= \frac{1}{2(3)(8)}\int_0^8\int_0^3\int_0^2 30\, dt_1\, dt_2\, dt_3 \\
&= 30
\end{aligned}$$

so we see that the mean is preserved by averaging (as expected). That is, $\mu_{X_V} = \mu_X = 30$.

To obtain the variance we write, for $V = 2(3)(8)$,

$$\begin{aligned}
\mathrm{Var}[X_V] &= \mathrm{E}\left[(X_V - \mu_{X_V})^2\right] \\
&= \mathrm{E}\left[\left(\frac{1}{V}\int_0^8\int_0^3\int_0^2 (X(t_1, t_2, t_3) - \mu_X)\, dt_1\, dt_2\, dt_3\right)^2\right] \\
&= \mathrm{E}\left[\frac{1}{V^2}\int_0^8\int_0^3\int_0^2\int_0^8\int_0^3\int_0^2 (X(t_1, t_2, t_3) - \mu_X) \times (X(s_1, s_2, s_3) - \mu_X)\, dt_1\, dt_2\, dt_3\, ds_1\, ds_2\, ds_3\right]
\end{aligned}$$

$$= \frac{1}{V^2} \int_0^8 \int_0^3 \int_0^2 \int_0^8 \int_0^3 \int_0^2 \mathrm{E}\left[(X(t_1,t_2,t_3) - \mu_X) \right.$$
$$\left. \times (X(s_1,s_2,s_3) - \mu_X)\right] \, dt_1\, dt_2\, dt_3\, ds_1\, ds_2\, ds_3$$
$$= \frac{1}{V^2} \int_0^8 \int_0^3 \int_0^2 \int_0^8 \int_0^3 \int_0^2 \mathrm{Cov}\left[X(t_1,t_2,t_3), \right.$$
$$\left. X(s_1,s_2,s_3)\right] dt_1\, dt_2\, dt_3\, ds_1\, ds_2\, ds_3$$
$$= \frac{\sigma_X^2}{V^2} \int_0^8 \int_0^3 \int_0^2 \int_0^8 \int_0^3 \int_0^2 \rho(t_1 - s_1, t_2 - s_2, t_3 - s_3)$$
$$\times\, dt_1\, dt_2\, dt_3\, ds_1\, ds_2\, ds_3$$
$$= \sigma_X^2 \gamma(2,3,8)$$

(Aside: The last three expressions would also have been obtained if we had first assumed $\mu_{X_V} = \mu_X = 0$, which would have made the earlier expressions somewhat simpler; however, this is only a trick for computing variance and must be used with care, i.e., the mean is not actually zero, but it may be set to zero for the purposes of this calculation.)

The variance function $\gamma(T_1,T_2,T_3)$ is nominally defined, as above, by a sixfold integration. Since the correlation function $\rho(t_1 - s_1, t_2 - s_2, t_3 - s_3)$ is constant along diagonal lines where $t_1 - s_1$, $t_2 - s_2$, and $t_3 - s_3$ are constants, the sixfold integration can be reduced to a threefold integration (see, e.g., Eq. 3.112):

$$\gamma(T_1,T_2,T_3)$$
$$= \frac{1}{[T_1T_2T_3]^2} \int_{-T_3}^{T_3} \int_{-T_2}^{T_2} \int_{-T_1}^{T_1} (|T_1| - |\tau_1|)(|T_2| - |\tau_2|)$$
$$\times (|T_3| - |\tau_3|)\rho(\tau_1,\tau_2,\tau_3)\, d\tau_1\, d\tau_2\, d\tau_3 \tag{3.156}$$

Since the given correlation function, Eq. 3.154, is quadrant symmetric, that is, since $\rho(\tau_1,\tau_2,\tau_3) = \rho(-\tau_1,\tau_2,\tau_3) = \rho(\tau_1,-\tau_2,\tau_3) = \cdots = \rho(-\tau_1,-\tau_2,-\tau_3)$, the variance function can be further simplified to

$$\gamma(T_1,T_2,T_3)$$
$$= \frac{8}{[T_1T_2T_3]^2} \int_0^{T_3} \int_0^{T_2} \int_0^{T_1} (|T_1| - \tau_1)(|T_2| - \tau_2)$$
$$\times (|T_3| - \tau_3)\rho(\tau_1,\tau_2,\tau_3)\, d\tau_1\, d\tau_2\, d\tau_3 \tag{3.157}$$

Thus, to find the variance of X_V, we must evaluate Eq. 3.157 for $T_1 = 2$, $T_2 = 3$, and $T_3 = 8$. For this, we will use Gaussian quadrature [see Griffiths and Smith (2006) or Press et al. (1997) and Appendices B and C],

$$\gamma(T_1,T_2,T_3) \simeq \frac{1}{T_1T_2T_3}$$
$$\times \sum_{k=1}^{n_g} w_k \left[\sum_{j=1}^{n_g} w_j \left\{ \sum_{i=1}^{n_g} w_i f(\tau_{1i},\tau_{2j},\tau_{3k}) \right\} \right] \tag{3.158}$$

where

$$f(\tau_{1i},\tau_{2j},\tau_{3k}) = (|T_1| - \tau_{1i})(|T_2| - \tau_{2j})(|T_3| - \tau_{3k})$$
$$\times \rho(\tau_{1i},\tau_{2j},\tau_{3k})$$
$$\tau_{1i} = \frac{T_1}{2}(1 + z_i)$$
$$\tau_{2j} = \frac{T_2}{2}(1 + z_j)$$
$$\tau_{3k} = \frac{T_3}{2}(1 + z_k)$$

and where w_i and z_i are the weights and evaluation points of Gaussian quadrature and n_g is the number of evaluation points to use. The accuracy of Gaussian quadrature is about the same as obtained by fitting a $(2n)$th-order polynomial to the integrand. The weights and evaluation points are provided in Appendix B for a variety of n_g values. Using $n_g = 20$, we get

$$\gamma(2,3,8) \simeq 0.878$$

Note that when $n_g = 5$ the Gaussian quadrature approximation gives $\gamma(2,3,8) \simeq 0.911$, a 4% relative error.

The variance of the $2 \times 3 \times 8$ average is thus

$$\mathrm{Var}\,[X_V] = \sigma_X^2 \gamma(2,3,8) \simeq (6)^2(0.878) = 31.6$$

so that $\sigma_{X_V} = \sqrt{31.6} = 5.6$.

3.7.10.3 Gaussian Correlation Function The Gaussian correlation function, in higher dimensions, is both separable (thus quadrant symmetric) and ellipsoidal,

$$\rho(\boldsymbol{\tau}) = \exp\left\{ -\pi \left[\left(\frac{\tau_1}{\theta_1}\right)^2 + \cdots + \left(\frac{\tau_n}{\theta_n}\right)^2 \right] \right\}$$
$$= \exp\left\{ -\frac{\pi\tau_1^2}{\theta_1^2} \right\} \cdots \exp\left\{ -\frac{\pi\tau_n^2}{\theta_n^2} \right\} \tag{3.159}$$

If all of the directional correlation lengths are equal, then the field is isotropic. Because the Gaussian model is separable, the higher dimensional spectral density and variance functions are simply products of their one-dimensional forms:

$$G(\boldsymbol{\omega}) = \sigma_X^2 \left(\frac{\theta_1\theta_2\cdots\theta_n}{\pi^n} \right)$$
$$\times \exp\left\{ -\frac{1}{4\pi} \left(\theta_1^2\omega_1^2 + \cdots + \theta_n^2\omega_n^2\right) \right\} \tag{3.160}$$

$$\gamma(\mathbf{T}) = \gamma_1(T_1)\gamma_2(T_2)\cdots\gamma_n(T_n) \tag{3.161}$$

where the one-dimensional variance functions are given by Eq. 3.92,

$$\gamma_i(T_i) = \frac{\theta_i^2}{\pi T_i^2} \left[\frac{\pi|T_i|}{\theta_i} \mathrm{erf}\left\{ \frac{\sqrt{\pi}|T_i|}{\theta_i} \right\} + \exp\left\{ -\frac{\pi T_i^2}{\theta_i^2} \right\} - 1 \right]$$

CHAPTER 4

Best Estimates, Excursions, and Averages

4.1 BEST LINEAR UNBIASED ESTIMATION

We often want some way of best estimating "future" events given "past" observations or, perhaps more importantly, of estimating unobserved locations given observed locations. Suppose that we have observed $X_1, X_2, \ldots, X_n$ and we want to estimate the optimal (in some sense) value for X_{n+1} using this information. For example, we could have observed the capacities of a series of piles and want to estimate the capacity of the next pile. One possibility is to write our estimate of X_{n+1} as a linear combination of our observations:

$$\hat{X}_{n+1} = \mu_{n+1} + \sum_{k=1}^{n} \beta_k (X_k - \mu_k) \tag{4.1}$$

where the hat indicates that this is an estimate of X_{n+1} and μ_k is the mean of X_k (the mean may vary with position). Note that we need to know the means in order to form this estimate. Equation 4.1 is referred to as the best linear unbiased estimator (BLUE) for reasons we shall soon see.

The question now is, what is the optimal vector of coefficients, $\boldsymbol{\beta}$? We can define "optimal" to be that which produces the minimum expected error between our estimate $\hat{X}_{n+1}$ and the true (but unknown) X_{n+1}. This *estimator error* is given by

$$X_{n+1} - \hat{X}_{n+1} = X_{n+1} - \mu_{n+1} - \sum_{k=1}^{n} \beta_k (X_k - \mu_k) \tag{4.2}$$

To make this error as small as possible, its mean should be zero and its variance minimized. The first criterion is automatically satisfied by the above formulation since

$$\begin{aligned}
\mathrm{E}\left[X_{n+1} - \hat{X}_{n+1}\right] &= \mathrm{E}\left[X_{n+1} - \mu_{n+1} - \sum_{k=1}^{n} \beta_k (X_k - \mu_k)\right] \\
&= \mu_{n+1} - \mu_{n+1} - \sum_{k=1}^{n} \beta_k \,\mathrm{E}\left[X_k - \mu_k\right] \\
&= -\sum_{k=1}^{n} \beta_k (\mu_k - \mu_k) \\
&= 0
\end{aligned}$$

We say that the estimator, Eq. 4.1, is *unbiased* because its mean is the same as the quantity being estimated.

Now we want to minimize the estimator's variance. Since the mean estimator error is zero, the variance is just the expectation of the squared estimator error,

$$\begin{aligned}
\mathrm{Var}\left[X_{n+1} - \hat{X}_{n+1}\right] &= \mathrm{E}\left[\left(X_{n+1} - \hat{X}_{n+1}\right)^2\right] \\
&= \mathrm{E}\left[X_{n+1}^2 - 2X_{n+1}\hat{X}_{n+1} + \hat{X}_{n+1}^2\right]
\end{aligned}$$

To simplify the following algebra, we will assume that $\mu_i = 0$ for $i = 1, 2, \ldots, n+1$. The final results, expressed in terms of covariances, will be the same even if the means are nonzero. For zero means, our estimator simplifies to

$$\hat{X}_{n+1} = \sum_{k=1}^{n} \beta_k X_k \tag{4.3}$$

and the estimator error variance becomes

$$\begin{aligned}
\mathrm{Var}\left[X_{n+1} - \hat{X}_{n+1}\right] &= \mathrm{E}\left[X_{n+1}^2\right] - 2\sum_{k=1}^{n} \beta_k \,\mathrm{E}\left[X_{n+1}X_k\right] \\
&\quad + \sum_{k=1}^{n}\sum_{j=1}^{n} \beta_k \beta_j \,\mathrm{E}\left[X_k X_j\right]
\end{aligned} \tag{4.4}$$

To minimize this with respect to our unknown coefficients $\beta_1, \beta_2, \ldots, \beta_n$, we set the following derivatives to zero:

$$\frac{\partial}{\partial \beta_\ell} \mathrm{Var}\left[X_{n+1} - \hat{X}_{n+1}\right] = 0 \quad \text{for } \ell = 1, 2, \ldots, n$$

which gives us n equations in n unknowns. Now

$$\frac{\partial}{\partial \beta_\ell} \mathrm{E}\left[X_{n+1}^2\right] = 0$$

$$\frac{\partial}{\partial \beta_\ell} \sum_{k=1}^{n} \beta_k \,\mathrm{E}\left[X_{n+1}X_k\right] = \mathrm{E}\left[X_{n+1}X_\ell\right]$$

$$\frac{\partial}{\partial \beta_\ell} \sum_{k=1}^{n}\sum_{j=1}^{n} \beta_k \beta_j \,\mathrm{E}\left[X_k X_j\right] = 2\sum_{k=1}^{n} \beta_k \,\mathrm{E}\left[X_\ell X_k\right]$$

which gives us

$$\frac{\partial}{\partial \beta_\ell} \text{Var}\left[X_{n+1} - \hat{X}_{n+1}\right] = -2\,\text{E}\left[X_{n+1}X_\ell\right] + 2\sum_{k=1}^{n} \beta_k\, \text{E}\left[X_\ell X_k\right] = 0$$

This means that

$$\text{E}\left[X_{n+1}X_\ell\right] = \sum_{k=1}^{n} \beta_k\, \text{E}\left[X_\ell X_k\right] \tag{4.5}$$

for $\ell = 1, 2, \dots, n$. If we define the matrix and vector components

$$C_{\ell k} = \text{E}\left[X_\ell X_k\right] = \text{Cov}\left[X_\ell, X_k\right]$$
$$b_\ell = \text{E}\left[X_\ell X_{n+1}\right] = \text{Cov}\left[X_\ell, X_{n+1}\right]$$

then Eq. 4.5 can be written as

$$b_\ell = \sum_{k=1}^{n} C_{\ell k}\beta_k$$

or, in matrix notation,

$$\mathbf{b} = \boldsymbol{C}\boldsymbol{\beta} \tag{4.6}$$

which has solution

$$\boldsymbol{\beta} = \boldsymbol{C}^{-1}\mathbf{b} \tag{4.7}$$

These are the so-called Yule–Walker equations and they can be solved by, for example, Gaussian elimination. Notice that $\boldsymbol{\beta}$ does not depend on spatial position, as a linear regression would. It is computed strictly from covariances. It is better to use covariances, if they are known, since this reflects not only distance but also the effects of differing geologic units. For example, two observation points may be physically close together, but if they are in different and largely independent soil layers, then their covariance will be small. Using only distances to evaluate the weights ($\boldsymbol{\beta}$) would miss this effect.

As the above discussion suggests, there is some similarity between best linear unbiased estimate and regression analysis. The primary difference is that regression ignores correlations between data points. However, the primary drawback to BLUE is that the means and covariances *must* be known a priori.

Example 4.1 Suppose that ground-penetrating radar suggests that the mean depth to bedrock μ in meters shows a slow increase with distance s in meters along the proposed line of a roadway, as illustrated in Figure 4.1, that is,

$$\mu(s) = 20 + 0.3s$$

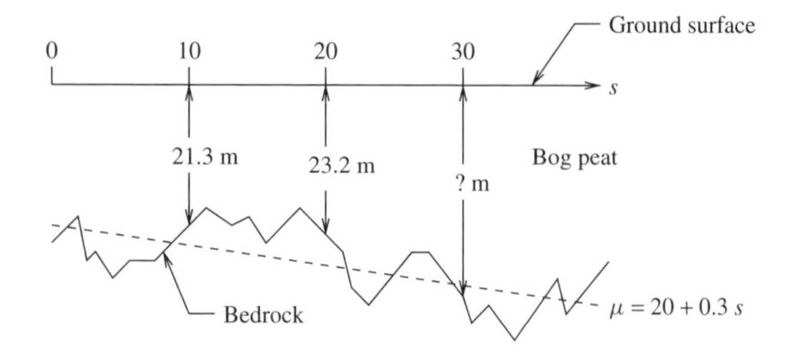

Figure 4.1 Depth to bedrock.

Furthermore suppose that a statistical analysis of bedrock depth at a similar site has given the following covariance function which is assumed to also hold at the current site,

$$C(\tau) = \sigma_X^2 \exp\left\{-\frac{|\tau|}{40}\right\}$$

where $\sigma_X = 5$ m and where τ is the separation distance between points. We want to estimate the bedrock depth X_3 at $s = 30$ m given the following observations of X_1 and X_2, at $s = 10$ m and $s = 20$ m, respectively,

$$x_1 = 21.3 \text{ m} \quad \text{at } s = 10$$
$$x_2 = 23.2 \text{ m} \quad \text{at } s = 20$$

SOLUTION We start by finding the components of the covariance matrix and vector;

$$\mathbf{b} = \begin{Bmatrix} \text{Cov}\left[X_1, X_3\right] \\ \text{Cov}\left[X_2, X_3\right] \end{Bmatrix} = \sigma_X^2 \begin{Bmatrix} e^{-20/40} \\ e^{-10/40} \end{Bmatrix}$$

$$\boldsymbol{C} = \begin{bmatrix} \text{Cov}\left[X_1, X_1\right] & \text{Cov}\left[X_1, X_2\right] \\ \text{Cov}\left[X_2, X_1\right] & \text{Cov}\left[X_2, X_2\right] \end{bmatrix} = \sigma_X^2 \begin{bmatrix} 1 & e^{-10/40} \\ e^{-10/40} & 1 \end{bmatrix}$$

Substituting these into Eq. 4.6 gives

$$\sigma_X^2 \begin{bmatrix} 1 & e^{-10/40} \\ e^{-10/40} & 1 \end{bmatrix} \begin{Bmatrix} \beta_1 \\ \beta_2 \end{Bmatrix} = \sigma_X^2 \begin{Bmatrix} e^{-20/40} \\ e^{-10/40} \end{Bmatrix}$$

Notice that the variance cancels out, which is typical when the variance is constant with position. We now get

$$\begin{Bmatrix} \beta_1 \\ \beta_2 \end{Bmatrix} = \begin{bmatrix} 1 & e^{-10/40} \\ e^{-10/40} & 1 \end{bmatrix}^{-1} \begin{Bmatrix} e^{-20/40} \\ e^{-10/40} \end{Bmatrix} = \begin{Bmatrix} 0 \\ e^{-10/40} \end{Bmatrix}$$

Thus, the optimal linear estimate of X_3 is

$$\begin{aligned}\hat{x}_3 &= \mu(30) + e^{-10/40}[x_2 - \mu(20)] \\ &= [20.0 + 0.3(30)] + e^{-10/40}[23.2 - 20.0 - 0.3(20)] \\ &= 29.0 - 2.8e^{-10/40} \\ &= 26.8 \text{ m}\end{aligned}$$

Notice that, because of the Markovian nature of the covariance function used in this example, the prediction of the future depends only on the most recent past. The prediction is independent of observations further in the past. This is typical of the Markov correlation function in one dimension (in higher dimensions, it is not so straightforward).

4.1.1 Estimator Error

Once the best linear unbiased estimate has been determined, it is of interest to ask how confident are we in our estimate? Can we assess the variability of our estimator? To investigate these questions, let us again consider a zero-mean process so that our estimator can be simply written as

$$\hat{X}_{n+1} = \sum_{k=1}^{n} \beta_k X_k \tag{4.8}$$

In this case, the variance is simply determined as

$$\begin{aligned}\text{Var}\left[\hat{X}_{n+1}\right] &= \text{Var}\left[\sum_{k=1}^{n} \beta_k X_k\right] \\ &= \text{Var}\left[\beta_1 X_1 + \beta_2 X_2 + \cdots + \beta_n X_n\right]\end{aligned} \tag{4.9}$$

The variance of a sum is the sum of the variances *only if the terms are independent.* In this case, the X's are not independent, so the variance of a sum becomes a double sum of all of the possible covariance pairs (see Section 1.7.2),

$$\text{Var}\left[\hat{X}_{n+1}\right] = \sigma_{\hat{X}}^2 = \sum_{k=1}^{n}\sum_{j=1}^{n} \beta_k \beta_j \text{ Cov}\left[X_k, X_j\right] = \boldsymbol{\beta}^{\text{T}} \boldsymbol{C} \boldsymbol{\beta} \tag{4.10}$$

where T means transpose.

However, the above estimator variance is often of limited value. We are typically more interested in asking questions such as: What is the probability that the true value of X_{n+1} exceeds our estimate, $\hat{X}_{n+1}$, by a certain amount. For example, we may want to compute

$$\text{P}\left[X_{n+1} > \hat{X}_{n+1} + b\right] = \text{P}\left[X_{n+1} - \hat{X}_{n+1} > b\right]$$

where b is some constant. Evidently, this would involve finding the distribution of the estimator error $E = (X_{n+1} - \hat{X}_{n+1})$. The variance of the estimator error can be found from Eq. 4.4 as follows:

$$\begin{aligned}\sigma_E^2 &= \text{Var}\left[X_{n+1} - \hat{X}_{n+1}\right] \\ &= \text{E}\left[X_{n+1}^2\right] - 2\sum_{k=1}^{n} \beta_k \text{ E}\left[X_{n+1}X_k\right] \\ &\quad + \sum_{k=1}^{n}\sum_{j=1}^{n} \beta_k \beta_j \text{ E}\left[X_k X_j\right] \\ &= \sigma_X^2 + \boldsymbol{\beta}^{\text{T}}\boldsymbol{C}\boldsymbol{\beta} - 2\boldsymbol{\beta}^{\text{T}}\mathbf{b} \quad \text{(rearranging terms)} \\ &= \sigma_X^2 + \sigma_{\hat{X}}^2 - 2\boldsymbol{\beta}^{\text{T}}\mathbf{b}\end{aligned} \tag{4.11}$$

So we see that the variance of the estimator error (often referred to directly as the estimator error) is the sum of the variance in X and the variance in $\hat{X}$ less a term which depends on the degree of correlation between X_{n+1} and the observations. As the correlation between the observations and the point being estimated increases, it becomes less and less likely that the true value of X_{n+1} will stray very far from its estimate. So for high correlations between the observations and the estimated point, the estimator error becomes small. This can be seen more clearly if we simplify the estimator error equation. To do this, we note that $\boldsymbol{\beta}$ has been determined such that $\boldsymbol{C}\boldsymbol{\beta} = \mathbf{b}$, or, putting it another way, $\boldsymbol{C}\boldsymbol{\beta} - \mathbf{b} = \mathbf{0}$ (where $\mathbf{0}$ is a vector of zeros). Now we write

$$\begin{aligned}\sigma_E^2 &= \sigma_X^2 + \boldsymbol{\beta}^{\text{T}}\boldsymbol{C}\boldsymbol{\beta} - 2\boldsymbol{\beta}^{\text{T}}\mathbf{b} \\ &= \sigma_X^2 + \boldsymbol{\beta}^{\text{T}}\boldsymbol{C}\boldsymbol{\beta} - \boldsymbol{\beta}^{\text{T}}\mathbf{b} - \boldsymbol{\beta}^{\text{T}}\mathbf{b} \\ &= \sigma_X^2 + \boldsymbol{\beta}^{\text{T}}(\boldsymbol{C}\boldsymbol{\beta} - \mathbf{b}) - \boldsymbol{\beta}^{\text{T}}\mathbf{b} \\ &= \sigma_X^2 - \boldsymbol{\beta}^{\text{T}}\mathbf{b}\end{aligned} \tag{4.12}$$

which is a much simpler way of computing σ_E^2 and more clearly demonstrates the variance reduction due to correlation with observations.

The estimator $\hat{X}_{n+1}$ is also the conditional mean of X_{n+1} given the observations. That is,

$$\text{E}\left[X_{n+1} \mid X_1, X_2, \ldots, X_n\right] = \hat{X}_{n+1} \tag{4.13}$$

The conditional variance of X_{n+1} is σ_E^2,

$$\text{Var}\left[X_{n+1} \mid X_1, X_2, \ldots, X_n\right] = \sigma_E^2 \tag{4.14}$$

Generally questions regarding the probability that the true X_{n+1} lies in some region should employ the conditional mean and variance of X_{n+1}, since this would then make use of all of the information at hand.

Example 4.2 Consider again Example 4.1. What is the variance of the estimator and the estimator error? Estimate the probability that X_3 exceeds $\hat{X}_3$ by more than 4 m.

SOLUTION We had

$$\boldsymbol{C} = \sigma_X^2 \begin{bmatrix} 1 & e^{-10/40} \\ e^{-10/40} & 1 \end{bmatrix} = (5)^2 \begin{bmatrix} 1 & e^{-10/40} \\ e^{-10/40} & 1 \end{bmatrix}$$

and

$$\boldsymbol{\beta} = \begin{Bmatrix} 0 \\ e^{-10/40} \end{Bmatrix}$$

so that

$$\sigma_{\hat{X}}^2 = \text{Var}\left[\hat{X}_3\right] = (5)^2 \left\{0 \quad e^{-10/40}\right\} \begin{bmatrix} 1 & e^{-10/40} \\ e^{-10/40} & 1 \end{bmatrix} \times \begin{Bmatrix} 0 \\ e^{-10/40} \end{Bmatrix} = (5)^2 \, e^{-20/40}$$

which gives $\sigma_{\hat{X}} = 5e^{-10/40} = 3.894$ m.

For the covariance vector found in Example 4.1,

$$\mathbf{b} = \sigma_X^2 \begin{Bmatrix} e^{-20/40} \\ e^{-10/40} \end{Bmatrix}$$

the estimator error is computed as

$$\begin{aligned} \sigma_E^2 = \text{Var}\left[X_3 - \hat{X}_3\right] &= \sigma_X^2 - \boldsymbol{\beta}^{\mathrm{T}} \mathbf{b} \\ &= \sigma_X^2 - \sigma_X^2 \{0 \; e^{-10/40}\} \begin{Bmatrix} e^{-20/40} \\ e^{-10/40} \end{Bmatrix} \\ &= (5)^2 \left(1 - e^{-20/40}\right) \end{aligned}$$

The standard deviation of the estimator error is thus $\sigma_E = 5\sqrt{1 - e^{-20/40}} = 3.136$ m. Note that this is less than the variability of the estimator itself and significantly less than the variability of X, due to the restraining effect of correlation between points.

To compute the required probability, we need to assume a distribution for the random variable $(X_3 - \hat{X}_3)$. Let us suppose that X is normally distributed. Since the estimate $\hat{X}$ is simply a sum of X's, it too must be normally distributed, which in turn implies that the quantity $X_3 - \hat{X}_3$ is normally distributed. We need only specify its mean and standard deviation, then, to fully describe its distribution.

We saw above that $\hat{X}_3$ is an unbiased estimate of X_3,

$$\text{E}\left[X_3 - \hat{X}_3\right] = 0$$

so that $\mu_E = 0$. We have just computed the standard deviation of $X_3 - \hat{X}_3$ as $\sigma_E = 3.136$ m. Thus,

$$\begin{aligned} \text{P}\left[X_3 - \hat{X}_3 > 4\right] &= \text{P}\left[Z > \frac{4 - 0}{3.136}\right] \\ &= 1 - \Phi(1.28) = 0.1003 \end{aligned}$$

4.1.2 Geostatistics: Kriging

Danie G. Krige's empirical work to evaluate mineral resources (1951) was formalize by Matheron (1962) into a statistical approach now commonly referred to as "Kriging" and normally used in geostatistics. Kriging is basically best linear unbiased estimation with the added ability to estimate certain aspects of the mean trend. We will give the theory for Kriging in this section, recognizing that some concepts will be repeated from best linear unbiased estimation. The application will be to a settlement problem in geotechnical engineering.

The purpose of Kriging is to provide a best estimate of a random field between known data. The basic idea is to estimate $X(\mathbf{x})$ at any point using a weighted linear combination of the values of X at each observation point. Suppose that $X_1, X_2, \ldots, X_n$ are observations of the random field $X(\mathbf{x})$ at the points $\mathbf{x}_1, \mathbf{x}_2, \ldots, \mathbf{x}_n$, that is, $X_k = X(\mathbf{x}_k)$. Then the Kriged estimated of $X(\mathbf{x})$ at $\mathbf{x}$ is given by

$$\hat{X}(\mathbf{x}) = \sum_{k=1}^{n} \beta_k X_k \tag{4.15}$$

where the n unknown weights β_k are to be determined to find the best estimate at the point $\mathbf{x}$. It seems reasonable that if the point $\mathbf{x}$ is particularly close to one of the observations, say X_j, then the weight β_j associated with X_j would be high. However, if $X(\mathbf{x})$ and X_j are in different (independent) soil layers, for example, then perhaps β_j should be small. Rather than using distance to determine the weights in Eq. 4.15, it is better to use covariance (or correlation) between the two points since this reflects not only distance but also, for example, the effects of differing geologic units.

In Kriging, it is assumed that the mean can be expressed as in a regression analysis,

$$\mu_X(\mathbf{x}) = \sum_{i=1}^{m} a_i g_i(\mathbf{x}) \tag{4.16}$$

where a_i is an unknown coefficient (which, as it turns out, need never be estimated) and $g_i(\mathbf{x})$ is a specified function of $\mathbf{x}$. Usually $g_1(x) = 1$, $g_2(x) = x$, $g_3(x) = x^2$, and so on in one dimension—similarly in higher dimensions. As in a regression analysis, the functions $g_1(\mathbf{x})$, $g_2(\mathbf{x}), \cdots$ should be (largely) linearly independent over the domain of the regression (i.e., the site domain). In order for the estimator (Eq. 4.15) to be unbiased, we require that the mean difference between the estimate $\hat{X}(\mathbf{x})$ and the true (but random) value $X(\mathbf{x})$ be zero,

$$\text{E}\left[\hat{X}(\mathbf{x}) - X(\mathbf{x})\right] = \text{E}\left[\hat{X}(\mathbf{x})\right] - \text{E}\left[X(\mathbf{x})\right] = 0 \tag{4.17}$$

where

$$\mathrm{E}\left[\hat{X}(\mathbf{x})\right] = \mathrm{E}\left[\sum_{k=1}^{n} \beta_k X_k\right] = \sum_{k=1}^{n} \beta_k \left(\sum_{i=1}^{m} a_i g_i(\mathbf{x}_k)\right)$$

$$\mathrm{E}\left[X(\mathbf{x})\right] = \sum_{i=1}^{m} a_i g_i(\mathbf{x})$$

The unbiased condition of Eq. 4.17 becomes

$$\sum_{i=1}^{m} a_i \left\{\sum_{k=1}^{n} \beta_k g_i(\mathbf{x}_k) - g_i(\mathbf{x})\right\} = 0 \qquad (4.18)$$

Since this must be true for any coefficients a_i, the unbiased condition reduces to

$$\sum_{k=1}^{n} \beta_k g_i(\mathbf{x}_k) = g_i(\mathbf{x}) \qquad (4.19)$$

which is independent of the unknown regression weights a_i.

The unknown Kriging weights $\boldsymbol{\beta}$ are obtained by minimizing the variance of the error, $E = \left(X(\mathbf{x}) - \hat{X}(\mathbf{x})\right)$, which reduces the solution to the matrix equation

$$\boldsymbol{K\beta} = \mathbf{M} \qquad (4.20)$$

where $\boldsymbol{K}$ and $\mathbf{M}$ depend on the covariance structure,

$$\boldsymbol{K} = \begin{bmatrix} C_{11} & C_{12} & \cdots & C_{1n} & g_1(\mathbf{x}_1) & g_2(\mathbf{x}_1) & \cdots & g_m(\mathbf{x}_1) \\ C_{21} & C_{22} & \cdots & C_{2n} & g_1(\mathbf{x}_2) & g_2(\mathbf{x}_2) & \cdots & g_m(\mathbf{x}_2) \\ \cdot & \cdot & & \cdot & \cdot & \cdot & & \cdot \\ \cdot & \cdot & & \cdot & \cdot & \cdot & & \cdot \\ \cdot & \cdot & & \cdot & \cdot & \cdot & & \cdot \\ C_{n1} & C_{n2} & \cdots & C_{nn} & g_1(\mathbf{x}_n) & g_2(\mathbf{x}_n) & \cdots & g_m(\mathbf{x}_n) \\ g_1(\mathbf{x}_1) & g_1(\mathbf{x}_2) & \cdots & g_1(\mathbf{x}_n) & 0 & 0 & \cdots & 0 \\ g_2(\mathbf{x}_1) & g_2(\mathbf{x}_2) & \cdots & g_2(\mathbf{x}_n) & 0 & 0 & \cdots & 0 \\ \cdot & \cdot & & \cdot & \cdot & \cdot & & \cdot \\ \cdot & \cdot & & \cdot & \cdot & \cdot & & \cdot \\ \cdot & \cdot & & \cdot & \cdot & \cdot & & \cdot \\ g_m(\mathbf{x}_1) & g_m(\mathbf{x}_2) & \cdots & g_m(\mathbf{x}_n) & 0 & 0 & \cdots & 0 \end{bmatrix}$$

in which C_{ij} is the covariance between X_i and X_j and

$$\boldsymbol{\beta} = \begin{Bmatrix} \beta_1 \\ \beta_2 \\ \cdot \\ \cdot \\ \cdot \\ \beta_n \\ -\eta_1 \\ -\eta_2 \\ \cdot \\ \cdot \\ \cdot \\ -\eta_m \end{Bmatrix}, \qquad \mathbf{M} = \begin{Bmatrix} C_{1x} \\ C_{2x} \\ \cdot \\ \cdot \\ \cdot \\ C_{nx} \\ g_1(\mathbf{x}) \\ g_2(\mathbf{x}) \\ \cdot \\ \cdot \\ \cdot \\ g_m(\mathbf{x}) \end{Bmatrix}$$

The quantities η_i are a set of Lagrangian parameters used to solve the variance minimization problem subject to the unbiased conditions of Eq. 4.19. Beyond allowing for a solution to the above system of equations, their actual values can be ignored. The covariance C_{ix} appearing in the vector on the right-hand side (RHS), $\mathbf{M}$, is the covariance between the ith observation point and the point $\mathbf{x}$ at which the best estimate is to be calculated.

Note that the matrix $\boldsymbol{K}$ is purely a function of the observation point locations and their covariances; thus it can be inverted once and then Eqs. 4.20 and 4.15 used repeatedly at different spatial points to build up the field of best estimates (for each spatial point, the RHS vector $\mathbf{M}$ changes, as does the vector of weights, $\boldsymbol{\beta}$).

The Kriging method depends upon two things: (1) knowledge of how the mean varies functionally with position, that is, $g_1, g_2, \ldots$ need to be specified, and (2) knowledge of the covariance structure of the field. Usually, assuming a mean which is either constant ($m = 1$, $g_1(\mathbf{x}) = 1$, $a_1 = \mu_X$) or linearly varying is sufficient. The correct form of the mean trend can be determined by

1. plotting the results and visually checking the mean trend,
2. performing a regression analysis, or
3. performing a more complex structural analysis; see, for example, Journel and Huijbregts (1978) for more details.

The covariance structure can be estimated by the methods discussed in Chapter 5 if sufficient data are available and used directly in Eq. 4.20 to define $\boldsymbol{K}$ and $\mathbf{M}$ (with, perhaps some interpolation for covariances not directly estimated). In the absence of sufficient data, a simple functional form for the covariance function is often assumed. A typical model is the Markovian in which the covariance decays exponentially with separation distance $\tau_{ij} = |\mathbf{x}_i - \mathbf{x}_j|$:

$$C_{ij} = \sigma_X^2 \, \exp\left\{-\frac{2|\tau_{ij}|}{\theta}\right\}$$

As mentioned in Chapter 3, the parameter θ is called the *correlation length*. Such a model now requires only the estimation of two parameters, σ_X and θ, but assumes that the field is *isotropic* and *statistically stationary*. Nonisotropic models are readily available and often appropriate for soils which display layering.

4.1.2.1 Estimator Error Associated with any estimate of a random process derived from a finite number of observations is an estimator error. This error can be used to assess the accuracy of the estimate.

The Kriging estimate is unbiased, so that

$$\mu_{\hat{X}}(\mathbf{x}) = \mathrm{E}\left[\hat{X}(\mathbf{x})\right] = \mathrm{E}\left[X(\mathbf{x})\right] = \mu_X(\mathbf{x})$$

Defining the error as the difference between the estimate $\hat{X}(\mathbf{x})$ and its true (but unknown and random) value $X(\mathbf{x})$, $E = X(\mathbf{x}) - \hat{X}(\mathbf{x})$, the mean and variance of the estimator error are given by

$$\mu_E = \mathrm{E}\left[X(\mathbf{x}) - \hat{X}(\mathbf{x})\right] = 0 \tag{4.21a}$$

$$\sigma_E^2 = \mathrm{E}\left[\left(X(\mathbf{x}) - \hat{X}(\mathbf{x})\right)^2\right] = \sigma_X^2 + \boldsymbol{\beta}_n^{\mathrm{T}}(\boldsymbol{K}_{n\times n}\boldsymbol{\beta}_n - 2\mathbf{M}_n) \tag{4.21b}$$

where $\boldsymbol{\beta}_n$ and $\mathbf{M}_n$ are the first n elements of $\boldsymbol{\beta}$ and $\mathbf{M}$ and $\boldsymbol{K}_{n\times n}$ is the $n \times n$ upper left submatrix of $\boldsymbol{K}$ containing the covariances. Note that $\hat{X}(\mathbf{x})$ can also be viewed as the conditional mean of $X(\mathbf{x})$ at the point $\mathbf{x}$. The conditional variance at the point $\mathbf{x}$ would then be σ_E^2.

Example 4.3 Foundation Consolidation Settlement Consider the estimation of consolidation settlement under a footing at a certain location given that soil samples/tests have been obtained at four neighboring locations. Figure 4.2 shows a plan view of the footing and sample locations. The samples and local stratigraphy are used to estimate the soil parameters C_c, e_o, H, and p_o appearing in the consolidation settlement equation

$$S = N\left(\frac{C_c}{1+e_o}\right) H \ \log_{10}\left(\frac{p_o + \Delta p}{p_o}\right) \tag{4.22}$$

Each of these four parameters is then treated as spatially varying and random between observation points. It is assumed that the estimation error in obtaining the parameters from the samples is negligible compared to field variability, and so this source of uncertainty will be ignored. The model error parameter N is assumed an ordinary random variable (not a random field) with mean 1.0 and standard deviation 0.1. The increase in pressure at middepth of the clay layer, Δp, depends on the load applied to the footing. We will assume that $\mathrm{E}\left[\Delta p\right] = 25$ kPa with standard deviation 5 kPa.

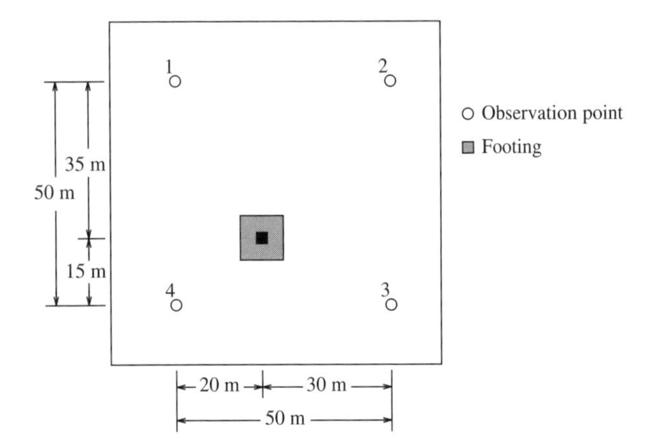

Figure 4.2 Consolidation settlement plan view with sample points.

The task now is to estimate the mean and standard deviation of C_c, e_o, H, and p_o at the footing location using the neighboring observations. Table 4.1 lists the soil settlement properties obtained at each of the four sample points.

In Table 4.1, we have assumed that all four random fields are stationary, with spatially constant mean and variance, the limited data not clearly indicating otherwise. In order to obtain a Kriging estimate at the footing location, we need to establish a covariance structure for the field. Obviously four sample points are far too few to yield even a rough approximation of the variance and covariance between samples, especially in two dimensions. We have assumed that experience with similar sites and similar materials leads us to estimate the coefficients of variation, v, shown in the table and a correlation length of about 60 m using an exponentially decaying correlation function. That is, we assume that the correlation structure is reasonably well approximated by

$$\rho(\mathbf{x}_i, \mathbf{x}_j) = \exp\left\{-\tfrac{2}{60}|\mathbf{x}_i - \mathbf{x}_j|\right\}$$

In so doing, we are assuming that the clay layer is horizontally isotropic, also a reasonable assumption. This yields the following correlation matrix between sample points:

$$\boldsymbol{\rho} = \begin{bmatrix} 1.000 & 0.189 & 0.095 & 0.189 \\ 0.189 & 1.000 & 0.189 & 0.095 \\ 0.095 & 0.189 & 1.000 & 0.189 \\ 0.189 & 0.095 & 0.189 & 1.000 \end{bmatrix}$$

Furthermore, it is reasonable to assume that the same correlation length applies to all four soil properties. Thus, the covariance matrix associated with the property C_c between

Table 4.1 Derived Soil Sample Settlement Properties

Sample Point	C_c	e_o	H (m)	p_o (kPa)
1	0.473	1.42	4.19	186.7
2	0.328	1.08	4.04	181.0
3	0.489	1.02	4.55	165.7
4	0.295	1.24	4.29	179.1
μ	0.396	1.19	4.27	178.1
v	0.25	0.15	0.05	0.05

sample points is just $\sigma^2_{C_c}\boldsymbol{\rho} = (0.25 \times 0.396)^2\boldsymbol{\rho}$. Similarly, the covariance matrix associated with e_o is its variance $[\sigma^2_{e_o} = (0.15 \times 1.19)^2 = 0.03186]$ times the correlation matrix, and so on.

In the following, we will obtain Kriging estimates from each of the four random fields [$C_c(\mathbf{x})$, $e_o(\mathbf{x})$, $H(\mathbf{x})$, and $p_o(\mathbf{x})$] independently. Note that this does not imply that the estimates will be independent, since if the sample properties are themselves correlated, which they most likely are, then the estimates will also be correlated. It is believed that this is a reasonably good approximation given the level of available data. If more complicated cross-correlation structures are known to exist and have been estimated, the method of co-Kriging can be applied; this essentially amounts to the use of a much larger covariance (Kriging) matrix and the consideration of all four fields simultaneously. Co-Kriging also has the advantage of also ensuring that the error variance is properly minimized. However, co-Kriging is not implemented here, since the separate Kriging preserves reasonably well any existing pointwise cross-correlation between the fields and since little is generally known about the actual cross-correlation structure.

The Kriging matrix associated with the clay layer thickness H is then obtained by multiplying $\sigma^2_H = (0.05 \times 4.27)^2$ by $\boldsymbol{\rho}$:

$$\boldsymbol{K}_H = \begin{bmatrix} 0.04558 & 0.00861 & 0.00432 & 0.00861 & 1 \\ 0.00861 & 0.04558 & 0.00861 & 0.00432 & 1 \\ 0.00432 & 0.00861 & 0.04558 & 0.00861 & 1 \\ 0.00861 & 0.00432 & 0.00861 & 0.04558 & 1 \\ 1 & 1 & 1 & 1 & 0 \end{bmatrix}$$

where, since we assumed stationarity, $m = 1$ and $g_1(\mathbf{x}) = 1$ in Eq. 4.16. Placing the coordinate axis origin at sample location 4 gives the footing coordinates $\mathbf{x} = (20, 15)$. Thus, the RHS vector $\mathbf{M}$ is

$$\mathbf{M}_H = \begin{Bmatrix} \sigma^2_H \rho(\mathbf{x}_1, \mathbf{x}) \\ \sigma^2_H \rho(\mathbf{x}_2, \mathbf{x}) \\ \sigma^2_H \rho(\mathbf{x}_3, \mathbf{x}) \\ \sigma^2_H \rho(\mathbf{x}_4, \mathbf{x}) \\ 1 \end{Bmatrix} = \begin{Bmatrix} (0.04558)(0.2609) \\ (0.04558)(0.2151) \\ (0.04558)(0.3269) \\ (0.04558)(0.4346) \\ 1 \end{Bmatrix} = \begin{Bmatrix} 0.01189 \\ 0.00981 \\ 0.01490 \\ 0.01981 \\ 1 \end{Bmatrix}$$

Solving the matrix equation $\boldsymbol{K}_H \boldsymbol{\beta}_H = \mathbf{M}_H$ gives the following four weights (ignoring the Lagrange parameter):

$$\boldsymbol{\beta}_H = \begin{Bmatrix} 0.192 \\ 0.150 \\ 0.265 \\ 0.393 \end{Bmatrix}$$

in which we can see that the samples which are closest to the footing are most heavily weighted (more specifically, the samples which are most highly correlated with the footing location are the most heavily weighted), as would be expected.

Since the underlying correlation matrix is identical for all four soil properties, the weights will be identical for all four properties; thus the best estimates at the footing are

$$\begin{aligned} \hat{C}_c &= (0.192)(0.473) + (0.150)(0.328) \\ &\quad + (0.265)(0.489) + (0.393)(0.295) = 0.386 \\ \hat{e}_o &= (0.192)(1.42) + (0.150)(1.08) \\ &\quad + (0.265)(1.02) + (0.393)(1.24) = 1.19 \\ \hat{H} &= (0.192)(4.19) + (0.150)(4.04) \\ &\quad + (0.265)(4.55) + (0.393)(4.29) = 4.30 \\ \hat{p}_o &= (0.192)(186.7) + (0.150)(181.0) \\ &\quad + (0.265)(165.7) + (0.393)(179.1) = 177.3 \end{aligned}$$

The estimation errors are given by the equation

$$\sigma^2_E = \sigma^2_X + \boldsymbol{\beta}_n^{\mathrm{T}}(\boldsymbol{K}_{n\times n}\boldsymbol{\beta}_n - 2\mathbf{M}_n)$$

Since the $n \times n$ submatrix of $\boldsymbol{K}$ is just the correlation matrix times the appropriate variance, and similarly $\mathbf{M}_n$ is the correlation vector (between samples and footing) times the appropriate variance, the error can be rewritten as

$$\sigma^2_E = \sigma^2_X\left(1 + \boldsymbol{\beta}_n^{\mathrm{T}}(\boldsymbol{\rho}\boldsymbol{\beta}_n - 2\boldsymbol{\rho}_x)\right)$$

where $\boldsymbol{\rho}_x$ is the vector of correlation coefficients between the samples and the footing (see the calculation of $\mathbf{M}_H$ above). For the Kriging weights and given correlation structure, this yields

$$\sigma^2_E = \sigma^2_X(0.719)$$

which gives the following individual estimation errors:

$$\begin{aligned} \sigma^2_{C_c} &= (0.009801)(0.719) = 0.00705 &\rightarrow \sigma_{C_c} &= 0.0839 \\ \sigma^2_{e_o} &= (0.03204)(0.719) = 0.0230 &\rightarrow \sigma_{e_o} &= 0.152 \\ \sigma^2_H &= (0.04558)(0.719) = 0.0328 &\rightarrow \sigma_H &= 0.181 \\ \sigma^2_{p_o} &= (79.31)(0.719) = 57.02 &\rightarrow \sigma_{p_o} &= 7.55 \end{aligned}$$

In summary, then, the variables entering the consolidation settlement formula have the following statistics based on the preceding Kriged estimates:

Variable	Mean	Standard Deviation (SD)	v
N	1.0	0.1	0.1
C_c	0.386	0.0839	0.217
e_o	1.19	0.152	0.128
H (m)	4.30	0.181	0.042
p_o (kPa)	177.3	7.55	0.043
Δp (kPa)	25.0	5.0	0.20

where v is the coefficient of variation.

A first-order approximation to the settlement, via Eq. 4.22, is thus

$$\mu_S = (1.0)\left(\frac{0.386}{1+1.19}\right)(4.30)\log_{10}\left(\frac{177.3+25}{177.3}\right)$$
$$= 0.0429 \text{ m}$$

To estimate the settlement variance, a first-order approximation yields

$$\sigma_S^2 = \sum_{j=1}^{m}\left(\frac{\partial S}{\partial X_j}\sigma_{X_j}\right)_\mu^2$$

where the subscript μ on the derivative implies that it is evaluated at the mean of all random variables and the variable X_j is replaced by each of N, C_c, ... in turn. Evaluation of the derivatives at the mean leads to the following table:

X_j	μ_{X_j}	$(\partial S/\partial X_j)_\mu$	σ_{X_j}	$[(\partial S/\partial X_j)\sigma_{X_j}]_\mu^2$
N	1.000	0.04342	0.1000	1.885×10^{-5}
C_c	0.386	0.11248	0.0889	8.906×10^{-5}
e_o	1.19	−0.01983	0.1520	0.908×10^{-5}
H	4.30	0.01010	0.1810	0.334×10^{-5}
p_o	177.3	−0.00023	7.5500	0.300×10^{-5}
Δp	25.0	0.00163	5.0000	6.618×10^{-5}

so that

$$\sigma_S^2 = \sum_{j=1}^{m}\left(\frac{\partial S}{\partial X_j}\sigma_{X_j}\right)_\mu^2 = 18.952\times10^{-5}\text{ m}^2$$

Hence $\sigma_S = 0.0138$ and the coefficient of variation of the settlement at the footing is $v_S = 0.0138/0.0429 = 0.322$. This is roughly a 10% decrease from the coefficient of variation of settlement obtained without the benefit of any neighboring observations (0.351). Although this does not seem significant in light of the increased complexity of the above calculations, it needs to be remembered that the contribution to overall uncertainty coming from N and Δp amounts to over 40%. Thus, the coefficient of variation v_S will decrease toward its minimum (barring improved information about N and/or Δp) of 0.213 as more observations are used and/or observations are taken closer to the footing. For example, if a fifth sample were taken midway between the other four samples (at the center of Figure 4.2), then the variance of each estimator decreases by a factor of 0.46 from the point variance (rather than the factor of 0.719 found above) and the settlement v_S becomes 0.285. Note that the reduction in variance can be found prior to actually performing the sampling since the estimator variance depends only on the covariance structure and the assumed functional form for the mean. Thus, the Kriging technique can also be used to plan an optimal sampling scheme—sample points are selected so as to minimize the estimator error.

Once the random-field model has been defined for a site, there are ways of analytically obtaining probabilities associated with design criteria, such as the probability of failure. For example, by assuming a normal or lognormal distribution for the footing settlement in this example, one can easily estimate the probability that the footing will exceed a certain settlement given its mean and standard deviation. Assuming the footing settlement to be normally distributed with mean 0.0429 m and standard deviation 0.0138, then the probability that the settlement will exceed 0.075 m is

$$\text{P}[S > 0.075] = 1 - \Phi\left(\frac{0.075 - 0.0429}{0.0138}\right)$$
$$= 1 - \Phi(2.33) = 0.01$$

4.2 THRESHOLD EXCURSIONS IN ONE DIMENSION

In both design and analysis contexts, the extremes of random processes are typically of considerable interest. Many reliability problems are defined in terms of threshold excursions—for example, when load exceeds a safe threshold (e.g., the strength). Most theories governing extremal statistics of random fields deal with excursion regions, regions in which the process X exceeds some threshold, and the few exact results that exist usually only apply asymptotically when the threshold level approaches infinity. A large class of random functions are not amenable to existing extrema theory at all, and for such processes the analysis of a sequence of realizations is currently the only way to obtain their extrema statistics. In this section we will investigate the basic theory of threshold excursions for one-dimensional processes. Since the statistics of threshold excursions depend heavily on the slope variance, we will begin by looking at the derivative, or slope, process.

4.2.1 Derivative Process

Consider a stationary random field $X(t)$. Its derivative is

$$\dot{X}(t) = \frac{dX(t)}{dt} = \lim_{\Delta t \to 0} \frac{X(t+\Delta t) - X(t)}{\Delta t} \tag{4.23}$$

We will concentrate on the finite-difference form of the derivative and write

$$\dot{X}(t) = \frac{X(t+\Delta t) - X(t)}{\Delta t} \tag{4.24}$$

with the limit being understood. The mean of the derivative process can be obtained by taking expectations of Eq. 4.24,

$$\mathrm{E}\left[\dot{X}(t)\right] = \mu_{\dot{X}} = \frac{\mathrm{E}\left[X(t+\Delta t)\right] - \mathrm{E}\left[X(t)\right]}{\Delta t} = 0 \tag{4.25}$$

since $\mathrm{E}\left[X(t+\Delta t)\right] = \mathrm{E}\left[X(t)\right]$ due to stationarity. Before computing the variance of the derivative process, it is useful to note that the (centered) finite-difference form of the second derivative of the covariance function of X, $C_X(\tau)$, at $\tau = 0$ is

$$\left.\frac{d^2 C_X(\tau)}{d\tau^2}\right|_{\tau=0} = \ddot{C}_X(0) = \frac{C_X(\Delta\tau) - 2C_X(0) + C_X(-\Delta\tau)}{\Delta\tau^2} \tag{4.26}$$

The variance of the derivative process, $\dot{X}(t)$, is thus obtained as

$$\begin{aligned}
\sigma_{\dot{X}}^2 = \mathrm{E}\left[\dot{X}^2\right] &= \frac{1}{\Delta t^2}\left\{2\mathrm{E}\left[X^2(t)\right] - 2\,\mathrm{E}\left[X(t)X(t+\Delta t)\right]\right\} \\
&= \frac{2\left[C_X(0) - C_X(\Delta t)\right]}{\Delta t^2} \\
&= -\frac{C_X(\Delta t) - 2C_X(0) + C_X(-\Delta t)}{\Delta t^2} \\
&= -\left.\frac{d^2 C_X(\tau)}{d\tau^2}\right|_{\tau=0}
\end{aligned} \tag{4.27}$$

where, due to stationarity, $\mathrm{E}\left[X^2(t+\Delta t)\right] = \mathrm{E}\left[X^2(t)\right]$ and, due to symmetry in the covariance function, we can write $2C_X(\Delta t) = C_X(\Delta t) + C_X(-\Delta t)$. From this we see that the derivative process will exist (i.e., will have finite variance) if the second derivative of $C_X(\tau)$ is finite at $\tau = 0$. A necessary and sufficient condition for $X(t)$ to be *mean square differentiable* (i.e., for the derivative process to have finite variance) is that the first derivative of $C_X(\tau)$ at the origin be equal to zero,

$$\left.\frac{dC_X(\tau)}{d\tau}\right|_{\tau=0} = \dot{C}_X(0) = 0 \tag{4.28}$$

If $\dot{C}_X(0)$ exists, it must be zero due to the symmetry in $C_X(\tau)$. Equation 4.28 is then equivalent to saying that $\dot{C}_X(0)$ exists if the equation is satisfied. In turn, if $\dot{C}_X(0) = 0$, then, because $C_X(\tau) \le C_X(0)$ so that $C_X(0)$ is a maximum, the second derivative, $\ddot{C}_X(0)$, must be finite and negative. This leads to a finite and positive derivative variance, $\sigma_{\dot{X}}^2$, according to Eq. 4.27.

For simplicity we will now assume that $\mathrm{E}\left[X(t)\right] = 0$, so that the covariance function becomes

$$C_X(\tau) = \mathrm{E}\left[X(t)X(t+\tau)\right] \tag{4.29}$$

There is no loss in generality by assuming $\mathrm{E}\left[X(t)\right] = 0$. A nonzero mean does not affect the covariance since the basic definition of covariance subtracts the mean in any case; see Eq. 1.29a or 3.4. The zero-mean assumption just simplifies the algebra. Differentiating $C_X(\tau)$ with respect to τ gives (the derivative of an expectation is the expectation of the derivative, just as the derivative of a sum is the sum of the derivatives)

$$\dot{C}_X(\tau) = \mathrm{E}\left[X(t)\dot{X}(t+\tau)\right]$$

Since $X(t)$ is stationary, we can replace t by $(t-\tau)$ (i.e., the statistics of X are the same at any point in time), which now gives

$$\dot{C}_X(\tau) = \mathrm{E}\left[X(t-\tau)\dot{X}(t)\right]$$

Differentiating yet again with respect to τ gives

$$\ddot{C}_X(\tau) = -\mathrm{E}\left[\dot{X}(t-\tau)\dot{X}(t)\right] = -C_{\dot{X}}(\tau)$$

In other words, the covariance function of the slope, $\dot{X}(t)$, is just equal to the negative second derivative of the covariance function of $X(t)$,

$$C_{\dot{X}}(\tau) = -\ddot{C}_X(\tau) \tag{4.30}$$

This result can also be used to find the variance of $\dot{X}(t)$,

$$C_{\dot{X}}(0) = -\ddot{C}_X(0)$$

which agrees with Eq. 4.27.

The cross-covariance between $X(t)$ and its derivative, $\dot{X}(t)$, can be obtained by considering (assuming, without loss in generality, that $\mu_X = 0$)

$$\begin{aligned}
\mathrm{Cov}\left[X,\dot{X}\right] = \mathrm{E}\left[X\dot{X}\right] &= \mathrm{E}\left[X(t)\left(\frac{X(t+\Delta t) - X(t)}{\Delta t}\right)\right] \\
&= \mathrm{E}\left[\left(\frac{X(t)X(t+\Delta t) - X^2(t)}{\Delta t}\right)\right] \\
&= \frac{C_X(\Delta t) - C_X(0)}{\Delta t} \\
&= \dot{C}_X(0)
\end{aligned}$$

Thus, if $\dot{X}$ exists [i.e., $\dot{C}_X(0) = 0$], it will be uncorrelated with X.

A perhaps more physical understanding of why some processes are not mean square differentiable comes if we consider the Brownian motion problem, whose solution possesses the Markov correlation function (the *Ornstein–Uhlenbeck* process—see Section 3.6.5). The idea in the Brownian motion model is that the motion of a particle changes randomly in time due to impulsive impacts by other (perhaps smaller) particles. At the instant of the impact, the particle velocity changes, and these changes are discontinuous. Thus, at the instant of each impact, the velocity derivative becomes infinite and so its variance becomes infinite.

Example 4.4 Show that a process having a Markov covariance function is not mean square differentiable, whereas a process having a Gaussian covariance function is.

SOLUTION The Markov covariance function is given by

$$C(\tau) = \sigma^2 \exp\left\{-\frac{2|\tau|}{\theta}\right\} \tag{4.31}$$

which is shown in Figure 4.3. The derivative of $C(\tau)$ is

$$\dot{C}(\tau) = \frac{d}{d\tau}C(\tau) = \begin{cases} \left(\dfrac{2\sigma^2}{\theta}\right) \exp\left\{\dfrac{2\tau}{\theta}\right\} & \text{if } \tau < 0 \\ -\left(\dfrac{2\sigma^2}{\theta}\right) \exp\left\{-\dfrac{2\tau}{\theta}\right\} & \text{if } \tau > 0 \end{cases} \tag{4.32}$$

which is undefined at $\tau = 0$. This is clearly evident in Figure 4.3. Thus, since $\dot{C}(0) \neq 0$, the Markov process is not mean square differentiable.

The Gaussian covariance function is

$$C(\tau) = \sigma^2 \exp\left\{-\frac{\pi\tau^2}{\theta^2}\right\} \tag{4.33}$$

which is shown in Figure 4.4. The derivative of $C(\tau)$ is now

$$\dot{C}(\tau) = -2\tau\left(\frac{\pi}{\theta^2}\right)\sigma^2 \exp\left\{-\frac{\pi\tau^2}{\theta^2}\right\} \tag{4.34}$$

and since $\dot{C}(0) = 0$, as can be seen in Figure 4.4, a process having Gaussian covariance function is mean square differentiable.

Vanmarcke (1984) shows that even a small amount of local averaging will convert a non–mean square differentiable process into one which is mean square differentiable (i.e., which possesses a finite variance derivative). In fact, all local average processes will be mean square differentiable. Suppose we define the local arithmetic average process, as usual, to be

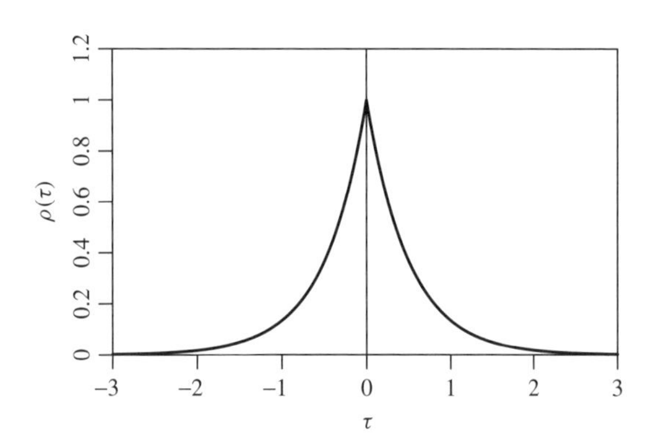

Figure 4.3 Typical Markov correlation function.

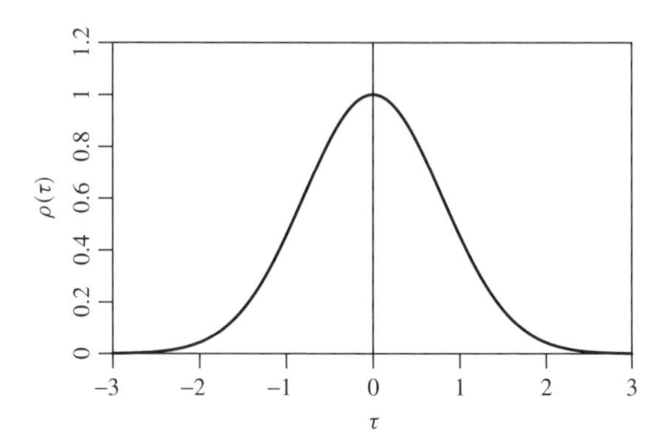

Figure 4.4 Typical Gaussian correlation function.

$$X_T(t) = \frac{1}{T}\int_{t-T/2}^{t+T/2} X(\xi)\, d\xi \tag{4.35}$$

The covariance function of $X_T(t)$ is [where we assume $X(t)$ has mean zero for simplicity in the interim step]

$$\begin{aligned} C_{X_T}(\tau) &= \mathrm{E}\left[X_T(t)X_T(t+\tau)\right] \\ &= \frac{1}{T^2}\int_0^T \int_\tau^{\tau+T} C_X(\xi - \eta)\, d\eta\, d\xi \end{aligned} \tag{4.36}$$

where $C_X(\tau)$ is the covariance function of $X(t)$.

If $X_T(t)$ is mean square differentiable, then $\dot{C}_{X_T}(0) = 0$. We can show this will be true for any averaging region $T > 0$ as follows:

$$\dot{C}_{X_T}(\tau) = \frac{d}{d\tau}C_{X_T}(\tau) = \frac{1}{T^2}\int_0^T \frac{d}{d\tau}\int_\tau^{\tau+T} C_X(\xi - \eta)\, d\eta\, d\xi \tag{4.37}$$

Noting that

$$\frac{d}{d\tau}\int_\tau^{\tau+T} g(\eta)\, d\eta = g(\tau + T) - g(\tau) \tag{4.38}$$

we see that

$$\dot{C}_{X_T}(\tau) = \frac{1}{T^2}\int_0^T \left[C_X(\xi - \tau - T) - C_X(\xi - \tau)\right] d\xi \tag{4.39}$$

At $\tau = 0$, where we make use of the fact that $C_X(-\xi) = C_X(\xi)$,

$$\begin{aligned} \dot{C}_{X_T}(0) &= \frac{1}{T^2}\int_0^T \left[C_X(\xi - T) - C_X(\xi)\right] d\xi \\ &= \frac{1}{T^2}\left\{\int_{-T}^0 C_X(\xi)\, d\xi - \int_0^T C_X(\xi)\, d\xi\right\} \end{aligned}$$

$$= \frac{1}{T^2} \int_0^T [C_X(-\xi) - C_X(\xi)] \, d\xi$$
$$= \frac{0}{T^2}$$
$$= 0$$

so that $X_T(t)$ is mean square differentiable according to the condition given by Eq. 4.28.

Applying Eq. 4.38 to Eq. 4.35 gives

$$\dot{X}_T(t) = \frac{1}{T} \frac{d}{dt} \int_{t-T/2}^{t+T/2} X(\xi) \, d\xi$$
$$= \frac{X(t+T/2) - X(t-T/2)}{T} \tag{4.40}$$

For stationary $X(t)$, the mean of $\dot{X}_T(t)$ is zero and its variance can be found as follows:

$$\sigma^2_{\dot{X}_T} = \mathrm{E}\left[\dot{X}^2_T(t)\right] = \frac{1}{T^2}\mathrm{E}\left[\left(X\left(t+\frac{T}{2}\right) - X\left(t-\frac{T}{2}\right)\right)^2\right]$$
$$= \frac{1}{T^2}\left\{2\mathrm{E}\left[X^2\right] - 2\mathrm{E}\left[X\left(t+\frac{T}{2}\right)X\left(t-\frac{T}{2}\right)\right]\right\}$$
$$= \frac{2\sigma_X^2}{T^2}[1 - \rho_X(T)] \tag{4.41}$$

since $\mathrm{E}\left[X^2(t+T/2)\right] = \mathrm{E}\left[X^2(t-T/2)\right] = \mathrm{E}\left[X^2(t)\right]$ due to stationarity.

In summary, if $\dot{C}_X(0) = 0$, then:

1. The derivative process $\dot{X}(t)$ has finite variance.
2. The derivative process $\dot{X}(t)$ is uncorrelated with $X(t)$.
3. The covariance function of $\dot{X}(t)$ is equal to $-\ddot{C}_X(\tau)$.

If $\dot{C}_X(0) = 0$ we say that $X(t)$ is mean square differentiable.

If $X(t)$ is not mean square differentiable, that is, $\dot{C}_X(0) \neq 0$, then any amount of local averaging will result in a process $X_T(t)$ which is mean square differentiable. The derivative of $X_T(t)$ has the following properties:

1. $\dot{X}_T(t) = \dfrac{X(t+T/2) - X(t-T/2)}{T}$
2. $\mu_{\dot{X}_T} = \mathrm{E}\left[\dot{X}_T(t)\right] = 0$
3. $\sigma^2_{\dot{X}_T} = \mathrm{Var}\left[\dot{X}_T(t)\right] = \dfrac{2\sigma_X^2}{T^2}[1 - \rho_X(T)]$

A possibly fundamental difficulty with locally averaging in order to render a process mean square differentiable is that $\sigma_{\dot{X}_T}$ then depends on the averaging size T and any standard deviation can be obtained simply by adjusting T. Since the above equations give no guidance on how T should be selected, its size must come from physical considerations of the process being modeled and the matching of the variance $\sigma_T^2 = \sigma_X^2\gamma(T)$ to what is observed. For example, CPT measurements represent a local average of soil properties over some "deformation region" that the cone imposes on the surrounding soil. This region might have radius of about 0.2 m (we have no reference for this estimate, this is just an engineering judgment regarding the amount of material displaced in the vicinity of a cone being forced through a soil).

4.2.2 Threshold Excursion Rate

The *mean rate* ν_b at which a stationary random process $X(t)$ crosses a threshold b was determined by Rice (1954) to be

$$\nu_b = \int_{-\infty}^{\infty} |\dot{x}| f_{X\dot{X}}(b, \dot{x}) \, d\dot{x} \tag{4.42}$$

where $f_{X\dot{X}}(x, \dot{x})$ is the joint probability density function of $X(t)$ and its derivative $\dot{X}(t)$. As we saw in the last section, if $X(t)$ is stationary, then $X(t)$ and $\dot{X}(t)$ are uncorrelated. If $X(t)$ is normally distributed, then, since $\dot{X}(t) = (X(t+\Delta t) - X(t))/\Delta t$ is just a sum of normals, $\dot{X}(t)$ must also be normally distributed. Since uncorrelated normally distributed random variables are also independent, this means that $X(t)$ and $\dot{X}(t)$ are independent and their joint distribution can be written as a product of their marginal distributions,

$$f_{X\dot{X}}(b, \dot{x}) = f_X(b) f_{\dot{X}}(\dot{x})$$

in which case Eq. 4.42 becomes

$$\nu_b = \int_{-\infty}^{\infty} |\dot{x}| f_X(b) f_{\dot{X}}(\dot{x}) \, d\dot{x} = f_X(b) \int_{-\infty}^{\infty} |\dot{x}| f_{\dot{X}}(\dot{x}) \, d\dot{x}$$
$$= f_X(b) \mathrm{E}\left[|\dot{X}|\right] \tag{4.43}$$

If X is normally distributed, with mean zero, then the mean of its absolute value is $\mathrm{E}[|X|] = \sigma_X\sqrt{2/\pi}$. Since $\dot{X}$ is normally distributed with mean zero, then $\mathrm{E}\left[|\dot{X}|\right] = \sigma_{\dot{X}}\sqrt{2/\pi}$, and we get

$$\nu_b = f_X(b)\sigma_{\dot{X}}\sqrt{\frac{2}{\pi}} = \frac{1}{\pi}\frac{\sigma_{\dot{X}}}{\sigma_X}\exp\left\{-\frac{b^2}{2\sigma_X^2}\right\} \tag{4.44}$$

where we substituted in the normal distribution for $f_X(b)$.

We are often only interested in the upcrossings of the threshold b [i.e., where $X(t)$ crosses the threshold b with positive slope]. Since every upcrossing is followed by a downcrossing, then the *mean upcrossing rate* ν_b^+ is equal to the *mean downcrossing rate* ν_b^-,

$$\nu_b^+ = \nu_b^- = \frac{\nu_b}{2} = \frac{1}{2\pi}\frac{\sigma_{\dot{X}}}{\sigma_X}\exp\left\{-\frac{b^2}{2\sigma_X^2}\right\} \tag{4.45}$$

4.2.3 Time to First Upcrossing: System Reliability

A classic engineering reliability problem is that of assessing the probability that the time to system failure exceeds some target lifetime. For example, if $X(t)$ is the load on a system and b is the system resistance, then the system will fail when $X(t)$ first exceeds b. If this occurs at some time prior to the design system lifetime, then the design has failed. The objective in design, then, is to produce a system having resistance b so that the time to failure T_f has sufficiently small probability of being less than the design lifetime.

We can use the mean upcrossing rates to solve this problem if we can determine the distribution of the time between upcrossings. Note that the previous section only gave the mean upcrossing rate, not the full distribution of times between upcrossings. Cramer (1966) showed that if $X(t)$ is stationary and Gaussian, then the time between upcrossings tends to a Poisson process for large thresholds ($b >> \sigma_X$). Let N_t be the number of upcrossings in time interval t and let T_f be the time to the first upcrossing. If N_t is a Poisson process, then it is parameterized by the mean upcrossing rate ν_b^+. Using the results of Section 1.9.5, we know that the probability that T_f exceeds some prescribed time t is

$$\mathrm{P}\left[T_f > t\right] = \mathrm{P}\left[N_t = 0\right] = \exp\{-\nu_b^+ t\} \tag{4.46}$$

4.2.4 Extremes

The largest or smallest values in a random sequence are also of considerable interest in engineering. For example, it is well known that failure tends to initiate at the lowest strength regions of a material. The tensile strength of a chain is a classic example. In geotechnical engineering, we know that shear failures (e.g., bearing capacity, slope stability) will tend to occur along surfaces which pass through regions where the ratio of shear strength to developed shear stress is a minimum.

The classic treatment of extremes (see Section 1.11) assumes that the random variables from which the extreme is being selected are mutually independent. When the set of random variables, $X(t)$, is correlated with correlation function $\rho_X(\tau)$, then the distribution of the extreme becomes considerably more complicated.

For example, if $\rho_X(\tau) = 1$ for all τ and $X(t)$ is stationary, then $X(t) = X$ for all t. That is, the random process becomes equal to a single random variable at all points in time—each realization of $X(t)$ is completely uniform. If we observe a realization of $X(t)$ at a sequence of times $X_1, X_2, \ldots, X_n$, then we will observe that all $X_1, X_2, \ldots, X_n$ are identical and equal to X. In this case, $Y_n = \max_{i=1}^n X_i = X$, and the distribution of the maximum, Y_n, is just equal to the distribution of X,

$$F_{Y_n}(y) = F_X(y) \tag{4.47}$$

Contrast this result with that obtained when $X_1, X_2, \ldots, X_n$ are independent, where, according to Eq. 1.199,

$$F_{Y_n}(y) = [F_X(y)]^n$$

Apparently, in the case of a correlated sequence of X_i's, the distribution of the maximum could be written as

$$F_{Y_n}(y) = [F_X(y)]^{n_{\text{eff}}} \tag{4.48}$$

where n_{eff} is the *effective* number of independent X_i's. When the X_i's are independent, $n_{\text{eff}} = n$. When the X_i's are completely correlated, $n_{\text{eff}} = 1$. The problem is determining the value of n_{eff} for intermediate magnitudes of correlation. Although determining n_{eff} remains an unsolved problem at the time of writing, Eqs. 4.47 and 4.48 form useful bounds; they also provide some guidelines, given knowledge about the correlation, for the judgmental selection of n_{eff}.

Consider a stationary Gaussian process $X(t)$ and let Y be the maximum value that $X(t)$ takes over some time interval $[0, t_1]$. Davenport (1964) gives the mean and standard deviation of Y to be [see also Leadbetter et al. (1983) and Berman (1992)]

$$\mu_Y = \mu_X + \sigma_X\left(a + \frac{\gamma}{a}\right) \tag{4.49a}$$

$$\sigma_Y = \sigma_X \frac{\pi}{6a} \tag{4.49b}$$

where $\gamma = 0.577216$ is Euler's number, and for time interval $[0, t_1]$,

$$a = \sqrt{2 \ln \nu_0^+ t_1} \tag{4.50}$$

and where ν_0^+ is the mean upcrossing rate of the threshold $b = 0$,

$$\nu_0^+ = \frac{1}{2\pi}\frac{\sigma_{\dot{X}}}{\sigma_X} \tag{4.51}$$

If Y is the minimum value that $X(t)$ takes over time interval $[0, t_1]$, then the only thing which changes is the sign in Eq. 4.49a,

$$\mu_Y = \mu_X - \sigma_X\left(a + \frac{\gamma}{a}\right) \tag{4.52}$$

Although these formulas do not give the entire distribution, they are often useful for first- or second-order Taylor series approximations. It should also be noted that they are only accurate for large $\nu_0^+ t_1 >> 1$. In particular Davenport's results assume asymptotic independence between values of $X(t)$, that is, it is assumed that $t_1 >> \theta$.

4.3 THRESHOLD EXCURSIONS IN TWO DIMENSIONS

In two and higher dimensions, we are often interested in asking questions regarding aspects such as the total area of a random field which exceeds some threshold, the number of excursion regions, and how clustered the excursion regions are. Unfortunately, theoretical results are not well advanced in two and higher dimensions for thresholds of practical interest (i.e., not of infinite height). In this section, some of the existing theory is presented along with some simulation-based estimates of the statistics of threshold excursions and extrema. The treatment herein is limited to the two-dimensional case, although the procedure is easily extended to higher dimensions. Seven quantities having to do with threshold excursions and extrema of two-dimensional random fields are examined:

1. The number of isolated excursion regions (N_b)
2. The area of isolated excursion regions (A_e)
3. The total area of excursion regions within a given domain ($A_b = \sum_{i=1}^{N_b} A_{e_i}$)
4. The number of holes appearing in excursion regions (N_h)
5. An integral geometric characteristic defined by Adler (1981) (Γ)
6. A measure of "clustering" defined herein (Ψ)
7. The distribution of the global maxima

These quantities will be estimated for a single class of random functions, namely Gaussian processes with Markovian covariance structure (Gauss–Markov processes), over a range of correlation lengths and threshold heights. In the following, the threshold height is expressed as $b\sigma_X$, that is, b is now in units of the standard deviation of the random field.

Within a given domain $\mathcal{V} = [0, T_1] \times [0, T_2]$ of area A_T, the total excursion area A_b can be defined by

$$A_b = \int_{\mathcal{V}} I_{\mathcal{V}}\Big(X(\mathbf{t}) - b\sigma_X\Big)\, d\mathbf{t} \tag{4.53}$$

where $b\sigma_X$ is the threshold of interest, σ_X^2 being the variance of the random field, and $I_{\mathcal{V}}(\cdot)$ is an indicator function defined on $\mathcal{V}$,

$$I_{\mathcal{V}}(t) = \begin{cases} 1 & \text{if } t \geq 0 \\ 0 & \text{if } t < 0 \end{cases} \tag{4.54}$$

For a stationary process, the expected value of A_b is simply

$$\mathrm{E}[A_b] = A_T \,\mathrm{P}[X(\mathbf{0}) \geq b\sigma_X] \tag{4.55}$$

which, for a zero-mean Gaussian process, yields

$$\mathrm{E}[A_b] = A_T\,[1 - \Phi(b)] \tag{4.56}$$

where Φ is the standard normal distribution function. The total excursion area A_b is made up of the areas of isolated (disjoint) excursions A_e as follows:

$$A_b = \sum_{i=1}^{N_b} A_{ei} \tag{4.57}$$

for which the isolated excursion regions can be defined using a point set representation:

$$\begin{aligned} \mathcal{A}_{ei} &= \{\mathbf{t} \in \mathcal{V} \;:\; X(\mathbf{t}) \geq b\sigma_X,\; \mathbf{t} \notin \mathcal{A}_{ej}\; \forall j \neq i\} \\ A_{ei} &= \mathcal{L}(\mathcal{A}_{ei}) \end{aligned} \tag{4.58}$$

where $\mathcal{L}(\mathcal{A}_{ei})$ denotes the Lebesque measure (area) of the point set $\mathcal{A}_{ei}$. Given this definition, Vanmarcke (1984) expresses the expected area of isolated excursions as a function of the second-order spectral moments

$$\mathrm{E}[A_{ei}] = 2\pi \left(\frac{F^c(b\sigma_X)}{f(b\sigma_X)}\right)^2 |\mathbf{\Lambda}_{11}|^{-1/2} \tag{4.59}$$

in which F^c is the complementary distribution function [for a Gaussian process, $F^c(b\sigma_X) = 1 - \Phi(b)$], f is the corresponding probability density function, and $\mathbf{\Lambda}_{11}$ is the matrix of second-order spectral moments with determinant $|\mathbf{\Lambda}_{11}|$, where

$$\mathbf{\Lambda}_{11} = \begin{bmatrix} \lambda_{20} & \lambda_{11} \\ \lambda_{11} & \lambda_{02} \end{bmatrix} \tag{4.60}$$

Equation 4.59 assumes that the threshold is sufficiently high so that the pattern of occurrence of excursions tends toward a two-dimensional Poisson point process. The joint spectral moments $\lambda_{k\ell}$ can be obtained either by integrating the spectral density function,

$$\lambda_{k\ell} = \int_{\infty}^{\infty}\int_{\infty}^{\infty} \omega_1^k \omega_2^\ell\, S_X(\omega_1, \omega_2)\, d\omega_1\, d\omega_2 \tag{4.61}$$

or through the partial derivatives of the covariance function evaluated at the origin,

$$\lambda_{k\ell} = -\left[\frac{\partial^{k+\ell} C_X(\boldsymbol{\tau})}{\partial\tau_1^k\, \partial\tau_2^\ell}\right]_{\boldsymbol{\tau}=\mathbf{0}} \tag{4.62}$$

The above relations presume the existence of the second-order spectral moments of $X(\mathbf{t})$, which is a feature of a mean square differentiable process. A necessary and sufficient condition for mean square differentiability is (see Section 4.2.1)

$$\left[\frac{\partial C_X(\boldsymbol{\tau})}{\partial\tau_1}\right]_{\boldsymbol{\tau}=\mathbf{0}} = \left[\frac{\partial C(\boldsymbol{\tau})}{\partial\tau_2}\right]_{\boldsymbol{\tau}=\mathbf{0}} = 0 \tag{4.63}$$

A quick check of the Gauss–Markov process whose covariance function is given by

$$B(\boldsymbol{\tau}) = \sigma^2 \exp\left\{-\frac{2}{\theta}|\boldsymbol{\tau}|\right\} \tag{4.64}$$

verifies that it is not mean square differentiable. Most of the existing theories governing extrema or excursion regions of random fields depend on this property. Other popular models which are not mean square differentiable and so remain intractable in this respect are:

1. Ideal white noise process
2. Moving average of ideal white noise
3. Fractal processes

4.3.1 Local Average Processes

One of the major motivations for the development of local average theory for random processes is to convert random functions which are not mean square differentiable into processes which are. Vanmarcke (1984) shows that even a very small amount of local averaging will produce finite covariances of the derivative process. For a two-dimensional local average process $X_D(\mathbf{t})$ formed by averaging $X(\mathbf{t})$ over $D = T_1 \times T_2$, Vanmarcke presents the following relationships for the variance of the derivative process $\dot{X}_D$ in the two coordinate directions:

$$\text{Var}\left[\dot{X}_D^{(1)}\right] = \frac{2}{T_1^2}\sigma^2\gamma(T_2)[1 - \rho(T_1|T_2)] \tag{4.65}$$

$$\text{Var}\left[\dot{X}_D^{(2)}\right] = \frac{2}{T_2^2}\sigma^2\gamma(T_1)[1 - \rho(T_2|T_1)] \tag{4.66}$$

where,

$$\dot{X}_D^{(i)} = \frac{\partial}{\partial t_i}X_D(\mathbf{t}), \qquad \gamma(T_1) = \gamma(T_1, 0),$$

$$\gamma(T_2) = \gamma(0, T_2)$$

$$\rho(T_i|T_j) = \frac{1}{T_j^2\sigma^2\gamma(T_j)}\int_{-T_j}^{T_j}(T_j - |\tau_j|)\,C_X(T_i, \tau_j)\,d\tau_j \tag{4.67}$$

Furthermore, Vanmarcke shows that the joint second-order spectral moment of the local average process is always zero for $D > 0$, that is,

$$\text{Cov}\left[\dot{X}_D^{(1)}, \dot{X}_D^{(2)}\right] = 0, \qquad \forall D > 0 \tag{4.68}$$

This result implies that the determinant of the second-order spectral moment matrix for the local average process can be expressed as the product of the two directional derivative process variances,

$$|\mathbf{\Lambda}_{11,D}|^{1/2} = \sigma_{\dot{X}_D}^2 = \left(\text{Var}\left[\dot{X}_D^{(1)}\right]\ \text{Var}\left[\dot{X}_D^{(2)}\right]\right)^{1/2} \tag{4.69}$$

Since the theory governing statistics of threshold excursions and extrema for mean square differentiable random functions is reasonably well established for high thresholds [see, e.g., Cramer and Leadbetter (1967), Adler (1981), and Vanmarcke (1984)], attention will now be focused on an empirical and theoretical determination of similar measures for processes which are not mean square differentiable. This will be accomplished through the use of a small amount of local averaging employing the results just stated. In particular, the seven quantities specified at the beginning of this section will be evaluated for the two-dimensional isotropic Markov process

$$C_X(\tau_1, \tau_2) = \sigma_X^2\ \exp\left\{-\frac{2}{\theta}\sqrt{\tau_1^2 + \tau_2^2}\right\} \tag{4.70}$$

realizations of which will be generated using the two-dimensional local average subdivision (LAS) method described in Section 6.4.6. Since the LAS approach automatically involves local averaging of the non–mean square differentiable point process (4.70), the realizations will in fact be drawn from a mean square differentiable process. The subscript D will be used to stress the fact that the results will be for the local average process and $\mathcal{Z}_D$ denotes a realization of the local average process.

4.3.2 Analysis of Realizations

Two-dimensional LAS-generated realizations of stationary, zero-mean, isotropic, Gaussian processes are to be analyzed individually to determine various properties of the discrete binary field, Y, defined by

$$Y_{jk,D} = I_{\mathcal{V}}\left(\mathcal{Z}_{jk,D} - b\sigma_D\right) \tag{4.71}$$

where subscripts j and k indicate spatial position $(t_{1j}, t_{2k}) = (j\Delta t_1, k\Delta t_2)$ and σ_D is the standard deviation of the local average process. The indicator function $I_{\mathcal{V}}$ is given by (4.54) and so $Y_D(\mathbf{x})$ has value 1 where the function $\mathcal{Z}_D$ exceeds the threshold and 0 elsewhere. In the following, each discrete value of $Y_{jk,D}$ will be referred to as a pixel which is "on" if $Y_{jk,D} = 1$ and "off" if $Y_{jk,D} = 0$. A space-filling algorithm was devised and implemented to both determine the area of each simply connected isolated excursion region, $A_{ei,D}$, according to (4.58), and find the number of "holes" in these regions. In this case, the Lebesque measure is simply

$$A_{ei,D} = \mathcal{L}(\mathcal{A}_{ei,D}) = \sum \Delta A_{ei,D} \tag{4.72}$$

where

$$\Delta A_{ei,D} = I_{\mathcal{A}_{ei,D}}\left(\mathcal{Z}_D(\mathbf{x}) - b\sigma_D\right)\ \Delta A \tag{4.73}$$

is just the incremental area of each pixel which is on within the discrete set of points $\mathcal{A}_{ei,D}$ constituting the ith simply connected region. In practice, the sum is performed only over those pixels which are elements of the set $\mathcal{A}_{ei,D}$. Note that the area determined in this fashion is typically slightly less than that obtained by computing the area within a smooth contour obtained by linear interpolation. The difference, however, is expected to be minor at a suitably fine level of resolution.

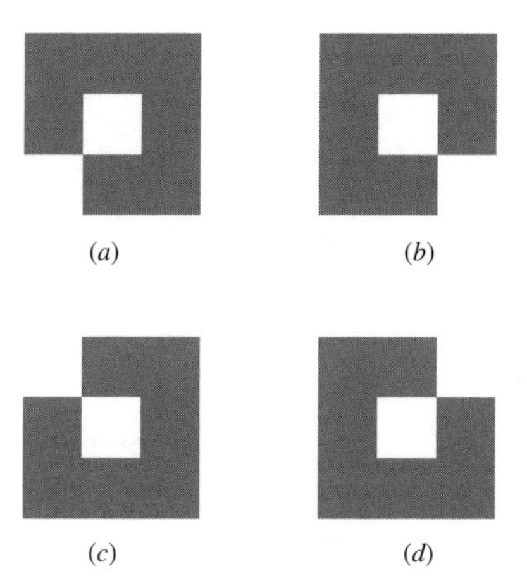

Figure 4.5 Examples of weakly surrounded holes: (*a*) and (*b*) are found to be holes while (*c*) and (*d*) are not.

A hole is defined as a set of one or more contiguous off pixels which are surrounded by on pixels. With reference to Figure 4.5, it can be seen that situations arise in which the hole is only "weakly" surrounded by on pixels. The algorithm was devised in such a way that only about half of these weakly surrounded regions are determined to be holes. In addition, if an off region intersects with the boundary of the domain, then it is not classified as a hole even if it is surrounded on all other sides by on regions. The fields to be generated will have resolution 128 × 128 and physical size 5 × 5. This gives a fairly small averaging domain having edge sizes of $T_1 = T_2 = \frac{5}{128}$ for which the variance function corresponding to Eq. 4.70 ranges in value from 0.971 to 0.999 for $\theta = \frac{1}{2}$ to $\theta = 4$. In all cases, $\sigma_X^2 = 1$ in Eq. 4.70 so that σ_D^2 equals the variance function.

Figure 4.6 shows a typical realization of the binary field Y obtained by determining the $b = 1$ excursion regions of Z for a correlation length $\theta = \frac{1}{2}$. Also shown in Figure 4.6 are the $b = 1$ contours which follow very closely the on regions. The centroid of each excursion is marked with a darker pixel.

In the sections to follow, trial functions are matched to the observed data and their parameters estimated. All curve fitting was performed by visual matching since it was found that existing least squares techniques for fitting complex nonlinear functions were in general unsatisfactory. In most cases the statistics were obtained as averages from 400 realizations.

4.3.3 Total Area of Excursion Regions

Since an exact relationship for the expected total area of excursion regions within a given domain, (4.56), is known for a Gaussian process, an estimation of this quantity from

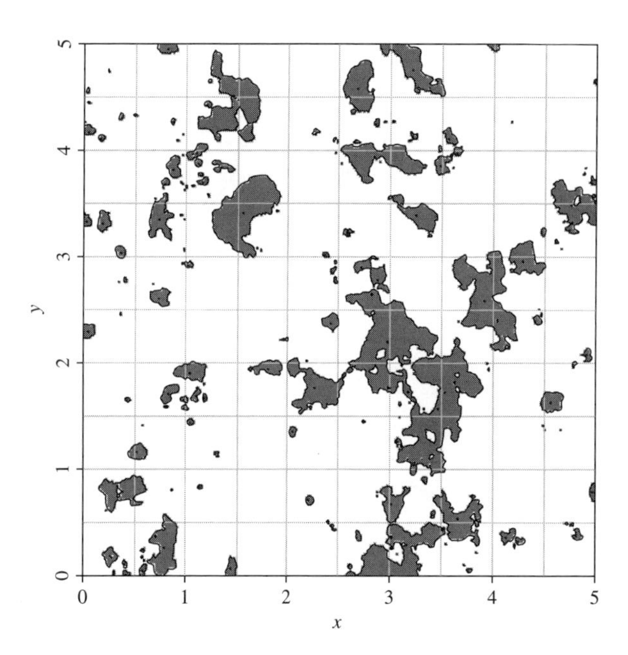

Figure 4.6 Sample function of binary field Y (Eq. 4.71). Regions shown in gray represent regions of $\mathcal{Z}$ which exceed the threshold $b = 1\sigma_D$, where $\mathcal{Z}$ is generated via the two-dimensional LAS algorithm according to Eq. 4.70 with $\theta = \frac{1}{2}$. Since Z is normally distributed, the gray regions on average occupy about 16% of the field.

a series of realizations represents a further check on the accuracy of the simulation method. Figure 4.7 shows the normalized average total area of excursions, $\overline{A}_{b,D}/A_T$, for $A_T = 25$. Here and to follow, the overbar denotes the quantity obtained by averaging over the realizations. The estimated area ratios show excellent agreement with the exact relationship.

4.3.4 Expected Number of Isolated Excursions

Figure 4.8 shows the average number of isolated excursion regions observed within the domain, $\overline{N}_{b,D}$, as a function of scale and threshold. Here the word "observed" will be used to denote the average number of excursion regions seen in the individual realizations. A similar definition will apply to other quantities of interest in the remainder of the chapter. The observed $\overline{N}_{b,D}$ is seen in Figure 4.8 to be a relatively smooth function defined all the way out to thresholds in excess of $3\sigma_D$.

An attempt can be made to fit the theoretical results which describe the mean number of excursions of a local average process above a relatively high threshold to the data shown in Figure 4.8. As we shall see, the theory for high thresholds is really only appropriate for high thresholds, as expected, and does not match well the results at lower

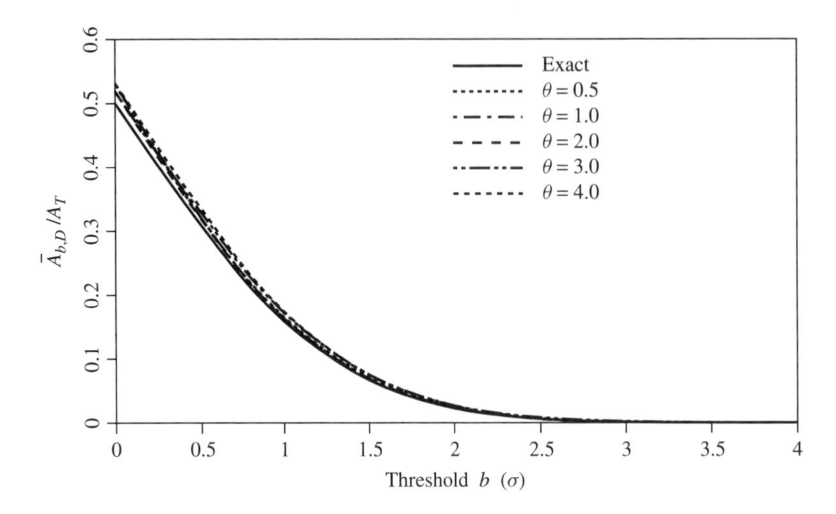

Figure 4.7 Average total area of excursion ratio, $\overline{A}_{b,D}/A_T$, as a function of threshold b.

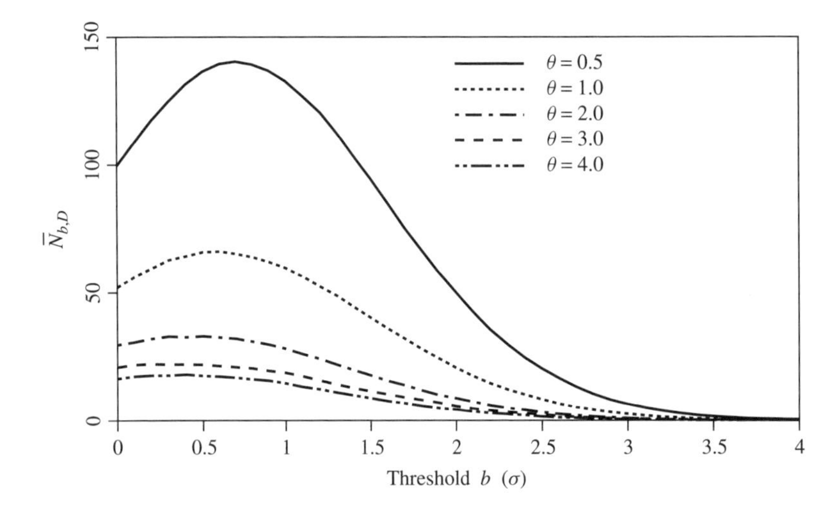

Figure 4.8 Average number of isolated excursions, $\overline{N}_{b,D}$, estimated from 400 realizations of the locally averaged two-dimensional Gauss–Markov process (Eq. 4.70).

thresholds. At high thresholds, the expected number of excursions is predicted by (Vanmarcke, 1984)

$$\mathrm{E}\left[N_{b,D}\right] = \frac{A_T f_D^2(b\sigma_D)}{2\pi\, F_D^c(b\sigma_D)}\, \sigma_{\dot{Z}_D}^2 \tag{4.74}$$

in which f_D and F_D^c are the pdf and complementary cdf of the local average process, respectively, and $\sigma_{\dot{Z}_D}^2$ is the geometric average of the directional variances of the derivative process as defined by Eq. 4.69. For the Gaussian process, Eq. 4.74 becomes

$$\mathrm{E}\left[N_{b,D}\right] = \frac{A_T\, e^{-b^2}}{4\pi^2 \sigma_D^2\,[1-\Phi(b)]}\, \sigma_{\dot{Z}_D}^2 \tag{4.75}$$

To determine $\sigma_{\dot{Z}_D}^2$ the functions $\rho(T_1|T_2)$ and $\rho(T_2|T_1)$ must first be calculated using Eq. 4.67. Consider $\rho(T_1|T_2)$ for the quadrant-symmetric Gauss–Markov process

$$\rho(T_1|T_2) = \frac{2}{T_2^2\sigma^2\gamma(T_2)} \int_0^{T_2} (T_2-\tau_2)\, B(T_1,\tau_2)\, d\tau_2$$

$$= \frac{2}{T_2^2\gamma(T_2)} \int_0^{T_2} (T_2-\tau_2)\, \exp\{-\tfrac{2}{\theta}\sqrt{T_1^2+\tau_2^2}\}\, d\tau_2$$

Making the substitution $r^2 = T_1^2 + \tau_2^2$ gives

$$\rho(T_1|T_2) = \frac{2}{T_2^2\sigma^2\gamma(T_2)} \int_{T_1}^{\sqrt{T_1^2+T_2^2}} \left[\frac{T_2\, re^{-2r/\theta}}{\sqrt{r^2-T_1^2}} - re^{-2r/\theta}\right] dr$$

To avoid trying to numerically integrate a function with a singularity at its lower bound, the first term in the integrand can be evaluated as follows:

$$\int_{T_1}^{\sqrt{T_1^2+T_2^2}} \frac{T_2 re^{-2r/\theta}}{\sqrt{r^2 - T_1^2}} dr$$

$$= \int_{T_1}^{\infty} \frac{T_2 re^{-2r/\theta}}{\sqrt{r^2 - T_1^2}} dr - \int_{\sqrt{T_1^2+T_2^2}}^{\infty} \frac{T_2 re^{-2r/\theta}}{\sqrt{r^2 - T_1^2}} dr$$

$$= T_2 T_1 K_1 \left(\frac{2T_1}{\theta}\right) - \int_{\sqrt{T_1^2+T_2^2}}^{a} \frac{T_2 re^{-2r/\theta}}{\sqrt{r^2 - T_1^2}} dr$$

$$- \int_{a}^{\infty} \frac{T_2 re^{-2r/\theta}}{\sqrt{r^2 - T_1^2}} dr$$

The second integral on the RHS can now be evaluated numerically, and for a chosen sufficiently large, the last integral has the simple approximation $\frac{1}{2}\theta T_2 \exp\{-2a/\theta\}$. The function K_1 is the modified Bessel function of order 1. Unfortunately, for small T_1, the evaluation of this integral is extremely delicate as it involves the small differences of very large numbers. An error of only 0.1% in the estimation of either K_1 or the integrals on the RHS can result in a drastic change in the value of $\sigma^2_{\dot{Z}_D}$, particularly at larger correlation lengths. The results in Table 4.2 were obtained using $T_1 = T_2 = \frac{5}{128}$, for which $\rho(T_1|T_2) = \rho(T_2|T_1)$, and a 20-point Gaussian quadrature integration scheme.

Using these variances, Eq. 4.74 was plotted against the observed $\overline{N}_{b,D}$ in Figure 4.9. The relatively poor agreement achieved may be as a result of the combination of the difficulty in accurately determining $\sigma^2_{\dot{Z}_D}$ for small averaging dimensions and the fact that Eq. 4.74 is an asymptotic relationship, valid only for $b \to \infty$. A much better fit in the tails ($b > 1.5$) was obtained using the empirically determined values of $\sigma^2_{\dot{Z}_D}$ shown in Table 4.3 (see page 146), which are typically about one-half to one-third those shown in Table 4.2. Using these values, the fit is still relatively poor at lower thresholds.

Table 4.2 Computed Variances of Local Average Derivative Process

Scale	$\rho(T_1\|T_2)$	$\sigma^2_{\dot{Z}_D}$
0.5	0.8482	196.18
1.0	0.9193	105.18
2.0	0.9592	53.32
3.0	0.9741	33.95
4.0	0.9822	23.30

An alternative approach to the description of $\overline{N}_{b,D}$ involves selecting a trial function and determining its parameters. A trial function of the form

$$\overline{N}_{b,D} \simeq A_T (a_1 + a_2 b) \exp\{-\tfrac{1}{2} b^2\} \tag{4.76}$$

where the symbol $\simeq$ is used to denote an empirical relationship, was chosen, and a much closer fit to the observed data, as shown in Figure 4.10, was obtained using the coefficients shown in Table 4.3. The functional form of Eq. 4.76 was chosen so that it exhibits the correct trends beyond the range of thresholds for which its coefficients were derived.

4.3.5 Expected Area of Isolated Excursions

Within each realization, the average area of isolated excursions, $\overline{A}_{e,D}$, is obtained by dividing the total excursion area by the number of isolated areas. Further averaging over the 400 realizations leads to the mean excursion areas shown in Figure 4.11 which are again referred to as the observed results. The empirical relationship of the previous section, Eq. 4.76, can be used along with the theoretically expected total excursion area (Eq. 4.56) to obtain the semiempirical relationship

$$\overline{A}_{e,D} \simeq \frac{[1 - \Phi(b)]\, e^{\frac{1}{2} b^2}}{a_1 + a_2 b} \tag{4.77}$$

which is compared to the observed data in Figure 4.12 and is seen to show very good agreement. For relatively high thresholds, dividing (4.56) by (4.75) and assuming independence between the number of regions and their total size yield the expected area to be

$$\mathrm{E}\left[A_{e,D}\right] = 4\pi^2[1 - \Phi(b)]^2 e^{b^2} \left(\frac{\sigma_D^2}{\sigma^2_{\dot{Z}_D}}\right) \tag{4.78}$$

Again the use of $\sigma^2_{\dot{Z}_D}$, as calculated from Eq. 4.69, gives a rather poor fit. Using the empirically derived variances shown in Table 4.3 improves the fit in the tails, as shown in Figure 4.13, but loses accuracy at lower thresholds for most scales.

4.3.6 Expected Number of Holes Appearing in Excursion Regions

In problems such as liquefaction or slope stability, we might be interested in determining how many strong regions appear in the site to help prevent global failure (see, e.g., Chapter 16). If excursion regions correspond to soil failure (in some sense), then holes in the excursion field would correspond to higher strength soil regions which do not fail and which help resist global failure. In this section, we look

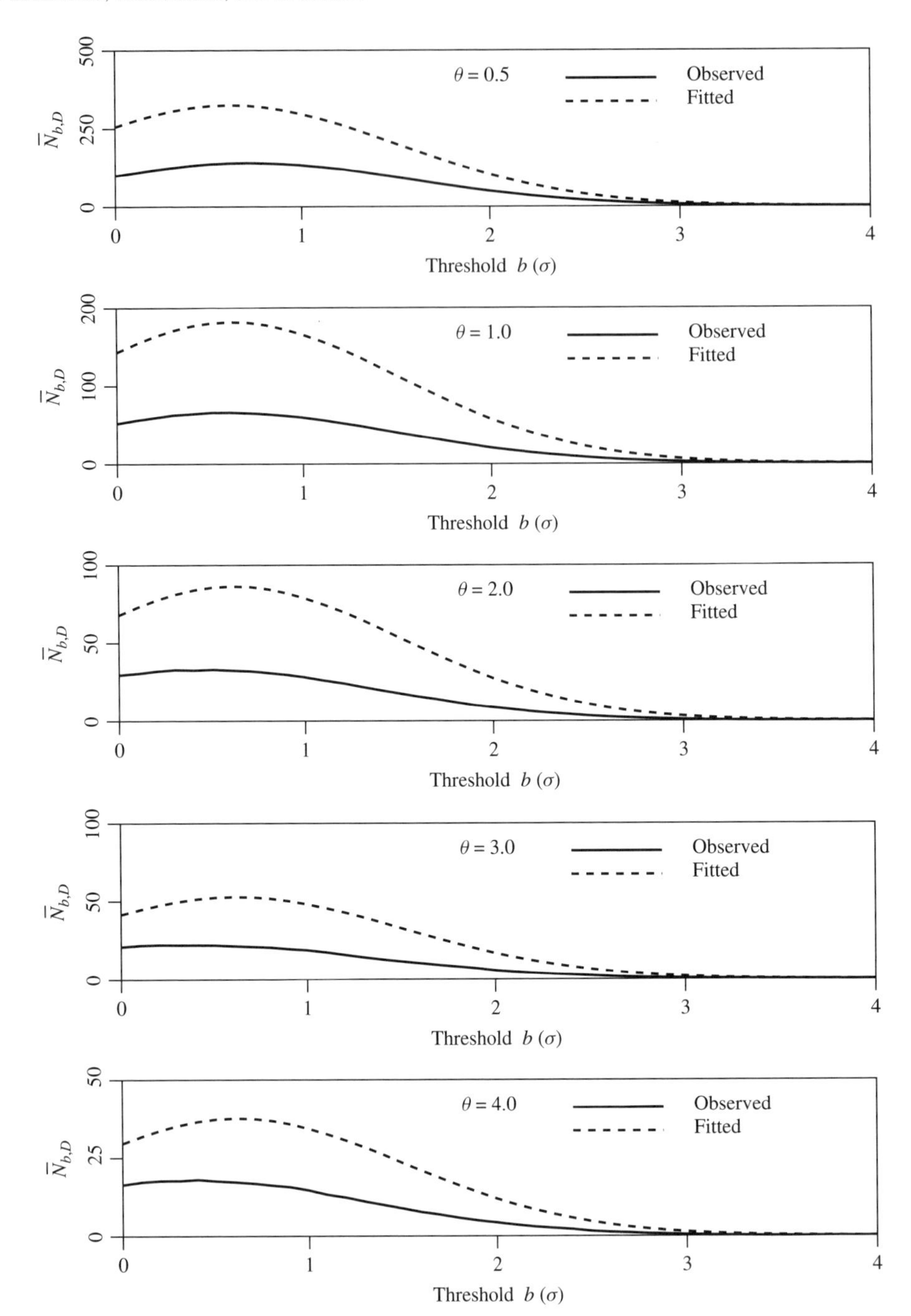

Figure 4.9 Comparison of theoretical fit by Eq. 4.75 with the observed average number of isolated excursions obtained by simulation.

at the number of holes (off regions surrounded by on regions) appearing in excursion regions. Since the data are being gathered via simulation, an empirical measure relating the average number of holes, $\overline{N}_{h,D}$, with the threshold height and the correlation length is derived here. The estimated $\overline{N}_{h,D}$ curves, obtained by finding the number of holes in each realization and averaging over 400 realizations, are shown in Figure 4.14. The empirical model used to fit these curves is

$$\overline{N}_{h,D} \simeq A_T(h_1 + h_2\, b)[1 - \Phi(b)] \tag{4.79}$$

where the parameters giving the best fit are shown in Table 4.4 and the comparison is made in Figure 4.15.

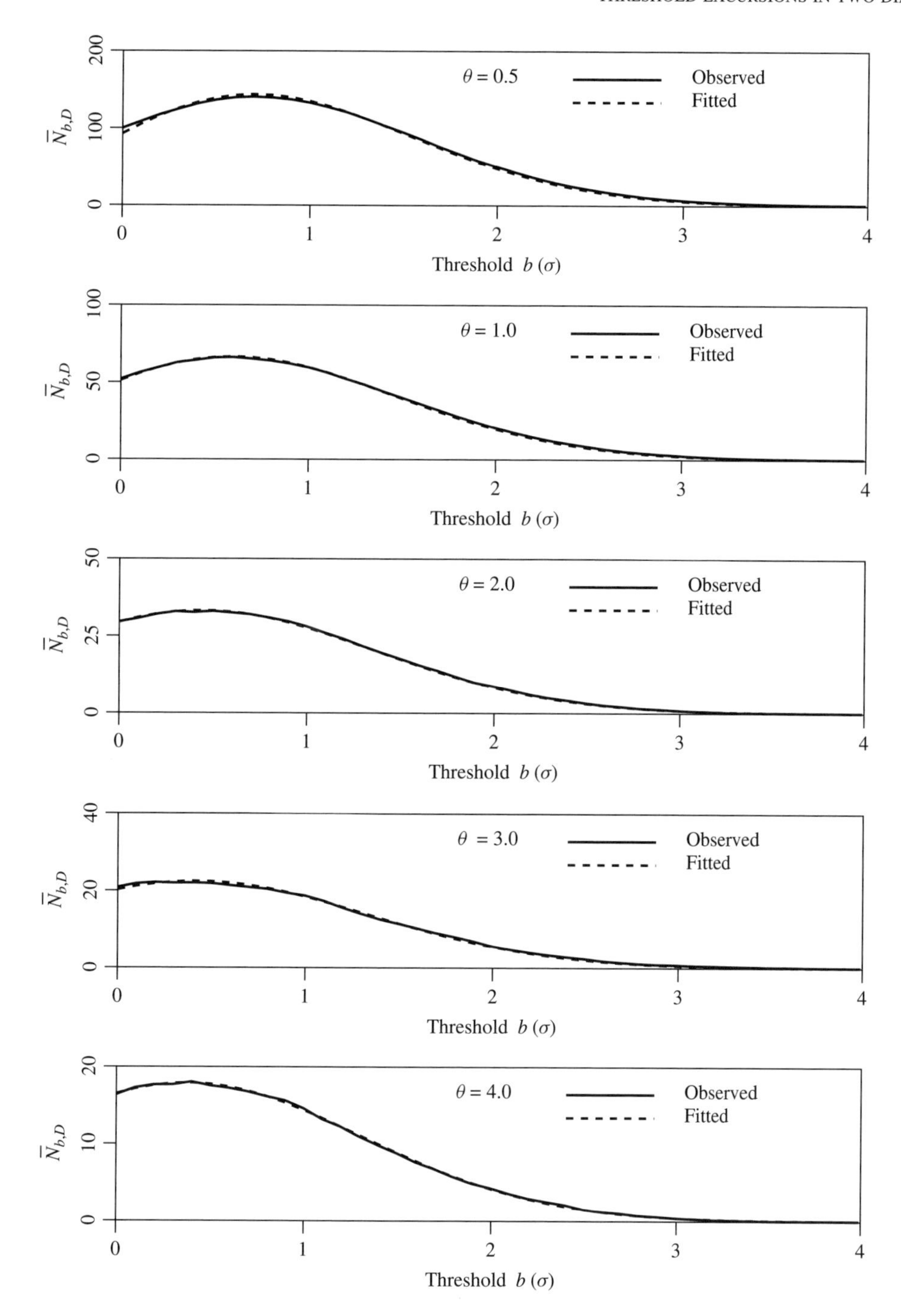

Figure 4.10 Comparison of empirical fit by Eq. 4.76 with the observed average number of isolated excursions obtained by simulation.

4.3.7 Integral Geometric Characteristic of Two-Dimensional Random Fields

In his thorough treatment of the geometric properties of random fields, Adler (1981) developed a so-called integral geometric (IG) characteristic $\Gamma(\mathcal{A}_{b,D})$ as a statistical measure of two-dimensional random fields. The definition of $\Gamma(\mathcal{A}_{b,D})$ will be shown here specifically for the two-dimensional case, although a much more general definition is given by Adler. First, using a point set representation, the excursion set $\mathcal{A}_{b,D}$ can be defined as the set of points in $\mathcal{V} = [0, T_1] \times [0, T_2]$ for which $Z_D(\mathbf{x}) \geq b\sigma_D$,

$$\mathcal{A}_{b,D} = \{\mathbf{t} \in \mathcal{V} \; : \; Z_D(\mathbf{t}) \geq b\sigma_D\} \tag{4.80}$$

Table 4.3 Empirically Determined Parameters of Eq. 4.76 and Variances of Derivative Process

Scale	a_1	a_2	$\sigma^2_{\dot{Z}_D}$
0.5	3.70	5.20	90.0
1.0	2.05	1.90	40.0
2.0	1.18	0.65	17.5
3.0	0.81	0.41	11.3
4.0	0.66	0.29	8.5

The Hadwiger characteristic of $\mathcal{A}_{b,D}$, $\varphi(\mathcal{A}_{b,D})$, is equal to the number of connected components of $\mathcal{A}_{b,D}$ (the number of isolated excursion regions) minus the number of holes in $\mathcal{A}_{b,D}$. Finally, if $\hat{\mathcal{V}}$ is defined as the edges of $\mathcal{V}$ which pass through the origin (the coordinate axes), then the IG characteristic is formally defined as

$$\Gamma(\mathcal{A}_{b,D}) = \varphi(\mathcal{A}_{b,D}) - \varphi(\mathcal{A}_{b,D} \cap \hat{\mathcal{V}}) \tag{4.81}$$

Essentially, $\Gamma(\mathcal{A}_{b,D})$ is equal to the number of isolated excursion areas which do not intersect the coordinate axes minus the number of holes in them. Figure 4.16 shows the average value of the IG characteristic, $\overline{\Gamma}(\mathcal{A}_{b,D})$, obtained from the locally averaged Gauss–Markov process realizations.

Adler presented an analytic result for the expected value of $\Gamma(\mathcal{A}_{b,D})$ which has been modified here to account for local averaging of a Gaussian process,

$$\mathrm{E}\left[\Gamma(\mathcal{A}_{b,D})\right] = \frac{bA_T}{(2\pi)^{3/2}\sigma_D^2}\exp\left\{-\frac{1}{2}b^2\right\}\sigma^2_{\dot{Z}_D} \tag{4.82}$$

Figure 4.17 shows the comparison between Eq. 4.82 and the observed data using the empirically estimated variances $\sigma^2_{\dot{Z}_D}$ shown in Table 4.3. The fit at higher thresholds appears to be quite reasonable. Using a function of the same form as Eq. 4.76,

$$\overline{\Gamma}(\mathcal{A}_{b,D}) \simeq A_T(g_1 + g_2 b)\exp\{-\tfrac{1}{2}b^2\}, \tag{4.83}$$

yields a much closer fit over the entire range of thresholds by using the empirically determined parameters shown in Table 4.5. Figure 4.18 illustrates the comparison.

4.3.8 Clustering of Excursion Regions

Once the total area of an excursion and the number of components which make it up have been determined, a natural question to ask is how the components are distributed: Do they tend to be clustered together or are they more uniformly distributed throughout the domain? When liquefiable soil pockets tend to occur well separated by stronger soil regions, the risk of global failure is reduced. However, if the liquefiable regions are clustered together, the likelihood of a large soil region liquefying is increased. Similarly, weak zones in a soil slope or under a footing do not necessarily represent a problem if they are evenly distributed throughout a stronger soil matrix; however, if the weak zones are clustered together, then they could easily lead to a failure mechanism.

It would be useful to define a measure, herein called Ψ, which varies from 0 to 1 and denotes the degree of clustering, 0 corresponding to a uniform distribution and larger values corresponding to denser clustering. The determination of such a measure involves first defining a reference domain within which the measure will be calculated. This is necessary since a stationary process over infinite space always has excursion regions throughout the space. On such a scale, the regions will always appear uniformly distributed (unless the correlation length also

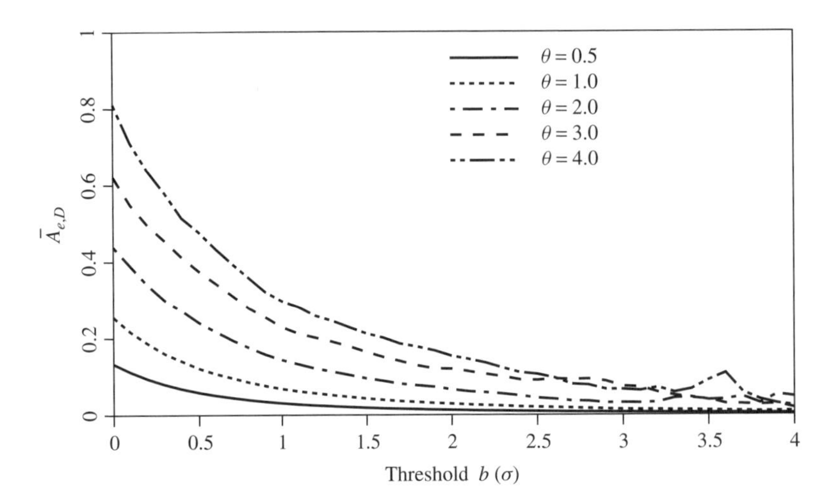

Figure 4.11 Average area of isolated excursion regions estimated from 400 realizations of the locally averaged two-dimensional Gauss–Markov process.

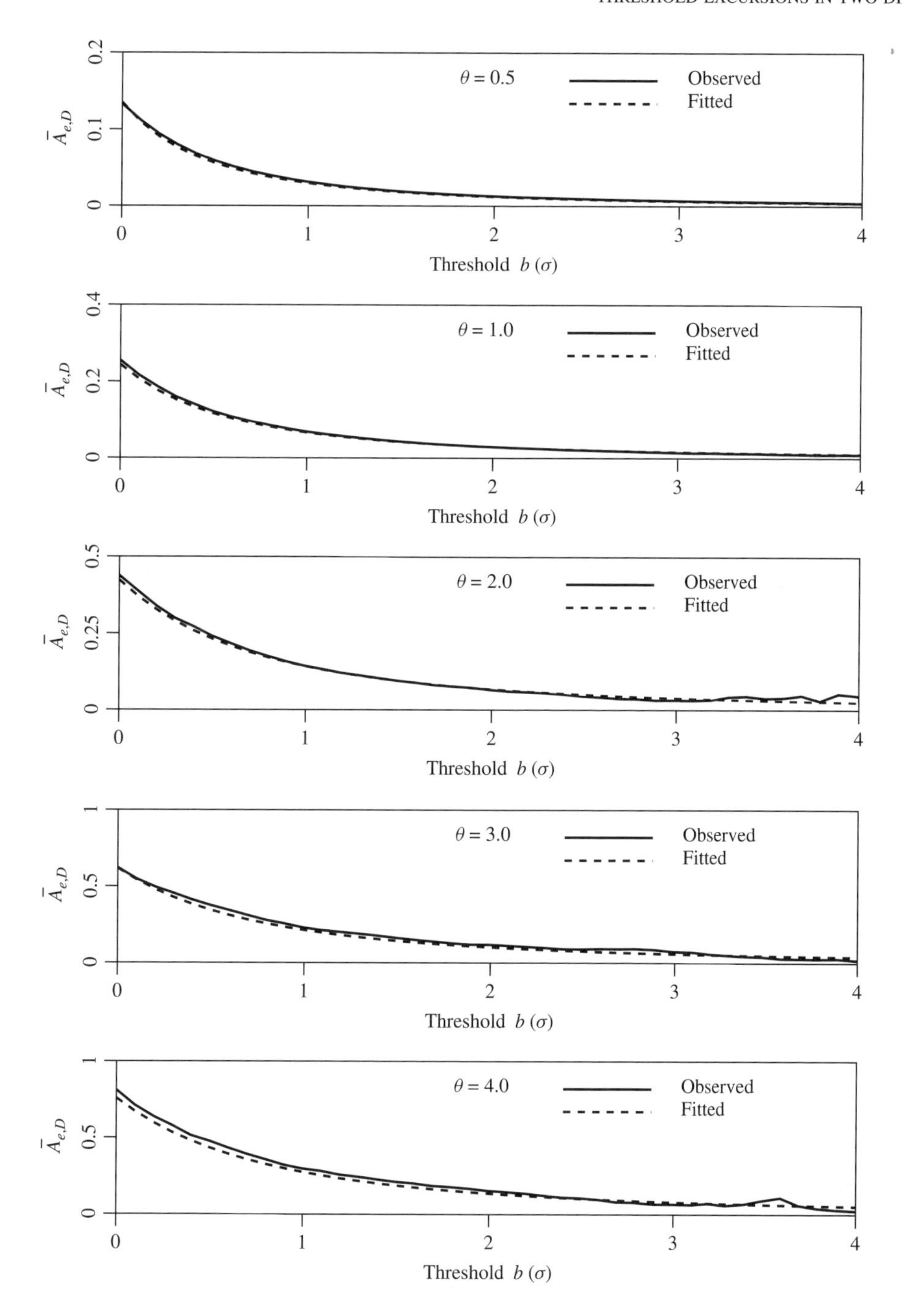

Figure 4.12 Comparison of semiempirical fit by Eq. 4.77 with the observed average area of isolated excursions obtained by simulation.

approaches infinity). For example, at scales approaching the boundaries of the known universe, the distribution of galaxies appears very uniform. It is only when attention is restricted to smaller volumes of space that one begins to see the local clustering of stars. Thus an examination of the tendency of excursions to occur in groups must involve a comparison within the reference domain of the existing pattern of excursions against the two extremes of uniform distribution and perfect clustering.

A definition for Ψ which satisfies these criteria can be stated as

$$\Psi = \frac{J_u - J_b}{J_u - J_c} \tag{4.84}$$

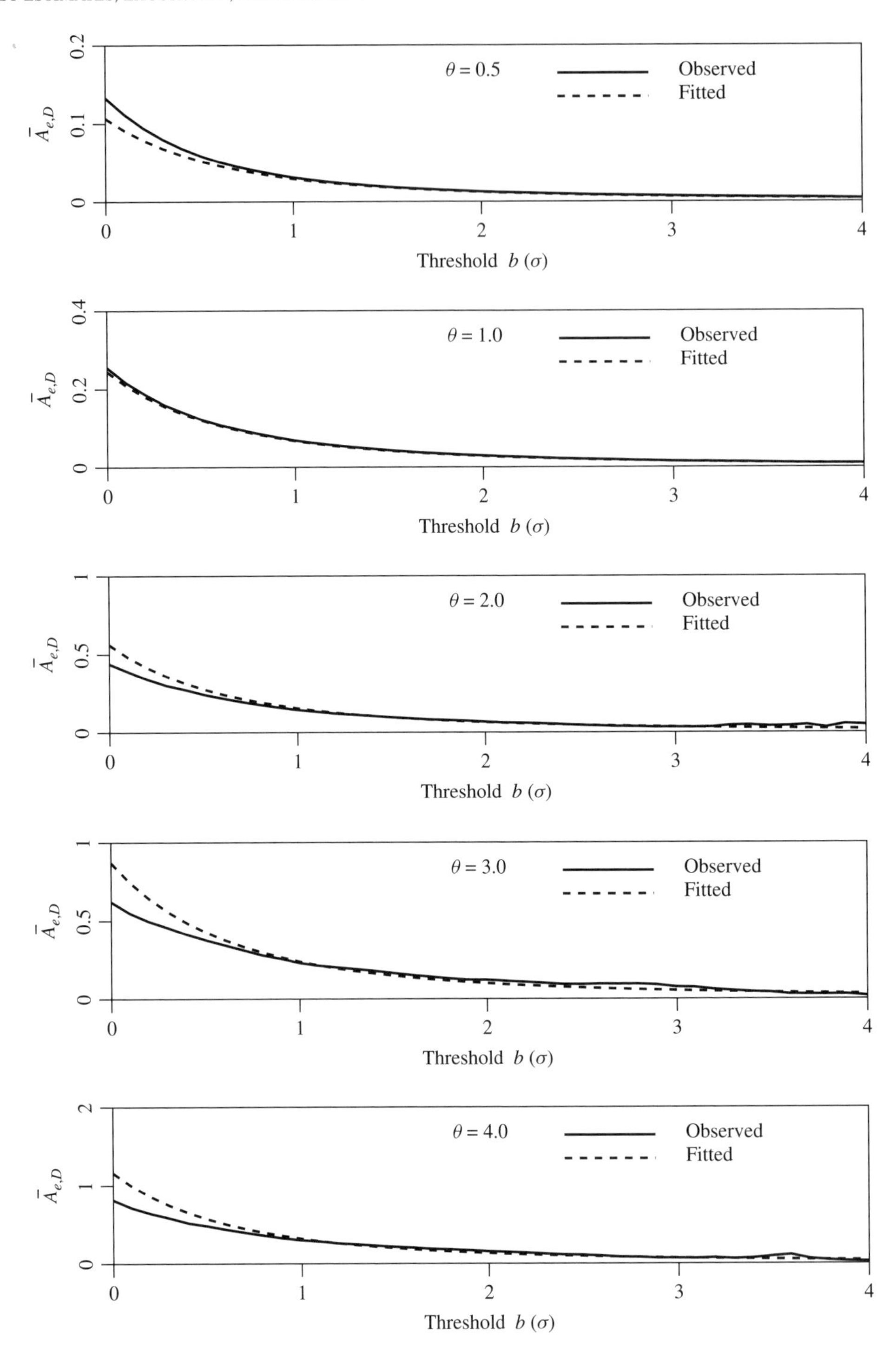

Figure 4.13 Comparison of fit by Eq. 4.78 using empirically derived variances $\sigma^2_{Z_D}$ with observed average area of isolated excursions obtained by simulation.

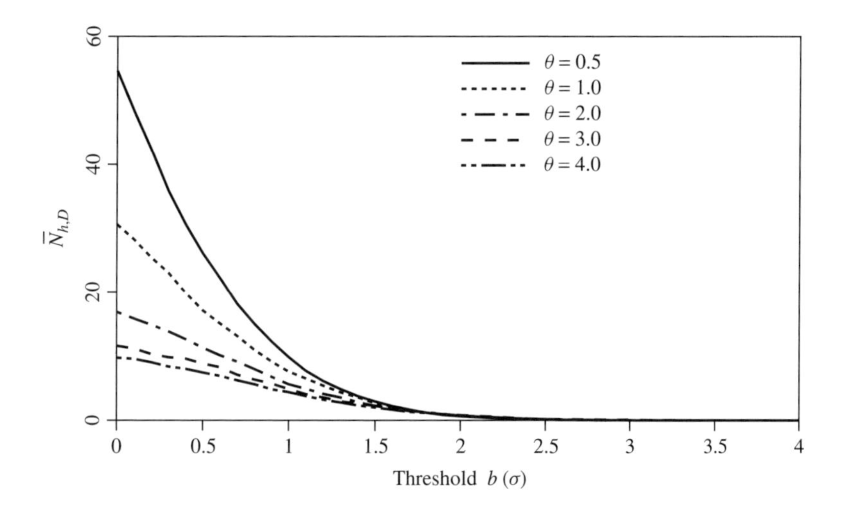

Figure 4.14 Average number of holes appearing in excursion regions.

where J_b is the polar moment of inertia of the excursion areas about their combined centroid, J_c is the polar moment of inertia of all the excursion areas concentrated within a circle, and J_u is the polar moment of inertia about the same centroid if the excursion area were distributed uniformly throughout the domain. Specifically

$$J_b = \sum_{i}^{N_{b,D}} J_{ei} + A_{ei,D} |\bar{\mathbf{x}}_b - \bar{\mathbf{x}}_i|^2 \tag{4.85}$$

$$J_{ei} = \sum_{j} \Delta A_{ei,D} |\bar{\mathbf{x}}_i - \mathbf{x}_j|^2 \tag{4.86}$$

$$J_u = \frac{A_{b,D}}{A_T} \int_{\mathcal{V}} |\bar{\mathbf{x}}_b - \mathbf{x}|^2 \, d\mathbf{x} \tag{4.87}$$

$$J_c = \frac{A_{b,D}^2}{2\pi} \tag{4.88}$$

where J_{ei} is the polar moment of inertia of the ith excursion region of area A_{ei} about its own centroid, $\bar{\mathbf{x}}_i$; $\Delta A_{ei,D}$ is as defined by Eq. 4.73 and $\bar{\mathbf{x}}_b$ is the centroid of all the excursion regions. The second moment of area was used in the definition since it is invariant under rotations. It can be easily seen that this definition will result in $\Psi = 0$ when the excursion regions are uniformly distributed over the space ($J_b \rightarrow J_u$) and $\Psi \rightarrow 1$ when the excursion regions are clustered within a small region ($J_b \rightarrow J_c$). It is also possible for Ψ to take negative values, indicating the occurrence of two local clusters at opposite sides of the domain. This information is just as valuable as positive values for Ψ but in practice has not been observed to occur on average.

All that remains is to define Ψ in the limiting cases. Equation 4.84 ensures that Ψ will be quite close to 1 in the case of only a single excursion region. It seems natural then to take $\Psi = 1$ if no excursions occur. At the other extreme, as $A_{b,D} \rightarrow A_T$, both the denominator and numerator of Eq. 4.84 become very small. Although the limit for noncircular domains is zero, it appears that the measure becomes somewhat unstable as $A_{b,D} \rightarrow A_T$. This situation is of limited interest since the cluster measure of a domain which entirely exceeds a threshold has little meaning. It is primarily a measure of the scatter of isolated excursions.

Individual realizations were analyzed to determine the cluster measure Ψ and then averaged over 200 realizations to obtain the results shown in Figure 4.19. Definite, relatively smooth trends both with correlation length and threshold height are evident, indicating that the measure might be useful in categorizing the degree of clustering.

4.3.9 Extremes in Two Dimensions

Extracting the maximum value from each realization of the random field, $\mathcal{Z}_D$, allows the estimation of its corresponding probability density function (or equivalently the cumulative distribution) with reasonable accuracy given a sufficient number of realizations. A total of 2200 realizations of the locally averaged Gauss–Markov process were generated for each correlation length considered. Conceptually it is not unreasonable to expect the cumulative distribution of the global maximum $F_{\max}(b)$ to have the form of an extreme-value distribution for a Gaussian process

$$F_{\max}(b) = [\Phi(b)]^{n_{\text{eff}}} \tag{4.89}$$

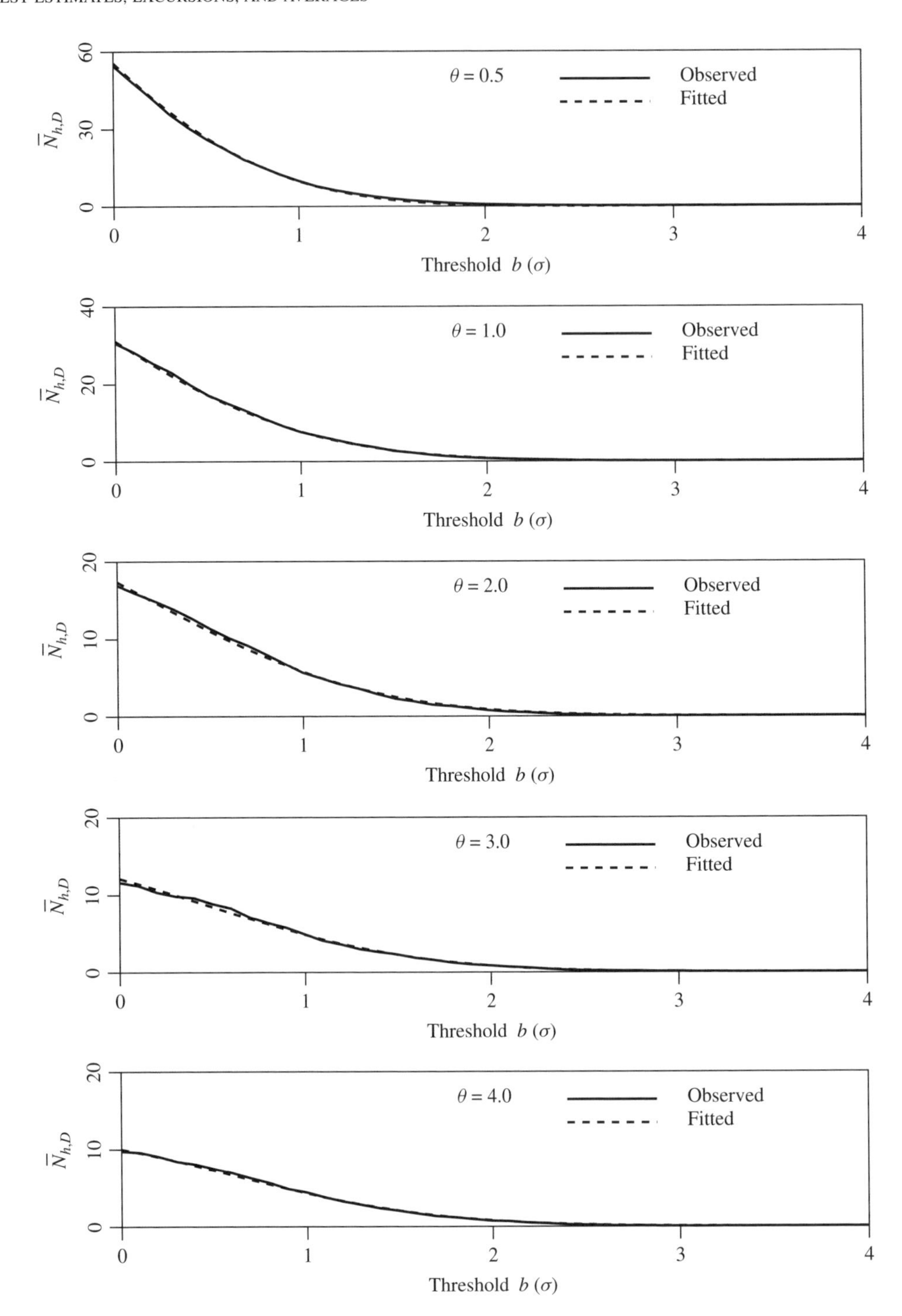

Figure 4.15 Comparison of empirical fit by Eq. 4.79 with observed average number of holes obtained by simulation.

where n_{eff} is the *effective* number of independent samples in each realization estimated by fitting Eq. 4.89 to the empirical cumulative distribution function at its midpoint. As the correlation length approaches zero, n_{eff} should approach the total number of field points (128×128), and as the scale becomes much larger than the field size, n_{eff} is expected to approach 1 (when the field becomes totally correlated). Except at the shortest correlation length considered, $\theta = 0.5$, the function defined by Eq. 4.89 was disappointing in its match with the cdf obtained from the

Table 4.4 Empirically Determined Parameters of Eq. 4.79 Based on Observed Average Number of Holes Obtained by Simulation

Correlation Length	h_1	h_2
0.5	4.45	−2.00
1.0	2.49	−0.55
2.0	1.39	0.06
3.0	0.97	0.25
4.0	0.80	0.28

realizations. Figure 4.20 illustrates the comparison for the empirically determined values of n_{eff} shown in Table 4.6. The better fit at the smallest correlation length is to be expected since at very small scales the field consists of a set of (almost) independent random variables and thus satisfies the conditions under which Eq. 4.89 theoretically applies. Not surprisingly, an improved match is obtained using a two-parameter type I extreme-value distribution having the double-exponential form

$$F_{\max}(b) = \exp\{-e^{-\alpha(b-\mu)}\} \tag{4.90}$$

where the parameters α and μ, estimated by an order statistics method developed by Leiblein (1954) using the simulation data, are presented in Table 4.6 for each correlation length. The comparison between the simulation-based cumulative distribution and that predicted by the type I extreme-value distribution is shown in Figure 4.21.

4.4 AVERAGES

We often wish to characterize random fields by averaging them over certain domains. For example, when arriving at characteristic soil properties for use in design (see Chapter 7), we usually collect field data and then use some sort of (possibly factored) average of the data as the representative value in the design process. The representative value has traditionally been based on the arithmetic average. However, two other types of averages have importance in geotechnical engineering: geometric and harmonic averages. All three averages are discussed next.

4.4.1 Arithmetic Average

The classical estimate of the central tendency of a random process is the *arithmetic average*, which is defined as

$$X_A = \begin{cases} \dfrac{1}{n}\displaystyle\sum_{i=1}^{n} X_i & \text{(discrete data)} \\ \dfrac{1}{T}\displaystyle\int_T X(\mathbf{x})\, d\mathbf{x} & \text{(continuous data)} \end{cases} \tag{4.91}$$

where T is the domain over which the continuous data are collected. The arithmetic average has the following properties:

1. X_A is an unbiased estimate of the true mean, μ_X. That is, $\text{E}[X_A] = \mu_X$.
2. X_A tends to have a normal distribution by the central limit theorem (see Section 1.10.8.1).
3. All observations are weighted equally, that is, are assumed to be equi-likely. Note that the true mean

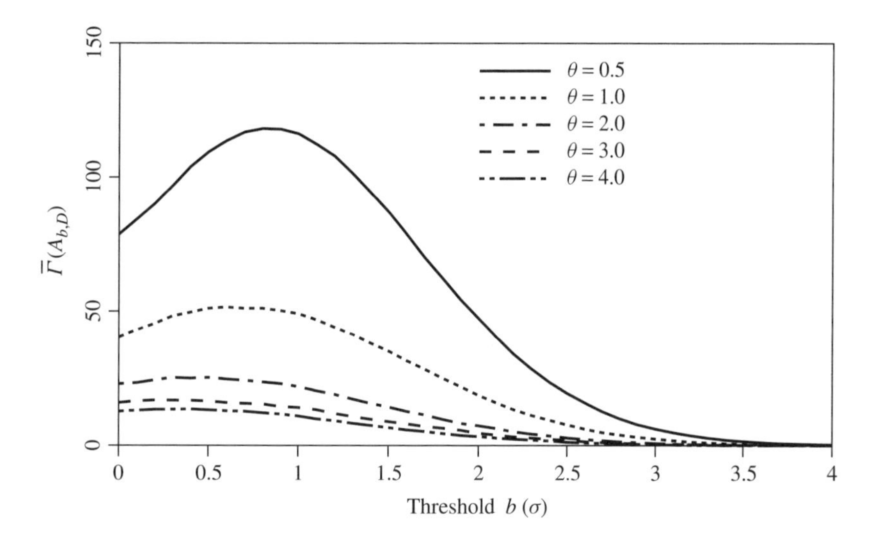

Figure 4.16 Average values of Adler's IG characteristic $\overline{\Gamma}$ obtained from 400 realizations of the locally averaged Gauss–Markov process.

Table 4.5 Empirically Determined Parameters of Eq. 4.83 Based on Observed Average IG Characteristic $\overline{\Gamma}$ Obtained by Simulation

Scale	g_1	g_2
0.5	2.70	5.10
1.0	1.50	1.80
2.0	0.87	0.58
3.0	0.61	0.32
4.0	0.50	0.22

is defined as a weighted average

$$\mu_X = \int_{\text{all } x} x f_X(x)\, dx$$

so that X_A is simply saying that the true distribution is unknown and assumed to be uniform. This assumption also means that low and high values are weighted equally and tend to cancel one another out (which is why X_A is an unbiased estimate of μ_X).

4. The variance of X_A depends on the degree of correlation between all the $X(\mathbf{x})$ values going into the average. As discussed in Section 3.4, the variance of X_A can be expressed as

$$\text{Var}[X_A] = \sigma_X^2 \gamma(T)$$

where $\gamma(T)$ is the variance reduction function defined by Eqs. 3.40–3.42.

4.4.2 Geometric Average

The *geometric average* is defined as the nth root of the product of n (nonnegative) random variables. Using this definition, the discrete set of random variables $X_1, X_2, \ldots, X_n$ has geometric average

$$X_G = (X_1 X_2 \cdots X_n)^{1/n} \tag{4.92}$$

This average is not well defined if the X's can be negative since the sign then becomes dependent on the number of negative values in the product, which may also be random. In this case, the geometric average may become imaginary. Thus, its use should be restricted to nonnegative random fields, as are most geotechnical properties.

The natural logarithm of X_G is

$$\ln X_G = \frac{1}{n} \sum_{i=1}^{n} \ln X_i \tag{4.93}$$

which is the average of the ln X values. Taking expectations gives the mean of ln X_G to be

$$\text{E}[\ln X_G] = \mu_{\ln X_G} = \mu_{\ln X}$$

In other words, the geometric average preserves the mean of ln X (just as the arithmetic average preserves the mean of X).

If Eq. 4.93 is made a power of e, we get an alternative way of computing the geometric average,

$$X_G = \exp\left\{\frac{1}{n} \sum_{i=1}^{n} \ln X_i\right\} \tag{4.94}$$

This latter expression is useful if X is a continuously varying spatial (and/or temporal) random field being averaged over some domain T, in which case the geometric average becomes its continuous equivalent,

$$X_G = \exp\left\{\frac{1}{T} \int_T \ln X(\mathbf{x})\, d\mathbf{x}\right\} \tag{4.95}$$

Some properties of the geometric average are as follows:

1. X_G weights low values more heavily than high values (low value dominated). This can be seen by considering what happens to the geometric average, see Eq. 4.92, if even a single X_i value is zero—X_G will become zero. Notice that X_A would be only slightly affected by a zero value. This property of being low-value dominated makes the geometric average useful in situations where the system behavior is dominated by low-strength regions in a soil (e.g., settlement, bearing capacity, seepage).
2. X_G tends to a lognormal distribution by the central limit theorem. To see this, notice that ln X_G is a sum of random variables, as seen in Eq. 4.93, which the central limit theorem tells us will tend to a normal distribution. If ln X_G is (at least approximately) normally distributed, then X_G is (at least approximately) lognormally distributed.
3. if X is lognormally distributed, then its geometric average X_G is also lognormally distributed with the same median.

The second property is important since it says that low-strength-dominated geotechnical problems, which can be characterized using a geometric average, will tend to follow a lognormal distribution. This may explain the general success of the lognormal distribution in modeling soil properties.

If X_G is lognormally distributed, its mean and variance are found by first finding the mean and variance of ln X_G, where in the continuous case

$$\ln X_G = \frac{1}{T} \int_T \ln X(\mathbf{x})\, d\mathbf{x} \tag{4.96}$$

Assuming that $X(\mathbf{x})$ is stationary, then taking expectations of both sides of the above equation leads to

$$\mu_{\ln X_G} = \mu_{\ln X} \tag{4.97}$$

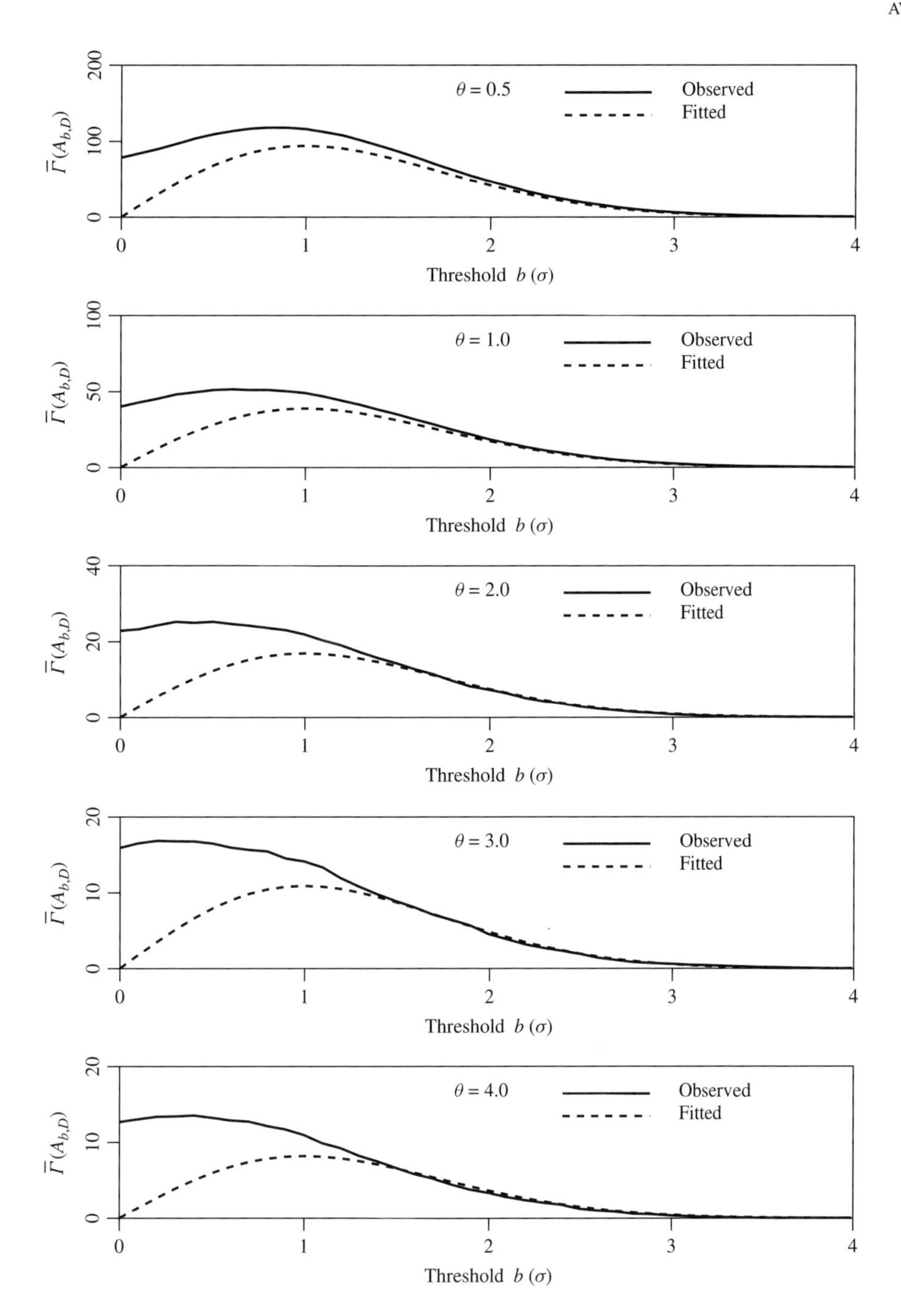

Figure 4.17 Comparison of theoretically predicted IG characteristic (Eq. 4.82) with observed average values obtained by simulation.

We note that since the median of a lognormally distributed random variable, X, is $\exp\{\mu_{\ln X}\}$, we see that the median of X_G is equal to the median of X. In other words, geometric averaging of a lognormally distributed random field, X, preserves both the type of the distribution and its median (this is analogous to arithmetic averaging of a normally distributed random field; the result is also normally distributed with the mean preserved). The preservation of the median of X is equivalent to the preservation of the mean of $\ln X$.

The variance of $\ln X_G$ is given by

$$\sigma^2_{\ln X_G} = \sigma^2_{\ln X} \gamma(T) \tag{4.98}$$

where $\gamma(T)$ is the variance reduction function defined for the $\ln X$ random field when arithmetically averaged over

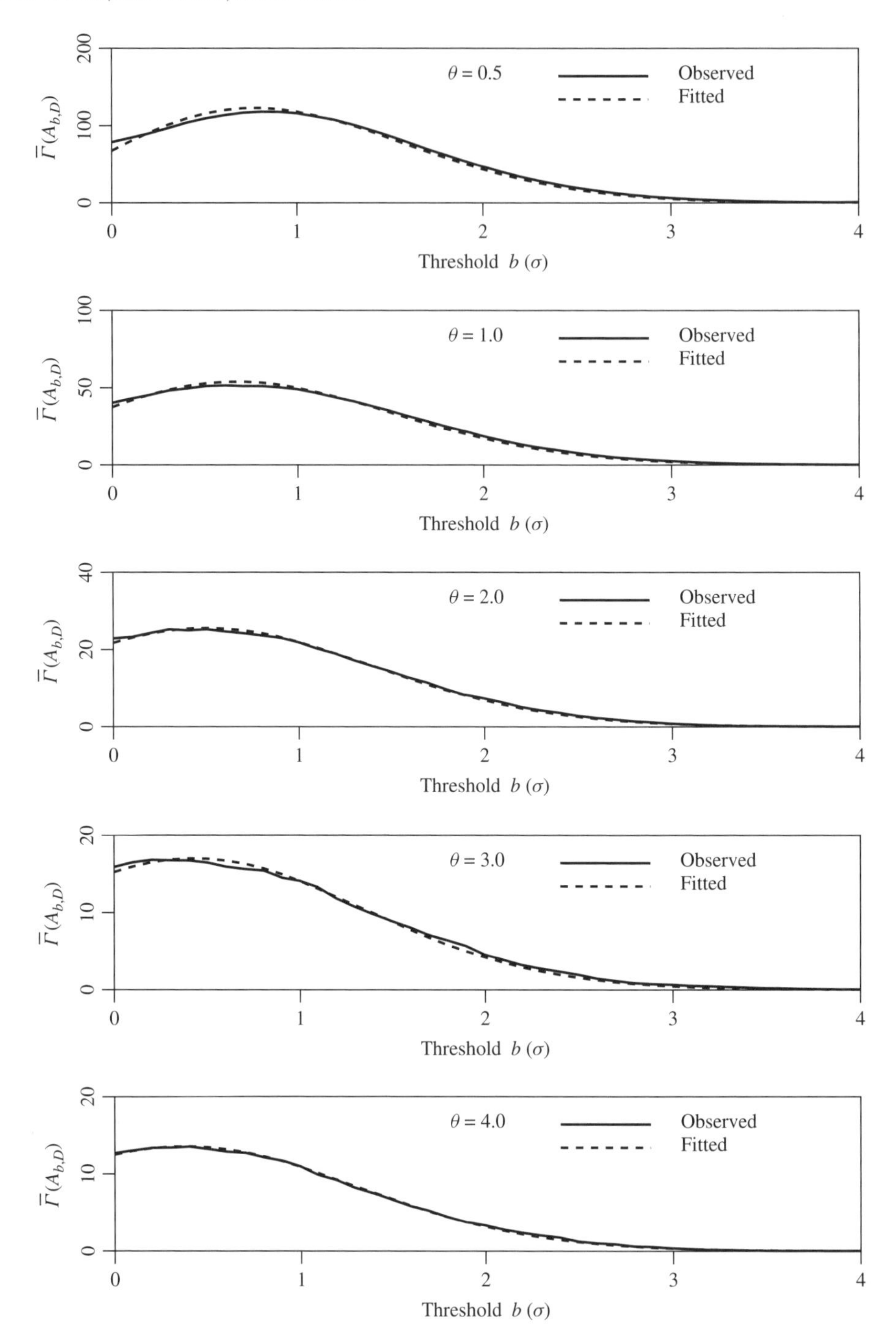

Figure 4.18 Comparison of empirically predicted IG characteristic (Eq. 4.83) with observed average values obtained by simulation.

the domain T. For example, if $\theta_{\ln X}$ is the correlation length of the ln X field and $\rho_{\ln X}(\boldsymbol{\tau}; \theta_{\ln X})$ is its correlation structure, then, from Eq. 3.40, we get

$$\gamma(T) = \frac{1}{|T|^2} \int_T \int_T \rho_{\ln X}(\boldsymbol{\xi} - \boldsymbol{\eta}; \theta_{\ln X}) \, d\boldsymbol{\xi} \, d\boldsymbol{\eta} \qquad (4.99)$$

where T may be a multidimensional domain and $|T|$ is its volume. The correlation length $\theta_{\ln X}$ can be estimated from observations $X_1, X_2, \ldots, X_n$ taken from the random field $X(\mathbf{x})$ simply by first converting all of the observations to $\ln X_1, \ln X_2, \ldots, \ln X_n$ and performing the required statistical analyses (see Chapter 5) on the converted data set.

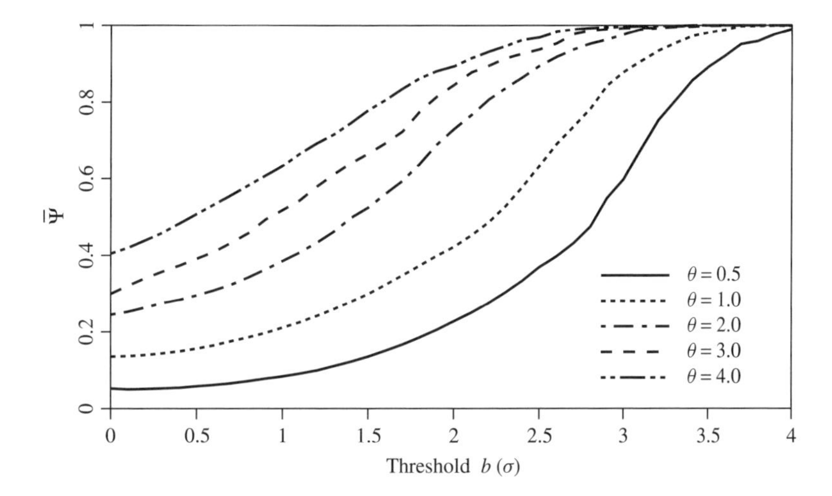

Figure 4.19 Average values of cluster measure $\overline{\Psi}$ estimated from 200 realizations of locally averaged Gauss–Markov process.

Table 4.6 Empirically Determined Effective Number of Independent Samples n_{eff} and Parameters of Type I Extreme Distribution (Eq. 4.90)

Scale	n_{eff}	α	μ
0.5	2900	3.14	3.41
1.0	900	2.49	3.05
2.0	180	2.05	2.52
3.0	70	1.78	2.15
4.0	35	1.62	1.86

Finally, the correlation function in logarithmic space can be converted to a correlation function in real space using Eq. 3.3,

$$\rho_X(\boldsymbol{\tau}) = \frac{\exp\left\{\sigma_{\ln X}^2 \rho_{\ln X}(\boldsymbol{\tau})\right\} - 1}{\exp\left\{\sigma_{\ln X}^2\right\} - 1} \tag{4.100}$$

For most random fields, the two correlation functions are quite similar and $\theta_X \simeq \theta_{\ln X}$.

Once the mean and variance of ln X_G have been computed, using Eqs. 4.97 and 4.98, lognormal transformations (Eq. 1.175) can be used to find the mean and variance of X_G:

$$\mu_{X_G} = \exp\left\{\mu_{\ln X} + \tfrac{1}{2}\sigma_{\ln X}^2 \gamma(T)\right\} = \frac{\mu_X}{\sqrt{\left(1 + v_X^2\right)^{1-\gamma(T)}}} \tag{4.101a}$$

$$\sigma_{X_G}^2 = \mu_{X_G}^2 \left[\exp\left\{\sigma_{\ln X}^2 \gamma(T)\right\} - 1\right] = \mu_{X_G}^2 \left[\left(1 + v_X^2\right)^{\gamma(T)}\right] \tag{4.101b}$$

where $v_X = \sigma_X/\mu_X$ is the coefficient of variation of X. Notice that the mean of the geometric average decreases as v_X increases. As the correlation length $\theta_{\ln X}$ increases, relative to the size of the averaging domain, T, the value of $\gamma(T)$ increases towards 1 (there is less independence between random variables in domain T, so there is less variance reduction). For strongly correlated random fields, then, the mean of the geometric average tends toward the global mean μ_X. At the other end of the scale, for poorly correlated random fields, where $\theta_{\ln X} << T$, the variance reduction function $\gamma(T) \to 0$ and the mean of the geometric average tends towards the median.

4.4.3 Harmonic Average

The harmonic average is particularly important in geotechnical engineering because it can be shown to be the exact average to use for several common geotechnical problems. Examples are (a) the settlement of a perfectly horizontally layered soil mass subject to uniform surface loading and (b) one-dimensional seepage through a soil. The harmonic average is defined by

$$X_H = \begin{cases} \left[\dfrac{1}{n}\displaystyle\sum_{i=1}^{n} \dfrac{1}{X_i}\right]^{-1} & \text{(discrete case)} \quad (4.102\text{a}) \\[2ex] \left[\dfrac{1}{T}\displaystyle\int_T \dfrac{d\mathbf{x}}{X(\mathbf{x})}\right]^{-1} & \text{(continuous case)} \quad (4.102\text{b}) \end{cases}$$

Example 4.5 Consider the layered soil shown in Figure 4.22 subjected to a surface stress σ. The elastic modulus of the ith layer is E_i. If the total settlement δ is expressed as

$$\delta = \frac{\sigma H}{E_{\mathit{eff}}}$$

derive E_{eff}.

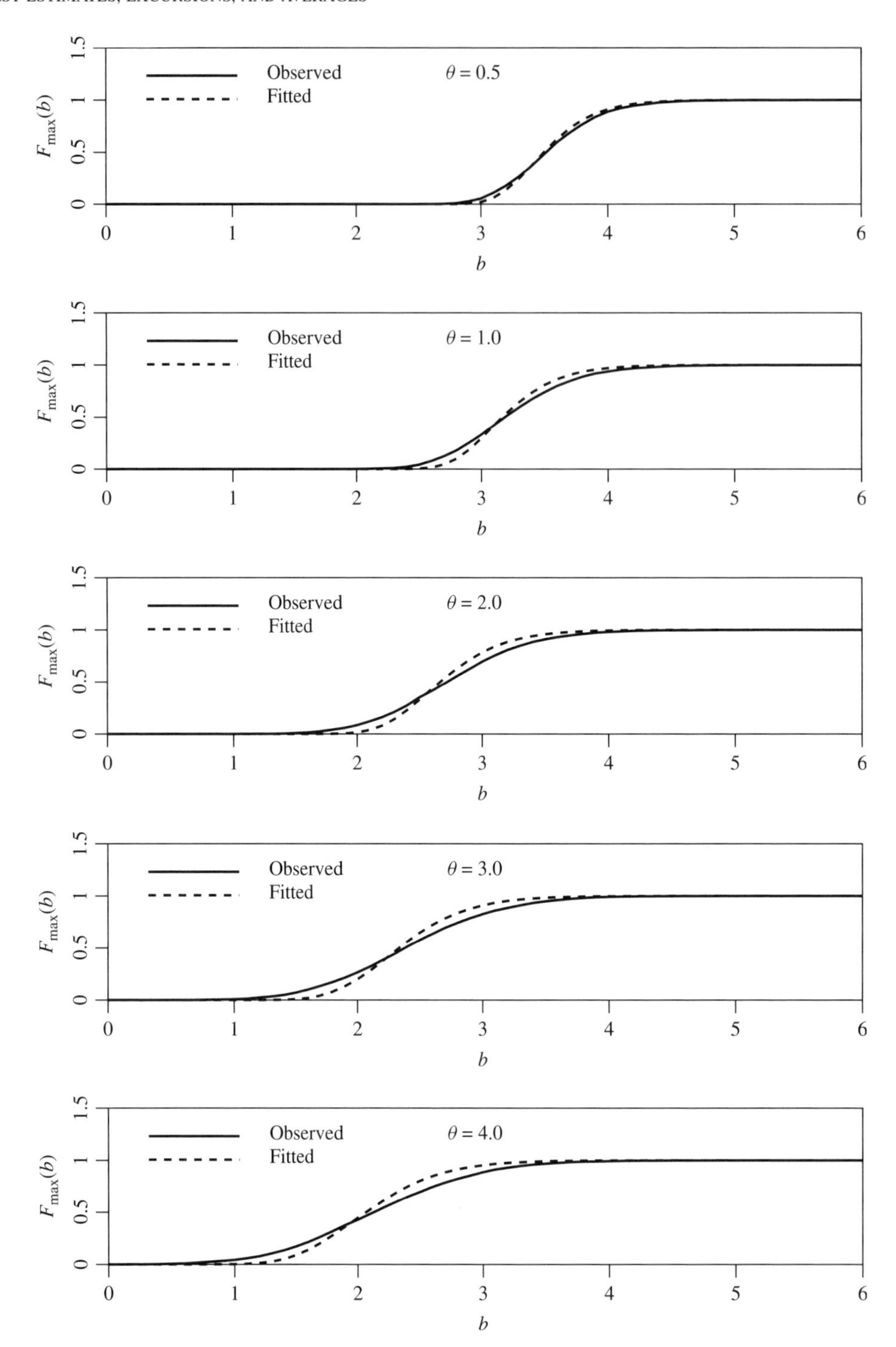

Figure 4.20 Observed cumulative distribution of global maximum of each realization compared to one-parameter extreme-value distribution given by Eq. 4.89.

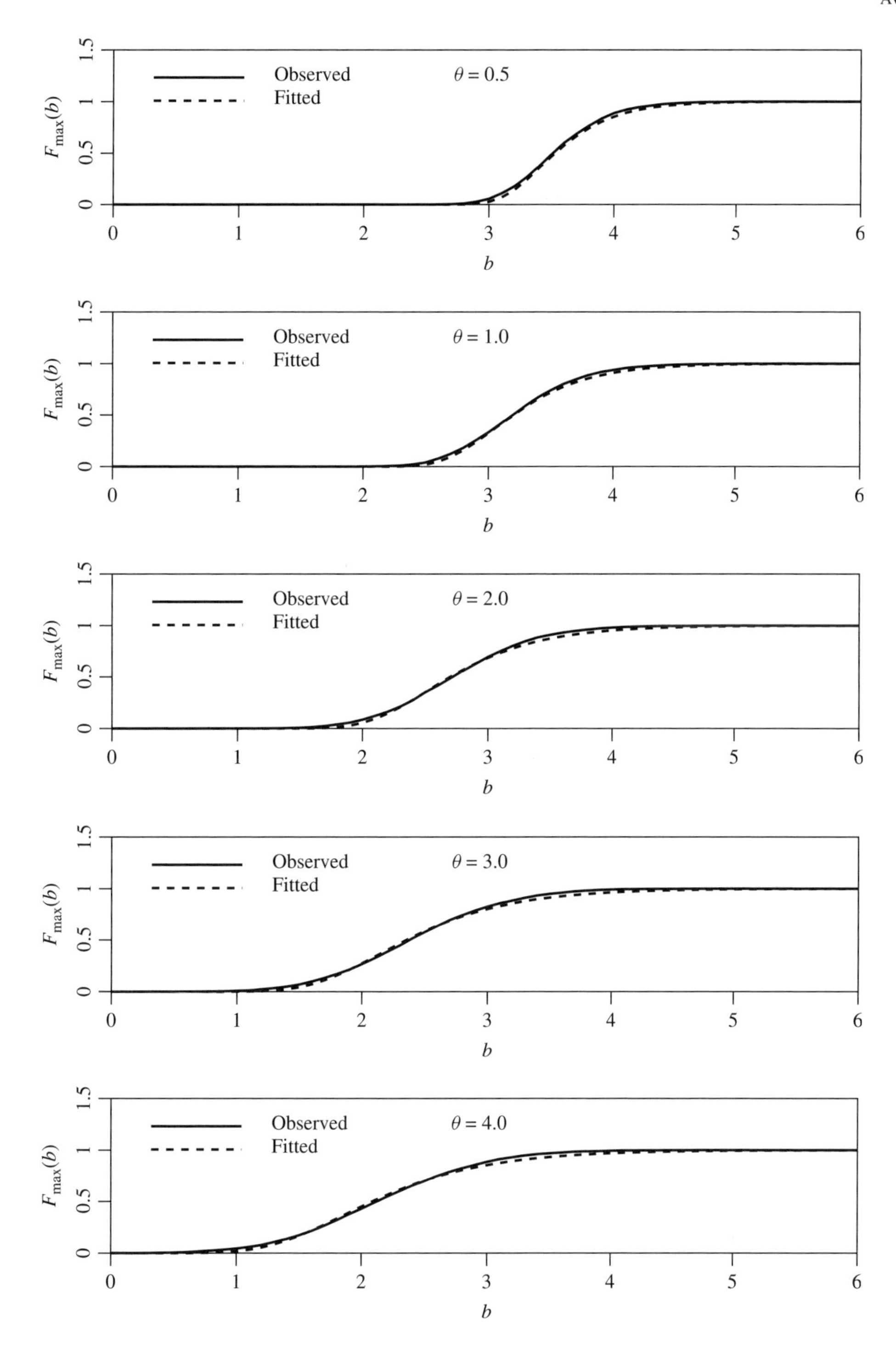

Figure 4.21 Observed cumulative distribution of global maximum of each realization compared to type I distribution given by Eq. 4.90.

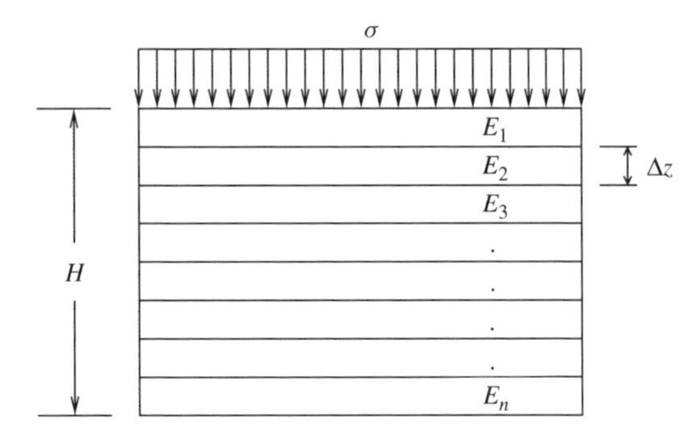

Figure 4.22 Settlement of perfectly horizontally layered soil.

SOLUTION We will assume that the load is uniformly distributed over a much wider area than seen in Figure 4.22, so that we can assume that the stress remains constant with depth. In this case, the settlement of the ith layer is

$$\delta_i = \frac{\sigma \ \Delta z_i}{E_i}$$

The total settlement is the sum of individual layer settlements,

$$\delta = \sum_{i=1}^{n} \frac{\sigma \ \Delta z_i}{E_i}$$

Equating this to $\sigma H / E_{\text{eff}}$ gives us

$$E_{eff} = \left[\frac{1}{H} \sum_{i=1}^{n} \frac{\Delta z_i}{E_i} \right]^{-1}$$

Finally, if the layer thicknesses are equal, then $H = n \ \Delta z$ and

$$E_{eff} = \left[\frac{1}{n} \sum_{i=1}^{n} \frac{1}{E_i} \right]^{-1}$$

which is the harmonic average (discrete case) defined above.

We can see from Eq. 4.102 that if any of the X values are zero, the harmonic average becomes zero. In fact, the harmonic average is even more strongly low-value dominated than is the geometric average. Unfortunately, the harmonic average is difficult to deal with from a probabilistic point of view since it has no known limiting distribution and its mean and variance are difficult to compute. When X is lognormally distributed, the lognormal distribution has been found to provide a reasonably good fit to the harmonic average for common types of random fields. Figure 4.23 illustrates the agreement between the frequency density plot of realizations of the harmonic average and a fitted lognormal distribution.

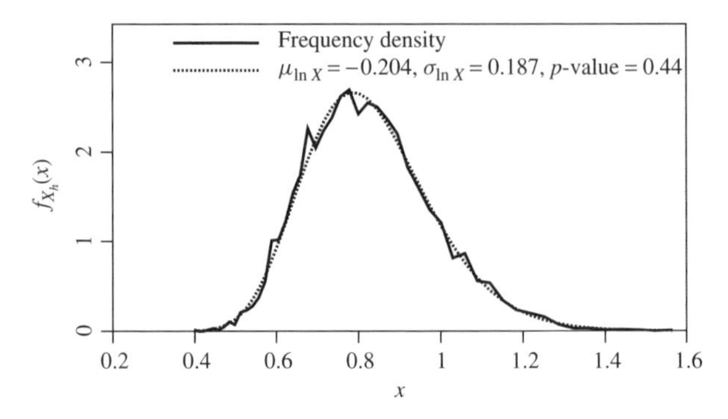

Figure 4.23 Frequency density plot of harmonic averages over area T^2 for $v_X = 0.5$ and $\theta_{\ln X} = 0.5T$ along with fitted lognormal distribution.

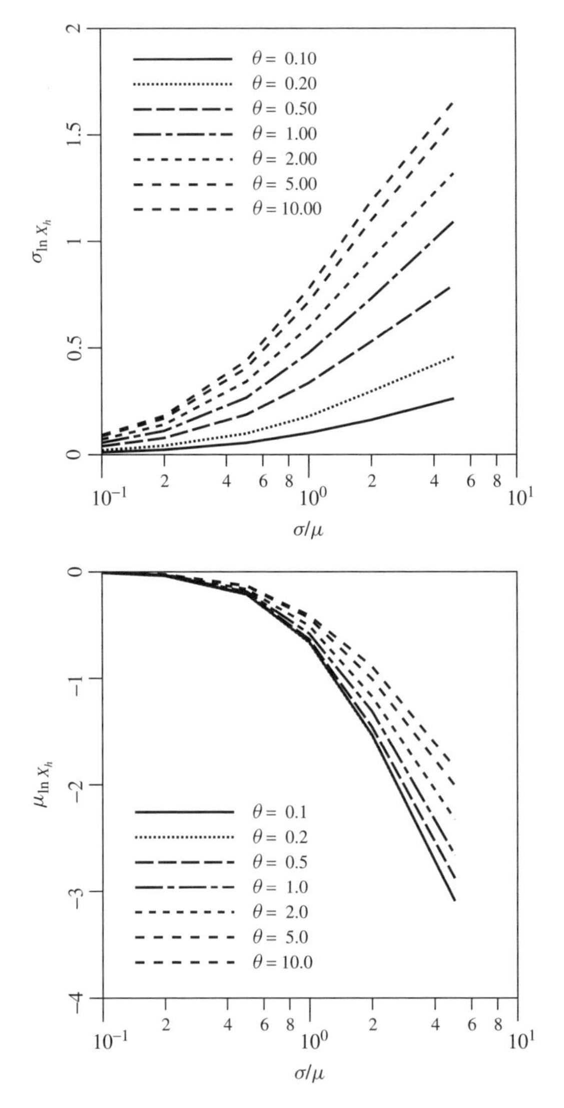

Figure 4.24 Mean and standard deviation of log-harmonic averages estimated from 5000 realizations ($\mu_X = 1.0$).

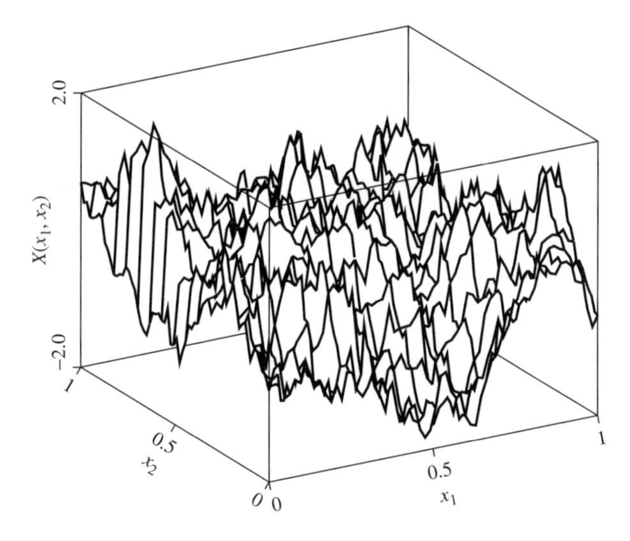

Figure 4.25 Sample two-dimensional random field having mean 1.0 and correlation length $\theta = 0.2$

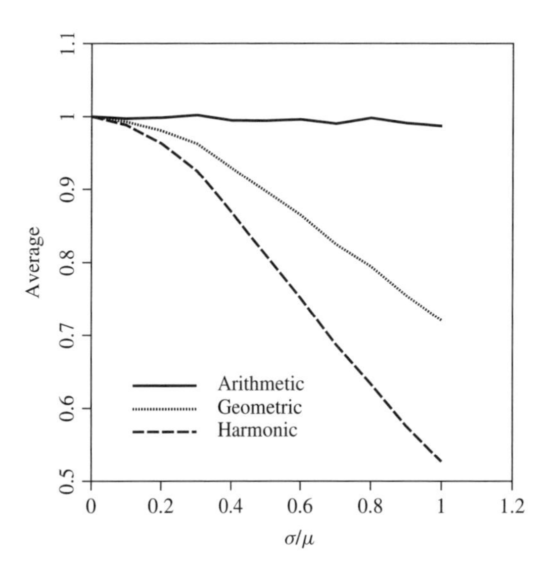

Figure 4.26 Comparison of arithmetic, geometric, and harmonic averages taken over a 1×1 random field with $\theta = 0.2$

Assuming that the harmonic average is at least approximately lognormally distributed, all that remains is to determine its two parameters, $\mu_{\ln X_H}$ and $\sigma_{\ln X_H}$. The authors have not found analytical approaches (beyond first order) to determine these quantities. Figure 4.24 illustrates how the mean and standard deviation of the harmonic average change with changing coefficient of variation and correlation length. These plots were obtained by simulating 5000 realizations of random fields of dimension $T \times T$ and computing their harmonic averages. Notice how the mean drops rapidly with increasing variability.

4.4.4 Comparison

Consider a two-dimensional random field of size 1×1, as illustrated in Figure 4.25. If we compute the average of the field shown in Figure 4.25 using the arithmetic, geometric, and harmonic averages, we will get three different values. As the variability of the random field is increased, the typical distance between the highs and lows increases. In turn, both the geometric and harmonic averages will decrease, since these are dominated by the low values.

Figure 4.26 illustrates how the means of the arithmetic, geometric, and harmonic averages change as the variance of the random field is increased. Both the geometric and harmonic averages fall quite rapidly, particularly when the coefficient of variation rises above about 0.2.

The harmonic average is the most strongly dominated by low values in the sample being averaged and the geometric average lies between the harmonic and arithmetic averages.

We conclude by commenting on the fact that these are not the only possible averages that can be used to characterize the behavior of soil masses. Other particularly important possibilities include averages which are distance weighted (i.e., soil properties close to a footing are more heavily weighted than those far away). Even better characterizations are obtained using averages which are weighted by correlations (i.e., soil properties more strongly correlated to the area of interest are weighted more heavily). The latter type of averaging is captured by a methodology called best linear unbiased estimate, a variant of which is called Kriging, as discussed in Section 4.1.

CHAPTER 5

Estimation

5.1 INTRODUCTION

The reliability assessment of geotechnical projects has been receiving increased attention from regulatory bodies in recent years. In order to provide a rational reliability analysis of a geotechnical system, there is a need for realistic random soil models which can then be used to assess probabilities relating to the design. Unfortunately, little research on the nature of soil spatial variability is available, and this renders reliability analyses using spatial variability suspect. In an attempt to remedy this situation, this chapter lays out the theory and discusses the analysis and estimation tools needed to analyze spatially distributed soil data statistically. Because of the complexity of the problem, the concentration herein is largely on the one-dimensional case. That is, the overall goal is to establish reasonable models for variability along a line. In order to achieve this goal, existing tools and estimators need to be critically reviewed to assess their performance for both large and small geotechnical data sets. The concentration on the one-dimensional case is reasonable, even when applied in a three-dimensional environment, because the directional "linear" statistics can be used to characterize separable models in higher dimensions.

Conceptually, at least, soils are basically deterministic. We could excavate an entire site and establish fairly closely the engineering properties of the soil throughout the site. Although such an undertaking would relieve us of having to deal with uncertainty, we would also be left with nothing upon which to found our structure, not to mention the cost. We must therefore live with uncertainty and attempt to quantify it rationally.

Although, traditionally, estimates of only the mean and variance have been largely sufficient for reliability estimates (via single random variable models), clients are now demanding full reliability studies. These studies require more sophisticated spatially variable models which involve knowledge of the spatial correlation structure of a soil. We now are asking questions such as: Knowing that soil properties are spatially correlated, what is a reasonable correlation function to use? Are soils best represented using fractal models or finite-scale models? What is the difference? How can this question be answered? Once a correlation function has been decided upon, how can its parameters be estimated? These are questions that this chapter addresses by looking at a number of tools which aid in selecting appropriate stochastic models. These tools include the sample covariance, spectral density, variance function, variogram, and wavelet variance functions. Common correlation functions, corresponding to finite-scale and fractal models, are investigated and estimation techniques discussed.

In general, statistical analyses can be separated into two areas which can be thought of as *descriptive* and *inferential* in nature. In the former, the goal is to best describe a particular data set with a view toward interpolating within the data set. For example, this commonly occurs when geotechnical data are obtained at a site for which a design is destined. Common descriptive statistics are the classic mean and variance estimates. The more advanced descriptive techniques most often used are those of regression using an appropriate polynomial which explains most of the variability, or best linear unbiased estimation (BLUE). Regression is purely geometry and observation based, while BLUE incorporates also the covariance structure between the data. Thus, the BLUE techniques require an a priori estimate of the covariance function governing the soil's spatial variability; this is often obtained by *inference* from other sites since it generally requires a very large data set to estimate reliably.

In general, inference occurs whenever one estimates properties at any unobserved spatial location. Here, the word inference will be taken to mean the estimation of stochastic model parameters which allow one to make probabilistic statements about an entire site for which data are limited or not available. This may be necessary, for example, in preliminary designs, in designs involving a future state, or in designs where a large site is to be characterized on the basis of a small test region. This chapter discusses inferential statistics for several reasons: (1) descriptive statistics are already reasonably well established and understood, (2) statistical results quoted in the literature must be inferential (unless you happen to be the author), (3) a priori knowledge of the second-moment (covariance) structure of soil properties is essential for BLUE estimators and Bayesian updating, and (4) site investigations are often not complete enough to even begin a spatial covariance estimation with any accuracy at all. Thus, most reliability-based designs will benefit from a database of inferred second-order soil statistics.

The difference between inferential and descriptive statistics becomes a critical issue in the interpretation of the estimates. This distinction can perhaps be best seen by asking yourself the following two questions, only one of which will be commonly true, prior to the statistical analysis of a data set:

1. Are the estimates being used to characterize the site at which the data are obtained (descriptive statistics)? If so, then estimator errors *decrease* with increasing correlation between experimental observations. That is, when a random field is highly correlated, only a few observations may be required to accurately characterize the field. For most practicing engineers, the answer to this question will be yes.
2. Are the estimates being used to characterize the soil "population" (inference)? That is, are the collected data being used to say something about all (similar?) soil sites? If so, then estimator errors *increase* with increasing correlation between observations. For example, when a random field is highly correlated, it does not matter how many observations one takes of the field, one will only see a small fraction of the distribution of possible values. The estimate could be quite in error (e.g., trying to make comments about the natural variability of cohesion by taking many samples in a gravel pit). For code developers, researchers, or anyone attempting to make statements about general a priori soil property statistics, the answer to this question will be yes.

Most practitioners are able to answer yes to question 1. However, most researchers publishing in the literature and anyone working on the development of a reliability-based geotechnical design code must answer yes to question 2. The site characterization problem (question 1) tends to be significantly simpler than the population characterization problem (question 2).

This chapter begins by looking at the problem of selecting a marginal distribution (i.e., a point distribution) and testing how well it fits the data. Then the simpler classical estimators of the mean, variance, and covariance structure, largely in the context of geotechnical engineering, are investigated. Some of the basic concepts of random-field theory, initially seen in Chapter 3, are elaborated on here for clarification. Where appropriate, we will distinguish between how the estimates apply to the characterization of a site (descriptive) and of the population (inference). The latter part of this chapter looks at more advanced methods of estimating the second-moment structure (i.e., the covariance structure) of a random field. Only the one-dimensional case is considered.

5.2 CHOOSING A DISTRIBUTION

When data have been collected on an input random variable of interest, the data can be used in one of three ways to specify a distribution:

1. The data values themselves are used directly in the simulation. This is sometimes called *trace-driven simulation*. This is the *least preferable* way to use the data since, in this case, the simulation can only reproduce what has happened historically and/or at the observation locations. There is seldom enough data to capture all the *possible future*, and/or spatial *variability*. This approach is most commonly reserved for earthquake ground motion simulation, where past recorded motions are used as system input to assess seismic response.
2. The data values are used to define an *empirical* distribution function directly. Random simulation then involves random samples drawn from the empirical distribution function. This is usually referred to as *sampling from the empirical distribution*. This method is better than the first as it is not *constrained* in the amount of data that can be simulated; however it still has drawbacks. Most notably, only observations in the *range* of the observed data can be simulated. This does not allow for the *extremes* which often control a design.
3. A reasonable distribution is *fitted* to the data. Now, random samples can be drawn from the fitted distribution in, for example, a *Monte Carlo simulation*. If a theoretical distribution that fits the observed data reasonably well can be found, then this is usually the preferred method. There are numerous advantages, most notably:
   (a) The "irregularities" in the empirical distribution are *smoothed* out with a fitted distribution. Since the irregularities are almost certainly due to the fact that only a finite sample is used, this is a desirable feature of the fitted distribution.
   (b) The fitted distribution can generate values outside the range of the observed sample. This means that *extremes* can be represented in a reasonable way.
   (c) Sometimes there are compelling physical reasons to have a given distributional form.
   (d) This is a more compact way to represent data. That is, most fitted distributions will have one or two parameters whereas the empirical distribution requires the storage of $2n$ values (n locations and n corresponding cumulative probabilities).

We will concentrate only on the fitting of a distribution to the data and we start by considering how to select the distribution governing the random field. Commonly, this will be a marginal distribution governing the distribution of a stationary random field at any *point* in the field, $f_X(x)$. We will rarely have enough information to prescribe a full joint distribution $f_{X_1X_2\dots}(x_1, x_2, \dots)$, except in the case of the normal distribution, so will not dwell on how to do this for the general case.

The first step in choosing a distribution is to consider what is physically reasonable for the soil property you are trying to model. The normal distribution is a very popular choice. This is particularly true when the soil property is a random field, since the full joint normal distribution is completely specified by only the mean and covariance structure. The one major disadvantage to the normal distribution is that its range is from $-\infty$ to $+\infty$. For most soil properties, for example, cohesion or elastic modulus, negative values do not have a physical meaning. Thus, for nonnegative soil properties, the normal distribution *cannot* be the true distribution, and other nonnegative distributions should be considered (e.g., lognormal, gamma, Weibull, or one of the extreme-value distributions). However, if the probability of obtaining a negative property value is small enough, the normal distribution is a reasonable approximation. For example, if the coefficient of variation, $v = \sigma/\mu$, of the soil property is less than about 30%, then the probability of obtaining a negative soil property value is only

$$\mathrm{P}[X < 0] = \mathrm{P}\left[Z < \frac{0-1}{\sigma/\mu}\right] = \mathrm{P}\left[Z < \frac{-1}{0.3}\right]$$
$$= \Phi(-3.33) = 0.0004$$

The difference between $v = \sigma/\mu$ of 0.3 and 1.0 is illustrated in Figure 5.1. Clearly, if v is as large as 1.0, then a fairly large proportion of possible realizations of X will be negative.

Given the advantages of the normal distributions (e.g., its ease of use and simple multivariate form), it may be desirable to use it when the coefficient of variation is acceptably small. However, in the example given above, if the target failure probability for a design is around 0.001 and if failure tends to occur in low-strength regions, then a model error as large as 0.0004 may not be acceptable. Overall, it is probably best to use a distribution which is physically reasonable where possible.

In geotechnical engineering, there are a number of soil properties which are bounded both above and below. This is another physical attribute of the property which should be considered when selecting a distribution. For example, friction angle (0–90°), porosity (0–1), degree of saturation (0–1), and relative density (0–1) are all bounded both above and below. Arguments may be made for bounding other soil properties, such as unit weight, both above and below. In the case of unit weight, we know that it cannot exceed the unit weight of the heaviest element but, practically speaking, is unlikely to exceed the unit weight of silicon, or perhaps calcium, or maybe even iron. Unfortunately, as with many soil properties that one might expect to have an upper bound, the upper bound is often *arbitrarily selected* and not precisely known. For example, an upper bound on a soil's unit weight might be assumed to be 26 kN/m^3 and this may yield a quite reasonable distribution. However, can it be said, with absolute certainty, that the unit weight will never exceed 26 kN/m^3? If not, then perhaps an unbounded distribution is more physically correct, so long as the likelihood of exceeding, say, 26 kN/m^3, is sufficiently small. We will never have enough information to state precisely which distribution is the true distribution for any soil property. Distributions should be selected that best satisfy the following guidelines:

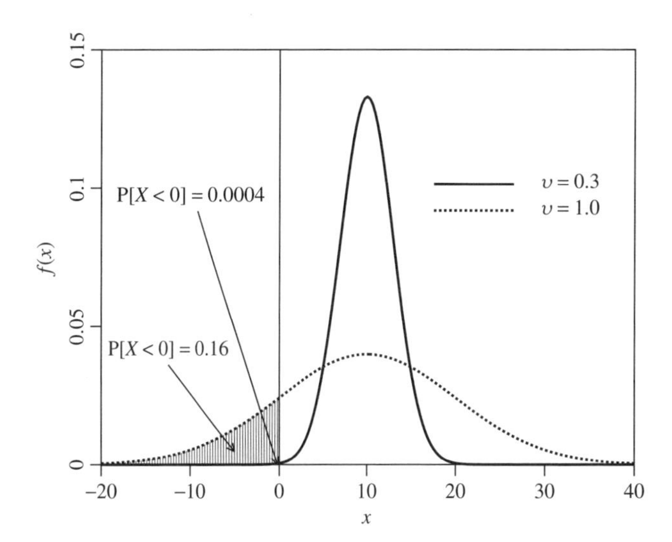

Figure 5.1 Comparison of probability that $X < 0$, where X is normally distributed, for two different coefficients of variation, $v = \sigma/\mu$.

1. If sufficient data are available, select the distribution which best fits the histogram of the data (see next section).
2. Ensure that the distribution is (at least approximately) physically reasonable. That is, if the soil property is strictly nonnegative, such as elastic modulus, then the normal distribution is not physically reasonable since it allows negative values. It may, however, be approximately reasonable if the probability of negative values is sufficiently small.

The second step in choosing a distribution is to read the literature; find out what other people have found to be successful probabilistic models for the soil/rock property in question.

The third step is to take whatever data are available, estimate the distribution parameters (see next section), and then see how well the distribution(s) you have selected actually match the data. If sufficient data are available (generally, at least 20 observations), the most common way to compare the data to the assumed distribution is by using a histogram. Histograms act as graphical estimates of the density function, as will be discussed shortly.

Finally, the selected distribution should be as simple as possible while still reflecting the basic nature of the variability. Distributions which involve more than two parameters are generally difficult to justify because we rarely have enough data to estimate even two parameters with any accuracy. There is little point in trying to match, say, a six-parameter distribution to the detailed erratic fluctuations in a histogram, even if the histogram is based on a large number of data. Many of the detailed fluctuations in a histogram will change if a different data set is collected or even if the histogram interval sizes are changed. Only the average (smoothed) behavior of the histogram should be matched in a fitted distribution.

5.2.1 Estimating Distribution Parameters

Once we have decided on a distribution to fit to the data that we have collected, the next step is to estimate the parameters of the distribution from the data. If we look at Sections 1.9 and 1.10, we will see that each distribution is characterized by one or more *parameters*. For example, the exponential distribution has one parameter, λ, while the lognormal distribution has two parameters, $\mu_{\ln X}$ and $\sigma_{\ln X}$.

In this section, we look at how these distribution parameters are estimated from our data. We call our parameter estimates *point estimates* because they consist of a single "best" value rather than a range. For example, we might say that a point estimate of the mean is 10.2. We obtain our point estimate by using an estimator, such as $\bar{x}$ (see Eq. 1.25).

There are basically two ways of obtaining point estimates in common usage. The simplest is the *method of moments* (MM) and this is probably the most common approach. A somewhat more difficult approach is the *maximum-likelihood* (ML) method. The ML estimates have some desirable statistical properties that sometimes make the extra effort well worth it.

In either case, when we consider our parameter estimates, there are four measures by which we judge their performance:

1. *Unbiasedness* $\longrightarrow$ E [estimator] = parameter
2. *Consistency* $\longrightarrow$ $\lim_{n\to\infty}$ estimator = parameter
3. *Efficiency* $\longrightarrow$ Var [estimator] large or small?
4. *Sufficiency* $\longrightarrow$ utilizes all pertinent information?

5.2.1.1 *Method of Moments*

For many of the distributions we have looked at, the parameters of the distribution are simply related to the moments; for example, for a normal distribution we have

$$\mathrm{E}[X] = \mu, \qquad \mathrm{E}\left[X^2\right] = \sigma^2 + \mu^2$$

while for the exponential distribution

$$\mathrm{E}[X] = \frac{1}{\lambda}, \qquad \mathrm{E}\left[X^2\right] = \frac{2}{\lambda^2}$$

(for the latter, we need only the first moment to find λ). We can use *sample moments* to replace the left-hand sides in the above and then solve for the parameter on the right. Sample moments are obtained by approximating the expectation integrals by a summation over equi-likely samples, each weighted by $1/n$ rather than $f(x)\,dx$,

$$\mathrm{E}[X] \simeq \bar{X} = \frac{1}{n}\sum_{i=1}^{n} X_i$$

$$\Longrightarrow \qquad \bar{x} = \frac{1}{n}\sum_{i=1}^{n} x_i \qquad \text{(sample mean)}$$

$$\mathrm{Var}[X] \simeq S^2 = \frac{1}{n}\sum_{i=1}^{n}(X_i - \bar{X})^2$$

$$\Longrightarrow \qquad s^2 = \frac{1}{n}\sum_{i=1}^{n}(x_i - \bar{x})^2 \qquad \text{(sample variance)}$$

Assuming that the sample X_i comes from the selected distribution, it follows that $\mathrm{E}[X_i] = \mu$ and that $\mathrm{E}\left[X_i^2\right] - \mathrm{E}^2[X_i] = \sigma^2$. The *bias* associated with the above sample moments can be determined by checking to see if the expectations of $\bar{X}$ and S^2 are in fact equal to μ_X and σ_X^2:

$$\mathrm{E}\left[\bar{X}\right] = \frac{1}{n}\sum_{i=1}^{n}\mathrm{E}[X_i] = \mu \quad \text{(OK, unbiased)}$$

$$\mathrm{E}\left[S^2\right] = \frac{1}{n}\sum_{i=1}^{n}\mathrm{E}\left[(X_i - \bar{X})^2\right]$$

$$= \frac{1}{n}\sum_{i=1}^{n}\mathrm{E}\left[X_i^2 - 2X_i\bar{X} + \bar{X}^2\right]$$

$$= \frac{1}{n}\sum_{i=1}^{n} \mathrm{E}\left[X_i^2 - 2X_i\left(\frac{1}{n}\sum_{j=1}^{n} X_j\right) + \frac{1}{n^2}\sum_{j=1}^{n}\sum_{k=1}^{n} X_j X_k\right]$$

$$= \frac{1}{n}\sum_{i=1}^{n}\left\{\mathrm{E}\left[X_i^2\right] - \frac{2}{n}\sum_{j=1}^{n}\mathrm{E}\left[X_i X_j\right] + \frac{1}{n^2}\sum_{j=1}^{n}\sum_{k=1}^{n}\mathrm{E}\left[X_j X_k\right]\right\}$$

Due to independence between samples, the expectation $\mathrm{E}\left[X_i X_j\right]$ becomes

$$\mathrm{E}\left[X_i X_j\right] = \begin{cases} \mathrm{E}\left[X_i\right]\mathrm{E}\left[X_j\right] & \text{if } i \neq j \\ \mathrm{E}\left[X_i^2\right] & \text{if } i = j \end{cases}$$

so that we get

$$\mathrm{E}\left[S^2\right] = \frac{1}{n}\sum_{i=1}^{n}\left\{\mathrm{E}\left[X^2\right] - \frac{2}{n}\left(\mathrm{E}\left[X^2\right] + (n-1)\mathrm{E}^2[X]\right) + \frac{1}{n^2}\left(n\mathrm{E}\left[X\right] + (n^2 - n)\mathrm{E}^2[X]\right)\right\}$$

$$= \frac{1}{n}\cdot n\left\{\left(1 - \frac{1}{n}\right)\mathrm{E}\left[X^2\right] - \left(\frac{n-1}{n}\right)\mathrm{E}^2[X]\right\}$$

$$= \left(\frac{n-1}{n}\right)\left(\mathrm{E}\left[X^2\right] - \mathrm{E}^2[X]\right)$$

$$= \left(\frac{n-1}{n}\right)\sigma^2$$

Since $\mathrm{E}\left[S^2\right] \neq \sigma^2$, the estimator for σ^2, as given above, is a *biased* estimator. To eliminate this bias, we must write

$$S^2 = \frac{1}{n-1}\sum_{i=1}^{n}(X_i - \bar{X})^2$$

which is unbiased since now $\mathrm{E}\left[S^2\right] = \sigma^2$. This is the common form of the estimator S^2 of σ^2.

The steps taken to compute point estimates of a distribution's parameters by the MM are as follows:

1. Decide on a distribution and identify its parameters.
2. Using the distribution's theoretical pdf, compute as many moments as there are unknown parameters (these moments are generally in terms of the parameters). Call these the *model* moments.
3. Compute the same number of sample moments.
4. Equate model and sample moments to find the parameter estimates.

Comments

- The MM can lead to large errors, especially when higher moments are involved. This is exacerbated if there are *outliers* in the data.
- The MM sometimes leads to biased estimators.

Example 5.1 Suppose we have the following set of 20 measurements on the lengths of fissures in a rock mass:

0.808, 1.005, 0.806, 0.661, 6.681, 0.057, 0.123, 9.372, 0.902, 0.764, 0.286, 0.764, 0.558, 1.813, 2.025, 6.559, 1.600, 3.014, 3.814, 4.503

Assuming that this data follows an exponential distribution, use the MM to fit the distribution to the data.

SOLUTION

1. We have been told what distribution we are to be fitting. The exponential pdf has the form (see Section 1.10.1)
$$f_X(x) = \lambda e^{-\lambda x} \tag{5.1}$$
where X is the fissure length and λ is the distribution parameter that we must find by the MM.
2. We have one unknown parameter, λ, so we must calculate the first moment from the theoretical distribution,
$$\mathrm{E}\left[X\right] = \frac{1}{\lambda} \tag{5.2}$$
3. Compute the first sample moment:
$$\bar{x} = \frac{1}{n}\sum_{i=1}^{n} = \frac{1}{20}(0.808 + 1.005 + \cdots + 4.503)$$
$$= 2.306 \tag{5.3}$$
4. Equate model and sample moments:
$$\frac{1}{\hat{\lambda}} = \bar{x} \qquad \Longrightarrow \qquad \hat{\lambda} = \frac{1}{\bar{x}} = \frac{1}{2.306} = 0.434 \tag{5.4}$$
where $\hat{\lambda} = 0.434$ is an estimate of the true distribution parameter λ (which remains unknown).

Note the use of the caret to denote the fact that this is just an estimate of the parameter and is not the parameter itself. This is an important distinction since estimates are in fact random variables, changing from sample set to sample set, whereas the parameter itself is deterministic. In general, the parameter is only known exactly when an infinite number of samples is taken.

Example 5.2 Say we have a set of data (observations) on the settlement of piles at various sites throughout a

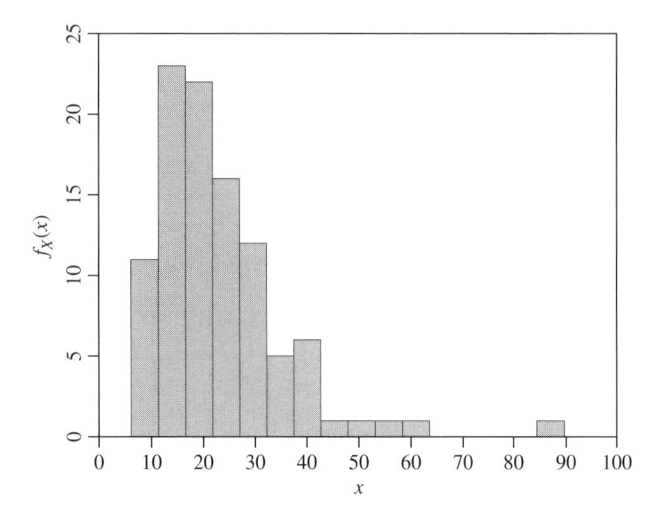

Figure 5.2 Histogram of pile settlements.

city, $x_1, x_2, \ldots, x_n$. A histogram of the data appears as in Figure 5.2. Use the MM to fit a distribution to the data.

SOLUTION

1. We first have to decide on an appropriate distribution to fit the data. A variety of common distributions have the positive right-skewed shape seen in Figure 5.2; in particular, any of the lognormal, gamma, Rayleigh, or chi-square distributions could be used. Suppose that we choose to fit a gamma distribution to this data set (we shall see later how we can formally compare the fits arising from various possible distributions). The gamma distribution has the general form

$$f_X(x) = \frac{\lambda\,(\lambda x)^{k-1}}{\Gamma(k)} e^{-\lambda x}, \qquad x \geq 0 \tag{5.5}$$

which has two parameters, λ and k (see Section 1.10.2).

2. Since the assumed distribution (gamma) has two unknown parameters, we need to compute the first two moments from the theoretical distribution. Section 1.10.2 gives

$$\mathrm{E}[X] = \frac{k}{\lambda} \tag{5.6a}$$

$$\mathrm{Var}[X] = \frac{k}{\lambda^2} \tag{5.6b}$$

3. We need two sample moments:

$$\bar{x} = \frac{1}{n}\sum_{i=1}^{n} x_i \tag{5.7a}$$

$$s^2 = \frac{1}{n-1}\sum_{i=1}^{n}(x_i - \bar{x})^2 \tag{5.7b}$$

4. Equate model and sample moments:

$$\left.\begin{aligned} \frac{\hat{k}}{\hat{\lambda}} &= \bar{x} \\ \frac{\hat{k}}{\hat{\lambda}^2} &= s^2 \end{aligned}\right\} \iff \left\{\begin{aligned} \hat{k} &= \frac{\bar{x}^2}{s^2} \\ \hat{\lambda} &= \frac{\bar{x}}{s^2} \end{aligned}\right.$$

Whether or not the gamma distribution is really the best choice of model for this particular data set must be determined by further testing. In general, the above procedure would be repeated for each candidate distribution. Then the various fitted distributions superimposed on the histogram and the best fit could be determined visually. This and a number of quantitative tests will be discussed shortly.

5.2.1.2 Maximum-Likelihood Estimators Another approach to estimating the parameters of a distribution is to find those parameters which yield the highest likelihood of actually observing the data. For example, suppose we have observed three values of X to be 1.2, 1.8, and 0.6. The probability of seeing the observations $\{1.2, 1.8, 0.6\}$ is vanishingly small if $\mu_X = 300$ and $\sigma_X = 1$. On the other hand, the probability of seeing $\{1.2, 1.8, 0.6\}$ is reasonably high if $\mu_X = 1.2$ and $\sigma_X = 0.4$. It is clear that there exists a combination of μ_X and σ_X which yields the highest likelihood of seeing the data. These optimum parameters are called *maximum-likelihood estimators* (MLEs). In general, maximum-likelihood estimators are preferred over method of moment estimators simply because they maximize the probability of observing the data. They also tend to yield estimators having minimum variance. One potential drawback to MLEs is that they can be biased. For example, the MLE of σ^2 is

$$S^2 = \frac{1}{n}\sum_{i=1}^{n}(X_i - \bar{X})^2$$

which is biased since

$$\mathrm{E}\left[S^2\right] = \left(\frac{n-1}{n}\right)\sigma^2$$

as was shown in the previous section.

The maximum likelihood (ML) method finds the distribution parameters that best explain the data collected. That is, the method finds estimators which *maximize* the likelihood of observing the data given the distribution. The ML technique generally has better properties that the MM (i.e., less bias, higher efficiency, etc.) but is somewhat less intuitive.

To illustrate the method, let us start with a simple example. Suppose that we have a random sample $X_1, X_2,$

$\ldots, X_n$ which comes from a population governed by $f_X = \lambda e^{-\lambda x}$, the exponential distribution. Since this is a random sample, the X's are independent. The probability of observing any particular x_i is

$$\mathrm{P}[X_i = x_i] = f_X(x_i)\ dx = \lambda e^{-\lambda x_i}\ dx \tag{5.8}$$

[note that we are abusing calculus slightly in the above by replacing the event $(x_i < X_i \le x_i + dx)$ with the event $(X_i = x_i)$ for simplicity] so that the probability of observing the set of values $\mathbf{x} = \{x_1, x_2, \ldots, x_n\}$ is the joint intersection,

$$\begin{aligned}
\mathrm{P}[X_1 = x_1 \cap X_2 = x_2 \cap \cdots \cap X_n = x_n] \\
&= \left(f_X(x_1)\ dx\right)\left(f_X(x_2)\ dx\right)\cdots\left(f_X(x_n)\ dx\right) \\
&= \prod_{i=1}^{n} f_X(x_i)\ dx \\
&= \left(\prod_{i=1}^{n} \lambda e^{-\lambda x_i}\right)(dx)^n
\end{aligned} \tag{5.9}$$

where we made use of independence to write the probability as a product. Ignoring the constant $(dx)^n$ (which is *very* small), we can define the *relative-likelihood function* $L(\lambda \mid \mathbf{x})$ to be

$$\begin{aligned}
L(\lambda \mid \mathbf{x}) = \prod_{i=1}^{n} f_X(x_i) &= \prod_{i=1}^{n} \lambda e^{-\lambda x_i} \\
&= \lambda^n \exp\left\{-\lambda \sum_{i=1}^{n} x_i\right\}
\end{aligned} \tag{5.10}$$

This is the relative likelihood of observing $\mathbf{x} = \{x_1, x_2, \ldots, x_n\}$ as a function of the unknown parameter λ. To find the maximum likelihood, we maximize this function with respect to λ, that is, we set

$$\frac{\partial L(\lambda \mid \mathbf{x})}{\partial \lambda} = 0 \tag{5.11}$$

Now

$$\frac{\partial L(\lambda \mid \mathbf{x})}{\partial \lambda} = \left(n\lambda^{n-1} - \lambda^n \sum_{i=1}^{n} x_i\right) e^{-\lambda \sum_{i=1}^{n} x_i}$$

so that

$$n\lambda^{n-1} - \lambda^n \sum_{i=1}^{n} x_i = 0 \quad \Longrightarrow \quad \hat{\lambda} = \frac{n}{\sum_{i=1}^{n} x_i} = \frac{1}{\bar{x}}$$

Note that in this case the ML and MM estimators are the same. This is not always true.

Since $L(\cdot)$ is always positive, the maximum of $L(\cdot)$ will occur in exactly the same location as the maximum of $\ln L(\cdot)$. Often it is easier to compute the derivative of $\ln L$ than it is to compute the derivative of L itself. In the above example,

$$\ln\left(L(\lambda \mid \mathbf{x})\right) = n \ln(\lambda) - \lambda \sum_{i=1}^{n} x_i \tag{5.12}$$

and setting the derivative of this to zero gives

$$\frac{\partial}{\partial \lambda} \ln\left(L(\lambda \mid \mathbf{x})\right) = \frac{n}{\lambda} - \sum_{i=1}^{n} x_i = 0 \tag{5.13}$$

yielding the same result, $\hat{\lambda} = 1/\bar{x}$, with somewhat less effort.

Example 5.3 Let X be a discrete random variable with a geometric distribution having parameter p (see Section 1.9.3). Find the MLE of p based on a random sample of size n.

SOLUTION If we have a random sample of size n, then we have observed independently $\mathbf{x} = \{x_1, x_2, \ldots, x_n\}$. We now want a MLE of p. The probability of observing x_i comes from the geometric distribution (see Section 1.9.3),

$$\mathrm{P}[X_i = x_i] = pq^{x_i - 1}, \qquad i = 1, 2, \ldots n \tag{5.14}$$

where $q = 1 - p$. The likelihood function is thus

$$\begin{aligned}
L(p \mid \mathbf{x}) &= \mathrm{P}[X_1 = x_1 \cap X_2 = x_2 \cap \cdots \cap X_n = x_n] \\
&= \prod_{i=1}^{n} pq^{x_i - 1} \\
&= p^n \prod_{i=1}^{n} (1 - p)^{x_i - 1}
\end{aligned} \tag{5.15}$$

while its logarithm is

$$\begin{aligned}
\ln(L) &= n \ln p + \sum_{i=1}^{n} (x_i - 1) \ln(1 - p) \\
&= n \ln p + \ln(1 - p) \sum_{i=1}^{n} (x_i - 1)
\end{aligned}$$

We now set the derivative of $\ln(L)$ to zero:

$$\frac{\partial \ln L}{\partial p} = \frac{n}{p} - \frac{\sum_{i=1}^{n} (x_i - 1)}{1 - p} = 0$$

Solving this for p, which we will now call $\hat{p}$, gives us

$$\hat{p} = \frac{n}{\sum_{i=1}^{n} x_i} = \frac{1}{\bar{x}}$$

where $\bar{x}$ is just the average of our observations.

Chapter 1 includes the MLEs for most of the common distributions covered (see also Law and Kelton, 2000).

5.2.2 Goodness of Fit

Once a distribution has been selected and then fit (by estimating its parameters) to the collected data, the *fit* must be assessed. That is, how well does the fitted distribution represent the true underlying distribution for our data? There are two commonly used approaches to answering this question:

1. Heuristic procedures
2. Goodness-of-fit tests

First, note that no fitted distribution will be an *exact* fit. But is the fitted distribution *reasonable* enough for its intended purpose? This issue is considered in the following sections.

5.2.2.1 Heuristic Procedures

Frequency Comparisons Frequency comparisons are made by plotting the observed frequency of occurrence against that predicted by the fitted theoretical distribution. A plot of the observed occurrence frequency is called a *histogram*. If the random variable is discrete, then the histogram is drawn as a bar graph, where the height of each bar is just equal to the number of times each discrete value occurs. When comparing the bar graph to the fitted distribution, the bar graph must be normalized, so that the sum of normalized frequencies becomes equal to 1.0, to agree with the sum of probabilities being unity. The normalization is performed by dividing the height of each bar by the total number of observations times the bar width, which, in the discrete case, is 1.0.

Example 5.4 Suppose that, just after construction, a series of 50 randomly selected 1-km-long sections of highway through a hilly region were selected to evaluate the annual probability of slope failure under the existing design code. The number of years until an observable slope failure occurred within each 1-km length, t_i, was recorded, with the following results:

3, 2, 8, 9, 10, 4, 4, 2, 7, 7, 1, 14, 2, 1, 8, 3, 4, 5, 4, 2, 10, 2, 1, 7, 8, 4, 3, 3, 21, 1, 3, 9, 1, 4, 5, 1, 4, 1, 4, 3, 5, 3, 1, 9, 1, 6, 3, 5, 12, 11

An analysis of similar data suggests that the annual probability of observable slope failure in each 1-km section of highway is 0.2. Assuming that sections fail independently and that each year constitutes an independent trial, how reasonable does this hypothesis appear to be?

SOLUTION If sections fail independently, and each year is also independent, then we have 50 independent observations of the "number of trials" (i.e., years) to first failure of a 1-km section. Under the given assumptions, the number of trials "to first failure" follows a geometric distribution. According to Section 1.9.3, the mean of the geometric distribution is

$$\mathrm{E}[T_1] = \frac{1}{p} \qquad \text{or} \qquad p = \frac{1}{\mathrm{E}[T_1]} \tag{5.16}$$

where T_1 is the number of trials (years) until first failure in a 1-km section. By the MM, where the arithmetic average of the observations, $\bar{t}$, is assumed to approximate $\mathrm{E}[T_1]$, we get the estimate $\hat{p}$ of the failure probability p to be

$$\hat{p} = \frac{1}{\bar{t}} \tag{5.17}$$

For our particular observations,

$$\bar{t} = \frac{1}{50}\sum_{t_i} = \frac{3+2+\cdots+11}{50} = 5.02$$

so that

$$\hat{p} = \frac{1}{5.02} = 0.199$$

Apparently the estimate of the annual probability of slope failure of 0.2 is in very close agreement with this data set.

The histogram is most easily formed by first sorting the observations from smallest to largest:

1, 1, 1, 1, 1, 1, 1, 1, 1, 2, 2, 2, 2, 2, 3, 3, 3, 3, 3, 3, 3, 3, 4, 4, 4, 4, 4, 4, 4, 4, 5, 5, 5, 5, 6, 7, 7, 7, 8, 8, 8, 9, 9, 9, 10, 10, 11, 12, 14, 21

from which we see that we have nine 1-km sections which had slope failures in the first year, five in the second year, and so on. Dividing each of these frequencies by $n = 50$ gives us a normalized frequency of 0.18 for the first year, 0.1 for the second year, and so on. This leads to the normalized histogram in Figure 5.3.

Figure 5.3 compares the histogram of the observed failure times, normalized to have unit area, with the fitted geometric distribution using $\hat{p} = 0.199$. At least visually, the fit appears to be reasonable. The observed histogram is somewhat erratic, but this is to be expected from such a small data set. If we were able to repeat the set of observations at another similar length of highway, we would presumably see a similarly shaped histogram but having different details. It is possible that, on average, the two histograms would agree.

Since the estimated value of p agrees well with that hypothesized and since the histogram appears to be reasonably well fit by the geometric distribution, we conclude that the distribution hypothesis appears reasonable. One of the primary assumptions underlying the assumed distribution is that sections fail independently and whether this is true or not has not been tested in this solution; in order to do so, we would typically need both spatial location information as well as more data. From an intuitive point

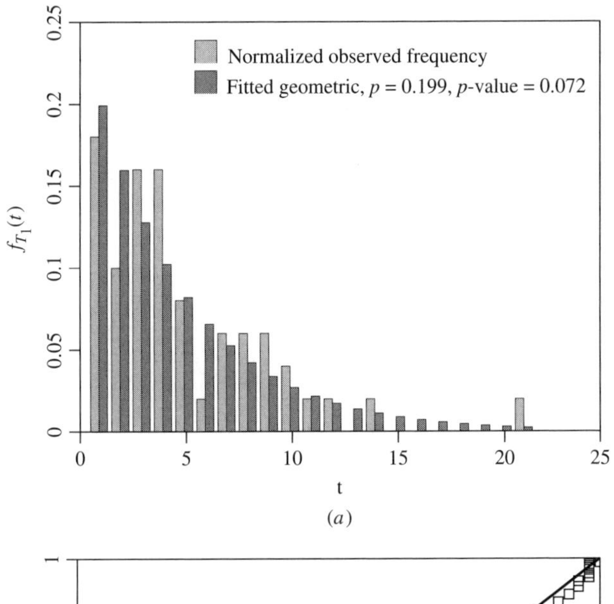

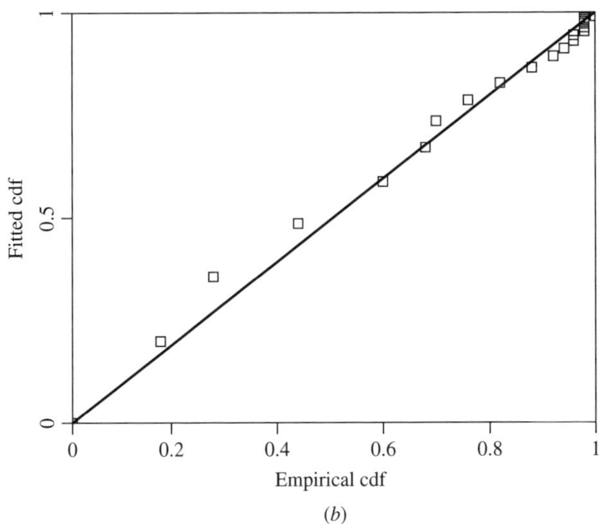

Figure 5.3 (*a*) Normalized histogram of discrete failure time data and associated fitted geometric distribution using $\hat{p} = 0.199$. (*b*) The probability–probability plot compares the empirical cumulative distribution with the fitted cumulative geometric distribution.

of view, it appears unlikely that sections will actually be completely independent. For one thing, ground tremors, if they occur, might encourage several sections to fail simultaneously—similarly with intense periods of rainfall in a region. However, a fully cross-correlated random field of slope failure times would be difficult to characterize using just these 50 observations (although an *a priori* model based on other information could be partially validated by these data). In general the assumption of independence is made by necessity and the resulting model must be viewed with caution. Actual failure probabilities would lie somewhere between those obtained using this independent model and a completely dependent model (which would be a single random variable).

Although the comparison between the observed and fitted histograms appears reasonable in Figure 5.3, we wonder if there is some better, or at least alternative, way to test the agreement. There are a number of quantitative goodness-of-fit tests available and one of the easiest of these to implement is the chi-square test, which will be discussed shortly. Another way to visually assess the appropriateness of a fitted distribution is to plot the empirical cumulative distribution against the fitted cumulative distribution. This type of comparison was shown in the lower plot of Figure 5.3 and will be discussed in more detail shortly.

When the data come from a continuous distribution, the histogram is obtained by breaking up the range of values covered by the data into k intervals $[a_0, a_1), \ldots, [a_{k-1}, a_k)$ and counting the frequency of occurrence of the data within each interval. The frequency count is then plotted. The main difficulty is in choosing a "good" number of intervals, which may involve some trial and error. Unfortunately, at least for small data sets, different numbers of intervals may yield significantly different distribution shapes. The authors commonly take k to be about 5–20% of the number of observations, n. We normally aim to have at least five observations occurring in each interval (as we shall see later, the chi-square goodness-of-fit test requires this), but there is a trade-off between obtaining a sufficient number of observations occurring in each interval and discerning the distribution shape. For example, if $n = 20$, we will probably have at least five observations in each interval if only $k = 2$ intervals are selected. However, it will be very difficult to discern the distribution shape with only two intervals. Admittedly, with only 20 observations, it is unlikely that we will be able to decide on a distribution in any case. However, the authors suggest a minimum of five intervals if the histogram is to be of any use in selecting a distribution.

For continuous data, the normalized histogram can again be compared with the fitted distribution. The histogram is normalized by dividing each frequency value by $n\,\Delta x$, where Δx is the interval width associated with each frequency. This results in a unit area lying under the histogram. The resulting plot is also often called a *frequency density plot*.

Example 5.5 Suppose that 50 clay samples have been tested and the measured cohesion of each has been recorded as follows (in kilopascals):

16.24, 16.13, 39.11, 30.40, 23.52, 9.07, 9.49, 25.92, 17.35, 5.83, 10.06, 39.25, 12.49, 8.71, 16.99, 20.74,

17.82, 32.96, 20.83, 18.10, 9.09, 22.94, 36.87, 17.11, 18.88, 14.89, 27.38, 16.59, 19.51, 29.82, 15.93, 21.27, 16.07, 42.06, 18.74, 16.00, 21.92, 13.33, 10.27, 27.30, 17.85, 18.15, 14.45, 14.82, 29.45, 17.76, 14.08, 14.55, 15.19, 36.15

In that cohesion is a nonnegative soil property, it is suggested that an appropriate distribution for cohesion is the lognormal, since the lognormal is also nonnegative (see Section 1.10.9). Is this suggestion supported by the above data?

SOLUTION The best way to determine the form of a distribution suggested by some data set is to plot a histogram. For this, it is useful to sort the data from smallest to largest:

5.83, 8.71, 9.07, 9.09, 9.49, 10.06, 10.27, 12.49, 13.33, 14.08, 14.45, 14.55, 14.82, 14.89, 15.19, 15.93, 16.00, 16.07, 16.13, 16.24, 16.59, 16.99, 17.11, 17.35, 17.76, 17.82, 17.85, 18.10, 18.15, 18.74, 18.88, 19.51, 20.74, 20.83, 21.27, 21.92, 22.94, 23.52, 25.92, 27.30, 27.38, 29.45, 29.82, 30.40, 32.96, 36.15, 36.87, 39.11, 39.25, 42.06

Suppose that we choose $k = 10$ intervals. We could subdivide the observed range evenly into 10 intervals between 5.83 and 42.06. This yields the histogram in Figure 5.4, whose shape is reasonably similar to that of a lognormal distribution.

Because a lognormal distribution is skewed to the right, observed values from this distribution will sometimes be quite large. When an exceptionally large observation occurs in a data set, subdividing the histogram evenly between $x_{\min}$ and $x_{\max}$ may not produce a very useful histogram, since many of the rightmost intervals (except the last) may be empty. To avoid this, it is generally better to subdivide the histogram range in the space of $\ln(x)$, henceforth called log space, rather than x itself. That is, we now choose to have 10 intervals evenly dividing the interval between $\ln(5.83) = 1.763$ and $\ln(42.06) = 3.739$. Each interval is of width $(3.739 - 1.763)/10 = 0.198$ in log space. To count the number of occurrences in each interval, we need to first transform our observations by taking their natural logarithms:

1.763, 2.164, 2.205, 2.207, 2.250, 2.309, 2.329, 2.525, 2.590, 2.645, 2.671, 2.677, 2.696, 2.701, 2.721, 2.768, 2.773, 2.777, 2.781, 2.788, 2.809, 2.833, 2.839, 2.854, 2.877, 2.880, 2.882, 2.896, 2.899, 2.930, 2.938, 2.971, 3.032, 3.036, 3.057, 3.088, 3.133, 3.158, 3.255, 3.307, 3.310, 3.383, 3.395, 3.415, 3.495, 3.587, 3.607, 3.666, 3.670, 3.739

The first histogram interval goes from 1.763 to $1.763 + 0.198 = 1.961$ and contains just one observation, $\ln(5.83) = 1.763$. The second interval goes from 1.961 to $1.961 + 0.198 = 2.159$ and contains zero observations. The third interval goes from 2.159 to 2.357 and contains six observations, and so on.

To plot the resulting histogram in real space, we must convert the intervals back to real space. Thus, the first interval goes from $e^{1.763} = 5.83$ to $e^{1.961} = 7.11$, the second interval goes from 7.11 to $e^{2.159} = 8.66$, and so on.

The only remaining complication occurs if we want to normalize the histogram so that it encloses unit area. In this case we must divide each frequency count by $n\,\Delta x_i$, where $n = 50$ and $\Delta x_1 = 7.11 - 5.83 = 1.28$ for the first interval. Thus, the normalized height of the first interval is $1/(50 \times 1.28) = 0.0156$. The second interval has width $\Delta x_2 = 8.66 - 7.11 = 1.55$, while the third interval has width $\Delta x_3 = e^{2.357} - e^{2.159} = 10.56 - 8.66 = 1.90$. Notice that the intervals get gradually wider as we move to the right. The normalized height of the third interval is therefore $6/(50 \times 1.90) = 0.0632$. The resulting histogram appears as in Figure 5.5.

First of all we notice that Figures 5.4 and 5.5 are quite different. This emphasizes the fact that the shape of a histogram is often very dependent on the number and size of the selected intervals. In the case of this particular data set, the even spacing of intervals in real space appears to give a smoother and more lognormal-looking histogram.

Although the fairly erratic histogram shown in Figure 5.5 does not appear strongly lognormally shaped, we shall see later that the Anderson–Darling goodness-of-fit test does

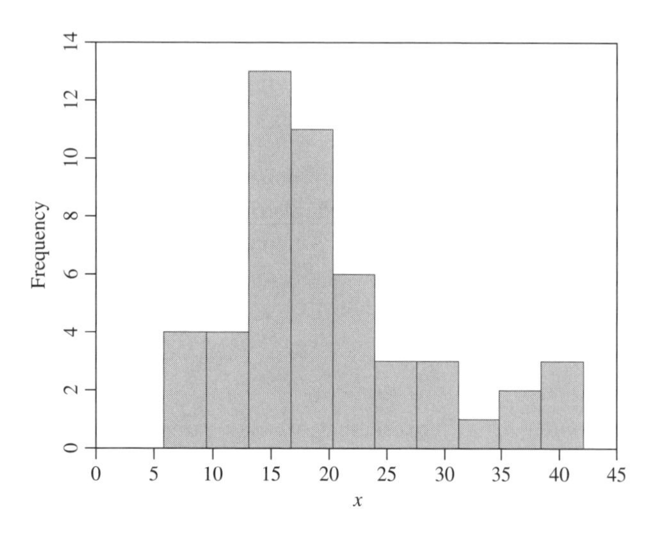

Figure 5.4 Histogram of cohesion data using 10 intervals between $x_{\min} = 5.83$ and $x_{\max} = 42.06$.

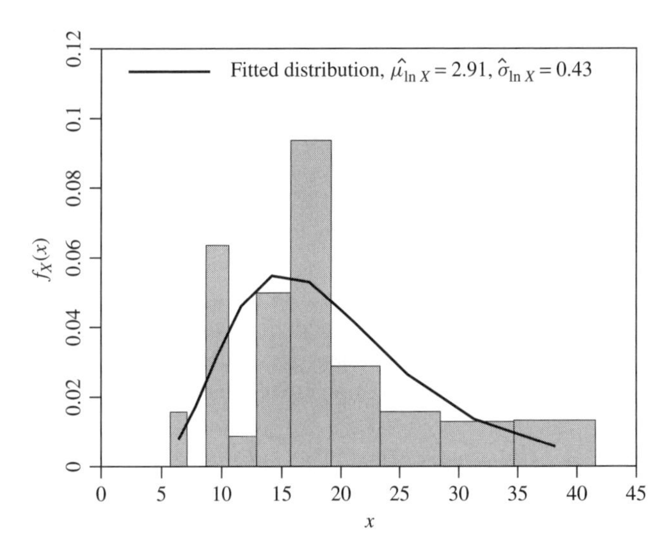

Figure 5.5 Frequency density plot of the 50 observed cohesion values with fitted lognormal distribution.

not reject the hypothesis that it does in fact come from a lognormal distribution. To illustrate the comparison, we have superimposed a lognormal distribution curve (solid line) having estimated parameters

$$\begin{aligned}\hat{\mu}_{\ln X} &= \frac{1}{n}\sum_{i=1}^{n} \ln x_i \\ &= \frac{1}{50}(1.763 + 2.164 + \cdots + 3.739) \\ &= 2.91 \\ \hat{\sigma}_{\ln X} &= \sqrt{\frac{1}{n-1}\sum_{i=1}^{n}\left(\ln x_i - \hat{\mu}_{\ln X}\right)^2} \\ &= 0.43\end{aligned}$$

In either the continuous or discrete case, there should be good agreement between the *empirical* (i.e., observed) and the *fitted* functions. The *degree* of agreement between the functions is sometimes difficult to assess, particularly for small n, but the following types of probability plots should help.

Probability Plots The idea with probability plots is to compare the *empirical* with the fitted cdf. Recall that the cdf is defined as $F(x) = \text{P}[X \le x]$. The fitted or hypothesized cdf will be denoted as $\hat{F}(x)$, since it is based on estimated parameters, such as $\bar{x}$, s^2, $\hat{\lambda}$, ...

Let $X_{(i)}$ be the ith smallest of the X_j's, that is, $X_{(1)}$ is the smallest, $X_{(2)}$ is the second smallest, and so on; then $X_{(i)}$ is called the *ith order statistic* of the X_j's. The empirical cdf, denoted $\tilde{F}_n(x)$, is then defined as

$$\tilde{F}_n(X_{(i)}) = \frac{i - 0.5}{n} \qquad \text{for } i = 1, \ldots, n$$

Note that this particular form of the empirical cdf, which is shifted by a factor of $\frac{1}{2}$, allows some probability for the random variable X to lie outside the range of the data. For example, suppose we observe $n = 4$ samples and they are $\{4.3, 5.6, 7.2, 8.3\}$. If we had defined $\tilde{F}_n(X_{(i)}) = i/n$, then $\tilde{F}_n(X_{(4)}) = \tilde{F}_n(8.3) = \frac{4}{4} = 1$ and we would essentially be saying $\text{P}[X > 8.3] = 0$, which does not seem likely. Similarly, if we defined $\tilde{F}_n(X_{(i)}) = (i-1)/n$, then $\tilde{F}_n(X_{(1)}) = \frac{0}{4} = 0$ and we would be saying $\text{P}[X \le 4.3] = 0$. With the current definition, we are admitting $\text{P}[X < 4.3] = 1/2n$ and $\text{P}[X > 8.3] = 1/2n$, which is reasonable.

To compare $\tilde{F}_n(x)$ and $\hat{F}(x)$, one could plot both on a graph *versus* x. However, looking for differences or similarities in two S-shaped curves is quite difficult. There are two more commonly used ways to compare the two cumulative distribution functions:

1. **Quantile–Quantile Plots:** Let $q_i = (i - 0.5)/n$, $i = 1, \ldots, n$. If we plot $\hat{x} = \hat{F}^{-1}(q_i)$ against $\tilde{F}_n^{-1}(q_i) = X_{(i)}$, then we get what we call a *quantile–quantile* plot, or just QQ plot for short. The value $\hat{x} = \hat{F}^{-1}(q_i)$ is the quantile corresponding to cumulative probability q_i, that is, $\hat{x}$ is the value such that $\text{P}[X \le \hat{x}] = q_i$.

   If $\hat{F}(x)$ and $\tilde{F}(x)$ both approximate the true underlying $F(x)$, then they should be approximately equal. If this is the case, then the QQ plot will be approximately a straight line with an intercept of 0 and a slope 1. For small sample sizes, departures from a straight line can be expected, but one would hope that the plot would approach a diagonal line *on average*.

   The construction of a QQ plot requires the calculation of the inverse, $\hat{F}^{-1}(q_i)$. Unfortunately, the inverse cdf is not easily computed for some distributions. For example, there is no closed-form inverse for the normal distribution and numerical procedures must be used. For this reason, the probability–probability plots are typically more popular, as discussed next. If one intends to write a program to produce QQ plots for mathematically more difficult distributions, special numerical procedures are available (see, e.g., Odeh and Evans, 1975).
2. **Probability–Probability Plots:** Plotting the fitted cdf, $\hat{F}(X_{(i)})$, *versus* the empirical cdf, $\tilde{F}_n(X_{(i)}) = q_i$, for $i = 1, \ldots, n$, yields a *probability–probability* plot, or PP plot for short. Again, if $\hat{F}(x)$ and $\tilde{F}(x)$ both approximate the underlying $F(x)$, and if the sample size n is large, then the PP plot will be approximately a straight diagonal line with an intercept of 0 and a slope 1. Figure 5.3 illustrated a PP plot.

Each plot has its own strengths. The QQ plot will amplify differences that exist between the tails of the model distribution function, $\hat{F}(x)$, and the tails of the sample distribution function, $\tilde{F}_n(x)$, whereas the PP plot will amplify differences between the middle of $\hat{F}(x)$ and the middle of $\tilde{F}_n(x)$.

These plots are built into most popular statistical packages, for example, Minitab, so that constructing both QQ and PP plots is straightforward using these packages.

5.2.2.2 Goodness-of-Fit Tests Suppose a set of observations $X_1, X_2, \ldots, X_n$ are to be taken at a site and a distribution fit is to be made to these observations. A goodness-of-fit test is a test of the following hypotheses:

H_o: the $X_1, X_2, \ldots, X_n$'s are governed by the fitted distribution function $\hat{F}$.

H_a: they are not.

Typical of any hypothesis test, the null or default hypothesis H_o is only rejected if the data are "sufficiently far" from H_o. For example, suppose that H_o states that the data follow a normal distribution with mean 10 and standard deviation 3. This hypothesis will probably be rejected if the collected data are $\{121, 182, 173\}$ since the data are so far away from the null. Alternatively, if the collected data are $\{11, 13, 12\}$, then the null hypothesis will probably not be rejected, since these values are quite likely to be seen under the null distribution. The fact that all three observations are above 10 is slightly suspicious but not that uncommon for only three observations. If the next 20 observations are all above 10, then it is more likely that H_o will be rejected.

One shortcoming of quantitative goodness-of-fit tests is that H_o tends to be rejected for large sample sizes. That is, the distribution assumed under H_o is almost certainly not the true distribution. For example, the true mean might be 10.0001, rather than 10, or the distribution shape might be slightly different than assumed, and so on. When the sample size is very large, these small discrepancies become significant and H_o is rejected. In this case, when H_o is rejected, the test is saying little about how reasonable the distribution is. For this reason, it is generally good practice to combine quantitative tests with a simple visual comparison of the distribution fit and employ engineering judgment. That is, goodness-of-fit tests are often overly "critical" and offer no advice on what might be a better fit. Since we are usually mostly interested in a distribution which is reasonable, but which might not fit every variation in the empirical distribution, it is important to exercise a healthy skepticism in the interpretation of these tests. They are tools in the selection process and are often best used in a comparative sense to compare a variety of candidate distributions.

It should also be emphasized that failure to reject H_o does not mean that H_o has been proven. For example, if H_o states that the mean is 10 and standard deviation is 3 and our observations are $\{11, 13, 12\}$, then we may not reject H_o, but we certainly cannot say that these three observations prove that the mean is 10 and standard deviation is 3. They just are not far enough away from a mean of 10 (where "distance" is measured in standard deviations) to warrant rejecting H_o. Failure to reject H_o simply means that the assumed (default) distribution is reasonable.

5.2.2.3 Chi-Square Test The *chi-square test* is essentially a *numerical comparison* of the observed histogram and the predicted histogram. We accomplish this by first constructing a histogram: the range over which the data lie is divided into k adjacent intervals $[a_0, a_1), [a_1, a_2), \ldots, [a_{k-1}, a_k)$. The idea is to then compare the number of observations falling within each interval to that which is *expected* under the fitted distribution.

Suppose that $X_1, X_2, \ldots, X_n$ are our observations. Let

$$N_j = \text{number of } X_i\text{'s in the } j\text{th interval } [a_{j-1}, a_j)$$

for $j = 1, 2, \ldots, k$. Thus, N_j is the height of the jth box in a histogram of the observations. We then compute the expected proportion p_j of the X_i's that would fall in the jth interval if a sample was drawn from the fitted distribution $\hat{f}(x)$:

- In the continuous case,

$$p_j = \int_{a_{j-1}}^{a_j} \hat{f}(x)\, dx \tag{5.18}$$

where $\hat{f}$ is the fitted pdf.

- For discrete data,

$$p_j = \sum_{a_{j-1} \le x_j \le a_j} \hat{p}(x_i) \tag{5.19}$$

where $\hat{p}$ is the fitted probability mass function.

Finally, we compute the test statistic

$$\chi^2 = \sum_{j=1}^{k} \frac{(N_j - np_j)^2}{np_j} \tag{5.20}$$

and reject H_o if χ^2 is too large. How large is too large? That depends on the number of intervals, k and the number of parameters in the fitted distribution that required estimation from the data. If m parameters were estimated from the data, then Chernoff and Lehmann (1954) showed that the rejection region lies asymptotically somewhere between $\chi^2_{\alpha,k-1}$ and $\chi^2_{\alpha,k-m-1}$. The precise rejection point is difficult to find, and so m is typically set to 0 and H_o rejected any time that

$$\chi^2 > \chi^2_{\alpha,k-1} \tag{5.21}$$

since this is conservative (the actual probability of rejecting H_o when it is true is at least as small as that claimed by the value of α). The critical values $\chi^2_{\alpha,\nu}$ for given type I error probability α and number of *degrees of freedom* $\nu = k - 1$ are shown in Table A.3 (Appendix A).

Unfortunately, the chi-square test tends to be overly sensitive to the number (and size) of the intervals selected; in fact, the test can fail to reject H_o for one choice of k and yet reject H_o for a different choice of k. No general prescription for the choice of intervals exists, but the following two guidelines should be adhered to when possible:

- Try to choose intervals such that $p_1 = p_2 = \cdots = p_k$. This is called the *equiprobable approach*.
- Seek $k \geq 3$ and expected frequency $np_j \geq 5$ for all j.

Despite this drawback of being sensitive to the interval choice, the chi-square test is very popular for at least two reasons: It is very simple to implement and understand and it can be applied to any hypothesized distribution. As we shall see, this is not true of all goodness-of-fit tests.

Example 5.6 A geotechnical company owns 12 CPT test rigs. On any given day, N_{12} of the test rigs will be out in the field and it is believed that N_{12} follows a binomial distribution. By randomly selecting 500 days over the last 5 years, the company has compiled the following data on the test rig usage frequency:

Number in Field	Frequency
0	37
1	101
2	141
3	124
4	57
5	27
6	11
7	2
8	0
9	0
10	0
11	0
12	0

Test the hypothesis that N_{12} follows a binomial distribution and report a p-value.

SOLUTION Clearly some of the intervals will have to be combined since they have very small expectation (recall, we want each interval to have expected frequency of at least 5). First of all, we can compute the estimate of p by using the fact that $\mathrm{E}[N_n] = np$, so that $\hat{p} = \bar{N}_n/n$ (this is also the MLE of p; see Section 1.9.2). By noting that the above table indicates that N_{12} was equal to 0 on 37 out of the 500 days and equal to 1 on 101 of the 500 days, and so on, then the value of $\bar{N}_{12}$ can be computed as

$$\bar{N}_{12} = \frac{1}{500}\sum_{i=1}^{500} N_{12_i}$$

$$= \frac{0(37) + 1(101) + 2(141) + \cdots + 12(0)}{500} = 2.396$$

so that

$$\hat{p} = \frac{2.396}{12} = 0.1997$$

Using the probability mass function for the binomial,

$$\mathrm{P}[N_n = k] = \binom{n}{k} p^k (1-p)^{n-k}$$

we can develop the following table:

k	Observed	Probability	Expected	$(N_j - np_j)^2/np_j$
0	37	$6.9064e-02$	34.53	$1.7640e-01$
1	101	$2.0676e-01$	103.38	$5.4794e-02$
2	141	$2.8370e-01$	141.85	$5.1119e-03$
3	124	$2.3593e-01$	117.96	$3.0891e-01$
4	57	$1.3243e-01$	66.21	$1.2828e+00$
5	27	$5.2863e-02$	26.43	$1.2234e-02$
6	11	$1.5386e-02$	7.69	$1.4215e+00$
7	2	$3.2902e-03$	1.64	$7.6570e-02$
8	0	$5.1302e-04$	0.25	$2.5651e-01$
9	0	$5.6883e-05$	0.028	$2.8442e-02$
10	0	$4.2574e-06$	0.002	$2.1287e-03$
11	0	$1.9311e-07$	0.0^31	$9.6557e-05$
12	0	$4.0148e-09$	0.0^52	$2.0074e-06$

We must combine the last seven intervals to achieve a single interval with expected number of 9.635 (greater than 5), leaving us with $k = 7$ intervals altogether. The corresponding number of observations in the last "combined" interval is 13. This gives us a chi-square statistic of

$$\chi^2 = 0.1764 + 0.05479 + 0.005112 + 0.3089 + 1.283$$

$$+ 0.01223 + \frac{(13 - 9.6253)^2}{9.6253}$$

$$= 3.02$$

At the $\alpha = 0.05$ significance level, our critical chi-square value is seen from Table A.3 to be

$$\chi^2_{0.05,7-1} = \chi^2_{0.05,6} = 12.592$$

and since 3.02 does not exceed the critical value, we cannot reject the hypothesis that the data follow a binomial distribution with $p = 0.1997$. That is, the assumed binomial distribution is reasonable.

The p-value (which is not to be confused with the parameter p) is the smallest value of α at which H_o would be rejected, so if the selected value of α is greater than the p-value, then H_o is rejected. The p-value for the above goodness-of-fit test is about 0.8 (obtained by reading along the chi-square table at six degrees-of-freedom until we find something near 3.02), which indicates that there is very little evidence in the sample against the null hypothesis (large p-values support the null hypothesis).

5.2.2.4 Kolmogorov–Smirnov Test The Kolmogorov–Smirnov (KS) test is essentially a numerical test of the empirical cumulative distribution function $\tilde{F}$ against the fitted cumulative distribution function $\hat{F}$. The KS test has the following advantages and disadvantages:

Advantages

- It does not require any grouping of the data (i.e., no information is lost and interval specification is not required).
- The test is valid (exactly) for any sample size n in the all-parameters-*known* case.
- The test tends to be more powerful than chi-Square tests against many alternatives (i.e., when the true distribution is some other theoretical distribution than that which is hypothesized under H_o, the KS test is better at identifying the difference).

Disadvantages

- It has a limited range of applicability since tables of critical values have only been developed for certain distributions.
- The critical values are not readily available for discrete data.
- The original form of the KS test is valid only if *all* the parameters of the hypothesized distribution are known a priori. When parameters are estimated from the data, this "extended" KS test is conservative.

The KS test seeks to see how close the empirical distribution $\tilde{F}$ is to $\hat{F}$ (after all, if the fitted distribution $\hat{F}$ is good, the two distributions should be very similar). Thus, the KS statistic D_n is simply the *largest* (vertical) distance between $\tilde{F}_n(x)$ and $\hat{F}(x)$ across *all values of* x and is defined as

$$D_n = \max\{D_n^+, D_n^-\}$$

where

$$D_n^+ = \max_{1 \le i \le n} \left\{ \frac{i}{n} - \hat{F}(X_{(i)}) \right\},$$

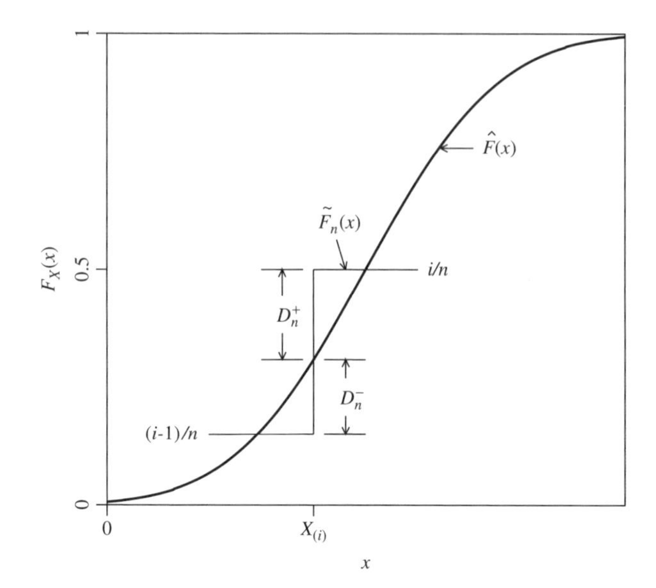

Figure 5.6 Illustration of D_n^+ and D_n^-.

$$D_n^- = \max_{1 \le i \le n} \left\{ \hat{F}(X_{(i)}) - \frac{i-1}{n} \right\}$$

In the KS test, the empirical distribution is taken as $\tilde{F}_n(X_{(i)}) = i/n$. At each $X_{(i)}$, the empirical distribution jumps from $(i-1)/n$ to i/n. Thus, the maximum difference between $\hat{F}(x)$ and $\tilde{F}(x)$ at the point $x = X_{(i)}$ is obtained by looking "backward" to the previous level of $\tilde{F}(x) = (i-1)/n$ and "forward" to the next level of $\tilde{F}(x) = i/n$ and choosing the largest. This forward and backward looking is performed in the calculation of D_n^+ and D_n^- and is illustrated in Figure 5.6. If D_n is excessively large, then the null hypothesis H_o that the data come from the fitted distribution $\hat{F}$ is rejected.

Prior to comparing D_n to the critical rejection value, a correction factor is applied to D_n to account for the behavior of different distributions. These scaled versions of D_n are referred to as the *adjusted* KS test statistics. Critical values for the adjusted KS test statistics for the all-parameters-known, the normal, and the exponential cases are given in Table 5.1. Critical values for the Weibull are given in Table 5.2. If the computed adjusted KS statistic is greater than the critical value given in the table, then the null hypothesis (that the distribution fits the data) is rejected, and an alternative distribution may be more appropriate.

5.2.2.5 Anderson–Darling Test The idea behind the Anderson–Darling (AD) test is basically the same as that behind the KS, but the AD is designed to better detect discrepancies in the tails and has higher power than the KS test (i.e., it is better able to discern differences between the hypothesized distribution and the actual distribution).

Table 5.1 Critical values for Adjusted KS Test Statistic

Case	Adjusted Test Statistic	1 − α: 0.850	0.900	0.950	0.975	0.990
All parameters known	$\left(\sqrt{n}+0.12+\frac{0.11}{\sqrt{n}}\right)D_n$	1.138	1.224	1.358	1.480	1.628
$N(\bar{X}(n), S^2(n))$	$\left(\sqrt{n}-0.01+\frac{0.85}{\sqrt{n}}\right)D_n$	0.775	0.819	0.895	0.955	1.035
exponential($\bar{X}(n)$)	$\left(D_n-\frac{0.2}{n}\right)\left(\sqrt{n}+0.26+\frac{0.5}{\sqrt{n}}\right)$	0.926	0.990	1.094	1.190	1.308

Table 5.2 Critical Values for Adjusted KS Test Statistic $\sqrt{n}D_n$ for Weibull Distribution

n	1 − α: 0.900	0.950	0.975	0.990
10	0.760	0.819	0.880	0.944
20	0.779	0.843	0.907	0.973
50	0.790	0.856	0.922	0.988
∞	0.803	0.874	0.939	1.007

The AD statistic A_n^2 has formal definition

$$A_n^2 = n\int_{\infty}^{\infty}\left[\tilde{F}_n(x)-\hat{F}(x)\right]^2\psi(x)\hat{f}(x)\,dx \tag{5.22}$$

where

$$\psi(x)=\frac{1}{\hat{F}(x)[1-\hat{F}(x)]}$$

is a *weight function* such that both tails are more heavily weighted than the middle of the distribution. In practice, the statistic A_n^2 is calculated as

$$A_n^2=\left(-\frac{1}{n}\left\{\sum_{i=1}^{n}(2i-1)[\ln Z_i+\ln(1-Z_{n+1-i})]\right\}\right)-n \tag{5.23}$$

where $Z_i=\hat{F}(X_{(i)})$ for $i=1,2,..,n$. Since A_n^2 is a (weighted) distance between cumulatives (as in the KS test), the null hypothesis will again be rejected if the statistic is too large. Again, just like with the KS test, a scaling is required to form the actual test statistic. The scaled A_n^2 is called the adjusted test statistic, and critical values for the all-parameters-known, the normal, the exponential, and the Weibull are given in Table 5.3. When it can be applied, the AS statistic is preferred over the KS.

Example 5.7 Suppose that the permeability of a set of 100 small-scale soil samples randomly selected from a site are tested and yield the following results (in units of 10^{-6} cm/s):

0	0	0	0	0	1	1	3	4	4
4	6	7	7	9	9	11	12	13	13
15	16	17	19	21	21	21	24	25	26
29	30	30	31	32	33	33	33	33	33
34	35	36	36	37	37	38	39	39	42
42	43	45	45	45	48	49	50	54	54
55	56	56	58	59	60	62	64	71	73
76	78	80	81	81	84	100	105	105	108
108	110	120	125	134	136	139	146	147	150
161	171	175	182	184	200	211	229	256	900

The data have been sorted from smallest to largest for convenience. Permeabilities of zero correspond to samples which are essentially rock, having extremely low permeabilities. As part of the distribution-fitting exercise, we want

Table 5.3 Critical Values for Adjusted AD Test Statistic

Case	Adjusted Test Statistic	1 − α: 0.900	0.950	0.975	0.990
All parameters known	A_n^2 for $n \geq 5$	1.933	2.492	3.070	3.857
$N(\bar{X}(n), S^2(n))$	$\left(1+\frac{4}{n}-\frac{25}{n^2}\right)A_n^2$	0.632	0.751	0.870	1.029
exponential($\bar{X}(n)$)	$\left(1+\frac{0.6}{n}\right)A_n^2$	1.070	1.326	1.587	1.943
Weibull($\hat{\alpha}, \hat{\beta}$)	$\left(1+\frac{0.2}{\sqrt{n}}\right)A_n^2$	0.637	0.757	0.877	1.038

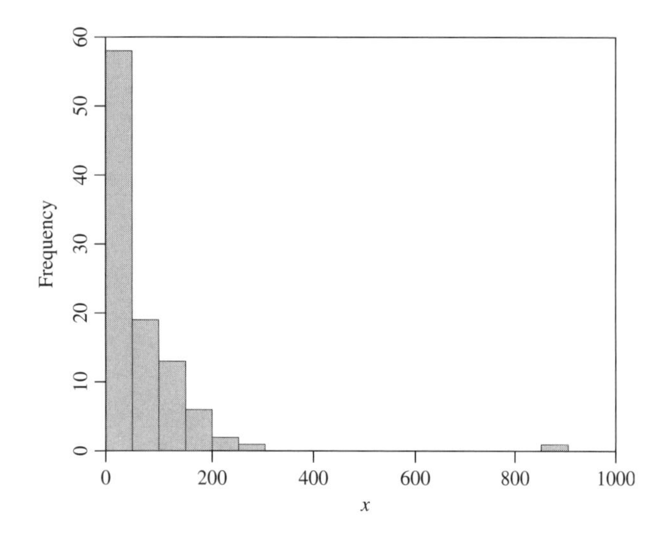

Figure 5.7 Histogram of permeability data using 18 intervals between $x_{\min} = 0$ and $x_{\max} = 900$.

to investigate how well the exponential distribution fits this particular data set. Test for goodness of fit of the exponential distribution using the chi-square, KS, and AD tests at a significance level of $\alpha = 5\%$.

SOLUTION We can start by computing some basic statistics. Let X be the permeability of a sample. Then the sample mean and variance are

$$\bar{x} = \frac{1}{100}(0 + 0 + \cdots + 900) = 69.7$$

$$s = \sqrt{\frac{1}{99}\Big((0 - 69.7)^2 + \cdots + (900 - 69.7)^2} = 101.8$$

For the exponential distribution, the mean and standard deviation are the same. While $\bar{x}$ and s are not extremely different, they are different enough that one would not immediately guess an exponential distribution on the basis of this information only.

It is instructive to look at a histogram of the data, as shown in Figure 5.7. Judging by the shape of the histogram, the bulk of the data appear to be quite well modeled by an exponential distribution. The suspicious thing about this data set is the single extremely large permeability of 900. Perhaps this is not part of the same population. For example, if someone misrecorded a measurement, writing down 900 instead of 90, then this one value is an error, rather than an observed permeability. Such a value is called an outlier. Special care needs to be taken to review outliers to ensure that they are actually from the population being studied. A common approach taken when the source of outliers is unknown is to consider the data set both with and without the outliers, to see how much influence the outliers have on the

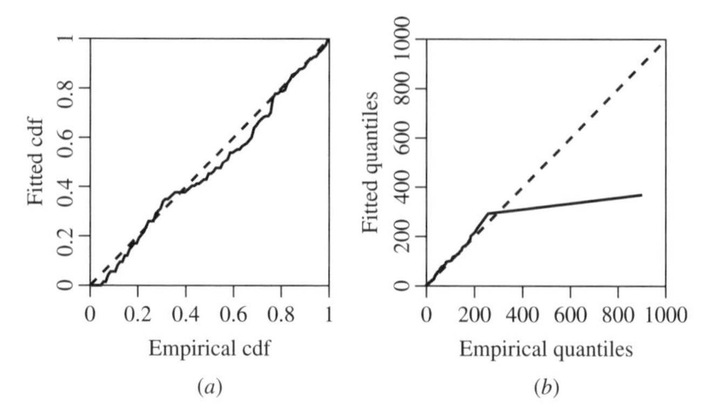

Figure 5.8 (*a*) PP and (*b*) QQ plots of exponential distribution fitted to 100 permeability observations.

fitted distribution. For example, if we recompute the sample mean and variance using just the first 99 observations, that is, without the largest 900 observation, we get

$$\bar{x} = \frac{1}{99}(0 + 0 + \cdots + 256) = 61.3$$

$$s = \sqrt{\frac{1}{98}\Big((0 - 69.7)^2 + \cdots + (256 - 69.7)^2} = 58.1$$

Now the mean and standard deviation are almost identical, as would be expected for the exponential distribution.

Figure 5.8 shows the PP and QQ plots for the original data set. The PP plot indicates that the exponential fit is pretty good for the bulk of the data. Recall that the PP emphasizes differences near the middle of the distribution, and even there, the PP plot stays pretty close to the target diagonal line. The QQ plot, on the other hand, emphasizes differences near the tails of the distribution. For this data set, the single observation (900) at the extreme right tail of the distribution clearly does not belong to this fitted exponential distribution. However, up to that single point, the QQ plot remains reasonably close to the target diagonal, indicating again that the exponential is reasonable for the bulk of the data.

To illustrate the effect that the single outlier (900) has on the QQ plot, it is repeated in Figure 5.9 for just the lower 99 observations. Evidently, for these 99 observations, the exponential distribution is appropriate. Now, let's see how our goodness-of-fit tests fare with the original 100 observations. For this data set, $\bar{x} = 69.7$, so our fitted exponential distribution has estimated parameter $\hat{\lambda} = 1/69.7$. Thus, the hypotheses that will be tested in each of the goodness-of-fit tests are as follows:

H_o : the $X_1, X_2, \ldots, X_{100}$'s are random variables with cdf $\hat{F}(x) = 1 - e^{-x/69.7}$.

H_a : they are not.

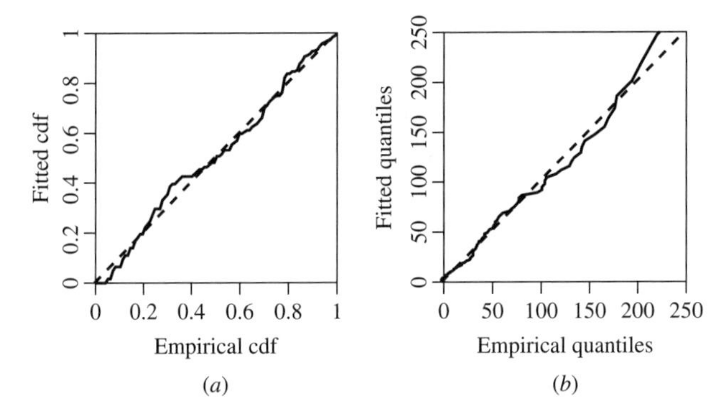

Figure 5.9 (*a*) PP and (*b*) QQ plots of exponential distribution fitted to lower 99 permeability observations.

Chi-Square Test

We want each interval to contain at least five observations, so, for $n = 100$, we should have at most $k \leq 20$ intervals. We try $k = 10$ intervals. We will select the interval widths so that each encloses an equal probability of 0.1 under the fitted distribution (see Figure 5.10). Our fitted distribution is $\hat{F}(x) = 1 - e^{-x/69.7}$, so our first interval boundary is where $\hat{F}(x) = 0.1$:

$$1 - e^{-x/69.7} = 0.1 \qquad \Longrightarrow \qquad x = -69.7\ \ln(1 - 0.1)$$

In general, our interval boundaries are at

$$x = -69.7\ \ln(1 - p), \qquad p = 0.1, 0.2, \ldots, 0.9, 1.0 \tag{5.24}$$

Note that the last interval goes from $x = -69.7\ \ln(1 - 0.9) = 160.49$ to ∞. Using Eq. 5.24, and counting the number of permeability observations occurring in each interval yield the following table:

Interval	Observed (N_j)	Expected (np_j)	$\frac{(N_j - np_j)^2}{np_j}$
0–7.34	14	10	1.6
7.34–15.55	7	10	0.9
15.55–24.86	7	10	0.9
24.86–35.60	14	10	1.6
35.60–48.31	14	10	1.6
48.31–63.87	11	10	0.1
63.87–83.92	8	10	0.4
83.92–112.18	7	10	0.9
112.18–160.49	8	10	0.4
160.49–∞	10	10	0.0
			$\sum = 8.4$

We reject H_o if $\chi^2 = 8.4$ exceeds $\chi^2_{\alpha,k-1} = \chi^2_{0.05,9} = 16.92$. Since $8.4 < 16.92$ we fail to reject H_o. Thus, the chi-square test does not reject the hypothesis that the data follow an exponential distribution with $\lambda = 1/69.7$.

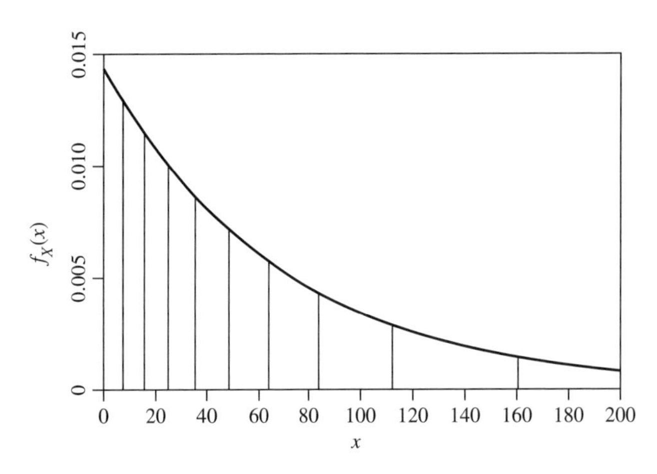

Figure 5.10 Dividing the fitted exponential distribution up into equal probability intervals. Each interval has probability 0.1.

Kolmolgorov–Smirnov Test

Figure 5.11 compares $\hat{F}(x) = 1 - e^{-x/69.7}$ and $\tilde{F}_n(x) = i/n$. The largest difference is shown. In detail

$$D_n^+ = 0.083, \qquad D_n^- = 0.040$$

and so $D_n = 0.083$. The adjusted D_n value is

$$D_{n,\text{adj}} = \left(D_n - \frac{0.2}{n}\right)\left(\sqrt{n} + 0.26 + \frac{0.5}{\sqrt{n}}\right)$$

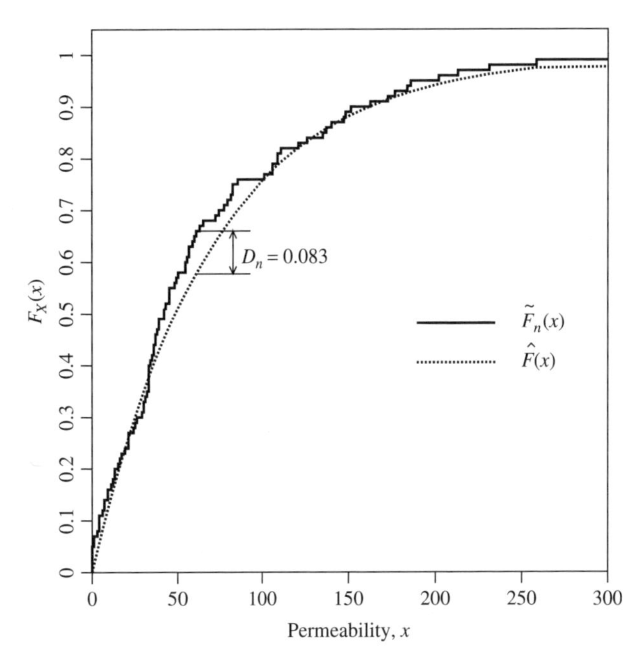

Figure 5.11 Empirical and fitted cumulative distribution functions of permeability data.

$$= \left(0.0828 - \frac{0.2}{100}\right)\left(\sqrt{100} + 0.26 + \frac{0.5}{\sqrt{100}}\right)$$
$$= 0.835$$

For $\alpha = 0.05$, the critical value is seen in Table 5.1 to be 1.094. Since $D_{n,\text{adj}}$ is 0.835, which is less than 1.094, we cannot reject the null hypothesis that the data follow an exponential distribution with $\lambda = 1/69.7$.

Anderson–Darling Test

We compute the AD test statistic from Eq. 5.23,

$$A_n^2 = \left(-\frac{1}{n}\left\{\sum_{i=1}^{n}(2i-1)[\ln Z_i + \ln(1 - Z_{n+1-i})]\right\}\right) - n$$

where $Z_i = \hat{F}(X_{(i)})$ for $i = 1, 2, \ldots, n$. Since the AD emphasizes discrepancies in the tails of the distribution, we expect that our outlier of 900 will significantly affect the test. However, we see that in fact it is at the other end of the distribution where we really run into trouble. We observed several zero permeability values, or, rather, we recorded several zero permeability values, possibly because we rounded the data or did not run the test carefully enough to determine a more precise value of permeability. The problem with this is that $\hat{F}(0) = 0$ and $\ln(0) = -\infty$. That is, the AD test statistic will turn out to be infinite due to our sloppy experimental/recording practices! We note that if, in fact, some permeabilities were zero, then they cannot follow an exponential distribution. Since the exponential distribution has $F(0) = \text{P}[X \le 0] = 0$, the probability of observing a zero value under the exponential distribution is zero. In other words, if zero values really were observed, the AD statistic would be quite correctly infinite, and the null hypothesis that the data follow an exponential distribution would be resoundingly rejected.

Common practice in such a situation is to either repeat the experiment for the low-permeability cases or estimate their actual permeabilities from judgment. Suppose we decide that the five zero permeabilities we recorded actually correspond to permeabilities $\{0.001, 0.01, 0.1, 0.2, 0.4\}$, then the AD statistic becomes

$$A_n^2 = 1.386$$

and the adjusted AD statistic is

$$A_{n,\text{adj}}^2 = \left(1 + \frac{0.6}{n}\right)A_n^2 = \left(1 + \frac{0.6}{100}\right)(1.386) = 1.394$$

From Table 5.3, we see that our critical statistic is 1.326, and since $A_{n,\text{adj}}^2 > 1.326$, we reject H_o and conclude, on the basis of this test, that the data do not follow an exponential distribution with $\lambda = 1/69.7$. This result does not agree with the chi-Square and KS tests but is due to the fact that the AD test is sensitive to discrepancies in the tails. If we repeat the test without the outlying 900, we get a significant reduction in the test statistic:

$$A_n^2 = 1.021, \qquad A_{n,\text{adj}}^2 = 1.027$$

and since $1.027 < 1.326$, we would now deem the exponential distribution to be reasonable. This result is still sensitive to our decision about what to do with the recorded zeros. If we remove both the zeros and the outlying 900, so that $n = 94$ now, we get

$$A_n^2 = 0.643, \qquad A_{n,\text{adj}}^2 = 0.647$$

In conclusion, the AD test suggests that the exponential distribution is quite reasonable for the bulk of the data but is not appropriate if the zeros and the 900 are to be acceptably modeled.

5.3 ESTIMATION IN PRESENCE OF CORRELATION

The classical estimators given in the previous sections all assume a *random sample*, that is, that all observations are statistically independent. When observations are dependent, as is typically the case at a single site, the estimation problem becomes much more complicated—all estimators known to the authors depend on *asymptotic independence*. Without independence between a significant portion of the sample, any estimate we make is doomed to be biased to an *unknown* extent. Another way of putting it is if our sample is composed of dependent observations, our estimates may say very little about the nature of the global population. This is a fundamental problem to which the only solution is to collect more data.

Correlation between data can, however, be beneficial if one is only trying to characterize the site from which the data are obtained (as is often the case). It all depends on what one is using the estimates for. To illustrate the two aspects of this issue, suppose that the soil friction angle at a site has been carefully measured along a line 10 km in length, as shown in Figure 5.12. Clearly there is a high degree of correlation between adjacent friction angle measurements. That is, if a measurement has a certain value, the next measurement will likely be similar—the "random field" is constrained against changing too rapidly by the high degree of spatial correlation.

Forget for the moment that we know the friction angle variation along our 10-km line and imagine that we have hired a geotechnical engineer to estimate the average friction angle for the entire site. The engineer may begin at $x = 0$ taking samples and accumulating friction angle data. After the engineer has traveled approximately 0.75 km, taking samples every few meters, the engineer might notice that the measurements are quite stable, with friction angles

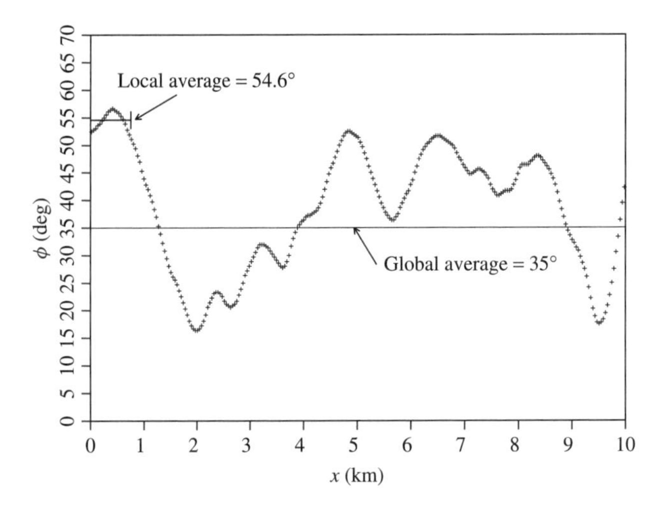

Figure 5.12 Friction angles measured at a site along a 10-km line.

lying consistently between 52° and 56°. Because of this stability, the engineer may decide that these measurements are representative and discontinue testing. If this is the case, the engineer will obtain a "sitewide" average friction angle estimate of 54.6° with standard deviation 1.5°.

While 54.6° is actually a very good estimate of the friction angle over the first 0.75 km, it is a very poor estimate of the sitewide average friction angle, which is almost 20° below the estimate. If we look at the actual friction angle variation shown in Figure 5.12, we see that there are only three locations where we could take samples over 0.75 km and get a good estimate of the global average (these would be in the regions where the trend intersects the average line). If we pick our sampling region randomly, we are almost certainly going to get a poor estimate of the global average unless we sample over a much larger distance (e.g., same number of samples, but much more spread out). It is also to be noted that the 10-km average shown in Figure 5.12 might be quite different from the 100-km, or 1000-km, average, and so on.

In general, the estimated sitewide average will be inaccurate when the sampling domain is too small. Even worse, the estimated standard deviation can be vastly in error, and unconservatively so. Positive correlation between samples reduces the standard deviation estimate, leading to a false sense of security at the global scale. For example, the friction angle standard deviation estimated over the first 0.75 km is only 1.5°, whereas the standard deviation over the 10 km is 11°, almost an order of magnitude different.

Why do we care about the global average and variance? If we are only designing foundations in the first 0.75 km of our site, then we do not care about the global statistics; in fact, using the global values would be an error. The estimated mean (54.6°) and standard deviation (1.5°) give us an accurate description of the 0.75-km site. In this case, the high correlation reduces the chance that we will get any excessively weak zones in the 0.75-km region. Alternatively, if we are trying to assess the risk of slope failure along a long run of highway and we only have access to data in a relatively small domain, then correlation between data may lead to significant errors in our predictions. For example, the high friction angle with low variability estimated over the first 0.75 km of Figure 5.12 will not reflect the considerably larger risk of slope failure in those regions where the friction angle descends below 20°. In this case, the correlation between data tends to hide the statistical nature of the soil that we are interested in.

In contrast, consider what happens when soil properties are largely independent, as is typically assumed under classical estimation theory. An example of such a field is shown in Figure 5.13, which might represent friction angles along a 10-km line where the soil properties at one point are largely independent of the soil properties at all other points. In this case, the average of observations over the first 0.75 km is equal to 37.2°. Thus, while the 0.75-km site is much more variable when correlation is not present, statistics taken over this site are much more representative of the global site, both in the mean and variance.

In summary, strong correlation tends to be beneficial if we only want to describe the site at which the data were taken (interpolation). Conversely, strong correlation is detrimental if we wish to describe the random nature of a much bigger site than that over which the data were gathered (extrapolation).

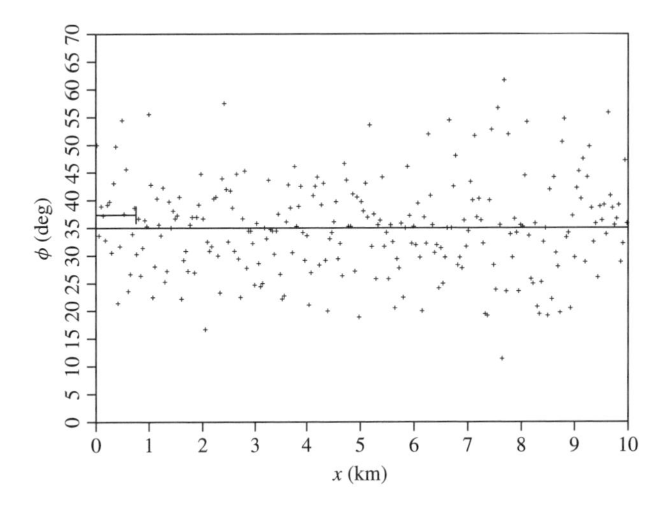

Figure 5.13 Friction angles measured at a site along a 10-km line where soil properties are largely spatially independent.

5.3.1 Ergodicity and Stationarity

As mentioned above, classical statistics are generally defined assuming independent observations. One way of achieving this in practice is to design an experiment that can be repeated over and over such that each outcome is independent of all others. The resulting observations can be assumed to be independent and used to accurately estimate the mean, variance, and so on, of the assumed distribution. Coming up with statistics in this fashion is called *averaging over the ensemble* of realizations. An *ensemble* is a collection of realizations of the random experiment and a *realization* is a particular single outcome of an experiment.

In geotechnical engineering, we generally only have one "experiment"—our planet as we see it. Rather than showing variability over an ensemble, our one realization shows variability over space. So the question is, how do we use classical statistics to estimate distribution parameters when we only have one spatially variable realization (experiment) from which to draw our observation?

Probabilists have answered this question by coming up with a concept called *ergodicity*. Ergodicity essentially says that, under certain conditions, averaging over the ensemble can be replaced by *averaging over space*. That is, if a stationary random process is ergodic, then its mean and covariance function can be found from a single realization of *infinite* extent,

$$\mu_X = \mathrm{E}[X(\mathbf{t})] = \lim_{|D|\to\infty} \frac{1}{|D|} \int_D X(\mathbf{t})\, d\mathbf{t} \tag{5.25}$$

$$\begin{aligned} C_X(\boldsymbol{\tau}) + \mu_X^2 &= \mathrm{E}[X(\mathbf{t}+\boldsymbol{\tau})X(\mathbf{t})] \\ &= \lim_{|D|\to\infty} \frac{1}{|D|} \int_D X(\mathbf{t}+\boldsymbol{\tau})X(\mathbf{t})\, d\mathbf{t} \end{aligned} \tag{5.26}$$

where D is the size of the domain over which our observations have been drawn and $C_X(\boldsymbol{\tau})$ is the covariance function of X. In order to guarantee the validity of the above relationships, two conditions must be imposed on the stationary random field $X(\mathbf{t})$. For Gaussian processes these conditions are

$$\lim_{|D|\to\infty} \frac{1}{|D|} \int_D C_X(\boldsymbol{\tau})\, d\boldsymbol{\tau} = 0 \tag{5.27a}$$

$$\lim_{|D|\to\infty} \frac{1}{|D|} \int_D |C_X(\boldsymbol{\tau})|^2\, d\boldsymbol{\tau} = 0 \tag{5.27b}$$

which are clearly met if

$$\lim_{\boldsymbol{\tau}\to\infty} C_X(\boldsymbol{\tau}) = 0 \tag{5.28}$$

Thus, ergodicity implies that the correlation coefficients between points separated by a large distances are negligible. In turn, this implies that the correlation length is much less than the observation domain, $\theta \ll D$.

A realization obtained from a particular algorithm is said to be ergodic if the desired mean and correlation structure can be obtained using Eqs. 5.25 and 5.26, respectively. Of course, realizations of infinite extent are never produced and so one cannot expect a finite realization to be ergodic (the word loses meaning in this context) any more than one can expect the average of a set of n independent observations to precisely equal the population mean. In fact, for finite-domain realizations, averaging must be performed over an ensemble of realizations in order to exactly calculate μ_X and $C_X(\boldsymbol{\tau})$. Although some algorithms may produce realizations which more closely approximate the desired statistics when averaged over a fixed (small) number of realizations than others, this becomes a matter of judgment. There is also the argument that since most natural processes are generally far from ergodic over a finite scale, why should a simulation of the process over a similar scale be ergodic?

The property of ergodicity *cannot* be proved or disproved from a single realization of finite extent. Typically, if a strong trend is seen in the data, this is indicative that the soil site is nonergodic. On the other hand, if the same strong trend is seen at every (similar) site, then the soil site may very well be ergodic. Since we never have soil property measurements over an infinite extent, we cannot say for sure whether our soil property realization is ergodic. Ergodicity in practice, then, is an assumption which allows us to carry on and compute statistics with an *assumed* confidence.

Consider a site from which a reasonably large set of soil samples has been gathered. Assume that the goal is to make statements using this data set about the stochastic nature of the soil at a different, although presumed similar, site. The collected data are a sample extracted over some sampling domain of extent D from a continuously varying soil property field. An example may be seen in Figure 5.14, where the solid line could represent the known, sampled, undrained shear strength of the soil and the dashed lines represent just two possibilities that the unknown shear strengths may take outside the sampling domain.

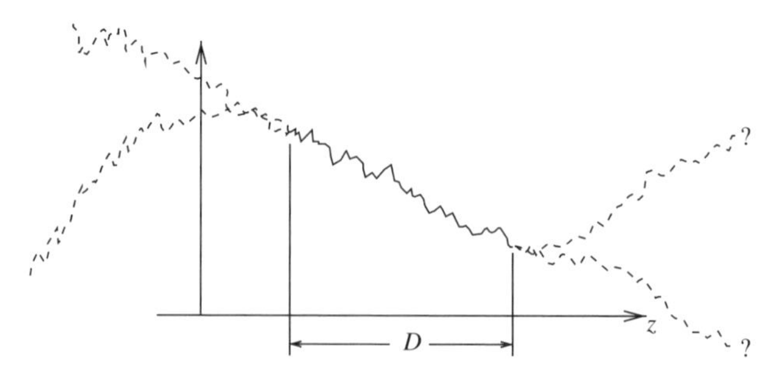

Figure 5.14 Soil sampled over a finite sampling domain.

Clearly this sample exhibits a strong spatial trend and would classically be represented by an equation of the form

$$S_u(z) = m(z) + \epsilon(z) \tag{5.29}$$

where $m(z)$ is a deterministic function giving the mean soil property at z and $\epsilon(z)$ is a random residual. If the goal were purely descriptive, then $m(z)$ would likely be selected so as to allow optimally accurate (minimum-variance) interpolation of S_u between observations. This generally involves letting $m(z)$ be a polynomial trend in z with coefficients selected to render ϵ mean zero with small variance.

However, if the data shown in Figure 5.14 are to be used to characterize another site, then the trend must be viewed with considerable caution. In particular one must ask if a similar trend is *expected* to be seen at the site being characterized and, if so, where the axis origin is to be located. In some cases, where soil properties vary predictably with depth, the answer to this question is affirmative. For example, undrained shear strength is commonly thought to increase with depth (but not always!). In cases where the same trend is *not* likely to reappear at the other site, then removal of the trend from the data and dealing with just the residual $\epsilon(z)$ has the following implications:

1. The covariance structure of $\epsilon(z)$ is typically drastically different than that of $S_u(z)$— it shows more spatial independence and has reduced variance.
2. The reintroduction of the trend to predict the deterministic part of the soil properties at the *target* site may be grossly in error.
3. The use of only the residual process, $\epsilon(z)$, at the other site will considerably underestimate the soil variability—the reported statistics will be *unconservative*. In fact, the more variability accounted for by $m(z)$, the less variable is $\epsilon(z)$.

From these considerations, it is easily seen that trends which are not likely to reappear, that is, trends which are not *physically (or empirically) based* and *predictable*, must not be removed prior to performing an inferential statistical analysis (e.g., to be used at other, similar, sites). The trend itself is part of the uncertainty to be characterized and removing it leads to unconservative reported statistics.

It should be pointed out at this time that one of the reasons for "detrending" the data is precisely to render the residual process largely spatially independent (removal of trends removes correlation). This is desirable because virtually all classical statistics are based on the idea that samples are composed of independent and identically distributed observations. Alternatively, when observations are dependent, the distributions of the estimators become very difficult to establish. This is compounded by the fact that the actual dependence structure is unknown. Only a limited number of asymptotic results are available to provide insight into the spatially dependent problem (Beran, 1994); simulation techniques are proving very useful in this regard (Cressie, 1993).

Another issue to be considered is that of the level of information available at the target site. Generally, a design does not proceed in the complete absence of site information. The ideal case involves gathering enough data to allow the main characteristics, say the mean and variance, of the soil property to be established with reasonable confidence. Then, inferred statistics regarding the spatial correlation (where correlation means correlation coefficient) structure can be used to complete the uncertainty picture and allow a reasonable reliability analysis or internal linear estimation. Under this reasoning, it makes sense to concentrate on statistics relating to the spatial correlation structure of a soil. Although the mean and variance will be obtained along the way as part of the estimation process, these results tend to be specifically related to, and affected by, the soil type—this is true particularly of the mean. The correlation structure is believed to be more related to the formation process of the soil; that is, the correlation between soil properties at two disjoint points will be related to where the materials making up the soil at the two points originated, to the common weathering processes experienced at the two points, geological deposition processes, and so on. Thus, the major factors influencing a soil's correlation structure can be thought of as being "external," that is, related to transport and weathering rather than to chemical and mechanical details of the soil particles themselves, common to most soil properties and types.

While it is undoubtedly true that many exceptions to the idea of an externally created correlation structure exist, the idea nevertheless gives some possible generality to estimates of the correlation obtained from a particular site and soil property. That is, it allows the correlation structure derived from a random field of a specific soil property to be used without change as a reasonable a priori correlation structure for other soil properties of interest and for other similar sites (although changes in the mean and possibly variance may, of course, be necessary).

Fundamental to the following statistical analysis is the assumption that the soil is spatially statistically homogeneous. This means that the mean, variance, correlation structure, and higher order moments are independent of position (and thus are the same from any reference origin). In the general case, isotropy is *not* usually assumed. That is, the vertical and horizontal correlation structures are allowed to be quite different. However, this is not an issue in this chapter since we will be restricting our attention to estimates of the correlation structure along a line in space, that is, in one dimension.

The assumption of spatial homogeneity does *not* imply that the process is relatively uniform over any finite domain. It allows *apparent* trends in the mean, variance, and higher order moments as long as those trends are just part of a larger scale random fluctuation; that is, the mean, variance, and so on, need only be constant over infinite space, not when viewed over a finite distance (Figure 5.14 may be viewed as an example of a homogeneous random field which appears nonstationary when viewed locally). Thus, this assumption does not preclude large-scale variations, such as often found in natural soils, although the statistics relating to the large scale fluctuations are generally harder to estimate reliably from a finite sampling domain.

The assumption of spatial homogeneity does, however, seem to imply that the site over which the measurements are taken is fairly uniform in geological makeup (or soil type). Again, this assumption relates to the level of uncertainty about the site for which the random model is aimed. Even changing geological units may be viewed as simply part of the overall randomness or uncertainty, which is to be characterized by the random model. The more that is known about a site, the less random the site model should be. However, the initial model that is used before significant amounts of data are explicitly gathered should be consistent with the level of uncertainty at the target site at the time the model is applied. Bayesian updating can be used to improve a prior model under additional site data.

With these thoughts in place, an appropriate inferential analysis proceeds as follows:

1. An initial regression analysis may be performed to determine if a statistically significant spatial trend is present. Since a trend with, for example, depth may have some physical basis and may be expected to occur identically at other sites, it may make sense to predict this trend and assume it to hold at the target site. If so, the remainder of the analysis is performed on the detrended data, $\epsilon(\mathbf{x}) = S_u(\mathbf{x}) - m(\mathbf{x})$, for example, and the trend and residual statistics must *both* be reported for use at the target site since they are intimately linked and cannot be considered separately. Using just the residual statistics leads to a stochastic model which is likely to be grossly in error.
2. Establish the second-moment behavior of the data set over space. Here interest may specifically focus on whether the soil is best modeled by a finite-scale stochastic model having limited spatial correlation or by a fractal model having significant lingering correlation over very large distances. These terms are discussed in more detail later.
3. For a selected spatial correlation function, estimate any required parameters from the data set.

Once the parameters of the random field model have been estimated, how the random field can be used depends on the questions being asked and the type of data available. In particular, the issue of whether or not data are available at the site being investigated has a significant impact on how the random-field model is defined and used. Two possible scenarios are as follows:

1. Data are gathered at the site in question over its entire domain:
   - A random field is being modeled whose values are known at the data site locations and no attempt will be made to extrapolate the field beyond the range of the data.
   - A representative random field model can always be estimated; estimates for μ_X, σ_X, and correlation structure are "local" and can be considered to be reasonably accurate for the purposes of modeling the site.
   - Best estimates of the random field between points at which data have been collected should be obtained using *best linear unbiased estimation* or *kriging*.
   - Probability estimates should be obtained using the conditioned random field. One possible approach is to use conditional simulation (all realizations pass through the known data but are random between the data sites).
2. Data are gathered at a similar site or over a limited portion of the site to be modeled:
   - There is much greater uncertainty in applying the statistics obtained from one site to that of another or in extending the results to a larger domain. Typically some assumptions need to be made about the "representativeness" of the sample. This situation typically arises in the preliminary phases of a design problem, before the site has been cleared, for example.
   - If the statistics can be considered representative, probability estimates can be made either analytically or through Monte Carlo simulations. BLUE or Kriging are not options since data are not available over the domain in question.
   - The treatment of trends in the data needs to be more carefully considered. If the trend seems to have some physical basis (such as an increase in certain soil properties with depth), then it may be reasonable to assume that the same trend exists at the site in question. However, if the trend has no particular physical basis, then it is entirely possible

that quite a different trend will be seen at the site in question. The random-field model should be able to accommodate this uncertainty.

5.3.2 Point versus Local Average Statistics

Random fields are characterized by their *point statistics*, that is, the mean, variance, marginal distribution, and so on, at each *point* in the random field. However, soil properties are rarely measured at a point. For example, porosity is ill-defined at the point level: At a particular point, the porosity is either 100% if the point happens to lie in a void or 0% if the point happens to lie inside a soil particle. The very definition of porosity (ratio of volume of voids to volume of soil) implies an average over the volume under consideration. Similarly, elastic modulus, friction angle, Poisson's ratio, consolidation ratio, and shear modulus are all ill-defined at the point scale (a point is one dimensional, so how can Poisson's ratio be defined?). Soil property measurements are generally averages over a volume (or possibly an area in the case of a shear box test).

So how do we relate local (possibly geometric) average measurements to the point-scale characteristics of our theoretical random field? The simple answer is that in practice so far we do not. Little work has been done in this area, and as we shall see, the theoretical backfiguring from the local average measurement to the point-scale statistic depends on knowledge of the *pointwise* correlation structure. In addition, the random-field models generally considered in this book are also *continuum models*, where the random field varies continuously. At the point scale, such models are not realistic, since soils are actually highly discontinuous at the microscale (solid to void or void to solid occurs at an interface not over some extended semisolid region).

Nevertheless, the continuum random-field models are useful so long as values derived from them for use in our soil models are local averages of some sort. To derive the point statistics associated with local average measurements, we need to know the following:

1. The size of the sample over which the measurement represents an average. For laboratory samples, this may be relatively easy to estimate depending on the test: For porosity, elastic, and hydraulic parameter tests, the size will be the laboratory sample volume. For laboratory shear tests, the "size" will probably be the shear plane area. For in situ tests, such as CPT, shear vane, and so on, one would have to estimate the volume of soil involved in the measurement; that is, a CPT cone may be averaging the soil resistance in a bulb of size about 100–200 mm radius in the vicinity of the cone.
2. The correlation coefficient between all points in the idealized continuum model. This is usually specified as a function of distance between points.
3. The type of averaging that the observations represent. Arithmetic averaging is appropriate if the quantity being measured is not dominated by low values. Porosity might be an example of a property which is simply an arithmetic average (sum of pore volumes divided by the total volume). Geometric averaging is appropriate for soil properties which are dominated by low values. Reasonable examples are hydraulic conductivity, elastic modulus, cohesion, and friction angle. Harmonic averaging is appropriate for soil properties which are strongly dominated by low values. Examples are the elastic modulus of horizontally layered soils and hydraulic conductivity in one-dimensional flow (i.e., through a pipe).

Assuming that each soil sample observation corresponds to an average over a volume of approximately D and that the correlation function is known, the relationship between the statistics of the observations (which will be discussed shortly) and the ideal random-field point statistics are as follows: Suppose that the following series of *independent* observations of X_D have been taken: $X_{D1}, X_{D2}, \ldots, X_{Dn}$. The sample mean of X_D is

$$\hat{\mu}_{X_D} = \frac{1}{n}\sum_{i=1}^{n} X_{D_i} \tag{5.30}$$

and the sample variance is

$$\hat{\sigma}_{X_D} = \frac{1}{n-1}\sum_{i=1}^{n}(X_{D_i} - \hat{\mu}_{X_D})^2 \tag{5.31}$$

If each soil sample is deemed to be an arithmetic average, X_D, of the idealized continuous soil property over volume D, that is,

$$X_D = \frac{1}{D}\int_D X(\mathbf{x})\, d\mathbf{x} \tag{5.32}$$

then the sample point mean and variance of the random field $X(\mathbf{x})$ are obtained from the sample mean and variance of X_D as follows:

$$\hat{\mu}_X = \hat{\mu}_{X_D} \tag{5.33a}$$

$$\hat{\sigma}_X^2 = \frac{\hat{\sigma}_{X_D}}{\gamma_X(D)} \tag{5.33b}$$

where $\gamma_X(D)$ is the variance reduction function (see Section 3.4) corresponding to the continuous random field, $X(\mathbf{x})$.

However, if each soil sample is deemed to be a geometric average, X_D, of the idealized continuous soil property over volume D, that is,

$$X_D = \exp\left\{\frac{1}{D}\int_D \ln X(\mathbf{x})\, d\mathbf{x}\right\} \tag{5.34}$$

then the sample point mean and variance of the random field $X(\mathbf{x})$ are obtained from the sample mean and variance of X_D as follows:

$$\hat{\mu}_X = \hat{\mu}_{X_D} \exp\left\{\ln\left(1+\hat{v}_{X_D}^2\right)\left[\frac{1-\gamma_{\ln X}(D)}{2\gamma_{\ln X}(D)}\right]\right\} \tag{5.35a}$$

$$\hat{\sigma}_X^2 = \hat{\mu}_X^2\left[\exp\left\{\frac{\ln\left(1+\hat{v}_{X_D}^2\right)}{\gamma_{\ln X}(D)}\right\} - 1\right] \tag{5.35b}$$

where $\hat{v}_{X_D} = \hat{\sigma}_{X_D}/\hat{\mu}_{X_D}$ is the sample coefficient of variation of X_D and $\gamma_{\ln X}(D)$ is the variance reduction function (see Section 3.4) corresponding to the continuous random field $\ln X(\mathbf{x})$.

If the soil sample represents a harmonic average of the random field $X(\mathbf{x})$, the relationship between the point statistics and harmonic average statistics will have to be determined by simulation on a case by case basis. See Section 4.4.3 for some guidance.

5.3.3 Estimating the Mean

Consider the classical sample estimate of the mean:

$$\hat{\mu}_X = \frac{1}{n}\sum_{i=1}^{n} X_i \tag{5.36}$$

If the field can be considered stationary, so that each X_i has the same mean, then $\mathrm{E}[\hat{\mu}_X] = \mu_X$ and this estimator is considered to be *unbiased* (i.e., it is "aimed" at the quantity to be estimated). It should be recognized that if a new set of observations of X is collected, the estimated mean will change. That is, $\hat{\mu}_X$ is itself a random variable. If the X_i's are independent, then the variance of $\hat{\mu}_X$ decreases as n increases. Specifically,

$$\mathrm{Var}[\hat{\mu}_X] = \frac{\sigma_X^2}{n}$$

which goes to zero as the number of independent observations, n, goes to infinity.

Now consider what happens to our estimate when the X_i's are completely correlated. In this case, $X_1 = X_2 = \cdots = X_n$ for a stationary process and

$$\hat{\mu}_X = \frac{1}{n}\sum_{i=1}^{n} X_i = X_1$$

and $\mathrm{Var}[\hat{\mu}_X] = \sigma_X^2$, that is, there is no reduction in the variability of the estimator $\hat{\mu}_X$ as n increases. This means that $\hat{\mu}_X$ will be a poor estimate of μ_X if the observations are highly correlated.

The true variance of the estimator $\hat{\mu}_X$ will lie somewhere between σ_X^2 and σ_X^2/n. In detail

$$\mathrm{Var}[\hat{\mu}_X] = \frac{1}{n^2}\sum_{i=1}^{n}\sum_{j=1}^{n} \mathrm{Cov}\left[X_i, X_j\right]$$

$$= \left[\frac{1}{n^2}\sum_{i=1}^{n}\sum_{j=1}^{n} \rho_{ij}\right]\sigma_X^2 \simeq \gamma(D)\sigma_X^2$$

where ρ_{ij} is the correlation coefficient between X_i and X_j and $\gamma(D)$ is call the *variance function* (see Section 3.4). The variance function lies between 0 and 1 and gives the amount of variance reduction that takes place when X is averaged over the sampling domain $D = n\,\Delta x$. For highly correlated fields, the variance function tends to remain close to 1, while for poorly correlated fields, the variance function tends toward $\Delta x/D = 1/n$. Figure 5.15 shows examples of a process $X(t)$ superimposed by its average over a width $D = 0.2$, $X_D(t)$, for poorly and highly correlated processes. When the process is poorly correlated, the variability of the average, $X_D(t)$, tends to be much smaller than that of the original $X(t)$, while if the process is highly correlated,

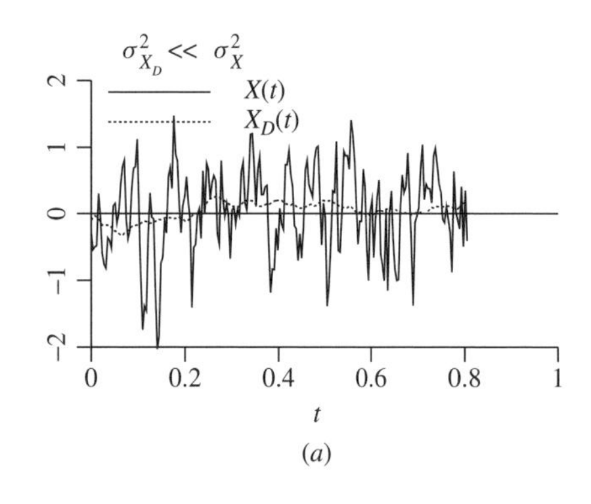

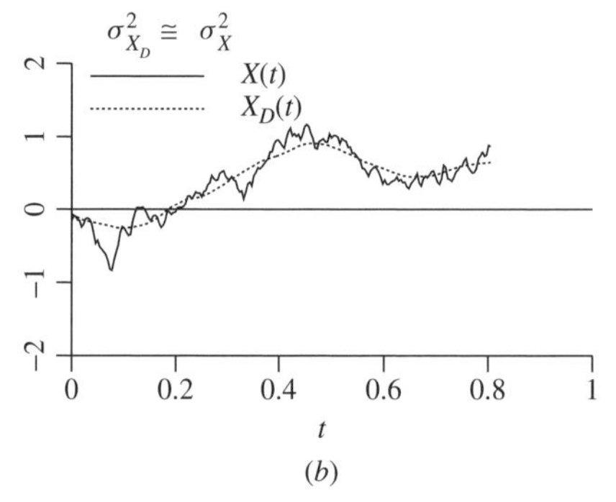

Figure 5.15 Effect of averaging on variance: (*a*) poorly correlated field; (*b*) highly correlated field.

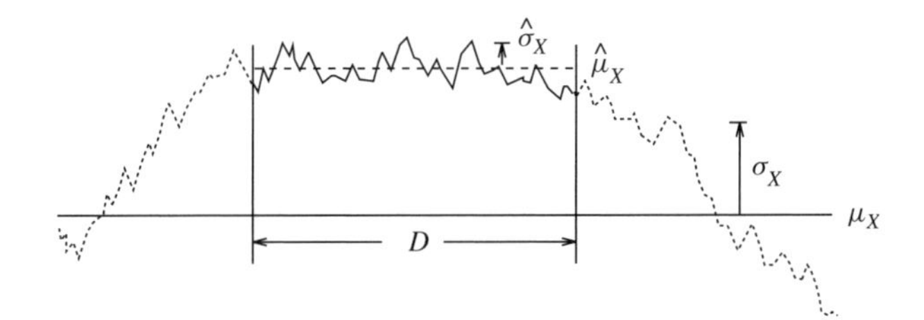

Figure 5.16 Local estimates of mean and standard deviation over sampling domain T.

the average $X_D(t)$ tends to follow $X(t)$ closely with little variance reduction.

The implications of this discussion are as follows: While the mean is typically estimated using Eq. 5.36, it is important to remember that, in the case of random fields with significant spatial correlation, this estimator may itself be highly variable and increasing the number of samples within a fixed domain may not decrease its variability (it would be better to increase the sampling domain size). See, for example, Figure 5.16.

On the other hand, the fact that the mean estimator may remain highly variable is really only important when the estimator is going to be used to model a random soil property at another site. If the data are being gathered at the site in question, then increasing the number of samples *does* reduce the uncertainty at the site, even if the true mean of the soil property in general remains questionable.

5.3.4 Estimating the Variance

Consider the traditional estimator of the variance:

$$\hat{\sigma}_X^2 = \frac{1}{n-1}\sum_{i=1}^{n}(X_i - \hat{\mu}_X)^2 \tag{5.37}$$

If the observations X_i are independent, then this estimator is unbiased with

$$\mathrm{E}\left[\hat{\sigma}_X^2\right] = \sigma_X^2 \tag{5.38}$$

If the observations X_i are (at least approximately) normally distributed, then the quantity $(n-1)\hat{\sigma}_X^2/\sigma_X^2$ follows a chi-square distribution with $n-1$ degrees of freedom. It follows from the fact that a chi-square distributed random variable with $n-1$ degrees of freedom has variance $2(n-1)$ (see Section 1.10.7) that the variance of $\hat{\sigma}_X^2$ is

$$\mathrm{Var}\left[\hat{\sigma}_X^2\right] = \frac{2\sigma_X^4}{n-1} \tag{5.39}$$

When the observations X_i are correlated, it can be shown that the MLE

$$\hat{\sigma}_X^2 = \frac{1}{n}\sum_{i=1}^{n}(X_i - \hat{\mu}_X)^2 \tag{5.40}$$

is a biased estimator with expectation

$$\mathrm{E}\left[\hat{\sigma}_X^2\right] = \sigma_X^2\Big(1 - \gamma(D)\Big) \tag{5.41}$$

In the presence of correlation, $\hat{\sigma}_X^2 < \sigma_X^2$, on average, since $\gamma(D)$ lies between 0 and 1. In fact, $\hat{\sigma}_X^2 \to 0$ as the field becomes increasingly correlated [since $\gamma(D) \to 1$ in this case]. This situation is illustrated in Figure 5.16 where a slowly varying (highly correlated) soil property is sampled over a relatively short distance D. In this case, the estimated variance is much smaller than the true variance and the estimated mean is considerably different than the true mean. In the case where the X_i's are independent, $\gamma(D)$ tends towards $1/n$ so that Eq. 5.40 is seen to be still biased. In most cases the unbiased estimator given by Eq. 5.37 is preferred when the data are independent.

It can be seen that the estimate given by Eq. 5.40 tends to become quite uncertain as the field becomes increasingly correlated. However, this is again only important if a good estimate of the true variance is being sought—if D denotes the site in question, then the data will accurately reflect that site (but cannot be used to extrapolate). In general the only way to get a good estimate of the variance is to increase the size of the sampling domain—the sampling domain size should be many times the correlation length θ. The difficulty with such a statement is that θ is generally unknown and might very well be infinite (as in fractal fields).

5.3.5 Trend Analysis

In the preceding sections, a stationary random field was implicitly assumed, having spatially constant mean and variance. In many cases this is not so, at least not apparently so, over the sampling domain. Often distinct trends in the mean can be seen, and sometimes the variance also clearly changes with position. We reiterate that if such trends are not physically based, that is, if there is no reason to suspect that identical trends would be repeated at another site, then their direct estimation depends on whether the data are being used to characterize this site or another. If the data are collected at the site to be estimated, then the direct estimation of the trends is worthwhile, otherwise probably not. If unexplainable trends are encountered during an exploration and the results are to be used to characterize another site, then probably a larger sampling domain needs to be considered.

Assume that the data are collected at the site to be characterized. In such a case, the task is to obtain estimates of $\mu_X(\mathbf{x})$ and $\sigma_X^2(\mathbf{x})$, both as functions of position. Trends in the variance typically require significant amounts of data to estimate accurately. The sampling domain is subdivided into small regions within each of which the variance is assumed spatially constant. This allows a "blockwise" estimation of the variance which may then be used to estimate a trend. Thus the estimation of a nonstationary variance is simply a reiteration of the stationary variance estimation procedure discussed earlier. Since often there are insufficient data to allow such a sophisticated analysis, the variance is usually assumed globally stationary.

Trends in the mean, in the case of stationary variance, can be obtained by least squares regression techniques. Here it is assumed that the mean can be described by a function of the form

$$\hat{\mu}_X(\mathbf{x}) = \sum_{k=1}^{M} a_k g_k(\mathbf{x}) \tag{5.42}$$

where a_k are the unknown coefficients to be solved for and $g_k(\mathbf{x})$ are prespecified functions of spatial position $\mathbf{x}$. In that complicated functions are often unjustifiable, usually the mean trend is taken to be linear so that, in one dimension, $g_1(x) = 1$, $g_2(x) = x$, and $M = 2$. In two dimensions, the corresponding mean function would be bilinear, with $g_1(\mathbf{x}) = 1$, $g_2(\mathbf{x}) = x_1$, $g_3(\mathbf{x}) = x_2$, and $g_4(\mathbf{x}) = x_1 x_2$. The coefficients a_k may be obtained by solving the so-called *normal* equations:

$$\boldsymbol{G}^{\mathrm{T}}\boldsymbol{G}\mathbf{a} = \boldsymbol{G}^{\mathrm{T}}\mathbf{y} \tag{5.43}$$

where $\mathbf{y}$ is the vector of observations (the measured values of the soil property in question), $\mathbf{a}$ is the vector of unknown coefficients in Eq. 5.42 and $\boldsymbol{G}$ is a matrix made up of the specified functions $g_k(\mathbf{x}_i)$ evaluated at each of the observation locations $\mathbf{x}_i$:

$$\boldsymbol{G} = \begin{bmatrix} g_1(\mathbf{x}_1) & g_2(\mathbf{x}_1) & \cdot & \cdot & \cdot & g_M(\mathbf{x}_1) \\ g_1(\mathbf{x}_2) & g_2(\mathbf{x}_2) & \cdot & \cdot & \cdot & g_M(\mathbf{x}_2) \\ \cdot & \cdot & \cdot & & & \cdot \\ \cdot & \cdot & & \cdot & & \cdot \\ \cdot & \cdot & & & \cdot & \cdot \\ g_1(\mathbf{x}_n) & g_2(\mathbf{x}_n) & \cdot & \cdot & \cdot & g_M(\mathbf{x}_n) \end{bmatrix}$$

Although the matrix $\boldsymbol{G}$ is of size $n \times M$, the normal equations boil down to just M equations in the M unknown coefficients of $\mathbf{a}$. As a word of caution, however, the normal equations often tend to be nearly singular, particularly for large data sets, and more advanced regression techniques may be required [such as the QR algorithm (see, e.g., Press et al., 1997)].

With this estimate of the mean, the process $X(\mathbf{x})$ can be converted into a mean stationary process $X'(\mathbf{x}) = X(\mathbf{x}) - \hat{\mu}_X(\mathbf{x})$. The deviation or residual process X' is now approximately mean zero. If a plot of $X'(\mathbf{x})$ over space seems to indicate a nonstationary variance, then the variance $\sigma_X^2(\mathbf{x})$ can be estimated by subdividing the sampling domain into small regions as discussed above. Otherwise an unbiased estimate of the stationary variance is

$$\hat{\sigma}_X^2 = \frac{1}{n-M} \sum_{i=1}^{n} \left(X_i - \hat{\mu}_X(\mathbf{x}_i)\right)^2$$

where M is the number of terms in Eq. 5.42.

If a nonstationary variance is detected and estimated, an approximately stationary field in both mean and variance can be produced through the transformation

$$X'(\mathbf{x}) = \frac{X(\mathbf{x}) - \hat{\mu}_X(\mathbf{x})}{\hat{\sigma}_X(\mathbf{x})}$$

In addition, such a transformation implies that X' has zero mean and unit variance (at least in approximation).

5.3.6 Estimating the Correlation Structure

An estimator of the correlation structure of a one-dimensional random field will be developed here. The extension to the multidimensional case is only slightly more complicated.

Consider the sequence of random variables $\{X_1, X_2, \ldots, X_n\}$ sampled from $X(x)$ at a sequence of locations separated by distance Δx. For the following estimator, it is essential that the data be equispaced. An unbiased estimator for the covariance, $C(j\,\Delta x)$, between any two random variables along x separated by the distance $j\,\Delta x$ for $j = 0, 1, \ldots, n - M - 1$ is given by

$$\hat{C}(j\,\Delta x) = \frac{1}{n-M-j} \times \sum_{i=1}^{n-j} \left(X_i - \hat{\mu}_X(x_i)\right)\left(X_{i+j} - \hat{\mu}_X(x_{i+j})\right)$$

where M is the number of unknowns used to estimate $\mu_X(x)$. The correlation coefficient is then estimated as

$$\hat{\rho}_X(j\,\Delta x) = \frac{\hat{C}(j\,\Delta x)}{\hat{\sigma}_X^2}$$

where $\hat{\sigma}_X^2 = \hat{C}(0)$ is the estimated variance.

In two dimensions, the estimator for the covariance at lag $\boldsymbol{\tau} = \{j\,\Delta x_1, k\,\Delta x_2\}$ involves a sum over all data pairs separated by the lag $\boldsymbol{\tau}$—similarly in higher dimensions. The normalizing factor $1/(n - M - j)$ becomes $1/(N_\tau - M)$, where N_τ is the total number of data pairs separated by $\boldsymbol{\tau}$ in the data set.

5.3.7 Example: Statistical Analysis of Permeability Data

Consider a set of permeability measurements made by an infiltrometer on 0.5 m × 0.5 m cells extracted from a rectangular test pad of poorly compacted clay, as shown in

Table 5.4 Permeability Data over 4-m Square Clay Test Pad

x_2	x_1 (m)							
(m)	0.25	0.75	1.25	1.75	2.25	2.75	3.25	3.75
0.25	53.69	61.94	82.38	65.49	49.71	17.85	42.83	14.71
0.75	98.42	46.87	109.41	99.40	7.01	16.71	20.70	1.88
1.25	41.81	6.32	20.75	31.51	6.11	26.88	33.71	13.48
1.75	149.19	11.47	0.63	14.88	8.84	73.17	40.83	29.96
2.25	140.93	30.31	1.04	0.92	2.81	34.85	3.31	0.24
2.75	105.74	1.27	10.58	0.21	0.04	0.57	2.92	7.09
3.25	99.05	12.11	0.12	0.97	5.09	6.90	0.65	1.29
3.75	164.42	7.38	13.35	10.88	8.53	2.22	3.26	0.73

Table 5.4. The test pad is of dimension 4 m × 4 m and the (x_1, x_2) coordinates shown in the table correspond to the center of each 0.5-m square cell. All values are in units of 10^{-7} cm/s. A quick review of the data reveals first that it is highly variable with $K_{\max}/K_{\min} > 4000$ and second that it tends from very high values at the left edge ($x_1 = 1$) to small values as x_1 increases. There also appears to be a similar but somewhat less pronounced trend in the x_2 direction, at least for larger values of x_1.

The high variability is typical of permeability data, since a boulder will have permeability approaching zero while an airspace will have permeability approaching infinity—soils typically contain both at some scale. Since permeability is bounded below by zero, a natural distribution to use in a random model of permeability is the lognormal. If K is lognormally distributed, then $\ln K$ will be normally distributed. In fact, the parameters of the lognormal distribution are just the mean and variance of $\ln K$ (see Section 1.10.9). Adopting the lognormal hypothesis, it is appropriate, before proceeding, to convert the data listed in Table 5.4 into $\ln K$ data, as shown in Table 5.5.

Two cases will be considered in this example:

1. The data are to be used to characterize other similar clay deposits. This is the more likely scenario for this particular sampling program.
2. The site to be characterized is the 4-m^2 test area (which may be somewhat hypothetical since it has been largely removed for laboratory testing).

Starting with case 1, any apparent trends in the data are treated as simply part of a longer scale fluctuation—the field is assumed to be stationary in mean and variance. Using Eqs. 5.36 and 5.37 the mean and variance are estimated as

$$\hat{\mu}_{\ln K} = -13.86, \qquad \hat{\sigma}^2_{\ln K} = 3.72$$

To estimate the correlation structure, a number of assumptions can be made:

(a) Assume that the clay bed is isotropic, which appears physically reasonable. Hence an isotropic correlation structure would be adopted which can be estimated by averaging over the lag τ in any direction. For example, when $\tau = 0.5$ m the correlation can be estimated by averaging over all samples separated by 0.5 m in any direction.

(b) Assume that the principal axes of anisotropy are aligned with the x_1 and x_2 coordinate axes and that the correlation function is *separable*. Now $\hat{\rho}_{\ln K}(\tau_1, \tau_2) = \hat{\rho}_{\ln K}(\tau_1)\hat{\rho}_{\ln K}(\tau_2)$ is obtained by averaging in the two coordinate directions separately and lag vectors not aligned with the coordinates need not be considered.

Table 5.5 Log Permeability Data over 4-m Square Clay Test Pad

x_2	x_1 (m)							
(m)	0.25	0.75	1.25	1.75	2.25	2.75	3.25	3.75
0.25	−12.13	−11.99	−11.71	−11.94	−12.21	−13.24	−12.36	−13.43
0.75	−11.53	−12.27	−11.42	−11.52	−14.17	−13.30	−13.09	−15.49
1.25	−12.38	−14.27	−13.09	−12.67	−14.31	−12.83	−12.60	−13.52
1.75	−11.11	−13.68	−16.58	−13.42	−13.94	−11.83	−12.41	−12.72
2.25	−11.17	−12.71	−16.08	−16.20	−15.08	−12.57	−14.92	−17.55
2.75	−11.46	−15.88	−13.76	−17.68	−19.34	−16.68	−15.05	−14.16
3.25	−11.52	−13.62	−18.24	−16.15	−14.49	−14.19	−16.55	−15.86
3.75	−11.02	−14.12	−13.53	−13.73	−13.97	−15.32	−14.94	−16.43

Because of the reduced number of samples contributing to each estimate, the estimates themselves will be more variable.

(c) Assume that the correlation structure is more generally anisotropic. Lags in any direction must be considered separately and certain directions and lags will have very few data pairs from which to derive an estimate. This typically requires a large amount of data.

Assumption (a) is preferred, but (b) will also be examined to judge the applicability of the first assumption. In assumption (b), the directional estimators are given by

$$\hat{\rho}_{\ln K}(j\,\Delta\tau_1) = \frac{1}{\hat{\sigma}^2_{\ln K}(n_2(n_1-j)-1)} \times \sum_{k=1}^{n_2}\sum_{i=1}^{n_1-j}(X'_{ik})(X'_{i+j,k}), \qquad j=0,1,\ldots,n_1-1$$

$$\hat{\rho}_{\ln K}(j\,\Delta\tau_2) = \frac{1}{\hat{\sigma}^2_{\ln K}(n_1(n_2-j)-1)} \times \sum_{k=1}^{n_1}\sum_{i=1}^{n_2-j}(X'_{ki})(X'_{k,i+j}), \qquad j=0,1,\ldots,n_2-1$$

where $X'_{ik} = \ln K_{ik} - \hat{\mu}_{\ln K}$ is the deviation in $\ln K$ about the mean, n_1 and n_2 are the number of samples in the x_1 and x_2 directions, respectively, and $\Delta\tau_1 = \Delta\tau_2 = 0.5$ m in this example. The subscripts on X' or $\ln K$ index first the x_1 direction and second the x_2 direction. The isotropic correlation estimator of assumption (a) is obtained using

$$\hat{\rho}_{\ln K}(j\ \Delta\tau) = \frac{1}{\hat{\sigma}^2_{\ln K}(n_2(n_1-j)+n_1(n_2-j)-1)} \times \left\{ \sum_{k=1}^{n_2}\sum_{i=1}^{n_1-j}(X'_{ik})(X'_{i+j,k}) + \sum_{k=1}^{n_1}\sum_{i=1}^{n_2-j}(X'_{ki})(X'_{k,i+j}) \right\},$$
$$j=0,1,\ldots,\max(n_1,n_2)-1$$

in which, if $n_1 \neq n_2$, then the $n_i - j$ appearing in the denominator must be treated specially. Specifically for any $j > n_i$, the $n_i - j$ term is set to zero.

Figure 5.17 shows the estimated directional and isotropic correlation functions for the $\ln K$ data. Note that at higher lags the curves become quite erratic. This is typical since they are based on fewer sample pairs as the lag increases. Also shown on the plot is a fitted exponentially decaying correlation function. The correlation length θ is estimated to be about $\hat{\theta} = 1.3$ m in this case. This was obtained simply by finding $\hat{\theta}$ which resulted in the fitted correlation function passing through the estimated correlation(s) at lag $\tau = 0.5$ m.

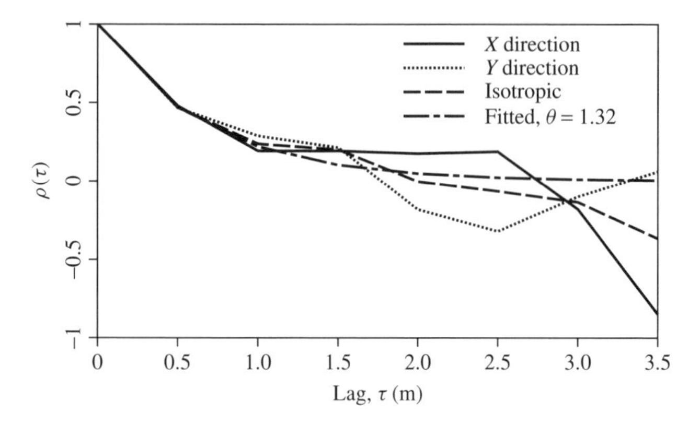

Figure 5.17 Estimated and fitted correlation function for $\ln K$ data.

It is important to point out that the estimated scale is quite sensitive to the mean. For example, if the mean of $\ln K$ is known to be -12.0 rather than -13.86, then the estimated scale jumps to $\hat{\theta} = 3.75$ m. In effect, the estimated scale is quite uncertain; it is best used to characterize the site at which the data were taken. Unfortunately, significantly better scale estimators have yet to be developed.

For case 2, where the data are being used to characterize the site from which it was sampled, the task is to estimate the trend in the mean. This can be done in a series of steps starting with simple functions for the mean (i.e., constant) and progressing to more complicated functions (e.g., bilinear, biquadratic), monitoring the residual variance for each assumed form. The form which accounts for a significant portion of the variance without being overly complex would be preferable.

Performing a least squares regression with a bilinear mean function on the data in Table 5.5 gives

$$\hat{\mu}_{\ln K}(\mathbf{x}) = -11.88 - 0.058x_1 - 0.102x_2 - 0.011x_1x_2$$

with corresponding residual variance of 2.58 (was 3.72 for the constant mean case). If a biquadratic mean function is considered, the regression yields

$$\hat{\mu}_{\ln K}(\mathbf{x}) = -12.51 + 0.643x_1 + 0.167x_2 - 0.285x_1x_2 - 0.0501x_1^2 - 0.00604x_2^2 + 0.0194x_1^2x_2 + 0.0131x_1x_2^2 - 0.000965x_1^2x_2^2$$

with a residual variance of 2.18. Since there is not much of a reduction in variance using the more complicated biquadratic function, the bilinear form is selected. For simplicity, only two functional forms were compared here. In general, one might want to consider all the possible combinations of monomials to select the best form.

Table 5.6 Log Permeability Residuals

x_2 (m)	x_1 (m) 0.25	0.75	1.25	1.75	2.25	2.75	3.25	3.75
0.25	−0.077	0.203	0.623	0.533	0.403	−0.487	0.533	−0.397
0.75	0.749	0.192	1.225	1.308	−1.159	−0.106	0.288	−1.929
1.25	0.124	−1.539	−0.133	0.513	−0.900	0.806	1.262	0.569
1.75	1.620	−0.681	−3.311	0.118	−0.132	2.247	1.937	1.896
2.25	1.785	0.558	−2.499	−2.307	−0.874	1.949	−0.088	−2.406
2.75	1.721	−2.343	0.133	−3.432	−4.736	−1.720	0.266	1.512
3.25	1.886	0.185	−4.036	−1.547	0.513	1.212	−0.749	0.340
3.75	2.612	−0.046	0.986	1.229	1.431	0.523	1.346	0.298

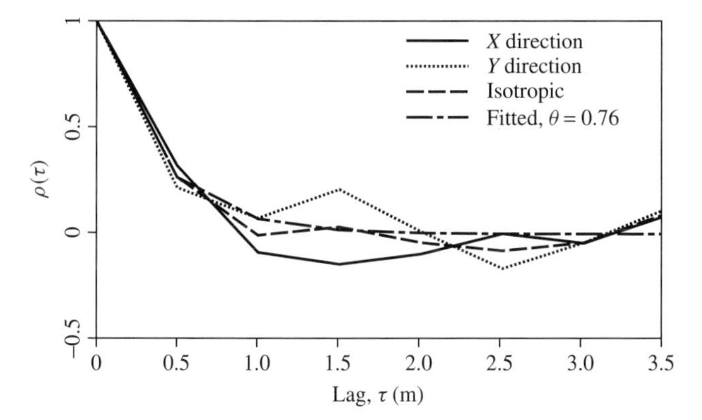

Figure 5.18 Estimated and fitted correlation function for $\ln K - \hat{\mu}_{\ln K}$ data.

Adopting the bilinear mean function, the residuals $\ln K' = \ln K - \hat{\mu}_{\ln K}$ are shown in Table 5.6. Figure 5.18 illustrates the estimated correlation structure of the residuals. Notice that the fitted correlation length has decreased to about 0.76 m. This reduction is typical since subtracting the mean tends to reduce the correlation between residuals. The estimated mean, variance, and correlation function (in particular the correlation length) can now be used confidently to represent the random field of log permeabilities at the site.

Comments The use of random-field models is not without its difficulties. This is particularly evident when estimating their parameters since random-field parameters must often be derived from a single realization (the site being explored). The interpretation of trends in the data as true trends in the mean or simply as large-scale fluctuations is a question which currently can only be answered by engineering judgment. The science of estimation in the presence of correlation between samples is not at all well developed.

As a result, the statistical parameters used to model a random field are generally uncertain and statements regarding probabilities are equally uncertain. That is, because of the uncertainty in estimates of mean properties, statements regarding the probability of failure of a slope, for example, cannot be regarded as absolute. However, they often yield reasonable approximations based on a very rational approach to the problem. In addition, probabilities can be used effectively in a relative sense; for example, the probability of failure of design A is less than that of design B. Since relative probabilities are less sensitive to changes in the underlying random-field parameters, they can be more confidently used in making design choices.

5.4 ADVANCED ESTIMATION TECHNIQUES

This section takes a *qualitative* look at a number of tools which tell something about the second-moment behavior of a one-dimensional random process. The intent in this section is to evaluate these tools with respect to their ability to discern between finite-scale and fractal behavior. In the following section, various maximum-likelihood approaches to the estimation of the parameters for the finite-scale and fractal models are given. Finally the results are summarized with a view toward their use in developing a priori soil statistics from a large geotechnical database.

5.4.1 Second-Order Structural Analysis

Attention is now turned to the stochastic characterization of the spatial correlation structure. In the following it is assumed that the data, x_i, $i = 1, 2, \ldots, n$, are collected at a sequence of equispaced points along a line and that the best stochastic model along that line is to be found. Note that the x_i may be some suitable transformation of the actual data derived from the samples. In the following, x_i is an observation of the random process $X_i = X(z_i)$, where z is an index (commonly depth) and $z_i = (i-1)\,\Delta z$, $i = 1, 2, \ldots, n$. A variety of tools will be considered in this section and their ability to identify the most appropriate stochastic model for X_i will be discussed. In particular, interest focuses on whether the process $X(z)$ is finite scaled or fractal in nature. The performance of the various tools in answering this question will be evaluated via simulation employing 2000 realizations of finite scale (Markov, see Section 3.6.5) and fractal (see Section 3.6.7) processes. Each simulation is of length 20.48 with $\Delta z = 0.02$, so that $n = 1024$ and realizations are produced via covariance matrix decomposition (see Section 6.4.2), a method which follows from a Cholesky decomposition. Unfortunately, large covariance matrices are often nearly singular and so are hard to decompose correctly. Since the covariance matrix for a one-dimensional equispaced random field is symmetric and Toeplitz (the entire matrix is known if only the first column is known—all elements along each diagonal are equal), the decomposition is done using the numerically more accurate Levinson–Durbin algorithm [see Marple (1987) and Brockwell and Davis (1987)].

5.4.1.1 Sample Correlation Function The classical sample average of x_i is computed as

$$\hat{\mu}_X = \frac{1}{n} \sum_{i=1}^{n} X_i \tag{5.44}$$

and the sample variance as

$$\hat{\sigma}_X^2 = \frac{1}{n} \sum_{i=1}^{n} (X_i - \hat{\mu}_X)^2 \tag{5.45}$$

The variance estimator is biased, since it is not divided by $n - 1$ as is usually seen. (A biased estimator is one whose expected value is not equal to the parameter it purports to estimate.) The use of a biased estimator here is for three reasons:

1. The expected error variance is smaller than that for the biased case (slightly).
2. The biased estimator, when estimating covariances, leads to a tractable nonnegative definite covariance matrix.
3. It is currently the most popular variance estimator in time series analysis (Priestley, 1981).

Probably the main reason for its popularity is due to its nonnegative definiteness. The covariance $C(\tau)$ between $X(z)$ and $X(z+\tau)$ is estimated, using a biased estimator for reasons discussed above, as

$$\hat{C}(\tau_j) = \frac{1}{n} \sum_{i=1}^{n-j+1} (X_i - \hat{\mu}_X)(X_{i+j-1} - \hat{\mu}_X), \qquad j = 1, 2, \ldots, n \tag{5.46}$$

where the lag $\tau_j = (j-1)\,\Delta z$. Notice that $\hat{C}(0)$ is the same as the estimated variance $\hat{\sigma}_X^2$. The sample correlation is obtained by normalizing,

$$\hat{\rho}(\tau_j) = \frac{\hat{C}(\tau_j)}{\hat{C}(0)} \tag{5.47}$$

One of the major difficulties with the sample correlation function resides in the fact that it is heavily dependent on the estimated mean $\hat{\mu}_X$. When the soil shows significant long-scale dependence, characterized by long-scale fluctuations (see, e.g., Figure 5.14), $\hat{\mu}_X$ is almost always a poor estimate of the true mean. In fact, it is not too difficult to show that although the mean estimator (Eq. 5.44) is unbiased, its variance is given by

$$\mathrm{Var}\left[\hat{\mu}_X\right] = \left[\frac{1}{n^2} \sum_{i=1}^{n} \sum_{j=1}^{n} \rho(\tau_{i-j})\right] \sigma_X^2 = \gamma_n \sigma_X^2 \simeq \gamma(D)\sigma_X^2 \tag{5.48}$$

In this equation, $D = (n-1)\,\Delta z$ is the sampling domain size (interpreted as the region defined by n equisized "cells," each of width Δz centered on an observation) and $\gamma(D)$ is the so-called variance function (Vanmarcke, 1984) which gives the variance reduction due to averaging over the length D,

$$\begin{aligned} \gamma(D) &= \frac{1}{D^2} \int_0^D \int_0^D \rho(\tau - s)\, d\tau\, ds \\ &= \frac{2}{D^2} \int_0^D (T - \tau)\,\rho(\tau)\, d\tau \end{aligned} \tag{5.49}$$

The discrete approximation to the variance function, denoted γ_n above, approaches $\gamma(D)$ as n becomes large. For highly correlated soil samples (over the sampling domain), $\gamma(D)$ remains close to 1.0, so that $\hat{\mu}_X$ remains highly variable, almost as variable as $X(z)$ itself. Notice that the variance of $\hat{\mu}_X$ is unknown since it depends on the unknown correlation structure of the process.

In addition, it can be shown that $\hat{C}(\tau_j)$ is biased according to (Vanmarcke, 1984)

$$\mathrm{E}\left[\hat{C}(\tau_j)\right] \simeq \sigma_X^2 \left(\frac{n-j+1}{n}\right) \left[\rho(\tau_j) - \gamma(D)\right] \tag{5.50}$$

where, again, the approximation improves as n increases. From this it can be seen that

$$\mathrm{E}\left[\hat{\rho}(\tau_j)\right] \simeq \frac{\mathrm{E}\left[\hat{C}(\tau_j)\right]}{\mathrm{E}\left[\hat{C}(0)\right]} \simeq \left(\frac{n-j+1}{n}\right)\left(\frac{\rho(\tau_j) - \gamma(D)}{1 - \gamma(D)}\right) \tag{5.51}$$

using a first-order approximation. For soil samples which show considerable serial correlation, $\gamma(D)$ may remain close to 1 and generally the term $[\rho(\tau_j) - \gamma(D)]$ will become negative for all $j^* \leq j \leq n$ for some $j^* < n$. What this means is that the estimator $\hat{\rho}(\tau)$ will typically dip below zero even when the field is actually highly positively correlated.

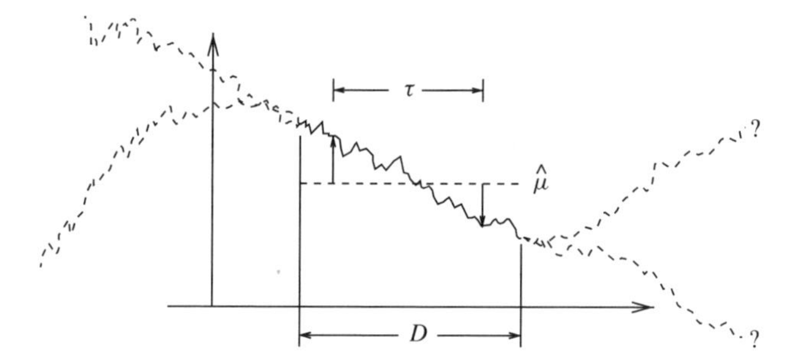

Figure 5.19 Covariance estimator on strongly dependent finite sample.

Another way of looking at this problem is as follows: Consider again the sample shown in Figure 5.14 and assume that the apparent trend is in fact just part of a long-scale fluctuation. Clearly Figure 5.14 is then a process with a very large correlation length compared to D. The local average $\hat{\mu}_X$ is estimated and shown by the dashed line in Figure 5.19. Now, for any τ greater than about half of the sampling domain, the product of the deviations from $\hat{\mu}_X$ in Eq. 5.46 will be negative. This means that the sample correlation function will decrease rapidly and become negative somewhere before $\tau = \frac{1}{2}D$ even though the true correlation function may remain much closer to 1.0 throughout the sample.

It should be noted that if the sample does in fact come from a short-scale process, with $\theta \ll D$, the variability of Eq. 5.48 and the bias of Eq. 5.51 largely disappear because $\gamma(D) \simeq 0$. This means that the sample correlation function is a good estimator of short-scale processes as long as $\theta \ll D$. However, if the process does in fact have long-scale dependence, then the correlation function cannot identify this and in fact continues to illustrate short-scale behavior. In essence, the estimator is analogous to a self-fulfilling prophecy: It always appears to justify its own assumptions.

Figures 5.20 and 5.21 illustrate the situation graphically using simulations from finite-scale and fractal processes. The max/min lines show the maximum and minimum correlations observed at each lag over the 2000 realizations. The finite-scale ($\theta = 3$) simulation shows reasonable agreement between $\hat{\rho}(\tau)$ and the true correlation because $\theta \ll D \simeq 20$. However, for the fractal process ($H = 0.95$) there is a very large discrepancy between the estimated average and true correlation functions. Clearly the sample correlation function fails to provide any useful information about large-scale or fractal processes.

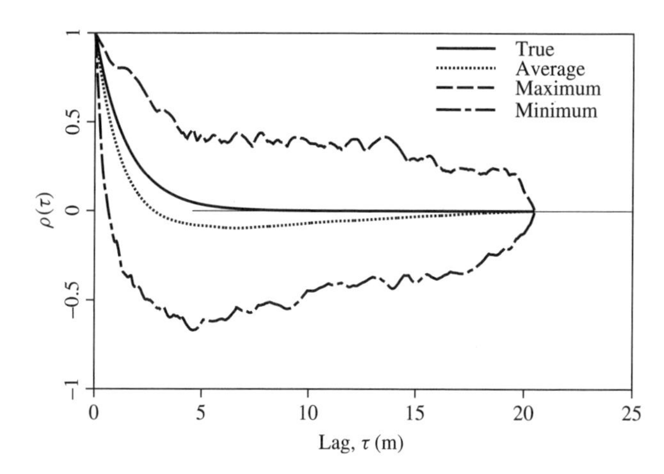

Figure 5.20 Correlation function estimates from a finite-scale process ($\theta = 3$).

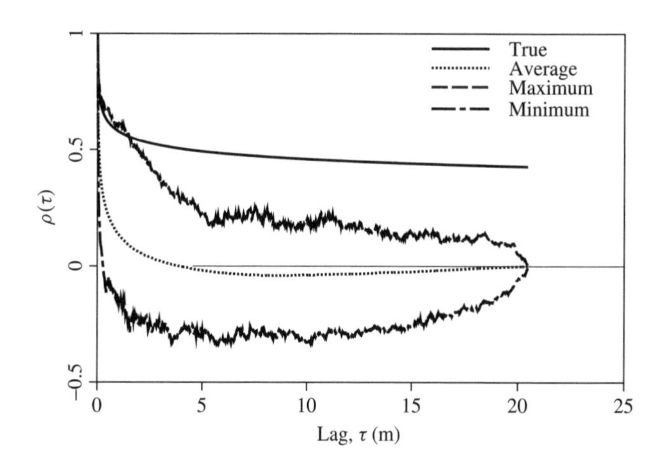

Figure 5.21 Correlation function estimates from a fractal process ($H = 0.95$).

5.4.1.2 Sample Semivariogram The semivariogram, $V(\tau)$, which is one-half of the variogram, as defined by Matheron (1962), gives essentially the same information as the correlation function since, for stationary processes, they are related according to

$$V(\tau_j) = \tfrac{1}{2}\mathrm{E}\left[(X_{i+j} - X_i)^2\right] = \sigma_X^2\Big(1 - \rho(\tau_j)\Big) \tag{5.52}$$

The sample semivariogram is defined by

$$\hat{V}(\tau_j) = \frac{1}{2(n-j)}\sum_{i=1}^{n-j}(X_{i+j} - X_i)^2, \qquad j = 0, 1, \ldots, n-1 \tag{5.53}$$

The major difference between $\hat{V}(\tau_j)$ and $\hat{\rho}(\tau_j)$ is that the semivariogram does not depend on $\hat{\mu}_X$. This is a clear advantage since many of the troubles of the correlation function relate to this dependence. In fact, it is easily shown that the semivariogram is an unbiased estimator with $\mathrm{E}\left[\hat{V}(\tau_j)\right] = \frac{1}{2}\mathrm{E}\left[(X_{i+j} - X_i)^2\right]$. Figures 5.22 and 5.23 show

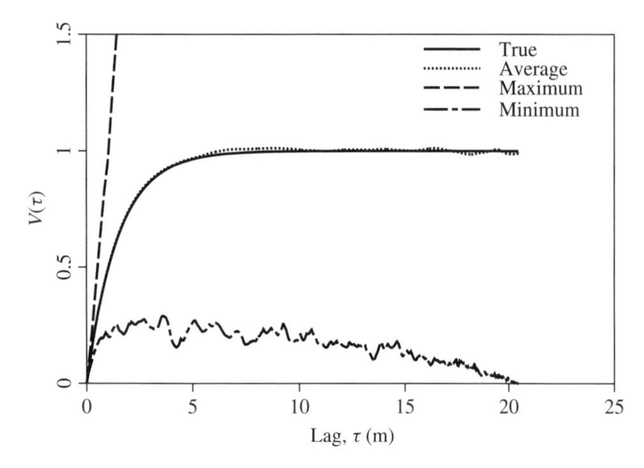

Figure 5.22 Semivariogram estimates from a finite-scale process ($\theta = 3$).

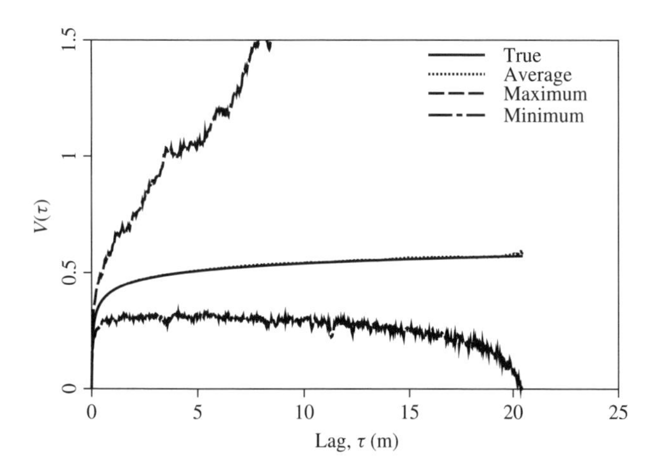

Figure 5.23 Semivariogram estimates from a fractal process ($H = 0.95$).

how this estimator behaves for finite-scale and fractal processes. Notice that the finite-scale semivariogram rapidly increases to its limiting value (the variance) and then flattens out whereas the fractal process leads to a semivariogram that continues to increase gradually throughout. This behavior can indicate the underlying process type and allow identification of a suitable correlation model. Note, however, the very wide range between the observed minimum and maximum (the maximum going off the plot, but having maximum values in the range from 5 to 10 in both cases). The high variability in the semivariogram may hinder its use in discerning between model types unless sufficient averaging can be performed.

The semivariogram finds its primary use in mining geostatistics applications [see, e.g., Journel and Huijbregts (1978)]. Cressie (1993) discusses some of its distributional characteristics along with robust estimation issues, but little is known about the distribution of the semivariogram when $X(z)$ is spatially dependent. Without the estimator distribution, the semivariogram cannot easily be used to test rigorously between competing model types (as in fractal vs. finite scale), nor can it be used to fit model parameters using the maximum-likelihood method.

5.4.1.3 *Sample Variance Function*

The variance function measures the decrease in the variance of an average as an increasing number of sequential random variables are included in the average. If the local average of a random process X_D is defined by

$$X_D = \frac{1}{D}\int_0^D X(z)\,dz \tag{5.54}$$

then the variance of X_D is just $\gamma(D)\sigma_X^2$. In the discrete case, which will be used here, this becomes

$$\bar{X} = \hat{\mu}_X = \frac{1}{n}\sum_{i=1}^{n} X(z_i) = \frac{1}{n}\sum_{i=1}^{n} X_i \tag{5.55}$$

where $\text{Var}\left[\bar{X}\right] = \gamma_n\sigma_X^2$ and γ_n, defined by Eq. 5.48, is the discrete approximation of $\gamma(D)$. If the X_i values are independent and identically distributed, then $\gamma_n = 1/n$, while if $X_1 = X_2 = \cdots = X_n$, then the X's are completely correlated and $\gamma_n = 1$ so that averaging does not lead to any variance reduction. In general, for correlation functions which remain nonnegative, $1/n \le \gamma_n \le 1$.

Conceptually, the rate at which the variance of an average decreases with averaging size tells about the spatial correlation structure. In fact, these are equivalent since in the one-dimensional (continuous) case

$$\gamma(D) = \frac{2}{D^2}\int_0^D (D-\tau)\rho(\tau)\,d\tau \quad \Longleftrightarrow \quad \rho(\tau) = \frac{1}{2}\frac{\partial^2}{\partial\tau^2}[\tau^2\gamma(\tau)] \tag{5.56}$$

Given a sequence of n equispaced observations over a sampling domain of size $D = (n-1)\ \Delta z$, the sample (discrete) variance function is estimated to be

$$\hat{\gamma}_i = \frac{1}{\hat{\sigma}_X^2(n-i+1)}\sum_{j=1}^{n-i+1}(X_{i,j} - \hat{\mu}_X)^2, \qquad i = 1, 2, \ldots, n \tag{5.57}$$

where $X_{i,j}$ is the local average

$$X_{i,j} = \frac{1}{i}\sum_{k=j}^{j+i-1} X_k, \qquad j = 1, 2, \ldots n-i+1 \tag{5.58}$$

Note that $\hat{\gamma}_1 = 1$ since the sum in Eq. 5.57 is the same as that used to find $\hat{\sigma}_X^2$ when $i = 1$. Also when $i = n$ the sample variance function $\hat{\gamma}_n = 0$ since $X_{n,j} = X_{n,1} = \hat{\mu}_X$. Thus, the sample variance function always connects the points $\hat{\gamma}_1 = 1$ and $\hat{\gamma}_n = 0$.

Unfortunately, the sample variance function is biased and its bias depends on the degree of correlation between observations. Specifically it can be shown that

$$\mathrm{E}\left[\hat{\gamma}_i\right] \simeq \frac{\gamma_i - \gamma_n}{1 - \gamma_n} \tag{5.59}$$

using a first-order approximation. This becomes unbiased as $n \to \infty$ only if $D = (n-1)\,\Delta z \to \infty$ *and* $\gamma(D) \to 0$ as well. What this means is that we need both the averaging region to grow large and the correlation function to decrease sufficiently rapidly within the averaging region in order for the sample variance function to become unbiased.

Figures 5.24 and 5.25 show sample variance functions averaged over 2000 simulations of finite-scale and fractal

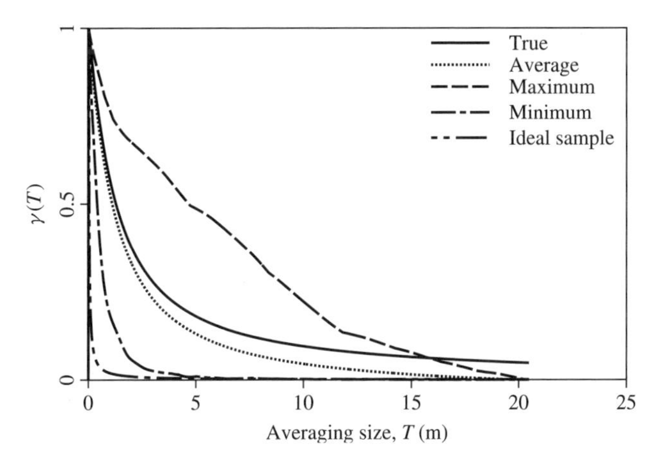

Figure 5.24 Variance function estimates from a finite-scale process ($\theta = 1$).

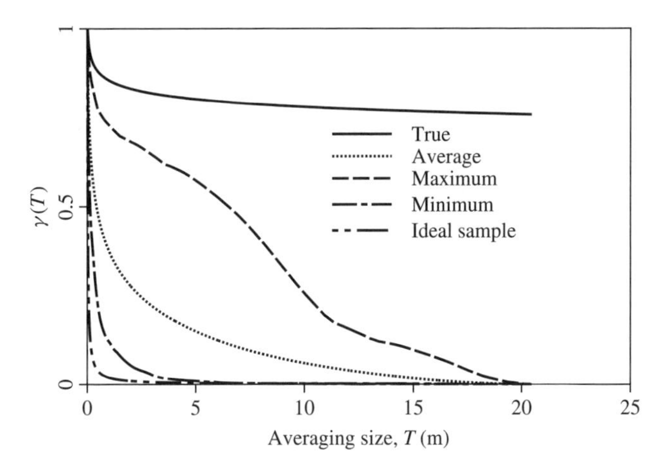

Figure 5.25 Variance function estimates from a fractal process ($H = 0.98$).

random processes. There is very little difference between the estimated variance function in the two plots, despite the fact that they come from quite different processes. Clearly, the estimate of the variance function in the fractal case is highly biased. Thus, the variance function plot appears not to be a good identification tool and is really only a useful estimate of second-moment behavior for finite-scale process with $\theta \ll D$. In the plots $T = i\,\Delta z$ for $i = 1, 2, \ldots n$.

5.4.1.4 *Wavelet Coefficient Variance* The wavelet basis has attracted much attention in recent years in areas of signal analysis, image compression, and, among other things, fractal process modeling (Wornell, 1996). It can basically be viewed as an alternative to Fourier decomposition except that sinusoids are replaced by "wavelets" which act only over a limited domain. In one dimension, wavelets are usually defined as translations along the real axis and dilations (scalings) of a "mother wavelet," as in

$$\psi_j^m(z) = 2^{m/2}\psi(2^m z - j) \tag{5.60}$$

where m and j are dilation and translation indices, respectively. The appeal to using wavelets to model fractal processes is that they are both self-similar in nature—as with fractal processes, all wavelets look the same when viewed at the appropriate scale (which, in the above definition, is some power of 2). The random process $X(z)$ is then expressed as a linear combination of various scalings, translations, and dilations of a common "shape." Specifically,

$$X(z) = \sum_m \sum_j X_j^m \psi_j^m(z) \tag{5.61}$$

If the wavelets are suitably selected so as to be orthonormal, then the coefficients can be found through the inversion

$$X_j^m = \int_{-\infty}^{\infty} X(z)\psi_j^m(z)\,dz \tag{5.62}$$

for which highly efficient numerical solution algorithms exist. The details of the wavelet decomposition will not be discussed here. The interested reader should see, for example, Strang and Nguyen (1996).

A theorem by Wornell (1996) states that, under reasonably general conditions, if the coefficients X_j^m are mutually uncorrelated, zero-mean random variables with variances

$$\sigma_m^2 = \mathrm{Var}\left[X_j^m\right] = \sigma^2 2^{-\gamma m} \tag{5.63}$$

then $X(z)$ obtained through Eq. 5.61 will have a spectrum which is very nearly fractal. Furthermore, Wornell makes theoretical and simulation-based arguments showing that the converse is also approximately true; namely that if $X(z)$ is fractal with spectral density proportional to $\omega^{-\gamma}$, then the coefficients X_j^m will be approximately uncorrelated with variance given by Eq. 5.63. If this is the case, then a plot of $\ln(\text{Var}\left[X_j^m\right])$ versus the scale index m will be a straight line.

Using a fifth-order Daubechies wavelet basis, Figures 5.26 and 5.27 show plots of the estimated wavelet coefficient variances $\hat{\sigma}_m^2$, where

$$\hat{\sigma}_m^2 = \frac{1}{2^{m-1}} \sum_{j=1}^{2^{m-1}} (x_j^m)^2 \tag{5.64}$$

against the scale index m for the finite-scale and fractal simulation cases. In Eq. 5.64, x_j^m is an estimate of X_j^m, obtained using observations in Eq. 5.62. The fractal simulations in Figure 5.27 yield a straight line, as expected, while the finite-scale simulations in Figure 5.26 show a slight flattening of the variance at lower values of m (larger scales). The lowest value of $m = 1$ is not plotted because the variance of this estimate is very large and it appears to suffer from

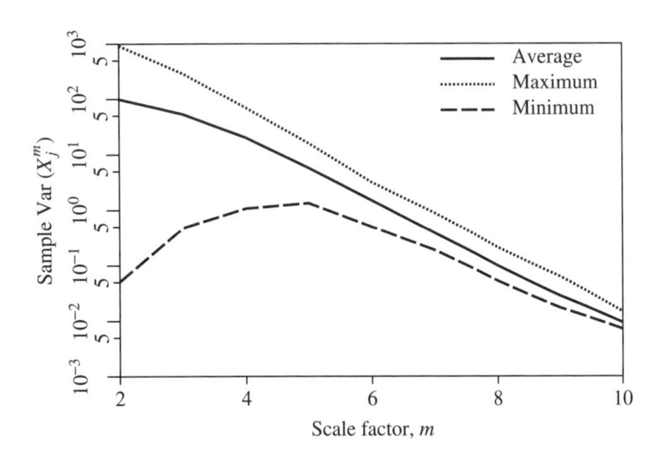

Figure 5.26 Wavelet coefficient variance estimates from a finite-scale process ($\theta = 3$).

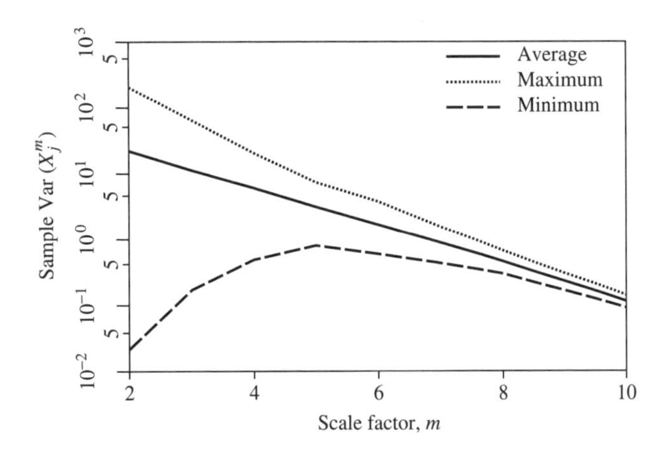

Figure 5.27 Wavelet coefficient variance estimates from a fractal process ($H = 0.95$).

the same bias as estimates of the spectral density function at $\omega = 0$ (as discussed later). On the basis of Figures 5.26 and 5.27, it appears that the wavelet coefficient variance plot may have some potential in identifying an appropriate stochastic model, although the difference in the plots is really quite small. Confident conclusions will require a large data set.

If it turns out that $X(z)$ is fractal and Gaussian, then the coefficients X_j^m are also Gaussian and (largely) uncorrelated as discussed above. This means that a maximum-likelihood estimation can be performed to evaluate the spectral exponent γ by looking at the likelihood of the computed set of coefficients x_j^m.

5.4.1.5 Sample Spectral Density Function The sample spectral density function, referred to here also as the periodogram despite its slightly nonstandard form, is obtained by first computing the Fourier transform of the data,

$$\chi(\omega_j) = \frac{1}{n} \sum_{k=0}^{n-1} X_{k+1} e^{-i\omega_j k} \tag{5.65}$$

for each Fourier frequency $\omega_j = 2\pi j/D, j = 0, 1, \ldots, (n-1)/2$. This is efficiently achieved using the fast Fourier transform. The periodogram is then given by the squared magnitude of the complex Fourier coefficients according to

$$\hat{G}(\omega_j) = \frac{D}{\pi} |\chi(\omega_j)|^2 \tag{5.66}$$

where $D = n\,\Delta z$ (note the slight change in definition for D, which is now equal to the *period* of the sequence). For stationary processes with finite variance, the periodogram estimates as defined here are independent and exponentially distributed with means equal to the true one-sided spectral density $G(\omega_j)$ (see Beran, 1994). Vanmarcke (1984) shows that the periodogram itself has a nonzero correlation length when $D = n\,\Delta z$ is finite, equal to $2\pi/D$. This suggests the presence of serial correlation between periodogram estimators. However, because the periodogram estimates at Fourier frequencies are separated by $2\pi/D$, they are approximately independent, according to the physical interpretation of the correlation length distance. The independence and distribution have also been shown by Yajima (1989) to hold for both fractal and finite-scale processes. Armed with this distribution on the periodogram estimates, one can perform maximum-likelihood estimation as well as (conceptually) hypothesis tests. If the periodogram is smoothed using some sort of smoothing window, as discussed by Priestley (1981), the smoothing may lead to loss of independence between estimates at sequential Fourier frequencies so that likelihood approaches become complicated. In this sense, it is best to smooth the periodogram (which is notoriously rough) by averaging over an ensemble of periodogram estimates taken from a sequence of realizations of the random process, where available.

Note that the periodogram estimate at $\omega = 0$ is *not* a good estimator of $G(0)$ and so it should not be included in the periodogram plot. In fact, the periodogram estimate at $\omega = 0$ is biased with $\mathrm{E}\left[\hat{G}(0)\right] = G(0) + n\mu_X^2/(2\pi)$ (Brockwell and Davis, 1987). Recalling that μ_X is unknown and its estimate is highly variable when strong correlation exists, the estimate $\hat{G}(0)$ should not be trusted. In addition, its distribution is no longer a simple exponential.

Certainly the easiest way to determine whether the data are fractal in nature is to look directly at a plot of the periodogram. Fractal processes have spectral density functions of the form $G(\omega) \propto \omega^{-\gamma}$ for $\gamma > 0$. Thus, $\ln G(\omega) = c - \gamma \ln \omega$ for some constant c, so that a log-log plot of the sample spectral density function of a fractal process will be a straight line with slope $-\gamma$. Figures 5.28 and 5.29 illustrate how the periodogram behaves when averaged over

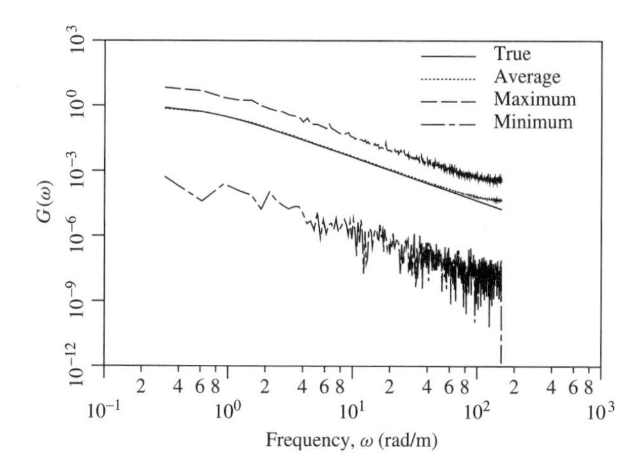

Figure 5.28 Periodogram estimates from a finite-scale process ($\theta = 3$).

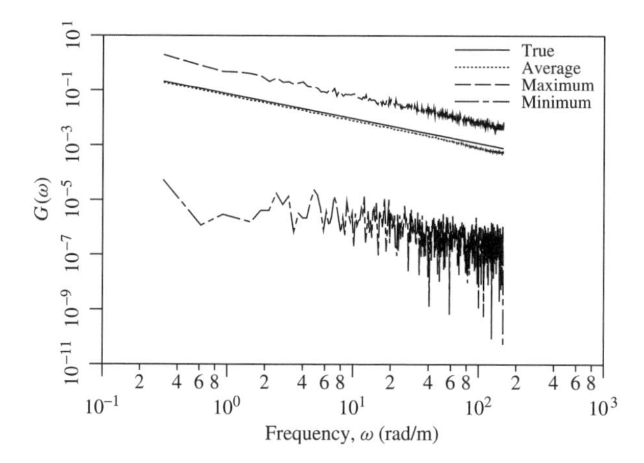

Figure 5.29 Periodogram estimates from a fractal process ($H = 0.95$).

both finite-scale and fractal simulations. The periodogram is a straight line with negative slope in the fractal case and becomes more flattened at the origin in the finite-scale case, as was observed for the wavelet variance plot. Again, the difference is only slight, so that a fairly large data set is required in order to decide on a model with any degree of confidence.

5.4.2 Estimation of First- and Second-Order Statistical Parameters

Upon deciding on whether a finite-scale or fractal model is more appropriate in representing the soil data, the next step is to estimate the pertinent parameters. In the case of the finite-scale model, the parameter of interest is the correlation length θ. For fractal models, the parameter of interest is the spectral exponent γ or, equivalently for $0 \le \gamma < 1$, the self-similarity parameter $H = \frac{1}{2}(\gamma + 1)$.

5.4.2.1 Finite-Scale Model If the process is deemed to be finite scale in nature, then a variety of techniques are available to estimate θ:

1. Directly compute the area under the sample correlation function. This is a nonparametric approach, although it assumes that the scale is finite and that the correlation function is monotonic. The area is usually taken to be the area up to when the function first becomes negative (the correlation length is not well defined for oscillatory correlation functions, other parameters may be more appropriate if this is the case). Also note that correlation estimates lying within the band $\pm 2n^{-1/2}$ are commonly deemed to be not significantly different than zero (see Priestley, 1981, p. 340; Brockwell and Davis, 1987, Chapter 7).
2. Use regression to fit a correlation function to $\hat{\rho}(\tau)$ or a semivariogram to $\hat{V}(\tau)$. For certain assumed correlation or semivariogram functions, this regression may be nonlinear in θ.
3. If the sampling domain D is deemed to be much larger than the correlation length, then the scale can be estimated from the variance function using an iterative technique such as that suggested by Vanmarcke (1984, p. 337).
4. Assuming a joint distribution for $X(t_j)$ with corresponding correlation function model, estimate unknown parameters (μ_X, σ_X^2, and correlation function parameters) using ML in the space domain.
5. Using the established results regarding the joint distribution for periodogram estimates at the set of Fourier frequencies, an assumed spectral density function can be "fit" to the periodogram using ML.

Because of the reasonably high bias in the sample correlation (or covariance) function estimates, even for finite-scale processes, using the sample correlation directly to estimate θ will not be pursued here. The variance function techniques have not been found by the authors to be particularly advantageous over, for example, the ML approaches and are also prone to error due to their high bias in long-scale or fractal cases. Here, the MLE in the space domain will be discussed briefly. The MLE in the frequency domain will be considered later.

It will be assumed that the data are normally distributed or have been transformed from their raw state into something that is at least approximately normally distributed. For example, many soil properties are commonly modeled using the lognormal distribution, often primarily because this distribution is strictly nonnegative. To convert lognormally distributed data to a normal distribution, it is sufficient merely to take the natural logarithm of the data prior to further statistical analysis. It should be noted that the normal model is commonly used for at least two reasons: It is analytically tractable in many ways and it is completely defined through knowledge of the first two moments, namely the mean and covariance structure. Since other distributions commonly require higher moments and since higher moments are generally quite difficult to estimate accurately, particularly in the case of geotechnical samples which are typically limited in size, the use of other distributions is often difficult to justify. The normal assumption can be thought of as a minimum-knowledge assumption which succinctly expresses the first two moments of the random field, where it is hoped that even these can be estimated with some confidence.

Since the data are assumed to be jointly normally distributed, the space domain MLEs are obtained by maximizing the likelihood of observing the spatial data under the assumed joint distribution. The likelihood of observing the sequence of observations $\mathbf{x}^{\mathrm{T}} = \{x_1, x_2, \ldots, x_n\}$ (superscript T denotes the vector or matrix transpose) given the distributional parameters $\boldsymbol{\phi}^{\mathrm{T}} = \{\mu_X, \sigma_X^2, \theta\}$ is

$$L(\mathbf{x}|\boldsymbol{\phi}) = \frac{1}{(2\pi)^{n/2}|\boldsymbol{C}|^{1/2}} \exp\left\{-\tfrac{1}{2}(\mathbf{x}-\boldsymbol{\mu})^{\mathrm{T}}\boldsymbol{C}^{-1}(\mathbf{x}-\boldsymbol{\mu})\right\} \tag{5.67}$$

where $\boldsymbol{C}$ is the covariance matrix between the observations, $\boldsymbol{C}_{ij} = \mathrm{E}\left[(X_i - \mu_i)(X_j - \mu_j)\right]$, $|\boldsymbol{C}|$ is the determinant of $\boldsymbol{C}$, and $\boldsymbol{\mu}$ is the vector of means corresponding to each observation location. In the following, the data are assumed to be modeled by a stationary random field, so that the mean is spatially constant and $\boldsymbol{\mu} = \mu_X \mathbf{1}$ where $\mathbf{1}$ is a vector of 1's. Again, if this assumption is not deemed warranted, a deterministic trend in the mean and variance can be removed from the data prior to the following statistical analysis through the transformation

$$x'(z) = \frac{x(z) - m(z)}{s(z)}$$

where $m(z)$ and $s(z)$ are deterministic spatial trends in the mean and standard deviation, possibly obtained by regression analysis of the data. Recall, however, that this is generally only warranted if the same trends are expected at the target site.

Also due to the stationarity assumption, the covariance matrix can be written in terms of the correlation matrix $\boldsymbol{\rho}$ as

$$\boldsymbol{C} = \sigma_X^2 \boldsymbol{\rho} \tag{5.68}$$

where $\boldsymbol{\rho}$ is a function only of the unknown correlation function parameter θ. If the correlation function has more than one parameter, then θ is treated as a vector of unknown parameters and the ML will generally be found via a gradient or grid search in these parameters. With Eq. 5.68, the likelihood function of Eq. 5.67 can be written as

$$L(\mathbf{x}|\boldsymbol{\phi}) = \frac{1}{(2\pi\sigma_X^2)^{n/2}|\boldsymbol{\rho}|^{1/2}} \exp\left\{-\frac{(\mathbf{x}-\boldsymbol{\mu})^{\mathrm{T}}\boldsymbol{\rho}^{-1}(\mathbf{x}-\boldsymbol{\mu})}{2\sigma_X^2}\right\} \tag{5.69}$$

Since the likelihood function is strictly nonnegative, maximizing $L(\mathbf{x}|\boldsymbol{\phi})$ is equivalent to maximizing its logarithm, which, ignoring constants, is given by

$$\mathcal{L}(\mathbf{x}|\boldsymbol{\phi}) = -\frac{1}{2}\ln\sigma_X^2 - \frac{1}{2}\ln|\boldsymbol{\rho}| - \frac{(\mathbf{x}-\boldsymbol{\mu})^{\mathrm{T}}\boldsymbol{\rho}^{-1}(\mathbf{x}-\boldsymbol{\mu})}{2\sigma_X^2} \tag{5.70}$$

The maximum of Eq. 5.70 can in principle be found by differentiating with respect to each unknown parameter μ_X, σ_X^2, and θ in turn and setting the results to zero. This gives three equations in three unknowns. The partial derivative of $\mathcal{L}$ with respect to μ_X, when set equal to zero, leads to the following estimator for the mean:

$$\hat{\mu}_X = \frac{\mathbf{1}^{\mathrm{T}}\boldsymbol{\rho}^{-1}\mathbf{x}}{\mathbf{1}^{\mathrm{T}}\boldsymbol{\rho}^{-1}\mathbf{1}} \tag{5.71}$$

Since this estimator still involves the unknown correlation matrix, it should be viewed as the value of μ_X which maximizes the likelihood function for a given value of the correlation parameter θ. If the two vectors $\mathbf{r}$ and $\mathbf{s}$ are solutions of the two systems of equations

$$\boldsymbol{\rho}\,\mathbf{r} = \mathbf{x} \tag{5.72a}$$

$$\boldsymbol{\rho}\,\mathbf{s} = \mathbf{1} \tag{5.72b}$$

then the estimator for the mean can be written as

$$\hat{\mu}_X = \frac{\mathbf{1}^{\mathrm{T}}\mathbf{r}}{\mathbf{1}^{\mathrm{T}}\mathbf{s}} \tag{5.73}$$

Note that this estimator is generally not very different from the usual estimator obtained by simply averaging the observations (Beran, 1994). Note also that $\mathbf{1}^T\mathbf{r} = \mathbf{x}^T\mathbf{s}$ so that really only $\mathbf{s}$ needs to be found in order to compute $\hat{\mu}_X$, However, $\mathbf{r}$ will be needed in the following, so it should be found anyhow.

The partial derivative of $\mathcal{L}$ with respect to σ_X^2, when set equal to zero, leads to the following estimator for σ_X^2:

$$\hat{\sigma}_X^2 = \frac{1}{n}(\mathbf{x} - \hat{\mu}_X \mathbf{1})^T \mathbf{r} \tag{5.74}$$

which is also implicitly dependent on the correlation function parameter through $\hat{\mu}_X$ and $\mathbf{r}$.

Thus, both the mean and variance estimators can be expressed in terms of the unknown parameter θ. Using these results, the maximization problem simplifies to finding the maximum of

$$\mathcal{L}(\mathbf{x}|\boldsymbol{\phi}) = -\tfrac{1}{2}\ln \hat{\sigma}_X^2 - \tfrac{1}{2}\ln|\boldsymbol{\rho}| \tag{5.75}$$

where the last term in Eq. 5.70 became simply $n/2$ and was dropped from Eq. 5.75 since it does not affect the location of the maximum.

In principle, Eq. 5.75 need now only be differentiated with respect to θ and the result set to zero to yield the optimal estimate $\hat{\theta}$ and subsequently $\hat{\sigma}_X^2$ and $\hat{\mu}_X$. Unfortunately, this involves differentiating the determinant of the correlation matrix, and closed-form solutions do not always exist. Solution of the MLEs may therefore proceed by iteration as follows:

1. Guess at an initial value for θ.
2. Compute the corresponding correlation matrix $\rho_{ij} = \rho(|z_i - z_j|)$, which in the current equispaced one-dimensional case is both symmetric and Toeplitz (elements along each diagonal are equal).
3. Solve Eqs. 5.72 for vectors $\mathbf{r}$ and $\mathbf{s}$.
4. Solve for the determinant of $\boldsymbol{\rho}$ (because this is often vanishing, it is usually better to compute the log determinant directly to avoid numerical underflow).
5. Compute the mean and variance estimates using Eqs. 5.74 and 5.73.
6. Compute the log-likelihood value $\mathcal{L}$ using Eq. 5.75.
7. Guess at a new value for θ and repeat steps 2–7 until the global maximum value of $\mathcal{L}$ is found.

Guesses for θ can be arrived at simply by stepping discretely through a likely range (and the speed of modern computers make this a reasonable approach), increasing the resolution in the region of located maxima. Alternatively more sophisticated techniques may be employed which look also at the magnitude of the likelihood in previous guesses. One advantage to the brute-force approach of stepping along at predefined increments is that it is more likely to find the global maximum in the event that multiple local maxima are present. With the speed of modern computers, this approach has been found to be acceptably fast for a low number of unknown parameters, less than, say, four, and where bounds on the parameters are approximately known.

For large samples, the correlation matrix $\boldsymbol{\rho}$ can become nearly singular, so that numerical calculations become unstable. In the one-dimensional case, Durbin–Levinson recursion, taking full advantage of the Toeplitz character of $\boldsymbol{\rho}$, yields a faster and more accurate decomposition and allows the direct computation of the log determinant as part of the solution (see, e.g., Marple, 1987).

One finite-scale model which has a particularly simple ML formulation is the jointly normal distribution with the Markov correlation function (see Section 3.6.5),

$$\rho(\tau) = \exp\left\{-\frac{2|\tau|}{\theta}\right\} \tag{5.76}$$

When observations are equispaced, the correlation matrix has a simple closed-form determinant and a tridiagonal inverse,

$$|\boldsymbol{\rho}| = (1 - q^2)^{n-1} \tag{5.77}$$

$$\boldsymbol{\rho}^{-1} = \left(\frac{1}{1-q^2}\right) \times \begin{bmatrix} 1 & -q & 0 & 0 & \cdot & \cdot & \cdot & 0 & 0 \\ -q & 1+q^2 & -q & 0 & \cdot & \cdot & \cdot & 0 & 0 \\ 0 & -q & 1+q^2 & -q & \cdot & \cdot & \cdot & 0 & 0 \\ 0 & 0 & -q & 1+q^2 & \cdot & \cdot & \cdot & 0 & 0 \\ \cdot & \cdot & \cdot & \cdot & \cdot & & & \cdot & \cdot \\ \cdot & \cdot & \cdot & \cdot & & \cdot & & \cdot & \cdot \\ \cdot & \cdot & \cdot & \cdot & & & \cdot & \cdot & \cdot \\ 0 & 0 & 0 & 0 & \cdot & \cdot & \cdot & 1+q^2 & -q \\ 0 & 0 & 0 & 0 & \cdot & \cdot & \cdot & -q & 1 \end{bmatrix} \tag{5.78}$$

where $q = \exp\{-2\ \Delta z/\theta\}$ for observations spaced Δz apart. Using these results, the ML estimation of q reduces to finding the root of the cubic equation,

$$f(q) = b_0 + b_1 q + b_2 q^2 + b_3 q^3 = 0 \tag{5.79}$$

on the interval $q \in (0, 1)$, where

$$b_0 = nR_1 \tag{5.80a}$$

$$b_1 = -(R_0 + nR_0') \tag{5.80b}$$

$$b_2 = -(n-2)R_1 \tag{5.80c}$$

$$b_3 = (n-1)R_0' \tag{5.80d}$$

$$R_0 = \sum_{i=1}^{n}(x_i - \hat{\mu}_X)^2 \tag{5.80e}$$

$$R_0' = R_0 - (x_1 - \hat{\mu}_X)^2 - (x_n - \hat{\mu}_X)^2 \tag{5.80f}$$

$$R_1 = \sum_{i=1}^{n-1} (x_i - \hat{\mu}_X)(x_{i+1} - \hat{\mu}_X) \tag{5.80g}$$

For given q, the corresponding MLEs of μ_X and σ_X^2 are

$$\hat{\mu}_X = \frac{Q_n - q(Q_n + Q'_n) + q^2 Q'_n}{n - 2q(n-1) + q^2(n-2)} \tag{5.81a}$$

$$\hat{\sigma}_X^2 = \frac{R_0 - 2qR_1 + q^2 R'_0}{n(1-q^2)} \tag{5.81b}$$

where

$$Q_n = \sum_{i=1}^{n} x_i \tag{5.82a}$$

$$Q'_n = Q_n - x_1 - x_n \tag{5.82b}$$

According to Anderson (1971), Eq. 5.79 will have one root between 0 and 1 (for positive R_1, which is n times the lag 1 covariance and should be positive under this model) and two roots outside the interval $(-1, 1)$. The root of interest is the one lying between 0 and 1 and it can be efficiently found using Newton–Raphson iterations with starting point $q = R_1/R'_0$ as long as that starting point lies within $(0, 1)$ (if not, use starting point $q = 0.5$).

Since the coefficients of the cubic depend on $\hat{\mu}_X$, which in turn depends on q, the procedure actually involves a global iteration outside the Newton–Raphson root-finding iterations. However, $\hat{\mu}_X$ changes only slightly with changing q, so global convergence is rapid if it is bothered with at all. Once the root q of Eq. 5.79 has been determined, the MLE of the correlation length is determined from

$$\hat{\theta} = -\frac{2\ \Delta z}{\ln q} \tag{5.83}$$

In general, estimates of the variances of the MLEs derived above are also desirable. One of the features of the ML approach is that asymptotic bounds on the covariance matrix $\boldsymbol{C}_{\hat{\theta}}$ between the estimators $\hat{\boldsymbol{\theta}}^{\mathrm{T}} = \{\hat{\mu}_X, \hat{\sigma}_X^2, \hat{\theta}\}$ can be found. This covariance matrix is called the Cramer–Rao bound, and the bound has been shown to hold asymptotically for both finite-scale and fractal processes (Dahlhaus, 1989; Beran, 1994)—in both cases for both n *and* the domain size going to infinity. If we let $\boldsymbol{\theta}^{\mathrm{T}} = \{\mu_X, \sigma_X^2, \theta\}$ be the vector of unknown parameters and define the vector

$$\boldsymbol{\mathcal{L}}' = \frac{\partial}{\partial \theta_j} \mathcal{L} \tag{5.84}$$

where $\mathcal{L}$ is the log-likelihood function defined by Eq. 5.70, then the matrix $\boldsymbol{C}_{\hat{\theta}}$ is given by the inverse of the Fisher information matrix,

$$\boldsymbol{C}_{\hat{\theta}}^{-1} = \mathrm{E}\left[[\boldsymbol{\mathcal{L}}'][\boldsymbol{\mathcal{L}}']^{\mathrm{T}} \right] \tag{5.85}$$

where the superscript T denotes the transpose and the expectation is over all possible values of $\mathbf{X}$ using its joint distribution (see Eq. 5.67) with parameters $\hat{\boldsymbol{\theta}}$. The above expectation is generally computed numerically since it is quite complex analytically. For the Gauss–Markov model the vector $\boldsymbol{\mathcal{L}}'$ is given by

$$\boldsymbol{\mathcal{L}}' = \left\{ \begin{array}{l} \frac{1}{\sigma_X^2} \mathbf{1}^{\mathrm{T}} \boldsymbol{\rho}^{-1} (\mathbf{X} - \boldsymbol{\mu}) \\ \frac{1}{2\sigma_X^4} (\mathbf{X} - \boldsymbol{\mu})^{\mathrm{T}} \boldsymbol{\rho}^{-1} (\mathbf{X} - \boldsymbol{\mu}) - \frac{n}{2\sigma_X^2} \\ \frac{2\ \Delta z (n-1) q^2}{\theta^2 (1-q^2)} - \frac{1}{2\sigma_X^2} (\mathbf{X} - \boldsymbol{\mu})^{\mathrm{T}} \boldsymbol{R}\, (\mathbf{X} - \boldsymbol{\mu}) \end{array} \right\} \tag{5.86}$$

where $\boldsymbol{R}$ is the partial derivative of $\boldsymbol{\rho}^{-1}$ with respect to the correlation length θ.

5.4.2.2 *Fractal Model*

The use of a fractal model is considerably more delicate than that of the finite-scale model. This is because the fractal model, with $G(\omega) \propto \omega^{-\gamma}$, has infinite variance. When $0 \leq \gamma < 1$, the infinite-variance contribution comes from the high frequencies so that the process is stationary but physically unrealizable. Alternatively, when $\gamma > 1$, the infinite variance comes from the low frequencies which yields a nonstationary (fractional Brownian motion) process. In the latter case, the infinite variance basically arises from the gradual meandering of the process over increasingly large distances as one looks over increasingly large scales. While nonstationarity is an interesting mathematical concept, it is not particularly useful or practical in soil characterization. It does, however, emphasize the dependence of the overall soil variation on the size of the region considered. This explicit emphasis on domain size is an important feature of the fractal model.

To render the fractal model physically useful for the case when $0 \leq \gamma < 1$, Mandelbrot and van Ness (1968) introduced a distance δ over which the fractal process is averaged to smooth out the high-frequencies and eliminate the high-frequency (infinite) variance contribution. The resulting correlation function is given by (Section 3.6.7)

$$\rho(\tau) = \frac{1}{2\delta^{2H}} \left[|\tau + \delta|^{2H} - 2|\tau|^{2H} + |\tau - \delta|^{2H} \right] \tag{5.87}$$

Unfortunately, the rather arbitrary nature of δ renders Mandelbrot's model of questionable practical value, particularly from an estimation point of view. If δ is treated as known, then one finds that the parameter H can be estimated to be any value desired simply by manipulating the size of δ. Alternatively, if both δ and H are estimated simultaneously via ML in the space domain (see previous section), then one finds that the likelihood surface has *many* local maxima, making it difficult to find the global maximum. Even when it has been found with reasonable confidence, it is the authors' experience that the global maximum tends to correspond to unreasonably large values of δ corresponding to overaveraging and thus far too smooth a process. Why this is so is yet to be determined.

A better approach to the fractal model is to employ the spectral representation, with $G(\omega) = G_o/\omega^\gamma$, and apply an upper frequency cutoff in the event that $0 \le \hat{\gamma} < 1$ or a lower frequency cutoff in the event that $\hat{\gamma} > 1$. Both of these approaches render the process both stationary and having finite variance. When $\gamma = 1$, the infinite-variance contribution appears at both ends of the spectrum and both an upper and lower cutoff are needed. The appropriate cutoff frequencies should be selected on the basis of the following:

1. A minimum descriptive scale, in the case $0 \le \hat{\gamma} < 1$, below which details of the process are of no interest. For example, if the random process is intended to be soil permeability, then the minimum scale of interest might correspond to the laboratory sample scale d at which permeability tests are carried out. The upper frequency cutoff might then be selected such that this laboratory scale corresponds to, say, one wavelength, $\omega_u = 2\pi/d$.
2. For the case $\hat{\gamma} > 1$, the lower bound cutoff frequency must be selected on the basis of the dimension of the site under consideration. Since the local mean will be almost certainly estimated by collecting some observations at the site, one can eliminate frequencies with wavelengths large compared to the site dimension. This issue is more delicate than that of an upper frequency cutoff discussed above because there is no natural lower frequency bound corresponding to a certain finite scale (whereas there is an upper bound corresponding to a certain finite-sampling resolution). If the frequency bound is made to be too high, then the resulting process may be missing the apparent long-scale trends seen in the original data set. As a tentative recommendation, the authors suggest using a lower cutoff frequency equal to the least nonzero Fourier frequency $\omega_o = 2\pi/D$, where D is the site dimension in the direction of the model.

The parameter γ is perhaps best estimated directly in the frequency domain via ML. There are at least two possible approaches, but here only the wavelet and periodogram MLEs will be discussed.

In the case of the wavelet basis representation, the approach is as follows (from Wornell, 1996). For an observed Gaussian process, $x(t_i)$, $i = 1, 2, \ldots, n$, the wavelet coefficients x_j^m, $m = 1, 2, \ldots, M$, $j = 1, 2, \ldots, 2^{m-1}$, where $n = 2^M$, can be obtained via Eq. 5.62 (preferably using an efficient wavelet decomposition algorithm). Since the input is assumed to be Gaussian, the wavelet coefficients will also be Gaussian. Wornell has shown that they are mean zero and largely independent if $X(z)$ comes from a fractal process so that the likelihood of obtaining the set of coefficients x_j^m is given by

$$L(\mathbf{x}; \gamma) = \prod_{m=1}^{M} \prod_{j=1}^{2^{m-1}} \frac{1}{\sqrt{2\pi\sigma_m^2}} \exp\left\{-\frac{(x_j^m)^2}{2\sigma_m^2}\right\} \tag{5.88}$$

where $\sigma_m^2 = \sigma^2 2^{-\gamma m}$ is the model variance for some unknown intensity σ^2. The log-likelihood function is thus

$$\mathcal{L} = -\frac{1}{2}\sum_{m=1}^{M}(2^{m-1})\ \ln(\sigma_m^2) - \frac{1}{2}\sum_{m=1}^{M}\left(\frac{1}{\sigma_m^2}\right)\sum_{j=1}^{2^{m-1}}(x_j^m)^2 \tag{5.89}$$

(discarding constant terms) which must be maximized with respect to σ^2 and γ. See Wornell (1996, Section 4.3) for details on a relatively efficient algorithm to maximize $\mathcal{L}$.

An alternative approach to estimating γ is via the periodogram. In the authors' opinion, the periodogram approach is somewhat superior to that of the wavelet because the two methods have virtually the same ability to discern between finite-scale and fractal processes and the periodogram has a vast array of available theoretical results dealing with its use and interpretation. While the wavelet basis may warrant further detailed study, its use in geotechnical analysis seems unnecessarily complicated at this time.

Since the periodogram has been shown to consist of approximately independent exponentially distributed estimates at the Fourier frequencies for a wide variety of random processes, including fractal, it leads easily to a MLE for γ. In terms of the one-sided spectral density function $G(\omega)$, the fractal process is defined by

$$G(\omega) = \frac{G_o}{|\omega|^\gamma}, \qquad 0 \le \omega < \infty \tag{5.90}$$

The likelihood of seeing the periodogram estimates $\hat{G}_j = \hat{G}(\omega_j)$, $j = 1, 2, \ldots, k$, where $k = \frac{1}{2}(n-1)$ and $\omega_j = 2\pi j/D$, is just

$$L(\hat{\mathbf{G}}; \boldsymbol{\theta}) = \prod_{j=1}^{k}\left(\frac{\omega_j^\gamma}{G_o}\right) \exp\left\{-\frac{\omega_j^\gamma}{G_o}\hat{G}_j\right\} \tag{5.91}$$

and its logarithm is

$$\mathcal{L} = -k\ \ln G_o + \gamma\sum_{j=1}^{k}\ln\omega_j - \frac{1}{G_o}\sum_{j=1}^{k}\omega_j^\gamma\hat{G}_j \tag{5.92}$$

which is maximized with respect to G_o and γ. The spectral intensity parameter G_o is not necessarily of primary interest since it may have to be adjusted anyhow to ensure that the area under $G(\omega)$ is equal to $\hat{\sigma}_X^2$ after the cutoff frequency discussed above is employed (or, alternatively, the cutoff frequency adjusted for fixed G_o).

Differentiating Eq. 5.92 with respect to G_o and setting the result to zero yield

$$\hat{G}_o = \frac{1}{k}\sum_{j=1}^{k} \hat{G}_j \, \omega_j^{\gamma} \tag{5.93}$$

In turn, differentiating Eq. 5.92 with respect to γ and setting the result to zero lead to the following root-finding problem in γ:

$$\frac{\sum_{j=1}^{k} \hat{G}_j \omega_j^{\gamma} \ \ln \omega_j}{\sum_{j=1}^{k} \hat{G}_j \omega_j^{\gamma}} - \sum_{j=1}^{k} \ln \omega_j = 0 \tag{5.94}$$

For almost all common processes, $0 \le \gamma \le 3$, so that Eq. 5.94 can be solved efficiently via bisection.

The Fisher information matrix, and thus the covariance matrix between the unknown parameters G_o and γ, is especially simple to compute for the periodogram MLE since, asymptotically, the estimator $\hat{G}_o$ is independent of $\hat{\gamma}$. The estimated variances of each estimated parameter can be found from

$$\sigma^2_{\hat{G}_o} \simeq \frac{\hat{G}_o^2}{\left(\sum_{i=1}^{k} \omega_i^{1-\hat{\gamma}} - k\right)^2 - \sum_{i=1}^{k} \omega_i^{2(1-\hat{\gamma})}} \tag{5.95}$$

$$\sigma^2_{\hat{\gamma}} \simeq \frac{1}{\left(\sum_{i=1}^{k} \ln \omega_i - k\right)^2 + k} \tag{5.96}$$

However, it should be pointed out that these variances are only achieved asymptotically as the sample length increases to infinity. In practice, this is not possible, so that estimates of γ and G_o will show much greater variability from sample to sample than suggested by the above bounds. What this means is that the uncertainty in the estimates of γ and G_o are best obtained via simulation or by considering a large number of sampling domains.

5.4.3 Summary

When attempting to identify which of a suite of stochastic models is best suited to represent a soil property, a variety of data transforms are available. Most commonly these are the sample correlation or covariance function, the semivariogram, the sample variance function, the sample wavelet coefficient variance function, and the periodogram. In trying to determine whether the soil property best follows a finite-scale model or a fractal $1/f$-type noise, the periodogram, wavelet variance, and semivariogram plots were found to be the most discriminating. In this sense, the periodogram is perhaps preferable due to the fact that it has been extensively studied, particularly in the context of time series analysis, and because it has a nice physical interpretation relating to the distribution of power to various component frequencies.

It is recognized that these data transforms have been evaluated by averaging over an ensemble of realizations. In many real situations only one or a few data sets will be available. Judging from the min/max curves shown on the plots of Figures 5.20–5.29, it may be difficult to assess the true nature of the process if little or no averaging can be performed. This is, unfortunately, a fundamental issue in all statistical analyses—confidence in the results decreases as the number of independent observations decreases. All that can really be said about the tools discussed above, for example, the periodogram, is that on average it shows a straight line with negative slope for fractal processes and flattens out at the origin for finite -scale processes. Because the periodogram ordinates are exponentially distributed and the ability to distinguish between process types depends on just the first few (low-frequency) ordinates, the use of only a single sample set may not lead to a firm conclusion. For this, special large-scale soil investigations, yielding a large number of soil samples, may be necessary.

The sample correlation or covariance functions are acceptable measures of second-moment structure when the correlation length of the process is small relative to the sampling domain, implying that many of the observations in the sample are effectively independent. However, these sample functions become severely biased when the sample shows strong dependence, preventing them from being useful to discern between finite-scale and fractal-type processes. Since the level of dependence is generally not known a priori, inferences based on the sample covariance and correlation functions are not generally reliable. Likewise, the sample variance function is heavily biased in the presence of strong dependence, rendering its use questionable unless the soil property is known to be finite scale with $\theta \ll D$.

Once a class of stochastic models has been determined using the periodogram, the periodogram can again be used to estimate the parameters of an assumed spectral density function via maximum likelihood. This method can be applied to either finite-scale or fractal processes, requiring only an assumption on the functional form of the spectral density function. The ML approach is preferred over other estimation techniques such as regression because of the many available results dealing with the distribution of ML estimates (see, e.g., Beran, 1994; DeGroot and Baecher, 1993).

If the resulting class of models is deemed to be fractal with $0 \le \hat{\gamma} < 1$, then the Mandelbrot model of Eq. 5.87 can also be fitted using ML in the space domain (preferably with δ taken to be some assumed small averaging length, below which details of the random soil property process are of no interest). For $\gamma > 1$, the fitted spectral density function $G(\omega) = G_o/\omega^{\gamma}$ is still of limited use because it corresponds to infinite variance. It must be truncated at some appropriate upper or lower bound (depending on whether γ is below or above 1.0) to render the model physically useful. The choice of truncation point needs additional investigation, although some rough guidelines were suggested above.

In the finite-scale case, generally only a single parameter, the correlation length, needs to be estimated to provide a completely usable stationary stochastic model. However, indications are that soil properties are fractal in nature, exhibiting significant correlations over very large distances. This proposition is reasonable if one thinks about the formation processes leading to soil deposits—the transport of soil particles by water, ice, or air often takes place over hundreds if not thousands of kilometers. There may, however, still be a place for finite-scale models in soil models. The major strength of the fractal model lies in its emphasis on the relationship between the soil variability and the size of the domain being considered. Once a site has been established, however, there may be little difference between a properly selected finite-scale model and the real fractal model over the finite domain. The relationship between such an "effective" finite-scale model and the true but finite-domain fractal model can be readily established via simulation.

CHAPTER 6

Simulation

6.1 INTRODUCTION

Stochastic problems are often very complicated, requiring overly simplistic assumptions in order to obtain closed-form (or exact) solutions. This is particularly true of many geotechnical problems where we do not even have exact analytical solutions to the deterministic problem. For example, multidimensional seepage problems, settlement under rigid footings, and pile capacity problems often lack exact analytical solutions, and discussion is ongoing about the various approximations which have been developed over the years. Needless to say, when spatial randomness is added to the problem, even the approximate solutions are often unwieldy, if they can be found at all. For example, one of the simpler problems in geotechnical engineering is that of Darcy's law seepage through a clay barrier. If the barrier has a large area, relative to its thickness, and flow is through the thickness, then a one-dimensional seepage model is appropriate. In this case, a closed-form analytical solution to the seepage problem is available. However, if the clay barrier has spatially variable permeability, then the one-dimensional model is no longer appropriate (flow lines avoid low-permeability regions), and even the deterministic problem no longer has a simple closed-form solution. Problems of this type, and most other geotechnical problems, are best tackled through simulation. *Simulation* is the process of producing reasonable replications of the real world in order to study the probabilistic nature of the response to the real world. In particular, simulations allow the investigation of more realistic geotechnical problems, potentially yielding entire probability distributions related to the output quantities of interest. A simulation basically proceeds by the following steps:

1. By taking as many observations from the "real world" as are feasible, the stochastic nature of the real-world problem can be estimated. From the raw data, histogram(s), statistical estimators, and goodness-of-fit tests, a distribution with which to model the problem is decided upon. Pertinent parameters, such as the mean, variance, correlation length, occurrence rate, and so on, may be of interest in characterizing the randomness (see Chapter 5).
2. A random variable or field, following the distribution decided upon in the previous step, is defined.
3. A *realization* of the random variable/field is generated using a *pseudo-random-number generator* or a *random-field generator*.
4. The response of the system to the random input generated in the previous step is evaluated.
5. The above algorithm is repeated from step 3 for as many times as are feasible, recording the responses and/or counting the number of occurrences of a particular response observed along the way.

This process is called *Monte Carlo simulation*, after the famed randomness of the gambling houses of Monte Carlo. The probability of any particular system response can now be estimated by dividing the number of occurrences of that particular system response by the total number of simulations. In fact, if all of the responses are retained in numerical form, then a histogram of the responses forms an estimate of the probability distribution of the system response. Thus, Monte Carlo simulations are a powerful means of obtaining probability distribution estimates for very complex problems. Only the response of the system to a known, *deterministic*, input needs to be computed at each step during the simulation. In addition, the above methodology is easily extended to multiple independent random variables or fields—in this case the distribution of each random variable or field needs to be determined in step 1 and a realization for each generated in step 3. If the multiple random variables or fields are not independent, then the process is slightly more complicated and will be considered in the context of random fields in the second part of this chapter.

Monte Carlo simulations essentially replicate the experimental process and are representative of the experimental results. The accuracy of the representation depends entirely on how accurately the fitted distribution matches the experimental process (e.g., how well the distribution matches the random field of soil properties). The outcomes of the simulations can be treated statistically, just as any set of observations can be treated. As with any statistic, the accuracy of the method generally increases as the number of simulations increases.

In theory, simulation methods can be applied to large and complex systems, and often the rigid idealizations and/or simplifications necessary for analytical solutions can be removed, resulting in more realistic models. However, in practice, Monte Carlo simulations may be limited by constraints of economy and computer capability. Moreover, solutions obtained from simulations may not be amenable to generalization or extrapolation. Therefore, as a general rule, Monte Carlo methods should be used only as a last resort: that is, when and if analytical solution methods are not available or are ineffective (e.g., because of gross idealizations). Monte Carlo solutions are also often a means of verifying or validating approximate analytical solution methods.

One of the main tasks in Monte Carlo simulation is the generation of random numbers having a prescribed probability distribution. Uniformly distributed random-number generation will be studied in Section 6.2. Some techniques for generating random variates from other distributions will be seen in Section 6.3. Techniques of generating random fields are considered starting in Section 6.4, and Section 6.6 elaborates in more detail about Monte Carlo simulation.

6.2 RANDOM-NUMBER GENERATORS

6.2.1 Common Generators

Recall that the $U(0, 1)$ distribution is a continuous uniform distribution on the interval from zero to one. Any one number in the range is just as likely to turn up as any other number in the range. For this reason, the continuous uniform distribution is the simplest of all continuous distributions. While techniques exist to generate random variates from other distributions, they all employ $U(0, 1)$ random variates. Thus, if a good uniform random-number generator can be devised, its output can also be used to generate random numbers from other distributions (e.g., exponential, Poisson, normal, etc.), which can be accomplished by an appropriate *transformation* of the uniformly distributed random numbers.

Most of the best and most commonly used uniform random-number generators are so-called *arithmetic generators*. These employ sequential methods where each number is determined by one or several of its predecessors according to a fixed mathematical formula. If carefully designed, such generators can produce numbers that *appear* to be independent random variates from the $U(0, 1)$ distribution, in that they pass a series of statistical tests (to be discussed shortly). In the sense that sequential numbers are not truly random, being derived from previous numbers in some deterministic fashion, these generators are often called *pseudo-random-number generators*.

A "good" arithmetic uniform random-number generator should possess several properties:

1. The numbers generated should appear to be independent and uniformly distributed.
2. The generator should be fast and not require large amounts of storage.
3. The generator should have the ability to reproduce a given stream of random numbers exactly.
4. The generator should have a very long period.

The ability to reproduce a given stream of random numbers is useful when attempting to compare the responses of two different systems (or designs) to random input. If the input to the two systems is not the same, then their responses will be naturally different, and it is more difficult to determine how the two systems actually differ. Being able to "feed" the two systems the same stream of random numbers allows the system differences to be directly studied.

The most popular arithmetic generators are *linear congruential generators* (LCGs) first introduced by Lehmer (1951). In this method, a sequence of integers $Z_1, Z_2, \ldots$ are defined by the recursive formula

$$Z_i = (aZ_{i-1} + c)(\text{mod } m) \tag{6.1}$$

where mod m means the whole remainder of $aZ_{i-1} + c$ after dividing it by m. For example, (15)(mod 4) is 3 and (17)(mod 4) is 1. In Eq. 6.1 m is the modulus, a is a multiplier, c is an increment, and all three parameters are positive integers. The result of Eq. 6.1 is an integer between 0 and $m - 1$ inclusive. The sequence starts by computing Z_1 using Z_0, where Z_0 is a positive integer *seed* or starting value. Since the resulting Z_i must be a number from 0 to $m - 1$, we can obtain a $[0, 1)$ uniformly distributed U_i by setting $U_i = Z_i/m$. The $[0, 1)$ notation means that U_i can be 0 but cannot be 1. The largest value that U_i can take is $(m - 1)/m$, which can be quite close to 1 if m is large. Also, because Z_i can only take on m different possible values, U_i can only take on m possible values between 0 and 1. Namely, U_i can have values $0, 1/m, 2/m, \ldots, (m - 1)/m$. In order for U_i to appear continuously uniformly distributed on $[0, 1)$, then, m should be selected to be a large number. In addition a, c, and Z_0 should all be less than m.

One sees immediately from Eq. 6.1 that the sequence of Z_i are completely dependent; Z_1 is obtained from Z_0, Z_2 is obtained from Z_1, and so on. For fixed values of a, c, and m, the same sequence of Z_i values will always be produced for the same starting seed, Z_0. Thus, Eq. 6.1 can reproduce a given stream of pseudo–random numbers exactly, so long as the starting seed is known. It also turns out that the sequence of U_i values will appear to be independent and uniformly distributed if the parameters a, c, and m are correctly selected.

One may also notice that if $Z_0 = 3$ produces $Z_1 = 746$, then whenever $Z_{i-1} = 3$, the next generated value will be $Z_i = 746$. This property results in a very undesirable phenomenon called *periodicity* that quite a number of rather common random-number generators suffer from. Suppose that you were unlucky enough to pick a starting seed, say $Z_0 = 83$ on one of these poor random-number generators that just happened to yield remainder 83 when $83a + c$ is divided by m. Then $Z_1 = 83$. In fact, the resulting sequence of "random" numbers will be $\{83, 83, 83, 83, \ldots\}$. We say that this particular stream of random variates has periodicity equal to one.

Why is periodicity to be avoided? To answer this question, let us suppose we are estimating an average system response by simulation. The simulated random inputs $U_1, U_2, \ldots, U_n$ result in system responses $X_1, X_2, \ldots, X_n$. The average system response is then given by

$$\bar{X} = \frac{1}{n}\sum_{i=1}^{n} X_i \tag{6.2}$$

and statistical theory tell us that the standard error on this estimate ($\pm$ one standard deviation) is

$$s_{\bar{X}} = \frac{s}{\sqrt{n}} \tag{6.3}$$

where

$$s^2 = \frac{1}{n-1}\sum_{i=1}^{n}(X_i - \bar{X})^2 \tag{6.4}$$

From this, we see that the standard error (Eq. 6.3) reduces toward zero as n increases, so long as the X_i's are independent. Now, suppose that we set $n = 1{,}000{,}000$ and pick a starting seed $Z_0 = 261$. Suppose further that this particular seed results in Z_i, $i = 1, 2, \ldots 10^6$, being the sequence $\{94, 4832, 325, 94, 4832, 325, \ldots\}$ with periodicity 3. Then, instead of 1,000,000 independent input values, as assumed, we actually only have 3 "independent" values, each repeated 333,333 times. Not only have we wasted a lot of computer time, but our estimate of the average system response might be very much in error – we assume that its standard error is $s/\sqrt{10^6} = 0.001s$, whereas it is actually $s/\sqrt{3} = 0.6s$, 600 times less accurate than we had thought!

Example 6.1 What are the first three random numbers produced by the LCG

$$Z_i = (25Z_{i-1} + 55)(\text{mod } 96)$$

for starting seed $Z_0 = 21$?

SOLUTION Since the modulus is 96, the interval $[0, 1)$ will be subdivided into at most 96 possible random values. Normally, the modulus is taken to be much larger to give a fairly fine resolution on the unit interval. However, with $Z_0 = 21$ we get

$$\begin{aligned}
Z_1 &= [25(21) + 55](\text{mod } 96)\\
&= 580(\text{mod } 96)\\
&= 4\\
Z_2 &= [25(4) + 55](\text{mod } 96)\\
&= 155(\text{mod } 96)\\
&= 59\\
Z_3 &= [25(59) + 55](\text{mod } 96)\\
&= 1530(\text{mod } 96)\\
&= 90
\end{aligned}$$

so that $U_1 = \frac{4}{96} = 0.042$, $U_2 = \frac{59}{96} = 0.615$, and $U_3 = \frac{90}{96} = 0.938$.

The maximum periodicity an LCG such as Eq. 6.1 can have is m, and this will occur only if a, c, and m are selected very carefully. We say that a generator has *full period* if its period is m. A generator which is full period will produce exactly one of each possible value, $\{0, 1, \ldots, m-1\}$, in each cycle. If the generator is good, all of these possible values will appear to occur in random order.

To help us choose the values of m, a, and c so that the generator has full period, the following theorem, proved by Hull and Dobell (1962), is valuable.

Theorem 6.1 The LCG defined by Eq. 6.1 has full period if and only if the following three conditions hold:

(a) The only positive integer that exactly divides both m and c is 1.
(b) If q is a prime number (divisible only by itself and 1) that exactly divides m, then q exactly divides $a - 1$.
(c) If 4 exactly divides m, then 4 exactly divides $a - 1$.

Condition (b) must be true of all prime factors of m. For example, $m = 96$ has two prime factors, 2 and 3, not counting 1. If $a = 25$, then $a - 1 = 24$ is divisible by both 2 and 3, so that condition (b) is satisfied. In fact, it is easily shown that the LCG $Z_i = (25Z_{i-1} + 55)(\text{mod } 96)$ used in the previous example is a full-period generator.

Park and Miller (1988) proposed a "minimal standard" (MS) generator with constants

$$\begin{aligned}
a &= 7^5 = 16{,}807,\\
c &= 0, \quad m = 2^{31} - 1 = 2{,}147{,}483{,}647
\end{aligned}$$

which has a periodicity of $m - 1$ or about 2×10^9. The only requirement is that the seed 0 must never be used.

This form of the LCG, that is, having $c = 0$, is called a *multiplicative* LCG:

$$Z_{i+1} = aZ_i (\text{mod } m) \tag{6.5}$$

which has a small efficiency advantage over the general LCG of Eq. 6.1 since the addition of c is no longer needed. However, most modern CPUs are able to do a vector multiply and add simultaneously, so this efficiency advantage is probably nonexistent. Multiplicative LCGs can no longer be full period because m now exactly divides both m and $c = 0$. However, a careful choice of a and m can lead to a period of $m - 1$, and only zero is excluded from the set of possible Z_i values—in fact, if zero is not excluded from the set of possible results of Eq. 6.5, then the generator will eventually just return zeroes. That is, once $Z_i = 0$ in Eq. 6.5, it remains zero forever. The constants selected by Park and Miller (1988) for the MS generator achieves a period of $m - 1$ and excludes zero. Possible values for U_i using the MS generator are $\{1/m, 2/m, \ldots, (m-1)/m\}$ and so both of the endpoints, 0 and 1, are excluded. Excluding the endpoints is useful for the generation of random variates from those other distributions which involve taking the logarithm of U or $1 - U$ [since $\ln(0) = -\infty$].

When implementing the MS generator on computers using 32-bit integers, the product aZ_i will generally result in an integer overflow. In their RAN0 function, Press et al. (1997) provide a 32-bit integer implementation of the MS generator using a technique developed by Schrage (1979).

One of the main drawbacks to the MS generator is that there is some correlation between successive values. For example, when Z_i is very small, the product aZ_i will still be very small (relative to m). Thus, very small values are always followed by small values. For example, if $Z_i = 1$, then $Z_{i+1} = 16{,}807$, $Z_{i+2} = 282{,}475{,}249$. The corresponding sequence of U_i is 4.7×10^{-10}, 7.8×10^{-6}, and 0.132. Any time that U_i is less than 1×10^{-6}, the next value will be less than 0.0168.

To remove the serial correlation in the MS generator along with this problem of small values following small values, a technique suggested by Bays and Durham and reported by Knuth (1981) is to use two LCGs; one an MS generator and the second to randomly shuffle the output from the first. In this way, U_{i+1} is not returned by the algorithm immediately after U_i but rather at some random time in the future. This effectively removes the problem of serial correlation. In their second edition of *Numerical Recipes*, Press et al. (1997) present a further improvement, due to L'Ecuyer (1988), which involves combining two different pseudorandom sequences, with different periods, as well as applying the random shuffle. The resulting sequence has a period which is the least common multiple of the two periods, which in Press et al.'s implementation is about 2.3×10^{18}. See Press et al.'s RAN2 function, which is what the authors of this book use as their basic random-number generator.

6.2.2 Testing Random-Number Generators

Most computers have a "canned" random-number generator as part of the available software. Before such a generator is actually used in simulation, it is strongly recommended that one identify exactly what kind of generator it is and what its numerical properties are. Typically, you should choose a generator that is identified (and tested) somewhere in the literature as being good (e.g., Press et al., 1997; the random-number generators given in the first edition of *Numerical Recipes* are not recommended by the authors, however). Before using other generators, such as those provided with computer packages (e.g., compilers), they should be subject to (at least) the empirical tests discussed below.

Theoretical Tests The best known theoretical tests are based on the rather upsetting observation by Marsaglia (1968) that LCG "random numbers fall mainly in the planes." That is, if U_1, U_2, ... is a sequence of random numbers generated by an LCG, the overlapping d-tuples $(U_1, U_2, \ldots, U_d)$, $(U_2, U_3, \ldots, U_{d+1})$, ... will all fall on a relatively small number of $(d-1)$-dimensional hyperplanes passing through the d-dimensional unit hypercube $[0, 1]^d$. For example, if $d = 2$, the pairs $(U_1, U_2), (U_2, U_3), \ldots$ will be arranged in lattice fashion along several families of parallel lines going through the unit square. The main problem with this is that it indicates that there will be regions within the hypercube where points will never occur. This can lead to bias or incomplete coverage in a simulation study.

Figure 6.1 illustrates what happens when the pairs (U_i, U_{i+1}) are plotted for the simple LCG of Example 6.1, $Z_{i+1} = (25Z_i + 55)(\text{mod } 96)$ on the left and Press et al.'s RAN2 generator on the right. The planes along which the pairs lie are clearly evident for the simpler LCG, and there are obviously large regions in the unit square that the generated pairs will never occupy. The RAN2 generator, on the other hand, shows much more random and complete coverage of the unit square, which is obviously superior.

Thus, one generator test would be to plot the $d = 2$ pairs of sequential pairs of generated values and look for obvious "holes" in the coverage. For higher values of d, the basic idea is to test the algorithm for gaps in $[0, 1]^d$ that cannot contain any d-tuples. The theory for such a test is difficult. However, if there is evidence of gaps, then the generator being tested exhibits poor behavior, at least in d dimensions. Usually, these tests are applied separately for each dimension from $d = 2$ to as high as $d = 10$.

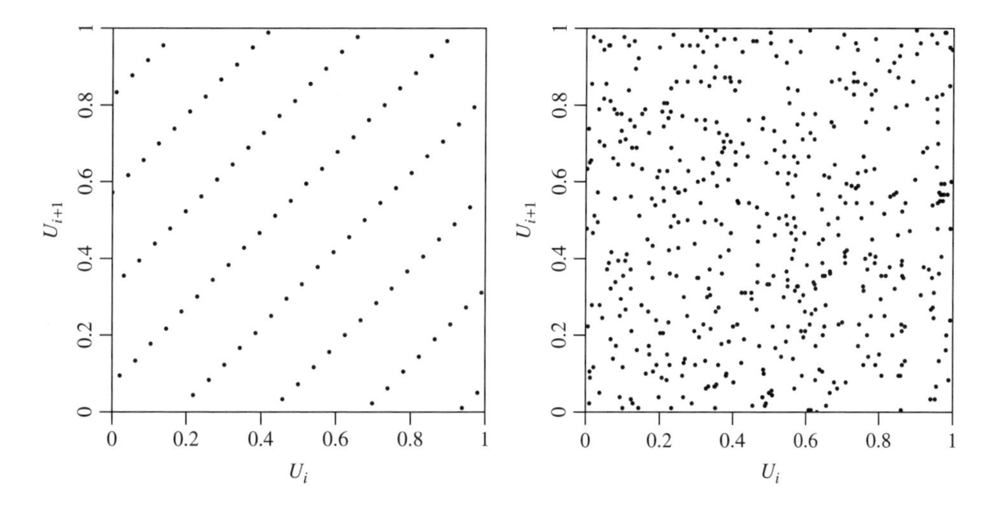

Figure 6.1 Plots of $d = 2$ pairs of generated values using simple LCG of Example 6.1 on left and Press et al.'s RAN2 generator on right.

Empirical Tests Empirically, it can be said that the generator performs adequately if it gives no evidence that generated random variates are not $U(0, 1)$. That is, the following hypotheses are tested:

H_o: $X_1, X_2, \ldots$ are independent random variables uniformly distributed on $(0, 1)$

H_a: $X_1, X_2, \ldots$ are not independent random variables uniformly distributed on $(0, 1)$

and the goodness-of-fit methods discussed in Section 5.2.2 can be applied to complete the test. Since the null hypothesis assumes independence, and this is a desirable feature of the random variates (at least, so far as classical statistical estimates are concerned), this should also be checked.

A direct test of the independence between random variates is the *runs test* which proceeds as follows:

1. We examine our sequence of n U_i's for subsequences in which the U_i's continue to increase (or decrease—we shall concentrate our attention here on *runs up*, which are the increasing subsequences). For example, suppose we generate $U_1, U_2, \ldots, U_{10}$ and get the sequence 0.29, 0.25, 0.09, 0.61, 0.90, 0.20, 0.46, 0.94, 0.13, 0.42. Then our runs up are as follows:

   0.29 is a run up of length 1.
   0.25 is a run up of length 1.
   0.09, 0.61, 0.90 is a run up of length 3.
   0.20, 0.46, 0.94 is a run up of length 3.
   0.13, 0.42 is a run up of length 2.

2. Count the number of runs up of length 1, 2, 3, 4, 5, and 6 or more and define

$$r_i = \begin{cases} \text{number of runs up of length } i & \text{for } i = 1, 2, 3, 4, 5 \\ \text{number of runs up of length} \geq 6 & \text{for } i = 6 \end{cases} \tag{6.6}$$

   For the 10 generated U_i values given above, we have $r_1 = 2$, $r_2 = 1$, $r_3 = 2$, and $r_4 = r_5 = r_6 = 0$.

3. Compute the test statistic

$$R = \frac{1}{n} \sum_{i=1}^{6} \sum_{j=1}^{6} a_{ij}(r_i - nb_i)(r_j - nb_j) \tag{6.7}$$

   where a_{ij} is the (i,j)th element (e.g., ith row, jth column) of the symmetric matrix (Knuth, 1981)

$$\begin{bmatrix} 4{,}529.4 & 9{,}044.9 & 13{,}568 & 18{,}091 & 22{,}615 & 27{,}892 \\ 9{,}044.0 & 18{,}097 & 27{,}139 & 36{,}187 & 45{,}234 & 55{,}789 \\ 13{,}568 & 27{,}139 & 40{,}721 & 54{,}281 & 67{,}852 & 83{,}685 \\ 18{,}091 & 36{,}187 & 54{,}281 & 72{,}414 & 90{,}470 & 111{,}580 \\ 22{,}615 & 45{,}234 & 67{,}852 & 90{,}470 & 113{,}262 & 139{,}476 \\ 27{,}892 & 55{,}789 & 83{,}685 & 111{,}580 & 139{,}476 & 172{,}860 \end{bmatrix}$$

   and

$$\{b_1, b_2, \ldots, b_6\} = \left\{\frac{1}{6}, \frac{5}{24}, \frac{11}{120}, \frac{19}{720}, \frac{29}{5040}, \frac{1}{840}\right\}$$

4. For large n (Knuth recommends $n \geq 4000$), R is approximately chi-square distributed with 6 DOFs. Thus, we would reject the null hypothesis that the U_i's are independent if R exceeds the critical value $\chi^2_{\alpha,6}$, for some assumed significance level α.

One potential disadvantage of empirical tests is that they are only *local*; that is, only that segment of a cycle that was actually used to generate the U_i's for the test is examined. Thus, the tests cannot say anything about how the generator might perform in other segments of the cycle. A big advantage, however, is that the actual random numbers that will be later used can be tested.

Note: Recall how statistical tests work. One would expect that even a "perfect" random-number generator would occasionally produce an "unacceptable" test statistic. In fact, unacceptable results *should* occur with probability α (which is just the type I error). Thus, it can be argued that hand-picking of segments to avoid "bad" ones is in fact a poor idea.

6.3 GENERATING NONUNIFORM RANDOM VARIABLES

6.3.1 Introduction

The basic ingredient needed for all common methods of generating random variates or random processes (which are sequences of random variables) from any distribution is a sequence of $U(0,1)$ random variates. It is thus important that the basic random-number generator be good. This issue was covered in the previous section, and standard "good" generators are readily available.

For most common distributions, efficient and exact generation algorithms exist that have been thoroughly tested and used over the years. Less common distributions may have several alternative algorithms available. For these, there are a number of issues that should be considered before choosing the best algorithm:

1. **Exactness:** Unless there is a significant sacrifice in execution time, methods which reproduce the desired distribution exactly, in the limit as $n \to \infty$, are preferable. When only approximate algorithms are available, those which are accurate over the largest range of parameter values are preferable.
2. **Execution Time:** With modern computers, setup time, storage, and time to generate each variate are not generally a great concern. However, if the number of realizations is to be very large, execution time may be a factor which should be considered.
3. **Simplicity:** Algorithms which are difficult to understand and implement generally involve significant debug time and should be avoided. All other factors being similar, the simplest algorithm is preferable.

Here, the most important general approaches for the generation of random variates from arbitrary distributions will be examined. A few examples will be presented and the relative merits of the various approaches will be discussed.

6.3.2 Methods of Generation

The most common methods used to generate random variates are:

1. Inverse transform
2. Convolution
3. Acceptance–rejection

Of these, the inverse transform and convolution methods are exact, while the acceptance–rejection method is approximate.

6.3.2.1 Inverse Transform Method Consider a continuous random variable X that has cumulative distribution function $F_X(x)$ that is strictly increasing. Most common continous distributions have strictly increasing $F_X(x)$ (e.g., uniform, exponential, Weibull, Rayleigh, normal, lognormal, etc.) with increasing x. This assumption is invoked to ensure that there is only one value of x for each $F_X(x)$. In this case, the inverse transform method generates a random variate from F as follows:

1. Generate $u \sim U(0,1)$.
2. Return $x = F^{-1}(u)$.

Note that $F^{-1}(u)$ will always be defined under the above assumptions since u lies between 0 and 1. Figure 6.2 illustrates the idea graphically. Since a randomly generated value of U, in this case 0.78, always lies between 0 and 1, the cdf plot can be entered on the vertical axis, read across to where it intersects $F(x)$, then read down to obtain the appropriate value of x, in this case 0.8. Repetition of this process results in x being returned in proportion to its density, since more "hits" are obtained where the cdf is the steepest (highest derivative, and, hence, highest density).

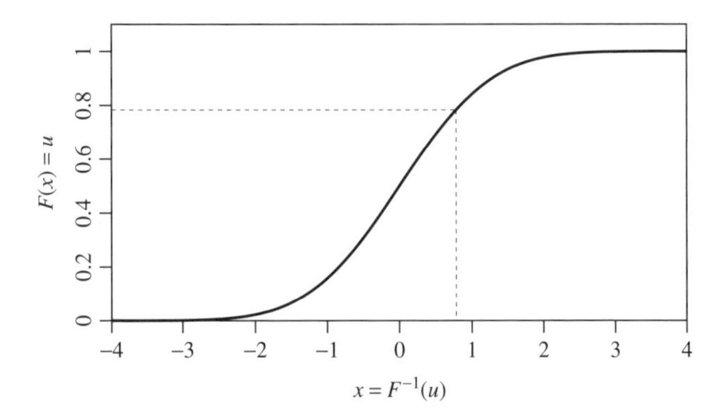

Figure 6.2 Inverse transform random-number generation.

The inverse transform method is the best method when the cumulative of the distribution function for generation can be easily "inverted." This includes a number of common distributions, such as the uniform, exponential, Weibull, and Rayleigh (unfortunately, the normal distribution has no closed-form inverse).

Example 6.2 Suppose that undersea slopes in the Baltic Sea fail at a mean rate of one every 400 years. Suppose also that times between failures are exponentially distributed and independent. Generate randomly two possible times between slope failures from this distribution.

SOLUTION We start by generating randomly two realizations of a uniformly distributed random variable on the interval (0, 1): say $u_1 = 0.27$, and $u_2 = 0.64$. Now, we know that for the exponential distribution,

$$F(x) = 1 - e^{-\lambda x}$$

where $\lambda = 1/400$ in this case. Setting $u = F(x)$ and inverting this relationship gives

$$x = -\frac{\ln(1-u)}{\lambda}$$

Note that since $1 - U$ is distributed identically to U, then this can be simplified to

$$x = -\frac{\ln(u)}{\lambda}$$

Admittedly, this leads to a different set of values in the realization, but the *ensemble* of realizations has the same distribution, and that is all that is important. This formulation may also be slightly more efficient since one operation has been eliminated. However, which form should be used also depends on the nature of the pseudo-random-number generator. Most generators omit either the 0 or the 1, at one of the endpoints of the distribution. Some generators omit both. However, if a generator allows a 0 to occur occasionally, then the form with $\ln(1-u)$ should be used to avoid numerical exceptions [$\ln(0) = -\infty$]. Similarly, if a generator allows 1 to occur occasionally, then $\ln(u)$ should be used. If both can appear, then the algorithm should specifically guard against an error using if-statements.

Using the $\ln(u)$ form gives the first two realizations of interfailure times to be

$$x_1 = -\frac{\ln(u_1)}{\lambda} = -400 \ln(0.27) = 523 \quad \text{years}$$

$$x_2 = -\frac{\ln(u_2)}{\lambda} = -400 \ln(0.64) = 179 \quad \text{years}$$

The inverse transform approach can also be used on *discrete* random variates, but with a slightly modified algorithm:

1. Generate u from the distribution $U(0,1)$.
2. Determine the smallest x_i such that $F(x_i) \geq u$, and return $x = x_i$.

Another way of stating this algorithm is as follows: Since the random variable is discrete, the unit interval can be split up into adjacent subintervals, the first having width equal to $\mathrm{P}[X = x_1]$, the second having width $\mathrm{P}[X = x_2]$, and so on. Then assign x according to whichever of these subintervals contains the generated u. There is a computational issue of how to *look* for the subinterval that contains a given u, and some approaches are better than others. In particular if x_j, $j = 1, 2, \ldots, m$, are equi-likely outcomes, then $i = \text{int}(1.0 + mu)$, where $\text{int}(\cdot)$ means integer part. This also assumes u can never quite equal 1.0, that is, the generator excludes 1.0. If 1.0 is possible, then add 0.999999 instead of 1.0 to mu. Now the discrete realization is $x = x_i$.

Both the continuous and discrete versions of the inverse transform method can be combined, at least formally, to deal with distributions which are *mixed*, that is, having both continuous and discrete components, as well as for continuous distribution functions with flat spots.

Over and above its intuitive appeal, there are three other main advantages to the inverse transform method:

1. It can easily be modified to generate from truncated distributions.
2. It can be modified to generate order statistics (useful in reliability, or lifetime, applications).
3. It facilitates variance–reduction techniques (where portions of the cdf are "polled" more heavily than others, usually in the tails of the distribution, and then resulting statistics corrected to account for the biased polling).

The inverse transform method requires a formula for F^{-1}. However, closed-form expressions for the inverse are not known for some distributions, such as the normal, the lognormal, the gamma, and the beta. For such distributions, numerical methods are required to return the inverse. This is the main disadvantage of the inverse transform method. There are other techniques specifically designed for some of these distributions, which will be discussed in the following. In particular, the gamma distribution is often handled by convolution (see next section), whereas a simple trigonometric transformation can be used to generate normally distributed variates (and further raising e to the power of the normal variate produces a lognormally distributed random variate).

6.3.2.2 Convolution The method of *convolution* can be applied when the random variable of interest can be expressed as a sum of other random variables. This is the case for many important distributions—most notably, recall that the gamma distribution, with integer k, can be expressed as the sum of k exponentially distributed and independent random variables.

For the convolution method, it is assumed that there are i.i.d. random variables $Y_1, Y_2, \ldots, Y_k$ (for fixed k), each with distribution $F(y)$ such that $Y_1 + Y_2 + \cdots + Y_k$ has the same distribution as X. Hence, X can be expressed as

$$X = Y_1 + Y_2 + \cdots + Y_k$$

For the method to work efficiently, it is further assumed that random variates for the Y_j's can be generated more readily than X itself directly (otherwise one would not bother with this approach). The convolution algorithm is then quite intuitive:

1. Generate $Y_1, Y_2, \ldots, Y_k$ i.i.d. each with distribution $F_Y(y)$.
2. Return $X = Y_1 + \cdots + Y_k$.

Note that some other generation method, for example, inverse transform, is required to execute step 1.

6.3.2.3 Acceptance–Rejection The acceptance–rejection method is less direct (and less intuitive) than the two previous methods; however, it can be useful when the other (direct) methods fail or are inefficient.

The acceptance–rejection method requires that a function $t(x)$ be specified that *majorizes* the density $f(x)$; that is, $t(x) \geq f(x)$ for all x. Now, $t(x)$ will not, in general, be a density since

$$c = \int_{-\infty}^{\infty} t(x)\, dx \geq \int_{-\infty}^{\infty} f(x)\, dx = 1$$

but the function $r(x) = t(x)/c$ clearly is a density [since the area under $r(x)$ is now 1.0]. Assume $t(x)$ must be such that $c < \infty$ and, in fact, efficiency is improved as c approaches 1.0. Now, the function t is selected arbitrarily but so that random variables, say Y, having density function $r(y) = t(y)/c$ are easily simulated [e.g., $R(y)$ is easily inverted]. In this case, the general acceptance–rejection algorithm for simulating X having density $f(x)$ is as follows:

1. Generate y having density $r(y)$.
2. Generate $u \sim U(0, 1)$ independently of Y.
3. If $u \leq f(y)/t(y)$, return $x = y$. Otherwise, go back to step 1 and try again.

There are two main things to consider in this algorithm:

1. Finding a suitable function $r(y)$, so that Y is simple to generate, may not be an easy task.
2. The probability of acceptance in step 3 can be shown to be $1/c$. This means that the method becomes very inefficient as c increases. For example, if $c = 100$, then only about 1 in 100 realizations are retained. This, combined with the fact that two random numbers must be generated for each trial random variate, makes the method quite inefficient under many circumstances.

6.3.3 Generating Common Continuous Random Variates

Uniform on (a, b) Solving $u = F(x)$ for x yields, for $0 \leq u \leq 1$,

$$x = F^{-1}(u) = a + (b - a)u$$

and the inverse transform method can be applied as follows:

1. Generate $u \sim U(0, 1)$.
2. Return $x = a + (b - a)u$.

Exponential Solving $u = F(x)$ for x yields, for $0 \leq u \leq 1$,

$$x = F^{-1}(u) = -\frac{\ln(1-u)}{\lambda} \stackrel{d}{=} -\frac{\ln(u)}{\lambda}$$

where $\stackrel{d}{=}$ implies equivalence in distribution. Now the inverse transform method can be applied as follows:

1. Generate $u \sim U(0, 1)$.
2. Return $x = -\ln(u)/\lambda$.

Gamma Considering the particular form of the gamma distribution discussed in Section 1.10.3,

$$f_{T_k}(t) = \frac{\lambda\,(\lambda t)^{k-1}}{(k-1)!} e^{-\lambda t}, \qquad t \geq 0 \tag{6.8}$$

where T_k is the sum of k-independent exponentially distributed random variables, each with mean rate λ. In this case, the generation of random values of T_k proceeds as follows:

1. Generate k-independent exponentially distributed random variables, $X_1, X_2, \ldots, X_k$, using the algorithm given above.
2. Return $T_k = X_1 + X_2 + \cdots + X_k$.

For the more general gamma distribution, where k is not integer, the interested reader is referred to Law and Kelton (2000).

Weibull Solving $u = F(x)$ for a Weibull distribution yields, for $0 \le u \le 1$,

$$x = F^{-1}(u) = \frac{1}{\lambda}[-\ln(1-u)]^{1/\beta} \overset{d}{=} \frac{1}{\lambda}(-\ln u)^{1/\beta}$$

and the inverse transform method can be applied to give

1. Generate $u \sim U(0,1)$.
2. Return $x = (-\ln u)^{1/\beta}/\lambda$.

Normal Since neither the normal distribution function nor its inverse has a simple closed-form expression, one must use a laborious numerical method to apply the inverse transform method. However, the following *radial transformation* method suggested by Box and Muller (1958) is exact, simple to use, and thus much more popular.

If X is normally distributed with mean μ_X and standard deviation σ_X, then realizations of X can be generated as follows:

1. Generate $u_1 \sim U(0,1)$ and $u_2 \sim U(0,1)$.
2. Form $g_1 = \sqrt{-2\ln u_1}\cos(2\pi u_2)$ and $g_2 = \sqrt{-2\ln u_1}\sin(2\pi u_2)$.
3. Form $x_1 = \mu_X + \sigma_X g_1$ and $x_2 = \mu_X + \sigma_X g_2$.
4. Return x_1 on this call to the algorithm and x_2 on the next call (so that the whole algorithm is run only on every second call).

The above method generates realizations of X_1 and X_2, which are *independent* $N(\mu_X, \sigma_X^2)$ random variates.

Example 6.3 Generate two independent realizations of a normally distributed random variable X having mean $\mu_X = 12$ and standard deviation $\sigma_X = 4$.

SOLUTION Using a random-number generator, such as Press et al.'s (1997) RAN2 routine (most spreadsheet programs also include random-number generators), two random numbers uniformly distributed between 0 and 1 are generated. The following are just two possibilities:

$$u_1 = 0.89362, \qquad u_2 = 0.42681$$

First, compute g_1 and g_2, which are realizations of a standard normal random variable (having mean 0 and standard deviation 1):

$$\begin{aligned} g_1 &= \sqrt{-2\ln u_1}\ \cos(2\pi u_2) \\ &= \sqrt{-2\ln(0.89362)}\ \cos[2\pi(0.42681)] = -0.42502 \\ g_2 &= \sqrt{-2\ln u_1}\ \sin(2\pi u_2) \\ &= \sqrt{-2\ln(0.89362)}\ \sin[2\pi(0.42681)] = 0.21050 \end{aligned}$$

Now compute the desired realizations of X:

$$\begin{aligned} x_1 &= \mu_X + \sigma_X g_1 = 12 + 4(-0.42502) = 10.29992 \\ x_2 &= \mu_X + \sigma_X g_1 = 12 + 4(0.21050) = 12.84200 \end{aligned}$$

Lognormal If X is lognormally distributed with mean μ_X and standard deviation σ_X, then $\ln X$ is normally distributed with mean $\mu_{\ln X}$ and standard deviation $\sigma_{\ln X}$. The generation of lognormally distributed X proceeds by first generating a normally distributed $\ln X$ as follows:

1. Generate normally distributed $\ln X$ with mean $\mu_{\ln X}$ and variance $\sigma^2_{\ln X}$ (see previous algorithm).
2. Return $X = e^{\ln X}$.

Empirical Sometimes a theoretical distribution that fits the data cannot be found. In this case, the observed data may be used directly to specify (in some sense) a usable distribution called an empirical distribution.

For continuous random variables, the type of empirical distribution that can be defined depends on whether the actual values of the individual original observations $x_1, x_2, \ldots, x_n$ are available or only the *number* of x_i's that fall into each of several specified intervals. We will consider the case where all of the original data are available.

Using all of the available observations, a continuous, piecewise-linear distribution function F can be defined by first sorting the x_i's from smallest to largest. Let $x_{(i)}$ denote the ith smallest of the x_j's, so that $x_{(1)} \le x_{(2)} \le \cdots \le x_{(n)}$. Then F is defined by

$$F(x) = \begin{cases} 0 & \text{if } x < x_{(1)} \\ \dfrac{i-1}{n-1} + \dfrac{x - x_{(i)}}{(n-1)(x_{(i+1)} - x_{(i)})} & \text{if } x_{(i)} \le x < x_{(i+1)} \\ 1 & \text{if } x_{(n)} \le x \end{cases}$$

for $i = 1, 2, \ldots, n-1$

Since the function $F(x)$ is a series of steps of height 0, $1/(n-1), 2/(n-1), \ldots, (n-2)/(n-1)$, 1 the generation conceptually involves generating $u \sim U(0,1)$, figuring out the index i of the step closest to u, and returning $x_{(i)}$. We will interpolate between the step below u and the step above. The following algorithm results:

1. Generate $u \sim U(0,1)$, let $r = (n-1)u$, and let $i = \text{int}(r) + 1$ where $\text{int}(\cdot)$ means integer part.
2. Return $x = x_{(i)} + (r - i + 1)(x_{(i+1)} - x_{(i)})$.

6.3.3.1 Generating Discrete Random Variates The discrete inverse transform methods may also be applied to generate random variables from the more common discrete probability distributions. The fact that these methods use the inverse transform is not always evident; however, in most cases they do.

Bernoulli If the probability of "success" is p, then:

1. Generate $u \sim U(0, 1)$.
2. If $u \leq p$, return $x = 1$. Otherwise, return $x = 0$.

Discrete Uniform

1. Generate $u \sim U(0, 1)$.
2. Return $x = i + \text{int}\big((j - i + 1)u\big)$, where i and j are the upper and lower discrete bounds and $\text{int}(\cdot)$ means the integer part.

Binomial To generate a binomial distributed random variate with parameters n and p:

1. Generate $y_1, y_2, \ldots, y_n$ independent Bernoulli random variates, each with parameter p.
2. Return $x = y_1 + y_2 + \cdots + y_n$.

Geometric

1. Generate $u \sim U(0, 1)$.
2. Return $x = \text{int}\,(\ln u / \ln(1 - p))$.

Negative Binomial If T_m is the number of trials until the m'th success, and T_m follows a negative binomial distribution with parameter p, then T_m can be written as the sum of m geometric distributed random variables. The generation thus proceeds by convolution:

1. Generate $y_1, y_2, \ldots, y_m$ independent geometric random variates, each with parameter p.
2. Return $T_m = y_1 + y_2 + \cdots + y_m$.

Poisson If N_t follows a Poisson distribution with parameter $r = \lambda t$, then N_t is the number of "arrivals" in time interval of length t, where arrivals arrive with mean rate λ. Since interarrival times are independent and exponentially distributed for a Poisson process, we could proceed by generating a series of k exponentially distributed random variables, each with parameter λ, until their sum just exceeds t. Then the realization of N_t is $k - 1$; that is, $k - 1$ arrivals occurred within time t, the kth arrival was after time t.

An equivalent and more efficient algorithm was derived by Law and Kelton (2000) by essentially working in the logarithm space to be as follows:

1. Let $a = e^{-r}$, $b = 1$, and $i = 0$, where $r = \lambda t$.
2. Generate $u_{i+1} \sim U(0, 1)$ and replace b by bu_{i+1}. If $b < a$, return $N_t = i$.
3. Replace i by $i + 1$ and go back to step 2.

6.3.3.2 Generating Arrival Process Times

Poisson Process Arrival Times The stationary Poisson process with rate $\lambda > 0$ has the property that the interarrival times, say $T_i = t_i - t_{i-1}$ for $i = 1, 2, \ldots$ are independent exponentially distributed random variables with common rate λ. Thus, the t_i's can be generated recursively as follows:

1. Generate $u \sim U(0, 1)$:
2. Return $t_i = t_{i-1} - (\ln\ u)/\lambda$.

Usually, t_0 is taken as zero.

6.3.4 Queueing Process Simulation

Most queueing simulations proceed by generating two streams of random variates: one for the interarrival times and the other for the service times. For common queueing models, these times have exponential distributions with means $1/\lambda$ and $1/\mu$ for the interarrival and service times, respectively, and are readily simulated using the results of the previous section. Using these quantities, the time of arrival of each customer, the time each spends in the queue, and the time being served, can be constructed. The algorithm requires a certain amount of bookkeeping, but it is reasonably straightforward.

Algorithms to simulate the arrivals and departures of customers in M/M/1 and M/M/k queueing systems are given in the following discussion. These algorithms may be used with any distribution of interarrival times and any distribution of service times, not just exponential distributions. Thus, the algorithms easily allow for simulating more general queueing processes than the M/M/k queue.

Both algorithms provide for the statistical estimation of the mean time in the system (W) and the mean time waiting in the queue W_q. They must be modified, however, if one wishes to estimate the fraction of time spent in a particular state.

Simulation of an M/M/1 Queueing Process From Higgins and Keller-McNulty (1995).

Variables

n = customer number
N_{max} = maximum number of arriving customers
$A(n)$ = interarrival time of the nth customer, the time lapse between the $(n - 1)$th and the nth customers

$S(n)$ = time it takes to service the nth customer
$T(n)$ = time that the nth customer arrives
$B(n)$ = time that nth customer begins being served
$D(n)$ = time that the nth customer leaves the system
$W(n)$ = time that the nth customer spends in the system
$W_q(n)$ = time that the nth customer spends waiting in the queue

Initialization

$n = 0$
$T(0) = 0$
$D(0) = 0$
N_{max} = determined from the user
λ = mean arrival rate, determined from the user, λ must be less than μ
μ = mean service rate, determined from the user

Algorithm Repeat the following steps for $n = 1$ to N_{max}:

1. Generate values for $A(n)$ and $S(n)$, that is, simulate exponentially distributed random values for interarrival and service times.
2. Set $T(n) = T(n-1) + A(n)$. That is, the arrival time of the nth customer is the arrival time of the $(n-1)$th customer plus the nth customer's interarrival time.
3. Set $B(n) = \max(D(n-1), T(n))$. That is, if the arrival time of the nth customer occurs before the departure time of customer $(n-1)$, the service time for the nth customer begins when the previous customer departs; otherwise, the nth customer begins service at the time of arrival.
4. Set $D(n) = B(n) + S(n)$. That is, add the service time to the time service begins to determine the time of departure of the nth customer.
5. Set $W_q(n) = B(n) - T(n)$. That is, the time spent in the queue is the difference between the time service begins and the arrival time.
6. Set $W(n) = D(n) - T(n)$. That is, the time spent in the system is the difference between the departure and arrival times.

Statistical Analysis The above algorithm really only supports the statistical estimation of the mean waiting time in the queue:

$$\bar{W}_q = \frac{1}{n}\sum_{n=1}^{N_{max}} W_q(n)$$

and the mean time in the system:

$$\bar{W} = \frac{1}{n}\sum_{n=1}^{N_{max}} W(n)$$

Be sure to run the simulation for a long enough period that the estimates are reasonably accurate (N_{max} must be increased as λ approaches μ). To test if you have selected a large enough N_{max}, try a number of different values (say $N_{max} = 1000$ and $N_{max} = 5000$, and see if the estimated mean times change significantly—if so, you need to choose an even larger N_{max}).

Simulation of an M/M/k Queueing Process From Higgins and Keller-McNulty (1995).

Variables

n = customer number
N_{max} = maximum number of arriving customers
$A(n)$ = interarrival time of the nth customer, the time lapse between the $(n-1)$th and the nth customers
$S(n)$ = time it takes to service the nth customer
$T(n)$ = time that the nth customer arrives
j = server number, $j = 1, 2, \ldots, k$
$F(j)$ = departure time of the customer most recently served by the jth server
J_{min} = server for which $F(j)$ is smallest, that is, the server with the earliest of the most recent departure times (e.g., if all servers are occupied, this is the server that will become free first)
$B(n)$ = time that nth customer begins being served
$D(n)$ = time that the nth customer leaves the system
$W(n)$ = time that the nth customer spends in the system
$W_q(n)$ = time that the nth customer spends waiting in the queue

Initialization

$n = 0$
$T(0) = 0$
$D(0) = 0$
$F(j) = 0$ for each $j = 1, 2, \ldots, k$
k = number of servers, determined from the user
N_{max} = determined from the user
λ = mean arrival rate, determined from the user
μ = mean service rate, determined from the user

Note that λ must be less than $k\mu$.

Algorithm Repeat the following steps for $n = 1$ to N_{max}:

1. Generate values for $A(n)$ and $S(n)$, that is, simulate exponentially distributed random values for interarrival and service times.
2. Set $T(n) = T(n-1) + A(n)$. That is, the arrival time of the nth customer is the arrival time of the $(n-1)$th customer plus the nth customer's interarrival time.
3. Find J_{min}, that is, find the smallest of the $F(j)$ values and set J_{min} to its index (j). In case of a tie, choose

$J_{\min}$ to be the smallest of the tying indices. For example, if $F(2)$ and $F(4)$ are equal and both the smallest out of the other $F(j)$'s, then choose $J_{\min} = 2$.
4. Set $B(n) = \max(F(J_{\min}), T(n))$. That is, if the arrival time of the nth customer occurs before any of the servers are free ($T(n) < F(J_{\min})$, then the service time for the nth customer begins when the ($J_{\min}$)th server becomes free; otherwise, if $T(n) > F(J_{\min})$, then a server is free and the nth customer begins service at the time of arrival.
5. Set $D(n) = B(n) + S(n)$. That is, add the service time to the time service begins to determine the time of departure of the nth customer.
6. Set $F(J_{\min}) = D(n)$. That is, the departure time for the server which handles the nth customer is updated.
7. Set $W_q(n) = B(n) - T(n)$. That is, the time spent in the queue is the difference between the time service begins and the arrival time.
8. Set $W(n) = D(n) - T(n)$. That is, the time spent in the system is the difference between the departure and arrival times.

Statistical Analysis The above algorithm really only supports the statistical estimation of the mean waiting time in the queue:

$$\bar{W}_q = \frac{1}{n} \sum_{n=1}^{N_{\max}} W_q(n)$$

and the mean time in the system:

$$\bar{W} = \frac{1}{n} \sum_{n=1}^{N_{\max}} W(n)$$

Be sure to run the simulation for a long enough period that the estimates are reasonably accurate ($N_{\max}$ must be increased as λ approaches μ). To test if you have selected a large enough $N_{\max}$, try a number of different values (say $N_{\max} = 1000$ and $N_{\max} = 5000$, and see if the estimated mean times change significantly—if so, you need to choose an even larger $N_{\max}$).

6.4 GENERATING RANDOM FIELDS

Random-field models of complex engineering systems having spatially variable properties are becoming increasingly common. This trend is motivated by the widespread acceptance of reliability methods in engineering design and is made possible by the increasing power of personal computers. It is no longer sufficient to base designs on best estimate or mean values alone. Information quantifying uncertainty and variability in the system must also be incorporated to allow the calculation of failure probabilities associated with various limit state criteria. To accomplish this, a probabilistic model is required. In that most engineering systems involve loads and materials spread over some spatial extent, their properties are appropriately represented by random fields. For example, to estimate the failure probability of a highway bridge, a designer may represent both concrete strength and input earthquake ground motion using independent random fields, the latter time varying. Subsequent analysis using a Monte Carlo approach and a dynamic finite-element package would lead to the desired statistics.

In the remainder of this chapter, a number of different algorithms which can be used to produce scalar multidimensional random fields are evaluated in light of their accuracy, efficiency, ease of implementation, and ease of use. Many different random-field generator algorithms are available of which the following are perhaps the most common:

1. Moving-average (MA) methods
2. Covariance matrix decomposition
3. Discrete Fourier transform (DFT) method
4. Fast Fourier transform (FFT) method
5. Turning-bands method (TBM)
6. Local average subdivision (LAS) method

In all of these methods, only the first two moments of the target field may be specified, namely the mean and covariance structure. Since this completely characterizes a Gaussian field, attention will be restricted in the following to such fields. Non-Gaussian fields may be created through nonlinear transformations of Gaussian fields; however, some care must be taken since the mean and covariance structure will also be transformed. In addition, only weakly homogeneous fields, whose first two moments are independent of spatial position, will be considered here.

The FFT, TBM, and LAS methods are typically much more efficient than the first three methods discussed above. However, the gains in efficiency do not always come without some loss in accuracy, as is typical in numerical methods. In the next few sections, implementation strategies for these methods are presented, and the types of errors associated with each method and ways to avoid them will be discussed in some detail. Finally, the methods will be compared and guidelines as to their use suggested.

6.4.1 Moving-Average Method

The *moving-average* technique of simulating random processes is a well-known approach involving the expression of the process as an average of an underlying white noise process. Formally, if $Z(\mathbf{x})$ is the desired zero mean process (a nonzero mean can always be added on later), then

$$Z(\mathbf{x}) = \int_{-\infty}^{\infty} f(\boldsymbol{\xi})\, dW(\mathbf{x} + \boldsymbol{\xi}) \qquad (6.9a)$$

or, equivalently,

$$Z(\mathbf{x}) = \int_{-\infty}^{\infty} f(\boldsymbol{\xi} - \mathbf{x})\, dW(\boldsymbol{\xi}) \tag{6.9b}$$

in which $dW(\boldsymbol{\xi})$ is the incremental white noise process at the location $\boldsymbol{\xi}$ with statistical properties:

$$\begin{aligned} \mathrm{E}\left[dW(\boldsymbol{\xi})\right] &= 0 \\ \mathrm{E}\left[dW(\boldsymbol{\xi})^2\right] &= d\boldsymbol{\xi} \\ \mathrm{E}\left[dW(\boldsymbol{\xi})\, dW(\boldsymbol{\xi}')\right] &= 0 \qquad \text{if } \boldsymbol{\xi} \neq \boldsymbol{\xi}', \end{aligned} \tag{6.10}$$

and $f(\boldsymbol{\xi})$ is a weighting function determined from the desired second-order statistics of $Z(\mathbf{x})$:

$$\begin{aligned} \mathrm{E}\left[Z(\mathbf{x})Z(\mathbf{x}+\boldsymbol{\tau})\right] &= \int_{-\infty}^{\infty}\int_{-\infty}^{\infty} f(\boldsymbol{\xi} - \mathbf{x})f(\boldsymbol{\xi}' - \mathbf{x} - \boldsymbol{\tau}) \\ &\quad \times \mathrm{E}\left[dW(\boldsymbol{\xi})\, dW(\boldsymbol{\xi}')\right] \\ &= \int_{-\infty}^{\infty} f(\boldsymbol{\xi} - \mathbf{x})f(\boldsymbol{\xi} - \mathbf{x} - \boldsymbol{\tau})\, d\boldsymbol{\xi} \end{aligned} \tag{6.11}$$

If $Z(\mathbf{x})$ is homogeneous, then the dependence on $\mathbf{x}$ disappears, and Eq. 6.11 can be written in terms of the covariance function (note by Eq. 6.10 that $\mathrm{E}\left[Z(\mathbf{x})\right] = 0$)

$$C(\boldsymbol{\tau}) = \int_{-\infty}^{\infty} f(\boldsymbol{\xi})f(\boldsymbol{\xi} - \boldsymbol{\tau})\, d\boldsymbol{\xi} \tag{6.12}$$

Defining the Fourier transform pair corresponding to $f(\boldsymbol{\xi})$ in n dimensions to be

$$F(\boldsymbol{\omega}) = \frac{1}{(2\pi)^n}\int_{-\infty}^{\infty} f(\boldsymbol{\xi})e^{-i\boldsymbol{\omega}\cdot\boldsymbol{\xi}}\, d\boldsymbol{\xi} \tag{6.13a}$$

$$f(\boldsymbol{\xi}) = \int_{-\infty}^{\infty} F(\boldsymbol{\omega})e^{i\boldsymbol{\omega}\cdot\boldsymbol{\xi}}\, d\boldsymbol{\omega} \tag{6.13b}$$

then by the convolution theorem Eq. 6.12 can be expressed as

$$C(\boldsymbol{\tau}) = (2\pi)^n \int_{-\infty}^{\infty} F(\boldsymbol{\omega})F(-\boldsymbol{\omega})\, e^{-i\boldsymbol{\omega}\cdot\boldsymbol{\tau}}\, d\boldsymbol{\omega} \tag{6.14}$$

from which a solution can be obtained from the Fourier transform of $C(\boldsymbol{\tau})$

$$F(\boldsymbol{\omega})F(-\boldsymbol{\omega}) = \frac{1}{(2\pi)^{2n}}\int_{-\infty}^{\infty} C(\boldsymbol{\tau})\, e^{-i\boldsymbol{\omega}\cdot\boldsymbol{\tau}}\, d\boldsymbol{\tau} \tag{6.15}$$

Note that the symmetry in the left-hand side of Eq. 6.15 comes about due to the symmetry $C(\boldsymbol{\tau}) = C(-\boldsymbol{\tau})$. It is still necessary to assume something about the relationship between $F(\boldsymbol{\omega})$ and $F(-\boldsymbol{\omega})$ in order to arrive at a final solution through the inverse transform. Usually, the function $F(\boldsymbol{\omega})$ is assumed to be either even or odd.

Weighting functions corresponding to several common one-dimensional covariance functions have been determined by a number of authors, notably Journel and Huijbregts (1978) and Mantoglou and Wilson (1981). In higher dimensions, the calculation of weighting functions becomes quite complex and is often done numerically using FFTs. The nonuniqueness of the weighting function and the difficulty in finding it, particularly in higher dimensions, renders this method of questionable value to the user who wishes to be able to handle arbitrary covariance functions.

Leaving this issue for the moment, the implementation of the MA method is itself a rather delicate problem. For a discrete process in one dimension, Eq. 6.9a can be written as

$$Z_i = \sum_{j=-\infty}^{\infty} f_j\, W_{i,j} \tag{6.16}$$

where $W_{i,j}$ is a discrete white noise process taken to have zero mean and unit variance. To implement this in practice, the sum must be restricted to some range p, usually chosen such that $f_{\pm p}$ is negligible:

$$Z_i = \sum_{j=-p}^{p} f_j\, W_{i,j} \tag{6.17}$$

The next concern is how to discretize the underlying white noise process. If Δx is the increment of the physical process such that $Z_i = Z((i-1)\Delta x)$ and Δu is the incremental distance between points of the underlying white noise process, such that

$$W_{i,j} = W((i-1)\Delta x + j\,\Delta u) \tag{6.18}$$

then $f_j = f(j\,\Delta u)$ and Δu should be chosen such that the quotient $r = \Delta x / \Delta u$ is an integer for simplicity. Figure 6.3 illustrates the relationship between Z_i and the discrete white noise process. For finite Δu, the discrete approximation (Eq. 6.17) will introduce some error into the estimated covariance of the realization. This error can often be removed through a multiplicative correction factor, as shown by Journel and Huijbregts (1978), but in general is reduced by taking Δu as small as practically possible (and thus p as large as possible).

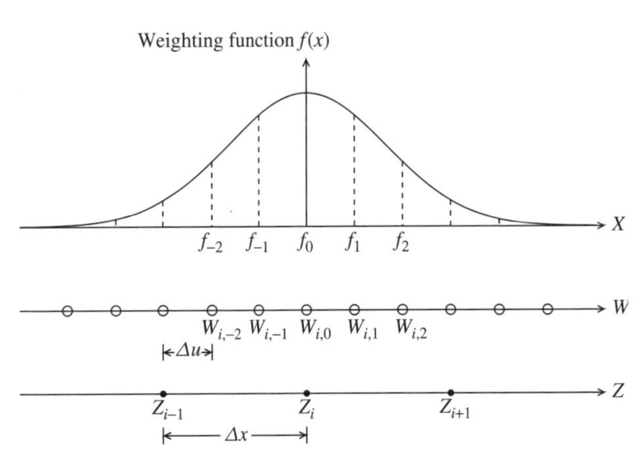

Figure 6.3 Moving-average process in one dimension.

Once the discretization of the underlying white noise process and the range p have been determined, the implementation of Eq. 6.17 in one dimension is quite straightforward and usually quite efficient for reasonable values of p. In higher dimensions, the method rapidly becomes cumbersome. Figure 6.4 shows a typical portion of a two-dimensional discrete process Z_{ij}, marked by $\times$'s, and the underlying white noise field, marked by dots. The entire figure represents the upper right corner of a two-dimensional field. The process Z_{ij} is now formed by the double summation

$$Z_{ij} = \sum_{k=-p_1}^{p_1} \sum_{\ell=-p_2}^{p_2} f_{k\ell}\, W_{i,j,k,\ell} \tag{6.19}$$

where $f_{k\ell}$ is the two-dimensional weighting function and $W_{i,j,k,\ell}$ is the discrete white noise process centered at the same position as Z_{ij}. The i and j subscripts on W are for bookkeeping purposes so that the sum is performed over a centered neighborhood of discrete white noise values.

In the typical example illustrated in Figure 6.4, the discretization of the white noise process is such that $r = \Delta u/\Delta x = 3$, and a relatively short correlation length was used so that $p = 6$. This means that if a $K_1 \times K_2$ field is to be simulated, the total number of white noise realizations to be generated must be

$$N_W = \left[1 + 2p_1 + r_1(K_1 - 1)\right]\left[1 + 2p_2 + r_2(K_2 - 1)\right] \tag{6.20}$$

or about $(rK)^2$ for a square field. This can be contrasted immediately with the FFT approach which requires the

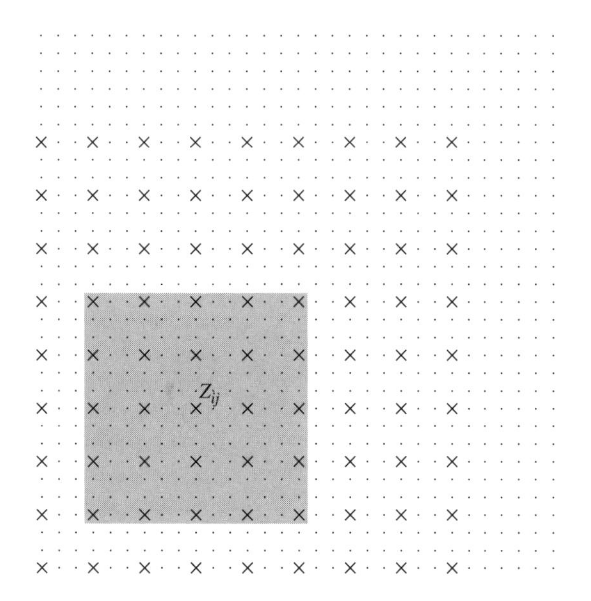

Figure 6.4 Two-dimensional MA process; Z_{ij} is formed by summing the contributions from the underlying white noise process in the shaded region.

generation of about $\frac{1}{2}K^2$ random values for a quadrant symmetric process (note that the factor of one-half is a consequence of the periodicity of the generated field). When $r = 3$, some 18 times as many white noise realizations must be generated for the MA algorithm as for the FFT method. Also the construction of each field point requires a total of $(2p + 1)^2$ additions and multiplications which, for the not unreasonable example given above, is $13^2 = 169$. This means that the entire field will be generated using $K^2(2p + 1)^2$ or about 11 million additions and multiplications for a 200×200 field. Again this can be contrasted to the two-dimensional FFT method (radix-2, row–column algorithm) which requires some $4K^2 \log_2 K$ or about 2 million multiply–adds. In most cases, the moving-average approach in two dimensions was found to run at least 10 times slower than the FFT approach. In three dimensions, the MA method used to generate a $64 \times 64 \times 64$ field with $p = 6$ was estimated to run over 100 times slower than the corresponding FFT approach. For this reason, and since the weighting function is generally difficult to find, the moving-average method as a general method of producing realizations of multidimensional random fields is only useful when the MA representation is particularly desired.

It can be noted in passing that the two-dimensional autoregressive MA (ARMA) model suggested by Naganum et al. (1987) requires about 50–150 multiply–adds (depending on the type of covariance structure modeled) for each field point. This is about 2–6 times slower than the FFT approach. While this is quite competitive for certain covariance functions, the corresponding run speeds for three-dimensional processes are estimated to be 15–80 times slower than the FFT approach depending on the choice of parameters p and r. Also, in a sequence of two studies, Mignolet and Spanos (1992) and Spanos and Mignolet (1992) discuss in considerable detail the MA, autoregressive (AR), and ARMA approaches to simulating two-dimensional random fields. In their examples, they obtain accurate results at the expense of running about 10 or more times slower than the fastest of the methods to be considered later in this chapter.

6.4.2 Covariance Matrix Decomposition

Covariance matrix decomposition is a direct method of producing a homogeneous random field with prescribed covariance structure $C(\mathbf{x}_i - \mathbf{x}_j) = C(\boldsymbol{\tau}_{ij})$, where $\mathbf{x}_i$, $i = 1, 2, \ldots, n$, are discrete points in the field and $\boldsymbol{\tau}_{ij}$ is the lag vector between the points $\mathbf{x}_i$ and $\mathbf{x}_j$. If $\boldsymbol{C}$ is a positive-definite covariance matrix with elements $C_{ij} = C(\boldsymbol{\tau}_{ij})$, then a mean zero discrete process $Z_i = Z(\mathbf{x}_i)$ can be produced (using vector notation) according to

$$\mathbf{Z} = \boldsymbol{L}\mathbf{U} \tag{6.21}$$

where $\boldsymbol{L}$ is a lower triangular matrix satisfying $\boldsymbol{L}\boldsymbol{L}^{\mathrm{T}} = \boldsymbol{C}$ (typically obtained using Cholesky decomposition), and $\mathbf{U}$ is a vector of n-independent mean zero, unit-variance Gaussian random variables. Although appealing in its simplicity and accuracy, this method is only useful for small fields. In two dimensions, the covariance matrix of a 128 × 128 field would be of size 16,384 × 16,384, and the Cholesky decomposition of such a matrix would be both time consuming and prone to considerable round-off error (particularly since covariance matrices are often poorly conditioned and easily become numerically singular).

6.4.3 Discrete Fourier Transform Method

The *discrete Fourier transform* method is based on the spectral representation of homogeneous mean square continuous random fields $Z(\mathbf{x})$, which can be expressed as (Yaglom, 1962)

$$Z(\mathbf{x}) = \int_{-\infty}^{\infty} e^{i\mathbf{x}\cdot\boldsymbol{\omega}}\, W(d\boldsymbol{\omega}) \tag{6.22}$$

where $W(d\boldsymbol{\omega})$ is an interval white noise process with mean zero and variance $S(\boldsymbol{\omega})\, d\boldsymbol{\omega}$. This representation is in terms of the physically meaningful spectral density function $S(\boldsymbol{\omega})$, and so is intuitively attractive. In practice, the n-dimensional integral becomes an n-dimensional sum which is evaluated separately at each point $\mathbf{x}$. Although potentially accurate, the method is computationally slow for reasonable field sizes and typical spectral density functions—the DFT is generally about as efficient as the MA discussed above. Its major advantage over the MA approach is that the spectral density function is estimated in practice using standard techniques.

In n dimensions, for real $Z(\mathbf{x})$, the DFT can be written as

$$Z(\mathbf{x}) = \sum_{k_1=-N_1}^{N_1} \sum_{k_2=-N_2}^{N_2} \cdots \sum_{k_n=-N_n}^{N_n} C_{k_1k_2\ldots k_n} \times \cos\left(\omega_{k_1}x_1 + \omega_{k_2}x_2 + \cdots + \omega_{k_n}x_n + \Phi_{k_1k_2\ldots k_n}\right)$$

where $\Phi_{k_1k_2\ldots k_n}$ is a random phase angle uniformly distributed on $[0, 2\pi]$, and $C_{k_1k_2\ldots k_n}$ is a random amplitude having Rayleigh distribution if Z is Gaussian. An alternative way of writing the DFT is

$$Z(\mathbf{x}) = \sum_{k_1=-N_1}^{N_1} \sum_{k_2=-N_2}^{N_2} \cdots \sum_{k_n=-N_n}^{N_n} A_{k_1k_2\ldots k_n} \times \cos(\omega_{k_1}x_1 + \omega_{k_2}x_2 + \cdots + \omega_{k_n}x_n) + B_{k_1k_2\ldots k_n} \sin(\omega_{k_1}x_1 + \omega_{k_2}x_2 + \cdots + \omega_{k_n}x_n)$$

where, for a stationary normally distributed $Z(\mathbf{x})$, the A and B coefficients are mutually independent and normally distributed with zero means and variances given by

$$\mathrm{E}\left[A_{k_1k_2\ldots k_n}^2\right] = \mathrm{E}\left[B_{k_1k_2\ldots k_n}^2\right] = S(\boldsymbol{\omega}_\mathbf{k})\ \Delta\boldsymbol{\omega}$$

In this equation, $\boldsymbol{\omega}_\mathbf{k} = \{\omega_{k_1}, \omega_{k_2}, \ldots, \omega_{k_n}\}$, and $S(\boldsymbol{\omega}_\mathbf{k})\ \Delta\boldsymbol{\omega}$ is the area under the spectral density function in an incremental region centered on $\boldsymbol{\omega}_\mathbf{k}$.

As mentioned above, the sum is composed of $(2N+1)^n$ terms (if $N_1 = N_2 = \cdots = N$), where $2N+1$ is the number of discrete frequencies taken in each dimension. Depending on the shape of the spectral density function, N might easily be of the order of 100, so that in three dimensions roughly 8 million terms must be summed for each spatial position desired in the generated field (thus, in three dimensions, a 20 × 20 × 20 random field would involve roughly 128 billion evaluations of sin or cosine).

This approach is really only computationally practical in one dimension where the DFT reduces to

$$Z(x) = \sum_{k=-N}^{N} A_k \cos(\omega_k x) + B_k \sin(\omega_k x)$$

where

$$\mathrm{E}[A_k] = \mathrm{E}[B_k] = 0$$

$$\mathrm{E}\left[A_k^2\right] = \mathrm{E}\left[B_k^2\right] = S(\omega_k)\ \Delta\omega$$

and where the A and B coefficients are mutually independent of all other A's and B's. If the symmetry in the spectral density function is taken advantage of, namely that $S(\omega) = S(-\omega)$, then the sum can be written

$$Z(x) = \sum_{k=0}^{N} A_k \cos(\omega_k x) + B_k \sin(\omega_k x) \tag{6.23}$$

where now the variances of the A and B coefficients are expressed in terms of the one-sided spectral density function

$$\mathrm{E}\left[A_k^2\right] = \mathrm{E}\left[B_k^2\right] = G(\omega_k)\Delta\omega_k \tag{6.24}$$

and where $\Delta\omega_0 = \frac{1}{2}(\omega_1 - \omega_0)$ and $\Delta\omega_k = \frac{1}{2}(\omega_{k+1} - \omega_{k-1})$.

Simulation proceeds as follows:

1. Decide on how to discretize the spectral density (i.e., on N and $\Delta\omega$).
2. Generate mean zero, normally distributed, realizations of A_k and B_k for $k = 0, 1, \ldots, N$ each having variance given by Eq. 6.24.
3. For each value of x desired in the final random process, compute the sum given by Eq. 6.23.

6.4.4 Fast Fourier Transform Method

If both space and frequency are discretized into a series of equispaced points, then the *fast Fourier transform* method developed by Cooley and Tukey (1965) can be used to compute Eq. 6.22. The FFT is much more computationally efficient than the DFT. For example, in one dimension the DFT requires N^2 operations, whereas the FFT requires only $N \log_2 N$ operations. If $N = 2^{15} = 32{,}768$, then the FFT will be approximately 2000 times faster than the DFT. For the purposes of this development, only the one-dimensional case will be considered and multidimensional results will be stated subsequently. For real and discrete $Z(x_j)$, $j = 1, 2, \ldots, N$, Eq. 6.22 becomes

$$\begin{aligned} Z(x_j) &= \int_{-\pi}^{\pi} e^{ix_j\omega}\, W(d\omega) \\ &= \lim_{K\to\infty} \sum_{k=-K}^{K} e^{ix_j\omega_k}\, W(\Delta\omega_k) \\ &= \lim_{K\to\infty} \sum_{k=-K}^{K} \Big\{ A(\Delta\omega_k)\cos(x_j\omega_k) \\ &\quad + B(\Delta\omega_k)\, \sin(x_j\omega_k) \Big\} \end{aligned} \tag{6.25}$$

where $\omega_k = k\pi/K$, $\Delta\omega_k$ is an interval of length π/K centered at ω_k, and the last step in Eq. 6.25 follows from the fact that Z is real. The functions $A(\Delta\omega_k)$ and $B(\Delta\omega_k)$ are i.i.d. random interval functions with mean zero and $\mathrm{E}[A(\Delta\omega_k)A(\Delta\omega_m)] = \mathrm{E}[B(\Delta\omega_k)B(\Delta\omega_m)] = 0$ for all $k \neq m$ in the limit as $\Delta\omega \to 0$. At this point, the simulation involves generating realizations of $A_k = A(\Delta\omega_k)$ and $B_k = B(\Delta\omega_k)$ and evaluating Eq. 6.25. Since the process is real, $S(\omega) = S(-\omega)$, and the variances of A_k and B_k can be expressed in terms of the one-sided spectral density function $G(\omega) = 2S(\omega)$, $\omega \geq 0$. This means that the sum in Eq. 6.25 can have lower bound $k = 0$. Note that an equivalent way of writing Eq. 6.25 is

$$Z(x_j) = \sum_{k=0}^{K} C_k \cos(x_j\omega_k + \Phi_k) \tag{6.26}$$

where Φ_k is a random phase angle uniformly distributed on $[0, 2\pi]$ and C_k follows a Rayleigh distribution. Shinozuka and Jan (1972) take $C_k = \sqrt{G(\omega_k)\ \Delta\omega}$ to be deterministic, an approach not followed here since it gives an upper bound on Z over the space of outcomes of $Z \leq \sum_{k=0}^{K} \sqrt{G(\omega_k)\ \Delta\omega}$, which may be an unrealistic restriction, particularly in reliability calculations which could very well depend on extremes.

Next, the process $Z_j = Z(x_j)$ is assumed to be periodic, $Z_j = Z_{K+j}$, with the same number of spatial and frequency discretization points ($N = K$). As will be shown later, the periodicity assumption leads to a symmetric covariance structure which is perhaps the major disadvantage to the DFT and FFT approaches. If the physical length of the one-dimensional process under consideration is D and the space and frequency domains are discretized according to

$$x_j = j\,\Delta x = \frac{jD}{K-1} \tag{6.27}$$

$$\omega_j = j\,\Delta\omega = \frac{2\pi j(K-1)}{KD} \tag{6.28}$$

for $j = 0, 1, \ldots, K-1$, then the Fourier transform

$$Z_j = \sum_{k=0}^{K} \mathcal{X}_k\, e^{i(2\pi jk/K)} \tag{6.29}$$

can be evaluated using the FFT algorithm. The Fourier coefficients, $\mathcal{X}_k = A_k - iB_k$, have the following symmetries due to the fact that Z is real:

$$A_k = \frac{1}{K}\sum_{j=0}^{K-1} Z_j \cos 2\pi\frac{jk}{K} = A_{K-k} \tag{6.30}$$

$$B_k = \frac{1}{K}\sum_{j=0}^{K-1} Z_j \sin 2\pi\frac{jk}{K} = -B_{K-k} \tag{6.31}$$

which means that A_k and B_k need only be generated randomly for $k = 0, 1, \ldots, K/2$ and that $B_0 = B_{K/2} = 0$. Note that if the coefficients at $K - k$ are produced independently of the coefficients at k, the resulting field will display aliasing (see Section 3.3.3). Thus, there is no advantage to taking Z to be complex, generating all the Fourier coefficients randomly, and attempting to produce two independent fields simultaneously (the real and imaginary parts), or in just ignoring the imaginary part.

As far as the simulation is concerned, all that remains is to specify the statistics of A_k and B_k so that they can be generated randomly. If Z is a Gaussian mean zero process, then so are A_k and B_k. The variance of A_k can be computed in a consistent fashion by evaluating $\mathrm{E}\left[A_k^2\right]$ using Eq. 6.30:

$$\mathrm{E}\left[A_k^2\right] = \frac{1}{K^2}\sum_{j=0}^{K-1}\sum_{\ell=0}^{K-1} \mathrm{E}\left[Z_j\, Z_\ell\right] \cos 2\pi\frac{jk}{K} \cos 2\pi\frac{lk}{K} \tag{6.32}$$

This result suggests using the covariance function directly to evaluate the variance of A_k; however, the implementation is complex and no particular advantage in accuracy is attained. A simpler approach involves the discrete approximation to the Wiener–Khinchine relationship:

$$\mathrm{E}\left[Z_j\, Z_\ell\right] \simeq \Delta\omega \sum_{m=0}^{K-1} G(\omega_m) \cos 2\pi\frac{m(j-l)}{K} \tag{6.33}$$

which when substituted into Eq. 6.32 leads to

$$\mathrm{E}\left[A_k^2\right] = \frac{\Delta\omega}{K^2}\sum_{j=0}^{K-1}\sum_{\ell=0}^{K-1}\sum_{m=0}^{K-1} G(\omega_m)\ \cos 2\pi\frac{m(j-\ell)}{K} C_{kj}C_{k\ell}$$

$$= \frac{\Delta\omega}{K^2}\sum_{m=0}^{K-1} G(\omega_m)\sum_{j=0}^{K-1} C_{mj}C_{kj}\sum_{\ell=0}^{K-1} C_{m\ell}C_{k\ell}$$

$$+\frac{\Delta\omega}{K^2}\sum_{m=0}^{K-1} G(\omega_m)\sum_{j=0}^{K-1} S_{mj}C_{kj}\sum_{\ell=0}^{K-1} S_{m\ell}C_{k\ell}, \quad (6.34)$$

where $C_{kj} = \cos 2\pi(kj/K)$ and $S_{kj} = \sin 2\pi(kj/K)$.

To reduce Eq. 6.34 further, use is made of the following two identities:

1. $\displaystyle\sum_{k=0}^{K-1} \sin 2\pi\frac{mk}{K}\ \cos 2\pi\frac{jk}{K} = 0$
2. $\displaystyle\sum_{k=0}^{K-1} \cos 2\pi\frac{mk}{K}\ \cos 2\pi\frac{jk}{K} = \begin{cases} 0 & \text{if } m \neq j \\ \frac{1}{2}K & \text{if } m = j,\ K-j \\ K & \text{if } m = j = 0, \text{ or } \frac{1}{2}K \end{cases}$

By identity 1, the second term of Eq. 6.34 is zero. The first term is also zero, except when $m = k$ or $m = K - k$, leading to the results

$$\mathrm{E}\left[A_k^2\right] = \begin{cases} \frac{1}{2}G(\omega_k)\ \Delta\omega & \text{if } k = 0 \\ \frac{1}{4}\left\{G(\omega_k) + G(\omega_{K-k})\right\}\ \Delta\omega & \text{if } k = 1,\ldots,\frac{1}{2}K-1 \\ G(\omega_k)\ \Delta\omega & \text{if } k = \frac{1}{2}K \end{cases} \quad (6.35)$$

remembering that for $k = 0$ the frequency interval is $\frac{1}{2}\ \Delta\omega$. An entirely similar calculation leads to

$$\mathrm{E}\left[B_k\right]^2 = \begin{cases} 0 & \text{if } k = 0,\ \frac{1}{2}K \\ \frac{1}{4}\left\{G(\omega_k) + G(\omega_{K-k})\right\}\ \Delta\omega & \text{if } k = 1,\ldots,\frac{1}{2}K-1 \end{cases} \quad (6.36)$$

Thus the simulation process is as follows:

1. Generate independent normally distributed realizations of A_k and B_k having mean zero and variance given by Eqs. 6.35 and 6.36 for $k = 0, 1, \ldots, K/2$ and set $B_0 = B_{K/2} = 0$.
2. Use the symmetry relationships, Eqs. 6.30 and 6.31, to produce the remaining Fourier coefficients for $k = 1 + K/2, \ldots, K-1$.
3. Produce the field realization by FFT using Eq. 6.29.

In higher dimensions a similar approach can be taken. To compute the Fourier sum over nonnegative frequencies only, the spectral density function $S(\boldsymbol{\omega})$ is assumed to be even in all components of $\boldsymbol{\omega}$ (quadrant symmetric) so that the "one-sided" spectral density function, $G(\boldsymbol{\omega}) = 2^n S(\boldsymbol{\omega})$ $\forall\omega_i \geq 0$, in n-dimensional space, can be employed. Using $L = K_1 - \ell$, $M = K_2 - m$, and $N = K_3 - n$ to denote the symmetric points in fields of size $K_1 \times K_2$ in two dimensions or $K_1 \times K_2 \times K_3$ in three dimensions, the Fourier coefficients yielding a real two-dimensional process must satisfy

$$\begin{aligned} A_{LM} &= A_{\ell m}, & B_{LM} &= -B_{\ell m} \\ A_{\ell M} &= A_{Lm}, & B_{\ell M} &= -B_{Lm} \end{aligned} \quad (6.37)$$

for $\ell, m = 0, 1, \ldots, \frac{1}{2}K_\alpha$ where K_α is either K_1 or K_2 appropriately. Note that these relationships are applied modulo K_α, so that $A_{K_1-0,m} \equiv A_{0,m}$, for example. In two dimensions, the Fourier coefficients must be generated over two adjacent quadrants of the field, the rest of the coefficients obtained using the symmetry relations. In three dimensions, the symmetry relationships are

$$\begin{aligned} A_{LMN} &= A_{\ell mn}, & B_{LMN} &= -B_{\ell mn} \\ A_{\ell MN} &= A_{Lmn}, & B_{\ell MN} &= -B_{Lmn} \\ A_{LmN} &= A_{\ell Mn}, & B_{LmN} &= -B_{\ell Mn} \\ A_{\ell mN} &= A_{LMn}, & B_{\ell mN} &= -B_{LMn} \end{aligned} \quad (6.38)$$

for $\ell, m, n = 0, 1, \ldots, \frac{1}{2}K_\alpha$. Again, only half the Fourier coefficients are to be generated randomly.

The variances of the Fourier coefficients are found in a manner analogous to the one-dimensional case, resulting in

$$\mathrm{E}\left[A_{\ell m}^2\right] = \tfrac{1}{8}\delta_{\ell m}^A\ \Delta\boldsymbol{\omega}\left(G_{\ell m}^d + G_{\ell N}^d + G_{Ln}^d + G_{LN}^d\right) \quad (6.39)$$

$$\mathrm{E}\left[B_{\ell m}^2\right] = \tfrac{1}{8}\delta_{\ell m}^B\ \Delta\boldsymbol{\omega}\left(G_{\ell m}^d + G_{\ell N}^d + G_{Ln}^d + G_{LN}^d\right) \quad (6.40)$$

for two dimensions and

$$\mathrm{E}\left[A_{\ell mn}\right]^2 = \tfrac{1}{16}\,\delta_{\ell mn}^A\ \Delta\boldsymbol{\omega}\left(G_{\ell mn}^d + G_{\ell mN}^d + G_{\ell Mn}^d + G_{Lmn}^d + G_{\ell MN}^d + G_{LmN}^d + G_{LMn}^d + G_{LMN}^d\right) \quad (6.41)$$

$$\mathrm{E}\left[B_{\ell mn}\right]^2 = \tfrac{1}{16}\,\delta_{\ell mn}^B\ \Delta\boldsymbol{\omega}\left(G_{\ell mn}^d + G_{\ell mN}^d + G_{\ell Mn}^d + G_{Lmn}^d + G_{\ell MN}^d + G_{LmN}^d + G_{LMn}^d + G_{LMN}^d\right) \quad (6.42)$$

in three dimensions, where for three dimensions

$$\Delta\boldsymbol{\omega} = \prod_{i=1}^{3} \Delta\omega_i \quad (6.43)$$

$$G_{lmn}^d = \frac{G(\omega_l, \omega_m, \omega_n)}{2^d} \quad (6.44)$$

and d is the number of components of $\boldsymbol{\omega} = (\omega_1, \omega_2, \omega_3)$ which are equal to zero. The factors $\delta^A_{\ell mn}$ and $\delta^B_{\ell mn}$ are given by

$$\delta^A_{\ell mn} = \begin{cases} 2 & \text{if } l = 0 \text{ or } \frac{1}{2}K_1, m = 0 \text{ or } \frac{1}{2}K_2, \\ & \quad n = 0 \text{ or } \frac{1}{2}K_3 \\ 1 & \text{otherwise} \end{cases} \tag{6.45}$$

$$\delta^B_{\ell mn} = \begin{cases} 0 & \text{if } l = 0 \text{ or } \frac{1}{2}K_1, m = 0 \text{ or } \frac{1}{2}K_2, \\ & \quad n = 0 \text{ or } \frac{1}{2}K_3 \\ 1 & \text{otherwise} \end{cases} \tag{6.46}$$

(ignoring the index n in the case of two dimensions). Thus, in higher dimensions, the simulation procedure is almost identical to that followed in the one-dimensional case—the only difference being that the coefficients are generated randomly over the half plane (two-dimensional) or the half volume (three-dimensional) rather than the half line of the one-dimensional formulation.

It is appropriate at this time to investigate some of the shortcomings of the method. First of all, can be shown that regardless of the desired target covariance function, the covariance function $\hat{C}_k = \hat{C}(k\,\Delta x)$ of the real FFT process is always symmetric about the midpoint of the field. In one dimension, the covariance function is given by (using complex notation for the time being)

$$\begin{aligned} \hat{C}_k &= \mathrm{E}\left[Z_{\ell+k}\overline{Z_\ell}\right] \\ &= \mathrm{E}\left[\sum_{j=0}^{K-1} \mathcal{X}_j \, \exp\left\{i\left(\tfrac{2\pi(\ell+k)j}{K}\right)\right\}\right. \\ &\qquad \left. \times \sum_{m=0}^{K-1} \overline{\mathcal{X}_m} \times \exp\left\{-i\left(\tfrac{2\pi \ell m}{K}\right)\right\}\right] \\ &= \sum_{j=0}^{K-1} \mathrm{E}\left[\mathcal{X}_j\overline{\mathcal{X}_j}\right] \, \exp\left\{i\left(\tfrac{2\pi jk}{K}\right)\right\} \end{aligned} \tag{6.47}$$

where use was made of the fact that $\mathrm{E}\left[\mathcal{X}_j\overline{\mathcal{X}_m}\right] = 0$ for $j \neq m$ (overbar denotes the complex conjugate). Similarly one can derive

$$\begin{aligned} \hat{C}_{K-k} &= \sum_{j=0}^{K-1} \mathrm{E}\left[\mathcal{X}_j\overline{\mathcal{X}_j}\right] \, \exp\left\{-i\left(\tfrac{2\pi jk}{K}\right)\right\} \\ &= \overline{\hat{C}_k} \end{aligned} \tag{6.48}$$

since $\mathrm{E}\left[\mathcal{X}_j\overline{\mathcal{X}_j}\right]$ is real. The covariance function of a real, process is also real in which case (6.48) becomes simply

$$\hat{C}_{K-k} = \hat{C}_k \tag{6.49}$$

In one dimension, this symmetry is illustrated by Figure 6.5. Similar results are observed in higher dimensions. In general, this deficiency can be overcome by generating a field twice as long as required in each coordinate direction and keeping only the first quadrant of the field. Figure 6.5 also compares the covariance, mean, and variance fields of the LAS method to that of the FFT method (the TBM method is not defined in one dimension). The two methods give satisfactory performance with respect to the variance and mean fields, while the LAS method shows superior performance with respect to the covariance structure.

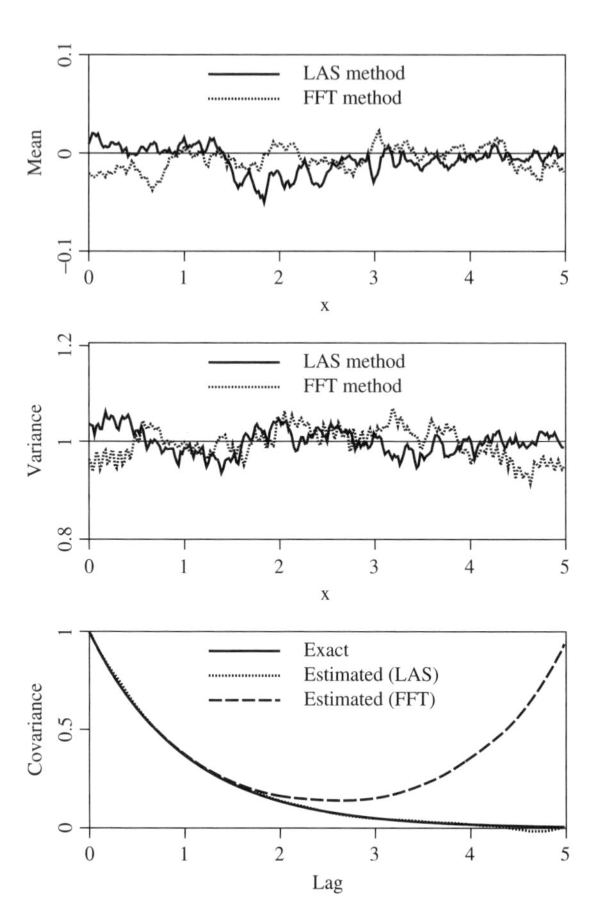

Figure 6.5 Mean, variance, and covariance of one-dimensional 128-point Gauss–Markov process estimated over ensemble of 2000 realizations generated by FFT.

The second problem with the FFT method relates primarily to its ease of use. Because of the close relationship between the spatial and frequency discretization, considerable care must be exercised when initially defining the spatial field and its discretization. First of all, the physical length of the field D must be large enough that the frequency increment $\Delta\omega = 2\pi(K-1)/KD \simeq 2\pi/D$ is sufficiently small. This is necessary if the sequence $\frac{1}{2}G(\omega_0)\Delta\omega, G(\omega_1)\Delta\omega, \ldots$ is to adequately approximate the target spectral density function. Figure 6.6 shows an example where the frequency discretization is overly coarse. Second, the physical resolution Δx must be selected

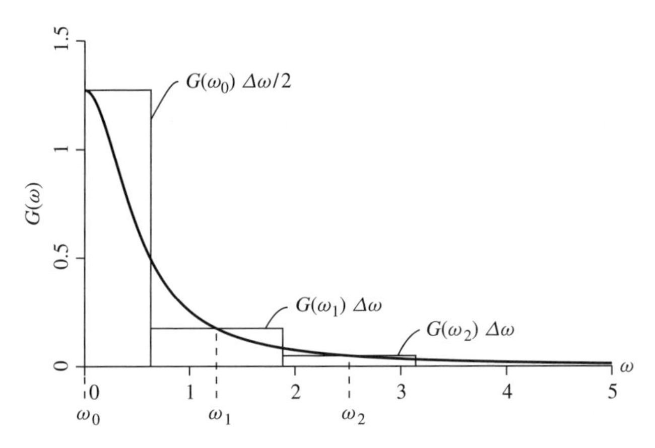

Figure 6.6 Example of overly coarse frequency discretization resulting in poor estimation of point variance ($D = 5$, $\theta = 4$).

so that the spectral density above the frequency $2\pi/\Delta x$ is negligible. Failure to do so will result in an underestimation of the total variance of the process. In fact the FFT formulation given above folds the power corresponding to frequencies between $\pi/\Delta x$ and $2\pi/\Delta x$ into the power at frequencies below the Nyquist limit $\pi/\Delta x$. This results in the point variance of the simulation being more accurate than if the power above the Nyquist limit were ignored; however, it leads to a nonuniqueness in that a family of spectral density functions, all having the same value of $G(\omega_k) + G(\omega_{K-k})$, yield the same process. In general, it is best to choose Δx so that the power above the Nyquist limit is negligible. The second term involving the symmetric frequency $G(\omega_{K-k})$ is included here because the point variance is the most important second-order characteristic.

Unfortunately, many applications dictate the size and discretization of the field a priori or the user may want to have the freedom to easily consider other geometries or spectral density functions. Without careful thought and analysis, the FFT approach can easily yield highly erroneous results.

A major advantage of the FFT method is that it can easily handle anisotropic fields with no sacrifice in efficiency. The field need not be square, although many implementations of the FFT require the number of points in the field in any coordinate direction to be a power of 2. Regarding efficiency, it should be pointed out that the time to generate the first realization of the field is generally much longer than that required to generate subsequent realizations. This is because the statistics of the Fourier coefficients must be calculated only once (see Eqs. 6.35 and 6.36).

The FFT method is useful for the generation of fractal processes, which are most naturally represented by the spectral density function. In fact the covariance function does not exist since the variance of a fractal process is ideally infinite. In practice, for such a process, the spectral density is truncated above and below to render a finite variance realization.

6.4.5 Turning-Bands Method

The *turning-bands* method, as originally suggested by Matheron (1973), involves the simulation of random fields in two- or higher dimensional space by using a sequence of one-dimensional processes along lines crossing the domain. With reference to Figure 6.7, the algorithm can be described as follows:

1. Choose an arbitrary origin within or near the domain of the field to be generated.
2. Select a line i crossing the domain having a direction given by the unit vector $\mathbf{u}_i$, which may be chosen either randomly or from some fixed set.
3. Generate a realization of a one-dimensional process, $Z_i(\xi_i)$, along the line i having zero mean and covariance function $C_1(\tau_i)$ where ξ_i and τ_i are measured along line i.
4. Orthogonally project each field point $\mathbf{x}_k$ onto the line i to define the coordinate ξ_{ki} ($\xi_{ki} = \mathbf{x}_k \cdot \mathbf{u}_i$ in the case of a common origin) of the one-dimensional process value $Z_i(\xi_{ki})$.
5. Add the component $Z_i(\xi_{ki})$ to the field value $Z(\mathbf{x}_k)$ for each $\mathbf{x}_k$.

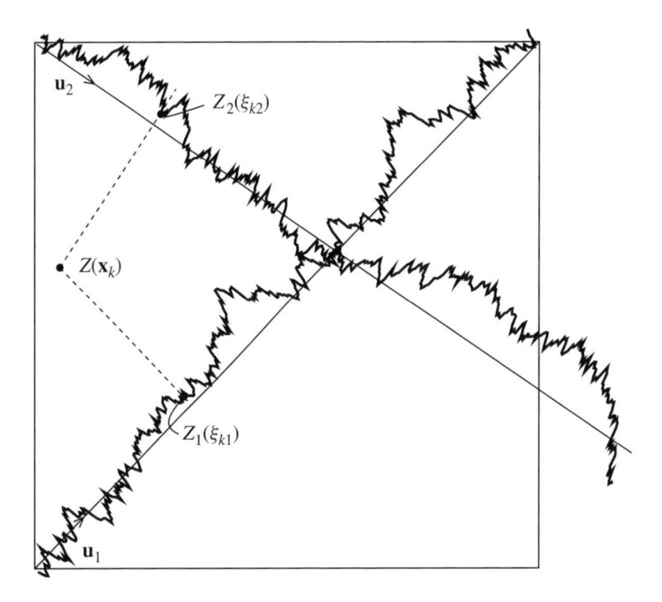

Figure 6.7 Turning-bands method. Contributions from the line process $Z_i(\xi_i)$ at the closest point are summed into the field process $Z(\mathbf{x})$ at $\mathbf{x}_k$.

6. Return to step 2 and generate a new one-dimensional process along a subsequent line until L lines have been produced.
7. Normalize the field $Z(\mathbf{x}_k)$ by dividing through by the factor $\sqrt{L}$.

Essentially, the generating equation for the zero-mean process $Z(\mathbf{x})$ is given by

$$Z(\mathbf{x}_k) = \frac{1}{\sqrt{L}} \sum_{i=1}^{L} Z_i(\mathbf{x}_k \cdot \mathbf{u}_i) \tag{6.50}$$

where, if the origins of the lines and space are not common, the dot product must be replaced by some suitable transform. This formulation depends on knowledge of the one-dimensional covariance function, $C_1(\tau)$. Once this is known, the line processes can be produced using some efficient one-dimensional algorithm. The last point means that the TBM is not a fundamental generator—it requires an existing one-dimensional generator (e.g., FFT or LAS, to be discussed next).

The covariance function $C_1(\tau)$ is chosen such that the multidimensional covariance structure $C_n(\boldsymbol{\tau})$ in n-dimensional space is reflected over the ensemble. For two-dimensional isotropic processes, Mantoglou and Wilson (1981) give the following relationship between $C_2(\boldsymbol{\tau})$ and $C_1(\eta)$ for $r = |\boldsymbol{\tau}|$:

$$C_2(r) = \frac{2}{\pi} \int_0^r \frac{C_1(\eta)}{\sqrt{r^2 - \eta^2}}\, d\eta \tag{6.51}$$

which is an integral equation to be solved for $C_1(\eta)$. In three dimensions, the relationship between the isotropic $C_3(r)$ and $C_1(\eta)$ is particularly simple:

$$C_1(\eta) = \frac{d}{d\eta}\left[\eta\, C_3(\eta)\right] \tag{6.52}$$

Mantoglou and Wilson supply explicit solutions for either the equivalent one-dimensional covariance function or the equivalent one-dimensional spectral density function for a variety of common multidimensional covariance structures.

In this implementation of the TBM, the line processes were constructed using a one-dimensional FFT algorithm, as discussed in the previous section. The LAS method was not used for this purpose because the local averaging introduced by the method would complicate the resulting covariance function of Eg. 6.51. Line lengths were chosen to be twice that of the field diagonal to avoid the symmetric covariance problem inherent with the FFT method. To reduce errors arising due to overly coarse discretization of the lines, the ratio between the incremental distance along the lines, $\Delta\xi$, and the minimum incremental distance in the field along any coordinate, Δx, was selected to be $\Delta\xi/\Delta x = \frac{1}{2}$.

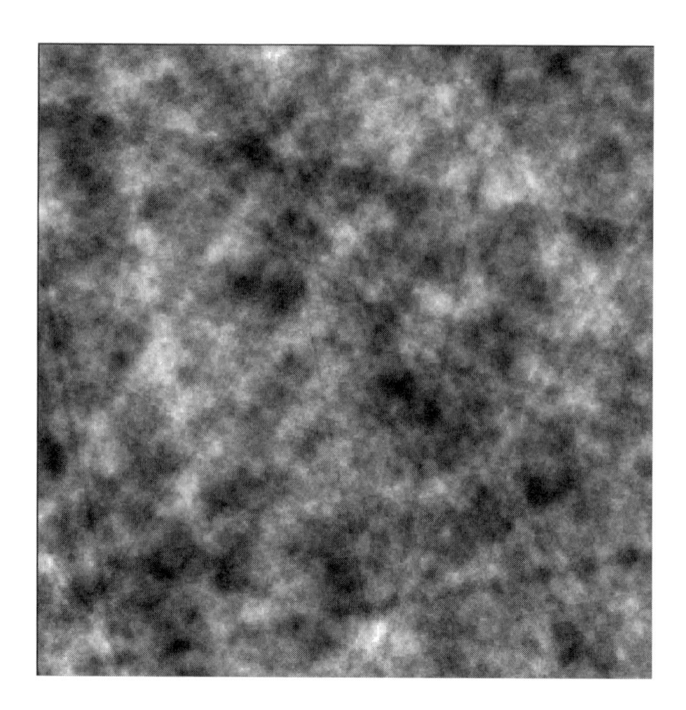

Figure 6.8 Sample function of two-dimensional field via TBM using 16 lines.

Figure 6.8 represents a realization of a two-dimensional process. The finite number of lines used, in this case 16, results in a streaked appearance of the realization. A number of origin locations were experimented with to mitigate the streaking, the best appearing to be the use of all four corners as illustrated in Figure 6.7 and as used in Figure 6.8. The corner selected as an origin depends on which quadrant the unit vector $\mathbf{u}_i$ points into. If one considers the spectral representation of the one-dimensional random processes along each line (see Eq. 6.22), it is apparent that the streaks are a result of constructive/destructive interference between randomly oriented traveling plane waves. The effect will be more pronounced for narrow-band processes and for a small number of lines. For this particular covariance function (Markov), the streaks are still visible when 32 lines are used, but, as shown in Figure 6.9, are negligible when using 64 lines (the use of number of lines which are powers of 2 is arbitrary). While the 16-line case runs at about the same speed as the two-dimensional LAS approach, the elimination of the streaks in the realization comes at a price of running about four times slower. The streaks are only evident in an average over the ensemble if nonrandom line orientations are used, although they still appear in individual realizations in either case. Thus, with respect to each realization, there is no particular advantage to using random versus nonrandom line orientations.

Since the streaks are present in the field itself, this type of error is generally more serious than errors in the

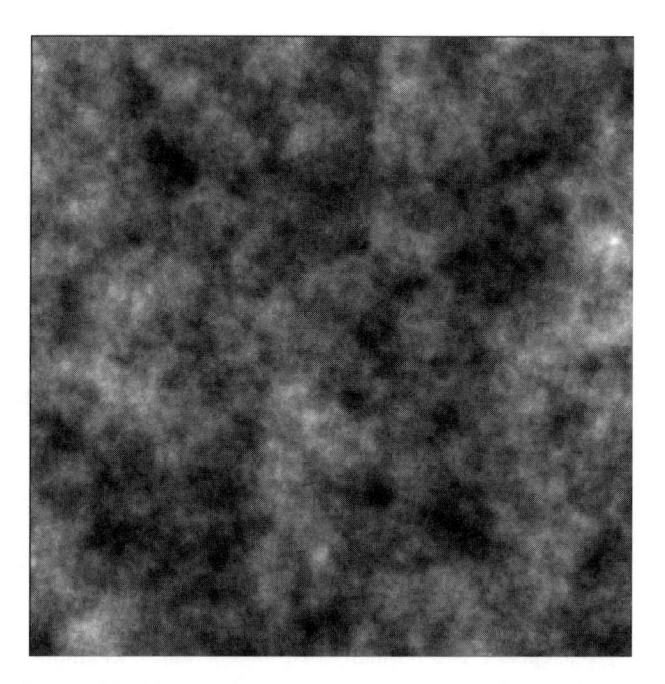

Figure 6.9 Sample function of two-dimensional field via TBM using 64 lines.

variance or covariance field. For example, if the field is being used to represent soil permeability, then the streaks could represent paths of reduced resistance to flow, a feature which may not be desirable in a particular study. Crack propagation studies may also be very sensitive to such linear correlations in the field. For applications such as these, the TBM should only be used with a sufficiently large number of lines. This may require some preliminary investigation for arbitrary covariance functions. In addition, the minimum number of lines in three and higher dimensions is difficult to determine due to visualization problems.

Note that the TBM does not suffer from the symmetric covariance structure that is inherent in the FFT approach. The variance field and covariance structure are also well preserved. However, the necessity of finding an equivalent one-dimensional covariance or spectral density function through an integral equation along with the streaked appearance of the realization when an insufficient number of lines are used makes the method less attractive. Using a larger number of lines, TBM is probably the most accurate of the three methods considered, at the expense of decreased efficiency, as long as the one-dimensional generator is accurate. TBM can be extended to anisotropic fields, although there is an additional efficiency penalty associated with such an extension since the one-dimensional process statistics must be recalculated for each new line orientation (see Mantoglou and Wilson, 1981, for details).

6.4.6 Local Average Subdivision Method

Of the three approximate methods considered, the *local average subdivision* (LAS) method (Fenton and Vanmarcke, 1990) is probably the most difficult to implement but the easiest to use. The local average subdivision method is a fast and generally accurate method of producing realizations of a discrete "local average" random process. The motivation for the method arose out of a need to properly account for the fact that most engineering measurements are actually local averages of the property in question. For example, soil porosity is ill-defined at the microscale—it is measured in practice using samples of finite volume, and the measured value is an average of the porosity through the sample. The same can be said of strength measurements, say triaxial tests on laboratory volumes, or CPT measurements which record the effects of deforming a bulb of soil around the cone. The variance of the average is strongly affected by the size of the sample. Depending on the distribution of the property being measured, the mean of the average may also be affected by the sample size—this is sometimes called the *scale effect*. These effects are relatively easily incorporated into a properly defined random local average process.

Another advantage to using local averages is that they are ideally suited to stochastic finite-element modeling using efficient, low-order, interpolation functions. Each discrete local average given by a realization becomes the average property within each discrete element. As the element size is changed, the statistics of the random property mapped to the element will also change in a statistically consistent fashion. This gives finite-element modelers the freedom to change mesh resolution without losing stochastic accuracy.

The concept behind the LAS approach derived from the stochastic subdivision algorithm described by Carpenter (1980) and Fournier et al. (1982). Their method was limited to modeling power spectra having a $\omega^{-\beta}$ form and suffered from problems with aliasing and "creasing." Lewis (1987) generalized the approach to allow the modeling of arbitrary power spectra without eliminating the aliasing. The stochastic subdivision is a midpoint displacement algorithm involving recursively subdividing the domain by generating new midpoint values randomly selected according to some distribution. Once chosen, the value at a point remains fixed, and at each stage in the subdivision only half the points in the process are determined (the others created in previous iterations). Aliasing arises because the power spectral density is not modified at each stage to reflect the increasing Nyquist frequency associated with each increase in resolution. Voss (in Peitgen and Saupe, 1988, Chapter 1) attempted to eliminate this problem with considerable success by adding randomness to all points at each stage in the subdivision in a method called *successive random additions*. However, the internal consistency easily achieved by the midpoint displacement methods (their ability to return to previous states while decreasing resolution through decimation) is largely lost with the successive random additions technique. The property of internal consistency in the midpoint displacement approaches implies that certain points retain their value throughout the subdivision, and other points are created to remain consistent with them with respect to correlation. In the LAS approach, internal consistency implies that the local average is maintained throughout the subdivision.

The LAS method solves the problems associated with the stochastic subdivision methods and incorporates into it concepts of local averaging theory. The general concept and procedure is presented first for a one-dimensional stationary process characterized by its second-order statistics. The algorithm is illustrated by a Markov process, having a simple exponential correlation function (see Section 3.6.5), as well as by a fractional Gaussian noise process as defined by Mandelbrot and van Ness (1968)—see Section 3.6.7. The simulation procedure in two and three dimensions is then described. Finally, some comments concerning the accuracy and efficiency of the method are made.

6.4.6.1 One-Dimensional Local Average Subdivision

The construction of a local average process via LAS essentially proceeds in a top-down recursive fashion as illustrated in Figure 6.10. In stage 0, a global average is generated for the process. In stage 1, the domain is subdivided into two regions whose "local" averages must in turn average to the global (or parent) value. Subsequent stages are obtained by subdividing each "parent" cell and generating values for the resulting two regions while preserving upwards averaging. Note that the global average remains constant throughout the subdivision, a property that is ensured merely by requiring that the average of each pair generated is equivalent to the parent cell value. This is also a property of any cell being subdivided. We note that the local average subdivision can be applied to any existing local average field. For example, the stage 0 shown in Figure 6.10 might simply be one local average cell in a much larger field. The algorithm proceeds as follows:

1. Generate a normally distributed global average (labeled Z_1^0 in Figure 6.10) with mean zero and variance obtained from local averaging theory (see Section 3.4).
2. Subdivide the field into two equal parts.
3. Generate two normally distributed values, Z_1^1 and Z_2^1, whose means and variances are selected so as to satisfy three criteria:
   (a) they show the correct variance according to local averaging theory
   (b) they are properly correlated with one another
   (c) they average to the parent value, $\frac{1}{2}(Z_1^1 + Z_2^1) = Z_1^0$

   That is, the distributions of Z_1^1 and Z_2^1 are conditioned on the value of Z_1^0.
4. Subdivide each cell in stage 1 into two equal parts.
5. Generate two normally distributed values, Z_1^2 and Z_2^2, whose means and variances are selected so as to satisfy four criteria:
   (a) they show the correct variance according to local averaging theory
   (b) they are properly correlated with one another
   (c) they average to the parent value, $\frac{1}{2}(Z_1^2 + Z_2^2) = Z_1^1$
   (d) they are properly correlated with Z_3^2 and Z_4^2

   The third criterion implies conditioning of the distributions of Z_1^2 and Z_2^2 on the value of Z_1^1. The fourth criterion will only be satisfied approximately by conditioning their distributions also on Z_2^1.

And so on in this fashion. The approximations in the algorithm come about in two ways: First, the correlation with adjacent cells across parent boundaries is accomplished through the parent values (which are already known having been previously generated). Second, the range of parent cells on which to condition the distributions will be limited to some neighborhood. Much of the remainder of this section is devoted to the determination of these conditional Gaussian distributions at each stage in the subdivision and to an estimation of the algorithmic errors. In the following, the term "parent cell" refers to the previous stage cell being subdivided, and "within-cell" means within the region defined by the parent cell.

To determine the mean and variance of the stage 0 value, Z_1^0, consider first a continuous stationary scalar random function $Z(t)$ in one dimension, a sample of which may appear as shown in Figure 6.11, and define a domain of interest $(0, D]$ within which a realization is to be produced. Two comments should be made at this point: First, as it

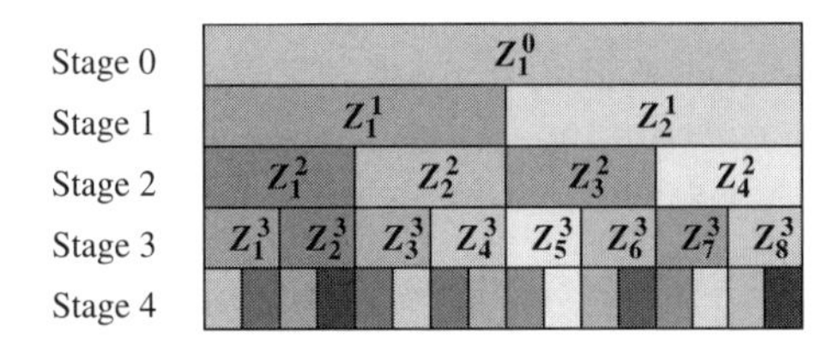

Figure 6.10 Top-down approach to LAS construction of local average random process.

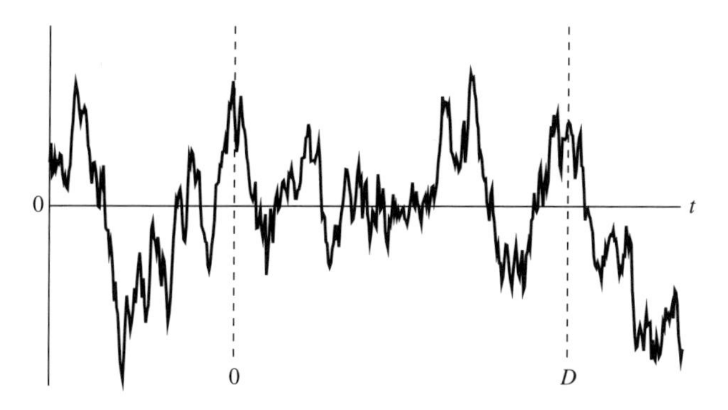

Figure 6.11 Realization of continuous random function Z with domain of interest $(0, D]$ shown.

is currently implemented, the LAS method is restricted to stationary processes fully described by their second-order statistics (mean, variance, and correlation function or, equivalently, spectral density function). This is not a severe restriction since it leaves a sufficiently broad class of functions to model most natural phenomena (Lewis, 1987); also, there is often insufficient data to substantiate more complex probabilistic models. Besides, a nonstationary mean and variance can be easily added to a stationary process. For example, $Y(t) = \mu(t) + \sigma(t)X(t)$ will produce a nonstationary $Y(t)$ from stationary $X(t)$ if $\mu(t)$ and/or $\sigma(t)$ vary with t (e.g., CPT soundings often show increases in both μ and σ with depth). Second, the subdivision procedure depends on the physical size of the domain being defined since the dimension over which local averaging is to be performed must be known. The process Z beyond the domain $(0, D]$ is ignored.

The average of $Z(t)$ over the domain $(0, D]$ is given by

$$Z_1^0 = \frac{1}{D}\int_0^D Z(\xi)\, d\xi \tag{6.53}$$

where Z_1^0 is a random variable whose statistics

$$\mathrm{E}\left[Z_1^0\right] = \mathrm{E}\left[Z\right] \tag{6.54}$$

$$\begin{aligned}\mathrm{E}\left[(Z_1^0)^2\right] &= \frac{1}{D^2}\int_0^D \int_0^D \mathrm{E}\left[Z(\xi)Z(\xi')\right]\, d\xi\, d\xi' \\ &= \mathrm{E}\left[Z\right]^2 + \frac{2}{D^2}\int_0^D (D-\tau)C(\tau)\, d\tau \end{aligned}\tag{6.55}$$

can be found by making use of stationarity and the fact that $C(\tau)$, the covariance function of $Z(t)$, is an even function of lag τ. Without loss in generality, $\mathrm{E}[Z]$ will henceforth be taken as zero. If $Z(t)$ is a Gaussian random function, Eqs. 6.54 and 6.55 give sufficient information to generate a realization of Z_1^0, which becomes stage 0 in the LAS method. If $Z(t)$ is not Gaussian, then the complete probability distribution function for Z_1^0 must be determined and a realization generated according to such a distribution. We will restrict our attention to Gaussian processes.

Consider now the general case where stage i is known and stage $i+1$ is to be generated. In the following the superscript i denotes the stage under consideration. Define

$$D^i = \frac{D}{2^i}, \qquad i = 0, 1, 2, \ldots, L \tag{6.56}$$

where the desired number of subintervals in the final realization is $N = 2^L$, and define Z_k^i to be the average of $Z(t)$ over the interval $(k-1)D^i < t \le kD^i$ centered at $t_k = (k - \frac{1}{2})D^i$, that is,

$$Z_k^i = \frac{1}{D^i}\int_{(k-1)D^i}^{kD^i} Z(\xi)\, d\xi \tag{6.57}$$

where $\mathrm{E}\left[Z_k^i\right] = \mathrm{E}\left[Z\right] = 0$. The target covariance between local averages separated by lag mD^i between centers is

$$\begin{aligned}\mathrm{E}\left[Z_k^i Z_{k+m}^i\right] &= \mathrm{E}\left[\left(\frac{1}{D^i}\right)^2 \int_{(k-1)D^i}^{kD^i} \int_{(k+m-1)D^i}^{(k+m)D^i} Z(\xi)Z(\xi')\, d\xi\, d\xi'\right] \\ &= \left(\frac{1}{D^i}\right)^2 \int_0^{D^i} \int_{mD^i}^{(m+1)D^i} C(\xi - \xi')\, d\xi\, d\xi' \\ &= \left(\frac{1}{D^i}\right)^2 \int_{(m-1)D^i}^{mD^i} \left[\xi - (m-1)D^i\right] C(\xi)\, d\xi \\ &\quad + \left(\frac{1}{D^i}\right)^2 \int_{mD^i}^{(m+1)D^i} \left[(m+1)D^i - \xi\right] B(\xi)\, d\xi \end{aligned}\tag{6.58}$$

which can be evaluated relatively simply using Gaussian quadrature as

$$\mathrm{E}\left[Z_k^i Z_{k+m}^i\right] \simeq \frac{1}{4}\sum_{\nu=1}^{n_g} w_\nu \left[(1+z_\nu))C(r_\nu) + (1-z_\nu)C(s_\nu)\right] \tag{6.59}$$

where $r_\nu = D^i\left(m - \frac{1}{2}(1-z_\nu)\right)$, $s_\nu = D^i\left(m + \frac{1}{2}(1+z_\nu)\right)$, and the weights, w_ν, and positions z_ν can be found in Appendix B for n_g Gauss points.

With reference to Figure 6.12, the construction of stage $i+1$ given stage i is obtained by estimating a mean for Z_{2j}^{i+1} and adding a zero mean discrete white noise $c^{i+1}U_j^{i+1}$ having variance $(c^{i+1})^2$

$$Z_{2j}^{i+1} = M_{2j}^{i+1} + c^{i+1}U_j^{i+1} \tag{6.60}$$

The best linear estimate for the mean M_{2j}^{i+1} can be determined by a linear combination of stage i (parent) values in

	j		$j+1$		
	$2j-1$	$2j$	$2j+1$	$2j+2$	

Figure 6.12 One-dimensional LAS indexing for stage i (top) and stage $i+1$ (bottom).

some neighborhood $j-n, \ldots, j+n$,

$$M_{2j}^{i+1} = \sum_{k=j-n}^{j+n} a_{k-j}^i Z_k^i \tag{6.61}$$

Multiplying Eq. 6.60 through by Z_m^i, taking expectations, and using the fact that U_j^{i+1} is uncorrelated with the stage i values allows the determination of the coefficients a in terms of the desired covariances,

$$\mathrm{E}\left[Z_{2j}^{i+1} Z_m^i\right] = \sum_{k=j-n}^{j+n} a_{k-j}^i \, \mathrm{E}\left[Z_k^i Z_m^i\right] \tag{6.62}$$

a system of equations ($m = j-n, \ldots, j+n$) from which the coefficients a_ℓ^i, $\ell = -n, \ldots, n$, can be solved. The covariance matrix multiplying the vector $\{a_\ell^i\}$ is both symmetric and Toeplitz (elements along each diagonal are equal). For $U_j^{i+1} \sim N(0,1)$ the variance of the noise term is $(c^{i+1})^2$ which can be obtained by squaring Eq. 6.60, taking expectations and employing the results of Eq. 6.62:

$$(c^{i+1})^2 = \mathrm{E}\left[(Z_{2j}^{i+1})^2\right] - \sum_{k=j-n}^{j+n} a_{k-j}^i \, \mathrm{E}\left[Z_{2j}^{i+1} Z_k^i\right] \tag{6.63}$$

The adjacent cell, Z_{2j-1}^{i+1}, is determined by ensuring that upwards averaging is preserved—that the average of each stage $i+1$ pair equals the value of the stage i parent:

$$Z_{2j-1}^{i+1} = 2Z_j^i - Z_{2j}^{i+1} \tag{6.64}$$

which incidentally gives a means of evaluating the cross-stage covariances:

$$\mathrm{E}\left[Z_{2j}^{i+1} Z_m^i\right] = \tfrac{1}{2}\,\mathrm{E}\left[Z_{2j}^{i+1} Z_{2m-1}^{i+1}\right] + \tfrac{1}{2}\,\mathrm{E}\left[Z_{2j}^{i+1} Z_{2m}^{i+1}\right] \tag{6.65}$$

which are needed in Eq. 6.62. All the expectations in Eqs. 6.62–6.65 are evaluated using Eq. 6.58 or 6.59 at the appropriate stage.

For stationary processes, the set of coefficients $\{a_\ell^i\}$ and c^i are independent of position since the expectations in Eqs. 6.62 and 6.63 are just dependent on lags. The generation procedure can be restated as follows:

1. For $i = 0, 1, 2, \ldots, L$ compute the coefficients $\{a_\ell^i\}$, $\ell = -n, \ldots, n$ using Eq. 6.62 and c^{i+1} using Eq. 6.63.
2. Starting with $i = 0$, generate a realization for the global mean using Eqs. 6.54 and 6.55.
3. Subdivide the domain.
4. For each $j = 1, 2, 3, \ldots, 2^i$, generate realizations for Z_{2j}^{i+1} and Z_{2j-1}^{i+1} using Eqs. 6.60 and 6.64.
5. Increment i and, if not greater than L, return to step 3.

Notice that subsequent realizations of the process need only start at step 2, and so the overhead involved with setting up the coefficients becomes rapidly negligible.

Because the LAS procedure is recursive, obtaining stage $i+1$ values using the previous stage, it is relatively easy to condition the field by specifying the values of the local averages at a particular stage. So, for example, if the global mean of a process is known a priori, then the stage 0 value can be set to this mean and the LAS procedure started at stage 1. Similarly, if the resolution is to be refined in a certain region, then the values in that region become the starting values and the subdivision resumed at the next stage.

Although the LAS method yields a local average process, when the discretization size becomes small enough, it is virtually indistinguishable from the limiting continuous process. Thus, the method can be used to approximate continuous functions as well.

Accuracy It is instructive to investigate how closely the algorithm approximates the target statistics of the process. Changing notation slightly, denote the stage $i+1$ algorithmic values, given the stage i values, as

$$\hat{Z}_{2j}^{i+1} = c^{i+1} U_j^{i+1} + \sum_{k=j-n}^{j+n} a_{k-j}^i Z_k^i \tag{6.66}$$

$$\hat{Z}_{2j-1}^{i+1} = 2Z_j^i - \hat{Z}_{2j}^{i+1} \tag{6.67}$$

It is easy to see that the expectation of $\hat{Z}$ is still zero, as desired, while the variance is

$$\begin{aligned}
\mathrm{E}\left[(\hat{Z}_{2j}^{i+1})^2\right] &= \mathrm{E}\left[\left(c^{i+1} U_j^{i+1} + \sum_{k=j-n}^{j+n} a_{k-j}^i Z_k^i\right)^2\right] \\
&= (c^{i+1})^2 + \sum_{k=j-n}^{j+n} a_{k-j}^i \sum_{\ell=j-n}^{j+n} a_{\ell-j}^i \, \mathrm{E}\left[Z_k^i Z_\ell^i\right] \\
&= \mathrm{E}\left[(Z_{2j}^{i+1})^2\right] - \sum_{k=j-n}^{j+n} a_{k-j}^i \, \mathrm{E}\left[Z_{2j}^{i+1} Z_k^i\right] \\
&\quad + \sum_{k=j-n}^{j+n} a_{k-j}^i \, \mathrm{E}\left[Z_{2j}^{i+1} Z_k^i\right] \\
&= \mathrm{E}\left[(Z_{2j}^{i+1})^2\right]
\end{aligned} \tag{6.68}$$

in which the coefficients c^{i+1} and a^i_ℓ were calculated using Eqs. 6.62 and 6.63 as before. Similarly, the within-cell covariance at lag D^{i+1} is

$$\begin{aligned}
\mathrm{E}\left[\hat{Z}^{i+1}_{2j-1}\hat{Z}^{i+1}_{2j}\right] &= \mathrm{E}\left[\left(2Z^i_j - c^{i+1}U^{i+1}_j - \sum_{k=j-n}^{j+n} a^i_{k-j} Z^i_k\right)\right. \\
&\quad \times \left.\left(c^{i+1}U^{i+1}_j + \sum_{\ell=j-n}^{j+n} a^i_{\ell-j} Z^i_\ell\right)\right] \\
&= 2\sum_{\ell=j-n}^{j+n} a^i_{\ell-j}\,\mathrm{E}\left[Z^i_\ell Z^i_j\right] - \mathrm{E}\left[(Z^{i+1}_{2j})^2\right] \\
&= 2\,\mathrm{E}\left[Z^{i+1}_{2j} Z^i_j\right] - \mathrm{E}\left[(Z^{i+1}_{2j})^2\right] \\
&= \mathrm{E}\left[Z^{i+1}_{2j-1} Z^{i+1}_{2j}\right]
\end{aligned} \tag{6.69}$$

using the results of Eq. 6.68 along with Eq. 6.65. Thus, the covariance structure within a cell is preserved *exactly* by the subdivision algorithm. Some approximation does occur across cell boundaries as can be seen by considering

$$\begin{aligned}
&\mathrm{E}\left[\hat{Z}^{i+1}_{2j}\hat{Z}^{i+1}_{2j+1}\right] \\
&= \mathrm{E}\left[\left(c^{i+1}U^{i+1}_j + \sum_{k=j-n}^{j+n} a^i_{k-j} Z^i_k\right)\right. \\
&\quad \times \left.\left(2Z^i_{j+1} - c^{i+1}U^{i+1}_{j+1} - \sum_{\ell=j-n+1}^{j+n+1} a^i_{\ell-j-1} Z^i_\ell\right)\right] \\
&= 2\sum_{k=j-n}^{j+n} a^i_{k-j}\,\mathrm{E}\left[Z^i_k Z^i_{j+1}\right] \\
&\quad - \sum_{\ell=j-n+1}^{j+n+1} a^i_{\ell-j-1} \sum_{k=j-n}^{j+n} a^i_{k-j}\,\mathrm{E}\left[Z^i_k Z^i_\ell\right] \\
&= \mathrm{E}\left[Z^{i+1}_{2j} Z^{i+1}_{2j+1}\right] + \mathrm{E}\left[Z^{i+1}_{2j} Z^{i+1}_{2j+2}\right] \\
&\quad - \sum_{\ell=j-n+1}^{j+n+1} a^i_{\ell-j-1}\,\mathrm{E}\left[Z^{i+1}_{2j} Z^i_\ell\right]
\end{aligned} \tag{6.70}$$

The algorithmic error in this covariance comes from the last two terms. The discrepancy between Eq. 6.70 and the exact covariance is illustrated numerically in Figure 6.13 for a zero mean Markov process having covariance and

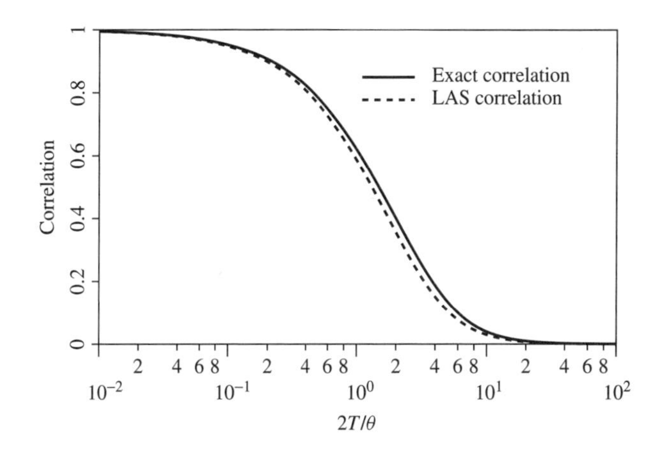

Figure 6.13 Comparison of algorithmic and exact correlation between adjacent cells across a parent cell boundary for varying effective cell dimension $2T/\theta$.

variance functions:

$$C(\tau) = \sigma^2 \exp\left\{-\frac{2|\tau|}{\theta}\right\} \tag{6.71}$$

$$\gamma(T) = \frac{\theta^2}{2T^2}\left[\frac{2|T|}{\theta} + \exp\left\{\frac{-2|T|}{\theta}\right\} - 1\right] \tag{6.72}$$

where T is the averaging dimension (in Figure 6.13, $T = D^{i+1}$) and θ is the correlation length of the process. The exact covariance is determined by Eq. 6.58 (for $m = 1$) using the variance function in Eq. 6.72. Although Figure 6.13 shows a wide range in the effective cell sizes, $2T/\theta$, the error is typically very small.

To address the issue of errors at larger lags and the possibility of errors accumulating from stage to stage, it is useful to look at the exact versus estimated statistics of the entire process. Figure 6.14 illustrates this comparison for the Markov process. It can be seen from this example, and from the fractional Gaussian noise example to come, that the errors are self-correcting, and the algorithmic correlation structure tends to the exact correlation function when averaged over several realizations. Spectral analysis of realizations obtained from the LAS method show equally good agreement between estimated and exact covariance functions (Fenton, 1990). The within-cell rate of convergence of the estimated statistic to the exact is $1/n_{\text{sim}}$, where n_{sim} is the number of realizations. The overall rate of convergence is about the same.

Boundary Conditions and Neighborhood Size When the neighborhood size $2n + 1$ is greater than 1 ($n > 0$), the construction of values near the boundary may require values from the previous stage which lie outside the boundary. This problem is handled by assuming that what happens

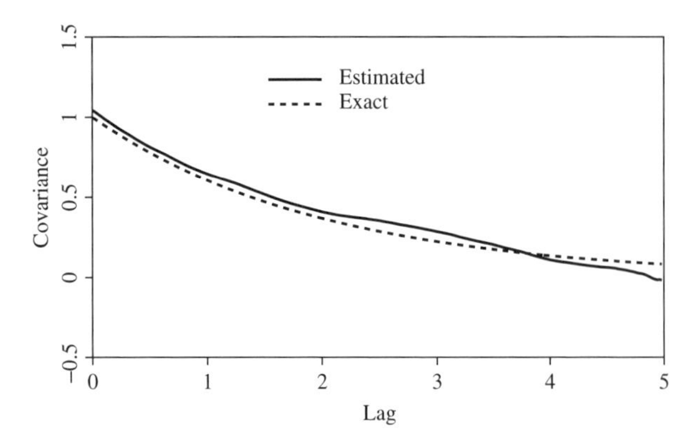

Figure 6.14 Comparison of exact and estimated covariance functions (averaged over 200 realizations) of Markov process with $\sigma = 1$ and $\theta = 4$.

outside the domain $(0, D]$ is of no interest and uncorrelated with what happens within the domain. The generating relationship (Eq. 6.60) near either boundary becomes

$$Z_{2j}^{i+1} = c^{i+1} U_j^{i+1} + \sum_{k=j-p}^{j+q} a_{k-j}^i Z_k^i \qquad (6.73)$$

where $p = \min(n, j-1)$, $q = \min(n, 2^i - j)$, and the coefficients a_ℓ^i need only be determined for $\ell = -p, \ldots, q$. The periodic boundary conditions mentioned by Lewis (1987) are not appropriate if the target covariance structure is to be preserved since they lead to a covariance which is symmetric about lag $D/2$ (unless the desired covariance is also symmetric about this lag).

In the implementation described in this section, a neighborhood size of 3 was used ($n = 1$), the parent cell plus its two adjacent cells. Because of the top-down approach, there seems to be little justification to using a larger neighborhood for processes with covariance functions which decrease monotonically or which are relatively smooth. When the covariance function is oscillatory, a larger neighborhood is required in order to successfully approximate the function. In Figure 6.15 the exact and estimated covariances are shown for a damped oscillatory process with

$$C(\tau) = \sigma^2 \cos(\omega\tau)\, e^{-2\tau/\theta} \qquad (6.74)$$

Considerable improvement in the model is obtained when a neighborhood size of 5 is used ($n = 2$). This improvement comes at the expense of taking about twice as long to generate the realizations. Many practical models of natural phenomena employ monotonically decreasing covariance functions, often for simplicity, and so the $n = 1$ implementation is usually preferable.

Fractional Gaussian Noise As a further demonstration of the LAS method, a self-similar process called fractional

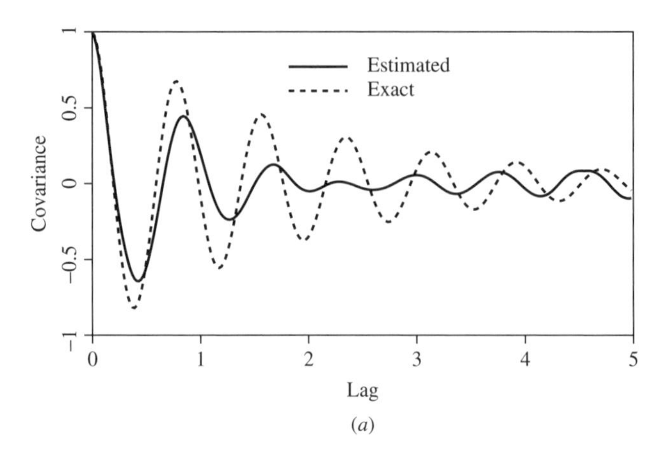

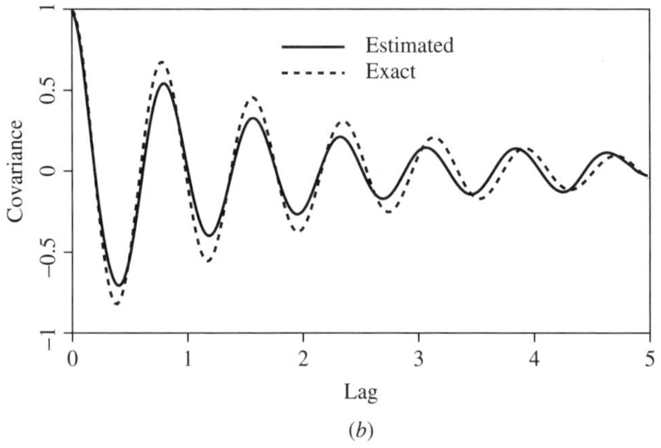

Figure 6.15 Effect of neighborhood size for (*a*) $n = 1$ and (*b*) $n = 2$ for damped oscillatory noise (Eq. 6.74).

Gaussian noise (see Section 3.6.7) was simulated, as shown in Figure 6.16. Fractional Gaussian noise (fGn) is defined by Mandelbrot and Van Ness (1968) to be the derivative of fractional Brownian motion (fBm) and is obtained by averaging the fBm over a small interval δ.

The resulting process has covariance and variance functions

$$C(\tau) = \frac{\sigma^2}{2\delta^{2H}}\left[|\tau+\delta|^{2H} - 2|\tau|^{2H} + |\tau-\delta|^{2H}\right] \qquad (6.75)$$

$$\gamma(T) = \frac{|T+\delta|^{2H+2} - 2|T|^{2H+2} + |T-\delta|^{2H+2} - 2\delta^{2H+2}}{T^2(2H+1)(2H+2)\delta^{2H}} \qquad (6.76)$$

defined for $0 < H < 1$. The case $H = 0.5$ corresponds to white noise and $H \to 1$ gives perfect correlation. In practice, δ is taken to be equal to the smallest lag between field points ($\delta = D/2^L$) to ensure that when $H = 0.5$ (white noise), $C(\tau)$ becomes zero for all $\tau \geq D/2^L$. A sample function and its corresponding ensemble statistics are

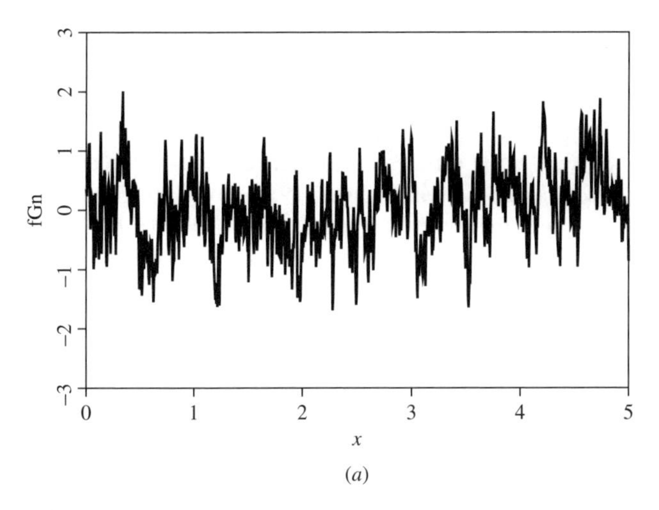

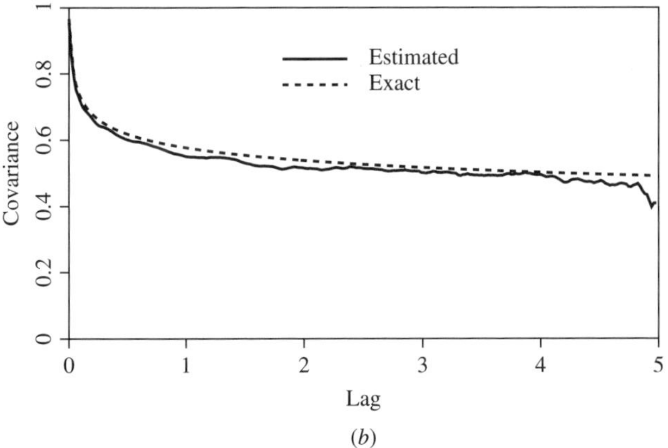

Figure 6.16 (*a*) LAS-generated sample function of fractional Gaussian noise for $H = 0.95$. (*b*) Corresponding estimated (averaged over 200 realizations) and exact covariance functions.

shown in Figure 6.16 for fGn with $H = 0.95$. The self-similar-type processes have been demonstrated by Mandelbrot (1982), Voss (1985), and many others [Mohr (1981), Peitgen and Saupe (1988), Whittle (1956), to name a few] to be representative of a large variety of natural forms and patterns, for example, music, terrains, crop yields, and chaotic systems. Fenton (1999b) demonstrated the presence of fractal behavior in CPT logs taken in Norway.

6.4.6.2 Multidimensional Local Average Subdivision

The two-dimensional LAS method involves a subdivision process in which a parent cell is divided into four equal sized cells. In Figure 6.17, the parent cells are denoted Z_l^i, $l = 1, 2, \ldots$, and the subdivided, or child cells, are denoted Z_j^{i+1}, $j = 1, 2, 3, 4$. Although each parent cell is eventually subdivided in the LAS process, only Z_5^i is subdivided in Figure 6.17 for simplicity. Using vector notation, the values of the column vector $\mathbf{Z}^{i+1} = \{Z_1^{i+1}, Z_2^{i+1}, Z_3^{i+1}, Z_4^{i+1}\}$ are obtained by adding a mean term to a random component. The mean term derives from a best linear unbiased estimate using a 3×3 neighborhood of the parent values, in this case the column vector $\mathbf{Z}^i = \{Z_1^i, \ldots, Z_9^i\}$. Specifically

Figure 6.17 Local average subdivision in two dimensions.

$$\mathbf{Z}^{i+1} = \boldsymbol{A}^{\mathrm{T}} \mathbf{Z}^i + \boldsymbol{L}\mathbf{U} \tag{6.77}$$

where $\mathbf{U}$ is a random vector with independent $N(0, 1)$ elements. This is essentially an ARMA model in which the "past" is represented by the previous coarser resolution stages. Defining the covariance matrices

$$\boldsymbol{R} = \mathrm{E}\left[\mathbf{Z}^i \mathbf{Z}^{i^{\mathrm{T}}}\right] \tag{6.78a}$$

$$\boldsymbol{S} = \mathrm{E}\left[\mathbf{Z}^i \mathbf{Z}^{i+1^{\mathrm{T}}}\right] \tag{6.78b}$$

$$\boldsymbol{B} = \mathrm{E}\left[\mathbf{Z}^{i+1} \mathbf{Z}^{i+1^{\mathrm{T}}}\right] \tag{6.78c}$$

then the matrix $\boldsymbol{A}$ is determined by

$$\boldsymbol{A} = \boldsymbol{R}^{-1} \boldsymbol{S} \tag{6.79}$$

while the lower triangular matrix $\boldsymbol{L}$ satisfies

$$\boldsymbol{L}\boldsymbol{L}^{\mathrm{T}} = \boldsymbol{B} - \boldsymbol{S}^{\mathrm{T}} \boldsymbol{A} \tag{6.80}$$

The covariance matrices $\boldsymbol{R}$, $\boldsymbol{S}$ and $\boldsymbol{B}$ must be computed as the covariances between local averages over the domains of the parent and child cells. This can be done using the variance function, although direct Gaussian quadrature of the covariance function has been found to give better numerical results. See Appendices B and C.

Note that the matrix on the right-hand side of Eq. 6.80 is only rank 3, so that the 4×4 matrix $\boldsymbol{L}$ has a special form with columns summing to zero (thus $L_{44} = 0$). While this results from the fact that all the expectations used in Eqs. 6.78 are derived using local average theory over the cell domains, the physical interpretation is that upwards averaging is preserved, that is, that $P_5 = \frac{1}{4}(Q_1 + Q_2 + Q_3 + Q_4)$. This means that one of the elements of $\mathbf{Q}$ is explicitly determined once the other three are known. In detail, Eq. 6.77 is carried out as follows:

$$Z_1^{i+1} = \sum_{l=1}^{9} A_{l1} Z_l^i + L_{11} U_1 \quad (6.81a)$$

$$Z_2^{i+1} = \sum_{l=1}^{9} A_{l2} Z_l^i + L_{21} U_1 + L_{22} U_2 \quad (6.81b)$$

$$Z_3^{i+1} = \sum_{l=1}^{9} A_{l3} Z_l^i + L_{31} U_1 + L_{32} U_2 + L_{33} U_3 \quad (6.81c)$$

$$Z_4^{i+1} = 4Z_5^i - Z_1^{i+1} - Z_2^{i+1} - Z_3^{i+1} \quad (6.81d)$$

where U_i are a set of three independent standard normally distributed random variables. Subdivisions taking place near the field boundaries are handled in much the same manner as in the one-dimensional case by assuming that conditions outside the field are uncorrelated with those inside the field.

The assumption of homogeneity vastly decreases the number of coefficients that need to be calculated and stored since the matrices $\boldsymbol{A}$ and $\boldsymbol{L}$ become independent of position. As in the one-dimensional case, the coefficients need only be calculated prior to the first realization—they can be reused in subsequent realizations reducing the effective cost of their calculation.

A sample function of a Markov process having isotropic covariance function

$$C(\tau_1, \tau_2) = \sigma^2 \exp\left\{-\frac{2}{\theta}\sqrt{\tau_1^2 + \tau_2^2}\right\} \quad (6.82)$$

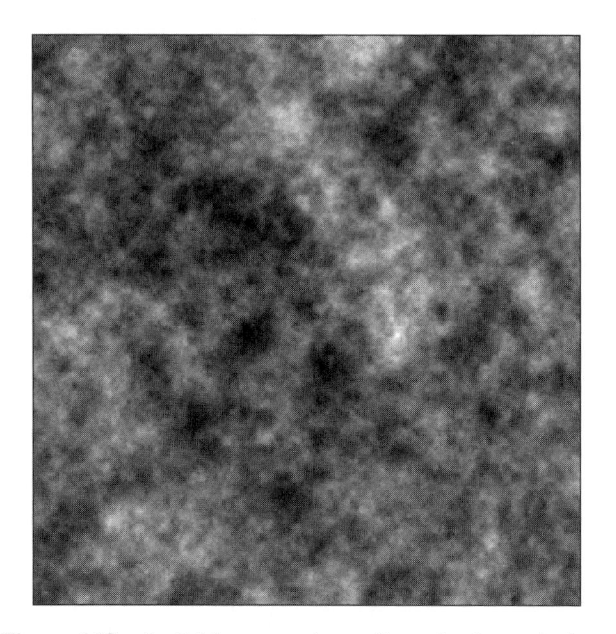

Figure 6.18 An LAS-generated two-dimensional sample function with $\theta = 0.5$. The field shown is 5×5 in size.

was generated using the two-dimensional LAS algorithm and is shown in Figure 6.18. The field, which is of dimension 5×5, was subdivided eight times to obtain a 256×256 resolution giving relatively small cells of size $\frac{5}{256} \times \frac{5}{256}$. The estimated covariances along three different directions are seen in Figure 6.19 to show very good agreement with the exact. The agreement improves (as $1/n_{\text{sim}}$) when the statistics are averaged over a larger number of simulations. Notice that the horizontal axis on Figure 6.19 extends beyond a lag of 5 to accommodate the estimation of the covariance along the diagonal (which has length $5\sqrt{2}$).

In three dimensions, the LAS method involves recursively subdividing rectangular parallelepipeds into eight equal volumes at each stage. The generating relationships are essentially the same as in the two-dimensional case except now seven random noises are used in the subdivision of each parent volume at each stage

$$Z_s^{i+1} = \sum_{l=1}^{27} A_{ls} Z_l^i + \sum_{r=1}^{s} L_{sr} U_r, \qquad s = 1, 2, \ldots, 7 \quad (6.83)$$

$$Z_8^{i+1} = 8Z_{14}^i - \sum_{s=1}^{7} Z_s^{i+1} \quad (6.84)$$

in which Z_s^{i+1} denotes a particular octant of the subdivided cell centered at Z_{14}^i. Equation 6.83 assumes a neighborhood size of $3 \times 3 \times 3$, and the subdivided cell is Z_{14}^i at the center of the neighborhood.

Figure 6.20 compares the estimated and exact covariance of a three-dimensional first-order Markov process having isotropic covariance

$$C(\tau_1, \tau_2, \tau_3) = \sigma^2 \exp\left\{-\frac{2}{\theta}\sqrt{\tau_1^2 + \tau_2^2 + \tau_3^2}\right\} \quad (6.85)$$

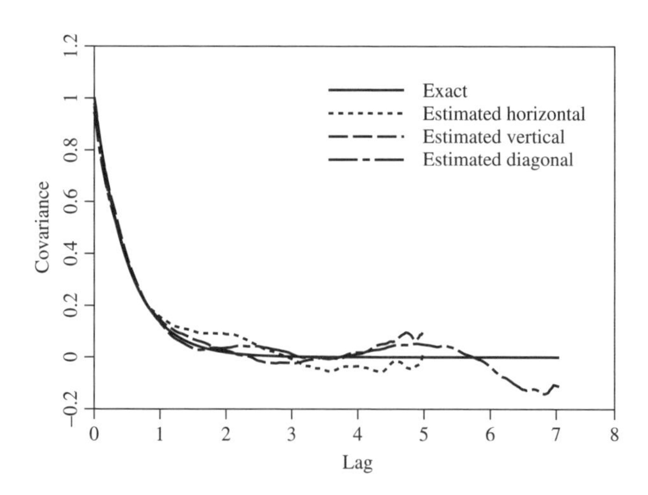

Figure 6.19 Comparison of exact and estimated covariance functions (averaged over 100 realizations) of two-dimensional isotropic Markov process with $\sigma = 1$ and $\theta = 0.5$.

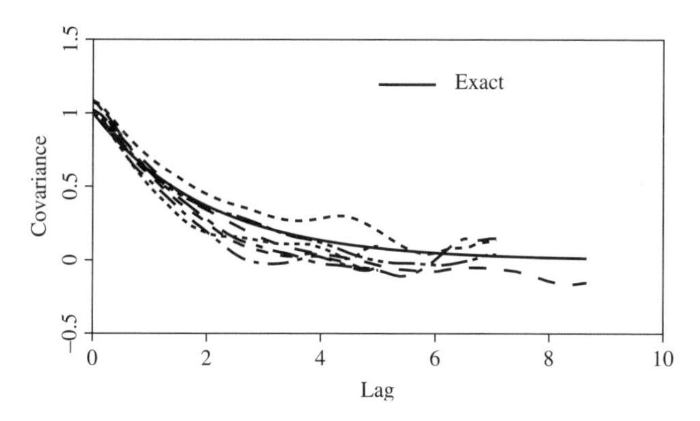

Figure 6.20 Comparison of exact and estimated covariance functions (averaged over 50 realizations) of three-dimensional isotropic Markov process with $\sigma = 1$ and $\theta = 0.5$. Dashed lines show covariance estimates in various directions through the field.

The physical field size of $5 \times 5 \times 5$ was subdivided six times to obtain a resolution of $64 \times 64 \times 64$ and the covariance estimates were averaged over 50 realizations.

6.4.6.3 Implementation and Accuracy To calculate stage $i + 1$ values, the values at stage i must be known. This implies that in the one-dimensional case, storage must be provided for at least $1.5N$ values where $N = 2^L$ is the desired number of intervals of the process. If rapid "zooming out" of the field is desired, it is useful to store all previous stages. This results in a storage requirement of $2N - 1$ in one dimension, $\frac{4}{3}(N \times N)$ in two dimensions, and $\frac{8}{7}(N \times N \times N)$ in three dimensions. The coefficients $\boldsymbol{A}$ and the lower triangular elements of $\boldsymbol{L}$, which must also be stored, can be efficiently calculated using Gaussian elimination and Cholesky decomposition, respectively.

In two and higher dimensions, the LAS method, as presented above with a neighborhood size of 3 in each direction, is incapable of preserving anisotropy in the covariance structure. The directional correlation lengths tend toward the minimum for the field. To overcome this problem, the LAS method can be mixed with the covariance matrix decomposition (CMD) method (see Eq. 6.21). As mentioned in Section 6.4.2, the CMD method requires large amounts of storage and is prone to numerical error when the field to be simulated is not small. However, the first several stages of the local average field could be produced directly by the CMD method and then refined by LAS in subsequent stages until the desired field resolution is obtained. The resulting field would have anisotropy preserved at the large scale.

Specifically, in the one-dimensional case, a positive integer k_1 is found so that the total number of cells, N_1, desired in the final field can be expressed as

$$N_1 = k_1(2^m) \tag{6.86}$$

where m is the number of subdivisions to perform and k_1 is as large as possible with $k_1 \leq k_{\max}$. The choice of the upper bound $k_{\max}$ depends on how large the initial covariance matrix used in Eq. 6.21 can be. If $k_{\max}$ is too large, the Cholesky decomposition of the initial covariance matrix will be prone to numerical errors and algorithmic nonpositive definiteness (which means that the Cholesky decomposition will fail). The authors suggest $k_{\max} \leq 256$.

In two dimensions, two positive integers k_1 and k_2 are found such that $k_1 k_2 \leq k_{\max}$ and the field dimensions can be expressed as

$$N_1 = k_1(2^m) \tag{6.87a}$$

$$N_2 = k_2(2^m) \tag{6.87b}$$

from which the first $k_1 \times k_2$ lattice of cell values are simulated directly using covariance matrix decomposition (Eg. 6.21). Since the number of subdivisions, m, is common to the two parameters, one is not entirely free to choose N_1 and N_2 arbitrarily. It docs, however, give a reasonable amount of discretion in generating nonsquare fields, as is also possible with both the FFT and TBM methods.

Although Figure 6.5 illustrates the superior performance of the LAS method over the FFT method in one dimension with respect to the covariance, a systematic bias in the variance field is observed in two dimensions (Fenton, 1994). Figure 6.21 shows a gray scale image of the

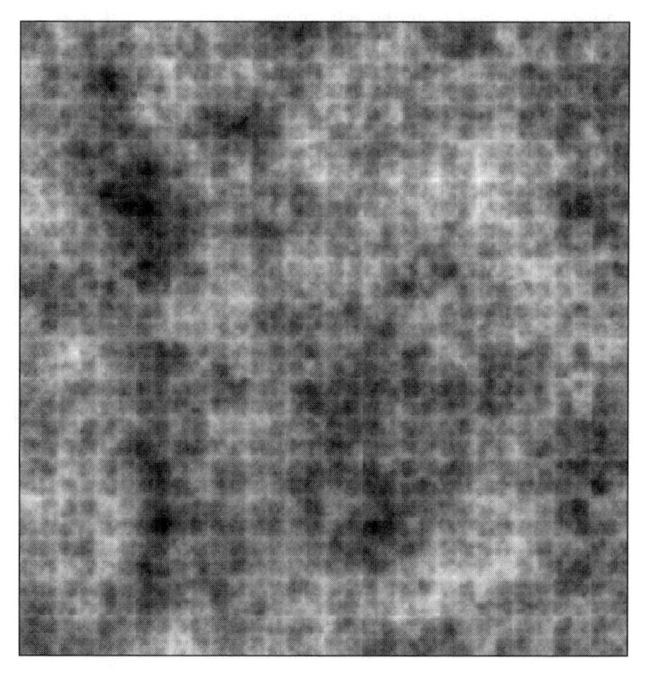

Figure 6.21 Two-dimensional LAS-generated variance field (averaged over 200 realizations).

estimated cell variance in a two-dimensional field obtained by averaging over the ensemble. There is a pattern in the variance field—the variance tends to be lower near the major cell divisions, that is, at the $\frac{1}{2}$, $\frac{1}{4}$, $\frac{1}{8}$, ... points of the field. This is because the actual diagonal, or variance, terms of the 4×4 covariance matrix corresponding to a subdivided cell are affected by the truncation of the parent cell influence to a 3×3 neighborhood. The error in the variance is compounded at each subdivision stage and cells close to "older" cell divisions show more error than do "interior" cells. The magnitude of this error varies with the number of subdivisions, the correlation length, and type of covariance function governing the process.

Figure 6.22 depicts the estimated variances along a line through the plane for both the LAS and TBM methods. Along any given line, the pattern in the LAS estimated variance seen in Figure 6.21 is not particularly noticeable, and the values are about what would be expected for an estimate over the ensemble. Figure 6.23 compares the estimated covariance structure in the vertical and horizontal directions, again for the TBM (64 lines) and LAS methods. In this respect, both the LAS and the TBM methods are reasonably accurate.

Figure 6.24 illustrates how well the LAS method combined with CMD preserves anisotropy in the covariance structure. In this figure the horizontal correlation length is $\theta_x = 10$ while the vertical correlation length is $\theta_y = 1$. As mentioned earlier the LAS algorithm, using a neighborhood size of 3, is incapable of preserving anisotropy. The anisotropy seen in Figure 6.24 is due to the initial CMD. The loss of anisotropy at very small lags (at the smaller

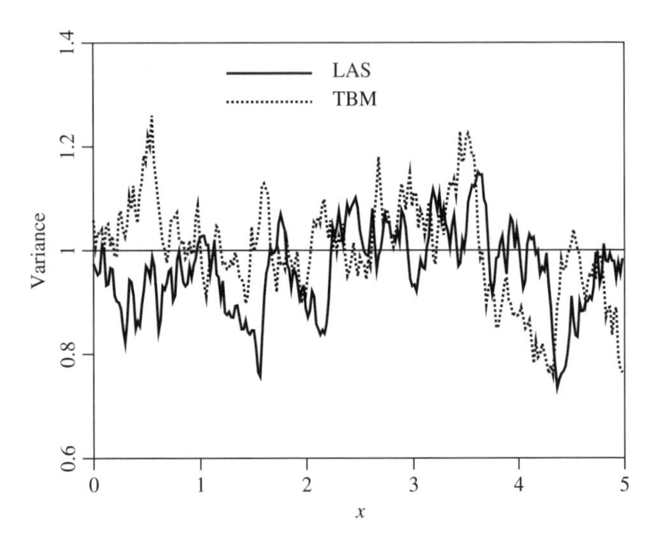

Figure 6.22 Variance along horizontal line through two-dimensional LAS and TBM fields estimated over 200 realizations.

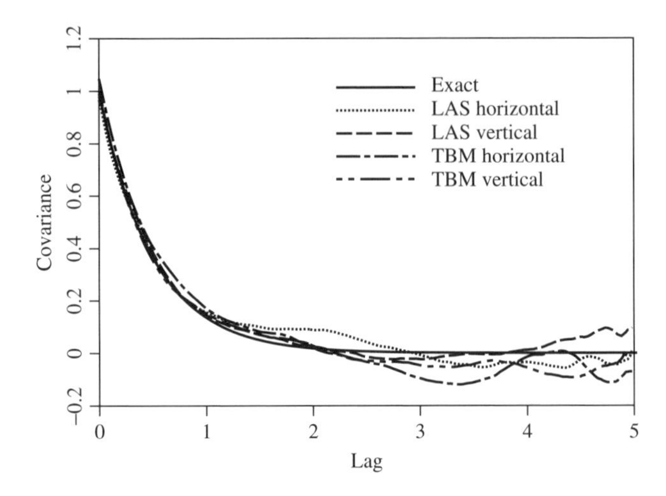

Figure 6.23 Covariance structure of LAS and TBM two-dimensional random fields estimated over 200 realizations.

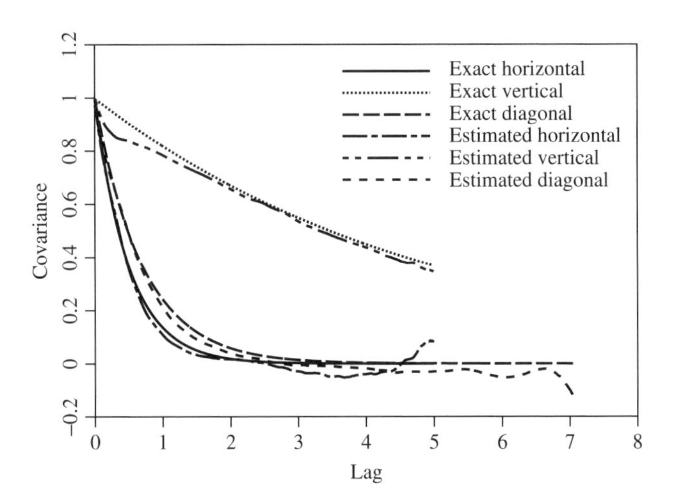

Figure 6.24 Exact and estimated covariance structure of anisotropic LAS-produced field with $\theta_x = 10$ and $\theta_y = 1$. The estimation is over 500 realizations.

scales where the subdivision is taking place) can be seen in the figure—that is, the estimated horizontal covariance initially drops too rapidly at small lags.

It may be possible to improve the LAS covariance approximations by extending the size of the parent cell neighborhood. A 3×3 neighborhood is used in the current implementation of the two-dimensional LAS algorithm, as shown in Figure 6.17, but any odd-sized neighborhood could be used to condition the statistics of the subdivided cells. Larger neighborhoods have not been tested in two and higher dimensions, although in one dimension increasing the neighborhood size to five cells resulted in a more accurate covariance function representation, as would be expected.

6.4.7 Comparison of Methods

The choice of a random-field generator to be used for a particular problem or in general depends on many issues. Table 6.1 shows the relative run times of the three algorithms to produce identically sized fields. The times have been normalized with respect to the FFT method so that a value of 2 indicates that the method took twice as long as did the FFT. If efficiency alone were the selection criteria, then either the TBM with a small number of lines or the LAS methods would be selected, with probably the LAS a better choice if streaking is not desired. However, efficiency of the random-field generator is often not an overriding concern—in many applications, the time taken to generate the field is dwarfed by the time taken to subsequently process or analyze the field by, for example, using the finite-element method. Substantial changes in generator efficiency may be hardly noticed by the user.

As a further comparison of the accuracy of the FFT, TBM, and LAS methods, a set of 200 realizations of a 128×128 random field were generated using the Markov covariance function with a correlation length $\theta = 2$ and a physical field size of 5×5. The mean and variance fields were calculated by estimating these quantities at each point in the field (averaging over the ensemble) for each algorithm. The upper and lower 90th percentiles are listed in Table 6.2 along with those predicted by theory under a normal distribution. To obtain these numbers, the mean and variance fields were first estimated, then upper and lower bounds were found such that 5% of the field exceeded the bounds above and below, respectively. Thus, 90% of the field is observed to lie between the bounds. It can be seen that all three methods yield very good results with respect to the expected mean and variance quantiles. The TBM results were obtained using 64 lines. Although these results are strictly only valid for the particular covariance function used, they are believed to be generally true over a wider variety of covariance functions and correlation lengths.

Table 6.1 Comparison of Run Times of FFT, TBM, and LAS Algorithms in One and Two Dimensions

			TBM	
Dimension	FFT	LAS	16 Lines	64 Lines
One	1.0	0.70	–	–
Two	1.0	0.55	0.64	2.6

Table 6.2 Upper and Lower 90th Percentiles of Estimated Mean and Variance Fields for FFT, TBM, and LAS Methods (200 realizations)

Algorithm	Mean	Variance
FFT	$(-0.06, 0.12)$	$(0.87, 1.19)$
TBM	$(-0.11, 0.06)$	$(0.83, 1.14)$
LAS	$(-0.12, 0.09)$	$(0.82, 1.13)$
Theory	$(-0.12, 0.12)$	$(0.84, 1.17)$

Purely on the basis of accuracy in the mean, variance, and covariance structures, the best algorithm of those considered here is probably the TBM method using a large number of lines. The TBM method is also one of the easiest to implement once an accurate one-dimensional generator has been implemented. Unfortunately, there is no clear rule regarding the minimum number of lines to be used to avoid streaking. In two dimensions using the Markov covariance function, it appears that at least 50 lines should be employed. However, as mentioned, narrow-band processes may require more. In three dimensions, no such statements can be made due to the difficulty in studying the streaking phenomena off a plane. Presumably one could use a "density" of lines similar to that used in the two-dimensional case, perhaps subtending similar angles, as a guide. The TBM method is reasonably easy to use in practice as long as the equivalent one-dimensional covariance or spectral density function can be found.

The FFT method suffers from symmetry in the covariance structure of the realizations. This can be overcome by generating fields twice as large as required in each coordinate direction and ignoring the surplus. This correction results in slower run times (a factor of 2 in one dimension, 4 in two dimensions, etc.). The FFT method is also relatively easy to implement and the algorithm is similar in any dimension. Its ability to easily handle anisotropic fields makes it the best choice for such problems. Care must be taken when selecting the physical field dimension and discretization interval to ensure that the spectral density function is adequately approximated. This latter issue makes the method more difficult to use in practice. However, the fact that the FFT approach employs the spectral density function directly makes it an intuitively attractive method, particularly in time-dependent applications.

The LAS method has a systematic bias in the variance field, in two and higher dimensions, which is not solvable without increasing the parent neighborhood size. However, the error does not result in values of variance that lie outside what would be expected from theory—it is primarily the pattern of the variance field which is of concern. Of the three methods considered, the LAS method is the most difficult to implement. It is, however, one of the easiest to use once coded since it requires no decisions regarding its parameters, and it is generally the most efficient. If the problem at hand requires or would benefit from a local average representation, then the LAS method is the logical choice.

6.5 CONDITIONAL SIMULATION OF RANDOM FIELDS

When simulation is being used to investigate the probabilistic nature of a particular site, we often have experimental information available at that site that should be reflected in the simulations. For example, suppose we are investigating the response of a soil site to loading from a structure, and we have four soil property measurements taken at spatial locations $\mathbf{x}_1, \mathbf{x}_2, \mathbf{x}_3$, and $\mathbf{x}_4$. Since we now know the soil properties at these four locations, it makes sense for any simulated soil property field to have the known values at these points in every simulation. The soil properties should only be random between the measured locations, becoming increasingly random with distance from the measured locations.

A random field which takes on certain known values at specific points in the field is called a *conditional random field*. This section seeks to produce simulations of a conditional random field, $Z_c(\mathbf{x})$, which takes on specific values $z(\mathbf{x}_\alpha)$ at the measurement locations $\mathbf{x}_\alpha$, $\alpha = 1, 2, \ldots, n_k$, where n_k is the number of measurement locations. Mathematically,

$$Z_c(\mathbf{x}) = \{Z(\mathbf{x}) \mid z(\mathbf{x}_\alpha), \alpha = 1, 2, \ldots, n_k\} \tag{6.88}$$

To accomplish the conditional simulation, the random field will be separated into two parts spatially: (1) $\mathbf{x}_\alpha$, $\alpha = 1, 2, \ldots, n_k$, being those points at which measurements have been taken, and at which the random field takes on deterministic values $z(\mathbf{x}_\alpha)$, and (2) $\mathbf{x}_\eta$, $\eta = 1, 2, \ldots, N - n_k$, being those points at which the random field is still random and at which we wish to simulate realizations of their possible random values. That is, the subscript α will denote known values, while the subscript η will denote unknown values which are to be simulated. N is the total number of points in the field to be simulated.

The conditional random field is simply formed from three components:

$$Z_c(\mathbf{x}) = Z_u(\mathbf{x}) + [Z_k(\mathbf{x}) - Z_s(\mathbf{x})] \tag{6.89}$$

where,

$Z_c(\mathbf{x})$ = desired conditional simulation

$Z_u(\mathbf{x})$ = unconditional simulation

$Z_k(\mathbf{x})$ = best linear unbiased estimate of field based on known (measured) values at $\mathbf{x}_\alpha$

$Z_s(\mathbf{x})$ = best linear unbiased estimate of field based on unconditional simulation values at $\mathbf{x}_\alpha$

The BLUE is discussed in more detail in Section 4.1. However, the best estimate at the measurement points, $\mathbf{x}_\alpha$, is just equal to the value at the measurement points. In other words, at each $\mathbf{x}_\alpha$, $Z_k(\mathbf{x}_\alpha) = z(\mathbf{x}_\alpha)$, while $Z_s(\mathbf{x}_\alpha) = Z_u(\mathbf{x}_\alpha)$. Thus, at each measurement point, $\mathbf{x}_\alpha$, Eq. 6.89 becomes

$$\begin{aligned} Z_c(\mathbf{x}_\alpha) &= Z_u(\mathbf{x}_\alpha) + [Z_k(\mathbf{x}_\alpha) - Z_s(\mathbf{x}_\alpha)] \\ &= Z_u(\mathbf{x}_\alpha) + [z(\mathbf{x}_\alpha) - Z_u(\mathbf{x}_\alpha)] = z(\mathbf{x}_\alpha) \end{aligned} \tag{6.90}$$

which is the measured value, as desired.

The unconditional simulation Z_u can be produced using one of the methods discussed in the previous sections. The BLUE of the field is obtained using the methodology presented in Section 4.1. In particular, the BLUE field based on the measured values, $Z_k(\mathbf{x})$, is determined by

$$Z_k(\mathbf{x}_\eta) = \mu_\eta + \sum_{\alpha=1}^{n_k} \beta_\alpha \Big(z(\mathbf{x}_\alpha) - \mu_\alpha \Big) \tag{6.91}$$

for $\eta = 1, 2, \ldots, N - n_k$, where μ_η is the unconditional field mean at $\mathbf{x}_\eta$, μ_α is the unconditional field mean at $\mathbf{x}_\alpha$, $z(\mathbf{x}_\alpha)$ is the measured value at $\mathbf{x}_\alpha$, and β_α is a weighting coefficient to be discussed shortly.

Similarly, the BLUE field of the simulation, $Z_s(\mathbf{x})$, is determined by

$$Z_s(\mathbf{x}_\eta) = \mu_\eta + \sum_{\alpha=1}^{n_k} \beta_\alpha \Big(Z_u(\mathbf{x}_\alpha) - \mu_\alpha \Big) \tag{6.92}$$

for $\eta = 1, 2, \ldots, N - n_k$. The only substantial difference between $Z_k(\mathbf{x})$ and $Z_s(\mathbf{x})$ is that the former is based on observed values, $z(\mathbf{x}_\alpha)$, while the latter is based on unconditional simulation values at the same locations, $Z_u(\mathbf{x}_\alpha)$. The difference appearing in Eq. 6.89 can be computed more efficiently and directly as

$$Z_k(\mathbf{x}_\eta) - Z_s(\mathbf{x}_\eta) = \sum_{\alpha=1}^{n_k} \beta_\alpha \Big(z(\mathbf{x}_\alpha) - Z_u(\mathbf{x}_\alpha) \Big) \tag{6.93}$$

for $\eta = 1, 2, \ldots, N - n_k$.

The weighting coefficients, β_α, are determined from

$$\beta = \boldsymbol{C}^{-1}\mathbf{b} \tag{6.94}$$

where β is the vector of weighting coefficients, β_α, and $\boldsymbol{C}$ is the $n_k \times n_k$ matrix of covariances between the unconditional random field values at the known points. The matrix $\boldsymbol{C}$ has components

$$C_{ij} = \text{Cov}\left[Z_u(\mathbf{x}_i), Z_u(\mathbf{x}_j)\right] \tag{6.95}$$

for $i, j = 1, 2, \ldots, n_k$. Finally, $\mathbf{b}$ is a vector of length n_k containing the covariances between the unconditional random field values at the known points and the prediction point, $Z_u(\mathbf{x}_\eta)$. It has components

$$b_\alpha = \text{Cov}\left[Z_u(\mathbf{x}_\eta), Z_u(\mathbf{x}_\alpha)\right] \tag{6.96}$$

for $\alpha = 1, 2, \ldots, n_k$.

Since $\boldsymbol{C}$ is dependent on the covariances between the known points, it only needs to be inverted once and can be used repeatedly in Eq. 6.94 to produce the vector of weights β for each of the $N - n_k$ best linear unbiased estimates (Eq. 6.93).

The conditional simulation of a random field proceeds in the following steps:

1. Partition the field into the known (α) and unknown (η) points.
2. Form the covariance matrix $\boldsymbol{C}$ between the known points (Eq. 6.95) and invert it to determine $\boldsymbol{C}^{-1}$ (in practice, this will more likely be an LU decomposition, rather than a full inversion),
3. Simulate the unconditional random field, $Z_u(\mathbf{x})$, at all points in the field.
4. for each unknown point, $\eta = 1, 2, \ldots, N - n_k$:
   (a) Form the vector $\mathbf{b}$ of covariances between the target point, $\mathbf{x}_\eta$ and each of the known points (Eq. 6.96).
   (b) Solve Eq. 6.94 for the weighting coefficients β_α, $\alpha = 1, 2, \ldots, n_k$.
   (c) Compute the difference $Z_k(\mathbf{x}_\eta) - Z_s(\mathbf{x}_\eta)$ by Eq. 6.93.
   (d) Form the conditioned random field by Eq. 6.89 at each $\mathbf{x}_\eta$.

6.6 MONTE CARLO SIMULATION

One of the primary goals of simulation is to estimate means, variances, and probabilities associated with the response of complex systems to random inputs. While it is generally preferable to evaluate these response statistics and/or probabilities analytically, where possible, we are often interested in systems which defy analytical solutions. For such systems, simulation techniques are ideal since they are simple and lead to direct results. The main disadvantage of simulation-derived moments or probabilities is that they do not lead to an understanding of how the probabilities or moments will change with changes in the system or input parameters. If the system is changed, the simulation must be repeated in order to determine the effect on response statistics and probabilities.

Consider the problem of determining the probability of failure of a system which has two random inputs, X_1 and X_2. The response of the system to these inputs is some function $g(X_1, X_2)$ which is also random because the inputs are random. For example, X_1 could be live load acting on a footing, X_2 could be dead load, and $g(X_1, X_2)$ would be the amount that the footing settles under these loads (in this example, we are assuming that the soil properties are nonrandom, which is unlikely—more likely that g is a function of a large number of random variables including the soil).

Now assume that system failure will occur whenever $g(X_1, X_2) > g_{\text{crit}}$. In the space of (X_1, X_2) values, there will be some region in which $g(X_1, X_2) > g_{\text{crit}}$, as illustrated in Figure 6.25, and the problem boils down to assessing the probability that the particular (X_1, X_2) which actually occurs will fall into the failure region. Mathematically, we are trying to determine the probability p_f, where

$$p_f = \mathrm{P}\left[g(X_1, X_2) > g_{\text{crit}}\right] \tag{6.97}$$

Let us further suppose, for example, that X_1 and X_2 follow a bivariate lognormal distribution (see Section 1.10.9.1) with mean well within the safe region and correlation coefficient between X_1 and X_2 of $\rho = -0.6$ (a negative correlation implies that as X_1 increases, X_2 tends to decrease—this is just an example). The distribution is illustrated in Figure 6.26. In terms of this joint distribution, the probability of failure

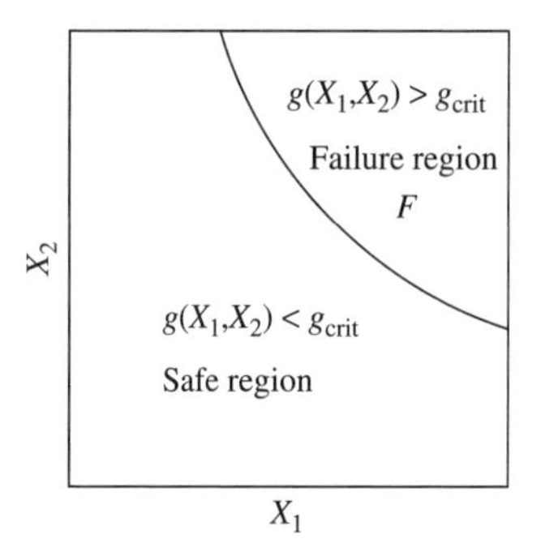

Figure 6.25 Failure and safe regions on the (X_1, X_2) plane.

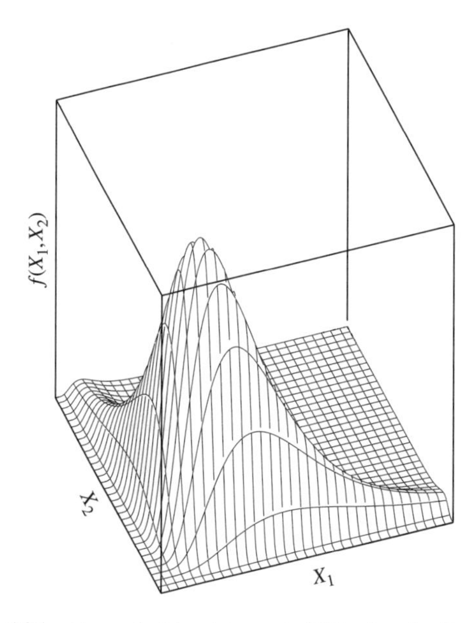

Figure 6.26 Example bivariate probability density function of X_1 and X_2.

can be expressed in terms of the joint probability density function, $f_{X_1X_2}(x_1, x_2)$:

$$p_f = \int_{x_2 \in F} \int_{x_1 \in F} f_{X_1X_2}(x_1, x_2)\, dx_1\, dx_2 \tag{6.98}$$

in which F denotes the failure region. Unfortunately, the lognormal distribution has no closed-form integral and so Eq. 6.98 must be evaluated numerically. One approach is to use some sort of numerical integration rule (such as Gaussian quadrature, see Appendix B). For nonrectangular failure regions, numerical integration algorithms may be quite difficult to implement.

An alternative and quite simple approach to evaluating Eq. 6.98 is to randomly simulate a sequence of realizations of X_1 and X_2, evaluate $g(X_1, X_2)$ for each, and check to see if $g(X_1, X_2)$ is greater than g_{crit} or not. This is called a *Monte Carlo simulation*. In detail, if x_{1i} and x_{2i} are the ith realizations of X_1 and X_2, respectively, for $i = 1, 2, \ldots, n$, and we define

$$I_i = \begin{cases} 1 & \text{if } g(x_{1i}, x_{2i}) > g_{\text{crit}} \\ 0 & \text{otherwise} \end{cases} \tag{6.99}$$

for each i, then our estimate of p_f is simply

$$\hat{p}_f = \frac{1}{n} \sum_{i=1}^{n} I_i \tag{6.100}$$

Or, in other words, the estimated probability of failure is equal to the number of realizations which failed divided by the total number of realizations.

Figure 6.27 illustrates the Monte Carlo simulation concept. Each circle represents a particular realization of (X_1, X_2) simulated from its joint distribution (Figure 6.26). Each plot shows 1000 realizations of (X_1, X_2). If a histogram of these realizations were to be constructed, one would obtain an estimate of the joint pdf of (X_1, X_2) shown in Figure 6.26. That is, Monte Carlo simulation allows the entire pdf to be estimated, not just the mean, variance, and exceedance probabilities.

Since none of the realizations in the left plot of Figure 6.27 lead to $g(x_1, x_2) > g_{\text{crit}}$ (i.e., fall in the failure region F), our estimate of p_f from this particular set of 1000 realizations is

$$\hat{p}_f = \frac{0}{1000} = 0$$

We know that this result cannot be correct since the lognormal distribution is unbounded in the positive direction (i.e., there will always be some nonzero probability that X will be greater than any number if X is lognormally distributed). In other words, sooner or later one or more realizations of (X_1, X_2) will appear in the failure region.

In the right plot of Figure 6.27, which is another 1000 realizations using a different starting seed, two of the realizations do fall in the failure region. In this case

$$\hat{p}_f = \frac{2}{1000} = 0.002$$

Figure 6.27 illustrates a fundamental issue relating to the accuracy of a probability estimate obtained from a Monte Carlo simulation. From these two plots, the true probability of failure could be anywhere between $p_f = 0$ to somewhat in excess of $p_f = 0.002$. If the target probability of failure is $p_f = 1/10{,}000 = 0.0001$, then clearly 1000 realizations are not sufficient to resolve this probability since one is unlikely to see a single failure from among the 1000 realizations. The question is: How many realizations should be performed in order to estimate p_f to within some acceptable accuracy? This question is reasonably easily answered by recognizing that I_i is a Bernoulli random variable so that we can make use of the statistical theory presented in Chapter 1. The standard deviation of $\hat{p}_f$ is, approximately,

$$\sigma_{\hat{p}_f} \simeq \sqrt{\frac{\hat{p}_f \hat{q}_f}{n}} \tag{6.101}$$

where the estimate of p_f is used (since p_f is unknown) and $\hat{q}_f = 1 - \hat{p}_f$. The two-sided $(1 - \alpha)$ confidence interval,

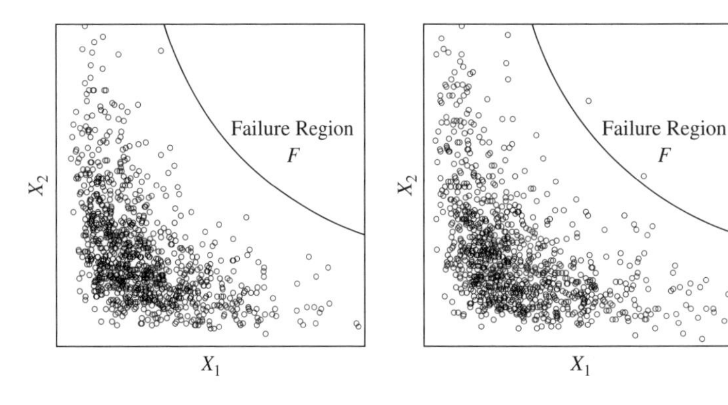

Figure 6.27 Two typical 1000-realization Monte Carlo simulations of (X_1, X_2). Points appearing within the failure region correspond to system failure.

$[L, U]$, on p_f is

$$[L, U]_{1-\alpha} = \hat{p}_f \pm z_{\alpha/2}\sqrt{\frac{\hat{p}_f \hat{q}_f}{n}} \tag{6.102}$$

where $z_{\alpha/2}$ is the point on the standard normal distribution satisfying $P[Z > z_{\alpha/2}] = \alpha/2$ (see the last line of Table A.2) and L and U are the lower and upper bounds of the confidence interval, respectively. This confidence interval makes use of the normal approximation to the binomial and is only valid if np_f and nq_f are large enough (see Section 1.10.8.2).

For example, if $\alpha = 0.05$, and $\hat{p}_f = 0.002$, based on the right plot of Figure 6.27 where $n = 1000$, then a 95% two-sided confidence interval on the true p_f is

$$\begin{aligned}
[L, U]_{0.95} &= 0.002 \pm z_{0.05}\sqrt{\frac{0.002(0.998)}{1000}} \\
&= 0.002 \pm (1.960)(0.001413) \\
&= [-0.00077, 0.0048]
\end{aligned}$$

This is a pretty wide confidence interval, which suggests that 1000 realizations is insufficient to properly resolve the true probability of failure in this case. Note also that the confidence interval includes a negative lower bound, which implies also that n is not large enough for this confidence interval to be properly found using a normal approximation to the binomial (the probability of failure cannot be negative).

The confidence interval idea can be used to prescribe the required number of realizations to attain a certain accuracy at a certain confidence level. For example, suppose we wish to estimate p_f to within 0.0005 with confidence 90%. The confidence interval $[L, U]_{1-\alpha}$ basically says that we are $(1 - \alpha)$ confident that the true value of p_f lies between L and U. Since L and U are centered on $\hat{p}_f$, another way of putting this is that we are $(1 - \alpha)$ confident that the true p_f is within $z_{\alpha/2}\sigma_{\hat{p}_f}$ of $\hat{p}_f$, since $z_{\alpha/2}\sigma_{\hat{p}_f}$ is half the confidence interval width.

We can use this interpretation to solve for the required value of n: Our desired maximum *error* on p_f is 0.0005 at confidence 90% ($\alpha = 0.10$), so we solve

$$0.0005 = z_{\alpha/2}\sigma_{\hat{p}_f} = z_{\alpha/2}\sqrt{\frac{\hat{p}_f \hat{q}_f}{n}}$$

for n. This gives

$$\begin{aligned}
n &= \hat{p}_f \hat{q}_f \left(\frac{z_{\alpha/2}}{0.0005}\right)^2 = 0.002(0.998)\left(\frac{1.645}{0.0005}\right)^2 \\
&= 21{,}604
\end{aligned}$$

which, as expected, is much larger than the 1000 realizations used in Figure 6.27.

In general, if the maximum error on p_f is e at confidence $1 - \alpha$, then the required number of realizations is

$$n = \hat{p}_f \hat{q}_f \left(\frac{z_{\alpha/2}}{e}\right)^2 \tag{6.103}$$

Figure 6.28 illustrates 100,000 realizations of (X_1, X_2). We now see that 49 of those realizations fell into the failure region, so that our improved estimate of p_f is

$$\hat{p}_f = \frac{49}{100{,}000} = 0.00049$$

which is quite different than suggested by the other attempts in Figure 6.27. If we want to refine this estimate and calculate it to within an error of 0.0001 (i.e., having a confidence interval [0.00039, 0.00059]) at confidence level 90%, we will need

$$n = (0.00049)(0.99951)\left(\frac{1.645}{0.0001}\right)^2 = 132{,}530$$

In other words, the 100,000 realizations used in Figure 6.28 gives us slightly less than 0.0001 accuracy on p_f with 90% confidence.

We note that we are often interested in estimating very small failure probabilities—most civil engineering works have target failure probabilities between 1/1000 and 1/100,000. As we saw above, estimating failure probabilities accurately in this range typically requires a very large number of realizations. Since the system response $g(X_1, X_2, \ldots)$ sometimes takes a long time to compute for each combination of $(X_1, X_2, \ldots)$, for example, when g involves a nonlinear finite-element analysis, large numbers of realizations may not be practical.

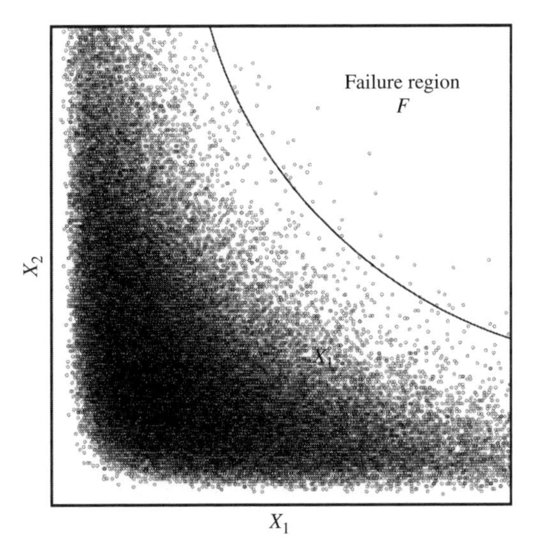

Figure 6.28 100,000 Monte Carlo simulations of (X_1, X_2). Points appearing within the failure region correspond to system failure.

There are at least three possible solutions when a large number (e.g., hundreds of thousands or millions) of realizations are impractical:

1. Perform as many realizations as practical, form a histogram of the response, and fit a distribution to the histogram. The fitted distribution is then used to predict failure probabilities. The assumption here is that the distribution of the system response continues to be modeled by the fitted distribution in the tails of the distribution. This is often believed to be a reasonable assumption. In order to produce a reasonably accurate histogram, the number of realizations should still be reasonably large (e.g., 500 or more).
2. Develop an analytical model for the probability of failure by determining the distribution of $g(X_1, X_2, \ldots)$; see Section 1.8. If the analytical model involves approximations, as they often do, some simulations should be performed to validate the model. The analytical model is then used to predict failure probabilities.
3. Employ variance reduction techniques to reduce the required number of realizations to achieve a desired accuracy. In the context of random fields, these techniques tend to be difficult to implement and will not be pursued further in this book. The interested reader is referred to Law and Kelton (2000) or Lewis and Orav (1989).

Monte Carlo simulations can also be used to estimate the moments of the response, $g(X_1, X_2, \ldots)$. If $(X_{1i}, X_{2i}, \ldots)$ is the ith realization, then the mean of g is estimated from

$$\hat{\mu}_g = \frac{1}{n}\sum_{i=1}^{n} g(X_{1i}, X_{2i}, \ldots) \tag{6.104}$$

and the variance of g is estimated from

$$\hat{\sigma}_g^2 = \frac{1}{n-1}\sum_{i=1}^{n}\left(g(X_{1i}, X_{2i}, \ldots) - \hat{\mu}_g\right)^2 \tag{6.105}$$

The error in the estimate of the mean, $\hat{\mu}_g$, decreases as the number of simulations n increases. The standard deviation of the estimate of the mean, sometimes called the *standard error*, is

$$\sigma_{\hat{\mu}_g} = \frac{\sigma_g}{\sqrt{n}} \simeq \frac{\hat{\sigma}_g}{\sqrt{n}} \tag{6.106}$$

and a $1-\alpha$ confidence interval on the true mean, μ_g, is [assuming $g(X_1, X_2, \ldots)$ is at least approximately normal]:

$$[L, U]_{1-\alpha} = \hat{\mu}_g \pm t_{\alpha/2,n-1}\frac{\hat{\sigma}_g}{\sqrt{n}} \tag{6.107}$$

where $t_{\alpha/2,n-1}$ is a percentile of the Student t-distribution; see Appendix A.2 for t-values using $\nu = n-1$. If the number of degrees of freedom, $\nu = n-1$, is large enough (e.g., larger than 100), the t-value can be replaced by $z_{\alpha/2}$, in which case the confidence interval can be used to determine the required number of simulations to achieve a certain level of accuracy on the mean estimate. For example, if we want to estimate the mean to within an error of e with confidence $(1-\alpha)$ we solve

$$n \simeq \left(\frac{z_{\alpha/2}\hat{\sigma}_g}{e}\right)^2 \tag{6.108}$$

The standard deviation (standard error) of the estimate of the variance, $\hat{\sigma}_g^2$, is

$$\sigma_{\hat{\sigma}_g^2} \simeq \hat{\sigma}_g^2\sqrt{\frac{2}{n-1}} \tag{6.109}$$

which assumes that $g(X_1, X_2, \ldots)$ is at least approximately normally distributed. Under the same assumption, the confidence interval on σ_g^2 is

$$[L, U]_{1-\alpha} = \left[\frac{(n-1)\hat{\sigma}_g^2}{\chi^2_{\alpha/2,n-1}}, \frac{(n-1)\hat{\sigma}_g^2}{\chi^2_{1-\alpha/2,n-1}}\right] \tag{6.110}$$

where $\chi^2_{\alpha,\nu}$ are quantiles of the chi-square distribution. See Appendix A.3 for values.

CHAPTER 7

Reliability-Based Design

7.1 ACCEPTABLE RISK

Before we talk about reliability-based design in geotechnical engineering, it is worth investigating the levels of risk that a reliability-based design is aiming to achieve. In many areas of design, particularly in civil engineering, the design is evaluated strictly in terms of the probability of failure, rather than by assessing both the probability of failure and the cost or consequences of failure. This is probably mostly due to the fact that the value of human life is largely undefined and a subject of considerable political and social controversy. For example, the failure of a bridge or structure may result in loss of life. How is this loss quantified? If we are to define risk as the probability of failure times the failure loss (as is commonly done), then we need a number to represent failure loss. In that this is difficult to quantify, many engineering designs proceed by considering *only* the probability of failure directly. Consequences are considered as an "add-on." For example, the National Building Code of Canada has an *importance factor* which can be set to 0.8 if the collapse of a structure is not likely to cause injury or serious consequence (such as might be the case for an unstaffed weather station located in the arctic)—in all other cases, the importance factor is taken as 1.0.

We note that the differing definitions of risk as either

1. the probability of failure or
2. the product of the probability of failure and the cost of failure

is a significant source of confusion and complicates the determination of what is an acceptable risk. We will consider "acceptable risk" here to mean "acceptable probability of failure," up to Section 7.6, but will bear in mind that this acceptable probability of failure will change with the severity of the consequences.

Most civil engineering structures are currently designed so that individual elements making up the structure have a "nominal" probability of failure of about 1 in 1000, and the same might be said about an individual geotechnical element such as a footing or pile. More specifically, we might say that for a random load L on an element with random resistance R we design such that

$$\mathrm{P}[L > R] \simeq \frac{1}{1000}$$

In fact, building codes are a bit vague on the issue of acceptable risk, partly because of the difficulty in assessing overall failure probabilities for systems as complex as entire buildings. The above failure probability is based on the loss of load-carrying capacity of a single building element, such as a beam or pile, but the codes also ensure a much lower probability of collapse by:

1. Ensuring that the system has many redundancies (if one element fails, its load is picked up by other elements)
2. Erring on the safe side in parameter estimates entering the probability estimate

So, in general, the number of failures resulting in loss of life is a good deal less than 1 in 1000 (perhaps ignoring those failures caused by deliberate sabotage or acts of war, which buildings are not generally designed against).

Another problem in determining acceptable risk lies in defining precisely what is meant by "failure"? Is this unacceptable deformations, which are unsightly, or complete destruction resulting in possible loss of life? Although the target probability of failure of about 1/1000 per element is deemed an acceptable risk, presumably this acceptable risk should change with the severity of the consequences. In other words, it is difficult to separate acceptable risk and consequence, nor should we.

In order to decide if a design is adequate, some idea of acceptable risk is necessary. Unfortunately, acceptable risk is also related to perceived risk. Consider the following two examples:

1. A staircase at an art gallery is suspended from two 25-mm cables, which are more than adequate to support the staircase. The staircase has a very small probability of failure. However, patrons are unwilling to use the staircase and so its utility is lost. Why? The patrons view the cables as being unsubstantial. Solution: Enclose the cables with fake pillars.
2. Safety standards for air travel are *much* higher than for travel by car, so annual loss of life in car accidents far

exceed that in airplane accidents. Why is there such a difference in acceptable risk levels? Presumably people are just more concerned about being suspended several thousand meters in the air with no visible means of support.

Acceptable risk has, at least, the following components:

1. **Public Opinion:** This is generally felt as public pressure on politicians, who in turn influence regulatory bodies, who in turn write the design codes.
2. **Failure Probability versus Cost Trade-off:** We accept higher risks in automobile travel partly because the alternatives are enormously expensive (at this time) and difficult to implement. Improved safety features such as air bags, antilock brakes, and roll bars are, however, an indication that our risk tolerance is decreasing. Unfortunately, *cost* is not always easy to determine (e.g., what is the value of human life?) so that this trade-off is sometimes somewhat irrational.
3. **Perceived Risk:** Some things just look like a disaster waiting to happen (as in our staircase example above), and we become unwilling to risk them despite their actual safety. The strict safety measures imposed on the airline industry may have a lot to do with our inherent fear of heights, as suggested above.

As Whipple (1986, p. 30) notes:

Traditionally, acceptable risk has been judged in engineering by whether good engineering practice has been followed, both in the application of appropriate design standards and in the analysis that results in engineering decisions where no standards apply precisely. Similarly, the courts rely on tradition-based standards in tort law to define a risk maker's responsibility to avert risk and a risk bearer's right to be free from significant risk impositions. A substantially different perspective holds in welfare economics, where risk is viewed as a social cost, and where acceptability depends to a significant degree on the costs of avoiding risk. Behind these professional perspectives is an evolving public opinion about which risks are too high and which are of little concern. One needs only note the ongoing toughening of laws concerning drunk driving and smoking in public places to see that the public definition of acceptable risk is dynamic.

One way to determined whether a risk is acceptable or not is to compare it to other common risks. For example, "the risk is the same as that of having a car accident while driving home today." As Whipple (1986) notes, the use of comparisons for judging the acceptability of risk is controversial. One criticism concerns the comparison of dissimilar risks, as Smith (1980 quoted by Whipple, 1986, p. 33) states:

A risk assessment procedure must demonstrate the relevance of the comparison. If tonsillectomies, for illustration, are less dangerous per hour than open-heart operations, it doesn't necessarily mean that the latter are too risky and that hospitals should be encouraged to remove more tonsils and to open fewer hearts. Nor does it mean that a particular energy system is acceptable merely because it is less dangerous than a tonsillectomy. The social benefits of these activities are so different that direct comparisons of their risks are nearly meaningless.

Nevertheless, comparisons are valuable, particularly if one is concentrating only on the probability of failure, and not on consequences and/or benefits. Some acceptable risk levels as suggested by Whipple (1986) are as follows:

Short-term risks, for example, recreational activities, $< 10^{-6}$/h
Occupational risks, $< 10^{-3}$/year, for example:
Logging, 1.4×10^{-3}/year
Coal mining, 6.4×10^{-4}/year
Heavy construction, 4.2×10^{-4}/year
All occupations, 1.1×10^{-4}/year
Safe occupations, 5×10^{-5}/year
Public risks, for example, living below a dam and involuntary exposure, $< 10^{-4}$/year

Risks are frequently ignored (and thus "accepted") when individual risks fall below 10^{-6}–10^{-7} per year.

One needs to be careful comparing risks in the above list. For example, some recreational activities which are quite risky on an hourly basis (e.g., scuba diving or parachuting) do not amount to a significant risk annually if few hours are spent over the year pursuing these activities.

Some individual risks per year in the United States are as follows (Whipple, 1986):

Accident death rate: 5×10^{-4}/year
Motor vehicle death rate: 2×10^{-4}/year
Fire and burns accident death rate: 4×10^{-5}/year
Accidental deaths from electric current: 5×10^{-6}/year
Accidental deaths from lightning, tornadoes, and hurricanes: 1×10^{-6}/year

Whipple compares human-caused disasters and natural disasters in Figures 7.1 and 7.2. Note that the occurrence frequency falls off with the severity of the disaster, as they should.

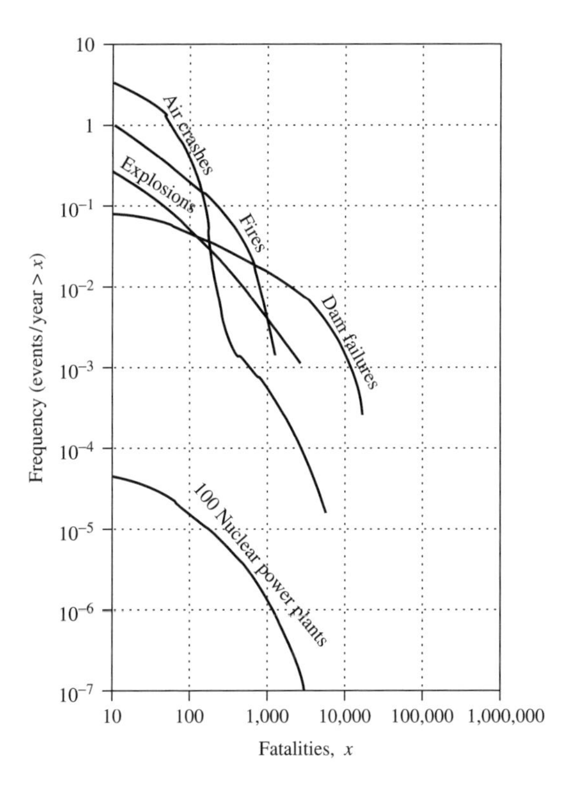

Figure 7.1 Frequency of human-caused events resulting in fatalities (Whipple, 1986).

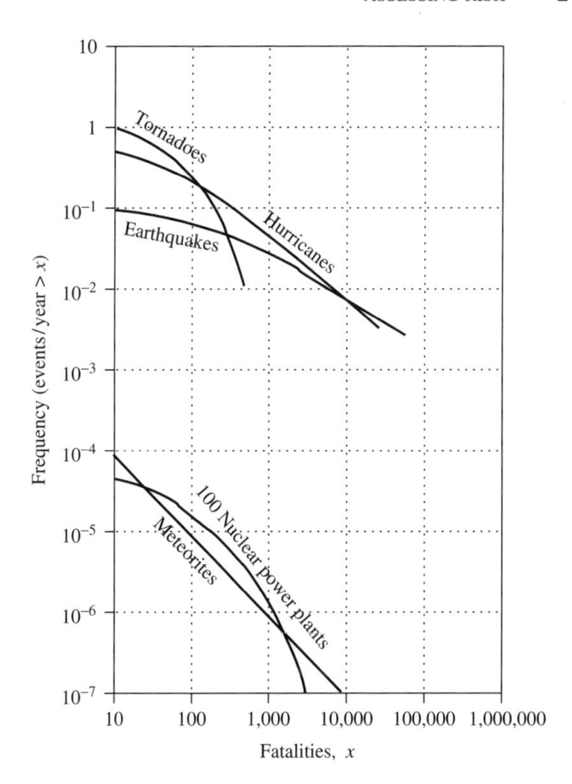

Figure 7.2 Frequency of natural disasters resulting in fatalities (Whipple, 1986).

7.2 ASSESSING RISK

Once an acceptable risk has been associated with a design, the next step is to assess the risk of failure. Many alternative approaches exist, ranging from exact analytical formulations (see Chapter 1), to approximate analytical methods (see FOSM in Section 1.8.4), to simulation methods. Chapter 6 was devoted to simulation methods. Here we will briefly discuss two popular approaches to estimating the probability of failure of a design. The first is called the Hasover–Lind first-order reliability method (FORM), and the second is the point estimate method (PEM).

7.2.1 Hasofer–Lind First-Order Reliability Method

The major drawback to the FOSM method (Section 1.8.4) when used to compute probabilities relating to failure, as pointed out by Ditlevson (1973), is that it can give different failure probabilities for the same problem when stated in equivalent, but different, ways. See also Madsen et al. (1986) and Baecher and Christian (2003) for detailed comparisons of the FOSM and FORM methods. A short discussion of the nonuniqueness of FOSM in the computation of failure probability is worth giving here, since it is this nonuniqueness that motivated Hasofer and Lind (1974) to develop an improved approach.

A key quantity of interest following an analysis using FOSM or FORM is the determination of the reliability index β for a given safety margin M. One classical civil engineering safety margin is

$$M = R - L \tag{7.1}$$

where R is the resistance and L is the load. Failure occurs when $R < L$, or, equivalently, when $M < 0$. The reliability index β, as defined by Cornell (1969), is

$$\beta = \frac{\mathrm{E}[M]}{\sqrt{\mathrm{Var}[M]}} \tag{7.2}$$

which measures how far the mean of the safety margin M is from zero (assumed to be the failure point) in units of number of standard deviations. Interest focuses on the probability that failure, $M < 0$, occurs. Since owners and politicians do not like to hear about probabilities of failure, this probability is often codified using the rather more obscure *reliability index*. There is, however, a unique relationship between the reliability index (β) and the probability of failure (p_f) given by

$$p_f = 1 - \Phi(\beta) \tag{7.3}$$

where Φ is the standard normal cumulative distribution function. Equation 7.3 assumes that M is normally distributed.

The point, line, or surface in higher dimensions, defined by $M = 0$, is generally called the *failure surface*. A similar concept was discussed in Section 6.6 where the line separating the "failure" and "safe" regions was used in a Monte Carlo analysis.

Consider the safety margin, $M = R - L$. If R is independent of L, then the FOSM method gives (see Eqs. 1.79 and 1.82)

$$\mathrm{E}[M] = \mathrm{E}[R] - \mathrm{E}[L] = \mu_R - \mu_L \tag{7.4}$$

and

$$\begin{aligned}\mathrm{Var}[M] &= \left(\frac{\partial M}{\partial R}\right)^2 \mathrm{Var}[R] + \left(\frac{\partial M}{\partial L}\right)^2 \mathrm{Var}[L] \\ &= \mathrm{Var}[R] + \mathrm{Var}[L] = \sigma_R^2 + \sigma_L^2 \end{aligned} \tag{7.5}$$

(note that because the safety margin is linear in this case, the first-order mean and variance of M are exact) so that

$$\beta = \frac{\mu_R - \mu_L}{\sqrt{\sigma_R^2 + \sigma_L^2}} \tag{7.6}$$

For nonnegative resistance and loads, as is typically the case in civil engineering, the safety margin could alternatively be defined as

$$M = \ln\left(\frac{R}{L}\right) = \ln(R) - \ln(L) \tag{7.7}$$

so that failure occurs if $M < 0$, as before. In this case, to first order,

$$\mathrm{E}[M] \simeq \ln(\mu_R) - \ln(\mu_L)$$

which is clearly no longer the same as before, and

$$\begin{aligned}\mathrm{Var}[M] &\simeq \left(\frac{\partial M}{\partial R}\right)^2 \mathrm{Var}[R] + \left(\frac{\partial M}{\partial L}\right)^2 \mathrm{Var}[L] \\ &= \frac{\mathrm{Var}[R]}{\mu_R^2} + \frac{\mathrm{Var}[L]}{\mu_L^2} = v_R^2 + v_L^2 \end{aligned} \tag{7.8}$$

where the derivatives are evaluated at the means and where v_R and v_L are the coefficients of variation of R and L, respectively. This gives a different reliability index:

$$\beta = \frac{\ln(\mu_R) - \ln(\mu_L)}{\sqrt{v_R^2 + v_L^2}} \tag{7.9}$$

The nonuniqueness of the FOSM method is due to the fact that different functional representations may have different mean estimates and different first derivatives. What the FOSM method is doing is computing the distance from the mean point to the failure surface *in the direction of the gradient at the mean point*. Hasofer and Lind (1974) solved the nonuniqueness problem by looking for the overall minimum distance between the mean point and the failure surface, rather than looking just along the gradient direction.

In the general case, suppose that the safety margin M is a function of a sequence of random variables $\mathbf{X}^T = \{X_1, X_2, \ldots\}$, that is,

$$M = f(X_1, X_2, \ldots) \tag{7.10}$$

and that the random variables $X_1, X_2, \ldots$ have covariance matrix $\boldsymbol{C}$. Then the Hasofer–Lind reliability index is defined by

$$\beta = \min_{M=0} \sqrt{(\mathbf{x} - \mathrm{E}[\mathbf{X}])^T \boldsymbol{C}^{-1} (\mathbf{x} - \mathrm{E}[\mathbf{X}])} \tag{7.11}$$

which is the minimum distance between the failure surface ($M = 0$) and the mean point ($\mathrm{E}[\mathbf{X}]$) in units of number of standard deviations—for example, if $M = f(X)$, then Eq. 7.11 simplifies to $\beta = \min_x (x - \mu_X)/\sigma_X$. Finding β under this definition is iterative; choose a value of $\mathbf{x}_0$ which lies on the curve $M = 0$ and compute β_0, choose another point $\mathbf{x}_1$ on $M = 0$ and compute β_1, and so on. The Hasofer–Lind reliability index is the minimum of all such possible values of β_i.

In practice, there are a number of sophisticated optimization algorithms, generally involving the gradient of M, which find the point where the failure surface is perpendicular to the line to the origin. The distance between these two points is β. Many spreadsheet programs now include such algorithms, and the user need only specify the minimization equation (see above) and the constraints on the solution (i.e., that $\mathbf{x}$ is selected from the curve $M = 0$ in this case). Unfortunately, nonlinear failure surfaces can sometimes have multiple local minima, with respect to the mean point, which further complicates the problem. In this case, techniques such as simulated annealing (see, e.g., Press et al, 1997) may be necessary, but which still do not guarantee finding the global minimum. Monte Carlo simulation is an alternative means of computing failure probabilities which is simple in concept, which is not limited to first order, and which can be extended easily to very difficult failure problems with only a penalty in computing time to achieve a high level of accuracy (see Section 6.6).

7.2.2 Point Estimate Method

The PEM is a simple, approximate way of determining the first three moments (the mean μ, variance σ^2, and skewness ν) of a variable that depends on one or more random input variables. Like FOSM (see Section 1.8.4) and FORM (see previous section), PEM does not require knowledge of the particular form of the probability density function of the input, nor does it typically explicitly account for spatial correlation.

The PEM is essentially a weighted average method reminiscent of numerical integration formulas involving "sampling points" and "weighting parameters." The PEM reviewed here will be the two point estimate method developed by Rosenblueth (1975, 1981) and also described by Harr (1987).

The PEM seeks to replace a continuous probability density function with a discrete function having the same first three central moments.

Steps for Implementing PEM

1. Determine the relationship between the dependent variable, W, and random input variables, $X, Y, \ldots$,

$$W = f(X, Y, \ldots) \tag{7.12}$$

2. Compute the locations of the two sampling points for each input variable. For a single random variable X with skewness ν_X the sampling points are given by

$$\xi_{X_+} = \tfrac{1}{2}\nu_X + \sqrt{1 + \left(\tfrac{1}{2}\nu_X\right)^2} \tag{7.13}$$

and

$$\xi_{X_-} = \xi_{X_+} - \nu_X \tag{7.14}$$

where ξ_{X_+} and ξ_{X_-} are standard deviation units giving the locations of the sampling points to the right and left of the mean, respectively. Figure 7.3 shows these sampling points located at $\mu_X + \xi_{X_+}\sigma_X$ and $\mu_X - \xi_{X_-}\sigma_X$. If the function depends on n variables, there will be 2^n sampling points corresponding to all combinations of the two sampling points for each variable. Figure 7.4 shows the locations of sampling points for a distribution of two random variables X

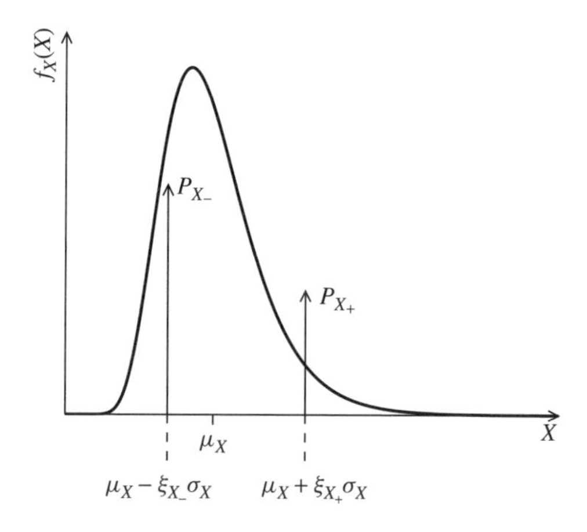

Figure 7.3 New PEM distribution.

and Y. Since $n = 2$, there are four sampling points given by

$$(\mu_X + \xi_{X_+}\sigma_X, \mu_Y + \xi_{Y_+}\sigma_Y)$$
$$(\mu_X + \xi_{X_+}\sigma_X, \mu_Y - \xi_{Y_-}\sigma_Y)$$
$$(\mu_X - \xi_{X_-}\sigma_X, \mu_Y + \xi_{Y_+}\sigma_Y)$$
$$(\mu_X - \xi_{X_-}\sigma_X, \mu_Y - \xi_{Y_-}\sigma_Y)$$

If skewness is ignored or assumed to equal zero, from Eqs. 7.13 and 7.14,

$$\xi_{X_+} = \xi_{X_-} = \xi_{Y_+} = \xi_{Y_-} = 1 \tag{7.15}$$

Each random variable then has point locations that are plus and minus one standard deviation from the mean.

3. Determine the weights P_i to give each of the 2^n point estimates. Just as a probability density function encloses an "area" of unity, so the probability weights must also sum to unity. The weights can also take into account correlation between two or more random variables. For a single random variable X, the weights are given by

$$P_{X_+} = \frac{\xi_{X_-}}{\xi_{X_+} + \xi_{X_-}} \tag{7.16a}$$

$$P_{X_-} = 1 - P_{X_+} \tag{7.16b}$$

For n random variables with no skewness, Christian and Baecher (1999) have presented a general expression for finding the weights, which takes into account the correlation coefficient ρ_{ij} between the ith and jth variables as follows:

$$P_{s_1 s_2,\ldots,s_n} = \frac{1}{2^n}\left[1 + \sum_{i=1}^{n-1}\sum_{j=i+1}^{n} s_i s_j \rho_{ij}\right] \tag{7.17}$$

where ρ_{ij} is the correlation coefficient between X_i and X_j and where $s_i = +1$ for points greater than the mean, and $s_i = -1$ for points smaller than the mean. The subscripts of the weight P indicate the location of the point that is being weighted. For example, for a point evaluated at $(x_1, y_1) = (\mu_X + \sigma_X, \mu_Y - \sigma_Y)$, $s_1 = +1$ and $s_2 = -1$ resulting in a negative product with a weight denoted by P_{+-}. For multiple random variables where skewness cannot be disregarded, the computation of weights is significantly more complicated. Rosenblueth (1981) presents the weights for the case of $n = 2$ to be the following:

$$P_{s_1 s_2} = P_{X s_1} P_{Y s_2} + s_1 s_2 \left[\frac{\rho_{XY}}{\sqrt{\left(1 + (\nu_X/2)^2\right)\left(1 + (\nu_Y/2)^2\right)}}\right] \tag{7.18}$$

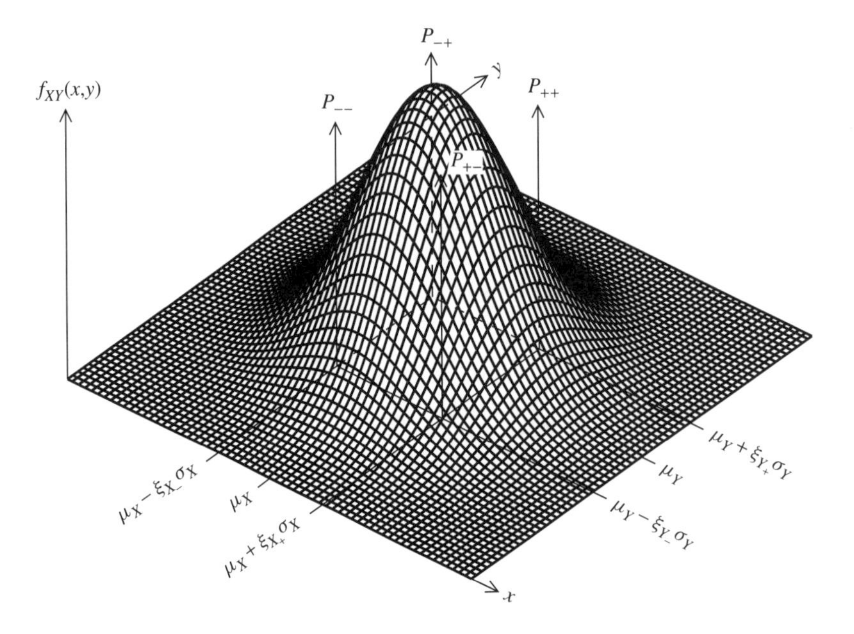

Figure 7.4 Point estimates for two random variables.

The notation is the same as for the previous equation with P_{Xs_i} and P_{Ys_j} being the weights for variables X and Y, respectively (see Eqs. 7.16). Here, ν_X is the skewness coefficient of X and ν_Y is the skewness coefficient of Y. For a lognormal distribution, the skewness coefficient ν can be calculated from the coefficient of variation v as follows (e.g., Benjamin and Cornell 1970):

$$\nu = 3 * v + v^3 \tag{7.19}$$

4. Determine the value of the dependent variable at each point. Let these values be denoted by $W_{X(+\text{ or }-)}$, $Y_{(+\text{ or }-),\ldots}$, depending upon the point at which W is being evaluated. For n random input variables, W is evaluated at 2^n points.
5. In general, the PEM enables us to estimate the expected values of the first three moments of the dependent variable using the following summations. Here, the P_i and W_i are the weight and the value of the dependent variable associated with some point location i where i ranges from 1 to 2^n. P_i is some P_{s_i,s_j} calculated in step 3 and W_i is the W_{s_i,s_j} value of the dependent variable evaluated at the specified location from step 4 above.

First moment:

$$\mu_W = \mathrm{E}[W] \simeq \sum_{i=1}^{2^n} P_i W_i \tag{7.20}$$

Second moment:

$$\sigma_W^2 = \mathrm{E}\left[(W - \mu_W)^2\right] \simeq \sum_{i=1}^{2^n} P_i (W_i - \mu_W)^2$$

$$= \sum_{i=1}^{2^n} P_i W_i^2 - \mu_W^2 \tag{7.21}$$

Third moment:

$$\nu_W = \frac{\mathrm{E}\left[(W - \mu_W)^3\right]}{\sigma_W^3} \simeq \frac{1}{\sigma_W^3} \sum_{i=1}^{2^n} P_i (W_i - \mu_W)^3$$

$$= \frac{1}{\sigma_W^3} \sum_{i=1}^{2^n} P_i W_i^3 - 3\mu_W P_i W_i^2 + 2\mu_W^3 \tag{7.22}$$

Example 7.1 Unconfined Triaxial Compression of a c', ϕ' Soil ($n = 2$) The unconfined ($\sigma_3' = 0$) compressive strength of a drained c', ϕ' soil is given from the Mohr–Coulomb equation as

$$q_u = 2c' \tan(45 + \tfrac{1}{2}\phi') \tag{7.23}$$

Considering the classical Coulomb shear strength law:

$$\tau_f = \sigma' \tan\phi' + c' \tag{7.24}$$

it is more fundamental to deal with $\tan\phi'$ (rather than ϕ') as the random variable. Thus, Eq. 7.23 can be rearranged as

$$q_u = 2c'[\tan\phi' + (1 + \tan^2\phi')^{1/2}] \tag{7.25}$$

Assuming, $\mu_{c'} = 100$ kPa, $\mu_{\tan\phi'} = \tan 30^\circ = 0.577$, and $v_{c'} = v_{\tan\phi'} = 0.5$, find the mean, variance, and skewness coefficient of q_u.

Following the steps discussed above for implementing PEM:

1. The function to be evaluated is Eq. 7.25.
2. It is assumed that the random shear strength variables c' and $\tan\phi'$ are uncorrelated and lognormally distributed. Thus, from Eqs. 7.13 and 7.14,

$$\xi_{c'_+} = \xi_{\tan\phi'_+} = 2.10, \qquad \xi_{c'_-} = \xi_{\tan\phi'_-} = 0.48$$

3. The weights are determined for the four sampling points from Eq. 7.18 using Eqs. 7.16 as follows:

$$P_{c'_+} = P_{\tan\phi'_+} = 0.185, \qquad P_{c'_-} = P_{\tan\phi'_-} = 0.815$$

Therefore, from Eq. 7.18 with $\rho_{ij} = 0$, the sampling point weights are

$$P_{++} = 0.034$$
$$P_{+-} = P_{-+} = 0.151$$
$$P_{-} = 0.665$$

4. The dependent variable q_u is evaluated at each of the points. Table 7.1 summarizes the values of the weights, the sampling points, and q_u for this case:
5. The first three moments of q_u can now be evaluated from Eqs. 7.20, 7.21, and 7.22 as follows:

$$\begin{aligned}\mu_{q_u} &= 0.034(1121.0) + 0.151(628.5) + 0.151(416.6) \\ &\quad + 0.665(233.5) \\ &= 350.9 \text{ kPa}\end{aligned}$$

$$\begin{aligned}\sigma^2_{q_u} &= 0.034(1121.0 - \mu_{q_u})^2 + 0.151(628.5 - \mu_{q_u})^2 \\ &\quad + 0.151(416.6 - \mu_{q_u})^2 + 0.665(233.5 - \mu_{q_u})^2 \\ &= 41657.0 \text{ kPa}^2\end{aligned}$$

Table 7.1 Weights, Sampling Points, and q_u Values for PEM

$P_{\pm\pm}$	c' (kPa)	$\tan\phi'$	$q_{u\pm\pm}$ (kPa)
0.034	205.0	1.184	1121.0
0.151	205.0	0.440	628.5
0.151	76.2	1.184	416.6
0.665	76.2	0.440	233.5

Table 7.2 Statistics of q_u Predicted Using PEM

$v_{c',\tan\phi'}$	ν_{q_u}	σ_{q_u} (kPa)	μ_{q_u} (kPa)	v_{q_u}
0.1	0.351	38.8	346.6	0.11
0.3	1.115	118.5	348.2	0.34
0.5	2.092	204.1	350.9	0.58
0.7	3.530	298.8	353.6	0.85
0.9	5.868	405.0	355.5	1.14

$$\begin{aligned}\nu_{q_u} &= \frac{1}{\sigma^3_{q_u}}\Big[0.034(1121.0 - \mu_{q_u})^3 \\ &\quad + 0.151(628.5 - \mu_{q_u})^3 + 0.151(416.6 - \mu_{q_u})^3 \\ &\quad + 0.665(233.5 - \mu_{q_u})^3)\Big] \\ &= 2.092\end{aligned}$$

Rosenblueth (1981) notes that for the multiple random variable case, skewness can only be reliably calculated if the variables are independent.

A summary of results for different coefficients of variation of c' and $\tan\phi'$, $v_{c',\tan\phi'}$, is presented in Table 7.2. For this example problem, FOSM and PEM give essentially the same results.

7.3 BACKGROUND TO DESIGN METHODOLOGIES

For over 100 years, *working stress design* (WSD), also referred to as *allowable stress design* (ASD), has been the traditional basis for geotechnical design relating to settlements or failure conditions. Essentially, WSD ensures that the *characteristic load* acting on a foundation or structure does not exceed some allowable limit. Characteristic values of either loads or soil properties are also commonly referred to as *nominal*, *working*, or *design* values. We will stick to the word *characteristic* to avoid confusion.

In WSD, the allowable limit is often based on a simple elastic analysis. Uncertainty in loads, soil strength, construction quality, and model accuracy is taken into account through a *nominal factor of safety* F_s, defined as the ratio of the characteristic resistance to the characteristic load:

$$F_s = \frac{\text{characteristic resistance}}{\text{characteristic load}} = \frac{\hat{R}}{\hat{L}} = \frac{\hat{R}}{\sum_{i=1}^{n} \hat{L}_i} \tag{7.26}$$

In general, the *characteristic resistance* $\hat{R}$ is computed by geotechnical formulas using conservative estimates of the soil properties while the *characteristic load* $\hat{L}$ is the sum of conservative unfactored estimates of characteristic load actions, $\hat{L}_i$, acting on the system (see Section 7.4.2 for further definitions of both terms). The load $\hat{L}$ is sometimes

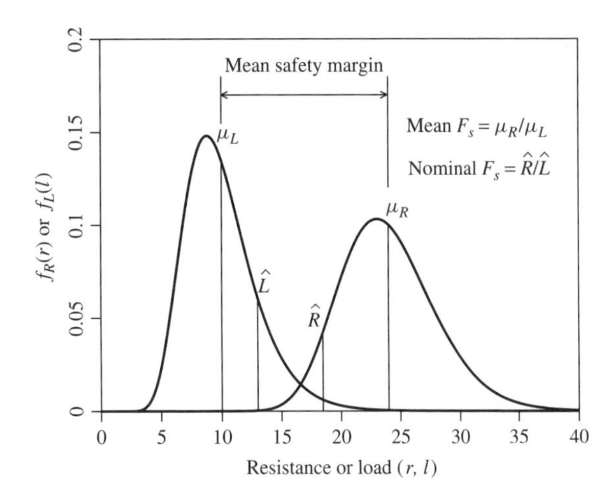

Figure 7.5 Load and resistance distributions.

taken as an upper percentile (i.e., a load only exceeded by a certain small percentage of loads in any one year), as illustrated in Figure 7.5, while $\hat{R}$ is sometimes taken as a cautious estimate of the mean resistance.

A geotechnical design proceeds by solving Eq. 7.26 for the characteristic resistance, leading to the following design requirement:

$$\hat{R} = F_s \sum_i \hat{L}_i \tag{7.27}$$

where $\hat{L}_i$ is the ith characteristic load effect. For example, $\hat{L}_1$ might be the characteristic dead load, $\hat{L}_2$ might be the characteristic live load, $\hat{L}_3$ might be the characteristic earthquake load, and so on. Although Eq. 7.26 is the formal definition of F_s, F_s is typically selected using engineering judgment and experience and then used in Eq. 7.27 to determine the required characteristic resistance (e.g., footing dimension).

There are a number of well-known problems with the WSD approach:

1. All uncertainty is lumped into the single factor of safety F_s.
2. The choice of F_s, although guided to some extent by geotechnical handbooks and codes, is left largely up to the engineer doing the design. Since engineering judgment is an essential component of design (see, e.g., Vick, 2002), the freedom to make this judgment is quite appropriate. However, just stating that the factor of safety should lie between 2 and 3 does not provide any guidance available from current research into the effects of spatial variability and level of site understanding on the probability of failure. The state of current knowledge should be available to designers.
3. The classic argument made against the use of a single factor of safety is that two soils with the same characteristic strength and characteristic load will have the same F_s value regardless of the actual variabilities in load and strength. This is true when the characteristic values are equal to the means, that is, when the factor of safety is defined in terms of the means, for example,

$$F_s = \frac{\text{mean resistance}}{\text{mean load}} \tag{7.28}$$

as it commonly is. The mean F_s was illustrated in Figure 7.5. Figure 7.6 shows how different geotechnical systems, having the same mean factor of safety,

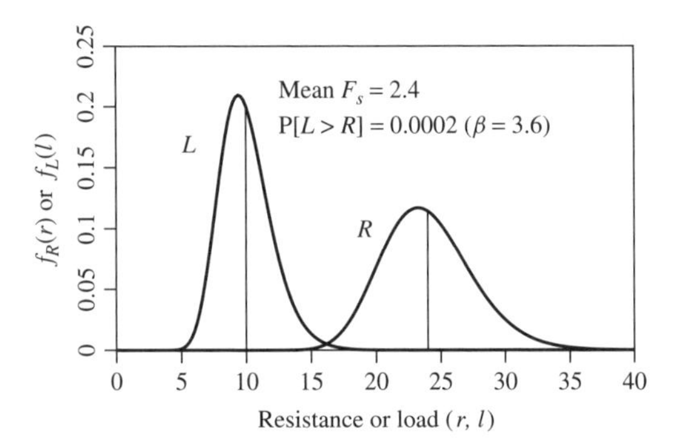

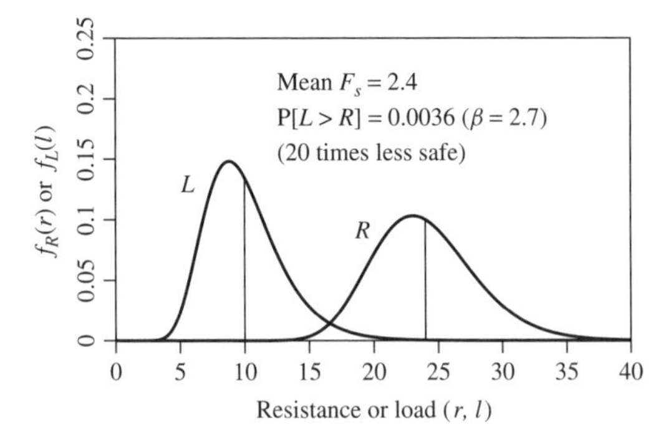

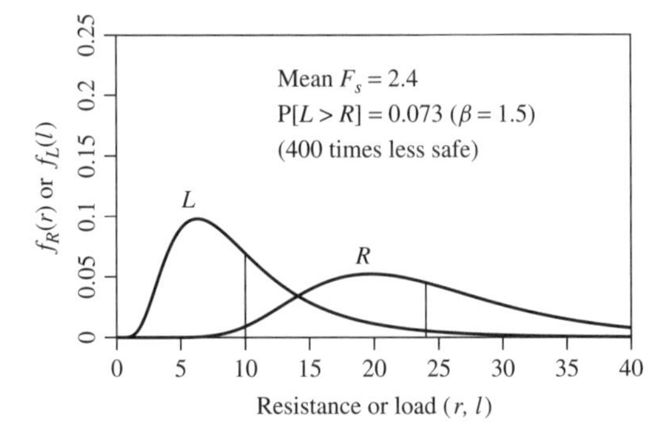

Figure 7.6 Three geotechnical problems can have precisely the same mean factor of safety and yet vastly different probabilities of failure, P$[L > R]$.

can have vastly different probabilities of failure. In other words, the mean factor of safety does not adequately reflect the actual design safety.

When the factor of safety is defined in terms of characteristic values, as is suggested in Eq. 7.26, and the characteristic values are taken as percentiles of the load and resistance distributions (e.g., L_c is that load exceeded only 2% of the time, $\hat{R}$ is that resistance exceeded 90% of the time), then changes in the variability of the load and resistance will result in a change in the factor of safety. However, in practice the use of clearly defined percentiles for characteristic values has rarely been done historically. This is to a great extent due to the fact that the distribution of the geotechnical "resistance" is different at every site and is rarely known. Usually, only enough samples to estimate the mean resistance are taken, and so the characteristic resistance is generally taken to be a "cautious estimate of the mean" [this is currently just one of the definitions of a characteristic value given by *Eurocode 7: Geotechnical Design* (EN 1997–1, 2003), Clause 2.4.3(6)]. In turn, this means that the mean factor of safety (perhaps using a cautious estimate of the mean in practice) has been traditionally used, despite the deficiencies illustrated in Figure 7.6.

It should also be noted that the evolution from WSD to more advanced reliability-based design methodologies is entirely natural. For at least the first half of the 20th century little was understood about geotechnical loads and resistances beyond their most important characteristics—their means. So it was appropriate to define a design code largely in terms of means and some single global factor of safety. In more recent years, as our understanding of the load and resistance distributions improve, it makes sense to turn our attention to somewhat more sophisticated design methodologies which incorporate these distributions.

The working stress approach to geotechnical design has nevertheless been quite successful and has led to many years of empirical experience. The primary impetus to moving away from WSD toward reliability-based design is to allow a better feel for the actual reliability of a system and to harmonize with structural codes which have been reliability based for some time now.

Most current reliability-based design codes start with an approach called *limit states design*. The "limit states" are those conditions in which the system ceases to fulfill the function for which it was designed. Those states concerning safety are called *ultimate limit states*, which include exceeding the load-carrying capacity (e.g., bearing failure), overturning, sliding, and loss of stability. Those states which restrict the intended use of the system are called *serviceability limit states*, which include deflection, permanent deformation, and cracking.

In 1943, Terzaghi's classic book *Theoretical Soil Mechanics* divided geotechnical design into two problems: *stability*, which is an ultimate limit state, and *elasticity*, which is a serviceability limit state. As a result, geotechnical engineers have led the civil engineering profession in limit states design. The basic idea is that any geotechnical system must satisfy at least two design criteria—the system must be designed against serviceability failure (e.g., excessive settlement) as well as against ultimate failure (e.g., bearing capacity failure).

At the ultimate limit state, the factor of safety assumes a slightly different definition:

$$F_s = \frac{\text{ultimate resistance}}{\text{characteristic load}} \tag{7.29}$$

However, the ultimate resistance has traditionally been found using conservative or cautious estimates of the mean soil properties so that Eq. 7.29 is still essentially a mean factor of safety.

Typical factors of safety for ultimate limit states are shown in Table 7.3. Notice that failures that involve *weakest-link* mechanisms, where the failure follows the weakest path through the soil (e.g., bearing capacity, which is a shearing-type failure of foundations, and piping) have the highest factors of safety. Failures that involve average soil properties (e.g., earthworks, retaining walls, deep foundations, and uplift) have lower factors of safety due to the reduction in variance that averaging results in.

In the late 1940s and early 1950s, several researchers (Taylor, 1948; Freudenthal, 1956; Hansen, 1953, 1956) began suggesting that the single factor of safety F_s be replaced by a set of *partial safety factors* acting on individual components of resistance and load. The basic idea was to attempt to account for the different degrees of uncertainty that exist for various load types and material properties (as we began to understand them). In principle, this allows for better quantification of the various sources of uncertainty in a design. The partial safety factors are selected so as to achieve a target reliability of the constructed system. That is, the safety factors for the components of load and resistance are

Table 7.3 Typical Factors of Safety in Geotechnical Design

Failure Type	Item	Factor of Safety
Shearing	Earthworks	1.3–1.5
	Retaining walls	1.5–2.0
	Foundations	2.0–3.0
Seepage	Uplift, heave	1.5–2.0
	Gradient, piping	3.0–5.0
Ultimate pile	Load tests	1.5–2.0
loads	Dynamic formula	3.0

Source: Terzaghi and Peck (1967).

selected by considering the distributions of the load and resistance components in such a way that the probability of system failure becomes acceptably small. This basic ideology led to the load and resistance factor design approach currently popular in most civil engineering codes, discussed in more detail in the next section.

7.4 LOAD AND RESISTANCE FACTOR DESIGN

Once the limit states have been defined for a particular problem, the next step is to develop design relationships for each of the limit states. The selected relationships should yield a constructed system having a target reliability or, conversely, an acceptably low probability of failure. A methodology which at least approximately accomplishes this goal and which has gained acceptance among the engineering community is the *load and resistance factor design* (LRFD) approach. In its simplest form, the load and resistance factor design for any limit state can be expressed as follows: Design the system such that its characteristic resistance $\hat{R}$ satisfies the following inequality:

$$\phi_g \hat{R} \geq \gamma \hat{L} \tag{7.30}$$

where ϕ_g is a resistance factor acting on the (geotechnical) characteristic resistance $\hat{R}$ and γ is a load factor acting on the characteristic load $\hat{L}$. Typically, the resistance factor ϕ_g is less than 1.0—it acts to reduce the characteristic resistance to a less likely *factored resistance*, having a suitably small probability of occurrence. Since, due to uncertainty, this smaller resistance may nevertheless occur in some small fraction of all similar design situations, it is the resistance assumed to exist in the design process. Similarly, the load factor γ is typically greater than 1.0 (unless the load acts in favor of the resistance). It increases the characteristic load to a *factored load*, which may occur in some (very) small fraction of similar design situations. It is this higher, albeit unlikely, load which must be designed against.

A somewhat more general form for the LRFD relationship appears as follows:

$$\phi_g \hat{R} \geq \eta \sum_{i=1}^{m} \gamma_i \hat{L}_i \tag{7.31}$$

where we apply separate load factors, γ_i, to each of m types of characteristic loads $\hat{L}_i$. For example, $\hat{L}_1$ might be the sustained or *dead* load, $\hat{L}_2$ might be the maximum lifetime dynamic or *live* load, $\hat{L}_3$ might be a load due to thermal expansion, and so on. Each of these load types will have their own distribution, and so their corresponding load factors can be adjusted to match their variability. The parameter η is an importance factor which is increased for important structures (e.g., structures which provide essential services after a disaster, such as hospitals). Some building codes, such as the National Building Code of Canada (National Research Council, 2005) adjust the load factors individually to reflect building importance, rather than use a single global importance factor.

The load, resistance, and importance factors are derived and adjusted to account for:

- Variability in load and material properties
- Variability in construction
- Model error (e.g., approximations in design relationships, failure to consider three dimensions and spatial variability, etc.)
- Failure consequences (e.g., the failure of a dam upstream of a community has much higher failure consequences than does a backyard retaining wall)

In geotechnical engineering, there are two common resistance factor implementations:

1. **Total Resistance Factor:** A single resistance factor is applied to the final computed soil resistance. This is the form that Eqs. 7.30 and 7.31 take.
2. **Partial Resistance Factors:** Multiple resistance factors are applied to the components of soil strength (e.g. $\tan\phi'$ and c') separately. This is also known as *factored strength* and *multiple resistance factor design* (MRFD).

There are advantages and disadvantages to both approaches and design codes are about equally divided on the choice of approach (some codes, such as the Eurocode 7, allow both approaches). The following comments can be made regarding the use of partial resistance factors:

1. Since the different components of soil strength will likely have different distributions, the individual resistance factors can be tailored to reflect the uncertainty in each strength component. For example, friction angle is generally determined more accurately than cohesion, and this can be reflected by using different resistance factors.
2. The partial factors only explicitly consider uncertainties due to material strength parameters—they do not include construction uncertainties, model error, and failure consequences. Additional factors would be required to consider these other sources of uncertainty.
3. When the failure mechanism is sensitive to changes in material strengths, then adjusting the material properties may lead to a different failure mechanism than expected.
4. The use of myriad partial factors in order to account separately for all sources of uncertainty can lead to confusion and loss of the understanding of the real soil

behavior. In addition, the estimation and calibration of multiple resistance factors is difficult and prone to significant statistical error. We may not be currently at a sufficient level of understanding of geotechnical variability to properly implement the partial resistance factor approach.

The following comments can be made on the use of a total resistance factor:

1. The geotechnical resistance is computed as in the WSD approach. Not only does this lead to a better representation of the actual failure mechanism, it also involves no fundamental change in how the practicing geotechnical engineer understands and computes the soil resistance. The engineer works with "real" numbers until the last step where the result is factored and the last factoring step is very similar to applying a factor of safety, except that the factor is specifically applied to the resistance. The total resistance factor approach allows for a smoother transition from WSD to LRFD. The only change is that loads are now separately factored.
2. The single soil resistance factor is consistent with structural codes, where each material has its own single resistance factor and soil is viewed as an engineering material. For example, concrete has a single resistance factor (ϕ_c), as does steel (ϕ_s). In this approach, soil will have a single "geotechnical" resistance factor (ϕ_g). Unlike concrete and steel, however, the variability (and understanding) of soil can be quite different at different sites, so the value of ϕ_g should depend on how well the site is understood.
3. The single resistance factor is used to account for all sources of uncertainty in the resistance. These include uncertainties in material properties (natural ground variability), site investigation, model errors (e.g., method of analysis and design), and construction errors.
4. A single resistance factor is much simpler to estimate from real data, from simulations, or to calibrate from existing design approaches (e.g., WSD).

Since there are little data on the true distributions of the individual components of geotechnical resistance, it makes sense at this time to keep the approach as simple as possible, while still harmonizing with the structural codes. At the moment, the simplest and easiest approach to implement is the total resistance factor. As more experience is gained with geotechnical uncertainties in the years to come, the multiple resistance factor approach may become more accurate.

7.4.1 Calibration of Load and Resistance Factors

All early attempts at producing geotechnical LRFD codes do so by calibration with WSD codes. This is perfectly reasonable since WSD codes capture over 100 years of experience and represent the current status of what society sees as *acceptable risk*. If the total resistance factor approach is used, then one factor of safety, F_s, becomes a set of load factors and a single resistance factor. Since the load factors are typically dictated by structural codes, the factor of safety can be simply translated into a single resistance factor to at least assure that the WSD level of safety is translated into the LRFD implementation. It is to be noted, however, that calibration of LRFD from WSD in this way leads to a code whose only advantage over WSD is that it is now consistent with structural LRFD codes. In order to achieve some of the other benefits of a reliability-based design, the resistance factor(s) must be based on more advanced statistical and probabilistic analyses.

The calibration of a single resistance factor from WSD is straightforward if the load factors are known a priori, for example, as given by associated structural codes or by statistical analysis of loads. Consider the WSD and LRFD design criteria:

$$\hat{R} \geq F_s \sum_{i=1}^{m} \hat{L}_i \qquad \text{(WSD)} \qquad (7.32a)$$

$$\phi_g \hat{R} \geq \sum_{i=1}^{m} \gamma_i \hat{L}_i \qquad \text{(LRFD)} \qquad (7.32b)$$

Solving Eq. 7.32b using the equality and substituting in Eq. 7.32a, also at the equality, gives the calibrated geotechnical resistance factor:

$$\phi_g = \frac{\sum_{i=1}^{m} \gamma_i \hat{L}_i}{F_s \sum_{i=1}^{m} \hat{L}_i} \qquad (7.33)$$

Notice that the resistance factor is dependent on the choice of load factors—one must ensure in any design that compatible factors are used. Clearly, if any of the factors are arbitrarily changed, then the resulting design will not have the same safety as under WSD.

Once the resistance factor has been obtained by calibration from WSD, it can then be used in Eq. 7.32b to produce the design. However, both Eqs. 7.32 are defined in terms of characteristic (nominal) loads ($\hat{L}_i$) and resistance ($\hat{R}$). As we shall see later, when we study the resistance factor from a more theoretical viewpoint, the precise definition of the characteristic values also affects the values of the load and resistance factors. This dependence can also be seen in Eqs. 7.32. For example, a designer is likely to choose a larger factor of safety, F_s, if the characteristic resistance is selected at the mean than if it is selected at the 5th percentile. A larger F_s value corresponds to a lower resistance factor, ϕ_g, so when the characteristic resistance is evaluated at the mean, we would expect the resistance factor to be lower.

Table 7.4 lists the load and resistance factors used in a variety of geotechnical design codes from around the world. The table is by no means complete—it is merely an attempt to show the differences between the codes for some common cases. Where the code suggests several different or a range of factors, only the range is presented (this would occur, e.g., when the factor used depends on the degree of uncertainty)—the actual load and resistance factors tend to be more intricately specified than suggested by the table. The table simply presents a range when several values are given by the code but does not attempt to explain the circumstances leading to the range. For this level of detail, the reader will have to consult the original code. This table is to be used mainly to get a general idea of where the various codes stand relative to one another and to see how the LRFD provisions are implemented (e.g., total resistance factor versus partial resistance factors).

To assess the relative conservatism of the various codes, the required area of a spread footing designed against bearing failure (ultimate limit state) using a dead load of 3700 kN, a live load of 1000 kN, $c' = 100$ kN/m^2, and $\phi' = 30°$ is computed using each code and shown in the rightmost column. The codes are ranked from the most conservative (requiring the largest area) at the top to the least conservative (smallest area) at the bottom. The load and resistance factors specified by the code were used in the computation of the required footing area. In cases where a range in load or resistance factors is given by the code, the midpoint of the range was used in the design computations. Again, it needs to be pointed out that this assignment of conservatism will not apply to all aspects of the codes—this is just a rough comparison.

Note that the codes either specify the partial factors acting separately on $\tan\phi'$ and c' or they specify the total resistance factor acting on the ultimate bearing (or sliding) capacity in Table 7.4. The codes are about equally split on how to implement the resistance factors. For example, the two Canadian codes listed—*Canadian Foundation Engineering Manual* [CFEM; Canadian Geotechnical Society (CGS), 1992] and *Canadian Highway Bridge Design Code* [CHBDC; Canadian Standards Association (CSA), 2000a]—implement the resistance factors in different ways (the 2006 CFEM is now in agreement with CHBDC). The same is true of the two Australian standards listed, AS 5100 and AS 4678. The Eurocode 7 actually has three models, but only two different models for the cases considered in Table 7.4 and these two cases correspond to the partial factor and total factor approaches.

7.4.2 Characteristic Values

Becker (1996a) notes that while there has been considerable attention paid in the literature to determining appropriate load and resistance factors, little has been paid to

Table 7.4 Comparative Values of Load and Resistance Factors

Code		Dead Load	Live Load	$\tan\phi'$	c'	Bearing	Sliding	Area
CFEM	1992	1.25	1.5	0.8	0.5–0.65	—	—	5.217
NCHRP 343	1991	1.3	2.17	—	—	0.35–0.6	0.8–0.9	4.876
NCHRP12–55	2004	1.25	1.75	—	—	0.45	0.8	4.700
Denmark	1965	1.0	1.5	0.8	0.57	—	—	4.468
B. Hansen	1956	1.0	1.5	0.83	0.59	—	—	4.145
CHBDC	2000	1.25	1.5	—	—	0.5	0.8	4.064
AS 5100	2004	1.2	1.5	—	—	0.35–0.65	0.35–0.65	3.942
AS 4678	2002	1.25	1.5	0.75–0.95	0.5–0.9	—	—	3.892
Eurocode 7	Model 1	1.0	1.3	0.8	0.8	—	—	3.061
Eurocode 7	Model 2	1.35	1.5	—	—	0.71	0.91	3.035
ANSI A58	1980	1.2–1.4	1.6	—	—	0.67–0.83	—	2.836

Notes:
CFEM = *Canadian Foundation Engineering Manual*
NCHRP = National Cooperative Highway Research Program
CHBDC = *Canadian Highway Bridge Design Code*
AS = Australian Standard
ANSI = American National Standards Institute

defining the characteristic loads and resistances used in the design, despite the fact that how the characteristic values are defined is of critical importance to the overall reliability. For example, using the same load and resistance factors with characteristic values defined at the 5th percentile or with characteristic values defined at the mean will yield designs with drastically differing failure probabilities.

In any reliability-based design, uncertain quantities such as load and resistance are represented by random variables having some distribution. Distributions are usually characterized by their mean, standard deviation, and some shape (e.g., normal or lognormal). In some cases, the characteristic values used in design are defined to be the means, but they can be more generally defined in terms of the means as (see Figure 7.5)

$$\hat{L} = k_L \mu_L \tag{7.34a}$$

$$\hat{R} = k_R \mu_R \tag{7.34b}$$

where k_L is the ratio of the characteristic to the mean load, $\hat{L}/\mu_L$, and k_R is the ratio of the characteristic to the mean resistance, $\hat{R}/\mu_R$. Normally, k_L is selected to be greater than or equal to 1.0, while k_R is selected to be less than or equal to 1.0.

The load acting on a foundation is typically composed of dead loads, which are largely static, and live loads, which are largely dynamic. Dead loads are relatively well defined and can be computed by multiplying volumes by characteristic unit weights. The mean and variance of dead loads are reasonably well known. Dynamic, or live, loads, on the other hand are more difficult to characterize probabilistically. A typical definition of a live load is the extreme dynamic load (e.g., bookshelves, wind loads, vehicle loads, etc.) that a structure will experience during its design life. We will denote this load as L_{L_e}, the subscript e implying an extreme (maximum) load over some time span. This definition implies that the live load distribution will change with the target design life so that both μ_{L_e} and its corresponding k_{L_e} become dependent on the design life.

Most geotechnical design codes recommend that the characteristic resistance be based on "a cautious estimate of the mean soil properties." In geotechnical practice, the choice of a characteristic value depends very much on the experience and risk tolerance of the engineer performing the design. An experienced engineer may pick a value based on the mean. Engineers more willing to take risks may use a value larger than the mean, trusting in the resistance factor to "make up the difference," while a risk-averse engineer may choose a resistance based on the minimum suggested by soil samples. Obviously, the resulting designs will have highly different reliabilities and costs.

Eurocode 7 has the following definitions for the characteristic value:

Clause 2.4.3(5): The characteristic value of a soil or rock parameter shall be selected as a cautious estimate of the value affecting the occurrence of the limit state.

Clause 2.4.3(6): The governing parameter is often a mean value over a certain surface or volume of the ground. The characteristic value is a cautious estimate of this mean value.

Clause 2.4.3(7): Characteristic values may be lower values, which are less than the most probable values, or upper values, which are greater. For each calculation, the most unfavorable combination of lower and upper values for independent parameters shall be used.

Lower values are typically used for the resistance (e.g., for cohesion) while upper values are typically used for load effects (e.g., live load). Unfortunately, the word *cautious* can mean quite different things to different engineers. For example, suppose that the true mean friction angle of a soil is 30°, and that a series of 10 soil samples result in estimated friction angles of (in descending order) 46°, 40°, 35°, 33°, 30°, 28°, 26°, 22°, 20°, and 18°. An estimate of the characteristic friction angle might be the actual average of these observations, 29.8°, perhaps rounded to 30°. A "cautious" estimate of the characteristic friction angle might be the median, which is $(30 + 28)/2 = 29°$. However, there are obviously some low strength regions in this soil, and a more cautious engineer might choose a characteristic value of 20° or 18°. The codes give very little guidance on this choice, even though the choice makes a large difference to the reliability of the final design.

The interpretations of "lower" and "upper" in clause 2.4.3(7) are not defined. To add to the confusion, Eurocode 1 (Basis of Design and Actions on Structures) has the following definition under clause 5 (Material Properties):

Unless otherwise stated in ENVs 1992 to 1999, the characteristic values should be defined as the 5% fractile for strength parameters and as the mean value for stiffness parameters.

For strength parameters (e.g., c and ϕ) a 5th percentile is very cautious. In the above example, the empirical 5th percentile would lie between 18° and 20°. A design based on this percentile could be considerably more conservative than a design based on, say, the median (29°), especially if the same resistance factors were used. The *User's Guide* to the National Building Code of Canada [National Research Council (NRC), 2006] states that characteristic geotechnical design values shall be based on the results of field and laboratory tests and take into account factors such as geological information, inherent variabilities, extent of zone of influence, quality of workmanship, construction effects, presence of fissures, time effects, and degree of ductility. Commentary K, Clause 12, states: "In essence, the characteristic value corresponds to the geotechnical engineer's best estimate of the most appropriate likely value for geotechnical properties relevant for the limit states investigated. A cautious estimate of the mean value for the affected ground (zone of influence) is generally considered as a logical value to use as the characteristic value."

A clearly defined characteristic value is a critical component to a successful load and resistance factor design. Without knowing where the characteristic value is in the resistance distribution, one cannot assess the reliability of a design, and so one cannot develop resistance factors based on a target system reliability. In other words, if a reliability-based geotechnical design code is to be developed, a clear definition of characteristic values is essential. As Becker (1996a) comments, it is not logical to apply resistance factors to poorly defined characteristic resistance values.

The use of the median as the characteristic geotechnical value may be reasonable since the median has a number of attractive probabilistic features:

1. When the geotechnical property being estimated is lognormally distributed, the median is less than the mean. This implies that the sample median can be viewed as a "cautious estimate of the mean."
2. The sample median can be estimated either by the central value of an ordered set of observations (see Section 1.6.2) or by computing the geometric average, X_G, of the observations, $X_1, X_2, \ldots, X_n$ (so long as all observations are positive):

$$X_G = [X_1 X_2, \ldots, X_n]^{1/n} = \exp\left\{\frac{1}{n}\sum_{i=1}^{n} \ln X_i\right\} \tag{7.35}$$

3. If the sample median is estimated by Eq. 7.35, and all observations are positive, then the sample median tends to have a lognormal distribution by the central limit theorem. If the observations come from a lognormal distribution, then the sample median will also be lognormally distributed. This result means that probabilities of events (e.g., failure) which are defined in terms of the sample median are relatively easily calculated.

There is some concern in the geotechnical engineering profession that defining the characteristic value will result in the loss of the ability of engineers to apply their judgment and experience to designs. As Mortensen (1983) notes, the explicit definition of characteristic values is not intended to undermine the importance of engineering judgment. Engineering judgment will still be essential in deciding which data are appropriate in determining the characteristic value (e.g., choice of soils in the zone of influence, weaker layers, etc.), as it has always been. The only difference is that all geotechnical engineers would now be aiming for a characteristic value at the same point in the resistance distribution. Ideally, this would mean that given the same site, investigation results, and design problem, two different engineers would arrive at very similar designs, and that both designs would have an acceptable reliability. The end result of a more clear definition should be improved guidance to and consistency among the profession, particularly among junior engineers.

As mentioned above, the resistance factor is intimately linked to the definition of the characteristic value. What was not mentioned is the fact that the resistance factor is also intimately linked to the uncertainty associated with the characteristic value. For example, if a single soil sample is taken at a site and the characteristic value is set equal to the strength property derived from that sample, then obviously there is a great deal of uncertainty about that characteristic value. Alternatively, if 100 soil samples are taken within the zone of influence of a foundation being designed, then the characteristic value used in the design will be known with a great deal of certainty. The resistance factor employed clearly depends on the level of certainty one has about the characteristic value. This is another area where engineering judgment and experience plays an essential role—namely in assessing what resistance factor should be used, which is analogous to the selection of a factor of safety.

The dependence of the resistance factor on how well the characteristic value is known is also reflected in structural codes. For example, the resistance factor for steel reinforcement is higher than that for concrete because steel properties are both better known and more easily controlled. Quality control of the materials commonly employed in structural engineering allows resistance factors that remain relatively constant. That is, a 30-MPa concrete ordered in London will be very similar in distribution to a 30-MPa concrete ordered in Ottawa, and so similar resistance factors can be used in both locations.

As an example of the dependence of the resistance factor on the certainty of the characteristic value, the Australian standard 5100.3–2004, Bridge Design, Part 3 (2004), defines ranges on the geotechnical resistance factor as given in Tables 7.5 and 7.6. The Australian standard further provides guidance on how to choose the resistance factor from within the range given in Table 7.5. Essentially, the choice of resistance factor value depends on how well the site is understood, on the design sophistication, level of

Table 7.5 Range of Values of Geotechnical Resistance Factor (ϕ_g) for Shallow Footings According to Australian Standard 5100.3–2004, Bridge Design, Part 3 (2004), Table 10.3.3(A)

Assessment Method of Ultimate Geotechnical Strength	Range of ϕ_g
Analysis using geotechnical parameters based on appropriate advanced in situ tests	0.50–0.65
Analysis using geotechnical parameters from appropriate advanced laboratory tests	0.45–0.60
Analysis using CPT tests	0.40–0.50
Analysis using SPT tests	0.35–0.40

construction control, and failure consequences, as specified in the Table 7.6.

Example 7.2 Suppose that we are performing a preliminary design of a strip footing, as shown in Figure 7.7, against bearing capacity failure (ultimate limit state). For simplicity, the soil is assumed to be weightless so that the ultimate bearing stress capacity, q_u, is predicted to be

$$q_u = cN_c \tag{7.36}$$

where c is the cohesion and N_c is the bearing capacity factor:

$$N_c = \frac{\tan^2(\pi/4 + \phi/2)\ \exp\{\pi \tan\phi\} - 1}{\tan\phi} \tag{7.37}$$

Table 7.6 Guide for Choice of Geotechnical Resistance Factor (ϕ_g) for Shallow Footings According to Australian Standard 5100.3–2004, Bridge Design, Part 3 (2004), Table 10.3.3(B)

Lower End of Range	Upper End of Range
Limited site investigation	Comprehensive site investigation
Simple methods of calculation	More sophisticated design method
Limited construction control	Rigorous construction control
Severe failure consequences	Less severe failure consequences
Significant cyclic loading	Mainly static loading
Foundations for permanent structures	Foundations for temporary structures
Use of published correlations for design parameters	Use of site-specific correlations for design parameters

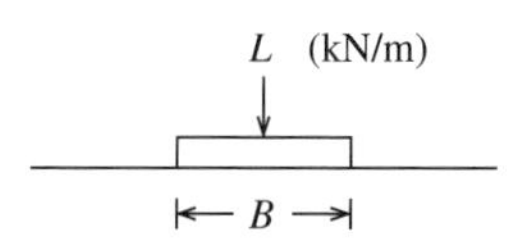

Figure 7.7 Strip footing of width B founded on a weightless soil.

for ϕ measured in radians. The footing is to support random live and dead loads having means $\mu_{L_e} = 300$ kN/m and $\mu_D = 900$ kN/m, respectively. Suppose further that three soil samples are available, yielding the results shown in Table 7.7. Determine the design footing width, B, using a traditional WSD approach as well as a LRFD approach based on a total resistance factor, ϕ_g.

SOLUTION The first step is to determine characteristic values of the cohesion and internal friction angle to be used in the design. The arithmetic averages of the soil samples are

$$\bar{c} = \frac{1}{n}\sum_{i=1}^{n} c_i = \tfrac{1}{3}(91.3 + 101.5 + 113.2) = 102.0 \quad \text{kN/m}^2$$

$$\bar{\phi} = \frac{1}{n}\sum_{i=1}^{n} \phi_i = \tfrac{1}{3}(21.8 + 25.6 + 29.1) = 25.5^\circ$$

The geometric averages of the soil samples are

$$c_G = \left[\prod_{i=1}^{n} c_i\right]^{1/n} = [91.3 \times 101.5 \times 113.2]^{1/3}$$
$$= 101.6 \quad \text{kN/m}^2$$

$$\phi_G = \left[\prod_{i=1}^{n} c_i\right]^{1/n} = [21.8 \times 25.6 \times 29.1]^{1/3} = 25.3^\circ$$

These averages are all shown in Table 7.7. The geometric average is an estimate of the median in cases where the

Table 7.7 Cohesion and Internal Friction Angle Estimates from Three Soil Samples

Soil Sample	c (kN/m^2)	ϕ (deg)
1	91.3	21.8
2	101.5	25.6
3	113.2	29.1
Arithmetic average	102.0	25.5
Geometric average	101.6	25.3
Characteristic value	100.0	25.0

soil property is (at least approximately) lognormally distributed. In the case of the cohesion, which is often taken to be lognormally distributed, the geometric average and the sample median almost coincide. The internal friction angle is not lognormally distributed (it possesses an upper bound), although the lognormal might be a reasonable approximation, and the geometric average is slightly lower than the sample median (25.3 vs. 25.6). The difference is negligible and is to be expected even if the property is lognormally distributed because of the differences in the way the median is estimated (one by using the central value, the other by using a geometric average of all of the values).

The characteristic value is conveniently rounded down from the geometric average slightly to a reasonable whole number, so that we will use $\hat{c} = 100$ kN/m^2 and $\hat{\phi} = 25^\circ$ as our characteristic design values.

Using $\hat{\phi} = 25^\circ = 0.4363$ rad in Eq. 7.37 gives us $N_c = 20.7$ so that

$$q_u = cN_c = (100)(20.7) = 2070 \quad \text{kN/m}^2$$

The third edition of the Canadian Foundation Engineering Manual (CGS, 1992) recommends a factor of safety, F_s, of 3. Using this, the allowable WSD bearing stress is

$$q_a = \frac{q_u}{F_s} = \frac{2070}{3} = 690 \quad \text{kN/m}^2$$

In order to find the required footing width using WSD, we need to know the characteristic load. It is probably reasonable to take the characteristic dead load equal to the mean dead load, $\mu_D = 900$ kN/m, since dead load is typically computed by multiplying mean unit weights times volumes, for example. Live loads are more difficult to define since they are time varying by definition. For example, the commentaries to the National Building Code of Canada (NRC, 2006) specifies that snow and wind loads be taken as those with only a 1-in-50 probability of exceedance in any one year, while the exceedance probability of characteristic use and occupancy loads is not defined. We shall assume that the live load mean of $\mu_{L_e} = 300$ kN/m (the subscript e denotes extreme) specified in this problem is actually the mean of the 50-year maximum live load (where the 50 years is the design life span of the supported structure). In other words, if we were to observe the maximum live load seen over 50 years at a large number of similar structures, the mean of those observations would be 300 kN/m (and coefficient of variation would be 30%). Our characteristic live load is thus taken to equal the mean, in this case 300 kN/m.

We can now compute the required design footing width as

$$B = \frac{\mu_{L_e} + \mu_D}{q_a} = \frac{300 + 900}{690} = 1.74 \quad \text{m}$$

If we have more certainty about the soil properties, which may be the case if we had more soil samples, and loads, then a reduced factor of safety may be sometimes used. For illustrative purposes, and for comparison to the LRFD computations to come, Table 7.8 lists the required design footing widths for three different factors of safety. The load and resistance factor design of the same footing proceeds as follows:

1. Define the characteristic load: Allen (1975) and Becker (1996a) suggest that the characteristic load values are obtained as multiples of the mean values (see Section 11.2):

$$\hat{L}_L = k_{L_e}\mu_{L_e} = 1.41(300) = 423 \quad \text{kN/m}$$

$$\hat{L}_D = k_D\mu_D = 1.18(900) = 1062 \quad \text{kN/m}$$

2. Choose a resistance factor. The following table is an example of the choices that might be available:

Level of Understanding	Resistance Factor, ϕ_g
Soil samples taken directly under footing	0.7
Soil samples taken near footing (<4 m)	0.6
Soil properties inferred from similar sites or by experience	0.54

3. Compute the ultimate soil capacity: We will again use the geometric average (or median) soil properties as the characteristic soil properties, rounded down to convenient whole numbers: $\hat{c} = 100$ kN/m^2 and $\hat{\phi} = 25^\circ$. This gives us (as before)

$$q_u = cN_c = (100)(20.7) = 2070 \quad \text{kN/m}^2$$

and our characteristic resistance is

$$\hat{R} = Bq_u$$

Table 7.8 Required Working Stress Design Footing Widths for Three Factors of Safety

Factor of Safety	Design Footing Width (m)
3.0	1.74
2.5	1.45
2.0	1.16

Note that we are taking our characteristic resistance at the median rather than as some multiple of the mean (as in Eq. 7.34b).

4. Solve the LRFD equation

$$\phi_g \hat{R} = \gamma_D \hat{L}_D + \gamma_L \hat{L}_L$$

for the required design footing width:

$$B = \frac{\gamma_D \hat{L}_D + \gamma_L \hat{L}_L}{\phi_g q_u}$$

5. If the soil samples are taken directly under the footing, so that we have a high level of understanding of the soil supporting the footing, then $\phi_g = 0.7$, according to the above example table, and we get the required design footing width to be

$$B = \frac{1.25(1062) + 1.5(429)}{0.7(2070)} = 1.36 \text{ m} \qquad (F_s \simeq 2.3)$$

where the F_s value is the factor of safety one would have to have used in the WSD approach to get the same footing width.

6. If the soil samples are taken near the footing, within about 4 m, so that we have an "average" level of site understanding according to the above table, then $\phi_g = 0.6$ and

$$B = \frac{1.25(1062) + 1.5(429)}{0.6(2070)} = 1.59 \text{ m} \qquad (F_s \simeq 2.7)$$

7. If the soil properties are inferred by samples from similar sites, or through experience, then the above table suggests that $\phi_g = 0.54$ and

$$B = \frac{1.25(1062) + 1.5(429)}{0.54(2070)} = 1.76 \text{ m} \qquad (F_s \simeq 3.0)$$

8. If the 2006 Canadian Foundation Engineering Manual (CGS, 2006) is used, then $\phi_g = 0.5$, which gives

$$B = \frac{1.25(1062) + 1.5(429)}{0.5(2070)} = 1.90 \text{ m} \qquad (F_s \simeq 3.3)$$

7.5 GOING BEYOND CALIBRATION

Reliability-based design is typically performed at one of three levels:

Level I: A semiprobabilistic approach where the design is based on factored load and resistance values. The load and resistance factors are determined by calibration (assuming past designs are achieving an acceptable risk level) or by the more advanced methods described next. This level is the basis of LRFD.

Level II: An approximate probabilistic analysis, which takes into account the probability distributions of loads and resistances but with the following simplifying assumptions:

- Loads and resistances are assumed to be independent.
- The components of the load and the resistance are assumed to be captured by single random variables, rather than by spatially and/or time-varying random fields.

Level III: A sophisticated probabilistic analysis in which soil resistance is modeled using cross-correlated space–time varying random fields. Depending on the scope of the geotechnical design, loads may also be space–time varying random fields (e.g., wind, earthquake, vehicle, and other dynamic loads are typically time varying). This level of analysis is typically very complex and usually only solvable via Monte Carlo simulation combined with sophisticated nonlinear multidimension finite-element models.

This section concentrates on level II analysis. It addresses the question as to how we can use simple probabilistic models to improve on load and resistance factors which have been calibrated from traditional working stress design. In this analysis we will again express the characteristic load and resistance as some multiples of the means, according to Eq. 7.34:

$$\hat{L} = k_L \mu_L, \qquad \hat{R} = k_R \mu_R$$

Note that if the characteristic resistance is the median resistance $\tilde{R}$, which is computed according to

$$\hat{R} = \tilde{R} = \frac{\mu_R}{\sqrt{1 + v_R^2}} \tag{7.38}$$

where v_R is the coefficient of variation of the resistance, then

$$k_R = \frac{1}{\sqrt{1 + v_R^2}} \tag{7.39}$$

So, for example, if the coefficient of variation of the resistance is 20%, then $k_R = 0.98$.

The simplest form of the design requirement under LRFD is as given by Eq. 7.30,

$$\phi_g \hat{R} \geq \gamma \hat{L}$$

and failure will occur if the actual resistance of the as-constructed design, R, is less than the actual load, L. That is,

$$\text{P}[\text{failure}] = \text{P}[R < L] = \text{P}\left[\frac{R}{L} < 1\right] \tag{7.40}$$

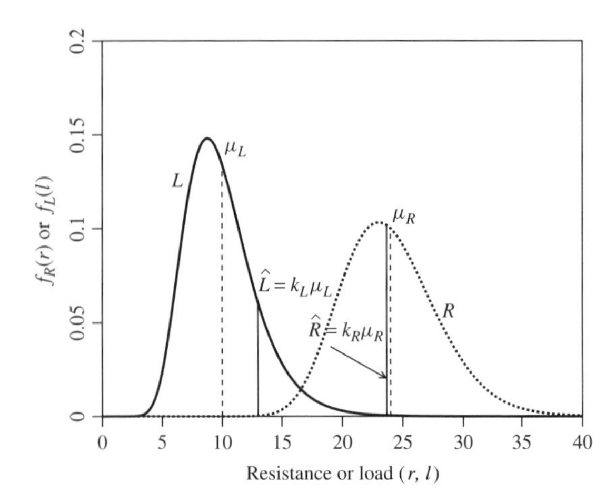

Figure 7.8 Distributions of load L and resistance R showing locations of characteristic load $\hat{L}$ and resistance $\hat{R}$.

The R/L form of this probability is appropriate if R and L are assumed to be lognormally distributed, as is commonly done (since they are both nonnegative).

In Figure 7.8, the left distribution is that of the actual load L, and the right distribution is that of the actual resistance R, and these distributions express the relative likelihoods that L or R take on any particular value. If a particular "as-constructed" geotechnical system happens to have a value of L which exceeds R, then the system will fail. Note that the probability that this occurs is *not* equal to the area of the overlap between the two distributions, although the failure probability does increase as this area gets larger (at least for the lognormal distribution).

Now let us define the *safety margin* M to be

$$M = \ln\left(\frac{R}{L}\right) \tag{7.41}$$

Clearly, if $M < 0$, the load exceeds the resistance and failure will have occurred. We are interested in computing the probability of failure in order to produce a design with acceptably small risk. If R and L are lognormally distributed, then M is normally distributed with mean and variance:

$$\mu_M = \mu_{\ln R} - \mu_{\ln L} \tag{7.42a}$$

$$\sigma_M = \sqrt{\sigma^2_{\ln R} + \sigma^2_{\ln L}} \tag{7.42b}$$

where we assumed the load and resistance to be independent in Eq. 7.42b. The probability of failure is thus

$$\mathrm{P}[M < 0] = \mathrm{P}\left[Z < \frac{0 - \mu_M}{\sigma_M}\right]$$
$$= \Phi\left(-\frac{\mu_{\ln R} - \mu_{\ln L}}{\sqrt{\sigma^2_{\ln R} + \sigma^2_{\ln L}}}\right) = \Phi(-\beta) \tag{7.43}$$

where Z is the standard normal, Φ is the standard normal cumulative distribution function, and β is the *reliability index*. Figure 7.9 illustrates the probability of failure (shaded region in both plots). It can be seen from the right plot, which is the distribution of M, that the reliability index, β, is the number of standard deviations that μ_M is away from the failure region, $M < 0$. In other words, β is defined as

$$\beta = \frac{\mu_M}{\sigma_M} = \frac{\mu_{\ln R} - \mu_{\ln L}}{\sqrt{\sigma^2_{\ln R} + \sigma^2_{\ln L}}} \tag{7.44}$$

The importance of the reliability index is historical. Society, and in particular politicians and government bodies, were not comfortable with expressions of probability of failure. Election platforms are not helped by any admission that their is some chance of loss of life in an engineered structure or when following a design code. The reliability index couches the potential for damage or loss of life in a nice "positive" way. Thus, most design codes are aimed at exceeding a target reliability index, rather than falling below an "acceptable" failure probability. Nevertheless, the overall aim of design codes is to reduce the probability of failure down to socially acceptable levels.

The left plot of Figure 7.9 shows the actual distribution of R/L, which is lognormally distributed since both R and L are lognormally distributed, as assumed. The failure region in the left plot is where $R/L < 1$. Taking the natural logarithm of R/L gives the right plot, which is a normal distribution. The shaded regions in both plots have the same area.

The failure probability given by Eq. 7.43 is exact so long as R and L are lognormally distributed and independent. If R and L are not independent, then σ_M will involve the covariance between R and L. Because the shear strength, τ, of soil is given by

$$\tau = c + \sigma \tan(\phi) \tag{7.45}$$

where σ is the normal stress (in this equation), then clearly the load and soil resistance are not independent since stress is, of course, a function of the load. Equation 7.45 suggests a strong positive correlation between τ and σ (i.e., as one increases, the other increases). Unfortunately, at this time, we are not aware of any studies characterizing the magnitude of the covariance between load and soil resistance over the multitude of geotechnical problems. We do note, however, that the assumption of independence between load and resistance is at least generally conservative, in that the assessed probability of failure will be higher under this assumption (the variance of M is larger under independence than under a positive correlation—another way of looking at this is if the strength almost always increases with load, then this would lead to a lower probability that load

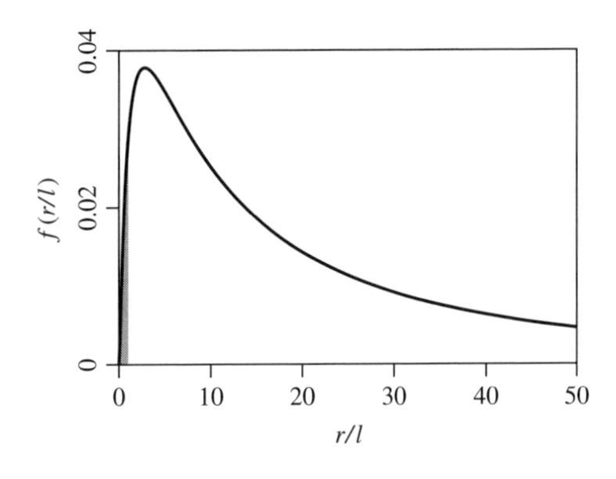

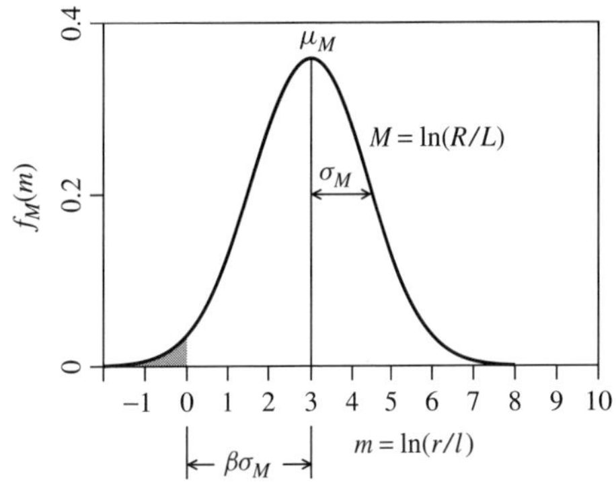

Figure 7.9 Distributions of R/L (left) and $\ln(R/L)$ (right) showing how reliability index β is defined.

will exceed resistance). We will therefore proceed in this section assuming that the load and resistance are independent. In subsequent chapters, where we use simulation to assess failure probabilities, we no longer need to assume independence and so can conceptually more accurately assess design requirements.

A reliability-based design typically proceeds by first selecting a target reliability index β and then designing the resistance such that

$$\mathrm{P}\left[\frac{R}{L} < 1\right] = \Phi(-\beta) \tag{7.46}$$

According to Eq. 7.44, the reliability index can be expressed as

$$\beta = \frac{\mu_M}{\sigma_M} = \frac{\mu_{\ln R} - \mu_{\ln L}}{\sqrt{\sigma_{\ln R}^2 + \sigma_{\ln L}^2}}$$

which can be solved for $\mu_{\ln R}$,

$$\begin{aligned}\mu_{\ln R} &= \mu_{\ln L} + \beta\sqrt{\sigma_{\ln R}^2 + \sigma_{\ln L}^2} \\ &\simeq \mu_{\ln L} + 0.71\beta\left(\sigma_{\ln R} + \sigma_{\ln L}\right)\end{aligned} \tag{7.47}$$

where the approximation $0.71(\sigma_{\ln R} + \sigma_{\ln L}) \simeq \sqrt{\sigma_{\ln R}^2 + \sigma_{\ln L}^2}$ is exact (insofar as $1/\sqrt{2} = 0.71$) when $v_R = v_L$ and is fairly accurate for common values of v_R and v_L otherwise. Since

$$\begin{aligned}\mu_{\ln L} &= \ln\left(\mu_L\right) - \tfrac{1}{2}\sigma_{\ln L}^2 \\ \mu_R &= \exp\left(\mu_{\ln R} + \tfrac{1}{2}\sigma_{\ln R}^2\right)\end{aligned}$$

we get

$$\begin{aligned}\mu_R = \mu_L \big[&\exp\left(\tfrac{1}{2}\sigma_{\ln R}^2 + 0.71\beta\sigma_{\ln R}\right) \\ &\times \exp\left(-\tfrac{1}{2}\sigma_{\ln L}^2 + 0.71\beta\sigma_{\ln L}\right)\big]\end{aligned}$$

or, rearranging,

$$\begin{aligned}&\exp\left(-\tfrac{1}{2}\sigma_{\ln R}^2 - 0.71\beta\sigma_{\ln R}\right)\mu_R \\ &\quad = \exp\left(-\tfrac{1}{2}\sigma_{\ln L}^2 + 0.71\beta\sigma_{\ln L}\right)\mu_L\end{aligned} \tag{7.48}$$

In turn, this relationship can be expressed in terms of the characteristic load and resistance,

$$\hat{R} = k_R\mu_R, \qquad \hat{L} = k_L\mu_L$$

as

$$\begin{aligned}&\left[\frac{\exp\left(-\tfrac{1}{2}\sigma_{\ln R}^2 - 0.71\beta\sigma_{\ln R}\right)}{k_R}\right]\hat{R} \\ &\quad = \left[\frac{\exp\left(-\tfrac{1}{2}\sigma_{\ln L}^2 + 0.71\beta\sigma_{\ln L}\right)}{k_L}\right]\hat{L}\end{aligned} \tag{7.49}$$

Recalling that the simplest form of the LRFD relationship is $\phi_g\hat{R} = \gamma\hat{L}$ implies that

$$\phi_g = \frac{\exp\left(-\tfrac{1}{2}\sigma_{\ln R}^2 - 0.71\beta\sigma_{\ln R}\right)}{k_R} \tag{7.50a}$$

$$\gamma = \frac{\exp\left(-\tfrac{1}{2}\sigma_{\ln L}^2 + 0.71\beta\sigma_{\ln L}\right)}{k_L} \tag{7.50b}$$

Equations 7.50 give us analytical means of evaluating the load and resistance factors if we know about the variability in the load and resistance and have selected a suitable reliability index, β.

If the load factors are known (e.g., from structural codes) and we are just interested in determining the resistance factor, ϕ_g, then we can deal with a more complicated (and more usual) LRFD equation:

$$\phi_g\hat{R} = \phi_g k_R\mu_R = \sum_i \gamma_i\hat{L}_i$$

The design proceeds by specifying μ_R such that

$$\mu_R = \frac{\sum_i \gamma_i \hat{L}_i}{\phi_g k_R} \tag{7.51}$$

where ϕ_g is found such that the design has a failure probability of no more than

$$\mathrm{P}\left[\frac{R}{L} < 1\right] = \Phi\left(-\frac{\mu_{\ln R} - \mu_{\ln L}}{\sqrt{\sigma_{\ln R}^2 + \sigma_{\ln L}^2}}\right) = \Phi(-\beta)$$

In order for this to be satisfied, we must have

$$\frac{\mu_{\ln R} - \mu_{\ln L}}{\sqrt{\sigma_{\ln R}^2 + \sigma_{\ln L}^2}} = \beta \tag{7.52}$$

or, in other words,

$$\mu_{\ln R} - \mu_{\ln L} = \beta\sqrt{\sigma_{\ln R}^2 + \sigma_{\ln L}^2}$$

Using the transformations

$$\mu_{\ln R} = \ln\left(\frac{\mu_R}{\sqrt{1 + v_R^2}}\right)$$

$$\mu_{\ln L} = \ln\left(\frac{\mu_L}{\sqrt{1 + v_L^2}}\right)$$

where v_R and v_L are the coefficients of variation (σ/μ) of the resistance and load, respectively (we shall see shortly how v_L is actually computed), gives

$$\ln\left(\frac{\mu_R\sqrt{1 + v_L^2}}{\mu_L\sqrt{1 + v_R^2}}\right) = \beta\sqrt{\sigma_{\ln R}^2 + \sigma_{\ln L}^2}$$

Now, replacing μ_R by Eq. 7.51 and raising e to the power of both sides yield

$$\frac{\sum_i \gamma_i \hat{L}_i}{\phi_g k_R \mu_L}\left(\frac{\sqrt{1 + v_L^2}}{\sqrt{1 + v_R^2}}\right) = \exp\left\{\beta\sqrt{\sigma_{\ln R}^2 + \sigma_{\ln L}^2}\right\}$$

which can now be solved for ϕ_g, giving

$$\phi_g = \frac{\sum_i \gamma_i \hat{L}_i}{k_R \mu_L}\left(\frac{\sqrt{1 + v_L^2}}{\sqrt{1 + v_R^2}}\right)\exp\left\{-\beta\sqrt{\sigma_{\ln R}^2 + \sigma_{\ln L}^2}\right\} \tag{7.53}$$

Thus, Eq. 7.53 can be used to determine the resistance factor ϕ_g so long as:

1. The load factors γ_i are known.
2. The characteristic loads $\hat{L}_i = k_{L_i}\mu_{L_i}$ are known, where k_{L_i} is the ratio of the ith characteristic load effect $\hat{L}_i$ to its mean μ_{L_i}.
3. The ratio k_R of the characteristic resistance $\hat{R}$ to its mean μ_R is known.
4. The total mean load μ_L computed as

$$\mu_L = \sum_i \mu_{L_i}$$

is known.
5. The coefficient of variation of the resistance, $v_R = \sigma_R/\mu_R$, is known.
6. The coefficient of variation of the load, v_L, where, assuming that the various load effects are statistically independent,

$$v_L = \frac{1}{\mu_L^2}\sum_i v_{L_i}^2 \mu_{L_i}^2 = \frac{1}{\mu_L^2}\sum_i v_{L_i}^2\left(\frac{\hat{L}_i}{k_{L_i}}\right)^2$$

is known.

7.5.1 Level III Determination of Resistance Factors

Generally speaking, a level III analysis, in which the soil is treated as a spatially (and possibly temporally) varying random field, can be used to determine required resistance factors. Because of the complexity of this approach, it usually proceeds using a method called *Monte Carlo simulation* (see Section 6.6). Monte Carlo simulation involves simulating a possible realization of the supporting (or forcing) soil mass and then analyzing the response of the soil to load (or restraint) using some sophisticated method which takes spatial variability into account—usually the finite-element method.

The main advantages of a level III analyses are as follows:

1. The soil is modeled more realistically.
2. The assumption of independence between load and resistance need no longer be made (so long as the analysis method is sufficient to properly model the interdependence between load and shear strength).
3. The entire distribution of the geotechnical response to the applied random load can be estimated, including probability of failure.
4. Complex problems can be studied.

The remainder of this book is devoted to examples in which a level III probabilistic analysis of soils is applied to a variety of common geotechnical problems such as settlement and bearing capacity of shallow footings, seepage problems, earth pressure, and slope stability.

7.6 RISK-BASED DECISION MAKING

One of the main concerns in design is how to make the best design decisions in the face of often considerable uncertainty. One approach is to express loss (or utility), usually in monetary terms, of the system being designed as a function of the design variables, which may be random. Some measure of this function is then minimized (or maximized), for example, expected value, to determine the best set of design variables. The definition of "best" depends on one's priorities, but usually one is looking for minimum cost and maximum reliability.

Decision making in the face of uncertainty involves choosing the best between a variety of possible design alternatives. In order to determine which is best, when outcomes are uncertain, the probabilities associated with design consequences must be estimated and the decision process follows from a risk assessment methodology. Risk assessments are generally related to *hazards* which refer to possible negative effects on structures, people, crops, or the environment. Possible hazards include flooding, earthquake, overloading, liquefaction, and the like.

It is assumed that design alternatives will be hazard specific. For example, one design alternative might be to build a levee of height 2.1 m to avoid damages incurred by flood hazard. This alternative will, in general, be assumed to not help with damages caused by other hazards, such as earthquake. This simplifying assumption allows hazards to be considered separately.

The hazard under consideration will be denoted with the letter H. Let H_i be the event that the hazard H reaches level i (e.g., 2.0-m flood, 2.1-m flood), having probability of occurrence $\mathrm{P}[H_i]$.

Let D_{jk} be the event that damage level j occurs in a structure (or area, person, etc.) k. This damage level will have some probability of occurrence that depends on the hazard level reached, H_i, that is, $\mathrm{P}\left[D_{jk} \mid H_i\right]$. This probability may be expressed in words as the probability that damage level j will occur in structure (area, person, etc.) k given that hazard level H_i has occurred. In the following, reference will be made to "structure k," but note that k counts any object (including people, housing, bridges, crops, areas, environments, etc.) that could sustain hazard-induced damage.

If event D_{jk} occurs, the ensuing cost will be denoted E_{jk}. The overall damage cost for structure k will be denoted E_k.

Let A_l be the lth design (or adaptation) alternative, for example, build levee to height 2.1 m, having fixed cost B_l (A_0 is commonly set to be the "do nothing" alternative, having cost $B_0 = 0$). The total cost of alternative A_l, including possible damage costs, will be C_l.

Given these definitions, the following quantities can be computed:

1. The probability that damage level j occurs in structure k is

$$\mathrm{P}\left[D_{jk}\right] = \sum_i \mathrm{P}\left[D_{jk} \mid H_i\right] \mathrm{P}[H_i] \qquad (7.54)$$

where the sum is over all possible hazard levels, $i = 1, 2, \ldots$.

2. The expected damage cost in structure k (this is the *impact* on structure k) is

$$\begin{aligned} \mathrm{E}[E_k] &= \sum_j E_{jk} \mathrm{P}\left[D_{jk}\right] \\ &= \sum_j E_{jk} \sum_i \mathrm{P}\left[D_{jk} \mid H_i\right] \mathrm{P}[H_i] \end{aligned} \qquad (7.55)$$

where the first sum is over all possible damage levels, $j = 0, 1, \ldots$, for example, from no damage to complete destruction.

3. The expected cost of design alternative A_l is

$$\begin{aligned} \mathrm{E}[C_l] &= B_l + \sum_k \mathrm{E}[E_k] = B_l \\ &\quad + \sum_k \sum_j E_{jk} \sum_i \mathrm{P}\left[D_{jk} \mid H_i\right] \mathrm{P}[H_i] \end{aligned} \qquad (7.56)$$

where the first sum is over all structures, $k = 1, 2, \ldots$.

The risk-based decision-making goal is to find the design alternative, A_l, having the minimum expected cost $\mathrm{E}[C_l]$.

Example 7.3 Flood Risk Assessment Due to ocean level rise and crustal subsidence a small town on the Atlantic coast will be increasingly susceptible to flooding in the years to come. Four alternative strategies that the town can adopt in light of the risk of flooding will be considered:

1. Do nothing. Spend no additional money protecting the town against floods, but pay for flood damages as they occur.
2. Floodproof individual buildings threatened by the flood. Floodproofing involves raising houses and small buildings and the closure/sealing of basements and ground floors of larger buildings.
3. Construct a levee around the town to protect against a 5.4-m flood.
4. Construct a levee around the town to protect against a 6.5-m flood.

It will be assumed that all strategies are aimed at a design life of 100 years, after which time a similar analysis can be repeated. The time value of money will be ignored in the following analysis, partly because it is not known when the flood damage will occur (although this uncertainty could be handled properly in a more detailed probabilistic analysis).

Two projected flood levels will be considered for the town over the coming century:

1. 5.4-m flood: This includes the projected 100-year return period storm surge combined with high tide (4.6 m) plus climate change effects (ocean rise, 0.5 m/century) plus crustal subsidence (this coastline is gradually sinking, 0.3 m/century). The occurrence of these floods will be modeled as a Poisson process. It will also be assumed that the storm surge duration is of the order of 12 h, so that the storm surge is sure to coincide with a high tide. Note that the recurrence period of the 5.4-m flood really only becomes 100 years toward the end of this century, that is, after the projected ocean level rise and crustal subsidence have taken place. A conservative approach is taken here by assuming that the recurrence period is already 100 years and will remain at that rate throughout the coming century. In other words, it is assumed that the mean rate at which the 5.4-m flood occurs is $\lambda = \frac{1}{100}$ per year throughout the coming century.
2. 6.5-m flood: This includes all of the components of the 5.4-m flood except that the storm surge now corresponds with the highest astronomical tide (HAT). In the region of the town, the HAT has a return period of about 1.5 years. If the storm surge lasts for 12 h (a conservative estimate), then the probability that the storm surge does coincide with the HAT is equal to 1 over the number of 12-h intervals between average occurrences of the HAT. Thus, the mean rate of occurrence of the 6.5-m flood per year is

$$\lambda = \left(\frac{1}{100}\right)\left(\frac{1}{1.5 \times 365.25 \times 2}\right) = 9 \times 10^{-6}$$

which will be assumed to be the rate of occurrence of a 6.5-m flood throughout the coming century.

Table 7.9 Flood Levels and Potential Impacts

Attribute	100-Year with Climate Change and Coastal Subsidence	100-Year with Climate Change, Coastal Subsidence, and HAT
Flood level	5.4 m	6.5 m
Number of structures in floodplain	42	116
m^2 of structures in floodplain	12,672	23,182
High-priority structures	K–12 school	K–12 school and fire station
Roads flooded	1300 m	2500 m

It will be assumed that the flooding of a structure in this town will incur an average structural repair cost of $240/m^2 and that the total flood cost will include $80/m^2 for damage to chattel (movable goods). The total flood repair cost is thus $320/m^2 over all flooded structures.

Costs

1. **Cost of repairs to damages**
   - 5.4 m: Existing properties in floodplain = 12,672 m^2; estimated flood damage cost = $240/m^2; total damage costs = **$3,041,280**
     Estimated damage costs for chattel (movable goods) = $80/m^2. Total chattel costs = **$1,003,622**
   - 6.5 m: Existing properties in floodplain = 23,182 m^2. Estimated flood damage cost = $240/m^2. Total damage costs = **$5,563,680**
     Estimated damage costs for chattel (movable goods) = $80/m^2. Total chattel costs = **$2,740,320**
2. **Costs of floodproofing small buildings** (up to 150 m^2 in plan, floodproofing type = elevation on bearing walls)
   - 5.4 m: Floodproofing cost = 31% of total property value
     Total property value of small buildings in floodplain = $1,642,000
     Floodproofing cost = 0.31(1,642,000) = **$509,020**
   - 6.5 m: Floodproofing cost = 31% of total property value
     Total property value of small buildings in floodplain = $5,029,900
     Floodproofing cost = 0.31(5,029,900) = **$1,559,269**
3. **Costs of floodproofing medium-sized buildings** (between 150 and 2000 m^2 in plan, floodproofing type = closure and seal)
   - 5.4 m: Floodproofing cost = 26% of total property value
     Total property value of medium-sized buildings in floodplain = $2,831,800
     Floodproofing cost = 0.26(2,831,800) = **$736,268**
   - 6.5 m: Floodproofing cost = 26% of total property value
     Total property value of medium-sized buildings in floodplain = $5,411,500
     Floodproofing cost = 0.26(5,411,500) = **$1,406,990**
4. **Costs of floodproofing large buildings** (over 2000 m^2 in plan, floodproofing type = closure and seal)

5.4 m Floodproofing cost = 26% of total property value
Total property value of school in floodplain = \$5,000,000
Floodproofing cost = 0.26(5,000,000) = **\$1,300,000**

6.5 m: Floodproofing cost = 26% of total property value
Total property value of school in floodplain = \$5,000,000
Floodproofing cost = 0.26(5,000,000) = **\$1,300,000**

5. **Cost of constructing a levee**
Regarding the construction of levees, we shall assume that the levee will have a 3-m-wide crest (to accommodate heavy equipment) and 3.5H:1V side slopes. We shall also provide a 3-m freeboard above the required flood elevation to avoid overtopping (and subsequent erosion) by wave actions, which are sure to accompany a storm surge. The average elevation of the land around the town on which the levee would be placed is about 5.0 m, and the levee would need to be approximately 900 m in length. The details of the two levees being considered are as follows;
5.4 m: For the 5.4-m flood, a 3-m freeboard brings the upper surface of the levee to an elevation of 8.4 m. The constructed height of the levee is thus $H = 8.4 - 5.0 = 3.4$ m having a cost of about \$2600/m. The total cost of this levee will therefore be 900(2600) = **\$2,340,000**.
6. For the 6.5-m flood, a 3-m freeboard results in a constructed levee height of $H = 9.5 - 5.0 = 4.5$ m having a cost of about \$4200/m. The total cost of this levee will therefore be 900(4200) = **\$3,780,000**.

Risk Assessment

We shall first consider the expected failure costs, $\mathrm{E}\left[C_f\right]$, associated with flooding for the "do-nothing" case. This is just equal to the cost of flood damage times the number of floods expected to occur during the design life of 100 years.

The 5.4-m flood has a (conservative) mean recurrence rate of $\lambda = 1/100$, so the expected number of such floods in $t = 100$ years is $\lambda t = 1$. The expected cost of repairs for damages for the 5.4-m flood is thus

$$\mathrm{E}\left[C_f\right] = 1 \times (3{,}041{,}280 + 1{,}003{,}622) = \$4{,}044{,}902$$

The expected number of 6.5-m floods in $t = 100$ years is (conservatively) $\lambda t = (9 \times 10^{-6})(100) = 0.0009$. The expected cost of repairs for damages for the 6.5-m flood is thus

$$\mathrm{E}\left[C_f\right] = 0.0009 \times (5{,}563{,}680 + 2{,}740{,}320) = \$7474$$

We see immediately that the best option with respect to the 6.5-m flood is to do nothing. The chances of the storm surge occurring at the same time as the HAT is so small that the risk is negligible compared to the initial costs of floodproofing or constructing a levee.

For the 5.4-m flood, we compare the following expected total costs:

1. The expected total cost of the do-nothing option is \$4,044,902.
2. The expected total cost of the floodproofing option is the sum of the floodproofing costs and the expected failure cost of the 6.5-m flood (we assume that floodproofing will certainly protect against the 5.4-m flood), (509,020 + 736,268 + 1,300,000) + 7474 = \$2,552,762.
3. The expected total cost of the levee construction option is 2,340,000 + 7474 = \$2,347,474.

The lowest expected total cost is option 3. This suggests that the town should construct a levee to protect against the 5.4-m flood.

We note, however, that options 2 and 3 have very similar total expected costs, so the two options are quite competitive. Some detailed comments about the risk assessment are as follows:

- We assume that both the floodproofing and 8.4-m levee provide sure protection against the 5.4-m flood. Because of the 3-m freeboard, this is probably largely true of the levee, pending a more detailed study of likely wave heights accompanying the storm surge at this location. However, in the opinion of the authors, floodproofing provides less certain protection unless properly maintained and implemented (i.e., are all doors adequately sealed against water ingress at the time of the flood?). Thus, the actual expected total cost of the floodproofing option may still include a significant flood damage cost.
- Floodproofing can be applied on a building-by-building basis, which may lead to cost savings (e.g., only protect the important buildings, or those buildings which would be most heavily damaged by a flood—garages, for example, may suffer very little damage in a flood).

- The above analysis is simplified by assuming that only two flood levels are possible: 5.4 and 6.5 m. In actuality, of course, there is a continuous range of possible flood levels. A more detailed risk analysis would consider separately the expected number of floods from, say, 5.0 to 5.5 m, the expected number of floods from 5.5 to 6.0 m, and so on. However, the individual probabilities of occurrence of floods on such a detailed level would be quite difficult to estimate on the basis of past records and even more difficult to predict with any accuracy for the future. The refinement would probably not be justified.

PART 2

Practice

CHAPTER 8

Groundwater Modeling

8.1 INTRODUCTION

In this chapter steady-state seepage through a soil mass with spatially random permeability is considered. The goal of the chapter is to present approximate theory and results which allow the assessment of probabilities relating to quantities of interest such as steady-state flow rates, exit gradients, and uplift pressures. The theoretical results are compared to simulation-based results.

The equation of steady groundwater flow followed in this chapter is Laplace's equation,

$$\nabla \cdot [\boldsymbol{K} \ \nabla\phi] = 0 \tag{8.1}$$

where $\boldsymbol{K}$ is the permeability tensor and ϕ is the hydraulic head.

8.2 FINITE-ELEMENT MODEL

The form of Laplace's equation given in Eq. 8.1 which arises in geomechanics, for example, concerning two-dimensional groundwater flow beneath a water-retaining structure or in an aquifer (Muskat, 1937), is conveniently solved by the finite-element method. The theory behind the finite-element method is discussed in detail in numerous publications (see, e.g., Smith and Griffiths, 2004) but essentially consists of "discretizing," or breaking, a continuum into numerous small pieces called finite elements similar to that shown later in Figure 8.5. The governing differential equation 8.1 is then solved approximately over each element, after which the elements are "assembled" into a "mesh" with appropriate boundary conditions to give a solution to the problem as a whole.

In a two-dimensional setting, the finite elements might be square with four-nodes as shown in Figure 8.1, while in three dimensions the elements might be cubic with eight-nodes as shown in Figure 8.2. The nodes are the places where the hydraulic head will eventually be calculated. It should realized that quadrilateral or hexahedral elements in general do not need to be square or cubic, but they always are in the applications described in this text using the random finite-element method (RFEM) as developed by the authors.

The end product of a finite-element analysis of a steady-seepage problem governed by Eq. 8.1 is an estimate of the hydraulic head at numerous nodes across the solution domain. Generally speaking, the more elements used to discretize the domain of the problem, the greater the accuracy of the solution. As with all numerical methods, however, a trade-off is necessary between accuracy and computer time. One of the key ingredients of a finite-element analysis is to design a mesh that is "fine enough" to give acceptable accuracy.

One of the great benefits of the finite-element method is that it is easy to model problems with variable properties. For example, a given soil deposit may consist of layers having different permeability values in which rows of element may be assigned different properties. In the RFEM, this feature is taken to the limit by analyzing problems in which *every* element in the mesh has a different property based on some underlying statistical distribution.

The finite-element discretization process essentially reduces the differential equation (8.1) to a set of equilibrium-type matrix equations at the element level of the form

$$\boldsymbol{k}_c \boldsymbol{\phi} = \mathbf{q} \tag{8.2}$$

where $\boldsymbol{k}_c$ is a symmetrical *conductivity matrix*, $\boldsymbol{\phi}$ is a vector of nodal hydraulic head values, and $\mathbf{q}$ is a vector of nodal inflows/outflows. The matrix equation essentially describes the relationship between flow rates and hydraulic head across a single finite element. The size of the matrix $\boldsymbol{k}_c$ depends on the element used. For example, when using four-node elements, $\boldsymbol{k}_c$ would have four rows and four columns.

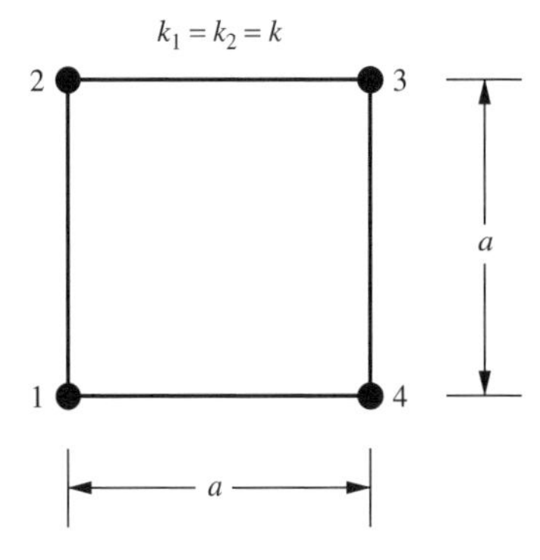

Figure 8.1 Two-dimensional isotropic square four-node finite element with node numbering.

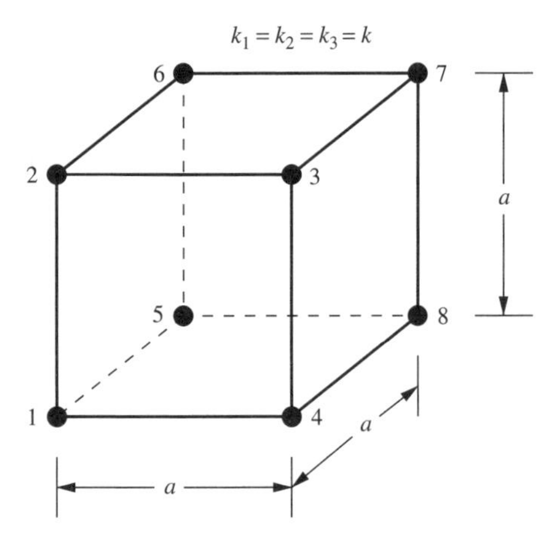

Figure 8.2 Three-dimensional isotropic cubic eight-node hexahedral finite element with node numbering.

At this stage the element conductivity matrices described in Eq. 8.2 are singular. After all the element matrices from the mesh have been assembled, however, with proper inclusion of boundary conditions, a system of symmetric linear simultaneous equations are obtained of the form

$$\boldsymbol{K}_c \boldsymbol{\Phi} = \mathbf{Q} \tag{8.3}$$

In Eq. 8.3, $\boldsymbol{K}_c$ represents the global conductivity matrix (of the entire mesh), $\boldsymbol{\Phi}$ holds the nodal values of the hydraulic head, and $\mathbf{Q}$ holds the nodal *net* flow rates. Since $\boldsymbol{K}_c$ is both symmetric and banded (all nonzero terms tend to be clustered near the diagonal), appropriate efficient storage strategies are essential.

Solution of Eq. 8.3 can be achieved using either direct or iterative solvers with the latter more likely to be used in very large problems typical of three-dimensional analysis where $\boldsymbol{K}_c$ cannot be fitted in the core memory of the computer.

Since the majority of internal nodes experience no net flow due to steady-state conditions, most components of $\mathbf{Q}$ equal zero. The nonzero terms in $\mathbf{Q}$, however, correspond to the boundary nodes at which the head is fixed. The steady-state flow rate through the system therefore is obtained by summing, separately, the positive and negative terms in $\mathbf{Q}$. These sums should of course be equal and opposite, with the positive sum representing the inflow upstream and the negative sum the outflow downstream.

As described later in this chapter, other quantities of engineering interest such as exit gradients and uplift forces can be obtained by further postprocessing of the nodal $\boldsymbol{\Phi}$ values. Exit gradients involve numerical differentiation of these values in the vicinity of the exit points immediately downstream of the structure, and uplift forces involve numerical integration of the pressure heads immediately beneath the structure.

8.2.1 Analytical Form of Finite-Element Conductivity Matrices

As mentioned previously, the RFEM formulations described in this text typically use square elements in two-dimensional or cubic elements in three dimensions. These simplified geometries enable the element conductivity matrices to be developed exactly for the flow problem and in closed form, in contrast to the approximate form typically obtained by numerical integration in general-purpose finite-element codes.

Assuming an isotropic square element of permeability k as shown in Figure 8.1 with an out-of-plane depth of one unit, the element conductivity matrix is given by

$$\boldsymbol{k}_c = \frac{k}{6}\begin{bmatrix} 4 & -1 & -2 & -1 \\ & 4 & -1 & -2 \\ & & 4 & -1 \\ & & & 4 \end{bmatrix} \tag{8.4}$$

Conductivity matrices of elements with different permeabilities as needed by the RFEM are easily obtained by simply changing the value of the permeability coefficient k in Eq. 8.4.

A similar approach can be used for the cubic isotropic element of permeability k shown in Figure 8.2. In this case the element conductivity matrix is given by

$$\boldsymbol{k}_c = \frac{ka}{12}\begin{bmatrix} 4 & 0 & -1 & 0 & 0 & -1 & -1 & -1 \\ & 4 & 0 & -1 & -1 & 0 & -1 & -1 \\ & & 4 & 0 & -1 & -1 & 0 & -1 \\ & & & 4 & -1 & -1 & -1 & 0 \\ & & & & 4 & 0 & -1 & 0 \\ & & & & & 4 & 0 & -1 \\ & & & & & & 4 & 0 \\ & & & & & & & 4 \end{bmatrix} \tag{8.5}$$

In both of Eqs. 8.4 and 8.5, only the upper triangular terms have been included for clarity, but the actual matrices are symmetrical.

8.3 ONE-DIMENSIONAL FLOW

One-dimensional flow occurs when all streamlines are parallel and straight. Low-permeability regions are not avoided by the fluid particles. If any portion of the soil mass through which a streamline passes is impermeable, then both the effective permeability and the flow rate become zero along that streamline. The *effective permeability* is defined as that uniform (everywhere the same) permeability which has the same flow rate as the actual spatially variable permeability. In general, the effective permeability is dominated by the low-permeability regions of the soil mass. This is particularly true in the one-dimensional flow case, where any "blockage" dominates the flow.

Example 8.1 Suppose that water flowing horizontally through a soil mass is idealized as being one-dimensional flow, and assume that the soil is composed of three units having different permeabilities. Figure 8.3 illustrates the flow regime. If we wish to express the flow rate Q as

$$Q = K_{\text{eff}} \left(\frac{A\,\Delta H}{L} \right)$$

where K_{eff} is the effective permeability that leads to the same flow rate as the true situation, A is the cross-sectional area perpendicular to the flow, ΔH is the head difference between left and right faces (ignoring gravity effects), and L is the flow path length, then what is K_{eff}?

SOLUTION In Figure 8.3, ΔH is the total head difference between right and left faces of the soil and ΔH_i is the head loss across the ith layer. The layers have equal thickness, $L/3$, and equal cross-sectional area, A.

We know that the total steady-state flow through each layer must be the same since there are no sinks or sources, that is, $Q_1 = Q_2 = Q_3 = Q$, so that

$$Q = K_1 \left(\frac{A\,\Delta H_1}{L/3} \right) = K_2 \left(\frac{A\,\Delta H_2}{L/3} \right) = K_3 \left(\frac{A\,\Delta H_3}{L/3} \right) \tag{8.6}$$

In other words, we must have

$$K_1\,\Delta H_1 = K_2\,\Delta H_2 = K_3 \Delta H_3$$

We can use this to express ΔH_2 and ΔH_3 in terms of ΔH_1,

$$\Delta H_2 = \frac{K_1}{K_2} \Delta H_1, \qquad \Delta H_3 = \frac{K_1}{K_3} \Delta H_1$$

We also know that the total head loss from right to left, ΔH, must just equal the sum of losses across each layer,

$$\begin{aligned} \Delta H &= \Delta H_1 + \Delta H_2 + \Delta H_3 \\ &= \Delta H_1 \left[1 + \frac{K_1}{K_2} + \frac{K_1}{K_3} \right] \end{aligned}$$

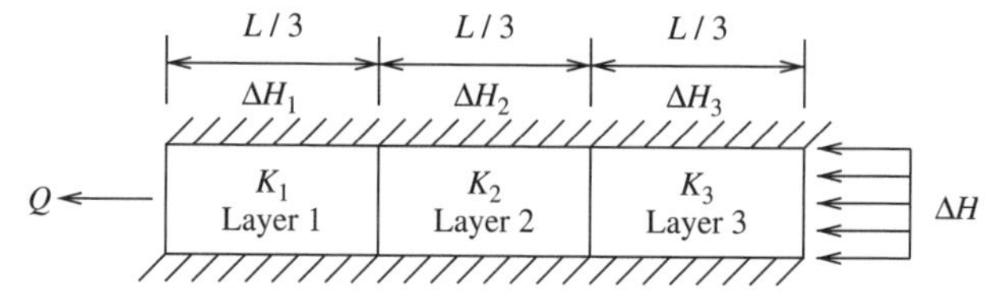

Figure 8.3 One-dimensional flow through three soil layers.

which we can solve for ΔH_1,

$$\Delta H_1 = \frac{\Delta H}{1 + K_1/K_2 + K_1/K_3}$$

$$\text{or} \qquad K_1\,\Delta H_1 = \frac{\Delta H}{1/K_1 + 1/K_2 + 1/K_3}$$

Finally, we can solve for Q using the first equality in Eq. 8.6,

$$Q = \frac{K_1\,\Delta H_1}{1/3} \frac{A}{L} = \frac{1}{(1/3)\,(1/K_1 + 1/K_2 + 1/K_3)} \times \left(\frac{A\,\Delta H}{L} \right)$$

from which we see that K_{eff} is the harmonic average,

$$K_{\text{eff}} = \frac{1}{(1/3)\,(1/K_1 + 1/K_2 + 1/K_3)}$$

The classic example of one-dimensional flow is flow through a pipe. The effective permeability is the harmonic average of the permeabilities along the flow path. If the flow path has length L, then the effective permeability is

$$K_{\text{eff}} = \left[\frac{1}{L} \int_0^L \frac{dx}{K(x)} \right]^{-1}$$

which is dominated by low values of $K(x)$ along the path. See Section 4.4 for a more detailed discussion of the harmonic average.

Example 8.2 Suppose that steady-state flow through a 2-m-thick compacted clay barrier is assumed to be one dimensional (Figure 8.4). This would occur, for example, if the permeability only varies through the thickness. Clay barriers which are made up of a series of parallel layers having different permeabilities might be reasonably assumed to behave this way. Suppose that the permeability of the clay is lognormally distributed with mean $\mu_K = 1 \times 10^{-8}$, standard deviation $\sigma_K = 0.5 \times 10^{-8}$, and correlation length $\theta_{\ln K} = 0.5$ m. Assume further that the correlation structure

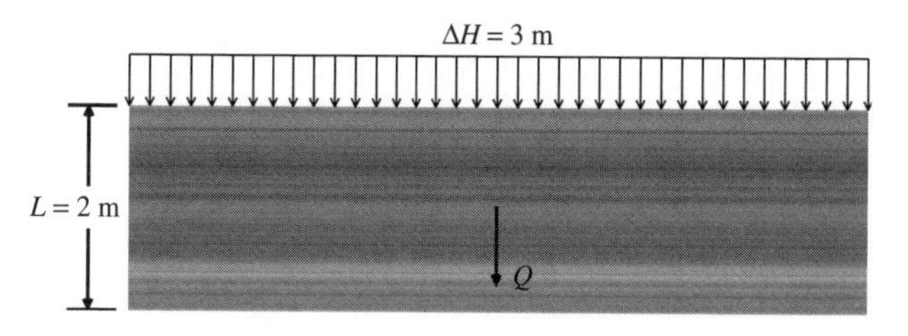

Figure 8.4 Flow through a compacted clay barrier idealized as one dimensional.

through the thickness is Markovian, so that

$$\rho_{\ln K}(\tau) = \exp\left\{-\frac{2\tau}{\theta_{\ln K}}\right\} = \exp\{-4\tau\} \qquad (8.7)$$

where τ is measured through the thickness. If the total head difference across the clay barrier is $\Delta H = 3$ m, then:

(a) Estimate the mean steady-state flow rate through a 1-m^2 area of the barrier.
(b) The design target is to restrict the steady-state flow through the barrier to less than 2×10^{-8} m^3/s per square meter of barrier. What is the probability that the actual steady-state flow rate exceeds this target?

SOLUTION We are told that $K(z)$ is lognormally distributed; thus its parameters are

$$\sigma^2_{\ln K} = \ln\left(1 + v_K^2\right) = \ln\left(1 + 0.5^2\right) = 0.223144$$
$$\mu_{\ln K} = \ln(\mu_K) - \tfrac{1}{2}\sigma^2_{\ln K} = \ln(1 \times 10^{-8}) - \tfrac{1}{2}(0.223144)$$
$$= -18.5323$$

where $v_K = \sigma_K/\mu_K$ is the coefficient of variation of K.

Because the flow is one dimensional, the total steady-state flow rate Q is given by

$$Q = K_{\text{eff}}\left(\frac{A\ \Delta H}{L}\right) \qquad (8.8)$$

where K_{eff} is the harmonic average

$$K_{\text{eff}} = \left[\frac{1}{L}\int_0^L \frac{dz}{K(z)}\right]^{-1} \qquad (8.9)$$

In order to answer parts (a) and (b), we need to determine the distribution of K_{eff}. Unfortunately, the harmonic average is the inverse of a sum of lognormally distributed random variables, which does not have a simple distribution. Let us start by defining

$$R(z) = \frac{1}{K(z)}$$

If $K(z)$ is lognormally distributed, then so is $R(z)$, with parameters

$$\mu_{\ln R} = -\mu_{\ln K} = 18.5323 \qquad (8.10a)$$
$$\sigma^2_{\ln R} = \sigma^2_{\ln K} = 0.223144 \qquad (8.10b)$$

Transforming these parameters back into real space, we get

$$\mu_R = \exp\left\{\mu_{\ln R} + \tfrac{1}{2}\sigma^2_{\ln R}\right\} = 1.2499969 \times 10^8$$
$$\sigma_R^2 = \mu_R^2\left(e^{\sigma^2_{\ln R}} - 1\right) = 3.906231 \times 10^{15}$$

At this point, it is worth making a couple of comments:

1. Because lognormal distribution transformations involve the sums and differences of large and small numbers, we generally keep quite a few significant digits in interim results.
2. The entire analysis can be scaled to make the mathematics more convenient. For example, if we start with $\mu_K = 1$ and $\sigma_K = 0.5$, then in part (a) we need only multiply the mean flow rate by 10^{-8} and in part (b) we need only compute the probability that the flow rate exceeds 2.

Now, let R_L be the arithmetic average of $R(z)$ over the depth L,

$$R_L = \frac{1}{L}\int_0^L R(z)\, dz \qquad (8.11)$$

so that $K_{\text{eff}} = [R_L]^{-1}$. Notice that if $\theta_{\ln K} \ll L$, then $R_L \simeq \mu_R$ since R_L is an arithmetic average, where

$$\mu_R = \exp\left\{\mu_{\ln R} + \tfrac{1}{2}\sigma^2_{\ln R}\right\} = \exp\left\{-\mu_{\ln K} + \tfrac{1}{2}\sigma^2_{\ln K}\right\}$$

which gives us

$$K_{\text{eff}} = \frac{1}{\mu_R} = \exp\left\{\mu_{\ln K} - \tfrac{1}{2}\sigma^2_{\ln K}\right\} \qquad (8.12)$$

That is, if the correlation length is small, so that the clay barrier consists of a large number of layers having mostly independent permeabilities, then the local average R_L will be a close estimate (with vanishing variance) of the true mean μ_R. In this case, the harmonic average can be computed by Eq. 8.12. Note also that in this case the harmonic average K_{eff} becomes deterministic.

Returning to our problem, where $\theta_{\ln K}$ is not vanishingly small, we can take expectations of Eq. 8.11 to find the mean and variance of R_L:

$$\mu_{R_L} = \mu_R = 1.2499969 \times 10^8$$
$$\sigma_{R_L} = \sigma_R^2\gamma(L)$$

where $\gamma(L)$ is the variance reduction function (see Section 3.4). To determine the value of $\gamma(L)$, which is defined in terms of $\rho_R(\tau)$, we must determine the nature of $\rho_R(\tau)$. If $K(z)$ is lognormally distributed, then $R(z) = 1/K(z)$ is also lognormally distributed and $\ln R = -\ln K$ is normally distributed with $\rho_{\ln R}(\tau) = \rho_{\ln K}(\tau)$ (see Eq. 8.7). Thus, the correlation structure of $\ln R$ is identical to that of $\ln K$.

Because the logarithm is a smoothly (and slowly) varying function of its argument (i.e., almost linear over ranges that typically include the mean plus or minus several standard deviations), it is quite reasonable to assume that $\rho_R(\tau) \simeq \rho_{\ln R}(\tau) = \rho_{\ln K}(\tau)$ with $\theta_R \simeq \theta_{\ln K} = 0.5$ m. With these approximations, Eq. 3.89 gives us

$$\gamma(L) = \frac{\theta_R^2}{2L^2}\left[\frac{2|L|}{\theta_R} + \exp\left\{-\frac{2|L|}{\theta_R}\right\} - 1\right]$$

$$= \frac{0.5^2}{2(2)^2}\left[\frac{2(2)}{0.5} + \exp\left\{\frac{-2(2)}{0.5}\right\} - 1\right]$$
$$= 0.21876$$

so that

$$\sigma_{R_L}^2 = \sigma_R^2 \gamma(L) = (3.906231 \times 10^{15})(0.21876)$$
$$= 8.54539 \times 10^{14}$$

Now we need only determine the form of the distribution for R_L. Since R_L is equal to a sum (Eq. 8.11) of lognormally distributed random variables, its true distribution is somewhere between a normal and lognormal distribution. As demonstrated by Fenton et al. (2003b), the distribution of R_L is closely approximated by a lognormal distribution with parameters

$$\sigma_{\ln R_L}^2 = \ln\left(1 + \frac{\sigma_{R_L}^2}{\mu_{R_L}^2}\right) = 0.053247$$

$$\mu_{\ln R_L} = \ln\left(\mu_{R_L}\right) - \tfrac{1}{2}\sigma_{\ln R_L}^2 = 18.617198$$

If R_L is lognormally distributed, then

$$K_{\text{eff}} = \frac{1}{R_L}$$

is also lognormally distributed with parameters

$$\mu_{\ln K_{\text{eff}}} = -\mu_{\ln R_L} = -18.617198 \tag{8.13a}$$

$$\sigma_{\ln K_{\text{eff}}}^2 = \sigma_{\ln R_L}^2 = 0.053247 \tag{8.13b}$$

We are now ready to answer our two questions, at least approximately:

(a) Taking expectations of Eq. 8.8 gives us, for $A = 1\ \text{m}^2$,

$$\mu_Q = \mu_{K_{\text{eff}}}\left(\frac{A\ \Delta H}{L}\right) = \mu_{K_{\text{eff}}}\left(\tfrac{3}{2}\right)$$

where, using Eqs. 8.13,

$$\mu_{K_{\text{eff}}} = \exp\left\{\mu_{\ln K_{\text{eff}}} + \tfrac{1}{2}\sigma_{\ln K_{\text{eff}}}^2\right\} = 0.843754 \times 10^{-8}$$

so that

$$\mu_Q = \left(0.843754 \times 10^{-8}\right)\left(\tfrac{3}{2}\right)$$
$$= 1.2656 \times 10^{-8} \quad \text{m}^3/\text{s}/\text{m}^2$$

(b) The probability we are looking for is

$$\text{P}\left[Q > 2 \times 10^{-8}\right]$$
$$= \text{P}\left[K_{\text{eff}} > \tfrac{4}{3} \times 10^{-8}\right]$$
$$= 1 - \Phi\left(\frac{\ln\left(\frac{4}{3} \times 10^{-8}\right) - \mu_{\ln K_{\text{eff}}}}{\sigma_{\ln K_{\text{eff}}}}\right)$$
$$= 1 - \Phi\left(\frac{\ln\left(\frac{4}{3} \times 10^{-8}\right) + 18.617198}{\sqrt{0.053247}}\right)$$
$$= 1 - \Phi(2.10)$$
$$= 0.0179$$

To check our approximations in this solution, we simulated a 2-m-thick clay barrier with parameters given in the introduction to this problem 100,000 times. For each realization, the flow rate through the barrier was computed according to Eqs. 8.8 and 8.9. The mean flow rate and the probability that the flow rate exceeded 2×10^{-8} were estimated from the 100,000 realizations with the following results:

$$\mu_Q \simeq 1.2631 \times 10^{-8}\ \text{m}^3/\text{s}/\text{m}^2$$
$$\text{P}\left[Q > 2 \times 10^{-8}\right] \simeq 0.0153$$

from which we see that the mean flow rate given by the above solution is within a 0.2% relative error from that determined by simulation. The probability estimate was not quite so close, which is not surprising given the fact that the distribution of K_{eff} is not actually lognormal, with a relative error of 17%. Nevertheless, the approximate solution has an accuracy which is quite acceptable given all other sources of uncertainty (e.g., μ_K, σ_K, and $\theta_{\ln K}$ are generally not well known).

8.4 SIMPLE TWO-DIMENSIONAL FLOW

Two-dimensional flow allows the fluid to avoid low-permeability zones simply by detouring around them. Although the streamlines, being restricted to lying in the plane, lack the complete directional freedom of a three-dimensional model (see next section), the low-permeability regions no longer govern the flow as strongly as they do in the one-dimensional case. The two-dimensional steady-state flow equation is given by Eq. 8.1, where the permeability K can be a tensor,

$$K = \begin{bmatrix} K_{11} & K_{12} \\ K_{21} & K_{22} \end{bmatrix}$$

in which K_{11} is the permeability in the x_1 direction, K_{22} is the permeability in the x_2 direction, and K_{12} and K_{21} are cross-terms usually assumed to be zero. In principle, all four components could be modeled as (possibly cross-correlated) random fields. However, since even the variance of even one of the components is rarely known with any accuracy, the usual assumption is that the permeability is isotropic. In this case, K becomes a scalar and the problem of estimating its statistical nature is vastly simplified. In this chapter, we will consider only the isotropic and stationary permeability case. This is usually sufficient to give us at least rough estimates of the probabilistic nature of seepage (Fenton and Griffiths, 1993).

A variety of boundary conditions are also possible, and we will start by looking at one of the simplest. Consider a soil mass which has no internal sources or sinks and has impervious upper and lower boundaries with constant head applied to the right edge, as illustrated in Figure 8.5, so that the mean flow is unidirectional.

Interest will be in determining the distribution of the total flow rate through Figure 8.5, bearing in mind that the permeability K is a spatially varying random field. To do this, let us define a quantity that will be referred to as *block permeability* (a term used in the water resources community), which is based on the total (random) flow rate Q. Specifically, for a particular realization of the spatially random permeability $K(\mathbf{x})$ on a given boundary value problem, the block permeability $\bar{K}$ is defined as

$$\bar{K} = \mu_K \left(\frac{Q}{Q_{\mu_K}} \right) \tag{8.14}$$

where $\mu_K = \text{E}[K]$ is the expected value of $K(\mathbf{x})$, Q is the total flow rate through the spatially random permeability field, and Q_{μ_K} is the deterministic total flow rate through the mean permeability field (having constant permeability μ_K throughout the domain). For the simple boundary value problem under consideration, Eq. 8.14 reduces to

$$\bar{K} = \left(\frac{X_L}{Y_L} \right) \left(\frac{Q}{\Delta H} \right) \tag{8.15}$$

where ΔH is the (deterministic) head difference between upstream and downstream faces. Since Q is dependent on $K(\mathbf{x})$, both Q and $\bar{K}$ are random variables and it is the distribution of $\bar{K}$ that is of interest. The definition of block

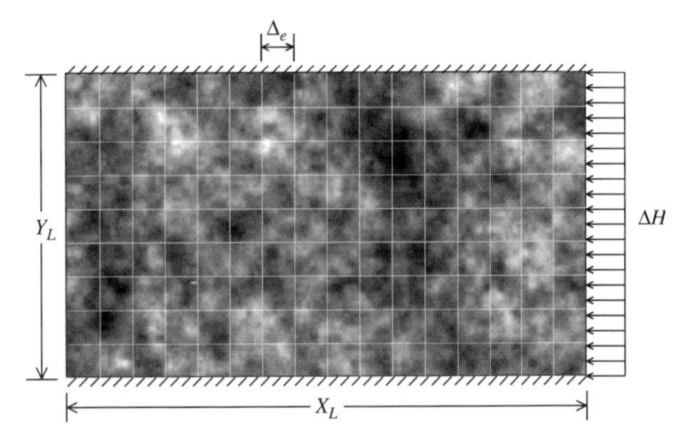

Figure 8.5 Two-dimensional finite-element model of soil mass having spatially random permeability. Top and bottom surfaces are impervious and a constant head is applied to the right face.

permeability used here essentially follows that of Rubin and Gómez-Hernández (1990) for a single block. Once the distribution of $\bar{K}$ is known, Eq. 8.15 is easily inverted to determine the distribution of Q for a specific geometry.

In the case of unbounded domains, considerable work has been done in the past to quantify a deterministic effective permeability measure K_{eff} as a function of the mean and variance of $\ln K$ at a point. In one dimension, the effective permeability is the harmonic mean (flow through a "series" of "resistors"), as discussed in the previous section, while in two dimensions the effective permeability is the geometric mean (Matheron, 1967). Indelman and Abramovich (1994) and Dagan (1993) develop techniques of estimating the effective permeability under the assumption that the domain is of infinite size and the mean flow is unidirectional. For three dimensions Gutjahr et al. (1978) and Gelhar and Axness (1983) used a small-perturbation method valid for small variances in an unbounded domain. Specifically they found that for n dimensions

$$K_{\text{eff}} = e^{\mu_{\ln K}} \left(1 - \tfrac{1}{2}\sigma^2_{\ln K}\right), \qquad n = 1 \tag{8.16a}$$

$$K_{\text{eff}} = e^{\mu_{\ln K}}, \qquad n = 2 \tag{8.16b}$$

$$K_{\text{eff}} \simeq e^{\mu_{\ln K}} \left(1 + \tfrac{1}{6}\sigma_{\ln K}\right), \qquad n = 3 \tag{8.16c}$$

where $\mu_{\ln K}$ and $\sigma^2_{\ln K}$ are the mean and variance of $\ln K$, respectively. Concerning higher order moments Dagan (1979) used a self-consistent model to estimate head and specific discharge variance for one-, two-, and three-dimensional flow in an infinite domain. Smith and Freeze (1979b) investigated head variability in a finite two-dimensional domain using Monte Carlo simulation.

Dykaar and Kitanidis (1992a,b) present a method for finding K_{eff} in a bounded domain using a spectral decomposition approach. The variance of block permeability is considered only briefly, through the use of simulations produced using FFT, to establish estimates of the averaging volume needed to reduce the variance to a negligible amount. No attempt was made to quantify the variance of block permeability. In perhaps the most pertinent work to this particular simple boundary value problem, Rubin and Gómez-Hernández (1990) obtained analytical expressions for the mean and variance of block permeability using perturbative expansions valid for small-permeability variance and based on some infinite-domain results. A first-order expansion was used to determine the block permeability covariance function.

In agreement with previous studies by Journel (1980), Freeze (1975), Smith and Freeze (1979b), Rubin and Gómez-Hernández (1990), and Dagan (1979, 1981, 1986), it is assumed here that $\ln K$ is an isotropic stationary Gaussian random field fully characterized by its mean $\mu_{\ln K}$, variance $\sigma^2_{\ln K}$, and correlation function $\rho_{\ln K}(\tau)$. The assumption regarding the distribution of K is basically predicated on the observation that field measurements of permeability are approximately lognormally distributed, as shown by Hoeksema and Kitanidis (1985) and Sudicky (1986). As argued in Section 4.4 this may be due to the fact that the geometric average, appropriate in two dimensions as indicated by Eq. 8.16b, tends to a lognormal distribution by the central limit theorem.

To solve Eq. 8.1 for the boundary value problem of Figure 8.5, the domain is discretized into elements of dimension $\Delta_1 \times \Delta_2$ (where $\Delta_1 = \Delta_2 = \Delta_e$ herein) and analyzed using the finite-element method (Smith and Griffiths, 2004). A realization of the random-field $\ln K(\mathbf{x})$ is then generated using the LAS method (Fenton, 1990) and permeabilities are assigned to individual elements using $K = e^{\ln K}$ (the permeability within each element is assumed constant). The total flow rate computed for this field can be used in Eq. 8.15 to yield a realization of the block permeability. Histograms of the block permeabilities are constructed by carrying out a large number of realizations for each case considered. The program used to carry out this simulation is called RFLOW2D and is available at `http://www.engmath.dal.ca/rfem`.

8.4.1 Parameters and Finite-Element Model

To investigate the effect of the form of the correlation function on the statistics of $\bar{K}$, three correlation functions were employed all having exponential forms:

$$\rho^a_{\ln K}(\boldsymbol{\tau}) = \exp\left\{-\frac{2}{\theta_{\ln K}}\sqrt{\tau_1^2 + \tau_2^2}\right\} \tag{8.17a}$$

$$\rho^b_{\ln K}(\boldsymbol{\tau}) = \exp\left\{-\frac{\pi}{\theta^2_{\ln K}}(\tau_1^2 + \tau_2^2)\right\} \tag{8.17b}$$

$$\rho^c_{\ln K}(\boldsymbol{\tau}) = \exp\left\{-\frac{2}{\theta_{\ln K}}(|\tau_1| + |\tau_2|)\right\} \tag{8.17c}$$

The first form is based on the findings of Hoeksema and Kitanidis (1985) but without the nugget effect which Dagan (1986) judges to have only a minor contribution when local averaging is involved. Notice that the second and third forms are separable and that the third form is not strictly isotropic.

It is well known that the block permeability ranges from the harmonic mean in the limit as the aspect ratio X_L/Y_L of the site tends to infinity and to the arithmetic mean as the aspect ratio reduces to zero. To investigate how the statistics of $\bar{K}$ change with practical aspect ratios, a study was conducted for ratios $X_L/Y_L \in \{\frac{1}{9}, 1, 9\}$. For an effective site dimension $D = \sqrt{X_L Y_L}$, the ratio $\theta_{\ln K}/D$ was varied over the interval $[0.008, 4]$. The very small ratios result in fields in which the permeabilities within each finite element are largely independent. Conceptually, when $\theta_{\ln K} = 0$ the permeability at all points become independent. This results in a white noise process which is physically unrealizable. In practice, correlation lengths less than the size of laboratory samples used to estimate permeability have little meaning since the permeability is measured at the laboratory scale. In this light, the concept of "permeability at a point" really means permeability in a representative volume (of the laboratory scale) centered at the point and correlation lengths much smaller than the volume over which permeability is measured have little meaning. In other words, while the RFEM assumes that the permeability field is continuously varying, and thus defined at the point level, it probably does not make sense to take the correlation length much smaller than the size of the volume used to estimate permeability.

Four different coefficients of variation were considered: $\sigma_K/\mu_K \in \{0.5, 1.0, 2.0, 4.0\}$ corresponding to $\sigma^2_{\ln K} \in \{0.22, 0.69, 1.61, 2.83\}$. It was felt that this represented enough of a range to establish trends without greatly compromising the accuracy of statistical estimates (as $\sigma^2_{\ln K}$ increases, more realizations would be required to attain a constant level of accuracy).

As mentioned, individual realizations were analyzed using the finite-element method (with four-node quadrilateral elements) to obtain the total flow rate through the domain. For each set of parameters mentioned above, 2000 realizations were generated and analyzed. No explicit attempt was made to track matrix condition numbers, but all critical calculations were performed in double precision and round-off errors were considered to be negligible.

8.4.2 Discussion of Results

The first task undertaken was to establish the form of the distribution of block permeability $\bar{K}$. Histograms were plotted for each combination of parameters discussed in the previous section and some typical examples are shown in Figure 8.6. To form the histograms, the permeability axis was divided into 50 equal intervals or "buckets." Computed block permeability values (Eq. 8.15) were normalized with respect to μ_K and the frequency of occurrence within each bucket was counted and then normalized with respect to the total number of realizations (2000) so that the area under the histogram becomes unity. A straight line was then drawn between the normalized frequency values centered on each bucket. Also shown on the plots are lognormal distributions fitted by estimating their parameters directly from the simulated data. A chi-squared goodness-of-fit test indicates that the lognormal distribution was acceptable

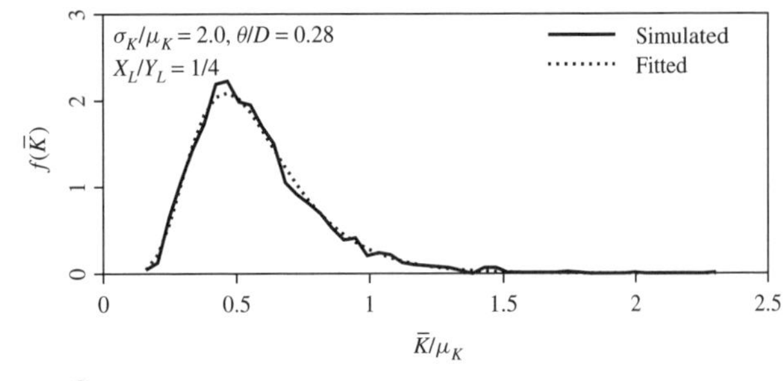

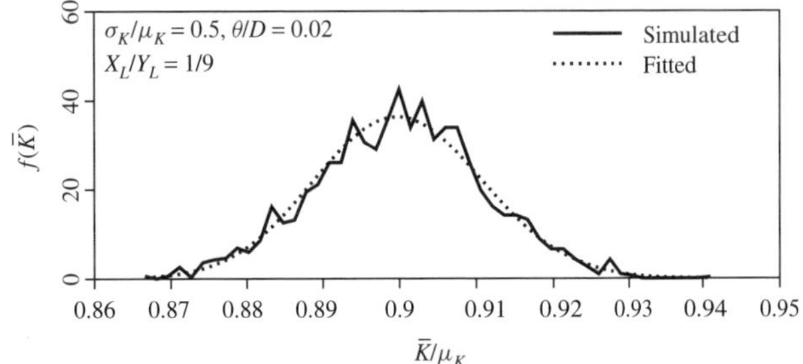

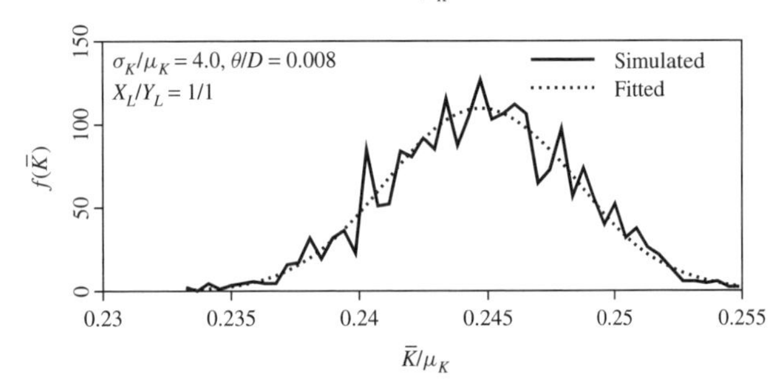

Figure 8.6 Typical histograms of block permeability for various block geometries and permeability statistics. The fitted distribution is lognormal with parameters estimated from the simulated data. The scale changes considerably from plot to plot.

93% of the time at the 5% significance level and 98% of the cases were acceptable at the 1% significance level (the significance level is the probability of mistakenly rejecting the lognormal hypothesis). Of those few cases rejected at the 1% significance level, no particular bias was observed, indicating that these were merely a result of the random nature of the analysis. The lowermost plot in Figure 8.6 illustrates one of the poorest fits which would be rejected at a significance level of 0.001%. Nevertheless, at least visually, the fit appears to be acceptable, demonstrating that the chi-squared test can be quite sensitive.

Upon accepting the lognormal model, the two parameters $m_{\ln \sigma_{\bar{K}}}$ and $s^2_{\ln \sigma_{\bar{K}}}$, representing the estimated mean and variance of $\ln(\bar{K})$, can be plotted as a function of the statistics of $\ln K$ and the geometry of the domain. Figures 8.7 and 8.8 illustrate the results obtained for the three correlation functions considered in Eqs. 8.17. One can see that for square domains, where $X_L/Y_L = \frac{1}{1}$, the statistics of $\bar{K}$ are closely approximated by

$$\mu_{\ln \bar{K}} \simeq \mu_{\ln K} \tag{8.18}$$

$$\sigma_{\ln \bar{K}} \simeq \sigma_{\ln K}\sqrt{\gamma_D} \tag{8.19}$$

where $\mu_{\ln K} = \ln \mu_K - \frac{1}{2}\sigma^2_{\ln K}$ and $\gamma_D = \gamma(D, D)$ (see Section 3.7.3). Note that these are just the statistics one would obtain by arithmetically averaging $\ln K(\mathbf{x})$ over the block [or, equivalently, by geometrically averaging $K(\mathbf{x})$]. Assuming the equality in Eqs. 8.18 and 8.19, the corresponding results in permeability space are

$$\mu_{\bar{K}} = \mu_K \ \exp\{-\tfrac{1}{2}\sigma^2_{\ln K}(1 - \gamma_D)\} \tag{8.20}$$

$$\sigma^2_{\bar{K}} = \mu^2_K \ \exp\{-\sigma^2_{\ln K}(1 - \gamma_D)\}\left[\exp\{\sigma^2_{\ln K}\gamma_D\} - 1\right] \tag{8.21}$$

If these expressions are extended beyond the range over which the simulations were carried out, then in the limit as $D \to 0$ they yield

$$\mu_{\bar{K}} \to \mu_K \tag{8.22a}$$

$$\sigma^2_{\bar{K}} \to \sigma^2_K \tag{8.22b}$$

since $\gamma_D \to 1$. This means that as the block size decreases, the statistics of the block permeability approach those of the point permeability, as expected. In the limit as $D \to \infty$, $\mu_{\bar{K}}$ approaches the geometric mean and $\sigma^2_{\bar{K}}$ approaches zero, which is to say that the block permeability approaches the effective permeability, again as expected.

If both γ_D and the product $\sigma^2_{\ln K}\gamma_D$ are small (e.g., when D is large), then first-order approximations to Eqs. 8.20 and 8.21 are

$$\mu_{\bar{K}} = \mu_K \exp\{-\tfrac{1}{2}\sigma^2_{\ln K}\} \tag{8.23}$$

$$\sigma^2_{\bar{K}} = \mu^2_{\bar{K}} \exp\{-\sigma^2_{\ln K}\}\left(\sigma^2_{\ln K}\gamma_D\right) \tag{8.24}$$

Rubin and Gómez-Hernández (1990) obtain Eq. 8.23 if only the first term in their expression for the mean is considered. In the limit as $D \to 0$, additional terms in their expression yield the result $\mu_{\bar{K}} \to \mu_K$, in agreement with the result given by Eq. 8.22a. With respect to the variance, Rubin and Gómez-Hernández generalize Eq. 8.24 using a first-order expansion to give the covariance between two equal-sized square blocks separated by a distance $\mathbf{h}$ as a function of the covariance between local averages of $\ln K$,

$$\text{Cov}\left[\bar{K}(\mathbf{x}),\ \bar{K}(\mathbf{x}+\mathbf{h})\right] = \mu^2_{\bar{K}} \ \exp\{-\sigma^2_{\ln K}\} \ \text{Cov}\left[K_D(\mathbf{x}), K_D(\mathbf{x}+\mathbf{h})\right] \tag{8.25}$$

where $\ln K_D(\mathbf{x})$ is the local average of $\ln K$ over the block centered at $\mathbf{x}$. Equation 8.25 reduces to Eq. 8.24 in the event that $\mathbf{h} = \mathbf{0}$ since $\text{Var}\left[\ln K_D(\mathbf{x})\right] = \sigma^2_{\ln K}\gamma_D$.

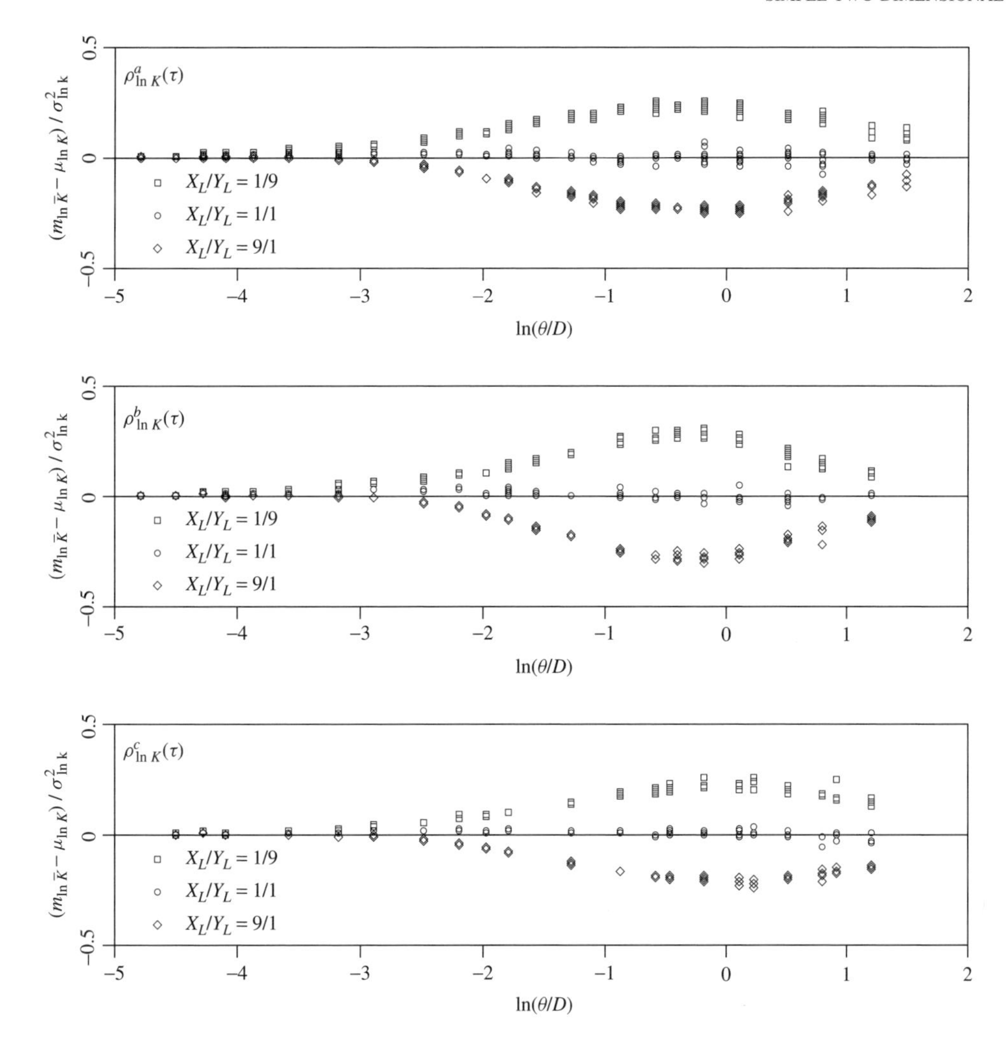

Figure 8.7 Estimated mean log-block permeability, $(m_{\ln \bar{K}} - \mu_{\ln K})/\sigma^2_{\ln K}$.

In many practical situations, neither γ_D nor $\sigma^2_{\ln K}\gamma_D$ is small, so that the approximations given by Eqs. 8.23, 8.24, and 8.25 can be greatly in error. To illustrate this, consider a hypothetical example in which $\mu_K = 1$ and $\sigma^2_K = 16$ (so that $\sigma^2_{\ln K} = 2.83$). It is expected that for a very small block the variance of the block permeability should be close to σ^2_K, namely $\sigma^2_{\bar{K}} \simeq 16$, as predicted by Eq. 8.22b. However, in this case, $\gamma \simeq 1$, and Eq. 8.24 yields a predicted variance of $\sigma^2_{\bar{K}} = 0.17$, roughly 100 times smaller than expected. Recall that Eq. 8.24 was derived on the basis of a first-order expansion and so is strictly only valid for $\sigma^2_{\ln K} \ll 1$.

For aspect ratios other than $\frac{1}{1}$, Figures 8.7 and 8.8 show clear trends in the mean and variance of $\ln \bar{K}$. At small aspect ratios in which the block permeability tends toward the arithmetic mean, $m_{\ln \sigma_{\bar{K}}}$ is larger than $\mu_{\ln K}$, reaching a peak at around $\theta_{\ln K} = D$. At large aspect ratios where the block permeability tends toward the harmonic mean, $m_{\ln \sigma_{\bar{K}}}$ is smaller than $\mu_{\ln K}$, again reaching a minimum around $\theta_{\ln K} = D$. Since the arithmetic and harmonic means bound the geometric mean above and below, respectively, the general form of the estimated results are as expected. While it appears that in the limit as $\theta_{\ln K} \to 0$ both the small and large aspect ratio mean results tend toward the geometric mean, this is actually due to the fact that the effective size of the domain $D/\theta_{\ln K}$ is tending to infinity so that the unbounded results of Eq. 8.16 apply. For such a situation, the small variances seen in Figure 8.8 are as expected. At the other extreme, as $\theta_{\ln K} \to \infty$, there appears again to be convergence to the geometric mean for all three aspect ratios considered. In this case, the field becomes perfectly correlated, so that all points have the same permeability and $\mu_{\bar{K}} = \mu_{\ln K}$ and $\sigma_{\bar{K}} = \sigma_{\ln K}$ for any aspect ratio.

Finally, we note that there is virtually no difference in the block permeability mean and variance arising from the

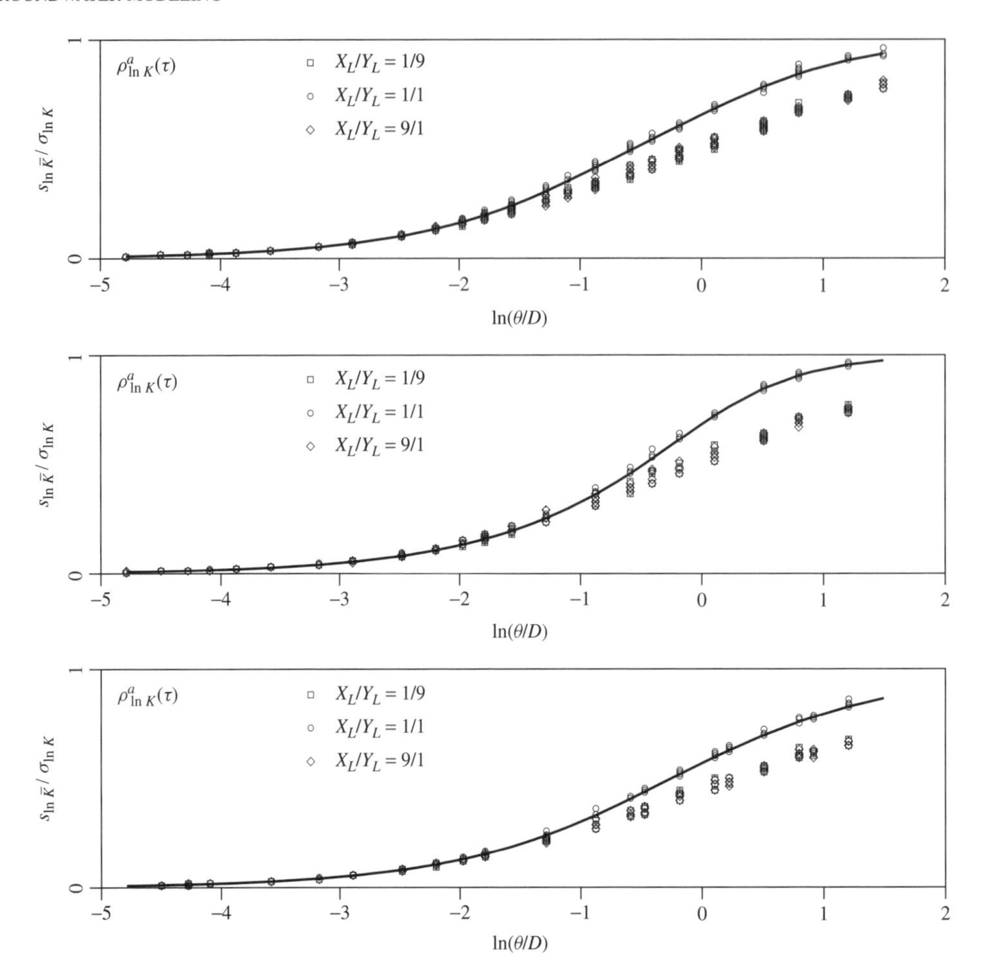

Figure 8.8 Estimated standard deviation of log-block permeability, $s_{\ln \bar{K}}/\sigma_{\ln K}$. Solid line corresponds to $s_{\ln \bar{K}}/\sigma_{\ln K} = \sqrt{\gamma(D, D)}$.

three correlation structures considered (see Eqs. 8.17). In other words, the probabilistic nature of flow through a random medium is insensitive to the form of the correlation function. In the remaining seepage problems considered in this chapter we will use the Markov correlation function, Eq. 8.17a.

8.5 TWO-DIMENSIONAL FLOW BENEATH WATER-RETAINING STRUCTURES

The finite-element method is an ideal vehicle for modeling flow beneath water-retaining structures (e.g., earth dams with or without cutoff walls) where the soil or rock properties are spatially variable (Griffiths and Fenton, 1993, 1998; Paice et al., 1994; and Griffiths et al.,1996). Finite-element methods which incorporate spatial variability represented as random fields generally fall into two camps: (1) the *stochastic finite-element method* in which the statistical properties are built directly into the finite-element equations themselves [see, e.g., Vanmarcke and Grigoriu (1983)] and (2) the RFEM which uses multiple analyses (i.e., *Monte Carlo* simulations) with each analysis stemming from a realization of the soil properties treated as a multidimensional random field. The main drawback to the stochastic finite-element method is that it is a first-order approach which loses accuracy as the variability increases. The main drawback to the RFEM is that it involves multiple finite-element analyses. However, with modern computers the Monte Carlo approach is now deemed to be both fast and accurate.

In this section, the RFEM has been used to examine two-dimensional confined seepage, with particular reference to flow under a water-retaining structure founded on a stochastic soil. In the study of seepage through soils beneath water-retaining structures, three important quantities need to be assessed by the designers (see Figure 8.9):

1. Seepage quantity
2. Exit gradients
3. Uplift forces

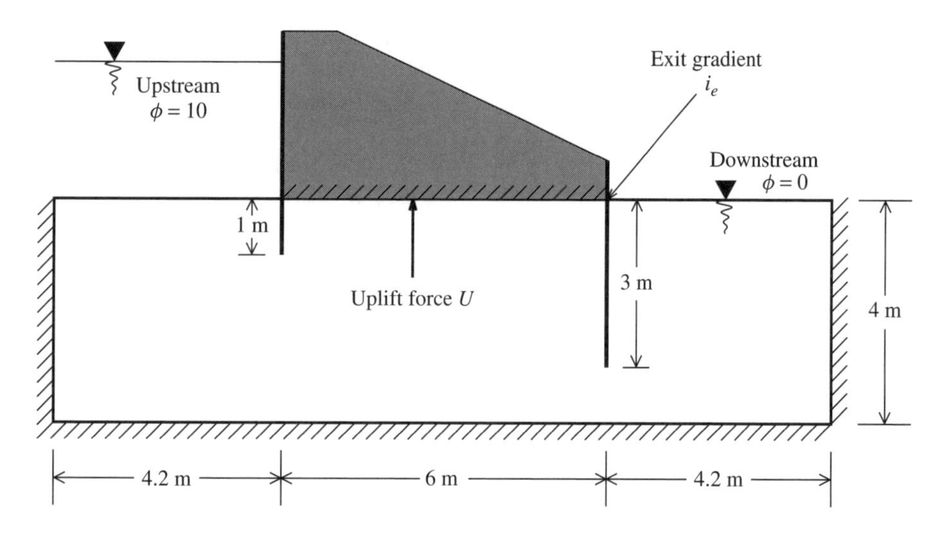

Figure 8.9 Confined seepage boundary value problem. The two vertical walls and the hashed boundaries are impermeable.

The classical approach used by civil engineers for estimating these quantities involves the use of carefully drawn flow nets (Casagrande, 1937; Cedergren, 1967; Verruijt, 1970). Various alternatives to flow nets are available for solving the seepage problem; however in order to perform quick parametric studies, for example, relating to the effect of cutoff wall length, powerful approximate techniques such as the method of fragments (Pavlovsky, 1933; Harr, 1962; Griffiths 1984) are increasingly employed. The conventional methods are deterministic, in that the soil permeability is assumed to be constant and homogeneous, although anisotropic properties and stratification can be taken into account.

A more rational approach to the modeling of soil is to assume the permeability of the soil underlying a structure, such as that shown in Figure 8.9, is random, that is, the soil is assumed to be a "random" field (e.g., Vanmarcke, 1984) characterized by a mean, standard deviation, and some correlation structure. While higher joint moments are possible, they are very rarely estimated with any accuracy, so generally just the first two moments (mean and covariance structure) are specified.

The stochastic flow problem posed in Figure 8.9 is far too difficult to contemplate solving analytically (and/or the required simplifying assumption would make the solution useless). The determination of probabilities associated with flow and exit gradients is conveniently done using Monte Carlo simulation. For this problem, we will use the LAS random-field generator (see Section 6.4.6) to generate realizations of the random permeability fields and then determine the resulting flow and head fields using the finite-element method for each realization. In detail, the simulated field of permeabilities is mapped onto a finite-element mesh, and potential and stream function boundary conditions are specified. The governing elliptic equation for steady flow (Laplace) leads to a system of linear "equilibrium" equations which are solved for the nodal potential values throughout the mesh using conventional Gaussian elimination within a finite-element framework. The program used to perform this study is called RFLOW2D and is available at `http://www.engmath.dal.ca/rfem`.

Only deterministic boundary conditions are considered in this analysis, the primary goal being to investigate the effects of randomly varying soil properties on the engineering quantities noted above. The method presented is nevertheless easily extended to random boundary conditions corresponding to uncertainties in upstream and downstream water levels, so long as these can be simulated.

The steady-flow problem is governed in two dimensions by Laplace's equation, in which the dependent variable ϕ is the piezometric head or potential at any point in the Cartesian field x–y:

$$K_{11}\frac{\partial^2\phi}{\partial x^2} + K_{22}\frac{\partial^2\phi}{\partial y^2} = 0 \qquad (8.26)$$

where K_{11} and K_{22} are the permeabilities in the x_1 and x_2 (horizontal and vertical) directions, respectively. The permeability field is assumed to be isotropic ($K_{11} = K_{22} = K$). While the method discussed in this section is simply extended to the anisotropic case (through the generation of a pair of correlated random fields), such site-specific extensions are left to the reader (the options in RFLOW2D easily allow this).

Note that Eq. 8.26 is strictly only valid for spatially constant K. In this analysis the permeability is taken to be constant within each element, its value being given by the local geometric average of the permeability field over the element domain. The geometric average was found to be appropriate in the previous section for square elements. From element to element, the value of K will vary, reflecting the random nature of the permeability. This approximation of the permeability field being made up of a sequence of local averages is consistent with the approximations made in the finite-element method and is superior to most traditional approaches in which the permeability of an element is taken to be simply the permeability at some point within the element.

The finite-element mesh used in this study is shown in Figure 8.10. It contains 1400 elements, and represents a model of two-dimensional flow beneath a dam which includes two cutoff walls. The upstream and downstream potential values are fixed at 10 and 0 m, respectively. The cutoff walls are assumed to have zero thickness, and the nodes along those walls have two potential values corresponding to the right and left sides of the wall.

For a given permeability field, the finite-element analysis computes nodal potential values which are then used to determine flow rates, uplift pressures, and exit gradients. The statistics of these quantities are assessed by producing multiple realizations of the random permeability field and analyzing each realization with the finite-element method to produce multiple realizations of the random flow rates, uplift pressures, and exit gradients. The random permeability field is characterized by three parameters defining its first two moments, namely the mean μ_K, the standard deviation σ_K, and the correlation length $\theta_{\ln K}$. In order to obtain reasonably accurate estimates of the output statistics, it was decided that each "run" would consist of the analysis of 1000 realizations.

8.5.1 Generation of Permeability Values

The permeability was assumed to be lognormally distributed and is obtained through the transformation

$$K_i = \exp\{\mu_{\ln K} + \sigma_{\ln K} G_i\} \tag{8.27}$$

in which K_i is the permeability assigned to the ith element, G_i is the local (arithmetic) average of a standard Gaussian random field $G(\mathbf{x})$ over the domain of the ith element, and $\mu_{\ln K}$ and $\sigma_{\ln K}$ are the mean and standard deviation of the logarithm of K (obtained from the "target" mean and standard deviation μ_K and σ_K).

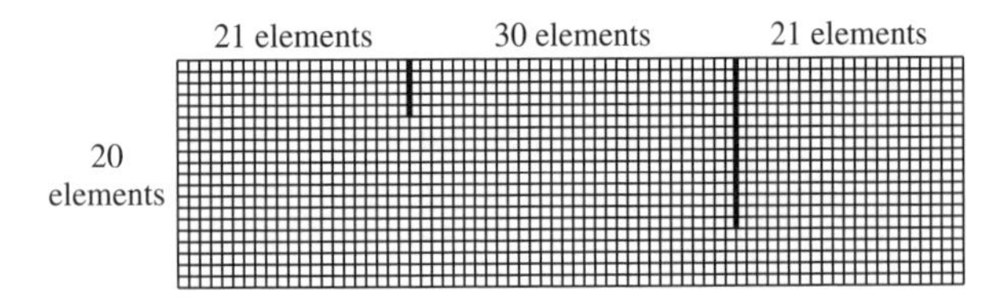

Figure 8.10 Finite-element mesh. All elements are 0.2-m × 0.2-m squares.

Realizations of the permeability field are produced using the LAS technique discussed in Section 6.4.6. The LAS technique renders realizations of the local averages G_i which are derived from the random field $G(\mathbf{x})$ having zero mean, unit variance, and a spatial correlation controlled by the correlation length. As the correlation length goes to infinity, G_i becomes equal to G_j for all elements i and j that is, the field of permeabilities tends to become uniform on each realization. At the other extreme, as the correlation length goes to zero, G_i and G_j become independent for all $i \neq j$—the soil permeability changes rapidly from point to point.

In the two-dimensional analyses presented in this section, the correlation lengths in the vertical and horizontal directions are taken to be equal (isotropic) for simplicity. Since actual soils are frequently layered, the correlation length horizontally is generally larger than it is vertically. However, the degree of layering is site specific and is left to the reader as a refinement. The results presented here are aimed at establishing the basic probabilistic behavior of flow under water-retaining structures. In addition, the two-dimensional model used herein implies that the out-of-plane correlation length is infinite—soil properties are constant in this direction—which is equivalent to specifying that the streamlines remain in the plane of the analysis. This is clearly a deficiency of the two-dimensional model; however, as we shall see in the next section, most of the characteristics of the random flow are nevertheless captured by the two-dimensional model.

8.5.2 Deterministic Solution

With regard to the seepage problem shown in Figure 8.10, a deterministic analysis was performed in which the permeability of all elements was assumed to be constant and equal to 1 m/s. This value was chosen as it was to be the mean value of subsequent stochastic analyses. Both the potential and the inverse streamline problems were solved, leading to the flow net shown in Figure 8.11.

All output quantities were computed in nondimensional form. In the case of the flow rate, the global flow vector $\mathbf{Q}$ was computed by forming the product of the potentials and the global conductivity matrix from Eq. 8.3. Assuming no sources or sinks in the flow regime, the only nonzero values in $\mathbf{Q}$ correspond to those freedoms on the upstream side at which the potentials were fixed equal to 10 m. These values were summed to give the total flow rate Q in cubic meters per second per meter, leading to a nondimensional flow rate $\bar{Q}$ defined by

$$\bar{Q} = \frac{Q}{\mu_K \Delta H} \tag{8.28}$$

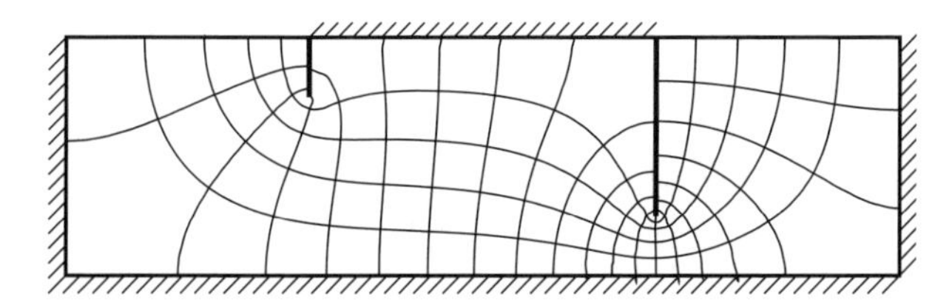

Figure 8.11 Deterministic flow net.

where μ_k is the (isotropic) mean permeability and ΔH is the total head difference between the up- and downstream sides. Here, $\bar{Q}$ is equivalent to the "shape factor" of the problem, namely the ratio of the number of flow channels divided by the number of equipotential drops (n_f/n_d) that would be observed in a carefully drawn flow net; alternatively, it is also equal to the reciprocal of the form factor utilized by the method of fragments.

In the following, the distribution of $\bar{Q}$ will be investigated. The actual flow rate is determined by inverting Eq. 8.28,

$$Q = \mu_K \Delta H \bar{Q} \tag{8.29}$$

which will have the same distribution as $\bar{Q}$ except with mean and standard deviation

$$\mu_Q = \mu_K \Delta H \mu_{\bar{Q}} \tag{8.30a}$$

$$\sigma_Q = \mu_K \Delta H \sigma_{\bar{Q}} \tag{8.30b}$$

The uplift force on the base of the dam, U, is computed by integrating the pressure distribution along the base of the dam between the cutoff walls. This quantity is easily deduced from the potential values at the nodes along this line together with a simple numerical integration scheme (e.g., repeated trapezium rule). A nondimensional uplift force $\bar{U}$ is defined as

$$\bar{U} = \frac{U}{\Delta H \gamma_w L} \tag{8.31}$$

where γ_w is the unit weight of water, L is the distance between the cutoff walls, and $\bar{U}$ is the uplift force expressed as a proportion of buoyancy force that would occur if the dam was submerged in water alone. The distribution of U is the same as the distribution of $\bar{U}$ except with mean and standard deviation

$$\mu_U = L\gamma_w \Delta H \mu_{\bar{U}} \tag{8.32a}$$

$$\sigma_U = L\gamma_w \Delta H \sigma_{\bar{U}} \tag{8.32b}$$

The exit gradient i_e is the rate of change of head at the exit point closest to the dam at the downstream end. This is calculated using a four-point backward-difference numerical differentiation formula of the form

$$i_e \approx \frac{1}{6b}(11\phi_0 - 18\phi_{-1} + 9\phi_{-2} - 2\phi_{-3}) \tag{8.33}$$

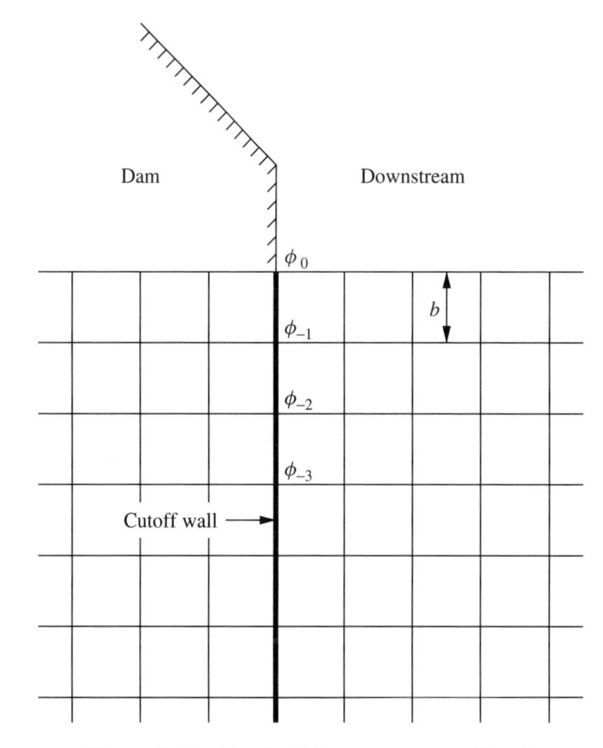

Figure 8.12 Detail of downstream cutoff wall.

where the ϕ_i values correspond to the piezometric head at the four nodes vertically below the exit point, as shown in Figure 8.12, and b is the constant vertical distance between nodes. It may be noted that the downstream potential head is fixed equal to zero, and thus $\phi_0 = 0.0$ m. The use of this four-point formula was arbitrary and was considered a compromise between the use of very low order formulas which would be too sensitive to random fluctuations in the potential and high-order formulas which would involve the use of correspondingly high-order interpolation polynomials which would be hard to justify physically.

Referring to Figures 8.9 and 8.10, the constants described above were given the following values:

$$\Delta H = 10 \text{ m}, \qquad \mu_K = 1 \text{ m/s}, \qquad \gamma_w = 9.81 \text{ kN/m}^3,$$
$$L = 15 \text{ m}, \qquad b = 0.2 \text{ m}$$

and a deterministic analysis using the mesh of Figure 8.10 led to the following output quantities:

$$\bar{Q} = 0.226, \qquad \bar{U} = 0.671, \qquad i_e = 0.688$$

This value of $i_e = 0.688$ would be considered unacceptable for design purposes as the critical hydraulic gradient for most soils is approximately unity and a factor of safety against piping of 4–5 is often recommended (see, e.g., Harr, 1962). However, the value of i_e is proportional to the head difference ΔH, which in this case, for simplicity and convenience of normalization, has been set to 10 m as noted previously.

These deterministic results will be compared with output from the stochastic analyses described in the next section.

8.5.3 Stochastic Analyses

In all the two-dimensional stochastic analyses that follow, the soil was assumed to be isotropic with a mean permeability $\mu_K = 1$ m/s. More specifically, the random fields were generated such that the target point mean permeability of each finite element was held constant at 1 m/s. Parametric studies were performed relating to the effect of varying the standard deviation (σ_K) and the correlation length ($\theta_{\ln K}$) of the permeability field. Following 1000 realizations, statistics relating to output quantities $\bar{Q}$, $\bar{U}$, and i_e were calculated.

8.5.3.1 Single Realization Before discussing the results from multiple realizations, an example of what a flow net might look like for a single realization is given in Figures 8.13*a* and *b* for permeability statistics $\mu_K = 1$ m/s, $\sigma_K = 1$ m/s, and $\theta_{\ln K} = 1.0$ m.

In Figure 8.13*a*, the flow net is superimposed on a gray scale which indicates the spatial distribution of the permeability values. Dark areas correspond to low permeability and light areas to high permeability. The streamlines clearly try to "avoid" the low-permeability zones, but this is not always possible as some realizations may generate a complete blockage of low-permeability material in certain parts of the flow regime. This type of blockage is most likely to occur where the flow route is compressed, such as under a cutoff wall. An example where this happens is shown in Figure 8.13*b*. Flow in these (dark) low-permeability zones is characterized by the streamlines moving further apart and the equipotentials moving closer together. Conversely, flow in the (light) high-permeability zones is characterized by the equipotentials moving further apart and the streamlines moving closer together. In both of these figures the contrast between stochastic flow and the flow through a deterministic field, such as that shown in Figure 8.11, is clear. In addition, the ability for the streamlines to avoid low-permeability zones means that the average permeability seen by the flow is higher than if the flow was constrained to pass through the low-permeability zones. This ability to circumnavigate the blockages is why the geometric average is a better model for two-dimensional flow than is the harmonic average.

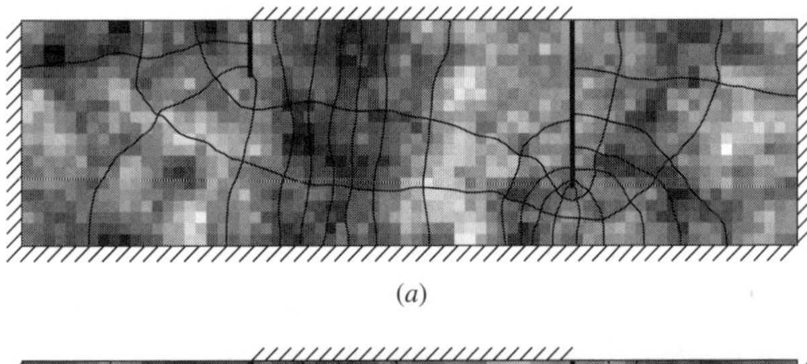

(*a*)

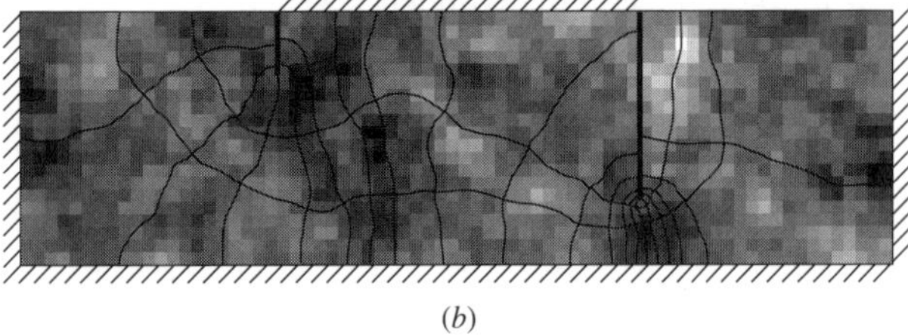

(*b*)

Figure 8.13 Stochastic flow net for two typical realizations.

Although *local* variations in the permeability have an obvious effect on the local paths taken by the water as it flows downstream, *globally* the stochastic and deterministic flow nets exhibit many similarities. The flow is predominantly in a downstream direction, with the fluid flowing down, under, and around the cutoff walls. For this reason the statistics of the output quantities might be expected to be rather insensitive to the geometry of the problem (e.g., length of walls) and qualitatively similar to the properties of a one-dimensional flow problem, aside from an average effective permeability, which is higher than in the one-dimensional case.

8.5.3.2 Statistics of Potential Field Figure 8.14 gives contours of the mean and standard deviation of the potential field following 1000 realizations for the case where $\theta_{\ln K} = 1.0$ and $v_K = \sigma_K/\mu_K = 1.0$. The mean potential values given in Figure 8.14 a are very similar to those obtained in the deterministic analysis summarized in the flow net of Figure 8.11. The standard deviation of the potentials given in Figure 8.14*b* indicate the zones in which the greatest uncertainty exists regarding the potential values. It should be recalled that the up- and downstream (boundary) potentials are deterministic, so the standard deviation of the potentials on these boundaries equals zero. The greatest values of standard deviation occur in the middle of the flow regime, which in the case considered here represents the zone beneath the dam and between the cutoff walls. The standard deviation is virtually constant in this zone. The statistics of the potential field are closely related to the statistics of the uplift force, as will be considered in the next section.

Other values of $\theta_{\ln K}$ and v_K led to the same mean contours as seen in Figure 8.14*a*. The potential standard deviations increase with increasing v_K, as expected, but tend towards zero as $\theta_{\ln K} \to 0$ or $\theta_{\ln K} \to \infty$. The potential standard deviations reach a maximum when $\theta_{\ln K} \simeq 6$ m or, more generally, when the correlation length is

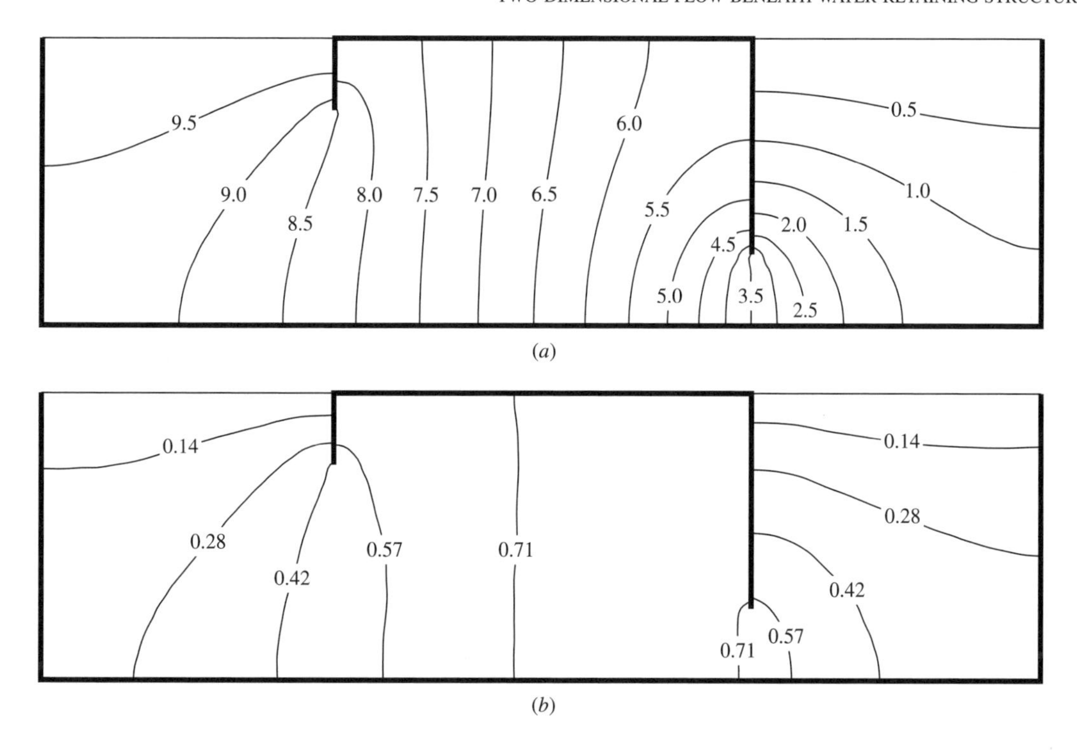

Figure 8.14 (*a*) Contours of potential mean and (*b*) potential standard deviation (both in meters) for $\sigma_K/\mu_K = 1.0$ and $\theta_{\ln K} = 1.0$ m.

approximately equal to the width of the water-retaining structure. This "worst-case" correlation length is commonly observed and can be used when the correlation length is unknown (which is almost always).

8.5.3.3 Flow Rate, Uplift, and Exit Gradient Parametric studies based on the mesh of Figure 8.10 were designed to show the effect of the permeability's standard deviation σ_K and correlation length $\theta_{\ln K}$ on the output quantities $\bar{Q}$, $\bar{U}$, and i_e. In all cases the mean permeability μ_K was maintained constant at 1 m/s.

Instead of plotting σ_K directly, the dimensionless coefficient of variation of permeability was used, and the following values were considered:

$$\frac{\sigma_K}{\mu_K} = 0.125, 0.25, 0.50, 1.0, 2.0, 4.0, 8.0, 16.0$$

together with correlation lengths

$$\theta_{\ln K} = 0.0, 1.0, 2.0, 4.0, 8.0, \infty \text{ m}$$

All permutations of these values were analyzed, and the results were summarized in Figures 8.15, 8.16, and 8.17 in the form of σ_K/μ_K versus the estimated means and standard deviations of $\bar{Q}$, $\bar{U}$, and i_e, denoted $(m_{\bar{Q}}, s_{\bar{Q}})$, $(m_{\bar{U}}, s_{\bar{U}})$, and (m_{i_e}, s_{i_e}), respectively.

Flow Rate Figure 8.15*a* shows a significant fall in $m_{\bar{Q}}$ (where $m_{\bar{Q}}$ is the simulation-based estimate of $\mu_{\bar{Q}}$) as σ_K/μ_K increases for $\theta_{\ln K} < 8$ m. As the correlation length approaches infinity, the expected value of $\bar{Q}$ approaches the constant 0.226. This curve is also shown in Figure 8.15*a*, although it should be noted it has been obtained through theory rather than simulation. In agreement with this result, the curve $\theta_{\ln K} = 8$ m shows a less marked reduction in $m_{\bar{Q}}$ with increasing coefficient of variation σ_K/μ_K. However, over typical correlation lengths, the effect on average flow rate is slight. The decrease in flow rate as a function of the variability of the soil mass is an important observation from the point of view of design. Traditional design practice may very well be relying on this variability to reduce flow rates on average. It also implies that ensuring higher uniformity in the substrate may be unwarranted unless the mean permeability is known to be substantially reduced and/or the permeability throughout the site is carefully measured. It may be noted that the deterministic result of $\bar{Q} = 0.226$ has been included in Figure 8.15*a*, and, as expected, the stochastic results converge on this value as σ_K/μ_K approaches zero.

Figure 8.15*b* shows the behavior of $s_{\bar{Q}}$, the estimate of $\sigma_{\bar{Q}}$, as a function of σ_K/μ_K. Of particular note is that $s_{\bar{Q}}$ reaches a maximum corresponding to σ_K/μ_K in the range 1.0–2.0 for finite $\theta_{\ln K}$. Clearly, when $\sigma_K = 0$, the permeability field will be deterministic and there will be no variability in the flow rate: $\sigma_{\bar{Q}}$ will be zero. What is not quite so obvious is that because the mean of $\bar{Q}$ falls

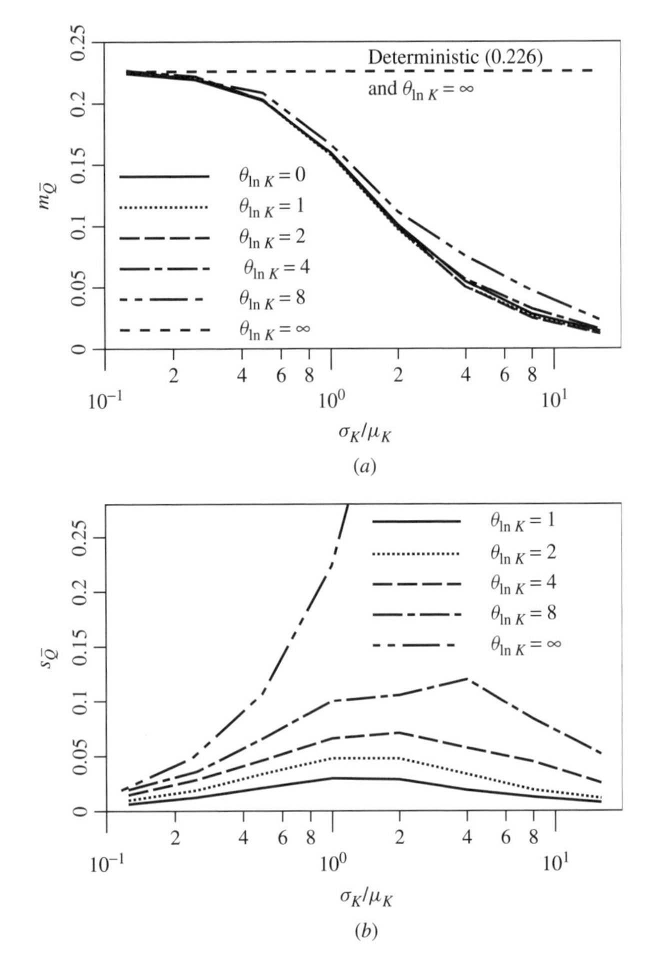

Figure 8.15 Effect of correlation length and coefficient of variation of permeability on (*a*) mean flow rate and (*b*) flow rate standard deviation.

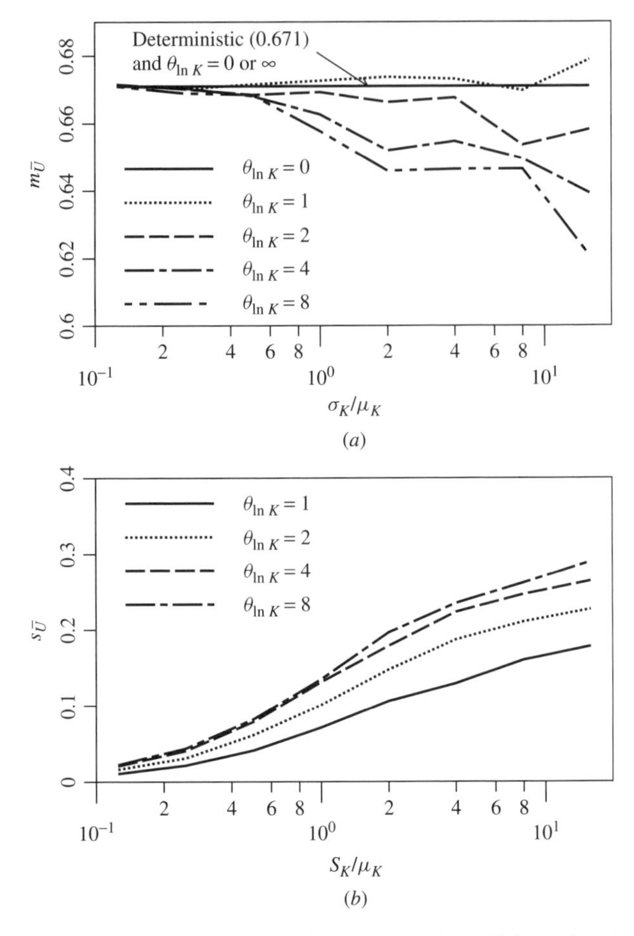

Figure 8.16 Effect of correlation length and coefficient of variation of permeability on (*a*) mean uplift force and (*b*) uplift force standard deviation.

to zero when $\sigma_K/\mu_K \to \infty$ for finite $\theta_{\ln K}$ (see Figure 8.15, where the curves go to zero as the permeability variability increases), the standard deviation of $\bar{Q}$ must also fall to zero since $\bar{Q}$ is nonnegative. Thus, $\sigma_{\bar{Q}} = 0$ when the permeability variance is both zero and infinite. It must, therefore, reach a maximum somewhere between these two bounds. The point at which the maximum occurs moves to the right as $\theta_{\ln K}$ increases.

In general, it appears that the greatest variability in $\bar{Q}$ occurs under rather typical conditions: correlation lengths between 1 and 4 m and coefficient of variation of permeability of around 1 or 2.

Uplift Force Figures 8.16*a* and *b* show the relationship between uplift force parameters $\mu_{\bar{U}}$ and $\sigma_{\bar{U}}$ and input permeability parameters σ_K/μ_K and $\theta_{\ln K}$. In the figure, $\mu_{\bar{U}}$ and $\sigma_{\bar{U}}$ are estimated from the simulation by $m_{\bar{U}}$ and $s_{\bar{U}}$, respectively. According to Figure 8.16*a*, $\mu_{\bar{U}}$ is relatively insensitive to the parametric changes. There is a gradual fall in $\mu_{\bar{U}}$ as both σ_K/μ_K and $\theta_{\ln K}$ increase, the greatest reduction being about 10% of the deterministic value of 0.671 when $\sigma_K/\mu_K = 16.0$ and $\theta_{\ln K} = 8.0$ m. The insensitivity of the uplift force to the permeability input statistics might have been predicted from Figures 8.11 and 8.14*a*, in which the contours of mean potential (piezometric head) are identical in both the deterministic and stochastic analyses.

Figure 8.16*b* shows that $\sigma_{\bar{U}}$ consistently rises as both σ_K/μ_K and $\theta_{\ln K}$ increase. It is known that in the limit as $\theta_{\ln K} \to \infty$, $\sigma_{\bar{U}} \to 0$ since under those conditions the permeability field becomes completely uniform. Some hint of this increase followed by a decrease is seen from Figure 8.16*b* in that the largest increases are for $\theta_{\ln K} = 0$ to $\theta_{\ln K} = 1$ while the increase from $\theta_{\ln K} = 4$ to $\theta_{\ln K} = 8$ is much smaller.

The actual value of $\sigma_{\bar{U}}$ for a given set of σ_K/μ_K and $\theta_{\ln K}$ could easily be deduced from the standard deviation of the potential values. Figure 8.14*b* gives contours of the standard deviation of the potential throughout the flow domain for the particular values $\sigma_K/\mu_K = 1.0$ and $\theta_{\ln K} = 1.0$. In Figure 8.14*b*, the potential standard deviation beneath the dam was approximately constant and equal to 0.71 m. After nondimensionalization by dividing by $\Delta H = 10$ m, these values closely agree with the corresponding value in the graph of Figure 8.16*b*.

The magnitude of the standard deviation of the uplift force given in Figure 8.16*b* across the range of parameters considered was not very great. The implication is that this quantity can be estimated with a reasonable degree of confidence. The explanation lies in the fact that the uplift force is calculated using potential values over quite a large number of nodes beneath the dam. This "averaging" process would tend to damp out fluctuations in the potential values that would be observed on a local scale, resulting in a variance reduction.

Exit Gradient The exit gradient is based on the first derivative of the potential (or piezometric head) with respect to distance at the exit point closest to the downstream end of the dam. It is well known that in a deterministic approach the largest value of i_e, and hence the most critical, lies at the exit point of the uppermost (and shortest) streamline. While for a single realization of a stochastic analysis this may not be the case, on average the location of the critical exit gradient is expected to occur at the "deterministic" location.

As i_e is based on a first derivative at a *particular* location within the mesh (see Figure 8.12), it can be expected to be the most susceptible to local variations generated by the stochastic approach. In order to average the calculation of i_e over a few nodes, it was decided to use a four-point (backward) finite-difference scheme as given previously in Eq. 8.33. This is equivalent to fitting a cubic polynomial over the potential values calculated at the four nodes closest to the exit point adjacent to the downstream cutoff wall. The cubic is then differentiated at the required point to estimate i_e. Note then that the gradient is estimated by studying the fluctuations over a length of 0.6 m vertically (the elements are 0.2 m by 0.2 m in size). This length will be referred to as the *differentiation length* in the following.

The variation of μ_{i_e} and σ_{i_e} over the range of parameters considered are given in Figures 8.17*a* and *b*. The figure shows the simulation-based estimates m_{i_e} and s_{i_e} of μ_{i_e} and σ_{i_e}, respectively. The sensitivity of i_e to σ_K/μ_K is clearly demonstrated. In Figure 8.17*a*, m_{i_e} agrees quite closely with the deterministic value of 0.688 for values of σ_K/μ_K in the range 0.0–1.0, but larger values start to show significant instability and divergence. It is interesting to note that for $\theta_{\ln K} \leq 1$ the tendency is for m_{i_e} to fall below the deterministic value of i_e as σ_K/μ_K is increased, whereas for larger values of $\theta_{\ln K}$ it tends to increase above the deterministic value. The scales 0 and 1 are less than and of the same magnitude as the differentiation length of 0.6 m used to estimate the exit gradient, respectively, while the scales 2, 4, and 8 are substantially greater. If this has some bearing on the divergence phenomena seen in Figure 8.17*a*, it calls into some question the use of a differentiation length to estimate the derivative at a point. Suffice to say that there may be some conflict between the numerical estimation method and random-field theory regarding the exit gradient that needs further investigation.

Figure 8.17*b* indicates the relatively large magnitude of σ_{i_e} which grows rapidly as σ_K/μ_K is increased. The influence of $\theta_{\ln K}$ in this case is not so great, with the results corresponding to $\theta_{\ln K}$ values of 1.0, 2.0, 4.0, and 8.0 m being quite closely grouped. It is noted that theoretically, as $\theta_{\ln K} \to \infty$, $\mu_{i_e} \to 0.688$ and $\sigma_{i_e} \to 0$. There appears to be some evidence of a reduction in σ_{i_e} as $\theta_{\ln K}$ increases, which is in agreement with the theoretical result. For correlation lengths negligible relative to the differentiation

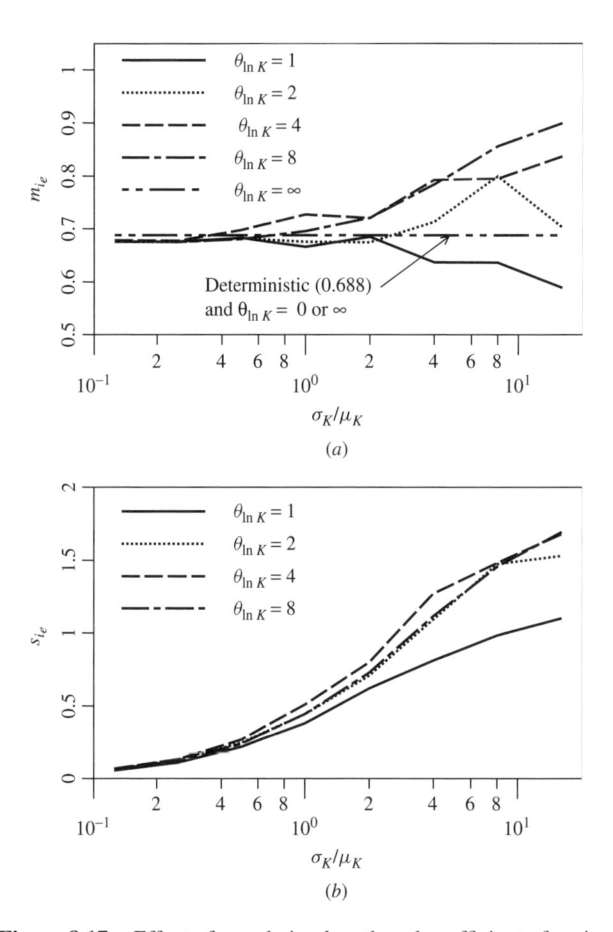

Figure 8.17 Effect of correlation length and coefficient of variation of permeability on (*a*) mean exit gradient and (*b*) exit gradient standard deviation.

length, that is, $\theta_{\ln K} = 0$, the variability in i_e is much higher than that for other scales at all but the highest permeability variance. This is perhaps to be expected, since $\theta_{\ln K} = 0$ yields large fluctuations in permeability within the differentiation length. It is also to be noted that the point correlation function employed in this study is Markovian (see Eq. 8.17a), which yields a random field which is nondifferentiable, its derivative having an infinite variance. Although local averaging does make the process differentiable, it is believed that this pointwise nondifferentiability may be partially to blame for the erratic behavior of the exit gradient.

8.5.4 Summary

This section presented a range of parametric studies which have been performed relating to flow beneath a water-retaining structure with two cutoff walls founded on a stochastic soil. Random-field concepts were used to generate permeability fields having predefined mean, standard deviation, and correlation structure. These values were mapped onto a finite-element mesh consisting of 1400 elements, and, for each set of parameters, 1000 realizations of the boundary value problem were analyzed. In all cases, the target mean permeability of each finite element was held constant and parametric studies were performed over a range of values of coefficient of variation and correlation length.

The three output quantities under scrutiny were the flow rate, the uplift force, and the exit gradient, the first two being nondimensionalized for convenience of presentation.

The mean flow rate was found to be relatively insensitive to typical correlation lengths but fell consistently as the variance of the permeability was increased. This observation may be of some importance in the design of such water-retaining structures. The standard deviation of the flow rate consistently increased with the correlation length but rose and then fell again as the coefficient of variation was increased.

The mean uplift force was rather insensitive to the parametric variations, falling by only about 10% in the worst case (high-permeability variance and $\theta_{\ln K} = 8$). The relatively small variability of uplift force was due to a "damping out" of local variations inherent in the random field by the averaging of potential values over the nodes along the full length of the base of the dam. Nevertheless, the standard deviation of the uplift force rose consistently with increasing correlation length and coefficient of variation, as was to be expected from the contour plots of the standard deviation of the potential values across the flow domain.

The mean exit gradient was much more sensitive to the statistics of the input field. Being based on a first derivative of piezometric head with respect to length at the exit point, this quantity is highly sensitive to local variations inherent in the potential values generated by the random field. Some local averaging was introduced both in the random-field simulation and by the use of the four-point numerical differentiation formula; however, the fluctuation in mean values was still considerable and the standard deviation values were high.

8.6 THREE-DIMENSIONAL FLOW

This section considers the stochastic three-dimensional boundary value problem of steady seepage, studying the influence of soil variability on "output" quantities such as flow rate (Griffiths and Fenton, 1995, 1997). The probabilistic results are contrasted with results obtained using an idealized two-dimensional model. For the computationally intensive three-dimensional finite-element analyses, strategies are described for optimizing the efficiency of computer code in relation to memory and CPU requirements. The program used to perform this analysis is RFLOW3D available at `http://www.engmath.dal.ca/rfem`.

The two-dimensional flow model used in the previous two sections rested on the assumption of perfect correlation in the out-of-plane direction. This assumption is no longer necessary with a three-dimensional model, and so the three-dimensional model is obviously more realistic. The soil permeability is simulated using the LAS method (Section 6.4.6) and the steady flow is determined using the finite-element method. The problem chosen for study is a simple boundary value problem of steady seepage beneath a single sheet pile wall penetrating a layer of soil. The variable soil property in this case is the soil permeability K, which is defined in the classical geotechnical sense as having units of length over time.

The overall dimensions of the problem to be solved are shown in Figures 8.18*a* and *b*. Figure 8.18*a* shows an isometric view of the three-dimensional flow regime, and Figure 8.18*b* shows an elevation which corresponds to the two-dimensional domain analyzed for comparison. In all results presented in this section, the dimensions L_x and L_y were held constant while the third dimension L_z was gradually increased to monitor the effects of three-dimensionality.

In all analyses presented in this section, a uniform mesh of cubic eight-node brick elements with a side length of 0.2 was used with 32 elements in the x direction ($L_x = 6.4$), 16 elements in the y direction ($L_y = 3.2$), and up to 16 elements in the z direction ($L_z = 0.8,\ 1.6, 3.2$).

The permeability field is assumed to be lognormally distributed and is obtained through the transformation

$$K_i = \exp\{\mu_{\ln K} + \sigma_{\ln K} G_i\} \tag{8.34}$$

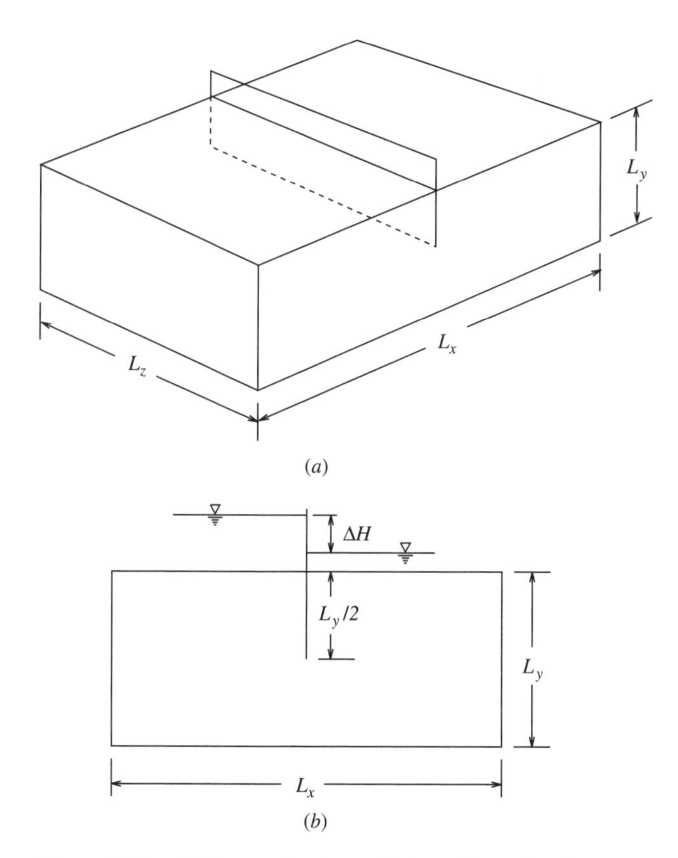

Figure 8.18 (*a*) Isometric view of three-dimensional seepage problem with (*b*) elevation.

in which K_i is the permeability assigned to the ith element, G_i is the local (arithmetic) average of a standard Gaussian random field $G(\mathbf{x})$ over the domain of the ith element, and $\mu_{\ln K}$ and $\sigma_{\ln K}$ are the mean and standard deviation of the logarithm of K, obtained from the prescribed mean and standard deviation μ_K and σ_K via the transformations of Eqs. 1.176. Realizations of the local average field G_i are generated by LAS (Section 6.4.6) using a Markov spatial correlation function (Section 3.7.10.2),

$$\rho(\tau) = \exp\left\{-\frac{2|\tau|}{\theta_{\ln K}}\right\} \tag{8.35}$$

where $|\tau|$ is the distance between points in the field and $\theta_{\ln K}$ is the correlation length.

In this three-dimensional analysis, the correlation lengths in all directions are taken to be equal (isotropic) for simplicity.

8.6.1 Simulation Results

A Monte Carlo approach to the seepage problem was adopted in which, for each set of input statistics ($\mu_K, \sigma_K, \theta_{\ln K}$) (or equivalently, $\mu_{\ln K}, \sigma_{\ln K}, \theta_{\ln K}$) and mesh geometry ($L_z$), 1000 realizations were performed. The main output quantities of interest from each realization in this problem are the total flow rate through the system Q and the exit gradient i_e.

We first focus on the flow rate. Following Monte Carlo simulations, the mean and standard deviation of Q were computed and presented in nondimensional form by representing Q in terms of a normalized flow rate $\bar{Q}$,

$$\bar{Q} = \frac{Q}{\Delta H\, \mu_K L_z} \tag{8.36}$$

where ΔH is the total head loss across the wall. In all the calculations performed in this study, ΔH was set to unity since it has a simple linear influence on the flow rate Q. Division by L_z has the effect of expressing the average flow rate over one unit of thickness in the z direction enabling a direct comparison to be made with the two-dimensional results. To get the true flow rate through the soil, Eq. 8.36 is inverted to give

$$Q = L_z \mu_K \Delta H \bar{Q} \tag{8.37}$$

which has the same distribution as $\bar{Q}$ except with mean and variance given by

$$\mu_Q = L_z \mu_K \Delta H \mu_{\bar{Q}} \tag{8.38a}$$

$$\sigma_Q = L_z \mu_K \Delta H \sigma_{\bar{Q}} \tag{8.38b}$$

The following parametric variations were implemented for fixed $\mu_K = 1$, $L_x = 6.4$, and $L_y = 3.2$:

$$\frac{\sigma_K}{\mu_K} = 0.125,\ 0.25,\ 0.5,\ 1,\ 2,\ 4,\ 8$$

$$\theta_{\ln K} = 1,\ 2,\ 4,\ 8,\ \infty \quad \text{(analytical)}$$

$$L_z = 0.8,\ 1.6,\ 3.2$$

and a selection of results will be presented here.

As the coefficient of variation of the input permeability ($v_K = \sigma_K/\mu_K$) was increased, the mean estimated normalized flow rate $m_{\bar{Q}}$ was observed to fall consistently from its deterministic value (assuming constant permeability throughout) of $\bar{Q}_{\text{det}} \approx 0.47$, as shown in Figure 8.19*a* for the case where $L_z/L_y = 1$. The fall in $m_{\bar{Q}}$ was steepest for small values of the correlation length $\theta_{\ln K}$; however, as $\theta_{\ln K}$ was increased, $m_{\bar{Q}}$ tended toward the deterministic result that would be expected for a strongly correlated permeability field ($\theta_{\ln K} \to \infty$).

The reduction in the expected flow rate with increased permeability variance but fixed mean has been described as "counterintuitive" by some observers. The explanation lies in the fact that in a continuous-flow regime such as the one modeled here, flow must be occurring in every region of the domain, so the greater the permeability variance, the greater the volume of low-permeability material that must be negotiated along any flow path. In an extreme case of "series" flow down a one-dimensional pipe of varying permeability cells, the effective permeability is given by the harmonic mean of the permeability values, which is heavily dependent on the lowest permeability encountered. The other extreme of "parallel" flow leads to the arithmetic mean. The three-dimensional example considered here is a complex combination of parallel and series flow which leads to an effective permeability more closely approximated by the geometric mean (Section 4.4.2), which is always smaller than the arithmetic mean (but not as small as the harmonic mean).

Figure 8.19*b* shows the estimated standard deviation of the normalized flow rate $s_{\bar{Q}}$ for the same geometry. For small $\theta_{\ln K}$ very little variation in $\bar{Q}$ was observed, even for high coefficients of variation. This is understandable if one thinks of the total flow through the domain as effectively an averaging process; high flow rates in some regions are offset by lower flow rates in other regions. It is well known in statistics that the variance of an average decreases linearly with the number of independent samples used in the average. In the random-field context, the "effective" number of independent samples increases as the correlation length decreases, and thus the decrease in variance in flow rate is to be expected. Conversely, when the correlation length is large, the variance in the flow rate is also expected to be larger—there is less averaging variance reduction within each realization. The maximum flow rate variance is obtained when the field becomes completely correlated ($\theta_{\ln K} = \infty$), in which case the permeability is uniform at each realization. Since the flow rate is proportional to the (uniform) permeability in this case, the flow rate variance exactly follows that of the permeability, and thus for $\theta_{\ln K} = \infty$,

$$\sigma_{\bar{Q}} = \frac{\sigma_K}{\mu_K} \bar{Q}_{\text{det}} \tag{8.39}$$

Notice also in Figure 8.19*b* that the standard deviation $s_{\bar{Q}}$ seems to reach a maximum at intermediate values of v_K for the smaller correlation lengths. This is because the mean $m_{\bar{Q}}$ is falling rapidly for smaller $\theta_{\ln K}$. In fact, in the limit as $v_K \to \infty$, the mean normalized flow rate will tend to zero, which implies that the standard deviation $s_{\bar{Q}}$ will also tend to zero (since $\bar{Q}$ is a nonnegative quantity). In other words, the standard deviation of the normalized flow rate will be zero when $v_K = 0$ or when $v_K = \infty$ and will reach a maximum somewhere in between for any $\theta_{\ln K} < \infty$, the maximum point being farther to the right for larger $\theta_{\ln K}$.

In order to assess the accuracy of these estimators in terms of their reproducibility, the standard deviation of the estimators can be related to the number of simulations of the Monte Carlo process (see Section 6.6). Assuming $\ln \bar{Q}$ is normally distributed, the standard deviations (standard errors) of the estimators are as follows (in this study $n = 1000$):

$$\sigma_{m_{\ln \bar{Q}}} = \frac{s_{\ln \bar{Q}}}{\sqrt{n}} \simeq 0.032 s_{\ln \bar{Q}} \tag{8.40a}$$

$$\sigma_{s^2_{\ln \bar{Q}}} = \sqrt{\frac{2}{n-1}}\, s^2_{\ln \bar{Q}} \simeq 0.045 s^2_{\ln \bar{Q}} \tag{8.40b}$$

Figures 8.20*a* and *b* show the influence of three-dimensionality on the estimated mean and standard deviation of $\bar{Q}$ by comparing results with gradually increasing numbers of elements in the z direction. Also included in these figures is the two-dimensional result which implies an infinite correlation length in the z direction and allows no flow out of the plane of the analysis. The particular cases shown correspond to a fixed correlation length $\theta_{\ln K} = 1$.

Compared with two-dimensional analysis, three dimensions allow the flow greater freedom to avoid the low permeability zones. This results in a less steep reduction in the expected flow rate with increasing v_K, as shown in

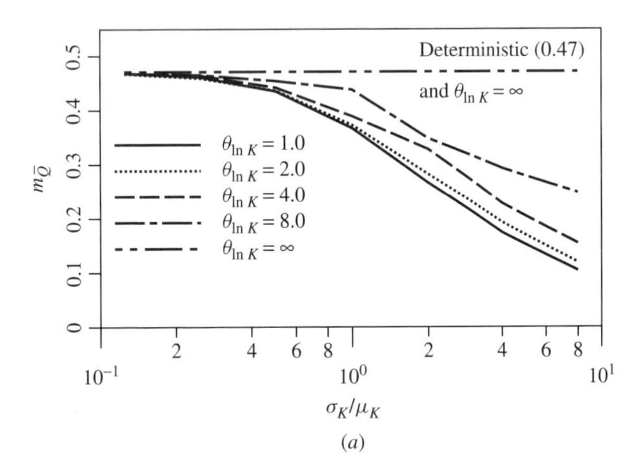

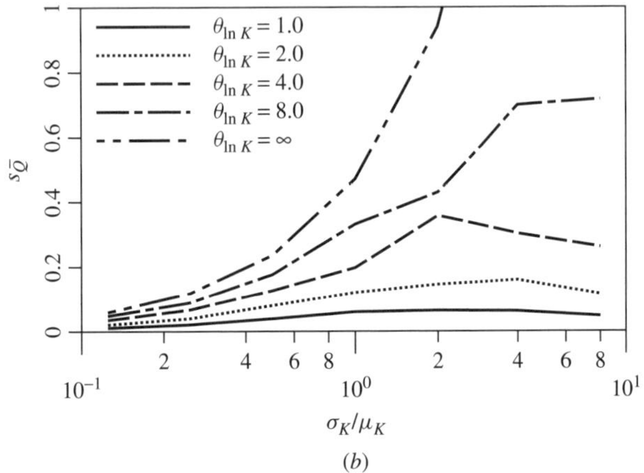

Figure 8.19 Influence of coefficient of variation v_K and correlation length $\theta_{\ln K}$ on (*a*) mean of and (*b*) standard deviation of normalized flow rate $\bar{Q}$. Plots are for $L_z/L_y = 1$.

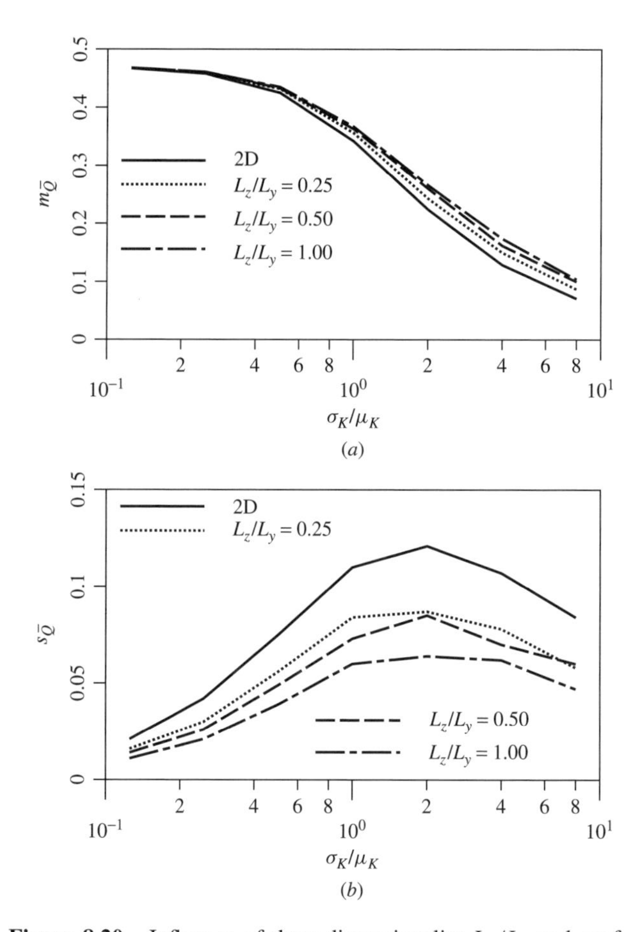

Figure 8.20 Influence of three-dimensionality L_z/L_y and coefficient of variation v_K on (*a*) mean, and (*b*) standard deviation of normalized flow rate $\bar{Q}$. Plots are for $\theta_{\ln K} = 1$; 2D = two dimensions.

Figure 8.20*a*. There is also a corresponding reduction in the variance of the expected flow rate as the third dimension is elongated as shown in Figure 8.20*b*. This additional variance reduction is due to the increased averaging one gets when one allows variability in the third dimension. The difference between the two- and three-dimensional results is not that great, however, and it could be argued that a two-dimensional analysis is a reasonable first approximation to the "true" behavior in this case. It should be noted that the two-dimensional approximation will tend to slightly underestimate the expected flow through the system, which is an unconservative result from the point of view of engineering design.

8.6.2 Reliability-Based Design

One of the main objectives of stochastic analyses such as those described in this section is to enable statements to be made relating to the probability of certain flow-related events occurring. Reliability-based design depends on this approach, so consider again the case of flow rate prediction beneath a water-retaining structure. Deterministic approaches using fixed values of permeability throughout the finite-element mesh will lead to a particular value of the flow rate which can then be factored (i.e., scaled up) as deemed appropriate by the designers. Provided this factored value is less than the maximum acceptable flow rate, the design is considered to be acceptable and in some sense "safe." Although the designer would accept that there is still a small possibility of failure, this is subjective and no attempt is made to quantify the risk.

On the other hand, the stochastic approach is more realistic in recognizing that even in a "well-designed" system, there will always be a possibility that the maximum acceptable flow rate could be exceeded if an unfortunate combination of soil properties should occur. The designers then have to make a quite different decision relating to how high a probability of "failure" would be acceptable.

Figure 8.21 shows typical histograms of the flow rate following 1000 realizations for the cases where $v_k = 0.50$

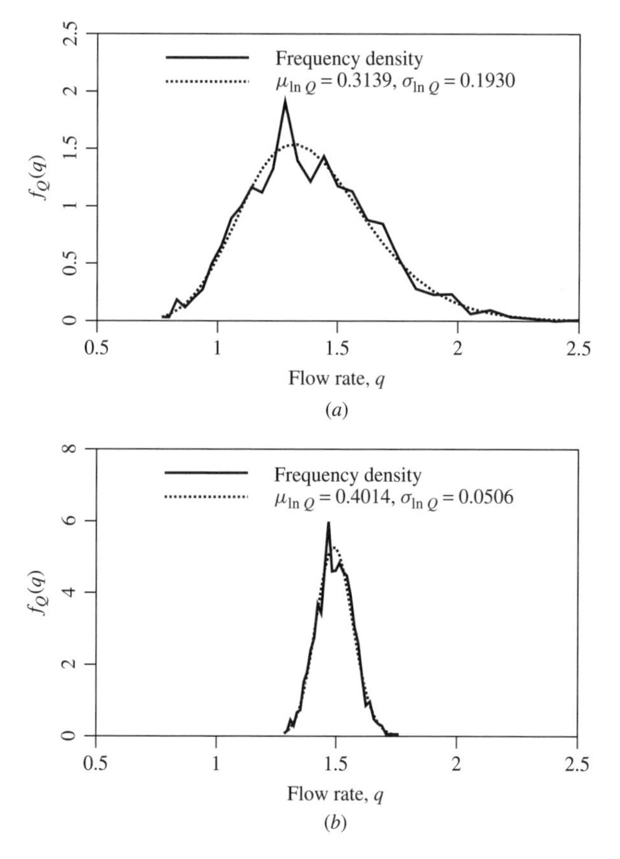

Figure 8.21 Histograms of simulated flow rates following 1000 realizations ($\theta_{\ln K} = 2.00$) along with fitted lognormal distribution for (*a*) $v_K = 0.5$ and (*b*) $v_K = 0.125$.

and $v_k = 0.125$. Both plots are for $L_z/L_y = 1.0$ and $\theta_{\ln K} = 2.0$ and fitted lognormal distributions are seen to match the histograms well. The parameters of the fitted distribution, $\mu_{\ln \bar{Q}}$ and $\sigma_{\ln \bar{Q}}$, are estimated from the suite of realizations and given in the plots. The histograms are normalized to produce a *frequency density plot* which has area beneath the curve of unity; in this way the histogram can be directly compared to the fitted distributions and allow the easy estimation of probabilities.

A chi-square goodness-of-fit hypothesis test was performed to assess the reasonableness of the lognormal distribution. The p-value of the test was 0.38 for Figure 8.21*a* and 0.89 for Figure 8.21*b*, indicating strong agreement between the histogram and the fitted distribution (i.e., we cannot reject the hypothesis that the distribution is lognormal).

The total deterministic flow rate through the three-dimensional geometry that would have occurred if the permeability was constant and equal to unity is given by

$$Q_{\text{det}} = L_z \mu_K \, \Delta H \, \bar{Q}_{\text{det}} = (3.2)(1)(1)0.47 = 1.50$$

where $L_z = 3.2$ is the width of the flow problem in the z direction and 0.47 represents the deterministic flow per unit width based on a two-dimensional analysis in the x–y plane (see Figure 8.18*b*). This value can be compared directly with the histograms in Figure 8.21 and it can be seen that the deterministic mean flow rate is quite close to the distribution mean when $v_K = 0.125$ (Figure 8.21*b*) but is shifted to the right of the distribution mean when $v_K = 0.50$ (Figure 8.21*a*). This shift is due to the falling mean flow rate as v_K increases (see Figure 8.19*a*).

For reliability-based design, a major objective of this type of analysis would be to estimate the probability that the deterministic flow rate underestimates the true flow rate. Such an underestimation would imply an "unsafe" design and should have an appropriately "low" probability. The actual value of an acceptable design probability of failure depends on a number of factors, including the importance of the water-retaining structure in relation to safety and infrastructure downstream.

Referring to the particular case shown in Figure 8.21*a*, the estimated probability that the deterministic flow rate underestimates the true flow rate is given by the following calculation, which assumes that Q is lognormally distributed (a reasonable assumption, as discussed above):

$$\text{P}[Q > Q_{\text{det}}] = 1 - \Phi\left(\frac{\ln\ 1.50 - 0.3139}{0.1930}\right) = 0.318$$

where $\mu_{\ln Q} = 0.3139$ and $\sigma_{\ln Q} = 0.1930$ are the parameters of the fitted distribution shown in Figure 8.21*a* and $\Phi(\cdot)$ is the standard normal cumulative distribution function. A similar calculation applied to the data in Figure 8.21*b* leads to

$$\text{P}[Q > Q_{\text{det}}] = 1 - \Phi\left(\frac{\ln\ 1.50 - 0.4014}{0.0506}\right) = 0.468$$

and for a range of different v_K values at constant correlation length $\theta_{\ln K} = 2.00$, the probability of an unsafe design has been plotted as a function of $\log_{10}\ v_K$ in Figure 8.22*a*.

Figure 8.22*a* shows that a deterministic calculation based on the mean permeability will always lead to a conservative estimate of the flow rate (i.e., $\text{P}[Q > Q_{\text{det}}] < 50\%$). As the coefficient of variation of the permeability increases, however, the probability that Q_{det} underestimates the flow rate decreases. For the range of v_K-values considered, the underestimation probability varied from less that 2% for $v_K = 8$ to a probability of 47% for $v_K = 0.125$. In the latter case, however, the standard deviation of the computed flow also becomes small, so the range of flow values more resembles a normal distribution than a lognormal one. In the limit as $v_K \to 0$, the random permeability field tends to its deterministic mean, but in probabilistic terms this implies an equal likelihood of the true flow rate falling on either side of the predicted Q_{det} value. Hence the curve in Figure 8.22*a* tends to a probability of 50% for small values of v_K. These

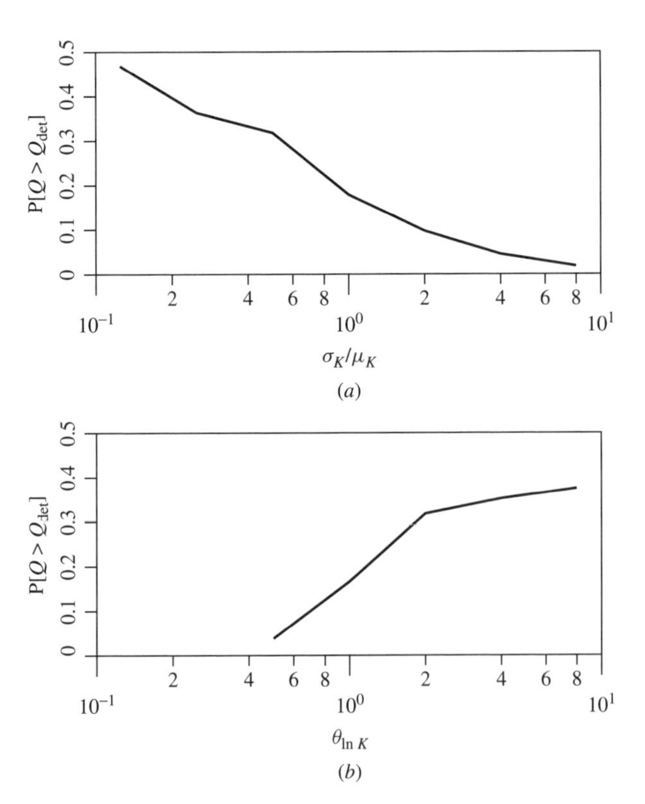

Figure 8.22 Probability of unsafe design, $\text{P}[Q > Q_{\text{det}}]$, plotted against (*a*) v_K with $\theta_{\ln K} = 2$ and (*b*) $\theta_{\ln K}$ with $v_K = 0.5$. Both plots are for $L_z/L_y = 1.00$.

results are reassuring from a design viewpoint because they indicate that the traditional approach leads to a conservative estimate of the flow rate—the more variable the soil, the more conservative the prediction. This observation is made with the knowledge that permeability is considered one of the most variable of soil properties with coefficients of variation ranging as high as 3 (see, e.g., Lee et al., 1983; Kulhawy et al., 1991; Phoon and Kulhawy, 1999).

The sensitivity of the probability $P[Q > Q_{det}]$ to variations in the correlation length $\theta_{\ln K}$ is shown in Figure 8.22*b*. The coefficient of variation of the soil permeability is maintained at a constant value given by $v_K = 0.50$ and the correlation length is varied in the range $0.5 < \theta_{\ln K} < 8$. This result shows that as the correlation length increases the probability of the true flow rate being greater than the deterministic value also increases, although its value is always less than 50%. In the limit, as $\theta_{\ln K} \to \infty$, each realization of the Monte Carlo process assumes a perfectly correlated field of permeability values. In this case, the flow rate distribution is identical to the permeability distribution (i.e., lognormal) with a mean equal to the flow rate that would have been computed using the mean permeability. The probability that the true flow rate exceeds the deterministic value therefore tends to the probability that the lognormally distributed random variable K exceeds its own mean when $\theta_{\ln K} = \infty$,

$$P[K > \mu_K] = 1 - \Phi\left(\tfrac{1}{2}\sigma_{\ln K}\right) = 1 - \Phi\left(\sqrt{\ln\left(1 + v_K^2\right)}\right) \tag{8.41}$$

which, for $v_K = 0.5$, gives $P[K > \mu_K] = 0.405$, which Figure 8.22*b* is clearly approaching as $\theta_{\ln K}$ increases.

As $\theta_{\ln K}$ is reduced, however, the probability of the true flow rate being greater than the deterministic value reduces quite steeply and approaches zero quite rapidly for $\theta_{\ln K} \leq 0.5$. The actual value of $\theta_{\ln K}$ in the field will not usually be well established (except perhaps in the vertical direction where sampling is continuous), so sensitivity studies help to give a feel for the importance of this parameter. For further information on soil property correlation, the interested reader is referred to Lumb (1966), Asaoka and Grivas (1982), DeGroot and Baecher (1993), and Marsily (1985).

The extrapolation of results in Figure 8.22*b* to very low probabilities (e.g., $P[Q > Q_{det}] < 0.01$) must be done cautiously, however, as more than the 1000 realizations of the Monte Carlo process used in this presentation would be needed for accuracy in this range. In addition, low probabilities can be significantly in error when estimated parameters are used to describe the distribution. Qualitatively, the fall in probability with decreasing $\theta_{\ln K}$ is shown well in the histogram of Figure 8.23, where for the case of $\theta_{\ln K} = 0.5$ the deterministic flow rate lies well toward the right-hand tail of the distribution leading to $P[Q > Q_{det}] \simeq 0.038$.

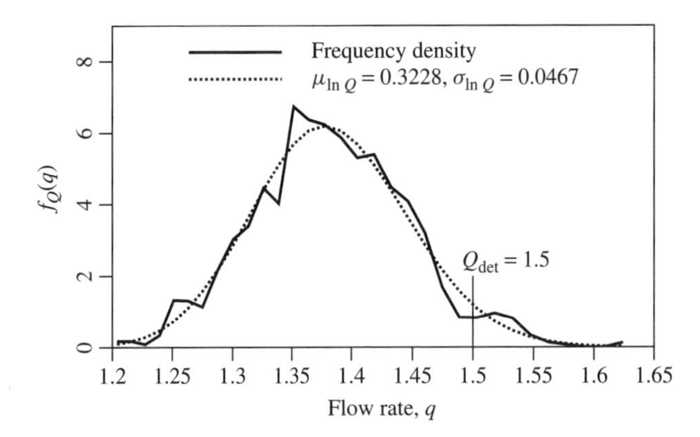

Figure 8.23 Histogram of simulated flow rates following 1000 realizations ($v_K = 0.5$, $\theta_{\ln K} = 0.5$) along with fitted lognormal distribution.

8.6.3 Summary

For low to moderate values of the correlation length ($\theta_{\ln K} < 8$), the expected value of the flow rate was found to fall consistently as the coefficient of variation of the permeability field was increased. The explanation lies in the fact that in a continuous-flow regime such as the one modeled here, the low-permeability zones cannot be entirely "avoided," so the greater the permeability variance, the greater the volume of low-permeability material that must be negotiated and the lower the flow rate. For higher values of the correlation length, the normalized flow rate mean tends to the deterministic value. The standard deviation of the flow rate was shown to consistently increase with the correlation length, staying within the bounds defined analytically for the limiting case of perfect correlation and to increase, then decrease, with the standard deviation of the input permeability for any finite correlation length.

The influence of three-dimensionality was to reduce the overall "randomness" of the results observed from one realization to the next. This had the effect of slightly increasing the expected flow rate and reducing the variance of the flow rate over those values observed from a two-dimensional analysis with the same input statistics. Although unconservative in the estimation of mean flow rates, there was not a great difference between the two- and three-dimensional results, suggesting that the simpler and less expensive two-dimensional approach may give acceptable accuracy for the cases considered.

Some of the results were reinterpreted from a reliability viewpoint, which indicated that if the flow rate was computed deterministically using the mean permeability, the probability of the true flow rate being greater would always be less than 50% (see also Eq. 8.41 when the correlation length goes to infinity). This probability fell to even smaller values as the variance of the input permeability was increased or the correlation length was reduced, implying that a deterministic prediction of flow rate based on the mean permeability would always be conservative on average.

8.7 THREE-DIMENSIONAL EXIT GRADIENT ANALYSIS

The traditional approach to estimating the exit gradient i_e downstream of water-retaining structures due to steady seepage is to assume homogeneous soil properties and proceed deterministically, perhaps using flow net techniques. Once the exit gradient is estimated, a large safety factor of 4–5 or even higher is then applied. The reason for this conservative approach is twofold. First, the consequence of piping and erosion brought about by i_e approaching the critical value i_c can be very severe, leading to complete and rapid failure of civil engineering structures with little advance warning. Second, the high safety factors reflect the designer's uncertainty in local variations of soil properties at the exit points and elsewhere within the flow domain. In this section, we present an alternative to the safety factor approach by expressing exit gradient predictions in the context of reliability-based design. Random-field theory and finite-element techniques are combined with Monte Carlo simulations to study the statistics of exit gradient predictions as a function of soil permeability variance and spatial correlation. The results for three dimensions are compared to those for two dimensions. The approach enables conclusions to be drawn about the probability of critical conditions being approached and hence failure occurring at a given site.

The aim of this section is to observe the influence of statistically variable soil permeability on the exit gradient i_e at the downstream side of a water-retaining structure in both two and three dimensions (Griffiths and Fenton, 1998). Smith and Freeze (1979a, b) were among the first to study the problem of confined flow through a stochastic medium using finite differences where flow between parallel plates and beneath a single sheet pile were presented.

The soil mass is assumed to have a randomly distributed permeability K, defined in the classical geotechnical sense as having units of length over time. The exit gradient is defined as the first derivative of the total head (or "potential"), itself a random variable, with respect to length at the exit points. To investigate the statistical behavior of exit gradients when the permeability field is random, a simple boundary value problem is considered—that of seepage beneath a single sheet pile wall penetrating to half the depth of a soil layer. Both two- and three-dimensional results are presented for comparison; Figures 8.24a and b show the meshes used for the two- and three-dimensional finite-element models, respectively. In two dimensions, it is assumed that all flow occurs in the plane of the analysis. More realistically, the three-dimensional model has no such restriction allowing flow to occur in any direction. This particular problem has been chosen because it is well understood, and a number of theoretical solutions exist for computing flow rates and exit gradients in the deterministic (constant-permeability) case (see, e.g., Harr, 1962; Verruijt, 1970; Lancellota, 1993).

The three-dimensional mesh has the same cross section in the x–y plane as the two-dimensional mesh (12.8×3.2) and extends by 3.2 units in the z direction. The two-dimensional mesh consists of square elements (0.2×0.2) and the three-dimensional mesh consists of cubic ($0.2 \times 0.2 \times 0.2$) elements. The boundary conditions are such that there is a deterministic fixed total head on the upstream and downstream sides of the wall. For simplicity the head difference across the wall is set to unity. The outer boundaries of the mesh are "no-flow" boundary conditions. In all cases, the sheet pile wall has a depth of 1.6 units, which is half the depth of the soil layer.

Figure 8.25a shows the classical smooth flow net for both the two- and three-dimensional cases, corresponding to a constant-permeability field, and Figure 8.25b shows a typical case in which the permeability is a spatially varying random field. In the latter case each element of the mesh has been assigned a different permeability value based on a statistical distribution. Note how the flow

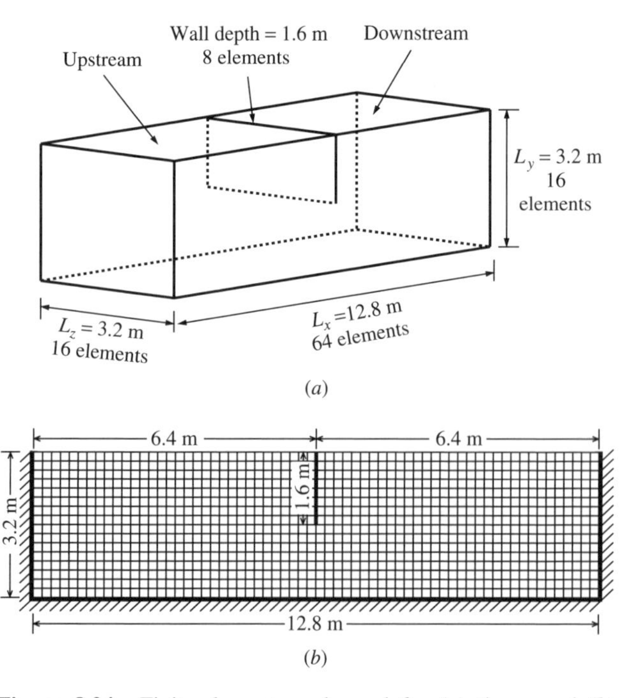

Figure 8.24 Finite-element mesh used for (a) three- and (b) two-dimensional seepage analyses.

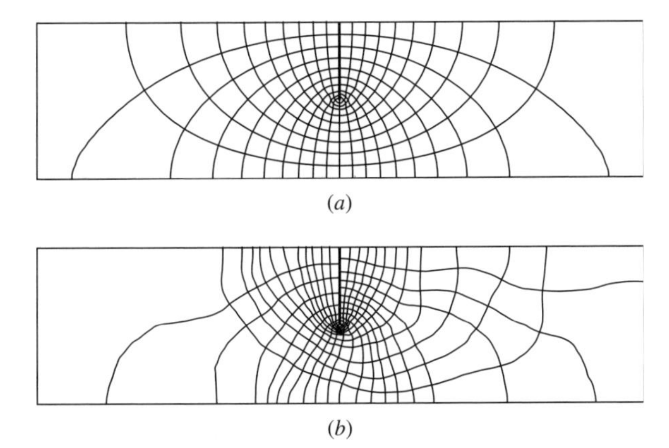

Figure 8.25 (*a*) Flow net for deterministic analysis (permeability equal to mean value everywhere) and (*b*) typical flow net when permeability is a spatially variable random field (with $\theta_{\ln K} = 2$ m and $v_K = 1$).

net becomes ragged as the flow attempts to avoid low-permeability zones.

The exit gradient against the downstream side of the wall is computed using a two-point numerical differentiation scheme as shown in Figure 8.26. The gradient is computed adjacent to the wall since this location always has the highest exit gradient in a constant-permeability field and will also give the highest expected value in a random-permeability field. In the three-dimensional analyses, the exit gradient was computed at all 17 downstream locations (there are 16 elements in the z direction, so there are 17 downstream nodes) although the two values computed at the center and edge of the wall have been the main focus of this study.

As in the previous sections, the permeability field will be assumed to be lognormally distributed with realizations produced by LAS (Section 6.4.6). The correlation

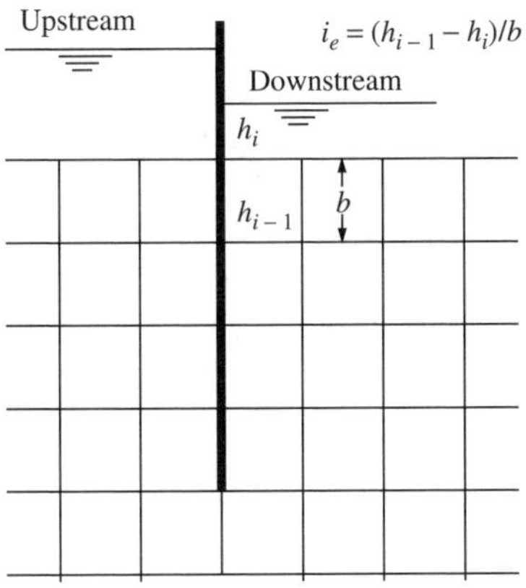

Figure 8.26 Numerical calculation of exit gradient i_e.

structure will be assumed isotropic, with site-specific anisotropic extensions being left to the reader. The computer programs used to produce the results presented in this section are RFLOW2D for the two-dimensional analyses and RFLOW3D for the three-dimensional analyses. Both programs are available at `http://www.engmath.dal.ca/rfem`.

The input to the random-field model comprises of the three parameters ($\mu_K, \sigma_K, \theta_{\ln K}$—a Markov correlation function is used, see Section 3.7.10.2). Based on these underlying statistics, each of the elements (1024 elements in the two-dimensional case and 16,384 in the three-dimensional case) is assigned a permeability from a realization of the permeability random field. A series of realizations are generated and the analysis of sequential realizations and the accumulation of results comprises a Monte Carlo process. In the current study, 2000 realizations were performed for each of the two-dimensional cases and 1000 in the three-dimensional cases. The reduced number of realizations in three dimensions was chosen to allow a greater number of parametric studies to be performed. Following Monte Carlo simulation of each parametric combination, 2000 (or 1000) values of the exit gradient i_e were obtained which were then analyzed statistically to give the mean, standard deviation, and hence probability of high values occurring that might lead to piping.

8.7.1 Simulation Results

The deterministic analysis of this seepage problem, with a constant-permeability throughout, gives an exit gradient of around $i_{\text{det}} = 0.193$, which agrees closely with the analytical solution for this problem (see, e.g., Lancellota, 1993).

Given that the critical exit gradient i_c (i.e., the value that would initiate piping) for a typical soil is approximately equal to unity, this deterministic value implies a factor of safety of around 5—a conservative value not untypical of those used in design of water-retaining structures (see, e.g., Harr, 1962; Holtz and Kovacs, 1981).

In all analyses the point mean permeability was fixed at $\mu_K = 1$ m/s while the point standard deviation and spatial correlation of permeability were varied in the ranges $0.03125 < \sigma_K/\mu_K < 32.0$ and $0.5 < \theta_{\ln K} < 16.0$ m. For each of these parametric combinations the Monte Carlo process led to *estimated* values of the mean and standard deviation of the exit gradient given by m_{i_e} and s_{i_e}, respectively.

8.7.1.1 Two-Dimensional Results Graphs of m_{i_e} versus v_K and m_{i_e} versus $\theta_{\ln K}$ for a range of values have been plotted in Figure 8.27. Figure 8.27*a* shows that as v_K tends to zero, the mean exit gradient tends, as expected, to the deterministic value of 0.193. For small

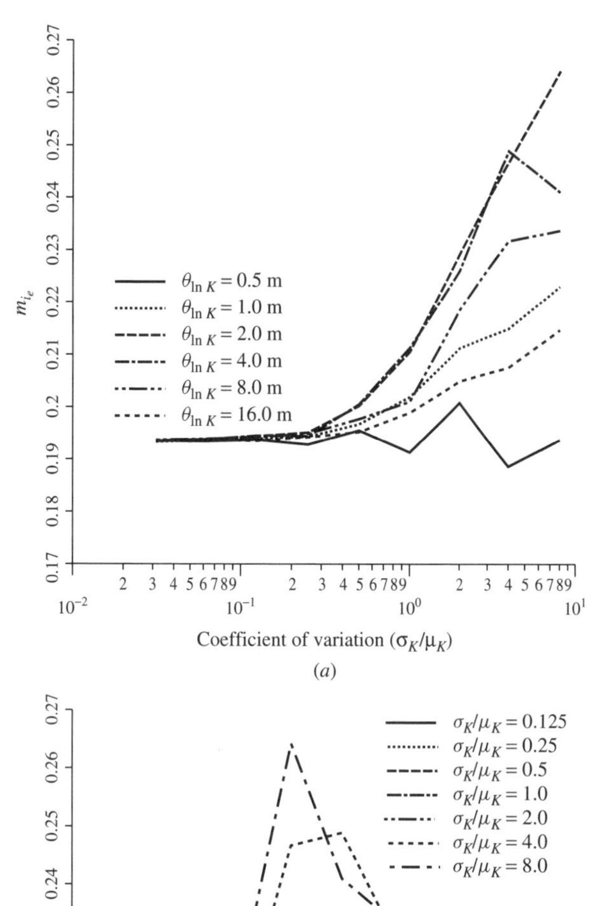

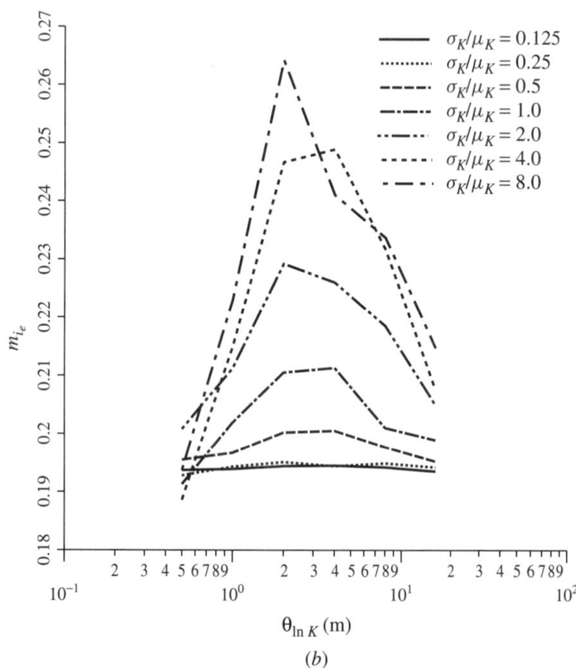

Figure 8.27 Estimated exit gradient mean m_{i_e} versus (*a*) coefficient of variation of permeability v_K and (*b*) correlation length $\theta_{\ln K}$ (two-dimensional analysis).

correlation lengths the mean exit gradient remains essentially constant as v_K is increased, but for higher correlation lengths, the mean exit gradient tends to rise. The amount by which the mean exit gradient increases, however, is dependent on $\theta_{\ln K}$ and appears to reach a maximum when $\theta_{\ln K} \approx 2$. This is shown more clearly in Figure 8.27*b* where the same results have been plotted with $\theta_{\ln K}$ along the abscissa. The maximum value of $m_{i_e} = 0.264$ recorded in this particular set of results corresponds to the case when $v_K = 8$ and represents an increase of 37% over the deterministic value.

The return to deterministic values as $\theta_{\ln K}$ increases is to be expected if one thinks of the limiting case where $\theta_{\ln K} = \infty$. In this case each realization would have a constant (although different) permeability, and thus the deterministic exit gradient would be obtained.

The standard error of the results shown in Figure 8.27 is $\sigma_{m_{i_e}} = s_{i_e}/\sqrt{n} \simeq 0.022 s_{i_e}$, where $n = 2000$ realizations. In other words, when $s_{i_e} \simeq 0.2$, the m_{i_e} are typically in error by less than about ± 0.0044, which basically agrees with the erratic behavior seen by the $\theta_{\ln K} = 0.5$ m line in Figure 8.27*a*.

Graphs of s_{i_e} versus v_K and s_{i_e} versus $\theta_{\ln K}$ for a same range of values have been plotted in Figure 8.28. Figure 8.28*a* shows that as v_K increases, the standard deviation of the exit gradient also increases. However, as was observed with the mean value of i_e, the standard deviation increases more substantially for some values of $\theta_{\ln K}$ than others. This is shown more clearly in Figure 8.28*b*. The peak in s_{i_e} again occurs around $\theta_{\ln K} \approx 2.0$.

It would appear therefore that there is a worst-case value of $\theta_{\ln K}$ from a reliability-based design viewpoint in which both the mean *and* the standard deviation of the exit gradient reach a local maximum at the same time. At this critical value of $\theta_{\ln K}$, the higher m_{i_e} implies that on the average i_e would be closer to the critical value i_c, and to make matters worse, the higher s_{i_e} implies greater uncertainty in trying to predict i_e.

The standard error of the square of the results shown in Figure 8.28 is $\sigma_{s_{i_e}^2} = \sqrt{2/(n-1)} s_{i_e}^2 \simeq 0.032 s_{i_e}^2$, where $n = 2000$ realizations. In other words, when $s_{i_e} \simeq 0.2$, the standard deviation of $s_{i_e}^2$ is approximately 0.0013 and so the standard error on s_{i_e} is roughly $\pm\sqrt{0.0013} = \pm 0.04$. The curves in Figure 8.27 show less error than this, as expected.

8.7.1.2 Three-Dimensional Results An identical set of parametric studies was performed using the three-dimensional mesh shown in Figure 8.24*b*. The flow is now free to meander in the z direction as it makes its primary journey beneath the wall from the upstream to the downstream side. Although the exit gradient can be computed at 17 nodal locations adjacent to the downstream side of the wall (16 elements), initial results are presented for the central location since this is considered to be the point where the effects of three-dimensionality will be greatest. Figures 8.29 and 8.30 are the three-dimensional counterparts of Figures 8.27 and 8.28 in two dimensions. Figure 8.29*a* shows the variation in m_{i_e} as a function of v_K for different $\theta_{\ln K}$ values. For low values of $\theta_{\ln K}$,

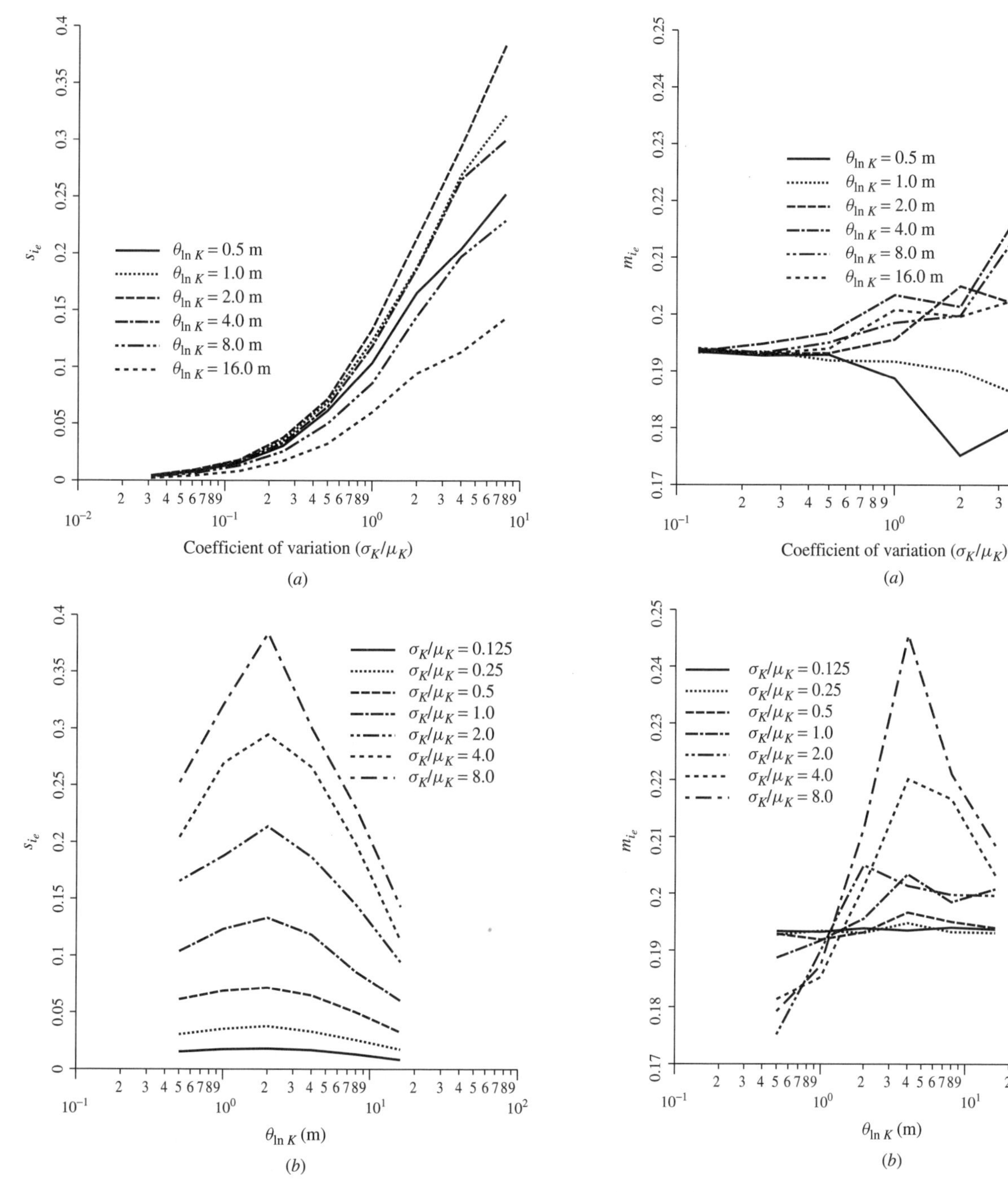

Figure 8.28 Estimated exit gradient standard deviation s_{i_e} versus (*a*) coefficient of variation of permeability v_K and (*b*) correlation length $\theta_{\ln K}$ (two-dimensional analysis).

Figure 8.29 Estimated exit gradient mean m_{i_e} versus (*a*) coefficient of variation of permeability v_K and (*b*) correlation length $\theta_{\ln K}$ (three-dimensional analysis).

the mean remains constant and even starts to fall as v_K is increased. For higher $\theta_{\ln K}$ values, the mean exit gradient starts to climb, and, as was observed in two dimensions (Figure 8.27*a*), there is a critical value of $\theta_{\ln K}$ for which the greatest values of m_{i_e} are observed. This is seen more clearly in Figure 8.29*b* in which m_{i_e} is plotted as a function of $\theta_{\ln K}$. The maxima in m_{i_e} are clearly seen and occur at higher values of $\theta_{\ln K} \approx 4$ than in two dimensions (Figure 8.27*b*), which gave maxima closer to $\theta_{\ln K} \approx 2$. The maximum value of $m_{i_e} = 0.243$ recorded in this particular

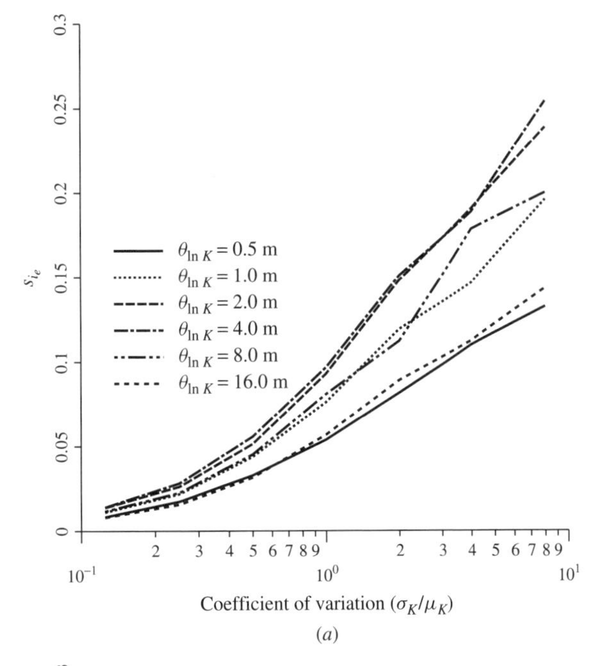

(*a*)

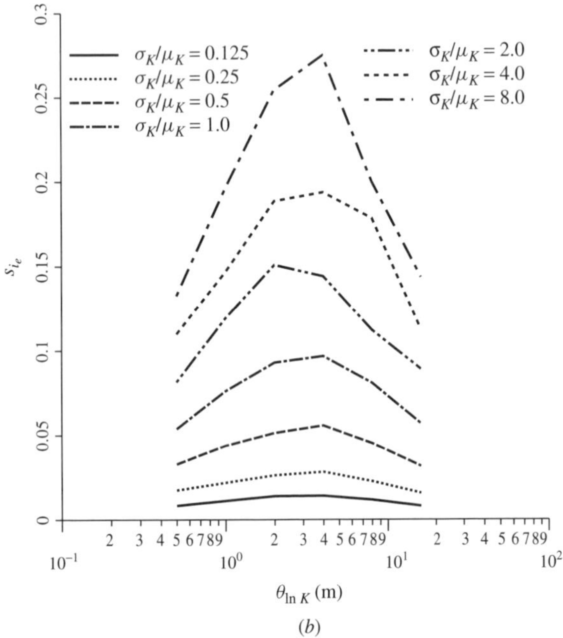

(*b*)

Figure 8.30 Estimated exit gradient standard deviation s_{i_e} versus (*a*) coefficient of variation of permeability v_K and (*b*) correlation length $\theta_{\ln K}$ (three-dimensional analysis).

set of results corresponds to the case $v_K = 8$, $\theta_{\ln K} = 4$ and represents an increase of 26% over the deterministic value. This should be compared with the 34% increase observed for the same case in two dimensions.

Figure 8.30 shows the behavior of s_{i_e} as a function of v_K and $\theta_{\ln K}$. Figure 8.30*a* indicates that the standard deviation of the exit gradient increases with v_K for all values of $\theta_{\ln K}$, but the extent of the increase is again dependent on the correlation length, as shown in Figure 8.30*b*, with the maxima occurring in the $\theta_{\ln K}$ range of 2–4.

8.7.2 Comparison of Two and Three Dimensions

Compared with two-dimensional analysis, three dimensions allows the flow greater freedom to avoid the low-permeability zones. The influence of three-dimensionality is therefore to reduce the overall randomness of the results observed from one realization to the next. This implies that the sensitivity of the output quantities to v_K will be reduced in three dimensions as compared with two dimensions. In the study of seepage quantities in the previous section three-dimensionality had the effect of slowing down the reduction in the expected flow rate as v_K was increased. Similarly, in this study of exit gradients, the *change* in m_{i_e} over its deterministic value with increasing v_K is less than it was in two dimensions.

For the case of $\theta_{\ln K} = 2$, Figure 8.31*a* presents results for m_{i_e} in both two and three dimensions. An additional three-dimensional result corresponding to the mean exit gradient at the edge of the wall is also included. A consistent pattern is observed in which the three-dimensional (center) result shows the smallest increase in m_{i_e} and the two-dimensional result shows the greatest increase. An intermediate result is obtained at the edge of the wall where the flow is restrained in one direction. The boundary conditions on this plane will ensure that the edge result lies between the two- and three-dimensional (center) results.

Figure 8.31*b* presents results for s_{i_e} for the same three cases. These results are much closer together, although, as expected, the three-dimensional (center) result gives the lowest values.

In summary, the effect of allowing flow in three dimensions is to increase the averaging effect discussed above within each realization. The difference between the two- and three-dimensional results is not that great, however, and it could be argued that a two-dimensional analysis is a reasonable first approximation to the true behavior. In relation to the prediction of exit gradients, it also appears that two dimensions is conservative, in that the increase in m_{i_e} with v_K observed for intermediate values of $\theta_{\ln K}$ is greater in two than in three dimensions.

8.7.3 Reliability-Based Design Interpretation

A factor of safety applied to a deterministic prediction is intended to eliminate any serious possibility of failure but without any objective attempt to quantify the risk.

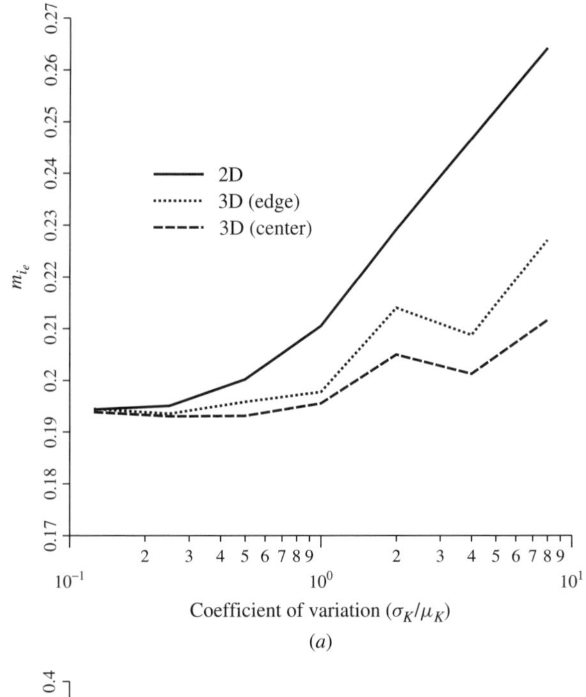

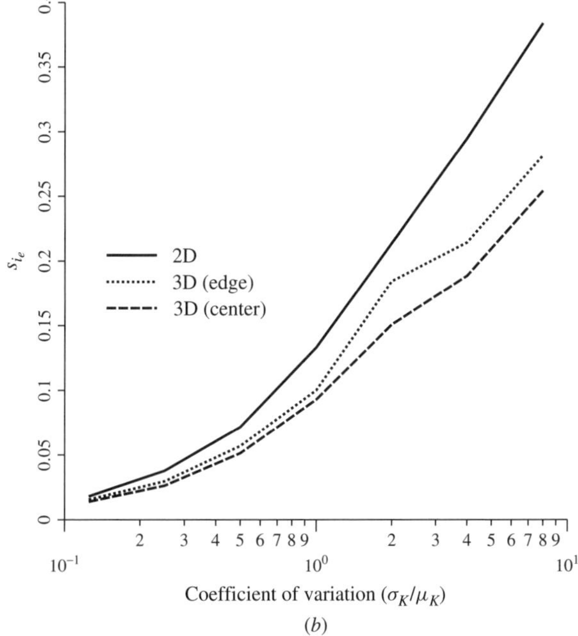

Figure 8.31 Effect of two (2D) and three (3D) dimensions on (*a*) m_{i_e} and (*b*) s_{i_e}. Both plots are for $\theta_{\ln K} = 2$.

Reliability-based design attempts to quantify risk by seeking answers to the following questions:

1. What is the probability that the actual exit gradient will exceed a deterministic prediction (based on a constant-properties throughout)?
2. What is the probability that the actual exit gradient will exceed the critical value, resulting in failure?

The Monte Carlo scheme described in this section enables probabilistic statements to be made. For example, if out of 1000 realizations, 50 gave an exit gradient $i_e \geq 1$, it could be concluded that the probability of piping or erosion was of the order of 50/1000, or 5%. In general, though, a histogram can be plotted, a distribution fitted, and the probabilities computed using the distribution.

8.7.3.1 Two Dimensions A typical histogram of exit gradient values corresponding to $\theta_{\ln K} = 2$ m and $v_K = 1$ for a two-dimensional analysis is shown in Figure 8.32. The ragged line comes from the frequency count obtained over the realizations and the smooth dotted line is based on a lognormal fit to those data. The good agreement suggests that the actual distribution of exit gradients is indeed lognormal. The mean and standard deviation of the underlying *normal* distribution of $\ln i_e$ are also printed on the figure. Since Figure 8.32 shows a fitted lognormal probability density function, probabilities can be deduced directly. For example, in the particular case shown, the probability that the actual exit gradient will exceed the deterministic value of $i_{\text{det}} = 0.193$ is approximated by

$$P[i_e > 0.193] = 1 - \Phi\left(\frac{\ln 0.193 + 1.7508}{0.6404}\right) \tag{8.42}$$

where $\Phi(\cdot)$ is the cumulative normal distribution function. In this case $\Phi(0.17) = 0.568$; thus

$$P[i_e > 0.193] = 0.43 \tag{8.43}$$

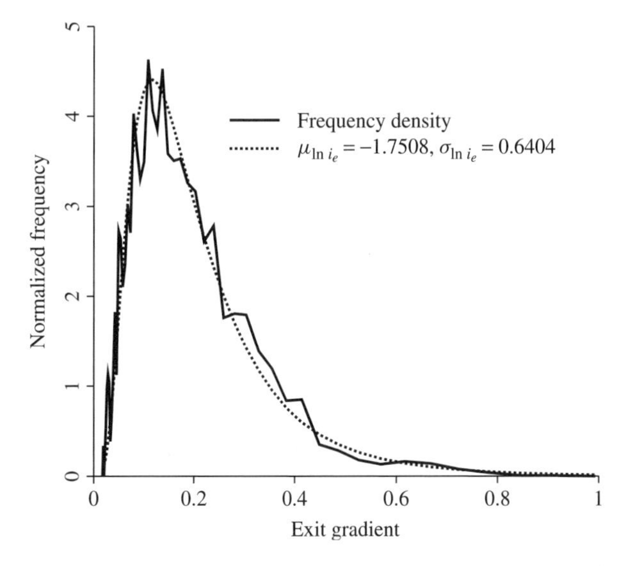

Figure 8.32 Histogram of exit gradients in two dimensions for the case $\theta_{\ln K} = 2$ and $v_K = 1$.

and there is a 43% probability that the deterministic prediction of $i_{\text{det}} = 0.193$ is unconservative.

A similar calculation has been performed for all the parametric variations considered in this study. In each case the following probability was calculated:

$$\mathrm{P}\left[i_e > \alpha i_{\text{det}}\right] \tag{8.44}$$

where α is a simple scaling factor on the deterministic exit gradient which serves the same purpose as the factor of safety. When $\alpha = 1$ (as in Eq. 8.43), the result is just the probability that the actual exit gradient will exceed the deterministic value. Larger values of α are interesting for design purposes where a prediction of the probability of failure is required. In the current example, the deterministic exit gradient is approximately equal to 0.2, so it would be of interest to know the probability of the actual exit gradient exceeding the critical hydraulic gradient $i_c \approx 1$. For this comparison therefore α would be set equal to 5.

A full range of probability values in two dimensions has been computed in this study and some selected results will now be described. A set of probabilities corresponding to $\theta_{\ln K} = 2$ is presented in Figure 8.33. The mean and standard deviation of the exit gradients reached a local maximum when $\theta_{\ln K} \simeq 2$ (see Figures 8.27*b* and 8.28*b*).

It should be noted that irrespective of the $\theta_{\ln K}$ or v_K, $\mathrm{P}\left[i_e > i_{\text{det}}\right]$ is always less than 50%. This is a reassuring result from a design standpoint. The probabilities which approach 50% correspond to a very low v_K and are somewhat misleading in that the computed exit gradients have a very low variance and are approaching the deterministic value. The 50% merely refers to an equal likelihood of the actual exit gradient lying on either side of an essentially normal distribution with a small variance. For small v_K, this is shown clearly by the sudden reduction to zero of the probability that i_e exceeds αi_{det} when $\alpha > 1$ (for example, when $\alpha = 1.1$).

As α is increased further, the probability consistently falls, although each curve exhibits a maximum probability corresponding to a different value of v_K. This interesting observation implies that there is a worst-case combination of $\theta_{\ln K}$ and v_K that gives the greatest likelihood of i_e exceeding i_{det}.

In consideration of failure conditions, the value of $P\left[i_e \geq 1\right]$, as indicated by the curve corresponding to $\alpha = 5$, is small but not insignificant, with probabilities approaching 10% for the highest v_K cases considered. In view of this result, it is not surprising that for highly variable soils a factor of safety against piping of up to 10 has been suggested by some commentators (see, e.g., Harr, 1987).

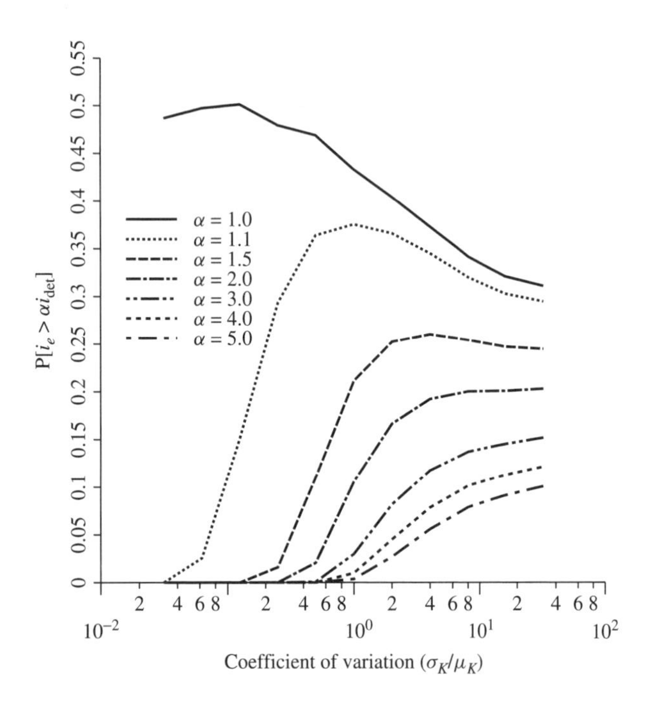

Figure 8.33 Probability that i_e exceeds αi_{det} versus v_K for the case $\theta_{\ln K} = 2$ m in two dimensions.

8.7.3.2 ***Three Dimensions*** An examination of the central exit gradients predicted by the three-dimensional analyses indicates that they are broadly similar to those obtained in two dimensions. Figure 8.34 shows a typical histogram of the central exit gradient value corresponding to $\theta_{\ln K} = 2$ m and $v_K = 1$ for a three-dimensional analysis. This is the same parametric combination given in Figure 8.32 for two dimensions. The fitted curve again indicates that the actual distribution of exit gradients is lognormal. The mean and standard deviation of the underlying *normal* distribution of $\ln i_e$ is also printed on the figure. In the case illustrated by Figure 8.34, the probability that the actual exit gradient

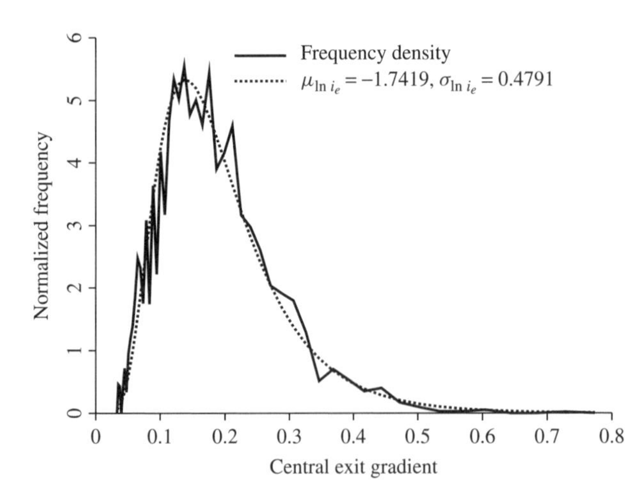

Figure 8.34 Histogram of central exit gradients in three dimensions for the case $\theta_{\ln K} = 2$ and $v_K = 1$.

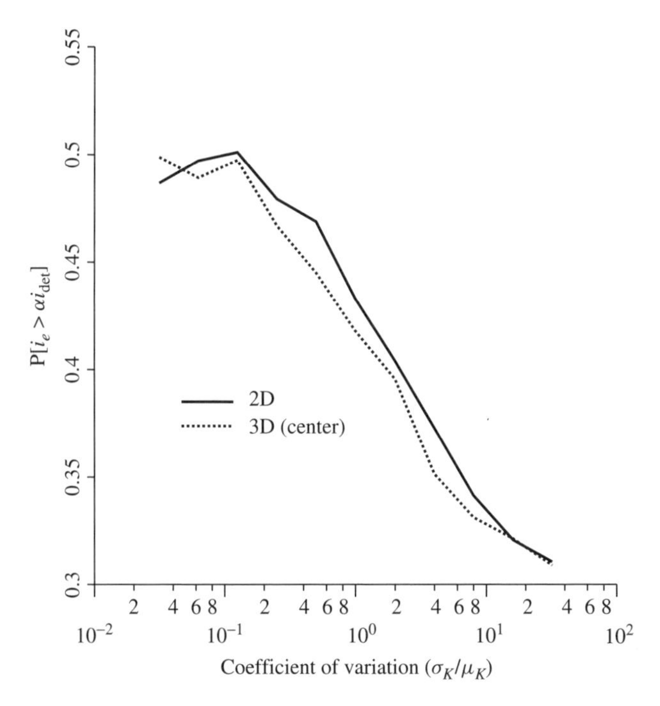

Figure 8.35 Probability that i_e exceeds i_{det} versus v_K for the case $\theta_{\ln K}$ =2 m in two (2D) and three (3D) dimensions.

will exceed the deterministic value of $i_{det} = 0.193$ is equal to 41% and is virtually the same as the 43% given in two dimensions.

The probability of an unconservative design based on three-dimensional studies of a full range of v_K values with $\theta_{\ln K} = 2$ is shown in Figure 8.35 together with the corresponding results in two dimensions ($\alpha = 1$). The three-dimensional results indicate a slight reduction in the probability that the deterministic value is unconservative. It appears that simpler and computationally less intensive two-dimensional analysis of exit gradients will generally give sufficiently accurate and conservative reliability estimates of exit gradients.

8.7.4 Concluding Remarks

Generally speaking, the computed variance of the exit gradient was considerably higher than other quantities of interest in the flow problem, such as the flow rate. This is hardly surprising when one considers that the exit gradient is a derivative quantity which is dependent on the total head value computed at a very specific location within the mesh at the downstream exit point.

An interesting result was that the computed exit gradient was found to reach a maximum for a particular value of the correlation length $\theta_{\ln K}$, somewhere between 2 and 4 m. The higher end of this range was observed in the three-dimensional studies and the lower end in two dimensions.

When the results were interpreted in the context of reliability-based design, conclusions could be reached about the probability of exit gradient values exceeding the deterministic value, or even reaching levels at which instability and piping could occur. In two dimensions and for the particular case of $\theta_{\ln K} = 2$ m, the probability of the actual exit gradient exceeding the deterministic value could be as high as 50% but generally lay in the 40% range for moderate values of v_K. The probability of an unconservative deterministic prediction was generally found to exhibit a maximum point corresponding to a particular combination of $\theta_{\ln K}$ and v_K. From a design point of view this could be considered a worst-case scenario leading to maximum uncertainty in the prediction of exit gradients.

With regard to the possibility of piping, erosion, and eventual failure of the system, a relationship was established between the traditional factor of safety (α) and the probability of failure. For the particular case mentioned above and assuming that the critical exit gradient is of the order $i_c \approx 1$, a factor of safety of $\alpha = 5$ could still imply a probability of failure as high as 10% if v_K is also high. This result suggests that factors of safety as high as 10 may not be unreasonable for critical structures founded in highly variable soil.

The three-dimensional studies were considerably more intensive computationally than their two-dimensional counterparts but had the modeling advantage of removing the requirement of planar flow. In three dimensions, the flow has greater freedom to avoid the low-permeability zones; thus there is less randomness associated with each realization. This was manifested in a reduced mean and standard deviation of the exit gradient as compared with two dimensions. The differences were not that great, however, and indicated that two-dimensional exit gradient studies in random soils will lead to conservative results while giving sufficient accuracy.

CHAPTER 9

Flow through Earth Dams

9.1 STATISTICS OF FLOW THROUGH EARTH DAMS

Many water-retaining structures in North America are earth dams and the prediction of flow through such structures is of interest to planners and designers. Although it is well known that soils exhibit highly variable hydraulic properties, the prediction of flow rates through earth dams is generally performed using deterministic models. In this section we consider a two-dimensional earth dam model and investigate the effects of spatially varying random hydraulic properties on two quantities of classical interest: (i) the total flow rate through the dam and (ii) the amount of drawdown of the free surface at the downstream face of the dam. The drawdown is defined as the elevation of the point on the downstream face of the dam at which the water first reaches the dam surface. Other issues which relate more to the structural reliability of an earth dam, such as failure by piping and flow along eroding fractures, are not addressed here. It is assumed that the permeability field is representable by a continuous random field and that interest is in the stable, steady-state, flow behavior. The study related here was performed by Fenton and Griffiths (1995, 1996) using the program RDAM2D, available at `http://www.engmath.dal.ca/rfem`.

The computation of flow through an earth dam is complicated by the fact that the location and profile of the free surface is not known a priori and must be determined iteratively. Nevertheless, the finite-element code required to perform such an analysis is really quite straightforward in concept, involving a simple Darcy flow model and iteratively adjusting the nodal elevations along the free surface to match their predicted potential heads (e.g., Smith and Griffiths, 2004). Lacy and Prevost (1987), among others, suggest a fixed-mesh approach where the elements allow for both saturated and unsaturated conditions. The approach suggested by Smith and Griffiths was selected here, due to its simplicity, with some modifications to improve convergence. See Fenton and Griffiths (1997a) for details on the convergence algorithm.

When the permeability is viewed as a spatially random field, the equations governing the flow become stochastic. Due to the nonlinear nature of these equations (i.e., the moving free surface), solving the stochastic problem using Monte Carlo simulations is appropriate. In this study a sequence of 1000 realizations of spatially varying soil properties with prescribed mean, variance, and spatial correlation structure are generated and then analyzed to obtain a sequence of flow rates and free-surface profiles. The mean and variance of the flow rate and drawdown statistics can then be estimated directly from the sequence of computed results. The number of realizations was selected so that the variance estimator of the logarithm of total flow rate had a coefficient of variation less than 5% (computed analytically under the assumption that log-flow rate is normally distributed; see Section 6.6).

Because the analysis is Monte Carlo in nature, the results are strictly only applicable to the particular earth dam geometries and boundary conditions studied; however, the general trends and observations may be extended to a range of earth dam boundary value problems. An empirical approach to the estimation of flow rate statistics and governing distribution is presented to allow these statistics to be easily approximated, that is, without the necessity of the full Monte Carlo analysis. This simplified procedure needs only a single finite-element analysis and knowledge of the variance reduction due to local averaging over the flow regime and will be discussed in detail in Section 9.1.3.

Figure 9.1 illustrates the earth dam geometries considered in this study, each shown for a realization of the soil permeability field. The square and rectangular dams were included since these are classical representations of the free-surface problem (Dupuit problem). The other two geometries are somewhat more realistic. The steep sloped dam, labeled dam 1 in Figure 9.1, can be thought of as a clay core held to its shape by highly permeable backfill having negligible influence on the flow rate (and thus the fill is not explicitly represented).

Figure 9.2 shows two possible realizations of dam 1. It can be seen that the free surface typically lies some distance below the top of the dam. Because the position of the surface is not known a priori, the flow analysis necessarily proceeds iteratively. Under the free surface, flow is assumed to be governed by Darcy's law characterized by an isotropic permeability $K(\mathbf{x})$, where $\mathbf{x}$ is the spatial location:

$$\nabla \cdot \mathbf{q} = 0, \qquad \mathbf{q} = -K(\mathbf{x})\, \nabla\phi \tag{9.1}$$

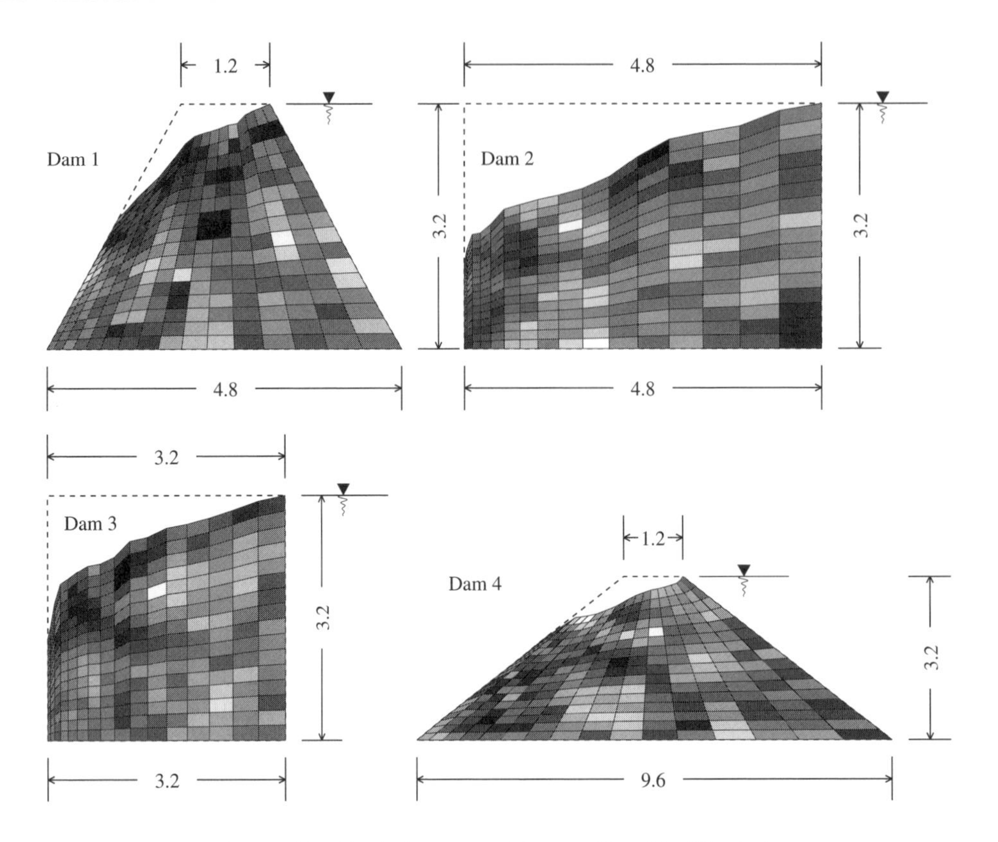

Figure 9.1 Earth dam geometries considered in this study.

where $\mathbf{q}$ is the specific discharge vector and ϕ is the hydraulic head.

As in Chapter 8, the permeability $K(\mathbf{x})$, is assumed to follow a lognormal distribution, with mean μ_K, variance σ_K^2, and parameters $\mu_{\ln K}$ and $\sigma_{\ln K}$ (see Eqs. 1.176). The correlation structure of the $\ln K(\mathbf{x})$ random field is assumed to be isotropic and Markovian with correlation function

$$\rho_{\ln K}(\tau) = \exp\left\{-\frac{2|\tau|}{\theta_{\ln K}}\right\} \tag{9.2}$$

where $\theta_{\ln K}$ is the correlation length.

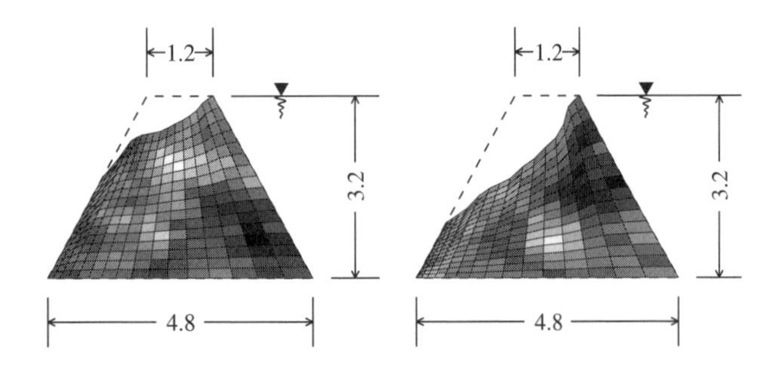

Figure 9.2 Finite-element discretization of dam 1 shown in Figure 9.1: two possible realizations.

Simulations of the soil permeability field proceeds in two steps: first an underlying Gaussian random field $G(\mathbf{x})$ is generated with mean zero, unit variance, and spatial correlation function (Eq. 9.2) using the LAS method. Next, since the permeability is assumed to be lognormally distributed, values of K_i, where i denotes the ith element, are obtained through the transformation

$$K_i = \exp\{\mu_{\ln K} + \sigma_{\ln K} G(\mathbf{x}_i)\} \tag{9.3}$$

where $\mathbf{x}_i$ is the centroid of the ith element and $G(\mathbf{x}_i)$ is the local average value generated by the LAS algorithm of the cell within which $\mathbf{x}_i$ falls. As will be discussed later, the finite-element mesh is deformed while iterating to find the free surface so that local average elements only approximately match the finite elements in area. Thus, for a given realization, the spatially "fixed" permeability field values are assigned to individual elements according to where the element is located on each free-surface iteration.

Both permeability and correlation length are assumed to be isotropic in this study. Although layered construction of an earth dam may lead to some anisotropy relating to the correlation length and permeability, this is not thought to be a major feature of the reconstituted soils typically used in earth dams. In contrast, however, natural soil deposits can exhibit quite distinct layering and stratification in which anisotropy cannot be ignored. Note that random fields with ellipsoidally anisotropic correlation functions, for example, of the form

$$\rho(\mathbf{t}) = \exp\left\{-2\sqrt{\frac{\tau_1^2}{\theta_1^2} + \frac{\tau_2^2}{\theta_2^2}}\right\} = \exp\left\{-\frac{2}{\theta_1}\sqrt{\tau_1^2 + \frac{\tau_2^2\theta_1^2}{\theta_2^2}}\right\} \tag{9.4}$$

where θ_1 and θ_2 are the directional correlation lengths, can always be transformed into isotropic forms by suitably scaling the coordinate axes. In this example, by using $x_2' = x_2(\theta_1/\theta_2)$, where x_2 is the space coordinate measured in the same direction as τ_2, Eq. 9.4 becomes isotropic with scale θ_1 and lag $\tau = \sqrt{\tau_1^2 + (\tau_2')^2}$, with τ_2' measured with respect to x_2'. Thus, if anisotropy is significant, such a transformation can be performed to allow the use of the results presented here, bearing in mind that it is the transformed geometry which must be used in the sequel.

The model itself is two dimensional, which is equivalent to assuming that the streamlines remain in the plane of analysis. This will occur if the dam ends are impervious and if the correlation length in the out-of-plane direction is infinite (implying that soil properties are constant in the out-of-plane direction). Clearly the latter condition will be false; however, a full three-dimensional analysis is beyond the scope of the present study. As was found for flow through bounded soil masses (e.g., around cutoff walls, see Section 8.7), it is believed that the two-dimensional analysis will still be reasonably accurate.

9.1.1 Random Finite-Element Method

For a given permeability field realization, the free-surface location and flow through the earth dam is computed using a two-dimensional iterative finite-element model derived from Smith and Griffiths (2004), program 7.3. The elements are four-node quadrilaterals and the mesh is deformed on each iteration until the total head along the free surface approaches its elevation head above a predefined horizontal datum. Convergence is obtained when the maximum relative change in the free-surface elevation at the surface nodes becomes less than 0.005. Figure 9.2 illustrates two possible free-surface profiles corresponding to different permeability field realizations with the same input statistics.

When the downstream face of the dam is inclined, the free surface tends to become tangent to the face, resulting in finite elements which can be severely skewed, leading in turn to inaccurate numerical results. This difficulty is overcome by proportionately shifting the mesh as the free surface descends to get a finer mesh near the top of the downstream face [the reader is referred to Fenton and Griffiths (1997a) for details]. Because of the mesh deformation taking place in each iteration along with the need to maintain the permeability realization as spatially fixed, the permeabilities assigned to each element are obtained by mapping the element centroids to the permeability field using Eq. 9.3. Thus the local average properties of the random field are only approximately reflected in the final mesh; some of the smaller elements may share the same permeability if adjacent elements fit inside a cell of the random field. This is believed to be an acceptable approximation, leading to only minor errors in the overall stochastic response of the system, as discussed next.

Preliminary tests performed for the study indicated that the response statistics only begin to show significant error when fewer than 5 elements were used in each of the two coordinate directions. In the current model 16 elements were used in each direction (256 elements in total). This ensures reasonable accuracy even in the event that some elements are mapped to the same random-field element. Because the elements are changing size during the iterative process, implying that the local average properties of the random-field generator are only approximately preserved in the final mesh, there is little advantage to selecting a local average random-field generator over a point process generator such as the FFT or TBM. The LAS algorithm was selected for use here primarily because it avoids the possible presence of artifacts (in the form of "streaks") in individual realizations arising in TBM realizations and the symmetric covariance structure inherent in the FFT algorithm (Fenton, 1994, see also Section 6.4). The LAS method is also much easier to use than the FFT approach.

Flow rate and drawdown statistics for the earth dam are evaluated over a range of the statistical parameters of K. Specifically, the estimated mean and standard deviation of the total flow rate, m_Q and s_Q, and of the drawdown, m_Y and s_Y, are computed for $\sigma_K/\mu_K = \{0.1, 0.5, 1.0, 2.0, 4.0, 8.0\}$ and $\theta_{\ln K} = \{0.1, 0.5, 1.0, 2.0, 4.0, 8.0\}$ by averaging over 1000 realizations for each (resulting in $6 \times 6 \times 1000 = 36,000$ realizations in total for each dam considered). An additional run using $\theta_{\ln K} = 16$ was performed for dam 1 to verify trends at large correlation lengths. The mean permeability μ_K is held fixed at 1.0. The drawdown elevations Y are normalized by expressing them as a fraction of the overall (original) dam height.

9.1.2 Simulation Results

On the basis of 1000 realizations, a frequency density plot of flow rates and drawdowns can be produced for each set of parameters of $K(\mathbf{x})$. Typical histograms are shown in Figure 9.3, with fitted lognormal and beta distributions superimposed on the flow rate and normalized drawdown histograms, respectively. The parameters of the fitted distributions are estimated by the method of moments from

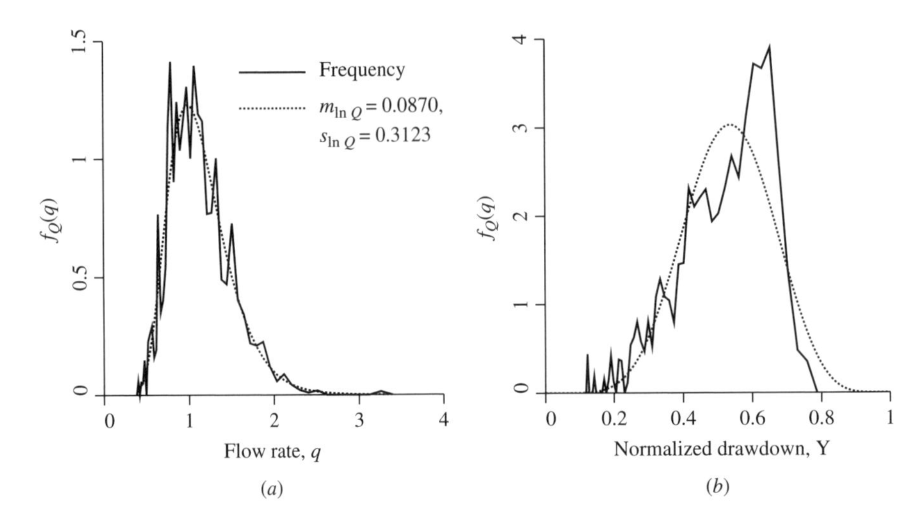

Figure 9.3 Frequency density plots of (*a*) flow rate and (*b*) normalized drawdown for dam 1 with $\sigma_K/\mu_K = 1$ and $\theta_{\ln K} = 1$.

the ensemble of realizations, which constitute a set of independent samples, using unbiased sample moments. For the lognormal distribution the estimators are

$$m_{\ln Q} = \frac{1}{n}\sum_{i=1}^{n} \ln Q_i \tag{9.5a}$$

$$s_{\ln Q}^2 = \frac{1}{n-1}\sum_{i=1}^{n}(\ln Q_i - m_{\ln Q})^2 \tag{9.5b}$$

where Q_i is the total flow rate through the ith realization. For $n = 1000$ realizations, the coefficients of variation of the estimators (assuming $\ln Q$ is approximately normally distributed) $m_{\ln Q}$ and $s_{\ln Q}^2$ are $0.032\sigma_{\ln K}/\mu_{\ln K}$ and 0.045, respectively.

It can be seen that the lognormal distribution fits the flow rate histogram reasonably well, as is typical; 60% of the cases considered (based on 1000 realizations each) satisfied the chi-square goodness-of-fit test at the 5% significance level. A review of the histograms corresponding to those cases not satisfying the test indicates that the lognormal distribution is still a reasonable approximation but the chi-square test is quite sensitive. For example, the histogram shown in Figure 9.3*a* fails the chi-square test at all significance levels down to 0.15%. From the point of view of probability estimates associated with flow rates, it is deemed appropriate therefore to assume that flow rates are well approximated by the lognormal distribution and all subsequent statistics of flow rates are determined from the fitted lognormal distribution.

Since the normalized drawdown is bound between 0 and 1, it was felt that perhaps a beta distribution might be an appropriate fit. Unfortunately the fit, obtained by the method of moments using unbiased sample moments of the raw data, was typically quite poor; the histogram shown in Figure 9.3*b* has sample mean and standard deviation 0.533 and 0.125, respectively, giving beta distribution parameters $\alpha = 7.91$ and $\beta = 6.93$. The fitted distribution fails to capture the skewness and upper tail behavior. Nevertheless, the drawdown mean and variance can be estimated reasonably accurately even though its actual distribution is unknown. For 1000 realizations, the estimators of the mean and variance of normalized drawdown have coefficients of variation of approximately $0.032s_Y/m_Y$ and 0.045 using a normal distribution approximation.

The estimated mean and variance of the total log-flow rate, denoted here as $m_{\ln Q}$ and $s_{\ln Q}^2$, respectively, are shown in Figure 9.4 as a function of the variance of log-permeability, $\sigma_{\ln K}^2 = \ln(1 + \sigma_K^2/\mu_K^2)$, and the correlation length, $\theta_{\ln K}$. These results are for dam 1 and are obtained from Eqs. 9.5. Clearly the mean log-flow rate tends to decrease from the deterministic value of $\ln(Q_{\mu_K}) = \ln(1.51) = 0.41$ (obtained by assuming $K = \mu_K = 1.0$ everywhere) as the permeability variance increases.

In terms of the actual flow rates which are assumed to be lognormally distributed, the transformations

$$m_Q = \exp\{m_{\ln Q} + \tfrac{1}{2}s_{\ln Q}^2\} \tag{9.6a}$$

$$s_Q^2 = m_Q^2(\exp\{s_{\ln Q}^2\} - 1) \tag{9.6b}$$

can be used to produce the mean flow rate plot shown in Figure 9.5. The apparent increase in variability of the estimators (see, e.g., the $\theta_{\ln K} = 16$ case) is due in part to the reduced vertical range but also partly to errors in the fit of the histogram to the lognormal distribution and the resulting differences between the raw data estimators and the log-data estimators.

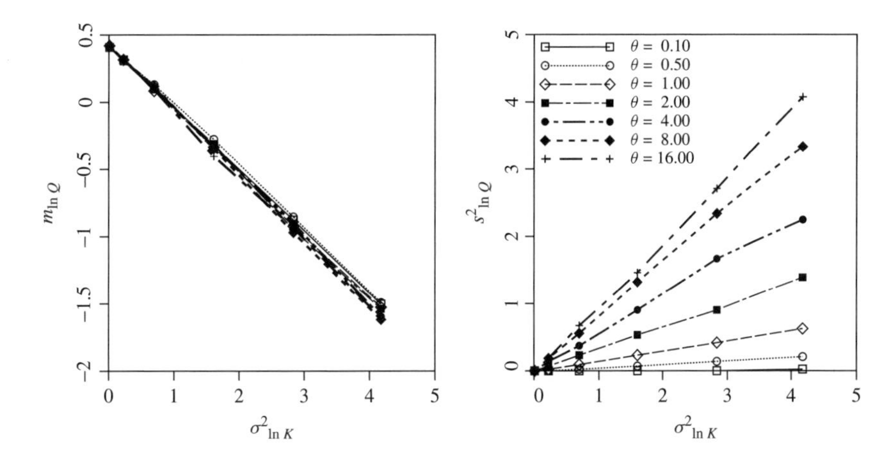

Figure 9.4 Estimated mean and standard deviation of log-flow rate through dam 1.

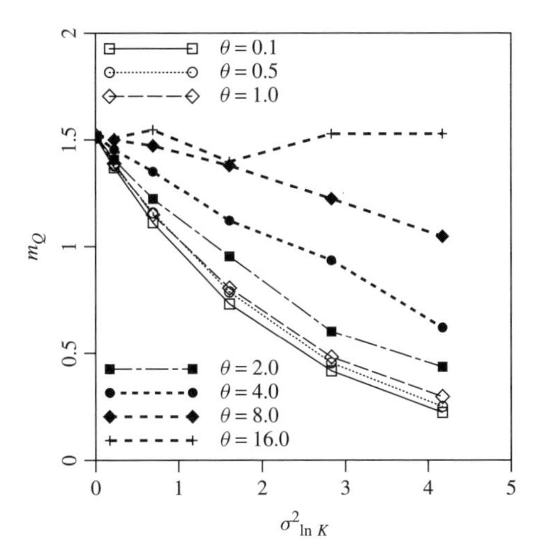

Figure 9.5 Estimated mean flow rate through dam 1.

It can be seen that the mean flow rate also reduces from the deterministic value $Q_{\mu_K} = 1.51$ with increasing $\sigma^2_{\ln K}$. The reduction is more pronounced for small correlation lengths but virtually disappears for correlation lengths considerably larger than the dam itself. It is known that as the correlation length becomes negligible compared to the size of the dam, the effective permeability approaches the geometric mean $K_G = \mu_K \exp\left\{-\frac{1}{2}\sigma^2_{\ln K}\right\}$ (Dagan, 1989), which for fixed μ_K illustrates the reduction in flow rate. Intuitively, one can think of this reduction in mean flow rate by first considering one-dimensional flow down a "pipe"—the total flow rate down the pipe is heavily dependent on the minimum permeability encountered along the way. As the variance of the permeability increases, and in the case of small correlation lengths, the chances of getting a small-permeability or "blocked" pipe also increases, resulting in a decreased mean flow rate. Similar, albeit less extreme, arguments can be made in the two-dimensional case, leading to the observed and predicted reduction in mean total flow rate as $\sigma^2_{\ln K}$ increases. As the correlation length increases to infinity, the mean flow rate m_Q becomes equal to Q_{μ_K} independent of $\sigma^2_{\ln K}$, as illustrated by the $\theta_{\ln K} = 16$ case in Figure 9.5. In this case, the random field is relatively uniform, and although individual realizations show considerable variability in total flow rate, the mean approaches the value predicted by $K = \mu_K$.

For very short correlation lengths, the variance of log-flow rate is very small, as evidenced by $s_{\ln Q}$ in Figure 9.4, increasing as the correlation length and $\sigma^2_{\ln K}$ increase. In the limit as $\theta_{\ln K} \to \infty$, it can be shown that $\sigma^2_{\ln Q} = \sigma^2_{\ln K}$ and $\mu_{\ln Q} = \ln(Q_{\mu_K}) - \sigma^2_{\ln K}/2$, trends which are seen in Figure 9.4 for $\theta_{\ln K} = 16$. Similar results were found for the other dam geometries.

Figure 9.6 shows the estimated mean and standard deviation of the normalized drawdown, m_Y and s_Y, respectively, again for the earth dam 1 shown in Figure 9.1. It can be seen that although some clear patterns exist for the mean drawdown with respect to the correlation length and $\sigma^2_{\ln K}$, the magnitude of the mean drawdown is little affected by these parameters and remains close to $Y = 0.58$ of the total dam height obtained in the deterministic case with $K = \mu_K = 1.0$. Note that for $\theta_{\ln K} = 4$, $\sigma^2_{\ln K} = 2.83$, the standard deviation of Y is estimated to be about 0.21, giving the standard deviation of the estimator m_Y to be about 0.0066. The 90% confidence interval on μ_Y is thus approximately [0.51, 0.53] for $m_Y = 0.52$. This observation easily explains the rather erratic behavior of m_Y observed in Figure 9.6.

The variability of the drawdown, estimated by s_Y, is significantly affected by $\theta_{\ln K}$ and $\sigma^2_{\ln K}$. For small correlation lengths relative to the dam size, the drawdown shows little variability even for high-permeability variance. This

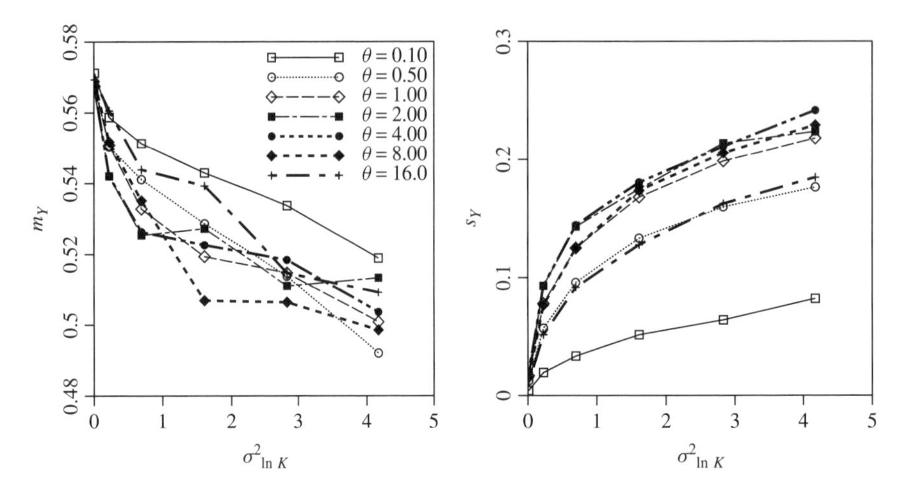

Figure 9.6 Estimated mean and standard deviation of normalized free-surface drawdown for dam 1.

suggests that, under these conditions, using a fixed free surface to model the dam may be acceptable. For larger correlation lengths, the drawdown shows more variability and the stochastic nature of the free-surface location should be included in an accurate analysis. Although it may seem that the drawdown variability continues to increase with increasing correlation length, it is known that this is not the case. There will be a worst-case scale at which the drawdown variability is maximized; at even larger scales, the drawdown variability will decrease since in the limit as $\theta_{\ln K} \to \infty$ the drawdown becomes equal to the deterministic result $Y = 0.58$ independent of the actual permeability. In other words, the drawdown becomes different from the deterministic result only in the presence of intermediate correlation lengths in the permeability field. To investigate this phenomenon, dam 1 was analyzed for the additional scale $\theta_{\ln K} = 16$ m, much greater than the earth dam dimension of around 3–4 m. It appears from Figure 9.6 that the drawdown variance is maximized for $\theta_{\ln K} = 4$ m, that is, for $\theta_{\ln K}$ of the order of the earth dam size.

9.1.3 Empirical Estimation of Flow Rate Statistics

For preliminary design reliability estimates, it is worth investigating approximate or empirical methods of estimating the mean and variance of flow through an earth dam. In the following, a semiempirical approach is adopted with the understanding that its accuracy in estimating flow statistics for problems other than those considered here is currently unknown. In practice the following results should be viewed as providing rough estimates, and more accurate estimates must currently be obtained via simulation.

The approach starts by noting that the mean $\mu_{\ln Q}$ and variance $\sigma^2_{\ln Q}$ of log-flow through a square two-dimensional domain with impervious top and bottom faces and constant head along both sides is accurately predicted by (on the basis of simulation studies, see Chapter 8)

$$\mu_{\ln Q} = \ln(Q_{\mu_K}) - \tfrac{1}{2}\sigma^2_{\ln K} \tag{9.7a}$$

$$\sigma^2_{\ln Q} = \sigma^2_{\ln K}\gamma(D, D) \tag{9.7b}$$

with equivalent results in real space (assuming that Q follows a lognormal distribution) given by

$$\mu_Q = Q_{\mu_k} \exp\{-\tfrac{1}{2}\sigma^2_{\ln K}(1 - \gamma(D, D))\} \tag{9.8a}$$

$$\sigma^2_Q = Q^2_{\mu_K} \exp\{-\sigma^2_{\ln K}(1 - \gamma(D, D))\} \times \left[\exp\{\sigma^2_{\ln K}\gamma(D, D)\} - 1\right] \tag{9.8b}$$

in which Q_{μ_K} is the flow rate obtained in a deterministic analysis of flow through a domain having permeability $K(\mathbf{x}) = \mu_K$ everywhere, D is the square root of the domain area (i.e., side length), and $\sigma^2_{\ln K}\gamma(D, D)$ is the variance of a local average of the random field $\ln(K)$ over the domain $D \times D$. In the event that $D >> \theta_{\ln K}$, so that $\gamma(D, D) \simeq 0$, Eqs. 9.7 become equal to that predicted using the geometric mean of permeability, that is, to the *effective* permeability defined by Dagan (1989) and Gelhar (1993). In the more general case, Rubin and Gómez-Hernández (1990) obtained similar results derived using a perturbation approach valid only when both $\gamma(D, D)$ and $\sigma^2_{\ln K}\gamma(D, D)$ are small. For values of $\gamma(D, D)$ and $\sigma^2_{\ln K}$ typical of this study, the perturbation approach can be considerably in error.

The parameter D characterizes the size of the averaging region. In the case of Eqs. 9.7, D refers to the side length of a two-dimensional *square* flow regime studied in Section 8.4; thus the flow is affected by the average permeability in a domain of size $D \times D$. The mean and variance of flow through such a two-dimensional domain is expected to depend in some way on the reduction in variance due to averaging over the domain, leading to the results given by Eqs. 9.7 and 9.8.

The shapes of the functions given by Eqs. 9.7 are very similar to those seen in Figure 9.4, suggesting that these functions can be used to predict the mean and variance of log-flow through an earth dam if an effective value of D can be found to characterize the flow through the dam. Thus, the task is to find the dimension D_{eff} of an equivalent two-dimensional square domain whose log-flow rate statistics (at least the mean and variance) are approximately the same as observed in the earth dam. One possible estimate of this effective dimension is

$$D_{\text{eff}} = \sqrt{\frac{A_{\text{wet}}}{\overline{Q}}} \tag{9.9}$$

where A_{wet} is the earth dam area (in-plane) under the free surface through which the flow takes place, that is, excluding the unsaturated soil above the free surface, and $\overline{Q}$ is the nondimensionalized flow rate through the earth dam obtained with $K(\mathbf{x}) = \mu_K$ everywhere, that is,

$$\overline{Q} = \frac{Q_{\mu_K}}{\mu_K H_{\text{eff}}\, z} \tag{9.10}$$

where H_{eff} is the effective fluid head and z is the out-of-plane thickness of the dam, which, for a two-dimensional analysis, is 1.0. Although it would appear reasonable to take H_{eff} as the average hydraulic head over the upstream face of the dam, it turns out to be better to take $H_{\text{eff}} = y_h/3$, the elevation of the centroid of the pressure distribution, where y_h is the upstream water head (and the overall height of the dam). Substitution of Eq. 9.10 into Eq. 9.9 along with this choice of H_{eff} gives

$$D_{\text{eff}} = \sqrt{\frac{A_{\text{wet}}\, \mu_K\, y_h\, z}{3Q_{\mu_K}}} \tag{9.11}$$

This equation can then be used in Eqs. 9.8 to estimate the desired flow rate statistics.

Figure 9.7 illustrates the agreement between the mean and standard deviation derived via simulation and predicted using Eq. 9.11 in Eqs. 9.7 for all four earth dam geometries shown in Figure 9.1. The dotted lines in Figure 9.7 denote the $\pm10\%$ relative error bounds. It can be seen that most of the predicted statistics match those obtained from simulation quite well in terms of absolute errors. A study of relative errors shows that 90% of the cases studied had relative errors less than 20% for the prediction of both the mean and standard deviation. There is no particular bias in the errors with respect to over- versus underestimation.

Admittedly, the effective dimension approach cannot properly reflect the correlation structure of the actual dam through a square-domain approximation—if the dam width is significantly greater than the dam height (as in dam 4), then the correlation between permeabilities at the top and bottom will generally be higher than from left edge to right edge. An "equivalent" square domain will not capture this. Thus, the effective dimension approach adopted here is expected to perform less well for long narrow flow regimes combined with correlation lengths approaching and exceeding the size of the dam. In fact, for the prediction of the mean, the simulation results belie this statement in that dam 4 performed much better than dams 1, 2, or 3. For the prediction of the standard deviation, dam 4 performed the least well, perhaps as expected. Nevertheless, overall the results are very encouraging. Thus, the effective dimension approach can be seen to give reasonable estimates of the mean and variance of log-flow rates through the dam in

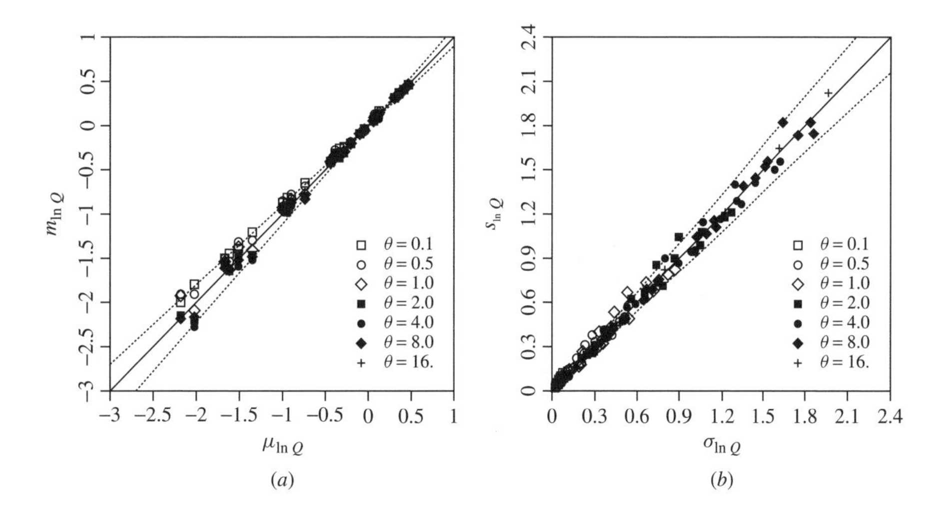

Figure 9.7 Comparison of (*a*) mean and (*b*) standard deviation statistics derived via simulation and as predicted by Eqs. 9.7.

most cases. To compute these estimates, the following steps must be performed:

1. Perform a single finite-element analysis using $K(\mathbf{x}) = \mu_K$ throughout the earth dam to determine Q_{μ_K} and the area of the dam below the free surface, A_{wet}.
2. Estimate the effective dam dimension using Eq. 9.11.
3. Compute the local average variance reduction factor $\gamma(D_{\text{eff}}, D_{\text{eff}})$ corresponding to the random field used to model the log-permeability field (see Section 3.4).
4. Estimate the mean and variance of log-flow through the dam using Eqs. 9.7. These values can be used directly in the lognormal distribution to compute probability estimates.

9.1.4 Summary

Although only a limited set of earth dam geometries are considered in this section, it should be noted that the stochastic response of a dam is dependent only on the ratio of the correlation length to the dam dimensions for given dam shape and type of random field. For example, consider two earth dams with the same overall shapes and permeability statistics μ_K and σ_K^2. If the second of the two dams is of twice the size and has twice the correlation length as the first, then the second will have twice the flow rate mean and standard deviation as the first, and they will have identical normalized drawdown statistics. Similarly, the results shown here are easily scaled for different values of μ_K; the important parameter as far as the stochastic response is concerned is the coefficient of variation σ_K/μ_K [or equivalently $\sigma_{\ln K}^2 = \ln(1 + \sigma_K^2/\mu_K^2)$]. These properties can be used to confidently employ the results of this section on earth dams of arbitrary dimension and mean permeability.

For correlation lengths which are *small* relative to the size of the dam, the simulation results indicate that:

1. The flow through the dam is well represented using only the estimated mean flow rate m_Q—the flow rate variance is small.
2. The mean flow rate falls rapidly as $\sigma_{\ln K}^2$ increases.
3. The free-surface profile will be relatively static and can be estimated confidently from a deterministic analysis. The simulation results imply that for both small and very large correlation lengths (relative to the dam size) the drawdown variability is small and the Monte Carlo analysis could proceed using a fixed free surface found from the deterministic analysis, avoiding the need to iterate on each realization.

As the correlation length becomes larger, the mean flow rate does not fall as rapidly with increasing $\sigma_{\ln K}^2$ while the variability of the flow rate from one realization to the next increases significantly. The variability in the free-surface location reaches a maximum for intermediate correlation lengths, apparently for scales of the order of the earth dam size.

The computation of estimates of the mean and variance of flow rates through an earth dam using Eqs. 9.7 allows designers and planners to avoid full-scale Monte Carlo simulations and can be used to approximately address issues regarding earth dam flow rate probabilities via a lognormal distribution. If more accurate estimates of these quantities are desired, particularly for correlation lengths approaching or greater than the dam size, then a full scale Monte Carlo simulation is currently the only viable choice (see RDAM2D at `http://www.engmath.dal.ca/rfem`). In that the mean, variance, and correlation length parameters of the permeability field, as estimated from the field, are themselves quite uncertain, the approximate estimate of the flow rate statistics may be quite appropriate in any case.

9.2 EXTREME HYDRAULIC GRADIENT STATISTICS

Earth dams fail from a variety of causes. Some, such as earthquake or overtopping, may be probabilistically quantifiable through statistical analysis. Others, such as internal erosion, are mechanically complex, depending on internal hydraulic gradients, soil gradation, settlement fracturing, drain and filter performance, and so on. In this section, the focus is on evaluating how internal hydraulic gradients are affected by spatial variability in soil permeability. The scope is limited to the study of simple but reasonable cases. Variability in internal hydraulic gradients is compared to traditional "deterministic" analyses in which the soil permeability is assumed constant throughout the earth dam cross section (Fenton and Griffiths, 1997b).

In a study of the internal stability of granular filters, Kenney and Lau (1985) state that grading stability depends on three factors: (1) size distribution of particles, (2) porosity, and (3) severity of seepage and vibration. Most soil stability tests proceed by subjecting soil samples to prescribed gradients (or fluxes or pressures) which are believed to be conservative, that is, which are considerably higher than believed "normal," as judged by current practice. For example, Kenney and Lau (1985) performed their soil tests using unit fluxes ranging from 0.48 to 1.67 cm/s. Lafleur et al. (1989) used gradients ranging from 2.5 to 8.0, while Sherard et al. (1984a, b) employed pressures ranging from 0.5 to 6 kg/cm^2 (corresponding to gradients up to 2000). Molenkamp et al. (1979) investigate the performance of filters under cyclically reversing hydraulic gradients. In all cases, the tests are performed under what are believed to be conservative conditions.

By considering the soil permeability to be a spatially random field with reasonable statistics, it is instructive to investigate just how variable the gradient (or flux or potential) can be at critical points in an earth dam. In this way, the extent of conservatism in the aforementioned tests can be assessed. Essentially this section addresses the question of whether uncertainty about the permeability field should be incorporated into the design process. Or put another way, are deterministic design procedures sufficiently safe as they stand? For example, if the internal gradient at a specific location in the dam has a reasonably high probability of exceeding the gradients under which soil stability tests were performed, then perhaps the use of the test results to form design criteria needs to be reassessed.

The study will concentrate on the two earth dam cross sections shown in Figure 9.8 with drains cross-hatched. The steeper sloped dam will be referred to here as dam A and the shallower cross section as dam B. The overall dimensions of the dams were arbitrarily selected since the results are scalable (Section 9.2.2), only the overall shape being of importance.

Sections 9.2.1 and 9.2.2 discuss the stochastic model, comprising random-field and finite-element models, used to represent the earth dam for the two geometries considered. In Section 9.2.3 the issue of how a simple internal drain, designed to avoid having the free surface exit on the downstream face of the dam above the drain, performs in the presence of spatially varying permeability. Successful drain performance is assumed to occur if the free surface remains contained within the drain at the downstream face with acceptably high probability.

Section 9.2.4 looks at the mean and standard deviation of internal hydraulic gradients. Gradients are defined here

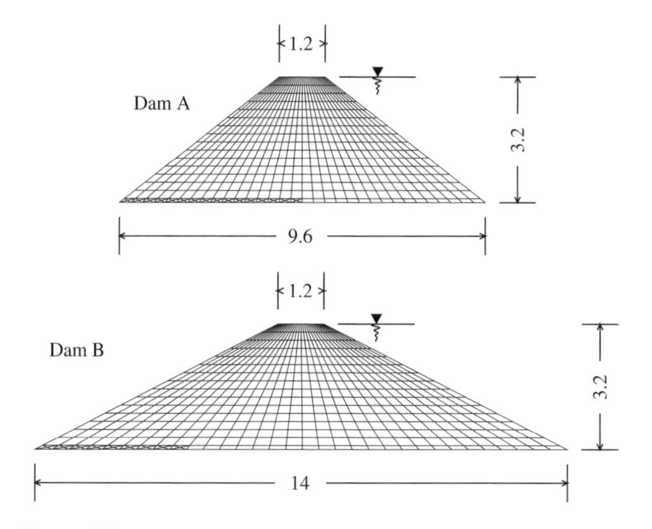

Figure 9.8 Two earth dam geometries considered in stochastic analysis.

strictly in magnitude—direction is ignored. Again the dams are considered to have a drain in place. Regions where the gradients are highest are identified and the distribution of these maximum gradients established via simulation. All simulation results were performed using the program RDAM2D available at `http://www.engmath.dal.ca/rfem`.

9.2.1 Stochastic Model

The stochastic model used to represent flow through an earth dam with free surface is an extension of the model developed in Section 9.1. When the permeability is viewed as a spatially random field, the equations governing the flow become stochastic. The random field characterizes uncertainty about the permeability at all points in the dam and from dam to dam. The flow through the dam will thus also be uncertain, and this uncertainty can be expressed by considering the probability distribution of various quantities related to flow.

The permeability $K(\mathbf{x})$ is assumed to follow a lognormal distribution, consistent with the findings of Freeze (1975), Hoeksema and Kitanidis (1985), and Sudicky (1986) and with the work of Griffiths and Fenton (1993), with mean μ_K and variance σ_K^2. Thus $\ln K$ is normally distributed (Gaussian) with mean $\mu_{\ln K}$ and variance $\sigma_{\ln K}^2$; see Eqs. 1.175 and 1.176.

Since $K(\mathbf{x})$ is a spatially varying random field, there will also be a degree of correlation between $K(\mathbf{x})$ and $K(\mathbf{x}')$, where $\mathbf{x}$ and $\mathbf{x}'$ are any two points in the field. Mathematically this concept is captured through the use of a spatial correlation function, which, in this study, is an exponentially decaying function of separation distance $\mathbf{t} = \mathbf{x} - \mathbf{x}'$ (this is a Markov model, see Section 3.7.10.2),

$$\rho(\mathbf{t}) = e^{-2|\mathbf{t}|/\theta_{\ln K}} \tag{9.12}$$

where $\theta_{\ln K}$ is the correlation length.

Simulation of the soil permeability field proceeds in two steps: first an underlying Gaussian random field $G(\mathbf{x})$ is generated with mean zero, unit variance, and spatial correlation function (Eq. 9.12) using the LAS method (Section 6.4.6). Next, since the permeability is assumed to be lognormally distributed, values of K_i, where i denotes the ith element, are obtained through the transformation

$$K_i = \exp\{\mu_{\ln K} + \sigma_{\ln K} G_i\} \tag{9.13}$$

where G_i is the local average of $G(\mathbf{x})$ over the domain of the ith element. The finite-element mesh is deformed while iterating to find the free surface so that local average elements only approximately match the finite elements in area. For a given realization, the spatially fixed permeability field values are assigned to individual elements according to where the element is located on each free-surface iteration.

Both permeability and correlation length are assumed to be isotropic in this study. Although layered construction of an earth dam may lead to some anisotropy relating to the correlation length and permeability, the isotropic model is employed for simplicity. In addition, the model itself is two dimensional, which is equivalent to assuming that streamlines remain in the plane of analysis. This will occur if the dam ends are impervious and if the correlation length in the out-of-plane direction is infinite (implying that soil properties are constant in the out-of-plane direction). Clearly the latter condition will be false; however, a full three-dimensional analysis is beyond the scope of the present study. It is believed that the two-dimensional analysis will still yield valuable insights to the problem, as indicated in Chapter 8.

Statistics of the output quantities of interest are obtained by Monte Carlo simulation employing 5000 realizations of the soil permeability field for each cross section considered. With this number of independent realizations, estimates of the mean and standard deviations of output quantities of interest have themselves standard deviations of approximately

$$s_{m_X} \simeq \sqrt{\frac{1}{n}} s_X = 0.014 s_X \tag{9.14a}$$

$$s_{s_X^2} \simeq \sqrt{\frac{2}{n-1}} s_X^2 = 0.02 s_X^2 \tag{9.14b}$$

where X is the output quantity of interest, m_X is its estimated mean, s_X is its estimated standard deviation, and s_{m_X} and $s_{s_X^2}$ are the estimated standard deviations of the estimators m_X and s_X, respectively (see Section 6.6).

Many of the statistical quantities discussed in the following are compared to the so-called deterministic case. The deterministic case corresponds to the traditional analysis approach in which the permeability is taken to be constant throughout the dam; here the deterministic permeability is equal to $\mu_K = 1.0$. For all stochastic analyses, the permeability coefficient of variation v_K is taken to be 0.50 and the correlation length is taken to be 1.0 (having the same units as lengths in Figures 9.8–9.10). These numbers are not excessive and are believed typical for a well-controlled earth dam fill.

9.2.2 Random Finite-Element Method

For a given permeability field realization, the free-surface location and flow through the earth dam are computed using a two-dimensional iterative finite-element model derived from Smith and Griffiths (2004), program 7.3. The elements are four-node quadrilaterals and the mesh is deformed on each iteration until the total head along the free surface approaches its elevation head above a predefined horizontal datum. Convergence is obtained when the maximum relative change in the free-surface elevation at the surface nodes becomes less than 0.005. Figure 9.9 illustrates two possible free-surface profiles for dam B corresponding to different permeability field realizations with the same input statistics. Lighter regions in the figure correspond to higher permeabilities. Along the base of the dam, from the downstream face, a drain is provided with fixed (nonrandom) permeability of 120 times the mean dam permeability. This permeability was selected to ensure that the free surface did not exit the downstream face above the drain for either cross section under deterministic conditions (constant permeability of $\mu_K = 1$ everywhere). It is assumed that this would be ensured in the normal course of a design. Notice that the drain itself is only approximately represented along its upper boundary because the elements are deforming during the iterations. This leads to some randomness in the drain behavior which, although not strictly quantifiable, may actually be quite realistic.

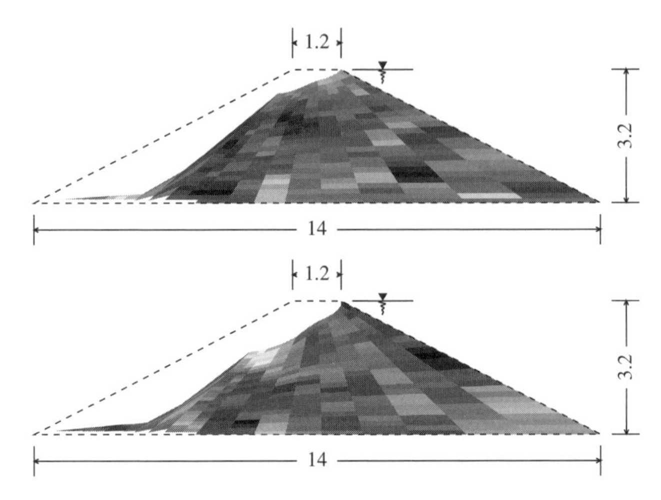

Figure 9.9 Two free-surface realizations for dam B.

In both cross sections, the free surface is seen to fall into the drain, although not as fast as classical free-surface profiles with drains would suggest. The difference here is that the drain has finite permeability; this leads to some backpressure causing the free surface to remain above it over some length. In that drains, which also act as filters, will not be infinitely permeable, these free surfaces are believed to be representative. Since the finite-element mesh is moving during the iterative analysis, the gradients must be calculated at fixed points rather than at the nodal locations. This means that gradients must be interpolated using the finite-element shape functions once the element enclosing an arbitrary fixed point is identified. Thus, two meshes are carried throughout the analysis—one fixed and one deforming according to the free-surface profile.

For each realization, the computer program computes the following quantities:

1. The free surface profile
2. The gradient, unit flux, and head at each point on a fixed-lattice discretization of the dam cross section. Points which lie above the free surface for this particular realization are assigned a gradient of zero. Gradients are computed as

$$g = \sqrt{\left(\frac{\partial \phi}{\partial x_1}\right)^2 + \left(\frac{\partial \phi}{\partial x_2}\right)^2}$$

   which is the absolute magnitude of the gradient vector. The vector direction is ignored.
3. Total flow rate through the cross section

All quantities form part of a statistical analysis by suitably averaging over the ensemble of realizations.

In Figure 9.8, the drains are denoted by cross-hatching and can be seen to lie along the downstream dam base. The dams are discretized into 32 × 16 elements and the drain has thickness of 0.1 in the original discretization. In dam A, the drain extends to the midpoint, while in dam B the drain has length 4, both measured from the downstream dam corner. The original, or undeformed, discretization shown in Figure 9.8 is also taken to be the fixed discretization over which gradients are obtained.

Elements falling within the domain of the drain during the analysis are assigned a permeability of $120\mu_K$, the remainder assigned random permeabilities as discussed in the previous section. As also mentioned previously, the elements in the deformed mesh are not rectangular so the drain is only approximated. Some elements lying above the drain have portions extending into the drain and vice-versa. The permeability-mapping algorithm has been devised to ensure that the drain itself is never blocked by portions of a low-permeability element extending into the drain. Given the uncertainty related to the infiltration of core fines into the drain, this model is deemed to be a reasonable approximation.

The overall dimensions of the dam and the assumed permeability statistics are scalable; that is, a dam having 10 times the dimensions of dam A or B will have 10 times the total flow rate, the same free-surface profile, and the same gradients if both have the same (space-scaled) permeability field. Output statistics are preserved if the prescribed correlation length is scaled by the same amount as the dam itself and the permeability mean and variance are unchanged. Regarding changes in the mean permeability, if the coefficient of variation ($v_K = \sigma_K/\mu_K$) remains fixed, then scaling μ_K results in a linear change in the flow rate (with unchanged v_Q), unchanged gradient, and free-surface profile statistics. The unit flux scales linearly with the permeability but is unaffected by changes in the dam dimension. The potential field scales linearly with the dam dimension but is unaffected by the permeability field (as long as the latter also scales with the dam dimension).

9.2.3 Downstream Free-Surface Exit Elevation

The drain is commonly provided to ensure that the free surface does not exit on the downstream face of the dam, resulting in its erosion. The lowering of the free surface by this means will be referred to herein as "drawdown." As long as the drawdown results in an exit point within the drain itself, the drain can be considered to be performing acceptably. Figure 9.10 shows the deterministic free-surface profiles for the two geometries considered. In both cases the free surface descends into the drain prior to reaching the downstream face.

Figure 9.11 shows histograms of free-surface exit elevations Y which are normalized with respect to the earth dam height. The dashed line is a normal distribution fit to the data, with parameters given in the line key. For the cases considered, the normalized dimension of the top of the drain is $0.1/3.2 = 0.031$, so there appears to be little danger of the free surface exiting above the drain when the soil is spatially variable, at least under the moderate levels of variability considered here. It is interesting to note that the normalized free-surface exit elevations obtained in the deterministic case (uniform permeability) are 0.024 for dam A and 0.015 for dam B. The mean values obtained from the simulation, as indicated in Figure 9.11, are 0.012 for dam A and 0.009 for dam B. Thus, the net effect of soil variability is to reduce the exit point elevation. Perhaps the major reason for this reduction arises from the length of the drain; in the presence of spatial variability there exists a higher

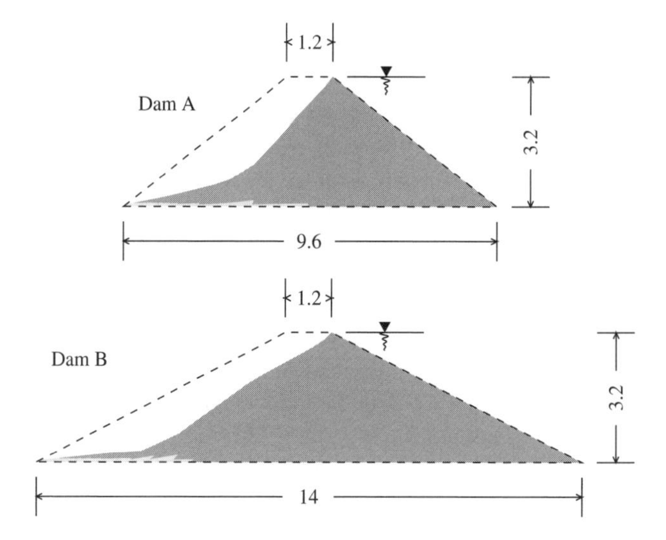

Figure 9.10 Deterministic free-surface profiles for two earth dam geometries considered.

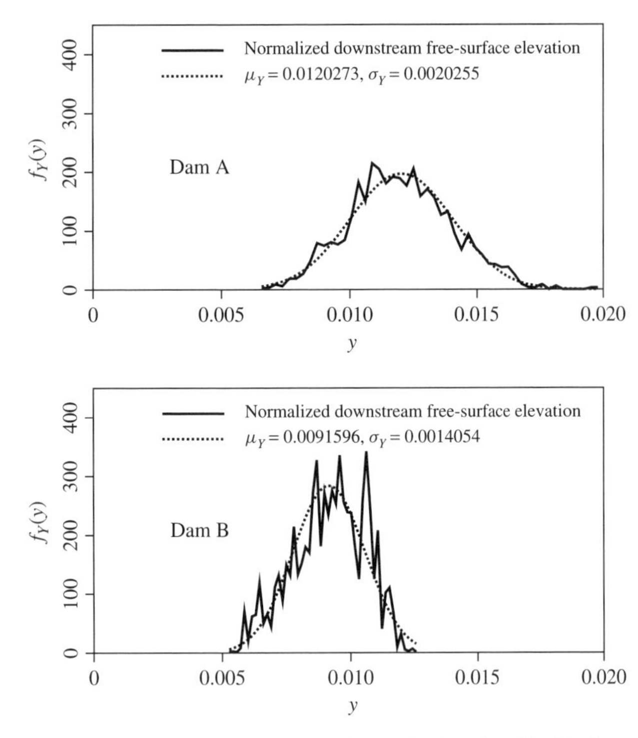

Figure 9.11 Normalized free-surface exit elevation distributions.

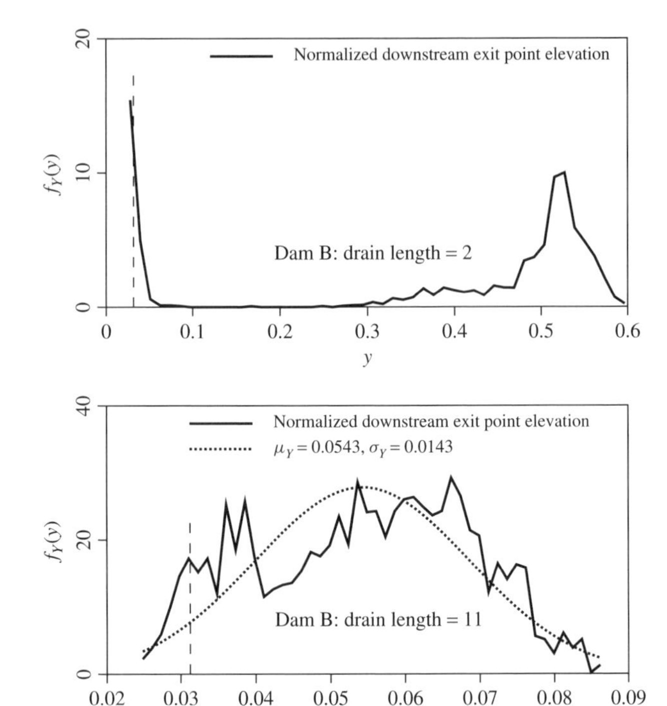

Figure 9.12 Effect of drain length on free-surface exit elevation.

probability that somewhere further along the drain the core permeability will be lower, tending to drive the flow into the drain. The deterministic flow rates (normalized with respect to μ_K) are 1.94 for dam A dam and 1.03 for dam B. The corresponding mean flow rates determined by simulation are somewhat reduced at 1.85 and 0.96, illustrating again that spatial variability tends to introduce blockages somewhere along the flow path. Since the coefficient of variation $v_Y = \sigma_Y/\mu_Y$ of the height of the free-surface exit point is only around 15–17% for both cross sections, permeability spatial variability primarily serves to lower the exit point height, reducing risk of surface erosion, and does not result in significant variability in the exit point elevation, at least under the input statistics assumed for these models. As will be seen later, the internal free-surface profile has somewhat more variability. However, this is not a design problem as it does not lead to emergence of the free surface on the downstream face of the core unless the drain is excessively short.

To investigate the effect of the drain length on the downstream exit elevation, the simulations for dam B (the shallower dam) were rerun for drain lengths of 2.0 and 11.0. The results are shown in Figure 9.12, which includes the location of the normalized drain height as a vertical dashed line. For the shorter drain length, the free-surface exit point distribution becomes bimodal, as perhaps expected. A deterministic analysis with the short drain predicts the free surface to exit at about the half height of the dam. Occasionally, realizations of the stochastic permeability field provide a high-permeability path to the drain and the free surface "jumps down" to exit through the drain rather than the downstream face. When the drain is extended a significant distance into the dam, the total flow rate increases significantly and the drain begins to approach its flow capacity. In this case, the response of the dam overlying the drain approaches that of the dam without a drain and the free-surface exit elevation rises (for a drain length of 11.0, only about 10% of realizations resulted in the free surface exiting within the drain). Again, this response is predicted by the deterministic analysis.

In summary, the free-surface exit elevation tends to have only a very small probability of exceeding the exit point elevation predicted by a deterministic analysis, implying that a deterministic analysis is conservative in this respect. However, if the drain becomes "plugged" due to infiltration of fines, perhaps effectively reducing its length, then the free surface may "jump" to the downstream face and in general has a higher probability of exiting somewhere (usually around midheight) on the downstream face. On the basis of these simulations, it appears that if the drain is behaving satisfactorily according to a deterministic analysis, then it will also behave satisfactorily in the presence of spatially random permeability, assuming that the permeability mean and variance are reasonably well approximated.

9.2.4 Internal Gradients

Figure 9.13 shows a gray-scale representation of the average internal gradients with dark regions corresponding to higher average gradients. Clearly the highest average gradients occur at the head of the drain (upper right corner of drain), as expected.

Approaching the downstream face of the dam, the average internal gradients tend to fade out slowly. This reflects two things: (1) the gradients near the free surface tend to be small and (2) the free surface changes location from realization to realization so that the average includes cases where the gradient is zero (above the free surface). The free surface itself sometimes comes quite close to the downstream face, but it always (with very high probability) descends to exit within the drain for the cases shown.

Figure 9.14 shows a gray-scale representation of the gradient standard deviation. Interestingly, larger standard deviations occur in the region of the mean free surface, as indicated by the dark band running along parallel to the downstream face, as well as near the head of the drain. Clearly, the area near the head of the drain is where the maximum gradients occur so this area has the largest potential for soil degradation and piping. The gradient distribution, as extracted from the finite-element program, at the drain head is shown in Figure 9.15. The gradients observed in dam A extend all the way up to about 3.5, which is in the range of the soil tests performed by Lafleur et al. (1989) on the filtration stability of broadly graded cohesionless soils. Thus it would appear that those test results, at least, cover the range of gradients observed in this dam. The wider dam B profile has a smaller range and lower gradients at the head of the drain and so should be safer with respect to piping failure.

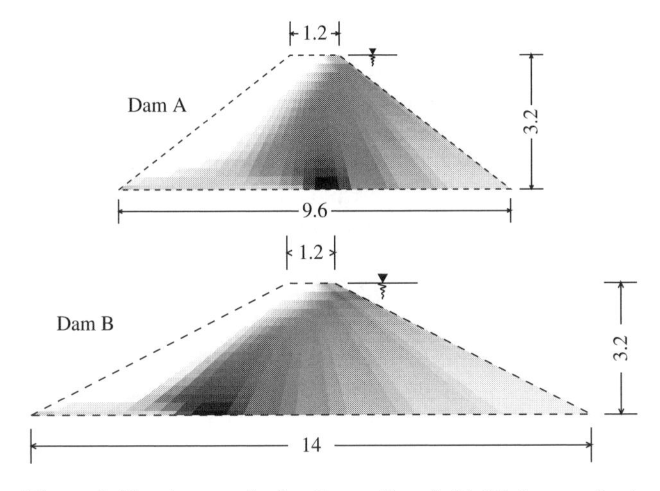

Figure 9.13 Average hydraulic gradient field. Higher gradients are dark.

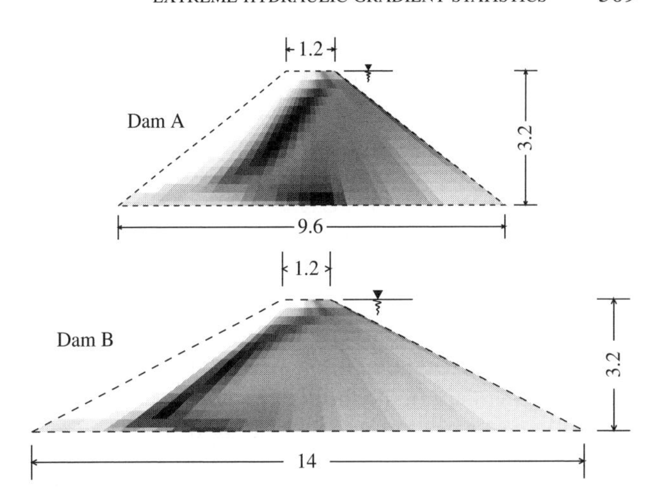

Figure 9.14 Hydraulic gradient standard deviation fields. Higher standard deviations are dark.

The deterministic gradients at the head of the drain were 1.85 for dam A and 1.08 for dam B. These values are very close to the mean gradients observed in Figure 9.15 but imply that the deterministic result is not a conservative measure of the gradients possible in the region of the drain. The coefficients of variation of the drain head gradients were 0.24 for dam A and 0.20 for dam B. Thus a rough

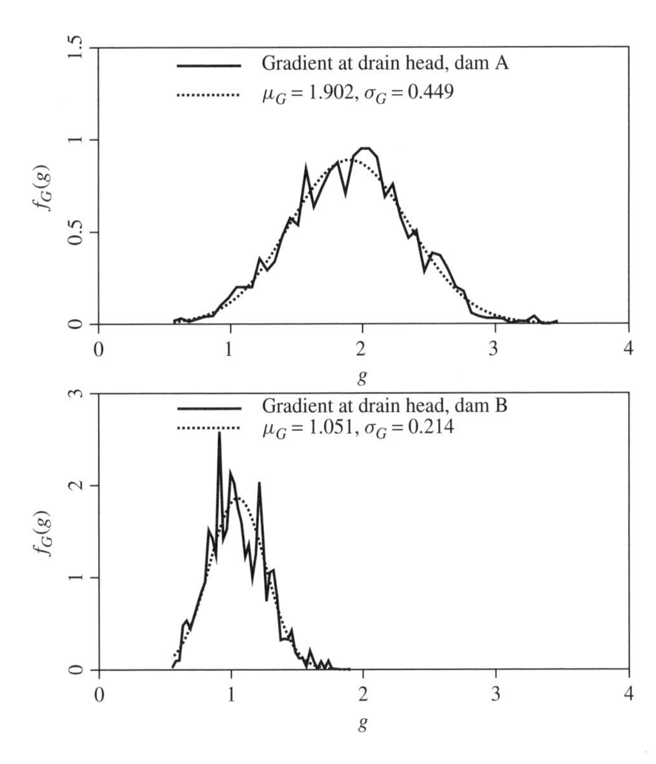

Figure 9.15 Hydraulic gradient distributions at drain heads (upper right corner of drain).

estimate of the maximum gradient distribution for dams such as these might be to take the deterministic gradient as a mean and apply a coefficient of variation of 25% under an assumed normal distribution. These results can be considered representative of any dam having one of these overall shapes with $\sigma_K/\mu_K = 0.5$ since the gradient is not affected by changing mean permeability or overall dam dimension.

For comparison, the unit flux near the head of the drain has a distribution similar to that of the gradient with mean $1.77\mu_K$ and coefficient of variation of 0.36 for dam A. For the shallower dam B, the mean flux at the head of the drain is $1.02\mu_K$ with coefficient of variation of 0.33. Although these results are not directly comparable with the stability tests carried out by Kenney and Lau (1985) under unit fluxes ranging from 0.48 to 1.67 cm/s (they did not report the corresponding permeability), it seems likely that the soil samples they were testing would have permeabilities much smaller than 1.0. In this case, the unit fluxes obtained in this simulation study are (probably) much smaller than those used in the test conditions, indicating again that the test conditions are conservative.

9.2.5 Summary

The primary conclusions derived from this study are as follows:

1. The downstream exit point elevation obtained using a deterministic analysis (constant permeability) is a conservative estimate. That is, the effect of spatial variability in the permeability field serves to lower the mean exit point elevation (as it does the mean flow rate).
2. The spatial variability of soil permeability does not significantly add variability to the free-surface location. The exception to this occurs when a sufficiently short drain is provided which keeps the free surface so close to the downstream dam face that it jumps to exit on the downstream face under slight changes of the permeability field in the region of the drain. In general, however, for a sufficiently long and clear drain (somewhere between one-fourth and one-half the base dimension), the free-surface profile is fairly stable. This observation has also been made in the absence of a drain (see previous section), even though the profile in that case is much higher.
3. A drain having permeability at least 120 times the mean permeability of the dam itself and having length between one-fourth and one-half of the base dimension was found to be successful in ensuring that the downstream free-surface exit point was consistently contained within the drain, despite variability in the permeability field. Specifically, the mean downstream exit point elevation was found to lie well within the drain. As noted above, the exit point elevation has relatively small standard deviation (coefficient of variation of less than 17%) so that the entire sample distribution also remained well within the drain.
4. Maximum internal hydraulic gradients occur near the head of the drain (upstream end), and although there is more variability in the gradient field than in the free-surface profile, the gradient distribution is not excessively wide. Coefficients of variation remain around 25% for both earth dam geometries considered (for an input coefficient of variation of 50% on the permeability).
5. There does not seem to be any significant probability that spatial variability in the permeability field will lead to hydraulic gradients exceeding those values used in tests leading to soil stability design criteria. Thus, design criteria based on published test results appear to be conservative, at least when considering only the influence of spatially varying permeability.
6. The hydraulic gradient distribution near the head of the drain has a mean very close to that predicted by a deterministic analysis with $K = \mu_K$ everywhere. A coefficient of variation of 25% can then be applied using a normal distribution to obtain a reasonable approximation to the gradient distribution.

Although these observations imply that existing design procedures based on "conservative" and deterministic tests do appear to be conservative and so can be used for drain design without regard to stochasticity, it must be emphasized that only one source of uncertainty was considered in this analysis under a single input coefficient of variation. A more complete (and complex) study would include soil particle distributions, differential settlements, and allow for particle movement, formation of preferential flow paths, drain blockage, and so on. The results of this study are, however, encouraging, in that the stability design of the soil only considers soil gradation issues under vibration and seepage and does not specifically account for larger scale factors such as differential settlement. What this means is that the results of this section are useful if the dam behaves as it was designed, without formation of large cracks due to settlement or preferential flow paths and without, for example, drain blockage, and suggest that such a design would be conservative without the need to explicitly consider stochastic variation in the soil permeability.

CHAPTER 10

Settlement of Shallow Foundations

10.1 INTRODUCTION

The settlement of structures founded on soil is a subject of considerable interest to practicing engineers since excessive settlements often lead to serviceability problems. In particular, unless the total settlements themselves are particularly large, it is usually differential settlements which lead to unsightly cracks in facades and structural elements, possibly even to structural failure, especially in unreinforced masonry elements. Existing code requirements limiting differential settlements to satisfy serviceability limit states [see building codes American Concrete Institute (ACI) 318-89, 1989, or Canadian Standards Association (CSA) A23.3-M84, 1984] specify maximum deflections ranging from $D/180$ to $D/480$, depending on the type of supported elements, where D is the center-to-center span of the structural element. Often, in practice, differential settlements between footings are generally controlled, not by considering the differential settlement itself, but by controlling the total settlement predicted by analysis using an estimate of the soil elasticity. This approach is largely based on correlations between total settlements and differential settlements observed experimentally (see, e.g., D'Appolonia et al., 1968) and leads to a limitation of 4–8 cm in total settlement under a footing, as specified in the *Canadian Foundation Engineering Manual* (CFEM), Part 2 (CGS 1978). Interestingly enough, the 1992 version of CFEM (CGS 1992) only specifies limitations on differential settlements, and so it is presumably assumed that by the early 1990s, geotechnical engineers have sufficient site information to assess differential settlements and/or are able to take spatial variability into account in order to assess the probability that differential settlements exceed a certain amount.

Because of the wide variety of soil types and possible loading conditions, experimental data on differential settlement of footings founded on soil are limited. With the aid of computers, it is now possible to probabilistically investigate differential settlements over a range of loading conditions and geometries. In this chapter, we investigate the distributions associated with settlement and differential settlement and present reasonably simple, approximate approaches to estimating probabilities associated with settlements and differential settlements.

In Section 4.4.3, we considered an example dealing with the settlement of a perfectly horizontally layered soil mass in which the equivalent elastic modulus (which may include the stress-induced effects of consolidation) varied randomly from layer to layer but was constant within each layer. The resulting global effective elastic modulus was the harmonic average of the layer moduli. We know, however, that soils will rarely be so perfectly layered, nor will their properties be perfectly uniform even within a layer. To illustrate what happens at the opposite extreme, where soil or rock properties vary only in the horizontal direction, consider the following example.

Example 10.1 Consider the situation in which a sequence of vertically oriented soil or rock layers are subjected to a rigid surface load, as illustrated in Figure 10.1. Assume that each layer has elastic modulus E_i which is constant in the vertical direction. If the total settlement δ is expressed as

$$\delta = \frac{\sigma H}{E_{\text{eff}}}$$

derive E_{eff}.

SOLUTION Because the load is assumed to be rigid, the settlement δ of all layers will be identical. Each layer picks up a portion of the total load which is proportional to its stiffness. If P_i is the load (per unit distance perpendicular to the plane of Figure 10.1) supported by the ith layer, then

$$P_i = \frac{\delta \Delta x E_i}{H}$$

The total load acting on Figure 10.1 is $P = (n\Delta x)\sigma$ which must be equal to the sum of layer loads;

$$(n\ \Delta x)\sigma = \sum_{i=1}^{n} P_i = \frac{\delta \Delta x}{H} \sum_{i=1}^{n} E_i$$

Solving for δ gives

$$\delta = \frac{\sigma H}{(1/n)\sum_{i=1}^{n} E_i}$$

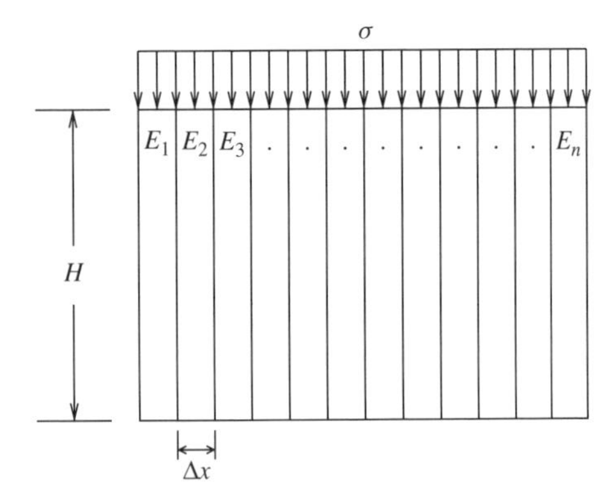

Figure 10.1 Settlement of perfectly vertically layered soil or rock.

from which we see that

$$E_{\text{eff}} = \frac{1}{n}\sum_{i=1}^{n} E_i$$

which is the arithmetic average.

Referring back to the example given in Section 4.4.3, we see that for a perfectly horizontally layered soil E_{eff} is the harmonic average, while for a perfectly vertically layered soil E_{eff} is the arithmetic average. A real soil will generally appear somewhere between these two extremes, as illustrated in Figure 10.2, which implies that E_{eff} will often lie somewhere between the harmonic and the arithmetic averages. Since the geometric average (see Section 4.4.2) does lie between the harmonic and arithmetic averages, it will often be an appropriate model for E_{eff}. Bear in mind, however, that if it is known that the soil or rock is strongly layered, one of the other averages may be more appropriate.

In the following two sections, random models for settlement of rigid shallow foundations are developed, first of all in two dimensions, then in three dimensions. The last two sections discuss how the probabilistic results can be used to develop a LRFD methodology for shallow-foundation serviceability limit states.

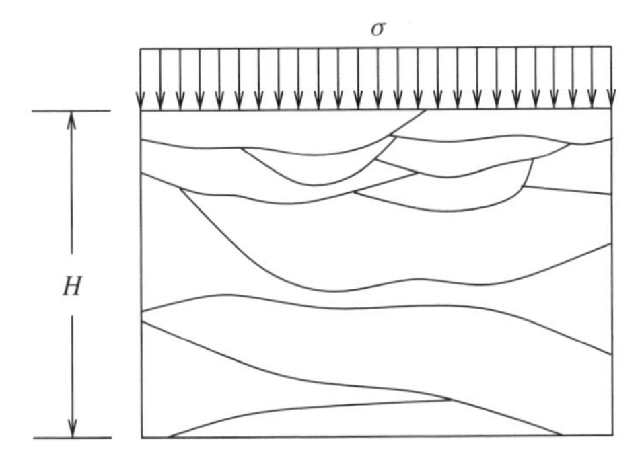

Figure 10.2 Layering of real soils is typically somewhere between perfect horizontal and vertical layering.

10.2 TWO-DIMENSIONAL PROBABILISTIC FOUNDATION SETTLEMENT

In this section, we first consider the distribution of settlements of a single footing, as shown in Figure 10.3*a*, and estimate the probability density function governing total settlement of the footing as a function of footing width for various statistics of the underlying soil. Only the soil elasticity is considered to be spatially random. Uncertainties arising from model and test procedures and in the loads are not considered. In addition, the soil is assumed to be isotropic; that is, the correlation structure is assumed to be the same in both the horizontal and vertical directions. Although soils generally exhibit a stronger correlation in the horizontal direction, due to their layered nature, the degree of anisotropy is site specific. In that this section is demonstrating the basic probabilistic behavior of settlement, anisotropy is left as a refinement for the reader. The program used to perform the study presented in this section is RSETL2D, available at `http://www.engmath.dal.ca/rfem`(Paice et al., 1994; Fenton et al., 1996; Fenton and Griffiths, 2002; Griffiths and Fenton, 2007).

In foundation engineering, both immediate and consolidation settlements are traditionally computed using elastic

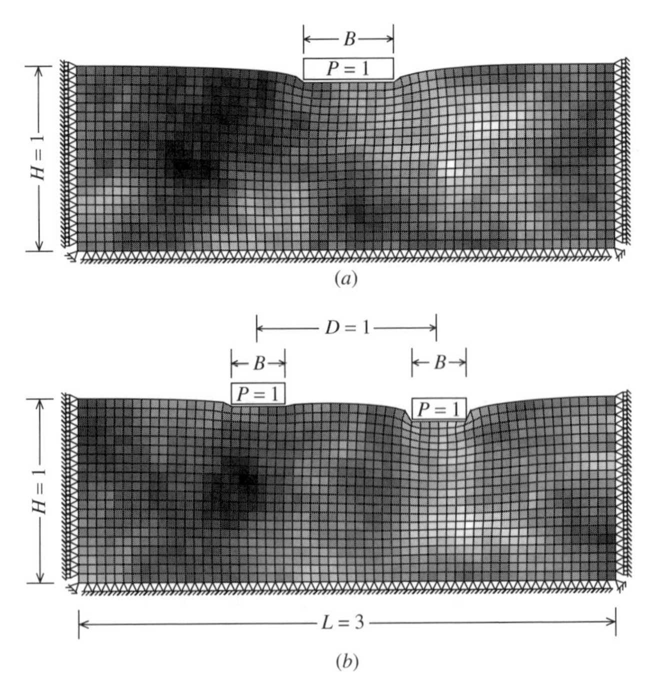

Figure 10.3 Random-field/finite-element representation of (*a*) single footing and (*b*) two footings founded on a soil layer.

theory. This section considers the elastic properties E that apply to either or both immediate and consolidation settlement as spatially random since these are usually the most important components of settlement.

The footings are assumed to be founded on a soil layer underlain by bedrock. The assumption of an underlying bedrock can be relaxed if a suitable averaging region is used. Guidelines to this effect are suggested below. The results are generalized to allow the estimation of probabilities associated with settlements under footings in many practical cases.

The second part of the section addresses the issue of differential settlements under a pair of footings, as shown in Figure 10.3*b*, again for the particular case of footings founded on a soil layer underlain by bedrock. The mean and standard deviation of differential settlements are estimated as functions of footing width for various input statistics of the underlying elastic modulus field. The probability distribution governing differential settlement is found to be conservatively estimated using a joint normal distribution with correlation predicted using local averages of the elastic modulus field under the two footings.

The physical problem is represented using a two-dimensional (plane-strain) model following the work of Paice et al. (1996), which is based on program 5.1 in Smith and Griffiths (2004). If the footings extend for a large distance in the out-of-plane direction z, then the two-dimensional elastic modulus field is interpreted either as an average over z or as having an infinite correlation length in the z direction. For footings of finite dimension, the two-dimensional model is admittedly just an approximation. However, the two-dimensional approximation is reasonable since the elastic modulus field is averaged by the foundation in the z direction in any case.

10.2.1 Random Finite-Element Method

Much discussion of the relative merits of various methods of representing random fields in finite-element analysis has been carried out in recent years (see, e.g., Li and Der Kiureghian, 1993). While the spatial averaging discretization of the random field used in this study is just one approach to the problem, it is appealing in the sense that it reflects the simplest idea of the finite-element representation of a continuum as well as the way that soil samples are typically taken and tested in practice, that is, as local averages. Regarding the discretization of random fields for use in finite-element analysis, Matthies et al. (1997) make the comment that "one way of making sure that the stochastic field has the required structure is to assume that it is a local averaging process," referring to the conversion of a nondifferentiable to a differentiable (smooth) stochastic process. They go on to say that the advantage of the local average representation of a random field is that it yields accurate results even for rather coarse meshes.

As illustrated in Figure 10.3, the soil mass is discretized into 60 four-noded quadrilateral elements in the horizontal direction by 20 elements in the vertical direction. Trial runs using 120×40 elements resulted in less than a 2.5% difference in settlements for the worst cases (narrowest footings) at a cost of more than 10 times the computing time, and so the 60×20 discretization was considered adequate. The overall dimensions of the soil model are held fixed at $L = 3$ and $H = 1$. No units will be used since the probabilistic properties of the soil domain are scaled by the correlation length, to be discussed shortly. The left and right faces of the finite-element model are constrained against horizontal displacement but are free to slide vertically while the nodes on the bottom boundary are spatially fixed. The footing(s) are assumed to be rigid, to not undergo any rotations, and to have a rough interface with the underlying soil (no-slip boundary). A fixed load $P = 1$ is applied to each footing—since settlement varies linearly with load, the results are easily scaled to different values of P.

To investigate the effect of the footing width B, the soil layer thickness H was held constant at 1.0 while the footing width was varied according to Table 10.1. Because the settlement problem is linear in many of its parameters, the following results can be scaled to arbitrary footing widths and soil layer thicknesses, so long as the ratio B/H is held fixed. For example, the settlement of a footing of width $B' = 2.0$ m on an $H' = 20$ m thick soil layer with load $P' = 1000$ kN and elastic modulus $E' = 60$ kN/m^2 corresponds to 0.06 times the settlement of a footing of width $B = 0.1$ m on an $H = 1.0$ m thick soil layer with $P = 1$ kN and elastic modulus $E = 1$ kN/m^2. The scaling factor from the unprimed to the primed case is $(P'/P)(E/E')$ as long as $B'/H' = B/H$. If B/H is not constant, a deterministic finite-element analysis will have to be performed to determine the scaling constant.

In the two-footing case, the distance between footing centers was held constant at 1.0, while the footing

Table 10.1 Input Parameters Varied in Study While Holding $H = 1$, $D = 1$, $P = 1$, $\mu_E = 1$, and $\nu = 0.25$ Constant

Parameter	Values Considered
σ_E	0.1, 0.5, 1.0, 2.0, 4.0
$\theta_{\ln E}$	0.01, 0.05, 0.1, 0.3, 0.5, 0.7, 1.0, 2.0, 5.0, 10.0, 50.0
B	0.1, 0.2, 0.5, 1.0 (single footing) 0.1, 0.3, 0.5 (two footings)

widths (assumed equal) were varied. Footings of width greater than 0.5 were not considered since this situation approaches that of a strip footing (the footings would be joined when $B = 1.0$). The soil has two properties of interest to the settlement problem: the (effective) elastic modulus $E(\mathbf{x})$ and Poisson's ratio $\nu(\mathbf{x})$, where $\mathbf{x}$ is spatial position. Only the elastic modulus is considered to be a spatially random soil property. Poisson's ratio was believed to have a smaller relative spatial variability and only a second-order importance to settlement statistics. It is held fixed at 0.25 over the entire soil mass for all simulations.

Figure 10.3 shows a gray-scale representation of two possible realizations of the elastic modulus field, along with the finite element mesh. Lighter areas denote smaller values of $E(\mathbf{x})$ so that the elastic modulus field shown in Figure 10.3*b* corresponds to a higher elastic modulus under the left footing than under the right—this leads to the substantial differential settlement indicated by the deformed mesh. This is just one possible realization of the elastic modulus field; the next realization could just as easily show the opposite trend.

The elastic modulus field is assumed to follow a lognormal distribution so that $\ln(E)$ is a Gaussian (normal) random field with mean $\mu_{\ln E}$ and variance $\sigma^2_{\ln E}$. The choice of a lognormal distribution is motivated by the fact that the elastic modulus is strictly nonnegative, a property of the lognormal distribution (but not the normal), while still having a simple relationship with the normal distribution. A Markovian spatial correlation function, which gives the correlation coefficient between log-elastic modulus values at points separated by distance τ, is used,

$$\rho_{\ln E}(\boldsymbol{\tau}) = \exp\left\{-\frac{2|\boldsymbol{\tau}|}{\theta_{\ln E}}\right\} \tag{10.1}$$

in which $\boldsymbol{\tau} = \mathbf{x} - \mathbf{x}'$ is the vector between spatial points $\mathbf{x}$ and $\mathbf{x}'$ and $|\boldsymbol{\tau}|$ is the absolute length of this vector (the lag distance). See Section 3.7.10.2 for more details. In this section, the word "correlation" refers to the correlation coefficient (normalized covariance). The correlation function decay rate is governed by the so-called correlation length $\theta_{\ln E}$, which, loosely speaking, is the distance over which log-elastic moduli are significantly correlated (when the separation distance $|\boldsymbol{\tau}|$ is greater than $\theta_{\ln E}$, the correlation between $\ln E(\mathbf{x})$ and $\ln E(\mathbf{x}')$ is less than 14%).

The assumption of isotropy is, admittedly, somewhat restrictive. In principle the methodology presented in the following is easily extended to anisotropic fields, although the accuracy of the proposed distribution parameter estimates would then need to be verified. For both the single- and two-footing problems, however, it is the horizontal correlation length which is more important. As will be seen, the settlement variance and covariance depend on the statistics of a local average of the log-elastic modulus field under the footing. If the vertical correlation length is less than the horizontal, this can be handled simply by reducing the vertical averaging dimension H to $H(\theta_{\ln E_h}/\theta_{\ln E_v})$. For very deep soil layers, the averaging depth H should probably be restricted to no more than about $10B$ since the stress under a footing falls off approximately according to $B/(B + H)$.

In practice, one approach to the estimation of $\theta_{\ln E}$ involves collecting elastic modulus data from a series of locations in space, estimating the correlations between the log-data as a function of separation distance, and then fitting Eq. 10.1 to the estimated correlations. As indicated in Sections 5.3.6 and 5.4.1.1, the estimation of $\theta_{\ln E}$ is not a simple problem since it tends to depend on the distance over which it is estimated. For example, sampling soil properties every 5 cm over 2 m will likely yield an estimated $\theta_{\ln E}$ of about 20 cm, while sampling every 1 km over 1000 km will likely yield an estimate of about 200 km. This is because soil properties vary at many scales; looked at closely, a soil can change significantly within a few meters relative to the few meters considered. However, soils are formed by weathering and glacial actions which can span thousands of kilometers, yielding soils which have much in common over large distances. Thus, soils can conceptually have lingering correlations over entire continents (even planets).

This lingering correlation in the spatial variability of soils implies that correlation lengths estimated in the literature should not just be used blindly. One should attempt to select a correlation length which has been estimated on a similar soil *over a domain of similar size* to the site being characterized. In addition, the level of detrending used to estimate the reported correlation length must be matched at the site being characterized. For example, if a correlation length, as reported in the literature, was estimated from data with a quadratic trend removed, then sufficient data must be gathered at the site being characterized to allow a quadratic trend to be fitted to the site data. The estimated correlation length then applies to the residual random variation around the trend. To facilitate this, researchers providing estimates of variance and correlation length in the literature should report (a) estimates with the trend removed, including the details of the trend itself, and (b) estimates without trend removal. The latter will typically yield significantly larger estimated variance and correlation length, giving a truer sense for actual soil variability.

In the case of two footings, the use of a correlation length equal to D is conservative in that it yields differential settlement variances which are at or close to their maximums, as will be seen shortly. In some cases, however, setting $\theta_{\ln E} = D$ may be unreasonably conservative.

If sufficient site sampling has been carried out to estimate the mean and variance of the soil properties at the site, then a significantly reduced correlation length is warranted. The literature should then be consulted to find a similar site on which a spatial statistical analysis has been carried out and an estimated correlation length reported.

In the case of a single footing, taking $\theta_{\ln E}$ large is conservative; in fact, the assumption that E is lognormally distributed and spatially constant leads to the largest variability (across realizations) in footing settlement. Thus, traditional approaches to randomness in footing settlement using a single random variable to characterize E are conservative—settlement will generally be less than predicted.

Throughout, the mean elastic modulus μ_E is held fixed at 1.0. Since settlement varies linearly with the soil elastic modulus, it is always possible to scale the settlement statistics to the actual mean elastic modulus. The standard deviation of the elastic modulus is varied from 0.1 to 4.0 to investigate the effects of elastic modulus variability on settlement variability. The parameters of the transformed $\ln(E)$ Gaussian random field may be obtained from Eqs. 1.176,

$$\sigma^2_{\ln E} = \ln(1 + v_E^2) \tag{10.2a}$$

$$\mu_{\ln E} = \ln(\mu_E) - \tfrac{1}{2}\sigma^2_{\ln E} \tag{10.2b}$$

where $v_E = \sigma_E/\mu_E$ is the coefficient of variation of the elastic modulus field. From Eq. 10.2a, it can be seen that $\sigma^2_{\ln E}$ varies from 0.01 to 2.83 in this study (note also that $\mu_{\ln E}$ depends on both μ_E and σ_E).

To investigate the effect of the correlation length $\theta_{\ln E}$ on the settlement statistics, $\theta_{\ln E}$ is varied from 0.01 (i.e., very much smaller than the soil model size) to 50.0 (i.e., substantially bigger than the soil model size) and up to 200 in the two-footing case. In the limit as $\theta_{\ln E} \to 0$, the elastic modulus field becomes a white noise field, with E values at any two distinct points independent. In terms of the finite elements themselves, values of $\theta_{\ln E}$ smaller than the elements result in a set of elements which are largely independent (increasingly independent as $\theta_{\ln E}$ decreases). Because of the averaging effect of the details of the elastic modulus field under a footing, the settlement in the limiting case $\theta_{\ln E} \to 0$ is expected to approach that obtained in the deterministic case, with $E = \tilde{\mu}_E$ (the median) everywhere, and has vanishing variance for finite $\sigma^2_{\ln E}$.

By similar reasoning the differential settlement (as in Figure 10.3*b*) as $\theta_{\ln E} \to 0$ is expected to go to zero. At the other extreme, as $\theta_{\ln E} \to \infty$, the elastic modulus field becomes the same everywhere. In this case, the settlement statistics are expected to approach those obtained by using a single lognormally distributed random variable E to model the soil, $E(\mathbf{x}) = E$. That is, since the settlement δ under a footing founded on a soil layer with uniform (but random) elastic modulus E is given by

$$\delta = \frac{\delta_{\text{det}}\mu_E}{E} \tag{10.3}$$

for δ_{det} the settlement when $E = \mu_E$ everywhere, then as $\theta_{\ln E} \to \infty$ the settlement assumes a lognormal distribution with parameters

$$\mu_{\ln \delta} = \ln(\delta_{\text{det}}) + \ln(\mu_E) - \mu_{\ln E} = \ln(\delta_{\text{det}}) + \tfrac{1}{2}\sigma^2_{\ln E} \tag{10.4a}$$

$$\sigma_{\ln \delta} = \sigma_{\ln E} \tag{10.4b}$$

where Eq. 10.2b was used in Eq. 10.4a. Also, since in this case the settlement under the two footings of Figure 10.3*b* becomes equal, the differential settlement becomes zero. Thus, the differential settlement is expected to approach zero at both very small and very large correlation lengths.

The Monte Carlo approach adopted here involves the simulation of a realization of the elastic modulus field and subsequent finite-element analysis (Smith and Griffiths, 2004) of that realization to yield a realization of the footing settlement(s). Repeating the process over an ensemble of realizations generates a set of possible settlements which can be plotted in the form of a histogram and from which distribution parameters can be estimated. In this study, 5000 realizations are performed for each input parameter set (σ_E, $\theta_{\ln E}$, and B). If it can be assumed that log-settlement is approximately normally distributed (which is seen later to be a reasonable assumption and is consistent with the distribution selected for E), and $m_{\ln \delta}$ and $s^2_{\ln \delta}$ are the estimators of the mean and variance of log-settlement, respectively, then the standard deviations of these estimators obtained from 5000 realizations are given by

$$\sigma_{m_{\ln \delta}} \simeq \frac{s_{\ln \delta}}{\sqrt{n}} = 0.014 s_{\ln \delta} \tag{10.5a}$$

$$\sigma_{s^2_{\ln \delta}} \simeq \sqrt{\frac{2}{n-1}}\, s^2_{\ln \delta} = 0.02 s^2_{\ln \delta} \tag{10.5b}$$

so that the estimator "errors" are negligible compared to the estimated variance (i.e., about 1 or 2% of the estimated standard deviation).

Realizations of the log-elastic modulus field are produced using the two-dimensional LAS technique (see Section 6.4.6). The elastic modulus value assigned to the ith element is

$$E_i = \exp\{\mu_{\ln E} + \sigma_{\ln E} G_i\} \tag{10.6}$$

where G_i is the local average over the ith element of a zero-mean, unit-variance Gaussian random field $G(\mathbf{x})$.

10.2.2 Single-Footing Case

A typical histogram of the settlement under a single footing, as estimated by 5000 realizations, is shown in Figure 10.4 for $B = 0.1$, $\sigma_E/\mu_E = 1$, and $\theta_{\ln E} = 0.1$. With the requirement that settlement be nonnegative, the shape of the histogram suggests a lognormal distribution, which was adopted in this study (see Eqs. 10.4). The histogram is normalized to produce a frequency density plot, where a straight line is drawn between the interval midpoints. Superimposed on the histogram is a fitted lognormal distribution with parameters given by $m_{\ln\delta}$ and $s_{\ln\delta}$ in the line key. At least visually, the fit appears reasonable. In fact, this is one of the worst cases out of all 220 parameter sets given in Table 10.1; a chi-square goodness-of-fit test yields a p-value of 8×10^{-10}. Large p-values support the lognormal hypothesis, so that this small value suggests that the data do not follow a lognormal distribution. Unfortunately, when the sample size is large ($n = 5000$ in this case), goodness-of-fit tests are quite sensitive to the "smoothness" of the histogram. They perhaps correctly indicate that the true distribution is not exactly as hypothesized but say little about the *reasonableness* of the assumed distribution. As can be seen from Figure 10.4, the lognormal distribution certainly appears reasonable.

Over the entire set of simulations performed for each parameter of interest (B, σ_E, and $\theta_{\ln E}$), 80% of the fits have p-values exceeding 5% and only 5% have p-values of less than 0.0001. This means that the lognormal distribution is generally a close approximation to the distribution of the simulated settlement data, typically at least as good as seen in Figure 10.4.

Accepting the lognormal distribution as a reasonable fit to the simulation results, the next task is to estimate the parameters of the fitted lognormal distributions as functions of the input parameters (B, σ_E, and $\theta_{\ln E}$). The lognormal distribution,

$$f_\delta(x) = \frac{1}{\sqrt{2\pi}\,\sigma_{\ln\delta}\, x} \exp\left\{-\frac{1}{2}\left(\frac{\ln x - \mu_{\ln\delta}}{\sigma_{\ln\delta}}\right)^2\right\}, \quad 0 \le x < \infty \tag{10.7}$$

has two parameters, $\mu_{\ln\delta}$ and $\sigma_{\ln\delta}$. Figure 10.5 shows how the estimator of $\mu_{\ln\delta}$, denoted $m_{\ln\delta}$, varies with $\sigma_{\ln E}$ for $B = 0.1$. All correlation lengths are drawn in the plot but are not individually labeled since they lie so close together. This observation implies that the mean log-settlement is largely independent of the correlation length $\theta_{\ln E}$. This is as expected since the correlation length does not affect the mean of a local average of a normally distributed process. Figure 10.5 suggests that the mean of log-settlement can be closely estimated by a straight line of the form (as suggested by Eq. 10.4a)

$$\mu_{\ln\delta} = \ln(\delta_{\text{det}}) + \tfrac{1}{2}\sigma^2_{\ln E} \tag{10.8}$$

where δ_{det} is the "deterministic" settlement obtained from a single finite-element analysis (or appropriate approximate calculation) of the problem using $E = \mu_E$ everywhere. This equation is also shown in Figure 10.5, and it can be seen that the agreement is very good. Even closer results were found for the other footing widths.

Estimates of the standard deviation of log-settlement, $s_{\ln\delta}$, are plotted in Figure 10.6 (as symbols) for the smallest and largest footing widths. Intermediate footing widths give similar results. In all cases, $s_{\ln\delta}$ increases to $\sigma_{\ln E}$ as $\theta_{\ln E}$ increases. The reduction in variance as $\theta_{\ln E}$ decreases is due to the local averaging variance reduction of the log-elastic modulus field under the footing (for smaller $\theta_{\ln E}$, there are more "independent" random field values, so that the variance reduces faster under averaging; see Section 3.4 for more details).

Following this reasoning, and assuming that local averaging of the area under the footing accounts for all of

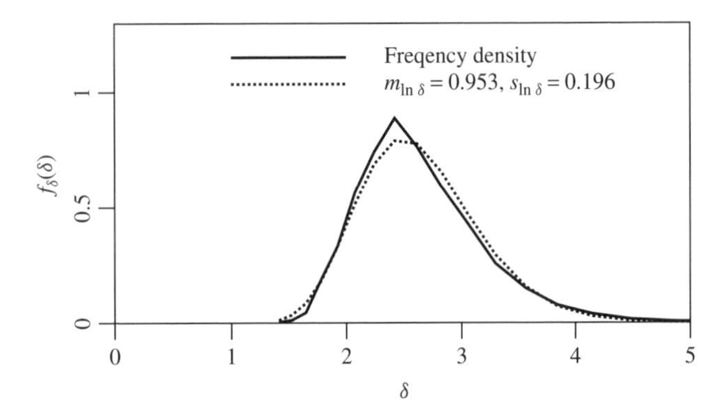

Figure 10.4 Typical frequency density plot and fitted lognormal distribution of settlement under a single footing.

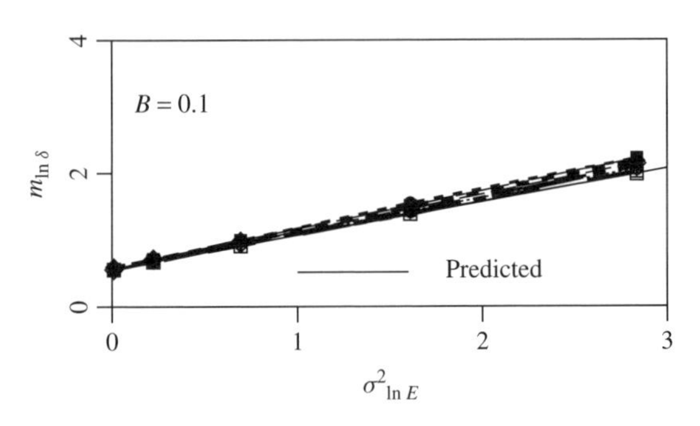

Figure 10.5 Estimated mean of log-settlement along with that predicted by Eq. 10.8.

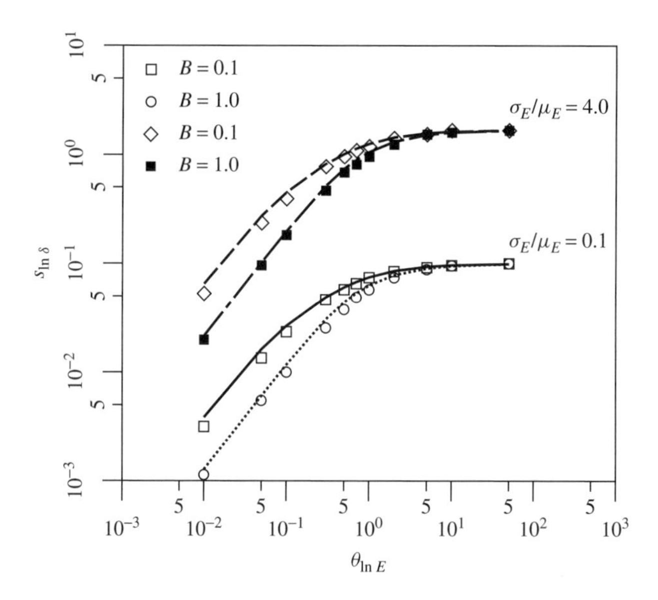

Figure 10.6 Comparison of simulated sample standard deviation of log-settlement, shown with symbols, with theoretical estimate via Eq. 10.9, shown with lines.

the variance reduction seen in Figure 10.6, the standard deviation of log-settlement is

$$\sigma_{\ln\delta} = \sqrt{\gamma(B,H)}\,\sigma_{\ln E} \tag{10.9}$$

where $\gamma(B,H)$ is the variance reduction function, which depends on the averaging region, $B \times H$, as well as on the correlation length, $\theta_{\ln E}$. Since $\sigma^2_{\ln E}$ is constant for each value of σ_E/μ_E (see Eq. 10.2a), Figure 10.6 is essentially a plot of the variance function, $\gamma(B,H)$, illustrating how the variance of a local average decreases as the correlation length decreases. See Appendix C for details on how to compute the variance reduction function.

Specifically, the variance function gives the amount that the log-elastic modulus variance is reduced when its random field is averaged over a region of size $B \times H$. The dependence of the variance function on H is apparently only valid for the geometries considered; if the footing is founded on a much deeper soil mass, one would not expect to average over the entire depth due to stress reduction with depth. As suggested in Section 10.2.1, H should be limited to no more than about $10B$. If in doubt, taking H to be relatively small (even zero) yields a conservative estimate of the settlement distribution, having large variability. This is equivalent to taking $\theta_{\ln E}$ large, as noted previously in Section 10.2.1. In practice, however, values of the normalized averaging area $BH/\theta^2_{\ln E}$ greater than about 5 yield values of $\sigma_{\ln\delta}$ less than about 15% of $\sigma_{\ln E}$ so that changes in H above this level have only a minor effect on the overall variance reduction. Predictions of $\sigma_{\ln\delta}$ using Eq. 10.9 are superimposed in Figure 10.6 using lines. The agreement is remarkable. Intermediate cases show similar, if not better, agreement with predictions.

An alternative physical interpretation of Eqs. 10.8 and 10.9 comes by generalizing the relationship given by Eq. 10.3 to the form

$$\delta = \frac{\delta_{\text{det}}\mu_E}{E_g} \tag{10.10}$$

where E_g is the geometric average of the elastic modulus values over the region of influence (see Section 4.4.2),

$$E_g = \exp\left\{\frac{1}{BH}\int_0^H \int_0^B \ln E(x,y)\,dx\,dy\right\} \tag{10.11}$$

Taking the logarithm of Eq. 10.10 and then computing its mean and variance lead to Eqs. 10.8, using Eq. 10.4a, and 10.9. The geometric mean is dominated by small values of elastic modulus, which means that the total settlement is dominated by low-elastic-modulus regions underlying the footing, as would be expected.

Example 10.2 Consider a single footing of width $B = 2.0$ m to be founded on a soil layer of depth 10.0 m and which will support a load $P = 1000$ kN. Suppose also that samples taken at the site have allowed the estimation of the elastic modulus mean and standard deviation at the site to be 40 and 40 MPa, respectively. Similarly, test results on a regular array at this or a similar site have resulted in an estimated correlation length, $\theta_{\ln E} = 3.0$ m. Assume also that Poisson's ratio is 0.25. Estimate the probability that the footing settlement will exceed 0.10 m.

SOLUTION The results from this section can be used as follows to estimate the probability that the settlement of the footing will exceed 0.10 m:

1. A finite-element analysis of the given problem with soil elastic modulus everywhere equal to $\mu_E = 40$ MPa gives a deterministic settlement of $\delta_{\text{det}} = 0.03531$ m.
2. Compute the variance of log-elastic modulus from Eq. 10.2, $\sigma^2_{\ln E} = \ln(2) = 0.69315$, so that $\sigma_{\ln E} = 0.83256$.
3. Compute the mean of log-settlement from Eq. 10.8, $\mu_{\ln\delta} = \ln(\delta_{\text{det}}) + 0.5\sigma^2_{\ln E} = -3.3437 + 0.5(0.69315) = -2.9971$.
4. Compute the standard deviation of log-settlement using Eq. 10.9, $\sigma_{\ln\delta} = \sqrt{\gamma(B,H)}\,\sigma_{\ln E} = \sqrt{0.22458}\,(0.83256) = 0.39455$. See Appendix C for an algorithm on the computation of the variance reduction function $\gamma(B,H)$.

As an aside, for $\mu_{\ln\delta} = -2.9971$ and $\sigma_{\ln\delta} = 0.39455$, the corresponding settlement mean and standard deviation are $\mu_\delta = \exp\{\mu_{\ln\delta} + \frac{1}{2}\sigma^2_{\ln\delta}\} = 0.0540$ m and $\sigma_\delta = \mu_\delta\sqrt{e^{\sigma^2_{\ln\delta}} - 1} = 0.0222$ m, respectively. A trial run of 5000 realizations for this problem gives $m_\delta = 0.0562$ and $s_\delta = 0.0201$ for relative differences between prediction and simulation of 3.9 and 10.4%, respectively. The estimated relative standard error on m_δ is approximately 0.5% for 5000 realizations.

5. Compute the desired probability using the lognormal distribution, $\text{P}[\delta > 0.10] = 1 - \Phi(1.7603) = 0.0392$, where $\Phi(\cdot)$ is the standard normal cumulative distribution.

A simulation run for this problem yielded 160 samples out of 5000 having settlement greater than 0.10 m. This gives a simulation-based estimate of the above probability of 0.032, which is in very good agreement with that predicted.

10.2.3 Two-Footing Case

Having established, with reasonable confidence, the distribution associated with settlement under a single footing founded on a soil layer, attention can now be turned to the more difficult problem of finding a suitable distribution to model differential settlement between footings. Analytically, if δ_1 is the settlement under the left footing shown in Figure 10.3*b* and δ_2 is the settlement of the right footing, then according to the results of the previous section, δ_1 and δ_2 will be jointly lognormally distributed random variables,

$$f_{\delta_1,\delta_2}(x,y) = \frac{1}{2\pi\sigma^2_{\ln\delta}rxy} \times \exp\left\{-\frac{1}{2r^2}\left[\Psi_x^2 - 2\rho_{\ln\delta}\Psi_x\Psi_y + \Psi_y^2\right]\right\}, \quad x \ge 0, \quad y \ge 0 \tag{10.12}$$

where $\Psi_x = (\ln x - \mu_{\ln\delta})/\sigma_{\ln\delta}$, $\Psi_y = (\ln y - \mu_{\ln\delta})/\sigma_{\ln\delta}$, $r^2 = 1 - \rho^2_{\ln\delta}$, and $\rho_{\ln\delta}$ is the correlation coefficient between the log-settlement of the two footings. It is assumed in the above that δ_1 and δ_2 have the same mean and variance, which, for the symmetric conditions shown in Figure 10.3*b* and stationary E field, will be true.

If the differential settlement between footings is defined by $\Delta = \delta_1 - \delta_2$, then the mean of Δ is zero if the elastic modulus field is statistically stationary. As indicated in Section 5.3.1, stationarity is a mathematical assumption that in practice depends on the level of knowledge that one has about the site. If a trend in the effective elastic modulus is known to exist at the site, then the following results can still be used by computing the deterministic differential settlement using the mean "trend" values in a deterministic analysis, then computing the probability of an *additional* differential settlement using the equations to follow. In this case the following probabilistic analysis would be performed with the trend removed from the elastic modulus field.

The exact distribution governing the differential settlement, assuming that Eq. 10.12 holds, is given as

$$f_\Delta(x) = \begin{cases} \int_0^\infty f_{\delta_1,\delta_2}(x+y,y)\,dy & \text{if } x \ge 0 \\ \int_{-x}^\infty f_{\delta_1,\delta_2}(x+y,y)\,dy & \text{if } x < 0 \end{cases} \tag{10.13}$$

which can be evaluated numerically but which has no analytical solution so far as the authors are aware. In the following a normal approximation to the distribution of Δ will be investigated.

Figure 10.7 shows two typical frequency density plots of differential settlement between two equal-sized footings ($B/D = 0.1$) with superimposed fitted normal distributions, where the fit was obtained by directly estimating the mean and standard deviation from the simulation. The normal distribution appears to be a reasonable fit in Figure 10.7*a*. Since a lognormal distribution begins to look very much like a normal distribution when $\sigma_{\ln\delta}$ is small, then for small $\sigma_{\ln\delta}$ both δ_1 and δ_2 will be approximately normally distributed. For small $\theta_{\ln E}$, therefore, since this leads to small $\sigma_{\ln\delta}$, the difference $\delta_1 - \delta_2$ will be very nearly normally distributed, as seen in Figure 10.7*a*.

For larger correlation lengths (and/or smaller D), the histogram of differential settlements becomes narrower than the normal, as seen in Figure 10.7*b*. What is less obvious in Figure 10.7*b* is that the histogram has much longer tails than predicted by the normal distribution. These long tails lead to a variance estimate which is larger than dictated by the central region of the histogram. Although the variance could be artificially reduced so that the fit is better near the origin, the result would be a significant underestimate of the probability of large differential settlements. This issue will be discussed at more length shortly when differential settlement probabilities are considered. Both plots are for $\sigma_E/\mu_E = 1.0$ and are typical of other coefficients of variation (COVs).

Assuming, therefore, that $\Delta = \delta_1 - \delta_2$ is (approximately) normally distributed, and that δ_1 and δ_2 are identically and lognormally distributed with correlation coefficient ρ_δ, differential settlement has parameters

$$\mu_\Delta = 0, \qquad \sigma^2_\Delta = 2(1-\rho_\delta)\sigma^2_\delta \tag{10.14}$$

Note that when $\theta_{\ln E}$ approaches zero, the settlement variance σ^2_δ also approaches zero. When $\theta_{\ln E}$ becomes very large, the correlation coefficient between settlements under

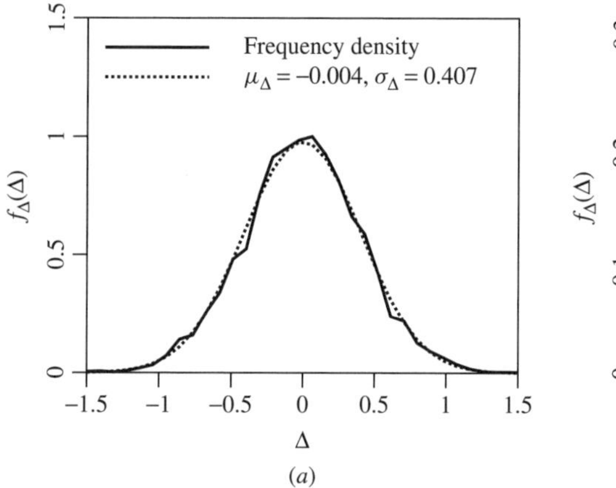

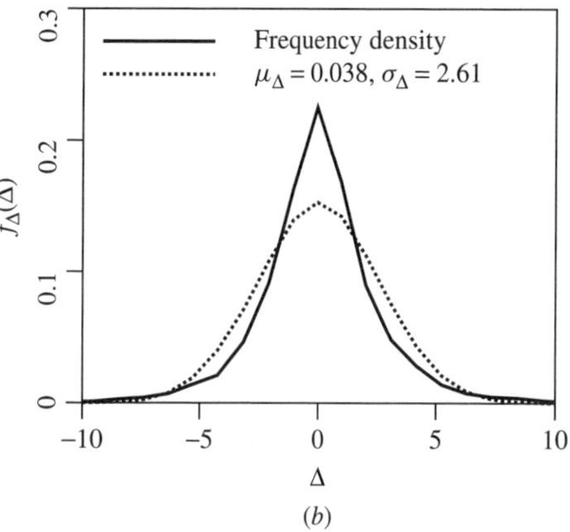

Figure 10.7 Frequency density and fitted distribution for differential settlement under two equal-sized footings with (*a*) $\theta_{\ln E}/D = 0.05$ and (*b*) $\theta_{\ln E}/D = 1.0$.

the two footings approaches 1. Thus, Eq. 10.14 is in agreement with the expectation that differential settlements will disappear for both very small and very large values of $\theta_{\ln E}$.

Since local averaging of the log-elastic modulus field under the footing was found to be useful in predicting the variance of log-settlement, it seems reasonable to suggest that the covariance between log-settlements under a pair of footings will be well predicted by the covariance between local averages of the log-elastic modulus field under each footing. For equal sized footings, the covariance between local averages of the log-elastic modulus field under two footings separated by distance D is given by

$$C_{\ln \delta} = \frac{\sigma^2_{\ln E}}{B^2H^2} \int_0^H \int_0^B \int_0^H \int_D^{D+B} \rho_{\ln E}(x_1 - x_2, y_1 - y_2) \times dx_2\, dy_2\, dx_1\, dy_1 \tag{10.15}$$

which can be evaluated reasonably accurately using a three-point Gaussian quadrature if $\rho_{\ln E}$ is smooth, as is Eq. 10.1. See Appendices B and C for details.

The correlation coefficient between settlements can now be obtained by transforming back from log-space (see Eq. 1.188),

$$\rho_\delta = \frac{\exp\{C_{\ln \delta}\} - 1}{\exp\{\sigma^2_{\ln \delta}\} - 1} \tag{10.16}$$

where $\sigma_{\ln \delta}$ is given by Eq. 10.9. The agreement between the correlation coefficient predicted by Eq. 10.16 and the correlation coefficient estimated from the simulations is shown in Figure 10.8. In order to extend the curve up to correlation coefficients close to 1, four additional correlation lengths were considered (now 15 correlation lengths are considered in total), all the way up to $\theta_{\ln E} = 200$. The general trends between prediction and simulation results are the same, although the simulations show more correlation

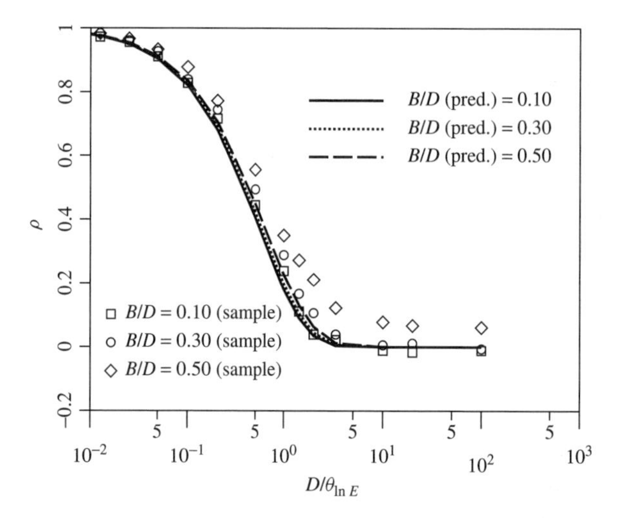

Figure 10.8 Predicted (pred.) and sample correlation coefficients between footing settlements for various relative separation distances between the footings and for $\sigma_E/\mu_E = 1$.

for larger footing widths than predicted by the above theory. For larger footing widths there is a physical interaction between the footings, where the stress under one footing begins to add to the stress under the other footing, so the true correlation is expected to be larger than that predicted purely on the basis of local averaging. The correlation predicted by Eq. 10.16, however, is at least conservative in that smaller correlations lead to larger probabilities of differential settlement.

Figure 10.9 shows the estimated standard deviation of Δ as a function of $\theta_{\ln E}/D$ for three footing widths and for $\sigma_E/\mu_E = 1$. Other values of σ_E/μ_E are of similar form. Superimposed on the sample standard deviations (shown as symbols) are the predicted standard deviations using

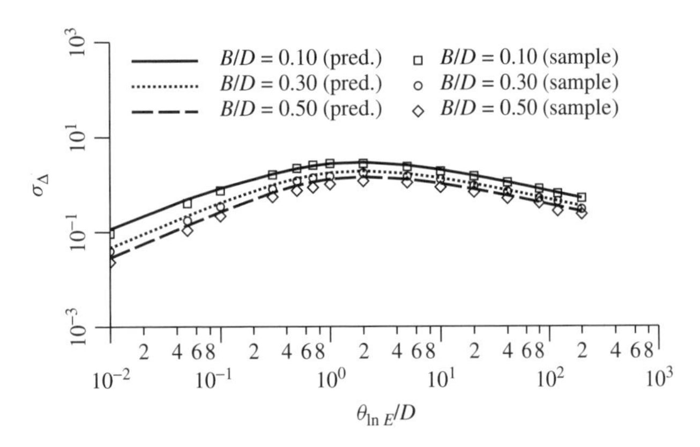

Figure 10.9 Predicted (pred.) and sample standard deviations of differential settlement for $\sigma_E/\mu_E = 1$.

Eq. 10.14 (shown as solid or dashed lines). The agreement is very good over the entire range of correlation lengths.

To test the ability of the assumed distribution to accurately estimate probabilities, the probability that the absolute value of Δ exceeds some threshold is compared to empirical probabilities derived from simulation. For generality, thresholds of $\alpha\mu_{|\Delta|}$ will be used, where $\mu_{|\Delta|}$ is the mean absolute differential settlement, which, if Δ is normally distributed, is given by

$$\mu_{|\Delta|} = \sqrt{\frac{2}{\pi}}\,\sigma_\Delta \tag{10.17}$$

Note that this relationship says that the mean absolute differential settlement is directly related to the standard deviation of Δ, which in turn is related to the correlation between the elastic moduli under the footings and the variability of the elastic moduli. In particular, this means that the mean absolute differential settlement is a function of just δ_{det}, $\sigma^2_{\ln E}$, and $\theta_{\ln E}$, increasing with δ_{det} and $\sigma^2_{\ln E}$, and reaching a maximum when $\theta_{\ln E}/D$ is near 1.0 (see Figure 10.9).

Figure 10.10 shows a plot of the probability

$$\mathrm{P}\left[|\Delta| > \alpha\mu_{|\Delta|}\right] = 2\Phi\left(\frac{-\alpha\mu_{|\Delta|} - \mu_\Delta}{\sigma_\Delta}\right) = 2\Phi\left(-\alpha\sqrt{\frac{2}{\pi}}\right) \tag{10.18}$$

for α varying from 0.5 to 4.0, shown as a solid line. The symbols show empirical probabilities that $|\Delta|$ is greater than $\alpha\mu_{|\Delta|}$ obtained via simulation (5000 realizations) for the 3 footing widths, 15 correlation lengths, and 5 elastic modulus COVs (thus, each column of symbols contains 225 points, 75 for each footing width).

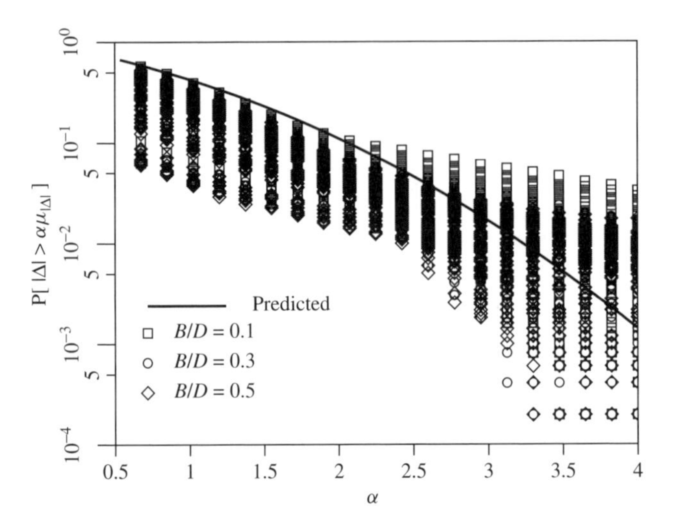

Figure 10.10 Simulation-based estimates of $\mathrm{P}\left[|\Delta| > \alpha\mu_{|\Delta|}\right]$ for all cases compared to that predicted by Eqs. 10.17 and 10.18.

It can be seen that the predicted probability is in very good agreement with average simulation results for large differential settlements, while being conservative (higher probabilities of exceedance) at lower differential settlements.

The normal distribution is considered to be a reasonable approximation for differential settlement in at least two ways; first of all it is a *conservative* approximation, that is, it *overestimates* the probability of differential settlement for the bulk of the data. Second, it is a *consistent* approximation in that it improves as the correlation length decreases, by virtue of the fact that the difference $\delta_1 - \delta_2$ approaches a normal distribution. Since the estimated correlation length decreases as a site is more thoroughly investigated and trends are removed, then the normal distribution becomes more accurate as more is known about the site. Conversely, if little is known about the site, the normal distribution properly reflects inherent uncertainty by generally predicting larger differential settlements.

Example 10.3 Consider two footings each of width $B = 2.0$ m separated by $D = 10$ m center to center. They are founded on a soil layer of depth 10 m and each supports a load $P = 1000$ kN. Assume also that $\mu_E = 40$ MPa, $\sigma_E = 40$ MPa, $\theta_{\ln E} = 1.0$ m, and Poisson's ratio is 0.25. If the footings support a floor or beam not attached to elements likely to be damaged by large deflection, then differential settlement is limited to $D/360 = 2.8$ cm. What is the probability that $|\Delta| > 2.8$ cm?

SOLUTION See the previous single-footing example for some of the earlier details; note, however, that the correlation length has changed in this example.

1. A deterministic finite-element analysis of this problem gives $\delta_{\det} = 0.03578$ under each footing (this number is very slightly different than that found in the single-footing case due to interactions between the two footings). For $\sigma^2_{\ln E} = 0.69315$, the log-settlement statistics under either footing are $\mu_{\ln\delta} = \ln(\delta_{\det}) + \frac{1}{2}\sigma^2_{\ln E} = -2.9838$ and $\sigma_{\ln\delta} = \sqrt{\gamma(B,H)\sigma^2_{\ln E}} = \sqrt{(0.055776)(0.69315)} = 0.19662$.
2. To calculate $C_{\ln\delta}$, a short program written to implement the approach presented in Appendix C gives $C_{\ln\delta} = 3.1356 \times 10^{-7}$.
3. In terms of actual settlement under each footing, the mean, standard deviation, and correlation coefficient are $\mu_\delta = \exp\{\mu_{\ln\delta} + \frac{1}{2}\sigma^2_{\ln\delta}\} = 0.051587$, $\sigma_\delta = \mu_\delta\sqrt{e^{\sigma^2_{\ln\delta}} - 1} = 0.010242$, and $\rho_\delta = e^{C_{\ln\delta}} - 1/e^{\sigma^2_{\ln\delta}} - 1 = 7.9547 \times 10^{-6}$, respectively. A 5000-realization simulation run for this problem gave estimates of settlement mean and standard deviation of 0.0530 and 0.0081, respectively, and an estimated correlation coefficient of -0.014 (where the negative correlation coefficient estimate is clearly due to a bias in the classical estimator; see Section 5.4.1.1 for a discussion of this issue).
4. The differential settlement Δ has parameters $\mu_\Delta = 0$ and $\sigma^2_\Delta = 2(1 - 7.9547 \times 10^{-6})(0.010242)^2 = 0.0002098$ and the mean absolute differential settlement in this case is predicted to be $\mu_{|\Delta|} = \sqrt{2(0.0002098)/\pi} = 0.011$. The simulation run estimated the mean absolute differential settlement to be 0.009.
5. The desired probability is predicted to be $P[|\Delta| > 0.028] = 2\Phi\left(-0.028/\sqrt{0.0002098}\right) = 2\Phi(-1.933) = 0.0532$. The empirical estimate of this probability from the simulation run is 0.0204.

The normal distribution approximation to Δ somewhat overestimates the probability that $|\Delta|$ will exceed $D/360$. This is therefore a conservative estimate. From a design point of view, if the probability derived in step 5 is deemed unacceptable, one solution is to widen the footing. This will result in a rapid decrease in $\mathrm{P}[|\Delta| > 0.028]$ in the case given above. In particular, if B is increased to 3.0 m, the empirical estimate of $\mathrm{P}[|\Delta| > 0.028]$ reduces by more than a factor of 10 to 0.0016.

The distribution of absolute differential settlement is highly dependent on the correlation length, primarily through the calculation of $\sigma_{\ln\delta}$. Unfortunately, the correlation length is a quantity that is very difficult to estimate and which is poorly understood for real soils, particularly in the horizontal direction, which is more important for differential settlement. If $\theta_{\ln E}$ is increased to 10.0 m in the above example, the empirical estimate of $\mathrm{P}[|\Delta| > 0.028]$ increases drastically to 0.44. From a design point of view, the problem is compounded since, for such a large correlation length, $\mathrm{P}[|\Delta| > 0.028]$ now decreases very slowly as the footing width is increased (holding the load constant). For example, a footing width of 5.0 m, with $\theta_{\ln E} = 10.0$ m, has $\mathrm{P}[|\Delta| > 0.028] = 0.21$. Thus, establishing the correlation length in the horizontal direction is a critical issue in differential settlement limit state design, and one which needs much more work.

10.2.4 Summary

On the basis of this simulation study, the following observations can be made. The settlement under a footing founded on a spatially random elastic modulus field of finite depth overlying bedrock is well represented by a lognormal distribution with parameters $\mu_{\ln\delta}$ and $\sigma^2_{\ln\delta}$ if E is also lognormally distributed. The first parameter, $\mu_{\ln\delta}$, is dependent on the mean and variance of the underlying log-elastic modulus field and may be closely approximated by considering limiting values of $\theta_{\ln E}$. One of the points made in this section is the observation that the second parameter, $\sigma^2_{\ln\delta}$, is very well approximated by the variance of a local average of the log-elastic modulus field in the region directly under the footing. This conclusion is motivated by the observation that settlement is inversely proportional to the geometric average of the elastic modulus field and gives the prediction of $\sigma^2_{\ln\delta}$ some generality that can be extended beyond the actual range of simulation results considered herein. For very deep soil layers underlying the footing, it is recommended that the depth of the averaging region not exceed about 10 times its width, due to stress reduction with depth. Once the statistics of the settlement, $\mu_{\ln\delta}$ and $\sigma^2_{\ln\delta}$, have been computed, using Eqs. 10.8 and 10.9, the estimation of probabilities associated with settlement involves little more than referring to a standard normal distribution table.

The differential settlement follows a more complicated distribution than that of settlement itself (see Eq. 10.13). This is seen also in the differential settlement histograms, which tend to be quite erratic with narrow peaks and long tails, particularly at large $\theta_{\ln E}/D$ ratios. Although the difference between two lognormally distributed random variables is not normally distributed, the normal approximation has nevertheless been found to be reasonable, yielding conservative estimates of probability over the bulk of the distribution. For a more accurate estimation of probability relating to differential settlement, where it can be assumed that footing settlement is lognormally distributed, Eq. 10.13 should be numerically integrated, and this approach will be investigated in the next section. Both the simplified normal approximation and the numerical integration of Eq. 10.13

depend upon a reasonable estimate of the covariance between footing settlements. Another observation made in this section is that this covariance is closely (and conservatively) estimated using the covariance between local averages of the log-elastic modulus field under the two footings. Discrepancies between the covariance predicted on this basis and simulation results are due to interactions between the footings when they are closely spaced; such interactions lead to higher correlations than predicted by local average theory, which leads to smaller differential settlements in practice than predicted. This is conservative. The recommendations regarding the maximum averaging depth made for the single-footing case would also apply here.

Example calculations are provided above to illustrate how the findings of the section may be used. These calculations are reasonably simple for hand calculations (except for the numerical integration in Eq. 10.15) and are also easily programmed. They allow probability estimates regarding settlement and differential settlement, which in turn allows the estimation of the risk associated with this particular limit state for a structure.

A critical unresolved issue in the risk assessment of differential settlement is the estimation of the correlation length, $\theta_{\ln E}$, since it significantly affects the differential settlement distribution. A tentative recommendation is to use a correlation length which is some fraction of the distance between footings, say $D/10$. There is, at this time, little justification for such a recommendation, aside from the fact that correlation lengths approaching D or bigger yield differential settlements which are felt to be unrealistic in a practical sense and, for example, not observed in the work of D'Appolonia et al. (1968).

10.3 THREE-DIMENSIONAL PROBABILISTIC FOUNDATION SETTLEMENT

This section presents a study of the probability distributions of settlement and differential settlement where the soil is modeled as a fully three-dimensional random field and footings have both length and breadth. The study is an extension of the previous section, which used a two-dimensional random soil to investigate the behavior of a strip footing of infinite length. The resulting two-dimensional probabilistic model is found to also apply in concept to the three-dimensional case here. Improved results are given for differential settlement by using a bivariate lognormal distribution, rather than the approximate univariate normal distribution used in the previous section. The program used to perform the study presented in this section is RSETL3D, available at `http://www.engmath.dal.ca/rfem` (Fenton and Griffiths, 2005b; Griffiths and Fenton, 2005).

The case of a single square, rigid pad footing is considered first, a cross section through which is shown in

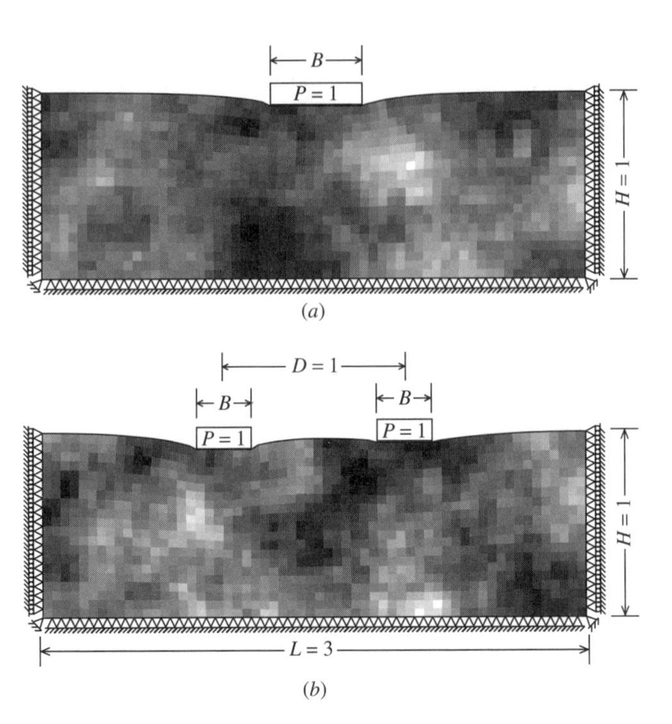

Figure 10.11 Slices through random-field/finite-element mesh of (*a*) single footing and (*b*) two footings founded on a soil layer.

Figure 10.11*a*, and the probability density function governing total settlement of the footing is estimated as a function of footing area for various statistics of the underlying soil. Only the soil elasticity is considered to be spatially random. Uncertainties arising from model and test procedures and in the loads are not considered. In addition, the soil is assumed to be isotropic; that is, the correlation structure is assumed to be the same in both the horizontal and vertical directions. Although soils generally exhibit a stronger correlation in the horizontal direction, due to their layered nature, the degree of anisotropy is site specific. In that this study is attempting to establish the basic probabilistic behavior of settlement, anisotropy is left as a refinement for site-specific investigations. The authors expect that the averaging model suggested in this section will drift from a geometric average to a harmonic average as the ratio of horizontal to vertical correlation lengths increases (see also Section 10.3.2). Only the geometric average model will be considered here since it allows for an approximate analytical solution. If the soil is known to be strongly layered, risk assessments should be performed by simulation using harmonic averages.

The footings are assumed to be founded on a soil layer underlain by bedrock. The assumption of an underlying bedrock can be relaxed if a suitable averaging region is used (recommendations about such an area are given in the previous section).

The second part of the section addresses the issue of differential settlements under a pair of footings, as shown in Figure 10.11*b*, again for the particular case of footings founded on a soil layer underlain by bedrock. The footing spacing is held constant at $D = 1$ while varying the footing size. Both footings are square and the same size. The mean and standard deviation of differential settlements are estimated as functions of footing size for various input statistics of the underlying elastic modulus field. The probability distribution governing differential settlement is found to be reasonably approximated using a joint lognormal distribution with correlation predicted using local geometric averages of the elastic modulus field under the two footings.

10.3.1 Random Finite-Element Method

The soil mass is discretized into 60 eight-node brick elements in each of the horizontal directions by 20 elements in the vertical direction. Each element is cubic with side length 0.05 giving a soil mass which has plan dimension 3×3 and depth 1. (*Note*: Length units are not used here since the results can be used with any consistent set of length and force units.) Figure 10.12 shows the finite-element mesh in three dimensions for the case of two footings.

The finite-element analysis uses a preconditioned conjugate gradient iterative solver (see Smith and Griffiths, 2004) that avoids the need to assemble the global stiffness matrix. Numerical experimentation indicates that the finite-element model gives excellent agreement with analytical results for a flexible footing. In the case of a rigid footing, doubling the number of elements in each direction resulted in a settlement which increased by about 3%, indicating that the

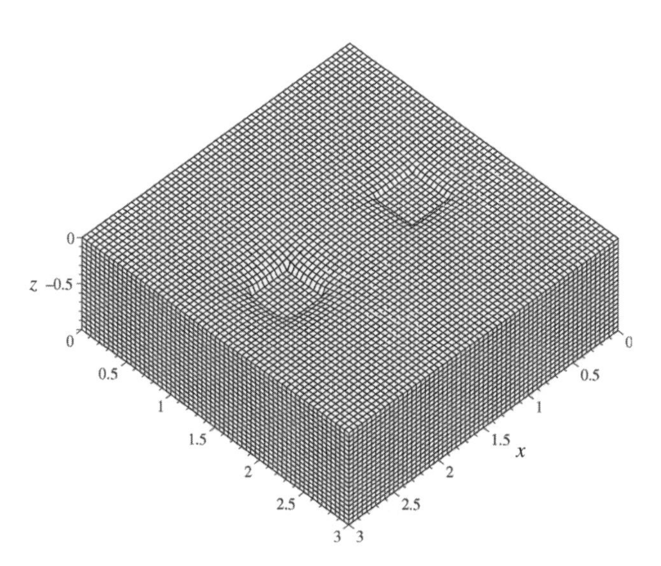

Figure 10.12 Finite-element mesh model of soil supporting two footings.

rigid-footing model may be slightly too stiff at the current resolution. However, the stochastic behavior will be unaffected by such slight shifts in total settlement (i.e., by the same fraction for each realization). The $60 \times 60 \times 20$ discretization was considered adequate to characterize the behavior of the settlement and differential settlement probability distributions.

The vertical side faces of the finite-element model are constrained against horizontal displacement but are free to slide vertically while the nodes on the bottom boundary are spatially fixed. The footing(s) are assumed to be rigid, to not undergo any rotations, and to have a rough interface with the underlying soil (no-slip boundary). A fixed load $P = 1$ is applied to each footing; since settlement varies linearly with load, the results are easily scaled to different values of P.

To investigate the effect of the square-footing area, the soil layer thickness H was held constant at 1.0 while the footing plan dimension B was varied according to Table 10.2. Because the settlement problem is linear in some of its parameters, the following results can be scaled to arbitrary square-footing areas so long as the ratio B/H is held fixed. For example, the settlement of a square footing of dimension $B' = 4.0$ m on an $H' = 20.0$ m thick soil layer with $P' = 1000$ kN and elastic modulus $E' = 60$ kN/m^2 corresponds to $\frac{5}{6}$ times the settlement of a footing of width $B = 0.2$ m on an $H = 1$ m thick soil layer with load $P' = 1$ kN and elastic modulus $E = 1$ kN/m^2. The scaling factor from the unprimed to the primed case is $(P'/P)(E/E')(B/B')$ as long as $B'/H' = B/H$. If B/H is not constant, a deterministic finite-element analysis will have to be performed to determine the scaling constant.

In the two-footing case, the distance between footing centers was held constant at $D = 1.0$, while the footing widths (assumed the same) were varied. Footings of width greater than 0.8 were not considered since this situation becomes basically that of a strip footing (the footings are joined when $B = 1.0$).

Table 10.2 Input Parameters Varied in Study While Holding $H = 1$, $D = 1$, $P = 1$, $\mu_E = 1$, and $\nu = 0.25$ Constant

Parameter	Values Considered
σ_E	0.1*, 0.5, 1.0*, 2.0, 4.0
$\theta_{\ln E}$	0.01, 0.1*, 0.5*, 1.0*, 5.0*, 10.0*
B	0.2, 0.4, 0.8, 1.6 (single footing)
	0.2*, 0.4*, 0.8* (two footings)

Note: Starred parameters were run with 1000 realizations in the two-footing case. The single-footing case and nonstarred parameters were run with 100 realizations.

The soil has two properties of interest to the settlement problem: the (effective) elastic modulus $E(\mathbf{x})$ and Poisson's ratio $\nu(\mathbf{x})$, where $\mathbf{x}$ is spatial position. Only the elastic modulus is considered to be a spatially random soil property. Poisson's ratio was believed to have a smaller relative spatial variability and only a second-order importance to settlement statistics. It is held fixed at 0.25 over the entire soil mass for all simulations.

Figure 10.11 shows a gray-scale representation of a possible realization of the elastic modulus field along a vertical plane through the soil mass cutting through the footing. Lighter areas denote smaller values of $E(\mathbf{x})$ so that the elastic modulus field shown in Figure 10.11*b* corresponds to a higher elastic modulus under the right footing than under the left; this leads to the substantial differential settlement indicated by the deformed mesh. This is just one possible realization of the elastic modulus field; the next realization could just as easily show the opposite trend (see, e.g., Figure 10.3).

The elastic modulus field is assumed to follow a lognormal distribution so that $\ln(E)$ is a Gaussian (normal) random field with mean $\mu_{\ln E}$ and variance $\sigma^2_{\ln E}$. The choice of a lognormal distribution is motivated in part by the fact that the elastic modulus is strictly nonnegative, a property of the lognormal distribution (but not the normal), while still having a simple relationship with the normal distribution. In addition, soil properties are generally measured as averages over some volume and these averages are often *low-strength dominated*, as may be expected. The authors have found in this and other studies that the geometric average well represents such low-strength-dominated soil properties. Since the distribution of a geometric average of positive quantities tends to the lognormal distribution by the central limit theorem, the lognormal distribution may very well be a natural distribution for many spatially varying soil properties.

A Markovian spatial correlation function, which gives the correlation coefficient between log-elastic modulus values at points separated by the distance τ, is used (see Section 3.7.10.2 for more details),

$$\rho_{\ln E}(\boldsymbol{\tau}) = \exp\left\{-\frac{2|\boldsymbol{\tau}|}{\theta_{\ln E}}\right\} \tag{10.19}$$

in which $\boldsymbol{\tau} = \mathbf{x} - \mathbf{x}'$ is the vector between spatial points $\mathbf{x}$ and $\mathbf{x}'$ and $|\boldsymbol{\tau}|$ is the absolute length of this vector (the lag distance). The results presented here are not particularly sensitive to the choice in functional form of the correlation; the Markov model is popular because of its simplicity.

As was found in the two-dimensional case for the two-footing case (see previous section), using a correlation length $\theta_{\ln E}$ equal to the footing spacing D is conservative in that it yields the largest probabilities of differential settlement. For total settlement of a single footing, taking $\theta_{\ln E}$ large is conservative since this leads to the largest variability of settlement (least variance reduction due to averaging of the soil properties under the footing).

To investigate the effect of the correlation length $\theta_{\ln E}$ on the settlement statistics, $\theta_{\ln E}$ is varied from 0.01 (i.e., very much smaller than the footing and/or footing spacing) to 10.0 (i.e., substantially larger than the footing and/or footing spacing). In the limit as $\theta_{\ln E} \to 0$, the elastic modulus field becomes a white noise field, with E values at any two distinct points independent. In terms of the finite elements themselves, values of $\theta_{\ln E}$ smaller than the elements results in a set of elements which are largely independent (increasingly independent as $\theta_{\ln E}$ decreases). But because the footing effectively averages the elastic modulus field on which it is founded, and since averages have decreased variance, the settlement in the limiting case $\theta_{\ln E} \to 0$ is expected to approach that obtained in the deterministic case, with E equal to its median everywhere (assuming geometric averaging), having vanishing variance for finite $\sigma^2_{\ln E}$.

At the other extreme, as $\theta_{\ln E} \to \infty$, the elastic modulus field becomes the same everywhere. In this case, the settlement statistics are expected to approach those obtained by using a single lognormally distributed random variable E to model the soil, $E(\mathbf{x}) = E$. That is, since the settlement δ under a footing founded on a soil layer with uniform (but random) elastic modulus E is given by

$$\delta = \frac{\delta_{\text{det}}\mu_E}{E} \tag{10.20}$$

for δ_{det} the settlement computed when $E = \mu_E$ everywhere, then as $\theta_{\ln E} \to \infty$ the settlement assumes a lognormal distribution with parameters

$$\mu_{\ln\delta} = \ln(\delta_{\text{det}}) + \ln(\mu_E) - \mu_{\ln E} = \ln(\delta_{\text{det}}) + \tfrac{1}{2}\sigma^2_{\ln E} \tag{10.21a}$$

$$\sigma_{\ln\delta} = \sigma_{\ln E} \tag{10.21b}$$

where Eq. 1.176b was used in Eq. 10.21a.

By similar reasoning the differential settlement between two footings (see Figure 10.11*b*) as $\theta_{\ln E} \to 0$ is expected to go to zero since the average elastic moduli seen by both footings approach the same value, namely the median (assuming geometric averaging). At the other extreme, as $\theta_{\ln E} \to \infty$ the differential settlement is also expected to approach zero, since the elastic modulus field becomes the same everywhere. Thus, the differential settlement approaches zero at both very small and very large correlation lengths—the largest differential settlements will occur at correlation lengths somewhere in between these two extremes. This "worst-case" correlation length has been observed by other researchers; see, for example, the work on a flexible foundation by Baecher and Ingra (1981).

The Monte Carlo approach adopted here involves the simulation of a realization of the elastic modulus field and subsequent finite-element analysis (e.g., Smith and Griffiths, 2004) of that realization to yield a realization of the footing settlement(s). Repeating the process over an ensemble of realizations generates a set of possible settlements which can be plotted in the form of a histogram and from which distribution parameters can be estimated.

If it can be assumed that log-settlement is approximately normally distributed (which is seen later to be a reasonable assumption and is consistent with the distribution selected for E), and $m_{\ln\delta}$ and $s^2_{\ln\delta}$ are the estimators of the mean and variance of log-settlement, respectively, then the standard deviations of these estimators obtained from the $n = 100$ realizations performed for the single-footing case are given by

$$\sigma_{m_{\ln\delta}} \simeq s_{\ln\delta}/\sqrt{n} = 0.1 s_{\ln\delta} \tag{10.22a}$$

$$\sigma_{s^2_{\ln\delta}} \simeq \sqrt{\frac{2}{n-1}}\, s^2_{\ln\delta} = 0.14 s^2_{\ln\delta} \tag{10.22b}$$

These estimator errors are not particularly small but, since the three-dimensional analysis is very time consuming, the number of realizations selected was deemed sufficient to verify that the geometric averaging model suggested in the two-dimensional case is also applicable in three dimensions. A subset of the cases considered in Table 10.2 (see starred quantities) was rerun using 1000 realizations to verify the probability distributions further out in the tails for the differential settlement problem.

Realizations of the log-elastic modulus field are produced using the three-dimensional LAS technique (Section 6.4.6). The elastic modulus value assigned to the ith element is

$$E_i = \exp\{\mu_{\ln E} + \sigma_{\ln E}\, G_i\} \tag{10.23}$$

where G_i is a local arithmetic average, over the element centered at $\mathbf{x}_i$, of a zero-mean, unit-variance Gaussian random field.

10.3.2 Single-Footing Case

A typical histogram of the settlement under a single footing is shown in Figure 10.13 for $B = 0.4$, $\sigma_E/\mu_E = 0.5$, and $\theta_{\ln E} = 1.0$ (1000 realizations were performed for this case to increase the resolution of the histogram). With the requirement that settlement be nonnegative, the shape of the histogram suggests a lognormal distribution, which was adopted in this study (see also Eqs. 10.21). The histogram is normalized to produce a frequency density plot, where a straight line is drawn between the interval midpoints. A chi-square goodness-of-fit test on the data of Figure 10.13 yielded a p-value of 0.54, indicating very good support for the lognormal hypothesis. The fitted lognormal distribution,

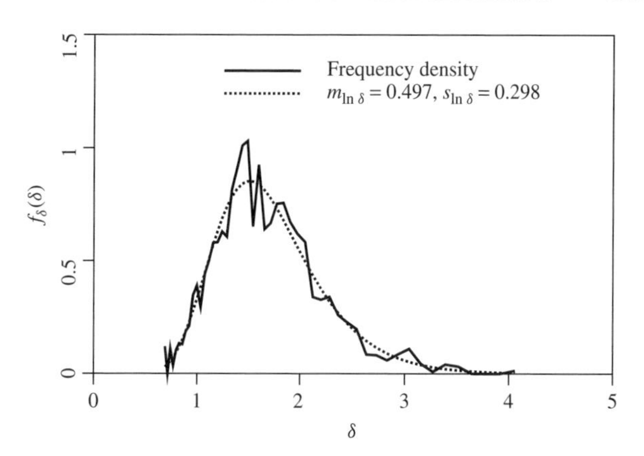

Figure 10.13 Typical frequency density plot and fitted lognormal distribution of settlement under single footing.

with parameters given by $m_{\ln\delta}$ and $s_{\ln\delta}$ shown in the line key, is superimposed on the plot.

Accepting the lognormal distribution as a reasonable fit to the simulation results, the next task is to estimate the parameters of the fitted lognormal distributions as functions of the input parameters (B, σ_E, and $\theta_{\ln E}$). The lognormal distribution,

$$f_\delta(x) = \frac{1}{\sqrt{2\pi}\,\sigma_{\ln\delta}\, x} \exp\left\{-\frac{1}{2}\left(\frac{\ln x - \mu_{\ln\delta}}{\sigma_{\ln\delta}}\right)^2\right\}, \quad 0 \le x < \infty \tag{10.24}$$

has two parameters, $\mu_{\ln\delta}$ and $\sigma_{\ln\delta}$. Figure 10.14 shows how the estimator of $\mu_{\ln\delta}$, denoted by $m_{\ln\delta}$, varies with $\sigma^2_{\ln E}$ for

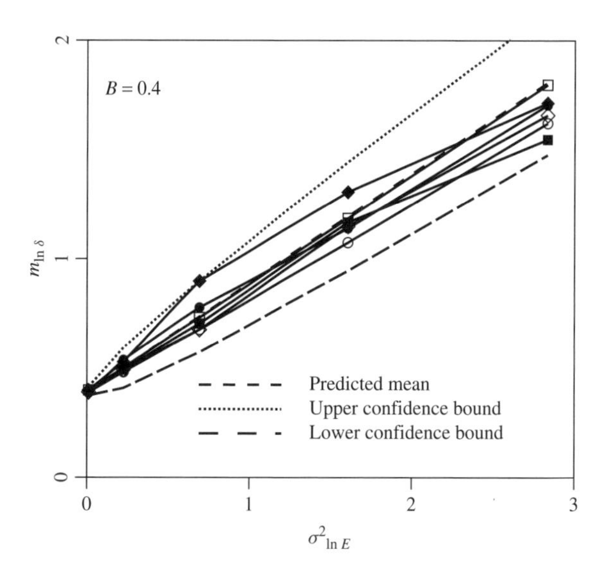

Figure 10.14 Estimated mean of log-settlement with prediction given by Eq. 10.21a.

$B = 0.4$ based on 100 realizations. All correlation lengths are drawn in the plot but are not individually labeled since they lie so close together. Also shown on the plot are the 95% confidence interval bounds on the true parameter, $\mu_{\ln\delta}$. As can be seen, all the estimators lie within these bounds indicating that the three-dimensional results are much the same as found using a two-dimensional model, namely that $\mu_{\ln\delta}$ is well predicted by Eq. 10.4a, which is identical to Eq. 10.21a. Similar results are observed for the other footing sizes considered.

Estimates of the standard deviation of log-settlement, $s_{\ln\delta}$, are plotted in Figure 10.15 (as symbols) for two footing sizes based on 100 realizations. The other footing sizes gave similar results. In all cases, $s_{\ln\delta}$ increases to $\sigma_{\ln E}$ as $\theta_{\ln E}$ increases, which is as expected since large correlation lengths give less variance reduction. Assuming that local geometric averaging of the volume directly under the footing accounts for all of the variance reduction seen in Figure 10.15, the standard deviation of log-settlement is predicted by

$$\sigma_{\ln\delta} = \sqrt{\gamma(B,B,H)}\,\sigma_{\ln E} \tag{10.25}$$

where $\gamma(B,B,H)$ is the three-dimensional variance reduction function (see Section 3.4), giving the amount that the variance is reduced due to averaging. The agreement between Eq. 10.25 and the estimated standard deviations is remarkable, indicating that a geometric average of the elastic modulus field under the footing is a good model for the effective elastic modulus seen by the footing. In this case,

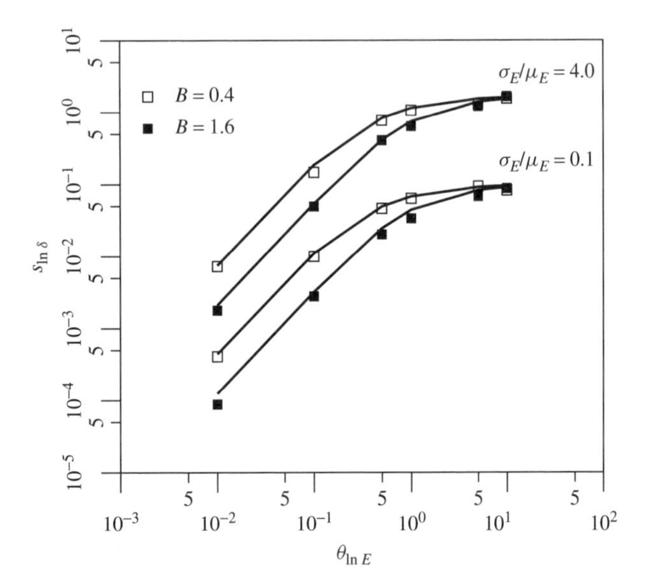

Figure 10.15 Comparison of simulated sample standard deviation of log-settlement, shown with symbols, to theoretical estimate via Eq. 10.25, shown with lines.

the geometric average E_g has the definition

$$E_g = \exp\left(\frac{1}{B^2H}\int_0^H\int_0^B\int_0^B \ln E(x,y,z)\,dx\,dy\,dz\right) \tag{10.26}$$

from which the settlement of the footing can be expressed as

$$\delta = \frac{\delta_{\det}\mu_E}{E_g} \tag{10.27}$$

Taking the logarithm of Eq. 10.27 and taking expectations lead to Eqs. 10.21a and 10.25. The practical implication of this result is that settlements are better predicted using an effective elastic modulus computed as a geometric average of the experimentally obtained moduli under the footing. For example, if n observations of the elastic modulus under a footing are taken, $E_1, E_2, \ldots, E_n$, then the footing settlement is best computed using the elastic modulus E_g computed as

$$E_g = \left(\prod_{i=1}^{n} E_i\right)^{1/n} = \exp\left(\frac{1}{n}\sum_{i=1}^{n}\ln E_i\right) \tag{10.28}$$

Once the parameters of the settlement distribution, $\mu_{\ln\delta}$ and $\sigma_{\ln\delta}$, have been calculated using Eqs. 10.21a and 10.25, probabilities associated with settlement are easily found,

$$\text{P}[\delta > x] = 1 - \Phi\left(\frac{\ln x - \mu_{\ln\delta}}{\sigma_{\ln\delta}}\right) \tag{10.29}$$

where Φ is the cumulative standard normal function. This computation assumes that settlement is lognormally distributed, as these studies clearly suggest.

10.3.3 Two-Footing Case

Consider now the case of two square footings each of plan dimension $B \times B$ and separated by center-to-center distance $D = 1$, as shown in Figure 10.11b. If the settlements δ_1 and δ_2 under each footing are lognormally distributed, as was found in the previous section, then the joint distribution between the two-footing settlements follows a bivariate lognormal distribution,

$$f_{\delta_1,\delta_2}(x,y) = \frac{1}{2\pi\sigma^2_{\ln\delta}rxy} \times \exp\left\{-\frac{1}{2r^2}\left[\Psi_1^2 - 2\rho_{\ln\delta}\Psi_1\Psi_2 + \Psi_2^2\right]\right\}, \quad x \geq 0, \quad y \geq 0 \tag{10.30}$$

where $\Psi_1 = (\ln x - \mu_{\ln\delta})/\sigma_{\ln\delta}$, $\Psi_2 = (\ln y - \mu_{\ln\delta})/\sigma_{\ln\delta}$, $r^2 = 1 - \rho^2_{\ln\delta}$, and $\rho_{\ln\delta}$ is the correlation coefficient between the log-settlement of the two footings. It is assumed in the above that δ_1 and δ_2 have the same mean and variance, which, for the symmetric conditions shown in Figure 10.11b, is a reasonable assumption.

Defining the differential settlement between footings to be $\Delta = \delta_1 - \delta_2$, the mean of Δ is zero if the elastic modulus field is statistically stationary, as assumed here (if the field is not stationary, then the differential settlement due to any trend in the mean must be handled separately). If Eq. 10.30 holds, then the exact distribution governing the differential settlement is given by

$$f_\Delta(x) = \int_0^\infty f_{\delta_1,\delta_2}(|x| + y, y)\, dy \tag{10.31}$$

and differential settlement probabilities can be computed as

$$\mathrm{P}[|\Delta| > x] = \mathrm{P}[\Delta < -x \,\cup\, \Delta > x] = 2\int_x^\infty f_\Delta(\xi)\, d\xi \tag{10.32}$$

Figure 10.16 shows typical frequency density plots of differential settlement for three different values of $\theta_{\ln E}$ between two equal-sized footings with $B = 0.4$ and $\sigma_E/\mu_E = 1.0$. Notice that the widest distribution occurs when $\theta_{\ln E}/D$ is equal to about 1.0, indicating that the worst case, when it comes to differential settlement, occurs when the correlation length is approximately equal to the distance between footings.

The distribution f_{δ_1,δ_2}, and thus also f_Δ, has three parameters, $\mu_{\ln\delta}$, $\sigma_{\ln\delta}$, and $\rho_{\ln\delta}$. The mean and standard deviation can be estimated using Eqs. 10.21a and 10.25. Since local averaging of the log-elastic modulus field under the footing was found to be an accurate predictor of the variance of log-settlement, it is reasonable to suggest that the covariance between log-settlements under a pair of footings can be predicted by the covariance between local averages of the log-elastic modulus field under each footing. As we shall see later, mechanical interaction between the footings (where the settlement of one footing causes some settlement in the adjacent footing) leads to higher covariances than suggested purely by consideration of the covariances between soil properties. However, when the footings are spaced sufficiently far apart, the mechanical interaction will be negligible. In this case, for equal-sized footings, the covariance between local averages of the log-elastic modulus field under two footings separated by distance D is given by

$$C_{\ln\delta} = \frac{\sigma_{\ln E}^2}{V_1 V_2}\int_{V_1}\int_{V_2} \rho_{\ln E}(\mathbf{x}_1 - \mathbf{x}_2)\, d\mathbf{x}_2\, d\mathbf{x}_1 \tag{10.33}$$

where V_1 is the $B \times B \times H$ volume under footing 1, V_2 is the equivalent volume under footing 2, and $\mathbf{x}$ is a spatial position. From this the correlation coefficient between the two local averages can be computed as

$$\rho_{\ln\delta} = \frac{C_{\ln\delta}}{\sigma_{\ln\delta}^2} \tag{10.34}$$

The predicted correlation can be compared to the simulation results by first transforming back from log-space,

$$\rho_\delta = \frac{\exp\{C_{\ln\delta}\} - 1}{\exp\{\sigma_{\ln\delta}^2\} - 1} \tag{10.35}$$

where $\sigma_{\ln\delta}$ is given by Eq. 10.25. The agreement between the correlation coefficient predicted by Eq. 10.35 and the correlation coefficient estimated from the simulations (1000 realizations) is shown in Figure 10.17. The agreement is reasonable, particularly for the smaller sized footings. For larger footings, the correlation is underpredicted, particularly at small $\theta_{\ln E}$. This is due to mechanical interaction between the larger footings, where the settlement of one footing induces some additional settlement in the adjacent footing on account of their relatively close proximity.

Armed with the relationships 10.21a, 10.25, and 10.34 the differential settlement distribution f_Δ can be computed using Eq. 10.31. The resulting predicted distributions have been superimposed on the frequency density plots of Figure 10.16 for $B = 0.4$. The agreement is very good for intermediate to large correlation lengths. At the smaller

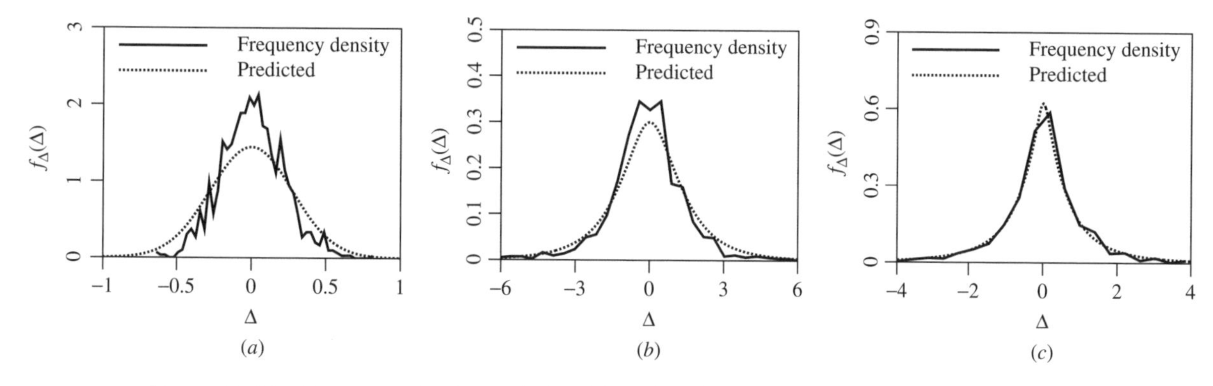

Figure 10.16 Frequency density with fitted bivariate lognormal distribution (Eq. 10.30) for differential settlement under two equal-sized footings with (*a*) $\theta_{\ln E}/D = 0.1$, (*b*) $\theta_{\ln E}/D = 1.0$, and (*c*) $\theta_{\ln E}/D = 10.0$.

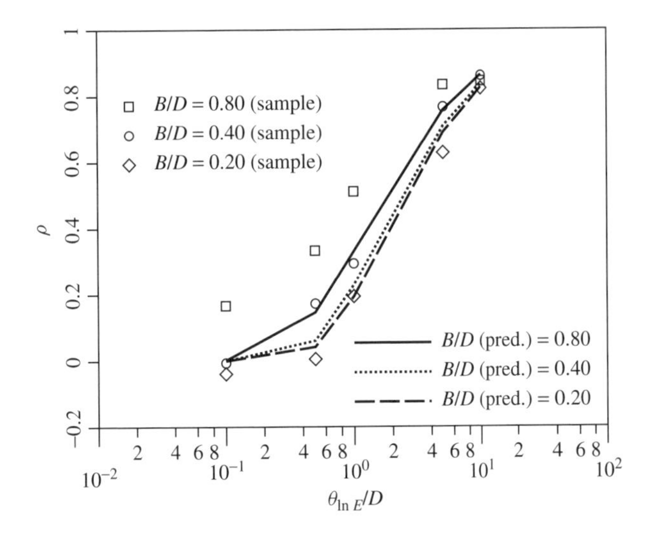

Figure 10.17 Predicted (pred.) and sample correlation coefficients between footing settlements for various relative separation distances between the footings and for $\sigma_E/\mu_E = 1$.

correlation lengths, Eq. 10.31 yields a distribution which is somewhat too wide—this is due to the underprediction of the correlation between footing settlements (Eq. 10.34) since, as the actual correlation between settlements increases, the differential settlement decreases and the distribution becomes narrower. However, an underprediction of correlation is at least conservative in that predicted differential settlement probabilities will tend to be too large.

To test the ability of the bivariate lognormal distribution to accurately estimate probabilities, the probability that the absolute value of Δ exceeds some threshold is compared to empirical probabilities derived from simulation. For generality, thresholds of $\alpha\mu_{|\Delta|}$ will be used, where $\mu_{|\Delta|}$ is the mean absolute differential settlement, which can be approximated as (this is exact if Δ is normally distributed)

$$\mu_{|\Delta|} \simeq \sqrt{\frac{2}{\pi}}\,\sigma_\Delta \tag{10.36}$$

where $\sigma_\Delta^2 = 2\sigma_\delta^2(1-\rho_\delta)$. Figure 10.18 shows a plot of the predicted (Eq. 10.32) versus empirical probabilities $\mathrm{P}\left[|\Delta| > \alpha\mu_{|\Delta|}\right]$ for α varying in 20 steps from 0.2 to 4.0. If the prediction is accurate, then the plotted points should lie on the diagonal line. The empirical probabilities are estimated by simulation. When the footings are well separated ($B/D = 0.2$, see Figure 10.18*a*) so that mechanical correlation is negligible, then the agreement between predicted and empirical probabilities is excellent. The two solid curved lines shown in the plot form a 95% confidence interval on the empirical probabilities, and it can be seen that most lie within these bounds. The few that lie outside are on the conservative side (predicted probability exceeds empirical probability).

As the footing size increases (see Figure 10.18*b*) so that their relative spacing decreases, the effect of mechanical correlation begins to be increasingly important and the resulting predicted probabilities increasingly conservative. A strictly empirical correction can be made to the correlation to account for the missing mechanical influences. If $\rho_{\ln\delta}$ is replaced by $(1 - B/2D)\rho_{\ln\delta} + B/2D$ for all B/D greater than about 0.3, which is an entirely empirical correction, the differential settlements are reduced and, as shown in Figure 10.19, the predicted probabilities become reasonably close to the empirical probabilities while still remaining slightly conservative. Until the complex interaction between two relatively closely spaced footings is

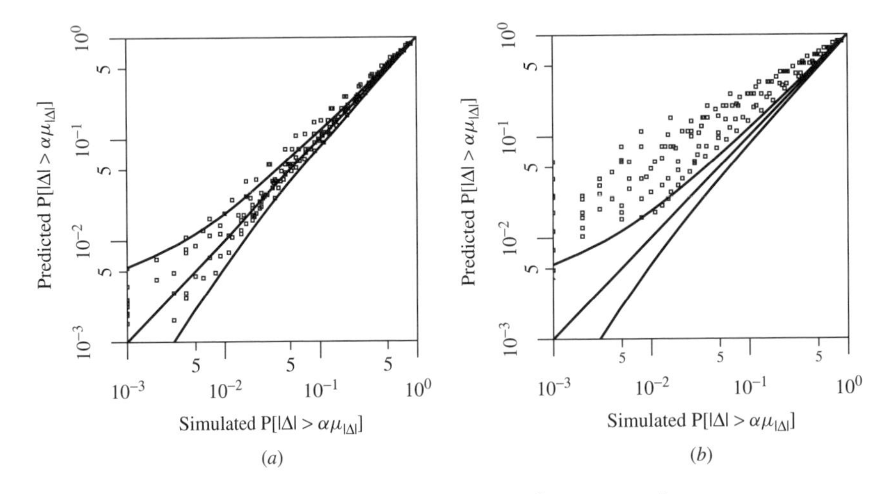

Figure 10.18 Predicted versus empirical probabilities $\mathrm{P}\left[|\Delta| > \alpha\mu_{|\Delta|}\right]$ for σ_E/μ_E values of 0.1 and 1.0, $\theta_{\ln E}$ varying from 0.1 to 10.0, and (*a*) $B/D = 0.2$ and (*b*) $B/D = 0.8$. Curved lines are 95% confidence intervals.

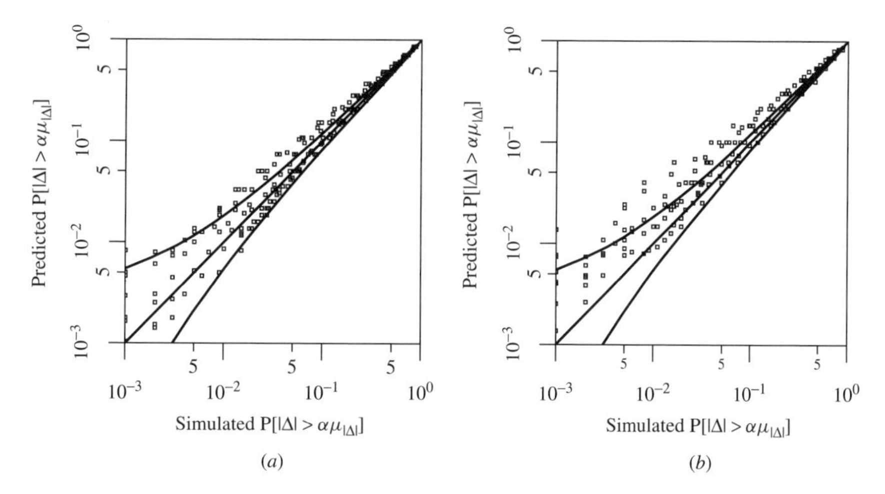

Figure 10.19 Predicted versus empirical probabilities $\mathrm{P}\left[|\Delta| > \alpha\mu_{|\Delta|}\right]$ corrected by empirically increasing $\rho_{\ln\delta}$ for (*a*) $B/D = 0.4$ and (*b*) $B/D = 0.8$.

fully characterized probabilistically, this simple empirical correction seems reasonable.

10.3.4 Summary

On the basis of this simulation study, the following observations can be made. As found in the two-dimensional case, the settlement under a footing founded on a three-dimensional spatially random elastic modulus field of finite depth overlying bedrock is well represented by a lognormal distribution with parameters $\mu_{\ln\delta}$ and $\sigma^2_{\ln\delta}$ if E is also lognormally distributed. The first parameter, $\mu_{\ln\delta}$, is dependent on the mean and variance of the underlying log-elastic modulus field and may be closely approximated by considering limiting values of $\theta_{\ln E}$. The second parameter, $\sigma^2_{\ln\delta}$, is very well represented by the variance of a local average of the log-elastic modulus field in the region directly under the footing. Once the parameters of the settlement, $\mu_{\ln\delta}$ and $\sigma^2_{\ln\delta}$, have been computed, using Eqs. 10.21a and 10.25, the estimation of probabilities associated with settlement involves little more than referring to a standard normal distribution table (see Eq. 10.29).

One of the implications of the findings for a single footing is that footing settlement is accurately predicted using a geometric average of the elastic modulus field in the volume under the footing. From a practical point of view, this finding implies that a geometric average of soil elastic modulus estimates made in the vicinity of the footing (e.g., by CPT soundings) should be used to represent the effective elastic modulus rather than an arithmetic average. The geometric average will generally be less than the arithmetic average, reflecting the stronger influence of weak soil zones on the total settlement.

Under the model of a lognormal distribution for the settlement of an individual footing, the bivariate lognormal distribution was found to closely represent the joint settlement of two footings when the footings are spaced sufficiently far apart (relative to their plan dimension) to avoid significant mechanical interaction. Using the bivariate lognormal model, probabilities associated with differential settlement are obtained that are in very good agreement with empirical probabilities obtained via simulation. The bivariate lognormal model is considerably superior to the approximate normal model developed in the two-dimensional case in the previous section, at the expense of a more complicated numerical integration (the normal approximation simply involved a table lookup).

When the footings become close enough that mechanical interaction becomes significant, the bivariate lognormal model developed here begins to overestimate the probabilities associated with differential settlement; that is, the differential settlements will be less than predicted. Although this is at least conservative, the reason is believed to be due to the fact that the stress field from one footing is affecting the soil under the other footing. This results in an increased correlation coefficient between the two-footing settlements that is not fully accounted for by the correlation between two local geometric averages alone. An empirical correction factor has been suggested in this section which yields more accurate probabilities and which should be employed if the conservatism without it is unacceptable.

10.4 STRIP FOOTING RISK ASSESSMENT

Foundation settlement, if excessive, can lead to unsightly cracking of structural and nonstructural elements of the supported building. For this reason most geotechnical design codes limit the settlement of footings to some reasonable amount, typically 25–50 mm (e.g., ASCE, 1994; CGS, 1992). Since the design of a footing is often governed by settlement, it would be useful to evaluate the reliability of typical "traditional" design methodologies.

In this section, the design of a strip footing against excessive settlement on a spatially random soil is studied and the reliability of the design assessed (Fenton et al., 2003b). The soil effective elastic modulus field $E(\mathbf{x})$, where $\mathbf{x}$ is spatial position, is modeled as a stationary spatially varying two-dimensional random field. Poisson's ratio is assumed deterministic and held constant at $\nu = 0.35$.

A two-dimensional analysis is performed on a strip footing assumed to be of infinite length out of plane. Spatial variation in the out-of-plane direction is ignored, which is equivalent to saying that the out-of-plane correlation length is infinite. This study provides a methodology to assessing the reliability of a traditional design method as well as to identify problems in doing so. A typical finite-element mesh showing a footing founded on a spatially random elastic modulus field, where light regions correspond to lower values of $E(\mathbf{x})$, is shown in Figure 10.20.

10.4.1 Settlement Design Methodology

The design method that will be studies is due to Janbu (1956), who expresses settlement under a strip footing in the form

$$\delta = \mu_0 \cdot \mu_1 \cdot \frac{qB}{E^*} \tag{10.37}$$

where q is the vertical stress in kilonewtons per square meter applied by the footing to the soil, B is the footing width, E^* is some equivalent measure of the soil elastic modulus underlying the footing, μ_0 is an influence factor for depth D of the footing below the ground surface, and μ_1 is an influence factor for the footing width B and depth of the soil layer H. A particular case study will be considered here for simplicity and clarity, rather than nondimensionalizing the problem. The particular case considered is of a footing founded at the surface of a soil layer ($\mu_0 = 1.0$) underlain by bedrock at a depth $H = 6$ m.

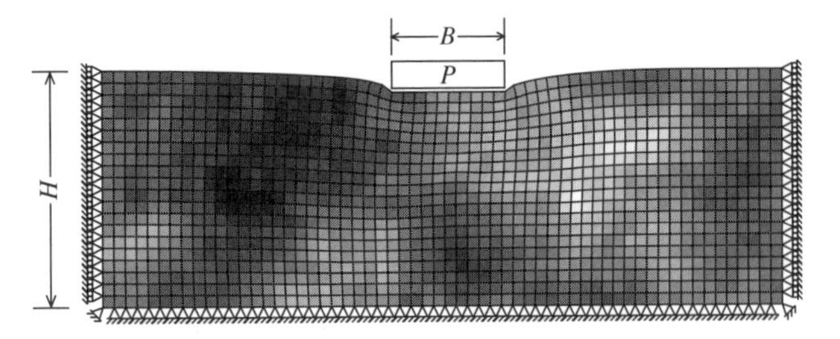

Figure 10.20 Deformed finite-element mesh with sample elastic modulus field.

The footing load is assumed to be deterministic and equal to $P = 1250$ kN per meter length of the footing in the out-of-plane direction. In terms of P, Eq. 10.37 can be rewritten as

$$\delta = \mu_0 \cdot \mu_1 \cdot \frac{P}{E^*} \tag{10.38}$$

In order to assess the design risk, we must compare Janbu's settlement predictions to settlements obtained by the RFEM. To properly compare the two, it is important to ensure that the two methods are in agreement when the soil is known and nonrandom. For this reason, it was decided to calibrate Janbu's relationship against the finite-element results obtained using deterministic and spatially constant elastic modulus $E^* = 30$ MPa for various ratios of H/B. Figure 10.21 illustrates how the influence factor μ_1 varies with $\ln(H/B)$. As can be seen, it is very nearly a straight line which is well approximated by

$$\mu_1 = a + b \ln\left(\frac{H}{B}\right) \tag{10.39}$$

where, for the case under consideration with a Poisson's ratio of 0.35, the line of best fit has $a = 0.4294$ and $b = 0.5071$, as shown fitted in Figure 10.21. The Janbu settlement prediction now can be written as

$$\delta = \mu_0 \left[a + b \ln\left(\frac{H}{B}\right)\right] \cdot \frac{P}{E^*} \tag{10.40}$$

The case where E^* is estimated by sampling the soil at a few locations below the footing is now considered. Letting $\hat{E}$ be the estimated elastic modulus, one possible estimator is

$$\hat{E} = \frac{H_1E_1 + H_2E_2 + \cdots + H_nE_n}{H} \tag{10.41}$$

where H_i is the thickness of the ith soil layer, E_i is the elastic modulus measured in the ith layer, and H is the total thickness of all layers. In this study individual layers are not considered directly, although spatial variability

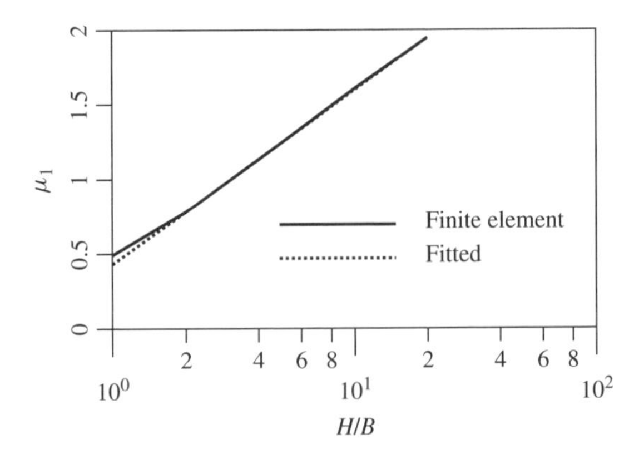

Figure 10.21 Effect of ratio H/B on settlement influence factor μ_1.

may lead to the appearance of layering. It will be assumed that n samples will be taken at equispaced distances over the depth H along a vertical line below the footing center. This sort of sampling might be obtained by using a CPT sounding, for example. In this case, the elastic modulus estimate would be computed as some sort of average of the observed values. We elect to use an arithmetic average of the observations, despite the recommendation in the previous two sections to use a geometric average, simply because the arithmetic average is more commonly used in practice. The estimate is, then, the classic formula

$$\hat{E} = \frac{1}{n} \sum_{i=1}^{n} E_i \tag{10.42}$$

No attempt is made in this study to account for measurement error. The goal here is to assess the foundation settlement variability when the design is based on actual observations of the elastic modulus at a few points.

Using the estimated elastic modulus, the settlement predicted by Janbu's method becomes

$$\delta_{\text{pred}} = \mu_0 \left[a + b \ln \left(\frac{H}{B} \right) \right] \cdot \frac{P}{\hat{E}} \tag{10.43}$$

If a maximum allowable settlement of 40 mm is to be designed for, then by setting $\delta_{\text{pred}} = \delta_{\max} = 0.04$ m, Eq. 10.43 can be solved for the required footing width B as

$$B = H \exp \left\{ -\frac{1}{b} \left(\frac{\hat{E}\,\delta_{\max}}{P\mu_0} - a \right) \right\} \tag{10.44}$$

Since the soil elastic modulus field $E(\mathbf{x})$ is random, the estimate $\hat{E}$ will also be random, which means that B is random. This is to be interpreted as follows: Consider a sequence of similar sites on each of which a footing is to be designed and placed to support the load P such that, for each, the settlement prediction is $\delta_{\max}$. Because the sites involve different realizations of the soil elastic modulus field, they will each have a different estimate $\hat{E}$ obtained by sampling. Thus, each site will have a different required footing width B.

The task now is to assess the distribution of the actual settlement experienced by each of these designed footings. If the prediction equation is accurate, then it is expected that approximately 50% of the footings will experience settlements in excess of $\delta_{\max}$ while the remaining 50% will experience less settlement. But, how much more or less? That is, what is the variability of settlement in this case? Note that this is a conditional probability problem. Namely, the random field $E(\mathbf{x})$ has been sampled at n points to obtain the design estimate $\hat{E}$. Given this estimate, B is obtained by Eq. 10.44. However, since the real field is spatially variable, $\hat{E}$ may or may not represent the actual elastic modulus as "seen" by the completed footing so that the actual settlement experienced by the footing will inevitably differ from the design target.

10.4.2 Probabilistic Assessment of Settlement Variability

The settlement variability will be assessed by Monte Carlo simulation. Details of the finite-element model and random-field simulator can be found in the previous two sections and in Section 6.4.6. The finite-element model used here is 60 elements wide by 40 elements in depth, with nominal element sizes $\Delta x = \Delta y = 0.15$ m, giving a soil regime of size 9 m wide by 6 m in depth. The Monte Carlo simulation consists of the following steps:

1. Generate a random field of elastic modulus local average cells using the LAS method which are then mapped onto the finite elements themselves.
2. "Virtually" sample the random field at 4 elements directly below the footing centerline (at depths 0, $H/3$, $2H/3$, and H). Then compute the estimated design elastic modulus $\hat{E}$ as the arithmetic average of these values.
3. Compute the required footing width B using Eq. 10.44.
4. Adjust both the (integer) number of elements n_W underlying the footing in the finite-element model and element width Δx such that $B = n_W\,\Delta x$. Note that the finite-element model assumes that the footing is a whole number of elements wide. Since B, as computed by Eq. 10.44, is continuously varying, some adjustment of Δx will be necessary. The final value of Δx is constrained to lie between (3/4)0.15 and (4/3)0.15 to avoid excessive element aspect ratios (Δy is held fixed at 0.15 m to maintain $H = 6$ m). Note also that the random field is *not* regenerated for the adjusted element size, so that some accuracy is lost with respect to the local average statistics. However, the approximation is deemed acceptable, given all other sources of uncertainty. Finally, the actual value of B used is constrained so that the footing is not less than 4 elements wide or more than 48 elements wide. This constraint is actually a more serious limitation, leading to some possible bias in the results, which is discussed further below.
5. Use the finite-element code to compute the simulated settlement δ_{sim}, which is interpreted as the settlement that the footing would actually experience on this particular realization of the spatially varying elastic modulus field.
6. Repeat from step 1 $n_{\text{sim}} = 2000$ times to yield 2000 realizations of δ_{sim}.

The sequence of realizations for δ_{sim} can then be statistically analyzed to determine its conditional probability density function (conditioned on $\hat{E}$).

The elastic modulus field is assumed to be lognormally distributed with parameters

$$\sigma^2_{\ln E} = \ln\left(1 + v_E^2\right), \qquad \mu_{\ln E} = \ln(\mu_E) - \tfrac{1}{2}\sigma^2_{\ln E} \quad (10.45)$$

where $v_E = \sigma_E/\mu_E$ is the coefficient of variation of the elastic modulus field. Since $E(\mathbf{x})$ is lognormally distributed, its logarithm is normally distributed and the elastic modulus value E_i assigned to the ith finite element can be obtained from a Gaussian random field through the transformation

$$E_i = \exp\{\mu_{\ln E} + \sigma_{\ln E} G_i\} \quad (10.46)$$

where G_i is the local average over the ith element of a zero-mean, unit-variance Gaussian random field $G(\mathbf{x})$, realizations of which are generated by the LAS method.

The Gaussian random field $G(\mathbf{x})$ is assumed to have a Markovian correlation structure, having correlation function

$$\rho(\tau) = \exp\left\{-\frac{2|\tau|}{\theta_{\ln E}}\right\} \quad (10.47)$$

where τ is the distance between any two points in the field and $\theta_{\ln E}$ is the correlation length. The random field has been assumed isotropic in this study, leaving the more site-specific anisotropic considerations for the reader to consider.

The simulation is performed for various statistics of the elastic modulus field. In particular, the mean elastic modulus μ_E is held fixed at 30 MPa, while the coefficient of variation v_E is varied from 0.1 to 1.0 and the correlation length $\theta_{\ln E}$ is varied from 0.1 to 15.

10.4.3 Prediction of Settlement Mean and Variance

It is hypothesized that if Janbu's relationship is sufficiently accurate for design purposes, it can also be used to predict the actual (simulated) settlement δ_{sim} reasonably accurately. That is, it is supposed that Eq. 10.40,

$$\delta = \mu_0 \left[a + b\ \ln\left(\frac{H}{B}\right)\right] \cdot \frac{P}{E^*}$$

will predict δ_{sim} for each realization if a suitable value of E^* can be found. In the previous two sections, it was shown found that settlement is very well predicted by setting E^* equal to the geometric average of the elastic modulus field over the region directly under the footing. This is what will be used here.

One difficulty is that the value of B in Eq. 10.40 is also derived from a sample of the random elastic modulus field. This means that δ is a function of both E^* and $\hat{E}$ where E^* is a local geometric average over a rectangle of *random* size $B \times H$. If Eq. 10.44 is substituted into Eq. 10.40, then δ can be expressed as

$$\delta = \frac{\hat{E}}{E^*}\delta_{\max} \quad (10.48)$$

Since E^* is a geometric average over a random area of size $B \times H$ of a lognormally distributed random field, then E^* is *conditionally* lognormally distributed with parameters

$$\mathrm{E}\left[\ln E^* \mid B\right] = \mu_{\ln E} \quad (10.49a)$$

$$\mathrm{Var}\left[\ln E^* \mid B\right] = \gamma(B,H)\sigma^2_{\ln E} \quad (10.49b)$$

where $\gamma(B,H)$ is the variance reduction function (see Section 3.4) defined in this case by the average correlation coefficient between every pair of points in the soil below the footing,

$$\gamma(B,H) = \frac{\int_0^B\int_0^B\int_0^H\int_0^H \rho(x_1 - x_2, y_1 - y_2)\, dy_1\, dy_2\, dx_1\, dx_2}{(HB)^2}$$

where, for the isotropic correlation function under consideration here, $\rho(x,y) = \rho(\sqrt{x^2+y^2}) = \rho(\tau)$; see Eq. 10.47. The variance function is determined numerically using Gaussian quadrature as discussed in Appendix C. The unconditional distribution parameters of ln E^* are obtained by taking expectations of Eqs. 10.49 with respect to B:

$$\mu_{\ln E^*} = \mu_{\ln E} \quad (10.50a)$$

$$\sigma^2_{\ln E^*} = \mathrm{E}\left[\gamma(B,H)\right]\sigma^2_{\ln E} \quad (10.50b)$$

A first-order approximation to $\mathrm{E}\left[\gamma(B,H)\right]$ is

$$\mathrm{E}\left[\gamma(B,H)\right] \simeq \gamma(\mu_B, H) \quad (10.51)$$

where μ_B is the mean footing width. Although a second-order approximation to $\mathrm{E}\left[\gamma(B,H)\right]$ could be considered, it was found to be only slightly different than the first-order approximation. It is recognized that the unconditional marginal distribution of E^* is probably no longer lognormal but histograms of E^* indicate that this is still a reasonable approximation.

The other random quantity appearing on the right-hand side of Eq. 10.48 is $\hat{E}$, which is an arithmetic average of a set of n observations,

$$\hat{E} = \frac{1}{n}\sum_{i=1}^{n} E_i$$

where E_i is the ith observed elastic modulus. It is assumed that elastic modulus samples are of approximately the same physical size as a finite element (e.g., a CPT cone measurement involves a "local average" bulb of deformed soil in the vicinity of the cone which might be on the order of the size of the elements used in this analysis). The first two moments of $\hat{E}$ are then

$$\mu_{\hat{E}} = \mu_E \tag{10.52a}$$

$$\sigma_{\hat{E}}^2 = \left(\frac{1}{n^2} \sum_{i=1}^{n} \sum_{j=1}^{n} \rho_{ij} \right) \sigma_E^2 \simeq \gamma(\Delta x, H)\sigma_E^2 \tag{10.52b}$$

where ρ_{ij} is the correlation coefficient between the ith and jth samples. The last approximation assumes that local averaging of $E(\mathbf{x})$ results in approximately the same level of variance reduction as does local averaging of $\ln E(\mathbf{x})$. This is not a bad approximation given all other sources of uncertainty.

If we can further assume that $\hat{E}$ is at least approximately lognormally distributed with parameters given by the transformations of Eq. 10.45, then δ in Eq. 10.48 will also be lognormally distributed with parameters

$$\mu_{\ln \delta} = \mu_{\ln \hat{E}} - \mu_{\ln E^*} + \ln(\delta_{\max}) \tag{10.53a}$$

$$\sigma_{\ln \delta}^2 = \sigma_{\ln \hat{E}}^2 + \sigma_{\ln E^*}^2 - 2 \operatorname{Cov}\left[\ln \hat{E},\ \ln E^*\right] \tag{10.53b}$$

The covariance term can be expressed as

$$\operatorname{Cov}\left[\ln \hat{E},\ \ln E^*\right] = \sigma_{\ln \hat{E}} \sigma_{\ln E^*} \rho_{\text{ave}} \tag{10.54}$$

where ρ_{ave} is the average correlation between every point in the domain defining E^* and every point in the domains defining $\hat{E}$. This can be expressed in integral form and solved numerically, but a simpler empirical approximation is suggested by observing that there will exist some "average" distance between the samples and the soil block under the footing, τ_{ave}, such that $\rho_{\text{ave}} = \rho(\tau_{\text{ave}})$. For the particular problem under consideration with $H = 6$ m, the best value of τ_{ave} was found by trial and error to be

$$\tau_{\text{ave}} = 0.1\mu_B \tag{10.55}$$

Finally, two of the results suggested above depend on the mean footing width μ_B. This can be obtained approximately as follows. First of all, taking the logarithm of Eq. 10.44 gives us

$$\ln B = \ln H - \frac{1}{b}\left(\frac{\delta_{\max} \hat{E}}{\mu_0 P} - a \right) \tag{10.56}$$

which has first two moments

$$\mu_{\ln B} = \ln H - \frac{1}{b}\left(\frac{\delta_{\max} \mu_{\hat{E}}}{\mu_0 P} - a \right) \tag{10.57a}$$

$$\sigma_{\ln B}^2 = \left(\frac{\delta_{\max}}{b\mu_0 P} \right)^2 \sigma_{\hat{E}}^2 \tag{10.57b}$$

and since B is nonnegative, it can be assumed to be at least approximately lognormally distributed (histogram plots of B indicate that this is a reasonable assumption) so that

$$\mu_B \simeq \exp\left\{ \mu_{\ln B} + \tfrac{1}{2}\sigma_{\ln B}^2 \right\}$$

With these results, the parameters of the assumed lognormally distributed settlement can be estimated using Eqs. 10.53 given the three parameters of the elastic modulus field, μ_E, σ_E, and $\theta_{\ln E}$.

10.4.4 Comparison of Predicted and Simulated Settlement Distribution

Before discussing the results, it is worth pointing out some of the difficulties with the comparison. First of all, as the coefficient of variation $v_E = \sigma_E/\mu_E$ increases, it becomes increasingly likely that the sample observations leading to $\hat{E}$ will be either very small or very large. If $\hat{E}$ is very small, then the resulting footing width, as predicted by Eq. 10.44, may be wider than the finite-element model (although, as discussed above, the footing width is arbitrarily restricted to being between 4 and 48 elements wide). It is recognized, however, that it is unlikely that a footing width in excess of 9 m would be the most economical solution. In fact, it is very likely that the designer would search for an alternative solution, such as a pile foundation, when faced with such a soft soil.

What this means is that it is difficult to evaluate the *unconditional* reliability of any single design solution since design solutions are rarely used in isolation; each is only one among a suite of solutions available to the designer and each has its own range of applicability (or, rather, economy). This implies that the reliability of a single design solution must be evaluated conditionally, that is, for the range of soil properties which make the solution economically optimal.

This conditional reliability problem is quite complex and beyond the scope of this study. Here the study is restricted to the unconditional reliability problem with the recognition that some of the simulation results at higher coefficients of variation are biased by restricting the "design" footing widths. In the worst case considered here, where $v_E = 1.0$, the fraction of footing widths found to be too wide, out of the 2000 realizations, ranged from 0% (for $\theta_{\ln E} = 0.1$) to 12% (for $\theta_{\ln E} = 15$).

The log-settlement mean, as predicted by Eq. 10.53a, is shown in Figure 10.22 along with the sample mean obtained from the simulation results for the minimum ($v_E = 0.1$) and maximum ($v_E = 1.0$) coefficients of variation considered. For small v_E, the agreement is excellent. For larger v_E, the maximum relative error is only about 7%, occurring at the smallest correlation length. Although the relative errors are minor, the small-scale behavior is not properly predicted by the analytical results and subsequent approximations built into Eq. 10.53a. It is believed that the major source of the discrepancies at small correlation lengths is due to the approximation of the second moment of $\hat{E}$ using the variance function $\gamma(\Delta x, H)$.

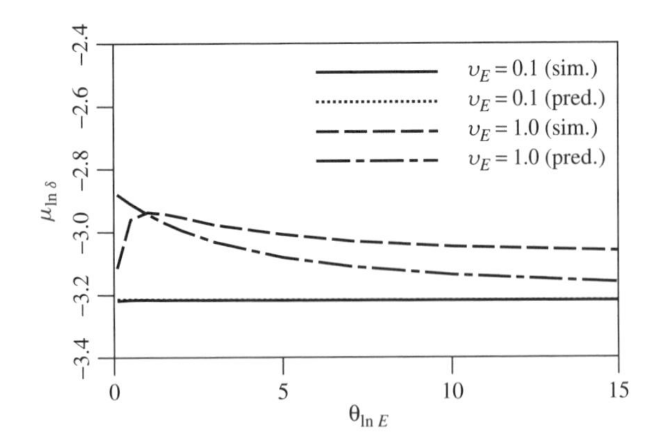

Figure 10.22 Comparison of predicted (pred.) and simulated (sim.) mean footing settlement.

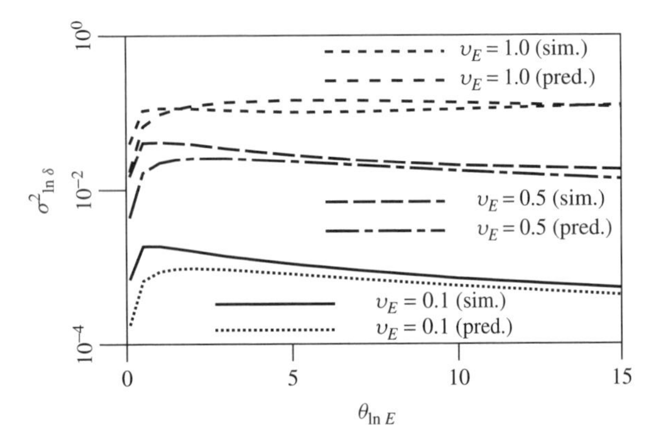

Figure 10.23 Comparison of predicted (pred.) and simulated (sim.) footing settlement variance.

The log-settlement variance, as predicted by Eq. 10.53b, is shown in Figure 10.23 along with the sample variance obtained from the simulation results for three different coefficients of variation v_E. Again, the agreement improves for increasing correlation lengths, but overall, the predicted variance is reasonably good and shows the same basic behavior as seen in the simulated results.

Figure 10.24 compares simulated versus predicted probability that the settlement exceeds some multiple of δ_{max} over all values of v_E and $\theta_{\ln E}$. The agreement is reasonable, tending to be slightly conservative with predicted "failure" probability somewhat exceeding simulated probability on average.

10.4.5 Summary

The results of Figure 10.24 indicate that the Janbu settlement prediction given by Eq. 10.37 has a reliability, when

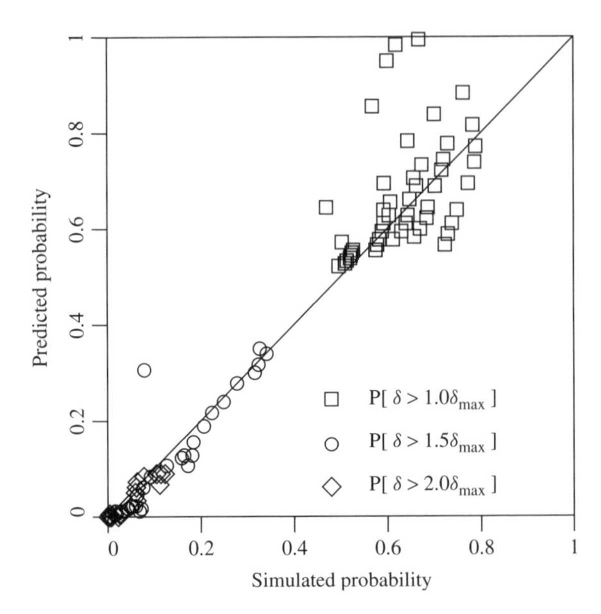

Figure 10.24 Comparison of predicted and simulated settlement probabilities.

used in design, that is reasonably well (and perhaps somewhat conservatively) estimated by Eqs. 10.53 so long as the basic statistics, μ_E, σ_E, and $\theta_{\ln E}$, of the elastic modulus field are known or estimated. Of these parameters, the most difficult to estimate is the correlation length $\theta_{\ln E}$ since its estimator requires extensive investigation. However, Figure 10.23 indicates that there is a worst case, in the sense of maximum variance, which occurs at about $\theta_{\ln E} \simeq 1$. Thus, if the correlation length is unknown, it should be conservative to use $\theta_{\ln E} \simeq 1$.

For a particular site, the reliability assessment of the footing design against excessive settlement proceeds as follows:

1. Sample the site at a number of locations and produce an estimate of the mean elastic modulus $\hat{E}$. In current practice this estimate seems to be an arithmetic average of the observed values. Although the results presented earlier in this chapter suggest that a geometric average would be more representative, the approach taken by current practice was adopted in this study.
2. Compute the required footing width B by Eq. 10.44. This constitutes the design phase.
3. Using the same data set collected in item 1, estimate $\mu_{\ln E}$ and $\sigma^2_{\ln E}$ by computing the sample mean and sample variance of the log-data. Assume that $\theta_{\ln E} \simeq 1$ unless a more sophisticated analysis is carried out.
4. Using Gaussian quadrature or some software package which numerically integrates a function, evaluate the

variance reduction functions $\gamma(B,H)$ and $\gamma(\Delta x,H)$. Note that the latter assumes that the data in step 1 were collected along a single vertical line below the footing.
5. Estimate $\mu_{\ln E^*}$ and $\sigma^2_{\ln E^*}$ using Eqs. 10.50 and 10.51.
6. Estimate $\mu_{\ln \hat{E}}$ and $\sigma^2_{\ln \hat{E}}$ using Eqs. 10.52 in the transformations of Eq. 10.45.
7. Compute τ_{ave} using Eq. 10.55 and then $\rho_{\text{ave}} = \rho(\tau_{\text{ave}})$ using Eq. 10.47. Compute the covariance of Eq. 10.54.
8. Compute the mean and variance of log-settlement using Eqs. 10.53.

Assuming that the settlement is lognormally distributed, probabilities relating to the actual settlement of the designed footing can now be computed as

$$\mathrm{P}\left[\delta > \delta_{\max}\right] = 1 - \Phi\left(\frac{\ln(\delta_{\max}) - \mu_{\ln\delta}}{\sigma_{\ln\delta}}\right) \qquad (10.58)$$

where Φ is the standard normal cumulative distribution function.

It is noted that this study involves a number of approximations and limitations, the most significant of which are the following:

1. Limiting the footing widths to some maximum upper value leads to some bias of the simulation results.
2. Janbu's influence factor μ_1 is approximated as a straight line. In fact, the curve flattens out for small values of H/B or large values of B. This approximation error could easily be contributing to the frequency of predicting excessively large footing widths for low $\hat{E}$.
3. Both E^* and $\hat{E}$ are assumed to be lognormally distributed, which is probably a reasonable assumption but which may lead to some discrepancies in extreme cases (such as for small correlation lengths). In addition, the variance of $\hat{E}$ is obtained using the variance function $\gamma(\Delta x,H)$. That is, a continuous local average over the height H in log-space is used to approximate the variance reduction of the average of a discrete set of observations in real space. The variance reduction is expected to be a reasonable estimate but not to be particularly accurate.
4. The covariance between $\ln E^*$ and $\ln \hat{E}$ is approximated by using an average correlation coefficient which is merely fitted by trial and error to the simulation results.

Perhaps one of the main results of the section, other than an approximate assessment of the reliability of a design methodology, is the recognition of the fact that the reliability assessment of design methodologies must be done conditionally. One task for the future is to determine how to specify the appropriate conditional soil property distributions as a function of design economies. Once this specification has been made, simulation can again be called upon to find the conditional reliabilities.

In addition, the results of this section do not particularly address sampling issues. For example, in the discussion above outlining how the reliability assessment would proceed, it was assumed that the same data used to estimate $\hat{E}$ would provide a reasonable estimate of both $\mu_{\hat{E}}$ and $\mu_{\ln E^*}$ (the latter using the logarithm of the data). Clearly, this introduces additional bias and uncertainty into the assessment that is not accounted for above.

10.5 RESISTANCE FACTORS FOR SHALLOW-FOUNDATION SETTLEMENT DESIGN

This section presents the results of a study in which a reliability-based settlement design approach is proposed and investigated via simulation using the RFEM. In particular, the effect of a soil's spatial variability and site investigation intensity on the resistance factors is quantified. The results of the section can be used to improve and generalize "calibrated" code provisions based purely on past experience (Fenton et al., 2005).

10.5.1 Random Finite-Element Method

A specific settlement design problem will be considered here in order to investigate the settlement probability distribution of footings designed against excessive settlement. The problem considered is that of a rigid, rough square-pad footing founded on the surface of a three-dimensional linearly elastic soil mass underlain by bedrock at depth H. Although only elastic settlement is specifically considered, the results can include consolidation settlement so long as the combined settlement can be adequately represented using an effective elastic modulus field. To the extent that the elastic modulus itself is a simplified representation of a soil's inverse compressibility, which is strain-level dependent, the extension of the approximation to include consolidation settlement is certainly reasonable and is as recommended, for example, in the *Canadian Highway Bridge Design Code Commentary* (CSA, 2000b).

The settlement of a rigid footing on a three-dimensional soil mass is estimated using a linear finite-element analysis. The mesh selected is 64 elements by 64 elements in plan by 32 elements in depth. Eight-node hexahedral elements each cubic with side length 0.15 m are used (note that metric units are used in this section, rather than making it nondimensional, since footing design will be based on a maximum tolerable settlement which is specified in meters) yielding a soil domain of size 9.6×9.6 m in plan by 4.8 m in depth. Because the stiffness matrix corresponding to a mesh of size $64 \times 64 \times 32$ occupies about 4 Gbytes of memory, a preconditioned conjugate gradient iterative solver (e.g., Smith and Griffiths, 2004), which avoids the need to assemble the global stiffness matrix, is employed in the finite-element code. A max-norm relative error tolerance of 0.005 is used to determine when the iterative solver has converged to a solution.

The finite-element model was tested in the deterministic case (uniform elastic soil properties) to validate its accuracy and was found to be about 20% stiffer (smaller settlements) than that derived by analytical approximations (see, e.g., Milovic, 1992). Using other techniques such as selectively reduced integration, nonconforming elements, and 20-node elements did not significantly affect the discrepancy between these results and Milovic's. The "problem" is that the finite elements truncate the singular stresses that occur along the edge of a rigid footing, leading to smaller settlements than predicted by theory. In this respect, Seyček (1991) compares real settlements to those predicted by theory and concluded that predicted settlements are usually considerably higher than real settlements. This is because the true stresses measured in the soil near the footing edge are finite and significantly less than the singular stresses predicted by theory. Seyček improves the settlement calculations by reducing the stresses below the footing. Thus, the finite-element results included here are apparently closer to actual settlements than those derived analytically, although a detailed comparison to Seyček's has not been performed by the authors. However, it is not believed that these possible discrepancies will make a significant difference to the probabilistic results of this section since the probability of failure (excessive settlement) involves a comparison between deterministic and random predictions arising from the same finite-element model, thus canceling out possible bias.

The rigid footing is assumed to have a rough interface with the underlying soil—no relative slip is permitted—and rotation of the footing is not permitted. Only square footings of dimension $B \times B$ are considered, where the required footing width B is determined during the design phase, to be discussed in the next section. Once the required footing width has been found, the design footing width must be increased to the next larger element boundary; this is because the finite-element mesh is fixed and footings must span an integer number of elements. For example, if the required footing width is 2.34 m and elements have dimension $\Delta x = \Delta y = 0.15$ m square, then the design footing width must be increased to 2.4 m (since this corresponds to 16 elements, rather than the 15.6 elements that 2.34 m would entail). This corresponds roughly to common design practice, where element dimensions are increased to an easily measured quantity.

Once the design footing width has been found, it must be checked to ensure that it is physically reasonable, both economically and within the finite-element model. First of all, there will be some minimum footing size. In this study the footings cannot be less than 4×4 elements in size—for one thing loaded areas smaller than this tend to have significant finite-element errors; for another they tend to be too small to construct. For example, if an element size of 0.15 m is used, then the minimum footing size is 0.6×0.6 m, which is not very big. French (1999) recommends a lower bound on footing size of 0.6 m and an upper economical bound of 3.7 m. If the design footing width is less than the minimum footing width, it is set equal to the minimum footing width. Second, there will be some maximum footing size. A spread footing bigger than about 4 m square would likely be replaced by some other foundation system (piles, mat, or raft). In this program, the maximum footing size is taken to be equal to two-thirds of the finite-element mesh width. This limit has been found to result in less than a 1% error relative to the same footing founded on a mesh twice as wide, so boundary conditions are not significantly influencing the results. If the design footing width exceeds the maximum footing width, then the probabilistic interpretation becomes somewhat complicated, since a different design solution would presumably be implemented. From the point of view of assessing the reliability of the "designed" spread footing, it is necessary to decide if this excessively large footing design would correspond to a success or to a failure. It is assumed in this study that the subsequent design of the alternative foundation would be a success, since it would have its own (high) reliability.

In all the simulations performed in this study, the lower limit on the footing size was never encountered, implying that for the choices of parameters selected in this study the probability of a design footing being less than 0.6×0.6 in dimension was very remote. Similarly, the maximum footing size was not exceeded in any but the most severe parameter case considered (minimum sampling, lowest resistance factor, highest coefficient of variation), where it was only exceeded in 2% of the possible realizations. Thus, the RFEM results presented here give reasonably accurate settlement predictions over the entire study.

The soil property of primary interest to settlement is elastic modulus E, which is taken to be spatially random and may represent both the initial elastic and consolidation behavior. Its distribution is assumed to be lognormal for two reasons: First, a geometric average tends to a lognormal distribution by the central limit theorem and the effective elastic modulus, as "seen" by a footing, was found to be closely represented by a geometric average in previous sections of this chapter and, second, the lognormal distribution is strictly nonnegative, which is physically reasonable for elastic modulus. The lognormal distribution has two parameters, $\mu_{\ln E}$ and $\sigma_{\ln E}$, which can be estimated by the sample mean and sample standard deviation of observations of $\ln(E)$. They can also be obtained from the mean and standard deviation of E using the transformations given by Eqs. 1.176.

A Markovian spatial correlation function, which gives the correlation coefficient between log-elastic modulus values at points separated by the lag vector $\boldsymbol{\tau}$ is used in this study,

$$\rho_{\ln E}(\boldsymbol{\tau}) = \exp\left\{-\frac{2|\boldsymbol{\tau}|}{\theta_{\ln E}}\right\} \tag{10.59}$$

in which $\boldsymbol{\tau} = \mathbf{x} - \mathbf{x}'$ is the vector between spatial points $\mathbf{x}$ and $\mathbf{x}'$ and $|\boldsymbol{\tau}|$ is the absolute length of this vector (the lag distance). The results presented here are not particularly sensitive to the choice in functional form of the correlation—the Markov model is popular because of its simplicity. The correlation function decay rate is governed by the correlation length $\theta_{\ln E}$, which, loosely speaking, is the distance over which log-elastic moduli are significantly correlated. The correlation structure is assumed to be isotropic in this study, which is appropriate for investigating the fundamental stochastic behavior of settlement. Anisotropic studies are more appropriate for site-specific analyses and for refinements to this study. In any case, anisotropy is not expected to have a large influence on the results of this section due to the averaging effect of the rigid footing on the properties it sees beneath it.

Poisson's ratio, having only a relatively minor influence on settlement, is assumed to be deterministic and is set equal to 0.3 in this study.

Realizations of the random elastic modulus field are produced using the LAS method (see Section 6.4.6). Local average subdivision produces a discrete grid of local averages, G_i, of a standard Gaussian random field, having correlation structure given by Eq. 10.59, where averaging is performed over the domain of the ith finite element. These local averages are then mapped to finite-element properties according to

$$E_i = \exp\{\mu_{\ln E} + \sigma_{\ln E} G_i\} \tag{10.60}$$

Figure 10.25 illustrates the finite-element mesh used in the study and Figure 10.26 shows a cross section through the soil mass under the footing for a typical realization of the soil's elastic modulus field. Figure 10.26 also illustrates the boundary conditions.

10.5.2 Reliability-Based Settlement Design

In this section we will investigate a reliability-based design methodology for the serviceability limit state of shallow

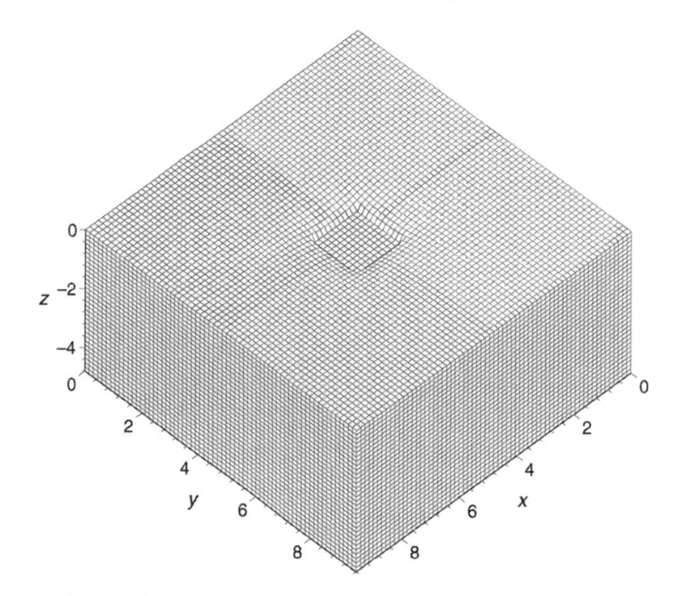

Figure 10.25 Finite-element mesh with one square footing.

footings. Footing settlement is predicted here using a modified Janbu et al. (1956) relationship, and this is the basis of design used in this section:

$$\delta_p = u_1 \frac{\hat{q}B}{\hat{E}} \tag{10.61}$$

where δ_p is the predicted footing settlement, $\hat{q} = \hat{P}/B^2$ is the characteristic stress applied to the soil by the characteristic load $\hat{P}$ acting over footing area $B \times B$, $\hat{E}$ is the estimate of elastic modulus underlying the footing, u_1 is an influence factor which includes the effect of Poisson's ratio ($\nu = 0.3$ in this study). The characteristic load $\hat{P}$ is often a nominal load computed from the supported live and dead loads (see Chapter 7), while the characteristic elastic modulus $\hat{E}$ is usually a cautious estimate of the mean elastic modulus under the footing obtained by taking laboratory samples or by in situ tests, such as CPT. In terms of the characteristic footing load $\hat{P}$, the settlement predictor thus becomes

$$\delta_p = u_1 \frac{\hat{P}}{B\hat{E}} \tag{10.62}$$

The relationship above is somewhat modified from that given by Janbu et al. (1956) and Christian and Carrier (1978) in that the influence factor u_1 is calibrated specifically for a square rough rigid footing founded on the surface of an elastic soil using the same finite-element model which is later used in the Monte Carlo simulations. This is done to remove bias (model) errors and concentrate specifically on the effect of spatial soil variability on required resistance factors. In practice, this means that the resistance factors given in this section are *upper bounds*, appropriate

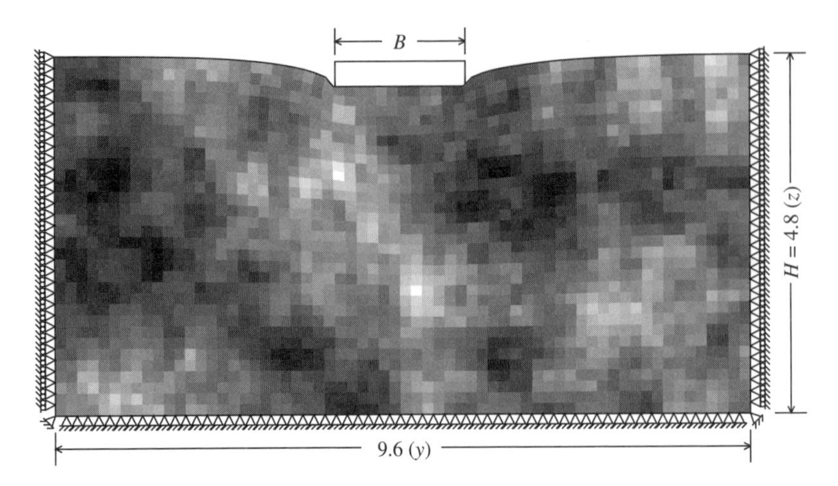

Figure 10.26 Cross section through realization of random soil underlying footing. Darker soils are stiffer.

for use when bias and measurement errors are known to be minimal.

The calibration of u_1 is done by computing, via the finite-element method, the deterministic settlement of a square rigid footing subjected to load $\hat{P}$ placed on a soil with elastic modulus $\hat{E}$ and Poisson's ratio ν. Once the settlement is obtained, Eq. 10.62 can be solved for u_1. Repeating this over a range of H/B ratios leads to the curve shown in Figure 10.27. This deterministic calibration was carried out over a larger range of mesh dimensions than indicated by Figure 10.25. A very close approximation to the finite-element results is given by the fitted relationship (obtained by consideration of the correct limiting form and by trial and error for the coefficients)

$$u_1 = 0.61\left(1 - e^{-1.18H/B}\right) \tag{10.63}$$

which is also shown in Figure 10.27.

Using Eq. 10.63 in Eq. 10.62 gives the following settlement prediction:

$$\delta_p = 0.61\left(1 - e^{-1.18H/B}\right)\left(\frac{\hat{P}}{B\hat{E}}\right) \tag{10.64}$$

The reliability-based design goal is to determine the footing width B such that the probability of exceeding a specified tolerable settlement $\delta_{\max}$ is acceptably small. That is, to find B such that

$$\text{P}[\delta > \delta_{\max}] = p_f = p_{\max} \tag{10.65}$$

where δ is the actual settlement of the footing "as placed" (which will be considered here to be the same as "as designed"). Design failure is assumed to have occurred if the actual footing settlement δ exceeds the maximum tolerable settlement $\delta_{\max}$. The probability of design failure is p_f and $p_{\max}$ is the maximum acceptable probability of design failure.

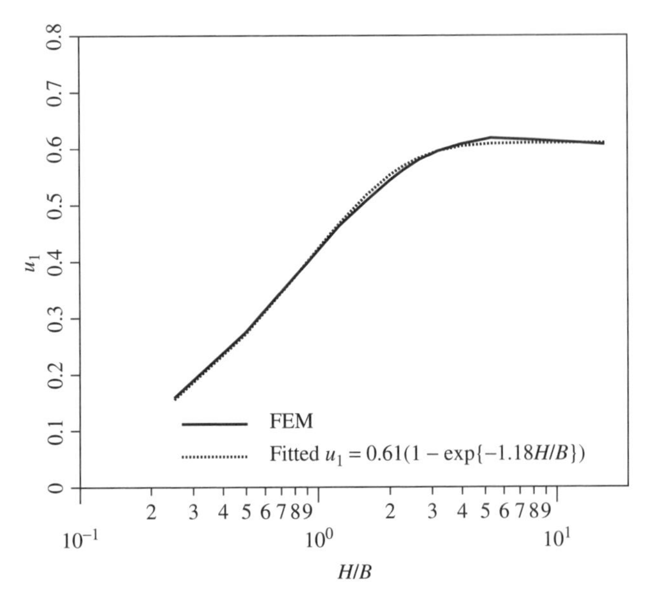

Figure 10.27 Calibration of u_1 using finite-element model (FEM).

A realization of the footing settlement δ is determined here using a finite-element analysis of a realization of the random soil. For u_1 calibrated to the finite-element results, δ can also be computed from

$$\delta = u_1 \frac{P}{BE_{\text{eff}}} \tag{10.66}$$

where P is the actual footing load and E_{eff} is the effective elastic modulus as seen by the footing (i.e., the uniform value of elastic modulus which would produce a settlement identical to the actual footing settlement). Both P and E_{eff} are random variables.

One way of achieving the desired design reliability is to introduce a load factor $\alpha \geq 1$ and a resistance factor $\phi_g \leq 1$ and then find B, α, and ϕ_g which satisfy both Eq. 10.65 and Eq. 10.62 with $\delta_p = \delta_{\max}$. In other words, find B and α/ϕ_g such that

$$\delta_{\max} = u_1 \left(\frac{\alpha \hat{P}}{B \phi_g \hat{E}} \right) \tag{10.67}$$

and

$$\mathrm{P}\left[u_1 \frac{P}{BE_{\text{eff}}} > u_1 \left(\frac{\alpha \hat{P}}{B \phi_g \hat{E}} \right) \right] = p_{\max} \tag{10.68}$$

In the above, we are assuming that the soil's elastic modulus E is the "resistance" to the load and that it is to be factored due to its significant uncertainty.

From these two equations, at most two unknowns can be found uniquely. For serviceability limit states, a load factor of 1.0 is commonly used, and $\alpha = 1$ will be used here. Note that only the ratio α/ϕ_g need actually be determined for the settlement problem.

Given α/ϕ_g, $\hat{P}$, $\hat{E}$, and H, Eq. 10.67 is relatively efficiently solved for B using a one-point iteration:

$$B_{i+1} = 0.61 \left(1 - e^{-1.18H/B_i}\right) \left(\frac{\alpha \hat{P}}{\delta_{\max} \phi_g \hat{E}} \right) \tag{10.69}$$

for $i = 0, 1, \ldots$ until successive estimates of B are sufficiently similar. A reasonable starting guess is $B_0 = 0.4(\alpha\hat{P})/(\delta_{\max}\phi_g\hat{E})$.

In Eq. 10.68, the random variables u_1 and B are common to both sides of the inequality and so can be canceled. It will also be assumed that the footing load is lognormally distributed and that the characteristic load $\hat{P}$ equals the (nonrandom) *median load*, that is,

$$\hat{P} = \exp\{\mu_{\ln P}\} \tag{10.70}$$

Setting the value of $\hat{P}$ to the median load considerably simplifies the theory in the sequel, but it should be noted that the definition of $\hat{P}$ will directly affect the magnitude of the estimated resistance factors. The lognormal distribution was selected because it results in loads which are strictly nonnegative (uplift problems should be dealt with separately and not handled via the tail end of a normal distribution assumption). The results to follow should be similar for any reasonable load distribution (e.g., gamma, chi square) having the same mean and variance.

Collecting all remaining random quantities leads to the simplified design probability

$$\mathrm{P}\left[P \frac{\hat{E}}{E_{\text{eff}}} > \frac{\alpha}{\phi_g} e^{\mu_{\ln P}} \right] = p_{\max} \tag{10.71}$$

The characteristic modulus $\hat{E}$ and the effective elastic modulus E_{eff} can also be reasonably assumed to be lognormally distributed. Under these assumptions, if Q is defined as

$$Q = P \frac{\hat{E}}{E_{\text{eff}}} \tag{10.72}$$

then Q is also lognormally distributed, and

$$\ln Q = \ln P + \ln \hat{E} - \ln E_{\text{eff}} \tag{10.73}$$

is normally distributed with mean

$$\mu_{\ln Q} = \mu_{\ln P} + \mu_{\ln \hat{E}} - \mu_{\ln E_{\text{eff}}} \tag{10.74}$$

It is assumed that the load distribution is known, so that $\mu_{\ln P}$, which is the mean of the logarithm of the total load, as well as its variance $\sigma^2_{\ln P}$ are known. The nature of the other two terms on the right-hand side will now be investigated.

Assume that $\hat{E}$ is estimated from a series of m soil samples that yield the observations $E_1^o, E_2^o, \ldots, E_m^o$. To investigate the nature of this estimate, it is first instructive to consider the effective elastic modulus E_{eff} as seen by the footing. Analogous to the estimate for $\hat{E}$, it can be imagined that the soil volume under the footing is partitioned into a large number of soil "samples" (although most of them, if not all, will remain unsampled) $E_1, E_2, \ldots, E_n$. If the soil is not strongly layered, the effective elastic modulus, as seen by the footing, E_{eff}, is a geometric average of the soil properties in the block under the footing, that is,

$$E_{\text{eff}} = \left(\prod_{i=1}^{n} E_i \right)^{1/n} = \exp\left\{ \frac{1}{n} \sum_{i=1}^{n} \ln E_i \right\} \tag{10.75}$$

If $\hat{E}$ is to be a good estimate of E_{eff}, which is desirable, then it should be similarly determined as a geometric average of the observed samples $E_1^o, E_2^o, \ldots, E_m^o$,

$$\hat{E} = \left(\prod_{j=1}^{m} E_j^o \right)^{1/m} = \exp\left\{ \frac{1}{m} \sum_{j=1}^{m} \ln E_j^o \right\} \tag{10.76}$$

since this estimate of E_{eff} is unbiased in the median; that is, the median of $\hat{E}$ is equal to the median of E_{eff}. This is a fairly simple estimator, and no attempt is made here to account for the location of samples relative to the footing. Note that if the soil is layered horizontally and it is desired to specifically capture the layer information, then Eqs. 10.75 and 10.76 can be applied to each layer individually—the final $\hat{E}$ and E_{eff} values are then computed as harmonic averages of the layer values. Although the distribution of a harmonic average is not simply defined, a lognormal approximation has been found to be often reasonable.

Under these definitions, the means of $\mu_{\ln \hat{E}}$ and $\mu_{\ln E_{\text{eff}}}$ are identical,

$$\mu_{\ln E_{\text{eff}}} = \mathrm{E}\left[\ln E_{\text{eff}}\right] = \mu_{\ln E} \tag{10.77}$$

$$\mu_{\ln \hat{E}} = \mathrm{E}\left[\ln \hat{E}\right] = \mu_{\ln E} \tag{10.78}$$

where $\mu_{\ln E}$ is the mean of the logarithm of elastic moduli of any sample. Thus, as long as Eqs. 10.75 and 10.76 hold, the mean of $\ln Q$ simplifies to

$$\mu_{\ln Q} = \mu_{\ln P} \tag{10.79}$$

Now, attention can be turned to the variance of $\ln Q$. If the variability in the load P is independent of the soil's elastic modulus field, which is entirely reasonable, then the variance of $\ln Q$ is

$$\sigma^2_{\ln Q} = \sigma^2_{\ln P} + \sigma^2_{\ln \hat{E}} + \sigma^2_{\ln E_{\text{eff}}} - 2\,\text{Cov}\left[\ln \hat{E}, \ln E_{\text{eff}}\right] \tag{10.80}$$

The variances of $\ln \hat{E}$ and $\ln E_{\text{eff}}$ can be expressed in terms of the variance of $\ln E$ using two variance reduction functions, γ^o and γ, defined as

$$\gamma^o(m) = \frac{1}{m^2} \sum_{i=1}^{m} \sum_{j=1}^{m} \rho^o_{ij} \tag{10.81a}$$

$$\gamma(n) = \frac{1}{n^2} \sum_{i=1}^{n} \sum_{j=1}^{n} \rho_{ij} \tag{10.81b}$$

where ρ^o_{ij} is the correlation coefficient between $\ln E^o_i$ and $\ln E^o_j$ and ρ_{ij} is the correlation coefficient between $\ln E_i$ and $\ln E_j$. These functions can be computed numerically once the locations of all soil samples are known. Both $\gamma^o(1)$ and $\gamma(1)$ have value 1.0 when only one sample is used to specify $\hat{E}$ or E_{eff}, respectively (when samples are "point" samples, then one sample corresponds to zero volume; however, it is assumed here that there is some representative sample volume from which the mean and variance of the elastic modulus field are estimated and this corresponds to the point measure). As the number of samples increases, the variance reduction function decreases toward zero at a rate inversely proportional to the total sample volume (see Vanmarcke, 1984). If the volume of the soil under the footing is $B \times B \times H$, then a reasonable approximation to $\gamma(n)$ is obtained by assuming a separable form:

$$\gamma(n) \simeq \gamma_1^2\left(\frac{2B}{\theta_{\ln E}}\right)\gamma_1\left(\frac{2H}{\theta_{\ln E}}\right) \tag{10.82}$$

where $\gamma_1(a)$ is the one-dimensional variance function corresponding to a Markov correlation function (see Section 3.6.5):

$$\gamma_1(a) = \frac{2}{a^2}\left[a + e^{-a} - 1\right] \tag{10.83}$$

An approximation to $\gamma^o(m)$ is somewhat complicated by the fact that samples for $\hat{E}$ are likely to be collected at separate locations. If the observations are sufficiently separated that they can be considered independent (e.g., separated by more than $\theta_{\ln E}$), then $\gamma^o(m) = 1/m$. If they are collected from within a contiguous volume V^o, then

$$\gamma^o(m) \simeq \gamma_1\left(\frac{2R}{\theta_{\ln E}}\right)\gamma_1\left(\frac{2R}{\theta_{\ln E}}\right)\gamma_1\left(\frac{2H}{\theta_{\ln E}}\right) \tag{10.84}$$

where the total plan area of soil sampled is $R \times R$ (e.g., a CPT sounding can probably be assumed to be sampling an effective area equal to about 0.2×0.2 m^2, so that $R = 0.2$ m for a single CPT). The true variance reduction function will be somewhere in between. In this study, the soil is sampled by examining one or more columns of the finite-element model, and so for an individual column, $R \times R$ becomes replaced by $\Delta x \times \Delta y$, which are the plan dimensions of the finite elements, and Eq. 10.84 can be used to obtain the variance reduction function for a single column. If more than one column is sampled, then

$$\gamma^o(m) \simeq \frac{\gamma_1(2\,\Delta x/\theta_{\ln E})\gamma_1(2\,\Delta y/\theta_{\ln E})\gamma_1(2H/\theta_{\ln E})}{n_{\text{eff}}} \tag{10.85}$$

where n_{eff} is the effective number of independent columns sampled. If the sampled columns are well separated (i.e., by more than the correlation length), then they could be considered independent, and n_{eff} would be equal to the number of columns sampled. If the columns are closely clustered (relative to the correlation length), then n_{eff} would decrease toward 1. The actual number is somewhere in between the number of columns sampled and 1 and should be estimated by judgment taking into account the distance between samples.

With these results,

$$\sigma^2_{\ln \hat{E}} = \gamma^o(m)\sigma^2_{\ln E} \tag{10.86a}$$

$$\sigma^2_{\ln E_{\text{eff}}} = \gamma(n)\sigma^2_{\ln E} \tag{10.86b}$$

The covariance term in Eq. 10.80 is computed from

$$\begin{aligned}\text{Cov}\left[\ln \hat{E}, \ln E_{\text{eff}}\right] &= \frac{1}{mn}\sum_{j=1}^{m}\sum_{i=1}^{n}\text{Cov}\left[\ln E^o_j, \ln E_i\right] \\ &= \sigma^2_{\ln E}\left[\frac{1}{mn}\sum_{j=1}^{m}\sum_{i=1}^{n}\rho'_{ij}\right] \\ &= \sigma^2_{\ln E}\rho'_{\text{ave}}\end{aligned} \tag{10.87}$$

where ρ'_{ij} is the correlation coefficient between $\ln E^o_j$ and $\ln E_i$ and ρ'_{ave} is the average of all these correlations. If the estimate $\ln \hat{E}$ is to be at all useful in a design, the value of ρ'_{ave} should be reasonably high. However, its magnitude depends on the degree of spatial correlation (measured by $\theta_{\ln E}$) and the distance between the observations E^o_i and the soil volume under the footing. The correlation function of Eq. 10.59 captures both of these effects. That is, there will exist an average distance τ'_{ave} such that

$$\rho'_{\text{ave}} = \exp\left\{\frac{-2\tau'_{\text{ave}}}{\theta_{\ln E}}\right\} \tag{10.88}$$

and the problem is to find a reasonable approximation to τ'_{ave} if the numerical calculation of Eq. 10.87 is to be avoided. The approximation considered in this study is that τ'_{ave} is defined as the average absolute distance between the E_i^o samples and a vertical line below the center of the footing, with a sample taken anywhere under the footing to be considered to be taken at the footing corner (e.g., at a distance $B/\sqrt{2}$ from the centerline); this latter restriction is taken to avoid a perfect correlation when a sample is taken directly at the footing centerline, which would be incorrect. A side study indicated that for all moderate correlation lengths ($\theta_{\ln E}$ of the order of the footing width) the true τ'_{ave} differed by less than about 10% from the approximation $B/\sqrt{2}$ for any sample taken under the footing.

Using these definitions, the variance of $\ln Q$ can be written as

$$\begin{aligned}\sigma^2_{\ln Q} &= \sigma^2_{\ln P} + \sigma^2_{\ln E}\left[\gamma^o(m) + \gamma(n) - 2\rho'_{\text{ave}}\right] \\ &\geq \sigma^2_{\ln P}\end{aligned} \tag{10.89}$$

The limitation $\sigma^2_{\ln Q} \geq \sigma^2_{\ln P}$ is introduced because it is possible, using the approximations suggested above, for the quantity inside the square brackets to become negative, which is physically inadmissable. It is assumed that if this happens the sampling has reduced the uncertainty in the elastic modulus field essentially to zero.

With these results in mind the design probability becomes

$$\begin{aligned}\text{P}\left[P\frac{\hat{E}}{E_{\text{eff}}} > \frac{\alpha}{\phi_g}e^{\mu_{\ln P}}\right] &= \text{P}\left[Q > \frac{\alpha}{\phi_g}e^{\mu_{\ln P}}\right] \\ &= \text{P}\left[\ln Q > \ln\alpha - \ln\phi_g + \mu_{\ln P}\right] \\ &= 1 - \Phi\left(\frac{-\ln\phi_g}{\sigma_{\ln Q}}\right) \\ &= p_{\max} \quad (\text{assuming } \alpha = 1)\end{aligned} \tag{10.90}$$

from which the required resistance factor ϕ_g can be found as

$$\phi_g = \exp\{-\beta\,\sigma_{\ln Q}\} \tag{10.91}$$

where β is the desired reliability index corresponding to $p_{\max}$. That is, $\Phi(\beta) = 1 - p_{\max}$. For example, if $p_{\max} = 0.05$, which will be assumed in the following, $\beta = 1.645$.

It is instructive at this point to consider a limiting case, namely where $\hat{E}$ is a perfect estimate of E_{eff}. In this case, $\hat{E} = E_{\text{eff}}$, which implies that $m = n$ and the observations $E_1^o, \ldots$ coincide identically with the samples $E_1, \ldots$. In this case, $\gamma^o = \rho'_{\text{ave}} = \gamma$, so that

$$\sigma^2_{\ln Q} = \sigma^2_{\ln P} \tag{10.92}$$

from which the required resistance factor can be calculated as

$$\phi_g = \exp\{-\beta\,\sigma_{\ln P}\} \tag{10.93}$$

For example, if $p_{\max} = 0.05$ and the coefficient of variation of the load is $v_P = 0.1$, then $\phi_g = 0.85$. Alternatively, for the same maximum acceptable failure probability, if $v_P = 0.3$, then ϕ_g decreases to 0.62.

One difficulty with the computation of $\sigma^2_{\ln E_{\text{eff}}}$ that is apparent in the approximation of Eq. 10.82 is that it depends on the footing dimension B. From the point of view of the design probability, Eq. 10.71, this means that B does not entirely disappear, and the equation is still interpreted as the probability that a footing of size $B \times B$ will fail to stay within the serviceability limit state. The major implication of this interpretation is that if Eq. 10.71 is used conditionally to determine ϕ_g, then the design resistance factor ϕ_g will have some dependence on the footing size; this is not convenient for a design code (imagine, for example, designing a concrete beam if ϕ_c varied with the beam dimension). Thus, strictly speaking, Eq. 10.71 should be used conditionally to determine the reliability of a footing against settlement failure once it has been designed. The determination of ϕ_g would then proceed by using the total probability theorem; that is, find ϕ_g such that

$$p_{\max} = \int_0^\infty \text{P}\left[Q > \frac{\alpha}{\phi_g}\hat{P} \,\middle|\, B\right] f_B(b)\, db \tag{10.94}$$

where f_B is the probability density function of the footing width B. The distribution of B is not easily obtained: It is a function of H, $\hat{P}$, $\delta_{\max}$, the parameters of $\hat{E}$, and the load and resistance factors α and ϕ_g—see Eq. 10.69—and so the value of ϕ_g is not easily determined using Eq. 10.94. One possible solution is to assume that changes in B do not have a great influence on the computed value of ϕ_g and to take $B = B_{\text{med}}$, where B_{med} is the (nonrandom) footing width required by design using the median elastic modulus along with a moderate resistance factor of $\phi_g = 0.5$ in Eq. 10.69. This approach will be adopted and will be validated by the simulation to be discussed next.

10.5.3 Design Simulations

As mentioned above, the resistance factor ϕ_g cannot be directly obtained by solving Eq. 10.71 for given B simultaneously with Eq. 10.67 since this would result in a resistance factor which depends on the footing dimension. To find the value of ϕ_g to be used for any footing size involves solving Eq. 10.94. Unfortunately, this is not feasible since the distribution of B is unknown (or at least very difficult to compute). A simple solution is to use Monte Carlo simulation to estimate the probability on the right-hand side of Eq. 10.94 and then use the simulation results to assess the validity of the simplifying assumption that B_{med} can be used to find ϕ_g using Eq. 10.71. The RFEM will be employed within a design context to perform the desired simulation. The approach is described as follows:

1. Decide on a maximum tolerable settlement δ_{max}. To illustrate the approach, we will select $\delta_{max} = 0.025$ m.
2. Estimate the characteristic footing load $\hat{P}$ to be the median load applied to the footing by the supported structure (it is assumed that the load distribution is known well enough to know its median, $\hat{P} = e^{\mu_{\ln P}}$).
3. Simulate an elastic modulus field $E(\mathbf{x})$ for the soil from a lognormal distribution with specified mean μ_E, coefficient of variation v_E, and correlation structure (e.g., Eq. 10.59) with correlation length $\theta_{\ln E}$. The field is simulated using the LAS method whose local average values are assigned to corresponding finite elements.
4. Virtually sample the soil to obtain an estimate $\hat{E}$ of its elastic modulus. In a real site investigation, the geotechnical engineer may estimate the soil's elastic modulus and depth to firm stratum by performing one or more CPT or SPT soundings. In this simulation, one or more vertical columns of the soil model are selected to yield the elastic modulus samples. That is, $\hat{E}$ is estimated using a geometric average, Eq. 10.76, where E_1^o is the elastic modulus of the top element of a column, E_2^o is the elastic modulus of the second to top element of the same column, and so on, to the base of the column. One or more columns may be included in the estimate, as will be discussed shortly, and measurement and model errors are not included in the estimate—the measurements are assumed to be precise.
5. Letting $\delta_p = \delta_{max}$ and for given factors α and ϕ_g, solve Eq. 10.69 for B. This constitutes the footing design. Note that design widths are normally rounded up to the next most easily measured dimension (e.g., 1684 mm would probably be rounded up to 1700 mm). In the same way, in this analysis the design value of B is rounded up to the next larger element boundary since the finite-element model assumes footings are a whole number of elements wide. (The finite-element model uses elements which are 0.15 m wide, so B is rounded up here to the next larger multiple of 0.15 m.)
6. Simulate a lognormally distributed footing load P having median $\hat{P}$ and variance σ_P^2.
7. Compute the "actual" settlement δ of a footing of width B under load P on a random elastic modulus field using the finite-element model. In this step, the virtually sampled random field generated in step 3 above is mapped to the finite-element mesh, the footing of width B (suitably rounded up to a whole number of elements wide) is placed on the surface, and the settlement is computed by finite-element analysis.
8. If $\delta > \delta_{max}$, the footing design is assumed to have failed.
9. Repeat from step 3 a large number of times ($n = 1000$, in this study), counting the number of footings n_f which experienced a design failure. The failure probability is then estimated as $\hat{p}_f = n_f/n$.

By repeating the entire process over a range of possible values of ϕ_g the resistance factor which leads to an acceptable probability of failure, $p_f = p_{max}$, can be selected. This "optimal" resistance factor will also depend on:

1. Number and locations of sampled columns (analogous to number and locations of CPT/SPT soundings)
2. Coefficient of variation of soil's elastic modulus, v_E
3. Correlation length $\theta_{\ln E}$

The simulation will be repeated over a range of values of these parameters to see how they affect ϕ_g.

Five different sampling schemes will be considered in this study, as illustrated in Figure 10.28 [see Jaksa et al. (2005) for a detailed study of the effectiveness of site investigations]. The outer solid line denotes the edge of the soil model and the interior dashed line the location of the footing. The small black squares show the plan locations where the site is virtually sampled. It is expected that the quality of the estimate of E_{eff} will improve for higher numbered sampling schemes. That is, the probability of design failure will decrease for higher numbered sampling schemes, everything else being held constant.

Table 10.3 lists the other parameters, aside from sampling schemes, varied in this study. In total 300 RFEM runs

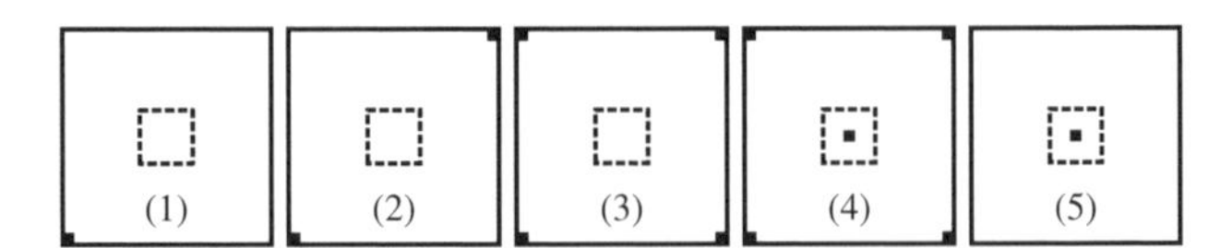

Figure 10.28 Sampling schemes considered in this study.

Table 10.3 Input Parameters Varied in Study While Holding $H = 4.8$ m, $D = 9.6$ m, $\mu_P = 1200$ kN, $v_P = 0.25$, $\mu_E = 20$ MPa, and $v = 0.3$ Constant

Parameter	Values Considered
v_E	0.1, 0.2, 0.5
$\theta_{\ln E}(m)$	0.1, 1.0 10.0, 100.0
ϕ_g	0.4, 0.5, 0.6, 0.7, 0.8

each involving 1000 realizations were performed. Based on 1000 independent realizations, the estimated failure probability $\hat{p}_f$ has standard error $\sqrt{\hat{p}_f(1-\hat{p}_f)/1000}$, which for a probability level of 5% is 0.7%. In other words, a true failure probability of 5% is estimated to within 0.7% with confidence 68% using 1000 observations.

10.5.4 Simulation Results

Figure 10.29 shows the effect of the correlation length on the probability of failure for sampling scheme 1 (a single sampled column at the corner of site) and for $v_E = 0.5$. The other sampling schemes and values of v_E displayed similarly shaped curves. Of particular note in Figure 10.29 is the fact that the probability of failure reaches a maximum for an intermediate correlation length, in this case when $\theta_{\ln E} \simeq 10$ m. This is as expected, since for stationary random fields the values of $\hat{E}$ and E_{eff} will coincide for both vanishingly small correlation lengths (where local averaging results in both becoming equal to the median) and for very large correlation lengths (where $\hat{E}$ and E_{eff} become perfectly correlated), and so the largest differences between $\hat{E}$ and E_{eff} will occur at intermediate correlation lengths. The true maximum could lie somewhere between $\theta_{\ln E} = 1$ m and $\theta_{\ln E} = 100$ m in this particular study.

Where the maximum correlation length occurs for arbitrary sampling patterns is still unknown. However, the authors expect that it is probably safe to say that taking $\theta_{\ln E}$ approximately equal to the average distance between sample locations and the footing center (but not less than the footing size) will yield suitably conservative failure probabilities. In the remainder of this study, the $\theta_{\ln E} = 10$ m

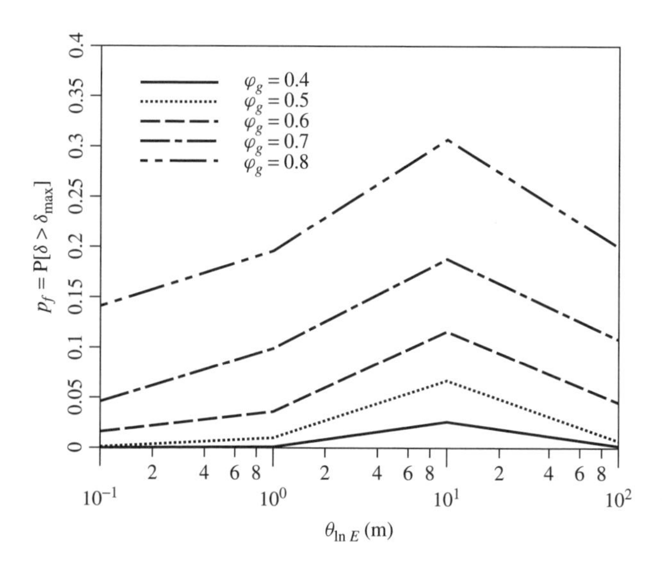

Figure 10.29 Effect of correlation length $\theta_{\ln E}$ on probability of settlement failure $p_f = \text{P}[\delta > \delta_{\max}]$.

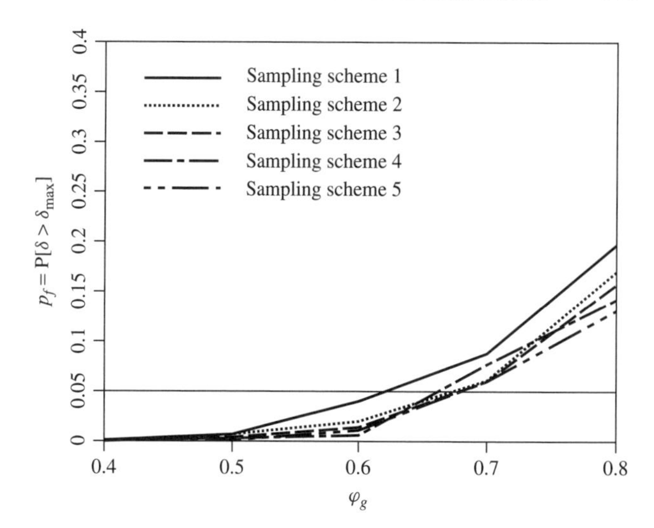

Figure 10.30 Effect of resistance factor ϕ_g on probability of failure $p_f = \text{P}[\delta > \delta_{\max}]$ for $v_E = 0.2$ and $\theta_{\ln E} = 10$ m.

results will be concentrated on since these yielded the most conservative designs.

Figure 10.30 shows how the estimated probability of failure varies with resistance factor for the five sampling schemes considered with $v_E = 0.2$ and $\theta_{\ln E} = 10$ m. This figure can be used for design by drawing a horizontal line across at the target probability $p_{\max}$—to illustrate this, a light line has been drawn across at $p_{\max} = 0.05$—and then reading off the required resistance factor for a given sampling scheme. For $p_{\max} = 0.05$, it can be seen that $\phi_g \simeq 0.62$ for the worst-case sampling scheme 1. For all the other sampling schemes considered, the required resistance factor is between about 0.67 and 0.69. Because the standard error of the estimated p_f values is 0.7% at this level, the relative positions of the lines tends to be somewhat erratic. What Figure 10.30 is saying, essentially, is that at low levels of variability increasing the number of samples does not greatly affect the probability of failure.

When the coefficient of variation v_E increases, the distinction between sampling schemes becomes more pronounced. Figure 10.31 shows the failure probability for the various sampling schemes at $v_E = 0.5$ and $\theta_{\ln E} = 10$ m. Improved sampling (i.e., improved understanding of the site) now makes a significant difference to the required value of ϕ_g, which ranges from $\phi_g \simeq 0.46$ for sampling scheme 1 to $\phi_g \simeq 0.65$ for sampling scheme 5, assuming a target probability of $p_{\max} = 0.05$. The implication of Figure 10.31 is that when soil variability is significant, considerable design/construction savings can be achieved when the sampling scheme is improved.

The approximation to the analytical expression for the failure probability can now be evaluated. For the case

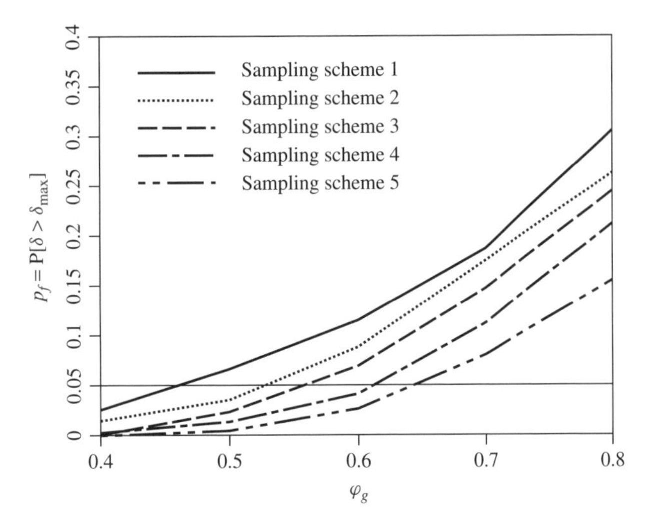

Figure 10.31 Effect of resistance factor ϕ_g on probability of failure $p_f = P[\delta > \delta_{max}]$ for $v_E = 0.5$ and $\theta_{\ln E} = 10$ m.

considered in Figure 10.31, $v_E = 0.5$ and $v_P = 0.25$, so that

$$\sigma^2_{\ln E} = \ln(1 + v_E^2) = 0.2231$$

$$\sigma^2_{\ln P} = \ln(1 + v_P^2) = 0.0606$$

To compute the variance reduction function $\gamma(n)$, the footing width corresponding to the median elastic modulus is needed. For this calculation, an initial value of ϕ_g is also needed, and the moderate value of $\phi_g = 0.5$ is recommended. For $\mu_E = 20{,}000$ kPa, the median elastic modulus $\tilde{E}$ is

$$\tilde{E} = \frac{\mu_E}{\sqrt{1+v_E^2}} = \frac{20{,}000}{\sqrt{1+0.5^2}} = 17{,}889 \quad \text{kPa}$$

and for $\mu_P = 1200$ kN, the median footing load is

$$\hat{P} = \frac{\mu_P}{\sqrt{1+v_P^2}} = \frac{1200}{\sqrt{1+0.25^2}} = 1164.2 \quad \text{kN}$$

Solving Eq. 10.69 iteratively gives $B_{med} = 2.766$ m. The corresponding variance reduction factors are

$$\gamma_1\left(\frac{2(4.8)}{10}\right) = \frac{2}{0.96^2}\left[0.96 + e^{-0.96} - 1\right] = 0.74413$$

$$\gamma_1\left(\frac{2(2.766)}{10}\right) = \frac{2}{0.5532^2}\left[0.5532 + e^{-0.5532} - 1\right]$$
$$= 0.83852$$

which gives

$$\gamma(n) \simeq (0.83852)^2(0.74413) = 0.5232$$

Now consider sampling scheme 1, which involves a single vertical sample with $R = \Delta x = 0.15$ m and corresponding variance reduction factor,

$$\gamma_1\left(\frac{2(0.15)}{10}\right) = \frac{2}{0.03^2}\left[0.03 + e^{-0.03} - 1\right] = 0.99007$$

$$\gamma^o(m) \simeq (0.99007)^2(0.74413) = 0.7294$$

For sampling scheme 1, $\tau'_{ave} \simeq \sqrt{2}(9.6/2) = 6.79$ m is the (approximate) distance from the sample point to the center of the footing. In this case,

$$\rho'_{ave} = \exp\left\{-\frac{2(6.79)}{10}\right\} = 0.2572$$

which gives us, using Eq. 10.89,

$$\sigma^2_{\ln Q} = 0.0606 + 0.2231\,[0.7294 + 0.5232 - 2(0.2572)]$$
$$= 0.2253$$

so that $\sigma_{\ln Q} = 0.4746$. For $\beta = 1.645$, the required resistance factor is determined by Eq. 10.91 to be

$$\phi_g = \exp\{-1.645(0.4746)\} = 0.46$$

The corresponding value on Figure 10.31 is also 0.46. Although this agreement is excellent, it must be remembered that this is an approximation, and the precise agreement may be due somewhat to mutually canceling errors and to chance, since the simulation estimates are themselves somewhat random. For example, if the more precise formulas of Eqs. 10.81a, 10.81b, and 10.87 are used, then $\gamma^o(m) = 0.7432$, $\gamma(n) = 0.6392$, and $\rho'_{ave} = 0.2498$, which gives

$$\sigma^2_{\ln Q} = 0.0606 + 0.2231\,[0.7432 + 0.6392 - 2(0.2498)]$$
$$= 0.2576$$

so that the "more precise" required resistance factor actually has poorer agreement with simulation:

$$\phi_g = \exp\{-1.645\sqrt{0.2576}\} = 0.43$$

It is also to be remembered that the more precise result above is still conditioned on $B = B_{med}$ and $\phi_g = 0.5$, whereas the simulation results are unconditional. Nevertheless, these results suggest that the approximations are insensitive to variations in B and ϕ_g and are thus reasonably general.

Sampling scheme 2 involves two sampled columns separated by more than $\theta_{\ln E} = 10$ m so that n_{eff} can be taken as 2. This means that $\gamma^o(m) \simeq 0.7294/2 = 0.3647$. The average distance from the footing centerline to the sampled columns is still about 6.79 m, so that $\rho'_{ave} = 0.2572$. Now

$$\sigma^2_{\ln Q} = 0.0606 + 0.2231\,[0.3647 + 0.5232 - 2(0.2572)]$$
$$= 0.1439$$

and the required resistance factor is

$$\phi_g = \exp\{-1.645\sqrt{0.1439}\} = 0.54$$

The corresponding value in Figure 10.31 is about 0.53.

Sampling scheme 3 involves four sampled columns separated by somewhat less than $\theta_{\ln E} = 10$ m. Due to the resulting correlation between columns, $n_{\text{eff}} \simeq 3$ is selected (i.e., somewhat less than the "independent" value of 4). This gives $\gamma^o(m) \simeq 0.7294/3 = 0.2431$. Since the average distance from the footing centerline to the sample columns is still about 6.79 m,

$$\sigma^2_{\ln Q} = 0.0606 + 0.2231\,[0.2431 + 0.5232 - 2(0.2572)] = 0.1268$$

The required resistance factor is

$$\phi_g = \exp\{-1.645\sqrt{0.1268}\} = 0.57$$

The corresponding value in Figure 10.31 is about 0.56.

Sampling scheme 4 involves five sampled columns also separated by somewhat less than $\theta_{\ln E} = 10$ m and $n_{\text{eff}} \simeq 4$ is selected to give $\gamma^o(m) \simeq 0.7294/4 = 0.1824$. One of the sampled columns lies below the footing, and so its distance to the footing centerline is taken to be $B_{\text{med}}/\sqrt{2} = 2.766/\sqrt{2} = 1.96$ m to avoid complete correlation. The average distance to sampling points is thus

$$\tau'_{\text{ave}} = \tfrac{4}{5}(6.79) + \tfrac{1}{5}(1.96) = 5.82$$

so that $\rho'_{\text{ave}} = 0.3120$. This gives

$$\sigma^2_{\ln Q} = 0.0606 + 0.2231\,[0.1824 + 0.5232 - 2(0.3120)] = 0.0788$$

The required resistance factor is

$$\phi_g = \exp\{-1.645\sqrt{0.0788}\} = 0.63$$

The corresponding value in Figure 10.31 is about 0.62.

For sampling scheme 5, the distance from the sample point to the center of the footing is zero, so τ'_{ave} is taken to equal the distance to the footing corner, $\tau'_{\text{ave}} = (2.766)/\sqrt{2} = 1.96$ m, as recommended earlier. This gives $\rho'_{\text{ave}} = 0.676$ and

$$\sigma^2_{\ln Q} = 0.0606 + 0.2231\,[0.7294 + 0.5232 - 2(0.676)] = 0.0606 + 0.2231\,[-0.0994] \rightarrow 0.0606$$

where approximation errors led to a negative variance contribution from the elastic modulus field which was ignored (i.e., set to zero). In this case, the sampled information is deemed sufficient to render uncertainties in the elastic modulus negligible, so that $\hat{E} \simeq E_{\text{eff}}$ and

$$\phi_g = \exp\{-1.645\sqrt{0.0606}\} = 0.67$$

The value of ϕ_g read from Figure 10.31 is about 0.65. If the more precise formulas for the variance reduction functions and covariance terms are used, then $\gamma^o(m) = 0.7432$, $\gamma(n) = 0.6392$, and $\rho'_{\text{ave}} = 0.6748$, which gives

$$\sigma^2_{\ln Q} = 0.0606 + 0.2231\,[0.7432 + 0.6392 - 2(0.6748)] = 0.0679$$

Notice that this is very similar to the approximate result obtained above, which suggests that the assumption that samples taken below the footing largely eliminate uncertainty in the effective elastic modulus is reasonable. For this more accurate result,

$$\phi_g = \exp\{-1.645\sqrt{0.0679}\} = 0.65$$

which is the same as the simulation results.

Perhaps surprisingly, sampling scheme 5 outperforms, in terms of failure probability and resistance factor, sampling scheme 4, even though sampling scheme 4 involves considerably more information. The reason for this is that in sampling scheme 4 the good information taken below the footing is diluted by poorer information taken from farther away. This implies that when a sample is taken below the footing, other samples taken from farther away should be downweighted. In other words, the simple averaging of data performed here should be replaced by distance-weighted averages.

The computations illustrated above for all five sampling schemes can be summarized as follows:

1. Decide on an acceptable maximum settlement $\delta_{\max}$. Since serviceability problems in a structure usually arise as a result of differential settlement, rather than settlement itself, the choice of an acceptable maximum settlement is usually made assuming that differential settlement will be less than the total settlement of any single footing [see, e.g., D'Appolonia et al. (1968) and the results of the first few sections of this chapter].
2. Choose statistical parameters of the elastic modulus field, μ_E, σ_E, and $\theta_{\ln E}$. The last can be the worst-case correlation length, suggested here to approximately equal the average distance between sample locations and the footing center, but not to be taken less than the median footing dimension. The values of μ_E and σ_E can be estimated from site samples (although the effect of using estimated values of μ_E and σ_E in these computations has not been investigated) or from the literature.
3. Use Eqs. 1.176 to compute the statistical parameters of $\ln E$ and then compute the median $\tilde{E} = \exp\{\mu_{\ln E}\}$.

4. Estimate statistical parameters for the load, μ_P and σ_P, and use these to compute the mean and variance of $\ln P$. Set $\hat{P} = \exp\{\mu_{\ln P}\}$.
5. Using a moderate resistance factor, $\phi_g = 0.5$, and the median elastic modulus $\tilde{E}$, compute the median value of B using the one-point iteration of Eq. 10.69. Call this B_{med}.
6. Compute $\gamma(n)$ using Eq. 10.82 (or Eq. 10.81b) with $B = B_{\text{med}}$.
7. Compute $\gamma^o(m)$ using Eq. 10.85 (or Eq. 10.81a).
8. Compute ρ'_{ave} using Eq. 10.88 (or Eq. 10.87) after selecting a suitable value for τ'_{ave} as the average absolute distance between the sample columns and the footing center (where distances are taken to be no less than the distance to the footing corner, $B_{\text{med}}/\sqrt{2}$).
9. Compute $\sigma_{\ln Q}$ using Eq. 10.89.
10. Compute the required resistance factor ϕ_g using Eq. 10.91.

10.5.5 Summary

The section presents approximate relationships based on random-field theory which can be used to estimate resistance factors appropriate for the LRFD settlement design of shallow foundations. Some specific comments arising from this research are as follows:

1. Two assumptions deemed to have the most influence on the resistance factors estimated in this study are (1) that the nominal load used for design, $\hat{P}$, is the median load and (2) that the load factor α is equal to 1.0. Changes in α result in a linear change in the resistance factor, for example, $\phi'_g = \alpha\phi_g$, where ϕ_g is the resistance factor found in this study and ϕ'_g is the resistance factor corresponding to an α which is not equal to 1.0. Changes in $\hat{P}$ (e.g., if $\hat{P}$ were taken as some other load exceedance percentile) would result in first-order linear changes to ϕ_g, but further study would be required to specify the actual effect on the resistance factor.
2. The resistance factors obtained in this study should be considered to be upper bounds since the additional uncertainties arising from measurement and model errors have not been considered. To some extent, these additional error sources can be accommodated here simply by using a value of v_E greater than would actually be true at a site. For example, if $v_E = 0.35$ at a site, the effects of measurement and model error might be accommodated by using $v_E = 0.5$ in the relationships presented here. This issue needs additional study, but Meyerhof's (1984, p. 6) comment that "in view of the uncertainty and great variability in in-situ soil-structure stiffnesses ... a partial factor of 0.7 should be used for an adequate reliability of serviceability estimates" suggests that the results presented here are reasonable (possibly a little conservative at the $v_E = 0.5$ level) for all sources of error.
3. The use of a median footing width B_{med} derived using a median elastic modulus and moderate $\phi_g = 0.5$ value, rather than by using the full B distribution in the computation of $\gamma(n)$, appears to be quite reasonable. This is validated by the agreement between the simulation results (where B varies with each realization) and the results obtained using the approximate relationships (see previous section).
4. The computation of a required resistance factor assumes that the uncertainty (e.g., v_E) is known. In fact, at a given site, all three parameters μ_E, v_E, and $\theta_{\ln E}$ will be unknown and only estimated to various levels of precision by sampled data. To establish a LRFD code, at least v_E and $\theta_{\ln E}$ need to be known a priori. One of the significant results of this research is that a worst-case correlation length exists which can be used in the development of a design code. While the value of v_E remains an outstanding issue, calibration with existing codes may very well allow its practical estimation.
5. At low uncertainty levels, that is, when $v_E \leq 0.2$ or so, there is not much advantage to be gained by taking more than two sampled columns (e.g., SPT or CPT borings) *in the vicinity of the footing*, as seen in Figure 10.30. This statement assumes that the soil is *stationary*. The assumption of stationarity implies that samples taken in one location are as good an estimator of the mean, variance, and so on, as samples taken elsewhere. Since this is rarely true of soils, the qualifier "in the vicinity" was added to the above statement.
6. Although sampling scheme 4 involved five sampled columns and sampling scheme 5 involved only one sampled column, sampling scheme 5 outperformed 4. This is because the distance to the samples was not considered in the calculation of $\hat{E}$. Thus, in sampling scheme 4 the good estimate taken under the footing was diluted by four poorer estimates taken some distance away. Whenever a soil is sampled directly under a footing, those sample results should be given much higher weighting than soil samples taken elsewhere. That is, the concepts of BLUE, which takes into account the correlation between estimate and observation, should be used (see Section 4.1). In this section a straightforward geometric average was used (arithmetic average of logarithms in log-space) for simplicity.

CHAPTER 11

Bearing Capacity

11.1 STRIP FOOTINGS ON c–ϕ SOILS

The design of a foundation involves the consideration of several limit states which can be separated into two groups: serviceability limit states, which generally translate into a maximum settlement or differential settlement, and ultimate limit states. The latter are concerned with the maximum load which can be placed on the footing just prior to a bearing capacity failure. This section looks at the ultimate bearing capacity of a smooth strip footing founded on a soil having spatially random properties [Fenton and Griffiths (2003); see also Fenton and Griffiths (2001), Griffiths and Fenton (2000b), Griffiths et al. (2002b), and Manoharan et al. (2001)]. The program used to perform the simulations reported here is called RBEAR2D and is available at `http://www.engmath.dal.ca/rfem`.

Most modern bearing capacity predictions involve a relationship of the form (Terzaghi, 1943; Meyerhof, 1951)

$$q_u = cN_c + \bar{q}N_q + \tfrac{1}{2}\gamma BN_\gamma \tag{11.1}$$

where q_u is the ultimate bearing stress, c is the cohesion, $\bar{q}$ is the overburden stress, γ is the unit soil weight, B is the footing width, and N_c, N_q, and N_γ are the bearing capacity factors which are functions of the friction angle ϕ. To simplify the analysis in this section, and to concentrate on the stochastic behavior of the most important term (at least as far as spatial variation is concerned), the soil is assumed weightless with no surcharge. Under this assumption, the bearing capacity equation simplifies to

$$q_u = cN_c \tag{11.2}$$

Bearing capacity predictions, involving specification of the N factors, are often based on plasticity theory (see, e.g., Prandtl, 1921; Terzaghi, 1943; Meyerhof, 1951; Sokolovski, 1965) of a rigid base punching into a softer material (Griffiths and Fenton, 2001). These theories assume a *uniform* soil underlying the footing—that is, the soil is assumed to have properties which are spatially constant. Under this assumption, most bearing capacity theories (e.g., Prandtl, 1921; Meyerhof, 1951, 1963) assume that the failure slip surface takes on a logarithmic spiral shape to give

$$N_c = \frac{e^{\pi \tan\phi}\, \tan^2(\pi/4 + \phi/2) - 1}{\tan\phi} \tag{11.3}$$

This relationship has been found to give reasonable agreement with test results (e.g., Bowles, 1996) under ideal conditions, and the displacement of the failed soil assumes the symmetry shown in Figure 11.1.

In practice, however, it is well known that the actual failure conditions will be somewhat more complicated than a simple logarithmic spiral. Due to spatial variation in soil properties, the failure surface under the footing will follow the weakest path through the soil, constrained by the stress field. For example, Figure 11.2 illustrates the bearing failure of a realistic soil with spatially varying properties. It can be seen that the failure surface only approximately follows a log-spiral on the right side and is certainly not symmetric. In this plot lighter regions represent weaker soil and darker regions indicate stronger soil. The weak (light) region near the ground surface to the right of the footing has triggered a nonsymmetric failure mechanism that is typically at a lower bearing load than predicted by traditional homogeneous and symmetric failure analysis.

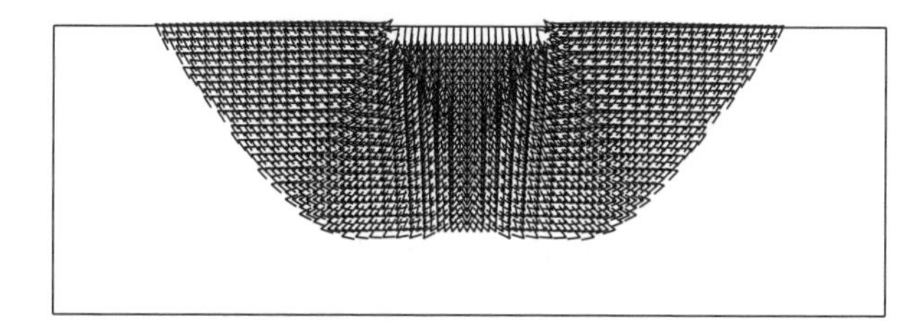

Figure 11.1 Displacement vector plot of bearing failure on uniform (spatially constant) soil.

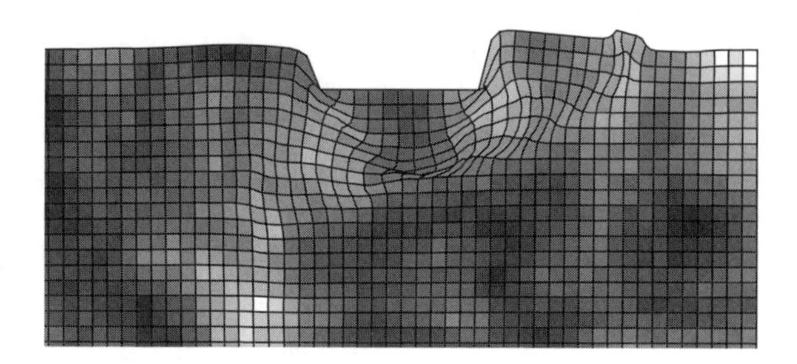

Figure 11.2 Typical deformed mesh at failure, where lighter regions indicate weaker soil.

The problem of finding the minimum strength failure slip surface through a soil mass is very similar in nature to the slope stability problem, and one which currently lacks a closed-form stochastic solution, so far as the authors are aware. In this section the traditional relationships shown above will be used as a starting point to this problem.

For a realistic soil, both c and ϕ are random, so that both quantities in the right-hand side of Eq. 11.2 are random. This equation can be nondimensionalized by dividing through by the cohesion mean:

$$M_c = \frac{q_u}{\mu_c} = \frac{c}{\mu_c} N_c \tag{11.4}$$

where μ_c is the mean cohesion and M_c is the stochastic equivalent of N_c, that is, $q_u = \mu_c M_c$. The stochastic problem is now boiled down to finding the distribution of M_c. A theoretical model for the first two moments (mean and variance) of M_c, based on geometric averaging, are given in the next section. Monte Carlo simulations are then performed to assess the quality of the predictions and determine the approximate form of the distribution of M_c. This is followed by an example illustrating how the results can be used to compute the probability of a bearing capacity failure. Finally, an overview of the results is given, including their limitations.

11.1.1 Random Finite-Element Method

In this study, the soil cohesion c is assumed to be lognormally distributed with mean μ_c, standard deviation σ_c, and spatial correlation length $\theta_{\ln c}$. The lognormal distribution is selected because it is commonly used to represent nonnegative soil properties and since it has a simple relationship with the normal. A lognormally distributed random field is obtained from a normally distributed random field $G_{\ln c}(\mathbf{x})$ having zero mean, unit variance, and spatial correlation length $\theta_{\ln c}$ through the transformation

$$c(\mathbf{x}) = \exp\{\mu_{\ln c} + \sigma_{\ln c} G_{\ln c}(\mathbf{x})\} \tag{11.5}$$

where $\mathbf{x}$ is the spatial position at which c is desired. The parameters $\mu_{\ln c}$ and $\sigma_{\ln c}$ are obtained from the specified cohesion mean and variance using the lognormal transformations of Eqs. 1.176.

The correlation coefficient between the log-cohesion at a point $\mathbf{x}_1$ and a second point $\mathbf{x}_2$ is specified by a correlation function, $\rho_{\ln c}(\tau)$, where $\tau = |\mathbf{x}_1 - \mathbf{x}_2|$ is the absolute distance between the two points. In this study, a simple exponentially decaying (Markovian) correlation function will be assumed, having the form (see also Section 3.7.10.2)

$$\rho_{\ln c}(\tau) = \exp\left(-\frac{2|\tau|}{\theta_{\ln c}}\right) \tag{11.6}$$

The spatial correlation length, $\theta_{\ln c}$, is loosely defined as the separation distance within which two values of $\ln c$ are significantly correlated, as discussed in Section 3.5. The correlation function, $\rho_{\ln c}$, acts between values of $\ln c$ since $\ln c$ is normally distributed, and a normally distributed random field is simply defined by its mean and covariance structure.

The random field is also assumed here to be statistically isotropic (the same correlation length in any direction through the soil). Although the horizontal correlation length is often greater than the vertical, due to soil layering, taking this into account is a site-specific refinement left to the reader. The main aspects of the stochastic behavior of bearing capacity for a relatively simple problem are presented here.

The friction angle ϕ is assumed to be bounded both above and below, so that neither normal nor lognormal distributions are appropriate. A beta distribution is often used for bounded random variables. Unfortunately, a beta-distributed random field has a complex joint distribution and simulation is cumbersome and numerically difficult. To keep things simple, the tanh transformation discussed in Section 1.10.10 is used. This transformation results in a bounded distribution which resembles a beta distribution but which arises as a simple transformation of a standard normal random field $G_\phi(\mathbf{x})$ according to

$$\phi(\mathbf{x}) = \phi_{\min} + \frac{1}{2}(\phi_{\max} - \phi_{\min})\left\{1 + \tanh\left(\frac{sG_\phi(\mathbf{x})}{2\pi}\right)\right\} \tag{11.7}$$

where $\phi_{\min}$ and $\phi_{\max}$ are the minimum and maximum friction angles, respectively, and s is a scale factor which governs the friction angle variability between its two bounds (see Figure 1.36).

The random field $G_\phi(\mathbf{x})$ has zero mean and unit variance, as does $G_{\ln c}(\mathbf{x})$. Conceivably, $G_\phi(\mathbf{x})$ could also have its own correlation length θ_ϕ distinct from $\theta_{\ln c}$. However, it seems reasonable to assume that if the spatial correlation structure is caused by changes in the constitutive nature of the soil over space, then both cohesion and friction angle would have similar correlation lengths. Thus, θ_ϕ is taken to be equal to $\theta_{\ln c}$ in this study. Both lengths will be referred to generically from now on simply as θ, remembering that this length reflects correlation between points in the underlying normally distributed random fields $G_{\ln c}(\mathbf{x})$ and $G_\phi(\mathbf{x})$ and not directly between points in the cohesion and friction fields. As mentioned above, both lengths can be estimated from data sets obtained over some spatial domain by statistically analyzing the suitably transformed data (inverses of Eqs. 11.5 and 11.7—see Eq. 1.191 for the inverse of Eq. 11.7). After transforming to the c and ϕ fields, the transformed correlation lengths will no longer be the same, but since both transformations are monotonic (i.e., larger values of $G_{\ln c}$ give larger values of c, etc.), the correlation lengths will be similar. For example, when $s = v = 1.0$, the difference in correlation lengths is less than 15% from each other and from the underlying

Gaussian field correlation length. In that all engineering soil properties are derived through various transformations of the physical soil behavior (e.g., cohesion is a complex function of electrostatic forces between soil particles), the final correlation lengths between engineering properties cannot be expected to be identical, only similar. For the purposes of a generic non-site-specific study, the above assumptions are believed reasonable.

The question as to whether the two parameters c and ϕ are correlated is still not clearly decided in the literature, and no doubt depends very much on the soil being studied. Cherubini (2000) quotes values of ρ ranging from -0.24 to -0.70, as does Wolff (1985) (see also Yuceman et al., 1973; Lumb, 1970; and Cherubini, 1997).

In that the correlation between c and ϕ is not certain, this section investigates the correlation extremes to determine if cross-correlation makes a significant difference. As will be seen, under the given assumptions regarding the distributions of c (lognormal) and ϕ (bounded), varying the cross-correlation ρ from -1 to $+1$ was found to have only a minor influence on the stochastic behavior of the bearing capacity.

11.1.2 Bearing Capacity Mean and Variance

The determination of the first two moments of the bearing capacity (mean and variance) requires first a failure model. Equations 11.2 and 11.3 assume that the soil properties are spatially uniform. When the soil properties are spatially varying, the slip surface no longer follows a smooth log-spiral and the failure becomes unsymmetric. The problem of finding the constrained path having the lowest total shear strength through the soil is mathematically difficult, especially since the constraints are supplied by the spatially varying stress field. A simpler approximate model will be considered here wherein geometric averages of c and ϕ, over some region under the footing, are used in Eqs. 11.2 and 11.3. The geometric average is proposed because it is dominated more by low strengths than is the arithmetic average. This is deemed reasonable since the failure slip surface preferentially travels through lower strength areas.

Consider a soil region of some size D discretized into a sequence of nonoverlapping rectangles, each centered on $\mathbf{x}_i$, $i = 1, 2, \ldots, n$. The geometric average of the cohesion c over the domain D may then be defined as

$$\begin{aligned}\bar{c} &= \left[\prod_{i=1}^{n} c(\mathbf{x}_i)\right]^{1/n} = \exp\left\{\frac{1}{n}\sum_{i=1}^{n} \ln c(\mathbf{x}_i)\right\} \\ &= \exp\left\{\mu_{\ln c} + \sigma_{\ln c}\bar{G}_{\ln c}\right\} \end{aligned} \tag{11.8}$$

where $\bar{G}_{\ln c}$ is the *arithmetic* average of $G_{\ln c}$ over the domain D. Note that an assumption is made in the above concerning $c(\mathbf{x}_i)$ being constant over each rectangle. In that cohesion is generally measured using some representative volume (e.g., a lab sample), the values of $c(\mathbf{x}_i)$ used above are deemed to be such measures.

In a similar way, the exact expression for the geometric average of ϕ over the domain D is

$$\bar{\phi} = \exp\left\{\frac{1}{n}\sum_{i=1}^{n} \ln \phi(\mathbf{x}_i)\right\} \tag{11.9}$$

where $\phi(\mathbf{x}_i)$ is evaluated using Eq. 11.7. A close approximation to the above geometric average, accurate for $s \le 2.0$, is

$$\bar{\phi} \simeq \phi_{\min} + \frac{1}{2}(\phi_{\max} - \phi_{\min})\left\{1 + \tanh\left(\frac{s\bar{G}_\phi}{2\pi}\right)\right\} \tag{11.10}$$

where $\bar{G}_\phi$ is the *arithmetic* average of G_ϕ over the domain D. For $\phi_{\min} = 5^\circ$, $\phi_{\max} = 45^\circ$, this expression has relative error of less than 5% for $n = 20$ independent samples. While the relative error rises to about 12%, on average, for $s = 5.0$, this is an extreme case, corresponding to an approximately uniformly distributed ϕ between the minimum and maximum values (Figure 1.36), which is felt to be unlikely to occur very often in practice. Thus, the above approximation is believed reasonable in most cases.

Using the latter result in Eq. 11.3 gives the "equivalent" value of N_c, $\bar{N}_c$, where the log-spiral model is assumed to be valid using a geometric average of soil properties within the failed region:

$$\bar{N}_c = \frac{e^{\pi \tan\bar{\phi}} \tan^2\left(\pi/4 + \bar{\phi}/2\right) - 1}{\tan\bar{\phi}} \tag{11.11}$$

so that, now

$$M_c = \frac{\bar{c}}{\mu_c}\bar{N}_c \tag{11.12}$$

If c is lognormally distributed, an inspection of Eq. 11.8 indicates that $\bar{c}$ is also lognormally distributed. If we can assume that $\bar{N}_c$ is at least approximately lognormally distributed, then M_c will also be at least approximately lognormally distributed (the central limit theorem helps out somewhat here). In this case, taking logarithms of Eq. 11.12 gives

$$\ln M_c = \ln\bar{c} + \ln\bar{N}_c - \ln\mu_c \tag{11.13}$$

so that, under the given assumptions, $\ln M_c$ is at least approximately normally distributed.

The task now is to find the mean and variance of $\ln M_c$. The mean is obtained by taking expectations of Eq. 11.13,

$$\mu_{\ln M_c} = \mu_{\ln\bar{c}} + \mu_{\ln\bar{N}_c} - \ln\mu_c \tag{11.14}$$

where

$$\begin{aligned}\mu_{\ln\bar{c}} &= \mathrm{E}\left[\mu_{\ln c} + \sigma_{\ln c}\bar{G}_{\ln c}\right] \\ &= \mu_{\ln c} + \sigma_{\ln c}\mathrm{E}\left[\bar{G}_{\ln c}\right]\end{aligned} \tag{11.15}$$

$$= \mu_{\ln c}$$

$$= \ln \mu_c - \frac{1}{2} \ln \left(1 + \frac{\sigma_c^2}{\mu_c^2}\right)$$

which used the fact that since $\bar{G}_{\ln c}$ is normally distributed, its arithmetic average has the same mean as $G_{\ln c}$, that is, $\mathrm{E}\left[\bar{G}_{\ln c}\right] = \mathrm{E}\left[G_{\ln c}\right] = 0$. The above result is as expected since the geometric average of a lognormally distributed random variable preserves the mean of the logarithm of the variable. Also Eq. 1.176b was used to express the mean in terms of the prescribed statistics of c.

A second-order approximation to the mean of the logarithm of Eq. 11.11, $\mu_{\ln \bar{N}_c}$, is

$$\mu_{\ln \bar{N}_c} \simeq \ln \bar{N}_c(\mu_{\bar{\phi}}) + \sigma_{\bar{\phi}}^2 \left(\frac{d^2 \ln \bar{N}_c}{d\bar{\phi}^2} \bigg|_{\mu_{\bar{\phi}}} \right) \tag{11.16}$$

where $\mu_{\bar{\phi}}$ is the mean of the geometric average of ϕ. Since $\bar{G}_\phi$ is an arithmetic average, its mean is equal to the mean of G_ϕ, which is zero. Thus, since the assumed distribution of ϕ is symmetric about its mean, $\mu_{\bar{\phi}} = \mu_\phi$ so that $\ln \bar{N}_c(\mu_{\bar{\phi}}) = \ln N_c(\mu_\phi)$.

A first-order approximation to $\sigma_{\bar{\phi}}^2$ is (note that this is a less accurate approximation than given by Eq. 1.196 and yet it still leads to quite accurate probability estimates, as will be seen)

$$\sigma_{\bar{\phi}}^2 = \left[\frac{s}{4\pi} (\phi_{\max} - \phi_{\min}) \sigma_{\bar{G}_\phi} \right]^2 \tag{11.17}$$

where, from local averaging theory (Vanmarcke, 1984), the variance of a local average over the domain D is given by (recalling that G_ϕ is normally distributed with zero mean and unit variance)

$$\sigma_{\bar{G}_\phi}^2 = \sigma_{G_\phi}^2 \gamma(D) = \gamma(D) \tag{11.18}$$

where $\gamma(D)$ is the "variance function" which reflects the amount that the variance is reduced due to local arithmetic averaging over the domain D (see Section 3.4). Note that in this study, $D = D_1 \times D_2$ is a two-dimensional rectangular domain so that $\gamma(D) = \gamma(D_1, D_2)$. The variance function can be obtained directly from the correlation function (see Appendix C).

The derivative in Eq. 11.16 is most easily obtained numerically using any reasonably accurate (N_c is quite smooth) approximation to the second derivative. See, for example, Press et al. (1997). If $\mu_{\bar{\phi}} = \mu_\phi = 25^\circ = 0.436$ rad (note that in all mathematical expressions, ϕ is assumed to be in radians), then

$$\frac{d^2 \ln \bar{N}_c}{d\bar{\phi}^2} \bigg|_{\mu_{\bar{\phi}}} = 5.2984 \text{ rad}^{-2} \tag{11.19}$$

Using these results with $\phi_{\max} = 45^\circ$ and $\phi_{\min} = 5^\circ$ so that $\mu_\phi = 25^\circ$ gives

$$\mu_{\ln \bar{N}_c} = \ln(20.72) + 0.0164 s^2 \gamma(D) \tag{11.20}$$

Some comments need to be made about this result: First of all, it increases with increasing variability in ϕ (increasing s). It seems doubtful that this increase would occur since increasing variability in ϕ would likely lead to more lower strength paths through the soil mass for moderate θ. Aside from ignoring the weakest path issue, some other sources of error in the above analysis follow:

1. The geometric average of ϕ given by Eq. 11.9 actually shows a slight decrease with s (about 12% less, relatively, when $s = 5$). Although the decrease is only slight, it at least is in the direction expected.
2. An error analysis of the second-order approximation in Eq. 11.16 and the first-order approximation in Eq. 11.17 has not been carried out. Given the rather arbitrary nature of the assumed distribution on ϕ, and the fact that this section is primarily aimed at establishing the approximate stochastic behavior, such refinements have been left for later work.

In light of these observations, a first-order approximation to $\mu_{\ln \bar{N}_c}$ may actually be more accurate. Namely,

$$\mu_{\ln \bar{N}_c} \simeq \ln \bar{N}_c(\mu_{\bar{\phi}}) \simeq \ln N_c(\mu_\phi) \tag{11.21}$$

Finally, combining Eqs. 11.15 and 11.21 into Eq. 11.14 gives

$$\mu_{\ln M_c} \simeq \ln N_c(\mu_\phi) - \frac{1}{2} \ln \left(1 + \frac{\sigma_c^2}{\mu_c^2}\right) \tag{11.22}$$

For independent c and ϕ, the variance of $\ln M_c$ is

$$\sigma_{\ln M_c}^2 = \sigma_{\ln \bar{c}}^2 + \sigma_{\ln \bar{N}_c}^2 \tag{11.23}$$

where

$$\sigma_{\ln \bar{c}}^2 = \gamma(D) \sigma_{\ln c}^2 = \gamma(D) \ln \left(1 + \frac{\sigma_c^2}{\mu_c^2}\right) \tag{11.24}$$

and, to first order,

$$\sigma_{\ln \bar{N}_c}^2 \simeq \sigma_{\bar{\phi}}^2 \left(\frac{d \ln \bar{N}_c}{d\bar{\phi}} \bigg|_{\mu_{\bar{\phi}}} \right)^2 \tag{11.25}$$

The derivative appearing in Eq. 11.25, which will be denoted as $\beta(\phi)$, is

$$\beta(\phi) = \frac{d \ln \bar{N}_c}{d\bar{\phi}} = \frac{d \ln N_c}{d\phi}$$

$$= \frac{bd}{bd^2 - 1} \left[\pi(1 + a^2)d + 1 + d^2 \right] - \frac{1 + a^2}{a} \tag{11.26}$$

where $a = \tan(\phi)$, $b = e^{\pi a}$, and $d = \tan(\pi/4 + \phi/2)$.

The variance of $\ln M_c$ is thus

$$\sigma^2_{\ln M_c} \simeq \gamma(D)\left\{\ln\left(1+\frac{\sigma_c^2}{\mu_c^2}\right) + \left[\left(\frac{s}{4\pi}\right)(\phi_{\max}-\phi_{\min})\beta(\mu_\phi)\right]^2\right\} \quad (11.27)$$

where ϕ is measured in radians.

11.1.3 Monte Carlo Simulation

A finite-element computer program based on program 6.1 in Smith and Griffiths (2004) was modified to compute the bearing capacity of a smooth rigid strip footing (plane strain) founded on a weightless soil with shear strength parameters c and ϕ represented by spatially varying and cross-correlated (pointwise) random fields, as discussed above. The bearing capacity analysis uses an elastic-perfectly plastic stress–strain law with a Mohr–Coulomb failure criterion. Plastic stress redistribution is accomplished using a viscoplastic algorithm. The program uses 8-node quadrilateral elements and reduced integration in both the stiffness and stress redistribution parts of the algorithm. The finite-element model incorporates five parameters: Young's modulus E, Poisson's ratio ν, dilation angle ψ, shear strength c, and friction angle ϕ. The program allows for random distributions of all five parameters; however, in the present study, E, ν, and ψ are held constant (at 100,000 kN/m^2, 0.3, and 0, respectively) while c and ϕ are randomized. The Young's modulus governs the initial elastic response of the soil but does not affect bearing capacity. Setting the dilation angle to zero means that there is no plastic dilation during yield of the soil. The finite-element mesh consists of 1000 elements, 50 elements wide by 20 elements deep. Each element is a square of side length 0.1 m and the strip footing occupies 10 elements, giving it a width of $B = 1$ m.

The random fields used in this study are generated using the LAS method (see Section 6.4.6). Cross-correlation between the two soil property fields (c and ϕ) is implemented via covariance matrix decomposition (see Section 6.4.2).

In the parametric studies that follow, the mean cohesion (μ_c) and mean friction angle (μ_ϕ) have been held constant at 100 kN/m^2 and 25° (with $\phi_{\min} = 5°$ and $\phi_{\max} = 45°$), respectively, while the coefficient of variation ($v = \sigma_c/\mu_c$), spatial correlation length (θ), and correlation coefficient, ρ, between $G_{\ln c}$ and G_ϕ are varied systematically according to Table 11.1.

Table 11.1 Random-Field Parameters Used in Study

θ	=	0.5	1.0	2.0	4.0	8.0	50.0
v	=	0.1	0.2	0.5	1.0	2.0	5.0
ρ	=	−1.0	0.0	1.0			

It will be noticed that coefficients of variation v up to 5.0 are considered in this study, which is an order of magnitude higher than generally reported in the literature (see, e.g., Phoon and Kulhawy, 1999). There are two considerations which complicate the problem of defining typical v's for soils that have not yet been clearly considered in the literature (Fenton, 1999a). The first has to do with the level of information known about a site. Prior to any site investigation, there will be plenty of uncertainty about soil properties, and an appropriate v comes by using v obtained from regional data over a much larger scale. Such a v value will typically be much greater than that found when soil properties are estimated over a much smaller scale, such as a specific site. As investigation proceeds at the site of interest, the v value drops. For example, a single sample at the site will reduce v slightly, but as the investigation intensifies, v drops toward zero, reaching zero when the entire site has been sampled (which, of course, is clearly impractical). The second consideration, which is actually closely tied to the first, has to do with scale. If one were to take soil samples every 10 km over 5000 km (macroscale), one will find that the v value of those samples will be very large. A value of v of 5.0 would not be unreasonable. Alternatively, suppose one were to concentrate one's attention on a single cubic meter of soil. If several 50-mm^3 samples were taken and sent to the laboratory, one would expect a fairly small v. On the other hand, if samples of size 0.1 μm^3 were taken and tested (assuming this was possible), the resulting v could be very large since some samples might consist of very hard rock particles, others of water, and others just of air (i.e., the sample location falls in a void). In such a situation, a v value of 5.0 could easily be on the low side. While the last scenario is only conceptual, it does serve to illustrate that v is highly dependent on the ratio between sample volume and sampling domain volume. This dependence is certainly pertinent to the study of bearing capacity since it is currently not known at what scale bearing capacity failure operates. Is the weakest path through a soil dependent on property variations at the microscale (having large v), or does the weakest path "smear" the small-scale variations and depend primarily on local average properties over, say, laboratory scales (small v)? Since laboratory scales are merely convenient for us, it is unlikely that nature has selected that particular scale to accommodate us. From the point of view of reliability estimates, where the failure mechanism might depend on microscale variations for failure initiation, the small v's reported in the literature might very well be dangerously unconservative. Much work is still required to establish the relationship between v, site investigation intensity, and scale. In the meantime, values of v over a fairly wide range are considered here since it is entirely possible that the higher values more truly reflect failure variability.

In addition, it is assumed that when the variability in the cohesion is large, the variability in the friction angle will also be large. Under this reasoning, the scale factor, s, used in Eq. 11.7 is set to $s = \sigma_c/\mu_c = v$. This choice is arbitrary but results in the friction angle varying from quite narrowly (when $v = 0.1$ and $s = 0.1$) to very widely (when $v = 5.0$ and $s = 5$) between its lower and upper bounds, 5° and 45°, as indicated in Figure 1.36.

For each set of assumed statistical properties given by Table 11.1, Monte Carlo simulations have been performed. These involve 1000 realizations of the soil property random fields and the subsequent finite-element analysis of bearing capacity. Each realization, therefore, has a different value of the bearing capacity and, after normalization by the *mean* cohesion, a different value of the bearing capacity factor:

$$M_{c_i} = \frac{q_{u_i}}{\mu_c}, \qquad i = 1, 2, \ldots, 1000$$

$$\hat{\mu}_{\ln M_c} = \frac{1}{1000} \sum_{i=1}^{1000} \ln M_{c_i} \tag{11.28}$$

where $\hat{\mu}_{\ln M_c}$ is the sample mean of $\ln M_c$ estimated over the ensemble of realizations. Figure 11.3 illustrates how the load–deformation curves determined by the finite-element analysis change from realization to realization.

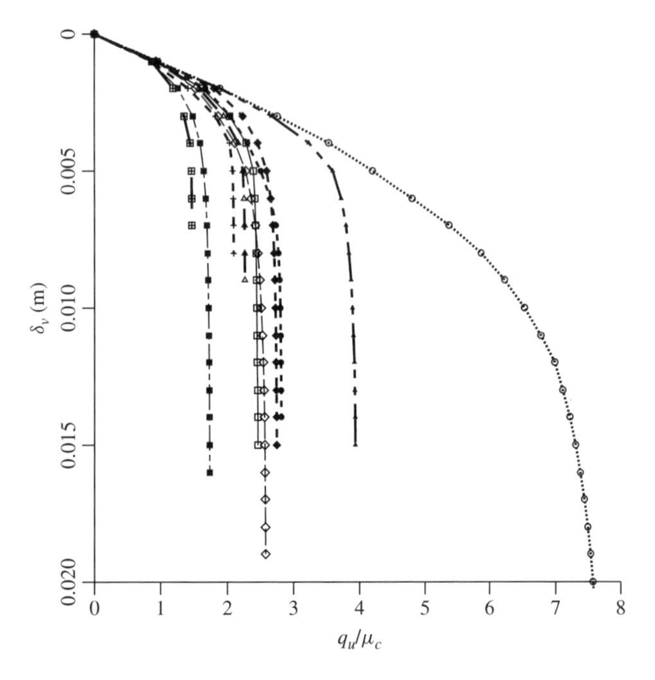

Figure 11.3 Typical load–deformation curves corresponding to different realizations of soil in bearing capacity analysis.

11.1.4 Simulation Results

Figure 11.4*a* shows how the sample mean log-bearing capacity factor, taken as the average over the 1000 realizations of $\ln M_{c_i}$, and referred to as $\hat{\mu}_{\ln M_c}$ in the figure, varies with correlation length, soil variability, and cross-correlation between c and ϕ. For small soil variability, $\hat{\mu}_{\ln M_c}$ tends toward the deterministic value of $\ln(20.72) = 3.03$, which is found when the soil takes on its mean properties everywhere. For increasing soil variability, the mean bearing capacity factor becomes quite significantly reduced from the traditional case. What this implies from a design standpoint is that the bearing capacity of a spatially variable soil will, on average, be *less* than the Prandtl solution based on the mean values alone. The greatest reduction from the Prandtl solution is observed for perfectly correlated c and ϕ ($\rho = +1$), the least reduction when c and ϕ are negatively correlated ($\rho = -1$), and the independent case ($\rho = 0$) lies between these two extremes. However, the effect of cross-correlation is seen to be not particularly large. If the negative cross-correlation indicated by both Cherubini (2000) and Wolff (1985) is correct, then the independent, $\rho = 0$, case is conservative, having mean bearing capacities consistently somewhat less than the $\rho = -1$ case.

The cross-correlation between c and ϕ is seen to have minimal effect on the sample standard deviation, $\hat{\sigma}_{\ln M_c}$, as shown in Figure 11.4*b*. The sample standard deviation is most strongly affected by the correlation length and somewhat less so by the soil property variability. A decreasing correlation length results in a decreasing $\hat{\sigma}_{\ln M_c}$. As suggested by Eq. 11.27, the function $\gamma(D)$ decays approximately with θ/D and so decreases with decreasing θ. This means that $\hat{\sigma}_{\ln M_c}$ should decrease as the correlation length decreases, which is as seen in Figure 11.4*b*.

Figure 11.4*a* also seems to show that the correlation length, θ, does not have a significant influence in that the $\theta = 0.1$ and $\theta = 8$ curves for $\rho = 0$ are virtually identical. However, the $\theta = 0.1$ and $\theta = 8$ curves are significantly lower than that predicted by Eq. 11.22 implying that the plot is somewhat misleading with respect to the dependence on θ. For example, when the correlation length goes to infinity, the soil properties become spatially constant, albeit still random from realization to realization. In this case, because the soil properties are spatially constant, the weakest path returns to the log-spiral and $\mu_{\ln M_c}$ will rise toward that given by Eq. 11.22, namely $\mu_{\ln M_c} = \ln(20.72) - \frac{1}{2}\ln(1 + \sigma_c^2/\mu_c^2)$, which is also shown on the plot. This limiting value holds because $\mu_{\ln N_c} \simeq \ln N_c(\mu_\phi)$, as discussed for Eq. 11.21, where for spatially constant properties $\bar{\phi} = \phi$.

Similarly, when $\theta \to 0$, the soil property field becomes infinitely "rough," in that all points in the field become independent. Any point at which the soil is weak will be surrounded by points where the soil is strong. A path

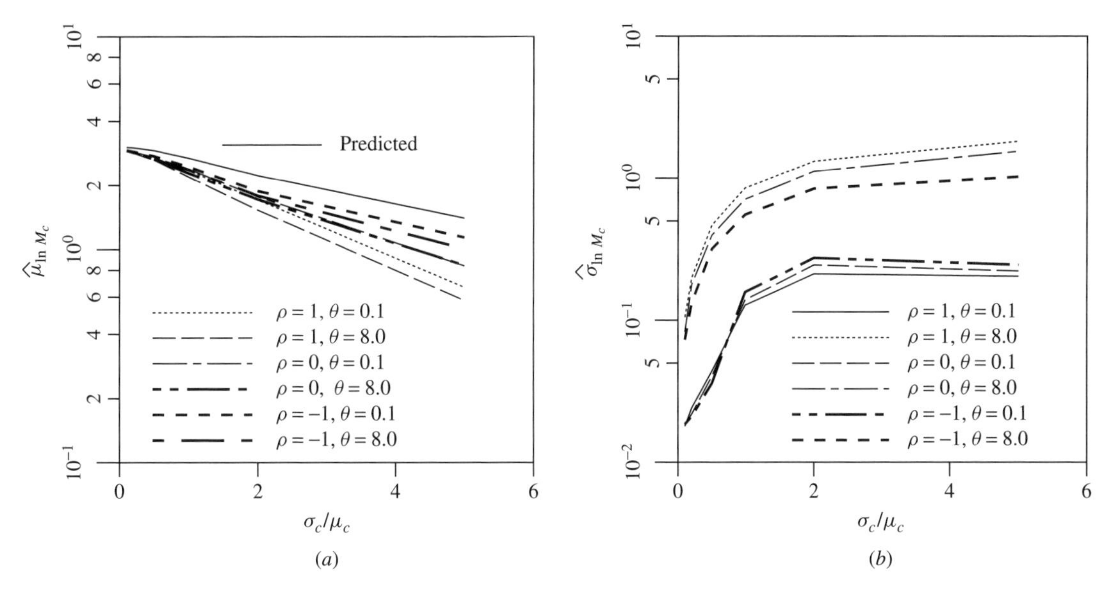

Figure 11.4 (a) Sample mean of log-bearing capacity factor, $\ln M_c$, along with its prediction by Eq. 11.22 and (b) its sample standard deviation.

through the weakest points in the soil might have very low average strength, but at the same time will become infinitely tortuous and thus infinitely long. This, combined with shear interlocking dictated by the stress field, implies that the weakest path should return to the traditional log-spiral with average shear strength along the spiral given by μ_ϕ and the median of c which is $\exp\{\mu_{\ln c}\}$. Again, in this case, $\mu_{\ln M_c}$ should rise to that given by Eq. 11.22.

The variation of $\mu_{\ln M_c}$ with respect to θ is more clearly seen in Figure 11.5. Over a range of values of σ_c/μ_c, the value of $\mu_{\ln M_c}$ rises toward that predicted by Eq. 11.22 at both high and low correlation lengths. At intermediate correlation lengths, the weakest path issue is seen to result in $\mu_{\ln M_c}$ being less than that predicted by Eq. 11.22 (see Figure 11.4a), the greatest reduction in $\mu_{\ln M_c}$ occurring when θ is of the same order as the footing width, B. It is hypothesized that $\theta \simeq B$ leads to the greatest reduction in $\mu_{\ln M_c}$ because it allows enough spatial variability for a failure surface which deviates somewhat from the log-spiral but which is not too long (as occurs when θ is too small) yet has significantly lower average strength than the $\theta \to \infty$ case. The apparent agreement between the $\theta = 0.1$ and $\theta = 8$ curves in Figure 11.4a is only because they are approximately equispaced on either side of the minimum at $\theta \simeq 1$.

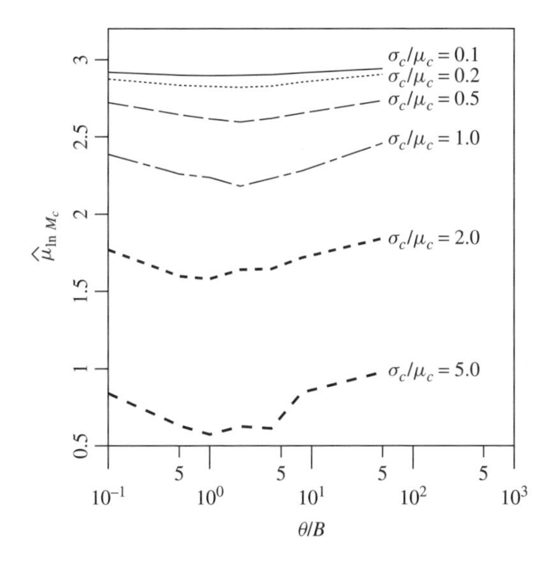

Figure 11.5 Sample mean of log-bearing capacity factor, $\ln M_c$, versus normalized correlation length θ/B.

As noted above, in the case where c and ϕ are independent ($\rho = 0$) the predicted mean, $\mu_{\ln M_c}$, given by Eq. 11.22 does not decrease as fast as observed in Figure 11.4a for intermediate correlation lengths. Nor does Eq. 11.22 account for changes in θ. Although an analytic prediction for the mean strength of the constrained weakest path through a spatially random soil has not yet been determined, Eq. 11.22 can be improved by making the following empirical corrections for the worst case ($\theta \simeq B$):

$$\mu_{\ln M_c} \simeq 0.92 \ln N_c(\mu_\phi) - 0.7 \ln\left(1 + \frac{\sigma_c^2}{\mu_c^2}\right) \tag{11.29}$$

where the overall reduction with σ_c/μ_c is assumed to follow the same form as predicted in Eq. 11.22. Some portion of the above correction may be due to finite-element model error (e.g., the finite-element model slightly underestimates the deterministic value of N_c, giving $N_c =$

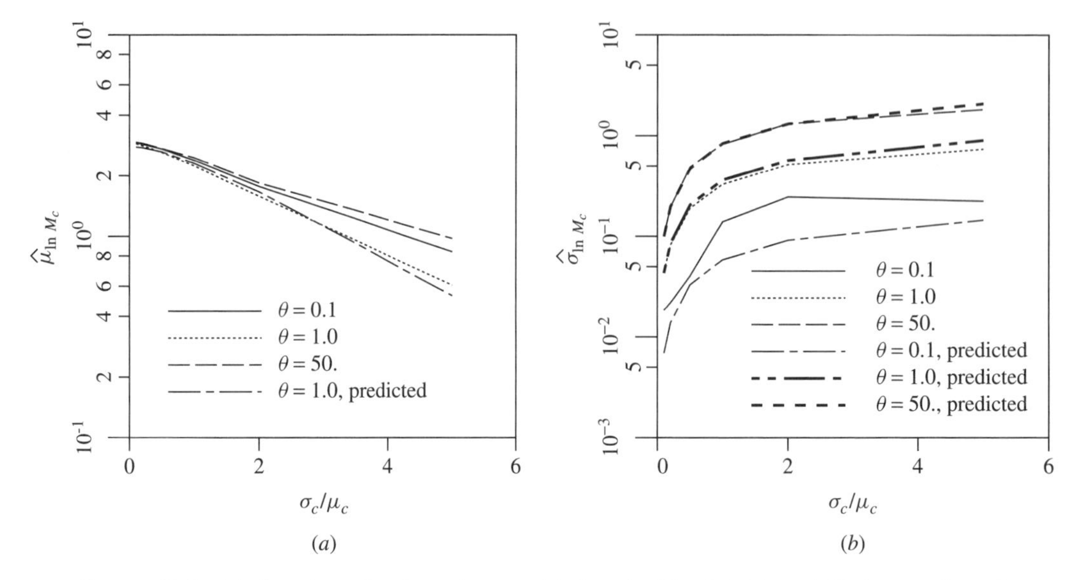

Figure 11.6 (*a*) Sample and estimated mean (via Eq. 11.29) of $\ln M_c$ and (*b*) its sample and estimated standard deviation (via Eq. 11.27).

19.6 instead of 20.7, a 2% relative error in $\ln N_c$), but most is attributed to the weakest path issue and model errors arising by relating a spatial geometric average to a failure which is actually taking place along a curve through the two-dimensional soil mass.

Figure 11.6 illustrates the agreement between the sample mean of $\ln M_c$ and that predicted by Eq. 11.29 and between the sample standard deviation of $\ln M_c$ and Eq. 11.27 for $\rho = 0$. The estimated mean is seen to be in quite good agreement with the sample mean for all θ when $\sigma_c/\mu_c < 2$, and with the worst case ($\theta = B$) for $\sigma_c/\mu_c > 2$.

The predicted standard deviation was obtained by assuming a geometric average over a region under the footing of depth equal to the mean wedge zone depth,

$$w \simeq \tfrac{1}{2}B \tan\left(\tfrac{1}{4}\pi + \tfrac{1}{2}\mu_\phi\right) \tag{11.30}$$

and width of about $5w$. This is a rough approximation to the area of the failure region within the mean log-spiral curve on either side of the footing. Thus, D used in the variance function of Eq. 11.27 is a region of size $5w \times w$, that is, $\gamma(D) = \gamma(5w, w)$.

Although Eq. 11.22 fails to reflect the effect of θ on the reduction in the mean log-bearing capacity factor with increasing soil variability, the sample standard deviation is extremely well predicted by Eq. 11.27 —being only somewhat underpredicted for very small correlation lengths. To some extent the overall agreement in variance is as expected since the variability along the weakest path will be similar to the variability along any nearby path through a statistically homogeneous medium.

The Monte Carlo simulation also allows the estimation of the probability density function of M_c. A chi-square goodness-of-fit test performed across all σ_c/μ_c, θ, and ρ parameter variations yields an average p-value of 33%. This is encouraging since large p-values indicate good agreement between the hypothesized distribution (lognormal) and the data. However, approximately 30% of the simulations had p-values less than 5%, indicating that a fair proportion of the runs had distributions that deviated from the lognormal to some extent. Some 10% of runs had p-values less than 0.01%. Figure 11.7*a* illustrates one of the better fits, with a p-value of 43% ($\sigma_c/\mu_c = 0.1$, $\theta = 4$, and $\rho = 0$), while Figure 11.7*b* illustrates one of the poorer fits, with a p-value of 0.01% ($\sigma_c/\mu_c = 5$, $\theta = 1$, and $\rho = 0$). It can be seen that even when the p-value is as low as 0.01%, the fit is still reasonable. There was no particular trend in degree of fit as far as the three parameters σ_c/μ_c, θ, and ρ was concerned. It appears, then, that M_c at least approximately follows a lognormal distribution. Note that if M_c does indeed arise from a geometric average of the underlying soil properties c and N_c, then M_c will tend to a lognormal distribution by the central limit theorem. It is also worth pointing out that this may be exactly why so many soil properties tend to follow a lognormal distribution.

11.1.5 Probabilistic Interpretation

The results of the previous section indicated that Prandtl's bearing capacity formula is still largely applicable in the case of spatially varying soil properties if geometrically averaged soil properties are used in the formula. The

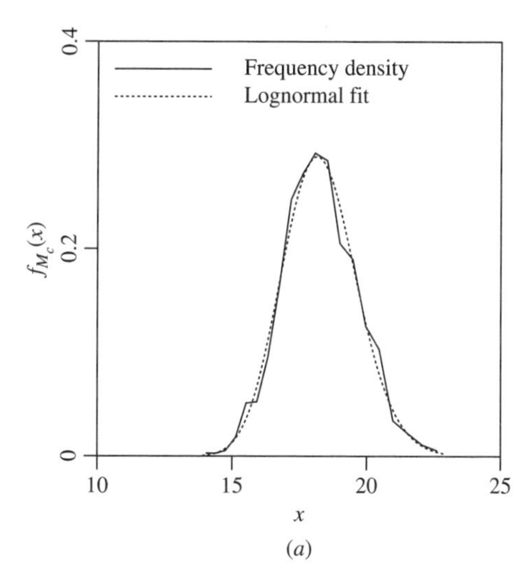

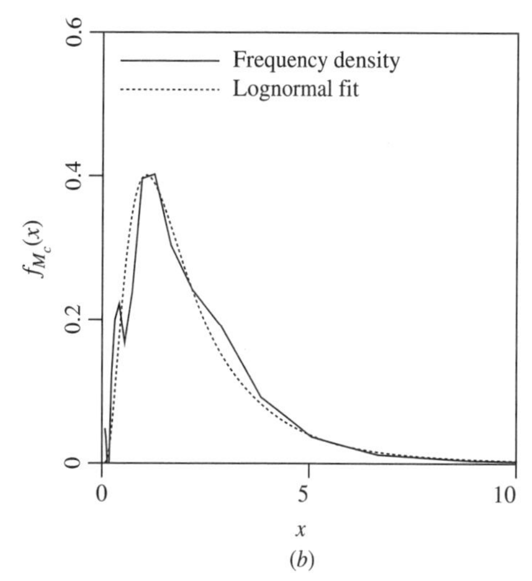

Figure 11.7 (*a*) Fitted lognormal distribution for $s = \sigma_c/\mu_c = 0.1$, $\theta = 4$, and $\rho = 0$ where the p-value is large (0.43) and (*b*) fitted lognormal distribution for $s = \sigma_c/\mu_c = 5$, $\theta = 1$, and $\rho = 0$ where the p-value is quite small (0.0001).

theoretical results presented above combined with the empirical correction to the mean proposed in the last section allows the approximate computation of probabilities associated with bearing capacity of a smooth strip footing. To illustrate this, consider an example strip footing of width $B = 2$ m founded on a weightless soil having $\mu_c = 75$ kPa, $\sigma_c = 50$ kPa, and $\theta = B = 2$ m (assuming the worst-case correlation length). Assume also that the friction angle ϕ is independent of c (conservative assumption) and ranges from 5° to 35°, with mean 20° and $s = 1$. In this case, the deterministic value of N_c, based purely on μ_ϕ is

$$N_c(\mu_\phi) = \frac{e^{\pi \tan \mu_\phi} \tan^2(\pi/4 + \mu_\phi/2) - 1}{\tan \mu_\phi} = 14.835 \tag{11.31}$$

so that, by Eq. 11.29,

$$\mu_{\ln \bar{M}_c} = 0.92 \ln(14.835) - 0.7 \ln\left(1 + \frac{50^2}{75^2}\right) = 2.2238 \tag{11.32}$$

For a footing width of $B = 2$, the wedge zone depth is

$$w = \tfrac{1}{2}B \tan\left(\frac{\pi}{4} + \frac{\mu_\phi}{2}\right) = \tan\left(\frac{\pi}{4} + \frac{20\pi}{360}\right) = 1.428 \tag{11.33}$$

Averaging over depth w by width $5w$ results in the variance reduction

$$\gamma(D) = \gamma(5w, w) = 0.1987$$

using the algorithm given in Appendix C for the Markov correlation function.

The slope of $\ln N_c$ at $\mu_\phi = 20°$ is 3.62779 (rad^{-1}), using Eq. 11.26. These results applied to Eq. 11.27 give

$$\sigma^2_{\ln \bar{M}_c} = 0.1987 \left\{ \ln\left(1 + \frac{50^2}{75^2}\right) + \left[\frac{s}{4\pi}(\phi_{\max} - \phi_{\min})\beta(\mu_\phi)\right]^2 \right\} = 0.07762 \tag{11.34}$$

so that $\sigma_{\ln \bar{M}_c} = 0.2778$.

The probability that M_c is less than half the deterministic value of N_c, based on μ_ϕ, is, then

$$\mathrm{P}\left[M_c \le \frac{14.835}{2}\right] = \Phi\left(\frac{\ln(14.835/2) - \mu_{\ln M_c}}{\sigma_{\ln M_c}}\right) = \Phi(-0.79) = 0.215 \tag{11.35}$$

where Φ is the cumulative distribution function for the standard normal and where M_c is assumed lognormally distributed, as was found to be reasonable above. A simulation of the above problem yields $\mathrm{P}\left[M_c \le 14.835/2\right] = 0.2155$. Although this amazing agreement seems too good to be true, this is, in fact, the first example problem that the authors considered. The caveat, however, is that predictions derived from the results of a finite-element program are being compared to the results of the same finite-element program, albeit at different parameter values. Nevertheless, the fact that the agreement here is so good is encouraging since it indicates that the theoretical results given above may have some overall generality—namely that Prandtl's bearing capacity solution is applicable to spatially variable soils if the soil properties are taken from geometric averages, suitably modified to reflect weakest path issues.

Inasmuch as the finite-element method represents the actual soil behavior, this observation seems reasonable.

11.1.6 Summary

Most soil properties are local averages of some sort and are derived from measurements of properties over some finite volume. In the case of the shear resistance of a soil sample, tests involve determining the average shear resistance over some surface through the soil sample. Since this surface will tend to avoid the high-strength areas in favor of low-strength areas, the average will be less than a strictly arithmetic mean over a flat plane. Of the various common types of averages—arithmetic, geometric, and harmonic—the one that generally shows the best agreement with "block" soil properties is the geometric average. The geometric average favors low-strength areas, although not as drastically as does a harmonic average, lying between the arithmetic and harmonic averages.

The bearing capacity factor of Prandtl (1921) has been observed in practice to give reasonable agreement with test results, particularly under controlled conditions. When soil properties become spatially random, the failure surface migrates from the log-spiral surface to some nearby surface which is weaker. The results presented in this section indicate that the statistics of the resulting surface are well represented by geometrically averaging the soil properties over a domain of about the size of the plastically deformed bearing failure region (taken to be $5w \times w$ in this study). That is, that Prandtl's formula can be used to predict the statistics of bearing capacity if the soil properties used in the formula are based on geometric averages, with some empirical adjustment for the mean.

In this sense, the weakest path through the soil is what governs the stochastic bearing capacity behavior. This means that the details of the distributions selected for c and ϕ are not particularly important, so long as they are physically reasonable, unimodal, and continuous. Although the lognormal distribution, for example, is mathematically convenient when dealing with geometric averages, very similar bearing capacity results are expected using other distributions, such as the normal distribution (suitably truncated to avoid negative strengths). The distribution selected for the friction angle basically resembles a truncated normal distribution over most values of s, but, for example, it is believed that a beta distribution could also have been used here without significantly affecting the results.

In the event that the soil is statistically anisotropic, that is, that the correlation lengths differ in the vertical and horizontal directions, it is felt that the above results can still be used with some accuracy by using the algorithm of Appendix C with differing vertical and horizontal correlation lengths. However, some additional study is necessary to establish whether the mean bearing capacity in the anisotropic case is at least conservatively represented by Eq. 11.29.

Some limitations to this study are as follows:

1. The simulations were performed using a finite-element analysis in which the values of the underlying normally distributed soil properties assigned to the elements are derived from arithmetic averages of the soil properties over each element domain. While this is believed to be a very realistic approach, intimately related to the soil property measurement process, it is nevertheless an approach where geometric averaging is being performed at the element scale (at least for the cohesion—note that arithmetic averaging of a normally distributed field corresponds to geometric averaging of the associated lognormally distribution random field) in a method which is demonstrating that geometric averaging is applicable over the site scale. Although it is felt that the fine-scale averaging assumptions should not significantly affect the large-scale results through the finite-element method, there is some possibility that there are effects that are not reflected in reality.
2. Model error has been entirely neglected in this analysis. That is, the ability of the finite-element method to reflect the actual behavior of an ideal soil, and the ability of Eq. 11.3 to do likewise have not been considered. It has been assumed that the finite- element method and Eq. 11.3 are sufficiently reasonable approximations to the behavior of soils to allow the investigation of the major features of stochastic soil behavior under loading from a smooth strip footing. Note that the model error associated with traditional usage of Eq. 11.3 may be due in large part precisely to spatial variation of soil properties, so that this study may effectively be reducing, or at least quantifying, model error (although whether this is really true or not will have to wait until sufficient experimental evidence has been gathered).

The geometric averaging model has been shown to be a reasonable approach to estimating the statistics of bearing capacity. This is particularly true of the standard deviation. Some adjustment was required to the mean since the geometric average was not able to completely account for the weakest path at intermediate correlation lengths. The proposed relationships for the mean and standard deviation, along with the simulation results indicating that the bearing capacity factor, M_c, is lognormally distributed, allow reasonably accurate calculations of probabilities associated with the bearing capacity. In the event that little is known about the cross-correlation of c and ϕ at a particular site, assuming that these properties are independent is deemed to be conservative (as long as the actual correlation is negative). In any case, the cross-correlation was not found to be a significant factor in the stochastic behavior of bearing capacity.

Perhaps more importantly, since little is generally known about the correlation length at a site, the results of this study indicate that there exists a worst-case correlation length of $\theta \simeq B$. Using this value, in the absence of improved information, allows conservative estimates of the probability of bearing failure. The estimate of the mean log-bearing capacity factor (Eq. 11.29) is based on this conservative case.

11.2 LOAD AND RESISTANCE FACTOR DESIGN OF SHALLOW FOUNDATIONS

The design of a shallow footing typically begins with a site investigation aimed at determining the strength of the founding soil or rock. Once this information has been gathered, the geotechnical engineer is in a position to determine the footing dimensions required to avoid entering various limit states. In so doing, it will be assumed here that the geotechnical engineer is in close communication with the structural engineer(s) and is aware of the loads that the footings are being designed to support. The limit states that are usually considered in the footing design are serviceability limit states (typically deformation) and ultimate limit states. The latter is concerned with safety and includes the load-carrying capacity, or *bearing capacity*, of the footing.

This section investigates a LRFD approach for shallow foundations designed against bearing capacity failure. The design goal is to determine the footing dimensions such that the *ultimate geotechnical resistance* based on characteristic soil properties, $\hat{R}_u$, satisfies

$$\phi_g \hat{R}_u \geq I \sum_i \alpha_i \hat{L}_i \tag{11.36}$$

where ϕ_g is the *geotechnical resistance factor*, I is an *importance factor*, α_i is the ith *load factor*, and $\hat{L}_i$ is the ith *characteristic load effect*. The relationship between ϕ_g and the probability that the designed footing will experience a bearing capacity failure will be summarized below (from Fenton et al., 2007a) followed by some results on resistance factors required to achieve certain target maximum acceptable failure probabilities (from Fenton et al., 2007b). The symbol ϕ is commonly used to denoted the resistance factor—see, for example, the National Building Code of Canada (NBCC) [National Research Council (NRC), 2005] and in Commentary K "Foundations" of the User's Guide—NBC 2005 Structural Commentaries (NRC, 2006). The authors are also adopting the common notation where the subscript denotes the material that the resistance factor governs. For example, where ϕ_c and ϕ_s are resistance factors governing concrete and steel, the letter g in ϕ_g will be taken to denote "geotechnical" or "ground."

The importance factor in Eq. 11.36, I, reflects the severity of the failure consequences and may be larger than 1.0 for important structures, such as hospitals, whose failure consequences are severe and whose target probabilities of failure are much less than for typical structures. Typical structures usually are designed using $I = 1$, which will be assumed in this section. Structures with low failure consequences (minimal risk of loss of life, injury, and/or economic impact) may have $I < 1$.

Only one load combination will be considered in this section, $\alpha_L \hat{L}_L + \alpha_D \hat{L}_D$, where $\hat{L}_L$ is the characteristic live load, $\hat{L}_D$ is the characteristic dead load, and α_L and α_D are the live- and dead-load factors, respectively. The load factors will be as specified by the *National Building Code of Canada* (NBCC; NRC, 2005); $\alpha_L = 1.5$ and $\alpha_D = 1.25$. The theory presented here, however, is easily extended to other load combinations and factors, so long as their (possibly time-dependent) distributions are known.

The characteristic loads will be assumed to be defined in terms of the means of the load components in the following fashion:

$$\hat{L}_L = k_{L_e} \mu_{L_e} \tag{11.37a}$$

$$\hat{L}_D = k_D \mu_D \tag{11.37b}$$

where μ_{L_e} and μ_D are the means of the live and dead loads, respectively, and k_{L_e} and k_D are live- and dead-load *bias* factors, respectively. The bias factors provide some degree of "comfort" by increasing the loads from the mean value to a value having a lesser chance of being exceeded. Since live loads are time varying, the value of μ_{L_e} is more specifically defined as the mean of the maximum live load experienced over a structure's lifetime (the subscript e denotes *extreme*). This definition has the following interpretation: If a series of similar structures, all with the same life span, is considered and the maximum live load experienced in each throughout its life span is recorded, then a histogram of this set of recorded maximum live loads could be plotted. This histogram then becomes an estimate of the distribution of these extreme live loads and the average of the observed set of maximum values is an estimate of μ_{L_e}. As an aside, the distribution of live load is really quite a bit more complicated than suggested by this explanation since it actually depends on both spatial position and time (e.g., regions near walls tend to experience higher live loads than seen near the center of rooms). However, historical estimates of live loads are quite appropriately based on spatial averages both conservatively and for simplicity, as discussed next.

For typical multistory office buildings, Allen (1975) estimates μ_{L_e} to be 1.7 kN/m^2, based on a 30-year lifetime. The corresponding characteristic live load given by the NBCC (NRC, 2005) is $\hat{L}_L = 2.4$ kN/m^2, which implies that $k_{L_e} = 2.4/1.7 = 1.41$. Allen further states that the mean live load *at any time* is approximately equal to the 30-year maximum mean averaged over an infinite area. The NBCC provides for a reduction in live loads with tributary area using the formula $0.3 + \sqrt{9.8/A}$, where A is the tributary area ($A > 20$ m^2). For $A \to \infty$, the mean live load at any time is thus approximately $\mu_L = 0.3(1.7) = 0.51$ kN/m^2. The bias factor which translates the instantaneous mean live load, μ_L to the characteristic live load, $\hat{L}_L$, is thus quite large having value $k_L = 2.4/0.51 = 4.7$.

Dead load, on the other hand, is largely static, and the time span considered (e.g., lifetime) has little effect on its distribution. Becker (1996b) estimates k_D to be 1.18. Figure 11.8 illustrates the relative locations of the mean and characteristic values for the three types of load distributions commonly considered.

The characteristic ultimate geotechnical resistance $\hat{R}_u$ is determined using characteristic soil properties, in this case characteristic values of the soil's cohesion, c, and friction angle, ϕ (note that although the primes are omitted from these quantities it should be recognized that the theoretical developments described in this paper are applicable to either total or effective strength parameters). To obtain the characteristic soil properties, the soil is assumed to be sampled over a single column somewhere in the vicinity of the footing, for example, a single CPT or SPT sounding near the footing. The sample is assumed to yield a sequence of m observed cohesion values $c_1^o, c_2^o, \ldots, c_m^o$, and m observed friction angle values $\phi_1^o, \phi_2^o, \ldots, \phi_m^o$. The superscript o denotes an observation. It is assumed here that the observations are error free, which is an *unconservative*

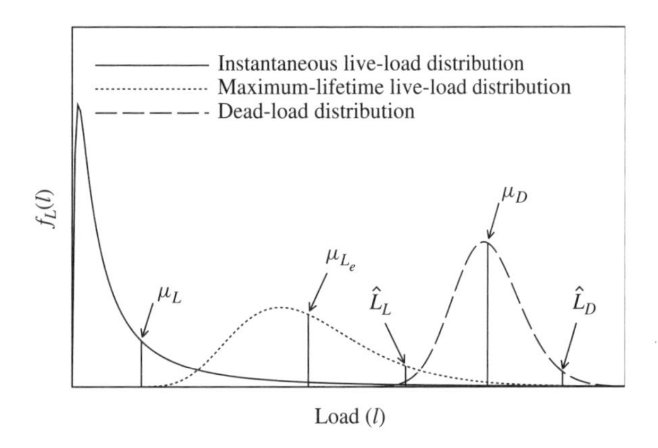

Figure 11.8 Characteristic and mean values of live and dead loads.

assumption. If the actual observations have considerable error, then the resistance factor used in the design should be reduced. This issue is discussed further in the summary.

The characteristic value of the cohesion, $\hat{c}$, is defined here as the median of the sampled observations, c_i^o, which, assuming c is lognormally distributed, can be computed using the geometric average:

$$\hat{c} = \left[\prod_{i=1}^{m} c_i^o\right]^{1/m} = \exp\left\{\frac{1}{m}\sum_{i=1}^{m} \ln c_i^o\right\} \tag{11.38}$$

The geometric average is used here because if c is lognormally distributed, as assumed, then $\hat{c}$ will also be lognormally distributed.

The characteristic value of the friction angle is computed as an arithmetic average:

$$\hat{\phi} = \frac{1}{m}\sum_{i=1}^{m} \phi_i^o \tag{11.39}$$

The arithmetic average is used here because ϕ is assumed to follow a symmetric bounded distribution and the arithmetic average preserves the mean. That is, the mean of $\hat{\phi}$ is the same as the mean of ϕ.

To determine the characteristic ultimate geotechnical resistance $\hat{R}_u$, it will first be assumed that the soil is weightless. This simplifies the calculation of the ultimate bearing stress q_u to

$$q_u = cN_c \tag{11.40}$$

The assumption of weightlessness is conservative since the soil weight contributes to the overall bearing capacity. This assumption also allows the analysis to explicitly concentrate on the role of cN_c on ultimate bearing capacity, since this is the only term that includes the effects of spatial variability relating to *both* shear strength parameters c and ϕ.

Bearing capacity predictions, involving specification of the N_c factor in this case, are generally based on plasticity theories (see, e.g., Prandtl, 1921; Terzaghi, 1943; Sokolovski, 1965) in which a rigid base is punched into a softer material. These theories assume that the soil underlying the footing has properties which are spatially constant (everywhere the same). This type of ideal soil will be referred to as a *uniform soil* henceforth. Under this assumption, most bearing capacity theories (e.g., Prandtl, 1921; Meyerhof, 1951, 1963) assume that the failure slip surface takes on a logarithmic spiral shape to give

$$N_c = \frac{e^{\pi \tan\phi}\tan^2(\pi/4 + \phi/2) - 1}{\tan\phi} \tag{11.41}$$

The theory is derived for the general case of a $c-\phi$ soil. One can always set $\phi = 0$ to obtain results for an undrained clayey soil.

Consistent with the theoretical results presented by Fenton et al. (2007b), this section will concentrate on the design of a strip footing. In this case, the characteristic ultimate geotechnical resistance $\hat{R}_u$ becomes

$$\hat{R}_u = B\hat{q}_u \tag{11.42}$$

where B is the footing width and $\hat{R}_u$ has units of load per unit length out-of-plane, that is, in the direction of the strip foot. The characteristic ultimate bearing stress $\hat{q}_u$ is defined by

$$\hat{q}_u = \hat{c}\hat{N}_c \tag{11.43}$$

where the characteristic N_c factor is determined using the characteristic friction angle in Eq. 11.41:

$$\hat{N}_c = \frac{e^{\pi \tan\hat{\phi}} \tan^2\left(\pi/4 + \hat{\phi}/2\right) - 1}{\tan\hat{\phi}} \tag{11.44}$$

For the strip footing and just the dead- and live-load combination, the LRFD equation becomes

$$\phi_g B\hat{q}_u = I\left[\alpha_L \hat{L}_L + \alpha_D \hat{L}_D\right]$$

$$\Longrightarrow \quad B = \frac{I\left[\alpha_L \hat{L}_L + \alpha_D \hat{L}_D\right]}{\phi_g \hat{q}_u} \tag{11.45}$$

To determine the resistance factor ϕ_g required to achieve a certain acceptable reliability of the constructed footing, it is necessary to estimate the probability of bearing capacity failure of a footing designed using Eq. 11.45. Once the probability of failure p_f for a certain design using a specific value for ϕ_g is known, this probability can be compared to the maximum acceptable failure probability p_m. If p_f exceeds p_m, then the resistance factor must be reduced and the footing redesigned. Similarly, if p_f is less than p_m, then the design is overconservative and the value of ϕ_g can be increased. A specific relationship between p_m and ϕ_g will be given below. Design curves will also be presented from which the value of ϕ_g required to achieve a maximum acceptable failure probability can be determined.

As suggested, the determination of the required resistance factor ϕ_g involves deciding on a maximum acceptable failure probability p_m. The choice of p_m derives from a consideration of acceptable risk and directly influences the size of ϕ_g. Different levels of p_m may be considered to reflect the "importance" of the supported structure—p_m may be much smaller for a hospital than for a storage warehouse.

The choice of a maximum failure probability p_m should consider the margin of safety implicit in current foundation designs and the levels of reliability for geotechnical design as reported in the literature. The values of p_m for foundation designs are nearly the same or somewhat less than those for concrete and steel structures because of the difficulties and high expense of foundation repairs. A literature review of the suggested acceptable probability of failure for foundations is listed in Table 11.2.

Table 11.2 Literature Review of Lifetime Probabilities of Failure of Foundations

Source	p_m
Meyerhof, 1970, 1993, 1995	10^{-2}–10^{-4}
Simpson et al., 1981	10^{-3}
NCHRP, 1991	10^{-2}–10^{-4}
Becker, 1996a	10^{-3}–10^{-4}

Meyerhof (1995, p. 132) was quite specific about acceptable risks: "The order of magnitude of lifetime probabilities of stability failure is about 10^{-2} for offshore foundation, about 10^{-3} for earthworks and earth retaining structures, and about 10^{-4} for foundations on land."

In this section three maximum lifetime failure probabilities, 10^{-2}, 10^{-3}, and 10^{-4} will be considered. In general, and without regard to the structural categorizations made by Meyerhof above, these probabilities are deemed by the authors to be appropriate for designs involving low, medium and high failure consequence structures, respectively. Resistance factors to achieve these target probabilities will be presented for the specific $c-\phi$ soil considered. These resistance factors are smaller than those the theory suggests for an undrained soil, since a $\phi = 0$ soil has only one source of uncertainty. In other words, the resistance factors based on a generalized $c-\phi$ soil are considered to be reasonably conservative.

We note that the effect of structural importance should actually be reflected in the importance factor, I, of Eq. 11.36 and not in the resistance factor. The resistance factor should be aimed at a medium, or common, structural importance level, and the importance factor should be varied above and below 1.0 to account for more and less important structures, respectively. However, since acceptable failure probabilities may not be simply connected to structural importance, we will assume $I = 1$ in the following. For code provisions, the factors recommended here should be considered to be the ratio ϕ_g/I.

11.2.1 Random Soil Model

The soil cohesion c is assumed to be lognormally distributed with mean μ_c, standard deviation σ_c, and spatial correlation length $\theta_{\ln c}$. A lognormally distributed random field is obtained from a normally distributed random field $G_{\ln c}(\mathbf{x})$ having zero mean, unit variance, and spatial correlation length $\theta_{\ln c}$ through the transformation

$$c(\mathbf{x}) = \exp\{\mu_{\ln c} + \sigma_{\ln c} G_{\ln c}(\mathbf{x})\} \tag{11.46}$$

where $\mathbf{x}$ is the spatial position at which c is desired, $\sigma^2_{\ln c} = \ln\left(1 + v_c^2\right)$, $\mu_{\ln c} = \ln(\mu_c) - \sigma^2_{\ln c}/2$, and $v_c = \sigma_c/\mu_c$ is the coefficient of variation.

The correlation coefficient between the log-cohesion at a point $\mathbf{x}_1$ and a second point $\mathbf{x}_2$ is specified by a correlation function $\rho_{\ln c}(\boldsymbol{\tau})$, where $\boldsymbol{\tau} = \mathbf{x}_1 - \mathbf{x}_2$ is the vector between the two points. In this section, a simple exponentially decaying (Markovian) correlation function will be assumed having the form

$$\rho_{\ln c}(\boldsymbol{\tau}) = \exp\left(-\frac{2|\boldsymbol{\tau}|}{\theta_{\ln c}}\right) \tag{11.47}$$

where $|\boldsymbol{\tau}| = \sqrt{\tau_1^2 + \tau_2^2}$ is the length of the vector $\boldsymbol{\tau}$. The spatial correlation length $\theta_{\ln c}$ is loosely defined as the separation distance within which two values of $\ln c$ are significantly correlated. Mathematically, $\theta_{\ln c}$ is defined as the area under the correlation function, $\rho_{\ln c}(\boldsymbol{\tau})$ (Vanmarcke, 1984).

The spatial correlation function $\rho_{\ln c}(\boldsymbol{\tau})$ has a corresponding variance reduction function $\gamma_{\ln c}(D)$, which specifies how the variance is reduced upon local averaging of $\ln c$ over some domain D. In the two-dimensional analysis considered here, $D = D_1 \times D_2$ is an area and the two-dimensional variance reduction function is defined by

$$\gamma_{\ln c}(D_1, D_2) = \frac{4}{(D_1 D_2)^2} \times \int_0^{D_1} \int_0^{D_2} (D_1 - \tau_1)(D_2 - \tau_2) \times \rho(\tau_1, \tau_2)\, d\tau_1\, d\tau_2 \tag{11.48}$$

which can be evaluated using Gaussian quadrature [see Fenton and Griffiths (2003), Griffiths and Smith (2006), and Appendix C for more details].

It should be emphasized that the correlation function selected above acts between values of $\ln c$. This is because $\ln c$ is normally distributed, and a normally distributed random field is simply defined by its mean and covariance structure. In practice, the correlation length $\theta_{\ln c}$ can be estimated by evaluating spatial statistics of the log-cohesion data directly (see, e.g., Fenton, 1999a). Unfortunately, such studies are scarce so that little is currently known about the spatial correlation structure of natural soils. For the problem considered here, it turns out that a worst-case correlation length exists which can be conservatively assumed in the absence of improved information.

The random field is also assumed here to be statistically isotropic (the same correlation length in any direction through the soil). Although the horizontal correlation length is often greater than the vertical, due to soil layering, taking this into account was deemed to be a site-specific refinement which does not lead to an increase in the general understanding of the probabilistic behavior of shallow foundations. The theoretical results presented here, however, apply also to anisotropic soils, so that the results are easily extended to specific sites. The authors have found that when the soil is sampled at some distance from the footing (i.e. not directly under the footing) that increasing the correlation length in the horizontal direction to values above the worst-case isotropic correlation length leads to a decreased failure probability, so that the isotropic case is also conservative for low to medium levels of site understanding. When the soil is sampled directly below the footing, the failure probability increases as the horizontal correlation length is increased above the worst case scale, which is unconservative.

The friction angle ϕ is assumed to be bounded both above and below, so that neither normal nor lognormal distributions are appropriate. A beta distribution is often used for bounded random variables. Unfortunately, a beta-distributed random field has a very complex joint distribution and simulation is cumbersome and numerically difficult. To keep things simple, a bounded distribution is selected which resembles a beta distribution but which arises as a simple transformation of a standard normal random field $G_\phi(\mathbf{x})$ according to

$$\phi(\mathbf{x}) = \phi_{\min} + \tfrac{1}{2}(\phi_{\max} - \phi_{\min}) \left\{1 + \tanh\left(\frac{sG_\phi(\mathbf{x})}{2\pi}\right)\right\} \tag{11.49}$$

where $\phi_{\min}$ and $\phi_{\max}$ are the minimum and maximum friction angles in radians, respectively, and s is a scale factor which governs the friction angle variability between its two bounds. See Section 1.10.10 for more details about this distribution. Figure 1.36 shows how the distribution of ϕ (normalized to the interval [0, 1]) changes as s changes, going from an almost uniform distribution at $s = 5$ to a very normal looking distribution for smaller s. Thus, varying s between about 0.1 and 5.0 leads to a wide range in the stochastic behavior of ϕ. In all cases, the distribution is symmetric so that the midpoint between $\phi_{\min}$ and $\phi_{\max}$ is the mean. Values of s greater than about 5 lead to a U-shaped distribution (higher at the boundaries), which is deemed unrealistic.

The following relationship between s and the variance of ϕ derives from a third-order Taylor series approximation to tanh and a first-order approximation to the final expectation,

$$\begin{aligned}\sigma_\phi^2 &= (0.5)^2(\phi_{\max} - \phi_{\min})^2\, \mathrm{E}\left[\tanh^2\left(\frac{sG_\phi}{2\pi}\right)\right] \\ &\simeq (0.5)^2(\phi_{\max} - \phi_{\min})^2\, \mathrm{E}\left[\frac{[sG_\phi/(2\pi)]^2}{1 + [sG_\phi/(2\pi)]^2}\right] \\ &\simeq (0.5)^2(\phi_{\max} - \phi_{\min})^2 \frac{s^2}{4\pi^2 + s^2}\end{aligned} \tag{11.50}$$

where $\mathrm{E}\left[G_\phi^2\right] = 1$ since G_ϕ is a standard normal random variable. Equation 11.50 slightly overestimates the true standard deviation of ϕ, from 0% when $s = 0$ to 11% when

$s = 5$. A much closer approximation over the entire range $0 \le s \le 5$ is obtained by slightly decreasing the 0.5 factor to 0.46 (this is an empirical adjustment):

$$\sigma_\phi \simeq \frac{0.46(\phi_{\max} - \phi_{\min})s}{\sqrt{4\pi^2 + s^2}} \tag{11.51}$$

The close agreement is illustrated in Figure 11.9.

Equation 11.50 can be generalized to yield the covariance between $\phi(\mathbf{x}_i)$ and $\phi(\mathbf{x}_j)$ for any two spatial points $\mathbf{x}_i$ and $\mathbf{x}_j$ as follows:

$$\begin{aligned}
\mathrm{Cov}\left[\phi(\mathbf{x}_i), \phi(\mathbf{x}_j)\right] &= (0.5)^2(\phi_{\max} - \phi_{\min})^2 \\
&\quad \times \mathrm{E}\left[\tanh\left(\frac{sG_\phi(\mathbf{x}_i)}{2\pi}\right) \tanh\left(\frac{sG_\phi(\mathbf{x}_j)}{2\pi}\right)\right] \\
&\simeq (0.5)^2(\phi_{\max} - \phi_{\min})^2 \\
&\quad \times \mathrm{E}\left[\frac{[sG_\phi(\mathbf{x}_i)/(2\pi)][sG_\phi(\mathbf{x}_j)/(2\pi)]}{1 + \frac{1}{2}\left\{[sG_\phi(\mathbf{x}_i)/(2\pi)]^2 + [sG_\phi(\mathbf{x}_j)/(2\pi)]^2\right\}}\right] \\
&\simeq (0.46)^2(\phi_{\max} - \phi_{\min})^2 \frac{s^2 \rho_\phi(\mathbf{x}_i - \mathbf{x}_j)}{4\pi^2 + s^2} \\
&= \sigma_\phi^2 \rho_\phi(\mathbf{x}_i - \mathbf{x}_j)
\end{aligned} \tag{11.52}$$

where the empirical correction found in Eq. 11.51 was introduced in the second to the last step.

It seems reasonable to assume that if the spatial correlation structure of a soil is caused by changes in the constitutive nature of the soil over space, then both cohesion and friction angle would have similar correlation lengths. Thus, θ_ϕ is taken to be equal to $\theta_{\ln c}$ in this study and ϕ is assumed to have the same correlation structure as c (Eq. 11.47), that is, $\rho_\phi(\boldsymbol{\tau}) = \rho_{\ln c}(\boldsymbol{\tau})$. Both correlation lengths will be referred to generically from now on simply as θ, and both correlation functions as $\rho(\boldsymbol{\tau})$, remembering that this length and correlation function reflects correlation between points in the underlying normally distributed random fields $G_{\ln c}(\mathbf{x})$ and $G_\phi(\mathbf{x})$ and not directly between points in the cohesion and friction fields (although the correlation lengths in the different spaces are quite similar). The correlation lengths can be estimated by statistically analyzing data generated by inverting Eqs. 11.46 and 11.49. Since both fields have the same correlation function, $\rho(\boldsymbol{\tau})$, they will also have the same variance reduction function, that is, $\gamma_{\ln c}(D) = \gamma_\phi(D) = \gamma(D)$, as defined by Eq. 11.48.

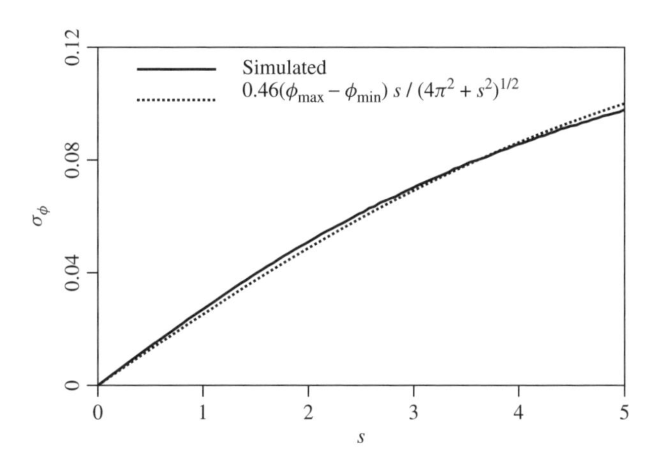

Figure 11.9 Relationship between σ_ϕ and s derived from simulation (100,000 realizations for each s) and the Taylor-series-derived approximation given by Eq. 11.51. The vertical scale corresponds to $\phi_{\max} - \phi_{\min} = 0.349$ rad (20°).

The two random fields, c and ϕ, are assumed to be independent. Nonzero correlations between c and ϕ were found by Fenton and Griffiths (2003) to have only a minor influence on the estimated probabilities of bearing capacity failure. Since the general consensus is that c and ϕ are negatively correlated (Cherubini, 2000; Wolff, 1985) and the mean bearing capacity for independent c and ϕ was slightly lower than for the negatively correlated case (see Section 11.1), the assumption of independence between c and ϕ is slightly conservative.

11.2.2 Analytical Approximation to Probability of Failure

In this section, an analytical approximation to the probability of bearing capacity failure of a strip footing is summarized. Equation 11.40 was developed assuming an ideal soil whose shear strength is everywhere the same (i.e., a *uniform soil*). When soil properties are spatially variable, as they are in reality, then the hypothesis made in this study is that Eq. 11.40 can be replaced by

$$q_u = \bar{c}\bar{N}_c \tag{11.53}$$

where $\bar{c}$ and $\bar{N}_c$ are the *equivalent cohesion* and *equivalent* N_c *factor*, defined as those *uniform* soil parameters which lead to the same bearing capacity as observed in the real, spatially varying, soil. In other words, it is proposed that equivalent soil properties, $\bar{c}$ and $\bar{\phi}$, exist such that a uniform soil having these properties will have the same bearing capacity as the actual spatially variable soil. The value of $\bar{N}_c$ is obtained by using the equivalent friction angle $\bar{\phi}$ in Eq. 11.41:

$$\bar{N}_c = \frac{e^{\pi \tan\bar{\phi}} \tan^2\left(\pi/4 + \bar{\phi}/2\right) - 1}{\tan\bar{\phi}} \tag{11.54}$$

In the design process, Eq. 11.53 is replaced by Eq. 11.43, and the design footing width B is obtained using Eq. 11.45, which, in terms of the characteristic design values, becomes

$$B = \frac{I\left[\alpha_L \hat{L}_L + \alpha_D \hat{L}_D\right]}{\phi_g \hat{c} \hat{N}_c} \tag{11.55}$$

The design philosophy proceeds as follows: Find the required footing width B such that the probability that the actual load L exceeds the actual resistance $q_u B$ is less than some small acceptable failure probability p_m. If p_f is the actual failure probability, then

$$p_f = \mathrm{P}\left[L > q_u B\right] = \mathrm{P}\left[L > \bar{c}\bar{N}_c B\right] \tag{11.56}$$

and a successful design methodology will have $p_f \le p_m$. Substituting Eq. 11.55 into Eq. 11.56 and collecting random terms to the left of the inequality leads to

$$p_f = \mathrm{P}\left[L \frac{\hat{c}\hat{N}_c}{\bar{c}\bar{N}_c} > \frac{I\left[\alpha_L \hat{L}_L + \alpha_D \hat{L}_D\right]}{\phi_g}\right] \tag{11.57}$$

Letting

$$Y = L \frac{\hat{c}\hat{N}_c}{\bar{c}\bar{N}_c} \tag{11.58}$$

means that

$$p_f = \mathrm{P}\left[Y > \frac{I\left[\alpha_L \hat{L}_L + \alpha_D \hat{L}_D\right]}{\phi_g}\right] \tag{11.59}$$

and the task is to find the distribution of Y. Assuming that Y is lognormally distributed [an assumption found to be reasonable by Fenton et al. (2007a) and which is also supported to some extent by the central limit theorem], then

$$\ln Y = \ln L + \ln \hat{c} + \ln \hat{N}_c - \ln \bar{c} - \ln \bar{N}_c \tag{11.60}$$

is normally distributed and p_f can be found once the mean and variance of $\ln Y$ are determined. The mean of $\ln Y$ is

$$\mu_{\ln Y} = \mu_{\ln L} + \mu_{\ln \hat{c}} + \mu_{\ln \hat{N}_c} - \mu_{\ln \bar{c}} - \mu_{\ln \bar{N}_c} \tag{11.61}$$

and the variance of $\ln Y$ is

$$\sigma^2_{\ln Y} = \sigma^2_{\ln L} + \sigma^2_{\ln \hat{c}} + \sigma^2_{\ln \bar{c}} + \sigma^2_{\ln \hat{N}_c} + \sigma^2_{\ln \bar{N}_c} - 2\,\mathrm{Cov}\left[\ln \bar{c}, \ln \hat{c}\right] - 2\,\mathrm{Cov}\left[\ln \bar{N}_c, \ln \hat{N}_c\right] \tag{11.62}$$

where the load L and soil properties c and ϕ have been assumed mutually independent.

To find the parameters in Eqs. 11.61 and 11.62, the following two assumptions are made:

1. The equivalent cohesion $\bar{c}$ is the geometric average of the cohesion field over some zone of influence D under the footing:

$$\bar{c} = \exp\left\{\frac{1}{D}\int_D \ln c(\mathbf{x})\, d\mathbf{x}\right\} \tag{11.63}$$

   Note that in this two-dimensional analysis D is an area and the above is a two-dimensional integration. If $c(\mathbf{x})$ is lognormally distributed, as assumed, then $\bar{c}$ is also lognormally distributed.
2. The equivalent friction angle, $\bar{\phi}$, is the arithmetic average of the friction angle over the zone of influence, D:

$$\bar{\phi} = \frac{1}{D}\int_D \phi(\mathbf{x})\, d\mathbf{x} \tag{11.64}$$

   This relationship also preserves the mean, that is, $\mu_{\bar{\phi}} = \mu_\phi$.

Probably the greatest source of uncertainty in this analysis involves the choice of the domain D over which the equivalent soil properties are averaged under the footing. The averaging domain was found by trial and error to be best approximated by $D = W \times W$, centered directly under the footing (see Figure 11.10). In this study, W is taken as 80% of the average mean depth of the wedge zone directly beneath the footing, as given by the classical Prandtl failure mechanism,

$$W = \frac{0.8}{2}\hat{\mu}_B \tan\left(\frac{\pi}{4} + \frac{\mu_\phi}{2}\right) \tag{11.65}$$

and where μ_ϕ is the mean friction angle (in radians), within the zone of influence of the footing, and $\hat{\mu}_B$ is an estimate of the mean footing width obtained by using mean soil properties (μ_c and μ_ϕ) in Eq. 11.45:

$$\hat{\mu}_B = \frac{I\left[\alpha_L \hat{L}_L + \alpha_D \hat{L}_D\right]}{\phi_g \mu_c \mu_{N_c}} \tag{11.66}$$

The footing shown on Figure 11.10 is just one possible realization since the footing width, B, is actually a random variable. The averaging area D with dimension W suggested by Eq. 11.65 is significantly smaller than that suggested in Section 11.1. In Section 11.1, it was assumed that the footing width was known, rather than designed, and

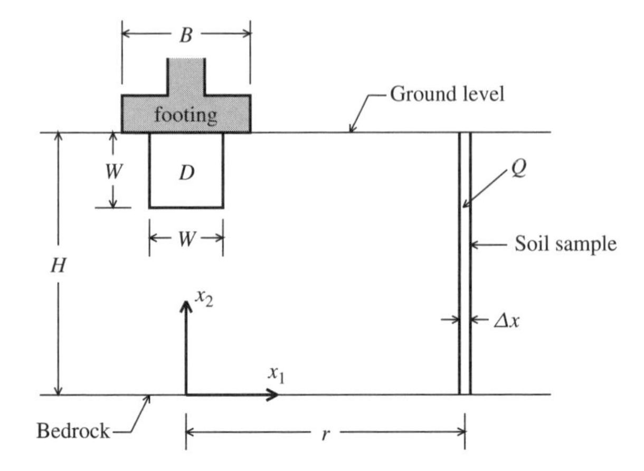

Figure 11.10 Averaging regions used to predict probability of bearing capacity failure.

recognized that the larger averaging region did not well represent the mean bearing capacity, which of course is the most important value in probability calculations. The smaller averaging region used in this study may be reasonable if one considers the actual quantity of soils involved in resisting the bearing failure along the failure surfaces. That is, D would be the area of soil which deforms during failure. Since this area will change, sometimes dramatically, from realization to realization, the above can only be considered a rough empirical approximation. The problem of deciding on an appropriate averaging region needs further study. In the simulations performed to validate the theory presented here, the soil depth is taken to be $H = 4.8$ m and $\Delta x = 0.15$ m, where Δx is the width of the columns of finite-elements used in the simulations (see, for example, Figure 11.2).

To first order, the mean of N_c is

$$\mu_{N_c} \simeq \frac{e^{\pi \tan \mu_\phi} \tan^2(\pi/4 + \mu_\phi/2) - 1}{\tan \mu_\phi} \tag{11.67}$$

Armed with the above information and assumptions, the components of Eqs. 11.61 and 11.62 can be computed as follows (given the basic statistical parameters of the loads, c, ϕ, the number and locations of the soil samples, and the averaging domain size D):

1. Assuming that the total load L is equal to the sum of the maximum live load L_{L_e} acting over the lifetime of the structure (this is a common, although rarely stated, definition of the live load) and the static dead load L_D, that is, $L = L_{L_e} + L_D$, both of which are random, then

$$\mu_{\ln L} = \ln(\mu_L) - \tfrac{1}{2}\ln\left(1 + v_L^2\right) \tag{11.68a}$$

$$\sigma_{\ln L}^2 = \ln\left(1 + v_L^2\right) \tag{11.68b}$$

where $\mu_L = \mu_{L_e} + \mu_D$ is the sum of the mean (max lifetime) live and (static) dead loads, and v_L is the coefficient of variation of the total load defined by

$$v_L^2 = \frac{\sigma_{L_e}^2 + \sigma_D^2}{\mu_{L_e} + \mu_D} \tag{11.69}$$

2. With reference to Eq. 11.38,

$$\mu_{\ln \hat{c}} = \mathrm{E}\left[\frac{1}{m}\sum_{i=1}^{m} \ln c_i^o\right] = \mu_{\ln c} \tag{11.70}$$

$$\sigma_{\ln \hat{c}}^2 \simeq \frac{\sigma_{\ln c}^2}{m^2}\sum_{i=1}^{m}\sum_{j=1}^{m} \rho(\mathbf{x}_i^o - \mathbf{x}_j^o) \tag{11.71}$$

where $\mathbf{x}_i^o$ is the spatial location of the center of the ith soil sample ($i = 1, 2, \ldots, m$) and ρ is the correlation function defined by Eq. 11.47. The approximation in the variance arises because correlation coefficients between the local averages associated with observations (in that all tests are performed on samples of some finite volume) are approximated by correlation coefficients between the local average centers. Assuming that $\ln \hat{c}$ actually represents a local average of $\ln c$ over a domain of size $\Delta x \times H$, where Δx is the horizontal dimension of the soil sample, which, for example, can be thought of as the horizontal zone of influence of a CPT sounding, and H is the depth over which the samples are taken, then $\sigma_{\ln \hat{c}}^2$ is probably more accurately computed as

$$\sigma_{\ln \hat{c}}^2 = \sigma_{\ln c}^2 \gamma(\Delta x, H) \tag{11.72}$$

3. With reference to Eq. 11.63,

$$\mu_{\ln \bar{c}} = \mathrm{E}\left[\frac{1}{D}\int_D \ln c(\mathbf{x})\, d\mathbf{x}\right] = \mu_{\ln c} \tag{11.73}$$

$$\sigma_{\ln \bar{c}}^2 = \sigma_{\ln c}^2 \gamma(D) \tag{11.74}$$

where $\gamma(D) = \gamma(W, W)$, as discussed above, is defined by Eq. 11.48.

4. Since $\mu_{\hat{\phi}} = \mu_\phi$ (see Eq. 11.39), the mean and variance of $\hat{N}_c$ can be obtained using first-order approximations to expectations of Eq. 11.44 (Fenton and Griffiths, 2003), as follows:

$$\mu_{\ln \hat{N}_c} = \mu_{\ln N_c} \simeq \ln \frac{e^{\pi \tan \mu_\phi} \tan^2(\pi/4 + \mu_\phi/2) - 1}{\tan \mu_\phi} \tag{11.75}$$

$$\sigma_{\ln \hat{N}_c}^2 \simeq \sigma_{\hat{\phi}}^2 \left(\left.\frac{d \ln \hat{N}_c}{d\hat{\phi}}\right|_{\mu_\phi}\right)^2 = \sigma_{\hat{\phi}}^2 \left[\frac{bd}{bd^2 - 1} \times \left[\pi(1 + a^2)d + 1 + d^2\right] - \frac{1 + a^2}{a}\right]^2 \tag{11.76}$$

where $a = \tan(\mu_\phi)$, $b = e^{\pi a}$, $d = \tan\left(\pi/4 + \mu_\phi/2\right)$. The variance of $\hat{\phi}$ can be obtained by making use of Eq. 11.52:

$$\sigma_{\hat{\phi}}^2 \simeq \frac{\sigma_\phi^2}{m^2}\sum_{i=1}^{m}\sum_{j=1}^{m} \rho(\mathbf{x}_i^o - \mathbf{x}_j^o) = \sigma_\phi^2 \gamma(\Delta x, H) \tag{11.77}$$

where $\mathbf{x}_i^o$ is the spatial location of the center of the ith soil observation ($i = 1, 2, \ldots, m$). See Eq. 11.51 for the definition of σ_ϕ. All angles are measured in radians, including those used in Eq. 11.51.

5. Since $\mu_{\bar{\phi}} = \mu_\phi$ (see Eq. 11.64), the mean and variance of $\bar{N}_c$ can be obtained in the same fashion as for $\hat{N}_c$ (in fact, they only differ due to differing local

averaging in the variance calculation). With reference to Eqs. 11.54 and 11.67,

$$\mu_{\ln \bar{N}_c} = \mu_{\ln \hat{N}_c} = \mu_{\ln N_c} \tag{11.78}$$

$$\sigma^2_{\ln \bar{N}_c} \simeq \sigma^2_{\bar{\phi}} \left(\frac{d \ln \bar{N}_c}{d\bar{\phi}} \bigg|_{\mu_\phi} \right)^2 = \sigma^2_{\bar{\phi}} \left[\frac{bd}{bd^2 - 1} \times \left[\pi(1+a^2)d + 1 + d^2 \right] - \frac{1+a^2}{a} \right]^2 \tag{11.79}$$

$$\sigma^2_{\bar{\phi}} = \sigma^2_{\phi} \gamma(D) = \sigma^2_{\phi} \gamma(W, W) \tag{11.80}$$

See previous item for definitions of a, b, and d. The variance reduction function, $\gamma(W, W)$ is defined for two dimension by Eq. 11.48, and Eq. 11.51 defines σ_ϕ.

6. The covariance between the observed cohesion values and the equivalent cohesion beneath the footing is obtained as follows for $D = W \times W$ and $Q = \Delta x \times H$:

$$\text{Cov}\left[\ln \bar{c}, \ln \hat{c}\right] \simeq \frac{\sigma^2_{\ln c}}{D^2 Q^2} \int_D \int_Q \rho(\mathbf{x}_1 - \mathbf{x}_2)\, d\mathbf{x}_1\, d\mathbf{x}_2 = \sigma^2_{\ln c} \gamma_{DQ} \tag{11.81}$$

where γ_{DQ} is the average correlation coefficient between the two areas D and Q. The area D denotes the averaging region below the footing over which equivalent properties are defined and the area Q denotes the region over which soil samples are gathered. These areas are illustrated in Figure 11.10. In detail, γ_{DQ} is defined by

$$\gamma_{DQ} = \frac{1}{(W^2\ \Delta x\ H)^2} \int_{-W/2}^{W/2} \int_{H-W}^{H} \int_{r-\Delta x/2}^{r+\Delta x/2} \int_0^H \times \rho(\xi_1 - x_1, \xi_2 - x_2)\, d\xi_2\, d\xi_1\, dx_2\, dx_1 \tag{11.82}$$

where r is the horizontal distance between the footing centerline and the centerline of the soil sample column. Equation 11.82 can be evaluated by Gaussian quadrature (see Appendices B and C).

7. The covariance between $\bar{N}_c$ and $\hat{N}_c$ is similarly approximated by

$$\text{Cov}\left[\ln \bar{N}_c, \ln \hat{N}_c\right] \simeq \sigma^2_{\ln N_c} \gamma_{DQ} \tag{11.83}$$

$$\sigma^2_{\ln N_c} \simeq \sigma^2_{\bar{\phi}} \left(\frac{d \ln N_c}{d\phi} \bigg|_{\mu_\phi} \right)^2 = \sigma^2_{\bar{\phi}} \left[\frac{bd}{bd^2 - 1} \left[\pi(1+a^2)d + 1 + d^2 \right] - \frac{1+a^2}{a} \right]^2 \tag{11.84}$$

Substituting these results into Eqs. 11.61 and 11.62 gives

$$\mu_{\ln Y} = \mu_{\ln L} \tag{11.85}$$

$$\sigma^2_{\ln Y} = \sigma^2_{\ln L} + \left[\sigma^2_{\ln c} + \sigma^2_{\ln N_c} \right] \times \left[\gamma(\Delta x, H) + \gamma(W, W) - 2\gamma_{DQ} \right] \tag{11.86}$$

which can now be used in Eq. 11.59 to produce estimates of p_f. Letting

$$q = I\left[\alpha_L \hat{L}_L + \alpha_D \hat{L}_D \right] \tag{11.87}$$

allows the probability of failure to be expressed as

$$p_f = \text{P}\left[Y > \frac{q}{\phi_g} \right] = \text{P}\left[\ln Y > \ln\left(\frac{q}{\phi_g} \right) \right] = 1 - \Phi\left(\frac{\ln(q/\phi_g) - \mu_{\ln Y}}{\sigma_{\ln Y}} \right) \tag{11.88}$$

where Φ is the standard normal cumulative distribution function.

Figure 11.11 illustrates the best and worst agreement between failure probabilities estimated via simulation and those computed using Eq. 11.88. The failure probabilities are slightly underestimated at the worst-case correlation lengths when the sample location is not directly below the footing. Given all the approximations made in the theory, the agreement is very good (within a 10% relative error), allowing the resistance factors to be computed with confidence even at probability levels which the simulation cannot estimate—the simulation involved only 2000 realizations and so cannot properly resolve probabilities much less than 0.001.

11.2.3 Required Resistance Factor

Equation 11.88 can be inverted to find a relationship between the acceptable probability of failure $p_f = p_m$ and the resistance factor ϕ_g required to achieve that acceptable failure probability,

$$\phi_g = \frac{I\left[\alpha_L \hat{L}_L + \alpha_D \hat{L}_D \right]}{\exp\left\{ \mu_{\ln Y} + \sigma_{\ln Y} \beta \right\}} \tag{11.89}$$

where β is the desired reliability index corresponding to p_m. That is $\Phi(\beta) = 1 - p_m$. For example, if $p_m = 0.001$, then $\beta = 3.09$.

The computation of $\sigma_{\ln Y}$ in Eq. 11.89 involves knowing the size of the averaging domain, D, under the footing. In turn, D depends on the average mean wedge zone depth (by assumption) under the footing, which depends on the mean footing width, $\hat{\mu}_B$. Unfortunately, the mean footing width given by Eq. 11.66 depends on ϕ_g, so solving Eq. 11.89 for ϕ_g is not entirely straightforward. One possibility is to iterate Eq. 11.89 until a stable solution is obtained. However, the authors have found that Eq. 11.89 is quite

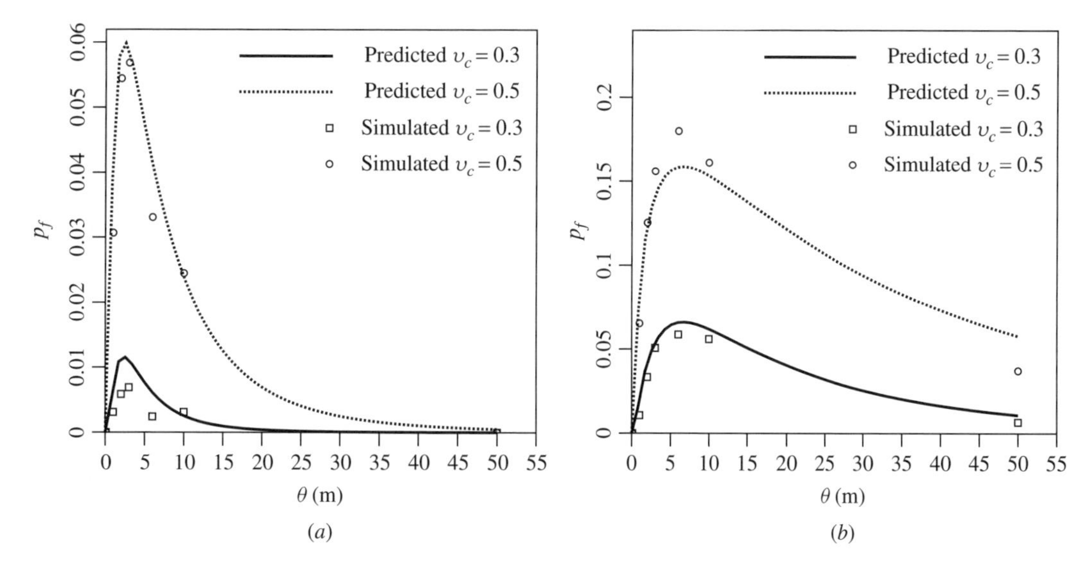

Figure 11.11 Comparison of failure probabilities estimated from simulation based on 2000 realizations and theoretical estimates using Eq. 11.88 with $\phi_g = 0.8$: Plot (*a*) probabilities when soil has been sampled directly under footing ($r = 0$ m), (*b*) probabilities when soil has been sampled 9 m from the footing centerline ($r = 9$ m). Note the change in the vertical scales—the probability of failure is much lower when samples are taken directly under the proposed footing.

insensitive to the initial size of D and using an "average" value of ϕ_g in Eq. 11.66 of 0.7 is quite sufficient. In other words, approximating

$$\hat{\mu}_B = \frac{I\left[\alpha_L \hat{L}_L + \alpha_D \hat{L}_D\right]}{0.7\mu_c \mu_{N_c}} \tag{11.90}$$

allows $\sigma_{\ln Y}$ to be suitably estimated for use in Eq. 11.89.

In the following, the value of ϕ_g required to achieve three target lifetime failure probability levels (10^{-2}, 10^{-3}, and 10^{-4}) for a specific case (a strip footing founded on a soil with specific statistic parameters) will be investigated. The results are to be viewed relatively. It is well known that the true probability of failure for any design will only be known once an infinite number of replications of that particular design have been observed over infinite time (and thus exposed to all possible loadings). One of the great advantages of probabilistic models is that it is possible to make probabilistic statements immediately, so long as we are willing to accept the fact that the probability estimates are only approximate. In that past history provides a wealth of designs which have been deemed by society to be acceptably reliable (or not, as the case may be), the results presented here need to be viewed *relative* to past designs so that the acceptable risk levels based on the past thousands of years of experience are incorporated. In other words, the results presented in the following, although rational and based on rigorous research, need to be moderated and adjusted by past experience.

The following parameters will be varied to investigate their effects on the resistance factor required to achieve a target failure probability p_m:

1. Three values of p_m are considered, 0.01, 0.001, and 0.0001, corresponding to reliability indices of approximately 2.3, 3.1, and 3.7, respectively.
2. The correlation length θ is varied from 0.0 to 50.0 m.
3. The mean cohesion was set to $\mu_c = 100$ kN/m^2. Four coefficients of variation for cohesion are considered, $v_c = 0.1, 0.2, 0.3$, and 0.5. The s factor for the friction angle distribution (see Figure 1.36) is set correspondingly to $s = 1, 2, 3$, and 5. That is, when $v_c = 0.2$, s is set to 2.0, and so on. The friction angle distribution is assumed to range from $\phi_{\min} = 0.1745$ radians (10°) to $\phi_{\max} = 0.5236$ rad (30°). The corresponding coefficients of variation for friction angle are $v_\phi = 0.07, 0.14, 0.20$, and 0.29.
4. Three sampling locations are considered: $r = 0, 4.5$, and 9.0 m from the footing centerline (see Figure 11.10 for the definition of r).

The design problem considered involves a strip footing supporting loads having means and standard deviations:

$$\mu_{L_e} = 200 \text{ kN/m}, \qquad \sigma_{L_e} = 60 \text{ kN/m} \tag{11.91a}$$

$$\mu_D = 600 \text{ kN/m}, \qquad \sigma_D = 90 \text{ kN/m} \tag{11.91b}$$

Assuming bias factors $k_D = 1.18$ (Becker, 1996b) and $k_{L_e} = 1.41$ (Allen, 1975) gives the characteristic loads

$$\hat{L}_L = 1.41(200) = 282 \quad \text{kN/m} \tag{11.92a}$$

$$\hat{L}_D = 1.18(600) = 708 \quad \text{kN/m} \tag{11.92b}$$

and the total factored design load (assuming $I = 1$) is

$$\begin{aligned} q &= I(\alpha_L \hat{L}_L + \alpha_D \hat{L}_D) \\ &= 1.5(282) + 1.25(708) = 1308 \quad \text{kN/m} \end{aligned} \tag{11.93}$$

So long as the ratio of dead to live load (assumed to be 3.0 in this study), the coefficients of variation of the load (assumed to be $v_{L_e} = 0.3$ and $v_D = 0.15$), and the characteristic bias factors k_{L_e} and k_D are unchanged, the results presented here are independent of the load applied to the strip footing. Minor changes in load ratios, coefficients of variation, and bias factors should not result in significant changes to the resistance factor.

Considering the slightly unconservative underestimation of the probability of failure in some cases (see Figure 11.11*b*), it is worthwhile first investigating to see how sensitive Eq. 11.89 is to changes in p_m of the same order as the errors in estimation of p_f. If p_m is replaced by $p_m/1.5$, then this corresponds to underestimating the failure probability by a factor of 1.5, which was well above the maximum difference seen between theory and simulation. It can be seen from Figure 11.12, which illustrates the effect of errors in the estimation of the failure probability, that the effect on ϕ_g is minor, especially considering all other sources of error in the analysis. Of the cases considered in this study, the ϕ_g values least affected by an underestimation of the probability occur when the soil is sampled under the footing ($r = 0$) and for small p_m, as seen in Figure 11.12*a*. The worst case is shown in Figure 11.12*b* and all other results (not shown) were seen to lie between these two plots. Even in the worst case of Figure 11.12*b*, the changes in ϕ_g due to errors in probability estimation are relatively small and will be ignored.

Figures 11.13, 11.14, and 11.15 show the resistance factors required for the cases where the soil is sampled directly under the footing, at a distance of 4.5 m, and at a distance of 9.0 m from the footing centerline, respectively, to achieve the three target failure probabilities. The worst-case correlation length is clearly between about 1 and 5 m, as evidenced by the fact that in all plots the lowest resistance factor occurs when $1 < \theta < 5$ m. This worst-case correlation length is of the same magnitude as the mean footing width ($\hat{\mu}_B = 1.26$ m) which can be explained as follows: If the random soil fields are stationary, then soil samples yield perfect information, regardless of their location, if the correlation length is either zero (assuming soil sampling involves some local averaging) or infinity. When the information is perfect, the probability of a bearing capacity failure goes to zero and $\phi_g \to 1.0$ (or possibly greater than 1.0 to compensate for the load bias factors). When the correlation length is zero, the soil sample will consist of an infinite number of independent "observations" whose average is equal to the true mean (or true median, if the average is a geometric average). Since the footing also averages the soil properties, the footing "sees" the same true mean (or true median) value predicted by the soil sample. When the correlation length goes to infinity, the soil becomes uniform, having the same value everywhere. In this case, any soil sample also perfectly predicts conditions under the footing.

At intermediate correlation lengths soil samples become imperfect estimators of conditions under the footing, and

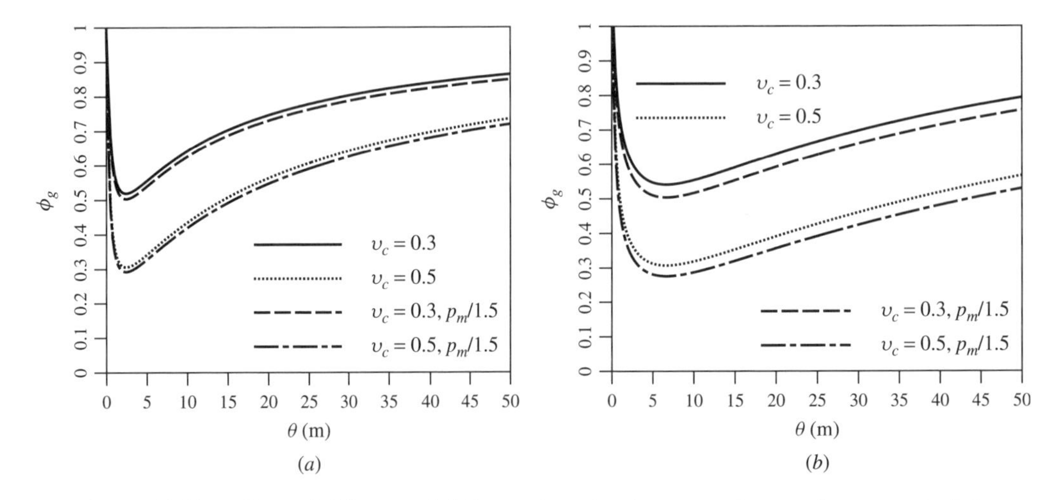

Figure 11.12 Effect of failure probability underestimation on resistance factor required by Eq. 11.89: (*a*) $r = 0$, $p_m = 0.001$; (*b*) $r = 9$, $p_m = 0.01$.

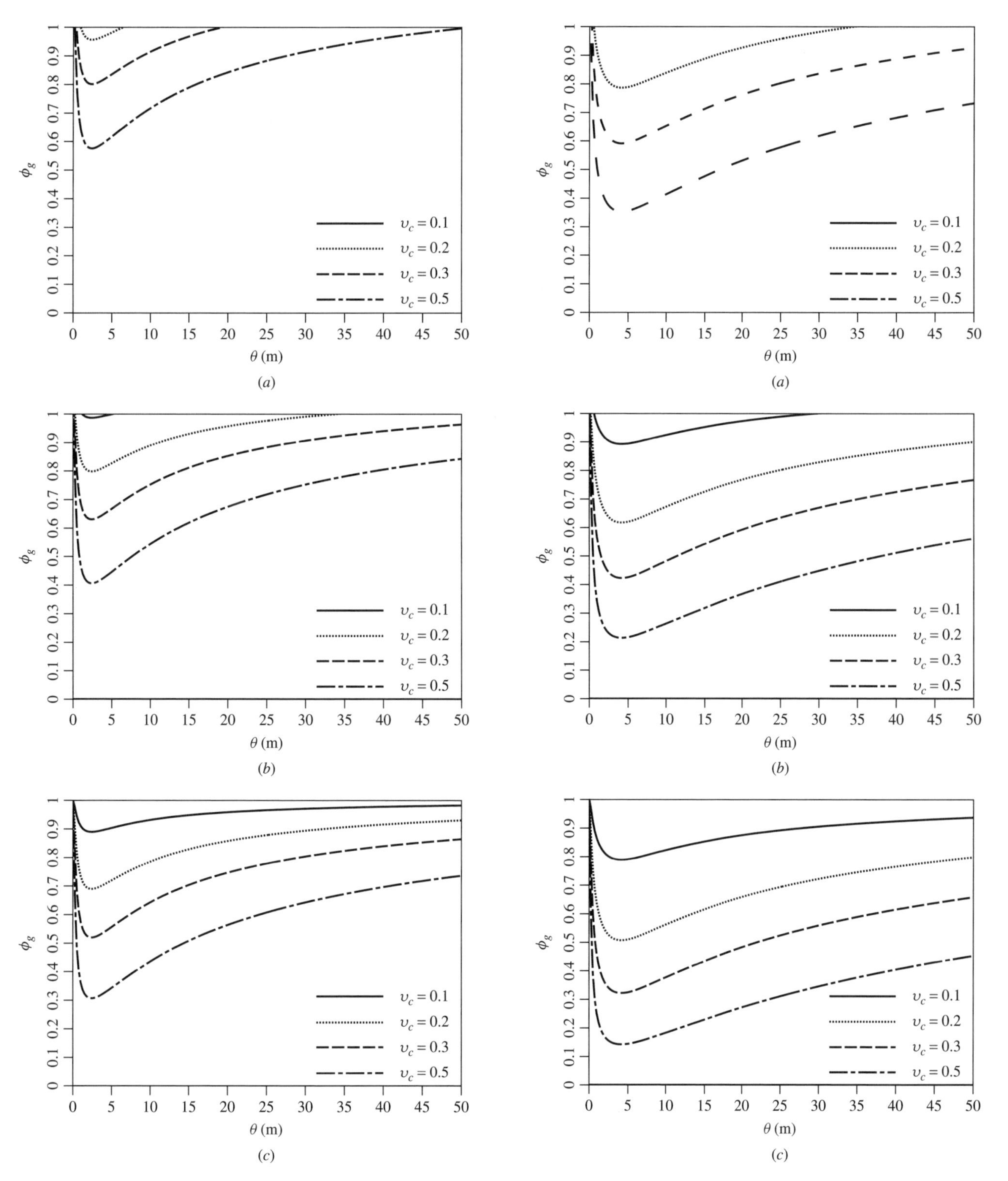

Figure 11.13 Resistance factors required to achieve acceptable failure probability p_m when soil is sampled directly under footing ($r = 0$): (*a*) $p_m = 0.01$; (*b*) $p_m = 0.001$; (*c*) $p_m = 0.0001$.

Figure 11.14 Resistance factors required to achieve acceptable failure probability p_m when soil is sampled at $r = 4.5$ m from footing centerline: (*a*) $p_m = 0.01$; (*b*) $p_m = 0.001$; (*c*) $p_m = 0.0001$.

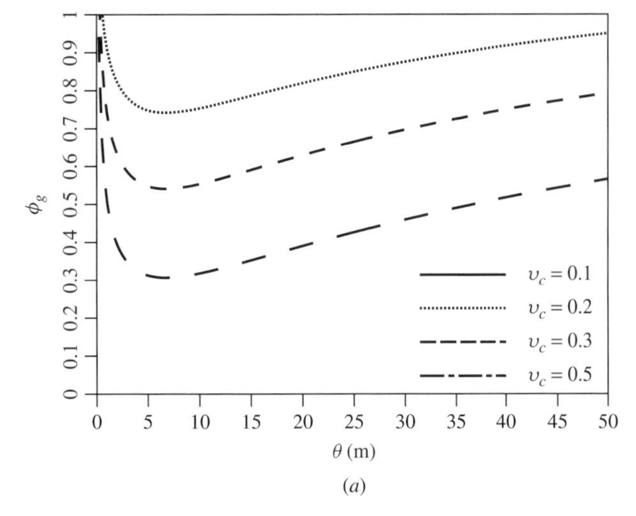

(*a*)

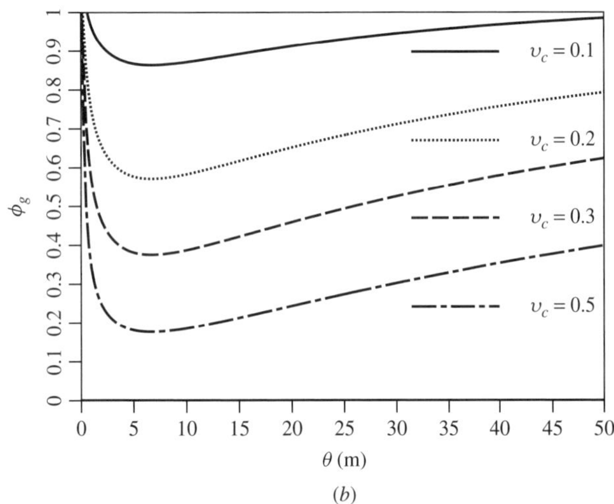

(*b*)

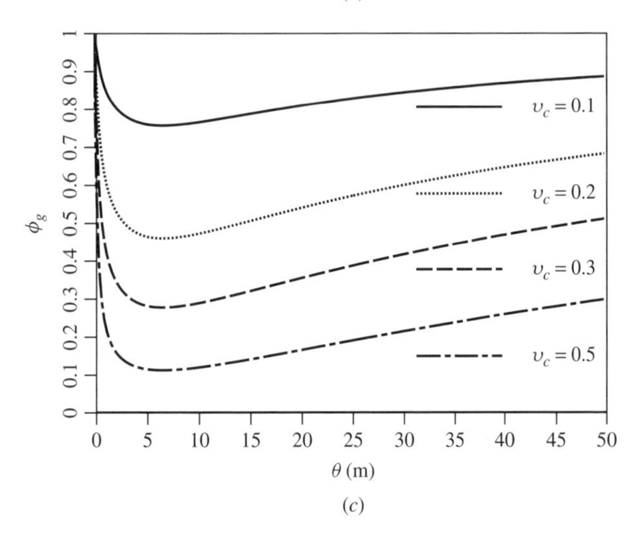

(*c*)

Figure 11.15 Resistance factors required to achieve acceptable failure probability p_m when soil is sampled at $r = 9.0$ m from footing centerline: (*a*) $p_m = 0.01$; (*b*) $p_m = 0.001$; (*c*) $p_m = 0.0001$.

so the probability of bearing capacity failure increases, or equivalently, the required resistance factor decreases. Thus, the minimum required resistance factor will occur at some correlation length between 0 and infinity. The precise value depends on the geometric characteristics of the problem under consideration, such as the footing width, depth to bedrock, length of soil sample, and/or the distance to the sample point. Notice in Figures 11.13, 11.14, and 11.15 that the worst-case point does show some increase as the distance to the sample location, r, increases.

As expected the smallest resistance factors correspond with the smallest acceptable failure probability considered, $p_m = 0.0001$, and with the poorest understanding of the soil properties under the footing (i.e., when the soil is sampled 9 m away from the footing centerline). When the cohesion coefficient of variation is relatively large, $v_c = 0.5$, with corresponding $v_\phi \simeq 0.29$, the worst-case values of ϕ_g dip almost down to 0.1 in order to achieve $p_m = 0.0001$. In other words, there will be a significant construction cost penalty if a high reliability footing is designed using a site investigation which is insufficient to reduce the residual variability to less than $v_c = 0.5$.

The simulation results can also be used to verify the theoretically determined resistance factors. This is done by using the simulation-based failure probabilities as values of p_m in the theory and comparing the resistance factor ϕ_g used in the simulation to that predicted by Eq. 11.89. The comparison is shown in Figure 11.16. For perfect agreement between theory and simulation, the points should align along the diagonal. The agreement is deemed to be very good and much of the discrepancy is due to failure probability estimator error, as discussed next. In general, however, the theory-based estimates of ϕ_g are seen to be conservative. That is, they are somewhat less than seen in the simulations on average.

Those simulations having less than 2 failures out of the 2000 realizations were omitted from the comparison in Figure 11.16, since the estimator error for such how probabilities is as big, or bigger, than the probability being estimated. For those simulations having 2 failures out of 2000 (included in Figure 11.16), the estimated probability of failure is 0.001 which has standard error $\sqrt{0.001(0.999)/2000} = 0.0007$. This error is almost as large as the probability being estimated, having a coefficient of variation of 70%. In fact most of the discrepancies in Figure 11.16 are easily attributable to estimator error in the simulation. The coefficient of variation of the estimator at the 0.01 probability level is 20%, which is still bigger than most of the relative errors seen in Figure 11.16 (the maximum relative error in Figure 11.16 is 0.28 at $\phi_g = 0.5$).

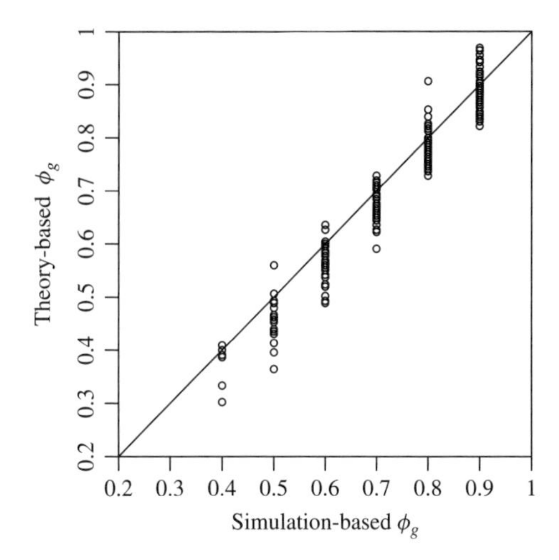

Figure 11.16 Required resistance factors, ϕ_g, based on simulation versus those based on Eq. 11.89. For perfect agreement, points would all lie on the diagonal.

The "worst-case" resistance factors required to achieve the indicated maximum acceptable failure probabilities, as seen in Figures 11.13–11.15, are summarized in Table 11.3. In the absence of better knowledge about the actual correlation length at the site in question, these factors are the largest values that should be used in the LRFD bearing capacity design of a strip footing founded on a $c - \phi$ soil.

It is noted, however, that the factors listed in Table 11.3 are sometimes quite conservative. For example, when $v_c = 0.3, r = 4.5$ m, and $p_m = 0.001$, Table 11.3 suggests that $\phi_g = 0.42$ for the $c-\phi$ soil considered here. However, if the soil is undrained, with $\phi = 0$ (all else being the same), then the only source of variability in the shear strength is the cohesion. In this case the above theory predicts a resistance factor of $\phi_g = 0.60$ which is considerably larger than suggested by Table 11.3.

To compare the resistance factors recommended in Table 11.3 to resistance factors recommended in the literature and to current geotechnical LRFD codes, changes in the load factors from code to code need to be taken into account. It will be assumed that all other sources define μ_{L_e}, μ_D, k_{L_e}, and k_D in the same way, which is unfortunately by no means certain. The easiest way to compare resistance factors is to compare the ratio of the resistance factor ϕ_g to the total load factor α. The total load factor, defined for fixed dead- to live-load ratio, is the single load factor which yields the same result as the individual live- and dead-load factors, that is, $\alpha\left(\hat{L}_L + \hat{L}_D\right) = \alpha_L \hat{L}_L + \alpha_D \hat{L}_D$. For mean dead- to live-load ratio $R_{D/L} = \mu_D/\mu_{L_e}$ and characteristic bias factors k_D and k_L,

$$\alpha = \frac{\alpha_L \hat{L}_L + \alpha_D \hat{L}_D}{\hat{L}_L + \hat{L}_D} = \frac{\alpha_L k_L \mu_{L_e} + \alpha_D k_D \mu_D}{k_L \mu_{L_e} + k_D \mu_D} = \frac{\alpha_L k_L + \alpha_D k_D R_{D/L}}{k_L + k_D R_{D/L}} \tag{11.94}$$

which, for $R_{D/L} = 3$, $k_L = 1.41$, $k_D = 1.18$, gives $\alpha = 1.32$. Table 11.4 compares the ratio of the resistance factors recommended in this study to total load factor with three other sources. The individual "current study" values correspond to the moderate case where $v_c = 0.3$ and acceptable failure

Table 11.3 Worst-Case Resistance Factors for Various Coefficients of Variation, v_c, Distance to Sampling Location, r, and Acceptable Failure Probabilities, p_m

	$r = 0.0$ m			$r = 4.5$ m			$r = 9.0$ m		
v_c	$p_m = 0.01$	0.001	0.0001	$p_m = 0.01$	0.001	0.0001	$p_m = 0.01$	0.001	0.0001
0.1	1.00	0.99	0.89	1.00	0.89	0.79	1.00	0.86	0.76
0.2	0.96	0.80	0.69	0.79	0.62	0.51	0.74	0.57	0.46
0.3	0.80	0.63	0.52	0.59	0.42	0.32	0.54	0.38	0.28
0.5	0.58	0.41	0.31	0.35	0.21	0.14	0.31	0.18	0.11

Table 11.4 Comparison of Resistance Factors Recommended in Study to Those Recommended by Three Other Sources

Source	Load Factors	ϕ_g	ϕ_g/α
Current study $r = 0$ m	$R_{D/L} = 3$, $\alpha_L = 1.5$, $\alpha_D = 1.25$	0.63	0.48
$r = 4.5$ m	$R_{D/L} = 3$, $\alpha_L = 1.5$, $\alpha_D = 1.25$	0.42	0.32
$r = 9.0$ m	$R_{D/L} = 3$, $\alpha_L = 1.5$, $\alpha_D = 1.25$	0.38	0.29
Foye et al., 2006	$R_{D/L} = 4$, $\alpha_L = 1.6$, $\alpha_D = 1.20$	0.70	0.54
CGS, (2006)	$R_{D/L} = 3$, $\alpha_L = 1.5$, $\alpha_D = 1.25$	0.50	0.38
Australian standard, 2004	$R_{D/L} = 3$, $\alpha_L = 1.8$, $\alpha_D = 1.20$	0.45	0.33

probability $p = 0.001$. The resistance factor derived from the Australian Standard (2004) on bridge foundations assumes a dead- to live-load ratio of 3.0 (not stated in the standard) and that the site investigation is based on CPT tests.

Apparently, the resistance factor recommended by Foye et al. (2006) assumes very good site understanding—they specify that the design assumes a CPT investigation which is presumably directly under the footing. Foye et al.'s recommended resistance factor is based on a reliability index of $\beta = 3$, which corresponds to $p_m = 0.0013$, which is very close to that used in Table 11.4 ($p_m = 0.001$). The small difference between the current study $r = 0$ result and Foye et al.'s may be due to differences in load bias factors—these are not specified by Foye et al.

The resistance factor specified by the Canadian Foundation Engineering Manual (CFEM, Canadian Geotechnical Society, 2006) is somewhere between that predicted here for the $r = 0$ and $r = 4.5$ m results. The CFEM resistance factor apparently presumes a reasonable, but not significant, understanding of the soil properties under the footing (e.g. $r \simeq 3$ m rather than $r = 0$ m). The corroboration of the rigorous theory proposed here by an experience-based code provision is, however, very encouraging. The authors also note that the CFEM is the only source listed in Table 11.4 for which the live- and dead-load bias factors used in this study can be reasonably assumed to also apply.

The Australian Standard AS 5100.3 (2004) resistance factor ratio is very close to that predicted here using $r = 4.5$ m. It is probably reasonable to assume that the Australian standard recommendations correspond to a moderate level of site understanding (e.g., $r = 4.5$ m) and an acceptable failure probability of about 0.0001.

11.3 SUMMARY

One of the main impediments to the practical use of these results is that they depend on a priori knowledge of the variance, and, to a lesser extent, since worst-case results are presented above, the correlation structure of the soil properties. However, assuming that at least one CPT sounding (or equivalent) is taken in the vicinity of the footing, it is probably reasonable to assume that the residual variability is reduced to a coefficient of variation of no more than about 0.3, and often considerably less (the results collected by other investigators, e.g. Phoon et al., 1999, suggest that this may be the case for "typical" site investigations). In this is so, the resistance factors recommended in Table 11.3 for $v_c = 0.3$ are probably reasonable for the load and bias factors assumed in this study.

The resistance factors recommended in Table 11.3 are conservative in (at least) the following ways:

1. It is unlikely that the correlation length of the residual random process at a site (after removal of any mean or mean trend estimated from the site investigation, assuming there is one) will actually equal the worst-case correlation length.
2. The soil is assumed weightless in this study. The addition of soil weight, which the authors feel to be generally less spatially variable than soil strength parameters, should reduce the failure probability and so result in higher resistance factors for fixed acceptable failure probability.
3. Sometimes more than one CPT or SPT is taken at the site in the footing region, so that the site understanding may exceed even the $r = 0$ m case considered here if trends and layering are carefully accounted for.
4. The parameters c and ϕ are assumed independent, rather than negatively correlated, which leads to a somewhat higher probability of failure and correspondingly lower resistance factor, and so somewhat conservative results. Since the effect of positive or negative correlation of c and ϕ was found by Fenton and Griffiths (2003) to be quite minor, this is not a major source of conservatism.

On the other hand, the resistance factors recommended in Table 11.3 are unconservative in (at least) the following ways:

1. Measurement and model errors are not considered in this study. The statistics of measurement errors are very difficult to determine since the true values need to be known. Similarly, model errors, which relate both the errors associated with translating measured values (e.g., CPT measurements to friction angle values) and the errors associated with predicting bearing capacity by an equation such as Eq. 11.40 are quite difficult to estimate simply because the true bearing capacity along with the true soil properties are rarely, if ever, known. In the authors' opinions this is the major source of unconservatism in the presented theory. When confidence in the measured soil properties or in the model used is low, the results presented here can still be employed by assuming that the soil samples were taken further away from the footing location than they actually were (e.g., if low-quality soil samples are taken directly under the footing, $r = 0$, the resistance factor corresponding to a larger value of r, say $r = 4.5$ m should be used).

2. The failure probabilities given by the above theory are slightly underpredicted when soil samples are taken at some distance from the footing at the worst-case correlation length. The effect of this underestimation on the recommended resistance factor has been shown to be relatively minor but nevertheless unconservative.

To some extent the conservative and unconservative factors listed above cancel one another out. Figure 11.16 suggests that the theory is generally conservative if measurement errors are assumed to be insignificant. The comparison of resistance factors presented in Table 11.4 demonstrates that the worst-case theoretical results presented in Table 11.3 agree quite well with current literature and LRFD code recommendations, assuming moderate variability and site understanding, suggesting that the theory is reasonably accurate. In any case, the theory provides an analytical basis to extend code provisions beyond mere calibration with the past.

The results presented in this section are for a $c-\phi$ soil in which both cohesion and friction contribute to the bearing capacity, and thus to the variability of the strength. If it is known that the soil is purely cohesive (e.g., "undrained clay"), then the strength variability comes from one source only. In this case, not only does Eq. 11.86 simplify since $\sigma^2_{\ln N_c} = 0$, but because of the loss of one source of variability, the resistance factors increase significantly. The net result is that the resistance factors presented in this paper are conservative when $\phi = 0$. Additional research is needed to investigate how the resistance factors should generally be increased for "undrained clays".

The effect of anisotropy in the correlation lengths has not been carefully considered in this study. It is known, however, that increasing the horizontal correlation length above the worst case length is conservative when the soil is not sampled directly below the footing. When the soil is sampled directly under the footing, weak spatially extended horizontal layers below the footing will obviously have to be explicitly handled by suitably adjusting the characteristic soil properties used in the design. If this is done, then the resistance factors suggested here should still be conservative. The theory presented in this section easily accomodates the anisotropic case.

One of the major advantages to a table such as 11.3 is that it provides geotechnical engineers with evidence that increased site investigation will lead to reduced construction costs and/or increased reliability. In other words, Table 11.3 is further evidence that you pay for a site investigation whether you have one or not (Institution of Civil Engineers, 1991).

CHAPTER 12

Deep Foundations

12.1 INTRODUCTION

Deep foundations, which are typically either piles or drilled shafts, will be hereafter collectively referred to as *piles* for simplicity in this chapter. Piles are provided to transfer load to the surrounding soil and/or to a firmer stratum, thereby providing vertical and lateral load-bearing capacity to a supported structure. In this chapter the random behavior of a pile subjected to a vertical load and supported by a spatially variable soil is investigated (Fenton and Griffiths, 2007). The program used to perform the simulations reported here is called RPILE1D, which is available at `http://www.engmath.dal.ca/rfem`.

The resistance, or bearing capacity, of a pile arises as a combination of side friction, where load is transmitted to the soil through friction along the sides of the pile, and end bearing, where load is transmitted to the soil (or rock) through the tip of the pile. As load is applied to the pile, the pile settles—the total settlement of the pile is due to both deformation of the pile itself and deformation of the surrounding soil and supporting stratum. The surrounding soil is, at least initially, assumed to be perfectly bonded to the pile shaft through friction and/or adhesion so that any displacement of the pile corresponds to an equivalent local displacement of the soil (the soil deformation reduces further away from the pile). In turn, the elastic nature of the soil means that this displacement is resisted by a force which is proportional to the soil's elastic modulus and the magnitude of the displacement. Thus, at least initially, the support imparted by the soil to the pile depends on the elastic properties of the surrounding soil. For example, Vesic (1977) states that the fraction of pile settlement due to deformation of the soil, δ_s, is a constant (dependent on Poisson's ratio and pile geometry) times Q/E_s, where Q is the applied load and E_s is the (effective) soil elastic modulus.

As the load on the pile is increased, the bond between the soil and the pile surface will at some point break down and the pile will both slip through the surrounding soil and plastically fail the soil under the pile tip. At this point, the ultimate bearing capacity of the pile has been reached. The force required to reach the point at which the pile slips through a sandy soil is conveniently captured using a soil–pile interface friction angle δ. The frictional resistance per unit area of the pile surface, f, can then be expressed as

$$f = \sigma_n' \tan\delta \tag{12.1}$$

where σ_n' is the effective stress exerted by the soil normal to the pile surface. In many cases, $\sigma_n' = K\sigma_o'$, where K is the earth pressure coefficient and σ_o' is the effective vertical stress at the depth under consideration. The total ultimate resistance supplied by the soil to an applied pile load is the sum of the end-bearing capacity (which can be estimated using the usual bearing capacity equation) and the integral of f over the embedded surface of the pile. For clays with zero friction angle, Vijayvergiya and Focht (1972) suggest that the average of f, denoted with an overbar, can be expressed in the form

$$\bar{f} = \lambda\left(\bar{\sigma}_o' + 2c_u\right) \tag{12.2}$$

where $\bar{\sigma}_o'$ is the average effective vertical stress over the entire embedment length, c_u is the undrained cohesion, and λ is a correction factor dependent on pile embedment length.

The limit state design of a pile involves checking the design at both the serviceability limit state and the ultimate limit state. The serviceability limit state is a limitation on pile settlement, which in effect involves computing the load beyond which settlements become intolerable. Pile settlement involves consideration of the elastic behavior of the pile and the elastic (e.g., E_s) and consolidation behavior of the surrounding soil.

The ultimate limit state involves computing the ultimate load that the pile can carry just prior to failure. Failure is assumed to occur when the pile slips through the soil (we are not considering structural failure of the pile itself), which can be estimated with the aid of Eq. 12.1 or 12.2, along with the end-bearing capacity equation. The ultimate pile capacity is a function of the soil's cohesion and friction angle parameters.

In this chapter, the soil's influence on the pile will be represented by bilinear springs (see, e.g., program 12 of Smith and Griffiths, 1982), as illustrated in Figure 12.1. The initial sloped portion of the load–displacement curve corresponds to the elastic (E_s) soil behavior, while the plateau corresponds to the ultimate shear strength of the pile–soil interface, which is a function of the soil's friction angle and cohesion. The next section discusses the finite-element and random-field models used to represent the pile

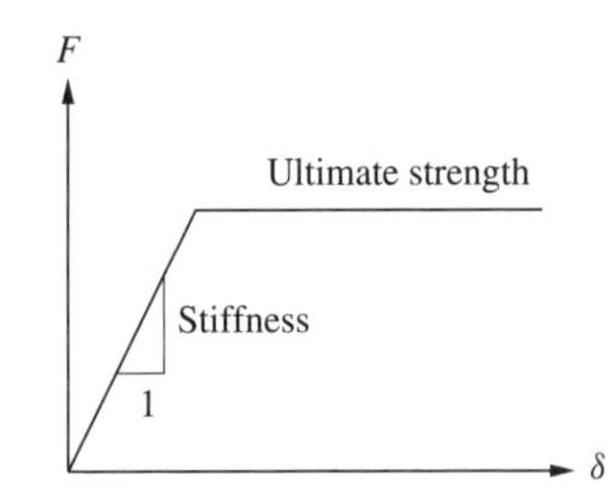

Figure 12.1 Bilinear load (F) versus displacement (δ) for soil springs.

and supporting soil in more detail. In the following section an analysis of the random behavior of a pile is described and presented. Only the effects of the spatial variability of the soil are investigated and not, for instance, those due to construction and placement variability. Finally, the results are evaluated and recommendations are made.

12.2 RANDOM FINITE-ELEMENT METHOD

The pile itself is divided into a series of elements, as illustrated in Figure 12.2. Each element has cross-sectional area A (assumed constant) and elastic modulus E_p, which can vary randomly along the pile. The stiffness assigned to the ith element is the geometric average of the product AE_p over the element domain.

As indicated in Figure 12.1, the ith soil spring is characterized by two parameters: its initial stiffness S_i and its ultimate strength U_i. The determination of these two parameters from the soil's elastic modulus, friction angle, and cohesion properties is discussed conceptually as follows:

1. The initial spring stiffness S_i is a function of the soil's spatially variable elastic modulus E_s. Since the strain induced in the surrounding soil due to displacement of the pile is complex, not least because the strain decreases nonlinearly with distance from the pile, the effective elastic modulus of the soil as seen by the pile at any point along the pile is currently unknown. The nature of the relationship between E_s and S_i remains a topic for further research. In this chapter, the spring stiffness contribution per unit length of the pile, $S(z)$, will be simulated directly as a lognormally distributed one-dimensional random process.
2. The ultimate strength of each spring is somewhat more easily specified so long as the pile–soil interface adhesion, friction angle, and normal stress are known. Assuming that soil properties vary only with depth z, the ultimate strength per unit pile length at depth z will have the general form (in the event that both adhesion and friction act simultaneously)

$$U(z) = p\left[\alpha c_u(z) + \sigma_n'(z)\tan\delta(z)\right] \qquad (12.3)$$

where $\alpha c_u(z)$ is the adhesion at depth z [see, e.g., Das (2000, p. 519) for estimates of the adhesion factor α], p is the pile perimeter length, $\sigma_n'(z)$ is the normal effective soil stress at depth z, and $\delta(z)$ is the interface friction angle at depth z. The normal stress is often taken as $K\sigma_o'$, where K is an earth pressure coefficient. Rather than simulate c_u and $\tan\,\delta$ and introduce the empirical and uncertain factors α and K, both of which could also be spatially variable, the ultimate strength per unit length, $U(z)$, will be simulated directly as a lognormally distributed one-dimensional random process.

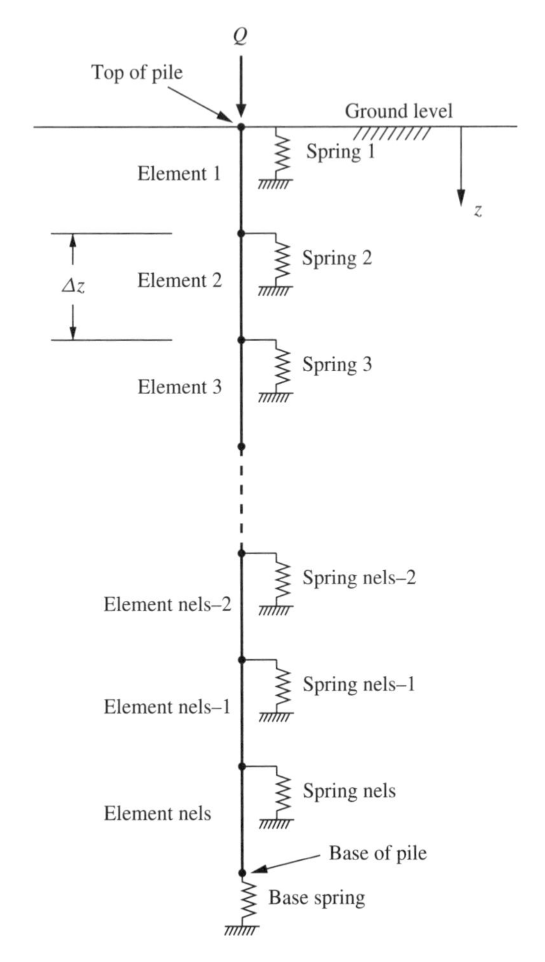

Figure 12.2 Finite-element representation of pile–soil system for a given number of elements (nels).

The RFEM thus consists of a sequence of pile elements joined by nodes, a sequence of spring elements attached to the nodes (see Figure 12.2), and three *independent* one-dimensional random processes described as follows:

- $S(z)$ and $U(z)$ are the spring stiffness and strength contributions from the soil per unit length along the pile and
- $E_p(z)$ is the elastic modulus of the pile.

It is assumed that the elastic modulus of the pile is a one-dimensional stationary lognormally distributed random process characterized by the mean pile stiffness μ_{AE_p}, standard deviation σ_{AE_p}, and correlation length $\theta_{\ln E_p}$, where A is the pile cross-sectional area. Note that, for simplicity, it is assumed that all three random processes have the same correlation lengths and all have the same correlation function (Markovian). While it may make sense for the correlation lengths associated with $S(z)$ and $U(z)$ to be similar, there is no reason that the correlation length of $E_p(z)$ should be the same as that in the soil. Keeping them the same merely simplifies the study while still allowing the study to assess whether a "worst-case" correlation length exists for the deep-foundation problem.

The elastic modulus assigned to each pile element will be some sort of average of $E_p(z)$ over the element length, and in this chapter the geometric average will be used:

$$E_{p_i} = \exp\left\{\frac{1}{\Delta z}\int_{z_i}^{z_i+\Delta z} \ln E_p(z)\, dz\right\} \tag{12.4}$$

where z_i is the depth to the top of the ith element. The geometric average is dominated by low stiffness values, which is appropriate for elastic deformation. It is to be noted that for a pile idealized using an elastic modulus varying only along the pile length, the true "effective" pile stiffness is the harmonic average

$$E_H = \left[\frac{1}{\Delta z}\int_{z_i}^{z_i+\Delta z} \frac{1}{E_p(z)}\, dz\right]^{-1}$$

which is even more strongly dominated by low stiffness values than the geometric average. However, the following justification can be argued about the use of the geometric average rather than the harmonic average over each element:

1. If the elements are approximately square (i.e., $\Delta z \simeq$ pile diameter), and the pile's true three-dimensional elastic modulus field is approximately isotropic (i.e., not strongly layered), then the effective elastic modulus of the element will be (at least closely approximated by) a geometric average. See, for example, Chapter 10, where this result was found for a soil block, which is a similar stochastic settlement problem to the pile element "block."
2. If the pile is subdivided into a reasonable number of elements along its length (say, 10 or more), then the overall response of the pile tends towards a harmonic average in any case since the finite element analysis will yield the exact "harmonic" result.

We are left now with the determination of the spring stiffness and strength values, S_i and U_i, from the one-dimensional random processes $S(z)$ and $U(z)$. Note that the spring parameters S_i and U_i have units of stiffness (kilonewtons per meter) and strength (kilonewtons), respectively, while $S(z)$ and $U(z)$ are the soil's contribution to the spring stiffness and strength *per unit length along the pile*. That is, $S(z)$ has units of kilonewtons per meter per meter and $U(z)$ has units of kilonewton per meter.

To determine the spring parameters S_i and U_i from the continuously varying $S(z)$ and $U(z)$, we need to think about the nature of the continuously varying processes and how they actually contribute to S_i and U_i. In the following we will discuss this only for the stiffness contribution S; the strength issue is entirely analogous and can be determined simply by substituting S with U in the following.

We will first subdivide each element into two equal parts, as shown in Figure 12.3, each of length $\Delta h = \Delta z/2$. The top of each subdivided cell will be at $t_j = (j-1)\,\Delta h$ for $j = 1, 2, \ldots, 2n+1$, where n is the number of elements. This subdivision is done so that the tributary lengths for each spring can be more easily defined: The stiffness for spring 1 is accumulated from the soil stiffness contribution

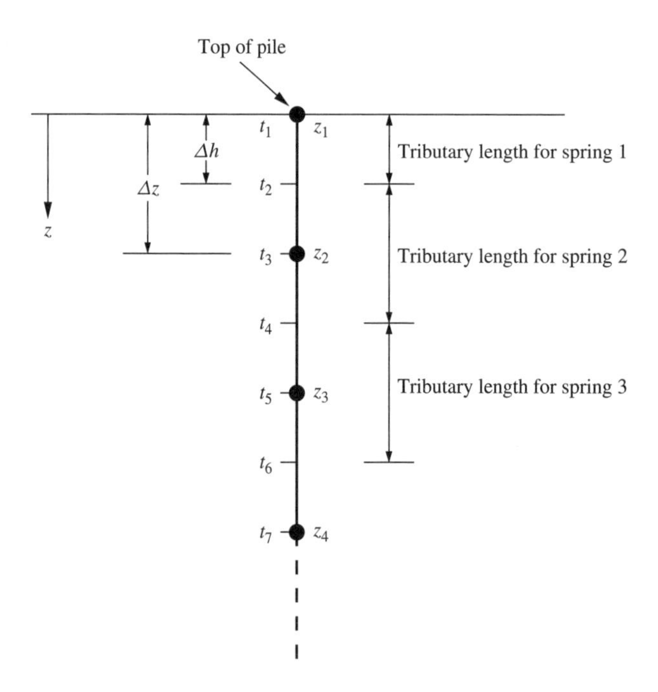

Figure 12.3 Subdivisions used to compute geometric averages.

$S(z)$ over the top cell from $z = t_1 = 0$ to $z = t_2 = \Delta h$. The stiffness for spring 2 is accumulated from the cell above spring 2 as well as from the cell below spring 2, that is, from $z = t_2 = \Delta h$ to $z = t_3 = 2\,\Delta h$ and from $z = t_3 = 2\,\Delta h$ to $z = t_4 = 3\,\Delta h$, and so on.

If the stiffness contributions $S(z)$ at each point z are independent (i.e., white noise), and if the pile stiffness is significantly larger than the soil stiffness, then S_i should be an arithmetic sum of $S(z)$ over the spring's tributary length,

$$S_i = \int_{z_i - \Delta z/2}^{z_i + \Delta z/2} S(z)\, dz \tag{12.5}$$

In other words, S_i should be an arithmetic average of $S(z)$ over the tributary length multiplied by the tributary length. However, $S(z)$ is not a white noise process—a low-stiffness region close to the pile will depress the stiffness contribution over a length of pile which will probably be significantly larger than the low-strength region itself. Thus, it makes sense to assume that S_i should be at least somewhat dominated by low-stiffness regions in the surrounding soil.

A compromise shall be made here: S_i will be an arithmetic sum of the two geometric averages over the ith spring's tributary areas (in the case of the top and bottom springs, only one tributary area is involved). The result is less strongly low-stiffness dominated than a pure geometric average, as might be expected by this sort of a problem where the strain imposed on the soil is relatively constant over the element lengths (i.e., the constant strain results in at least some arithmetic averaging). The exact nature of the required average is left for future research.

If the mean of $S(z)$ is allowed to vary linearly with depth z, then

$$\mu_S = \mathrm{E}[S(z)] = a + bz \tag{12.6}$$

If the stiffnesses per unit length at the top and bottom of the pile are s_{top} and s_{bot}, respectively, and we measure z downward from the top of the pile, then

$$a = s_{\text{top}} \tag{12.7a}$$

$$b = \frac{s_{\text{bot}} - s_{\text{top}}}{L} \tag{12.7b}$$

where L is the pile length.

It is assumed that $S(z)$ is lognormally distributed. It thus has parameters

$$\mu_{\ln S} = \ln(a + bz) - \tfrac{1}{2}\sigma_{\ln S}^2 \tag{12.8a}$$

$$\sigma_{\ln S}^2 = \ln(1 + v_S^2) \tag{12.8b}$$

where v_S is the coefficient of variation of $S(z)$. It will be assumed that v_S is constant with depth, so that $\sigma_{\ln S}$ is also constant with depth. Now $S(z)$ can be expressed in terms of the underlying zero-mean, unit-variance, normally distributed one-dimensional random process $G(z)$,

$$\begin{aligned} S(z) &= \exp\{\mu_{\ln S} + \sigma_{\ln S} G(z)\} \\ &= \exp\left\{\ln(a + bz) - \tfrac{1}{2}\sigma_{\ln S}^2 + \sigma_{\ln S} G(z)\right\} \end{aligned} \tag{12.9}$$

In other words

$$\ln S(z) = \ln(a + bz) - \tfrac{1}{2}\sigma_{\ln S}^2 + \sigma_{\ln S} G(z) \tag{12.10}$$

Now let S_{G_j} be the geometric average of the soil spring stiffness contribution $S(z)$ over the jth cell, that is, over a length of the pile from t_j to $t_j + \Delta h$, $j = 1, 2, \ldots, 2n$,

$$\begin{aligned} S_{G_j} &= \exp\left\{\frac{1}{\Delta h}\int_{t_j}^{t_j+\Delta h} \ln S(z)\, dz\right\} \\ &= \exp\left\{\frac{1}{\Delta h}\int_{t_j}^{t_j+\Delta h} \left[\ln(a + bz) - \tfrac{1}{2}\sigma_{\ln S}^2 \right.\right. \\ &\qquad \left.\left. + \sigma_{\ln S} G(z)\right] dz\right\} \\ &= \exp\left\{\frac{1}{\Delta h}\int_{t_j}^{t_j+\Delta h} \ln(a + bz)\, dz - \tfrac{1}{2}\sigma_{\ln S}^2 + \sigma_{\ln S} G_j\right\} \end{aligned} \tag{12.11}$$

where G_j is the *arithmetic* average of $G(z)$ from $z = t_j$ to $z = t_j + \Delta h$:

$$G_j = \frac{1}{\Delta h}\int_{t_j}^{t_j+\Delta h} G(z)\, dz \tag{12.12}$$

Now define

$$\begin{aligned} m_j &= \frac{1}{\Delta h}\int_{t_j}^{t_j+\Delta h} \ln(a + bz)\, dz - \tfrac{1}{2}\sigma_{\ln S}^2 \\ &= \frac{1}{b\,\Delta h}\left[a_1 \ln(a_1) - a_2 \ln(a_2)\right] - 1 - \tfrac{1}{2}\sigma_{\ln S}^2 \end{aligned} \tag{12.13}$$

where

$$a_1 = a + b(t_j + \Delta h) \tag{12.14a}$$

$$a_2 = a + bt_j \tag{12.14b}$$

If $b = 0$, that is, the soil stiffness contribution is constant with depth, then m_j simplifies to

$$m_j = \ln(s_{\text{top}}) - \tfrac{1}{2}\sigma_{\ln S}^2 \tag{12.15}$$

Using m_j, the geometric average becomes

$$S_{G_j} = \exp\left\{m_j + \sigma_{\ln S} G_j\right\} \tag{12.16}$$

Notice that m_j is the arithmetic average of $\mu_{\ln S}$ over the distance from $z = t_j$ to $z = t_j + \Delta z$.

The contribution to the spring stiffness is now $\Delta h\, S_{G_j}$. In particular, the top spring has contributing soil stiffness from $z = 0$ to $z = \Delta h$, so that $S_1 = \Delta h\, S_{G_1}$. Similarly, the next spring down has contributions from the soil from $z = \Delta h$

to $z = 2\ \Delta h$ as well as from $z = 2\ \Delta h$ to $z = 3\ \Delta h$, so that

$$S_2 = \Delta h \left[S_{G_2} + S_{G_3} \right] \tag{12.17}$$

and so on.

The finite-element analysis is displacement controlled. In other words, the load corresponding to the prescribed maximum tolerable serviceability settlement, δ_{max}, is determined by imposing a displacement of δ_{max} at the pile top. Because of the nonlinearity of the springs, the finite-element analysis involves an iteration to converge on the set of admissible spring forces which yield the prescribed settlement at the top of the pile. The relative maximum-error convergence tolerance is set to a very small value of 0.00001.

The pile capacity corresponding to the ultimate limit state is computed simply as the sum of the U_i values over all of the springs.

12.3 MONTE CARLO ESTIMATION OF PILE CAPACITY

To assess the probabilistic behavior of deep foundations, a series of Monte Carlo simulations, with 2000 realizations each, were performed and the distributions of the serviceability and ultimate limit state loads were estimated. The serviceability limit state was defined as being a settlement of $\delta_{max} = 25$ mm. Because the maximum tolerable settlement cannot easily be expressed in dimensionless form, the entire analysis will be performed for a particular case study; namely, a pile of length 10 m is divided into $n = 30$ elements with $\mu_{AE_p} = 1000$ kN, $\sigma_{AE_p} = 100$ kN, $\mu_S = 100$ kN/m/m, and $\mu_U = 10$ kN/m. The base of the pile is assumed to rest on a slightly firmer stratum, so the base spring has mean stiffness 200 kN/m and mean strength 20 kN (note that this is in addition to the soil contribution arising from the lowermost half-element). Coefficients of variation of spring stiffness and strength, v_S and v_U, taken to be equal and collectively referred to as v, ranged from 0.1 to 0.5. Correlation lengths $\theta_{\ln S}$, $\theta_{\ln E_p}$, and $\theta_{\ln U}$, all taken to be equal and referred to collectively simply as θ, ranged from 0.1 m to 100.0 m. The spring stiffness and strength parameters were assumed to be mutually independent as well as being independent of the pile elastic modulus.

The first task is to determine the nature of the distribution of the serviceability and ultimate pile loads. Figure 12.4*a* shows one of the best and Figure 12.4*b* one of the worst fits of a lognormal distribution to the serviceability pile load histogram with chi-square goodness-of-fit p-values of 0.84 and 0.0006, respectively (the null hypothesis being that the serviceability load follows a lognormal distribution). The plot in Figure 12.4*b* would result in the lognormal hypothesis being rejected for any significance level in excess of 0.06%. Nevertheless, a visual inspection of the plot suggests that the lognormal distribution is quite *reasonable*—in fact, it is hard to see why one fit is so much "better" than the other. It is well known, however, that when the number of simulations is large, goodness-of-fit tests tend to be very sensitive to small discrepancies in the fit, particularly in the tails.

Figure 12.5 shows similar results for the ultimate pile capacities, which are simply obtained by adding up the ultimate spring values. In both figures, the lognormal distribution appears to be a very reasonable fitted, despite the very low p-value of Figure 12.5*b*.

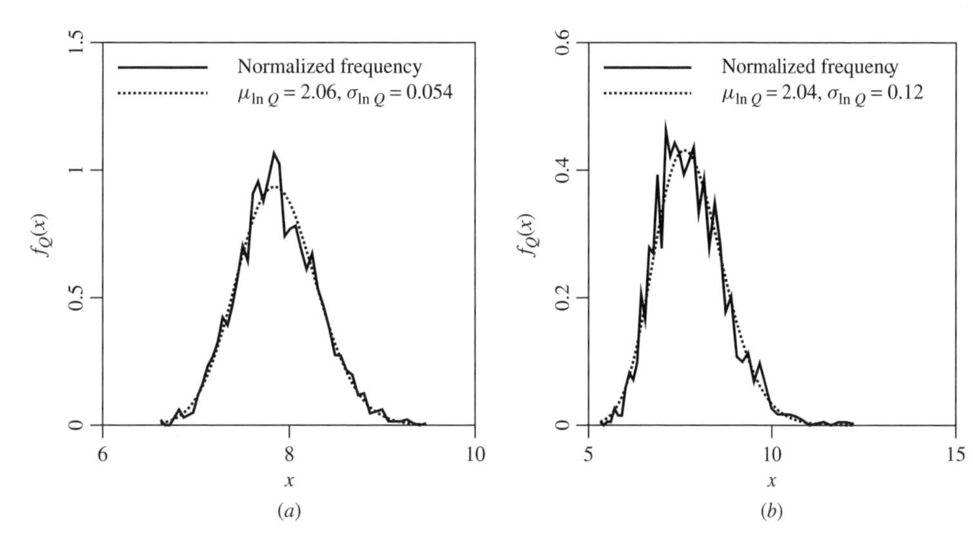

Figure 12.4 Estimated and fitted lognormal distributions of serviceability limit state loads Q for (*a*) $v = 0.2$ and $\theta = 1$ m (p-value 0.84) and (*b*) $v = 0.5$ and $\theta = 1.0$ m (p-value 0.00065).

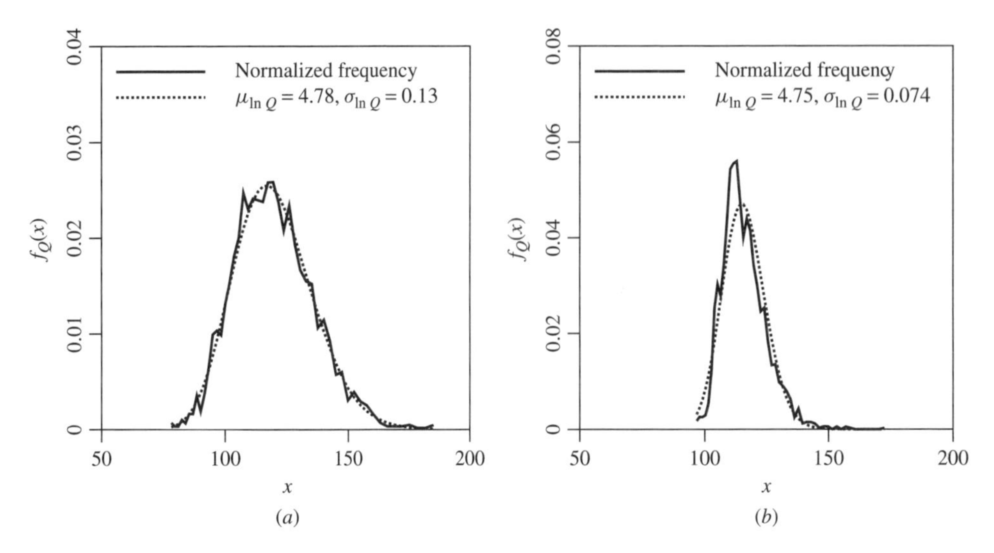

Figure 12.5 Estimated and fitted lognormal distributions of ultimate limit state loads Q for (*a*) $v = 0.2$ and $\theta = 10$ m (p-value 0.94) and (*b*) $v = 0.4$ and $\theta = 0.1$ m (p-value 8×10^{-11}).

If the pile capacities at both the serviceability and ultimate limit states are lognormally distributed, then the computation of the probability that the actual pile capacity Q is less than the design capacity Q_{des} proceeds as follows,

$$\mathrm{P}\left[Q < Q_{\text{des}}\right] = \Phi\left(\frac{\ln Q_{\text{des}} - \mu_{\ln Q}}{\sigma_{\ln Q}}\right) \tag{12.18}$$

where Φ is the standard normal cumulative distribution function. For this computation we need only know the mean and standard deviation of $\ln Q$. Figure 12.6 shows the estimated mean and variance of $\ln Q$ for the serviceability limit state, that is, those loads Q which produce the maximum tolerable pile settlement, which in this case is 25 mm. The estimate of $\mu_{\ln Q}$ is denoted $m_{\ln Q}$ while the estimate of $\sigma_{\ln Q}$ is denoted $s_{\ln Q}$. Similarly, Figure 12.7 shows the estimated mean and standard deviation of $\ln Q$ at the ultimate limit state, that is, at the point where the pile reaches failure and the capacity of all springs has been fully mobilized.

12.4 SUMMARY

Aside from the changes in the magnitudes of the means and standard deviations, the statistical behavior of the maximum loads at serviceability and ultimate limit states are very similar. First of all the mean loads are little affected by both the coefficient of variation (v) and the correlation length (θ);

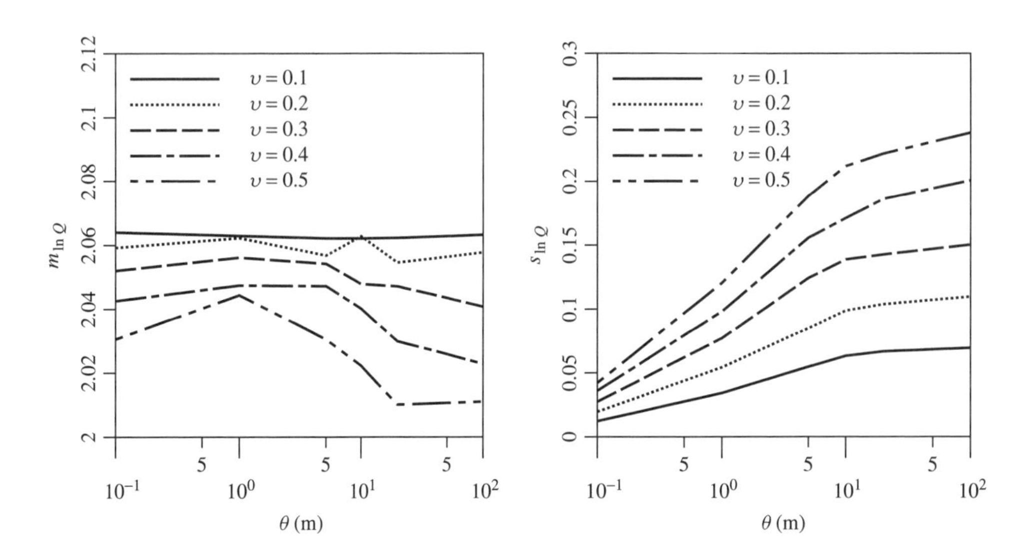

Figure 12.6 Estimated mean $m_{\ln Q}$ and standard deviation $s_{\ln Q}$ of maximum load Q at serviceability limit state.

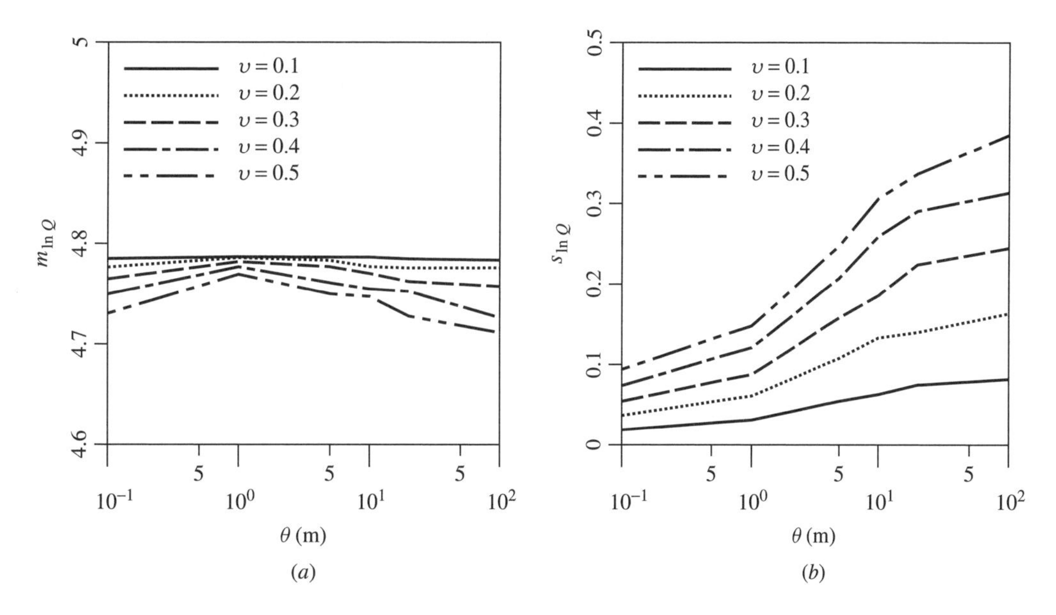

Figure 12.7 Estimated mean $m_{\ln Q}$ and standard deviation $s_{\ln Q}$ of maximum load Q at ultimate limit state.

note that the vertical axes for the plots in Figures 12.6*a* and 12.7*a* are over a fairly narrow range. The mean in Q and the mean in $\ln Q$ show similar behavior. There are only slight reductions in the mean for increasing v. This suggests that the pile is more strongly controlled by arithmetic averaging of the soil strengths, which is perhaps not surprising if the pile is much stiffer than the surrounding soil. In fact, it could be argued that some of the reduction in mean with v is due to the fact that geometric averaging was done over the half-element lengths. In other words, it is possible that only arithmetic averaging should be done in this pile model. This needs further study.

The load standard deviation (for both Q and $\ln Q$) increases monotonically for increasing coefficient of variation, as expected (i.e., as the soil becomes increasingly variable, one expects its ability to support the pile would also become increasingly variable). This behavior was also noted by Phoon et al. (1990). The standard deviation approaches zero as the correlation length goes to zero, which is also to be expected due to local averaging (geometric or otherwise). At the opposite extreme as $\theta \to \infty$, the standard deviation approaches that predicted if the soil is treated as a single lognormally distributed random variable (with an independent base spring variable). For example, when $\theta \to \infty$, $\sigma_{\ln Q}$ is expected to approach 0.407 for the ultimate limit state with $v = 0.5$. It is apparent in the plot of Figure 12.7*b* that the uppermost curve is approaching 0.407, as predicted.

The predicted value of $\sigma_{\ln Q}$ for the ultimate limit state (Figure 12.7) as $\theta \to \infty$ is obtained through the following reasoning: For the pile problem investigated here, the soil strength is assumed constant with depth, so that $b = 0$ and $a = u_{\text{top}} = 10$ kN/m and thus

$$U(z) = \exp\left\{\ln a - \tfrac{1}{2}\sigma_{\ln U}^2 + \sigma_{\ln U} G(z)\right\} \tag{12.19}$$

(see Eq. 12.9), where $\sigma_{\ln U}^2 = \ln(1 + v_U^2)$. Using Eqs. 12.15 and 12.16 for the ultimate limit state gives

$$\begin{aligned} U_{G_j} &= \exp\left\{\ln a - \tfrac{1}{2}\sigma_{\ln U}^2 + \sigma_{\ln U} G_j\right\} \\ &= \frac{a}{\sqrt{1 + v_U^2}} \exp\left\{\sigma_{\ln U} G_j\right\} \end{aligned} \tag{12.20}$$

The ultimate pile capacity is just the sum of these geometric average strengths plus the resistance provided at the base:

$$Q = U_b + \sum_{j=1}^{2n} U_{G_j}\ \Delta h \tag{12.21}$$

where U_b is the ultimate strength contributed by the soil under the pile base. As $\theta \to \infty$, the $G(z)$ random field becomes constant, $G(z) = G$, which means that each average G_j of $G(z)$ becomes the same; $G_1 = G_2 = \cdots = G$ and Eq. 12.21 simplifies to

$$Q = U_b + 2nU_{G_j}\ \Delta h \tag{12.22}$$

If the soil base strength was zero, so that $Q = 2nU_{G_j}\ \Delta h$, then it is a relatively easy matter to show that $\sigma_{\ln Q} = \sqrt{\ln\left(1 + v_U^2\right)}$. However, with the base strength present, we must first compute the mean and variance of Q; to do this, we note that the quantity $\exp\left\{\sigma_{\ln U} G_j\right\}$ appearing in Eq. 12.20 is lognormally distributed with mean $\sqrt{1 + v_U^2}$

and variance $(1 + v_U^2)v_U^2$. The mean and variance of Q are thus

$$\mu_Q = \mathrm{E}[U_b] + 2na\ \Delta h = 20 + 2(30)(10)(\tfrac{10}{60}) = 120\ \text{kN}$$

$$\sigma_Q^2 = \mathrm{Var}[U_b] + 4n^2(a^2 v_u^2)\ \Delta h$$

$$= [0.5(20)]^2 + 4(30)^2(10)^2(0.5)^2(\tfrac{10}{60})^2 = 2600\ \text{kN}^2$$

for our particular study, so that $\sigma_Q = 51$ kN. If Q is assumed to be at least approximately lognormally distributed, then for $v = 0.5$, we get

$$\sigma_{\ln Q} = \sqrt{\ln\left(1 + (\tfrac{51}{120})^2\right)} = 0.407$$

which is clearly what the $v = 0.5$ curve in the plot of Figure 12.7*a* is tending toward.

The mean shows somewhat of a maximum at correlation lengths of 1–10 m for $v > 0.1$. If the design load Q_{des} is less than the limit state load Q, then this maximum means that the nominal factor of safety, FS, reaches a maximum for values of θ around half the pile length. The reason for this maximum is currently being investigated more carefully. However, since the mean only changes slightly while the standard deviation increases significantly with increasing correlation length, the probability of design failure, that is, the probability that the actual pile capacity Q is less than the design capacity Q_{des}, will show a general increase with correlation length (assuming that $\ln Q_{\text{des}} < \mu_{\ln Q}$) to a limiting value when $\theta \to \infty$. In other words, from a reliability-based design point of view, the worst-case correlation length is when $\theta \to \infty$ and the soil acts as a single random variable.

This observation makes sense since variance reduction only occurs if independent random variables are averaged. That is, if the soil acts as a single random variable, then the variance remains unreduced and the failure probability is maximized. The implication of this worst case is that reliability-based pile design can conservatively ignore spatial variation in soil properties so long as end bearing is not a major component of the pile capacity (bearing capacity is significantly affected by spatial variability; see Chapter 11). For piles that depend mostly on skin friction, then, the reliability-based design at both serviceability and ultimate limit states can proceed using single random variables to represent the soil's elastic behavior (serviceability) and shear strength (ultimate).

CHAPTER 13

Slope Stability

13.1 INTRODUCTION

Slope stability analysis is a branch of geotechnical engineering that is highly amenable to probabilistic treatment and has received considerable attention in the literature. The earliest studies appeared in the 1970s (e.g., Matsuo and Kuroda, 1974; Alonso, 1976; Tang et al., 1976; Vanmarcke, 1977) and have continued steadily (e.g., D'Andrea and Sangrey, 1982; Li and Lumb, 1987; Mostyn and Li, 1993; Chowdhury and Tang, 1987; Whitman, 2000; Wolff, 1996; Lacasse, 1994; Christian et al., 1994; Christian, 1996; Lacasse and Nadim, 1996; Hassan and Wolff, 2000; Duncan, 2000; Szynakiewicz et al., 2002; El-Ramly et al., 2002; Griffiths and Fenton, 2004; Griffiths et al., 2006, 2007).

Two main observations can be made in relation to the existing body of work on this subject. First, the vast majority of probabilistic slope stability analyses, while using novel and sometimes quite sophisticated probabilistic methodologies, continue to use classical slope stability analysis techniques (e.g., Bishop, 1955) that have changed little in decades and were never intended for use with highly variable soil shear strength distributions. An obvious deficiency of the traditional slope stability approaches is that the shape of the failure surface (e.g., circular) is often fixed by the method; thus the failure mechanism is not allowed to "seek out" the most critical path through the soil. Second, while the importance of spatial correlation (or autocorrelation) and local averaging of statistical geotechnical properties has long been recognized by many investigators (e.g., Mostyn and Soo, 1992), it is still regularly omitted from many probabilistic slope stability analyses.

In recent years, the authors have been pursuing a more rigorous method of probabilistic geotechnical analysis (e.g., Griffiths and Fenton, 2000a; Paice, 1997), in which nonlinear finite-element methods (program 6.3 from Smith and Griffiths, 2004) are combined with random-field generation techniques. This method, called the random finite-element method (RFEM), fully accounts for spatial correlation and averaging and is also a powerful slope stability analysis tool that does not require a priori assumptions relating to the shape or location of the failure mechanism.

This chapter applies the RFEM to slope stability risk assessment. Although the authors have also considered c–ϕ slopes (Szynakiewicz et al., 2002), the next section considers a cohesive soil and investigates the general probabilistic nature of a slope. The final section develops a risk assessment model for slopes. Both sections employ the RFEM program called RSLOPE2D to perform the slope stability simulations. This program is available at `http://www.engmath.dal.ca/rfem`.

13.2 PROBABILISTIC SLOPE STABILITY ANALYSIS

In order to demonstrate some of the benefits of RFEM and to put it in context, this section investigates the probabilistic stability characteristics of a cohesive slope using both simple and more advanced methods. Initially, the slope is investigated using simple probabilistic concepts and classical slope stability techniques, followed by an investigation on the role of spatial correlation and local averaging. Finally, results are presented from a full RFEM approach. Where possible throughout this section, the probability of failure (p_f) is compared with the traditional factor of safety (F_S) that would be obtained from charts or classical limit equilibrium methods.

The slope under consideration, denoted the test problem, is shown in Figure 13.1 and consists of undrained clay, with shear strength parameters $\phi_u = 0$ and c_u. In this study, the slope inclination and dimensions given by β, H, and D and the saturated unit weight of the soil γ_{sat} are held constant, while the undrained shear strength c_u is assumed to be a random variable. In the interests of generality, the undrained shear strength will be expressed in dimensionless form c, where $c = c_u/(\gamma_{\text{sat}}H)$.

13.2.1 Probabilistic Description of Shear Strength

In this study, the shear strength c is assumed to be characterized statistically by a lognormal distribution defined by a mean μ_c and a standard deviation σ_c. Figure 13.2 shows the distribution of a lognormally distributed cohesion having mean $\mu_c = 1$ and standard deviation $\sigma_c = 0.5$. The probability of the strength dropping below a given value can be found from standard tables by first transforming the

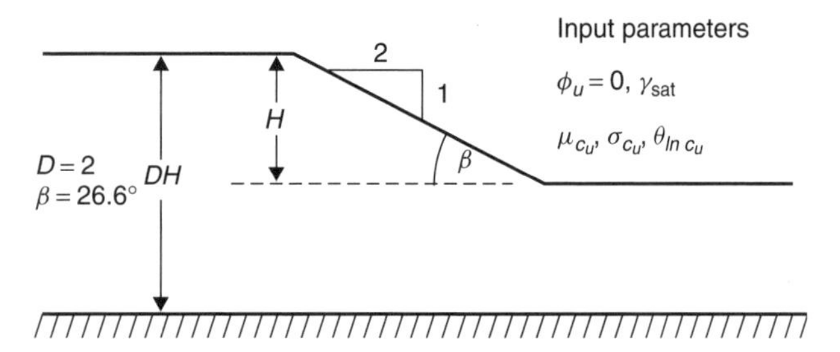

Figure 13.1 Cohesive slope test problem.

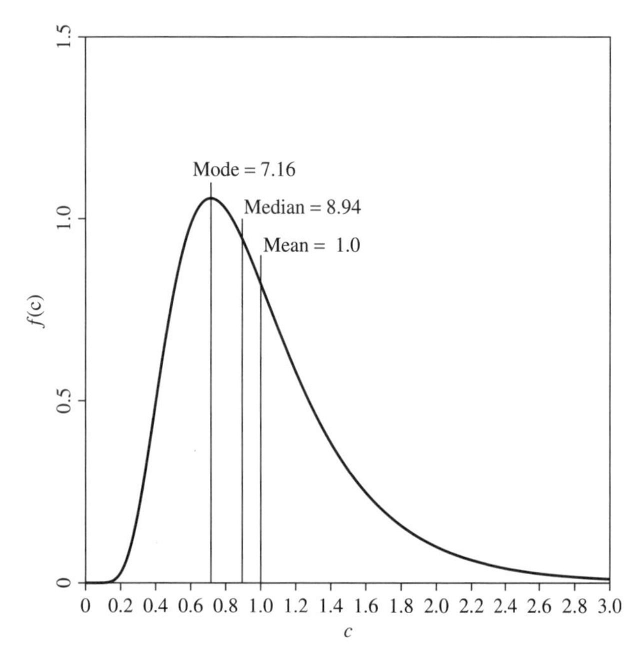

Figure 13.2 Lognormal distribution with mean 1 and standard deviation 0.5 ($v_c = 0.5$).

lognormal to the normal:

$$\begin{aligned} \mathrm{P}[c < a] &= \mathrm{P}[\ln c < \ln a] = \mathrm{P}\left[Z < \frac{\ln a - \mu_{\ln c}}{\sigma_{\ln c}}\right] \\ &= \Phi\left(\frac{\ln a - \mu_{\ln c}}{\sigma_{\ln c}}\right) \end{aligned} \tag{13.1}$$

as is discussed in Section 1.10.9. The lognormal parameters $\mu_{\ln c}$ and $\sigma_{\ln c}$ given μ_c and σ_c are obtained via the transformations

$$\sigma_{\ln c}^2 = \ln\left(1 + v_c^2\right) \tag{13.2a}$$

$$\mu_{\ln c} = \ln(\mu_c) - \tfrac{1}{2}\sigma_{\ln c}^2 \tag{13.2b}$$

in which the coefficient of variation of c, v_c, is defined as

$$v_c = \frac{\sigma_c}{\mu_c} \tag{13.3}$$

A third parameter, the spatial correlation length $\theta_{\ln c}$, will also be considered in this study. Since the actual undrained shear strength field is lognormally distributed, its logarithm yields an "underlying" normal distributed (or Gaussian) field. The spatial correlation length is measured with respect to this underlying field, that is, with respect to $\ln c$. In particular, the spatial correlation length ($\theta_{\ln c}$) describes the distance over which the spatially random values will tend to be significantly correlated in the underlying Gaussian field. Thus, a large value of $\theta_{\ln c}$ will imply a smoothly varying field, while a small value will imply a ragged field. The spatial correlation length can be estimated from a set of shear strength data taken over some spatial region simply by performing the statistical analyses on the log-data. In practice, however, $\theta_{\ln c}$ is not much different in magnitude from the correlation length in real space, and, for most purposes, θ_c and $\theta_{\ln c}$ are interchangeable given their inherent uncertainty in the first place. In the current study, the spatial correlation length has been nondimensionalized by dividing it by the height of the embankment H and will be expressed in the form

$$\Theta = \frac{\theta_{\ln c}}{H} \tag{13.4}$$

It has been suggested (see, e.g., Lee et al., 1983; Kulhawy et al., 1991) that typical v_c values for undrained shear strength lie in the range 0.1–0.5. The spatial correlation length, however, is less well documented and may well exhibit anisotropy, especially since soils are typically horizontally layered. While the advanced analysis tools used later in this study have the capability of modeling an anisotropic spatial correlation field, the spatial correlation, when considered, will be assumed to be isotropic. Anisotropic site-specific applications are left to the reader.

13.2.2 Preliminary Deterministic Study

To put the probabilistic analyses in context, an initial deterministic study has been performed assuming a *uniform soil*. By a uniform soil we mean that the soil properties are the same at all points through the soil mass. For the simple slope shown in Figure 13.1, the factor of safety F_S can readily be obtained from Taylor's (1937) charts or simple limit equilibrium methods to give Table 13.1.

Table 13.1 Factors of Safety for Uniform Soil

c	F_S
0.15	0.88
0.17	1.00
0.20	1.18
0.25	1.47
0.30	1.77

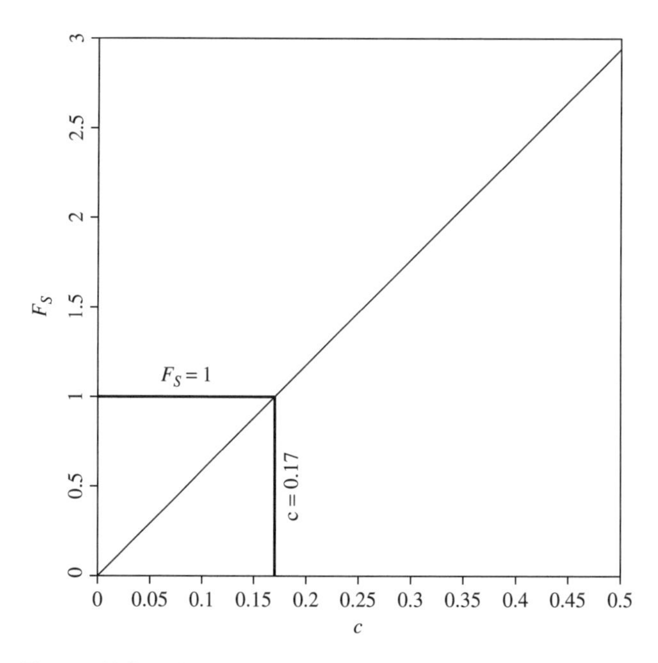

Figure 13.3 Linear relationship between F_S and c for uniform cohesive slope with slope angle $\beta = 26.57^\circ$ and depth ratio $D = 2$.

These results, shown plotted in Figure 13.3, indicate the linear relationship between c and F_S. The figure also shows that the test slope becomes unstable when the shear strength parameter falls below $c = 0.17$. The depth ratio mentioned in Figure 13.3 is defined in Figure 13.1.

13.2.3 Single-Random-Variable Approach

The first probabilistic analysis to be presented here investigates the influence of giving the shear strength c a lognormal probability density function similar to that shown in Figure 13.2, based on a mean μ_c and a standard deviation σ_c. The slope is assumed to be uniform, having the same value of c everywhere; however, the value of c is selected randomly from the lognormal distribution. Anticipating the random-field analyses to be described later in this section, this *single-random-variable* (SRV) approach implies a spatial correlation length of $\Theta = \infty$.

The probability of failure (p_f) in this case is simply equal to the probability that the shear strength parameter c will be less than 0.17. Quantitatively, this equals the area of the probability density function corresponding to $c \leq 0.17$.

For example, if $\mu_c = 0.25$ and $\sigma_c = 0.125$ ($v_c = 0.5$), Eqs. 1.176 state that the mean and standard deviation of the underlying *normal* distribution of the strength parameter are $\mu_{\ln c} = -1.498$ and $\sigma_{\ln c} = 0.472$.

The probability of failure is therefore given by

$$p_f = p[c < 0.17] = \Phi\left(\frac{\ln 0.17 - \mu_{\ln c}}{\sigma_{\ln c}}\right) = 0.281$$

where Φ is the cumulative standard normal distribution function (see Section 1.10.8).

This approach has been repeated for a range of μ_c and v_c values, for the slope under consideration, leading to Figure 13.4, which gives a direct relationship between the F_S and the probability of failure. It should be emphasized that the F_S in this plot is based on the value that would have been obtained if the slope had consisted of a uniform soil with a shear strength equal to the mean value μ_c from Figure 13.3. We shall refer to this as the *factor of safety based on the mean*.

From Figure 13.4, the probability of failure p_f clearly increases as the F_S decreases; however, it is also shown that for $F_S > 1$, the probability of failure increases as the v_c increases. The exception to this trend occurs when $F_S < 1$. As shown in Figure 13.4, the probability of failure in such cases is understandably high; however, the role of v_c is to have the opposite effect, with lower values of v_c tending to give the highest values of the probability of failure. This is explained by the "bunching up" of the shear strength distribution at low v_c rapidly excluding area to the right of the critical value of $c = 0.17$.

Figure 13.5 shows that the median (see Section 1.6.2), $\tilde{\mu}_c$ is the key to understanding how the probability of failure changes in this analysis. When $\tilde{\mu}_c < 0.17$, increasing v_c causes p_f to fall, whereas when $\tilde{\mu}_c > 0.17$, increasing v_c causes p_f to rise.

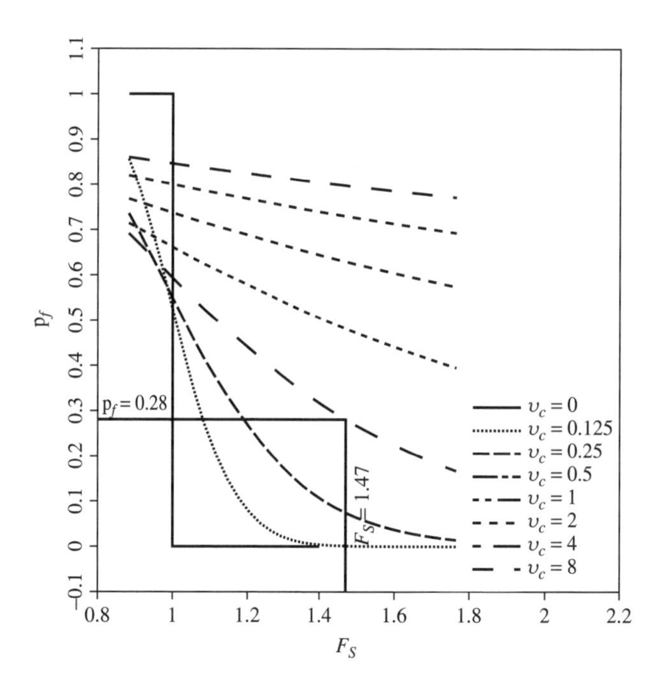

Figure 13.4 Probability of failure versus F_S (based on mean) in SRV approach.

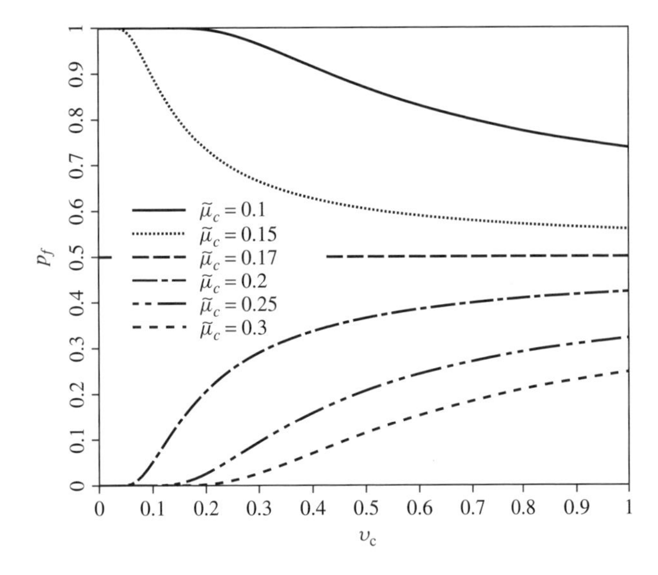

Figure 13.5 Probability of failure p_f versus coefficient of variation v_c for different medians of c, $\tilde{\mu}_c$.

While the SRV approach described in this section leads to simple calculations and useful qualitative comparisons between the probability of failure and the F_S, the quantitative value of the approach is more questionable. An important observation highlighted in Figure 13.4 is that a soil with a mean strength of $\mu_c = 0.25$ (implying $F_S = 1.47$) would give a probability of failure as high as $p_f = 0.28$ for a soil with $v_c = 0.5$. Practical experience indicates that slopes with an F_S as high as 1.47 rarely fail.

An implication of this result is that either the perfectly correlated SRV approach is entirely pessimistic in the prediction of the probability of failure, and/or it is unconservative to use the mean strength of a variable soil to estimate the F_S. Presented with a range of shear strengths at a given site, a geotechnical engineer would likely select a "pessimistic" or "lowest plausible" value for design, c_{des}, that would be lower than the mean. Assuming for the time being that the SRV approach is reasonable, Figure 13.6 shows the influence on the probability of failure of two strategies for factoring the mean strength μ_c prior to calculating the F_S for the test problem. In Figure 13.6*a*, a linear reduction in the design strength has been proposed using a strength reduction factor f_1, where

$$c_{\text{des}} = \mu_c(1 - f_1) \tag{13.5}$$

and in Figure 13.6*b*, the design strength has been reduced from the mean by a factor f_2 of the standard deviation, where

$$c_{\text{des}} = \mu_c - f_2\sigma_c \tag{13.6}$$

All the results shown in Figure 13.6 assume that after factorization, $c_{\text{des}} = 0.25$, implying an F_S of 1.47. The

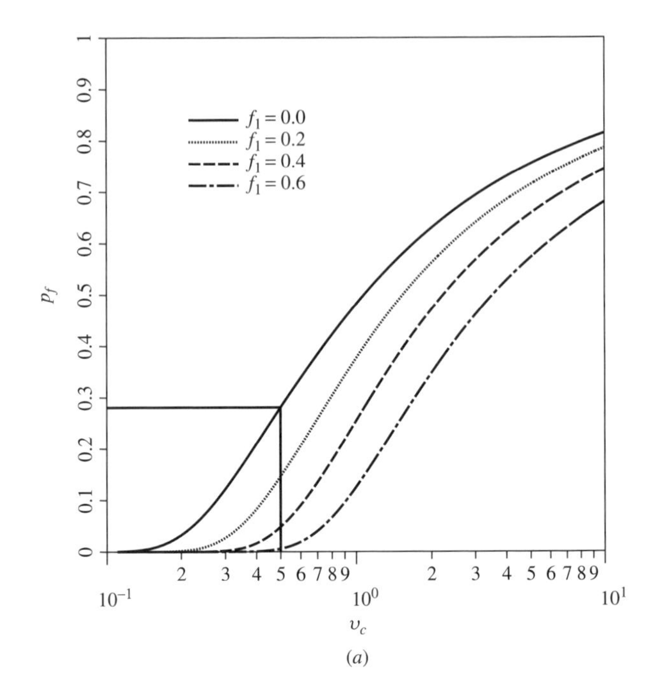

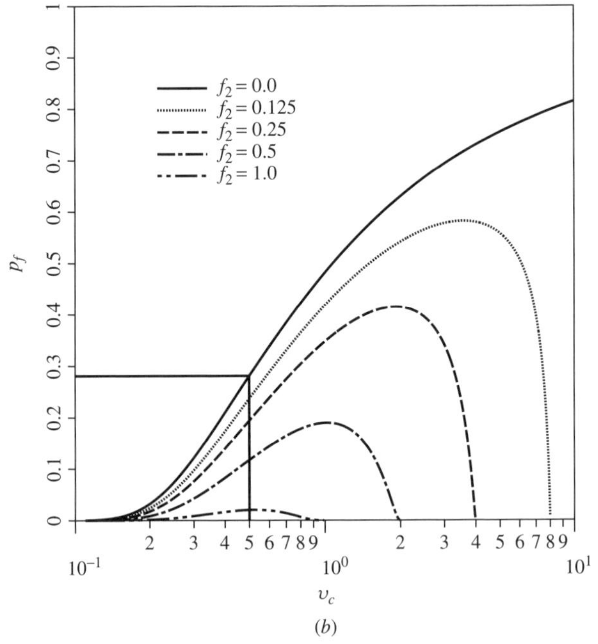

Figure 13.6 Influence of different design strength factoring strategies on probability of failure–F_S relationship: (*a*) linear factoring and (*b*) standard deviation factoring; all curves assume $F_S = 1.47$ (based on $c_{\text{des}} = 0.25$).

probability of failure of $p_f = 0.28$ with no strength factorization, $f_1 = f_2 = 0$, has also been highlighted for the case of $v_c = 0.5$. In both plots, an increase in the strength reduction factor reduces the probability of failure, which is to be expected; however, the nature of the two sets of reduction curves is quite different, especially for higher values of v_c. From the linear mean strength reduction (Eq. 13.5), $f_1 = 0.6$ would result in a probability of failure of about 0.6%. By comparison, a mean strength reduction of one standard deviation given by $f_2 = 1$ (Eq. 13.6) would result in a probability of failure of about 2%. Figure 13.6*a* shows a gradual reduction of the probability of failure as f_1 is increased; however, a quite different behavior is shown in Figure 13.6*b*, where standard deviation factoring results in a very rapid reduction in the probability of failure, especially for higher values of $v_c > 2$. This curious result is easily explained by the functional relationship between p_f and v_c, where the design strength can be written as

$$c_{\text{des}} = 0.25 = \mu_c - f_2\sigma_c = \mu_c(1 - f_2 v_c) \tag{13.7}$$

Hence as $v_c \to 1/f_2$, $\mu_c \to \infty$. With the mean strength so much greater than the critical value of 0.17, the probability of failure falls very rapidly toward zero.

13.2.4 Spatial Correlation

Implicit in the SRV approach described above is that the spatial correlation length is infinite. In other words only uniform soils are considered in which the single property assigned to the slope is taken at random from a lognormal distribution. A more realistic model would properly take account of smaller spatial correlation lengths in which the soil strength is allowed to vary spatially within the slope. The parameter that controls this behavior (at least under the simple spatial variability models considered here) is the spatial correlation length $\theta_{\ln c}$ as discussed previously. In this work, an exponentially decaying (Markovian) correlation function is used of the form

$$\rho(\tau) = e^{-2|\tau|/\theta_{\ln c}} \tag{13.8}$$

where $\rho(\tau)$ is the familiar correlation coefficient between two points in the soil mass which are separated by distance τ. A plot of this function is given in Figure 13.7 and indicates, for example, that the soil strength at two points separated by $\tau = \theta_{\ln c}$ ($\tau/\theta_{\ln c} = 1$) will have a correlation coefficient of $\rho = 0.135$. This correlation function is merely a way of representing the observation that soil samples taken close together are more likely to have similar properties than samples taken from far apart. There is also the issue of anisotropic spatial correlation in that soils are likely to have longer spatial correlation lengths in the horizontal direction than in the vertical, due to the depositional history. While the tools described in this section can take account of anisotropy, this refinement is left to the reader for site-specific refinements.

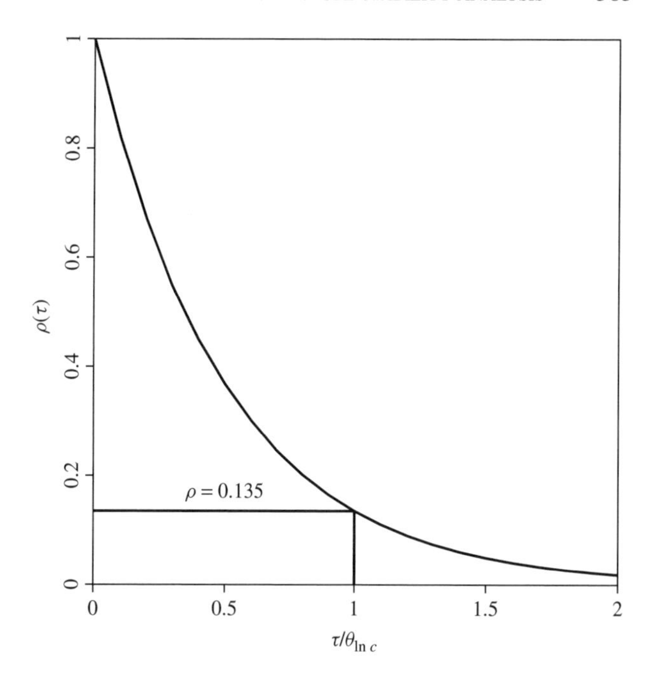

Figure 13.7 Markov correlation function.

13.2.5 Random Finite-Element Method

A powerful and general method of accounting for spatially random shear strength parameters and spatial correlation is the RFEM, which combines elasto-plastic finite-element analysis with random-field theory generated using the LAS method (Section 6.4.6). The methodology has been described in more detail in previous chapters, so only a brief description will be repeated here.

A typical finite-element mesh for the test problem considered in this section is shown in Figure 13.8. The majority of the elements are square; however, the elements adjacent to the slope are degenerated into triangles.

The code developed by the authors enables a random field of shear strength values to be generated and mapped onto the finite-element mesh, taking full account of element size in the local averaging process. In a random field, the value assigned to each cell (or finite element in this case) is itself a random variable; thus, the mesh of Figure 13.8, which

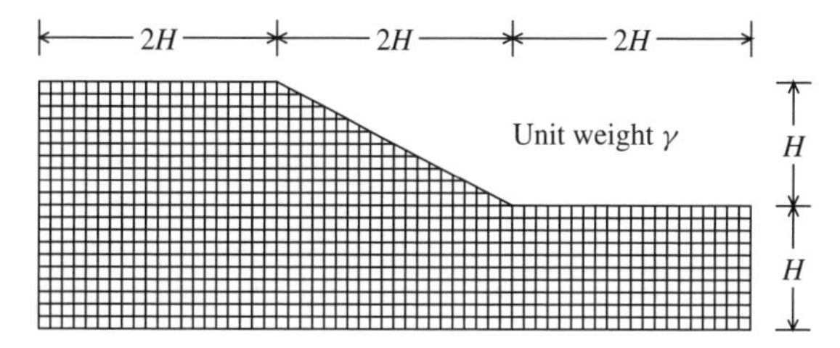

Figure 13.8 Mesh used for RFEM slope stability analysis.

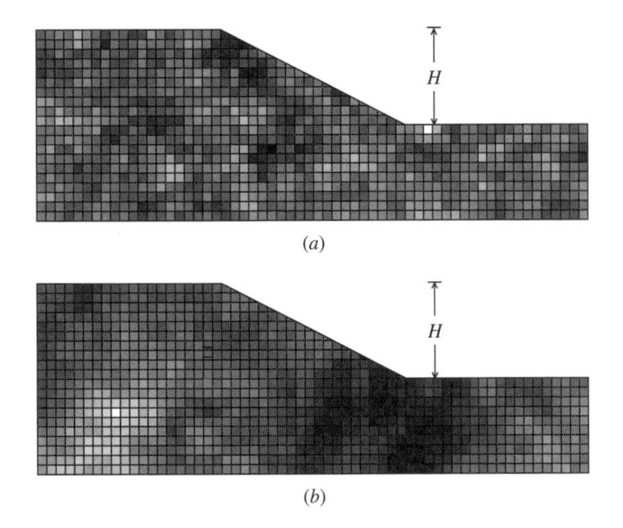

Figure 13.9 Influence of correlation length in RFEM analysis: (*a*) $\Theta = 0.2$; (*b*) $\Theta = 2.0$.

has 910 finite elements, contains 910 random variables. The random variables can be correlated to one another by controlling the spatial correlation length $\theta_{\ln c}$ as described previously; hence, the SRV approach discussed in the previous section, where the spatial correlation length is implicitly set to infinity, can now be viewed as a special case of a much more powerful analytical tool. Figures 13.9*a* and 13.9*b* show typical meshes corresponding to different spatial correlation lengths. Figure 13.9*a* shows a relatively low spatial correlation length of $\Theta = 0.2$, and Figure 13.9*b* shows a relatively high spatial correlation length of $\Theta = 2$. Light regions depict "weak" soil. It should be emphasized that both these shear strength distributions come from the same lognormal distribution, and it is only the spatial correlation length that is different.

In brief, the analyses involve the application of gravity loading and the monitoring of stresses at all the Gauss points. The slope stability analyses use an elastic-perfectly plastic stress–strain law with a Tresca failure criterion which is appropriate for "undrained clays." If the Tresca criterion is violated, the program attempts to redistribute excess stresses to neighboring elements that still have reserves of strength. This is an iterative process which continues until the Tresca criterion and global equilibrium are satisfied at all points within the mesh under quite strict tolerances.

Plastic stress redistribution is accomplished using a viscoplastic algorithm with 8-node quadrilateral elements and reduced integration in both the stiffness and stress redistribution parts of the algorithm. The theoretical basis of the method is described more fully in Chapter 6 of the text by Smith and Griffiths (2004), and for a detailed discussion of the method applied to slope stability analysis, the reader is referred to Griffiths and Lane (1999).

For a given set of input shear strength parameters (mean, standard deviation, and spatial correlation length), Monte Carlo simulations are performed. This means that the slope stability analysis is repeated many times until the statistics of the output quantities of interest become stable. Each "realization" of the Monte Carlo process differs in the locations at which the strong and weak zones are situated. For example, in one realization, weak soil may be situated in the locations where a critical failure mechanism develops causing the slope to fail, whereas in another, strong soil in those locations means that the slope remains stable.

In this study, it was determined that 1000 realizations of the Monte Carlo process for each parametric group was sufficient to give reliable and reproducible estimates of the probability of failure, which was simply defined as the proportion of the 1000 Monte Carlo slope stability analyses that failed.

In this study,"failure" was said to have occurred if, for any given realization, the algorithm was unable to converge within 500 iterations. While the choice of 500 as the iteration ceiling is subjective, Figure 13.10 confirms, for the case of $\mu_c = 0.25$ and $\Theta = 1$, that the probability of failure defined this way, is stable for iteration ceilings greater than about 200.

13.2.6 Local Averaging

The input parameters relating to the mean, standard deviation, and spatial correlation length of the undrained strength

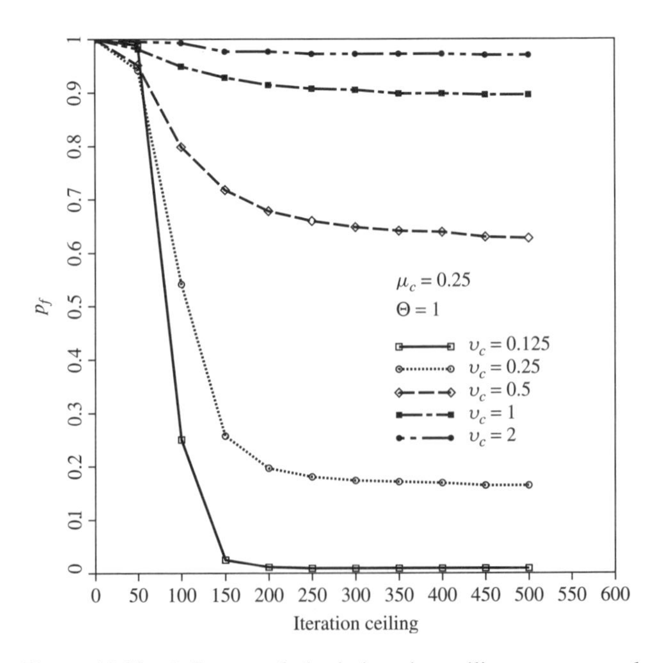

Figure 13.10 Influence of plastic iteration ceiling on computed probability of failure.

are assumed to be defined at the point level. While statistics at this resolution are obviously impossible to measure in practice, they represent a fundamental baseline of the inherent soil variability which can be corrected through local averaging to take account of the sample size.

In the context of the RFEM approach, each element is assigned a constant property at each realization of the Monte Carlo process. The assigned property represents an average over the area of each finite element used to discretize the slope. If the point distribution is normal, local arithmetic averaging is used which results in a reduced variance but the mean is unaffected. In a lognormal distribution, however, local geometric averaging is used (see Section 4.4.2), and both the mean and the standard deviation are reduced by this form of averaging as is appropriate for situations in which low-strength regions dominate the effective strength. The reduction in both the mean and standard deviation is because from Eqs. 1.175a and 1.175b, the mean of a lognormally random variable depends on both the mean *and* the variance of the underlying normal log-variable. Thus, the coarser the discretization of the slope stability problem and the larger the elements, the greater the influence of local averaging in the form of a reduced mean and standard deviation. These adjustments to the points statistics are fully accounted for in the RFEM and are implemented before the elasto-plastic finite-element slope stability analysis takes place.

13.2.7 Variance Reduction over Square Finite Element

In this section, the algorithm used to compute the locally averaged statistics applied to the mesh is described. A lognormal distribution of a random variable c, with point statistics given by a mean μ_c, a standard deviation σ_c, and spatial correlation length $\theta_{\ln c}$ is to be mapped onto a mesh of square finite elements. Each element will be assigned a single value of the undrained strength parameter.

The locally averaged statistics over the elements will be referred to here as the "area" statistics with the subscript A. Thus, with reference to the underlying normal distribution of $\ln c$, the mean, which is unaffected by local averaging, is given by $\mu_{\ln c_A}$, and the standard deviation, which is affected by local averaging is given by $\sigma_{\ln c_A}$.

The variance reduction factor due to local averaging γ is defined as (see also Section 3.4)

$$\gamma(A) = \left(\frac{\sigma_{\ln c_A}}{\sigma_{\ln c}}\right)^2 \tag{13.9}$$

and is a function of the element size, A, and the correlation function from Eq. 13.8, repeated here explicitly for the two-dimensional isotropic case (i.e., the correlation length is assumed the same in any direction for simplicity):

$$\rho(\tau_1, \tau_2) = \exp\left\{-\frac{2}{\theta_{\ln c}}\sqrt{\tau_1^2 + \tau_2^2}\right\} \tag{13.10}$$

where τ_1 is the difference between the x_1 coordinates of any two points in the random field, and τ_2 is the difference between the x_2 coordinates. We assume that x_1 is measured in the horizontal direction and x_2 is measured in the vertical direction.

For a square finite element of side length $\alpha\theta_{\ln c}$ as shown in Figure 13.11, so that $A = \alpha\theta_{\ln c} \times \alpha\theta_{\ln c}$, it can be shown (Vanmarcke, 1984) that for an isotropic spatial correlation field, the variance reduction factor is given by

$$\gamma(A) = \frac{4}{(\alpha\theta_{\ln c})^4}\int_0^{\alpha\theta_{\ln c}}\int_0^{\alpha\theta_{\ln c}} (\alpha\theta_{\ln c} - x_1)(\alpha\theta_{\ln c} - x_2) \times \exp\left\{-\frac{2}{\theta_{\ln c}}\sqrt{x^2 + y^2}\right\} dx_1\, dx_2 \tag{13.11}$$

Numerical integration of this function leads to the variance reduction values given in Table 13.2 and shown plotted in Figure 13.11.

Figure 13.11 indicates that elements that are small relative to the correlation length ($\alpha \to 0$) lead to very little variance reduction [$\gamma(A) \to 1$], whereas elements that are large relative to the correlation length can lead to very significant variance reduction [$\gamma(A) \to 0$].

The statistics of the underlying log-field, including local arithmetic averaging, are therefore given by

$$\sigma_{\ln c_A} = \sigma_{\ln c}\sqrt{\gamma(A)} \tag{13.12a}$$

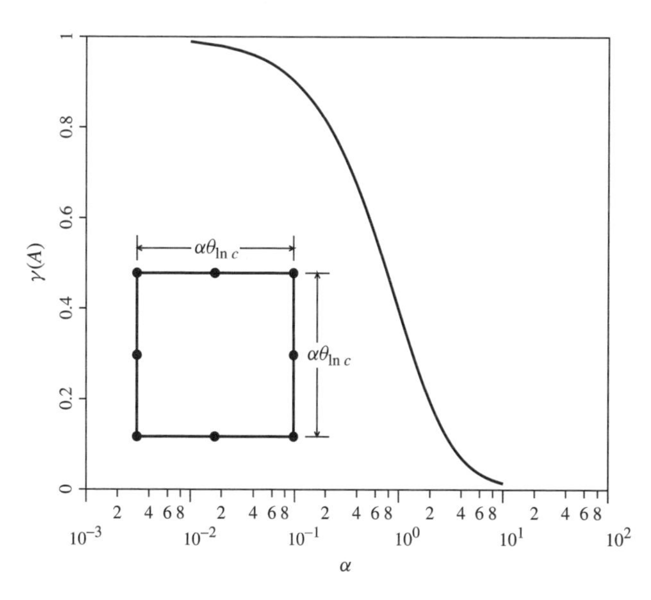

Figure 13.11 Variance reduction when arithmetically averaging over square element of side length $\alpha\theta_{\ln c}$ with Markov correlation function ($A = \alpha\theta_{\ln c} \times \alpha\theta_{\ln c}$).

Table 13.2 Variance Reduction due to Arithmetic Averaging over Square Element

α	$\gamma(A)$
0.01	0.9896
0.10	0.9021
1.00	0.3965
10.00	0.0138

and

$$\mu_{\ln c_A} = \mu_{\ln c} \tag{13.12b}$$

which leads to the following statistics of the lognormal field, including local geometric averaging, that is actually mapped onto the finite-element mesh (from Eqs. 1.175a and 1.175b)

$$\mu_{c_A} = \exp\left\{\mu_{\ln c} + \tfrac{1}{2}\sigma^2_{\ln c}\gamma(A)\right\} \tag{13.13a}$$

$$\sigma_{c_A} = \mu_{c_A}\sqrt{\exp\{\sigma^2_{\ln c}\gamma(A)\} - 1} \tag{13.13b}$$

from which it is easy to see that local geometric averaging affects both the mean and the standard deviation. Recall also that arithmetic averaging of $\ln c$ corresponds to geometric averaging of c (see Section 4.4.2 for more details).

It is instructive to consider the range of locally averaged statistics since this helps to explain the influence of the spatial correlation length $\Theta(= \theta_{\ln c}/H)$ on the probability of failure in the RFEM slope analyses described in the next section.

Expressing the mean and the coefficient of variation of the locally averaged variable as a proportion of the point values of these quantities leads to Figures 13.12*a* and 13.12*b*, respectively. In both cases, there is virtually no reduction due to local averaging for elements that are small relative to the spatial correlation length ($\alpha \to 0$). This is to be expected since the elements are able to model the point field quite accurately. For larger elements relative to the spatial correlation length, however, Figure 13.12*a* indicates that the average of the locally averaged field tends to a constant equal to the median, and Figure 13.12*b* indicates that the coefficient of variation of the locally averaged field tends to zero.

From Eqs. 13.12 and 13.13, the expression plotted in Figure 13.12*a* for the mean can be written as

$$\frac{\mu_{c_A}}{\mu_c} = \frac{1}{(1+v_c^2)^{[1-\gamma(A)]/2}} \tag{13.14}$$

which states that when $\gamma(A) \to 0$, $\mu_{c_A}/\mu_c \to 1/\sqrt{1+V_c^2}$, thus $\mu_{c_A} \to e^{\mu_{\ln c}} = \tilde{\mu}_c$, which is the median of c. Similarly, the expression plotted in Figure 13.12*b* for the coefficient of variation of the locally geometrically averaged variable can be written as

$$\frac{v_{c_A}}{v_c} = \frac{\sqrt{(1+v_c^2)^{\gamma(A)} - 1}}{v_c} \tag{13.15}$$

which states that when $\gamma(A) \to 0$, $v_{c_A}/v_c \to 0$, thus $v_{c_A} \to 0$.

Further examination of Eqs. 13.14 and 13.15 shows that for all values of $\gamma(A)$ the median of the geometric average equals the median of c:

$$\tilde{\mu}_{c_A} = \tilde{\mu}_c \tag{13.16}$$

Hence it can be concluded that:

1. Local geometric averaging reduces both the mean and the variance of a lognormal point distribution.
2. Local geometric averaging preserves the median of the point distribution.

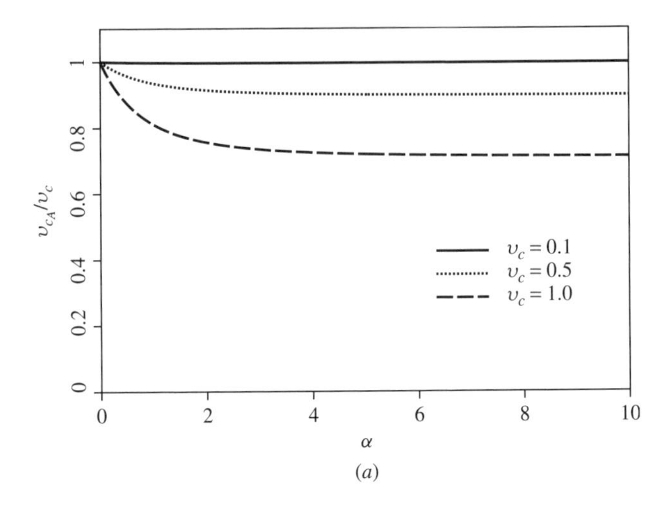

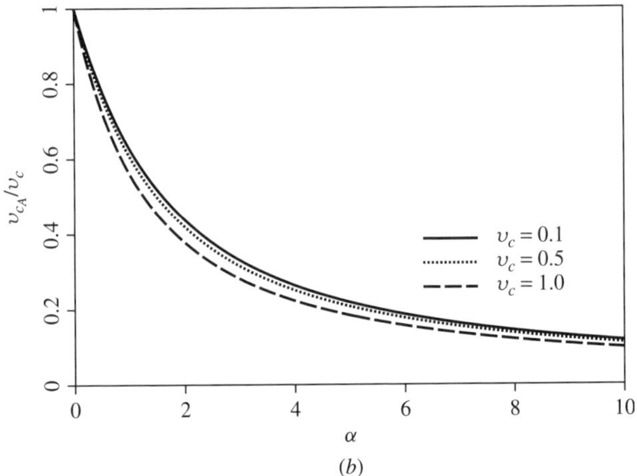

Figure 13.12 Influence of element size, expressed in the form of size parameter α, on statistics of local averages: influence on the (*a*) mean and (*b*) coefficient of variation.

3. In the limit as $A \to \infty$ and/or $\Theta \to 0$, local geometric averaging removes all variance, and the mean tends to the median.

13.2.8 Locally Averaged SRV Approach

In this section the probability of failure is reworked with the SRV approach using properties derived from local averaging over an individual finite element, termed finite-element locally averaged properties throughout the rest of this section. With reference to the mesh shown in Figure 13.8, the square elements have a side length of $0.1H$, thus $\Theta = 0.1/\alpha$. Figure 13.13 shows the probability of failure p_f as a function of Θ for a range of input point coefficients of variation, with the point mean fixed at $\mu_c = 0.25$. The probability of failure is defined, as before, by $p[c < 0.17]$, but this time the calculation is based on the finite-element locally averaged properties, μ_{c_A} and σ_{c_A} from Eqs. 13.13. The Figure clearly shows two tails to the results, with $p_f \to 1$ as $\Theta \to 0$ for all $v_c > 1.0783$, and $p_f \to 0$ as $\Theta \to 0$ for all $v_c < 1.0783$. The horizontal line at $p_f = 0.5$ is given by $v_c = 1.0783$, which is the special value of the coefficient of variation that causes the median of c to have value $\tilde{\mu}_c = 0.17$. Recalling Table 13.1, this is the critical value of c that would give $F_S = 1$ in the test slope. Higher values of v_c lead to $\tilde{\mu}_c < 0.17$ and a tendency for $p_f \to 1$ as $\Theta \to 0$. Conversely, lower values of v_c lead to $\tilde{\mu}_c > 0.17$ and a tendency for $p_f \to 0$. Figure 13.14 shows the same data plotted the other way round with v_c along the abscissa. This Figure clearly shows the full influence of spatial correlation in the range $0 \leq \Theta < \infty$. All the curves cross over at the critical value of $v_c = 1.0783$, and it is of interest to note the step function corresponding to $\Theta = 0$ when p_f changes suddenly from zero to unity.

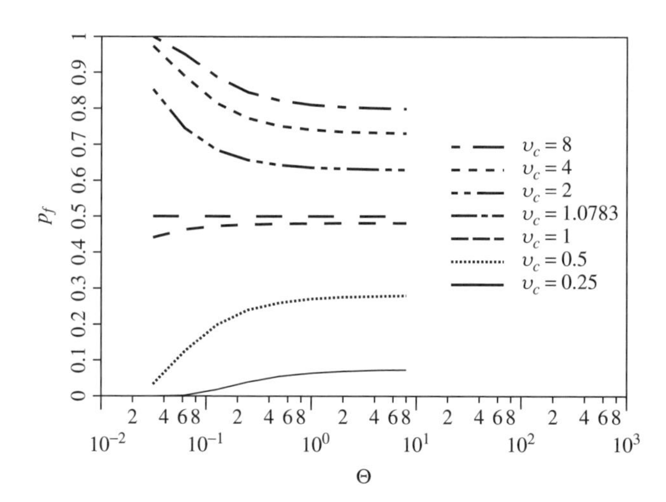

Figure 13.13 Probability of failure versus spatial correlation length based on finite-element locally geometrically averaged properties; the mean is fixed at $\mu_c = 0.25$.

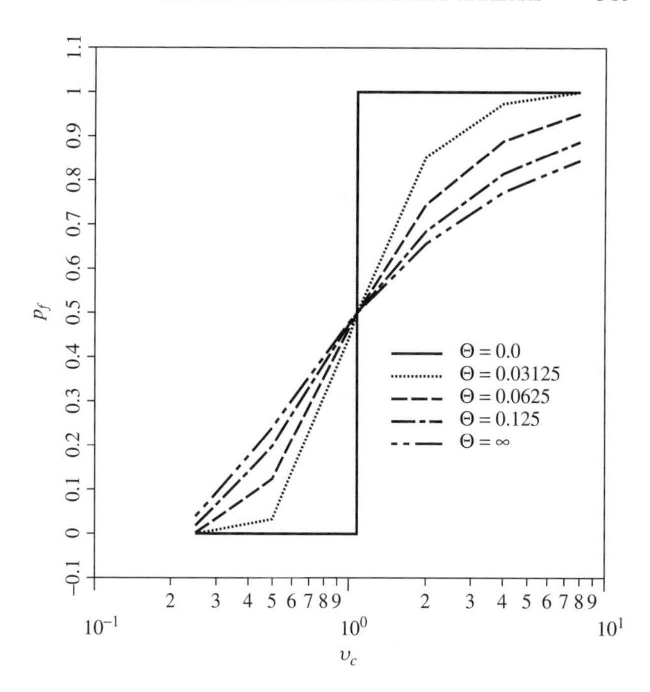

Figure 13.14 Probability of failure versus coefficient of variation based on finite-element locally geometrically averaged properties; the mean is fixed at $\mu_c = 0.25$.

It should be emphasized that the results presented in this section involved no finite-element analysis and were based solely on an SRV approach with statistical properties based on finite-element locally geometrically averaged properties based on a typical finite element of the mesh in Figure 13.8.

13.2.9 Results of RFEM Analyses

In this section, the results of full nonlinear RFEM analyses with Monte Carlo simulations are described, based on a range of parametric variations of μ_c, v_c, and Θ.

In the elasto-plastic RFEM approach, the failure mechanism is free to "seek out" the weakest path through the soil. Figure 13.15 shows two typical random field realizations and the associated failure mechanisms for slopes with $\Theta = 0.5$ and $\Theta = 2$. The convoluted nature of the failure mechanisms, especially when $\Theta = 0.5$, would defy analysis by conventional slope stability analysis tools. While the mechanism is attracted to the weaker zones within the slope, it will inevitably pass through elements assigned many different strength values. This weakest path determination, and the strength averaging that goes with it, occurs quite naturally in the finite-element slope stability method and represents a very significant improvement over traditional limit equilibrium approaches to probabilistic slope

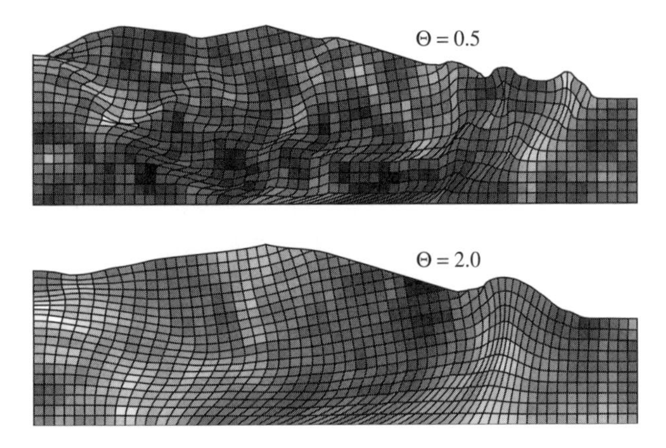

Figure 13.15 Typical random-field realizations and deformed mesh at slope failure for two different spatial correlation lengths. Light zones are weaker.

stability, in which local averaging, if included at all, has to be computed over a failure mechanism that is preset by the particular analysis method (e.g., a circular failure mechanism when using Bishop's method).

Fixing the point mean strength at $\mu_c = 0.25$, Figures 13.16 and 13.17 show the effect of the spatial correlation length Θ and the coefficient of variation v_c on the probability of failure for the test problem. Figure 13.16 clearly indicates two branches, with the probability of failure tending to unity or zero for higher and lower values of v_c, respectively. This behavior is qualitatively similar to that observed in Figure 13.13, in which an SRV approach was used to predict the probability of failure based solely on finite-element locally averaged properties. Figure 13.17 shows the same results as Figure 13.16, but plotted the other way round with the coefficient of variation along the abscissa. Figure 13.17 also demonstrates that when Θ becomes large, corresponding approximately to an SRV approach with no local averaging, the probability of failure is overestimated (conservative) when the coefficient of variation is relatively small and underestimated (unconservative) when the coefficient of variation is relatively high. Figure 13.17 also demonstrates that the SRV approach described earlier in the section, which gave $p_f = 0.28$ corresponding to $\mu_c = 0.25$ and $v_c = 0.5$ with no local averaging, is indeed pessimistic. The RFEM results show that the inclusion of spatial correlation and local averaging in this case will always lead to a smaller probability of failure.

Comparison of Figures 13.13 and 13.14 with Figures 13.16 and 13.17 highlights the influence of the finite-element approach to slope stability, where the failure mechanism is free to locate itself optimally within the mesh. From Figures 13.14 and 13.17, it is clear that the "weakest path" concept made possible by the RFEM approach

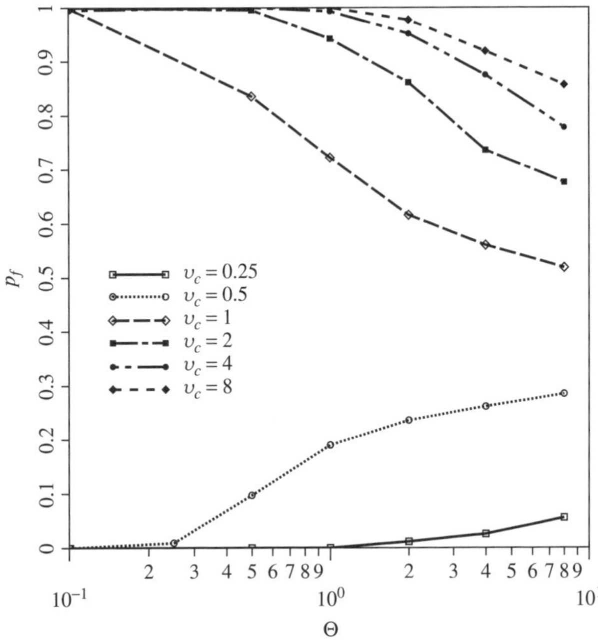

Figure 13.16 Probability of failure versus spatial correlation length from RFEM; the mean is fixed at $\mu_c = 0.25$.

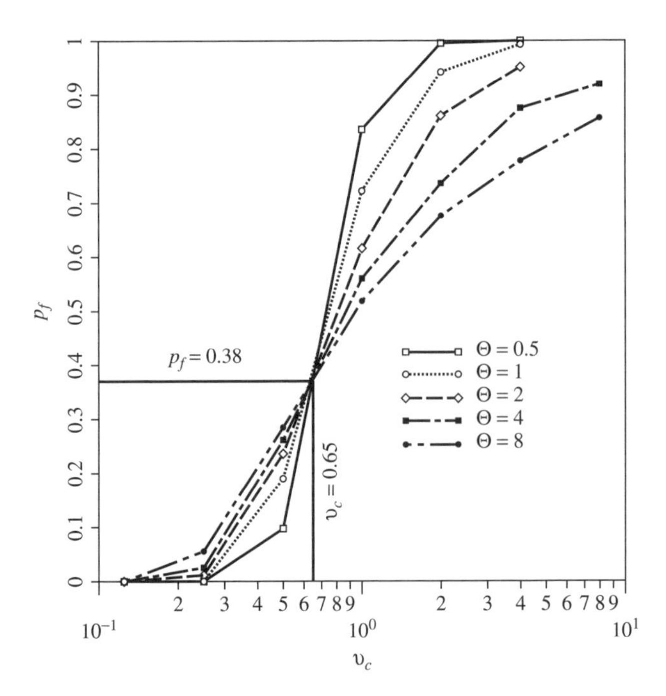

Figure 13.17 Probability of failure versus coefficient of variation from RFEM; the mean is fixed at $\mu_c = 0.25$.

has resulted in the crossover point falling to lower values of both v_c and p_f. With only finite-element local averaging, the crossover occurred at $v_c = 1.0783$, whereas by the RFEM it occurred at $v_c \approx 0.65$. In terms of the probability of failure with only finite-element local averaging, the crossover occurred at $p_f = 0.5$, whereas by the RFEM it occurred at $p_f \approx 0.38$. The RFEM solutions show that the SRV approach becomes unconservative over a wider range of v_c values than would be indicated by finite- element local averaging alone.

Figure 13.18 gives a direct comparison between Figures 13.13 and 13.16, indicating clearly that for higher values of v_c, RFEM always gives a higher probability of failure than when using finite- element local averaging alone. This is caused by the weaker elements in the distribution dominating the strength of the slope and the failure mechanism "seeking out" the weakest path through the soil.

At lower values of v_c, the locally averaged results tend to overestimate the probability of failure and give conservative results compared with RFEM. In this case the stronger elements of the slope are dominating the solution, and the higher median combined with the "bunching up" of the locally averaged solution at low values of Θ means that potential failure mechanisms cannot readily find a weak path through the soil.

In all cases, as Θ increases, the RFEM and the locally averaged solutions converge on the SRV solution corresponding to $\Theta = \infty$ with no local averaging. The $p_f = 0.28$ value, corresponding to $v_c = 0.5$, and discussed earlier in the section, is also indicated in Figure 13.18.

All of the above results and discussion in this section so far were applied to the test slope from Figure 13.1 with the mean strength fixed at $\mu_c = 0.25$ corresponding to a factor of safety (based on the mean) of 1.47. In the next set of results μ_c is varied while v_c is held constant at 0.5. Figure 13.19 shows the relationship between F_S (based on the mean) and p_f assuming finite-element local averaging only, and Figure 13.20 shows the same relationship as computed using RFEM.

Figure 13.19, based on finite-element local averaging only, shows the full range of behavior for $0 \leq \Theta < \infty$. The figure shows that Θ only starts to have a significant influence on the F_S vs. p_f relationship when the correlation length becomes significantly smaller than the slope height ($\Theta << 1$). The step function in which p_f jumps from zero to unity occurs when $\Theta = 0$ and corresponds to a local average having zero variance. In this limiting case, the local average of the soil is deterministic, yielding a constant strength everywhere in the slope. With $v_c = 0.5$, the critical value of mean shear strength that would give $\mu_{c_A} = \tilde{\mu}_c = 0.17$ is easily shown by Eq. 13.14 to be $\mu_c = 0.19$, which corresponds to an $F_S = 1.12$. For higher values of Θ,

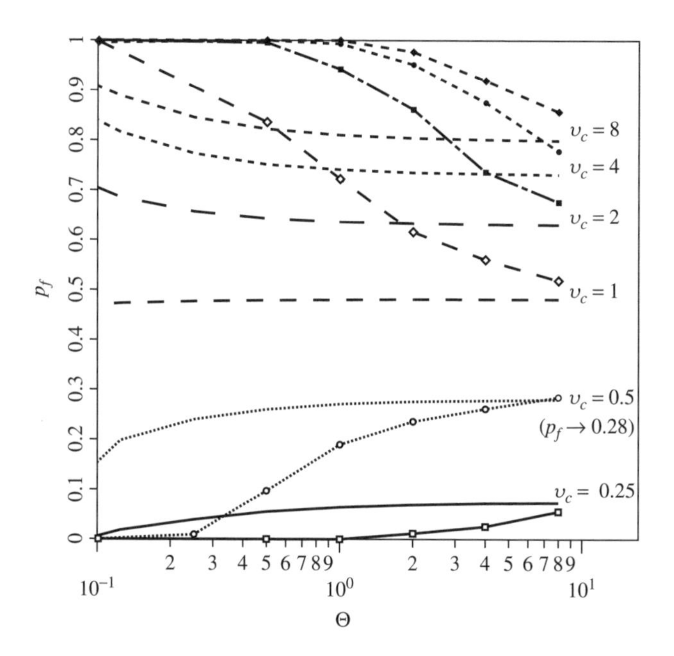

Figure 13.18 Comparison of the probabilities of failure predicted by RFEM and by finite-element local geometric averaging alone; the curves which include points come from the random finite-element method; the mean is fixed at $\mu_c = 0.25$.

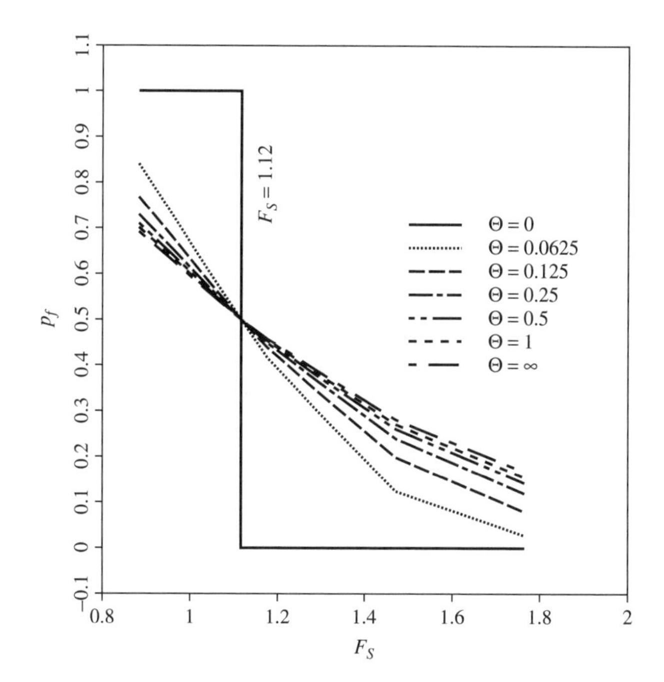

Figure 13.19 Probability of failure versus F_S (based on mean) using finite-element local geometric averaging only for test slope; the coefficient of variation is fixed at $v_c = 0.5$.

the relationship between F_S and p_f is quite bunched up and generally insensitive to Θ. For example, there is little difference between the curves corresponding to $\Theta = \infty$ and $\Theta = 0.5$. It should also be observed from Figure 13.19 that for $F_S > 1.12$, failure to account for local averaging by assuming $\Theta = \infty$ is conservative, in that the predicted p_f is higher than it should be. When $F_S < 1.12$, however, failure to account for local averaging is unconservative.

Figure 13.20 gives the same relationships as computed using RFEM. By comparison with Figure 13.19, the RFEM results are more spread out, implying that the probability of failure is more sensitive to the spatial correlation length Θ. Of greater significance is that the crossover point has again shifted by RFEM as it seeks out the weakest path through the slope. In Figure 13.20, the crossover occurs at $F_S \approx 1.37$, which is significantly higher and of greater practical significance than the crossover point of $F_S \approx 1.12$ by finite-element local geometric averaging alone. The theoretical line corresponding to $\Theta = \infty$ is also shown in this plot. From a practical viewpoint, the RFEM analysis indicates that failure to properly account for local averaging is unconservative over a wider range of factors of safety than would be the case by finite-element local averaging alone. To further highlight this difference, the particular results from Figures 13.19 and 13.20 corresponding to $\Theta = 0.5$ (spatial correlation length equal to half the embankment height) have been replotted in Figure 13.21.

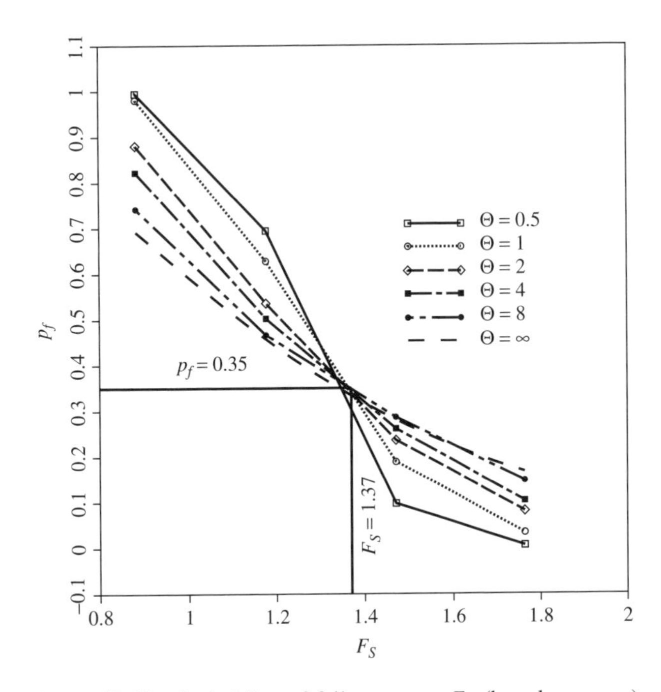

Figure 13.20 Probability of failure versus F_S (based on mean) using RFEM for test slope; the coefficient of variation is fixed at $v_c = 0.5$.

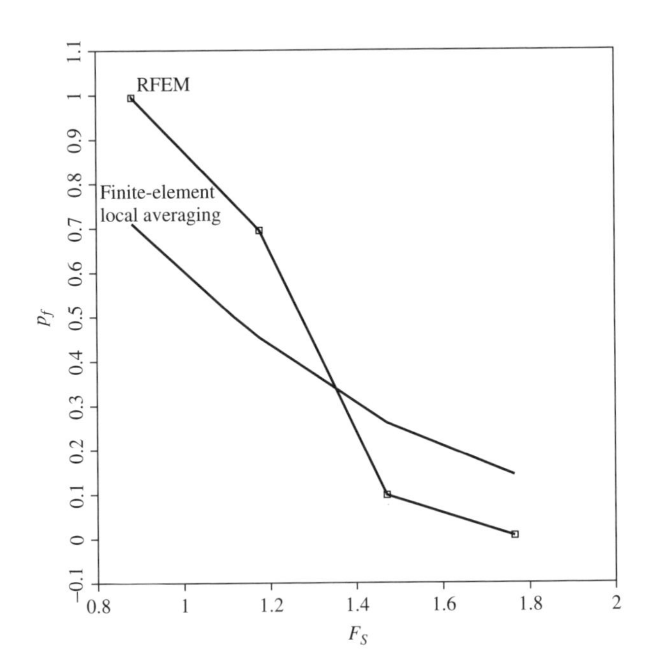

Figure 13.21 Comparison of probabilities of failure versus F_S (based on mean) using finite-element local geometric averaging alone with RFEM for test slope; $v_c = 0.5$ and $\Theta_{\ln c} = 0.5$.

13.2.10 Summary

The section has investigated the probability of failure of a cohesive slope using both simple and more advanced probabilistic analysis tools. The simple approach treated the strength of the entire slope as a single random variable, ignoring spatial correlation and local averaging. In the simple studies, the probability of failure was estimated as the probability that the shear strength would fall below a critical value based on a lognormal probability density function. These results led to a discussion on the appropriate choice of a design shear strength value suitable for deterministic analysis. Two factorization methods were proposed that were able to bring the probability of failure and the FS more into line with practical experience.

The second half of the section implemented the RFEM on the same test problem. The nonlinear elasto-plastic analyses with Monte Carlo simulation were able to take full account of spatial correlation and local averaging and observe their impact on the probability of failure using a parametric approach. The elasto-plastic finite-element slope stability method makes no a priori assumptions about the shape or location of the critical failure mechanism and, therefore, offers very significant benefits over traditional limit equilibrium methods in the analysis of spatially variable soils. In the elasto-plastic RFEM, the failure mechanism is free to seek out the weakest path through the soil, and it has been shown that this generality can lead to higher probabilities of failure than could be explained by finite-element local averaging alone.

In summary, simplified probabilistic analysis in which spatial variability is ignored by assuming perfect correlation can lead to unconservative estimates of the probability of failure. This effect is most pronounced at relatively low factors of safety (Figure 13.20) or when the coefficient of variation of the soil strength is relatively high (Figure 13.18).

13.3 SLOPE STABILITY RELIABILITY MODEL

The failure prediction of a soil slope has been a longstanding geotechnical problem and one which has attracted a wide variety of solutions. Traditional approaches to the problem generally involve assuming that the soil slope is homogeneous (spatially constant) or possibly layered, and techniques such as Taylor's (1937) stability coefficients for frictionless soils, the method of slices, and other more general methods involving arbitrary failure surfaces have been developed over the years. The main drawback to these methods is that they are not able to easily find the critical failure surface in the event that the soil properties are spatially varying.

In the realistic case where the soil properties vary randomly in space, the slope stability problem is best captured via a nonlinear finite-element model which has the distinct advantage of allowing the failure surface to seek out the path of least resistance, as pointed out in the previous section. In this section such a model is employed, which, when combined with a random-field simulator, allows the realistic probabilistic evaluation of slope stability (Fenton and Griffiths, 2005c). This work builds on the previous section, which looked in some detail at the probability of failure of a single slope geometry. Two slope geometries are considered in this section, one shallower with a 2 : 1 gradient and the other steeper with a 1 : 1 gradient. Both slopes are assumed to be composed of undrained clay, with $\phi_u = 0$, of height H with the slope resting on a foundation layer, also of depth H. The finite-element mesh for the 2 : 1 gradient slope is shown in Figure 13.22. The 1 : 1 slope is similar, except that the horizontal length of the slope is H rather than $2H$.

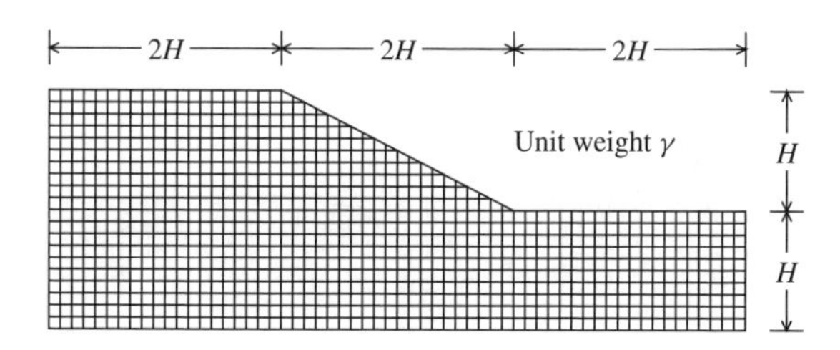

Figure 13.22 Mesh used for stability analysis of 2 : 1 gradient slope.

The soil is represented by a random spatially varying undrained cohesion field $c_u(\mathbf{x})$ which is assumed to be lognormally distributed, where $\mathbf{x}$ is the spatial position. The cohesion has mean μ_{c_u} and standard deviation σ_{c_u} and is assumed to have an exponentially decaying (Markovian) correlation structure:

$$\rho_{\ln c_u}(\tau) = e^{-2|\tau|/\theta_{\ln c_u}} \tag{13.17}$$

where τ is the distance between two points in the field. Note that the correlation structure has been assumed isotropic in this study. The use of an anisotropic correlation is straightforward, within the framework developed here, but is considered a site-specific extension. In this section it is desired to investigate the stochastic behavior of slope stability for the simpler isotropic case, leaving the effect of anisotropy for the reader.

The correlation function has a single parameter, $\theta_{\ln c_u}$, the correlation length. Because c_u is assumed to be lognormally distributed, its logarithm, $\ln c_u$, is normally distributed. In this study, the correlation function is measured relative to the underlying normally distributed field. Thus, $\rho_{\ln c_u}(\tau)$ gives the correlation coefficient between $\ln c_u(\mathbf{x})$ and $\ln c_u(\mathbf{x}')$ at two points in the field separated by the distance $\tau = |\mathbf{x} - \mathbf{x}'|$. In practice, the parameter $\theta_{\ln c_u}$ can be estimated from spatially distributed c_u samples by using the logarithm of the samples rather than the raw data themselves. If the actual correlation between points in the c_u field is desired, the following transformation can be used (Vanmarcke, 1984):

$$\rho_{c_u}(\tau) = \frac{\exp\{\rho_{\ln c_u}(\tau)\sigma^2_{\ln c_u}\} - 1}{\exp\{\sigma^2_{\ln c_u}\} - 1} \tag{13.18}$$

The spatial correlation length can be nondimensionalized by dividing it by the slope height H as was done in Eq. 13.4:

$$\Theta = \frac{\theta_{\ln c_u}}{H} \tag{13.19}$$

Thus, the results given here can be applied to any size problem, so long as it has the same slope and same overall bedrock depth–slope height ratio D. The standard deviation σ_{c_u} may also be expressed in terms of the dimensionless coefficient of variation

$$v_c = \frac{\sigma_{c_u}}{\mu_{c_u}} \tag{13.20}$$

If the mean and variance of the underlying $\ln c_u$ field are desired, they can be obtained through the transformations

$$\sigma^2_{\ln c_u} = \ln\left(1 + v_c^2\right), \qquad \mu_{\ln c_u} = \ln(\mu_{c_u}) - \tfrac{1}{2}\sigma^2_{\ln c_u} \tag{13.21}$$

By using Monte Carlo simulation, where the soil slope is simulated and analyzed by the finite-element method repeatedly, estimates of the probability of failure are obtained over a range of soil statistics. The failure probabilities are compared to those obtained using a harmonic average of the cohesion field employed in Taylor's stability coefficient method, and very good agreement is found. The study indicates that the stability of a spatially varying soil slope is well modeled using a harmonic average of the soil properties.

13.3.1 Random Finite-Element Method

The slope stability analyses use an elastic-perfectly plastic stress–strain law with a Tresca failure criterion. Plastic stress redistribution is accomplished using a viscoplastic algorithm which uses 8-node quadrilateral elements and reduced integration in both the stiffness and stress redistribution parts of the algorithm. The theoretical basis of the method is described more fully in Chapter 6 of the text by Smith and Griffiths (2004). The method is discussed in more detail in the previous section.

In brief, the analyses involve the application of gravity loading and the monitoring of stresses at all the Gauss points. If the Tresca criterion is violated, the program attempts to redistribute those stresses to neighboring elements that still have reserves of strength. This is an iterative process which continues until the Tresca criterion and global equilibrium are satisfied at all points within the mesh under quite strict tolerances.

In this study,"failure" is said to have occurred if, for any given realization, the algorithm is unable to converge within 500 iterations (see Figure 13.10). Following a set of 2000 realizations of the Monte Carlo process the probability of failure is simply defined as the proportion of these realizations that required 500 or more iterations to converge.

The RFEM combines the deterministic finite -element analysis with a random-field simulator, which, in this study, is the LAS discussed in Section 6.4.6. The LAS algorithm produces a field of random element values, each representing a local average of the random field over the element domain, which are then mapped directly to the finite elements. The random elements are local averages of the log-cohesion, $\ln c_u$, field. The resulting realizations of the log-cohesion field have correlation structure and variance correctly accounting for local averaging over each element. Much discussion of the relative merits of various methods of representing random fields in finite-element analysis has been carried out in recent years (see, e.g., Li and Der Kiureghian, 1993). While the spatial averaging discretization of the random field used in this study is just one approach to the problem, it is appealing in the sense that it reflects the simplest idea of the finite-element representation of a continuum as well as the way that soil samples are typically taken and tested in practice, that is, as local averages. Regarding the discretization of random fields for use in finite-element analysis, Matthies et al. (1997, p. 294) makes the following comment: "One way of making sure that the stochastic field has the required structure is to assume that it is a local averaging process," referring to the conversion of a nondifferentiable to a differentiable (smooth) stochastic process. Matthies further goes on to say that the advantage of the local average representation of a random field is that it yields accurate results even for rather coarse meshes.

Figure 13.23 illustrates two possible realizations arising from the RFEM for the 2 : 1 slope—similar results were observed for the 1 : 1 slope. In this figure, dark regions correspond to stronger soil. Notice how convoluted the failure region is, particularly at the smaller correlation length. It can be seen that the slope failure involves the plastic deformation of a region around a failure "surface" which undulates along the weakest path. Clearly, failure is more complex than just a rigid "circular" region sliding along a clearly defined interface, as is typically assumed.

13.3.2 Parametric Studies

To keep the study nondimensional, the soil strength is expressed in the form of a dimensionless shear strength:

$$c = \frac{c_u}{\gamma H} \tag{13.22}$$

which, if c_u is random, has mean

$$\mu_c = \frac{\mu_{c_u}}{\gamma H} \tag{13.23}$$

where γ is the unit weight of the soil, assumed in this study to be deterministic. In the 2 : 1 slope case where the cohesion field is assumed to be everywhere the same and

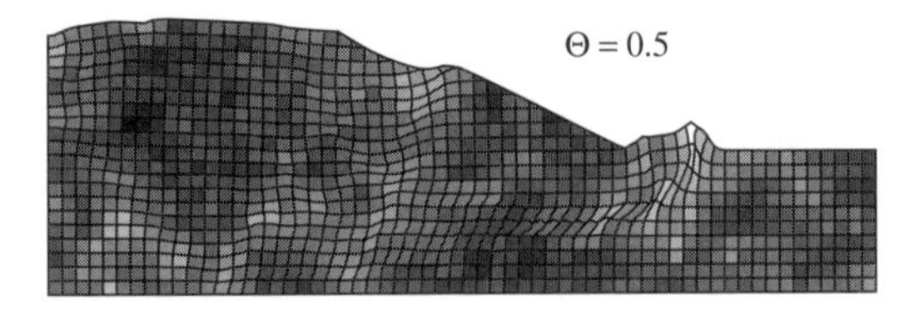

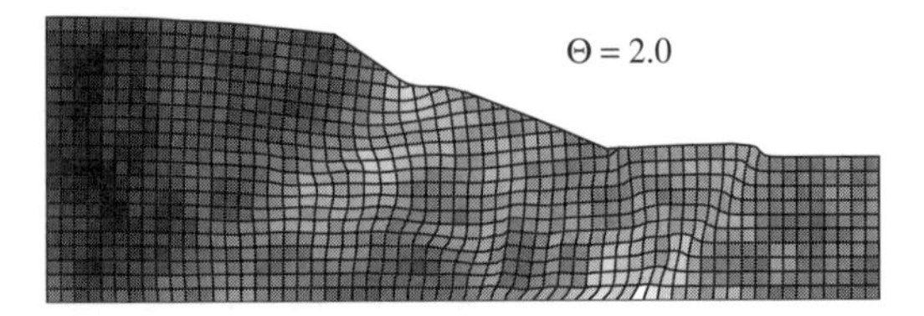

Figure 13.23 Two typical failed random-field realizations. Low-strength regions are light.

equal to μ_{c_u}, a value of $\mu_c = 0.173$ corresponds to a factor of safety $F_S = 1.0$, which is to say that the slope is on the verge of failure. For the 1 : 1 slope, $\mu_c = 0.184$ corresponds to a factor of safety $F_S = 1.0$. Both of these values were determined by finding the deterministic value of c_u needed to just achieve failure in the finite-element model, bearing in mind that the failure surface cannot descend below the base of the model. These values are almost identical to what would be identified using Taylor's charts (Taylor, 1937), although as will be seen later, small variations in the choice of the critical values of μ_c can result in significant changes in the estimated probability of slope failure, particularly for larger factors of safety.

This study considers the following values of the input statistics. For the 2 : 1 slope, μ_c is varied over the following values:

$$\mu_c = 0.15, 0.173, 0.20, 0.25, 0.30$$

and over

$$\mu_c = 0.15, 0.184, 0.20, 0.25, 0.30$$

for the 1 : 1 slope. For the normalized correlation length Θ and coefficient of variation v_c, the following ranges were investigated:

$$\Theta = 0.10, 0.20, 0.50, 1.00, 2.00, 5.00, 10.0$$

$$v_c = 0.10, 0.20, 0.50, 1.00, 2.00, 5.00$$

For each set of the above parameters, 2000 realizations of the soil field were simulated and analyzed, from which the probability of slope failure was estimated. This section concentrates on the development of a failure probability model, using a harmonic average of the soil, and compares the simulated probability estimates to those predicted by the harmonic average model.

13.3.3 Failure Probability Model

In Taylor's stability coefficient approach to slope stability (Taylor, 1937), the coefficient

$$c = \frac{c_u}{\gamma H} \tag{13.24}$$

assumes that the soil is completely uniform, having cohesion equal to c_u everywhere. This coefficient may then be compared to the critical coefficient obtained from Taylor's charts to determine if slope failure will occur or not. For the slope geometry studied here, slope failure will occur if $c < c_{\text{crit}}$ where $c_{\text{crit}} = 0.173$ for the 2 : 1 slope and $c_{\text{crit}} = 0.184$ for the 1 : 1 slope.

In the case where c_u is randomly varying in space, two issues present themselves. First of all Taylor's method cannot be used on a nonuniform soil, and, second, Eq. 13.24 now includes a random quantity on the right-hand side [namely, $c_u = c_u(\mathbf{x})$] so that c becomes random. The first issue can be solved by finding some representative or *equivalent* value of c_u, which will be referred to here as $\bar{c}_u$, such that the stability coefficient method still holds for the slope. That is, $\bar{c}_u$ would be the cohesion of a uniform soil such that it has the same factor of safety as the real spatially varying soil.

The question now is: How should this equivalent soil cohesion value be defined? First of all, each soil realization will have a different value of $\bar{c}_u$, so that Eq. 13.24 is still a function of a random quantity, namely

$$c = \frac{\bar{c}_u}{\gamma H} \tag{13.25}$$

If the distribution of $\bar{c}_u$ is found, the distribution of c can be derived. The failure probability of the slope then becomes equal to the probability that c is less than the Taylor critical value c_{crit}.

This line of reasoning suggests that $\bar{c}_u$ should be defined as some sort of average of c_u over the soil domain where failure is occurring. Three common types of averages present themselves, as discussed in Section 4.4:

1. *Arithmetic Average:* The arithmetic average over some domain, A, is defined as

$$X_a = \frac{1}{n}\sum_{i=1}^{n} c_{u_i} = \frac{1}{A}\int_A c_u(\mathbf{x})\, d\mathbf{x} \tag{13.26}$$

for the discrete and continuous cases, where the domain A is assumed to be divided up into n samples in the discrete case. The arithmetic average weights all of the values of c_u equally. In that the failure surface seeks a path through the weakest parts of the soil, this form of averaging is not deemed to be appropriate for this problem.

2. *Geometric Average:* The geometric average over some domain, A, is defined as

$$X_g = \left(\prod_{i=1}^{n} c_{u_i}\right)^{1/n} = \exp\left\{\frac{1}{A}\int_A \ln c_u(\mathbf{x})\, d\mathbf{x}\right\} \tag{13.27}$$

The geometric average is dominated by low values of c_u and, for a spatially varying cohesion field, will always be less than the arithmetic average. This average potentially reflects the reduced strength as seen along the failure path and has been found by the authors (Fenton and Griffiths, 2002, 2003) to well represent the bearing capacity and settlement of footings founded on spatially random soils. The geometric average is also a "natural" average of the lognormal distribution since an arithmetic average of the underlying normally distributed random variable, $\ln c_u$, leads to the geometric average when converted back to the lognormal distribution. Thus, if c_u is lognormally distributed, its geometric local average will also be lognormally distributed with the median preserved.

3. *Harmonic Average:* The harmonic average over some domain, A, is defined as

$$X_h = \left[\frac{1}{n} \sum_{i=1}^{n} \frac{1}{c_{u_i}} \right]^{-1} = \left[\frac{1}{A} \int_A \frac{d\mathbf{x}}{c_u(\mathbf{x})} \right]^{-1} \quad (13.28)$$

This average is even more strongly influenced by small values than is the geometric average. In general, for a spatially varying random field, the harmonic average will be smaller than the geometric average, which in turn is smaller than the arithmetic average. Unfortunately, the mean and variance of the harmonic average, for a spatially correlated random field, are not easily found.

Putting aside for the moment the issue of how to compute the equivalent undrained cohesion, $\bar{c}_u$, the size of the averaging domain must also be determined. This should approximately equal the area of the soil which fails during a slope subsidence. Since the value of $\bar{c}_u$ changes only slowly with changes in the averaging domain, only an approximate area need be determined. The area selected in this study is a parallelogram, as shown in Figure 13.24, having slope length equal to the length of the slope and horizontal surface length equal to H. For the purposes of computing the average, it is further assumed that this area can be approximated by a rectangle of dimension $w \times h$ (averages over rectangles are generally easier to compute). Thus, a rectangular $w \times h$ area is used to represent a roughly circular band (on average) within which the soil is failing in shear.

In this study, the values of w and h are taken to be

$$w = \frac{H}{\sin \beta}, \qquad h = H \sin \beta \qquad (13.29)$$

such that $w \times h = H^2$, where β is the slope angle (26.6° for the 2 : 1 slope and 45° for the 1 : 1 slope). It appears, when

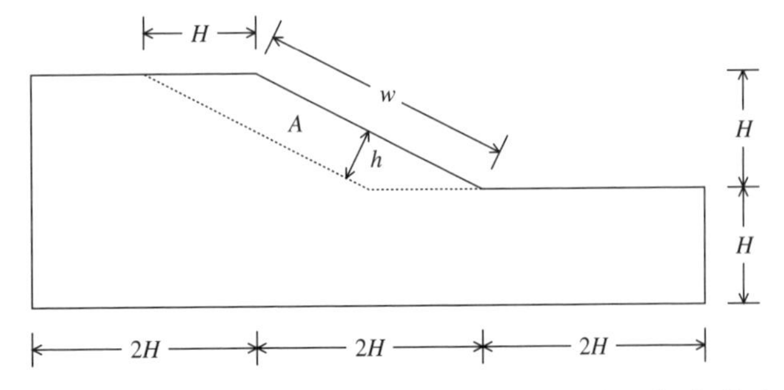

Figure 13.24 Assumed averaging domain (1 : 1 slope is similar).

comparing Figure 13.23 to 13.24, that the assumed averaging domain of Figure 13.24 is smaller than the deformed regions seen in Figure 13.23. A general prescription for the size of the averaging domain is not yet known, although it should capture the approximate area of the soil involved in resisting the slope deformation. The area assumed in Figure 13.24 is to be viewed as an initial approximation which, as will be seen, yields surprisingly good results. It is recognized that the true average will be of the minimum soil strengths within a roughly circular band—presumably the area of this band is on average approximated by the area shown in Figure 13.24.

With an assumed averaging domain, $A = w \times h$, the geometric average leads to the following definition for $\bar{c}_u$:

$$\bar{c}_u = X_g = \exp\left\{ \frac{1}{A} \int_A \ln c_u(\mathbf{x}) \, d\mathbf{x} \right\} \qquad (13.30)$$

which, if c_u is lognormally distributed, is also lognormally distributed. The resulting coefficient

$$c = \frac{\bar{c}_u}{\gamma H} \qquad (13.31)$$

is then also lognormally distributed with mean and variance

$$\mu_{\ln c} = \mu_{\ln c_u} - \ln(\gamma H) \qquad (13.32a)$$

$$\sigma^2_{\ln c} = \sigma^2_{\ln \bar{c}_u} = \gamma(w,h)\sigma^2_{\ln c_u} \qquad (13.32b)$$

The function $\gamma(w,h)$ is the so-called variance function, which lies between 0 and 1, and gives the amount that the variance of a local average is reduced from the point value. It is formally defined as the average of correlations between every pair of points in the averaging domain:

$$\gamma(w,h) = \frac{1}{A^2} \int_A \int_A \rho(\boldsymbol{\xi} - \boldsymbol{\eta}) \, d\boldsymbol{\xi} \, d\boldsymbol{\eta} \qquad (13.33)$$

Solutions to this integral, albeit sometimes approximate, exist for most common correlation functions. Alternatively, the integral can be calculated accurately using a numerical method such as Gauss quadrature. See Appendix C for more details.

The probability of failure p_f can now be computed by assuming that Taylor's stability coefficient method holds when using this equivalent value of cohesion, namely by computing

$$p_f = \mathrm{P}[c < c_{\text{crit}}] = \Phi\left(\frac{\ln c_{\text{crit}} - \mu_{\ln c}}{\sigma_{\ln c}} \right) \qquad (13.34)$$

where the critical stability coefficient for the 2 : 1 slope is $c_{\text{crit}} = 0.173$ and for the 1 : 1 slope is $c_{\text{crit}} = 0.184$; Φ is the cumulative distribution function for the standard normal. Unfortunately, the geometric average for $\bar{c}_u$ leads to predicted failure probabilities which significantly underestimate the probabilities determined via simulation, and changes in the averaging domain size does not particularly improve the prediction. This means that the soil strength as "seen" by the finite-element model is even lower, in general, than that predicted by the geometric average. Thus, the geometric average was abandoned as the correct measure for $\bar{c}_u$.

Since the harmonic average yields values which are even lower than the geometric average, the harmonic average over the same domain, $A = w \times h$, is now investigated as representative of $\bar{c}_u$, namely

$$\bar{c}_u = X_h = \left[\frac{1}{A}\int_A \frac{d\mathbf{x}}{c_u(\mathbf{x})}\right]^{-1} \qquad (13.35)$$

Unfortunately, so far as the authors are aware, no relatively simple expressions exist for the moments of $\bar{c}_u$, as defined above, for a spatially correlated random field. The authors are continuing research on this problem but, for the time being, these moments can be obtained by simulation. It may seem questionable to be developing a probabilistic model with the nominal goal of eliminating the necessity of simulation, when that model still requires simulation. However, the moments of the harmonic mean can be arrived at in a small fraction of the time taken to perform the nonlinear slope stability simulation.

In order to compute probabilities using the statistics of $\bar{c}_u$, it is necessary to know the distribution of $c = \bar{c}_u/(\gamma H)$. For lognormally distributed c_u, the distribution of the harmonic average is not simple. However, since $\bar{c}_u$ is strictly nonnegative ($c_u \geq 0$), it seems reasonable to suggest that $\bar{c}_u$ is at least approximately lognormal. A histogram of the harmonic averages obtained in the case where $v_c = 0.5$ and $\Theta = 0.5$ is shown in Figure 13.25, along with a fitted lognormal distribution. The p-value for the chi-Square goodness-of-fit test is 0.44, indicating that the lognormal distribution is very reasonable, as also indicated by the plot. Similar results were obtained for other parameter values.

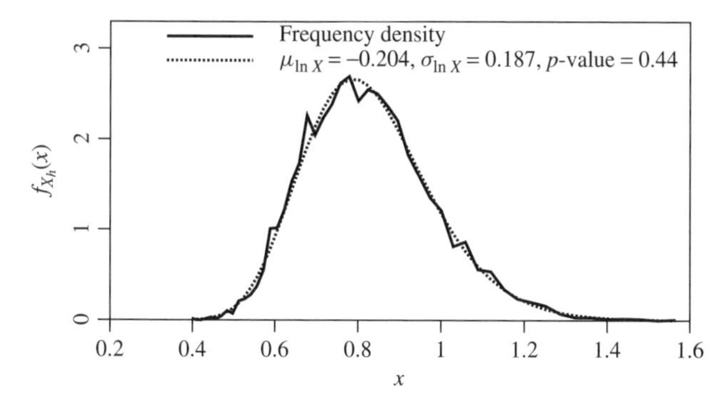

Figure 13.25 Histogram of harmonic averages along with fitted lognormal distribution.

The procedure to estimate the mean and variance of the harmonic average $\bar{c}_u$ for each parameter set (μ_c, v_c, and Θ) considered in this study involves (a) generating a large number of random cohesion fields, each of dimension $w \times h$, (b) computing the harmonic average of each using Eq. 13.28, and (c) estimating the mean and variance of the resulting set of harmonic averages. Using 5000 random-field realizations, the resulting estimates for the mean and standard deviation of $\ln X_h$ are shown in Figure 13.26 for random fields with mean 1.0. Since $\bar{c}_u$ is assumed to be (at least approximately) lognormally distributed, having parameters $\mu_{\ln \bar{c}_u}$ and $\sigma_{\ln \bar{c}_u}$, the mean and standard deviation of the logarithm of the harmonic averages are shown in Figure 13.26 for the two slopes considered. Of note in Figure 13.26 is the fact that there is virtually no difference in the mean and standard deviation for the 2 : 1 and 1 : 1 slopes, even though the averaging regions have quite different shapes. Admittedly the two averaging regions have the same area, but this only slow change in harmonic average statistics with averaging dimension has been found also to be true of changing areas. This implies that the accurate determination of the averaging area is not essential to the accuracy of failure probability predictions.

Given the results of Figure 13.26, the slope failure probability can now be computed as in Eq. 13.34:

$$p_f = \mathrm{P}[c < c_{\text{crit}}] = \Phi\left(\frac{\ln c_{\text{crit}} - \mu_{\ln c}}{\sigma_{\ln c}}\right) \qquad (13.36)$$

except that now the mean and standard deviation of $\ln c$ are computed using the harmonic mean results of Figure 13.26 suitably scaled for the actual value of $\mu_{c_u}/\gamma H$ as follows:

$$\mu_{\ln c} = \ln\left(\frac{\mu_{c_u}}{\gamma H}\right) + \mu_{\ln X_h} = \ln(\mu_c) + \mu_{\ln X_h} \qquad (13.37a)$$

$$\sigma_{\ln c} = \sigma_{\ln X_h} \qquad (13.37b)$$

where $\mu_{\ln X_h}$ and $\sigma_{\ln X_h}$ are read from Figure 13.26, given the correlation length and coefficient of variation.

Figure 13.27 shows the predicted failure probabilities versus the failure probabilities obtained via simulation over all parameter sets considered. The agreement is remarkably good, considering the fact that the averaging domain was rather arbitrarily selected, and there was no a priori evidence that the slope stability problem should be governed by a harmonic average. The results of Figure 13.27 indicate that the harmonic average gives a good probabilistic model of slope stability.

There are a few outliers in Figure 13.27 where the predicted failure probability considerably overestimates that obtained via simulation. For the 2 : 1 slope, these outliers correspond to the cases where (1) $\mu_c = 0.3$, $v_c = 1.0$,

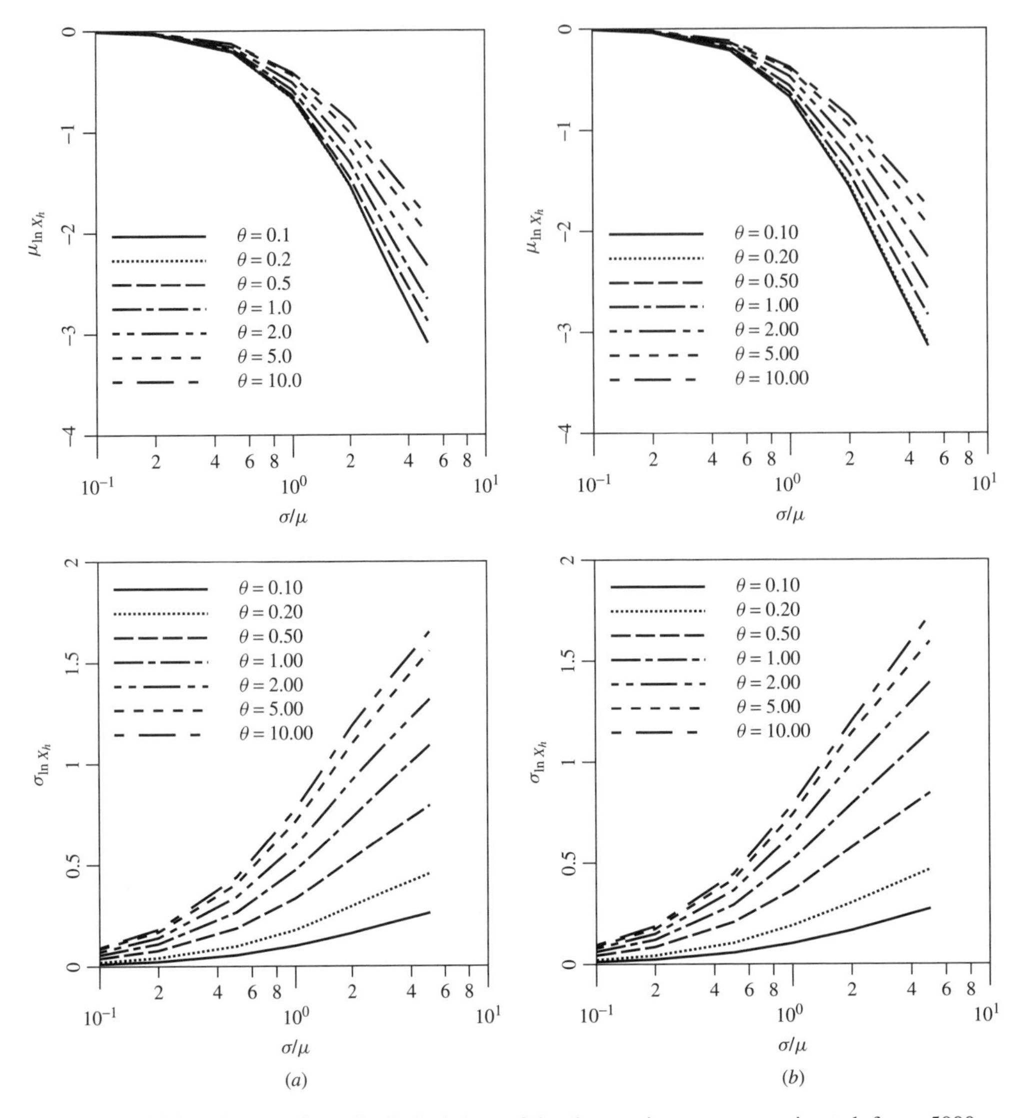

Figure 13.26 Mean and standard deviation of log-harmonic averages estimated from 5000 simulations: (*a*) 2 : 1 cohesive slope; (*b*) 1 : 1 cohesive slope.

and $\Theta = 0.1$ (simulated probability is 0.047 versus predicted probability of 0.86) and (2) $\mu_c = 0.3$, $v_c = 1.0$, and $\Theta = 0.2$ (simulated probability is 0.31 versus predicted probability of 0.74). Both cases correspond to the largest F_S considered in the study ($\mu_c = 0.3$ gives an $F_S = 1.77$ in the uniform soil case). Also the small correlation lengths yield the smallest values of $\sigma_{\ln c}$ which, in turn, implies that the cumulative distribution function of $\ln c$ increases very rapidly over a small range. Thus, slight errors in the estimate of $\mu_{\ln c}$ makes for large errors in the probability.

For example, the worst case seen in Figure 13.27*a* has predicted values of

$$\mu_{\ln c} = \ln(\mu_c) + \mu_{\ln X_h} = \ln(0.3) - 0.66 = -1.864$$

$$\sigma_{\ln c} = \sigma_{\ln X_h} = 0.10$$

The predicted failure probability is thus

$$\begin{aligned} \mathrm{P}\left[c < 0.173\right] &= \Phi\left(\frac{\ln 0.173 + 1.864}{0.10}\right) = \Phi(1.10) \\ &= 0.86 \end{aligned}$$

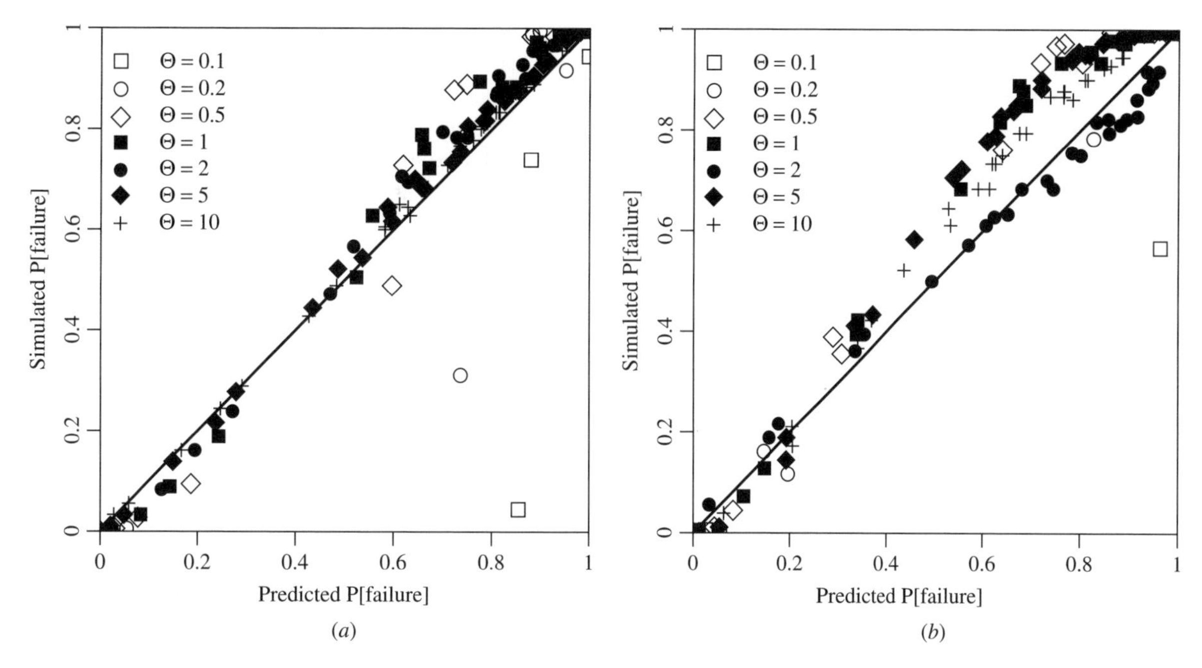

Figure 13.27 Simulated failure probabilities versus failure probabilities predicted using a harmonic average of c_u over domain $w \times h$: (*a*) 2 : 1 cohesive slope; (*b*) 1 : 1 cohesive slope.

As mentioned, a relatively small error in the estimation of $\mu_{\ln c}$ can lead to a large change in probability. For example, if $\mu_{\ln c}$ was -1.60 instead of -1.864, a 14% change, then the predicted failure probability changes significantly to

$$P[c < 0.173] = \Phi\left(\frac{\ln 0.173 + 1.6}{0.10}\right) = \Phi(-1.54)$$
$$= 0.062$$

which is about what was obtained via simulation. The conclusion drawn from this example is that small errors in the estimation of $\mu_{\ln c}$ or, equivalently, in c_{crit} can lead to large errors in the predicted slope failure probability if the standard deviation of $\ln c$ is small. The latter occurs for small correlation lengths, Θ. In most cases for small values of Θ the failure probability tends to be either close to zero ($v_c < 1.0$) or close to 1.0 ($v_c > 1.0$), in which case the predicted and simulated probabilities are in much better agreement. That is, the model shows very good agreement with simulation for all but the case where a large F_S is combined with a small correlation length and intermediate coefficient of variation ($v_c \simeq 1.0$). This means that the selected harmonic average model is not the best predictor in the region where the cumulative distribution is rapidly increasing. However, in these cases, the predicted failure probability is overestimated, which is at least conservative.

For all other results, especially where the F_S is closer to 1.0 ($\mu_c < 0.3$), the harmonic average model leads to very good estimates of failure probability with somewhat more scatter seen for the 1 : 1 slope. The increased scatter for the 1 : 1 slope is perhaps as expected since the steeper slope leads to a larger variety of critical failure surfaces. In general, for both slopes the predicted failure probability is seen to be conservative at small failure probabilities, slightly overestimating the failure probability.

13.3.4 Summary

This study investigates the failure probabilities of two undrained clay slopes, one with gradient 2 : 1 and the other with gradient 1 : 1. The basic idea of the section is that the Taylor stability coefficients are still useful if an "equivalent" soil property can be found to represent the spatially random soil. It was found that a harmonic average of the soil cohesion over a region of dimension $H^2(\sin\ \beta \times 1/\sin\ \beta) = H^2$ yields an equivalent stability number with an approximately lognormal distribution that quite well predicts the probability of slope failure. The harmonic average was selected because it is dominated by low-strength regions appearing in the soil slope, which agrees with how the failure surface will seek out the low-strength areas. The dimension of the averaging region was rather arbitrarily selected—the equivalent stability coefficient mean and variance is only slowly affected by changes in the averaging region dimension—but is believed to reasonably approximate the area of the 'average' slope failure band.

An important practical conclusion arising from the fact that soil slopes appear to be well characterized by a harmonic average of soil sample values, rather than by an arithmetic average, as is traditionally done, has to do with how soil samples are treated. In particular, the study suggests that the reliability of an existing slope is best estimated by sampling the soil at a number of locations and then using a harmonic average of the sample values to estimate the soil's equivalent cohesion. Most modern geotechnical codes suggest that soil design properties be taken as "cautious estimates of the mean"—the harmonic average, being governed by low-strength regions, is considered by the authors to be such a "cautious estimate" for slope stability calculations.

CHAPTER 14

Earth Pressure

14.1 INTRODUCTION

Traditional geotechnical analysis uses the factor-of-safety approach in one of two ways. In foundation analysis, for example, Terzaghi's bearing capacity equation leads to an estimate of the ultimate value, which is then divided by the factor of safety to give allowable loading levels for design. Alternatively, in slope stability analysis, the factor of safety represents the factor by which the shear strength parameters of the soil would need to be reduced to reach the limit state. Either way, the factor of safety represents a blanket factor that implicitly includes all sources of variability and uncertainty inherent in the geotechnical analysis.

The approaches described in this chapter attempt to include the effects of soil property variability in a more scientific way using statistical methods (Griffiths et al., 2005). If it is assumed that the soil parameters in question (e.g., friction angle, cohesion, compressibility, and permeability) are random variables that can be expressed in the form of a probability density function, then the issue becomes one of estimating the probability density function of some outcome that depends on the input random variables. The output can then be interpreted in terms of probabilities, leading to statements such as: "The design load on the foundation will give a probability of bearing capacity failure of $p_1\%$," "The embankment has a probability of slope failure of $p_2\%$," "The probability of the design settlement levels being exceeded is $p_3\%$," or "The probability that the earth pressure acting on a retaining wall exceeds the design value is $p_4\%$".

The effect of spatial variability on active and passive earth pressures is investigated in this chapter. The spatial variability is represented using random fields and the soil response computed by the finite-element method. This is another manifestation of the RFEM. The program used to determine many of the results in this chapter is called REARTH2D and is available at `http://www.engmath.dal.ca/rfem`.

The random fields are simulated using the LAS method (see Section 6.4.6) while the finite-element analysis is a nonlinear elasto-plastic algorithm which employs the Mohr–Coulomb failure criterion [see Griffiths and Fenton (2001) and Smith and Griffiths (2004) for more details].

14.2 PASSIVE EARTH PRESSURES

In this section we examine various ways of computing probabilities relating to passive earth pressures and we will start with an example which uses the FOSM method. The limiting horizontal passive earth force against a smooth wall of height H is given from the Rankine equation as

$$P_p = \tfrac{1}{2}\gamma H^2 K_p + 2c'H\sqrt{K_p} \tag{14.1}$$

where the passive earth pressure coefficient is written in this case as (Griffiths et al., 2002c)

$$K_p = \left[\tan\phi' + \sqrt{1+\tan^2\phi'}\right]^2 \tag{14.2}$$

a form which emphasizes the influence of the fundamental variable $\tan\phi'$.

In dimensionless form we can write

$$\frac{P_p}{\gamma H^2} = \tfrac{1}{2}K_p + 2\sqrt{K_p}\frac{c'}{\gamma H} \tag{14.3}$$

or

$$\bar{P}_p = \tfrac{1}{2}K_p + 2\bar{c}\sqrt{K_p} \tag{14.4}$$

where $\bar{P}_p$ is a dimensionless passive earth force and $\bar{c} = c'/\gamma H$ is a dimensionless cohesion.

Operating on Eq. 14.4 and treating $\tan\phi'$ and $\bar{c}$ as uncorrelated random variables, the first-order approximation to the mean of $\bar{P}_p$ is given by Eq. 1.79 to be

$$\mu_{\bar{P}_p} = \mathrm{E}\left[\bar{P}_p\right] = \tfrac{1}{2}\mu_{K_p} + 2\mu_{\bar{c}}\sqrt{\mu_{K_p}} \tag{14.5}$$

and from Eq. 1.83, the first-order approximation to the variance of $\bar{P}_p$ is

$$\sigma^2_{\bar{P}_p} = \mathrm{Var}\left[\bar{P}_p\right] \simeq \left(\frac{\partial \bar{P}_p}{\partial \bar{c}}\right)^2 \mathrm{Var}\left[\bar{c}\right] + \left(\frac{\partial \bar{P}_p}{\partial(\tan\phi')}\right)^2 \mathrm{Var}\left[\tan\phi'\right] \tag{14.6}$$

The required derivatives are computed analytically from Eq. 14.4 at the means as follows:

$$\frac{\partial \bar{P}_p}{\partial \bar{c}} = 2\sqrt{\mu_{K_p}} \tag{14.7a}$$

$$\frac{\partial \bar{P}_p}{\partial(\tan\phi')} = \frac{\mu_{K_p}}{\sqrt{1+\mu^2_{\tan\phi'}}} + 2\mu_{\bar{c}}\frac{\sqrt{\mu_{K_p}}}{\sqrt{1+\mu^2_{\tan\phi'}}} \tag{14.7b}$$

It is now possible to compute the mean and standard deviation of the horizontal earth force for a range of input soil property variances. In this example, we shall assume that the coefficient of variation (v) values for both $\bar{c}$ and $\tan\phi'$ are the same, that is,

$$v_{\bar{c},\tan\phi'} = \frac{\sigma_{\bar{c}}}{\mu_{\bar{c}}} = \frac{\sigma_{\tan\phi'}}{\mu_{\tan\phi'}} \tag{14.8}$$

Table 14.1 shows the influence of variable input on the passive force in the case of $\mu_{\bar{c}} = 5$ and $\mu_{\tan\phi'} = \tan 30^\circ = 0.577$. It can be seen that in this case the process results in a slight magnification of the coefficient of variation of the passive force over the input values. For example, $v_{\bar{c},\tan\phi'} = 0.5$ leads to $v_{\bar{P}_p} = 0.53$ and so on. The ratio of the output $v_{\bar{P}_p}$ to the input $v_{\bar{c},\tan\phi'}$ can also be obtained analytically from Eqs. 14.5 and 14.6 to give

$$\frac{v_{\bar{P}_p}}{v_{\bar{c},\tan\phi'}} \approx 2\frac{\sqrt{(\sqrt{\mu_{K_p}} + 2\mu_{\bar{c}})^2(\mu_{K_p} - 1)^2 + 4\mu_{\bar{c}}^2(\mu_{K_p} + 1)^2}}{(\sqrt{\mu_{K_p}} + 4\mu_{\bar{c}})(\mu_{K_p} + 1)} \tag{14.9}$$

This equation is plotted in Figure 14.1 for a range of $\mu_{\bar{c}}$ values. The graph indicates that in many cases the FOSM method causes the ratio given by Eq. 14.9 to be less than unity. In other words, the coefficient of variation of the output passive force is smaller than the coefficient of variation of the input strength parameters. For higher fiction angles, however, this trend is reversed.

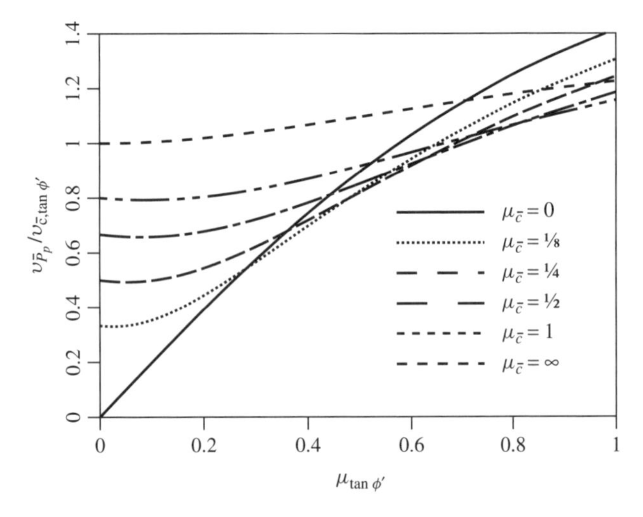

Figure 14.1 $v_{\bar{P}_p}/v_{\bar{c},\tan\phi'}$ versus $\mu_{\tan\phi'}$ for passive earth pressure analysis by FOSM.

14.2.1 Numerical Approach

An alternative approach evaluates the derivatives numerically using a central finite-difference formula. In this case, the dependent variable $\bar{P}_p$ is sampled across two standard deviations in one variable while keeping the other variable fixed at the mean. This large central difference interval encompasses about 68% of all values of the input parameters $\bar{c}$ and $\tan\phi'$, so the approximation is only reasonable if the function $\bar{P}_p$ from Eq. 14.4 does not exhibit much nonlinearity across this range. The finite-difference formulas take the form

$$\frac{\partial\bar{P}_p}{\partial\bar{c}} \approx \frac{\bar{P}_p(\mu_{\bar{c}} + \sigma_{\bar{c}}, \mu_{\tan\phi'}) - \bar{P}_p(\mu_{\bar{c}} - \sigma_{\bar{c}}, \mu_{\tan\phi'})}{2\sigma_{\bar{c}}} = \frac{\Delta P_{p(\bar{c})}}{2\sigma_{\bar{c}}} \tag{14.10}$$

and

$$\frac{\partial\bar{P}_p}{\partial(\tan\phi')} \approx \frac{\bar{P}_p(\mu_{\bar{c}}, \mu_{\tan\phi'} + \sigma_{\tan\phi'}) - \bar{P}_p(\mu_{\bar{c}}, \mu_{\tan\phi'} - \sigma_{\tan\phi'})}{2\sigma_{\tan\phi'}} = \frac{\Delta P_{p(\tan\phi')}}{2\sigma_{\tan\phi'}} \tag{14.11}$$

The main attraction of this approach is that, once the derivative terms are squared and substituted into Eq. 14.6, the variances of $\bar{c}$ and $\tan\phi'$ cancel out, leaving

$$\text{Var}\left[\bar{P}_p\right] \approx \left(\tfrac{1}{2}\Delta\bar{P}_{p(\bar{c})}\right)^2 + \left(\tfrac{1}{2}\Delta\bar{P}_{p(\tan\phi')}\right)^2 \tag{14.12}$$

In this case, $\bar{P}_p$ is a linear function of $\bar{c}$ and is slightly nonlinear with respect to $\tan\phi'$. It is clear from a comparison of Tables 14.1 and 14.2 that the numerical and analytical approaches in this case give essentially the same results.

Table 14.1 Statistics of $\bar{P}_p$ Predicted Using FOSM (Analytical Approach) with $\mu_{\bar{c}} = 5$ and $\mu_{\tan\phi'} = \tan 30^\circ = 0.577$

$v_{\bar{c},\tan\phi'}$	$\partial\bar{P}_p/\partial\bar{c}$	Var $[\bar{c}]$	$\partial\bar{P}_p/\partial(\tan\phi')$	Var $[\tan\phi']$	Var $[\bar{P}_p]$	$\sigma_{\bar{P}_p}$	$\mu_{\bar{P}_p}$	$v_{\bar{P}_p}$
0.1	3.46	0.25	17.60	0.0033	4.03	2.01	18.82	0.11
0.3	3.46	2.25	17.60	0.0300	36.29	6.02	18.82	0.32
0.5	3.46	6.25	17.60	0.0833	100.81	10.04	18.82	0.53
0.7	3.46	12.25	17.60	0.1633	197.59	14.06	18.82	0.75
0.9	3.46	20.25	17.60	0.2700	326.64	18.07	18.82	0.96

Table 14.2 Statistics of $\bar{P}_p$ Predicted Using FOSM (Numerical Approach) with $\mu_{\bar{c}} = 5$ and $\mu_{\tan\phi'} = \tan 30° = 0.577$

$v_{\bar{c},\tan\phi'}$	$\Delta\bar{P}_{p(\bar{c})}/2$	$\Delta\bar{P}_{p(\tan\phi')}/2$	$\text{Var}\left[\bar{P}_p\right]$	$\sigma_{\bar{P}_p}$	$\mu_{\bar{P}_p}$	$v_{\bar{P}_p}$
0.1	1.73	1.02	4.03	2.01	18.82	0.11
0.3	5.20	3.04	36.26	6.02	18.82	0.32
0.5	8.66	5.05	100.53	10.03	18.82	0.53
0.7	12.12	7.04	196.54	14.02	18.82	0.74
0.9	15.59	9.00	323.93	18.00	18.82	0.96

14.2.2 Refined Approach Including Second-Order Terms

In the above example, a first-order approximation was used to predict both the mean and variance of $\bar{P}_p$ from Eqs. 1.79 and 1.83. Since the variances of $\bar{c}$ and $\tan\phi'$ are both known, it is possible to refine the estimate of $\mu_{\bar{P}_p}$ by including second-order terms from Eq. 1.76a, leading to

$$\begin{aligned}\mu_{\bar{P}_p} &\approx \bar{P}_p(\mu_{\bar{c}}, \mu_{\tan\phi'}) + \tfrac{1}{2}\,\text{Var}\,[\bar{c}]\,\frac{\partial^2\bar{P}_p}{\partial\bar{c}^2} \\ &\quad + \tfrac{1}{2}\,\text{Var}\left[\tan\,\phi'\right]\frac{\partial^2\bar{P}_p}{\partial(\tan\,\phi')^2} \\ &\quad + \text{Cov}\left[\bar{c}, \tan\,\phi'\right]\frac{\partial^2\bar{P}_p}{\partial\bar{c}\;\partial(\tan\,\phi')}\end{aligned} \tag{14.13}$$

where all derivatives are evaluated at the mean. Noting that in this case $\partial^2\bar{P}_p/\partial\bar{c}^2 = 0$ and $\text{Cov}\left[\bar{c}, \tan\,\phi'\right] = 0$, the expression simplifies to

$$\mu_{\bar{P}_p} \approx \bar{P}_p(\mu_{\bar{c}}, \mu_{\tan\phi'}) + \tfrac{1}{2}\,\text{Var}\left[\tan\,\phi'\right]\frac{\partial^2\bar{P}_p}{\partial(\tan\,\phi')^2} \tag{14.14}$$

where the analytical form of the second derivative is given by

$$\begin{aligned}\frac{\partial^2\bar{P}_p}{\partial(\tan\,\phi')^2} &= \frac{2}{1+\mu^2_{\tan\phi'}}\left[\mu_{K_p} + \mu_{\bar{c}}\sqrt{\mu_{K_p}}\right] \\ &\quad - \frac{\mu_{\tan\phi'}}{(1+\mu^2_{\tan\phi'})^{3/2}}\left[\mu_{K_p} + 2\mu_{\bar{c}}\sqrt{\mu_{K_p}}\right]\end{aligned} \tag{14.15}$$

Combining Eqs. 14.14 and 14.15 for the particular case of $\mu_{\bar{c}} = 5$ and $\mu_{\tan\phi'} = 0.577$ leads to

$$\mu_{\bar{P}_p} = 18.82 + 4.94\;\text{Var}\left[\tan\,\phi'\right] \tag{14.16}$$

Table 14.3 shows a reworking of the analytical results from Table 14.1 including second-order terms in the estimation of $\mu_{\bar{P}_p}$. A comparison of the results from the two tables indicates that the second-order terms have marginally increased $\mu_{\bar{P}_p}$ and thus reduced $v_{\bar{P}_p}$. The differences introduced by the second-order terms are quite modest, however, indicating the essentially linear nature of this problem.

Table 14.3 Statistics of $\bar{P}_p$ Predicted Using FOSM (Analytical Approach Including Second-Order Terms) with $\mu_{\bar{c}} = 5$ and $\mu_{\tan\phi'} = \tan 30° = 0.577$

$v_{\bar{c},\tan\,\phi'}$	Var [tan ϕ]	$\sigma_{\bar{P}_p}$	$\mu_{\bar{P}_p}$	$v_{\bar{P}_p}$
0.1	0.0033	2.01	18.84	0.11
0.3	0.0300	6.02	18.97	0.32
0.5	0.1833	10.04	19.23	0.52
0.7	0.1633	14.06	19.63	0.72
0.9	0.2700	18.07	20.15	0.90

14.2.3 Random Finite-Element Method

For reasonably "linear" problems, the FOSM and FORM (see Section 7.2.1 for a discussion of the latter) are able to take account of soil property variability in a systematic way. These traditional methods, however, typically take no account of spatial correlation, which is the tendency for properties of soil elements "close together" to be correlated while soil elements "far apart" are uncorrelated. In soil failure problems such as passive earth pressure analysis, it is possible to account for local averaging and spatial correlation by prescribing a potential failure surface and averaging the soil strength parameters along it (e.g., El-Ramly et al., 2002; Peschl and Schweiger, 2003). A disadvantage of this approach is that the location of the potential failure surface must be anticipated in advance, which rather defeats the purpose of a general random soil model.

To address the correlation issue, the passive earth pressure problem has been reanalyzed using the RFEM via the program REARTH2D (available at `http://www.engmath.dal.ca/rfem`), enabling soil property variability and spatial correlation to be accounted for in a rigorous and general way. The methodology involves the generation and mapping of a random field of c' and $\tan\phi'$ properties onto a quite refined finite-element mesh. Full account is taken of local averaging and variance reduction (see Section 6.4.6) over each element, and an exponentially decaying spatial correlation function is incorporated. An elasto-plastic finite-element analysis is then performed using a Mohr–Coulomb failure criterion.

In a passive earth pressure analysis the nodes representing the rigid wall are translated horizontally into the mesh and the reaction forces back-figured from the developed stresses. The limiting passive resistance (P_p) is eventually reached and the analysis is repeated numerous times using Monte Carlo simulations. Each realization of the Monte Carlo process involves a random field with the same mean, standard deviation, and spatial correlation length. The spatial distribution of properties varies from one realization to the next, however, so that each simulation leads to a different value of P_p. The analysis has the option of including cross correlation between properties and anisotropic spatial correlation lengths (e.g., the spatial correlation length in a naturally occurring stratum of soil is often higher in the horizontal direction). Neither of these options has been investigated in the current study to facilitate comparisons with the FOSM.

Lognormal distributions of c' and $\tan\phi'$ have been used in the current study and mapped onto a mesh of eight-node, quadrilateral, plane-strain elements. Examples of different spatial correlation lengths are shown in Figure 14.2 in the form of a gray scale in which weaker regions are lighter and stronger regions are darker.

Examples of a relatively low spatial correlation length and a relatively high correlation length are shown. It should be emphasized that the mean and standard deviation of the random variable being portrayed are the same in both figures. The spatial correlation length (which has units of length) is defined with respect to the underlying normal distribution and denoted as $\theta_{\ln c', \ln\tan\phi'}$. Both c' and $\tan\phi'$ were assigned the same isotropic correlation length in this study. A convenient nondimensional form of the spatial correlation length can be achieved in the earth pressure analysis by dividing by the wall height H, thus $\Theta = \theta_{\ln c', \ln\tan\phi'}/H$.

14.2.4 Parametric Studies

Quite extensive parametric studies of the passive earth pressure problem by RFEM were performed by Tveten (2002). A few of these results are presented here in which the coefficients of variation of c' and $\tan\phi'$ and spatial correlation length Θ have been varied. In all cases, the mean strength parameters have been held constant at $\mu_{c'} = 100$ kPa and $\mu_{\tan\phi'} = \tan 30^\circ = 0.577$. In addition, the soil unit weight was fixed at 20 kN/m^3 and the wall height set to unity. Thus, the dimensionless cohesion described earlier in the paper is given by $\bar{c} = c'/(\gamma H) = 5$. The variation in the limiting mean passive earth pressure, μ_{P_p}, normalized with respect to the value that would be given by simply substituting the mean strength values, $P_p(\mu_{c'}, \mu_{\tan\phi'}) = 376.4$ kN/m, is shown in Figure 14.3.

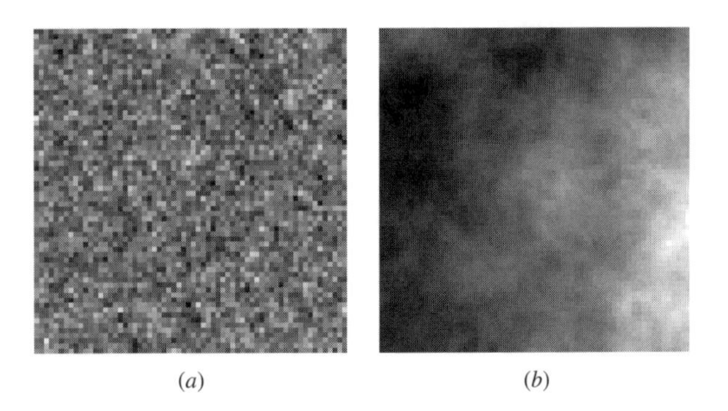

Figure 14.2 Typical random fields in RFEM approach: (*a*) low correlation length; (*b*) high correlation length.

The figure shows results for spatial correlation lengths in the range $0.01 < \Theta < 10$. At the lower end, the small spatial correlation lengths result in very significant local averaging over each finite element. In the limit as $\Theta \to 0$, local averaging causes the mean of the properties to tend to the median and the variance to tend to zero. For a typical random variable X, the properties of the lognormal distribution give that

$$\frac{\tilde{\mu}_X}{\mu_X} = \frac{1}{\sqrt{1 + v_X}} \tag{14.17}$$

With reference to Figure 14.3 and the curve corresponding to $v_{\bar{c},\tan\phi'} = 0.8$, the ratio given by Eq. 14.17 is 0.781. For a soil with $\mu_{c'} = 100$ kPa and $\mu_{\tan\phi'} = \tan 30^\circ = 0.577$, as $\Theta \to 0$, these properties tend to $\tilde{\mu}_{c'} = 78.1$ kPa and $\tilde{\mu}_{\tan\phi'} = 0.451$, respectively. The limiting passive earth pressure with these median values is 265.7 kN/m, which leads to a normalized value of 0.71, as indicated at the left side of Figure 14.3

At the other extreme, as $\Theta \to \infty$, each realization of the Monte Carlo leads to an analysis of a uniform soil. In this

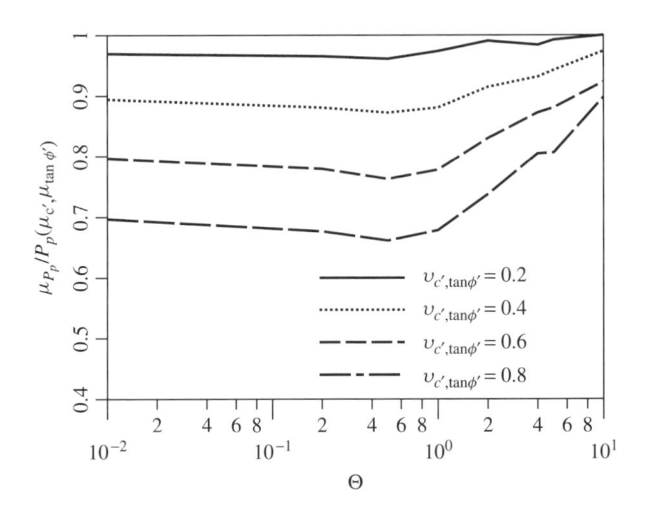

Figure 14.3 Influence of Θ on normalized μ_{P_p} for different $v_{\bar{c},\tan\phi'}$.

case there is no reduction of strength due to local averaging and the lines in Figure 14.3 all tend to unity on the right side. This is essentially the result indicated by the FOSM analysis.

All the lines indicate a slight minimum in the limiting passive resistance occurring close to or slightly lower than $\Theta \approx 1$. This value of Θ implies a spatial correlation length of the order of the height of the wall itself. Similar behavior was observed in Chapter 11 in relation to bearing capacity analysis. It is speculated that at this spatial correlation length there is a greater likelihood of weaker zones of soil aligning with each other, facilitating the formation of a failure mechanism.

The above discussion highlights the essential difference and benefits offered by the RFEM over conventional probabilistic methods. These can be summarized as follows:

1. The RFEM accounts for spatial correlation in a rigorous and objective way.
2. The RFEM does not require the user to anticipate the location or length of the failure mechanism. The mechanism forms naturally wherever the surface of least resistance happens to be.

Figure 14.4 shows the deformed mesh at failure from a typical realization of the Monte Carlo process. It can be seen that in this case the weaker light zone near the ground surface toward the center has triggered a quite localized mechanism that outcrops at this location.

Some other differences between FOSM and RFEM worth noting are as follows:

1. Figure 14.3 indicates that for intermediate values of Θ the RFEM results show a fall and even a minimum in the μ_{P_p} response as Θ is reduced, while FOSM gave essentially constant values. In fact, when second-order terms were included (Table 14.3) a slight *increase* in μ_{P_p} was observed.

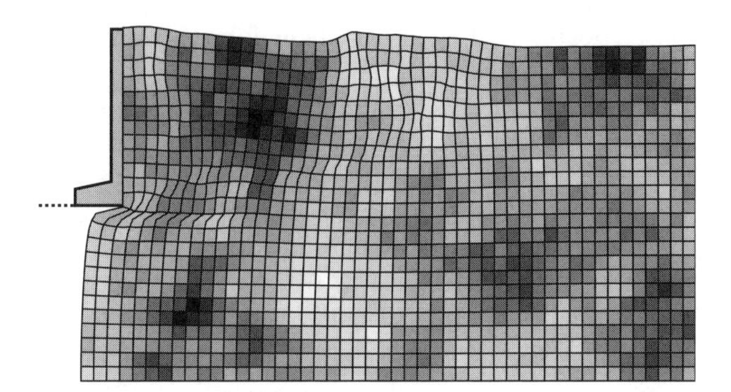

Figure 14.4 Typical passive failure mechanism. Light zones indicate weaker soil.

2. Using the FOSM Tables 14.1–14.3 indicated that the coefficient of variation of the passive earth force was similar to the coefficient of variation of the input shear strength parameters. Due to local averaging in the RFEM, on the other hand, the coefficient of variation of the passive earth force falls as Θ is reduced. As $\Theta \to 0$ in the RFEM approach, the coefficient of variation of the passive force also tends to zero.

14.2.5 Summary

The section has discussed two methods for implementing probabilistic concepts into geotechnical analysis of a simple problem of passive earth pressure. The "simple" method was the FOSM and the "sophisticated" method was the RFEM:

1. Probabilistic methods offer a more rational way of approaching geotechnical analysis, in which probabilities of design failure can be assessed. This is more meaningful than the abstract factor-of-safety approach. Being relatively new, however, probabilistic concepts can be quite difficult to digest, even in so-called simple methods.
2. The RFEM indicates a significant reduction in mean compressive strength due to the weaker zones dominating the overall strength at intermediate values of Θ. The observed reduction in the mean strength by RFEM is greater than could be explained by local averaging alone.
3. The study has shown that proper inclusion of spatial correlation, as used in the RFEM, is essential for quantitative predictions in probabilistic geotechnical analysis. While simpler methods such as FOSM (and FORM) are useful for giving guidance on the sensitivity of design outcomes to variations of input parameters, their inability to systematically include spatial correlation and local averaging limits their usefulness.
4. The study has shown that the RFEM is one of the very few methods available for modeling highly variable soils in a systematic way. In the analysis of soil masses, such as the passive earth pressure problem considered herein, a crucial advantage of RFEM is that it allows the failure mechanism to "seek out" the critical path through the soil.

14.3 ACTIVE EARTH PRESSURES: RETAINING WALL RELIABILITY

14.3.1 Introduction

Retaining wall design has long been carried out with the aid of either the Rankine or Coulomb theory of earth pressure.

To obtain a closed-form solution, these traditional earth pressure theories assume that the soil is uniform. The fact that soils are actually spatially variable leads, however, to two potential problems in design:

1. Do sampled soil properties adequately reflect the equivalent properties of the entire retained soil mass?
2. Does spatial variability of soil properties lead to active earth pressure effects that are significantly different than those predicted using traditional models?

This section combines nonlinear finite-element analysis with random-field simulation to investigate these two questions and assess just how safe current design practice is. The specific case investigated is a two-dimensional frictionless wall retaining a cohesionless drained backfill. The wall is designed against sliding using Rankine's earth pressure theory. The design friction angle and unit-weight values are obtained by sampling the simulated random soil field at one location and these sampled soil properties are then used as the equivalent soil properties in the Rankine model. Failure is defined as occurring when the Rankine predicted force acting on the retaining wall, modified by an appropriate factor of safety, is less than that computed by the RFEM employing the actual soil property (random) fields. Using Monte Carlo simulation, the probability of failure of the traditional design approach is assessed as a function of the factor of safety used and the spatial variability of the soil (Fenton and Griffiths, 2005a).

Retaining walls are, in most cases, designed to resist active earth pressures. The forces acting on the wall are typically determined using the Rankine or Coulomb theory of earth pressure after the retained soil properties have been estimated. This section compares the earth pressures predicted by Rankine's theory against those obtained via finite-element analysis in which the soil is assumed to be spatially random. The specific case of a two-dimensional cohesionless drained soil mass with a horizontal upper surface retained by a frictionless wall is examined. For a cohesionless soil the property of interest is the friction angle. The wall is assumed to be able to move away from the soil a sufficient distance to mobilize the frictional resistance of the soil.

The traditional theories of lateral active earth pressures are derived from equations of limit equilibrium along a planar surface passing through the soil mass. The soil is assumed to have a spatially constant friction angle. Under these conditions, and for the retaining problem considered herein, Rankine proposed the active earth pressure coefficient to be

$$K_a = \tan^2\left(\tfrac{\pi}{4} - \tfrac{1}{2}\phi'\right) \tag{14.18}$$

where ϕ' is the soil's drained friction angle (radians). Traditional theories assume that the unit weight γ is spatially constant also, so that the total lateral active earth force acting on a wall of height H, at height $H/3$, is given by

$$P_a = \tfrac{1}{2}\gamma H^2 K_a \tag{14.19}$$

The calculation of the lateral design load on a retaining wall involves estimating the friction angle ϕ' and the unit weight γ and then using Eqs. 14.18 and 14.19. To allow some margin for safety, the value of P_a may be adjusted by multiplying by a conservative factor of safety F_S.

Due to spatial variability, the failure surface is often more complicated than a simple plane and the resulting behavior cannot be expected to match that predicted by theory. Some work on reliability-based design of earth retaining walls has been carried out; see, for example, Basheer and Najjar (1996) and Chalermyanont and Benson (2004). However, these studies consider the soil to be spatially uniform; that is, each soil property is represented by a single random variable and every point in the soil is assigned the same property value. For example, a particular realization might have $\phi' = 32^\circ$, which would be assumed to apply to all points in the soil mass. The assumption that the soil is spatially uniform is convenient since most geotechnical predictive models are derived assuming spatially uniform properties (e.g., Rankine's earth pressure theory). These studies serve to help develop understanding of the underlying issues in reliability-based design of retaining walls but fail to include the effects of spatial variability. As will be seen, the failure surface can be significantly affected by spatial variability.

When spatial variability is included in the soil representation, alternative tractable solutions to the reliability issue must be found. For geotechnical problems which do not depend too strongly on extreme microscale soil structure, that is, which involve some local averaging, it can be argued that the behavior of the spatially random soil can be closely represented by a spatially uniform soil which is assigned the "equivalent" properties of the spatially random soil. The authors have been successful in the past with this equivalent property representation for a variety of geotechnical problems by defining the equivalent uniform soil as some sort of average of the random soil—generally the geometric average has been found to work well (see, e.g., Chapters 8, 9, 10, and 11). If the above argument holds, then it implies that the spatially random soil can be well modeled by equations such as 14.18 and 14.19, even though these equations are based on uniform soil properties—the problem becomes one of finding the appropriate equivalent soil properties.

In practice, the values of ϕ' and γ used in Eqs. 14.18 and 14.19 are obtained through site investigation. If the investigation is thorough enough to allow spatial variability to be characterized, an equivalent soil property can, in principle, be determined using random-field theory combined with simulation results. However, the level of site investigation required for such a characterization is unlikely to be worth carrying out for most retaining wall designs. In the more common case, the geotechnical engineer may base the design on a single estimate of the friction angle and unit weight. In this case, the accuracy of the prediction arising from Eqs. 14.18 and 14.19 depends very much on how well the single estimate approximates the equivalent value. This section addresses the above issues.

Figure 14.5 shows plots of what a typical retained soil might look like once the retaining wall has moved enough to mobilize the active soil behavior for two different possible realizations. The soil's spatially random friction angle is shown using a gray-scale representation, where light areas correspond to lower friction angles. Note that although the unit weight γ is also spatially random, its variability is not shown on the plots—its influence on the stochastic behavior of earth pressure was felt to be less important than that of the ϕ' field.

The wall is on the left-hand face and the deformed mesh plots of Figure 14.5 are obtained using the RFEM with eight-node square elements and an elastic, perfectly plastic constitutive model (see next section for more details). The wall is gradually moved away from the soil mass until plastic failure of the soil occurs and the deformed mesh at failure is then plotted. It is clear from these plots that the failure pattern is more complex than that found using traditional theories, such as Rankine's. Instead of a well-defined failure plane, the particular realization shown in the upper plot of Figure 14.5, for example, seems to have a failure wedge forming some distance from the wall in a region with higher friction angles. The formation of a failure surface can be viewed as the mechanism by which lateral loads stabilize to a constant value with increasing wall displacement.

Figure 14.5 also illustrates that choosing the correct location to sample the soil may be important to the accuracy of the prediction of the lateral active load. For example, in the lower plot of Figure 14.5, the soil sample, taken at the midpoint of the soil regime, results in a friction angle estimate which is considerably lower than the friction angle typically seen in the failure region (recall that white elements correspond to lower friction angles). The resulting predicted lateral active load, using Rankine's theory, is about 1.5 times that predicted by the RFEM, so that a wall designed using this soil sample would be overdesigned.

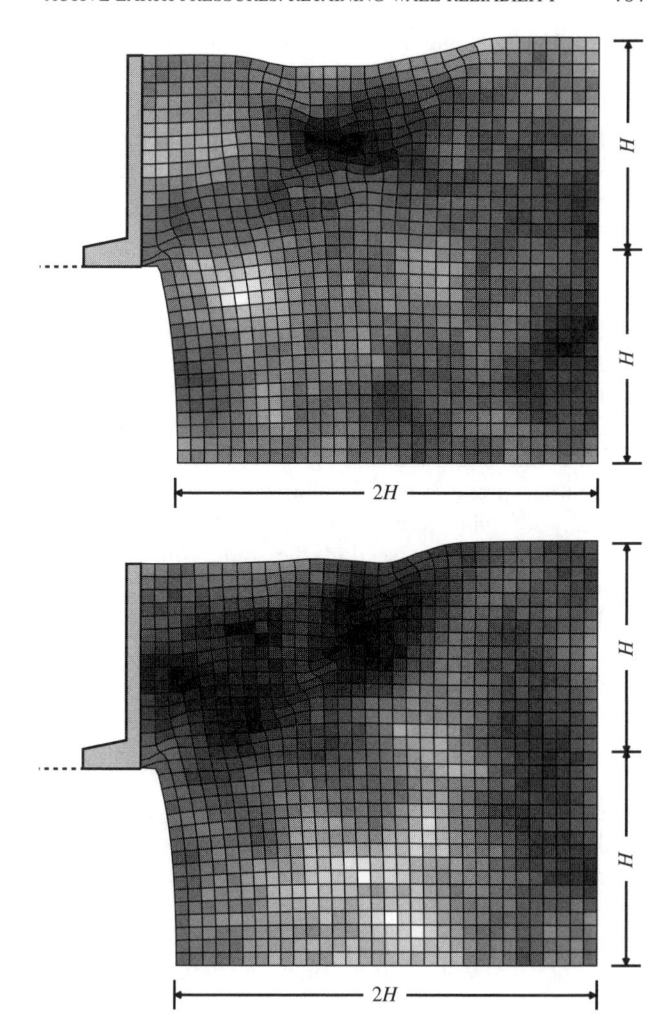

Figure 14.5 Active earth displacements for two different possible soil friction angle field realizations (both with $\theta_{\ln \tan\phi'}/H = 1$ and $v_{\tan\phi'} = 0.3$).

Quite the opposite is found for the more complex failure pattern in the upper plot of Figure 14.5, where the lateral active load found via the RFEM is more than two times that predicted using Rankine's theory and so a Rankine-based design would be unconservative. The higher RFEM load is attributed to the low-friction-angle material found in near proximity to the wall.

14.3.2 Random Finite-Element Method

The soil mass is discretized into 32 eight-noded square elements in the horizontal direction by 32 elements in the vertical direction. Each element has a side length of $H/16$, giving a soil block which is $2H$ in width by $2H$ in depth. (*Note*: Length units are not used here since the results can be used with any consistent set of length and force units.) The retaining wall extends to a depth H along the left face.

The finite-element earth pressure analysis uses an elastic, perfectly plastic Mohr–Coulomb constitutive model with stress redistribution achieved iteratively using an elasto-viscoplastic algorithm essentially similar to that described in the text by Smith and Griffiths (2004). The active wall considered in this study is modeled by translating the top 16 elements on the upper left side of the mesh uniformly horizontally and away from the soil. This translation is performed incrementally and models a rigid, smooth wall with no rotation.

The initial stress conditions in the mesh prior to translation of the nodes are that the vertical stresses equal the overburden pressure and the horizontal stresses are given by Jaky's (1944) formula in which $K_0 = 1 - \sin\phi'$. As described in the next section, the study will assume that $\tan\phi'$ is a lognormally distributed random field; hence K_0 will also be a random field (albeit fully determined by ϕ'), so that the initial stresses vary randomly down the wall face.

The boundary conditions are such that the right side of the mesh allows vertical but not horizontal movement, and the base of the mesh is fully restrained. The top and left sides of the mesh are unrestrained, with the exception of the nodes adjacent to the "wall," which have fixed horizontal components of displacement. The vertical components of these displaced nodes are free to move down, as active conditions are mobilized. These boundary conditions have been shown to work well for simple earth pressure analysis (see, e.g., Griffiths, 1980).

Following incremental displacement of the nodes, the viscoplastic algorithm monitors the stresses in all the elements (at the Gauss points) and compares them with the strength of the element based on the Mohr–Coulomb failure criterion. If the failure criterion is not violated, the element is assumed to remain elastic; however, if the criterion is violated, stress redistribution is initiated by the viscoplastic algorithm. The process is inherently iterative, and convergence is achieved when all stresses within the mesh satisfy the failure criterion and global stress equilibrium within quite tight tolerances.

At convergence following each increment of displacement, the mobilized active reaction force on the wall is computed by integrating the stresses in the elements attached to the displaced nodes. The finite-element analysis is terminated when the incremental displacements have resulted in the active reaction force reaching its minimum limiting value.

The cohesionless soil being studied here has two properties of primary interest to the active earth pressure problem: the friction angle $\phi'(\mathbf{x})$ and the unit weight $\gamma(\mathbf{x})$, where $\mathbf{x}$ is the spatial position. Both are considered to be spatially random fields. The finite-element model used in this study also includes the soil's dilation angle, taken to be zero, Poisson's ratio, taken to be 0.3, and Young's modulus, taken to be 1×10^5. These three properties are assumed to be spatially constant—this does not introduce significant error since these properties play only a minor role in the limiting active earth pressures.

The two properties which are considered to be spatially random, ϕ' and γ, are characterized by their means, their standard deviations, and their correlation lengths (which are measures of the degree of spatial correlation). The unit weight is assumed to have a lognormal distribution, primarily because of its simple relationship with the normal distribution, which is fully specified by the first two moments, and because it is nonnegative. The friction angle ϕ' is generally bounded, which means that its distribution is a complicated function with at least four parameters (see Section 1.10.10). However, $\tan\phi'$ varies between zero and infinity as ϕ' varies between zero and 90°. Thus, a possible distribution for $\tan\phi'$ is also the lognormal. This distribution will be assumed in this section; that is, the friction angle field will be represented by the lognormally distributed $\tan\phi'$ field.

The spatial correlation structure of both fields will be assumed to be the same. This is not only for simplicity, since it can be argued that the spatial correlation of a soil is governed largely by the spatial variability in a soil's source materials, weathering patterns, stress and formation history, and so on. That is, the material source, weathering, stress history, and so on, forming a soil at a point will be similar to those at a closely neighboring point, so one would expect that all the soil's properties will vary similarly between the two points (aside from deviations arising from differing nonlinear property response to current conditions).

With this argument in mind, the spatial correlation function for the $\ln(\gamma)$ and $\ln(\tan\phi')$ fields, both normally distributed, is assumed to be Markovian,

$$\rho(\boldsymbol{\tau}) = \exp\left\{\frac{-2|\boldsymbol{\tau}|}{\theta}\right\} \tag{14.20}$$

where θ is the correlation length beyond which two points in the field are largely uncorrelated, $\boldsymbol{\tau}$ is the vector between the two points, and $|\boldsymbol{\tau}|$ is its absolute length.

In this study, the two random fields γ and $\tan\phi'$ are first assumed to be independent. Thus, two independent standard normal random fields $G_1(\mathbf{x})$ and $G_2(\mathbf{x})$ are simulated using the LAS method (see Section 6.4.6) using the correlation structure given by Eq. 14.20. These fields are then transformed to the target fields through the relationships

$$\gamma(\mathbf{x}) = \exp\left\{\mu_{\ln\gamma} + \sigma_{\ln\gamma} G_1(\mathbf{x})\right\} \tag{14.21a}$$

$$\tan\,\phi'(\mathbf{x}) = \exp\left\{\mu_{\ln\tan\phi'} + \sigma_{\ln\tan\phi'} G_2(\mathbf{x})\right\} \tag{14.21b}$$

where μ and σ are the mean and standard deviation of the subscripted variable obtained using the transformations

$$\sigma^2_{\ln\gamma} = \ln\left(1 + v^2_\gamma\right) \tag{14.22a}$$

$$\mu_{\ln\gamma} = \ln(\mu_\gamma) - \tfrac{1}{2}\sigma^2_{\ln\gamma} \tag{14.22b}$$

and $v_\gamma = \sigma_\gamma/\mu_\gamma$ is the coefficient of variation of γ. A similar transformation can be applied for the mean and variance of $\tan\phi'$ by replacing γ with $\tan\phi'$ in the subscripts of Eq. 14.22.

Since the friction angle ϕ' and unit weight γ generally have a reasonably strong positive correlation, a second case will be considered in this study where the two fields are significantly correlated; specifically, a correlation coefficient of $\rho = 0.8$ will be assumed to act between $\ln(\gamma)$ and $\ln(\tan\phi')$ at each point $\mathbf{x}$ in the soil. Thus, when the friction angle is large, the unit weight will also tend to be large within their respective distributions. The correlation between the fields is implemented using the covariance matrix decomposition method (see Section 6.4.2).

Once realizations of the soil have been produced using LAS and the above transformations, the properties can be mapped to the elements and the soil mass analyzed by the finite-element method. See Figure 14.5 for two examples. Repeating this analysis over a sequence of realizations (Monte Carlo simulation, see Section 6.6) yields a sequence of computed responses, allowing the distribution of the response to be estimated.

14.3.3 Active Earth Pressure Design Reliability

As mentioned in Section 14.3.1, the design of a retaining wall involves two steps: (1) estimating the pertinent soil properties and (2) predicting the lateral load through, for example, Eq. 14.19. The reliability of the resulting design depends on the relationship between the predicted and actual lateral loads. Disregarding variability on the resistance side and assuming that the design wall resistance R satisfies

$$R = F_S P_a \tag{14.23}$$

where F_S is a factor of safety and P_a is the predicted active lateral earth load (Eq. 14.19), then the wall will survive if the true active lateral load P_t is less than $F_S P_a$. The true active lateral load will inevitably differ from that predicted because of errors in the estimation of the soil properties and because of the spatial variability present in a true soil which is not accounted for by classical theories, such as Eqs. 14.18 and 14.19. The probability of failure of the retaining system will be defined as the probability that the true lateral load P_t exceeds the factored resistance,

$$p_f = \mathrm{P}\left[P_t > R\right] = \mathrm{P}\left[P_t > F_S P_a\right] \tag{14.24}$$

This is the theoretical definition of the failure probability p_f. In the following section, the estimate of this failure probability $\hat{p}_f$ will be obtained by Monte Carlo simulation. The "true" (random) lateral load P_t will be assumed in this study to be closely approximated by the load computed in the finite-element analysis of each soil realization. That is, it is assumed that the finite-element analysis, which accounts for spatial variability, will produce a realistic assessment of the actual lateral active soil load for a given realization of soil properties.

The predicted lateral load P_a depends on an estimate of the soil properties. In this section, the soil properties γ and $\tan\phi'$ will be estimated using only a single "virtual sample" taken at a distance H in from the base of the retaining wall and a distance H down from the soil surface. The term virtual sample means that the properties are sampled from the random-field realizations assigned to the finite-element mesh. Specifically, virtual sampling means that for $\mathbf{x}_s$ being the coordinates of the sample point, the sampled soil properties $\hat{\gamma}$ and $\hat{\phi}'$ are obtained from each random-field realization as

$$\hat{\gamma} = \gamma(\mathbf{x}_s) \tag{14.25a}$$

$$\hat{\phi}' = \tan^{-1}\left(\tan(\phi'(\mathbf{x}_s))\right) \tag{14.25b}$$

Armed with these sample properties, the predicted lateral load becomes (for ϕ' in radians)

$$P_a = \tfrac{1}{2}\hat{\gamma}H^2\tan^2\left(\tfrac{\pi}{4} - \tfrac{1}{2}\hat{\phi}'\right) \tag{14.26}$$

No attempt is made to incorporate measurement error. The goal of this study is to assess the design risk arising from the spatial variability of the soil and not from other sources of variability.

Table 14.4 lists the statistical parameters varied in this study. The coefficient of variation $v = \sigma/\mu$ is changed for both the unit weight γ and the friction $\tan\phi'$ fields identically. That is, when the coefficient of variation of the unit weight field is 0.2, the coefficient of variation of the $\tan\phi'$ field is also 0.2, and so on. For each parameter set considered in Table 14.4, the factor of safety F_S, is varied from 1.5 to 3.0. This range is somewhat wider than the range of 1.5–2.0 recommended by the *Canadian Foundation Engineering Manual* (CFEM; CGS, 1992) for retaining wall systems.

Table 14.4 Parameters Varied in Study While Holding Retained Soil Dimension H and Soil Properties $\mu_{\tan\phi'} = \tan 30°$, $\mu_\gamma = 20$, $E = 1 \times 10^5$, and $\nu = 0.3$ Constant

Parameter	Values Considered
σ/μ	0.02, 0.05, 0.1, 0.2, 0.3, 0.5
θ/H	0.1, 0.2, 0.5, 1.0, 2.0, 5.0
ρ	0.0, 0.8

Note: For each parameter set, 1000 realizations were run.

The correlation length θ, which is normalized in Table 14.4 by expressing it as a fraction of the wall height θ/H, governs the degree of spatial variability. When θ/H is small, the random field is typically rough in appearance—points in the field are more independent. Conversely, when θ/H is large, the field is more strongly correlated so that it appears smoother with less variability in each realization. A large correlation length has two implications: First, the soil properties estimated by sampling the field at a single location will be more representative of the overall soil mass and, second, the reduced spatial variability means that the soil will behave more like that predicted by traditional theory. Thus, for larger correlation lengths, fewer "failures" are expected (where the actual lateral limit load exceeds the factored prediction) and the factor of safety can be reduced. For intermediate correlation lengths, however, the soil properties measured at one location may be quite different from those actually present at other locations. Thus, for intermediate correlation lengths, more failures are expected. When the correlation length becomes extremely small, much smaller than the soil property sample size, local averaging effects begin to take over and both the sample and overall soil mass return to being an effectively uniform soil (with properties approaching the median), accurately predicted by traditional theory using the sample estimate.

Following this reasoning, the maximum probability of failure of the design is expected to occur when the correlation length is some intermediate value. Evidence supporting this argument is found in the next section.

14.3.4 Monte Carlo Results

Both plots of Figure 14.5 indicate that it is the high-friction-angle regions which attract the failure surface in the active case. While this is not always the case for all realizations, it tends to be the most common behavior. Such a counterintuitive observation seems to be largely due to the interaction between the initial horizontal stress distribution, as dictated by the $K_o = 1 - \sin\phi'$ random field, and the friction angle field.

To explain this behavior, it is instructive to consider the Mohr's circles corresponding to $K_o = 1 - \sin\phi'$ (at rest, initial, conditions) and $K_a = (1 - \sin\phi')/(1 + \sin\phi')$ (active failure conditions). As ϕ' increases from zero, the distance between the initial and failure circles increases, reaching a maximum when $\phi' = \tan^{-1}(0.5\sqrt{2}\sqrt{\sqrt{2} - 1}) = 24.47°$. Beyond this point, the distance between the initial and failure circles decreases with increasing ϕ'. Since the average drained friction angle used in this study is 30° (to first order), the majority of realizations of ϕ' are in this region of decreasing distance between circles. This supports the observation that, under these conditions, the higher friction angle regions tend to reach active failure first. It can still be stated that failure is always attracted to the weakest zones, even if those weakest zones happen to have a higher friction angle. In this sense the gray scale shown in Figure 14.5 is only telling part of the story—it is really the Coulomb shear strength ($\sigma' \tan\phi'$) which is important.

The attraction of the failure surface to the high-friction-angle regions is due to the fact that the initial conditions vary with ϕ' according to Jaky's formula in this study. In a side investigation, it was found that if the value of K_o is held fixed, then the failure surface does pass through the lower friction angle regions. Figure 14.6 shows the effect that K_o has on the location of the failure surface. In Figure 14.6*a*, K_o is held spatially constant at 0.5 and, in this case, the failure surface clearly gravitates toward the low-friction-angle regions. In Figure 14.6*b*, K_o is set equal to $1 - \sin\phi'$, as in the rest of the section, and the failure surface clearly prefers the high-friction-angle regions. The authors also investigated the effect of spatially variable versus spatially constant unit weight and found that this had little effect on the failure surface location, at least for the levels of variability considered here. The location of the failure surface seems to be primarily governed by the nature of K_o (given random ϕ').

The migration of the failure surface through the weakest path means that, in general, the lateral wall load will be different than that predicted by a model based on uniform soil properties, such as Rankine's theory. Figure 14.7 shows the estimated probability of failure $\hat{p}_f$ that the actual lateral active load exceeds the factored predicted design load (see Eq. 14.24) for a moderate correlation length ($\theta/H = 1$) and for various coefficients of variation in the friction angle and unit weight. The estimates are obtained by counting the number of failures encountered in the simulation and dividing by the total number of realizations considered ($n = 1000$). In that this is an estimate of a proportion, its standard error (one standard deviation) is $\sqrt{p_f(1 - p_f)/n}$, which is about 1% when $p_f = 20\%$ and about 0.3% when $p_f = 1\%$. The figure shows two cases: (a) where the

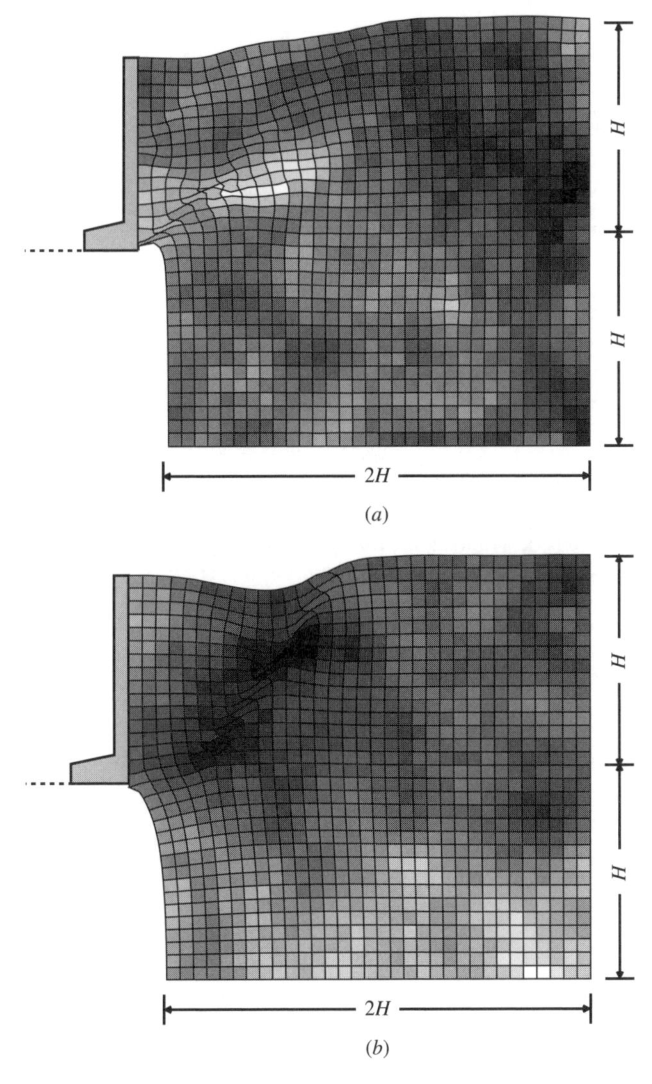

Figure 14.6 Active earth displacements for two different possible soil friction angle field realizations (both with $\theta/H = 1$ and $\sigma/\mu = 0.3$): (*a*) K_o held spatially constant at 0.5; (b) $K_o = 1 - \sin\phi'$ is a spatially random field derived from ϕ'.

friction angle and unit-weight fields are independent and (b) where there is a strong correlation between the two fields.

As expected, the probability of failure increases as the soil becomes increasingly variable. Figure 14.7 can be used to determine a required factor of safety corresponding to a target probability of failure. For example, if the fields are assumed to be independent (Figure 14.7*a*), with $v = 0.2$, and the soil properties are sampled as in this study, then a required factor of safety of about $F_S = 2$ is appropriate for a target probability of failure of 5%. The required factor of safety increases to 3 or more when $v \geq 0.3$. Recalling that only one sample is used in this study to characterize the soil and that the sample is well outside the expected failure zone (albeit without any measurement error), the required factor of safety may be reduced if more samples are taken or if the sample is taken closer to the wall resulting in a more accurate characterization of the soil.

Figure 14.7*b* shows the estimated probability of failure for the same conditions as in Figure 14.7*a*, except that now the friction angle and unit-weight fields are strongly correlated ($\rho = 0.8$). The main effects of introducing correlation between the two fields are (1) slightly reducing the average wall reaction and (2) significantly reducing the wall reaction variance (correlation between "input" parameters tends to reduce variance in the "output"). These two effects lead to a reduction in failure probability which leads in turn to a reduction in the required factor of safety for the same target failure probability. For example, the required factor of safety in the case of strongly correlated fields with $v \geq 0.3$ is only $F_S \geq 2$ for a probability of failure of 5%.

Figure 14.8 shows the estimated probability of failure $\hat{p}_f$ for $v = 0.2$ against the correlation length θ/H for the two cases of (a) independence between the friction angle and unit-weight fields and (b) strong correlation between the fields ($\rho = 0.8$). Notice that for the correlated fields of Figure 14.8*b*, the probability of failure is negligible for all $F_S \geq 2$ when $v = 0.2$.

As anticipated in Section 14.3.3 and shown in Figure 14.8, there is a worst-case correlation length, where the probability of failure reaches a maximum. A similar worst case is seen for all v values considered. This worst-case correlation length is typically of the order of the depth of the wall ($\theta = 0.5H$ to $\theta = H$). The importance of this observation is that this worst-case correlation length can be conservatively used for reliability analyses in the absence of improved information. Since the correlation length is quite difficult to estimate in practice, requiring substantial data, a methodology that does not require its estimation is preferable.

14.3.5 Summary

On the basis of this simulation study, the following observations can be made for a cohesionless backfill:

1. The behavior of a spatially variable soil mass is considerably more complex than suggested by the simple models of Rankine and Coulomb. The traditional approach to compensating for this model error is to appropriately factor the lateral load predicted by the model.
2. The failure mode of the soil in the active case suggests that the failure surface is controlled by high-friction-angle regions when K_o is defined according to Jaky's formula (and is thus spatially variable). When K_o

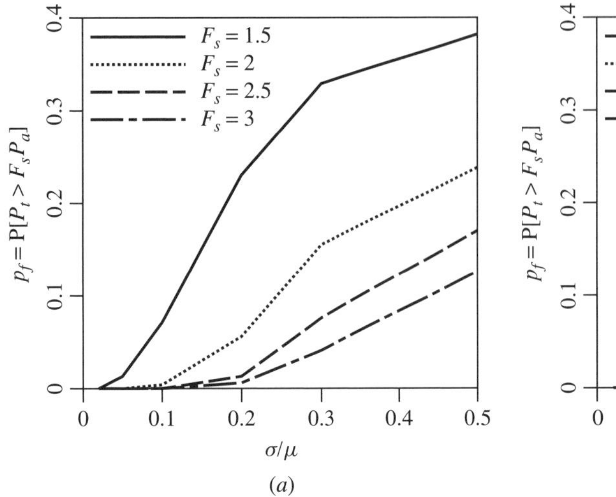

(*a*)

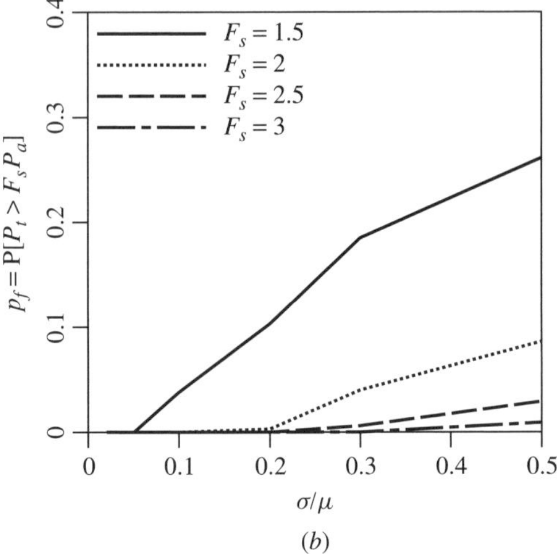

(*b*)

Figure 14.7 Estimated probability that actual load exceeds design load, $\hat{p}_f$, for $\theta/H = 1$: (*a*) ϕ' and γ fields are independent ($\rho = 0$); (*b*) two fields are strongly correlated ($\rho = 0.8$).

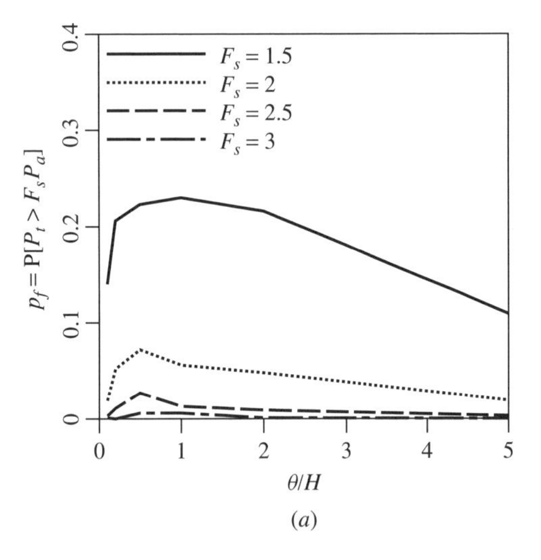

(*a*)

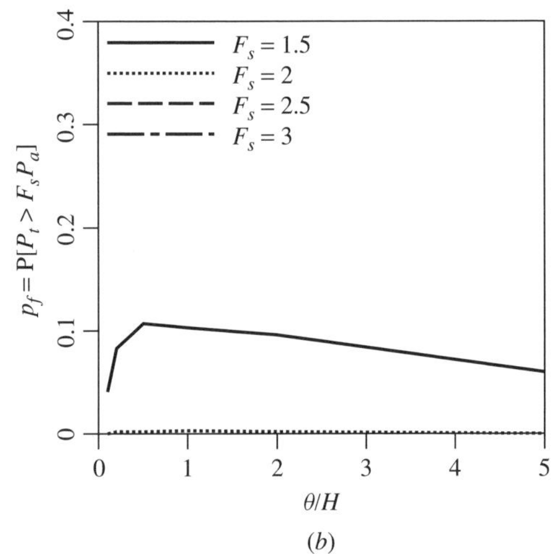

(*b*)

Figure 14.8 Estimated probability that actual load exceeds design load $\hat{p}_f$ for $\sigma/\mu = 0.2$: (*a*) ϕ' and γ fields are independent ($\rho = 0$); (*b*) two fields are strongly correlated ($\rho = 0.8$).

is held spatially constant, the failure surface tends to pass preferentially through the low-friction-angle regions.

3. Taking the friction angle and unit-weight fields to be independent is conservative in that it leads to higher estimated probabilities of failure.
4. In the case when the friction angle and unit-weight fields are taken to be independent and when the soil is sampled at a single point at a moderate distance from the wall, the probabilities of failure are quite high and a factor of safety of about 2.0—3.0 is required to maintain a reasonable reliability (95%) unless it is known that the coefficient of variation for the soil is less than about 20%. Since for larger coefficients of variation the required factors of safety are above those recommended by, say, the CFEM (CGS, 1992), the importance of a more than minimal site investigation is highlighted.
5. Assuming a strong correlation between the friction angle and unit-weight fields leads to factors of safety which are more in line with those recommended by CFEM. However, further research is required to determine if (and under what conditions) this strong correlation should be depended upon in a design.
6. As has been found for a number of different classical geotechnical problems (e.g., differential settlement and bearing capacity), a worst-case correlation length exists for the active earth pressure problem which is of

the order of the retaining wall height. The important implication of this observation is that the correlation length need not be estimated—the worst-case scale can be used to yield a conservative design at a target reliability. This is a practical advantage because the correlation length is generally difficult and expensive to estimate accurately, requiring a large number of samples.

In summary, there is much that still needs to be investigated to fully understand the probabilistic active behavior of retained soils. In particular, the effect of sampling intensity on design reliability and the type of sample average best suited to represent the equivalent soil property are two areas which must be investigated further using this study as a basis before a formal reliability-based design code can be developed.

CHAPTER 15

Mine Pillar Capacity

15.1 INTRODUCTION

In this chapter we investigate the effect of spatial variability on the overall strength of rock or coal pillars (Griffiths et al., 2001, 2002a). These pillars are commonly provided at various intervals to provide roof support in a mine. The probabilistic estimates of pillar capacity are produced using the program RPILL2D (or RPILL3D), which is available at `http://www.engmath.dal.ca/rfem`. The results of this study enable traditional approaches involving factors of safety to be reinterpreted as a *probability of failure* in the context of reliability-based design.

A review and assessment of existing design methods for estimating the factor of safety of coal pillars based on statistical approaches was covered by Salamon (1999). This chapter investigates in a rigorous way the influence of rock strength variability on the overall compressive strength of rock pillars typically used in mining and underground construction. The investigation merges elasto-plastic finite-element analysis (e.g., Smith and Griffiths, 2004) with random-field theory (e.g., Vanmarcke, 1984; Fenton 1990) within a Monte Carlo framework in an approach referred to as the random finite-element method (RFEM).

The rock strength is characterized by its unconfined compressive strength or *cohesion* c using an elastic, perfectly plastic Tresca failure criterion. The variable c is assumed to be lognormally distributed (so that $\ln c$ is normally distributed) with three parameters as shown in Table 15.1. The correlation length describes the distance over which the spatially random values are significantly correlated in the underlying Gaussian field. A large correlation length implies a smoothly varying field, while a small correlation length implies a highly variable field. In order to nondimensionalize the analysis, the rock strength variability is expressed in terms of its coefficient of variation:

$$v_c = \frac{\sigma_c}{\mu_c} \tag{15.1}$$

and the correlation length is normalized with respect to the pillar dimension B,

$$\Theta = \frac{\theta_{\ln c}}{B} \tag{15.2}$$

where B is the height (and width) of the pillar as illustrated in Figure 15.1.

The spatially varying rock strength field is simulated using the LAS method (see Section 6.4.6), which produces a sequence of normally distributed random values G_i, which represent local arithmetic averages of the standardized $\ln c$ field over each element $i = 1, 2, \ldots$. In turn, the ith element is assigned a random value, c_i, which is a local geometric average, over the element, of the continuously varying random field having point statistics derived from Table 15.1, according to

$$c_i = \exp\{\mu_{\ln c} + \sigma_{\ln c} G_i\} \tag{15.3}$$

(recall that the geometric average is the arithmetic average of the logarithms raised to the power e, see Section 4.4.2). The element values thus correctly reflect the variance reduction due to arithmetic averaging over the element as well as the correlation structure dictated by the correlation length, $\theta_{\ln c}$. In this study, an exponentially decaying (Markovian)

Table 15.1 Input Parameters for Rock Strength c

Parameters	Symbols	Units
Mean	μ_c	kN/m^2
Standard deviation	σ_c	kN/m^2
Correlation length	$\theta_{\ln c}$	m

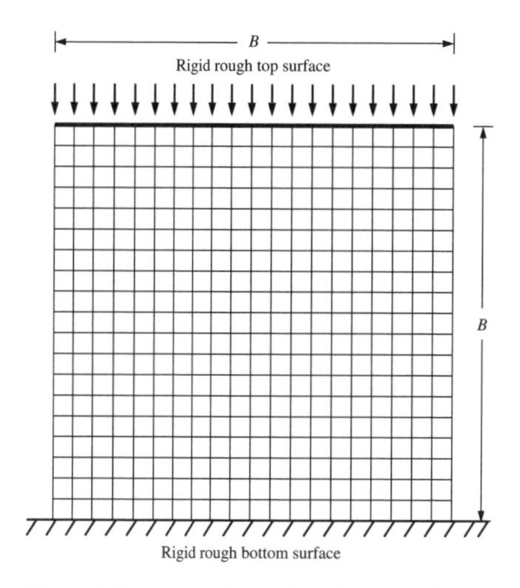

Figure 15.1 Mesh used for finite-element pillar analyses.

correlation function is assumed:

$$\rho(\tau) = \exp\left\{-\frac{2|\tau|}{\theta_{\ln c}}\right\} \tag{15.4}$$

where τ is the distance between any two points in the rock mass. Notice that the above correlation function is isotropic, which is to say two points separated by 0.2 m vertically have the same correlation coefficient as two points separated by 0.2 m horizontally. While it is unlikely that actual rock properties will have an isotropic correlation structure (e.g., due to layering), the basic probabilistic behavior of pillar failure can be established in the isotropic case and anisotropic site-specific refinements left to the reader. The methodologies and general trends will be similar to the results presented here.

The present study is confined to plane strain pillars with square dimensions. A typical finite-element mesh is shown in Figure 15.1 and consists of 400 eight-node plane strain quadrilateral elements. Each element is assigned a different c-value based on the underlying lognormal distribution, as discussed above. For each Monte Carlo simulation, the block is compressed by incrementally displacing the top surface vertically downward. At convergence following each displacement increment, the nodal reaction loads are summed and divided by the width of the block B to give the average axial stress. The maximum value of this axial stress q_u is then defined as the compressive strength of the block.

This study focuses on the dimensionless *bearing capacity factor* N_c defined for each of the n_{sim} Monte Carlo simulations as

$$N_{c_i} = \frac{q_{u_i}}{\mu_c}, \qquad i = 1, 2, \ldots, n_{\text{sim}} \tag{15.5}$$

It should be noted that N_{c_i} for each simulation is nondimensionalized by dividing q_u by the *mean* compressive strength μ_c. The N_{c_i} values are then analyzed statistically leading to a sample mean m_{N_c},

$$m_{N_c} = \frac{1}{n_{\text{sim}}} \sum_{i=1}^{n_{\text{sim}}} N_{c_i} \tag{15.6}$$

and sample standard deviation s_{N_c},

$$s_{N_c} = \sqrt{\frac{1}{n_{\text{sim}} - 1} \sum_{i=1}^{n_{\text{sim}}} (N_{c_i} - m_{N_c})^2} \tag{15.7}$$

These statistics, in turn, can be used to estimate probabilities concerning the compressive strength of the pillar.

A *uniform* rock, having spatially constant strength c, has an unconfined compressive strength from Mohr's circle given by $N_c = 2$; hence, for a uniform rock,

$$q_u = 2c \tag{15.8}$$

Of particular interest in this study, therefore, is to compare this deterministic value of 2 with m_{N_c} from the RFEM analyses.

15.2 LITERATURE

Although reliability-based approaches have not yet been widely implemented by geotechnical engineers in routine design, there has been a significant growth in interest in this area as an alternative to the more traditional factor of safety. A valid criticism of the factor of safety is that it does not give as much physical insight into the likelihood of design failure as a probabilistic measure (e.g., Singh, 1972). Even though a reliability-based analysis tells more about the safety of a design, engineers have tended to prefer the factor of safety approach since there is a perception that it takes less time to compute (e.g., Thorne and Quine, 1993). This perception is no doubt well based since factor of safety approaches are generally fairly simple, but the old addage—You get what you pay for—applies here. The understanding of the basic failure mechanism afforded by the consideration of spatial variation is well worth the effort. In addition to increasing understanding and safety, reliability-based design can also maximize cost efficiency (e.g., Call, 1985).

Both variability and correlation lengths of material properties can affect the reliability of geotechnical systems. While the variability of geotechnical properties are hard to determine since soil and rock properties can vary widely (e.g., Phoon and Kulhawy, 1999; Harr, 1987; Lumb, 1966; Lee et al., 1983), there is some consensus that v_c values for rock strength range from 0.30 to 0.50 (e.g., Hoek 1998; Savely, 1987; Hoek and Brown, 1997). This variability has been represented in the present study by a lognormal distribution that ensures nonnegative strength values. The correlation length can also affect system reliability (e.g., Mostyn and Li, 1993; Lacasse and Nadim, 1996; DeGroot, 1996; Wickremesinghe and Campanella, 1993; Cherubini, 2000). In mining applications, material variability is not usually accounted for directly; however, empirical formulas have been developed to adjust factors of safety accordingly (e.g., Salamon, 1999; Peng and Dutta, 1992; Scovazzo 1992).

Finite-element analysis has been used in the past to account for varying properties of geotechnical problems including pillar design (see, e.g., Park, 1992; Peng and Dutta, 1992; Tan et al., 1993; Mellah et al., 2000; Dai et al., 1993). In this chapter, elasto-plastic finite-element analysis has been combined with random-field theory to investigate the influence of material variability and correlation lengths on mine pillar stability. By using multiple simulations, the Monte-Carlo technique can be used to predict pillar reliability involving materials with high variances and spatial variability that would not be amenable to analysis by first-order second-moment methods.

15.3 PARAMETRIC STUDIES

Analyses were performed with input parameters within the following ranges:

$$0.01 < \Theta < 10, \qquad 0.05 < v_c < 1.6$$

For each pair of values of v_c and Θ, 2500 Monte Carlo simulations were performed, and from these, the estimated statistics of the bearing capacity factor N_c were computed leading to the sample mean m_{N_c} and sample standard deviation s_{N_c}.

In order to maintain reasonable accuracy and run-time efficiency, the sensitivity of results to mesh density and the number of Monte Carlo simulations was examined. Figure 15.2 shows the effect of varying the mesh size with all other variables held constant. Since there is little change from the 20×20 element mesh to the 40×40 element mesh, the 20×20 element mesh is deemed to give reasonable precision for the analysis. Figure 15.3 shows the convergence of m_{N_c} as the number of simulations increases. The figure displays five repeated analyses with identical properties and indicates that 2500 simulations give reasonable precision and reproducibility. Although higher precision could be achieved with greater mesh density and simulation counts, the use of a 20×20 element mesh with $n_{\text{sim}} = 2500$ simulations is considered to be accurate enough in view of the inherent uncertainty of the input statistics.

The accuracy of results obtained from Monte Carlo analyses can also be directly computed from the number of simulations. Estimated mean bearing capacities will have a standard error ($\pm$ one standard deviation) equal to the sample standard deviation times $1/\sqrt{n_{\text{sim}}} = 1/\sqrt{2500} = 0.020$ or about 2% of the sample standard deviation. Similarly, the estimated

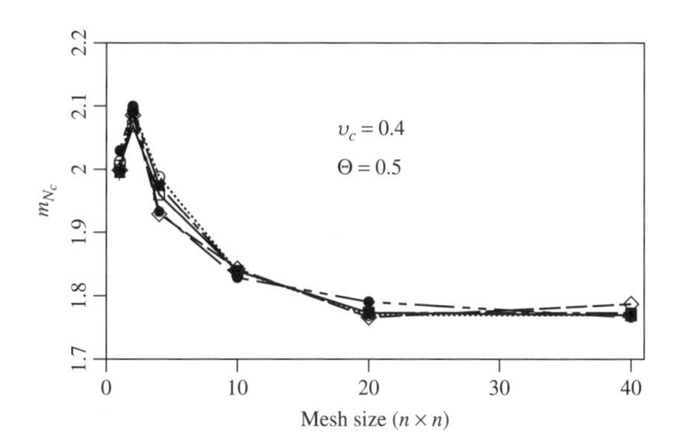

Figure 15.2 Influence of mesh density on accuracy of computed m_{N_c} with 2500 simulations.

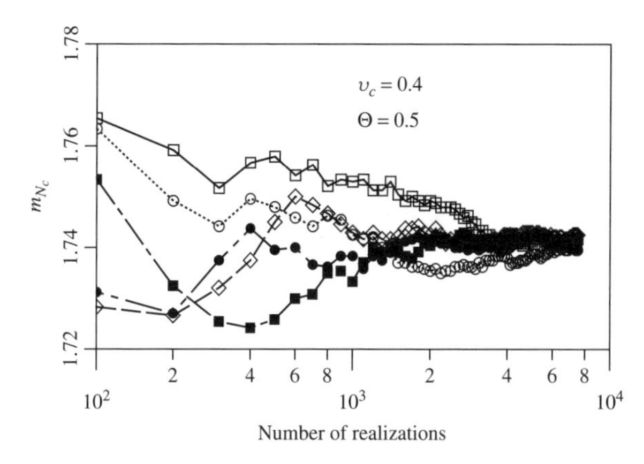

Figure 15.3 Influence of number of simulations on accuracy of computed m_{N_c}.

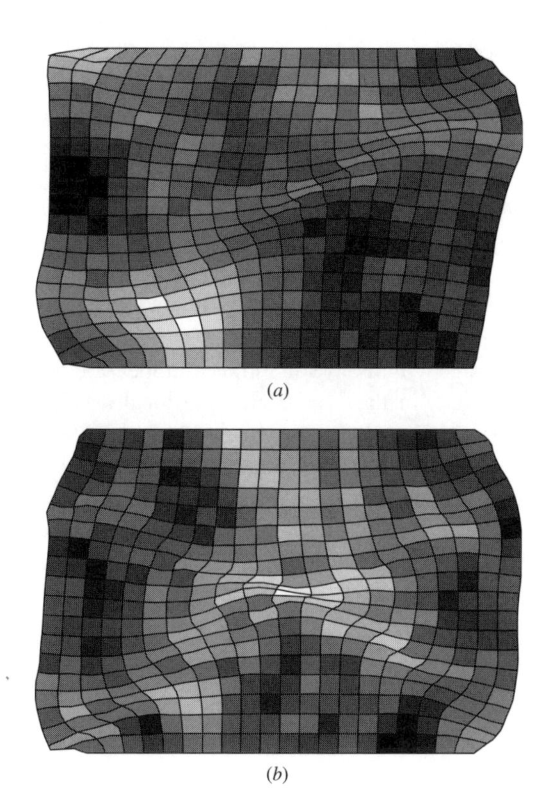

Figure 15.4 Typical deformed meshes and gray scales at failure: (*a*) $v_c = 0.5, \Theta = 0.4$; (*b*) $v_c = 0.5, \Theta = 0.2$. Lighter zones signify weaker rock.

variance will have a standard error equal to the sample variance times $\sqrt{(2/(n_{\text{sim}} - 1))} = \sqrt{(2/2499)} = 0.028$, or about 3% of the sample variance. This means that estimated quantities will generally be within about 4% of the true (i.e., finite element) quantities, statistically speaking.

Figures 15.4*a* and 15.4*b* show two typical deformed meshes at failure, corresponding to $\Theta = 0.4$ and $\Theta = 0.2$, respectively. Lighter regions in the plots indicate weaker rock, and darker regions indicate stronger rock. It is clear that the weak (dark) regions have triggered quite irregular failure mechanisms. In general, the mechanism is attracted to the weak zones and "avoids" the strong zones. This suggests that failure is not simply a function of the arithmetic average of rock strength—it is somewhat reduced due to the failure path seeking out weak materials.

15.3.1 Mean of N_c

A summary of the sample mean bearing capacity factor (m_{N_c}), computed using the values provided in Section 15.3, for each simulation is shown in Figure 15.5. The plots confirm that for low values of v_c, m_{N_c} tends to the deterministic value of 2. As the v_c of the rock increases, the mean bearing capacity factor falls quite rapidly, especially for smaller values of Θ. As shown in Figure 15.5*b*, however, m_{N_c} reaches a minimum at about $\Theta = 0.2$ and starts to climb again. In the limit as $\Theta \to 0$, there are no "preferential" weak paths the failure mechanism can follow, and the mean bearing capacity factors return to deterministic values dictated by the median (see, e.g., Eq. 14.17). For example, in Figure 15.5*b*, when $v_c = 1$, $m_{N_c} \to 2/\sqrt{2} = 1.41$ as $\Theta \to 0$. In principle, the $\Theta = 0$ case is somewhat delicate to investigate. Strictly speaking, any local average of a (finite variance) random $\ln c$ field having $\Theta = 0$ will have zero variance (since the local average will involve an infinite number of independent points). Thus, in the $\Theta = 0$ case the local average representation, that is, the finite-element method (as interpreted here), will necessarily return to the deterministic case. The detailed investigation of this trend is also complicated by the fact that rock properties are never determined at the "point" level—they are based on a local average over the rock sample volume. Thus, while recognizing the apparent trend with small Θ in this study, the theoretical and numerical verification of the limiting trend is left for further research. Also included in Figure 15.5*a* is the horizontal line corresponding to the solution that would be obtained for $\Theta = \infty$. This hypothetical case implies that each simulation of the Monte Carlo process produces a uniform soil, albeit with properties varying from one simulation to the next. In this case, the distribution of q_u will be statistically similar to the distribution of c but magnified by 2, thus $m_{N_c} = 2$ for all values of v_c.

15.3.2 Coefficient of Variation of N_c

Figure 15.6 shows the influence of Θ and v_c on the sample coefficient of variation of the estimated bearing capacity factor, $v_{N_c} = s_{N_c}/m_{N_c}$. The plots indicate that v_{N_c} is positively correlated with both v_c and Θ, with the limiting value of $\Theta = \infty$ giving the straight line $v_{N_c} = v_c$.

15.4 PROBABILISTIC INTERPRETATION

Following Monte Carlo simulations for each parametric combination of input parameters (Θ and v_c), the suite of computed bearing capacity factor values from Eq. 15.5 was plotted in the form of a histogram, and a "best-fit" lognormal distribution superimposed. An example of such a plot is shown in Figure 15.7 for the case where $\Theta = 0.5$ and $v_c = 0.4$.

Since the lognormal fit has been normalized to enclose an area of unity, areas under the curve can be directly related to probabilities. From a practical viewpoint, it would

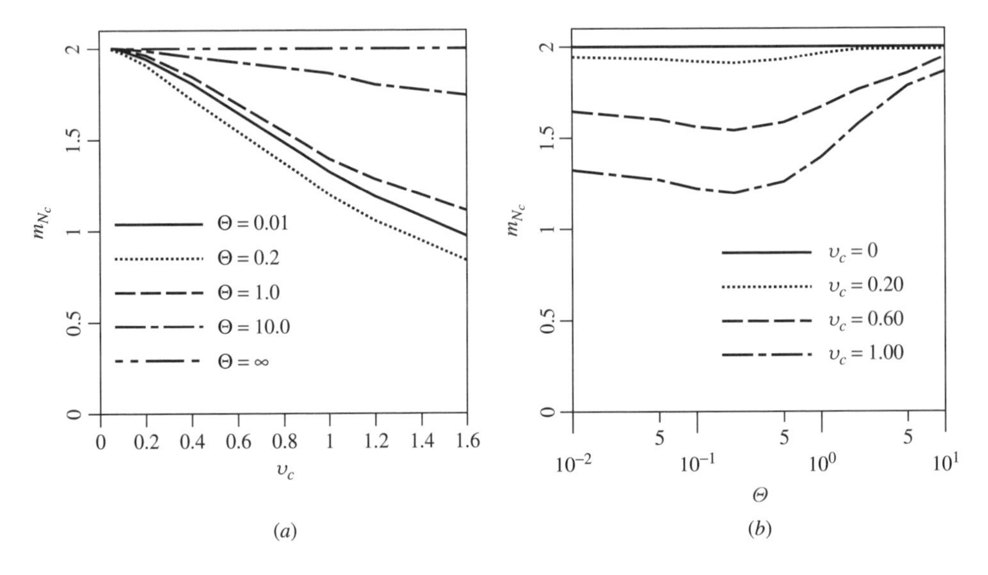

Figure 15.5 Variation of m_{N_c} with (*a*) coefficient of variation v_c and (*b*) correlation length Θ.

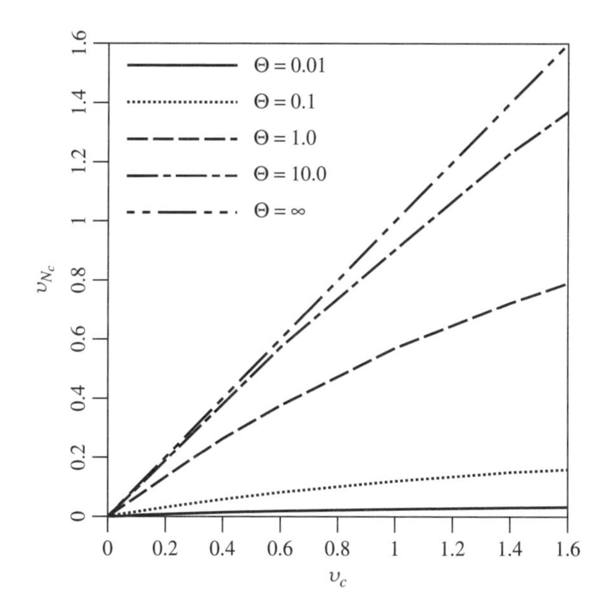

Figure 15.6 Effect of coefficient of variation in c, υ_c, on coefficient of variation in N_c, υ_{N_c}.

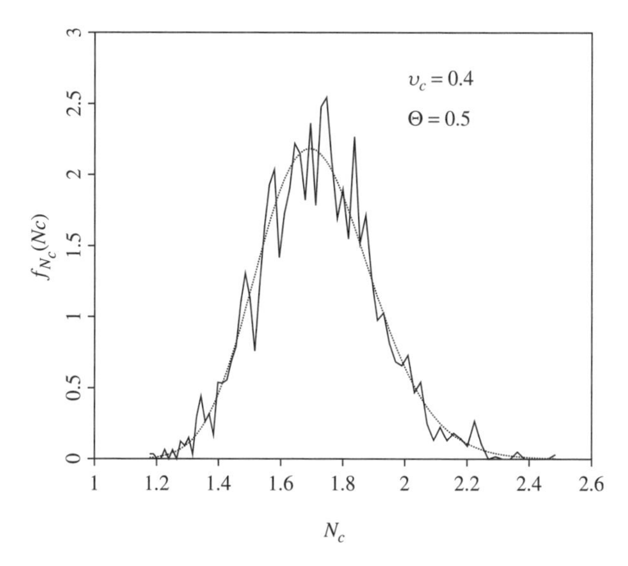

Figure 15.7 Histogram and lognormal fit for typical set of simulated N_c values.

be of interest to estimate the probability of *design failure*, defined here as occurring when the computed compressive strength is less than the deterministic value based on the mean strength divided by a *factor of safety* F_S, that is,

$$\text{Design failure if } q_u < \frac{2\mu_c}{F_S} \tag{15.9}$$

or alternatively,

$$\text{Design failure if } N_c < \frac{2}{F_S} \tag{15.10}$$

The probability of failure as defined in Eq. 15.10 can be expressed as the area under the probability density function to the left of a "target" design value $2/F_S$; hence, from the properties of the underlying normal distribution we get

$$\mathrm{P}\left[N_c < \frac{2}{F_S}\right] = \Phi\left(\frac{\ln(2/F_S) - \mu_{\ln N_c}}{\sigma_{\ln N_c}}\right) \tag{15.11}$$

where Φ is the cumulative standard normal distribution function.

For the particular case shown in Figure 15.7, the fitted lognormal distribution has the sample statistics $m_{N_c} = 1.721$ and $s_{N_c} = 0.185$. These values indicate a median given by $\tilde{\mu}_{N_c} = 1.711$ and a mode given by $\text{Mode}_{N_c} = 1.692$. Furthermore, the distribution of ln N_c has mean and standard deviation, using Eqs. 1.176, of $\mu_{\ln N_c} \simeq 0.537$ and $\sigma_{\ln N_c} \simeq 0.107$. For the particular case of $F_S = 1.5$, Eq. 15.11 gives $p(N_c < 2/1.5) = 0.01$, indicating a 1% probability of design failure as defined above. This implies a 99% reliability that the pillar will remain stable. It should be noted that for the relatively small standard deviation indicated in Figure 15.7, the lognormal distribution looks very similar to a normal distribution.

15.4.1 General Observations on Probability of Failure

While the probability of design failure is directly related to the estimated values of m_{N_c} and s_{N_c}, it is of interest to observe the separate influences of m_{N_c} and s_{N_c}. If s_{N_c} is held constant, increasing m_{N_c} clearly decreases the probability of failure as shown in Figure 15.8*a* since the curves move consistently to the right and the area to the left of any stationary target decreases. The situation is less clear if m_{N_c} is held constant and s_{N_c} is varied, as shown in Figure 15.8*b*.

Figure 15.9*a* shows how the probability of design failure, as defined in Eq. 15.11, varies as a function of υ_{N_c} and the ratio of the target value $2/F_S$ to the *mean* of the lognormal distribution m_{N_c}. If the target value is less than or equal to the mean, the probability of failure always increases as υ_{N_c} is increased. If the target value is larger than the mean, however, the probability of failure initially falls and then gradually rises.

A more fundamental parameter when estimating probabilities of lognormal distributions is the median, $\tilde{\mu}_{N_c}$, which represents the 50% probability location. Figure 15.9*b* shows how the probability of design failure varies as a function of υ_{N_c} and the ratio of the target value $2/F_S$ to the *median*. In this case the probabilistic interpretation is clearly defined. If the target is less than the median, the probability always increases as υ_{N_c} is increased, whereas if the target is greater than the median, the probability always decreases. If the target equals the median, the probability of failure is 50%, irrespective of the value of υ_{N_c}. It might also be noted in Figure 15.9*b* that while the rate of change of probability

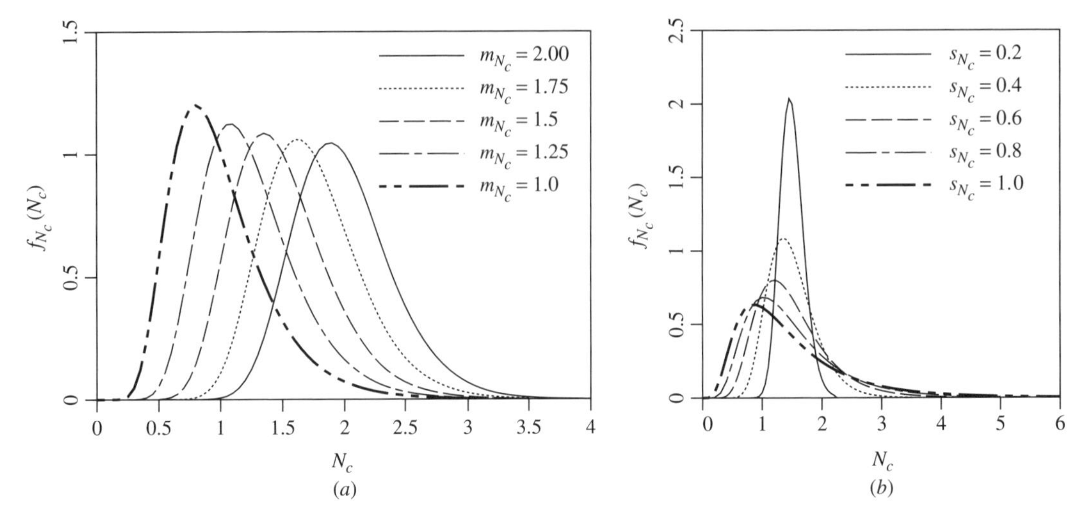

Figure 15.8 Lognormal distribution plots with (*a*) constant standard deviation ($s_{N_c} = 0.4$) and varying mean and (*b*) constant mean ($m_{N_c} = 1.5$) and varying standard deviation.

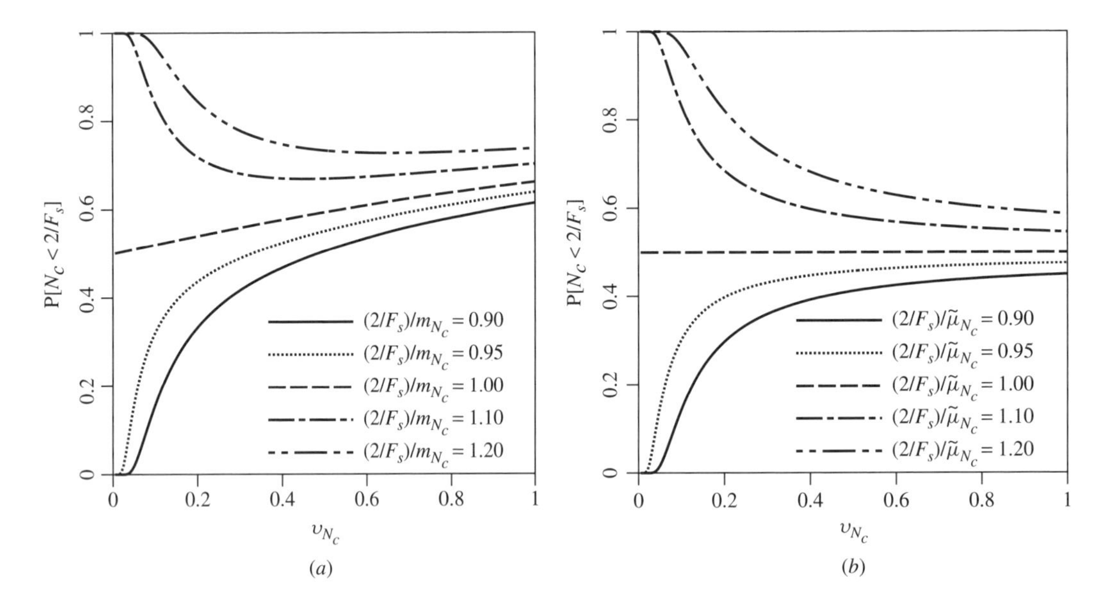

Figure 15.9 Probability of N_c being less than $2/F_S$ for (*a*) different $(2/F_S)/m_{N_c}$ values and (*b*) different $(2/F_S)/\tilde{\mu}_{N_c}$ values.

is quite high at low values of v_{N_c}, the curves tend to flatten out quite rapidly as v_{N_c} is increased.

15.4.2 Results from Pillar Analyses

The influence of these rather complex interactions on the pillar stability analyses can be seen in Figures 15.10, where the probability of design failure is shown as a function of the correlation length Θ for different values of v_c. Each of the four plots corresponds to a different value of the factor of safety, where $F_S = 1.5$, 2.0, 2.5, and 3.0, respectively. Consider in more detail the results shown in Figure 15.10*a* for the case of $F_S = 1.5$, where the target value is $2/F_S = 1.33$. To help with the interpretation, tabulated values of the statistics of N_c corresponding to different values of v_c are presented in Tables 15.2–15.5. Small values of $v_c \leq 0.20$ result in correspondingly small values of v_{N_c} and high values of $m_{N_c} \approx 2$, as shown in Table 15.2, leading to low probabilities of design failure for all Θ. For larger values of v_c, for example, $v_c = 0.4$, the mean m_{N_c} has fallen but is still always higher than the target value of 1.33, as shown in Table 15.3. With $1.33/m_{N_c} < 1$, Table 15.3 indicates that the increasing values of v_{N_c} result in a gradually increasing probability of design failure. This trend is also confirmed by Figure 15.9*a*. Consider now the

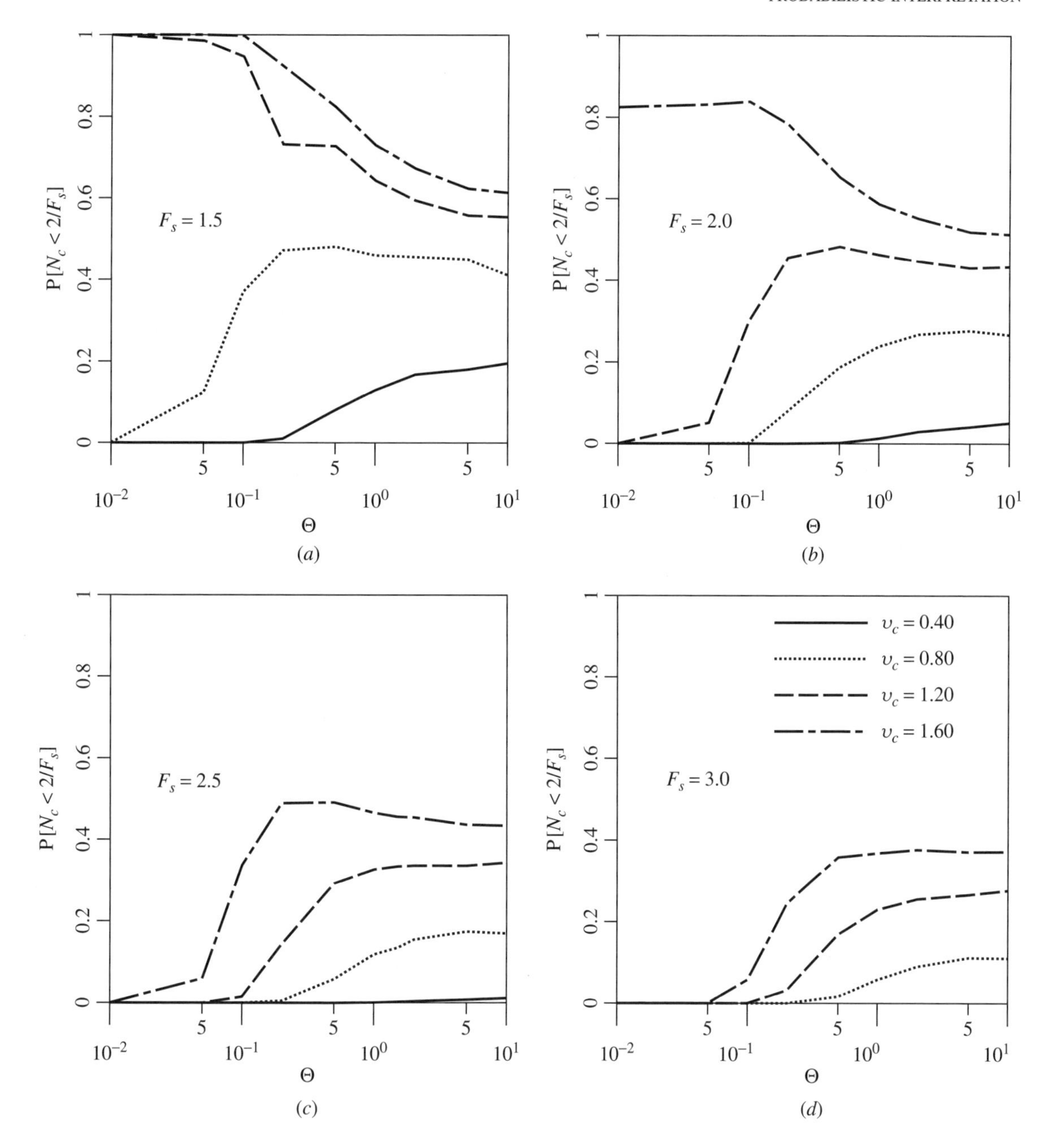

Figure 15.10 Probability of design failure as function of v_c and Θ for four different factors of safety, F_S.

behavior of the probabilities for rather high values of v_c, such as $v_c = 1.2$. From Table 15.4, the mean values of m_{N_c} have fallen quite significantly and are often smaller than the target value of 1.33. More significantly in this case, the median of N_c is *always* smaller than the target of 1.33. Small values of Θ imply small values of v_{N_c} and an almost certain probability of design failure (≈ 1). With $1.33/\tilde{\mu}_{N_c} > 1$, Table 15.4 indicates that the increasing values of v_{N_c} result in a falling probability of design failure. This trend is also confirmed by Figure 15.9*b*. For intermediate values of v_c, such as $v_c = 0.8$, the probability of design failure from Figure 15.10*a* is seen to rise and then fall. This interesting result implies a worst-case combination of v_c and Θ which would give a maximum probability of design failure.

The results tabulated in Table 15.5 indicate that at low values of Θ, the $\tilde{\mu}_{N_c}$ is slightly larger than the target, and this, combined with the low value of v_{N_c}, gives a negligible probability of failure. As Θ is increased, v_{N_c} increases and the $\tilde{\mu}_{N_c}$ decreases. Both of these effects cause the probability of failure to rise as confirmed by Figure 15.9*b*.

Table 15.2 Probability of Design Failure for $F_S = 1.5$ and $v_c = 0.2$

Θ	m_{N_c}	v_{N_c}	$1.33/\tilde{\mu}_{N_c}$	$P[N_c < 1.33]$
0.01	1.943	0.008	0.686	0.000
0.10	1.917	0.031	0.696	0.000
0.20	1.909	0.056	0.670	0.000
0.50	1.930	0.099	0.694	0.000
1.00	1.964	0.134	0.685	0.002
2.00	1.985	0.164	0.681	0.009
5.00	1.987	0.180	0.682	0.016
10.00	1.987	0.190	0.683	0.021
∞	2.000	0.200	0.680	0.026

Table 15.3 Probability of Design Failure for $F_S = 1.5$ and $v_c = 0.4$

Θ	m_{N_c}	v_{N_c}	$1.33/\tilde{\mu}_{N_c}$	$P[N_c < 1.33]$
0.01	1.809	0.014	0.737	0.000
0.10	1.747	0.058	0.764	0.000
0.20	1.721	0.107	0.779	0.010
0.50	1.770	0.193	0.767	0.083
1.00	1.847	0.264	0.747	0.130
2.00	1.880	0.310	0.743	0.163
5.00	1.944	0.358	0.728	0.181
10.00	1.953	0.380	0.730	0.196
∞	2.000	0.400	0.718	0.195

Table 15.4 Probability of Design Failure for $F_S = 1.5$ and $v_c = 1.2$

Θ	m_{N_c}	v_{N_c}	$1.33/\tilde{\mu}_{N_c}$	$P[N_c < 1.33]$
0.01	1.189	0.028	1.122	1.000
0.10	1.083	0.136	1.242	0.946
0.20	1.055	0.239	1.299	0.867
0.50	1.125	0.468	1.309	0.727
1.00	1.283	0.662	1.246	0.643
2.00	1.479	0.838	1.176	0.588
5.00	1.719	1.003	1.099	0.545
10.00	1.801	1.108	1.105	0.545
∞	2.000	1.200	1.041	0.517

At approximately $\Theta = 0.5$, the $\tilde{\mu}_{N_c}$ approaches the target, giving a maximum probability of design failure close to 0.5. As indicated in Table 15.5, further increase in Θ causes the $1.33/\tilde{\mu}_{N_c}$ ratio to fall quite consistently. Although v_{N_c} is still rising, the overall behavior is dominated by the falling $1.33/\tilde{\mu}_{N_c}$ ratio, and the probability of failure falls as implied in Figure 15.9*b*.

Table 15.5 Probability of Design Failure for $F_S = 1.5$ and $v_c = 0.8$

Θ	m_{N_c}	v_{N_c}	$1.33/\tilde{\mu}_{N_c}$	$P[N_c < 1.33]$
0.01	1.478	0.022	0.902	0.000
0.10	1.387	0.103	0.966	0.370
0.20	1.371	0.178	0.988	0.472
0.50	1.429	0.336	0.984	0.481
1.00	1.542	0.472	0.956	0.460
2.00	1.659	0.607	0.940	0.456
5.00	1.816	0.754	0.920	0.450
10.00	1.905	0.738	0.870	0.416
∞	2.000	0.800	0.854	0.411

Figures 15.10*b–d*, corresponding to higher factors of safety, display similar maxima in their probabilities; however, there is an overall trend that shows the expected reduction in the probability of failure as the factor of safety is increased. Figure 15.10*d*, corresponding to $F_S = 3$, indicates that for a reasonable upper-bound value of $v_{N_c} = 0.6$, the probability of design failure will be negligible for $\Theta < 1$.

The program that was used to produce the results in this chapter enables the reliability of rock pillars with varying compressive strength and spatial correlation to be assessed. In particular, a direct comparison can be made between the probability of failure and the more traditional factor of safety.

Table 15.6 shows the factor of safety and probability of failure for pillar strength as a function of Θ for the particular case of $v_c = 0.4$. When v_c and Θ are known, a factor of safety can be chosen to meet the desired probability of failure or acceptable risk. For instance, if a target probability of failure of 1% is desired for $v_c = 0.4$ and $\Theta = 0.2$, a factor of safety of at least $F_S = 1.5$ should be applied to the mean shear strength value. When Θ is not known, a conservative estimate should be made that would lead to the

Table 15.6 Probability of Pillar Failure for $v_c = 0.4$

	Θ				
F_S	0.10	0.20	1.00	2.00	10.00
1.00	0.99	0.93	0.67	0.64	0.60
1.25	0.07	0.27	0.34	0.36	0.36
1.50	0.00	0.01	0.13	0.13	0.20
1.75	0.00	0.00	0.04	0.07	0.10
2.00	0.00	0.00	0.01	0.03	0.05
2.25	0.00	0.00	0.00	0.01	0.02
2.50	0.00	0.00	0.00	0.00	0.01
2.75	0.00	0.00	0.00	0.00	0.01
3.00	0.00	0.00	0.00	0.00	0.00

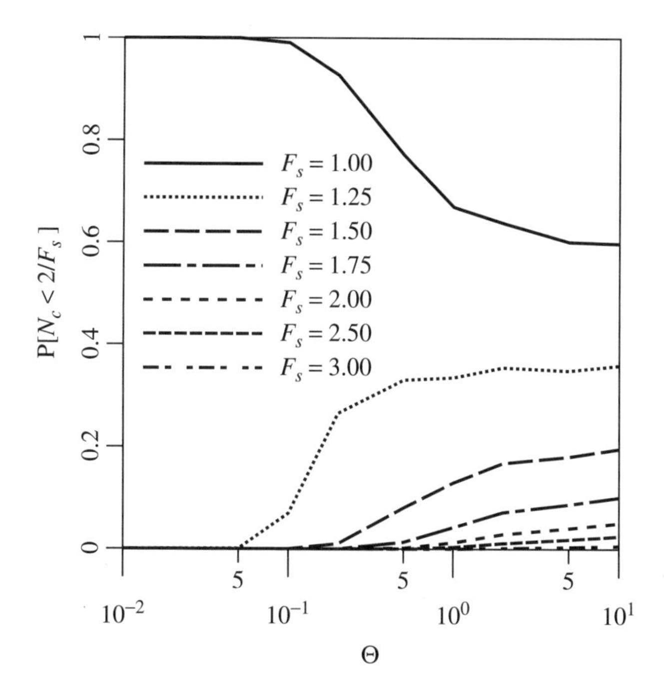

Figure 15.11 Probability of design failure as function of Θ for $v_c = 0.4$ and for different factors of safety, F_S.

most conservative prediction. For instance, if a 1% probability of failure is acceptable for $v_c = 0.4$ with unknown Θ, a factor of safety of at least $F_S = 2.75$ is called for.

Figure 15.11 shows a plot of the results from Table 15.6.

15.5 SUMMARY

The chapter has shown that rock strength variability in the form of a spatially varying lognormal distribution can significantly reduce the compressive strength of an axially loaded rock pillar.

The following more specific conclusions can be made:

1. As the coefficient of variation of the rock strength increases, the expected compressive strength decreases. For a given coefficient of variation, the expected mean compressive strength reaches a minimum corresponding to a critical value of the correlation length. In the absence of good information relating to the correlation length, this critical value should be used in design.
2. The coefficient of variation of the compressive strength is observed to be positively correlated with both the correlation length and the coefficient of variation of the rock strength.
3. The probability of failure is a function of m_{N_c}, s_{N_c}, and the target design value $2/F_S$. The chapter has shown that the interpretation of the probability of failure is most conveniently explained by comparing the target design value with the median of the lognormal distribution.
4. By interpreting the Monte Carlo simulations in a probabilistic context, a direct relationship between the factors of safety and probability of failure can be established.

CHAPTER 16

Liquefaction

16.1 INTRODUCTION

Consider a soil mass subjected to an earthquake. Upon shaking, the soil particles tend to want to settle into a more densely packed arrangement. This will occur with only some surface settlement if the soil is dry or only partially saturated with water. If the soil is fully saturated, then in order for the soil to become more densely packed, the water between the particles is forced to escape. However, if the water cannot easily escape, this can lead to a very dangerous situation in which the pore water pressures exceed the contact pressure between soil particles, and the soil effectively turns into a fluid.

In this chapter, we examine the effect of spatial variability on the extent and severity of earthquake-induced liquefaction (Fenton, 1990; Fenton and Vanmarcke, 1998). The analysis is largely Monte Carlo simulation based. The software used to model the response of a soil to earthquake input is called DYNA1D (Prevost, 1989).

Under earthquake shaking, liquefaction-induced failure can occur only if the resistance to seismic stresses is sufficiently low over a sufficiently large volume of foundation soil; high liquefaction potential need not be a problem if confined to small, isolated volumes, as demonstrated in a liquefaction stability analysis by Hryciw et al. (1990). Thus, studies of liquefaction risk at a site should consider not only the liquefaction potential at sample points, as traditionally done, but also the spatial variation of liquefaction potential over the entire site.

Because of the highly nonlinear nature of liquefaction response of a soil mass, the spatial distribution of liquefied regions can be most accurately obtained through multiple simulation runs, that is, through a Monte Carlo analysis. In Monte Carlo simulation, the challenge is to simulate sets of properly correlated finite-element local averages; each set of simulated values serves as input into (multiple) deterministic finite-element analysis. Sample statistics of response parameters (i.e., the occurrence and extent of liquefaction) can then be computed.

An appropriate soil model for stochastic finite-element analysis involves a partition of the soil volume into a set of finite elements. A vector of material properties, drawn from realizations of three-dimensional local average random fields, is then associated with each element. In the simulation, one must account for "point variance" reduction and correlation between elements, consistent with the dimensions of the finite elements and the correlation parameters of the underlying fields. The variance reduction function, which reflects the amount of variance reduction due to local averaging and depends on one or more correlation lengths, frugally captures the correlation structure and is well suited for the simulation of local averages (Vanmarcke, 1984).

Although the soil properties are being modeled by a three-dimensional random field, the liquefaction analysis will be carried out using only a one-dimensional finite-element program applied to one column of the soil mass at a time. This considerable simplification of the problem is necessitated by the enormous computational requirements of a nonlinear, time-stepping, stochastic Monte Carlo simulation analysis. In addition, and again partly due to computational time issues but also due to the one-dimensional sequential approximation to a three-dimensional problem, the chapter considers only the *initiation* of liquefaction. While it is well known that postevent pore pressure redistribution is important in liquefaction, it is not felt that a one-dimensional model will properly reflect this redistribution since in the one-dimensional model almost all shear wave motion is absorbed by the liquefied layer and the surface ceases to move. On the shorter time scale of the event itself, the initiation of liquefaction in the soil is believed to be modeled reasonably accurately via this one-dimensional approximation. The chapter concentrates on the spatial variation, over a horizontal plane, of the initial liquefaction state. A picture of the initial liquefaction state is built up by looking at horizontal cross sections through the collection of one-dimensional soil columns making up the soil mass.

16.2 MODEL SITE: SOIL LIQUEFACTION

An earthquake of magnitude $M_s = 6.0$, on April 26, 1981, in the Imperial Valley near Westmorland, California, caused significant damage, in many cases through liquefaction. This prompted a detailed geological survey of the valley, including the installation of accelerometers and piezometers to record ground motions and changes in pore water pressure during future earthquakes at the Wildlife Management Area. The Wildlife Management Area is located 3 km south of Calipatria in the Imperial Wildfowl Management Area, lying on the west side of the incised floodplain of the Alamo River.

The site was instrumented in 1982 with surface and down-hole (7.5-m-depth) accelerometers and six pore water pressure transducers (Bennett et al., 1984). The Superstition Hills event ($M_s = 6.6$), recorded in 1987 (Holzer et al., 1988), resulted in liquefaction at the site in the form of sand boils and limited lateral spreading and motivates this study—the following model is based on the Wildlife site.

Within the upper three geological units, a closer examination by Holzer et al. (1988) revealed five soil strata to the level of the down-hole accelerometer:

1. Layer 1 (0.0–1.2 m): very loose silt
2. Layer 2 (1.2–2.5 m): very loose silt
3. Layer 3 (2.5–3.5 m): very loose to loose sandy silt
4. Layer 4 (3.5–6.8 m): loose to medium dense silty sand
5. Layer 5 (6.8–7.5 m): medium to stiff clayey silt

The water table at a depth of 1.2 m forms the boundary between layers 1 and 2.

The random medium representation of the soil properties and deterministic finite-element program used to assess the spatial variation of liquefaction at the model site are described in the following sections. Recognizing that little information concerning spatial variability of the soil properties at the site is available, the model requires many parameters to be assumed using reasonable estimates. Since many of these statistical parameters were not verified at the Wildlife site, this example serves primarily to investigate the degree of spatial variability in liquefaction under reasonable assumptions and to investigate techniques of evaluating liquefaction risk in the presence of spatial variability. The intensity of the earthquake excitation and the correlation lengths of the soil properties were varied for the purpose of sensitivity analysis.

The soil volume to be modeled is 80 × 80 m laterally by 7.5 m in depth and is partitioned into a 16 × 16 × 32 set of finite elements. Thus, each element has dimensions 5 × 5 m laterally by 0.23 m vertically. Realizations of the random soil properties within each element are obtained by columnwise extraction from a set of three-dimensional local average simulations.

16.2.1 Stochastic Soil Model

For this study, the soil parameters expected to have the greatest impact on site response and liquefaction likelihood and selected to be modeled as three-dimensional random fields were permeability k, porosity n, modulus of elasticity (solid phase) E, Poisson's ratio (solid phase) ν, and dilation reference angle Φ. The ratio of Φ to the friction angle determines whether initial contraction is followed by dilation or contraction in the soil during shaking. All of these properties, and in particular the dilation reference angle, are generally found through laboratory tests on soil samples. These parameters are required as input to the finite-element analysis program to be discussed later. Their treatment and precise interpretation within the finite-element algorithm is discussed in detail by Prevost (1989). Permeability and, indirectly, porosity, are perhaps the most important parameters influencing liquefaction in sandy soils. Water trapped within the soil structure carries an increasing fraction of the stress as the soil tries to densify during shaking. Eventually the intergranular effective stresses may become so low that relative movement between particles becomes possible and the medium effectively liquefies.

Beyond CPT tests performed at a small number of locations, the published site information (Bennett et al., 1984; Holzer et al., 1988) contains barely enough data to establish mean parameters as estimated by Keane and Prevost (1989) and listed in Table 16.1 as a function of depth. Estimates of the statistical nature of the above parameters are based on a combination of engineering judgment and a review of the literature (Fenton, 1990). Assumed variances associated with each parameter are also shown in Table 16.1 as a function of depth.

In all cases the random material parameters are obtained by transforming a three-dimensional zero-mean, unit-variance homogeneous Gaussian field $Z(\mathbf{x})$, realizations of which are produced using the three-dimensional LAS method (Section 6.4.6). Letting $U_i(\mathbf{x})$ represent the value of the ith soil property at the spatial point $\mathbf{x} = \{x, y, z\}^T$, with z the depth below the surface,

$$U_i(\mathbf{x}) = \mathcal{T}_i\left(\mu_i(z) + \sigma_i(z) Z_i(\mathbf{x})\right) \tag{16.1}$$

where $\mu_i(z)$ is the mean, $\sigma_i(z)$ is the standard deviation, and $\mathcal{T}_i$ is a transformation taking the Gaussian process, $Z_i(\mathbf{x})$, into the marginal distribution appropriate for property i. Notice that the formulation allows trends in the mean and variance as a function of depth to be incorporated.

For the permeability, elastic modulus, and dilation reference angle, all assumed to be lognormally distributed, the transformation $\mathcal{T}_i$ is the exponential

$$U_i(\mathbf{x}) = \exp\{\mu_{\ln i}(z) + \sigma_{\ln i}(z) Z_i(\mathbf{x})\} \tag{16.2}$$

Porosity is related to both permeability and soil relative density, the latter of which is also related to the initial vertical stresses in the medium as well as the shear wave velocities. The porosity at the Wildlife site is assumed to have a constant mean of 0.42. Recognizing that n must be bounded, the following transformation $\mathcal{T}_n$ (see Eq. 16.1)

Table 16.1 Geotechnical Parameters at the Wildlife Site

Property	Statistic	0–1.2	1.2–2.5	2.5–3.5	3.5–6.8	6.8–7.5
Permeability, k (m/s)	Mean	1×10^{-5}	1×10^{-5}	1×10^{-5}	1×10^{-4}	1×10^{-6}
	$\mu_{\ln k}$	−11.7	−11.7	−11.9	−9.7	−14.1
	$\sigma^2_{\ln k}$	0.6	0.6	0.8	1.0	0.5
Porosity,[a] n	Mean	0.42	0.42	0.42	0.42	0.42
	$\mu_{n'}$	0	0	0	0	0
	$\sigma^2_{n'}$	1.0	1.0	1.0	1.0	1.0
Elastic modulus, E (N/m^2)	Mean	3.9×10^7	3.7×10^7	5.4×10^7	5.4×10^7	7.0×10^7
	$\mu_{\ln E}$	17.1	17.1	17.4	17.2	17.7
	$\sigma^2_{\ln E}$	0.8	0.6	0.8	1.2	0.8
Poisson's ratio,[b] ν	Mean	0.275	0.275	0.275	0.275	0.275
	$\mu_{\nu'}$	0	0	0	0	0
	$\sigma^2_{\nu'}$	1.0	1.0	1.0	1.0	1.0
Dilation reference angle, Φ	Mean	21.3°	20.0°	19.0°	18.0°	5.0°
	$\mu_{\ln \Phi}$	2.95	2.90	2.84	2.77	1.51
	$\sigma^2_{\ln \Phi}$	0.2	0.2	0.2	0.3	0.2

[a] See Eqs. 16.3 and 16.4.
[b] See Eqs. 16.6 and 16.7.

changes a normally distributed variate into a bounded distribution:

$$U_n = a + (b-a)\mathcal{T}_n(Y) = a + \frac{b-a}{2}\left\{1 + \tanh\left(\frac{Y}{2\pi}\right)\right\} \tag{16.3}$$

which is a one-to-one mapping of $Y \in (-\infty, \infty)$ into $U_n \in (a, b)$, where Y is obtained from the random field Z according to Eq. 16.1:

$$Y(\mathbf{x}) = \mu_{n'}(z) + \sigma_{n'}(z) Z_{n'}(\mathbf{x}) \tag{16.4}$$

where $\mu_{n'}$ and $\sigma_{n'}$ are the mean and standard deviation of Y, which can be obtained in practice by taking the first two moments of the inverse:

$$Y = \mathcal{T}_n^{-1}\left(\frac{U_n - a}{b-a}\right) = \pi \ \ln\left(\frac{U_n - a}{b - U_n}\right) \tag{16.5}$$

See Section 1.10.10 for more details on this bounded distribution. For the assumed value of $\sigma^2_{n'} = 1.0$ used herein (see Table 16.1), the distribution of U_n is bell shaped with mode at the midpoint, $\frac{1}{2}(b+a)$. In this case study, it is assumed that $n \in (0.22, 0.62)$ with mean 0.42. While this may seem to be a fairly wide range on the porosity, it should be noted that the distribution given by Eq. 1.194 implies that 90% of porosity realizations lie between 0.37 and 0.47. The solid phase (soil) mass density, ρ_s, was taken to be 2687 kg/m^3 (Keane and Prevost, 1989) giving a mean soil dry unit mass of (1 − 0.42)(2687) = 1558 kg/m^3.

Because it is well known that soil porosity is related to permeability, the underlying Gaussian fields $Z_{n'}$ and $Z_{\ln k}$ are generated so as to be mutually correlated on a point-by-point basis. This is accomplished by generating two independent random fields and then linearly combining them using the Cholesky decomposition of the 2 × 2 cross-correlation matrix to yield two properly correlated random fields (see Section 6.4.2). A correlation coefficient of 0.5 is assumed; however, it must be recognized that the true correlation between these properties is likely to be quite variable and site specific. Although the other random soil properties are also felt to be correlated with soil porosity, their degree of correlation is significantly less certain than in the case of permeability, which is already somewhat speculative. For this reason, the other random properties are assumed to be independent. Recalling that the introduction of correlation decreases the variability between pairs of random variables, the assumption of independence increases the overall variability contained in the model. Thus, it is deemed better to assume independence than to assume an erroneous correlation. The effect of cross-correlation between, say, porosity and the dilation reference angle on the spatial distribution of liquefaction is left an open question that may be better left until more is known about the statistical correlations between these properties.

Poisson's ratio is also chosen to be a bounded random variable, $\nu \in (0.075, 0.475)$, according to Eq. 16.3 with constant mean 0.275. Now Y is given by

$$Y(\mathbf{x}) = \mu_{\nu'}(z) + \sigma_{\nu'}(z) Z_{\nu'}(\mathbf{x}) \tag{16.6}$$

so that

$$U_\nu = 0.075 + 0.4\,\mathcal{T}_\nu(Y) \tag{16.7}$$

and the transformation $\mathcal{T}_\nu$ is the same as $\mathcal{T}_n$ in Eq. 16.3. Under this transformation, with $\sigma_{\nu'} = 1$, 90% of realizations of Poisson's ratio will lie between 0.24 and 0.31.

The relationship between the dilation reference angle Φ and the friction angle at failure, ϕ, as interpreted internally by the finite-element analysis program, determines whether the soil subsequently dilates or contracts upon shaking. If the ratio Φ/ϕ exceeds 1.0, then only contraction occurs, otherwise initial contraction is followed by dilation. Since contraction results in increasing pore water pressure, this ratio is of considerable importance in a liquefaction analysis. Rather than considering both the dilation and friction angles to be random, only the dilation angle was selected as random; the friction angle was prescribed deterministically with $\phi = 21^\circ$ in layer 1, 20° in layer 2, 22° in layers 3 and 4, and 35° in layer 5, as estimated by Keane and Prevost (1989). These assumptions still lead to the ratio Φ/ϕ being random.

The covariance function $C(\boldsymbol{\tau})$ used to model the spatial variability of all the random soil properties is of a simple exponential form parameterized by θ_v and θ_h, the correlation lengths in the vertical and horizontal directions, respectively,

$$C(\tau_1, \tau_2, \tau_3) = \sigma^2 \exp\left\{-\frac{2}{\theta_h}\left(\sqrt{\tau_1^2 + \tau_2^2}\right) - \frac{2|\tau_3|}{\theta_v}\right\} \quad (16.8)$$

where $\boldsymbol{\tau} = \{\tau_1, \tau_2, \tau_3\}^{\mathrm{T}} = \mathbf{x} - \mathbf{x}'$ denotes the separation distance between two spatial points, $\mathbf{x}$ and $\mathbf{x}'$. Note that Eq. 16.8 has a partially separable form in τ_3 (vertically). This covariance function governs the underlying Gaussian random fields; after transformation into the desired marginal distributions, the covariance structure is also transformed so that comparison between statistics derived from real data and Eq. 16.8 must be made with caution. From the point of view of estimation, the statistical parameters governing the underlying Gaussian fields can always be simply obtained by performing an inverse transformation on the data prior to estimating the statistics. For example, if the parameter is treated as a lognormally distributed random process by transforming a normally distributed random field using the relationship $U = \exp\{Y\}$, then the corresponding mean, variance, and correlation length of Y can be found from the raw data by taking the logarithm of the data prior to statistical analysis. In the absence of spatial data, the following discussion is derived from the literature and is assumed to apply to the underlying Gaussian fields directly.

In the vertical direction, Marsily (1985) proposes that the correlation length of soil permeability is of the order of 1 m, and so $\theta_v = 1$ m is adopted here. The horizontal correlation length, θ_h, is highly dependent on the horizontal extent and continuity of soil layers. The U.S. Geological Survey (Bennett et al., 1984) indicated that the layers at the Wildlife site are fairly uniform, and a ratio of horizontal to vertical correlation lengths $\theta_h/\theta_v \simeq 40$ was selected implying $\theta_h \simeq 40$ m; this is in the same range as Vanmarcke's (1977) estimate of 55 m for the compressibility index of a sand layer. Although compressibility and permeability are, of course, different engineering properties, one might argue that the correlation length depends largely on the geological processes of transport of raw materials, layer deposition, and common weathering rather than on the actual property studied. Based on this reasoning, all the random soil properties are modeled using the same correlation lengths as well as the same form of the covariance function.

The simulations are repeated using a larger vertical correlation length, $\theta_v = 4$ m, while holding the ratio of horizontal to vertical correlation lengths constant at 40. In the following, only the vertical correlation length is referred to when indicating the case studied.

16.2.2 Stochastic Earthquake Model

Earthquake ground motions vary from point to point in both time and space. Techniques have been developed to generate such fields of motion (Vanmarcke et al., 1993), while studies of earthquake motions over arrays of seismometers provide estimates of the space–time correlation structure of ground motion (e.g., Boissières and Vanmarcke, 1995). Input earthquake motions in this study, applied to the base of the soil model on a pointwise basis, are realizations of a space–time random field with the following assumed space–frequency correlation function:

$$\rho(\omega, \boldsymbol{\tau}) = \exp\left\{-\frac{\omega|\boldsymbol{\tau}|}{2\pi cs}\right\} \quad (16.9)$$

where $\boldsymbol{\tau} = \mathbf{x} - \mathbf{x}'$ is the lag vector between spatial points $\mathbf{x}$ and $\mathbf{x}'$, ω is the wave component frequency (radians per second), c is the shear wave velocity (taken as 130 m/s at the base of the soil model), and $s = 5.0$ is a dimensionless parameter controlling the correlation decay.

Only one component of motion is used, modeled after the north–south (NS) component of the Superstition Hills event. Analyses by Keane and Prevost (1989) indicate that including the east–west and vertical components makes little difference to the computed (deterministic) site response (the north–south component had dominant amplitudes), and using it alone, Keane obtains remarkably good agreement with the recorded site response.

The marginal spectral density function $G(\omega)$ governing the input motion spectral content was derived from the dominant NS component of earthquake acceleration, shown in Figure 16.1, recorded at the down-hole accelerometer for the Superstition Hills event. To reduce the number of time steps in the analysis, only 20.48 s of motion were generated—1024 time steps at 0.02 s each. Using the maximum entropy method, a pseudoevolutionary spectral density function was estimated in four consecutive time windows, starting at 7 s into the recorded acceleration as denoted by dashed lines in Figure 16.1. The derived

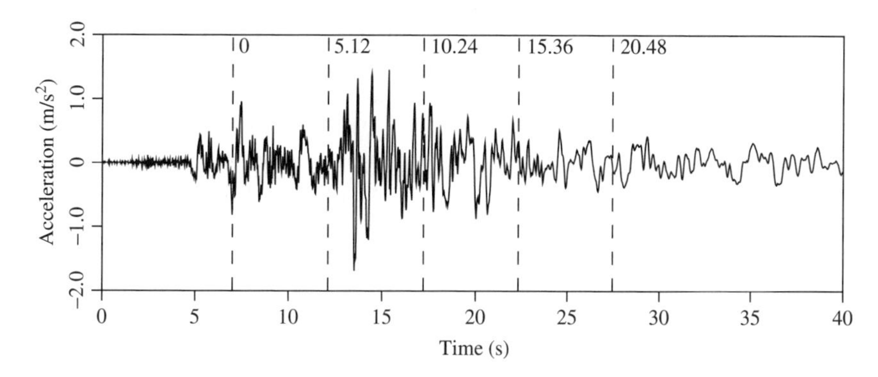

Figure 16.1 Recorded accelerogram at 7.5-m depth during Superstition Hills event (NS component).

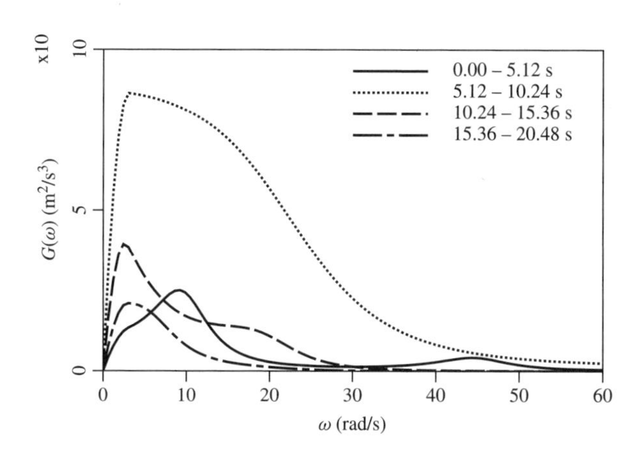

Figure 16.2 Pseudoevolutionary spectral density function estimated from Superstition Hills event (NS component) for four consecutive time windows.

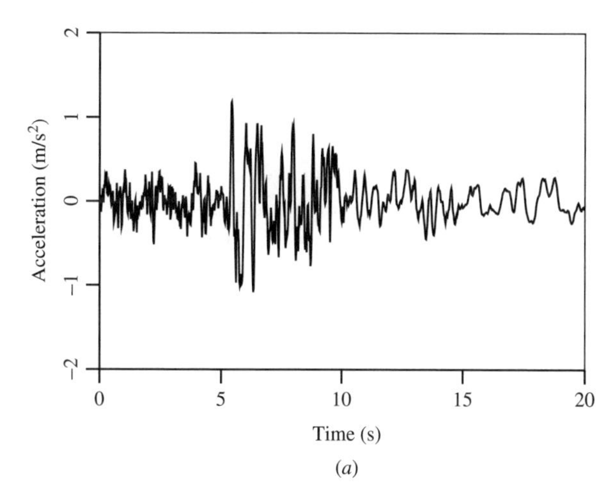

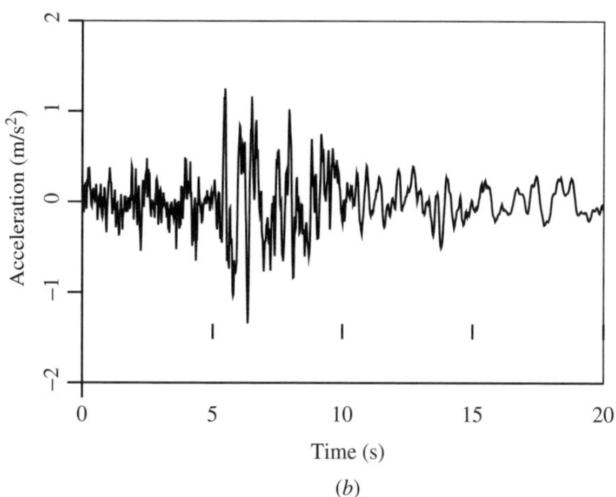

Figure 16.3 Sample acceleration records generated at two points, a and b, separated by 80 m.

spectral density functions shown in Figure 16.2, one for each time window, were then used to produce nonstationary earthquake acceleration realizations. The last $G(\omega)$ was actually based on the entire trailing portion of the recorded motion.

Admittedly, the down-hole motions include both upward propagating energy and downward propagating reflected energy, the latter of which is modified by existing material properties in the soil above the recording point. However, only the spectral density function of the down-hole motion is used to control the generated motions, not the detailed recordings themselves. The resulting simulations can be thought of as having a mean which includes the *mean* soil properties in the overlying field.

Figure 16.3 shows a realization of the input acceleration field sampled at two points separated by 80 m. Although the motions are quite similar, they are not identical and may be considered representative of the possible base motion at the site. The same marginal spectral density function was used at all spatial points over the base of the soil model, presumably reflecting the filtering of bedrock motion typical at the site. To partially assess the effect of earthquake intensity on the spatial distribution of liquefaction at the site, the study was repeated using the first set of artificial motions scaled by a factor of 0.7.

16.2.3 Finite-Element Model

The soil mass is divided into 256 columns arranged on a 16 × 16 grid, each column consisting of 32 elements (33 nodes) vertically. Realizations of the soil mass are excited by artificial earthquake motions applied at the base of each soil column and analyzed using DYNA1D (see Section 16.1). DYNA1D employs multiple yield level elasto-plastic constitutive theory to take into account the nonlinear, anisotropic, and hysteretic stress–strain behavior of the soil as well as the effects of the transient flow of pore water through the soil media and its contractive/dilative nature. Each finite element is assigned soil properties, either deterministic values or from realizations of random fields.

Soil columns are then analyzed individually, so that 256 invocations of the finite-element analysis are required for each realization of the soil mass. The column analyses are independent, and the only link between the soil columns is through their correlated properties. It is unknown how the coupling between columns in a fully three-dimensional dynamic analysis would affect the determination of global liquefaction potential; however, it is believed that the analysis proposed herein represents a reasonable approximation to the fully three-dimensional analysis at this time, particularly since the site is reasonably level and only liquefaction initiation is considered.

The surface response obtained from the analysis of a single column of soil is shown in Figure 16.4 along with a typical realization of the input motion acting at the column base. The soil at a depth of about 2.7 m began to liquefy after about 10 s of motion. This is characterized at the surface by a dramatic reduction in response as the liquefied layer absorbs the shear wave motion propagating from below.

Of particular interest in the evaluation of liquefaction potential at the site is the prediction of surface displacement and pore water pressure buildup while shaking lasts. As the global analysis consists of a series of one-dimensional column analyses, it was decided not to use the surface displacement predictions as indicators of liquefaction potential. Rather, the pore pressure ratio associated with each element was selected as the liquefaction potential measure to be studied. Redistribution of pore water pressure after the earthquake excitation, which could lead to further liquefaction of upper soil layers, was not considered in this initial study.

16.2.4 Measures of Liquefaction

The finite-element program calculates the excess pore water pressure, u_i, in each element i as a function of time. The

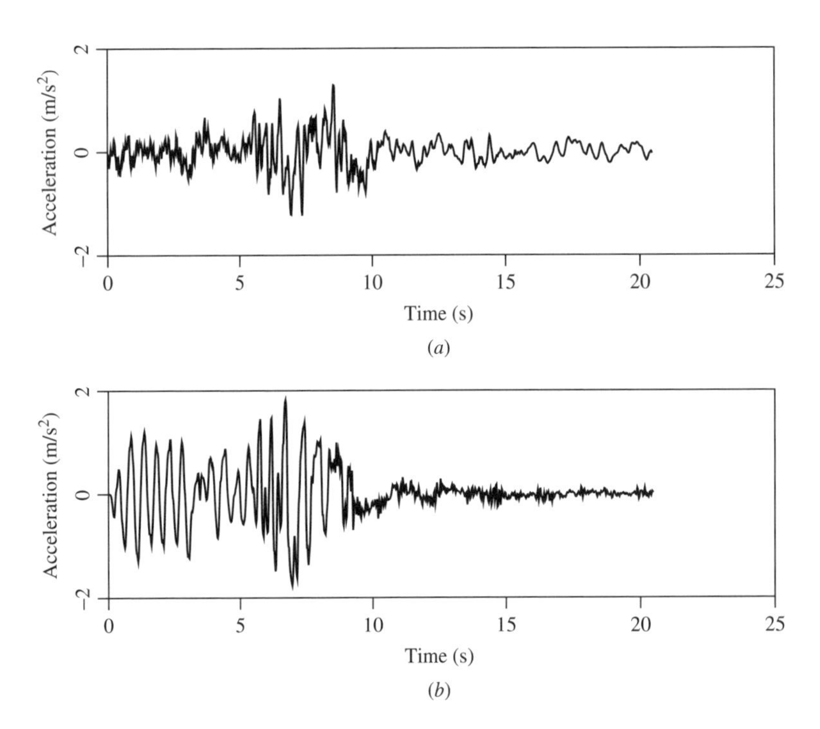

Figure 16.4 (*a*) Base input and (*b*) surface response computed by DYNA1D for a particular soil column realization.

ratio $q_i = u_i/\sigma'_{oi}$, where σ'_{oi} is the initial vertical effective stress in the ith element, is commonly thought of as the parameter measuring the occurrence of liquefaction (Seed, 1979) and will be referred to herein as the *liquefaction index*. Note that, owing to the one-dimensional nature of the finite-element analysis, the horizontal effective stress is ignored and liquefaction is based only on the initial vertical effective stress.

When q_i reaches a value of 1, the pore water is carrying the load so that soil particles become free to slip and liquefaction occurs. It is possible, however, for liquefaction to take place at values of q_i slightly less than 1, as it is only necessary that most of the lateral strength or bearing capacity is lost. Fardis and Veneziano (1982) suggest that the liquefied fraction of the ith element of soil, η_i, be calculated as

$$\eta_i = \mathrm{P}\left[\frac{u_i}{\sigma'_{oi}} \geq 0.96\right] \tag{16.10}$$

for undrained and partially drained effective stress models. The probability $\mathrm{P}[\cdot]$ on the right-hand side can be evaluated through a sequence of simulations. Fardis then goes on to evaluate the risk of liquefaction L as the probability that the maximum of $\eta(z)$ over the depth z is close to 1:

$$L = \mathrm{P}\left[\max_z(\eta(z)) \approx 1\right] \tag{16.11}$$

where now $\eta(z)$ is interpreted, not as a probability, but rather as the sample liquefied fraction.

For individual soil columns where interaction with adjacent soil is ignored, such an approach is reasonable since the occurrence of liquefaction at a given layer will result in the loss of lateral resistance at the surface. Shinozuka and Ohtomo (1989) have a slightly different approach involving summing the liquefaction indices q over depth to obtain the vertically averaged liquefaction index $\mathcal{Q}$:

$$\mathcal{Q} = \frac{1}{h}\int_0^h \frac{u(z)}{\sigma'_o(z)}\, dz \tag{16.12}$$

where h is the total depth of the modeled column. In this way the effect of the vertical extent of a liquefied region can be incorporated into a risk analysis. But how important is the vertical extent of liquefaction? While it certainly has bearing on liquefaction risk, it is easy to imagine a situation in which a thin layer some distance below the surface becomes completely liquefied while adjoining layers above and below remain stable. Such a condition could yield a relatively low value of $\mathcal{Q}$ even though lateral stability at the surface may be lost. On the other hand, the vertical extent of liquefied regions may be more important to the occurrence of sand boils and vertical settlement. In that the risk of sand boils and/or vertical settlement is quantifiable using point or vertically averaged approaches, whereas the loss of lateral stability resulting in spreading or lateral movement depends on the spatial distribution of liquefaction, this study concentrates on the latter issue.

In the three-dimensional situation, neither approach discussed above is deemed entirely suitable. If the term *global liquefaction* is used to denote the loss of lateral stability leading to large surface displacements at the site, then the occurrence of high q_i indices at an individual point (or small region) will not necessarily imply global liquefaction if adjacent regions retain sufficient strength. Likewise if a particular layer is found to have high q values over a significant lateral extent, then global liquefaction risk could be high even though the average for the site may be low. In this study, the lateral spatial extent of liquefied regions is assumed to be the more important factor in the determination of global liquefaction risk for a site. For each realization, the analysis proceeds as follows:

1. Compute the liquefaction index $q_{ij}(t_\ell) = u_i/\sigma'_{oi}$ for each element i in the jth column at each time step t_ℓ and repeat for all the columns.
2. Compute the sum

   $$Q_{i\ell} = \frac{1}{A}\sum_{j=1}^{n_c} q_{ij}(t_\ell)\, \Delta A_j$$

   where A is the total area of the site model, ΔA_j is the area of the jth column, and n_c is the number of columns; $Q_{i\ell}$ is the ith layer average liquefaction index at each time step t_ℓ.
3. Determine the indices i^* and ℓ^* which maximize $Q_{i\ell}$. The index i^* now represents the depth of the plane with the maximum likelihood of liquefying at the time t_{ℓ^*} and $q_{i^*j}(t_{\ell^*})$ is the corresponding two-dimensional field of liquefaction indices (indexed by j).
4. Determine the excursion area fraction defined by

   $$A_q = \frac{1}{A}\sum_{j=1}^{n_c} I_A\left(q_{i^*j}(t_{\ell^*}) - q\right)\, \Delta A_j$$

   for a variety of levels $q \in (0, 1)$. The indicator function $I_A(\cdot)$ has value 1 for positive arguments and 0 otherwise.

Repeating the above steps for a number of realizations allows the estimation of the spatial statistics of the liquefaction indices q on the horizontal plane of maximum liquefaction likelihood. In particular, the excursion area fractions A_q are evaluated for $q = \{0.1, 0.2, \ldots, 0.9\}$.

Liquefaction of a column is defined as occurring when the liquefaction index q_i exceeds 0.96 in some element; the analysis of that column is then discontinued to save considerable computational effort, and the liquefaction indices are subsequently held constant. In fact, numerical tests indicated that, at least under this one-dimensional model, the liquefied element absorbs most of the input motion (see Figure 16.4) so that little change was subsequently observed in the liquefaction indices of higher elements. The liquefied element can be considered the location of liquefaction initiation since postevent pore pressure redistribution is being ignored (and, in fact, is not accurately modeled with this one-dimensional simplification).

The horizontal plane having the highest average liquefaction index is found and the statistics of those indices determined. This plane will be referred to as the maximal plane. It is recognized that when liquefaction does take place it is not likely to be confined to a horizontal plane of a certain thickness. At the very least the plane could be inclined, but more likely liquefaction would follow an undulating surface. This level of sophistication is beyond the scope of this initial study, however, which is confined to the consideration of liquefaction occurring along horizontal planes arranged over depth.

Figure 16.5 illustrates two realizations of the maximal plane. Regions which have liquefied are shown in white. In both examples, a significant portion of the area has q indices exceeding 0.9 and there is clearly significant spatial variation. The gray scale representation was formed by linear interpolation from the 16×16 mesh of finite elements.

16.3 MONTE CARLO ANALYSIS AND RESULTS

The four cases considered are summarized in Table 16.2. In the following, the first set of simulated ground motions are referred to as event 1 and the ground motions scaled by a factor of 0.7 as event 2. The average depth at which the maximal plane occurs is about 2.7 m for cases 1 and 3 and about 3.0 m for cases 2 and 4. Thus, it appears that the larger correlation lengths result in somewhat lower maximal planes. These results are in basic agreement with the location of liquefied units observed by Holzer et al. (1989).

Table 16.2 Monte Carlo Cases

Case	Input Motion Scaling Factor	Vertical Correlation Length (m)	Number of Realizations
1	1.0 (event 1)	1.0	100
2	1.0 (event 1)	4.0	100
3	0.7 (event 2)	1.0	100
4	0.7 (event 2)	4.0	100

The average excursion area, expressed as a fraction of the total domain area, of the maximal plane exceeding a threshold liquefaction index q, $\bar{A}_q$, is shown in Figure 16.6. Excursion fraction $\bar{A}_q$ is obtained by averaging the A_q values over the 100 realizations for each case. The trend in Figure 16.6 is evident:

1. The correlation length has little effect on the average excursion fraction $\bar{A}_q$.
2. The intensity of the input motion has a significant effect on the excursion fractions, as expected. A 30% reduction in input motion intensity reduced the liquefaction index corresponding to $\bar{A}_q = 17\%$ from 0.9 to about 0.35, almost a threefold reduction.

According to Figure 16.6, only about 20% of the model site had liquefaction indices in excess of 0.9 under event 1.

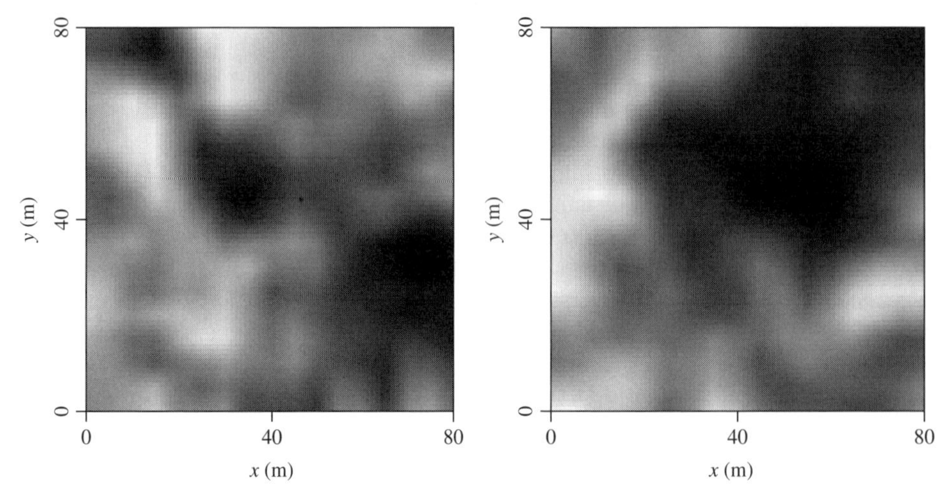

Figure 16.5 Gray-scale maps of the planes having the highest average liquefaction index q drawn from two realizations of the soil mass.

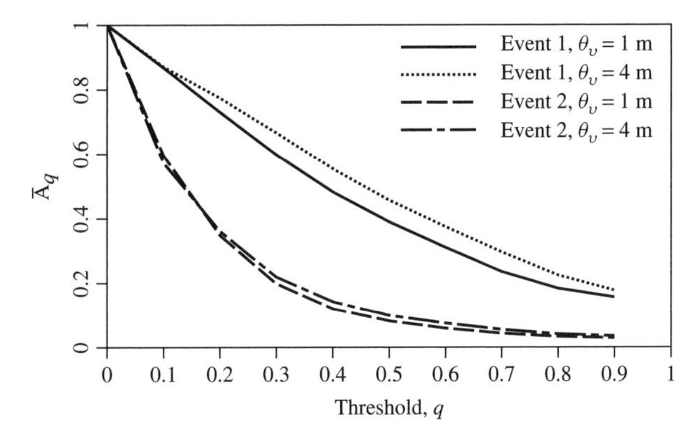

Figure 16.6 Average fraction of maximal plane, $\bar{A}_q$, having liquefaction indices in excess of indicated q thresholds.

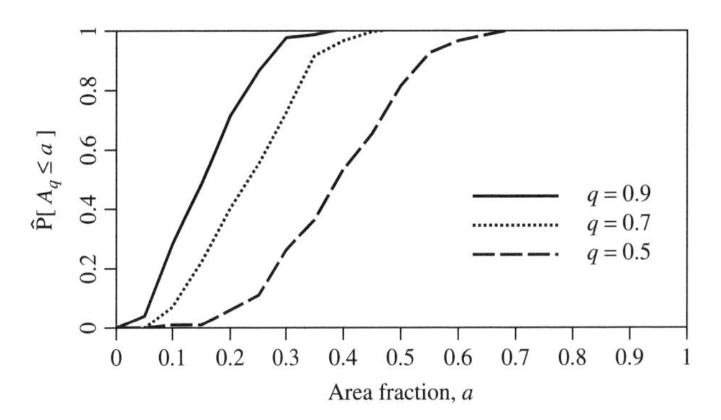

Figure 16.7 Estimated probability distribution of area fraction with liquefaction index greater than q (for event 1, $\theta_v = 1$ m).

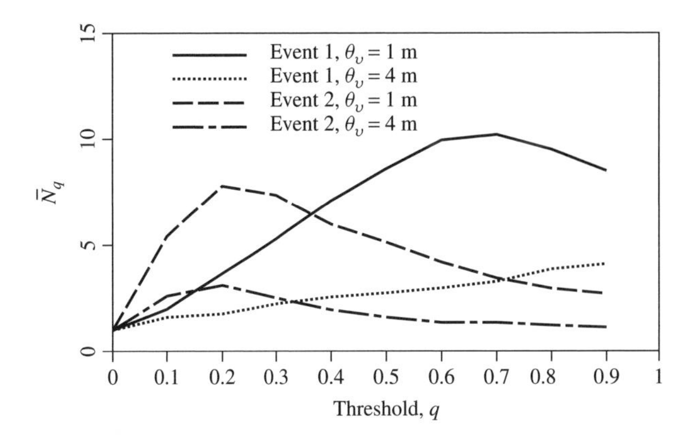

Figure 16.8 Average number of isolated excursion areas, $\bar{N}_q$, where liquefaction indices exceed the threshold q.

Since an event of this magnitude did result in sand boils and lateral spreading at the Wildlife site, the simulation results suggest that global liquefaction may occur even if only a relatively low percentage of the site is predicted to liquefy. This observation emphasizes the need to rationally quantify the spatial distribution of liquefaction and its effect on global liquefaction risk in future studies.

It appears that the likelihood of global liquefaction due to event 2 is quite low. To some extent, this is substantiated by the fact that the Wildlife site did not (globally) liquefy during the Elmore Ranch event ($M_s = 6.2$ compared to the Superstition Hills event, $M_s = 6.6$) (Keane and Prevost, 1989). Figure 16.6 suggests a possible approach to the evaluation of liquefaction risk using the knowledge that the Wildlife site is highly liquefiable: Determine the average area of the maximal planes which exceed a liquefaction index of 0.9—global liquefaction risk increases as this area increases. In this particular study (case 1 or 2) only some 15–20% of the total area liquefied under this criterion. It is unknown at this time if this proportion of liquefaction is generally sufficient to result in global liquefaction. Such a measure needs to be substantiated and verified through similar studies of other sites and other earthquakes. It nevertheless suggests that, under reasonable assumptions about the site and the earthquake, soil liquefaction can vary significantly over space, and only a small fraction need actually liquefy to cause structural damage.

The spatial variability of liquefaction can be quantified in a number of ways; the total area of excursions (exceeding some liquefaction index), the number of isolated excursions, and the degree to which the individual excursions are clustered. Figure 16.7 shows the estimated probability distribution, $\hat{P}$, of the area fraction having liquefaction indices greater than q for event 1, $\theta_v = 1.0$ m. From this plot it can be seen that, for example, $\hat{P}[A_{0.9} > 0.2] = 1 - 0.7 = 0.3$, that is, 30% of realizations have more than 20% of the maximal plane area with liquefaction indices higher than 0.9. Similarly, more than 10% of the maximal plane area effectively liquefies ($q_{ij} \geq 0.9$) with probability 72%.

Figure 16.8 shows the average number of isolated excursions above the liquefaction index q for each case study, and Figure 16.9 shows the corresponding cluster measure, Ψ, both averaged over 100 realizations. The cluster measure, as defined by Fenton and Vanmarcke (1992), reflects the degree to which excursion regions are clustered: Ψ has value 0 if the excursions are uniformly distributed through the domain and value 1 if they are clumped into a single region or excursion. Both figures exhibit much more pronounced effects due to changes in the correlation length. The correlation length $\theta_v = 4$ m ($\theta_h = 160$ m) substantially decreases the average number of excursions and substantially increases the cluster measure. This implies that for the same total area exceeding a certain index q, the regions show higher clustering at higher correlation lengths. In turn, higher clustering implies a higher likelihood of

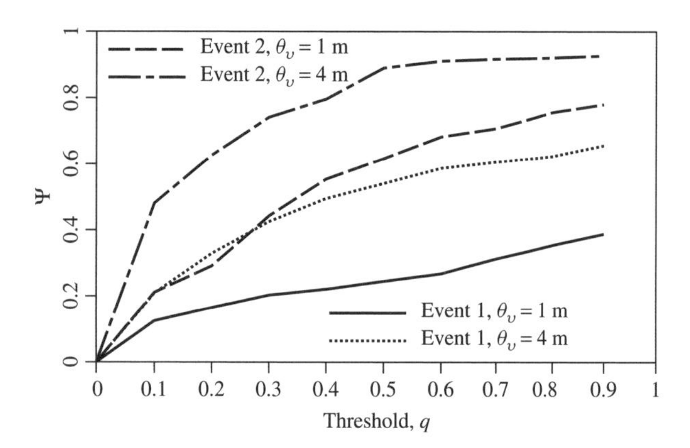

Figure 16.9 Average cluster measure Ψ of isolated excursions where liquefaction indices exceed the threshold q.

global liquefaction since there are fewer pockets of "resistance" within the excursion region. Notice that event 2 typically has higher mean values of Ψ since it has fewer excursions at high thresholds (a single excursion, or no excursions, corresponds to $\Psi \simeq 1$). The likelihood of liquefaction, thus, cannot depend on the cluster measure alone; it must also take into consideration the total excursion area above a high threshold.

16.4 SUMMARY

It is recognized that the one-dimensional finite-element analysis employed in this study cannot capture some of the details of spatial liquefaction, the connection between soil columns being only through their correlated properties and earthquake ground motion. However, the resulting analysis was tractable at this time (a fully three-dimensional analysis is still prohibitively computationally expensive), allowing the analysis of a sufficient number of realizations for reasonable statistics. It is believed that the major, large-scale, features of the spatial distribution of liquefaction initiation are nevertheless captured by the present analysis, allowing the following observations to be made.

Perhaps the major observation to be drawn from this study is that (predicted) soil liquefaction shows considerable spatial variability under reasonable assumptions regarding the site and its excitation. The recognition of this spatial variability may significantly advance our understanding and modeling of the phenomenon, allowing the probabilistic assessment of the spatial extent of liquefaction damage. The present study indicated that as little as 15–20% of a site which is known to have liquefied was actually predicted to liquefy during the event. Whether this was the actual fraction of liquefaction at the site is unknown. There is also a possibility of further postevent liquefaction.

Given the fact that the Wildlife site was known to have liquefied during the Superstition Hills event, the following summary of the results of this model study can be made;

1. The spatially random approach to liquefaction analysis enables quantifying the probability of effectively liquefied area fractions or excursions at the site. For example, on the basis of this study, more than 10% of the model site over a plane at about 2.7 m depth was predicted to effectively liquefy ($q \geq 0.9$) with probability 72% during event 1 ($\theta_v = 1$ m), which was modeled after the Superstition Hills event.
2. The likelihood of global liquefaction resulting in loss of lateral stability at the surface appears to be most easily quantified by the total area of the domain whose liquefaction indices exceed some threshold index q^*. In this case study if the threshold index is taken as 0.9, a high likelihood of global liquefaction might be associated with mean total excursion areas A_{q^*} in excess of about 15–20% of the total domain area. This measure incorporates the effect of earthquake intensity but needs to be calibrated through other studies and, in time, through fully three-dimensional models.
3. The likelihood of liquefaction can be modified by the cluster measure—as the cluster measure decreases, the liquefied regions become separated by pockets of resistance, and the likelihood of global liquefaction at the site decreases. This correction incorporates the effect of correlation lengths of the soil properties.

The recognition that liquefaction is a spatially varying phenomenon and the development of techniques to quantify this variability, along with its implications on risk, are important starts in the understanding of global liquefaction failure at a site. The study also illustrates the potential danger in assessing liquefaction risk at a site on the basis of, for example, CPT data collected at a single location. Data from several different locations should be considered so that liquefiable regions can be more closely identified and subsequently modeled in a dynamic structural evaluation.

REFERENCES

Abramowitz, M., and Stegun, I., Eds. (1970). *Handbook of Mathematical Functions*, 10th ed., Dover, New York.

Adler, R. J. (1981). *The Geometry of Random Fields*, Wiley, New York.

Allen, D. E. (1975). "Limit states design: A probabilistic study," *Can. J. Civ. Eng.*, **36**(2), 36–49.

Alonso, E. E. (1976). "Risk analysis of slopes and its application to slopes in Canadian sensitive clays," *Géotechnique*, **26**, 453–472.

American Concrete Institute (ACI) (1989). *ACI 318-89, Building Code Requirements for Reinforced Concrete*, ACI, Detroit, MI.

American Society of Civil Engineers (ASCE) (1994). *Settlement Analysis*, Technical Engineering and Design Guides, No. 9, adapted from U.S. Army Corps of Engineers, Reston, VA.

Anderson, T. W. (1971). *The Statistical Analysis of Time Series*, Wiley, New York.

Asaoka, A., and Grivas, D. A. (1982). "Spatial variability of the undrained strength of clays," *ASCE J. Geotech. Eng.*, **108**(5), 743–756.

Australian Standard (2004). *Bridge Design, Part 3: Foundations and Soil-Supporting Structures*, AS 5100.3–2004, Sydney, Australia.

Australian Standard (2002). *Earth-Retaining Structures*, AS 4678–2004, Sydney, Australia.

Baecher, G. B., and Ingra, T. S. (1981). "Stochastic FEM in settlement predictions," *ASCE J. Geotech. Eng.*, **107**(4), 449–464.

Baecher, G. B., and Christian, J. T. (2003). *Reliability and Statistics in Geotechnical Engineering*, Wiley, Chichester, United Kingdom.

Basheer, I. A., and Najjar, Y. M. (1996). "Reliability-based design of reinforced earth retaining walls," *Transport. Res. Rec.*, **1526**, 64–78.

Becker, D. E. (1996a). "Eighteenth Canadian Geotechnical Colloquium: Limit states design for foundations. Part I. An overview of the foundation design process," *Can. Geotech. J.*, **33**, 956–983.

Becker, D. E. (1996b). "Eighteenth Canadian Geotechnical Colloquium: Limit states design for foundations. Part II. Development for the National Building Code of Canada," *Can. Geotech. J.*, **33**, 984–1007.

Benjamin, J. R., and Cornell, C. A. (1970). *Probability, Statistics, and Decision for Civil Engineers*, McGraw-Hill, New York.

Bennett, M. J., McLaughlin, P. V., Sarmiento, J. S., and Youd, T. L. (1984). "Geotechnical investigation of liquefaction sites, Imperial County, California," open-file report 84-252, U.S. Department of the Interior Geological Survey, Menlo Park, CA.

Beran, J. (1994). *Statistics for Long-Memory Processes*, V. Isham *et al.*, Eds., Chapman & Hall, New York.

Berman, S. M. (1992). *Sojourns and Extremes of Stochastic Processes*, Wadsworth and Brooks/Cole, Pacific Grove, CA.

Bishop, A.W. (1955). "The use of the slip circle in the stability analysis of slopes," *Géotechnique*, **5**(1), 7–17.

Boissières, H. P., and Vanmarcke, E. H. (1995). "Spatial correlation of earthquake ground motion: non-parametric estimation," *Soil Dynam. Earthquake Eng.*, **14**, 23–31.

Bowles, J. E. (1996). *Foundation Analysis and Design*, 5th ed., McGraw-Hill, New York.

Box, G. E. P., and Mullef, M. E. (1958). "A note on the generation of random normal variates," *Ann. Math. Statist.*, **29**, 610–611.

Brockwell, P. J. and Davis, R. A. (1987). *Time Series: Theory and Methods*, P. Bickel *et al.*, Eds., Springer-Verlag, New York.

Call, R. D. (1985). "Probability of stability design of open pit slopes," in *Rock Masses: Modeling of Underground Openings/Probability of Slope Failure/Fracture of Intact Rock*, Proc. Symp. Geotech. Eng. Div., C.H. Dowding, Ed., American Society of Civil Engineers, Denver, CO, pp. 56–71.

Canadian Geotechnical Society (1978). *Canadian Foundation Engineering Manual*, CGS, 2nd ed., Montreal, Quebec.

Canadian Geotechnical Society (1992). *Canadian Foundation Engineering Manual*, CGS, 3rd ed., Montreal, Quebec.

Canadian Geotechnical Society (CGS) (2006). *Canadian Foundation Engineering Manual*, CGS, 4th ed., Montreal, Quebec.

Canadian Standards Association (CSA) (1984). *CAN3-A23.3-M84 Design of Concrete Structures for Buildings*, CSA, Toronto, Ontario.

Canadian Standards Association (CSA) (2000a). *Canadian Highway Bridge Design Code (CHBDC)*, CSA Standard S6-00, CSA, Rexdale, Ontario.

Canadian Standards Association (CSA) (2000b). *Commentary on CAN/CSA-S6-00 Canadian Highway Bridge Design Code*, CAN/CSA-S6.1-00, CSA, Mississauga, Ontario.

Carpenter, L. (1980). "Computer Rendering of fractal curves and surfaces," in *Association for Computing Machinery's Special*

Interest Group on Graphics and Interactive Techniques SIGGRAPH 80 Proceedings, Assoc. Computing Machinery, New York, pp. 108–120.

Casagrande, A. (1937). "Seepage through dams," *J. New. Engl. Water Works Assoc.*, **51**(2), 131–172.

Cedergren, H. R. (1967). *Seepage, Drainage and Flow Nets*, Wiley, Chichester, United Kingdom.

Chalermyanont, T., and Benson, C. H. (2004). "Reliability-based design for internal stability of mechanically stabilized earth walls," *ASCE J. Geotech. Geoenv. Eng.*, **130**(2), 163–173.

Chernoff, H., and Lehmann, E. L. (1954). "The use of maximum likelihood estimates in χ^2 tests for goodness of fit," *Ann. Math. Statist.*, **25**, 579–586.

Cherubini, C. (1997). "Data and considerations on the variability of geotechnical properties of soils," in *Proc. Int. Conf. Safety and Reliability, ESREL 97*, Vol. 2, Lisbon, Spain, pp. 1583–1591.

Cherubini, C. (2000). "Reliability evaluation of shallow foundation bearing capacity on c', ϕ' soils," *Can. Geotech. J.*, **37**, 264–269.

Chowdhury, R. N., and Tang, W. H. (1987). "Comparison of risk models for slopes," in *Proc. 5th Int. Conf. Applications of Statistics and Probability in Soil and Structural Engineering*, Vol. 2, pp. 863–869.

Christian, J. T. (1996). "Reliability methods for stability of existing slopes," in *Uncertainty in the geologic environment: From theory to practice*, Geotechnical Special Publication No. 58, C.D. Shackelford *et al.*, Eds., American Society of Civil Engineers, New York, pp. 409–418.

Christian, J. T. and Baecher, G. B. (1999). "Point-estimate method as numerical quadrature," *ASCE J. Geotech. Geoenv. Eng.*, **125**(9), 779–786.

Christian, J. T. and Carrier, W. D. (1978). "Janbu, Bjerrum and Kjaernsli's chart reinterpreted," *Can. Geotech. J.*, **15**, 123–128.

Christian, J. T., Ladd, C. C., and Baecher, G. B. (1994). "Reliability applied to slope stability analysis," *ASCE J. Geotech. Eng.*, **120**(12), 2180–2207.

Cinlar, E. (1975). *Introduction to Stochastic Processes*, Prentice-Hall, Englewood Cliffs, NJ.

Cooley, J. W. and Tukey, J. W. (1965). "An algorithm for the machine calculation of complex Fourier series," *Math. Comput.*, **19**(90), 297–301.

Cornell, C. A. (1969). "A probability-based structural code," *J. Amer. Concrete Inst.*, **66**(12), 974–985.

Craig, R. F. (2001). *Soil Mechanics*, 6th ed., Spon Press, New York.

Cramer, H. (1966). "On the intersections between the trajectories of a normal stationary stochastic process and a high level," *Arkiv för Matematik*, **6**, 337.

Cramer, H., and Leadbetter, M. R. (1967). *Stationary and Related Stochastic Processes*, Wiley, New York.

Cressie, N. A. C. (1993). *Statistics for Spatial Data*, 2nd ed., Wiley, New York.

D'Andrea, R. A., and Sangrey, D. A. (1982). "Safety factors for probabilistic slope design," *ASCE J. Geotech. Eng.*, **108**(GT9), 1101–1118.

D'Appolonia, D. J., D'Appolonia, E., and Brissette, R. F. (1968). "Settlement of spread footings on sand," *ASCE J. Soil Mech. Found. Div.*, **94**(SM3), 735–760.

Dagan, G. (1979). "Models of groundwater flow in statistically homogeneous porous formations," *Water Resourc. Res.*, **15**(1), 47–63.

Dagan, G. (1981). "Analysis of flow through heterogeneous random aquifers by the method of embedding matrix: 1, Steady flow," *Water Resourc. Res.*, **17**(1), 107–122.

Dagan, G. (1986). "Statistical theory of groundwater flow and transport: Pore to laboratory, laboratory to formation, and formation to regional scale," *Water Resourc. Res.*, **22**(9), 120S–134S.

Dagan, G. (1989). *Flow and Transport in Porous Formations*, Springer-Verlag, New York.

Dagan, G. (1993). "Higher-order correction of effective conductivity of heterogeneous formations of lognormal conductivity distribution," *Transp. Porous Media*, **12**, 279–290.

Dahlhaus, R. (1989). "Efficient parameter estimation for self-similar processes," *Ann. Statist.*, **17**, 1749–1766.

Dai, Y., Fredlund, D. G., and Stolte, W. J. (1993). "A probabilistic slope stability analysis using deterministic computer software," in *Probabilistic Methods in Geotechnical Engineering*, K.S. Li and S.-C.R. Lo, Eds., Balkema, Rotterdam, The Netherlands, pp. 267–274.

Das, B. M. (2000). *Fundamentals of Geotechnical Engineering*, Brooks/Cole, Pacific Grove, CA.

Davenport, A. G. (1964). "Note on the random distribution of the largest value of a random function with emphasis on gust loading," *Proc. Inst. Civ. Eng.*, **28**, 187–196.

DeGroot, D. J., and Baecher, G. B. (1993). "Estimating autocovariance of in-situ soil properties," *ASCE J. Geotech. Eng.*, **119**(1), 147–166.

DeGroot, D. J. (1996). "Analyzing spatial variability of in-situ properties," in *Uncertainty in the geologic environment: From theory to practice*, Geotechnical Special Publication No. 58, C.D. Shackelford *et al.*, Eds., American Society of Civil Engineers, New York, pp. 210–238.

Devore, J. L. (2003). *Probability and Statistics for Engineering and the Sciences*, 6th ed., Duxbury, New York.

Ditlevson, O. (1973). "Structural reliability and the invariance problem," Research Report No. 22, Solid Mechanics Division, University of Waterloo, Waterloo, Canada.

Duncan, J. M. (2000). "Factors of safety and reliability in geotechnical engineering," *ASCE J. Geotech. Geoenv. Eng.*, **126**(4), 307–316.

Dykaar, B. B., and Kitanidis, P. K. (1992a). "Determination of the effective hydraulic conductivity for heterogeneous porous media using a numerical spectral approach: 1, Method," *Water Resourc. Res.*, **28**(4), 1155–1166.

Dykaar, B. B., and Kitanidis, P. K. (1992b). "Determination of the effective hydraulic conductivity for heterogeneous porous media using a numerical spectral approach: 2, Results," *Water Resourc. Res.*, **28**(4), 1167–1178.

El-Ramly, H., Morgenstern, N. R., and Cruden, D. M. (2002). "Probabilistic slope stability analysis for practice," *Can. Geotech. J.*, **39**, 665–683.

EN 1997-1 (2003). *Eurocode 7 Geotechnical design—Part 1: General rules*, CEN (European Committee for Standardization), Brussels, Belgium.

ENV 1991-1 (1994). *Eurocode 1 Basis of design and actions on structures—Part 1: Basis of design*, CEN (European Committee for Standardization), Brussels, Belgium.

Ewing, J. A. (1969). "Some measurements of the directional wave spectrum," *J. Marine Res.*, **27**, 163–171.

Fardis, M. N., and Veneziano, D. (1982). "Probabilistic analysis of deposit liquefaction," *ASCE J. Geotech. Eng.*, **108**(3), 395–418.

Fenton, G. A. (1990). *Simulation and Analysis of Random Fields*, Ph.D. Thesis, Dept. Civil Eng. and Op. Res., Princeton University, Princeton, NJ.

Fenton, G. A. (1994). "Error evaluation of three random field generators," *ASCE J. Eng. Mech.*, **120**(12), 2478–2497.

Fenton, G. A. (1999a). "Estimation for stochastic soil models," *ASCE J. Geotech. Geoenv. Eng.*, **125**(6), 470–485.

Fenton, G. A. (1999b). "Random field modeling of CPT data," *ASCE J. Geotech. Geoenv. Eng.*, **125**(6), 486–498.

Fenton, G. A., and Griffiths, D. V. (1993). "Statistics of block conductivity through a simple bounded stochastic medium," *Water Resourc. Res.*, **29**(6), 1825–1830.

Fenton, G. A., and Griffiths, D. V. (1995). "Flow through earthdams with spatially random permeability," in *Proc. 10th ASCE Engineering Mechanics Conf.*, Boulder, CO, pp. 341–344.

Fenton, G. A., and Griffiths, D. V. (1996). "Statistics of free surface flow through a stochastic earth dam," *ASCE J. Geotech. Eng.*, **122**(6), 427–436.

Fenton, G. A., and Griffiths, D. V. (1997a). "A mesh deformation algorithm for free surface problems," *Int. J. Numer. Anal. Methods Geomech.*, **21**(12), 817–824.

Fenton, G. A., and Griffiths, D. V. (1997b). "Extreme hydraulic gradient statistics in a stochastic earth dam," *ASCE J. Geotech. Geoenv. Eng.*, **123**(11), 995–1000.

Fenton, G. A., and Griffiths, D. V. (2001). "Bearing capacity of spatially random $c - \phi$ soils," in *Proc. 10th Int. Conf. on Computer Methods and Advances in Geomechanics (IACMAG 01)*, Tucson, AZ, pp. 1411–1415.

Fenton, G. A., and Griffiths, D. V. (2002). "Probabilistic foundation settlement on spatially random soil," *ASCE J. Geotech. Geoenv. Eng.*, **128**(5), 381–390.

Fenton, G. A., and Griffiths, D. V. (2003). "Bearing capacity prediction of spatially random $c - \phi$ soils," *Can. Geotech. J.*, **40**(1), 54–65.

Fenton, G. A., and Griffiths, D. V. (2005a). "Reliability of traditional retaining wall design," *Géotechnique*, **55**(1), 55–62.

Fenton, G. A., and Griffiths, D. V. (2005b). "Three-dimensional probabilistic foundation settlement," *ASCE J. Geotech. Geoenv. Eng.*, **131**(2), 232–239.

Fenton, G. A., and Griffiths, D. V. (2005c). "A slope stability reliability model," *Proc. K.Y. Lo Symposium*, London, Ontario.

Fenton, G. A., and Griffiths, D. V. (2007). "Reliability-based deep foundation design," in *Probabilistic Applications in Geotechnical Engineering*, GSP No. 170, Proc. Geo-Denver 2007 Symposium, American Society of Civil Engineers, Denver, CO.

Fenton, G. A., and Vanmarcke, E. H. (1990). "Simulation of random fields via local average subdivision," *ASCE J. Eng. Mech.*, **116**(8), 1733–1749.

Fenton, G. A., and Vanmarcke, E. H. (1992). "Simulation-based excursion statistics," *ASCE J. Eng. Mech.*, **118**(6), 1129–1145.

Fenton, G. A., and Vanmarcke, E. H. (1998). "Spatial variation in liquefaction risk," *Géotechnique*, **48**(6), 819–831.

Fenton, G. A., Griffiths, D. V., and Cavers, W. (2005). "Resistance factors for settlement design," *Can. Geotech. J.*, **42**(5), 1422–1436.

Fenton, G. A., Paice, G. M., and Griffiths, D.V. (1996). "Probabilistic analysis of foundation settlement," in *Proc. ASCE Uncertainty'96 Conf.*, Madison, WI, pp. 651–665.

Fenton, G. A., Griffiths, D. V., and Urquhart, A. (2003a). "A slope stability model for spatially random soils," in *Proc. 9th Int. Conf. Applications of Statistics and Probability in Civil Engineering (ICASP9)*, A. Kiureghian *et al.*, Eds., Millpress, San Francisco, CA, pp. 1263–1269.

Fenton, G. A., Zhou, H., Jaksa, M. B., and Griffiths, D. V. (2003b). "Reliability analysis of a strip footing designed against settlement," in *Proc. 9th Int. Conf. Applications of Statistics and Probability in Civil Engineering (ICASP9)*, A. Kiureghian *et al.*, Eds., Millpress, San Francisco, CA, pp. 1271–1277.

Fenton, G. A., Zhang, X. Y., and Griffiths, D. V. (2007a). "Reliability of shallow foundations designed against bearing failure using LRFD," *Georisk*, **1**(4), 202–215.

Fenton, G. A., Zhang, X. Y., and Griffiths, D. V. (2007b). "Loadxsxse and resistance factor design of shallow foundations against bearing failure," *Can. Geotech. J.*, submitted for publication.

Fournier, A., Fussell, D., and Carpenter, L. (1982). "Computer rendering of stochastic models," *Commun. ACM*, **25**(6), 371-384.

Foye, K. C., Salgado, R., and Scott, B. (2006). "Resistance factors for use in shallow foundation LRFD," *ASCE J. Geotech. Geoenv. Eng.*, **132**(9), 1208–1218.

Freeze, R. A. (1975). "A stochastic-conceptual analysis of one-dimensional groundwater flow in nonuniform homogeneous media," *Water Resourc. Res.*, **11**(5), 725–741.

French, S. E. (1999). *Design of Shallow Foundations*, ASCE Press, Reston, VA.

Freudenthal, A. M. (1956). "Safety and the probability of structure failure," *Trans. ASCE*, **121**, 1337–1375.

Gelhar, L. W. (1993). *Stochastic Subsurface Hydrology*, Prentice Hall, Englewood Cliffs, NJ.

Gelhar, L. W., and Axness, C. L. (1983). "Three-dimensional stochastic analysis of macrodispersion in aquifers," *Water Resourc. Res.*, **19**(1), 161–180.

Gradshteyn, I. S., and Ryzhik, I. M. (1980). *Table of Integrals, Series, and Products*, 4th ed., Academic, Toronto, Ontario.

Griffiths, D. V. (1980). "Finite element analyses of walls, footings and slopes," in *Proc. Symp. on Comp. Applic. Geotech. Probs. in Highway Eng.*, PM Geotechnical Analysts, Cambridge, United Kingdom, pp. 122–146.

Griffiths, D. V. (1984). "Rationalised charts for the method of fragments applied to confined seepage," *Géotechnique*, **34**(2), 229–238.

Griffiths, D. V., and Fenton, G. A. (1993). "Seepage beneath water retaining structures founded on spatially random soil," *Géotechnique*, **43**(4), 577–587.

Griffiths, D. V., and Fenton, G. A. (1995). "Observations on two- and three-dimensional seepage through a spatially random soil," in *Proc. 7th Int. Conf. on Applications of Statistics and Probability in Civil Engineering*, Paris, France, pp. 65–70.

Griffiths, D. V., and Fenton, G. A. (1997). "Three-dimensional seepage through spatially random soil," *ASCE J. Geotech. Geoenv. Eng.*, **123**(2), 153–160.

Griffiths, D. V., and Fenton, G. A. (1998). "Probabilistic analysis of exit gradients due to steady seepage," *ASCE J. Geotech. Geoenv. Eng.*, **124**(9), 789–797.

Griffiths, D. V., and Fenton, G. A. (2000a). "Influence of soil strength spatial variability on the stability of an undrained clay slope by finite elements," in *Slope Stability 2000*, Geotechnical Special Publication No. 101, American Society of Civil Engineering, New York, pp. 184–193.

Griffiths, D. V., and Fenton, G. A. (2000b). "Bearing capacity of heterogeneous soils by finite elements," in *Proc. 5th Int. Congress on Numerical Methods in Engineering and*

Scientific Applications (CIMENICS'00), N. Troyani and M. Cerrolaza, Eds., Sociedad Venezolana de Métodos Numéricos en Ingeniería, pp. CI 27–37.

Griffiths, D. V., and Fenton, G. A. (2001). "Bearing capacity of spatially random soil: The undrained clay Prandtl problem revisited," *Géotechnique*, **51**(4), 351–359.

Griffiths, D. V., and Fenton, G. A. (2004). "Probabilistic slope stability analysis by finite elements," *ASCE J. Geotech. Geoenv. Eng.*, **130**(5), 507–518.

Griffiths, D. V., and Fenton, G. A. (2005). "Probabilistic settlement analysis of rectangular footings," in *Proc. XVI Int. Conf. Soil Mech. Geotech. Eng. (ICSMGE)*, Millpress Science, Osaka, Japan, pp. 1041–1044.

Griffiths, D. V., and Fenton, G. A. (2007). "Probabilistic settlement analysis by stochastic and Random Finite Element Methods," in *Proc. XIII PanAmerican Conf. on Soil Mechanics and Geotechnical Engineering*, Isla de Margarita, Venezuela, pp. 166–176.

Griffiths, D. V., and Lane, P. A. (1999). "Slope stability analysis by finite elements," *Géotechnique*, **49**(3), 387–403.

Griffiths, D. V., and Smith, I. M. (2006). *Numerical Methods for Engineers*, 2nd ed., Chapman & Hall/CRC Press, Boca Raton, FL.

Griffiths, D. V., Paice, G. M., and Fenton, G. A. (1994). "Finite element modeling of seepage beneath a sheet pile wall in spatially random soil," in *Proc. Int. Conf. of the Int. Assoc. Computer Methods and Advances in Geomechanics (IACMAG 94)*, H.J. Siriwardane and M.M. Zaman, Eds., Morgantown, WV, pp. 1205–1210.

Griffiths, D. V., Fenton, G. A., and Paice, G. M. (1996). "Reliability-based exit gradient design of water retaining structures," in *Proc. ASCE Uncertainty'96 Conference*, Madison, WI, pp. 518–534.

Griffiths, D. V., Fenton, G. A., and Lemons, C. B. (2001). "Underground pillar stability: A probabilistic approach," in *Proc. XV Int. Conf. Soil Mech. Geotech. Eng. (ICSMGE)*, Istanbul, Turkey, pp. 1343–1346.

Griffiths, D. V., Fenton, G. A., and Lemons, C. B. (2002a). "Probabilistic analysis of underground pillar stability," *Int. J. Numer. Anal. Methods Geomechan.*, **26**, 775–791.

Griffiths, D. V., Fenton, G. A., and Manoharan, N. (2002b). "Bearing capacity of a rough rigid strip footing on cohesive soil: A probabilistic study," *ASCE J. Geotech. Geoenv. Eng.*, **128**(9), 743–755.

Griffiths, D. V., Fenton, G. A., and Tveten, D. E. (2002c). "Probabilistic geotechnical analysis: How difficult does it need to be?," in *Proc. Int. Conf. on Probabilistics in Geotechnics: Technical and Economic Risk Estimation*, R. Pottler *et al.*, Eds., United Engineering Foundation, Graz, Austria, pp. 3–20.

Griffiths, D. V., Fenton, G. A., and Tveten, D. E. (2005). "Probabilistic earth pressure analysis by the random finite element method," in *Proc. 11th Int. Conf. on Computer Methods and Advances in Geomechanics (IACMAG 05)*, Vol. 4, G. Barla and M. Barla, Eds., Turin, Italy, pp. 235–249.

Griffiths, D. V., Fenton, G. A., and Ziemann, H. R. (2006). "The influence of strength variability in the analysis of slope failure risk," in *Geomechanics II, Proc. 2nd Japan-US Workshop on Testing, Modeling and Simulation*, P.V. Lade, and T. Nakai, Eds., Geotechnical Special Publication No. 156, American Society of Civil Engineers, Kyoto, Japan, pp. 113–123.

Griffiths, D. V., Fenton, G. A., and Denavit, M. D. (2007). "Traditional and advanced probabilistic slope stability analysis," in *Probabilistic Applications in Geotechnical Engineering*, Geotechnical Special Publication No. 170, Proc. Geo-Denver 2007 Symposium, American Society of Civil Engineers, Denver, CO.

Gumbel, E. (1958). *Statistics of Extremes*, Columbia University Press, New York.

Gutjahr, A. L., Gelhar, L. W., Bakr, A. A., and MacMillan, J. R. (1978). "Stochastic analysis of spatial variability in subsurface flows: 2, Evaluation and application," *Water Resourc. Res.*, **14**(5), 953–959.

Hansen, B. (1953). *Earth Pressure Calculation*, Danish Technical Press, Copenhagen, Denmark.

Hansen, B. (1956). *Limit Design and Safety Factors in Soil Mechanics*, Bulletin No. 1, Danish Geotechnical Institute, Copenhagen, Denmark.

Harr, M. E. (1987). *Reliability-Based Design in Civil Engineering*, McGraw-Hill, New York.

Harr, M. E. (1962). *Groundwater and Seepage*, McGraw-Hill, New York.

Hasofer, A. M., and Lind, N. C. (1974). "Exact and invariant second-moment code format," *ASCE J. Engl. Mech. Div.*, **100**, 111–121.

Hassan, A. M., and Wolff, T. F. (2000). "Effect of deterministic and probabilistic models on slope reliability index," in *Slope Stability 2000*, Geotechnical Special Publication No. 101, American Society of Civil Engineers, New York, pp. 194–208.

Higgins, J. J., and Keller-McNulty, S. (1995). *Concepts in Probability and Stochastic Modeling*, Duxbury, New York.

Hoek, E. (1998). "Reliability of Hoek-Brown estimates of rock mass properties and their impact on design," *Int. J. Rock Mechan., Min. Sci. Geomechan. Abstr.*, **34**(5), 63–68.

Hoek, E., and Brown, E.T. (1997). "Practical estimates of rock mass strength," *Int. J. Rock Mechan. Mining Sc.*, **34**(8), 1165.

Hoeksema, R. J., and Kitanidis, P. K. (1985). "Analysis of the spatial structure of properties of selected aquifers," *Water Resourc. Res.*, **21**(4), 563–572.

Holtz, R. D., and Kovacs, W. D. (1981). *An Introduction to Geotechnical Engineering*, Prentice-Hall, Englewood Cliffs, NJ.

Holtz, R. D., and Krizek, R. J. (1971). "Statistical evaluation of soils test data," in *1st Int. Conf. on Applications of Statistics and Probability to Soil and Structural Engineering*, Hong Kong University Press, Hong Kong, pp. 229–266.

Holzer, T. L., Youd, T. L., and Bennett, M. J. (1988). "In situ measurement of pore pressure build-up during liquefaction," in *Proc. 20th Joint Meeting of United States-Japan Panel on Wind and Seismic Effects*, National Institute of Standards and Technology, Gaithersburg, MD, pp. 118–130.

Holzer, T. L., Youd, T. L., and Hanks, T. C. (1989). "Dynamics of liquefaction during the 1987 Superstition Hills, California earthquake," *Science*, **244**, 56–59.

Hryciw, R. D., Vitton, S., and Thomann, T. G. (1990). "Liquefaction and flow failure during seismic exploration," *ASCE J. Geotech. Eng.*, **116**(12), 1881–1899.

Hull, T. E., and Dobell, A. R. (1962). "Random number generators," *SIAM Rev.*, **4**, 230–254.

Indelman, P., and Abramovich, B. (1994). "A higher-order approximation to effective conductivity in media of anisotropic random structure," *Water Resourc. Res.*, **30**(6), 1857–1864.

Institution of Civil Engineers (1991). *Inadequate Site Investigation*, Thomas Telford, London.

Jaksa, M. B., Goldsworthy, J. S., Fenton, G. A., Kaggwa, W. S., Griffiths, D. V., Kuo, Y. L., and Poulos, H. G. (2005).

"Towards reliable and effective site investigations," *Géotechnique*, **55**(2), 109–121.

Jaky, J. (1944). "The coefficient of earth pressure at rest," *J. Soc. Hung. Architects Eng.*, 355–358.

Janbu, N., Bjerrum, L., and Kjaernsli, B. (1956). "Veiledning ved losning av fundamenteringsoppgaver," Publication 16, Norwegian Geotechnical Institute, Oslo, pp. 30–32.

Journel, A. G. (1980). "The lognormal approach to predicting the local distribution of selective mining unit grades," *Math. Geol.*, **12**(4), 283–301.

Journel, A. G., and Huijbregts, Ch. J. (1978). *Mining Geostatistics*, Academic, New York.

Keane, C. M., and Prevost, J. H. (1989). "An analysis of earthquake data observed at the Wildlife Liquefaction Array Site, Imperial County, California," in *Proc. 2nd U.S.-Japan Workshop on Liquefaction, Large Ground Deformations and Their Effects on Lifelines*, Technical Report NCEER-89-0032, Buffalo, NY, pp. 176–192.

Kenney, T. C. and Lau, D. (1985). "Internal stability of granular filters," *Can. Geotech. J.*, **22**, 215–225.

Knuth, D. E. (1981). *Seminumerical algorithms*, Vol. 2 of *The Art of Computer Programming*, 2nd ed., Addison-Wesley, Reading, MA.

Krige, D. G. (1951). "A statistical approach to some basic mine valuation problems on the Witwatersrand," *J. Chem., Metal., Mining Soc. S. Afr.*, **52**(6), 119–139.

Kulhawy, F. H., Roth, M. J. S., and Grigoriu, M. D. (1991). "Some statistical evaluations of geotechnical properties," in *Proc. 6th Int. Conf. Appl. Statistical Problems in Civil Engineering (ICASP6)*, Mexico City, pp. 705–712.

Lacasse, S. (1994). "Reliability and probabilistic methods," in *Proc. 13th Int. Conf. on Soil Mechanics Foundation Engineering*, pp. 225–227.

Lacasse, S., and Nadim, F. (1996). "Uncertainties in characterising soil properties," in *ASCE Uncertainties'96 Conference Proceedings*, C.H. Benson, Ed., Madison, WI, pp. 49–75.

Lacy, S. J., and Prevost, J.H. (1987). "Flow through porous media: A procedure for locating the free surface," *Int. J. Numer. Anal. Methods Geomech.*, **11**(6), 585–601.

Lafleur, J., Mlynarek, J., and Rollin, A. L. (1989). "Filtration of broadly graded cohesionless soils," *ASCE J. Geotech. Eng.*, **115**(12), 1747–1768.

Lancellota, R. (1993). *Geotechnical Engineering*, Balkema, Rotterdam, The Netherlands.

Law, A. M. and Kelton, W. D. (1991). *Simulation Modeling and Analysis*, 2nd ed., McGraw-Hill, New York.

Law, A. M. and Kelton, W. D. (2000). *Simulation Modeling and Analysis*, 3rd ed., McGraw-Hill, New York.

Leadbetter, M. R., Lindgren, G., and Rootzen, H. (1983). *Extremes and Related Properties of Random Sequences and Processes*, Springer-Verlag, New York.

L'Ecuyer, P. (1988). "Efficient and portable combined random number generators," *Commun. ACM*, **31**, 742–749 and 774.

Lee, I. K., White, W., and Ingles, O. G. (1983). *Geotechnical Engineering*, Pitman, London.

Lehmer, D. H. (1951). "Mathematical methods in large-scale computing units," *Ann. Comput. Lab.*, **26**, 141–146.

Leiblein, J. (1954). "A new method of analysing extreme-value data," Technical Note 3053, National Advisory Committee for Aeronautics (NACA), Washington, DC.

Lewis, J. P. (1987). "Generalized stochastic subdivision," *ACM Trans. Graphics*, **6**(3), 167–190.

Lewis, P. A. W. and Orav, E. J. (1989). *Simulation Methodology for Statisticians, Operations Analysts, and Engineers*, Vol. 1, Wadsworth & Brooks, Pacific Grove, CA.

Li, C.-C. and Kiureghian, A. (1993). "Optimal discretization of random fields," *ASCE J. Eng. Mech.*, **119**(6), 1136–1154.

Li, K. S. and Lumb, P. (1987). "Probabilistic design of slopes," *Can. Geotech. J.*, **24**, 520–531.

Lumb, P. (1966). "The variability of natural soils," *ASCE J. Geotech. Eng.*, **3**(2), 74–97.

Lumb, P. (1970). "Safety factors and the probability distribution of soil strength," *Can. Geotech. J.*, **7**, 225–242.

Madsen, H. O., Krenk, S., and Lind, N. C. (1986). *Methods of Structural Safety*, Prentice-Hall, Englewood Cliffs, NJ.

Mandelbrot, B. B. (1982). *The Fractal Geometry of Nature*, W.H. Freeman, New York.

Mandelbrot, B. B., and Ness, J. W. (1968). "Fractional Brownian motions, fractional noises and applications," *SIAM Rev.*, **10**(4), 422–437.

Manoharan, N., Griffiths, D. V., and Fenton, G. A. (2001). "A probabilistic study of rough strip footing on cohesive soil," in *6th U.S. National Congress on Computational Mechanics (VI USACM)*, University of Michigan, Dearborn, MI, p. 257.

Mantoglou, A., and Wilson, J. L. (1981). "Simulation of random fields with the turning bands method," Report #264, Massachusetts Institute of Technology, Dept. Civil Eng., Cambridge, MA.

Marple, S. L., Jr. (1987). *Digital Spectral Analysis*, in *Signal Processing Series*, A.V. Oppenheim, Ed., Prentice-Hall, Englewood Cliffs, NJ.

Marsaglia, G. (1968). "Random numbers fall mainly in the planes," *Natl. Acad. Sci. Proc.*, **61**, 25–28.

Marsily, G. (1985). "Spatial variability of properties in porous media: A stochastic approach," in *Advances in Transport Phenomena in Porous Media*, J. Bear and M.Y. Corapcioglu, NATO Advanced Study Institute on Fundamentals of Transport Phenomena in Porous Media, Dordrecht, pp. 719–769.

Matern, B. (1960). "Spatial variation: Stochastic models and their application to some problems in forest surveys and other sampling investigations," *Swedish Forestry Res. Inst.*, **49**(5).

Matheron, G. (1962). "Traité de Géostatistique Appliquée, Tome I," *Memoires du Bureau de Recherches Geologiques et Minieres*, Vol. 14, Editions Technip, Paris.

Matheron, G. (1967). *Eléments Pour une Théorie des Milieux Poreux*, Masson et Cie, Paris.

Matheron, G. (1973). "The intrinsic random functions and their applications," *Adv. in Appl. Probab.*, **5**, 439–468.

Matsuo, M., and Kuroda, K. (1974). "Probabilistic approach to the design of embankments," *Soils Found.*, **14**(1), 1–17.

Matthies, H. G., Brenner, C. E., Bucher, C. G., and Soares, C. G. (1997). "Uncertainties in probabilistic numerical analysis of structures and solids–stochastic finite elements," *Struct. Safety*, **19**(3), 283–336.

Mellah, R., Auvinet, G., and Masrouri, F. (2000). "Stochastic finite element method applied to non-linear analysis of embankments," *Prob. Eng. Mech.*, **15**, 251–259.

Menon, M. V. (1963). "Estimation of the shape and scale parameters of the Weibull distribution," *Technometrics*, **5**, 175–182.

Meyerhof, G. G. (1951). "The ultimate bearing capacity of foundations," *Géotechnique*, **2**(4), 301–332.

Meyerhof, G. G. (1963). "Some recent research on the bearing capacity of foundations," *Can. Geotech. J.*, **1**(1), 16–26.

Meyerhof, G. G. (1970). "Safety factors in soil mechanics," *Can. Geotech. J.*, **7**, 349–355.

Meyerhof, G. G. (1984). "Safety factors and limit states analysis in geotechnical engineering," *Can. Geotech. J.*, **21**(1), 1–7.

Meyerhof, G. G. (1993). "Development of geotechnical limit state design," in *Proc. Int. Symp. on Limit State Design in Geotechnical Engineering*, Danish Geotechnical Society, Copenhagen, Denmark, pp. 1–12.

Meyerhof, G. G. (1995). "Development of geotechnical limit state design," *Can. Geotech. J.*, **32**, 128–136.

Mignolet, M. P., and Spanos, P. D. (1992). "Simulation of homogeneous two-dimensional random fields: Part I —AR and ARMA Models," *ASME J. Appl. Mech.*, **59**, S260–S269.

Milovic, D. (1992). "Stresses and displacements for shallow foundations," in *Developments in Geotechnical Engineering Series*, Vol. 70, Elsevier, Amsterdam.

Mohr, D. L. (1981). "Modeling data as a fractional Gaussian noise," Ph.D. Thesis, Princeton University, Dept. Stat., Princeton, NJ.

Molenkamp, F., Calle, E. O., Heusdens, J. J., and Koenders, M. A. (1979). "Cyclic filter tests in a triaxial cell," in *Proc. 7th European Conf. on Soil Mechanics and Foundation Engineering*, Brighton, England, pp. 97–101.

Mortensen, K. (1983). "Is limit state design a judgement killer?" Bulletin No. 35, Danish Geotechnical Institute, Copenhagen, Denmark.

Mostyn, G. R., and Li, K. S. (1993). "Probabilistic slope stability–State of play," in *Proc. Conf. on Probabilistic Methods in Geotechnical Engineering*, K.S. Li and S.-C.R. Lo, Eds, Balkema, Rotterdam, The Netherlands, pp. 89–110.

Mostyn, G. R. and Soo, S. (1992). "The effect of autocorrelation on the probability of failure of slopes," in *Proc. 6th Australia, New Zealand Conf. on Geomechanics: Geotechnical Risk*, pp. 542–546.

Muskat, M. (1937). *The Flow of Homogeneous Fluids through Porous Media*, McGraw-Hill, New York.

Naganum, T., Deodatis, G., and Shinozuka, M. (1987). "An ARMA model for two-dimensional processes," *ASCE J. Eng. Mech.*, **113**(2), 234–251

National Cooperative Highway Research Program (NCHRP) (1991). *Manuals for the Design of Bridge Foundations*, Report 343, NCHRP, Transportation Research Board, National Research Council, Washington, DC.

National Cooperative Highway Research Program (NCHRP) (2004). *Load and Resistance Factors for Earth Pressures on Bridge Substructures and Retaining Walls*, Report 12-55, NCHRP, Transportation Research Board, National Research Council, Washington, DC.

National Research Council (NRC) (2005). *National Building Code of Canada*, National Research Council of Canada, Ottawa.

National Research Council (NRC) (2006). *User's Guide–NBC 2005 Structural Commentaries (Part 4 of Division B)*, 2nd ed., National Research Council of Canada, Ottawa.

Odeh, R. E., and Evans, J. O. (1975). "The percentage points of the normal distribution," *Appl. Statist.*, **23**, 96–97.

Paice, G. M. (1997). "Finite element analysis of stochastic soils," Ph.D. Thesis, University of Manchester, Dept. Civil Engineering, Manchester, United Kingdom.

Paice, G. M., Griffiths, D. V., and Fenton, G. A. (1994). "Influence of spatially random soil stiffness on foundation settlements," in *ASCE Settlement'94 Conference*, A.T. Young and G.Y. Félio, Eds., American Society of Civil Engineers, Texas A&M University, pp. 628–639.

Paice, G. M., Griffiths, D. V., and Fenton, G. A. (1996). "Finite element modeling of settlements on spatially random soil," *ASCE J. Geotech. Eng.*, **122**(9), 777–779.

Papoulis, A. (1991). *Probability, Random Variables, and Stochastic Processes*, 3rd ed., McGraw-Hill, New York.

Park, D. (1992). "Numerical modeling as a tool for mine design," in *Proc. Workshop on Coal Pillar Mechanics and Design, 33rd US Symp. on Rock Mechanics*, U.S. Bureau of Mines, Sante Fe, NM, pp. 250–268.

Park, S. K., and Miller, K. W. (1988). "Random number generators: Good ones are hard to find," *Commun. ACM*, **31**, 1192–1201.

Pavlovsky, N. N. (1933). "Motion of water under dams," in *Proc. 1st Congress on Large Dams*, Stockholm, Sweden, pp. 179–192.

Peitgen, H-O., and Saupe, D., Eds. (1988). *The Science of Fractal Images*, Springer-Verlag, New York.

Peng, S. S., and Dutta, D. (1992). "Evaluation of various pillar design methods: A case study," in *Proc. Workshop on Coal Pillar Mechanics and Design, 33rd US Symp. on Rock Mechanics*, U.S. Bureau of Mines, Sante Fe, NM, pp. 269–276.

Peschl, G. M., and Schweiger, H. F. (2003). "Reliability analysis in geotechnics with finite elements. Comparison of probabilistic, stochastic and fuzzy set methods," in *3rd Int. Symp. Imprecise Probabilities and their Applications (ISIPTA'03)*, Lugano, Switzerland, pp. 437–451.

Phoon, K-K., and Kulhawy, F. H. (1999). "Characterization of geotechnical variability," *Can. Geotech. J.*, **36**, 612–624.

Phoon, K. K., Quek, S. T., Chow, Y. K., and Lee, S. L. (1990). "Reliability Analysis of pile settlement," *ASCE J. Geotech. Eng.*, **116**(11), 1717–1735.

Prandtl, L. (1921). "Uber die Eindringungsfestigkeit (Harte) plastischer Baustoffe und die Festigkeit von Schneiden," *Zeitschr. ang. Math. Mechan.*, **1**(1), 15–20.

Press, W. H., Teukolsky, S. A., Vetterling, W. T., and Flannery, B. P. (1997). *Numerical Recipes in C: The Art of Scientific Computing*, 2nd ed., Cambridge University Press, New York.

Prevost, J. H. (1989). "A computer program for nonlinear seismic response analysis," NCEER-89-0025, National Center for Earthquake Engineering Research, Buffalo, NY.

Priestley, M. B. (1981). *Spectral Analysis and Time Series*, Vol. 1: *Univariate Series*, Academic, New York.

Rice, S. O. (1954). "Mathematical analysis of random noise," in *Selected Papers on Noise and Stochastic Processes*, N. Wax, Ed., Dover, New York, pp. 133–294.

Rosenblueth, E. (1975). "Point estimates for probability moments," *Proc. Nat. Acad. Sci. USA*, **72**(10), 3812–3814.

Rosenblueth, E. (1981). "Two-point estimates in probabilities," *Appl. Math. Modelling*, **5**, 329–335.

Rubin, Y., and Gómez-Hernández, J. J. (1990). "A stochastic approach to the problem of upscaling of conductivity in disordered media: Theory and unconditional numerical simulations," *Water Resourc. Res.*, **26**(4), 691–701.

Salamon, M. D. G. (1999). "Strength of coal pillars from back-calculation," in *Proc. 37th US Rock Mechanics Symp. on Rock Mechanics for Industry*, Balkema, Rotterdam, The Netherlands pp. 29–36.

Savely, J. P. (1987). "Probabilistic analysis of intensely fractured rock masses," in *Proc. 6th Int. Congress on Rock Mechanics*, Montreal, Quebec, pp. 509–514.

Schrage, L. (1979). "A more portable random number generator," *Assoc. Comput. Mach. Trans. Math. Software*, **5**, 132–138.

Scovazzo, V. A. (1992). "A practitioner's approach to pillar design," in *Proc. Workshop on Coal Pillar Mechanics and Design, 33rd US Symp. on Rock Mechanics*, U.S. Bureau of Mines, Sante Fe, NM, pp. 277–282.

Seed, H. B. (1979). "Soil liquefaction and cyclic mobility evaluation for level ground during earthquakes," *ASCE J. Geotech. Eng.*, **105**(GT2), 201–255.

Seycek, J. (1991). "Settlement calculation limited to actual deformation zone," in *Deformations of Soils and Displacements of Structures, Proc 10th European Conf. Soil Mechan. Found. Eng., Florence, Italy*, Balkema, Rotterdam, The Netherlands pp. 543–548.

Sherard, J. L., Dunnigan, L. P., and Talbot, J. R. (1984a). "Basic properties of sand and gravel filters," *ASCE J. Geotech. Eng.*, **110**(6), 684–700.

Sherard, J. L., Dunnigan, L. P., and Talbot, J. R. (1984b). "Filters for silts and clays," *ASCE J. Geotech. Eng.*, **110**(6), 701–718.

Shinozuka, M., and Jan, C. M. (1972). "Digital simulation of random processes and its applications," *J. Sound Vibration*, **25**(1), 111–128.

Shinozuka, M., and Ohtomo, K. (1989). "Spatial severity of liquefaction," Technical Report NCEER-89-0032, in *Proc. 2nd U.S.-Japan Workshop on Liquefaction, Large Ground Deformations and Their Effects on Lifelines*, T. D. O'Rourke and M. Hamada, Eds.

Simpson, B., Pappin, J. W., and Croft, D. D. (1981). "An approach to limit state calculations in geotechnics," *Ground Eng.*, **14**(6), 21–28.

Singh, A. (1972). "How reliable is the factor of safety in foundation engineering?," in *Proc. 1st Int. Conf. Applications of Statistics and Probability to Soil and Structural Engineering*, P. Lumb, Ed., Hong Kong University Press, Hong Kong, pp. 390–409.

Smith, K. (1980). "Risk analysis: Toward a standard method," paper presented at the *American/European Nuclear Societies' Meeting on Thermal Reactor Safety*, April 8–11, Knoxville, TN.

Smith, L., and Freeze, R. A. (1979a). "Stochastic analysis of steady state groundwater flow in a bounded domain: 1. One-dimensional simulations," *Water Resourc. Res.*, **15**(3), 521–528.

Smith, L., and Freeze, R. A. (1979b). "Stochastic analysis of steady state groundwater flow in a bounded domain: 2) Two-dimensional simulations," *Water Resourc. Res.*, **15**(6), 1543–1559.

Smith, I. M., and Griffiths, D. V. (1982). *Programming the Finite Element Method*, 2nd ed., Wiley, New York.

Smith, I. M., and Griffiths, D. V. (2004). *Programming the Finite Element Method*, 4th ed., Wiley, New York.

Sokolovski, V. V. (1965). *Statics of Granular Media*, Pergamon, London.

Spanos, P. D., and Mignolet, M. P. (1992). "Simulation of homogeneous two-dimensional random fields: Part II—MA and ARMA Models," *ASME J. Appl. Mech.*, **59**, S270–S277.

Strang, G., and Nguyen, T. (1996). *Wavelets and Filter Banks*, Wellesley-Cambridge, New York.

Sudicky, E. A. (1986). "A natural gradient experiment on solute transport in a sand aquifer: Spatial variability of hydraulic conductivity and its role in the dispersion process," *Water Resourc. Res.*, **22**(13), 2069–2083.

Szynakiewicz, T., Griffiths, D. V., and Fenton, G. A. (2002). "A probabilistic investigation of c', ϕ' slope stability," in *Proc. 6th Int. Cong. Numerical Methods in Engineering and Scientific Applications, CIMENICS'02*, Sociedad Venezolana de Métodos Numéricos en Ingeniería, pp. 25–36.

Tan, C. P., Donald, I. B., and Melchers, R. E. (1993). "Probabilistic slip circle analysis of earth and rock fill dams," in *Prob. Methods in Geotechnical Engineering*, K. S. Li and S.-C. R. Lo, Eds, Balkema, Rotterdam, The Netherlands, pp. 233–239.

Tang, W. H., Yuceman, M. S., and Ang, A. H. S. (1976). "Probability-based short-term design of slopes," *Can. Geotech. J.*, **13**, 201–215.

Taylor, D. W. (1937). "Stability of earth slopes," *J. Boston Soc. Civil Eng.*, **24**(3), 337–386.

Taylor, D. W. (1948). *Fundamentals of Soil Mechanics*, Wiley, New York.

Terzaghi, K. (1943). *Theoretical Soil Mechanics*, Wiley, New York.

Terzaghi, K., and Peck, R. P. (1967). *Soil Mechanics in Engineering Practice*, 2nd ed., Wiley, New York.

Thoman, D. R., Bain, L. J., and Antle, C. E. (1969). "Inferences on the parameters of the Weibull distribution," *Technometrics*, **11**, 445–460.

Thorne, C. P., and Quine, M. P. (1993). "How reliable are reliability estimates and why soil engineers rarely use them," in *Probabilistic Methods in Geotechnical Engineering*, K.S. Li and S.-C. R. Lo, Eds., Balkema, Rotterdam, The Netherlands, 325–332.

Tveten, D. E. (2002). "Application of probabilistic methods to stability and earth pressure problems in geomechanics," Master's Thesis, Colorado School of Mines, Division of Engineering, Golden, CO.

Vanmarcke, E. H. (1977). "Probabilistic Modeling of Soil Profiles," *ASCE J. Geotech. Eng.*, **103**(GT11), 1227–1246.

Vanmarcke, E. H. (1984). *Random Fields: Analysis and Synthesis*, MIT Press, Cambridge, Massachusetts.

Vanmarcke, E. H., and Grigoriu, M. (1983). "Stochastic finite element analysis of simple beams," *ASCE J. Eng. Mech.*, **109**(5), 1203–1214.

Vanmarcke, E. H., Heredia-Zavoni, E., and Fenton, G. A. (1993). "Conditional simulation of spatially correlated earthquake ground motion," *ASCE J. Eng. Mech.*, **119**(11), 2333–2352.

Verruijt, A. (1970). *Theory of Groundwater Flow*, MacMillan, London.

Vesic, A. S. (1977). *Design of Pile Foundations*, in National Cooperative Highway Research Program Synthesis of Practice No. 42, Transportation Research Board, Washington, DC.

Vick, S. G. (2002). *Degrees of Belief: Subjective Probability and Engineering Judgement*, American Society of Civil Engineers, Reston, VA.

Vijayvergiya, V. N. and Focht, J. A. (1972). "A new way to predict capacity of piles in clay," presented at the Fourth Offshore Technology Conference, Houston, TX, Paper 1718.

Voss, R. (1985). "Random fractal forgeries," *Special Interest Group on Graphics (SIGGRAPH) Conference Tutorial Notes*, ACM, New York.

Whipple, C. (1986). "Approaches to acceptable risk," in *Proc. Eng. Found. Conf. Risk-Based Decision Making in Water Resources*, Y.Y. Haimes and E.Z. Stakhiv, Eds., pp. 30–45.

Whitman, R. V. (2000). "Organizing and evaluating uncertainty in geotechnical engineering," *ASCE J. Geotech. Geoenv. Eng.*, **126**(7), 583–593.

Whittle, P. (1956). "On the variation of yield variance with plot size," *Biometrika*, **43**, 337–343.

Wickremesinghe, D., and Campanella, R. G. (1993). "Scale of fluctuation as a descriptor of soil variability," in *Prob. Methods in Geotech. Eng.*, K.S. Li and S.-C.R. Lo, Eds., Balkema, Rotterdam, The Netherlands, pp. 233–239.

Wolff, T. F. (1996). "Probabilistic slope stability in theory and practice," in *Uncertainty in the Geologic Environment: From Theory to Practice*, Geotechnical Special Publication No. 58, C. D. Shackelford *et al.*, Eds., American Society of Civil Engineers, New York, pp. 419–433.

Wolff, T. H. (1985). "Analysis and design of embankment dam slopes: a probabilistic approach," Ph.D. Thesis, Purdue University, Lafayette, IN.

Wornell, G. W. (1996). *Signal Processing with Fractals: A Wavelet Based Approach*, in *Signal Processing Series*, A.V. Oppenheim, Ed., Prentice Hall, Englewood Cliffs, NJ.

Yaglom, A. M. (1962). *An Introduction to the Theory of Stationary Random Functions*, Dover, Mineola, NY.

Yajima, Y. (1989). "A central limit theorem of Fourier transforms of strongly dependent stationary processes," *J. Time Ser. Anal.*, **10**, 375–383.

Yuceman, M. S., Tang, W. H., and Ang, A. H. S. (1973). "A probabilistic study of safety and design of earth slopes," University of Illinois, Urbana, Civil Engineering Studies, Structural Research Series 402, Urbana-Champagne, IL.

PART 3

Appendixes

APPENDIX A

Probability Tables

A.1 NORMAL DISTRIBUTION: $\Phi(z) = \int_{-\infty}^{z} \frac{1}{\sqrt{2\pi}} e^{-\frac{1}{2}x^2}\, dx$

z	.00	.01	.02	.03	.04	.05	.06	.07	.08	.09
0.0	.50000	.50398	.50797	.51196	.51595	.51993	.52392	.52790	.53188	.53585
0.1	.53982	.54379	.54775	.55171	.55567	.55961	.56355	.56749	.57142	.57534
0.2	.57925	.58316	.58706	.59095	.59483	.59870	.60256	.60641	.61026	.61409
0.3	.61791	.62171	.62551	.62930	.63307	.63683	.64057	.64430	.64802	.65173
0.4	.65542	.65909	.66275	.66640	.67003	.67364	.67724	.68082	.68438	.68793
0.5	.69146	.69497	.69846	.70194	.70540	.70884	.71226	.71566	.71904	.72240
0.6	.72574	.72906	.73237	.73565	.73891	.74215	.74537	.74857	.75174	.75490
0.7	.75803	.76114	.76423	.76730	.77035	.77337	.77637	.77935	.78230	.78523
0.8	.78814	.79102	.79389	.79673	.79954	.80233	.80510	.80784	.81057	.81326
0.9	.81593	.81858	.82121	.82381	.82639	.82894	.83147	.83397	.83645	.83891
1.0	.84134	.84375	.84613	.84849	.85083	.85314	.85542	.85769	.85992	.86214
1.1	.86433	.86650	.86864	.87076	.87285	.87492	.87697	.87899	.88099	.88297
1.2	.88493	.88686	.88876	.89065	.89251	.89435	.89616	.89795	.89972	.90147
1.3	.90319	.90490	.90658	.90824	.90987	.91149	.91308	.91465	.91620	.91773
1.4	.91924	.92073	.92219	.92364	.92506	.92647	.92785	.92921	.93056	.93188
1.5	.93319	.93447	.93574	.93699	.93821	.93942	.94062	.94179	.94294	.94408
1.6	.94520	.94630	.94738	.94844	.94949	.95052	.95154	.95254	.95352	.95448
1.7	.95543	.95636	.95728	.95818	.95907	.95994	.96079	.96163	.96246	.96327
1.8	.96406	.96485	.96562	.96637	.96711	.96784	.96855	.96925	.96994	.97062
1.9	.97128	.97193	.97257	.97319	.97381	.97441	.97500	.97558	.97614	.97670
2.0	.97724	.97778	.97830	.97882	.97932	.97981	.98030	.98077	.98123	.98169
2.1	.98213	.98257	.98299	.98341	.98382	.98422	.98461	.98499	.98537	.98573
2.2	.98609	.98644	.98679	.98712	.98745	.98777	.98808	.98839	.98869	.98898
2.3	.98927	.98955	.98982	$.9^{2}0096$	$.9^{2}0358$	$.9^{2}0613$	$.9^{2}0862$	$.9^{2}1105$	$.9^{2}1343$	$.9^{2}1575$
2.4	$.9^{2}1802$	$.9^{2}2023$	$.9^{2}2239$	$.9^{2}2450$	$.9^{2}2656$	$.9^{2}2857$	$.9^{2}3053$	$.9^{2}3244$	$.9^{2}3430$	$.9^{2}3612$
2.5	$.9^{2}3790$	$.9^{2}3963$	$.9^{2}4132$	$.9^{2}4296$	$.9^{2}4457$	$.9^{2}4613$	$.9^{2}4766$	$.9^{2}4915$	$.9^{2}5059$	$.9^{2}5201$
2.6	$.9^{2}5338$	$.9^{2}5472$	$.9^{2}5603$	$.9^{2}5730$	$.9^{2}5854$	$.9^{2}5975$	$.9^{2}6092$	$.9^{2}6207$	$.9^{2}6318$	$.9^{2}6427$
2.7	$.9^{2}6533$	$.9^{2}6635$	$.9^{2}6735$	$.9^{2}6833$	$.9^{2}6928$	$.9^{2}7020$	$.9^{2}7109$	$.9^{2}7197$	$.9^{2}7282$	$.9^{2}7364$
2.8	$.9^{2}7444$	$.9^{2}7522$	$.9^{2}7598$	$.9^{2}7672$	$.9^{2}7744$	$.9^{2}7814$	$.9^{2}7881$	$.9^{2}7947$	$.9^{2}8011$	$.9^{2}8073$
2.9	$.9^{2}8134$	$.9^{2}8192$	$.9^{2}8249$	$.9^{2}8305$	$.9^{2}8358$	$.9^{2}8411$	$.9^{2}8461$	$.9^{2}8511$	$.9^{2}8558$	$.9^{2}8605$
3.0	$.9^{2}8650$	$.9^{2}8693$	$.9^{2}8736$	$.9^{2}8777$	$.9^{2}8817$	$.9^{2}8855$	$.9^{2}8893$	$.9^{2}8929$	$.9^{2}8964$	$.9^{2}8999$
3.1	$.9^{3}0323$	$.9^{3}0645$	$.9^{3}0957$	$.9^{3}1259$	$.9^{3}1552$	$.9^{3}1836$	$.9^{3}2111$	$.9^{3}2378$	$.9^{3}2636$	$.9^{3}2886$
3.2	$.9^{3}3128$	$.9^{3}3363$	$.9^{3}3590$	$.9^{3}3810$	$.9^{3}4023$	$.9^{3}4229$	$.9^{3}4429$	$.9^{3}4622$	$.9^{3}4809$	$.9^{3}4990$
3.3	$.9^{3}5165$	$.9^{3}5335$	$.9^{3}5499$	$.9^{3}5657$	$.9^{3}5811$	$.9^{3}5959$	$.9^{3}6102$	$.9^{3}6241$	$.9^{3}6375$	$.9^{3}6505$
3.4	$.9^{3}6630$	$.9^{3}6751$	$.9^{3}6868$	$.9^{3}6982$	$.9^{3}7091$	$.9^{3}7197$	$.9^{3}7299$	$.9^{3}7397$	$.9^{3}7492$	$.9^{3}7584$
3.5	$.9^{3}7673$	$.9^{3}7759$	$.9^{3}7842$	$.9^{3}7922$	$.9^{3}7999$	$.9^{3}8073$	$.9^{3}8145$	$.9^{3}8215$	$.9^{3}8282$	$.9^{3}8346$
3.6	$.9^{3}8408$	$.9^{3}8469$	$.9^{3}8526$	$.9^{3}8582$	$.9^{3}8636$	$.9^{3}8688$	$.9^{3}8738$	$.9^{3}8787$	$.9^{3}8833$	$.9^{3}8878$
3.7	$.9^{3}8922$	$.9^{3}8963$	$.9^{4}0038$	$.9^{4}0426$	$.9^{4}0799$	$.9^{4}1158$	$.9^{4}1504$	$.9^{4}1837$	$.9^{4}2158$	$.9^{4}2467$
3.8	$.9^{4}2765$	$.9^{4}3051$	$.9^{4}3327$	$.9^{4}3592$	$.9^{4}3848$	$.9^{4}4094$	$.9^{4}4330$	$.9^{4}4558$	$.9^{4}4777$	$.9^{4}4987$
3.9	$.9^{4}5190$	$.9^{4}5385$	$.9^{4}5572$	$.9^{4}5752$	$.9^{4}5925$	$.9^{4}6092$	$.9^{4}6252$	$.9^{4}6406$	$.9^{4}6554$	$.9^{4}6696$
4.0	$.9^{4}6832$	$.9^{4}6964$	$.9^{4}7090$	$.9^{4}7211$	$.9^{4}7327$	$.9^{4}7439$	$.9^{4}7546$	$.9^{4}7649$	$.9^{4}7748$	$.9^{4}7843$

Notes:

1. For $z = i.jk$, where i, j, and k are digits, enter table at line $i.j$ under column $.0k$.
2. $0.9^{4}7327$ is short for 0.99997327, etc.
3. $\Phi(-z) = 1 - \Phi(z)$.

A.2 INVERSE STUDENT t-DISTRIBUTION: $\alpha = \mathrm{P}\left[T > t_{\alpha,\nu}\right]$

	α									
ν	.40	.25	.10	.05	.025	.01	.005	.0025	.001	.0005
1	0.325	1.000	3.078	6.314	12.706	31.821	63.657	127.321	318.309	636.619
2	0.289	0.816	1.886	2.920	4.303	6.965	9.925	14.089	22.327	31.599
3	0.277	0.765	1.638	2.353	3.182	4.541	5.841	7.453	10.215	12.924
4	0.271	0.741	1.533	2.132	2.776	3.747	4.604	5.598	7.173	8.610
5	0.267	0.727	1.476	2.015	2.571	3.365	4.032	4.773	5.893	6.869
6	0.265	0.718	1.440	1.943	2.447	3.143	3.707	4.317	5.208	5.959
7	0.263	0.711	1.415	1.895	2.365	2.998	3.499	4.029	4.785	5.408
8	0.262	0.706	1.397	1.860	2.306	2.896	3.355	3.833	4.501	5.041
9	0.261	0.703	1.383	1.833	2.262	2.821	3.250	3.690	4.297	4.781
10	0.260	0.700	1.372	1.812	2.228	2.764	3.169	3.581	4.144	4.587
11	0.260	0.697	1.363	1.796	2.201	2.718	3.106	3.497	4.025	4.437
12	0.259	0.695	1.356	1.782	2.179	2.681	3.055	3.428	3.930	4.318
13	0.259	0.694	1.350	1.771	2.160	2.650	3.012	3.372	3.852	4.221
14	0.258	0.692	1.345	1.761	2.145	2.624	2.977	3.326	3.787	4.140
15	0.258	0.691	1.341	1.753	2.131	2.602	2.947	3.286	3.733	4.073
16	0.258	0.690	1.337	1.746	2.120	2.583	2.921	3.252	3.686	4.015
17	0.257	0.689	1.333	1.740	2.110	2.567	2.898	3.222	3.646	3.965
18	0.257	0.688	1.330	1.734	2.101	2.552	2.878	3.197	3.610	3.922
19	0.257	0.688	1.328	1.729	2.093	2.539	2.861	3.174	3.579	3.883
20	0.257	0.687	1.325	1.725	2.086	2.528	2.845	3.153	3.552	3.850
21	0.257	0.686	1.323	1.721	2.080	2.518	2.831	3.135	3.527	3.819
22	0.256	0.686	1.321	1.717	2.074	2.508	2.819	3.119	3.505	3.792
23	0.256	0.685	1.319	1.714	2.069	2.500	2.807	3.104	3.485	3.768
24	0.256	0.685	1.318	1.711	2.064	2.492	2.797	3.091	3.467	3.745
25	0.256	0.684	1.316	1.708	2.060	2.485	2.787	3.078	3.450	3.725
26	0.256	0.684	1.315	1.706	2.056	2.479	2.779	3.067	3.435	3.707
27	0.256	0.684	1.314	1.703	2.052	2.473	2.771	3.057	3.421	3.690
28	0.256	0.683	1.313	1.701	2.048	2.467	2.763	3.047	3.408	3.674
29	0.256	0.683	1.311	1.699	2.045	2.462	2.756	3.038	3.396	3.659
30	0.256	0.683	1.310	1.697	2.042	2.457	2.750	3.030	3.385	3.646
40	0.255	0.681	1.303	1.684	2.021	2.423	2.704	2.971	3.307	3.551
50	0.255	0.679	1.299	1.676	2.009	2.403	2.678	2.937	3.261	3.496
60	0.254	0.679	1.296	1.671	2.000	2.390	2.660	2.915	3.232	3.460
70	0.254	0.678	1.294	1.667	1.994	2.381	2.648	2.899	3.211	3.435
80	0.254	0.678	1.292	1.664	1.990	2.374	2.639	2.887	3.195	3.416
90	0.254	0.677	1.291	1.662	1.987	2.368	2.632	2.878	3.183	3.402
100	0.254	0.677	1.290	1.660	1.984	2.364	2.626	2.871	3.174	3.390
110	0.254	0.677	1.289	1.659	1.982	2.361	2.621	2.865	3.166	3.381
120	0.254	0.677	1.289	1.658	1.980	2.358	2.617	2.860	3.160	3.373
∞	0.253	0.674	1.282	1.645	1.960	2.326	2.576	2.807	3.090	3.291

A.3 INVERSE CHI-SQUARE DISTRIBUTION: $\alpha = P\left[\chi^2 > \chi^2_{\alpha,\nu}\right]$

	α										
ν	.995	.990	.975	.950	.900	.500	.100	.050	.025	.010	.005
1	0.00	0.00	0.00	0.00	0.02	0.45	2.71	3.84	5.02	6.63	7.88
2	0.01	0.02	0.05	0.10	0.21	1.39	4.61	5.99	7.38	9.21	10.60
3	0.07	0.11	0.22	0.35	0.58	2.37	6.25	7.81	9.35	11.34	12.84
4	0.21	0.30	0.48	0.71	1.06	3.36	7.78	9.49	11.14	13.28	14.86
5	0.41	0.55	0.83	1.15	1.61	4.35	9.24	11.07	12.83	15.09	16.75
6	0.68	0.87	1.24	1.64	2.20	5.35	10.64	12.59	14.45	16.81	18.55
7	0.99	1.24	1.69	2.17	2.83	6.35	12.02	14.07	16.01	18.48	20.28
8	1.34	1.65	2.18	2.73	3.49	7.34	13.36	15.51	17.53	20.09	21.95
9	1.73	2.09	2.70	3.33	4.17	8.34	14.68	16.92	19.02	21.67	23.59
10	2.16	2.56	3.25	3.94	4.87	9.34	15.99	18.31	20.48	23.21	25.19
11	2.60	3.05	3.82	4.57	5.58	10.34	17.28	19.68	21.92	24.72	26.76
12	3.07	3.57	4.40	5.23	6.30	11.34	18.55	21.03	23.34	26.22	28.30
13	3.57	4.11	5.01	5.89	7.04	12.34	19.81	22.36	24.74	27.69	29.82
14	4.07	4.66	5.63	6.57	7.79	13.34	21.06	23.68	26.12	29.14	31.32
15	4.60	5.23	6.26	7.26	8.55	14.34	22.31	25.00	27.49	30.58	32.80
16	5.14	5.81	6.91	7.96	9.31	15.34	23.54	26.30	28.85	32.00	34.27
17	5.70	6.41	7.56	8.67	10.09	16.34	24.77	27.59	30.19	33.41	35.72
18	6.26	7.01	8.23	9.39	10.86	17.34	25.99	28.87	31.53	34.81	37.16
19	6.84	7.63	8.91	10.12	11.65	18.34	27.20	30.14	32.85	36.19	38.58
20	7.43	8.26	9.59	10.85	12.44	19.34	28.41	31.41	34.17	37.57	40.00
21	8.03	8.90	10.28	11.59	13.24	20.34	29.62	32.67	35.48	38.93	41.40
22	8.64	9.54	10.98	12.34	14.04	21.34	30.81	33.92	36.78	40.29	42.80
23	9.26	10.20	11.69	13.09	14.85	22.34	32.01	35.17	38.08	41.64	44.18
24	9.89	10.86	12.40	13.85	15.66	23.34	33.20	36.42	39.36	42.98	45.56
25	10.52	11.52	13.12	14.61	16.47	24.34	34.38	37.65	40.65	44.31	46.93
26	11.16	12.20	13.84	15.38	17.29	25.34	35.56	38.89	41.92	45.64	48.29
27	11.81	12.88	14.57	16.15	18.11	26.34	36.74	40.11	43.19	46.96	49.64
28	12.46	13.56	15.31	16.93	18.94	27.34	37.92	41.34	44.46	48.28	50.99
29	13.12	14.26	16.05	17.71	19.77	28.34	39.09	42.56	45.72	49.59	52.34
30	13.79	14.95	16.79	18.49	20.60	29.34	40.26	43.77	46.98	50.89	53.67
40	20.71	22.16	24.43	26.51	29.05	39.34	51.81	55.76	59.34	63.69	66.77
50	27.99	29.71	32.36	34.76	37.69	49.33	63.17	67.50	71.42	76.15	79.49
60	35.53	37.48	40.48	43.19	46.46	59.33	74.40	79.08	83.30	88.38	91.95
70	43.28	45.44	48.76	51.74	55.33	69.33	85.53	90.53	95.02	100.43	104.21
80	51.17	53.54	57.15	60.39	64.28	79.33	96.58	101.88	106.63	112.33	116.32
90	59.20	61.75	65.65	69.13	73.29	89.33	107.57	113.15	118.14	124.12	128.30
100	67.33	70.06	74.22	77.93	82.36	99.33	118.50	124.34	129.56	135.81	140.17

APPENDIX B

Numerical Integration

B.1 GAUSSIAN QUADRATURE

The definite integral

$$I = \int_a^b f(x)\,dx$$

can be approximated by a weighted sum of $f(x)$ evaluated at a series of carefully selected locations between a and b. Gaussian quadrature involves the numerical approximation

$$I \simeq \frac{b-a}{2} \sum_{i=1}^{n_g} w_i f(\xi_i) \tag{B.1}$$

where n_g is the number of points at which $f(x)$ is evaluated, w_i are weighting factors, and ξ_i are locations

$$\xi_i = \tfrac{1}{2}(b+a) + \tfrac{1}{2}(b-a)z_i = a + \tfrac{1}{2}(b-a)(1+z_i) \tag{B.2}$$

The weights w_i and standardized locations z_i are given in Tables B.1–B.3.

The approximation is easily extended to higher dimensions. For example,

$$\int_{a_1}^{a_2} \int_{b_1}^{b_2} f(x_1, x_2)\,dx_2\,dx_1$$

$$\simeq \frac{a_2-a_1}{2}\frac{b_2-b_1}{2} \sum_{j=1}^{n_g} w_j \sum_{i=1}^{n_g} w_i f(\xi_i, \eta_j) \tag{B.3}$$

where

$$\begin{aligned} \xi_i &= \tfrac{1}{2}(a_2+a_1) + \tfrac{1}{2}(a_2-a_1)z_i \\ &= a_1 + \tfrac{1}{2}(a_2-a_1)(1+z_i) \end{aligned} \tag{B.4a}$$

$$\begin{aligned} \eta_i &= \tfrac{1}{2}(b_2+b_1) + \tfrac{1}{2}(b_2-b_1)z_i \\ &= b_1 + \tfrac{1}{2}(b_2-b_1)(1+z_i) \end{aligned} \tag{B.4b}$$

Table B.1 Weights w_i and Standardized Locations z_i for $n_g = 1, \ldots, 5$

n_g	w_i	z_i
$n_g = 1$	2.000000000000000	0.000000000000000
$n_g = 2$	1.000000000000000	−0.577350269189626
	1.000000000000000	0.577350269189626
$n_g = 3$	0.555555555555556	−0.774596669241483
	0.888888888888889	0.000000000000000
	0.555555555555556	0.774596669241483
$n_g = 4$	0.347854845137454	−0.861136311594053
	0.652145154862546	−0.339981043584856
	0.652145154862546	0.339981043584856
	0.347854845137454	0.861136311594053
$n_g = 5$	0.236926885056189	−0.906179845938664
	0.478628670499366	−0.538469310105683
	0.568888888888889	0.000000000000000
	0.478628670499366	0.538469310105683
	0.236926885056189	0.906179845938664

Table B.2 Weights w_i and Standardized Locations z_i for $n_g = 6, \ldots, 10$

n_g	w_i	z_i
$n_g = 6$	0.171324492379170	−0.932469514203152
	0.360761573048139	−0.661209386466265
	0.467913934572691	−0.238619186083197
	0.467913934572691	0.238619186083197
	0.360761573048139	0.661209386466265
	0.171324492379170	0.932469514203152
$n_g = 7$	0.129484966168870	−0.949107912342759
	0.279705391489277	−0.741531185599394
	0.381830050505119	−0.405845151377397
	0.417959183673469	0.000000000000000
	0.381830050505119	0.405845151377397
	0.279705391489277	0.741531185599394
	0.129484966168870	0.949107912342759
$n_g = 8$	0.101228536290376	−0.960289856497536
	0.222381034453374	−0.796666477413627
	0.313706645877887	−0.525532409916329
	0.362683783378362	−0.183434642495650
	0.362683783378362	0.183434642495650
	0.313706645877887	0.525532409916329
	0.222381034453374	0.796666477413627
	0.101228536290376	0.960289856497536
$n_g = 9$	0.081274388361574	−0.968160239507626
	0.180648160694857	−0.836031107326636
	0.260610696402935	−0.613371432700590
	0.312347077040003	−0.324253423403809
	0.330239355001260	0.000000000000000
	0.312347077040003	0.324253423403809
	0.260610696402935	0.613371432700590
	0.180648160694857	0.836031107326636
	0.081274388361574	0.968160239507626
$n_g = 10$	0.066671344308688	−0.973906528517172
	0.149451349150581	−0.865063366688985
	0.219086362515982	−0.679409568299024
	0.269266719309996	−0.433395394129247
	0.295524224714753	−0.148874338981632
	0.295524224714753	0.148874338981632
	0.269266719309996	0.433395394129247
	0.219086362515982	0.679409568299024
	0.149451349150581	0.865063366688985
	0.066671344308688	0.973906528517172

Table B.3 Weights w_i and Standardized Locations z_i for $n_g = 16, 20$

n_g	w_i	z_i
$n_g = 16$	0.02715245941175409485 2	−0.98940093499164993259 6
	0.06225352393864789286 3	−0.94457502307323257607 8
	0.09515851168249278481 0	−0.86563120238783174388 0
	0.12462897125553387205 2	−0.75540440835500303389 5
	0.14959598881657673208 1	−0.61787624440264374844 7
	0.16915651939500253818 9	−0.45801677765722738634 2
	0.18260341504492358886 7	−0.28160355077925891323 0
	0.18945061045506849628 5	−0.09501250983763744018 5
	0.18945061045506849628 5	0.09501250983763744018 5
	0.18260341504492358886 7	0.28160355077925891323 0
	0.16915651939500253818 9	0.45801677765722738634 2
	0.14959598881657673208 1	0.61787624440264374844 7
	0.12462897125553387205 2	0.75540440835500303389 5
	0.09515851168249278481 0	0.86563120238783174388 0
	0.06225352393864789286 3	0.94457502307323257607 8
	0.02715245941175409485 2	0.98940093499164993259 6
$n_g = 20$	0.01761400713915211831 2	−0.99312859918509492478 6
	0.04060142980038694133 1	−0.96397192727791379126 8
	0.06267204833410906357 0	−0.91223442825132590586 8
	0.08327674157670474872 5	−0.83911697182221882339 5
	0.10193011981724043503 7	−0.74633190646015079261 4
	0.11819453196151841731 2	−0.63605368072651502545 3
	0.13168863844917662689 8	−0.51086700195082709800 4
	0.14209610931838205132 9	−0.37370608871541956067 3
	0.14917298647260374678 8	−0.22778585114164507808 0
	0.15275338713072585069 8	−0.07652652113349733375 5
	0.15275338713072585069 8	0.07652652113349733375 5
	0.14917298647260374678 8	0.22778585114164507808 0
	0.14209610931838205132 9	0.37370608871541956067 3
	0.13168863844917662689 8	0.51086700195082709800 4
	0.11819453196151841731 2	0.63605368072651502545 3
	0.10193011981724043503 7	0.74633190646015079261 4
	0.08327674157670474872 5	0.83911697182221882339 5
	0.06267204833410906357 0	0.91223442825132590586 8
	0.04060142980038694133 1	0.96397192727791379126 8
	0.01761400713915211831 2	0.99312859918509492478 6

APPENDIX C

Computing Variances and Covariances of Local Averages

C.1 ONE-DIMENSIONAL CASE

Let X_A be the local arithmetic average of a stationary one-dimensional random process $X(x)$ over some length A, where

$$X_A = \frac{1}{A}\int_0^A X(x)\,dx$$

The variance of X_A is given by $\sigma_A^2 = \gamma(A)\sigma_X^2$, in which $\gamma(A)$ is the variance reduction function (see Section 3.4),

$$\begin{aligned}\gamma(A) &= \frac{1}{A^2}\int_0^A\int_0^A \rho_X(\xi-\eta)\,d\xi\,d\eta \\ &= \frac{2}{A^2}\int_0^A (A-\tau)\rho_X(\tau)\,d\tau \qquad \text{(C.1)}\end{aligned}$$

where $\rho_X(\tau)$ is the correlation function of $X(x)$. Equation C.1 can be efficiently (and accurately if ρ_X is relatively smooth between Gauss points) evaluated by Gaussian quadrature,

$$\gamma(A) \simeq \frac{1}{A}\sum_{i=1}^{n_g} w_i(A-x_i)\rho_X(x_i) \qquad \text{(C.2)}$$

where $x_i = \frac{1}{2}A(1+z_i)$ and the Gauss points z_i and weights w_i are given in Appendix B for various choices in the number of Gauss points, n_g.

Now consider two local averages, as illustrated in Figure C.1, defined as

$$X_A = \frac{1}{A}\int_{a_1}^{a_2} X(x)\,dx$$

$$X_B = \frac{1}{B}\int_{b_1}^{b_2} X(x)\,dx$$

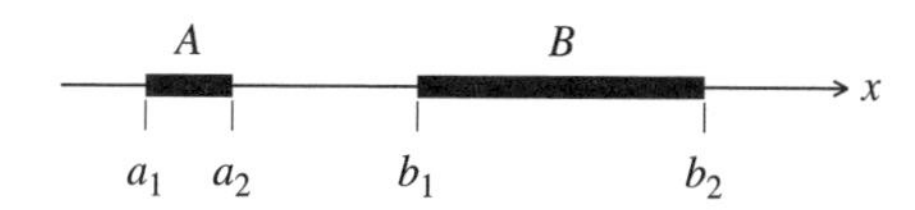

Figure C.1 Two local averages over lengths A and B, respectively.

where $A = a_2 - a_1$ and $B = b_2 - b_1$. The covariance between X_A and X_B is

$$\begin{aligned}\text{Cov}[X_A, X_B] &= \frac{\sigma_X^2}{AB}\int_{b_1}^{b_2}\int_{a_1}^{a_2} \rho(\xi-\eta)\,d\xi\,d\eta \\ &\simeq \frac{\sigma_X^2}{4}\sum_{i=1}^{n_g} w_i \sum_{j=1}^{ng} w_j\rho(\xi_i-\eta_j)\end{aligned}$$

where

$$\xi_i = a_1 + \tfrac{1}{2}A(1+z_i), \qquad \eta_j = b_1 + \tfrac{1}{2}B(1+z_j)$$

C.2 TWO-DIMENSIONAL CASE

Consider the two local averages shown in Figure C.2, X_A and X_B, where A and B are now areas. The local averages are defined as

$$X_A = \frac{1}{A}\int_{a_{1y}}^{a_{2y}}\int_{a_{1x}}^{a_{2x}} X(x,y)\,dx\,dy$$

$$X_B = \frac{1}{B}\int_{b_{1y}}^{b_{2y}}\int_{b_{1x}}^{b_{2x}} X(x,y)\,dx\,dy$$

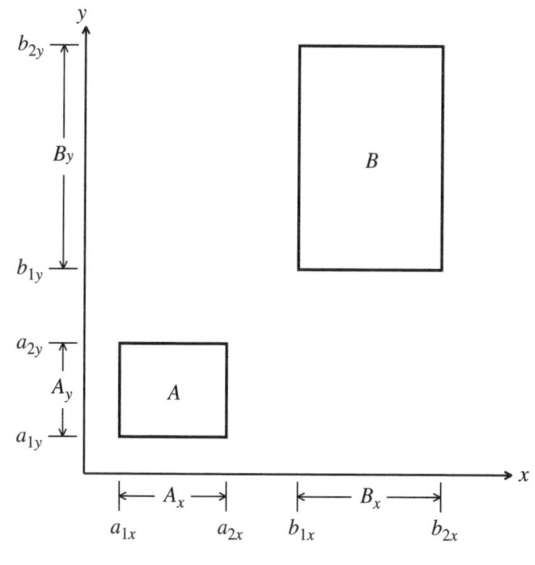

Figure C.2 Two local averages over areas A and B, respectively.

where

$$A = A_x A_y = (a_{2x} - a_{1x})(a_{2y} - a_{1y})$$
$$B = B_x B_y = (b_{2x} - b_{1x})(b_{2y} - b_{1y})$$

The variance of X_A is $\sigma_A^2 = \gamma(A)\sigma_X^2$, where $\gamma(A)$ is the variance reduction function (see Section 3.4) defined in two dimensions as

$$\begin{aligned}\gamma(A) &= \frac{1}{A^2}\int_0^{A_x}\int_0^{A_x}\int_0^{A_y}\int_0^{A_y} \\ &\quad\times \rho_X(\eta_1 - \xi_1, \eta_2 - \xi_2)\, d\xi_2\, d\eta_2\, d\xi_1\, d\eta_1 \\ &= \frac{4}{A^2}\int_0^{A_x}\int_0^{A_y} \\ &\quad\times (A_x - \tau_1)(A_y - \tau_2)\rho_X(\tau_1, \tau_2)\, d\tau_2\, d\tau_1 \end{aligned} \qquad \text{(C.3)}$$

where $\rho_X(\tau_1, \tau_2)$ is the correlation function of $X(x, y)$, which is assumed to be stationary and quadrant symmetric in the above (see Section 3.7.4). Equation C.3 can be approximated by Gaussian quadrature as

$$\begin{aligned}\gamma(A) &\simeq \frac{1}{A}\sum_{j=1}^{n_g} w_j \sum_{i=1}^{n_g} w_i (A_x - \xi_i)(A_y - \eta_j)\rho_X(\xi_i, \eta_j) \\ &= \frac{1}{4}\sum_{j=1}^{n_g} w_j (1 - z_j)\sum_{i=1}^{ng} w_i (1 - z_i)\rho_X(\xi_i, \eta_j)\end{aligned}$$

where

$$\xi_i = \tfrac{1}{2}A_x(1 + z_i), \qquad \xi_j = \tfrac{1}{2}A_y(1 + z_j)$$

and the weights w_i and Gauss points z_i are found in Appendix B.

The covariance between X_A and X_B is given by

$$\begin{aligned}\text{Cov}[X_A, X_B] &= \frac{\sigma_X^2}{AB}\int_{a_{1y}}^{a_{2y}}\int_{a_{1x}}^{a_{2x}}\int_{b_{1y}}^{b_{2y}}\int_{b_{1x}}^{b_{2x}} \\ &\quad\times \rho_X(x - \xi, y - \eta)\, d\xi\, d\eta\, dx\, dy \\ &\simeq \frac{\sigma_X^2}{16}\sum_{i=1}^{n_g} w_i \sum_{j=1}^{n_g} w_j \sum_{k=1}^{n_g} w_k \sum_{\ell=1}^{n_g} w_\ell \\ &\quad\times \rho_X(x_j - \xi_\ell, y_i - \eta_k)\end{aligned}$$

where

$$x_j = a_{1x} + \tfrac{1}{2}A_x(1 + z_j), \qquad \xi_\ell = b_{1x} + \tfrac{1}{2}B_x(1 + z_\ell)$$
$$y_i = a_{1y} + \tfrac{1}{2}A_y(1 + z_i), \qquad \eta_k = b_{1y} + \tfrac{1}{2}B_y(1 + z_k)$$

and the weights w_i and Gauss points z_i can be found in Appendix B.

C.3 THREE-DIMENSIONAL CASE

Consider the two local averages X_A and X_B where A and B are now volumes. The notation used in three dimensions basically follows that shown for two dimensions in Figure C.2, with the addition of the third z direction. The local averages are defined as

$$X_A = \frac{1}{A}\int_{a_{1z}}^{a_{2z}}\int_{a_{1y}}^{a_{2y}}\int_{a_{1x}}^{a_{2x}} X(x, y, z)\, dx\, dy\, dz$$
$$X_B = \frac{1}{B}\int_{b_{1z}}^{b_{2z}}\int_{b_{1y}}^{b_{2y}}\int_{b_{1x}}^{b_{2x}} X(x, y, z)\, dx\, dy\, dz$$

where

$$A = A_x A_y A_z = (a_{2x} - a_{1x})(a_{2y} - a_{1y})(a_{2z} - a_{1z})$$
$$B = B_x B_y B_z = (b_{2x} - b_{1x})(b_{2y} - b_{1y})(b_{2z} - b_{1z})$$

The variance of X_A is $\sigma_A^2 = \gamma(A)\sigma_X^2$, where $\gamma(A)$ is the variance reduction function (see Section 3.4) defined in three dimensions as

$$\begin{aligned}\gamma(A) &= \frac{1}{A^2}\int_0^{A_x}\int_0^{A_x}\int_0^{A_y}\int_0^{A_y}\int_0^{A_z}\int_0^{A_z} \\ &\quad\times \rho_X(\eta_1 - \xi_1, \eta_2 - \xi_2, \eta_3 - \xi_3)\, d\xi_3\, d\eta_3 \\ &\quad\times d\xi_2\, d\eta_2\, d\xi_1\, d\eta_1 \\ &= \frac{8}{A^2}\int_0^{A_x}\int_0^{A_y}\int_0^{A_z}(A_x - \tau_1)(A_y - \tau_2)(A_z - \tau_3) \\ &\quad\times \rho_X(\tau_1, \tau_2, \tau_3)\, d\tau_3\, d\tau_2\, d\tau_1\end{aligned} \qquad \text{(C.4)}$$

where $\rho_X(\tau_1, \tau_2, \tau_3)$ is the correlation function of $X(x, y, z)$, which is assumed to be stationary and quadrant symmetric in the above (see Section 3.7.4). Equation C.4 can be approximated by Gaussian quadrature as

$$\begin{aligned}\gamma(A) &\simeq \frac{1}{A}\sum_{k=1}^{n_g} w_k \sum_{j=1}^{n_g} w_j \sum_{i=1}^{n_g} w_i (A_x - \xi_i)(A_y - \eta_j)(A_z - \psi_k) \\ &\quad\times \rho_X(\xi_i, \eta_j, \psi_k) \\ &= \frac{1}{8}\sum_{k=1}^{n_g} w_k (1 - z_k)\sum_{j=1}^{n_g} w_j (1 - z_j)\sum_{i=1}^{ng} w_i (1 - z_i) \\ &\quad\times \rho_X(\xi_i, \eta_j, \psi_k)\end{aligned}$$

where

$$\xi_i = \tfrac{1}{2}A_x(1 + z_i), \quad \xi_j = \tfrac{1}{2}A_y(1 + z_j), \quad \psi_k = \tfrac{1}{2}A_z(1 + z_k)$$

and the weights w_i and Gauss points z_i are found in Appendix B.

The covariance between X_A and X_B is given by

$$\begin{aligned}
\text{Cov}[X_A, X_B] &= \frac{\sigma_X^2}{AB} \int_{a_{1z}}^{a_{2z}} \int_{a_{1y}}^{a_{2y}} \int_{a_{1x}}^{a_{2x}} \int_{b_{1z}}^{b_{2z}} \int_{b_{1y}}^{b_{2y}} \int_{b_{1x}}^{b_{2x}} \\
&\quad \times \rho_X(x - \xi, y - \eta, z - \psi) \\
&\quad \times \; d\xi \; d\eta \; d\psi \; dx \; dy \; dz \\
&\simeq \frac{\sigma_X^2}{64} \sum_{i=1}^{n_g} w_i \sum_{j=1}^{n_g} w_j \sum_{k=1}^{n_g} w_k \sum_{\ell=1}^{n_g} w_\ell \sum_{m=1}^{n_g} w_m \\
&\quad \times \sum_{n=1}^{n_g} w_n \rho_X(x_k - \xi_n, y_j - \eta_m, z_i - \psi_\ell)
\end{aligned}$$

where

$$\begin{aligned}
x_k &= a_{1x} + \tfrac{1}{2}A_x(1 + z_k), & \xi_n &= b_{1x} + \tfrac{1}{2}B_x(1 + z_n) \\
y_j &= a_{1y} + \tfrac{1}{2}A_y(1 + z_j), & \eta_m &= b_{1y} + \tfrac{1}{2}B_y(1 + z_m) \\
z_i &= a_{1z} + \tfrac{1}{2}A_z(1 + z_i), & \psi_\ell &= b_{1z} + \tfrac{1}{2}B_z(1 + z_\ell)
\end{aligned}$$

and the weights w_i and Gauss points z_i can be found in Appendix B.

INDEX

D

E

F

G

H

I

J

K

L

M

N

O

P

Q

R

S